KV-387-951

Adaptive Methods for Control System Design

OTHER IEEE PRESS BOOKS

Singular Perturbations in Systems and Control, *Edited by P. V. Kokotovic and H. K. Khalil*
Getting the Picture, *By S. B. Weinstein*
Space Science and Applications, *Edited by J. H. McElroy*
Medical Applications of Microwave Imaging, *Edited by L. Larsen and J. H. Jacobi*
Modern Spectrum Analysis, II, *Edited by S. B. Kesler*
The Calculus Tutoring Book, *By C. Ash and R. Ash*
Imaging Technology, *Edited by H. Lee and G. Wade*
Phase-Locked Loops, *Edited by W. C. Lindsey and Chak M. Chie*
VLSI Circuit Layout: Theory and Design, *Edited by T. C. Hu and E. S. Kuh*
Monolithic Microwave Integrated Circuits, *Edited by R. A. Pucel*
Next Generation Computers, *Edited by E. A. Torrero*
Kalman Filtering: Theory and Application, *Edited by H. W. Sorenson*
Spectrum Management and Engineering, *Edited by F. Matos*
Digital VLSI Systems, *Edited by M. T. Elmasry*
Introduction to Magnetic Recording, *Edited by R. M. White*
Insights Into Personal Computers, *Edited by A. Gupta and H. D. Toong*
Television Technology Today, *Edited by T. S. Rzeszewski*
The Space Station: An Idea Whose Time Has Come, *By T. R. Simpson*
Marketing Technical Ideas and Products Successfully! *Edited by L. K. Moore and D. L. Plung*
The Making of a Profession: A Century of Electrical Engineering in America, *By A. M. McMahon*
Power Transistors: Device Design and Applications, *Edited by B. J. Baliga and D. Y. Chen*
VLSI: Technology Design, *Edited by O. G. Folberth and W. D. Grobman*
General and Industrial Management, *By H. Fayol; revised by I. Gray*
A Century of Honors, *An IEEE Centennial Directory*
MOS Switched-Capacitor Filters: Analysis and Design, *Edited by G. S. Moschytz*
Distributed Computing: Concepts and Implementations, *Edited by P. L. McEntire, J. G. O'Reilly, and R. E. Larson*
Engineers and Electrons, *By J. D. Ryder and D. G. Fink*
Land-Mobile Communications Engineering, *Edited by D. Bodson, G. F. McClure, and S. R. McConoughey*
Frequency Stability: Fundamentals and Measurement, *Edited by V. F. Kroupa*
Electronic Displays, *Edited by H. I. Refioglu*
Spread-Spectrum Communications, *Edited by C. E. Cook, F. W. Ellersick, L. B. Milstein, and D. L. Schilling*
Color Television, *Edited by T. Rzeszewski*
Advanced Microprocessors, *Edited by A. Gupta and H. D. Toong*
Biological Effects of Electromagnetic Radiation, *Edited by J. M. Osepchuk*
Engineering Contributions to Biophysical Electrocardiography, *Edited by T. C. Pilkington and R. Plonsey*
The World of Large Scale Systems, *Edited by J. D. Palmer and R. Saeks*
Electronic Switching: Digital Central Systems of the World, *Edited by A. E. Joel, Jr.*
A Guide for Writing Better Technical Papers, *Edited by C. Harkins and D. L. Plung*
Low-Noise Microwave Transistors and Amplifiers, *Edited by H. Fukui*
Digital MOS Integrated Circuits, *Edited by M. I. Elmasry*
Geometric Theory of Diffraction, *Edited by R. C. Hansen*
Modern Active Filter Design, *Edited by R. Schaumann, M. A. Soderstrand, and K. B. Laker*
Adjustable Speed AC Drive Systems, *Edited by B. K. Bose*
Optical Fiber Technology, II, *Edited by C. K. Kao*
Protective Relaying for Power Systems, *Edited by S. H. Horowitz*
Analog MOS Integrated Circuits, *Edited by P. R. Gray, D. A. Hodges, and R. W. Broderson*
Interference Analysis of Communication Systems, *Edited by P. Stavroulakis*
Integrated Injection Logic, *Edited by J. E. Smith*
Sensory Aids for the Hearing Impaired, *Edited by H. Levitt, J. M. Pickett, and R. A. Houde*
Data Conversion Integrated Circuits, *Edited by D. J. Dooley*
Semiconductor Injection Lasers, *Edited by J. K. Butler*
Satellite Communications, *Edited by H. L. Van Trees*
Frequency-Response Methods in Control Systems, *Edited by A. G. J. MacFarlane*
Programs for Digital Signal Processing, *Edited by the Digital Signal Processing Committee, IEEE*
Automatic Speech & Speaker Recognition, *Edited by N. R. Dixon and T. B. Martin*
Speech Analysis, *Edited by R. W. Schafer and J. D. Markel*
The Engineer in Transition to Management, *By I. Gray*

Adaptive Methods for Control System Design

Edited by
Madan M. Gupta
Professor of Engineering
Cybernetics Research Laboratory
University of Saskatchewan

Associate Editor
Chi-Hau Chen
Professor of Electrical Engineering
Southeastern Massachusetts University

A volume in the IEEE PRESS Selected Reprint Series, prepared under the sponsorship of the IEEE Systems, Man, and Cybernetics Society.

IEEE
PRESS

The Institute of Electrical and Electronics Engineers, Inc., New York

IEEE PRESS

1986 Editorial Board

M. E. Van Valkenburg, *Editor in Chief*

J. K. Aggarwal, *Editor, Selected Reprint Series*

Glen Wade, *Editor, Special Issue Series*

James Aylor
J. E. Brittain
R. W. Brodersen
B. D. Carroll
R. F. Cotellessa
M. S. Dresselhaus
Thelma Estrin
L. H. Fink
S. K. Ghandhi
Irwin Gray
H. A. Haus
E. W. Herold
R. C. Jaeger
Shlomo Karni
J. O. Limb
R. W. Lucky
E. A. Marcatili
W. R. Perkins
A. C. Schell
M. I. Skolnik
D. L. Vines

W. R. Crone, *Managing Editor*

Hans P. Leander, *Technical Editor*

Teresa Abiuso, *Administrative Assistant*

David G. Boulanger, *Associate Editor*

Copyright © 1986 by
THE INSTITUTE OF ELECTRICAL AND ELECTRONICS ENGINEERS, INC.
345 East 47th Street, New York, NY 10017-2394
All rights reserved.

PRINTED IN THE UNITED STATES OF AMERICA

IEEE Order Number: PC01651

Library of Congress Cataloging-in-Publication Data

Adaptive methods for control system design.

(IEEE Press selected reprint series)
Includes index.
1. Adaptive control systems. I. Gupta, Madan M. II. Chen, C. H. (Chi-hau), 1937–
TJ217.A3215 1986 629.8′36 86-10562

ISBN 0-87942-207-6

Contents

Foreword ix

Preface xi

Acknowledgment xiii

Adaptive Methods for Control Systems Design: An Overview xv

Part I: Adaptive Control: General Introduction 1
Design of a Self-Optimizing Control System, *R. E. Kalman* (*Transactions of the ASME*, February 1958) 5
Adaptive System Design: A Genetic Approach, *K. De Jong* (*IEEE Transactions on Systems, Man, and Cybernetics*, September 1980) 16
Theory and Applications of Adaptive Control—A Survey, *K. J. Åström* (*Automatica*, September 1983) 25
Parameter Adaptive Control Algorithms—A Tutorial, *R. Isermann* (*Automatica* September 1982) 41
State-of-the-Art and Prospects of Adaptive Systems, *A. A. Voronov and V. Yu. Rutkovsky* (*Automatica*, September 1984) 57
Model Algorithmic Control (MAC); Basic Theoretical Properties, *R. Rouhani and R. K. Mehra* (*Automatica*, July 1982) 68
A New Solution to Adaptive Control, *J. M. Martin-Sanchez* (*Proceedings of the IEEE*, August 1976) 82
Optimality in Discrete Adaptive Systems, *Ya. Z. Tsypkin* (*Reprints of the 9th World Congress of the International Federation of Automatic Control*, 1984) 92
Global Stability of Adaptive Pole Placement Algorithms, *H. Elliott, R. Cristi, and M. Das* (*IEEE Transactions on Automatic Control*, April 1985) 96

Part II: Model Reference Adaptive Approach 105
Liapunov Redesign of Model Reference Adaptive Control Systems, *P. C. Parks* (*IEEE Transactions on Automatic Control*, July 1966) 109
Control of Unknown Plants in Reduced State Space, *P. N. Nikiforuk, M. M. Gupta, and H. H. Choe* (*IEEE Transactions on Automatic Control*, October 1969) 115
Comparative Studies of Model Reference Adaptive Control Systems, *C-C. Hang and P. C. Parks* (*IEEE Transactions on Automatic Control*, October 1973) 123
Model Reference Adaptive Control with an Augmented Error Signal, *R. V. Monopoli* (*IEEE Transactions on Automatic Control*, October 1974) 133
Unification of Discrete Time Explicit Model Reference Adaptive Control Designs, *I. D. Landau and R. Lozano* (*Automatica*, July 1981) 144
Stable Model Reference Adaptive Control in the Presence of Bounded Disturbances, *G. Kreisselmeier and K. S. Narendra* (*IEEE Transactions on Automatic Control*, December 1982) 163
Error Models for Stable Hybrid Adaptive Systems, *K. S. Narendra, I. H. Khalifa, and A. M. Annaswamy* (*IEEE Transactions on Automatic Control*, April 1985) 170

Part III: Self-Tuning Regulators 179
On Self Tuning Regulators, *K. J. Åström and B. Wittenmark* (*Automatica*, March 1973) 181
Self-Tuning Control, *D. W. Clarke and P. J. Gawthrop* (*Proceedings of the IEEE*, June 1979) 195
Self-Tuning Regulators for a Class of Multivariable Systems, *U. Borison* (*Automatica*, March 1979) 203
Self-Tuning Controllers Based on Pole-Zero Placement, *K. J. Åström and B. Wittenmark* (*Proceedings of the IEE*, May 1980) 210
Multivariable Pole-Assignment Self-Tuning Regulators, *D. L. Prager and P. E. Wellstead* (*Proceedings of the IEE*, vol. 128, Part D, pp. 9–18, Jan. 1981) 221
Predicator-Based Self-Tuning Control, *V. Peterka* (*Automatica*, January 1984) 231

Practical Issues in the Implementation of Self-Tuning Control, *B. Wittenmark and K. J. Åström* (*Automatica*, September 1984) 243

Part IV: Adaptive Control of Uncertain Plants 255
Dual-Control Theory. I, *A. A. Fel'dbaum* (*Avtomtika i Telemekhanika*, 1960) 257
Actively Adaptive Control for Nonlinear Stochastic Systems, *E. Tse and Y. Bar-Shalom* (*Proceedings of the IEEE*, August 1976) 264
Adaptive Stochastic Control for a Class of Linear Systems, *E. Tse and M. Athans* (*IEEE Transactions on Automatic Control*, February 1972) 274
Adaptive Control with the Stochastic Approximation Algorithm: Geometry and Convergence, *A. H. Becker, Jr., P. R. Kumar, and C-Z. Wei* (*IEEE Transactions on Automatic Control*, April, 1985) 288
The Strong Consistency of the Stochastic Gradient Algorithm of Adaptive Control, *H. F. Chen and P. E. Caines* (*Proceedings of the 23rd IEEE Conference on Decision and Control*, 1984) 297
Stochastic Adaptive Control for Exponentially Convergent Time-Varying Systems, *G. C. Goodwin, D. J. Hill, and X. Xianya* (*Proceedings of the 23rd IEEE Conference on Decision and Control*, 1984) 301

Part V: Applications to Aircraft Control Problems 307
Stability Analysis and Design for Aircraft Gust Alleviation Control, *P. N. Nikiforuk, M. M. Gupta, and K. Kanai* (*Automatica*, September 1974) 309
Gust-Alleviation Control Systems for Aircraft, *D. McLean* (*Proceedings of the IEE*, July 1978) 320
Design of a Two-Level Adaptive Flight Control System, *P. N. Nikiforuk, H. Ohta, and M. M. Gupta* (*Journal of Guidance and Control*, January/February 1979) 331
Command and Stability Systems for Aircraft: A New Digital Adaptive Approach, *U. Hartmann and V. Krebs* (*Automatica*, March 1980) 338

Part VI: Applications to Autopilots 351
Adaptive Autopilots for Tankers, *C. G. Källström, K. J. Åström, N. E. Thorell, J. Eriksson, and L. Sten* (*Automatica*, May 1979) 353
Adaptive Steering of Ships—A Model Reference Approach, *J. Van Amerongen* (*Automatica*, January 1984) 367

Part VII: Applications to Process Control, Robotics and Other Fields 379
Successful Adaptive Control of Paper Machines, *T. Cegrell and T. Hedqvist* (*Automatica*, January 1975) 381
Self-Tuning Control of an Ore Crusher, *U. Borison and R. Syding* (*Automatica*, January 1976) 388
Self-Tuning Control of a Titanium Dioxide Kiln, *G. A. Dumont and P. R. Bélanger* (*IEEE Transactions on Automatic Control*, August 1978) 395
Self-Tuning Adaptive Control of Cement Raw Material Blending, *L. Keviczky, J. Hetthéssy, M. Hilger, and J. Kolostori* (*Automatica*, November 1978) 402
Adaptive Identification and Control Algorithms for Nonlinear Bacterial Growth Systems, *D. Dochain and G. Bastin* (*Automatica*, September 1984) 410
Model Reference Adaptive Control Algorithms for Industrial Robots, *S. Nicosia and P. Tomei* (*Automatica*, September 1984) 424
Adaptive Locomotion of a Multilegged Robot over Rough Terrain, *R. B. McGhee and G. I. Iswandhi* (*IEEE Transactions on Systems, Man, and Cybernetics*, April 1979) 434
Adaptive Control of Left Ventricular Bypass Assist Devices, *B. C. McInnis, Z-W. Guo, P. C. Lu, and J-C. Wang* (*IEEE Transactions on Automatic Control*, April 1985) 441
Adaptive Load-Frequency Control of the Hungarian Power System, *I. Vajk, M. Vajta, L. Keviczky, R. Haber, J. Hetthéssy, and K. Kovács* (*Automatica* March 1985) 449

Author Index 459

Subject Index 461

Editor's Biography 463

Dedication

To My Grandparents and Parents
for Instilling Within Me
the Thirst for Knowledge,
and the Quest for Excellence;

AND

To My Teachers, Research Colleagues, and Students
Who Have All Taught and Inspired Me.

ओं कायेन वाचा मनसेन्द्रियैर्वा
बुद्ध्यात्मना वानुसृतः स्वभावात् ।
करोमि यद्यत्सकलं परस्मै
नारायणायेति समर्पयामि ॥

Oh Lord!

to you we dedicate humbly and with all the fervour of our hearts whatever we have accomplished, either consciously or unconsciously, through our bodies, minds and senses, indeed through our very souls.

—Madan M. Gupta

Foreword

EFFECTIVE application of automatic control principles is necessary to ensure that technological systems and processes function well. In addition, there is a need for reliable and maintainable systems, and their effective and efficient operation. In many cases such systems have to work in environments about which very little is known. Special types of control must then be employed to ascertain the desirable functional characteristics of reliability and maintainability.

The notion of adaptive control was introduced about three decades ago to provide effective control of systems in unknown environments. The idea of designing a system that can adjust its own structure and parameters to accomplish a specified purpose, without much prior knowledge of the environment in which the system will operate, was very appealing. Over the last three decades, a considerable amount of theoretical research has been carried out in various areas of adaptive control. As a result, not only are we now rich with theoretical constructs; there are several examples of noteworthy applications as well.

Most of the literature on adaptive control is scattered over numerous publications. In this reprint volume, the editors have gathered a rich variety of past and current research results in the area. The volume deals with most of the major adaptive control methods, and the reprints provide historical perspectives, overviews of theoretical and applied research, and detailed treatments of some important concepts, as well as references to the abundance of papers in the field.

The reprints are authored by professionals from 16 countries, and the coverage is nicely balanced, with 29 survey/theory papers and 15 application papers.

This book should be of interest to anyone within the field of adaptive control, including graduate students, researchers, and control system design engineers.

ANDREW P. SAGE

Preface

THE emergence of new and complex technological systems over the past two decades has demanded better control systems. This has led to a better appreciation of adaptive methods for control systems design. The growing interest in the field of adaptive control systems is evidenced by the fact that more than 6000 publications have appeared over the past three decades. Also many sessions have been devoted to this field at major international conferences and symposia in the area of control systems. For example, at the IFAC Congress held at Kyoto (August, 1981) about ten percent of all the papers were devoted to some aspects of adaptation and learning. Also, at the 1984 IEEE-CDC, ten percent of the sessions were devoted to the field of adaptive control systems.

A practical problem for researchers working in this area was to develop better control strategies for misbehaved systems encompassing uncertainties, nonlinearities, and distributiveness in their parameters. Such systems are hard to model. And this led to the development of various methodologies which exhibit learning and adaptive capabilities. These methodologies ranged from the earlier hill-climbing methods to more recent approaches such as model reference adaptive control systems and self-tuning regulators. The field has evolved progressively over the last three decades. Some researchers have used rigorous mathematical approaches to develop the area while others have blended a sound theoretical basis with the practical needs.

Although there has been a large number of research publications and over a dozen books written in the related areas of adaptive and learning systems, until now there has been no single resource to which students or researchers could refer to in learning about the major developments. Most of the books available in this area are largely devoted to the adaptive methods encompassing only the author's own research interests.

In designing the present volume, the editors' goal was to present a pedagogically sound reprint volume in adaptive control systems which could be useful as a supplementary or even as a main text for graduate students. Additionally, this collection of literature should have conceptual and theoretical information, with a comprehensive view of the general field of adaptive methods for control systems design that can stimulate the research interest of readers. It will also give a comprehensive view of the field to practicing engineers, academics, and students.

In order to meet these objectives, 44 of the most significant articles from over 300 articles published between 1955 and 1985 were chosen for this volume, giving the reader a wide perspective. The main sources for these articles are IEE, IFAC, IEEE, and ASME including their transactions and conference proceedings. This collection contains a wide breadth of classical papers, survey papers, and a variety of current papers giving both theory and a wide range of applications. Thus, this collection provides an 'instant library' from which the students, practicing engineers, and researchers may obtain an overall picture of the early and recent aspects of this important field of adaptive control.*

The field has advanced rapidly and no paper can be singled out as the seminal contribution. It was, therefore, a difficult task for the editor to select a relatively few articles from a large number of candidates. We are keenly aware of the excellent papers that had to be excluded in order to produce a balanced collection. The overview and additional citation given in each part will provide the readers with helpful pointers into the literature.

This collection of papers is divided into seven parts, progressing from fundamental problems to historical perspectives, and from theoretical developments to advanced applications:

Part I: Adaptive Control: General Introduction
Part II: Model Reference Adaptive Approach
Part III: Self-Tuning Regulators
Part IV: Adaptive Control of Uncertain Plants
Part V: Applications to Aircraft Control Problems
Part VI: Applications to Adaptive Autopilots
Part VII: Application to Process Control, Robotics, and other Fields

Part I contains nine articles which give historical perspectives and surveys. These papers point out the main problems associated with Hu adaptive systems and its theory and design methods. In Parts II and III, the theory for the design of model reference adaptive systems and self-tuning regulators is given. Part IV deals with adaptive control of uncertain plants using dual control and other related approaches. In Parts V, VI, and VII, we give applications oriented papers dealing with the adaptive control of aircraft, adaptive autopilots, and applications of adaptive control in process control, robotics, biomedical engineering and power systems control. In some of these papers, especially in the area of autopilots, process control and power systems some real-life applications are given. Additionally, each part of the volume precedes with editorial comments

* Most of the papers for this volume have been taken from IEE, IEEE, ASME and IFAC sources. The countrywise distribution of these (theoretical & applications) papers is as follows: Australia (1 + 0), Belgium (0 + 1), Canada (2 + 3), China (2 + 0), Czechoslovakia (1 + 0), Federal Republic of Germany (2 + 1), France (1 + 0), Hungary (0 + 2), Italy (0 + 1), Japan (0 + 2), Netherland (0 + 1), Spain (1 + 0), Sweden (5 + 3), U.K. (4 + 1), USA (9 + 2), and USSR (3 + 0), where the first and second figures correspond to the number of theoretical and applications papers, respectively. Some of these papers have authorship from more than one country.

and a list of selected bibliography; this will provide additional information to the readers.

In the following, we present a cross interaction matrix of the book, this will help the readers to follow a certain course during their studies. The rows of this matrix are divided into three headings: Surveys, Theory, and Applications. The columns have four classes: Model Reference Adaptive Control, Self-Tuning Regulators, Control of Uncertain Plants, and Miscellaneous Adaptive Control Methodologies. The readers will notice that the classes have fuzzy boundaries, however, this cognitive matrix will guide them through the book. The book is more or less self-contained, however, as a prequisite for a thorough study of the various theoretical concepts presented in this volume, the readers will need a sufficient knowledge of conventional control theory, tools of the modern control theory, various methods of stability analysis (Liapunov and Popov methods of stability, probability theory, and stochastic process). No attempt has been made to include these theoretical tools in this volume since a number of excellent references are available in these areas.

It is our hope that this volume will provide the reader not only with valuable conceptual and technical information but also with a comprehensive view of the general field of adaptive control systems design and its problems, accomplishments, and future potentials.

CROSS-INTERACTION MATRIX OF PAPER CONTENTS

	Model Reference Adaptive Control	Self-Tuning Regulators	Control of Uncertain Plants	Miscellaneous Adaptive Control Methodologies
Surveys	3, 5	1, 3	2, 6, 7	0, 1, 2, 3, 4, 5, 6, 7, 8, 9
Theory	10, 11, 12, 13, 14, 15, 16	17, 18, 19, 20 21, 22, 23	24, 25, 26, 27 28, 29	—
Applications	30, 31, 32, 33, 35, 41	34, 36, 37, 38, 39, 44	30, 36	31, 32, 33, 40, 42, 43

Nov. 5, 1985

MADAN M. GUPTA
University of Saskatchewan

Acknowledgment

LEARNING and adaptation are natural characteristics that are inherent in all living beings. This subject was formally introduced to me in 1961, and ever since the beauty and the strength of this subject has intrigued me. I have learned this subject through several phases since then: first as a student and then as a student advisor, teacher and researcher in the field. During these phases, I have learned from my professors (J. R. Handa, I. J. Nagrath, J. L. Douce, P. C. Parks, and K. C. Ng), research students (H. H. Choe, R. Hoffman, L. G. Mason, R. Weddige, L. Wagner, and N. Hori, in particular), research colleagues (Dr. P. N. Nikiforuk, Dr. K. Kanai, Dr. K. Tamura, Dr. Y. Yamane, Dr. H . Ohta, Dr. N. Minamide, Dr. Y. Mutoh, Dr. Y. Yamamoto, Dr. T. Mori, and Dr. J. B. Kiszka), and many other members of the learning and adaptive community who directly or indirectly have influenced my thought process in this field; Professor Ya. Z. Tsypkin, Professor R. V. Monopoli, Professor K. J. Astrom, Professor P. C. Parks, Professor A. P. Sage, Dr. I. D. Landau, Dr. G. C. Goodwin, Professor N. K. Sinha, Professor K. S. Narendra and Dr. P. R. Belanger are some of the names which are very important to me in the field.

This volume was conceived several years ago, the goal was to assemble and produce a comprehensive set of literature in this growing but scattered field of adaptive control systems. Several colleagues from the learning and adaptive community have provided me with useful feedback at various phases of this project. Discussions with Professor K. J. Astrom and Dr. B. Wittenmark have helped me in the selection process of these articles.

Indeed, it is a great pleasure to acknowledge the encouragement and adaptive feedback provided by Professor Andrew P. Sage during all the editorial phases, since this book's inception to its completion, of this project. Finally, I would like to thank my wife Suman, three sons (Anu, Ashu, and Amit), and my mother who let me finish this project and many other equally important assignments without many interruptions. Of course, the cheerful environment of the College of Engineering at the University of Saskatchewan has provided me with an ideal atmosphere to work in the adaptive control field.

Nov. 5, 1985 MADAN M. GUPTA

Adaptive Methods for Control Systems Design: An Overview

MADAN M. GUPTA

I. Why Adaptive Control?

TODAY'S technological systems are very complex and so are their control problems. During the past two decades there has been a demand for improved control systems, which has led to a better appreciation for the need for adaptive methods in control system design. Recently, the field of adaptive control has faced an exponential growth, and new theoretical concepts with many interesting applications are evolving.

We give a dictionary meaning for "*adapt*" and "*adaptation*".*

> **adapt: 1 a:** to make suitable or fit (as for a particular use, purpose or situation) ··· **b:** to make suitable (for a new or different use or situation) by means of changes or modifications ··· **2:** to adjust (oneself to particular conditions or ways bring (oneself) into harmony with a particular environment ··· to bring oneself or esp. one's acts, behavior, or mental state into harmony with changed conditions or environment

> **adaptation: 1 a:** the act or process of adapting, fitting or modifying ··· **b:** the state or condition of being adapted or adjusted or of adapting or adjusting oneself ··· **2:** adjusting to environmental conditions: as **a:** adjustment of a sense organ to the intensity or quality of stimulation prevailing at the moment ··· effected by changes in sensitivity and occurring as a heightened sensitivity or as a physical adjustment to meet change conditions or as decline or loss of sensitivity to a constant stimulus ··· **b:** modification of an organism or of its parts or organs fitting it more perfectly for existence under the conditions of its environment and resulting from the action of natural selection upon variation ··· **c:** the continuing process through which the organization of groups is modified to meet the requirements of their social and physical environment **3:** something that is adapted: a modification for a new use or an alteration or change in form or structure ···

The dictionary meaning of "*adapt*" and "*adaptation*" is expressed primarily in terms of biological adaptation to environmental changes. The same definition may be carried over to "human-made" or "artificial" adaptive control systems, the main topic of this book.

For our control systems problem, we would like to construct an adaptive mechanism which forces a given process to its optimum performance, according to some desired criterion, even under the changing environmental conditions. A simple example of adaptive feedback mechanism is the automatic gain control (AGC) used in a receiver system for maintaining a constant output intensity over a wide range of varying input signal conditions.

The purpose of the present volume is to present a broad prespective of adaptive methods in control systems design. For this purpose, we have assembled a wide breadth of classical papers, and survey papers, and a variety of contemporary papers presenting theory and a wide range of applications.

In an ideal situation, an adaptive control system should have the following characteristics:

- i) It should be able to adapt continually to bounded environmental variations and/or changing system requirements.
- ii) It should have some learning abilities.
- iii) An adaptive control system should be "reproductive" in the sense that for changing situations a controller will evolve with a new set of parameters and a new set of control strategies. Such a controller, for example, would be able to steer a ship or pilot an aircraft under varying, but bounded, environmental conditions.
- iv) It should have the capability of "self-repair" in case of internal parameter failures.
- v) The adaptive controller should be "robust" in the sense that the performance of the controlled dynamic system should be insensitive to changes in its environment and/or its parameters and modelling errors.

Such an ideal adaptive control system will naturally have the potential for much superior control capabilities and provide better performance characteristics for a plant operating over a wide range of environmental parameters compared to non-adaptive type of controllers. The design of such an ideal adaptive controller represents a significant challenge to control systems engineers.

II. Evolution of Feedback Control Theory

The theory of feedback control systems has gone through various phases of evolution. We classify this development of the theory into three main categories with increasing complex-

The author is with Cybernetics Research Laboratory, College of Engineering, University of Saskatchewan, Saskatoon, Saskatchewan, Canada S7N 0W0.

* Webster's Third New International Dictionary [1968].

ities: the deterministic control theory, the stochastic control theory, and the adaptive control theory.

In the deterministic control theory, one assumes complete knowledge of the plant under control. A wealth of theory was developed in this area especially for linear systems in which the superposition theorem (convolution integral) plays the main role. Mathematical tools such as the Laplace transform theory, the transfer function approach, and the state-space formulation of the systems are used extensively. A rich theory of optimal control, mainly based upon state-variable feedback, was developed. The theory is rich both in notion and its development, but it gives the illusion that the principal difficulties are overcome. These difficulties were observed as soon as uncertainties, nonlinearities, and other such factors were encountered in real life systems. For nonlinear systems, some other powerful tools of analysis and synthesis have evolved.

Stochastic control theory deals with uncertainties that are inherent to control systems [4]–[11]. Linear deterministic control theoretical tools were extended to deal with stochastic situations, and such powerful methods as the Wiener and Kalman filtering methods were developed. Extension of these methods were obtained for a certain class of nonlinear systems.

These methods, both deterministic and stochastic, are heavily dependent upon the assumption that a sufficient *a priori* knowledge of the system and its disturbances are available in either a deterministic or a statistical sense. We know that in real-life situations, especially while dealing with complex dynamic systems operating in a complex environment, neither the plant equations nor its environment are completely known, and that their characteristics cannot be measured in advance.

Such an uncertain situation, where *a priori* knowledge of the plant and its environment, is insufficient led to the introduction of the notion of "adaptivity." The problem of controlling a plant enveloped by uncertainties with almost no, or with a very little *a priori*, information can be solved by "learning" about the process during its operation, and resynthesizing the control strategy based upon this new information. In this way the plant control strategy can be adapted to initial uncertainty and a new changing environment.

The principle of optimality is one of the subjects in the field of automatic control theory that is discussed frequently. The notion of optimality was introduced to satisfy the need to control a process in some best manner. Although several ways of defining a performance index have evolved, the quadratic performance criterion was chosen as the most satisfactory formulation because of its attractive mathematical properties and, in certain cases, useful physical characteristics. Some of these control problems were solved using various optimal control theoretical tools such as Pontryagin's maximum principle and Bellman's dynamic programming method. These tools were initially developed for deterministic problems, but extension of them have been developed, with some successes, for stochastic and adaptive control problems as well.

In the adaptive control literature the terms "adaptive control", "self-adaptive control", and "learning and adaptive control" have been used interchangeably, since as a rule, they do not have a unique notion or interpretation. It has been shown in the literature that adaptation with learning capabilities should produce a robust adaptive control [34]. Questions relating to "explicit learning" and "implicit learning" now arise. Explicit learning implies obtaining knowledge about the system in terms of its explicit parameters, such as finding the coefficients of the differential equation describing the system. Such learning behavior can be implemented using such modules as identifiers, observers, filters, etc. Implicit learning methods, which are found to be much more useful, and perhaps less complex in design procedure, look at the system dynamic response and infer from it the dynamics of the system directly. Based upon this implicit information, the adaptive controller synthesizes the control strategy. Implicit and explicit methods of learning in adaptive control are also referred to in the literature as direct and indirect methods respectively.

III. Formulation of Adaptive Control Problems

Now, we define adaptive control problems for a general class of dynamic systems with varying parameters. For this purpose consider the following sets of equations describing the plant state evolution process and measurements with associated uncertainties.

Plant State-Evolution Process: The state evolution process of the plant can be described, in general, by the following nonlinear and time-varying vector differential equation:

$$\dot{x}(t)=f[x(t),\ u(t),\ w(t),\ t],\qquad t\geq t_0, \tag{1a}$$

where

$x(t) \in R^n$:	state vector of the plant,
$u(t) = u(x, t) \in R^m$:	a vector of state feedback control inputs,
$w(t) \in R^w$:	a vector of disturbances on the plant states and its parameters, and,
$f[\cdot] \in R^n$:	an unknown or partially known nonlinear and time-varying vector function of its arguments.

Plant Output Measurements: The measurements of the plant output are possibly contaminated by measurement noise $v(t)$ and, in general, can be described as

$$y(t)=g[x(t),\ u(t),\ v(t),\ t], \tag{1b}$$

where

$y(t) \in R^p$:	plant measurements,
$v(t) \in R^v$:	measurement noise, and
$g[\cdot] \in R^p$:	an unknown or partially known function of its arguments.

The Performance Index

The performance index J describing the operating quality (performance) of the system may be described as an integral of the scalar function (in some cases a vector function) given by

$$J=\int_{t_0}^{t} L[y(t),\ u(t),\ w(t),\ t]\ dt,$$

$$x(t) \in X,\ u(t) \in U, \tag{1c}$$

where $L[\cdot]$ is, in general, a nonlinear scalar function of the measurement $y(t)$ (or states $x(t)$) and control efforts $u(t)$. This nonlinear function is, in most cases, taken as a quadratic function of $x(t)$ and $u(t)$. The fact that $x(t) \in X$ and $u(t) \in U$ in (1c) provides some physical and economical constraints on the states $x(t)$ as well as on the control efforts $u(t)$. One particular conceptual realization of the implied control process described in the above equations is depicted in Fig. 1.

The Feedback Control Problem

A general feedback control problem for the dynamic process described by (1a) and (1b) is *"synthesize an on-line and real-time feedback control signal u[x(t), t] using some control strategy in order to optimize (maximize or minimize) the performaces index J, in (1c), whereas x(t) and u(t) are constrained by some allowable sets X and U, respectively".*

Philosophically, the control problem described above is very challenging, and one would like to find an analytical solution or a computer algorithm which can, in whole or part, provide a control $u(t) = u[x(t), t]$. The control $u(t)$ should force the plant so as to yield some desirable characteristics, whereas, the desirable performance is defined in (1c). Unfortunately, no analytical solution for such a general class of dynamic processes is available. Fortunately, such an analytical solution, except for a mathematical curiosity, is not needed. One may construct, however, a conceptual control algorithm based upon intuitive and intelligent human reasoning. This is the very reason that some researchers in the control systems community are striving to develop *"self-learning and adaptive control algorithms."*

A need for self-adaptive control algorithms, having some desirable learning capabilities, arises in the control of processes which are time-varying, nonlinear, and have unknown dynamics with unknown disturbances acting upon them. For such an inherently complex problem, no general analytical solution can be found. Although a potentially acceptable *a priori* control structure can be defined, it is generally not possible to specify, in advance, the parameters within this structure.

The description of the plant and its measurement as defined in (1a) and (1b) is very general. The control problem for such processes can be classified under the three main problem categories given below. This classification will depend upon the amount of *a priori* knowledge of the functions $f[\cdot]$ and $g[\cdot]$ and the parameters of their arguments.

i) *A black-box control problem:* If the function $f[\cdot]$ and $g[\cdot]$ are completely or largely unknown with unknown disturbances $w(t)$ and $v(t)$, the control problem can be categorized as that of a black box control problem. Such a control problem is generally unsolvable analytically.
ii) *A white-box control problem:* If the functions $f[\cdot]$ and $g[\cdot]$ are completely known with a complete *a priori* information (in a deterministic or a stochastic sense) on disturbances $w(t)$ and $v(t)$, the control problem can be categorized as a white-box control problem. If the white-box control problem deals with the control of a linear system with a quadratic performance index and Gaussian input and output disturbances, the problem is known as a Linear-Quadratic Gaussian or LQG control problem. Some elegant mathematical approaches exist for the solution of this type of control problem [4]–[6]. However an analytical solution may be extremely difficult to find or may not even exist for a general class of nonlinear and nonquadratic white-box control problems.
iii) *A grey-box control problem:* If the functions $f[\cdot]$ and $g[\cdot]$ are partially known, with a partial *a priori* information on their disturbances, we have a grey-box control problem. The domain of this control problem lies in between the white-box and the black-box control domains. The adaptive control methods deal primarily with the grey-box control problem domain, and are the subject of this volume.

A possible approach to the solution of the grey-box control problem is to accumulate dynamically, in a direct or indirect way, information about the system response and to simultaneously generate an acceptable control signal in an adaptive feedback manner.

Consider, for example, the control of a power generating unit. Invariably, performance depends on a number of unspecified and unknown parameters and environmental disturbances such as customer demand overtime. While there may be forecast, there will be a number of unpredictable components. This is a good example of the grey-box control problem. For this system, it is generally difficult to provide a fixed control strategy based upon a finite amount of *a priori* information.

Based upon our earlier definition of "adapt" and "adaptive," the adaptive control problem can be stated as follows therefore: *"Design a feedback controller having the "learning" and "adaptation" capabilities which forces an arbitrary (or partially known) dynamic process operating in an uncertain environment to respond in some "acceptable fashion."*

In 1958 R. E. Kalman [1] postulated building a machine which adjusts itself automatically to control an arbitrary dynamic process. The design of a self-regenerative type of adaptive controller may provide a real challenge to control system engineers [2]. For example, one can design an adaptive controller for a certain class of dynamic processes. In order to produce another adaptive controller for another class of dynamic processes (for example, a process having time constants of the order of hours rather than seconds), one has to change certain strategies rather than the whole structure of the adaptive controller. But the design of such a self-regenerating type adaptive controller provides a real challenge to the control engineers.

The performance index is defined in (1c). To incorporate this within an adaptive controller requires some on-line means for evaluating performance of the dynamic process and that of the adaptive control strategy. Intuitively, performance of the adaptive control strategy should depend upon the ability of the resulting adaptive system to respond quickly and efficiently to

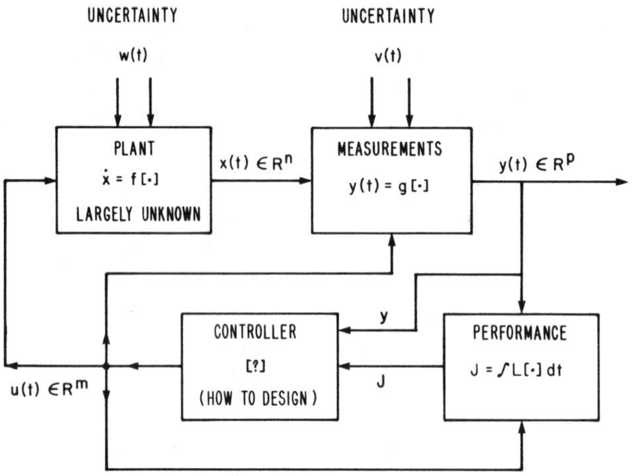

Fig. 1. Control problems for a general class of dynamic systems.

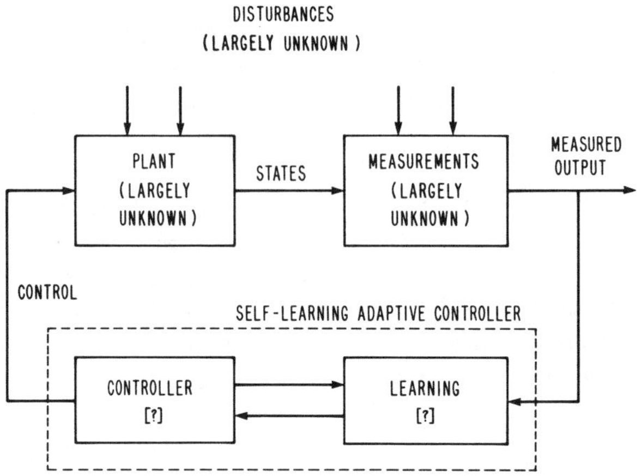

Fig. 2. A feedback control system with conceptual self-learning and adaptive capabilities.

problem at hand by generating control signals to optimize performance of the dynamic process under control. Also, good ''adaptive strategies'' should be robust in the sense that they should be insensitive to, and should respond optimally over, a whole range of unpredictable environmental disturbances to which the dynamic process might be subjected. Instances where such a problem may arise are in the control of a power generating unit subject to a wide range of load fluctuations, or in the control of an aircraft with wide variations in the physical plant parameters with speed and altitude. Of course, even ordinary feedback control loops inherently desensitize the response of a process for small changes in the parameters or slight changes in the environment. But, here we are postulating the design of a self-adaptive control strategy with inherent learning capabilities which can control a class of dynamic processes described in (1a) and (1b) over a very wide-range of environments and parameters.

In Part I of this volume, we have reproduced some excellent survey papers, which give an integrated view of the field of adaptive control systems. In this paper, without being repetitious, we present an overview of adaptive methods for control system design.

IV. A Brief Review of Adaptive Control Approaches

Prior to the last three decades, emphasis in the field of feedback control was largely on systems with known dynamics with constant or almost constant (slowly time-varying) parameters and with a known environment. Thus, the system designer was concerned primarily with whether or not a feedback control system would work at all, would be stable, would return a disturbed system to equilibrium within a reasonable time, would keep system errors within reasonable values, would not have transients with excessive over- and underswings, and would not give responses that are too oscillatory. The system designer did not try to attain (or maintain) optimum performance when the system parameters change or system is subjected to unknown disturbances. He usually confined himself only to linear systems and for nonlinear ones he approximated the system by a linear model. Most of the control mathematical tools used for studying nonlinear systems were meager, and since optimum performance can usually be obtained only by the use of nonlinear controllers he carefully avoided the issue of optimal performance for such systems. Most of the work on systems with stochastic disturbances used only trial and error methods as opposed to analytical design techniques which are available today.

In 1966, almost two decades ago, R. Oldenburger was one of the first to select a set of published research literature under the title "Optimal and Self-Optimizing Control" [3]. Most of the research that was reported by Oldenburger in this volume was conducted in the Soviet Union and the United States. The major research efforts during the mid-1960's to mid-1970's were in optimal-control theory, both deterministic and stochastic, with *a priori* known information, for example. However, the growth in modern technological systems needed robust controllers. Controllers based upon optimal control theory are inherently sensitive to modeling errors, and thus are not robust.

Some new notation and concepts like "*self-adaptive*" or "*adaptive*" control emerged. These tools provided self-optimization capabilities in feedback control systems where the process dynamics experience changes because of environmental changes. Some of the theoretical tools provided additional insight into the feedback control problems with some practical considerations. An earlier paper contemplating the philosophy and some practical considerations was due to Kalman [1]. This topic was enthusiastically perused by many researchers, one example being the design of self-adaptive autopilot control systems [12]. Most of the designs were based upon intuition but in some cases they had sound theoretical basis. Early adaptive control attempts faced implementation problems. However, they were largely responsible for laying the foundations of adaptive control systems. The results in the areas of identification, observeration, estimation, filtering and optimal control theory gave a basic understanding of adaptive control design methodologies. Some of these identification and estimation tools were directly transported in the design philosopy of adaptive feedback control systems. A considerable amount of effort was, and is being, spent in developing self-adaptive control approaches [27]. Some of these works represent a successful theoretical attempt with better operational characteristics and established theoretical basis for the design of adaptive controllers. The state of the recent theory providing a guideline for the design is progressing as evidenced by the 1984 IEEE-CDC [14].

In the foregoing discussion, a description was given of the adaptive control problem and some possible conceptual approaches to the solution of this problem that have been adopted by the researchers and reported in the literature. In the following paragraphs, an attempt is made to bring some of these approaches into perspective.

The two most important approaches to adaptive systems are model reference adaptive control (MRAC) and self-tuning regulators (STR). Although these approaches were developed almost independently, they share some common design features.

In the model reference adaptive control (MRAC), the performance of a desired system is described in terms of a reference model. The adaptive controller is designed to force the plant to behave like the reference model. This idea was introduced by Whitaker and his colleagues [15], [28], [29], [31], [32] and was further developed by Humphries and Sage [33], Parks [16], Gupta *et al.* [17], [18], Monopoli [19], Landau [20] and others. The book by Landau [21] and other survey papers in the area [16], [20], [27] give a comprehensive picture of these works.

An alternative approach to adaptive control, known as the self-tuning regulator, has attracted much interest during the past decade. The basic idea of a self-tuning regulator has evolved from the work of Kalman in 1958 on self-optimizing systems [1]. Between 1958 and 1973, several interesting expositions on these ideas have appeared [22]–[24]. In 1973, Astrom and Wittenmark developed the self-tuning regulator for the stochastic minimum variance control problems [25]. The approach is very flexible with respect to the underlying design method, and recently many different extensions of the approach have appeared in the literature. A description of self-tuning regulators is found in Astrom [26], [27]. Though some interesting applications of self-tuning regulators have appeared, some of these are reproduced in this volume, some of the basic theoretical issues, such as stability and optimization concerns still remain open questions.

Theoretical developments and application of these two major approaches will be appearing in much detail in this volume.

V. Some Implementation Aspects

As described in the previous sections, realization of adaptive control approaches need the measurement and estimation of information that is contained in the measureable outputs of the plant. Depending upon the model assumed by the designers, these output measurements are subject to various mathematical operations such as model, parameter, and state estimators. Estimation and filtering theories have been developed by control system engineers and communication scientists and, to some extent, have been applied to adaptive control problems. Filters have been used to ameliorate the effects of uncertainties. In the indirect approach, an estimation of the plant parameters is obtained by using on-line identifiers. Such identifiers are themselves composed of state and parameter estimators and therefore, increase the complexity of the composite adaptive feedback control system. These added features are intended to reduce the effects of uncertainties and, thereby, to increase the fidelity of the information which is used in the adaptive feedback mechanism. In some cases, however, these blocks, Figs. 3(a) and (3b), may introduce stability problems which may not be easily solvable [27].

The approaches and methodologies used in adaptive feedback mechanism are numerous. Some of them make explicit use of the information contained in the input signals and output measurements. Others use this information in an implicit way. The explicit use of information is possible through the effects of such subsystems as filters, estimators, observers, and identifiers. In the implicit information use no subsystems are used. The output of the plant is used to generate information, such as an error signal, which in turn is used to generate an appropriate adaptive feedback control signal (Fig. 4).

During the last two decades, many concepts and methodologies of adaptive feedback mechanisms have evolved. The most

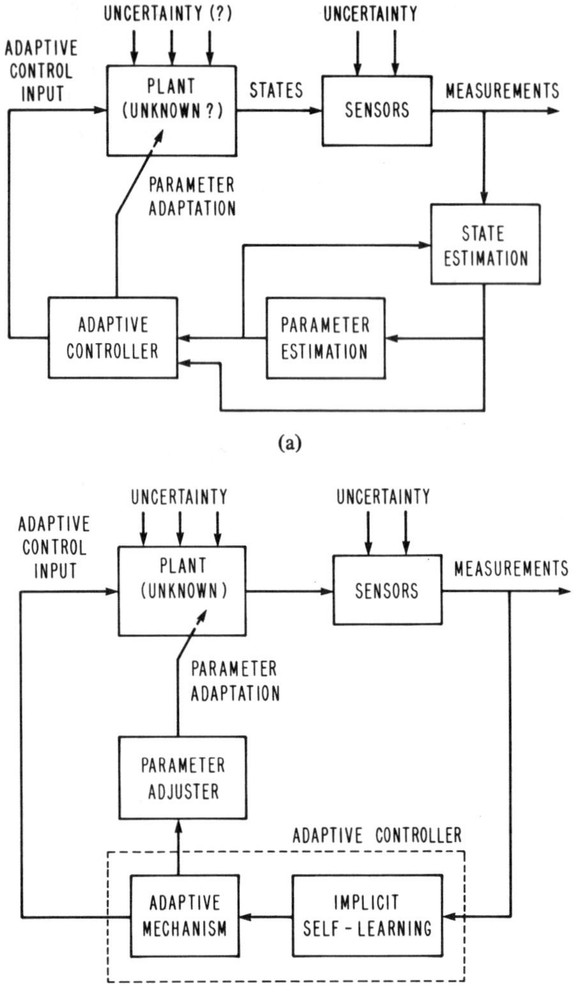

Fig. 3. (a) Feedback adaptive control systems with state and parameter estimators. (b) A conceptual feedback adaptive control system with implicit learning.

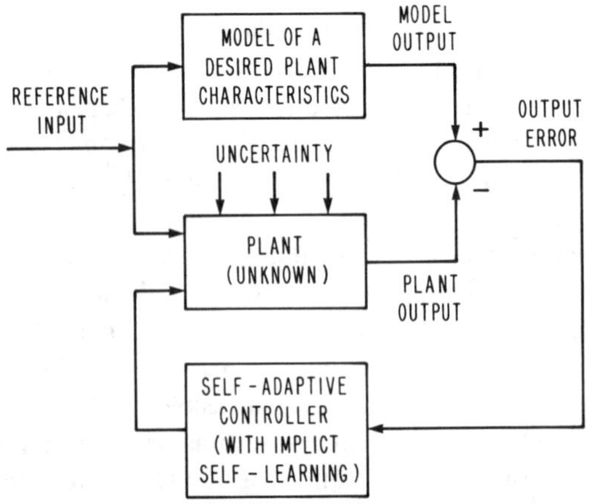

Fig. 4. An ideal model reference adaptive system (MRAS).

popular approaches of model reference adaptive technique and self-tuning regulator were described in the preceding section. Either of these conceptual approaches may be implemented using one of the following two techniques:

i) signal synthesis adaptive technique
ii) parameter adaptive technique.

The fundamental difference between these two realization approaches is that the first approach results in generation of a "signal" whereas the second is alternation of controller parameters. In the signal synthesis technique, an adaptive feedback signal is generated and this is used to modify the plant control inputs so as to force the plant outputs to yield some optimum or acceptable performance. In the parameter adaptive techniques, an adaptive signal is generated and used to modify the controller and/or plant parameters so as to force the plant outputs to yield some optimum or acceptable response.

Some research has been reported which make use of both signal synthesis and parameter adaptive mechanisms as illustrated in Fig. 3(b). It may be shown mathematically that, in general, both adaptive mechanisms yield the same equivalent system. It has been found, however, that from a practical viewpoint the signal synthesis adaptive technique may be much more attractive than that of the parameter adaptive mechanism since, in the former case, the requirement for the accessibility of the plant parameters is eliminated. Of course, controller parameters may be modified as these will generally be accessible.

Implementation of model reference adaptive systems provides a challenge to the designers innovation. In the implementation of this approach, the performance of the desired system is modeled using a mathematical model simulated on a computer. The plant outputs are compared with the model outputs and an error signal is generated. This error signal is used either to modify the plant control inputs, and/or to adjust its parameters. This type of adaptive controller ideally forces the error signal, representing the difference between model output and systems output, to zero with increasing time for a plant operating in an uncertain and unknown environment. An ideal scheme for implementing such an adaptive controller is shown in Fig. 4.

The adaptive feedback method based upon this model reference adaptive approach has proven to be among the more popular ones. Earlier methods used heuristical approaches in the design of the adaptive feedback loop, and often such schemes led to unstable adaptation for certain types of input signals. However, the pioneer works of Parks and others [16]–[21] used advanced stability methods viz. Liapunov stability technique and Popov's hyperstability method. This led to stable adaptation in a limited region of system operation. Now several design methods have emerged that are based upon stability approaches.

VI. Intrinsic Problems and Future Challenges

For industrial applications, a self-adaptive mechanism having self-learning capabilities is needed. In this mechanism the system must learn from the present and weighted past measurements of the plant and use this information in an adaptive mode so as to yield an optimum performance. But many unusual problems are encountered in practical control systems, these are

i) unknown and stochastic types of parameters;
ii) appreciable but unknown and variable dead time, for example, in chemical processes;
iii) nonlinear and distributive types of processes, for example, in power systems;
iv) stochastic disturbances, such as variation in ambient temperature or pressure of wind speed; and
v) nonminimum phase behavior.

These problems are intrinsic to the design of adaptive controllers for such applications as in aerospace vehicles and aircraft, autopilots for ships and tankers, chemical plants, power systems and robotics. In the past decade, studies in adaptive control theory were mainly concerned with either the application of linear systems or in some cases nonlinear systems. However, the design of adaptive feedback controllers for a class of nonlinear uncertain systems in the grey box category is still a challenge to researchers. In a recent editorial Ljung and Anderson [35] have commented on this widely recognized gaps in the field. Many successful applications have been reported, despite the magnitude of the difficulties encountered in attempting to apply some of the existing adaptive control theory to realistic situations. The comments of Ya. A. Tsypkin [34]: *"It is difficult to find more fashionable and attractive terms in the modern theory of automatic control then the terms of adaptation and learning. At the same time, it is not simple to find any other concepts which are less complex and more vague."*

We know intuitively what encompasses an adaptive system. We desire to design perhaps a nonlinear feedback controller with an adaptive philosophy or mode of operation. This controller must have some on-line learning capabilities, and should generate proper feedback control signal using the available measured noisy information. This feedback control signal must then be such as to control the output variables of a class of unknown plants in an adaptive mode, to follow the outputs of a desired model, Fig. 4.

In this volume, a small but representative sample from a variety and large amount of literature in the adaptive area is reprinted. Let our readers, both novice and experienced, discover for themselves "how much" we have achieved in the field of adaptive control.

REFERENCES

[1] R. E. Kalman, "Design of a self-optimizing control system," *Trans. ASME,* vol. 80, pp. 468–478, 1958.
[2] K. DeJong, "Adaptive system design: A genetic approach," *IEEE Trans. Systems, Man and Cybernetics,* vol. SMC-10, no. 9, pp. 566–574, 1980.
[3] R. Oldenburger, Ed., *Optimal and Self-Optimizing Control.* Cambridge, MA: MIT Press, 1966.
[4] A. P. Sage, *Optimum Systems Control.* Englewood Cliffs, NJ: Prentice-Hall, 1968.
[5] A. P. Sage and J. L. Melsa, *System Identification.* New York and London: Academic Press, 1971.
[6] E. A. Bryson and Y. C. Ho, *Applied Optimal Control.* Waltham, MA: Blaisdell Publishing Co., 1975.
[7] K. J. Astrom and B. Wittenmark, *Computer Control Systems—Theory and Design.* Englewood Cliffs, NJ: Prentice-Hall, 1984.
[8] K. J. Astrom, *Introduction to Stochastic Control Theory.* New York: Academic Press, 1970.
[9] B. D. O. Anderson and J. B. Moore, *Linear Optimal Control.* Englewood Cliffs, NJ: Prentice-Hall, 1971.
[10] ——, *Optimal Filtering.* Englewood Cliffs, NJ: Prentice-Hall, 1979.
[11] T. Kailath, *Linear Systems.* Englewood Cliffs, NJ: Prentice-Hall, 1980.
[12] J. Van Amerongen and A. J. Udink Ten Cate, "Model reference adaptive autopilots for ships," *Automatica,* vol. 11, pp. 441–449, 1976.
[13] G. C. Goodwin and K. S. Sin, *Adaptive Filtering Prediction and Control.* Englewood Cliffs, NJ: Prentice-Hall, 1984.
[14] Proceedings of the 1984, 23rd IEEE-CDC, Las Vegas.
[15] H. P. Whitaker, J. Yamron, and A. Kezcr, "Design of model reference adaptive control systems for aircraft," Rep. R-164, Instrumentation Laboratory, Massachusetts, Institute of Technology, Cambridge, MA, 1958.
[16] P. C. Parks, "Liapunov redesign of model reference adaptive control systems," *IEEE Trans. Autom. Control,* vol. AC-11, pp. 362–367, 1966.
[17] M. M. Gupta and P. N. Nikiforuk, "On the design of an adaptive controller for uncertain plants via Liapunov signal synthesis method," *J. Cybern.,* vol. 9, pp. 73–98, 1979.
[18] M. M. Gupta, P. N. Nikiforuk, and H. Ohta, "Design of a two-level adaptive flight control system," *J. Guidance and Contr.,* vol. 2, no. 1, pp. 79–85.
[19] R. V. Monopoli, "Model reference adaptive control with an augmented error signal," *IEEE Trans. Autom. Control,* vol. AC-19, pp. 474–482, 1974.
[20] I. D. Landua, "A survey of model reference adaptive techniques—Theory and applications," *Automatica,* vol. 10, pp. 353–379, 1974.
[21] ——, *Adaptive Control—The Model Reference Approach.* New York: Marcel Dekker, 1979.
[22] V. Peterka, "Adaptive digitial relgulation of noisy systems," present at the 2nd Prague IFACA Symposium Identification and Parameter Estimation, 1970.
[23] K. J. Astrom, "Self-tuning regulators—Design principles and applications," in *Applications of Adaptive Control,* K. S. Narendra and R. V. Monopoli, Eds. New York: Academic Press, 1980.
[24] K. J. Astrom and B. Wittenmark, "Self-tuning controllers based on pole-zero placement," *IEE Proc.,* vol. 127, pt. D, no. 3, pp. 120–130, 1980.
[25] ——, "On self-tuning regulators," *Automatica,* vol. 9, pp. 195–199, 1973.
[26] ——, "Analysis of a self-tuning regulator for nonminimum phase systems," presented at the IFAC Symp. Stoch. Control, Budapest, 1974.
[27] K. J. Astrom, "Theory and applications of adaptive control—A survey," *Automatica,* vol. 19, no. 5, pp. 471–486, 1983.
[28] P. V. Osburn, "Investigation of a method of adaptive control," Sc. D. Thesis, Department of Aeronautics and Astronautics, M.I.T. Instrumentation Laboratory rep. T-266, 1961.
[29] K. E. Hagen, "Dynamic behavior of parameter adaptive control systems," Ph.D. Thesis, Department of Aeronautics and Astronautics M.I.T. Instrumentation Laboratory rep. T-368, 1964.
[30] M. Margolis and C. T. Leondes, "A parameter tracking servo for adaptive control systems," *IEEE Trans. Automat. Contr.,* PGAC 4, no. 2, pp. 110–111, 1959.
[31] B. H. Willis, "Effect of parameter variation upon model reference adaptive control system performance," M.I.T. Instrumentation Laboratory rep. T-305, 1962.
[32] H. P. Whitaker, "Model reference adaptive control systems for large flexible boosters," M.I.T. Instrumentation Laboratory rep. E-1036, 1960.
[33] J. T. Humphries, R. E. Uhrig, and A. P. Sage, "A model reference adaptive control systems for a nuclear rocket engine," in *Proc. Joint Automatic Control Conf.,* pp. 492–499, 1966.
[34] Y. J. Tsypkin, *Adaptation and Learning in Automatic Systems.* New York: Academic Press, 1971.
[35] L. Ljung and B. D. O. Anderson, "Adaptive control, Where are we?" *Automatica,* vol. 20, no. 5, pp. 499–500, 1984.

BOOKS IN THE FIELD OF ADAPTIVE CONTROL SYSTEMS

R. Bellman, *Adaptive Processes—A Guided Tour.* Princeton, NJ: Princeton University Press, 1961.

E. Mishkin and L. B. Braun, Eds., *Adaptive Control Systems*. New York: McGraw-Hill, 1961.

D. Wilde, *Optimum Seeking Methods*. Englewood Cliffs, NJ: Prentice-Hall, 1964.

A. A. Feldbaum, *Optimal Control System*. New York: Academic Press, 1965.

D. Sworder, *Optimal Adaptive Control Systems*. New York: Academic Press, 1966.

W. Eveleigh, *Adaptive Control and Optimization Techniques*. New York: McGraw-Hill, 1967.

Y. Sawaragi, Y. Sunahara, and T. Nakamizo, *Statistical Decision Theory in Adaptive Control Systems*. New York: Academic Press, 1967.

S. L. Yakowitz, *Mathematics of Adaptive Control Processes*. New York: American Elsevier, 1969.

J. M. Mendel and K. S. Fu, *Adaptive, Learning, and Pattern Recognition Systems*. New York: Academic Press, 1970.

Y. Z. Tsypkin, *Adaptation and Learning in Automatic Systems*. New York: Academic Press, 1971.

——, *Foundations of the Theory of Learning Systems*. New York: Academic Press, 1973.

G. N. Saridis, *Self-Organizing Control of Stochastic Systems*. New York: Marcel Dekker, 1977.

I. D. Landua, *Adaptive Control—The Model Reference Approach*. New York: Marcel Dekker, 1979.

K. S. Narendra and R. V. Monopoli, Eds., *Application of Adaptive Control*. New York: Academic Press, 1980.

H. Unbehauen, Ed., *Methods and Applications in Adaptive Control*. Berlin: Springer-Verlag, 1980.

C. J. Harris and S. A. Billings, *Self-Tuning and Adaptive Control*. Theory and Applications, Stevenage: Peregrinus, 1981.

G. C. Goodwin and K. S. Sin, *Adaptive Filtering Prediction and Control*. Englewood Cliffs, NJ: Prentice Hall, 1984.

B. Widrow and S. D. Stearms, *Adaptive Signal Processing*. Englewood Cliffs, NJ: Prentice-Hall, 1985.

Part I
Adaptive Control: General Introduction

ONE of the fundamental tenets of modern control theory is that a plant cannot be well controlled until it can be modeled in quantitative terms, and its states are observed or reconstructed in a deterministic or stochastic sense. This premise has resulted in such notions as identification, parameter and state estimation, filtering and observers. One of the fundamental implicit concerns in the field of control theory the notion of "learning and self-adaptive control." Using some of these concepts and notions the field of adaptive control has evolved over the past three decades from some primitive hill-climbing methods to advanced adaptive control techniques.

Today adaptive control, though still in early stages of development, has potential applications for control of almost all technological systems in which the inherent complexities associated with explicit identification and parameter estimation do not permit the applications of conventional control tools.

In Part I of this volume, we give an overview of adaptive control system design methods that define basic adaptive control problems and some inherent complexities associated with such problems. We also give a list of important bibliographical material available in the field. This part contains a series of nine important articles written, from 1958 to April 1985, by some well-known researchers in the field. Some of the articles provide basic notions and motivations, while other give important surveys.

In the first paper in this part published in 1958, R. E. Kalman postulate the design of a machine which adjusts itself automatically to control an arbitrary dynamic process. Almost two and half decades ago, when "minis-" and "micros-" were not even dreamed of, he discussed some engineering features of a "small computer" which can control an arbitrary process in a stand-alone adaptive mode. He also discussed some possible applications to aircraft and chemical processes, where such "self-optimizing" controllers can be used. When we examine Kalman's optimization method, which was based upon a simple theory and a primitive machine (computer), and today's advances in the adaptive control and computer fields, we have to admit that a great deal has been achieved.

An unconventional but natural design approach to adaptive feedback control is presented by K. DeJong in the second paper entitled "Adaptive System Design: A Genetic Approach." In this approach the author used reproductive plans in which he attempts to provide a solution to adaptive control problems by dynamically accumulating information about the problem at hand and using this information to generate an acceptable solution. The author presents an example to provide an intuitive idea of how reproductive plans operate, and makes some observations concerning the applicability of this approach to various problems in adaptive system design for complex processes. Reproductive plans should be viewed as providing a powerful albeit unconventional tools for thc adaptive system designer, giving him the potential for solving highly complex industrial problems beyond the reach of existing techniques. We have selected this article for this volume with a hope that the recent advances in artificial intelligence, micro-electronics and computing power will provide an additional impetus to reorganize our thinking process in the design of adaptive control algorithms in a most natural way. Perhaps these concepts may lead to the new notion of "intelligent adaptive control."

Karl J. Astrom's survey is entitled "Theory and Applications of Adaptive Control: A Survey," in the third paper. This 1983 paper is an expanded version of the paper originally presented at the 1981 IFAC World Congress—Kyoto at a plenary session. In part, it indicates that when designed with adequate practical precautions, present adaptive techniques may be used successfully in a variety of applications even though many theoretical problems remain to be solved. This survey paper discusses some important issues and practical problems.

In the fourth paper, R. Isermann presents a tutorial on parameter adaptive control algorithms. He discusses theoretical investigations in the field and gives practical experiences. He gives an introduction to adaptive control algorithms with recursive parameter estimation; some of these techniques have obtained increasing attention in recent years, especially in the design of adaptive controllers. Both deterministic and stochastic cases have been dealt with. Though the recursive parameter estimation problem are very attractive theoretically, they present an inherent problem of stability, convergence, and performance when used in an adaptive feedback mode. Also, they create a problem when such algorithms are being employed for fast processes with time constants in the range of milliseconds. The author of this tutorial gives some indication of how to get around some of these problems.

In the fifth paper, our Soviet colleagues, A. A. Voronov and V. Yu. Rutkovsky, present the state-of-the-art as well as prospects of adaptive systems derived from the work of Soviet researchers. Most of the survey papers appearing in the West have neglected, to some extent, to report the Soviet research, but this paper recently appeared in *Automatica* and it provides some interesting insights into some adaptive algorithms, including optimal adaptive systems developed in the Soviet Union. In this paper, the authors clearly indicate that in spite of many theories in the area of adaptive control, there is a great need for unified terminology, design of better adaptation algorithms, and new computational tools. As stated in this paper, many adaptation algorithms have been synthesized under the condition of convergence with $t \to \infty$, others are too complicated and require a complete state vector for feedback control. The authors emphasize the use of adaptive methods for practical problems, and rightly point out the difficulties associated with the presently available methodologies.

Rouhani and Mehra present some basic theoretical properties of model algorithmic control (MAC) in the sixth paper. The time-domain and the frequency-domain analysis of model algorithmic control explains most of its heuristically observed properties, in particular, with respect to stability, robustness, and nonlinear inputs both for deterministic and stochastic environments. The authors emphasize the use of algorithmic-adaptive type of control with an emphasis on fast digital computer, with fast memory access and substantial information storage. This paper along with the second paper by DeJong illustrate some of the future directions in the area of adaptive feedback control methodologies.

The paper entitled "A New Solution to Adaptive Control" (the seventh paper) by J. M. Martin-Sanchez suggests a new method for linear time-varying multivariable processes with unknown parameters. The principle characteristics of this new proposed adaptive control as reported are:

i) Knowledge of the plant's parameters or its variation with time is not required. For practical applications, no *a priori* data are required.
ii) The system only uses the information given by plant inputs and outputs and does not require an estimation of the plant state variables.
iii) The adaptive mechanism generates matrices $A(k)$ and $B(k)$ by means of simple arithmetic operations and the calculation of the control signal is also extremely simple.
iv) When the control system is in its equilibrium point, the process output is equal to the desired output. In this way, the main control problem becomes the design of the desired output, in accord with the limitations of the plant and the desired specifications.
v) We can always modify the desired output, in real time, thus making it possible to take care of severe perturbations that may act on the system.

As reported by the author, the main advantages of this method over MRAS are:

a) no reference model is needed;
b) a priori knowledge of the plant parameters is not required;
c) it can work with structural difference;
d) it uses only I/O data, and for this reason no state estimation or adaptive observer is required.

Martin-Sanchez's proposal seems to be sound both for deterministic and stochastic systems and may be applied to many practical situations. Since the publication of this paper, Martin-Sanchez has published quite extensively using this basic philosophy, and the researchers will find this philosophy quite useful in their works.

In the eighth paper of this part, we present a paper by our Soviet colleague, Ya. Z. Tsypkin, entitled "Optimality in Discrete Adaptive Systems" presented at the 1984 IFAC Congress, Budapest. The author discusses adaptive algorithms for discrete adaptation involving identification algorithm for direct adaptive systems and prediction, and adaptation for indirect adaptive systems. This paper, and the fifth paper, are included to give some insight into the methodologies adopted in the USSR.

In the last article (ninth paper) of this part, we reproduce a recently published paper entitled "Global Stability of Adaptive Pole Placement Algorithms" by H. Elliott, R. Gristi, and M. Das. In this paper, the authors present direct and indirect control schemes for assigning the closed-loop poles of a single-input, single-output systems in both the continuous and discrete-time cases. The authors prove, in the presence of persistent excitation, the global stability for an arbitrary assignment of closed-loop poles. The reported results are valid for both minimum and nonminimum phase type of systems. Many researchers have investigated the adaptive pole-zero placement problem, and hopefully this paper along with the references cited in this paper will provide a deeper insight into the pole-placement problem to the readers. Also, see the twentieth paper on self-tuning regulators based on pole-zero placement.

This part of the volume should stimulate the interest of readers just by exploring certain ideas and philosophy in the field of adaptive control. The additional selected bibliography should help the reader to explore the areas further. In the following parts, we will treat some of these approaches in more detail.

Selected Bibliography

J. Adams, "Understanding adaptive control," *Automation,* vol. 17, pp. 108–113, 1970.

B. D. O. Anderson and R. R. Bitmead, "Stability theorems for the relaxation of the strictly positive real condition in hyberstable adaptive schemes," in *Proc. of the 23rd IEEE Conference on Decision and Control,* pp. 1284–1291, 1984.

R. B. Asher, I. I. Andrisani, and P. Dorato, "Biography on adaptive control systems," *Proc. IEEE,* vol. 64, no. 8, pp. 1226–1240, 1976.

K. J. Astrom and P. Eykhoff, "System identification—A survey," *Automatica,* no. 7, pp. 123–162, 1971.

K. J. Astrom, "Theory and applications of adaptive control," IFAC Congress, Kyoto, Japan, 1981.

K. G. Astrom, "Interactions between excitation and unmodeled dynamics in adaptive control," in *Proc. 23rd Conference on Decision and Control,* pp. 1276–1281, 1984.

M. Bodson and S. S. Sastry, "Exponential convergence and robustness margins in adaptive control," in *Proc. of the 23rd IEEE Conference on Decision and Control,* pp. 1282–1285, 1984.

R. E. Bowles, "Self-adaptive control system," U.K. Patent I 242 286, Oct. 14, 1968, Publ. Sept. 17, 1971.

R. L. Carroll and D. P. Lindorff, "An adaptive observer for single-input, single-output linear systems," *IEEE Trans. Automat. Contr.,* vol. AC-18, pp. 428–435, 1973.

R. L. Carroll, "New adaptive algorithms in Lyapunov synthesis," in *Proc. IEEE Conf. Dec. and Contr.,* pp. 282–287, 1974.

——, "New adaptive algorithms in Lyapunov synthesis," *IEEE Trans. Automat. Contr.,* vol. AC-21, pp. 246–249, 1976.

B. L. Deekshatulir and H. V. Gururaja, "Critically damped adaptive system," *Int. J. Contr.,* vol. 8, no. 1, pp. 23–32, 1968.

J. L. Douce, K. C. Ng, and M. M. Gupta, "Dynamics of the parameter perturbation process," *Proc. IEE,* vol. 113, no. 6, pp. 1077–1083, 1966.

B. Egardt, "Stability of adaptive controllers," in *Lecture Notes in Control and Information Sciences.* Berlin: Springer-Verlag, 1979.

V. W. Eveleigh and N. R. Powell, "Adaptive control of a reflective satellite communication system," presented at *Proc. IEEE 7th Symp. Adapt.*

Proc., Dec. and Contr., 1968.

A. A. Feldbaum, "Dual control theory, I-IV," *Automat. Remote Contr.,* vol. 21, pp. 874–880, pp. 1033–1039, 1961; vol. 22, pp. 1–2, pp. 109–121, 1962.

A. Feuer, B. R. Barmish, and A. S. Morse, "An unstable dynamical system associated with model adaptive control," *IEEE Trans. Automat. Contr.,* vol. AC-23, no. 3, pp. 499–500, 1978.

K. S. Fu, "Adaptive models of human controllers," in *Proc. IEEE Int. Conv. Dig.,* pp. 50–51, 1971.

G. C. Goodwin, P. J. Ramadge, and P. E. Caines, "Discrete time multivariable adaptive control," *IEEE Trans. Automat. Contr.,* vol. AC-25, no. 3, pp. 449–456, 1980.

P. C. Gregory, Ed., "Proceedings of the self adaptive flight control systems symposium," WADC Tech. Rep. 59-49, Wright Air Development Center, Wright-Patterson Air Force Base, Ohio, 1959.

M. P. Groover, "A new look at adaptive control," *Automation,* vol. 20, no. 4, pp. 60–63, 1973.

M. M. Gupta, "On Van-der Grinten technique of optimization," in *Advances of Automatic Controls, Proc. Inst. Mech. Eng.,* pp. 307–308, 1965.

——, "Dynamic sensitivity to step perturbation in linear system parameters," in *Proc. 2nd IFAC Symp. on Syst. Sens. and Adapt.,* Dubrovnik, Yugoslavia, pp. B.70–B.82, 1968.

——, "Simultaneous generation of the first and higher order gradient functions of a cost function," *IEEE Trans. Automat. Contr.,* vol. AC-13, no. 1, pp. 115–116, 1968.

——, "On the characteristics of the parameter perturbation process dynamics," *IEEE Trans. on Automat. Contr.,* vol. AC-14, no. 5, pp. 440–442, 1969.

M. M. Gupta and P. N. Nikiforuk, "Some further contributions to the dynamic sensitivity of the parameter perturbation processes," *Trans. ASME, J. of Dyn. Syst., Meas. and Contr.,* vol. 93, Series G, no. 3, pp. 173–179, 1971.

——, "On the design of an adaptive controller for uncertain plants via Lyapunov signal synthesis method," *J. Cybern.,* vol. 9, pp. 73–98, 1979.

M. M. Gupta, P. N. Nikiforuk, and Y. Yamane, "An indirect approach to adaptive control," *Proc. 1980 JACC,* pp. FP-2-F1 to FP-2-F10, 1980.

K. Harris and R. D. Bell, "Model-algorithmic adaptive control for nonlinear systems," in *Proc. 23rd IEEE Conf. on Decision and Control,* pp. 681–682, 1984.

I. M. Horowitz, "Plant adaptive vs. ordinary feedback systems," *IRE Trans. Automat. Contr.,* vol. AC-17, no. 1, pp. 48–56, 1962.

I. M. Horowitz, J. W. Smay, and A. Shapiro, "A synthesis theory for the externally excited adaptive systems," *IEEE Trans. Automat. Contr.,* vol. AC-19, no. 2, pp. 101–107, 1974.

A. G. Ivakhenenko, "On constructing an extremum controller without hunting oscillations," *IEEE Trans. Automat. Contr.,* vol. AC-12, no. 2, pp. 144–153, 1967.

H. G. Jacob, "Hill-climbing controller for plants with any dynamics and rapid drifts,"*Int. J. Contr.,* vol. 16, no. 1, pp. 17–35, 1972.

O. L. R. Jacobs and P. Saratchandran, "Comparison of adaptive controllers," *Automatica,* vol. 16, no. 1, pp. 89–97, 1980.

R. E. Kalman, "Design of a self-optimizing control system," *Trans. ASME,* vol. 80, pp. 468–478, 1958.

——, "On the general theory of control," in *Proc. 1st IFAC Congr.,* pp. 481–492, 1960.

R. E. Kalman and J. E. Bertram, "Control system analysis and design via the second method of Lyapunov: I. continuous time systems," *J. Basic Eng.,* vol. 82, ser. D, no. 2, 1960.

——, "Control system analysis and design via the second method of Lyapunov: II. discrete time systems," *Trans. ASME: J. Basic Eng.,* vol. 82, ser. D, pp. 394–399, 1960.

R. E. Kalman, "When is a linear control system optimal?" *J. Basic Eng., Trans. ASME,* vol. 86, pp. 51–60, 1964.

R. L. Kosut and B. D. O. Anderson, "Robust adaptive control: Conditions for local stability," in *Proc. 23rd IEEE Conf. on Decision and Control,* pp. 1002–1008, 1984.

A. S. Krasovskiy, "Optimality and adaptivity of control systems," *J. Cybern.,* vol. 3, no. 3, pp. 74–32, 1973.

G. Kreisselmeier, "Adaptive observers with experimental rate of convergence," *IEEE Trans. Automat. Contr.,* vol. AC-25, no. 4, pp. 717–722, 1977.

——, "Adaptive control via adaptive observation and asymptotic feedback matrix synthesis," *IEEE Trans. Automat. Contr.,* vol. AC-25, no. 4, *IEEE Trans. Automat. Contr.,* vol. AC-25, 1980.

I. N. Krutov, "Synthesis of searchless self-adjusting systems on the basis of the root-locus method II," *Automat. and Remote Contr.,* vol. 33, no. 12, pt. 1, pp. 1973–1981.

——, "Synthesis of searchless self-adjusting systems based on the root-locus method I," *Automat. and Remote Contr.,* vol. 33, no. 1, pt. 1, pp. 1641–1656, 1973.

R. Ku and M. Athans, "On the adaptive control of linear systems using the one-loop feedback optimal approach," *IEEE Trans. Automat. Contr.,* vol. AC-18, pp. 489–493, 1973.

P. Kudva and K. S. Narendra, "Synthesis of an adaptive observer using Lyapunov's direct method," *Int. J. Contr.,* vol. 18, pp. 1201–1210, 1973.

P. Kuduva and V. Gourishankar, "On the stability problem of multivariable model-following systems," *Int. J. Contr.,* vol. 24, no. 6, pp. 801–805, 1976.

T. Kurokawa and H. Tamura, "A self-organizing time-optimal controller," *Int. J. Contr.,* vol. 16, no. 2, pp. 225–241, 1972.

D. G. Lainiotis, "Partitioning: A unifying framework for adaptive systems I: Estimation," *Proc. IEEE,* vol. 64, no. 8, pp. 1126–1142, 1976.

——, "Partitioning: A unifying framework of adaptive systems II: Control," *Proc. IEEE,* vol. 64, no. 8, pp. 1182–1197, 1976.

I. D. Landau, "A survey of model reference adaptive techniques—Theory and applications," *Automatica,* vol. 10, pp. 353–379, 1974.

——, *Adaptive Control—The Model Reference Approach.* New York: Marcel Dekker, 1979.

——, "Combining model reference adaptive controllers and stochastic self-tuning regulators," presented at IFAC Congress, Kyoto, Japan, 1981.

W. E. Larimore, S. Mahmood, and R. K. Mehra, "Multivariable adaptive model algorithimic control," *Proc. of the 23rd IEEE Conference on Decision and Control,* pp. 675–680, 1984.

L. Ljung, "Strong convergence of a stochastic approximation algorithm," *Ann. Stat.,* vol. 6, no. 3, pp. 680–696, 1978.

L. Ljung and P. Caines, "Asymptotic normality of prediction error estimators for approximate system models," *Automatica,* vol. 3, pp. 29–46, 1979.

R. K. Mehra, "Approaches to adaptive filtering," *IEEE Trans. Automat. Contr.,* vol. AC-17, pp. 693–698, 1972.

J. Merkel, "An adaptive modeling technique using a class of Lyapunov functions," *Proc. 1972 Int. Conf. Cybern. and Soc.,* vol. 358–359, 1972.

N. Minamide, M. M. Gupta, and P. N. Nikiforuk, "Design of adaptive tracking systems for plants of unknown order," presented at *IFAC Workshop on Adaptive Systems in Control and Signal Processing,* 1983.

E. Mishkin and L. B. Braun, Eds. *Adaptive Control Systems.* New York: McGraw-Hill, 1961.

A. S. Morse, "Global stability of parameter—Adaptive control systems," *IEEE Trans. Automat. Contr.,* vol. AC-25, no. 3, pp. 433–439, 1980.

——, "New directions in parameter adaptive control," in *Proc. of the 23rd IEEE Conf. on Decision and Control,* pp. 1566–1568, 1984.

K. S. Narendra and J. H. Taylor, *Frequency Domain Criteria for Absolute Stability.* New York and London: Academic Press, 1973.

K. S. Narendra and R. V. Monopoli, Ed., *Applications of Adaptive Control.* New York and London: Academic Press, 1980.

P. N. Nikiforuk and M. M. Gupta, "A bibliography on the properties, generations, and control system applications of shift-register sequences," *Inter. J. Contr.,* vol. 9, no. 2, pp. 217–234, 1969.

P. N. Nikiforuk, M. M. Gupta, and H. H. Choe, "Control of unknown dynamic systems in reduced state space," *Proc. 1969 JACC,* pp. 15–22, 1969.

——, "Control of unknown dynamic systems in reduced state space," *IEEE Trans. Automat. Contr.,* vol. AC-14, no. 5, pp. 489–496, 1969.

P. N. Nikiforuk, M. M. Gupta, and K. Kanai, "Optimal control synthesis with respect to desired response characteristics and available states," *Proc. 1972 JACC,* pp. 1–9, 1972.

P. N. Nikiforuk, M. M. Gupta, K. Kania, and L. Wagner, "A deterministic type controller for a class of uncertain plants with partial identification," in *Proc. 3rd IFAC Symp. Ident. and Syst. Parameter Est.,* 1973.

P. N. Nikiforuk, M. M. Gupta, and K. Tamura, "A minimum tume adaptive observer for linear discrete systems," in *Proc. IFAC Symp. on Multivariable Syst.,* (MVTS), Fredericton, Canada, pp. 197–203, 1977.

P. N. Nikiforuk, N. Minamide, and M. M. Gupta, "Design of a continuous-time adaptive regulator," International Federation of Automatic Control, 8th Triennial World Congress, pp. VII.41–VII.46, 1983.

H. Ohta, M. M. Gupta, and P. N. Nikiforuk, "Development of equiobservable forms with applications to adaptive control," IX Triennial World Congress, International Federation of Automatic Control, vol. VIII, pp. 125–129, 1984.

——, "On the development of equiobservable forms for linear multivariable systems," *IEEE Trans. Automat. Contr.,* vol. AC-29, no. 8, pp.

730–733, 1984.
D. Orlicki, L. Valavani, M. Athans, and G. Stein, "Robust adaptive control with variable dead zones," in *Proc. 23rd IEEE Conf. Decision and Control,* 1984.
I. S. Parry and G. H. Houpis, "A parameter identification self-adaptive control system," *IEEE Trans. Automat. Contr.,* vol. AC-15, no. 4, pp. 462–468, 1970.
A. K. Petrov, "Determination of an optimal self-adaptive algorithm," *Eng. Cybern.,* vol. 9, no. 1, pp. 185–197, 1971.
V. M. Popov, "Absolute stability of nonlinear systems of automatic control," *Automat. Remote Contr.,* vol. 22, no. 8, 1962.
——, *Hyperstability of Automatic Control Systems.* New York: Springer-Verlag, 1973.
L. Praly, "Robust model reference adaptive controllers: Part 1: Stability analysis," in *Proc. 23rd Conf. Decision and Control,* 1984.
B. Riedle and P. Kolotivic, "A stability-instability boundary for disturbance-free slow adaptation and unmodeled dynamics," *Proc. 23rd Conf. on Decision and Control,* 1984.
A. P. Sage and B. R. Eisenberg, "Suboptimal adaptive control of a nonlinear plant," *IEEE Trans. Automat. Contr.,* vol. AC-11, no. 3, pp. 621–623, 1966.
G. N. Saridis, *Self-Organizing Control of Stochastic Systems.* New York: Marcel Dekker, 1977.
V. A. Serdyukov, "Synthesis of adaptive systems by Lyapunov's direct method," *Automat. and Remote Contr.,* vol. 33, no. 7, pp. 1138–1144, 1972.
N. K. Sinha, "Adaptive control with incomplete identification," *Proc. 3rd IFAC Symp. Sens., Adapt. and Opt.,* pp. 343–348, 1973.
N. K. Sinha, M. M. Gupta, and P. N. Nikiforuk, "Recent advances in adaptive control," *J. Cybern.,* vol. 6, no. 2, pp. 79–100, 1977.
J. Sternby, "Extremum control systems—An area for adaptive control?" in *Proc. 1981 JACC,* 1981.
R. L. Stratonovisch, "Does the synthesis theory of optimal adaptive self-teaching and self-adjusting systems exist?" *Automat. and Remote Contr.,* no. 1, pp. 83–92, 1968.
Ya. Z. Tsypkin, "Adaptation, training and self-organization in automatic systems," *Automat. and Remote Contr.,* vol. 27, pp. 16–51, 1966.
——, "Adaptation, learning and self-learning in control systems," presented at *Proc. 3rd IFAC Congr.,* 1966.
——, "Does the synthesis theory of optimal adaptive systems still exist?" *Automat. and Remote Contr.,* no. 1, pp. 93–98, 1968.
——, "Self-learning—What is it?" *IEEE Trans. Automat. Contr.,* vol. AC-13, no. 6, pp. 608–612, 1968.
——, *Adaptation and Learning in Automatic Systems.* New York and London: Academic Press, 1971.
——, "Dynamic adaptation algorithms," *Automat. and Remote Contr.,* vol. 33, no. 1, pt. 1, pp. 59–67, 1972.
——, "Adaptive system theory today and tomorrow," in *Proc. 3rd IFAC Symp. Sens., Adapt., and Opt.,* pp. 47–67, 1973.
——, *Foundations of the Theory of Learning Systems.* New York and London: Academic Press, 1973.
H. Unbehauen, Ed., *Methods and Applications in Adaptive Control.* Berlin: Springer-Verlag, 1980.
D. Williamson, "Observation of bilinear systems with application to biological control," *Automatica,* vol. 13, pp. 243–254, 1977.
Y. Yamane, M. M. Gupta, and P. N. Nikiforuk, "An indirect adaptive control for linear unknown discrete systems of unknown order," presented at Proc. 9th SICE Symp. Contr. Theory, 1980.
Y. Yamane, P. N. Nikiforuk, and M. M. Gupta, "Deadbeat adaptive control in feedback," presented at IFAC Workshop on Adaptive Systems in Control and Signal Processing, 1983.
P. C. Young, "Process parameter estimation and self adaptive control," in *Theory of Self-Adapting Control System,* P. H. Hammond, Ed. New York: Plenum Press, 1966.
V. I. Zudov, *Methods of A. M., Lyapunov and Their Applications.* Groningen, Netherlands: Noordhoff, 1964.

Design of a Self-Optimizing Control System

By R. E. KALMAN,[1] NEW YORK, N. Y.

This paper examines the problem of building a machine which adjusts itself automatically to control an arbitrary dynamic process. The design of a small computer which acts as such a machine is presented in detail. A complete set of equations describing the machine is derived and listed; engineering features of the computer are discussed briefly. This machine represents a new concept in the development of automatic control systems. It should find widespread application in the automation of complex systems such as aircraft or chemical processes, where present methods would be too expensive or time-consuming to apply.

Introduction

The art of the design of systems for the automatic control of dynamic processes of many different kinds (such as airplanes, chemical plants, military-weapon systems, and so on) has been reduced gradually to standard engineering practice during the years following World War II. In the simplest possible setting, the problem that the engineer faces in designing such automatic control systems is shown in Fig. 1. It is desired that the output of the process $c(t)$, which may be position, speed, temperature, pressure, flow rate, or the like, be as close as possible at all times to an arbitrarily given input $r(t)$ to the system. In other words, at all instants of time it is desired to keep the error $e(t) = r(t) - c(t)$ as small as possible. Control is accomplished by varying some physical quantity $m(t)$, called the control effort, which affects the output of the process.

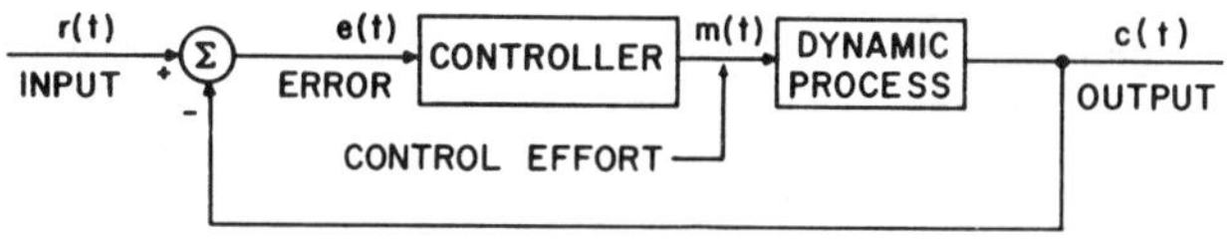

Fig. 1 Block Diagram of Simplest Control Problem

As long as the deviations from an equilibrium value of $r(t)$, $c(t)$, and therefore of $e(t)$ and $m(t)$, are small, the system can be regarded as approximately linear and there is a wealth of theoretical as well as practical information on which engineering design may be based. (When the system is not linear, present-day knowledge supplies only fragmentary suggestions for design; however, nonlinear effects are frequently of secondary importance.) It is generally agreed that the design of high-performance control systems is essentially a problem of matching the dynamic characteristics of a process by those of the controller. Practically speaking, this means that if the dynamic characteristics of the process are known with sufficient accuracy, then the characteristics of a controller necessary to give a certain desired type of performance can be specified. Usually, this amounts to writing down in quantitative terms the differential equations of the controller. Thus the design procedure can be divided roughly into the following distinct stages:

I Measure the dynamic characteristics of the process.
II Specify the desired characteristics of the controller.
III Put together a controller using standard elements (amplifiers, integrators, summers, electric networks, and so on) which has the required dynamic characteristics.

This subdivision of effort in designing a control system is oversimplified, but it will be a convenient starting point for the following discussion.

It has been pointed out by Bergen and Ragazzini (1)[2] that if a high degree of flexibility is desired in design stage (III), it is advantageous to use a sampled-data system. In principle, a sampled-data system is one where the controller is a digital computer. It is probably no exaggeration to say that, because of the great inherent flexibility of a digital computer, *any* desired controller characteristic is practically realizable. The use of a digital computer for the controller reduces stage (III) to a straightforward operation, like that of transcribing a handwritten manuscript by means of a typewriter.

Since the theory of linear control systems is well developed, stages (I–II) also can be made to consist of more-or-less standard procedures. Quick and convenient design even in stage (III) demands or at least suggests a digital computer; so the question arises whether or not stages (I–II) also can be reduced to completely mechanical operations which can be performed by a digital computer. Accordingly, the problem considered in this paper can be stated as follows:

To design a machine which, when inserted in the place of the controller in Fig. 1, will automatically perform steps (I–III), and set itself up as a controller which is optimum in some sense. The design of this machine is to be based on broad principles only. Its operation should require no direct human intervention but merely the measurements of $r(t)$ and $c(t)$.

In other words, such a machine, if it can be built, eliminates the lengthy, tedious, and costly procedure of engineering design—it is only necessary to connect the machine to *any* process. Thus the machine would seemingly eliminate the need for the control-systems engineer, but the latter can be reassured by the fact that the design of the machine itself is a far more ambitious and challenging undertaking than that of conventional control systems.

An even more decisive advantage of the machine over present-day design procedures is the following: In carrying out steps (I–III) it is generally taken for granted that the dynamic characteristics of the process will change only slightly under any operating conditions encountered during the lifetime of the control system. Such *slight* changes are foreseen and are usually counteracted by using feedback. Should the changes become large, the control equipment as originally designed may fail to meet performance specifications. Instances where difficulties of this type are encountered are:

(*a*) Changes of aircraft characteristics with speed.
(*b*) Chemical processes.
(*c*) Any large-scale control operation, where the nature of the system can be affected by uncontrolled and unforeseen factors.

By contrast, the machine can *repeat* steps (I–III) continually and thereby detect and make corrections in accordance with any

[1] Department of Electrical Engineering and Electronics Research Laboratories, Columbia University; formerly, Engineering Research Laboratory, E. I. du Pont de Nemours & Company, Wilmington, Del.

Presented at the Instruments and Regulators Division Conference, Evanston, Ill., April 8–10, 1957, of The American Society of Mechanical Engineers.

Note: Statements and opinions advanced in papers are to be understood as individual expressions of their authors and not those of the Society. Manuscript received at ASME Headquarters, January 14, 1957. Paper No. 57—IRD-12.

[2] Numbers in parentheses refer to the References at the end of the paper.

Reprinted with permission from *Trans. ASME*, vol. 80, pp. 468–478, Feb. 1958.

changes in the dynamic characteristics of a process which it controls. Such a control system operates always at or near some "optimum," provided only that changes in the dynamic characteristics of the controlled process do not occur very abruptly. It may be said that the machine adapts itself to changes in its surroundings—this may be regarded as an extension of the principle of feedback. The author prefers to call this property of the machine "self-optimization." The word "ultrastability" has been suggested also in a similar context by Ashby (2).

In the stated degree of generality, the problem is certainly not at a stage at present where any clear-cut ("unique") solution can be expected. Therefore this paper does not treat the general problem but presents a specific approach which leads to a practically satisfactory solution. This point is of considerable interest, since some earlier speculations relating to the problem were mostly of theoretical nature, without an attempt to appraise the difficulties (cost, complexity, and so on) of practical implementation (2–5). A machine based on the principles discussed in what follows actually has been built and will be described briefly in a later section.

It should be emphasized that the machine has been designed from a practical engineering point of view, rather than deduced from some law of physics or mathematics. The various single elements in the design of the machine are based on known principles. The choice between alternate possibilities in each stage of the design has been guided by efficiency and cost considerations. It is claimed that the over-all design uniting these principles in one machine is new and represents a major advance in regard to practicality over suggestions contained in the current literature.

General Design Considerations

From the technological point of view, it is clear that the machine discussed in the preceding section must be a computer. There are two possible choices, analog or digital computer. The latter choice is preferable. The reason is this. An analog computer is basically a method of simulating simple dynamic processes as they occur in the physical universe. The machine in question is required to simulate the actions of man, not of nature. This requires much greater flexibility and at the present state of computer technology such flexibility is provided only by digital computers.

The words "digital" and "analog" used here refer to the *external* characteristics of computers. Mathematically speaking, an analog computer performs the operations of analysis, such as differentiation, integration, computing logarithms, and so on, while a digital computer performs only arithmetic operations; namely, addition and multiplication. An analog computer operates on continuous functions (of time), the digital computer deals with discrete numbers. As far as the *internal* construction of these machines is concerned, it may happen that a computer which is called analog by its user contains discrete components (such as very fast counting circuits); and a computer which is called digital by its user may contain continuous components (such as potentiometers). Following these remarks, the computer that is described later may be called externally digital, internally analog.

In a digital computer, mathematical operations must be expressed (using approximations of various types) in numerical form. For instance, a function such as e^x must be computed by means of a series, which involves only repeated addition and multiplication. Another example is measuring the dynamic characteristics (transfer function or impulse response) of a process. Mathematically, this leads to the problem of solving an integral equation for which no satisfactory analog computing technique exists at present. On a digital computer the problem reduces to solving a set of simultaneous algebraic equations which is much simpler than solving an integral equation.

These considerations suggest the first fundamental design requirement:

(*A*) *The machine must be a digital computer.*

Recall now that the machine has a twofold job; namely, design and control. (i) It must measure the dynamic characteristics of the process and then determine the best form of the controller. (ii) It must control the process by providing the required control action $m(t)$. It is naturally desirable to keep these distinct functions independent. Therefore:

(*B*) *The operations necessary for designing a suitable controller must not be allowed to interact with the control action itself.*

It will be seen later that this requirement cannot be satisfied completely; the degree to which it must be relaxed to provide satisfactory operation is one of the unanswered questions at present.

Special Design Considerations

There are several practical requirements, all quite self-evident, which must be satisfied if the machine is to fulfill the expectations presented in the Introduction. All of these are related to design problem (I).

The functioning of the machine must not be critically dependent on obtaining measurements with high accuracy. Determination of the dynamic characteristics of the process is based on knowledge of $m(t)$ and $c(t)$. Since the first of these is actually produced by the machine itself, it may be assumed to be known with arbitrary accuracy; $c(t)$, however, corresponds to some physical quantity such as temperature, flow, and so on, whose determination is always accompanied by errors due to the imperfect operation of measuring equipment. These errors are called *measurement noise*. The standard method of reducing measurement noise is to take a large number of measurements. This leads to the requirement:

(*C*) *The determination of the dynamic characteristics of the process must be based on a large number of measurements so as to minimize the effects of measurement noise.*

As pointed out in the Introduction, one of the potential advantages of such a machine is that it can constantly repeat the entire design procedure and thereby adjust itself in a manner corresponding to any changes in process characteristics. But because of requirement (*C*), the determination of process characteristics requires a large number of measurements, taking a (possibly) long period of time. Since the system characteristics at the end of a series of measurements may be appreciably different from what they were at the beginning of the series of measurements, it is clear that older measurements ("obsolete data") should not be regarded as being as good as more recent measurements. This may be stated as:

(*D*) *Among any two measurements of $c(t)$, the more recent one should be given the higher weight: Measurements of $c(t)$ made infinitely long ago should be given zero weight.*

The cost, size, probability of breakdown, and so on of the machine is roughly proportional to the number of computations it has to perform per unit time. Therefore other things being equal, the number of computations should be as small as possible:

(*E*) *The methods of numerical computation to be used in the machine should be highly efficient.*

This last requirement will make it possible also to choose between alternative methods of computation.

Computation of Transfer Function From Measurements

Sampling. We now examine in detail the problem of measuring the dynamic characteristics of the process to be controlled. To do this, the functions $m(t)$ and $c(t)$ must be known. Since, according to requirement (*A*), the machine is to be a digital computer, it is necessary to replace $m(t)$ and $c(t)$, which are continuously varying functions of time, by sequences of numbers which are discretely varying functions of time. This process is known as *sampling*. The most common way of doing this is to perform measurements periodically. Let the sampling instants be $t = kT$, $k = 0, 1, 2, \ldots$, where T is called the sampling period. Then sampling replaces $m(t)$ and $c(t)$ by the sequences of numbers

$$\begin{matrix} m(0), & m(T), & m(2T), \ldots, & m(kT), \ldots \\ c(0), & c(T), & c(2T), \ldots, & c(kT), \ldots \end{matrix} \qquad k = 0, 1, \ldots \ldots [1]$$

In order to simplify the notation, we frequently will write $m_k = m(kT)$ and $c_k = c(kT)$ from now on. As a result of the sampling process, all experimental information about the functions $m(t)$ and $c(t)$ is contained in the numbers [1]. The sampling process is illustrated in Fig. 2.

The theory of linear control systems in which some of the controlled quantities are subject to sampling (the so-called sampled-data systems) is well developed. For further information, see Ragazzini and Zadeh (6) and Truxal (7).

FIG. 2 SAMPLING PROCESS

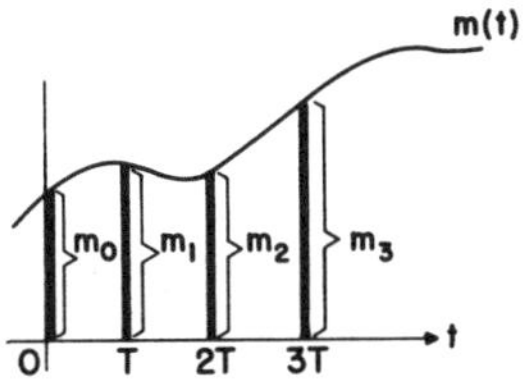

Step Response of the Process. If the process is *linear, time-invariant, and stable,* it is well known that $c(t)$ is related to $m(t)$ by the convolution integral

$$c(t) = \int_{-\infty}^{t} h(t-u)dm(u) \ldots\ldots\ldots\ldots [2]$$

where $h(t)$ is the step-function response of the process; $h(t) = 0$ when $t < 0$. Once $h(t)$ (or one of its equivalent forms, for instance, its Laplace transform) is known, the dynamic behavior of the process in question is completely characterized. But to find $h(t)$ given $m(t)$ and $c(t)$ by means of Equation [2] requires solving an integral equation which is a very difficult task.

If we consider now the closed-loop system shown in Fig. 1, it is clear that the input $m(t)$ to the process is the output of the self-optimizing controller. Therefore $m(t)$ must depend on the output of a digital computer; in other words, $m(t)$ must be a function of time which is completely determined by its values m_k at the sampling instants. To construct a function $m(t)$ from the series of numbers m_k which has a definite value at every instant of time calls for some method of interpolation. The simplest and practically most frequently used method (6, 7) is to hold the value of $m(t)$ constant after each sampling instant until the next sampling instant. In mathematical notation

$$m(t) = m_k, \qquad kT \leqslant t < (k+1)T \ldots\ldots\ldots [3]$$

Assuming that $m(t)$ is given by Equation [3], it is easy to show that the convolution integral Equation [2] reduces to the sum

$$c(t) = \sum_{l=-\infty}^{lT<t} h(t-lT)(m_l - m_{l-1}) \ldots\ldots\ldots [4]$$

Noting that $h(kT) = 0$ for all $k < 0$, and considering only sampled values of $c(t)$ and $h(t)$, Equation [4] can be rewritten in the simpler form

$$c_k = \sum_{l=-\infty}^{l=k} (h_{k-l} - h_{k-l-1})m_l = \sum_{l=-\infty}^{l=k} g_{k-l}m_l \ldots\ldots [5]$$

where the g_k's are recognized as the samples of the response of the system to a unit pulse. According to Equation [5], the dynamic behavior of the process is now represented by the sequence of numbers

$$g_0 = h(0), \quad g_1 = h(T) - h(0), \ldots, \quad g_k = h(kT) - h[(k-1)T], \ldots$$

Moreover, if the input-output sequences [1] are known after some sampling instant, say, $k = 0$, then the numbers g_k can be determined by solving an infinite set of simultaneous linear algebraic equations given by Equation [5]. Since $h_k \rightarrow$ const with $k \rightarrow \infty$ (otherwise the process would not be stable and therefore Equation [5] would not be valid at all) it can be assumed in practice that $h_k = h_N$ for all $k > N$ if N is sufficiently large. This assumption means that $g_k = 0$ for all $k > N$ so that only a *finite* set of linear algebraic equations has to be solved to get the g_k.

But even with this simplification it would be quite inefficient to represent the process by means of the g_k because this would require a large amount of storage in the digital computer. For instance, if the step response of the process is

$$h(t) = 1 - \exp(-t/\tau)$$

$$g_0 = 0, \qquad g_k = [\exp(T/\tau) - 1]\exp(-kT/\tau), \qquad k \geq 1$$

then approximately $N = 5\tau/T$ numbers are necessary if the error due to neglecting the terms g_k, $k > N$ is to be less than 1 per cent. If fast control is required, the time constant of the closed-loop system must be much less than τ; on the other hand, the response of the closed-loop system on the average cannot take place in less than T seconds. Thus τ/T must be large, which means that a large number of values of g_k must be stored. This and other practical considerations to be discussed later indicate that the numbers g_k do not represent the dynamic characteristics of a process efficiently.

Pulse Transfer Function. A different way to represent a dynamic process is to assume that there is a linear differential equation relating $m(t)$ to $c(t)$. Consequently, m_k and c_k may be assumed to be related by means of a linear difference equation

$$c_k + b_1c_{k-1} + \ldots + b_nc_{k-n} = a_0m_k + a_1m_{k-1} + \ldots + a_qm_{k-q} \ldots\ldots [6]$$

where the a_i and b_i are real constants and b_0 has been set arbitrarily equal to unity. If the differential equation relating $m(t)$ and $c(t)$ is known, the Difference Equation [6] can be derived readily using the theory of sampled-data systems. Such a derivation shows that in general $q = n$. By rearranging Equation [6], it follows that c_k can be expressed in terms of previous inputs and outputs

$$c_k = a_0m_k + a_1m_{k-1} + \ldots + a_nm_{k-n} - b_1c_{k-1} - \ldots - b_nc_{k-n} \ldots [6a]$$

Usually $a_0 = 0$, since most physical systems do not respond instantaneously. The theoretical difference between Equations [6a] and [4] is that in the latter case in principle all past inputs are needed to determine the present output while in the former case only a finite number of past inputs and outputs is needed. The practical difference is that when the system is known to be governed by a difference equation, much fewer a_i and b_i than g_k are needed to represent the system.

Using the notation $z^ic_k = c_{k+i}$ (where i is any integer), it is possible to write down the following basic relationship between the g_k defined by Equation [5] and the a_i and b_i defined by Equation [6]

$$G(z) = \frac{a_1 z^{-1} + \ldots + a_n z^{-n}}{1 + b_1 z^{-1} + \ldots + b_n z^{-n}}$$

$$= g_1 z^{-1} + g_2 z^{-2} + \ldots + g_k z^{-k} + \ldots \ldots \ldots [7]$$

where the right-hand term is obtained by the formal expansion of the rational fraction $G(z)$ by long division according to ascending powers of z^{-1}. The first term, g_0, is missing because it was assumed that $a_0 = 0$ which implies that $h_0 = g_0 = 0$. The function $G(z)$ is called the *pulse transfer function* of the process (6, 7). It has the same role in the analysis of linear sampled-data systems as the transfer function (Laplace transform of a differential equation) in the analysis of linear continuous systems.

The number of the a_i and b_i used to represent the process is based also on an assumption as to what the value of n should be. This is a matter of approximation; in other words, n should be chosen sufficiently large so that the a_i and b_i represent the process with some desired accuracy. But the characteristics of the process are not known in advance so that some initial guess must be made about n in setting up the machine. It is, of course, possible in principle to let the machine check the adequacy of this initial guess once experimental data about the process are available. For simplicity, however, the machine discussed in this paper was designed to operate with a fixed choice of n ($n = 2$).

Finally, it should be recalled that use of the numbers g_k is feasible only if the process is stable. No such restriction is inherent in the representation by Equation [6].

To summarize, the first step in the design of the machine is:

(i) *The dynamic characteristics of the process are to be represented in the form of Equation [6], the coefficients of which are to be computed from measurements. The number $n = q$ is assumed arbitrarily. In general, the higher n, the more accurate the representation of the process by the Difference Equation [6].*

Method of Determining Coefficients. According to design requirement (C), the coefficients in Equation [6] must be determined from a large number of measurements. This can be done as follows: Suppose we make a particular guess for the a_i and b_i at the Nth sampling instant. Let us denote these assumed values by $a_i(N)$ and $b_i(N)$, and compute all the past values of c_k using this particular set of coefficients and Equation [6a]. Denoting by $c_k^*(N)$ the values of the output computed in this way, we have

$$c_k^*(N) = -b_1(N)c_{k-1} - b_2(N)c_{k-2} - \ldots - b_n(N)c_{k-n}$$
$$+ a_1(N)m_{k-1} + a_2(N)m_{k-2} + \ldots + a_n(N)m_{k-n} \ldots \ldots [8]$$
$$k = 0, 1, \ldots, N$$

A convenient measure of how good this choice of coefficients, in the light of past measured data, is the mean squared error

$$\frac{1}{N} \sum_{k=0}^{k=N} \epsilon_k^2(N) = \frac{1}{N} \sum_{k=0}^{k=N} [c_k - c_k^*(N)]^2 \ldots \ldots [9]$$

where $\epsilon_k^2(N)$ represents the squared error between measured values c_k in the past and the predicted values $c_k^*(N)$ based on a certain choice of coefficients made at the Nth sampling instant; choosing the coefficients $a_i(N)$ and $b_i(N)$ in such a fashion that the mean squared error Equation [9] is a minimum called *least-squares filtering*. In general, any method for determining the $a_i(N)$ and $b_i(N)$ differs from least-squares filtering only in the form of the appropriate expression to be minimized. The advantage of least-squares filtering is that the computations can be carried out fairly simply (see Appendix), which is usually not the case if other types of error expression are used.

In view of design requirement (D), the more recent measurements should receive greater weight than very old ones, since the process dynamics may change with time. To meet this requirement, we proceed as follows: Let $W(t)$ be a continuous, monotonically decreasing function of time such that

$$\left.\begin{array}{c} W(0) = 1 \\ 0 < W(t) < 1, 0 < t < \infty \\ W(\infty) = 0 \\ \int_0^\infty W(t)dt < \infty \end{array}\right\} \ldots \ldots \ldots [10]$$

A function satisfying such conditions is called a *weighting function.* Writing W_k for $W(kT)$, the final criterion of determining the coefficients may be stated as follows: Choose $a_i(N)$, $b_i(N)$ in such a way that the expression

$$E(N) = \sum_{k=0}^{k=N} \epsilon_k^2(N) W_{N-k} \ldots \ldots \ldots [11]$$

is a minimum. In other words the errors which would have been committed with the present choice of the coefficients $N - k$ sampling periods ago are to be weighted by a number $0 < W_{N-k} < 1$. Practically speaking, this means that the coefficients are calculated by disregarding errors which would have been committed in predicting the output a very long time ago (when the process may have been different) but trying to keep errors in predicting recent outputs small. None of these considerations, however, determines the precise form of the function $W(t)$; this question will be settled later so that an efficient computation procedure is obtained. We now state the second step in the design of the machine:

(ii) *The coefficients a_i and b_i should be determined anew at each sampling instant so as to minimize the weighted mean-square error $E(N)$.*

Numerical Solution of Weighted Least-Squares Filtering Problem. The explicit process necessary to determine the $a_i(N)$ and $b_i(N)$ requires, even after numerous simplifications, lengthy and somewhat involved calculations. These are discussed and recorded in detail in the Appendix. Only a few remarks are given here:

1 It is necessary to compute a number of so-called pseudo-correlation functions in order to write the error expression $E(N)$ in a simple form. These pseudo-correlation functions embody all measurement data up to the Nth sampling instant which is necessary to compute $E(N)$. To compute $E(N + 1)$, it is necessary to modify the pseudo-correlation functions so as to include the data received at the $(N + 1)$st sampling instant. It turns out that this process can be carried out in a simple way only if W_k is the unit pulse response (cf. Equations [5] and [7]) of a linear system governed by a difference equation. Then computation of the pseudo-correlation functions is carried out by passing products of measured values of m_k and c_k through a linear low-pass filter.

2 In order to apply Equation [6] to characterize a process, it is necessary that m_k and c_k be measured with respect to two reference values m_r and c_r such that, if m_r is a constant input to the system, c_r is the output in the steady state. Since the correct choice of such reference levels is not known in general, they must not enter into the computations of the type of Equation [6a]. In practice, the reference levels are usually determined by extraneous considerations such as calibration and range of measuring instruments. One way of avoiding the effect of incorrect reference levels (so-called *bias errors*) is to pass m_k and c_k through identical high-pass filters. After a sufficiently long period of time the bias errors, which are equivalent to a constant input to the filter, will be attenuated by an arbitrarily large factor at the output of an appropriately designed high-pass filter.

After the pseudo-correlation functions have been obtained, the determination of the coefficients reduces to solving a set of simultaneous linear algebraic equations. To do this efficiently, an iteration procedure is used; it turns out that high-pass filtering m_k and c_k (which is equivalent approximately to subtracting the instantaneous mean value of these series of numbers) is a necessary requirement to insure the convergence of the iteration procedure.

The third step in the design is as follows:

(iii) *The calculations necessary for determining the coefficients consist of modifications of the classical least-squares filtering procedure and are given in the Appendix.*

Optimal Adjustment of Controller

Once the pulse-transfer function of the process to be controlled has been obtained, the synthesis of an "optimal" controller as a set of difference equations becomes a routine task (1, 8, 9).

It is not easy to agree, however, on what constitutes optimal control. The design of an optimal controller depends in general on two considerations:

(*a*) The nature of the input and disturbance signals to the system.

(*b*) The performance criterion used.

For instance, the inputs to the system may consist of step functions of various magnitudes; the performance criterion may be the length of time after the application of the step required by the control system to bring the error within prescribed limits. Or the input may consist of signals which are defined only in the statistical sense, in which case a reasonable performance criterion is the mean squared value of the error signal.

To include in the design of the machine means by which the machine can decide what class of input signals it is subjected to and what type of optimal controller should be used appears to be too ambitious a task at the present time. For this reason, in the practical realization of the machine (see the section Description of Computer), a prearranged method of optimizing the controller was used.

This method was described in a recent note by the author (8). The input signals are to consist of steps. The controller is to be designed in such a fashion that the error resulting from a step input becomes zero in minimum time and remains zero at all values of time thereafter. As a result of these assumptions the optimal controller is described by a difference equation whose coefficients are simple multiples of the coefficients of the pulse transfer function (see Equation [25] in the Appendix.)

We note the last step in the design:

(iv) *The choice of an optimal controller is largely arbitrary, depending on what aspect of system response is to be optimized. The determination of the coefficients in the describing equations of the controller is a routine matter if the coefficients of the pulse-transfer function are known.*

Summary of Machine Organization

Since the describing equations of the self-optimizing controller are somewhat involved, it is helpful to visualize the various computation processes as shown in Fig. 3.

Numbers in brackets indicate equations which characterize the particular operations performed. It should be remembered, of course, that there are many pseudo-correlation functions, coefficients, and so on, to be computed, some of which are indicated only in a schematic fashion.

It is perhaps worth while to emphasize that the closed-loop system consisting of the self-optimizing machine and the process is highly nonlinear. The principal nonlinear operations are:

(*a*) The multiplications before the input to low-pass filters whose outputs are the pseudo-correlation functions.

(*b*) The determination of controller coefficients.

These nonlinear operations have made it necessary to design the self-optimizing machine step by step. There exists at present no general theory for the design of nonlinear control systems of this type.

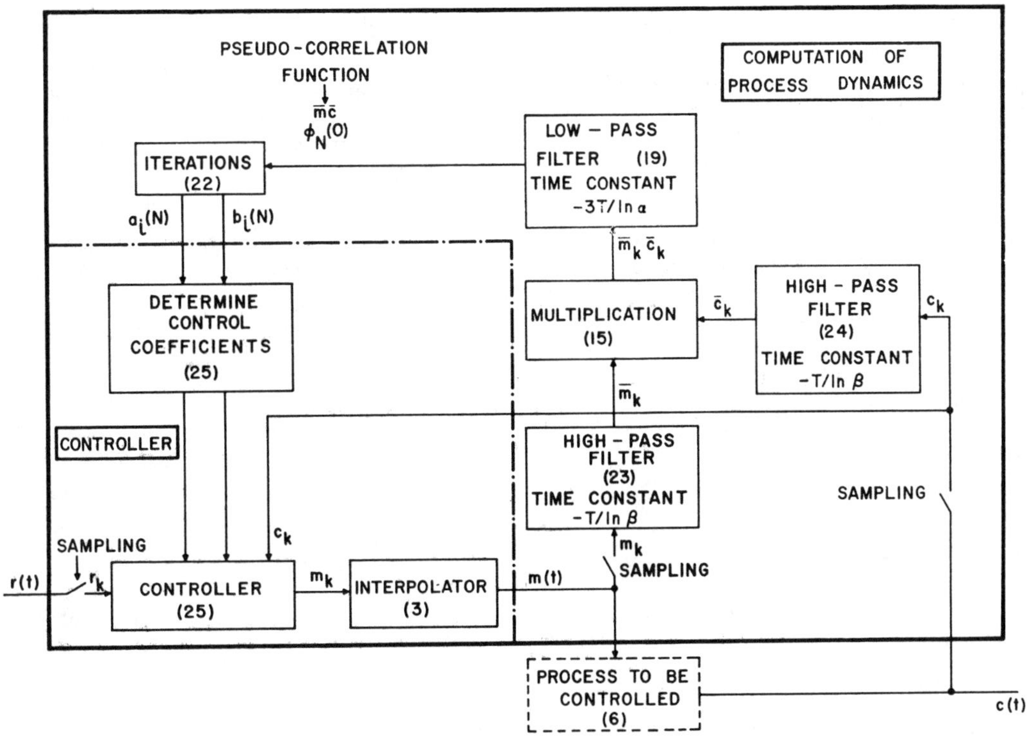

Fig. 3 Block Diagram of Computation Steps for Self-Optimizing Controller

Unsolved Questions

According to the preceding discussion, the operation of the self-optimizing system depends mainly on the accuracy of the computation of the pulse-transfer function from measurement data. Now suppose that the system is under very good control and that the input and disturbances to the system are nearly constant. In that case m_k and c_k will vary only very slightly about their equilibrium values. As a result, the numbers $\bar{m}_k$ and $\bar{c}_k$ (approximately the deviations of m_k and c_k from equilibrium) which are the inputs to the computation process determining the transfer function will be small and of roughly the same order of magnitude as the measurement noise. Under such circumstances, the transfer function cannot be computed very accurately. If the transfer function is not known accurately, then the controller cannot be set up accurately either and the system will not be operating optimally. But then the control will be less good and the deviations from the equilibrium values will increase. This, in turn, will improve the signal-to-noise ratio of the quantities $\bar{m}_k$ and $\bar{c}_k$; the computation of the transfer function will be more accurate, control action more nearly optimal, and so on. This shows that the operation of the system is limited basically by measurement noise. The fluctuations around the equilibrium condition must always be large enough to measure the transfer function with reasonable accuracy even in face of measurement noise. Thus the operation of the system depends on not being entirely at rest; if it were, it is impossible to say anything about the dynamic characteristics of the controlled process. A more precise answer to the problem involved here calls for further study.

Let us now examine qualitatively the effect of the choice α and β (cf. Fig. 3 and Appendix, Equations [19, 23, 24]) on this aspect of system performance. If α is very close to unity, the computation of the pulse-transfer function involves a large number of samples of $\bar{m}_k$ and $\bar{c}_k$ so that even if the system is at rest, i.e., $\bar{m}_k$ and $\bar{c}_k$ are practically zero, the computation of the pulse-transfer function is not affected for a long time, because the system "remembers" results of old measurements. On the other hand, if the process dynamics change rapidly in time, then α should be chosen fairly small because otherwise the computed transfer function will not be the actual transfer function. Thus α is a design parameter whose choice depends somewhat on the nature of a particular situation encountered. There is no reason, of course, why the system cannot adjust α also, but this is a problem beyond the scope of this paper.

The choice of β is guided by similar considerations. If the inputs to the system change slowly then β should be very close to unity for then the low-frequency components in m_k and c_k (slow "drift" about equilibrium point) will be very heavily attenuated. If the system is a more lively one, i.e., m_k and c_k fluctuate appreciably in time due to the effect of inputs or disturbances acting on the system, the β should be chosen smaller to improve the transient response of the high-pass filter. Thus β is another design parameter for the self-optimizing system.

Additional possibilities for improving these aspects of system operation should be considered in future work. More complicated weighting-functions and high-pass filters, suspending the operation of transfer-function computation when signal-to-noise levels become too low, putting in periodic test signals to check the operation of various parts of the computer, and the like, are some topics for future research.

Description of Computer

As soon as the operations discussed in the foregoing sections have been reduced to a set of numerical calculations (see Appendix) the machine has been synthesized in principle. This means that any general-purpose digital computer can be programmed to act as the self-optimizing machine.

In practical applications, however, a general-purpose digital computer is an expensive, bulky, extremely complex, and somewhat awkward piece of equipment. Moreover, the computational capabilities (speed, storage capacity, accuracy) of even the smaller commercially available general-purpose digital computers are considerably in excess of what is demanded in performing the computations listed in the Appendix.

For these reasons, a small special-purpose computer was constructed which could be called externally digital and internally analog according to the terminology in the section General Design Considerations. Briefly, this computer is organized as follows:

The computer operates on numbers whose absolute values do not exceed unity. Each number is represented by a 60-cycle-per-sec (cps) voltage. Numbers are stored on multiturn potentiometers, by positioning a given potentiometer by means of a servo arrangement in such a fashion that its output voltage (with unit excitation) is a 60-cps signal of the required magnitude and sign. Numbers are added by feeding corresponding voltages into electronic summing circuits. Two numbers a and b are multiplied by the following well-known method: If output of the potentiometer with unit excitation is b, then the output of the potentiometer with excitation a will be ab. The storage locations and summers can be interconnected in such a fashion that, in any one step of computation, the computer is capable of performing any one of the following types of operations

$$\left.\begin{aligned} & a_1b_1 + a_2b_2 + \ldots + a_7b_7 = x \\ \text{or} \quad & \\ & a_1b_1c_1d_1 + a_2b_2c_2 + a_3b_3c_3 = x \\ \text{or} \quad & \\ & a_1b_1c_1d_1e_1f_1g_1h_1 = x \end{aligned}\right\} \quad \ldots\ldots\ldots [12]$$

and so on

where each quantity appearing on the left-hand side of Equations [12] is an arbitrary number; x is the desired result of the computation. The fact that several additions and multiplications can be performed simultaneously is very convenient from the standpoint of programming the computer. Usually, each of Equations [12] must be broken up into several parts in programming them on a general-purpose computer.

The front view of the computer, which is roughly of the size of an average filing cabinet, is shown in Fig. 4. Only connections for input-output signals appear on the front panel. The programming of the computer is achieved by inserting wires into a "patch panel" on top of the computer which is shown in Fig. 5. Almost every signal voltage inside the computer is brought out to some contact on the patch panel. This arrangement makes it possible to interconnect the basic components of the computer in any manner desired and also facilitates troubleshooting and maintenance. The disadvantage of a patch-panel type of programming is that the change of program is a time-consuming operation; however, this is of minor significance since the machine is intended to operate with a fixed program in any typical application. The control panel shown in Fig. 5 also contains means for changing the sampling rate and reading numbers into any one of the storage locations in the computer.

The wiring necessary to connect computer components with the patch panel, together with associated relays, timing and checking circuits takes up approximately one third of the volume of the computer. Another one third of the volume is required for the electronic circuits performing summation and multiplication and

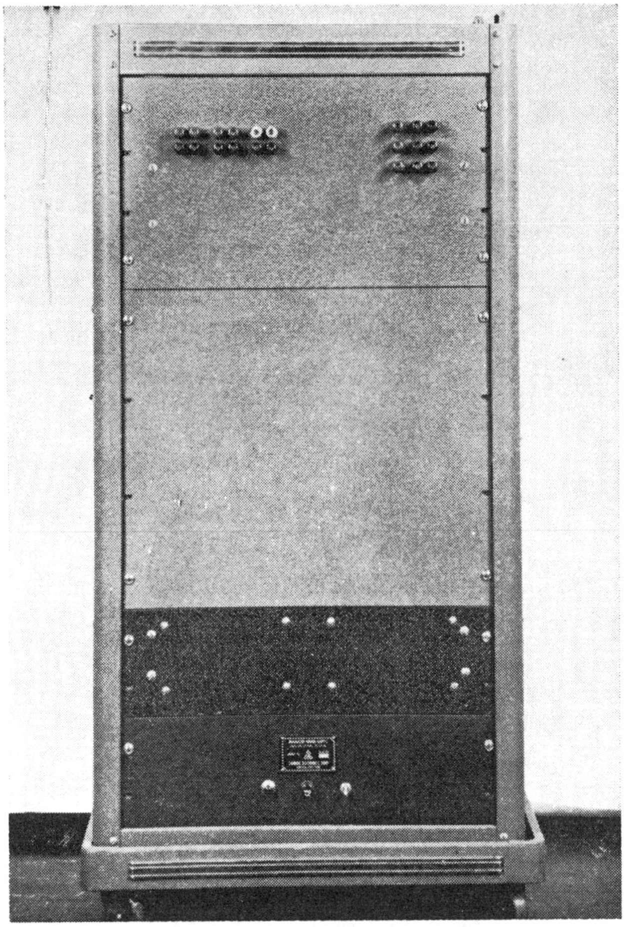

Fig. 4 Front View of Computer

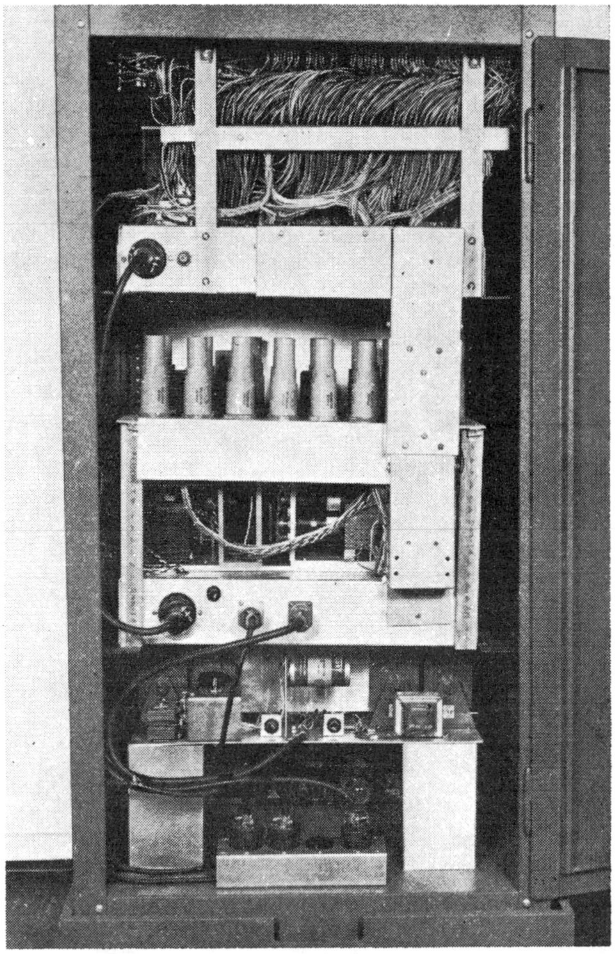

Fig. 6 Rear View of Computer

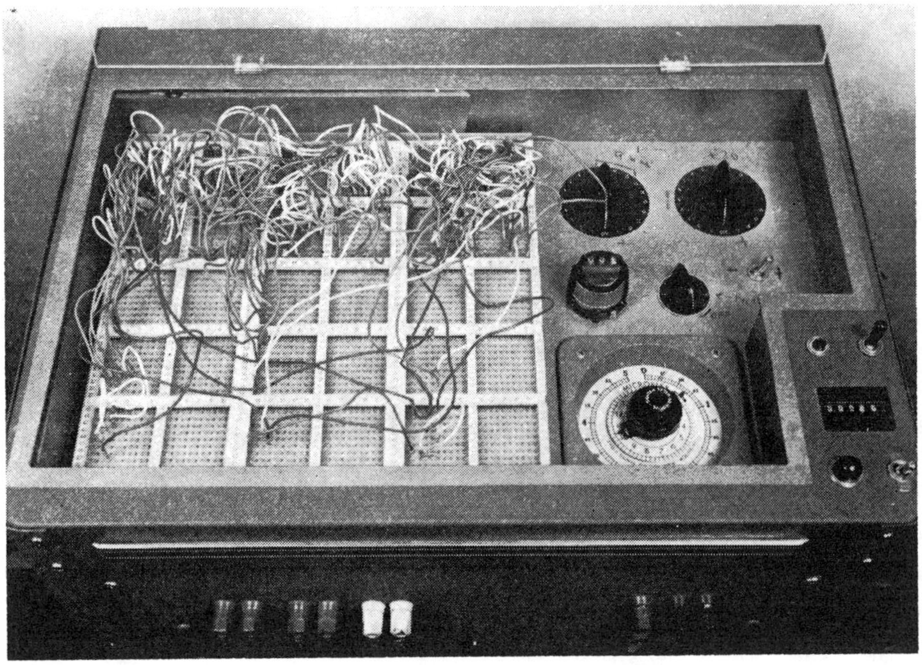

Fig. 5 Control Panel of Computer

the storage potentiometers. The remaining one third of space is taken up by power supplies. The internal arrangement of the computer is shown in the rear view of Fig. 6.

The computer described shows that the practical realization of a self-optimizing machine is well within the technological means available at the present time. Actually, the computer described was constructed in 1954/1955. The computer also represents savings in cost and complexity over currently available general purpose digital computers. On the other hand, when self-optimizing control of a large-scale installation is desired, in other words, when there are several dynamic processes to be controlled simultaneously and possibly in an interdependent fashion, then the general-purpose digital computer is much better matched to the problem both in terms of cost and computational capability.

Conclusions

This paper shows the feasibility of mechanizing much of the process by which automatic control systems for standard applications are being designed today. The amount of numerical computations necessary for accomplishing this is relatively modest (after the numerous simplifications discussed) and can be readily implemented in practice at moderate cost.

More importantly, however, the machine described here is an ideal controller since it needs merely to be interconnected with the process to be controlled to achieve optimum control after a short transitory period and hold it thereafter even if the process characteristics change with time. The task of the control engineer of the future will be not to design a specific system, but to improve the principles on which machines of the type described here will operate. Unlike his predecessor, the stock in trade of the new control-systems engineer will not be the graph paper, the slide rule, or even the analog computer but a firm and deep-seated understanding of the fundamental principles, physical and mathematical, on which automatic control is based. The drudgery of computing will be taken over by machines but the challenge of thinking remains.

Acknowledgments

The research reported here was supported by the Engineering Research Laboratory, E. I. du Pont de Nemours & Co., Wilmington, Del., to whom the author is indebted for permission to publish this paper. The author wishes also to thank various members of the Engineering Research Laboratory for their help and interest during the progress of this work, and to Dr. J. R. Ragazzini, Columbia University, for several stimulating discussions.

References

1 "Sampled-Data Processing Techniques for Feedback Control Systems," by A. R. Bergen and J. R. Ragazzini, Trans. AIEE, vol. 73, part II, 1954, pp. 236–247.

2 "Design for a Brain," by W. R. Ashby, John Wiley & Sons, Inc., New York, N. Y., 1952.

3 "Possibilities of a Two Time Scale Computing System for Control and Simulation of Dynamic Systems," by H. Ziebolz and H. M. Paynter, Proceedings of the National Electronics Conference, vol. 9, 1953, pp. 215–223.

4 "Determination of System Characteristics From Normal Operating Records," by T. P. Goodman and J. B. Reswick, Trans. ASME, vol. 77, 1955, pp. 259–268.

5 "Self-Optimizing Systems," by E. G. C. Burt, preprint for International Control Systems Conference, Heidelberg, Germany, September, 1956.

6 "The Analysis of Sampled-Data Systems," by J. R. Ragazzini and L. A. Zadeh, Trans. AIEE, vol. 71, part II, 1952, pp. 225–234.

7 "Automatic Feedback Control System Synthesis," by J. G. Truxal, McGraw-Hill Book Company, Inc., New York, N. Y., 1955.

8 R. E. Kalman, discussion of reference (1), Trans. AIEE, vol. 73, part II, 1954, pp. 245–246.

9 "Digital Controllers for Sampled-Data System," by J. E. Bertram, Trans. AIEE, vol. 75, part II, 1956, pp. 151–159.

10 "Introduction to Numerical Analysis," by F. B. Hildebrand, McGraw-Hill Book Company, Inc., New York, N. Y., 1956.

11 "Numerical Analysis," by W. E. Milne, Princeton University Press, Princeton, N. J., 1949.

Appendix

The following is the detailed derivation of the complete set of equations characterizing the self-optimizing controller in the special case when $n = 2$ in the Difference Equation [6]. Using these equations, any digital computer may be programmed to act as a self-optimizing controller. When $n > 2$, the required equations can be obtained similarly.

First of all, instead of performing the computations required to minimize Equation [11] at every sampling instant, they may be performed at every qth (where q is a positive integer) sampling instant. This does not affect the reasoning in the section Method of Determining Coefficients, and results in considerable simplification in the required computations. With this change, the error expression Equation [11] becomes

$$E(N) = \sum_{j=0}^{j=N/q} \epsilon_{qj}^2(N) W_{N-qj} \quad \ldots\ldots\ldots\ldots [13]$$

where $k = qj$ and N is a number divisible by q.

Now assume that $n = 2$ in Equation [6]. Using the recurrence relation Equation [8], $\epsilon_{qj}^2(N)$ can be written as

$$\left.\begin{aligned}
\epsilon_{qj}^2(N) &= [c_{qj} - c_{qj}^*(N)]^2 \\
&= c_{qj}^2 + b_1^2(N)c_{qj-1}^2 + b_2^2(N)c_{qj-2}^2 \\
&\quad + 2b_1(N)c_{qj}c_{qj-1} + 2b_2(N)c_{qj}c_{qj-2} \\
&\qquad + 2b_1(N)b_2(N)c_{qj-1}c_{qj-2} \\
&\quad - 2a_1(N)c_{qj}m_{qj-1} - 2a_2(N)c_{qj}m_{qj-2} \\
&\quad - 2b_1(N)a_1(N)c_{qj-1}m_{qj-1} \\
&\qquad - 2b_1(N)a_2(N)c_{qj-1}m_{qj-2} \\
&\quad - 2b_2(N)a_1(N)c_{qj-2}m_{qj-1} \\
&\qquad - 2b_2(N)a_2(N)c_{qj-2}m_{qj-2} \\
&\quad + a_1^2(N)m_{qj-1}^2 + a_2^2(N)m_{qj-2}^2 \\
&\qquad + 2a_1(N)a_2(N)m_{qj-1}m_{qj-2}
\end{aligned}\right\} \ldots [14]$$

The measured values of c and m occur in Equation [14] always in terms of the type

$$c_{qj-r}c_{qj-s} \qquad c_{qj-r}m_{qj-s} \qquad m_{qj-r}m_{qj-s} \ldots\ldots\ldots [15]$$

where $r, s = 0, 1, 2$. If we now let

$$q = n + 1 = 3$$

then it is clear that factors of the same type will be multiplied by the same coefficients in Equation [14], regardless of the value of j. This property does not arise when $q < 3$. Using the symmetry introduced by the particular choice of q, $E(N)$ can be put in a simpler form by defining the *pseudo-correlation functions*

$$\left.\begin{aligned}
\phi_{N-r}^{cc}(r - s) &= \sum_{j=1}^{j=N/3} c_{3j-r}c_{3j-s}W_{N-3j} \\
\phi_{N-r}^{cm}(r - s) &= \sum_{j=1}^{j=N/3} c_{3j-r}m_{3j-s}W_{N-3j} \\
\phi_{N-r}^{mm}(r - s) &= \sum_{j=1}^{j=N/3} m_{3j-r}m_{3j-s}W_{N-3j}
\end{aligned}\right\} \ldots [16]$$

With these definitions, $E(N)$ can be written as follows, arranging the terms in the same fashion as in Equation [14]

$$\left.\begin{aligned}
E(N) = {} & \phi_N^{cc}(0) + b_1^2(N)\phi_{N-1}^{cc}(0) + b_2^2(N)\phi_{N-2}^{cc}(0) \\
& + 2b_1(N)\phi_N^{cc}(-1) + 2b_2(N)\phi_N^{cc}(-2) \\
& \qquad + 2b_1(N)b_2(N)\phi_{N-1}^{cc}(-1) \\
& - 2a_1(N)\phi_N^{cm}(-1) - 2a_2(N)\phi_N^{cm}(-2) \\
& - 2b_1(N)a_1(N)\phi_{N-1}^{cm}(0) \\
& \qquad - 2b_1(N)a_2(N)\phi_{N-1}^{cm}(-1) \\
& - 2b_2(N)a_1(N)\phi_{N-2}^{cm}(1) - 2b_2(N)a_2(N)\phi_{N-2}^{cm}(0) \\
& + a_1^2(N)\phi_{N-1}^{mm}(0) + a_2^2(N)\phi_{N-2}^{mm}(0) \\
& \qquad + 2a_1(N)a_2(N)\phi_{N-1}^{mm}(-1)
\end{aligned}\right\} \quad \ldots [17]$$

Remark. The conventional definition of correlation functions is

$$\phi_N^{cc}(r) = \frac{1}{N}\sum_{k=0}^{k=N} c_k c_{k+r}$$

To evaluate this function iteratively, as is done in Equation [19] for pseudo-correlation functions, it would be necessary to compute

$$\phi_N^{cc}(r) = c_N c_{N+r}/N + (N-1)\phi_{N-1}^{cc}(r)/N$$

Since the factor $(N - 1)/N$ cannot be calculated accurately enough as $N \rightarrow \infty$, such an iterative calculation would be impractical.

The pseudo-correlation functions can be evaluated iteratively as follows: Suppose that, in addition to meeting Conditions [10], the weighting function W_k is a sequence of numbers such as the g_k given by Equation [7]. Then it follows that the pseudo-correlation functions can be regarded as the output of a linear system governed by a difference equation, whose input consists of products such as Equation [15]. In particular, if we let

$$W^{3j} = \alpha^j \qquad (0 < \alpha < 1) \ldots\ldots\ldots\ldots [18]$$

then every pseudo-correlation function satisfies a first-order difference equation of the type

$$\phi_{3j-r}^{cm}(r-s) - \alpha\phi_{3(j-1)-r}^{cm}(r-s) = c_{3j-r}m_{3j-s} \ldots [19]$$

According to Equation [17] the determination of the coefficients of the pulse-transfer function requires first that all input-output data (the measured values of c and m) be consolidated into the pseudo-correlation functions. Because of the recurrence relation Equation [19], the computation of the latter is quite simple, since to get the pseudo-correlation functions at the Nth sampling instant requires only the knowledge of the same functions at the end of the $(N - 3)$th sampling instant, plus the values of c_{N-2}, c_{N-1}, c_N, m_{N-2}, m_{N-1}. Once the new pseudo-correlation functions have been computed, the data measured during the preceding three sampling periods can be discarded and the system is ready to receive new data. Thus the use of the pseudo-correlation functions and the choice of a suitable weighting function greatly simplifies the implementation of mean-square filtering.

In order that $E(N)$ be a minimum with respect to the a_i and b_i, it is *necessary* that the partial derivatives

$$\frac{\partial E(N)}{\partial a_i} = 0 \qquad \frac{\partial E(N)}{\partial b_i} = 0 \quad (i = 1, 2, \ldots, n) \ldots [20]$$

vanish. The proof that these conditions are also *sufficient* to insure the existence of a minimum of $E(N)$ is quite difficult. Refer to Milne (11) for discussion of a closely related problem.

The Conditions [20] lead to four linear equations in the coefficients $a_1(N)$, $a_2(N)$, $b_1(N)$, $b_2(N)$ as follows

$$\left.\begin{aligned}
& a_1(N)\phi_{N-1}^{mm}(0) + a_2(N)\phi_{N-1}^{mm}(-1) - b_1(N)\phi_{N-1}^{cm}(0) \\
& \qquad - b_2(N)\phi_{N-2}^{cm}(1) = \phi_N^{cm}(-1) \\
& a_1(N)\phi_{N-1}^{mm}(-1) + a_2(N)\phi_{N-2}^{mm}(0) \\
& \qquad - b_1(N)\phi_{N-1}^{cm}(-1) - b_2(N)\phi_{N-2}^{cm}(0) = \phi_N^{cm}(-2) \\
& -a_1(N)\phi_{N-1}^{cm}(0) - a_2(N)\phi_{N-1}^{cm}(-1) \\
& \qquad + b_1(N)\phi_{N-1}^{cc}(0) + b_2(N)\phi_{N-1}^{cc}(-1) = -\phi_N^{cc}(-1) \\
& -a_1(N)\phi_{N-2}^{cm}(1) - a_2(N)\phi_{N-2}^{cm}(0) \\
& \qquad + b_1(N)\phi_{N-1}^{cc}(-1) + b_2(N)\phi_{N-2}^{cc}(0) = -\phi^{cc}(-2)
\end{aligned}\right\} \quad \ldots [21]$$

Any method for solving linear simultaneous equations can be used for finding the a_i and b_i from Equation [21]. However, the standard elimination methods (which, incidentally, are much more efficient than solving Equation [21] by Cramer's rule) require a rather large amount of storage and somewhat lengthy computations. These disadvantages become increasingly worse as n increases. However, an exact computation of a solution of Equation [21] is very wasteful in that, if a solution of Equation [21] at the $(N - 3)$th sampling instant is available, then that solution is also an excellent guess for the solution of Equation [21] at the Nth sampling instant since the correlation function can have changed only slightly, unless a very small value of α is used. This suggests an *iteration* procedure for solving Equation [21], of which the simplest is the so-called Gauss-Seidel method (10).

Applying the Gauss-Seidel method to Equation [21] leads to the equations

$$a_1(N) = \frac{-a_2(N-3)\phi_{N-1}^{mm}(-1) + b_1(N-3)\phi_{N-1}^{cm}(0) + b_2(N-3)\phi_{N-2}^{cm}(1) + \phi_N^{cm}(-1)}{\phi_{N-1}^{mm}(0)} \ldots [22a]$$

$$a_2(N) = \frac{-a_1(N)\phi_{N-1}^{mm}(-1) + b_1(N-3)\phi_{N-1}^{cm}(-1) + b_2(N-3)\phi_{N-2}^{cm}(0) + \phi_N^{cm}(-2)}{\phi_{N-2}^{mm}(0)} \ldots [22b]$$

$$b_1(N) = \frac{a_1(N)\phi_{N-1}^{cm}(0) + a_2(N)\phi_{N-1}^{cm}(-1) - b_2(N-3)\phi_{N-1}^{cc}(-1) - \phi_N^{cc}(-1)}{\phi_{N-1}^{cc}(0)} \ldots [22c]$$

$$b_2(N) = \frac{a_1(N)\phi_{N-1}^{cm}(1) + a_2(N)\phi_{N-2}^{cm}(0) - b_1(N)\phi_{N-1}^{cc}(-1) - \phi_N^{cc}(-2)}{\phi_{N-2}^{cc}(0)} \ldots [22d]$$

If desired, the cycle of iterations just written down can be repeated to obtain better accuracy.

A necessary and sufficient condition for the convergence of the iteration Equations [22] is that the diagonal coefficients in Equations [21], i.e., $\phi_{N-1}^{mm}(0)$, $\phi_{N-2}^{mm}(0)$, $\phi_{N-1}^{cc}(0)$, $\phi_{N-2}^{cc}(0)$ should be larger in absolute value than any of the other coefficients in the same equation. To insure rapid convergence, it is highly desirable that the diagonal coefficients be as large as possible compared to the off-diagonal coefficients.

A glance at Equation [19] shows that the pseudo-correlation

functions just mentioned are always the sum of positive numbers because the right-hand side of Equation [19] is always positive, being a square. To make the pseudo-correlation functions corresponding to the off-diagonal elements in Equations [21] smaller in absolute value than the diagonal elements, the right-hand side of Equation [19] for these functions must be alternatively positive and negative. This can be achieved by subtracting from each c_k and m_k the average (mean) values of these quantities over a long period of time. Unless this is done, c_k and m_k might vary only slightly about a large average value in which case all the correlation functions will be approximately equal and the iteration Equation [22] will not converge fast enough, if at all.

To estimate the mean of a time series in a very reliable way is not an easy problem. In the present case, however, sophisticated statistical methods are not required because the precise knowledge of the mean is not important. The simplest procedure then is to put both c_k and m_k through identical high-pass filters which remove the slowly varying components (i.e., the mean) of these quantities. When the mean is constant in time, it is equal to the zero frequency component of the signal. The simplest high-pass filter on numerical data is represented by the difference equation

$$c_k - c_{k-1} = \bar{c}_k - \beta\bar{c}_{k-1} \qquad (0 < \beta < 1) \ldots\ldots [23]$$

where $\bar{c}_k$ is approximately equal to c_k − mean (c_k). The closer β is to 1, the better the removal of the mean if the latter is constant. On the other hand, if the mean varies β should be somewhat smaller for best results. A similar equation holds for $\bar{m}_k$

$$m_k - m_{k-1} = \bar{m}_k - \beta\bar{m}_{k-1} \qquad (0 < \beta < 1) \ldots\ldots [24]$$

A simple substitution in Equation [6] shows that $\bar{c}_k$ and $\bar{m}_k$ are related by the same difference equation as c_k and m_k. This is because if two quantities are linearly related, the relationship remains undisturbed if both quantities are put through identical linear filters. Thus the removal of the mean represented by Equations [23] and [24] does not affect the computation of the pulse-transfer function of the process to be controlled, except for greatly improving the convergence of the iteration process Equations [22]. Hence all pseudo-correlation functions should be computed using the $\bar{c}_k$ and $\bar{m}_k$.

It remains to show how the equations of the controller can be obtained from the knowledge of the coefficients of the pulse-transfer function. As mentioned earlier, the controller is to be digital. Using a method of synthesis due to the author (8), which yields the optimum design if the closed-loop system is to respond to a unit step input in minimal time without overshoot (for a given fixed sampling period T), the numbers necessary to specify the controller are very simply related to the coefficients of the pulse-transfer function of the process which is to be controlled. In fact, the difference equation specifying the controller is

$$[a_1(N) + a_2(N)]m_k - a_1(N)m_{k-1} - a_2(N)m_{k-2}$$
$$= e_k + b_1(N)e_{k-1} + b_2(N)e_{k-2} \ldots\ldots [25]$$

where $\qquad e_k = r_k - c_k$

Equation [25] is valid for $N + 1 \leq k \leq N + 3$, after which a new set of coefficients must be used from the next determination of the pulse-transfer function. It should be noted that Equation [25] holds only if the (continuous) transfer function of the process is approximately $H(s) = K/(s + a)(s + b)$ with a, b, $K > 0$. If, for instance, $a = 0$, the form of Equation [25] is different. For methods of synthesizing digital controllers which are optimal in some other sense, see references (1, 9).

For convenience, the time sequence of computations to be performed during a cycle of $q = 3$ sampling periods is listed as follows.

$k = N - 2$

(1) Compute m_{N-2} using [25]
(2) Compute $\bar{c}_{N-2}$ using [23]
(3) Compute $\bar{m}_{N-2}$ using [24]
(4) Compute $\phi_{N-2}^{\bar{m}\bar{m}}(0)$, $\phi_{N-2}^{\bar{c}\bar{c}}(0)$, $\phi_{N-2}^{\bar{c}\bar{m}}(0)$ using Equation [19]

$k = N - 1$

(1) Compute m_{N-1} using [25]
(2) Compute $\bar{c}_{N-1}$ using [23]
(3) Compute $\bar{m}_{N-1}$ using [24]
(4) Compute $\phi_{N-1}^{\bar{m}\bar{m}}(0)$, $\phi_{N-1}^{\bar{m}\bar{m}}(-1)$, $\phi_{N-1}^{\bar{c}\bar{c}}(0)$, $\phi_{N-1}^{\bar{c}\bar{c}}(-1)$ $\phi_{N-1}^{\bar{c}\bar{m}}(0)$, $\phi_{N-1}^{\bar{c}\bar{m}}(-1)$, $\phi_{N-2}^{\bar{c}\bar{m}}(1)$ using Equation [19]

$k = N$

(1) Compute m_N using [25]
(2) Compute $\bar{c}_N$ using [23]
(3) Compute $\bar{m}_N$ using [24]
(4) Compute $\phi_N^{\bar{c}\bar{c}}(-1)$, $\phi_N^{\bar{c}\bar{c}}(-2)$, $\phi_N^{\bar{c}\bar{m}}(-1)$, $\phi_N^{\bar{c}\bar{m}}(-2)$ using Equation [19]
(5) Compute $a_1(N)$, $a_2(N)$, $b_1(N)$, $b_2(N)$ using [22]

Discussion

RANE L. CURL.[3] The author has presented with skill his proposal for a self-optimizing control system. He has also covered most of the limitations in both the theory and design of his machine. I will only mention perhaps one or two points that come to mind.

On the first stage of the author's procedure, *measure the dynamic characteristics of the process*, a difficulty would be met in most real processes of the regulatory type. The proposed method of determining the system characteristics is subject to error when the existence of an error signal is due to load disturbances entering between the *control effort* and the *output*. This error may be of two types. The first is from poor "response" information in the presence of noise, and is inherent in any method which does not use process response information over a very long time. The desire to make the self-optimizing machine respond to changes in process dynamics is anathema to obtaining a good measure of the transfer function in the presence of noise. The second type of error is inherent in *all* methods which determine process dynamics while the process is on closed loop control. The noise circulates in the loop and there exists a correlation between the *noise component* of the output $c(t)$, and the control effort $m(t)$.

The importance of the regulatory type of controller and the difficulty of obtaining good process dynamics when it is in use suggests a reason additional to that of the author as to why technological unemployment of control engineers will not result from this machine.

The author's use of $n = 2$, while a strict limitation, was, as the author correctly stated, a matter of convenience and not an inherent limitation. It would be of interest if the author would comment on the behavior of the machine described in his paper when used with systems having incompatible transfer functions, i.e., for processes for which Equation [25] does not represent the optimum controller.

The well known "optimizing" controller for adjusting a set point in order to maximize yield, profit, etc., introduces its own disturbance as a "tracer" on system performance. This is another possibility, in some cases, to computation suspension at low signal to noise ratios as in the author's machine.

I agree with the author that this machine does "represent. . . an advance. . . .in practicality over suggestions. . . .in the current literature." But I ask last the primary unanswered question: Does it work?

[3] Shell Development Company, Emeryville, Calif.

Author's Closure

Before taking up in detail the questions raised in Dr. Curl's discussion, the author wishes to answer his last and most important point, "Does it work?" The answer is, "Yes."

Dr. Curl's remarks on difficulties of determining the process transfer function amplify some of the matters discussed in the section, Unsolved Questions. As in any method of measurement based on statistical principles, the determination of the process dynamics depends on obtaining a large number of data with stationary statistical properties so that the effect of unwanted influences acting on the system can be averaged out. If the load disturbances have a nonzero mean value, then their effect on the plant appears as a shift in the operating point. The computation procedure determines the linear system dynamics for small deviations about this "phantom" operating point. Since the computation of the transfer function can take account of slow changes, shifts in the mean value of the load disturbances do not affect the operation of the system, provided that these shifts occur slowly relative to the sampling period. The accuracy of computation of the transfer function depends on the effective signal-to-noise ratio, that is, on the ratio of the mean-square value of the control effort $m(t)$ required under normal operating conditions to the mean-square value of the combined effect of load disturbances and measurement noise. When this ratio is too small, it may be improved by introducing special "test signals" into the plant, or the operation of the transfer-function computation may be temporarily suspended until the signal-to-noise ratio is improved.

The effect of circulating noise determines the maximum accuracy achievable by a self-optimizing system and can, in general, only be determined experimentally. If the effect is too large, more accurate instrumentation must be used. It should be borne in mind also that since the controller of a self-optimizing system is closely matched to the dynamics of the plant, any errors due to circulating noise can be rapidly corrected. In other words, measurement noise is not amplified by the system.

By way of illustration, it may be pointed out that measurements performed by the author using high-accuracy measuring equipment support the foregoing remarks. The computation of the transfer function of a crude third-order electrical analog (3 capacitors in cheap electronic circuity without voltage regulation) yielded the following experimental results, over about 500 sampling points:

Largest time constant $\tau_1 \cong$ constant $\pm$ 0.1 per cent
Next time constant $\tau_1/3 \cong$ constant $\pm$ 1.0 per cent
Smallest time constant $\tau_1/10 \cong$ constant $\pm$ 10 per cent

The high accuracy with which the dominant time constant τ_1 can be determined is quite remarkable. The variation is only slightly worse than the errors introduced by the measuring process. On the other hand, the large error in the determination of the smallest time constant is due to the combined effect of amplifier noise, temperature transients, and so forth. From this measurement, it may be concluded that the system may be regarded as effectively second-order. Indeed, the system could be controlled quite satisfactorily with a sampled-data controller with a fixed, second-order program. Conclusive results concerning the performance of the self-optimizing controller in an actual plant installation cannot be given here.

In conclusion, the author does not share Dr. Curl's pessimism that the presence of noise problems makes a self-optimizing system impractical. Probably the most serious practical difficulty barring better process control at the present time is the unavailability of accurate data on process dynamics. This difficulty can be circumvented in many cases by use of a self-optimizing controller. The author may not be unduly optimistic in expressing his feeling that (disregarding economic considerations) sufficient theoretical and technological know-how exists already to bring practical process control close to the best performance achievable in the light of the limitations imposed by physical measuring equipment.

Adaptive System Design: A Genetic Approach

KENNETH DE JONG

Abstract—**An unconventional approach to adaptive system design is presented which has considerable promise for complex automation problems. The heart of this approach is a sophisticated search technique called a reproductive plan which is based on information processing models from population genetics. Both simulation and analysis results are presented which suggest that reproductive plans out-perform existing techniques on complex process response surfaces. An example is presented to provide an intuitive idea of how reproductive plans operate, and some observations are made concerning the applicability of this approach to various problems in adaptive system design.**

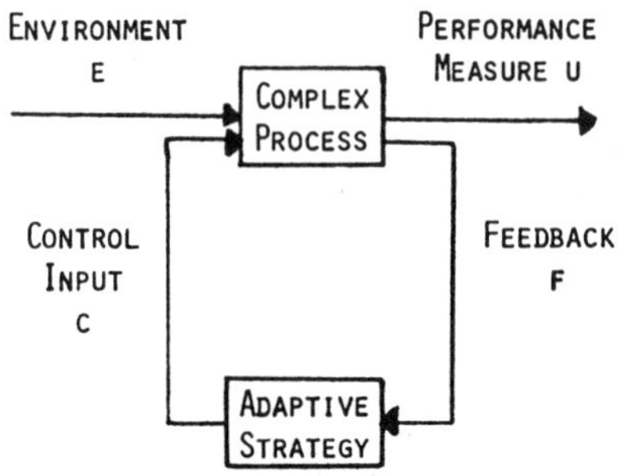

Fig. 1. Adaptive system model.

I. Introduction

THE NEED for an adaptive solution to a problem arises in a wide variety of contexts. Typically, the inherent complexity of a problem or the uncertainty surrounding it prevents one from specifying an acceptable *a priori* solution. Instead, an attempt is made to solve the problem adaptively by dynamically accumulating information about the problem at hand and using this information to generate an acceptable solution [18]. The author's interest in adaptive systems stems from the frequency with which problems of this type arise in the area of software design.

Consider, for example, the problem of choosing a good data structure for an information retrieval system. More often than not, critical information about the characteristics of the data to be stored as well as the relative frequencies of access operations are not known until application time. Whether systems based on *a priori* assumptions about these characteristics produce acceptable performance depends, of course, on the extent to which the assumptions hold. Adaptive solutions avoid this problem by modifying the underlying structures as the characteristics of a particular application become known (see, for example, [12] or [15]).

The difficulties encountered in specifying scheduling algorithms for time-sharing systems provide another context for adaptive system design. A standard approach is to manually "tune" the scheduling algorithms over a period of time to the average demand characteristics of the user community. After the initial tuning period, the system runs without manual intervention until an "unusual" situation arises which results in system degradation. Adaptive solutions avoid this problem by modifying the underlying algorithms dynamically in response to the particular demand characteristics at hand (for example, [1] or [14]).

As a final example from the software area, consider the problem of designing a game-playing program. Its performance invariably depends on the unspecified values of a fair number of parameters, presenting the designer with a difficult tuning process. An adaptive approach allows not only for the design of self-tuning programs but also for the possibility of strategy generating systems.

Because of this wide variety of contexts in which adaptive systems can occur both in the software area and elsewhere, the major theme of the research here has been to avoid the case study approach to adaptive system design and focus on the problem in a more general setting. That is, rather than consider special purpose techniques which are applicable to only specific situations, a more general class of adaptive techniques is sought which permit a broad range of applications while remaining computationally efficient.

II. The Problem of Adaptation

As illustrated in Fig. 1, we have chosen to represent adaptive systems abstractly as a standard feedback loop in classical control theory. This has two advantages. First, it immediately suggests a large body of control theory knowledge which can be brought to bear on the problem. Second, it admits to a relatively simple formal representation of the problem of adaptation, namely:

- a set E of environments in which a complex process must operate,
- a set C of allowable control inputs available to an

Manuscript received July 20, 1978; revised August 6, 1979.
The author is with the Department of Computer Science, University of Pittsburgh, 322 Alumni Hall, Pittsburgh, PA 15260.

Reprinted from *IEEE Trans. Syst., Man, Cybernetics*, vol. SMC-10, pp. 566–574, Sept. 1980.

adaptive strategy for modifying the behavior of the complex process,

- a process performance measure u: $E \times C \rightarrow R$ which specifies the performance of the process in environment e under control input c,
- a feedback function f: $E \times C \rightarrow R^n$ which provides the adaptive strategy with dynamic information about the process being controlled,
- a set S of adaptive strategies, each of which attempts to use the accumulating knowledge to improve the performance of the process via changes in the control inputs. More formally, each strategy is a map

$$s: (\text{traj}_t(c_t), \text{traj}_t(f(c_t))) \rightarrow C$$

from the set of time trajectories of C and f into the control set C.

Finally, some means of evaluating the performance of an adaptive strategy is required. Intuitively, an adaptive strategy should be evaluated on its ability to respond quickly to the problem at hand by generating control inputs which yield good performance with respect to the process being controlled. Moreover, good adaptive strategies should be robust in the sense that they should respond quickly across the whole range of environments to which the controlled process might be subjected. More formally, a map

$$X_E: S \rightarrow R$$

is required to evaluate and compare adaptive strategies.

With the general framework now specified, the problem of adaptation can be put more clearly in focus. Each environment e in E to which the controlled process might be subjected can be viewed as defining a *performance response surface* over the control input space C by considering the function u_e: $C \rightarrow R$ defined by

$$u_e(c) = u(c, e)$$

the restriction of the performance measure u to environment e. It is the response surface defined by u_e which must be explored by an adaptive strategy in order to generate good performance.

Viewing the problem of adaptation in this way immediately suggests a variety of techniques from such areas as sampling theory, function optimization, and control theory which might be incorporated into an adaptive strategy. However, it should be noted that each environmental performance measure u_e is really the composition of two functions: a behavioral function b_e: $C \rightarrow Q$ which specifies how the control inputs affect the state of the process, and a measurement function m: $Q \rightarrow R$ which yields the performance measure based on the state of the process. The function m is usually man-generated and, hence, of fairly simple form (e.g., a polynomial of degree 2). However, for processes of any interest, b_e is a complex "black box" function unavailable in closed form. As a consequence the composition u_e is generally an extremely complex, often high-dimensional, multimodal function which may be discontinuous, may contain noise, and is almost never available in closed form. These observations immediately rule out many of the techniques referred to above. Most function optimization and hill-climbing methods are local techniques in that they quickly find the nearest local optimum, but leave the global optimum of a multimodal surface undiscovered (see, for example, [11] or [19]). On the other hand, statistical sampling techniques often avoid this problem, but their computation time grows rapidly with the dimensionality of the problem (see, for example, [2] or [8]). When faced with the more difficult surfaces, the current state of the art is to revert to some form of random search. However, from an adaptive system point of view, to apply random search is to ignore the information about the response surface which is accumulating throughout the period of adaptation. The critical question, then, for adaptive system design, is whether there are ways of exploiting this accumulation of information in order to improve performance.

III. Current Assumptions

The research presented in the remainder of this paper deals with a class of adaptive strategies which fall within the general framework discussed above. However, to sharpen the focus of the research several assumptions have been made.

1) A discrete time scale is assumed, while the general framework avoids any such distinction.

2) No *a priori* information is available to an adaptive strategy about the type of response surfaces defined by the u_e. However, they are assumed to be stationary over the period of adaptation.

3) The only feedback available to an adaptive strategy is the value of the process performance measure u, i.e.,

$$f(e, c) = u(e, c).$$

This is usually referred to as "zero-order" feedback in the sense that the very least information required for adaptation is an indication of how well the controlled process is performing.

4) Two alternative methods of evaluating the performance of adaptive strategies are used. The first, referred to as "on-line" performance, is defined as

$$x_e(s) = \text{ave}_t(u_e(t)), \qquad t \in [0, T].$$

That is, an adaptive strategy is evaluated in terms of the average performance of the controlled process over time. Evaluation measures of this form are used in situations like operating system control when all experimentation by the adaptive strategy whether good or bad is included in the evaluation process. A second measure, referred to as "off-line" performance, is given by

$$x_e^*(s) = \text{ave}_t(u_e^*(t)), \qquad t \in [0, T]$$

where $u_e^*(t)$ is defined as the best performance achieved within the time interval $[0, t]$. Evaluation measures of this form are used when experimentation with new control inputs can be performed off-line (via a simulation model or during periods of inactivity) while the on-line system runs with the best control inputs generated at that time.

Both of these adaptive performance measures are local in the sense that they evaluate the performance of an adaptive strategy in a particular environment e. The corresponding global measures of robustness are defined as

$$X_E(s)=\text{ave}_e(x_e(s)), \qquad e\in E$$

and

$$X_E^*(s)=\text{ave}_e(x_e^*(s)), \qquad e\in E.$$

That is, robustness is defined in terms of the average performance of a strategy s over the entire set E of process environments.

IV. Artificial Genetic-Based Adaptive Systems

As noted earlier, the critical question for adaptive system design is whether or not there exist general yet efficient ways to exploit accumulating information about the behavior of the process being controlled. The research reported here is an outgrowth of an on-going project attempting to answer such questions [9]. The basic viewpoint of this research is that nature is a rich source of examples of sophisticated adaptive systems. The goal of this project is to understand and abstract from natural systems the mechanisms of adaptation in order to design artificial systems of comparable sophistication. Much of this research has centered around the design and analysis of artificial adaptive systems based on models drawn from the area of population genetics (see, for example, [3], [6], [7], or [10]). A class of genetic-based adaptive strategies, called reproductive plans, has evolved from this research and the performance characteristics of this class is the subject of the remainder of this paper.

Briefly, a reproductive plan operates as follows. Each point in the control space C is considered as an individual and represented uniquely within the system by a string generated from some alphabet (there is some evidence that the alphabet $\{0,1\}$, i.e., a binary representation, is in a certain sense optimal [9]). The strings play the role of "genetic material" with specific positions on the string (genes) taking on a variety of values (alleles) with the performance function u_e defining the concept of fitness for elements in the control space. At any particular point in time, a reproductive plan maintains a population $A(t)$ of strings which represent the current set of control inputs being evaluated for fitness. A new generation $A(t+1)$ of strings is generated by simulating the dynamics of population development as illustrated in Fig. 2. The process begins by randomly generating the initial population $A(0)$. Each individual in the current population is submitted for evaluation as a control input, saving its associated performance index. A selection probability is then defined as indicated over the current population $A(t)$. Finally, the next generation $A(t+1)$ is produced by selecting individuals via the selection probabilities to undergo "reproduction" via genetic operators.

To get a feeling for how the reproductive plans work, note that the selection probabilities are defined in such a

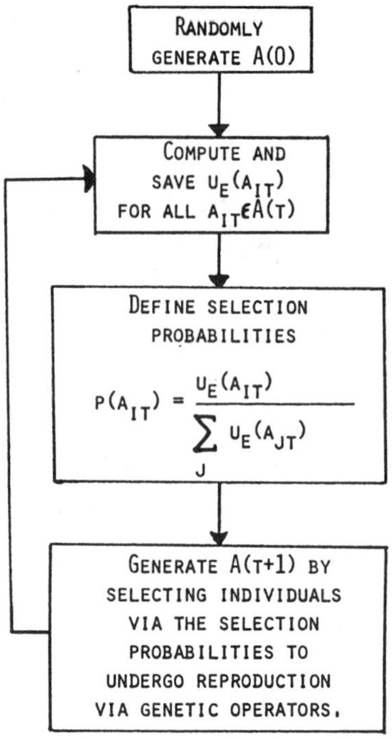

Fig. 2. Basic structure of reproductive plans.

way that the expected number of offspring produced by an individual is proportional to his associated performance value. This can be seen by considering the process of selecting individuals for reproduction as N samples from $A(t)$ with replacement using the selection probabilities. The expected number of offspring from individual a_{it} is given by

$$O(a_{it})=N*p(a_{it})$$
$$=N*\frac{u_e(a_{it})}{\sum_j u_e(a_{jt})}$$
$$=\frac{u_e(a_{it})}{\frac{1}{N}*u_e(A(t))}$$

which indicates that individuals with average performance ratings produce on the average one offspring while better individuals produce more than one and below average individuals less than one. Hence, with no other mechanism for adaptation, reproduction proportional to performance results in a sequence of generations $A(t)$ in which the best individual in $A(0)$ takes over a larger and larger proportion of the population.

However, in nature as well as in these artificial systems, offspring are almost never exact duplicates of a parent. It is the role of genetic operators to exploit the selection process by producing new individuals which have high-performance expectations. The choice of operators is motivated by the mechanisms of nature: crossover, mutation, inversion, and so on [4], [13]. The exact form taken by such operators depends upon the particular "genetic" representation chosen for points in C. In order to illustrate more clearly the role of genetic operators in these artificial genetic systems, a very basic subclass of reproductive plans is presented which exhibits surprising adaptive capabilities.

The simplest reproductive plans represent a point in C as a single string of length L taken from the alphabet $\{0,1\}$. In this context two basic genetic operators are used: crossover and mutation. The crossover operation works as follows. Whenever an individual a_i is selected from the current population $A(t)$ to undergo reproduction, a mate a_j is also selected from $A(t)$. Their offspring a_k is produced by concatenating an initial segment of a_i with a final segment from a_j. The segments are defined by selecting at random a crossover point from the $L-1$ possible crossover points. For example, if $a_i = 0010110$ and $a_j = 0101100$, then a crossover occurring between the third and fourth positions in the binary strings would generate $a_k = 0011100$. Thus, the strategy employed by crossover is to construct new individuals from existing high-performance individuals. Notice, however, that crossover will explore only those subspaces of the control space C which are already represented in $A(t)$. If, for example, every individual in $A(t)$ contains a zero in the first gene position, crossover will never generate a new individual with a 1 in that position. A subspace may not be represented in $A(t)$ for several reasons. It may have been deleted by selection due to associated poor performance. It may also be missing because of the limited size of $A(t)$. In the basic reproductive plans this problem is resolved via the second genetic operator: mutation.

The mutation operator generates a new individual by independently modifying one or more of the gene values of an existing individual. For example, a mutation occurring in individual $a_i = 0100110$ at the third position would generate $a_j = 0110110$. In nature as well as in these artificial systems, the probability of a gene undergoing a mutation is less than 0.001 suggesting that it is not a primary genetic operator. Rather, it serves to guarantee that the probability of searching a particular subspace of C is never zero.

V. Analysis of Reproductive Plans

Since the goal of an adaptive strategy is to rapidly generate and maintain good performance of a complex process in an unknown environment, our analysis of reproductive plans will focus on how they search the set C of allowable controls. Recall that elements of C are represented internally as strings drawn from some finite alphabet. For simplicity (and also because of some theoretical justification [9]), a binary representation space A is chosen in which each element in C is represented as a binary string of length L in A.

A Kth-order hyperplane of A can then be defined to be the $(L-K)$-dimensional subspace of A specified by fixing exactly K of the L positions in a string and leaving the remainder as "don't care" elements. These hyperplanes can be represented visually as follows:

$$0\text{-}\text{-}\text{-}\cdots\text{-} \overset{d}{=} \{a_i \in A : v_{i1} = 0\}$$

$$\text{-}\text{-}11\text{-}\cdots\text{-} \overset{d}{=} \{a_i \in A : v_{i2} = 1 \text{ and } v_{i3} = 1\}.$$

If we consider all possible hyperplanes which can be defined on a fixed set of K positions, this set $\{H_i\}$ of Kth-order hyperplanes forms a uniform partition of the space A. For example,

$$H_1 = 0\text{-}\text{-}\text{-}\cdots\text{-}$$

$$H_2 = 1\text{-}\text{-}\text{-}\cdots\text{-}$$

form a first-order partition of A with exactly half of the points falling into each hyperplane. Clearly this is just one of L possible first-order partitions of A, and in general there are $\binom{L}{j}$ jth-order partitions of A giving a grand total of

$$\sum_{j=0}^{L} \binom{L}{j} = 2^L$$

distinct hyperplane partitions of A.

The motivation for focusing on hyperplane partitions is as follows. If we consider the performance measure u_e: $A \to R$ restricted to any particular hyperplane H_i, the restriction u_{e_i} defines a random variable with a well-defined mean and variance which are, of course, unknown to an adaptive strategy. Hence, one can view the problem of searching A for good control points in terms of optimal sampling of the random variables defined by u_e on the various hyperplanes making up a partition of A. A good adaptive strategy will rapidly exploit its accumulating information about u_e to shift its sampling to those hyperplanes which have high expectation of good performance.

The knowledgeable reader will have already noted that this viewpoint puts us in direct contact with statistical decision theory and in particular with K-armed bandit problems. In its simplest form, one is presented with K slot-machines which have varying (unknown) rates of payoff and the problem is to minimize total losses over N trials. The optimal (but alas, nonrealizable) strategy is to play the best paying machine all the time. Lacking this *a priori* information, the problem becomes one of minimizing the expected number of trials to the lower-paying machines. Each trial yields more information about the relative performances of each machine. The goal is to exploit this information as quickly and efficiently as possible.

Note that a distinct K-armed bandit problem can be associated with each hyperplane partition of A. Hence, an adaptive strategy can be viewed as simultaneously solving 2^L K_j-armed bandit problems! The question we will continue to explore in this section is how well reproductive plans allocate trials to these K_j-armed bandits.

In order to accomplish this we fix our attention on a particular hyperplane partition $\{H_i\}$ in relationship to the

INSTITIÚID TEICNEOLAÍOCHTA AN LEABHARLANN LEITIR CEANAINN
S00 25060
629.836

population $A(t)$ of N individuals maintained by a reproductive plan. Sine $\{H_i\}$ is a partition of the space A, each a_{it} in $A(t)$ lies in some H_i. Let $M_i(t)$ represent the number of individuals from $A(t)$ which lie in H_i at time t. Because of the way in which the selection probabilities were defined for reproductive plans (Section IV), we know that the expected number of offspring $O(H_i)$ produced by individuals in H_i at time t is given by

$$O(H_i) = \sum_{j=1}^{M_i(t)} \frac{u_e(a_{jt})}{\bar{u}_e(t)}$$

$$= \frac{M_i(t)}{\bar{u}_e(t)} * \sum_{j=1}^{M_i(t)} \frac{u_e(a_{jt})}{M_i(t)}$$

$$= M_i(t) * \frac{\hat{u}_e(H_i(t))}{\bar{u}_e(t)}.$$

If in fact the offspring $O(H_i)$ themselves lie in H_i, then we have

$$M_i(t+1) = M_i(t) * \frac{\hat{u}_e(H_i(t))}{\bar{u}_e(t)}.$$

That is, the number of trials allocated to H_i varies from one time step to the next in proportion to its performance relative to the average. Hence, we see that the time-varying selection probability distributions provide a highly dynamic sampling bias so that hyperplanes (bandits) whose sampled mean remains consistently above average over a period of time will experience a rapid (exponential) increase in the number of trials allocated to them. This dynamic decisionmaking process which exploits accumulating information about the problem at hand has been analyzed in detail in [9] and [5] and, in particular, has been shown to be a near-optimal solution to the K-armed bandit problem.

This near-optimal behavior was predicated on the fact that the offspring $O(H_i)$ themselves lie in H_i. Whether or not this is true depends on the genetic operators used to construct them. In the basic reproductive plan (RP) there are two such operators: crossover and mutation. An offspring will lie in H_i only if the k positions which define H_i remain unchanged between parent and offspring. Intuitively, if crossover occurs within these defining positions, one or more of them will likely be changed. Hence, it is fairly easy to show that the probability of a parent in H_i producing an offspring outside of H_i is no greater than $\frac{d(H_i)-1}{L-1}$, where $d(H_i)$ is the "definition length" of H_i, namely, the length of the smallest segment containing all the defining positions of H_i. As a consequence we note that crossover has little effect on the allocation of trials to the bandits associated with short-definition hyperplanes (relative to L), while the allocation of trials to long-definition hyperplanes is considerably disrupted.

The probability of a parent in H_i producing an offspring outside H_i via mutation is just $P_m * \frac{k}{L}$, where P_m is the probability of a bit undergoing a mutation and k is the order of H_i. In nature and in reproductive plans $P_m < 0.001$. Hence, mutation has very little effect on the allocation of trials according to performance.

We are now in a position to summarize how reproductive plans search A for high-performance control points. Beginning with an initial random set of 50–100 trials, they generate a near-optimal sequence of trials to short-definition hyperplane partitions. As elements of high-performance hyperplanes begin to dominate $A(t)$, there is an effective reduction in the dimensionality of the search space and a corresponding reduction in the definition lengths of hyperplanes, providing for another cycle of near-optimal sampling.

VI. Examples of Reproductive Plans at Work

So far the discussion has centered around the description and analysis of reproductive plans. In this section we focus on the application of these genetic-based techniques to the problems of designing adaptive systems. The approach will be to illustrate how reproductive plans can be used in two broad areas: parameter optimization problems and control of dynamic systems.

A. Parameter Optimization Problems

Parameter optimization problems arise in a wide variety of contexts including such diverse areas as the design of responsive operating systems, efficient oil refineries, auditoriums with good acoustics, and airplanes with proper aerodynamics. The basic characteristics of such problems are a complex process whose behavior is affected by the values of a set of well-known parameters, a concept of performance of the complex process in a given environment, and long-term stable environments permitting optimization of the performance of the process in a given environment.

As discussed in Section II, the performance measure u_e of a complex process can be viewed as defining over the control space a nontrivial response surface the form of which is generally not known and frequently not amenable to standard optimization techniques. In such cases the problem is easily transformed into a problem amenable to the genetic-based procedures described in the previous sections and is accomplished as follows.

Let $X_1, X_2, \cdots, X_N$ represent the N control parameters of the given complex process. In any practical situation the X_i are subject at the very least to interval constraints of the form $a_i < X_i < b_i$, defining an N-dimensional hypercube of control points. (Note that control spaces with additional constraints can always be embedded in a hypercube via standard penalty function extension techniques.) Since reproductive plan RP works internally with binary strings, the values of X_i in the interval $[a_i, b_i]$ are represented by $L_i = \log_2[(b_i - a_i) * \Delta X_i]$ bits where ΔX_i represents the level of resolution desired. Hence points in the N-dimensional control space to be searched by the reproductive plans are represented by binary strings of length $L = \sum_{i=1}^{N} L_i$ each segment of which represents the value of one of the N parameter settings.

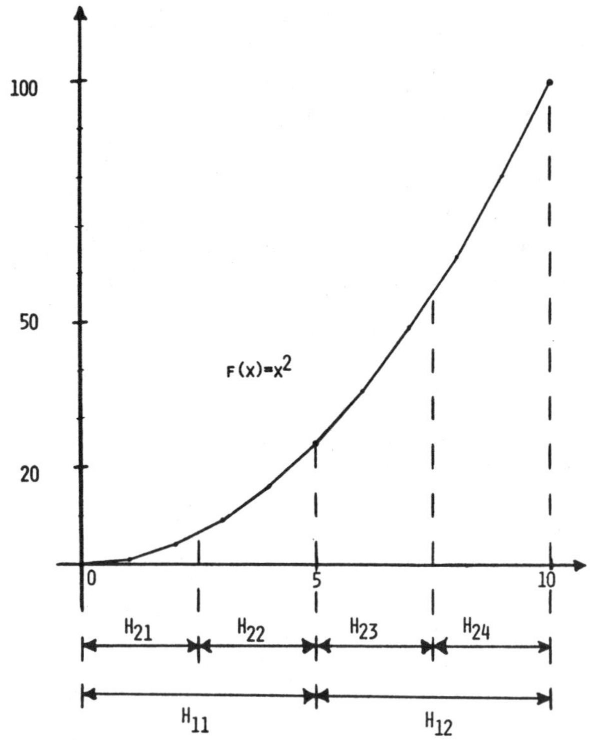

Fig. 3. Sample hyperplane partitions of the control space C.

To visualize how reproductive plans rapidly and efficiently search such spaces, consider a simple example. Suppose a single parameter X is to be optimized over the interval [0, 10] with the precision of two decimal places. Hence, $\log_2[(10-0)*10^2]=10$ bits are required and the correspondence between [0, 10] and the binary representation space A is given by $x=\frac{a_i}{100}$. Suppose further that associated with the complex process is a simple (but, of course, unknown) performance measure to be maximized of the form $f(x)=x^2$.

In order to see how RP searches A, we focus our attention initially on a first-order partition P_1 defined by

$$H_{11}: 0\text{---}\cdots\text{---}$$
$$H_{12}: 1\text{---}\cdots\text{---}.$$

P_1 simply divides the space in half as shown in Fig. 3. Since RP generates an initial population $A(0)$ randomly from a uniform distribution over A, we expect half of $A(0)$ to lie in H_{11} and half in H_{12}. Notice, however, that $\bar{f}(H_{11})<\bar{f}(H_{12})$. Since P_1 is a short-definition partition relative to $L=10$, RP will allocate trials to H_{11} and H_{12} according to the near optimal strategy described earlier. In other words, after relatively few generations the population $A(t)$ consists almost entirely of individuals from H_{12}.

We now consider a refinement P_2 of P_1 given by

$$H_{21}: 00\text{---}\cdots\text{---}$$
$$H_{22}: 01\text{---}\cdots\text{---}$$
$$H_{23}: 10\text{---}\cdots\text{---}$$
$$H_{24}: 11\text{---}\cdots\text{---}$$

P_2 simply divides the space in quarters as illustrated in Fig. 3. As we noted above, RP rapidly generates a population $A(t_1)$ in which most individuals begin with a one. This is in effect a reduction in the search space A to an $L-1$ dimensional space. Hence, after a few generations, P_2 effectively becomes a first-order partition of A_{L-1} to which RP now allocates a near-optimal sequence of trials. Since $\bar{f}(H_{23})<\bar{f}(H_{24})$, RP rapidly generates a population $A(t_2)$ which lies almost entirely in H_{24}, effecting yet another reduction in the search space.

The important thing to note here is that the same remarks hold for any other short-definition partitions, for example:

$$\text{---}\cdots\text{--}0$$
$$\text{---}\cdots\text{--}1.$$

While such partitions are harder to visualize, each is being sampled at a near-optimal rate simultaneously by RP. It is this parallelism which gives even simple reproductive plans like RP their surprising adaptive capabilities.

Because reproductive plans represent a significant shift away from traditional adaptive system design, it was important to collect performance data early particularly in comparison with the more standard techniques. In order to do this in an application-independent setting, a set of five performance measures to be minimized were carefully chosen to include instances of continuous, discontinuous, convex, nonconvex, unimodal, multimodal, low- and high-dimensional functions as well as functions with Gaussian noise.

The basic reproductive plan (RP) was implemented in PL/I and its performance evaluated over the selected set of response surfaces. In addition, three other strategies were evaluated on the same surfaces:

- Rand: a simple strategy consisting of pure uniform random search.
- Praxis: a sophisticated optimization technique based on conjugate direction methods [2].
- DFP: a sophisticated optimization technique based on variable-metric methods [11], [19].

Pure random search provides a lower bound on adaptive system performance in that it uses none of the accumulating information about a response surface to focus its exploration of the control space. Any adaptive system which claims to make use of accumulating information should obviously perform better. The two optimization techniques were chosen in order to provide a clear comparison between state-of-the-art optimization methods and genetic-based strategies. The precise details of the study can be found in [5]; for our purposes the following synopsis will suffice.

Reproductive plans consistently and significantly outperform pure random search. Figs. 4 and 5 give typical performance curves x_e and x_e^* for both techniques averaged over an ensemble of runs on response surface u_e. Recall that x_e and x_e^* themselves are defined in terms of the average values of the response surface u_e at the

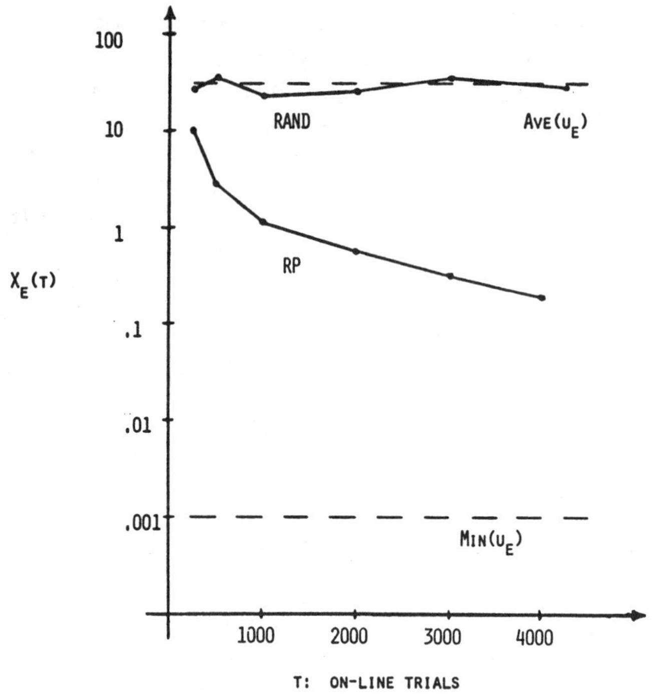

Fig. 4. On-line performance curves x_e for Rand and RP averaged over an ensemble of runs on response surface u_e.

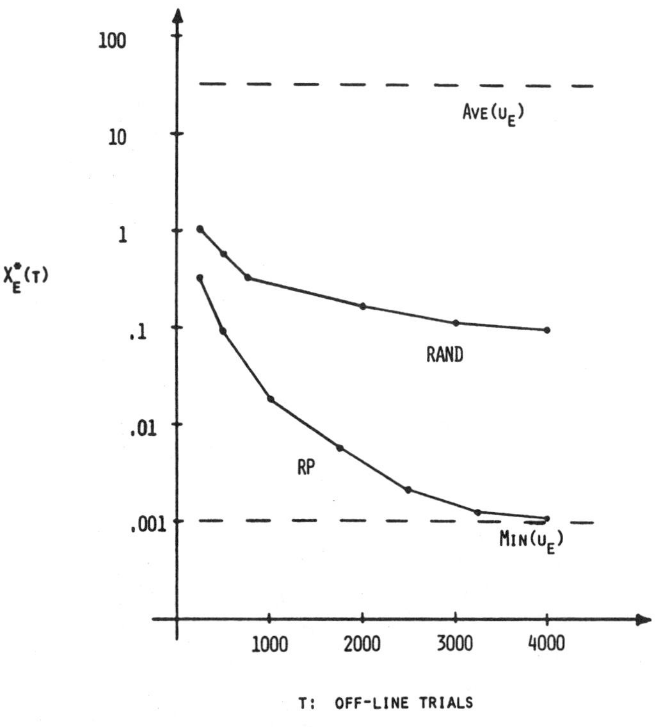

Fig. 5. Off-line performance curves x_e^* for Rand and RP averaged over an ensemble of runs on response surface u_e.

control points generated by Rand and RP over time. So, for example, x_e^* in Fig. 5 illustrates the convergence rate of the *average* value of the trials which, of course, is slower than the convergence rate of the trials themselves. Fig. 4 nicely captures the fact that random search is not using the accumulating information to focus its attention; rather each point in the space remains equally likely and, hence, $x_e(t)$ over an ensemble of runs approximates the average value of the response surface u_e.

Reproductive plans are insensitive to the form of the response surface. They work equally as well on discontinuous, noisy, high-dimensional, and multimodal surfaces. The curves in Fig. 4 are representative of the performance of RP on any one of the test surfaces.

Reproductive plans are efficient in the sense that excellent (but not necessarily optimal) performance is being generated by the time 1000 samples have been taken from large search spaces ($\gg 10^{30}$).

The comparison between reproductive plans and standard optimization techniques were mixed. As one might expect, the optimization routines outperform reproductive plans on those surfaces which satisfied a tight set of constraints, namely, convex, continuous, quadratic-like surfaces. However, on the more difficult discontinuous, noisy, multimodal surfaces reproductive plans were clearly superior.

These encouraging results suggest that reproductive plans hold a valid place parameter optimization strategies falling in between special techniques which work well on limited classes of response surfaces and the inefficient technique of pure random search.

B. *Controlling Dynamic Systems*

Up to this point we have been considering complex processes with response surfaces which are stationary with respect to time. That is, repeated applications of a particular control vector generate identical responses. In this section we consider the application of genetic-based strategies to the control of systems whose response surfaces vary with time. Changes in the response characteristics of a system can be due to environmental changes (e.g., changing traffic patterns relating to a computer-controlled traffic signal) or due to changes in state of the complex process itself (e.g., position in space, the contents of internal memory cells, etc.) and frequently is due to a complex combination of both. Notice, however, that the genetic strategies defined in the previous sections are insensitive to the *source* of these changes. Rather, the sensitivity is related to the rate at which the response surfaces change relative to the sampling rate available to the genetic algorithms for testing and evaluating new control vectors. With this observation in mind, we consider the applications of genetic strategies to three broad classes of changing response surfaces: first, slow long-term drifting changes; second, periods of stability with intermittant abrupt changes; and third, continuously changing surfaces.

1) Drifting Response Surfaces: Systems with response surfaces which drift slowly over time are frequently encountered in manufacturing environments where small variations in the quality of raw materials and/or minor changes to the manufacturing process itself causes some shifting and/or distending of the response surface over time without changing its global features significantly.

Changes of this type have little or no impact on the application of genetic strategies and can be applied as described in previous sections. One can view the long-term effects of mutation as a continuous low-level sampling of the neighborhood region surrounding the best areas of the control space as defined by the current response surface. Simulation studies have confirmed the ease with which reproductive plans "track" movement of this type.

2) Epochal Changes in Response Surfaces: Systems whose response surfaces are essentially stable for long periods of time (epochs) followed by a relatively sudden shift to a significantly new response surface which in turn is reasonably stable are also frequently encountered in adaptive system design. Attempting to design an adaptive traffic controller capable of responding to changing traffic conditions is a good example of this type of change. The various traffic patterns encountered over the period of a day can be viewed as defining a piecewise-stable environment relative to the rate at which sampling of control vectors occurs. Clearly, the genetic strategies will do well up to the first epochal change. However, as defined, they will respond very slowly to an abrupt significant change in the response surface. Fortunately, this problem can be handled rather easily in two quite distinct ways.

The first solution is to provide for the adaptive system a "system reset button" which is enabled by some external sensor (e.g., a clock, a meter, etc.) and essentially restarts the genetic algorithms with a new randomly generated population when a new epoch begins. This, of course, relies on one's ability to sense epochal changes (which, at least in the case of a traffic controller, is not too difficult), but has the virtue of using the genetic strategies as described in the previous sections.

An alternative approach is to rely on the genetic algorithms themselves to adapt not only to small changes in the response surface but also epochal changes. A relatively minor change to the reproductive plans, which was initially studied for totally different reasons, appears to be a solution to this problem. The change, simply put, is to allow the rate of mutation to undergo adaptation itself in parallel with the exploration of the control space. Intuitively, mutation can be viewed as a pressure to expand the scope of search, while crossover together with the selection probabilities defines a pressure to converge toward high performance hyperplanes. Higher mutation rates are desirable during transition into and at the beginning of epochs, while lower rates are desirable as adaptation to the current response surface occurs. Several ways of incorporating changing mutation rates have been/are being studied, the most elegant of which is to associate a mutation rate with individuals (e.g., an extra 8 bits on the chromosome) and allow its values to undergo changes via selection and genetic operators. Encouraging preliminary results using this approach indicate that with no external sensing device required the genetic strategies reduce, then increase and reduce again, the rate of mutation in response to epochal changes.

3) Continuously Changing Response Surfaces: Response surfaces which are continuously changing during the period of adaptation are generally associated with non-Markovian processes (i.e., processes with memory). Typically, the application of a particular control sequence causes an internal change of state (e.g., position in space, contents of a queue, etc.) which significantly changes the response characteristics of the system. Hence, using our earlier notation, the performance function u_e for a given (stable) environment e must be viewed as

$$u_e: C \times X \rightarrow R$$

a function of both the particular control c being selected as well as the current state x of the process. Clearly, to control systems of this type effectively, one must have not only performance feedback but state information as well.

Classical optimal control theory approaches problems of this type by assuming knowledge of the relationship between the application of a control and its effect on the state of the process. This usually takes the form of "equations of motion." For example,

$$\dot{x} = -\frac{1}{a} * x + c$$

describes a typical first-order linear system. By defining an appropriate performance measure (e.g., mean square error from some desired state) optimal control functions can frequently be derived from the "equations of motion." Notice, however, that the availability of such *a priori* information *precludes* the need for an adaptive system as defined in Section I. It is the lack of sufficient *a priori* knowledge about a system which requires a strategy for dynamically acquiring the information to construct an acceptable control function. Attempting to control the performance of complex software systems such as an operating system or a data base management system is a good example of problems of this type. Notice, however, that even having the "equations of motion" for a particular system is not necessarily sufficient *a priori* information. One's ability to solve systems of equations involving nonlinear and interacting terms is often quite limited leading quickly to nonoptimal approximations to the control functions. It is precisely in such situations when the necessary *a priori* information is unavailable that the genetic-based strategies described in the previous sections provide promising alternatives. We will briefly consider two such approaches.

One straightforward approach is to use the genetic algorithms over time to select an appropriate control function from a large (predefined) family of control functions. For example, one might assume (or require) a control function of the form:

$$c(x) = a_0 + a_1 x + \cdots + a_n x^n$$

and use the genetic strategies to rapidly search the n-dimensional space of coefficients for appropriate ones. The performance of a particular set of coefficients is simply defined in terms of the performance of the associated control function when applied to the complex process.

The advantage of such an approach is that it recasts the problem in the well-studied parameter optimization context discussed earlier in this section. The disadvantage is that *a priori* limits are placed on the kinds of control functions that can be "discovered" by genetic plans.

An intriguing alternative approach currently under study attempts to avoid such *a priori* limitations by evolving control functions which are expressed in a special programming language called production systems. Briefly, a production system program consists of a collection of "production rules" of the form

IF ⟨condition⟩ THEN ⟨action⟩

in which the conditions act as detectors, classifiers, etc. of the state space of the complex process being controlled and the actions specify the appropriate controls to be applied. The space to be searched is the set of all such production system programs (of bounded length). The genetic algorithms accomplish this by maintaining a population of such programs, evaluating their fitness as control functions, and then evolving new control programs using the reproductive plans described in the earlier sections. Preliminary results suggest that the genetic algorithms are capable of rapidly generating high-performance control functions (programs) in this form.

VII. Observations and Conclusions

Reproductive plans have been successfully applied to a variety of difficult optimization problems arising in such areas as file system design, data base design, and acoustical design. The motivation in each case was the total intractability of the problem using conventional techniques. The chief virtue of optimization problems is that the area is well-studied and provides a clear-cut testbed on which to evaluate reproductive plans.

The second phase is to study reproductive plans in several distinctly different situations. Currently projects are under way to apply reproductive plans in the areas of efficient approximation algorithms for NP-hard problems, adaptive data base design techniques, adaptive resource management techniques, and self-improving game-playing programs. As discussed in Section VI, preliminary results are encouraging. Because reproductive plans operate simultaneously on a large data base (population) of information, they are relatively insensitive to feedback functions which may vary over time and/or have noise, discontinuities, delays, and failures associated with them.

At the same time it is clear that reproductive plans should not be viewed as a replacement for existing techniques when applicable. Clearly techniques which are capable of exploiting *a priori* knowledge about a problem (e.g., linearity) will in general outperform a technique which makes no such assumptions. Rather, reproductive plans should be viewed as providing a powerful albeit unconventional tool for the adaptive system designer, giving him the potential for solving highly complex problems beyond the reach of existing techniques.

References

[1] M. J. Bauer, "A simulation approach to the design of dynamic feedback scheduling algorithms for time-shared computer systems," *Simuletter, Ass. Comput. Mach. SIGSIM Quart.*, vol. 514, pp. 23–31, 1974.

[2] R. P. Brent, "Algorithms for finding zeros and extrema of functions without calculating derivatives," Ph.D. thesis, Computer Science Dep. Stanford Univ., Stanford, CA, 1971.

[3] D. J. Cavicchio, Jr., "Adaptive search using simulated evolution," Ph.D. thesis, Dep. Computer and Communication Sciences, Univ. Michigan, Ann Arbor, 1970.

[4] J. F. Crow and M. Kimura, *An Introduction to Population Genetics Theory*. New York: Harper & Row, 1970.

[5] K. A. De Jong, "Analysis of the behavior of a class of genetic adaptive systems," Ph.D. thesis, Dep. Computer and Communication Sciences, Univ. Michigan, Ann Arbor, 1975.

[6] N. Y. Foo and J. L. Bosworth, "Algebraic, geometric, and stochastic aspects of genetic operators," Univ. Michigan, Ann Arbor, Tech. Rep. 003120-2-T, 1972.

[7] D. R. Frantz, "Non-linearities in genetic adaptive search," Ph.D. thesis, Dep. Computer and Communication Sciences, Univ. Michigan, Ann Arbor, 1972.

[8] J. D. Hill, "A search technique for multimodal surfaces," *IEEE Trans. Syst. Sci. Cybern.*, vol. SSC-3, no. 1, pp. 2–8, 1969.

[9] J. H. Holland, *Adaptation in Natural and Artificial Systems*. Ann Arbor, MI: Univ. Michigan, 1975.

[10] R. B. Hollstien, "Artificial genetic adaptation in computer control systems," Ph.D. thesis, Dep. Computer and Communication Sciences, Univ. Michigan, Ann Arbor, 1971.

[11] S. L. S. Jacoby, J. S. Kowalik, and J. T. Pizzo, *Iterative Methods for Non-linear Optimization Problems*. Englewood Cliffs, NJ: Prentice-Hall, 1972.

[12] K. Maruyama and S. E. Smith, "Optimal reorganization of distributed space disk files," *Commun. Ass. Comput. Mach.*, vol. 19, no. 11, pp. 634–642, 1976.

[13] L. E. Mettler and T. G. Gregg, *Population Genetics and Evolution*. Foundations of Modern Genetics Series. Englewood Cliffs, NJ: Prentice-Hall, 1969.

[14] D. Potier, E. Gelenbe, and J. Lenfart, "Adaptive allocation of Central Processing Unit Quanta," *J. Ass. Comput. Mach.*, vol. 23, no. 1, pp. 97–102, 1976.

[15] R. Rivest, "On self-organizing sequential search heuristics," Massachusetts Institute of Technology, *Commun. Ass. Comput. Mach.*, vol. 19, no. 2, pp. 63–67, 1976.

[16] H. H. Rosenbrock, "An automatic method for finding the greatest or least value of a function," *The Computer J.*, vol. 3, pp. 175–184, Oct. 1960.

[17] J. Shekel, "Test functions for multimodal search techniques," presented at Fifth Ann. Princeton Conf. Inform. Sci. Syst., Princeton, NJ, 1971.

[18] Y. Z. Tsypkin, *Adaptation and Learning in Automatic Systems*. New York: Academic, 1971.

[19] D. J. Wilde and C. S. Beightler, *Foundations of Optimization*. Englewood Cliffs, NJ: Prentice-Hall, 1969.

Theory and Applications of Adaptive Control—A Survey*

K. J. ÅSTRÖM†

A survey of adaptive control theory and its applications indicates that when designed with adequate practical precautions it may be used successfully in a variety of applications even though many theoretical problems remain to be solved.

Key Words—Adaptive control; model reference; self-tuning regulators; gain scheduling; stability analysis; stochastic control theory; dual control; auto-tuning.

Abstract—Progress in theory and applications of adaptive control is reviewed. Different approaches are discussed with particular emphasis on model reference adaptive systems and self-tuning regulators. Techniques for analysing adaptive systems are discussed. This includes stability and convergence analysis. It is shown that adaptive control laws can also be obtained from stochastic control theory. Issues of importance for applications are covered. This includes parameterization, tuning, and tracking, as well as different ways of using adaptive control. An overview of applications is given. This includes feasibility studies as well as products based on adaptive techniques.

1. INTRODUCTION

ACCORDING to Webster's dictionary to *adapt* means "to change (oneself) so that one's behavior will conform to new or changed circumstances". The words adaptive control have been used at least from the beginning of the 1950s. There is, for example, a patent on an daptive regulator by Caldwell (1950).

Over the years there have been many attempts to define adaptive control (Truxal, 1964; Saridis, Mendel and Nikolic, 1973). Intuitively an adaptive regulator can change its behaviour in response to changes in the dynamics of the process and the disturbances. Since ordinary feedback was introduced for the same purpose the question of the difference between feedback control and adaptive control immediately arises. A meaningful definition of adaptive control which makes it possible to look at a regulator and decide if it is adaptive or not is still missing. There appears, however, to be a consensus that a constant gain feedback is not an adaptive system. In this paper I will therefore take the pragmatic approach that adaptive control is simply a special type of nonlinear feedback control. Adaptive control often has the characteristic that the states of the process can be separated into two categories, which change at different rates. The slowly changing states are viewed as parameters.

Research on adaptive control was very active in the early 1950s. It was motivated by design of autopilots for high performance aircraft. Such aircraft operate over a wide range of speeds and altitudes. It was found that ordinary constant gain, linear feedback can work well in one operating condition. However, difficulties can be encountered when operating conditions change. A more sophisticated regulator which works well over a wide range of operating conditions is therefore needed. The work on adaptive flight control was characterized by a lot of enthusiasm, bad hardware and nonexisting theory. A presentation of the results is given in Gregory (1959) and Mishkin and Braun (1961). Interest in the area diminished due to lack of insight and a disaster in a flight test (see Taylor and Adkins, 1965).

In the 1960s there were many contributions to control theory, which were important for the development of adaptive control. State space and stability theory were introduced. There were also important results in stochastic control theory. Dynamic programming, introduced by Bellman (1957, 1961) and dual control theory introduced by Feldbaum (1960a, b, 1961a, b, 1965), increased the understanding of adaptive processes. Fundamental contributions were also made by Tsypkin (1971), who showed that many schemes for learning and adaptive control could be described in a common framework as recursive equations of the stochastic approximation type. There were also major developments in system identification and in parameter estimation (Åström and Eykhoff, 1971).

The interest in adaptive control was renewed in the 1970s. The progress in control theory during the previous decade contributed to an improved understanding of adaptive control. The rapid and revolutionary progress in microelectronics has made it possible to implement adaptive regulators simply and cheaply. There is now a vigorous development of the field both at universities and in industry.

There are several surveys on adaptive control. The early work was surveyed by Aseltine, Mancini and Sarture (1958); Stromer (1959) and Jacobs (1961). Surveys of special areas in the field are given by Landau (1974); Wittenmark (1975); Unbehauen and Schmidt (1975); Parks, Schaufelberger and Schmid (1980). The papers by Truxal (1964) and Tsypkin (1973) also given an enlightening perspective. An extensive bibliography which covers more than 700 papers is given by Asher, Andrisani and Dorato (1976). Three books, Narendra and Monopoli (1980); Unbehauen (1980); Harris and Billings (1981), contains representative collections of papers dealing with recent applications.

When selecting the material for this paper I deliberately choose to focus on the simplest types of adaptive regulators. The idea is to describe the principles behind those adaptive schemes which are now finding their way towards applications and products. This means that many interesting adaptive schemes are left out. Self-optimizing controls were recently surveyed by Sternby (1980). Other forms of adaptation that occurs in learning systems and in biological systems are described in Saradis (1977); Mendel and Fu (1970).

2. APPROACHES TO ADAPTIVE CONTROL

Three schemes for parameter adaptive control—gain scheduling, model reference control and self-tuning regulators are described in a common framework. The starting point is an ordinary feedback control loop with a process and a regulator with adjustable parameters. The key problem is to find a convenient way of changing the regulator parameters in response

*Received 2 December 1981; revised 4 April 1983; revised 2 May 1983. This paper is an expanded and updated version of a plenary lecture given at the 8th IFAC World Congress on Control Science and Technology for the Progress of Society which was held in Koyto, Japan during August 1981. The published proceedings of this IFAC meeting may be ordered from Pergamon Press Ltd, Headington Hill Hall, Oxford OX3 0BW, U.K. This paper was recommended for publication in revised form by editor G. Saridis.

†Department of Automatic Control, Lund Institute of Technology, Box 725, S-220 07 Lund, Sweden.

Reprinted with permission from *Automatica*, vol. 19, pp. 471–486, Sept. 1983.
Copyright © 1983, International Federation of Automatic Control IFAC. Published by Pergamon Press Ltd.

to changes in process and disturbance dynamics. The schemes differ only in the way the parameters of the regulator are adjusted.

Gain scheduling

It is sometimes possible to find auxiliary variables which correlate well with the changes in process dynamics. It is then possible to reduce the effects of parameter variations by changing the parameters of the regulator as functions of the auxiliary variables (Fig. 1). This approach is called *gain scheduling* because the scheme was originally used to accommodate changes in process gain only.

The concept of gain scheduling originated in connection with development of flight control systems. In this application the Mach-number and the dynamic pressure are measured by air data sensors and used as scheduling variables.

The key problem in the design of systems with gain scheduling is to find suitable scheduling variables. This is normally done based on knowledge of the physics of a system. For process control the production rate can often be chosen as a scheduling variable since time constants and time delays are often inversely proportional to production rate.

When scheduling variables have been obtained, the regulator parameters are determined at a number of operating conditions using some suitable design method. Stability and performance of the system are typically evaluated by simulation. Particular attention is given to the transition between different operating conditions. The number of operating conditions are increased if necessary.

It is sometimes possible to obtain gain schedules by introducing normalized dimension-free parameters in such a way that the normalized model does not depend on the operating conditions. The auxiliary measurements are used together with the process measurements to calculate the normalized measurement variables. The normalized control variable is calculated and retransformed before it is applied to the process. An example of this is given in Källström and co-workers (1979). The flow control scheme proposed by Niemi (1981) is also of this type. The regulator obtained can be regarded as composed of two nonlinear static systems with a linear regulator in between. Sometimes the calculation of the normalized variables is based on variables obtained by Kalman filtering. The system then becomes even more complex.

One drawback of gain scheduling is that it is an open-loop compensation. There is no feedback which compensates for an incorrect schedule. Gain scheduling can thus be viewed as feedback control system where the feedback gains are adjusted by feed forward compensation. Another drawback of gain scheduling is that the design is time consuming. The regulator parameters must be determined for many operating conditions. The performance must be checked by extensive simulations. This difficulty is partly avoided if scheduling is based on normalized variables. Gain scheduling has the advantage that the parameters can be changed very quickly in response to process changes. The limiting factors depend on how quickly the auxiliary measurements respond to process changes.

There is a controversy in nomenclature whether gain scheduling should be considered as an adaptive system or not because the parameters are changed in open loop. Irrespective of this discussion, gain scheduling is a very useful technique to reduce the effects of parameter variations. It is in fact the predominant method to handle parameter variations in flight control systems (Stein, 1980). There is a commercial regulator for process control Micro-Scan 1300 made by Taylor Instruments which is based on gain scheduling (Andreiev, 1977).

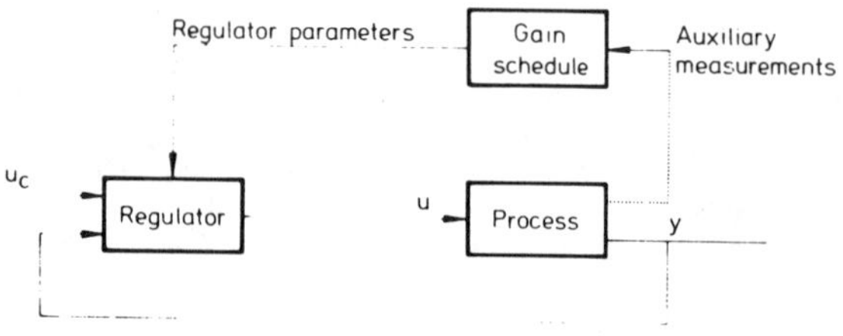

FIG. 1. Block diagram of a system where influences of parameter variations are reduced by gain scheduling.

Model reference adaptive systems MRAS

Another way to adjust the parameters of the regulator is shown in Fig. 2. This scheme was originally developed by Whitaker, Yamron and Kezer (1958) for the servo problem. The specifications are given in terms of a reference model which tells how the process output ideally should respond to the command signal. Notice that the reference model is part of the control system. The regulator can be thought of as consisting of two loops. The inner loop is an ordinary control loop composed of the process and the regulator. The parameters of the regulator are adjusted by the outer loop, in such a way that the error e between the model output y_m and the process output y becomes small. The outer loop is thus also a regulator loop. The key problem is to determine the adjustment mechanism so that a stable system which brings the error to zero is obtained. This problem is nontrivial. It is easy to show that it cannot be solved with a simple linear feedback from the error to the controller parameters.

The following parameter adjustment mechanism, called the 'MIT-rule', was used in the original MRAS

$$\frac{d\theta}{dt} = ke\,\text{grad}_\theta\, e. \tag{1}$$

In this equation e denotes the model error. The components of the vector θ are the adjustable regulator parameters. The components of the vector $\text{grad}_\theta\, e$ are the sensitivity derivatives of the error with respect to the adjustable parameters. The sensitivity derivatives can be generated as outputs of a linear system driven by process inputs and outputs. The number k is a parameter which determines the adaptation rate.

Whitaker and co-workers motivated the rule (1) as follows. Assume that the parameters θ change much slower than the other system variables. To make the square of the error e small it seems reasonable to change the parameters in the direction of the negative gradient of e^2.

If (1) is rewritten as

$$\theta(t) = -k \int^t e(s)\,\text{grad}_\theta\, e(s)\,ds$$

it is seen that the adjustment mechanism can be thought of as composed of three parts: a linear filter for computing the sensitivity derivatives from process inputs and outputs, a multiplier and an integrator. This configuration is typical for many adaptive systems.

The MIT-rule will perform well if the parameter k is small. The allowable size depends on the magnitude of the reference signal. Consequently it is not possible to give fixed limits which guarantee stability. The MIT-rule can thus give an unstable closed loop system. Modified adjustment rules can be obtained using stability theory. These rules are similar to the MIT-rule. The sensitivity derivatives in (1) will be replaced by other functions. This is discussed further in Section 3.

The MRAS was originally proposed by Whitaker and co-workers (1958). Further work was done by Parks (1966), Hang and Parks (1973), Monopoli (1973), Landau (1974) and Ionescu and Monopoli (1977). There has been a steady interest in the method (Hang and Parks, 1973). Landau's book (Landau, 1979)

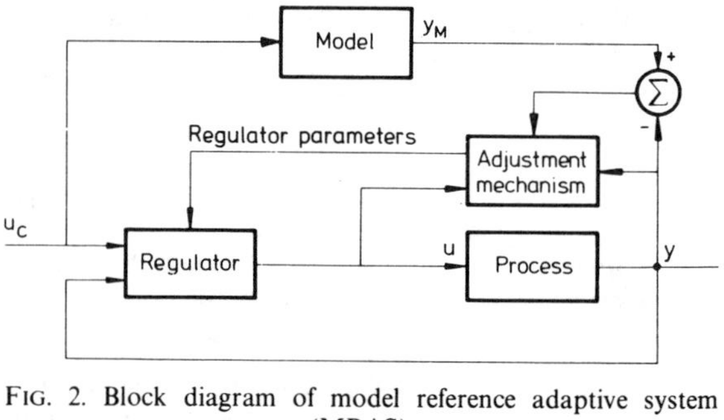

FIG. 2. Block diagram of model reference adaptive system (MRAS).

gives a comprehensive treatment of work up to 1978. It also includes many references. Recent contributions are discussed in Section 4.

The MRAS shown in Fig. 2 is called a *direct* scheme because the regulator parameters are updated directly. There are also other MRAS schemes where the regulator parameters are updated *indirectly* (Narendra and Valavani, 1979).

Self-tuning regulators STR

A third method for adjusting the parameters is to use the self-tuning regulator. Such a system is shown in Fig. 3. The regulator can be thought of as composed of two loops. The inner loop consists of the process and an ordinary linear feedback regulator. The parameters of the regulator are adjusted by the outer loop, which is composed of a recursive parameter estimator and a design calculation. To obtain good estimates it may also be necessary to introduce perturbation signals. This function is not shown in Fig. 3 in order to keep the figure simple.

Notice that the box labelled 'regulator design' in Fig. 3 represents an on-line solution to a design problem for a system with known parameters. This is called the *underlying design problem*. Such a problem can be associated with most adaptive control schemes. However, the problem is often given indirectly. To evaluate adaptive control schemes it is often useful to find the underlying design problem because it will give the characteristics of the system under the ideal conditions when the parameters are known exactly.

The self-tuning regulator was originally developed for the stochastic minimum variance control problem (Åström and Wittenmark, 1973). Since the approach is very flexible with respect to the underlying design method many different extensions have been made. Self-tuners based on phase and amplitude margins are discussed in Åström (1982). Pole-placement self-tuners have been investigated by many authors: Edmunds (1976); Wouters (1977); Wellstead and co-workers (1979); Wellstead, Prager and Zanker; Åström and Wittenmark (1980). Minimum variance self-tuners with different extensions are treated in Peterka (1970, 1982); Åström and Wittenmark (1973, 1974); Clarke and Gawthrop (1975, 1979); Gawthrop (1977). The LQG design method is the basis for the self-tuners presented in Peterka and Åström (1973); Åström (1974); Åström and Zhao-ying (1982); Menga and Mosca (1980).

The self-tuner also contains a recursive parameter estimator. Many different estimation schemes have been used, for example stochastic approximation, least squares, extended and generalized least squares, instrumental variables, extended Kalman filtering and the maximum likelihood method.

The self-tuning regulator was originally proposed by Kalman (1958), who built a special-purpose computer to implement the regulator. Several experimental investigations were carried out as digital computers became available. The self-tuning regulator has recently received considerable attention because it is flexible, easy to understand and easy to implement with microprocessors (Åström, 1980a; Kurz, Isermann and Schumann 1980; Isermann, 1980a).

The self-tuner shown in Fig. 3 is called an *explicit* STR or an STR based on estimation of an explicit process model. It is sometimes possible to re-parameterize the process so that it can be expressed in terms of the regulator parameters. This gives a significant simplification of the algorithm because the design calculations are eliminated. Such a self-tuner is called an *implicit* STR because it is based on estimation of an implicit process model.

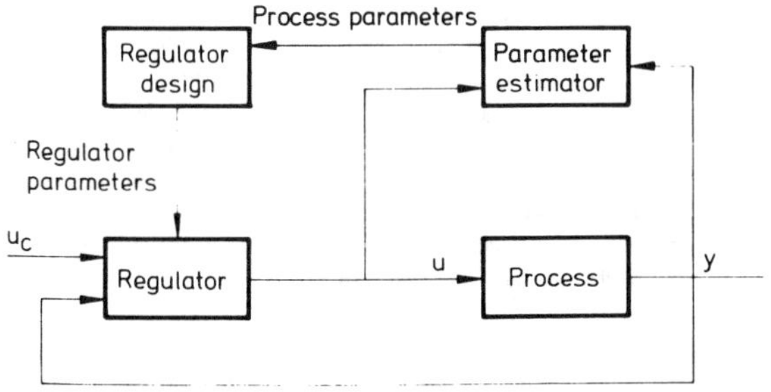

FIG. 3. Block diagram of a self-tuning regulator (STR).

Relations between MRAS and STR

The MRAS was obtained by considering a deterministic servo-problem and the STR by considering a stochastic regulation problem. In spite of the differences in their origin it is clear from Figs. 2 and 3 that the MRAS and the STR are closely related. Both systems have two feedback loops. The inner loop is an ordinary feedback loop with a process and a regulator. The regulator has adjustable parameters which are set by the outer loop. The adjustments are based on feedback from the process inputs and outputs. The methods for design of the inner loop and the techniques used to adjust the parameters in the outer loop may be different, however.

The direct MRAS is closely related to the implicit STR and the indirect MRAS to the explicit STR. The relations between STR and MRAS are further elaborated in the next Section. More details are found in Egardt (1979b, 1980a); Landau (1979); Åström (1980a).

3. AN EXAMPLE

To give a better insight into the different approaches the algorithms will be worked out in detail. The example also serves to introduce some notation needed for the theory in Section 4. A design problem for systems with known parameters is first described. Different adaptive control laws are then given. A pole-placement design is chosen as the underlying design problem. This is useful in order to discuss similarities and differences between self-tuners and model reference adaptive systems. It is also a convenient way to unify many algorithms. Additional details about the design procedure are given in Åström and Wittenmark (1984).

The underlying design problem for known systems

Consider a single input–single output, discrete time system described by

$$Ay = Bu \tag{2}$$

where u is the control signal and y the output signal. The symbols A and B denote relatively prime polynomials in the forward shift operator. Assume that it is desired to find a regulator such that the relation between the command signal and the desired output signal is given by

$$A_m y_m = B_m u_c \tag{3}$$

where A_m and B_m are polynomials.

A general linear control law which combines feedback and feedforward is given by

$$Ru = Tu_c - Sy \tag{4}$$

where R, S, and T are polynomials. This control law represents a negative feedback with the transfer function $-S/R$ and a feed forward with the transfer function T/R.

Elimination of u between (2) and (4) gives the following equation for the closed-loop system

$$(AR + BS)y = BTu_c. \tag{5}$$

The process zeros, given by $B = 0$, will thus be closed-loop zeros unless they are cancelled by corresponding closed-loop poles. Since unstable or poorly damped zeros cannot be cancelled the polynomial B is factored as

$$B = B^+B^- \tag{6}$$

where B^+ contains those factors which can be cancelled and B^- the remaining factors of B. The zeros of B^+ must of course be stable and well damped. To make the factorization unique it is also required that B^+ is monic.

It follows from (5) that the characteristic polynomial of the closed-loop system is $AR + BS$. This polynominal may be designed to have three types of factors: the cancelled process zeros, the desired model poles and the desired observer* poles. Let B^+, A_m and A_0 denote these factors then

$$AR + BS = B^+ A_m A_0. \quad (7)$$

Since B^+ divides B it follows from this equation that B^+ divides R. Hence

$$R = B^+ R_1. \quad (8)$$

Equation (7) can then be written as

$$AR_1 + B^- S = A_0 A_m. \quad (7')$$

Requiring that the relation (5) between the command signal and the process output should be equal to the desired relation (3) gives

$$B_m = B^- B_m^+ \quad (9)$$

$$T = A_0 B_m^+. \quad (10)$$

The specifications must thus be such that B^- is a factor of B_m otherwise there is no solution to the design problem.

To complete the solution of the problem it remains to give conditions which guarantee that there exist solutions to (7) which give a causal control law. The feedback transfer function S/R is causal if

$$\deg S \leq \deg R.$$

It is thus desirable to find a solution S to (7) which has as low degree as possible. There is always a solution with

$$\deg S \leq \deg A - 1.$$

It then follows from (7) that

$$\deg R = \deg A_0 + \deg A_m + \deg B^+ - \deg A.$$

The condition

$$\deg A_0 \geq 2 \deg A - \deg A_m - \deg B^+ - 1 \quad (11)$$

thus guarantees that the feedback transfer function S/R is causal. Similarly the inequality

$$\deg A_m - \deg B_m \geq \deg A - \deg B. \quad (12)$$

implies that the feed forward transfer function T/R is causal.

To solve the pole-placement design problem which give the selected poles A_m and A_0, with A and B given, the equation (7′) is first solved to obtain R_1 and S. The desired feedback is then given by (4) with R and T given by (8) and (10).

There may be several solutions to the Diophantine equation (7) which satisfy the causality conditions (11) and (12).

All solutions will give the same closed-loop transfer function. The different solutions give however different responses to disturbances and measurement errors.

It follows from (2), (3), (7′) and (8)–(10) that the control law (4) can be written as

$$u = G_m G_p^{-1} u_c - \frac{S}{R}[y - y_m]$$

where

$$G_p = B/A, G_m = B_m/A_m, \text{ and } y_m = G_m u_c.$$

*The introduction of the observer polynomial A_0 is guided by the state-space solution to the problem which is a combination of state feedback and an observer. For details see Åström and Wittenmark (1984).

This shows that the pole-placement design can be interpreted as model following. This is important in order to establish the relations between the STR and the MRAS. Equation (4) is, however, preferable in realizations.

Parameter estimation

The control law (4) is not realizable if the parameters of the model (2) are unknown. However, the parameters may be estimated. There are many ways to do this (Ljung and Söderström, 1983). Many estimators can be described by the recursive equation

$$\theta(t) = \theta(t-1) + P(t)\varphi(t)\varepsilon(t) \quad (13)$$

where the components of the vector θ are the estimated parameters, the vector φ is a regression vector, and ε is the prediction error. The quantities φ and ε depend on the identification method and the model structure. For example if the least-squares method is applied to the model (2) the prediction error is given by

$$\varepsilon(t) = q^{-\deg A}[A(q)y(t) - B(q)u(t)]$$
$$= y(t) - \varphi^T(t)\theta(t-1)$$

where

$$\varphi(t) = [-y(t-1), \ldots, -y(t - \deg A)u(t-d), \ldots, u(t - \deg A)]$$

and

$$d = \deg A - \deg B.$$

The elements of the vector φ are thus delayed values of the input u and the output y.

The quantity P in (13) depends on the particular estimation technique. It may be a constant which gives an updating formula similar to the MIT-rule (1).

Another method (Kaczmarz, 1937) can be viewed as a recursive solution of a set of linear equations. This method is described by (13) with P as the scalar

$$P(t) = \frac{1}{\varphi^T(t)\varphi(t)}. \quad (14)$$

In stochastic approximation methods P is a scalar given by

$$P(t) = \left[\sum_{k=1}^{t} \varphi^T(k)\varphi(k)\right]^{-1}. \quad (15)$$

The recursive least-squares method is given by (13) with

$$P(t) = \left[\sum_{k=1}^{t} \varphi(k)\varphi^T(k)\right]^{-1}. \quad (16)$$

Some minor modifications have to be made if the denominator in (14) and (15) are zero or if the matrix in (16) is singular.

The properties of the estimates depend on the model and the disturbances. In the deterministic case, when there are no disturbances, estimates converge to the correct value in a finite number of steps. The algorithms with P given by (16) have this property, for example. Algorithms with a constant P converge exponentially provided that there is persistent excitation. When data is generated by (2) with independent random variables added to the right-hand side it is necessary to have algorithms where $P(t)$ goes to zero for increasing t in order to get estimates which converge to the correct value. This is the case when P is given by (15) or (16). These algorithms are said to have decreasing gain. An algorithm with decreasing gain is, however, useless when the process parameters are changing. For such a case (14) can be used or (16) can be replaced by

$$P(t) = \left[\sum_{k=1}^{t} \lambda^{t-k}\varphi(k)\varphi^T(k)\right]^{-1} \quad (17)$$

where $0 \leq \lambda \leq 1$ ia a *forgetting factor or a discounting factor*. This choice of P corresponds to a least-squares estimate with an exponential discounting of past data.

An explicit self-tuner

An explicit self-tuner based on the pole-placement design can be expressed as follows.

Algorithm 1.

Step 1. Estimate the coefficients of the polynomials A and B in (2) recursively using (13) with (14), (15), (16), or (17).

Step 2. Substitute A and B by the estimates obtained in step 1 and solve (7′) to obtain R_1 and S. Calculate R by (8) and T by (10).

Step 3. Calculate the cntrol signal from (4).

Repeat the steps 1–3 at each sampling period. □

An implicit self-tuner

In the implicit self-tuner the design calculations are eliminated and the regulator parameters are updated directly. The algorithm can be derived as follows. One has

$$A_m A_0 y = AR_1 y + B^- Sy = BR_1 u + B^- Sy = B^- [Ru + Sy] \quad (18)$$

where the first equality follows from (7′), the second from (2), and the third from (8). Notice that (18) can be interpreted as a process model, which is parameterized in B^-, R, and S. An estimation of the parameters of the model (18) gives the regulator parameters directly. A solution to the bilinear estimation problem is given in Åström (1980c). In the special case of minimum-phase systems when $B^- = b_0$ the implicit algorithm can be expressed as follows.

Algorithm 2.

Step 1. Estimate the coefficients of the polynomials R, S in (18) recursively using (13) with

$$\varepsilon(t) = q^{-\deg A_0 A_m}[A_0 A_m y(t) - b_0\{Ru(t) + Sy(t)\}]$$

$$= q^{-\deg A_0 A_m} A_0 A_m y(t) - \varphi^T(t)\theta(t-1)$$

where

$$\varphi(t) = [-y(t-d), \ldots, -y(t-d-\deg S)\ b_0 u(t-d), \ldots, b_0 u(t-d-\deg R)]$$

with

$$d = \deg A - \deg B$$

and (14), (15), (16), or (17).

Step 2. Calculate the control signal from (4), with R and S substituted by their estimates obtained in step 1.

Repeat the steps 1 and 2 at each sampling period. □

The simple self-tuner in Åström and Wittenmark (1973) corresponds to this algorithm with P given by (17).

Other implicit self-tuners

Algorithm 2 is based on a re-parameterization of the process model (2). The re-parameterization is nontrivial in the sense that (18) has more parameters than (2). The parameterization (18) has the drawback that the model obtained is not linear in the parameters. This makes the parameter estimation more difficult. It is thus natural to investigate other parameterizations. One possibility is to write the model (18) as

$$A_0 A_m y = \mathcal{R}u + \mathcal{S}y \quad (19)$$

where

$$\mathcal{R} = B^- R$$

$$\mathcal{S} = B^- S.$$

The estimated polynomials will then have a common factor B^- which represents the unstable modes. To avoid cancellation of such modes it is then necessary to cancel the common factor before calculating the control law. The following control algorithm is then obtained.

Algorithm 3.

Step 1. Estimate the coefficients of the polynomials $\mathcal{R}$ and $\mathcal{S}$ in the model (19).

Step 2. Cancel possible factors in $\mathcal{R}$ and $\mathcal{S}$ to obtain R and S.

Step 3. Calculate the control signal from (4) where R and S are those obtained in step 2.

Repeat the steps 1–3 at each sampling period. □

This algorithm clearly avoids a nonlinear estimation problem. There are, however, more parameters to estimate than in Algorithm 2 because the parameters of the polynomial B^- are estimated twice.

There are several other possibilities. For the case $B^+ = $ const it is possible to proceed as follows. Write the model (2) as

$$Az = u$$
$$y = Bz. \quad (20)$$

If the polynomials A and B are coprime there exist two polynomials U and V such that

$$UA + VB = 1 \quad (21)$$

it follows from (10), (20), and (21) that

$$A_0 A_m z = A_0 A_m (UA + VB) z = (RA + SB) z.$$

Equation (20) gives

$$A_0 A_m Uu + A_0 A_m Vy - Ru - Sy = 0.$$

or

$$U(A_0 A_m u) + V(A_0 A_m y) - Ru - Sy = 0. \quad (22)$$

Notice that this equation is linear in the parameters. An adaptive algorithm similar to Algorithm 3 can be constructed based on (22). This was proposed by Elliott (1982). Difficulties with this algorithm have been reported by Johnson, Lawrence and Lyons (1982).

Relations to other algorithms

The simple example is useful because it allows a unification of many different adaptive algorithms. The simple self-tuner based on least-squares estimation and minimum variance control introduced in Åström and Wittenmark (1973) is the special case of Algorithm 2 with $A_0 A_m = z^m$ and $B^- = b_0$.

The model reference adaptive algorithm in Monopoli (1974) is the special case of Algorithm 2 with $B^- = b_0$ and an estimator (13) with P being a constant scalar. The self-tuning controller of Clarke and Gawthrop (1975, 1979) corresponds to the same design with least-squares estimation.

The self-tuning pole placement algorithms of Wellstead, Prager and Zanker (1979) are equivalent to Algorithm 1.

A class of algorithms called IDCOM have been introduced by Richalet and co-workers (1978). These algorithms are based on the idea of estimating an impulse response model of the process and using a control design which gives an exponential decay of disturbances. A simple version of IDCOM may be viewed as the special case of Algorithm 2 with $A_0 A_m = z^{m-1}(z-a)$ and $R = r_0 z^k$.

4. THEORY

The closed-loop systems obtained with adaptive control are nonlinear. This makes analysis difficult, particularly if there are random disturbances. Progress in theory has therefore been slow and painstaking. Current theory gives insight into some special problems. Much work still remains before a reasonably complete theory is available. Analysis of stability, convergence, and performance are key problems. Another purpose of the theory is to find out if control structures like those in Section 2 are reasonable, or if there are better ways to do adaptive control.

Stability

Stability is a basic requirement on a control system. Much effort has been devoted to stability analysis of adaptive systems. It is important to keep in mind that the stability concepts for nonlinear differential equations refer to stability of a particular solution. It may thus happen that one solution is stable and that another solution is unstable.

Stability analysis has not been applied to systems with gain scheduling. This is surprising since such systems are simpler than MRAS and STR.

The stability theories of Liapunov and Popov have been extensively applied to adaptive control. The major developments of MRAS were all inspired by the desire to construct adjustment mechanisms, which would give stable solutions. Parks (1966) applied Liapunov theory to the general MRAS problem for systems with state feedback and also output feedback for systems, whose transfer functions are strictly positive real. Landau (1979) applied hyperstability to a wide variety of MRAS configurations. The key observation in all these works is that the closed-loop system can be represented as shown in Fig. 4. The system can thus be viewed as composed of a linear system and a nonlinear passive system. If the linear system is strictly positive real, it follows from the passivity theorem that the error e goes to zero. See Desoer and Vidyasagar (1975), for example.

To obtain the desired representation it is necessary to parameterize the model so that it is *linear in the parameters.* The model should thus be of the form

$$y(t) = \varphi^{\mathrm{T}}(t)\theta.$$

This requirement strongly limits the algorithms that can be considered.

The general problem with output feedback poses additional problems, because it is not possible to obtain the desired representation by filtering the model error. Monopoli (1974) showed that it is necessary to augment the error by adding additional signals. For systems with output feedback the variable ε in Fig. 4 should thus be the augmented error.

There are some important details in the stability proofs based on Fig. 4. To ensure stability it must be shown that the vector φ is bounded. This is easy for systems which only have a variable gain, because the regression vector φ has only one component which is the command signal. The components of the vector φ are, however, in general functions of the process inputs and outputs. It is then a nontrivial problem to ensure that φ is bounded. It should also be noticed that it follows from the passivity theorem that ε goes to zero. The parameter error will not go to zero unless the matrix $\Sigma\varphi\varphi^{\mathrm{T}}/t$ is always larger than a positive definite matrix.

When the transfer function $G(s)$ is not positive real there is an additional difficulty because the signal ε is the augmented error. It thus remains to show that the model error also goes to zero.

Several of these difficulties remained unnoticed for many years. The difficulties were pointed out in Morgan and Narendra (1977); Feuer and Morse (1978). Stability proofs were given recently by Egardt (1979a); Fuchs (1979); Goodwin, Ramadge and Caines (1980); Gawthrop (1980); de Larminat (1979); Morse (1980); Narendra and Lin (1979); Narendra, Lin and Valavani (1980). The following result is due to Goodwin, Ramadge and Caines (1980).

Theorem 1. Let the system (2) be controlled by the adaptive Algorithm 2 with $A_0A_m = z^m$, $B^- = b_0$ and Kaczmarz method of parameter estimation, i.e. (14). Assume that

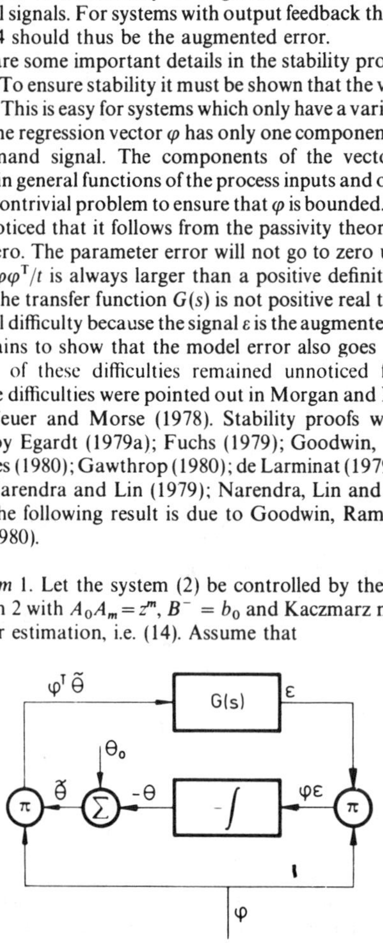

FIG. 4. Block diagram of a MRAS redrawn for the purpose of applying hyperstability theory. The variable ε is the (filtered) model error, φ is a vector of regression variables, θ are the adjustable parameters and θ_0 their true values.

(A1) the pole excess $d = \deg A - \deg B$ is known;

(A2) the estimated model is at least of the same order as the process;

(A3) the polynomial B has all zeros inside the unit disc.

The signals u and y are then bounded and $y(t)$ approaches the command signal $y_c(t)$ as time goes to infinity. □

The proof is not based on hyperstability theory. It is an analysis based on the particular structure of the problem. Notice that the theorem does not say that the parameter estimates converge to their true values.

Theorem 1 is important, because it is a simple and rigorous stability proof for a reasonable adaptive problem. The assumptions required are however very restrictive.

The Assumption A1 means for discrete systems that the time delay is known with a precision, which corresponds to a sampling period. This is not unreasonable. For continuous time systems the assumption means that the slope of the high frequency asymptote of the Bode diagram is known. If this is the case, it is possible to design a robust high gain regulator for the problem (Horowitz, 1963).

Assumption A2 is very restrictive, since it implies that the estimated model must be at least as complex as the true system, which may be nonlinear with distributed parameters. Almost all control systems are in fact designed based on strongly simplified models. High frequency dynamics are often neglected in the simplified models. It is therefore very important that a design method can cope with model uncertainty (Bode, 1945; Horowitz, 1963). The parameters obtained for low-order models depend critically on the frequency content of the input signal (Mannerfelt, 1981). In adaptive control based on low order models it is important that parameters are not updated unless the input signal has sufficient energy at the relevant frequencies.

Assumption A3 is also crucial. It arises from the necessity to have a model, which is linear in the parameters. It follows from (8) that this is possible only if $B^- = b_0$. In other words the underlying design method is based on cancellation of all process zeros. Such a design will not work even for systems with known constant parameters if the system has an unstable inverse.

Notice that Theorem 1 requires that there are no disturbances. The analysis by Egardt (1979a, 1980b, c) also applies to the case when there are disturbances. Egardt has given counterexamples which show that modifications of the algorithms or additional assumptions are necessary if there are disturbances. One possibility is to bound the parameter estimates *a priori* for example by introducing a saturation in the estimator. Another possibility is to introduce a dead zone in the estimator which keeps the estimates constant if the residuals are small. These results also hold for continuous time systems as has also been shown by Narendra and Peterson (1981).

Convergence analysis

Determination of convergence conditions, possible convergence points and convergence rates are the essential problems in convergence analysis.

Convergence analysis reduces to analysis of equations such as (13). Such problems are extensively discussed in the literature on system identification (Eykhoff, 1981). However, there are two complications in the adaptive case. Since the process input is generated by feedback the excitation of the process depends on the process disturbances. It is then difficult to show that the input is persistently exciting, a necessity for convergence of the estimate. The input is also correlated with the disturbances because it is generated by feedback, the regression vector φ in (13) will then also depend on past estimates. This means that (13) is not a simple state equation.

For the purpose of convergence analysis it is commonly assumed that the system is driven by disturbances. It is then possible to apply the powerful results from ergodic theory and martingale theory to establish convergence of some adaptive schemes. There are also some other methods which give more of a system theoretic insight.

Martingale theory

A very general proof for convergence of the least squares algorithm was given by Sternby (1977) by applying a martingale convergence theorem. The convergence condition is simply that $P(t) \to 0$ as $t \to \infty$. An extension of this result was applied to adaptive systems (Sternby and Rootzen, 1982). The results are limited to the model

$$Ay = Bu + e$$

where e is white noise. The approach is Bayesian which means that the parameters are assumed to be random variables. This poses some conceptual difficulties because nothing can be said about convergence for particular values of the parameters.

A convergence theorem for the simple self-tuner based on modified stochastic approximation estimation, and minimum variance control was given by Goodwin, Ramadge and Caines (1981). This case corresponds to the special case of Algorithm 2. A system described by the model

$$Ay = Bu + Ce \tag{23}$$

where e is white noise was investigated. Application of a martingale convergence theorem gave the following result.

Theorem 2. Let the process (23) be controlled by Algorithm 2 with $A_0 A_m = z^m$, $B^- = b_0$ and $d = 1$ about a modified stochastic approximation identification, i.e. (13) with

$$P(t) = a_0/t$$

Let Assumptions A1–A3 of Theorem 1 hold and assume moreover that the function

$$G(z) = C(z) - a_0/2 \tag{24}$$

is strictly positive real. Then the inputs and the outputs are mean-square bounded and the adaptive regulator converges to the minimum variance regulator for the system (23). □

Various extensions to larger pole excess and different modifications of the least-squares estimation have been given. A convergence proof for the general Algorithm 2 with least-squares estimation is, however, still not available.

Averaging methods

The algorithms in Section 2 are motivated by the assumption that the parameters change slower than the state variables of the system. The state variables of the closed-loop system can thus be separated into two groups. The variables θ and P in (13) will thus change slowly and the input u and the output y will change rapidly. It is then natural to try to describe the parameters approximately by some approximation of $P\varphi\varepsilon$ in (13). One possibility is to replace $P\varphi\varepsilon$ by its mean value, for example. Such ideas were used by Krylov and Bogoliubov (1937) in deterministic analysis of nonlinear oscillations.

The method of averaging was used by Åström and Wittenmark (1973) to determine possible equilibrium points for the parameters in a tuning problem. For the simple self-tuner based on minimum variance control and least squares or stochastic approximation it was shown that the equilibria are characterized by the equation

$$E\varphi\varepsilon = 0. \tag{25}$$

This equation implies

$$\begin{aligned} &\lim_{t\to\infty} \frac{1}{t} \sum_{k=1}^{t} y(k+\tau)\, y(k) = 0,\ \tau = d,\ d+1, \ldots, d + \deg S \\ &\lim_{t\to\infty} \frac{1}{t} \sum_{k=1}^{t} y(k+\tau) u(k) = 0,\ \tau = d,\ d+1, \ldots, d + \deg R \end{aligned} \tag{26}$$

where $d = \deg A - \deg B$. This characterizes the possible equilibria even if the process is nonlinear or of high order. Åström and Wittenmark (1973) also showed the surprising result that the minimum variance control is an equilibrium even when $C \neq 1$ in (23) and the least-squares estimates are biased.

The method of averaging was developed extensively by Ljung (1977a, b) for the tuning problem. Since the estimator gain $P(t)$ goes to zero as time goes to infinity it can be expected that the separation of fast and slow modes will be better and better as time increases. It was shown by Ljung that asymptotically as time goes to infinity the estimates are described by an ordinary differential equation.

For example, consider an algorithm based on stochastic approximation, i.e. (13), with P given by (15). Introduce the transformed time defined by

$$\tau(t) = c \sum_{k=1}^{t} |P(k)| \tag{27}$$

where $|\cdot|$ denotes some norm and c is a constant. Ljung (1977a) showed that the estimates will approximatively be described by the solutions to the ordinary differential equation

$$\frac{d\theta}{d\tau} = f(\theta) \tag{28}$$

where

$$f(\theta) = E\varphi\varepsilon \tag{29}$$

and E denotes the mean value calculated under the assumption that the parameter θ is constant. Ljung also showed that the estimates converge to the solution of (28) as time increases. The approach can be used for a wide class of problems. Since the estimates are approximately described by ordinary differential equations the technique is referred to as the 'ODE-approach'.

The method is useful for determination of possible equilibrium points and their local stability as well as for determination of convergence rates. The method is however difficult to apply because the function f in (28) is complicated even for simple problems. Another drawback is that the method is based on the assumption that the signals are bounded. It is thus necessary to establish boundedness by some other technique.

Ljung (1977b) applied the ODE-approach to the simple self-tuner based on stochastic approximation and minimum variance control. Theorem 2 was proven under the additional assumption of bounded signals. A similar analysis of the self-tuner based on minimum variance control and least squares estimation showed that the condition corresponding to (24) is that the function

$$G(z) = 1/C(z) - \tfrac{1}{2}$$

is strictly positive real.

Holst (1979) made a detailed analysis of the local behaviour of (28) for the self-tuner based on minimum variance control and least-squares estimation. He showed that the equilibrium corresponding to minimum variance control is locally stable if the function $C(z)$ is positive at the zeros of $B(z)$.

Ljung and Wittenmark (1974) have constructed a counter-example where (28) has a limit cycle. Because of the transformation (28), the estimates will oscillate with ever increasing period. Since the convergence of the parameters depend on the polynomial C, it is clear that convergence can be lost if the characteristics of the disturbances change.

Averaging techniques have also been applied to the tracking problem. In this case the parameters will not converge because the gain $P(t)$ of the estimator does not go to zero. Kushner (1977) and Kushner and Clark (1978) used weak convergence theory to approximate $P\varphi\varepsilon$ in (13) by its mean value and a random term which describes the fluctuations. Singular perturbation methods have also been used to investigate (13).

Stochastic control theory

Regulator structures like MRAS and STR are based on heuristic arguments. It would be appealing to obtain the regulators from a unified theoretical framework. This can be done using nonlinear stochastic control theory. The system and its environment are then described by a stochastic model. The criterion is formulated as to minimize the expected value of a loss function, which is a scalar function of states and controls.

The problem of finding a control, which minimizes the expected loss function, is difficult. Conditions for existence of optimal controls are not known. Under the assumption that a solution exists, a functional equation for the optimal loss function can be derived using dynamic programming. This equation, which is called the *Bellman equation*, can be solved numerically only in very simple cases. The structure of the optimal regulator obtained is shown in Fig. 5 (Bellman, 1961; Bertsekas, 1976). The controller can be thought of as composed of two parts: an estimator and a feedback regulator. The estimator generates the conditional probability distribution of the state from the measurements. This distribution is called the *hyperstate* of the problem. The feedback regulator is a nonlinear function, which maps the hyperstate into the space of control variables.

The structural simplicity of the solution is obtained at the price of introducing the hyperstate, which is a quantity of very high dimension. Notice that the structure of the regulator is similar to the STR in Fig. 3. The self-tuning regulator can be regarded as an approximation where the conditional probability distribution is replaced by a distribution with all mass at the conditional mean value. It will be shown by an example that many interesting properties are lost by such an approximation.

Notice that there is no distinction between the parameters and the other state variables in Fig. 5. This means that the regulator can handle very rapid parameter variations.

The optimal control law has an interesting property. The control attempts to drive the output to its desired value, but it will also introduce perturbations when the parameters are uncertain. This will improve the estimates and the future controls. The optimal control gives the correct balance between maintaining good control and small estimation errors. This property is called *dual control* (Feldbaum, 1965; Florentin, 1971; Jacobs and Patchell, 1972; Bar-Shalom and Tse, 1974, 1976; Tse and Bar-Shalom, 1975). Optimal stochastic control theory also offers other possibilities to obtain sophisticated adaptive algorithms (Saridis, 1977).

Example. Dual control of an integrator. A simple example is used for illustration. Consider a system described by

$$y(t+1) = y(t) + bu(t) + e(t) \tag{30}$$

where u is the control, y the output, e white noise normal $(0, \sigma_e)$. Let the criterion be to minimize the mean square deviation of the output y.

If the parameter b is assumed to be a random variable with a Gaussian prior distribution, the conditional distribution of b, given inputs and outputs up to time t, is Gaussian with mean $\hat{b}(t)$ and standard deviation $\sigma(t)$ (Åström and Wittenmark, 1971). The hyperstate can then be characterized by the triple $(y(t), \hat{b}(t), \sigma(t))$. The equations for updating the hyperstate are the same as the ordinary Kalman filtering equations (Åström, 1970).

Introduce the loss function

$$V_N = \min_u E\left\{\sum_{k=t+1}^{t+N} y^2(k) | Y_t\right\} \tag{31}$$

where Y_t denotes the data available at time t, i.e. $\{y(t), y(t-1), \ldots\}$. By introducing the normalized variables

$$\eta = y/\sigma_e, \qquad \beta = \hat{b}/\sigma, \qquad \mu = -u\hat{b}/y \tag{32}$$

it can be shown that V_N depends on η and β only. The Bellman equation for the problem can be written as

$$V_N(\eta, \beta) = \min_\mu \Big\{ 1 + \eta^2[(1-\mu)^2 + \mu^2\beta^{-2}] + \int_{-\infty}^{\infty} V_{N-1}\left(\eta(1+\mu) + \varepsilon\sqrt{(\beta^2 + \mu^2\eta^2)}/\beta, \sqrt{(\beta^2 + \mu^2\eta^2)} - \varepsilon\mu\eta/\beta\right)\varphi(\varepsilon)\, d\varepsilon \Big\} \tag{33}$$

where φ is the normal (0, 1) probability density (Åström, 1978). When the minimization is performed, the control law is obtained as the function $\mu_N(\eta, \beta)$. When the Bellman equation is solved numerically it turns out that the control law is independent of N for large N. A graph of the control law for $N = 30$ is shown in Fig. 6. With the accuracy of the graph there is no difference between the control laws obtained for $N = 20$ and 30.

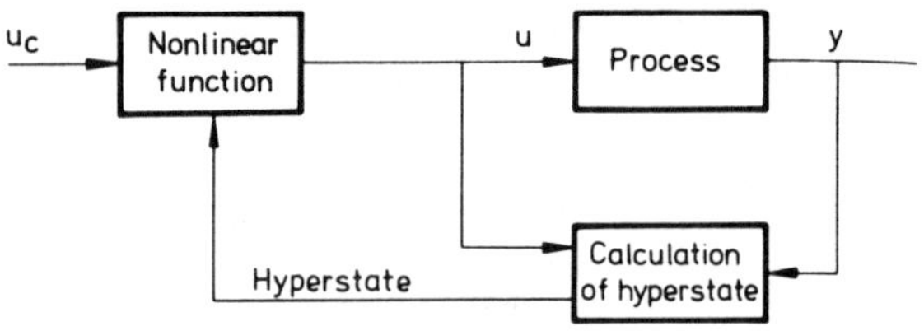

FIG. 5. Block diagram of an adaptive regulator obtained from stochastic control theory.

Some approximations to the optimal control law are also given. The *certainty equivalence control*

$$u(t) = -y(t)/\hat{b} \tag{34}$$

is obtained simply by solving the control problem in the case of known parameters and substituting the known parameters with their estimates. The self-tuning regulator can be interpreted as a certainty equivalence control. Using the normalized variables the control law becomes

$$\mu = 1. \tag{34'}$$

The control law

$$u(t) = -\frac{1}{\hat{b}(t)} \frac{\hat{b}^2(t)}{\hat{b}^2(t) + \sigma^2(t)} y(t) \tag{35}$$

is another approximation, which is called *cautious control*, because it hedges and uses lower gain when the estimates are uncertain. In normalized variables this control law can be expressed as

$$\mu = \frac{\beta^2}{1 + \beta^2}. \tag{35'}$$

Notice that all control laws are the same for large β, i.e. if the estimate is accurate. The optimal control law is also close to the cautious control for large control errors. For estimates with poor precision and moderate control errors the dual control gives larger control actions than the other control laws. This can be interpreted as the introduction of probing signals to achieve better estimates. Also notice that in a large region the certainty equivalence control is closer to the dual control law than the cautious control is. □

If the Bellman equation can be solved as in the example it is possible to obtain reasonable approximations to the dual control law. Notice, however, that cautious and certainty equivalence controls do not have the dual property.

Unfortunately the solution of the Bellman equation requires substantial computations. In the example the normalized states

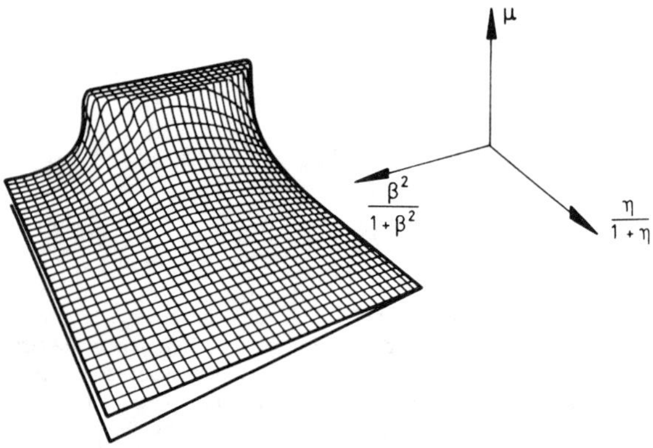

FIG. 6. Optimal dual control for the integrator with unknown gain. The nomalized control variable μ is shown as a function of the normalized control error η and the normalized parameter uncertainty β. The surface is truncated at $\mu = 3$.

were quantized in 64 levels. The solution of the problem for $N=30$ required 180 CPU-hours on the DEC VAX 11/780. It is not possible to do the calculations in reasonable time for a realistic problem with four or more parameters. To appreciate the difficulties consider the model (23) with $C=1$ and e white Gaussian noise. Since the conditional distribution of the parameters is Gaussian the triple (φ, θ, P) where φ is the regression vector, θ the estimates and P the conditional covariance matrix is a hyperstate. For a second order model φ and θ have four components and P has 10 different elements. Eighteen real numbers are thus needed to characterize the hyperstate. If each variable is quantized into 32 levels $2^{90} \approx 10^9$ memory locations are needed to store the loss function. When $C \neq 1$ in (23) the complexity increases drastically because the conditional distribution is no longer Gaussian and many more variables are required to characterize the distribution.

5. USES AND ABUSES OF ADAPTIVE TECHNIQUES

Before going into the details of applications of adaptive control some different ways to use adaptive techniques will be discussed.

Auto-tuning

It is possible to tune regulators with three to four parameters by hand if there is not too much interaction between adjustments of different parameters. For more complex regulators it is however necessary to have suitable tuning tools. Traditionally tuning of more complex regulators have followed the route of modelling or identification and regulator design. This is often a time-consuming and costly procedure which can only be applied to important loops or to systems which are made in large quantities.

Both the MRAS and the STR become constant gain feedback controls when the estimated parameters are constant. Compare Figs. 2 and 3. The adaptive loop can thus be used as a *tuner* for a control loop. In such applications the adaptation loop is simply switched on, perturbation signals may be added. The adaptive regulator is run until the performance is satisfactory. The adaptation loop is then disconnected and the system is left running with fixed regulator parameters. Auto-tuning can be considered as a convenient way to incorporate automatic modelling and design into a regulator. It widens the class of problems where systematic design methods can be used cost effectively. Auto-tuning is particularly useful for design methods like feed forward which critically depend on good models.

Automatic tuning can be applied to simple PID controllers as well as to more complicated regulators. It is useful to combine auto-tuning with diagnostic tools for checking the performance of the control loops. For minimum variance control the performance evaluation can be done simply by monitoring the covariances of the inputs and the outputs (Åström, 1970). Tuning can then be initiated when there are indications that a loop is badly tuned.

It is very convenient to introduce auto-tuning in a DDC-package. One tuning algorithm can then serve many loops. Since a good tuning algorithm only requires a few kbytes of memory in a control computer substantial benefits are obtained at marginal cost. Halme (1980) describes a tuning algorithm based on least squares estimation and minimum variance control which has been incorporated in a commercial DDC-package.

Auto-tuning can also be included in single-loop regulators. For example, it is possible to design regulators where the mode switch has three positions: manual, automatic, and tuning. The tuning algorithm represents however a major part of the software of a single loop regulator. Memory requirements will typically be more than doubled when auto-tuning is introduced. A single loop adaptive regulator has been announced by the Leeds & Northrup Co. (Andreiev, 1981).

Automatic construction of gain schedules

The adaptive control loop may also be used to build a gain schedule. The parameters obtained when the system is running in one operating condition are then stored in a table. The gain schedule is obtained when the process has operated at a range of operating conditions, which covers the operating range.

There are also other ways to combine gain scheduling with adaptation. A gain schedule can be used to quickly get the parameters into the correct region and adaptation can then be used for fine tuning.

Adaptive regulators

The adaptive techniques may of course also be used for genuine adaptive control of systems with time varying parameters. There are many ways to do this.

The operator interface is important, since adaptive regulators may also have parameters which must be chosen. It has been my experience that regulators without any externally adjusted parameters can be designed for specific applications, where the purpose of control can be stated *a priori*. The shipsteering autopilot discussed in Section 7 is such an example.

In many cases it is, however, not possible to specify the purpose of control *a priori*. It is at least necessary to tell the regulator what it is expected to do. This can be done by introducing dials that give the desired properties of the closed-loop system. Such dials are called *performance related*. New types of regulators can be designed using this concept. For example, it is possible to have a regulator with one dial, which is labelled with the desired closed-loop bandwidth. Another possibility would be to have a regulator with a dial, which is labelled with the weighting between state deviation and control action in a LQG problem. A third possibility would be to have a dial labelled with the phase margin or the amplitude margin. The characteristics of a regulator with performance related knobs are shown by the following example.

Example. The bandwidth self-tuner. The bandwidth self-tuner is an adaptive regulator which has one adjustable parameter on the front panel which is labelled with the *desired closed-loop bandwidth*. The particular implementation is in the form of a pole-placement self-tuner. The details of the algorithm are given in Åström (1979).

The response of the servo to a square wave command signal is shown in the Fig. 7. It is clear from the figure that the servo performs very well after the first command step. It is seen in Fig. 7(a) that the control signal gives a moderate 'kick' after a command step when the requested bandwidth is low (1.5 rad s^{-1}). The control signal then decreases gradually to a steady-state level. When a larger bandwidth (4.5 rad s^{-1}) is demanded as in Fig. 7(b) the value of the control signal immediately after the

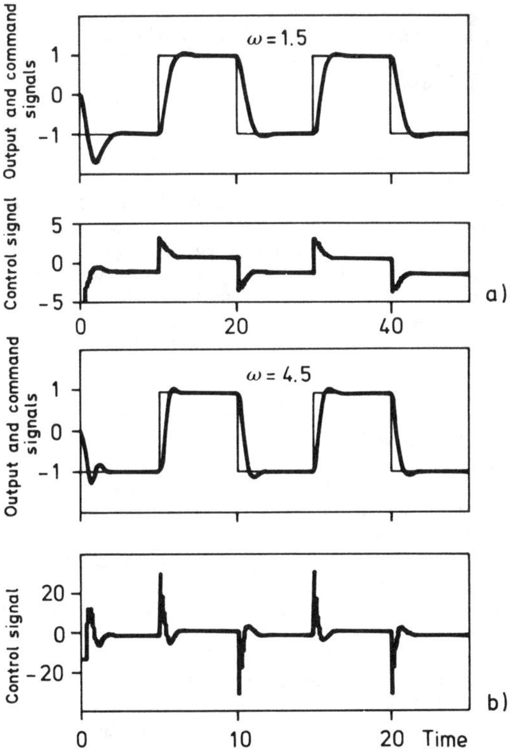

FIG. 7. Simulation of the bandwidth self-tuner. The process has the transfer function $1/(s+1)^2$. The requested bandwidth is 1.5 rad s^{-1} in (a) and 4.5 rad s^{-1} in (b).

step is more than 30 times larger than its steady-state value. A low bandwidth gives a slow response, small control signals and low sensitivity to measurement noise. A high bandwidth gives a fast response. The control actions will however be large and the system will be sensitive to noise. It is easy for an operator to determine the bandwidth which is suitable for a particular application by experimentation. Notice that apart from the specifications in terms of the bandwidth all necessary adjustments are handled automatically by the self-tuner. □

Abuses of adaptive control

An adaptive regulator, being inherently nonlinear, is more complicated than a fixed gain regulator. Before attempting to use adaptive control it is therefore important to first examine if the control problem cannot be solved by constant gain feedback. Problems of this type have only rarely been investigated. Two exceptions are Åström (1980b) and Jacobs (1980). In the vast literature on adaptive control there are many cases where a constant gain feedback can do as well as an adaptive regulator. A typical example is the very ambitious feasibility study of adaptive autopilots for aircrafts (IEEE, 1977). The aircraft used in the experiments could easily be controlled with conventional methods.

Notice that it is not possible to judge the need for adaptive control from the variations of the open loop dynamics over the operating range. Many cases are known where a constant gain feedback can cope well with considerable variations in system dynamics (Åström, 1980d). There are also design techniques for constant gain feedback that can cope with considerable gain variations (Horowitz, 1963).

6. PRACTICAL ASPECTS

Since there is no complete theory for adaptive control, there are in applications many problems which must be solved intuitively with support of simulation. This situation is not unique for adaptive control. Similar problems occur in many other areas.

Consider for example conventional PID control. The system obtained when a PID regulator is connected to a linear system is well understood and can be analysed with great precision. When implementing a PID regulator it is necessary, however, to consider many issues like hand/automatic transfer, bumpless parameter changes, reset windup and nonlinear output, etc. Several PID regulators may also be connected via logical selectors. The systems obtained are then nonlinear and the linear analysis of the ideal case is of limited value. Since the nonlinear modifications give substantial improvements in performance they are widely used although they are poorly understood theoretically and not widely publicised.

The situation is similar for adaptive control. The major difference is that the adaptive systems are inherently nonlinear and more complex. For example in an adaptive regulator windup can occur in the inner loop as well as in the outer loop. A few of these problems will be discussed in this Section. Additional details are found in Clarke and Gawthrop (1981); Schumann, Lachmann and Isermann (1982); Åström (1980d).

Parameter tracking

Since the key property of an adaptive regulator is its ability to track variations in process dynamics, the performance of the parameter estimator is crucial. A fundamental result of system identification theory is that the input signal to the process must be *persistently exciting* or *sufficiently rich* (Åström and Bohlin, 1966). In the adaptive systems the input signal is generated by feedback. Under such circumstances there is no guarantee that the process will be properly excited. On the contrary, good regulation may give a poor excitation. Consequently there are inherent limitations unless extra perturbation signals are introduced, as is suggested by dual control theory. If perturbation signals are not introduced the parameter updating should be switched off when the system is not properly excited. Examples of what happens if this is not down are given by Rohrs and co-workers (1982).

To track parameter variations it is necessary to discount old data. This will involve compromises. If data are discounted too fast the estimates will be uncertain even if the true parameters are constant. If old data are discounted slowly the estimates of constant parameters will be good but it is impossible to track rapid parameter variations.

Exponential discounting is a simple way to discard old data. In the case of least squares estimation this leads to the recursive formula (13) with P given by (17). If the time constant of the exponential discounting is T the forgetting factor is given by

$$\lambda = 1 - h/T$$

where h is the sampling period. The choice of a suitable T is a compromise between the precision of the estimate and the ability to track parameter variations.

The simple exponential discounting works very well if the process is properly excited. Practical experiences which report good results with this method are reported by Borisson and Syding (1976); Åström and co-workers (1977); Westerlund, Toivonen and Nyman (1980).

Estimator windup and bursts

There are severe problems with exponential discounting when the excitation of the process changes. A typical example is when the major excitation is caused by set point changes. There may then be long periods with no excitation at all. When the set point is kept constant the relevant data will be discounted even if no new data is obtained.

The effect can be seen analytically. It follows from (17) that

$$P(t) = [\varphi(t)\varphi^T(t) + \lambda P^{-1}(t-1)]^{-1}. \tag{34}$$

If the process is poorly excited the input u and the output y are small. Since the components of the new vector φ are delayed values of y and u the vector φ is also small. In the extreme case when the regression vector φ is zero it follows from (34) that the gain matrix $P(t)$ grows exponentially. When the gain becomes sufficiently large the estimator becomes unstable. Small residuals then give very large changes in the parameters, and the closed loop system may become unstable. The process will then be well excited, and the parameter estimates will quickly achieve good values. Looking at the process output there will be periods of good regulation followed by bursts (Morris, Fenton and Nazer, 1977; Fortescue, Kershenbaum and Ydstie, 1981). There will be bursts because the estimator with exponential forgetting is an open-loop unstable system. The phenomenon is therefore also called *estimator windup* in analogy with integrator windup in simple regulators. A typical case of estimator windup is shown in Fig. 8. Notice the nearly exponential growth of the diagonal element of P in the intervals when the command signal remains

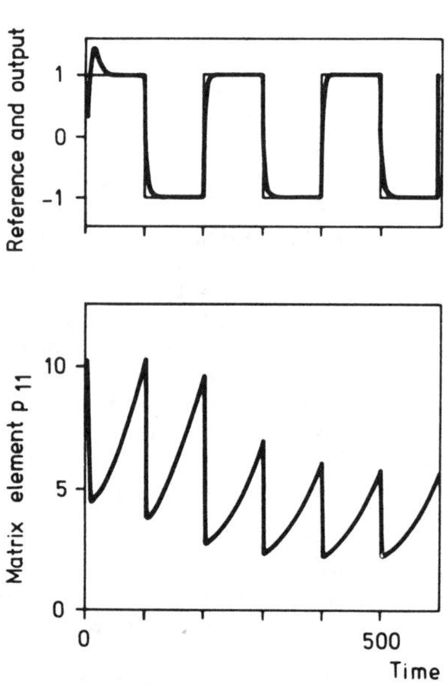

FIG. 8. Estimator windup. The reference signal, the output signal and one diagonal element of the matrix P are shown. Notice that P increases exponentially when the set point is constant.

constant. The variations in P have little effect on the regulation because they are not large enough. They will be much larger if the command changes less frequently. Without further excitation the adaptive loop may then even become unstable.

Notice that φ need not be zero for windup to occur. If φ lies in a given subspace for a period then P will grow exponentially in directions orthogonal to the subspace.

There are many ways to avoid estimator windup and bursts. Since the problem is caused by poor excitation in combination with discounting of old data there are two possibilities to avoid the difficulty, to make sure that the process is properly excited or to eliminate the discounting. The condition for persistent excitation can be monitored and perturbation can be introduced when the excitation is poor. This is in the spirit of dual control theory.

In some cases it is not feasible to introduce extra perturbations to obtain good excitation. Covariance windup can then be avoided by discounting old data only when there is proper excitation. Several ad hoc procedures have been proposed. It has been proposed to update the estimates only when $|\varphi|$ or $|P\varphi|$ are larger than a given quantity. Fortescue, Kershenbaum and Ydstie (1981) have proposed a method for varying the forgetting factor automatically. This decreases the probability for bursts but does not eliminate them. A simple scheme to limit the estimator gain has been proposed by Irving (1979). A forgetting factor $\lambda < 1$ is used if the trace of the matrix P is sufficiently small. The forgetting factor is set to unity if tr P exceeds a critical value. Hägglund (1983) has proposed to forget data only in the directions where new information is obtained.

Robustness

In practice it is necessary to make sure that the regulator works well under extreme conditions. There are many *ad hoc* features that have proven useful. It is often useful to limit the control signal and its rate of change (Wittenmark, 1973). To avoid that a single large measurement error upsets the recursive estimation it is useful to use robust estimation (Huber, 1981; Poljak and Tsypkin, 1976).

Equation (13) is then modified to

$$\theta(t) = \theta(t-1) + P(t)\varphi(t)f(\varepsilon(t)) \tag{35}$$

where the function f is a saturation function or some similar nonlinearity. The function f can be determined from the probability distribution of the disturbances.

Numerics and coding

It is very important that no numerical problems occur in any operating modes. There are potential problems both in the parameter estimation and in the control design. All estimation methods are poorly conditioned if the models are over-parameterized. The conditioning will also depend on the excitation of the process which may change considerably. The least-squares estimation problem is poorly conditioned for high signal to noise ratios. It is particularly harmful if estimation is based on signals which have a high superimposed d.c. level. In such cases it is useful to remove the d.c.-level and to use square root algorithms instead of the ordinary algorithms (Lawson and Hanson, 1974; Peterka, 1975; Bierman, 1977).

Many control design methods are also poorly conditioned for some model parameters. The singularities are typically associated with loss of observability or controllability of the model. Since parameter estimation algorithms may give any values it is important to guard against the difficulties that may arise in explicit or indirect schemes. This problem is discussed in Åström (1979).

Adaptive regulators are easily implemented using microcomputers (Bengtsson, 1979; Clarke and Gawthrop, 1981; Fjeld and Wilhelm, 1981; Glattfelder, Huguenin and Schaufelberger (1980). The simple self-tuner (Algorithm 2) is easily coded in less than 100 lines in a high-level programming language (much less if readabilty is sacrificed). Other adaptive algorithms may require considerably longer codes. The adaptive algorithms are in any case at least an order of magnitude more complex than fixed gain regulators.

Integral action

It is very important that a control system has the ability to maintain small errors in spite of large low frequency disturbances. For fixed gain regulators this is ensured by introducing integral action which gives a high-loop gain at low frequencies. There are several different possibilities to obtain similar effects in adaptive systems. Some adaptive schemes introduce integral action automatically when needed. Integral action can also be forced by choosing special regulator structures. It can also be introduced indirectly via the parameter estimation. A discussion of some alternatives and their pros and cons is given in Åström (1980d).

Supervisory loops

The adaptive systems in Figs. 2 and 3 can be regarded as hierarchical systems with two levels. The lower level is the ordinary feedback loop with the process and the regulator. The adaptation loop which adjusts the parameters represents the higher level. The adaptation loop in typical STR or MRAS requires parameters like the model order, the forgetting factor and the sampling period. A third layer can be added to the hierarchy to set these parameters. It has already been mentioned that the forgetting factor can be determined by monitoring the excitation of the process. By storing process inputs and outputs it is also possible to estimate models having different sampling periods, different orders and different structures. Experimentation with systems of this type are given by Saridis (1977); Schumann, Lachmann and Isermann (1982).

7. APPLICATIONS

There are over 1500 papers on adaptive control, several hundred simulations have been performed, and several hundred experiments on laboratory processes have been made. Adaptive techniques are also starting to be used in commercial products. An overview of the applications is given in this section.

Laboratory experiments

Over the past ten years there have been extensive laboratory experiments with adaptive control mostly in universities but also to an increasing extent in industrial companies. Schemes like gain scheduling, MRAS, and STR have been investigated. The goal of the experiments has been to understand the algorithms and to investigate many of the factors, which are not properly covered by theory.

Industrial feasibility studies

There have been a number of industrial feasibility studies of adaptive control. The following list is not exhaustive but it covers some of the major studies.

autopilots for aircrafts and missiles
- IEEE (1977)
- Young (1981)

autopilots for ships
- Källström and co-workers (1979)
- van Amerongen (1981, 1982)

cement mills (grinding and mixing of raw material)
- Csaki and co-workers (1978)
- Keviczky and co-workers (1978)
- Kunze and Salaba (1978)
- Westerlund, Toivonen and Nyman (1980)
- Westerlund (1981)

chemical reactors
- Harris, MacGregor and Wright (1980)
- Clarke and Gawthrop (1981)
- Buchholt, Clement and Bay Jorgensen (1979)

diesel engines
- Zanker (1980)

digesters
- Sastry (1979)

glass furnaces
- Haber and co-workers (1979)

heat exchangers
- Jensen and Hänsel (1974)
- Zanker (1980)
- Kurz, Isermann and Schumann (1980)

heating and ventilation
  Clarke and Gawthrop (1981)
  Jensen and Hänsel (1974)
  Schumann (1980)
motor drives
  Courtiol (1975)
  Courtiol and Landau (1975)
  Speth (1969)
  Negoesco, Courtiol and Francon (1978)
optical telescope
  Gilbart and Winston (1974)
ore crusher
  Borisson and Syding (1976)
paper machines
  Borisson and Wittenmark (1974)
  Cegrell and Hedqvist (1975)
  Fjeld and Wilhelm (1981)
pH-control
  Shinsky (1974)
  Buchholt and Kümmel (1979)
  Bergmann and Lachmann (1980)
  Jacobs, Hewkin and White (1980)
power systems
  Irving (1979)
  Irving and van Mien (1979)
  Isermann (1980b)
rolling mills
  Bengtsson (1979)
  Seyfried and Stöle (1975)
titanium oxide kiln
  Dumont and Bélanger (1978)

The references above and the recent books on applications of adaptive control, Narendra and Monopoli (1980) and Unbehauen (1980), contain more details and many additional references.

The feasibility studies have shown that there are indeed cases, where adaptive control is very useful. They have also shown that there are cases where the benefits are marginal.

Signal processing applications

Algorithms similar to the ones discussed in this paper have found extensive applications in the communications field. There has in fact been almost parallel development of adaptive algorithms in the control and communications communities. The signal processing problems are mostly concerned with adaptive filtering. Among the problems considered in communications we can mention adaptive prediction in speech coding and adaptive noise cancellation. Adaptive echo cancellation, based on VLSI technology with adjustment of over 100 parameters, is, for example, beginning to be introduced in telephone systems. A survey of these applications are given in Falconer (1980)

Industrial products

Adaptive control is now also finding its way into industrial products. There appears to be many different ways of using adaptive techniques.

Gain scheduling is the predominant technique for design of autopilots for high performance aircrafts. It is considered as a well established standard technology in the aerospace industry. Gain scheduling is also starting to be used in the process industry. A regulator for process control Micro Scan 1300 which uses gain scheduling has been announced by Taylor Instruments (Andreiev, 1977). It is also easy to implement gain scheduling using the modern hardware for distributed process control.

There are also products based on the STR and the MRAS, both general purpose self-tuners as stand alone systems and software in DDC-packages. A self-tuning PID regulator has been announced by Leeds & Northrup Co. The Swedish company ASEA AB makes a general purpose self-tuner NOVATUNE and a general purpose adaptive optimizer NOVAMAX. The Finnish cement factory Lohja Corporation has announced a cement kiln control system with a self-tuner. There are more commercial adaptive regulators which are on their way to the market. Two examples are chosen to illustrate typical industrial products.

Example. A ship steering autopilot. A ship has to operate in a varying environment: wind, waves, and currents may change considerably. The dynamics of a ship may also vary with trim, loading, and water depth. An adaptive autopilot based on recursive estimation and LQG control is manufactured by Kockumation AB in Sweden. Fig. 9 compares the performances of this autopilot with a conventional autopilot under comparable conditions. It is clear from the figure that the performance of the adaptive autopilot is superior to the conventional system. The reason for this is that the control law is more complex. If the operating conditions were known it would be possible to redesign the conventional autopilot so that it gives the same performance as the adaptive system. In the particular case this would, however, require a regulator with at least eight parameters. It would not be practical to tune such a regulator manually. □

The following example illustrates the benefits of self-tuning in a typical process control problem.

Example. Paper machine control. The feasibility of minimum variance control as a design method for control of basis weight and moisture content of a paper machine was established by Åström (1967). The application of a self-tuner for adjusting both feedback and feed forward parameters was suggested in Åström and Wittenmark (1973). The industrial feasibility of such a scheme was demonstrated by Borisson and Wittenmark (1974) and Cegrell and Hedqvist (1975). Five years later the concept was taken up by one of the manufacturers of paper machine controls (Fjeld and Wilhelm, 1981). The histogram in Fig. 10 shows a comparison between the self-tuner and a conventional system. The figure indicates that the ability of the self-tuner to adjust the parameters in response to a changing environment is beneficial. □

8. CONCLUSIONS

The adaptive technique is slowly emerging after 25 years of research and experimentation. Important theoretical results on

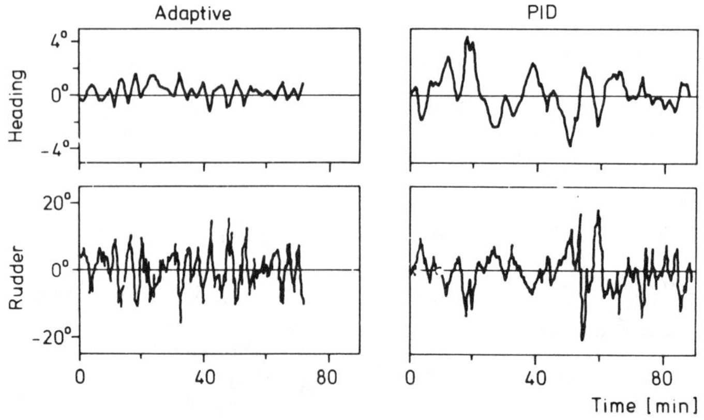

FIG. 9. Heading and rudder angles for a conventional and an adaptive autopilot. From Källström and co-workers (1979).

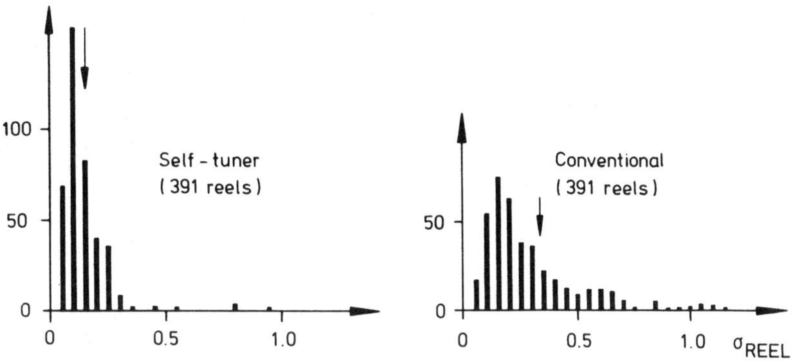

FIG. 10. Histograms for standard deviation in moisture for a self-tuner and a conventional fixed gain regulator. From Fjeld and Wilhelm (1981).

stability and structure have been established. Much theoretical work still remains to be done. The advent of microprocessors has been a strong driving force for the applications. Laboratory experiments and industrial feasibility studies have contributed to a better understanding of the practical aspects of adaptive control. There are also a number of adaptive regulators appearing on the market.

Acknowledgements—My research in adaptive control has for many years been supported by the Swedish Board of Technical Development (STU). The writing of this paper was made under the contract 82-3430. This support is gratefully acknowledged. Over the years I have also learned much from discussions with colleagues and graduate students.

REFERENCES

Andreiev, N. (1977). A process controller that adapts to signal and process conditions. *Control Engng.*, Dec., 38.

Andreiev, N. (1981). A new dimension: a self-tuning controller that continually optimizes PID constants. *Control Engng.*, Aug., 84.

Asher, R. B., D. Adrisani and P. Dorato (1976). Bibliography on adaptive control systems. *IEEE Proc.*, **64**, 1126.

Åström, K. J. (1967). Computer control of a paper machine—an application of linear stochastic control theory. *IBM J. Res. & Dev.*, **11**, 389.

Åström, K. J. (1970). *Introduction to Stochastic Control Theory.* Academic Press, New York.

Åström, K. J. (1974). A self-tuning regulator for non-minimum phase systems. Report TFRT-3113, Department of Automatic Control, Lund Institute of Technology.

Åström, K. J. (1978). Stochastic control problems. In W. A. Coppel (Ed.), *Mathematical Control Theory.* Lecture Notes in Mathematics. Springer, Berlin.

Åström, K. J. (1979). Simple self-tuners I. Report CODEN: LUTFD2/(TFRT-7184)/(1979), Department of Automatic Control, Lund Institute of Technology.

Åström, K. J. (1980a). Self-tuning regulators—design principles and applications. In K. S. Narendra and R. V. Monopoli (Eds.), *Applications of Adaptive Control.* Academic Press, New York.

Åström, K. J. (1980b). Why use adaptive techniques for steering large tankers? *Int. J. Control*, **32**, 689.

Åström, K. J. (1980c). Direct methods for non-minimum phase systems. *Proc. 19th CDC Conference*, Albuquerque, NM.

Åström, K. J. (1980d). Design principles for self-tuning regulators. In H. Unbehauen (Ed.), *Methods and Applications in Adaptive Control.* Springer, Berlin.

Åström, K. J. (1982). Ziegler-Nichols auto-tuners. Report CODEN: LUTFD2/(TFRT-3167)/01-025/(1982).

Åström, K. J. and T. Bohlin (1966). Numerical identification of linear dynamic systems from normal operating records. In P. H. Hammond (Ed.), *Theory of Self-adaptive Control Systems.* Plenum Press.

Åström, K. J. and P. Eykhoff (1971). System identification—a survey. *Automatica*, **7**, 123.

Åström, K. J. and B. Wittenmark (1971). Problems of identification and control. *J. Math. Anal. & Applic.*, **34**, 90.

Åström, K. J. and B. Wittenmark (1973). On self-tuning regulators. *Automatica*, **9**, 185.

Åström, K. J. and B. Wittenmark (1974). Analysis of a self-tuning regulator for nonminimum phase systems. *Preprints IFAC Symposium on Stochastic Control*, Budapest.

Åström, K. J. and B. Wittenmark (1980). Self-tuning controllers based on pole-zero placement. *IEE Proc.*, **127**, 120.

Åström, K. J. and B. Wittenmark (1984). *Computer Controlled Systems—Theory and Design.* Prentice-Hall.

Åström, K. J. and Z. Zhao-ying (1982). A linear quadratic Gaussian self-tuner. *Proc. Workshop on Adaptive Control*, Florence, Italy.

Åström, K. J., U. Borisson, L. Ljung and B. Wittenmark (1977). Theory and applications of self-tuning regulators. *Automatica*, **13**, 457.

Aseltine, J. A., A. R. Mancini and C. W. Sarture (1958). A survey of adaptive control systems. *IRE Trans*, **AC-6**, 102.

Bar-Shalom, Y. and E. Tse (1974). Dual effect, certainty equivalence and separation in stochastic control. *IEEE Trans Aut. Control*, **AC-19**, 494.

Bar-Shalom, Y. and E. Tse (1976). Caution, probing and the value of information in the control of uncertain systems. *Annals Econ. Soc. Measurement*, **5**, 323.

Bellman, R. (1957). *Dynamic Programming.* Princeton University Press.

Bellman, R. (1961). *Adaptive Processes—A Guided Tour.* Princeton University Press.

Bengtsson, G. (1979). Industrial Experience with a micro-computer based self-tuning regulator. *Preprints 18th CDC Conference*, Fort Lauderdale.

Bergmann, S. and K.-H. Lachmann (1980). Digital parameter adaptive control of a pH process. *Preprints JACC*, San Francisco.

Bertsekas, D. P. (1976). *Dynamic Programming and Stochastic Control.* Academic Press, New York.

Bierman, G. J. (1977). *Factorization Methods for Discrete Sequential Estimation.* Academic Press, New York.

Bode, H. W. (1945). *Network Analysis and Feedback Amplifier Design.* Van Nostrand.

Borisson, U. and R. Syding (1976). Self-tuning control of an ore crusher. *Automatica*, **12**, 1.

Borisson, U. and B. Wittenmark (1974). An industrial application of a self-tuning regulator. *Proc. IFAC Symposium on Digital Computer Applications to Process Control*, Zurich.

Buchholt, F. and Kümmel, M. (1979). Self-tuning control of a pH-neutralization process. *Automatica*, **15**, 665.

Buchholt, F., K. Clement and S. Bay Jorgensen (1979). Self-tuning control of a fixed bed chemical reactor. *Preprints 5th IFAC Symposium on Identification and System Parameter Estimation*, Darmstadt.

Caldwell, W. I. (1950). Control system with automatic response adjustment. American patent, 2,517,081. Filed 25 April 1947.

Cegrell, T. and T. Hedqvist (1975). Successful adaptive control of paper machines. *Automatica*, **11**, 53.

Clarke, D. W. and P. J. Gawthrop (1975). A self-tuning controller. *IEE Proc.*, **122**, 929.

Clarke, D. W. and P. J. Gawthrop (1979). Self-tuning control. *IEE Proc.*, **126**, 633.

Clarke, D. W. and P. J. Gawthrop (1981). Implementation and application of microprocessor-based self-tuners. *Automatica*, **17**, 233.

Courtiol, B. (1975). Applying model reference adaptive techniques for the control of electromechanical systems. *Proc. 6th IFAC World Congress*. Instrument Society of America, Pittsburgh, Pennsylvania.

Courtiol, B. and I. D. Landau (1975). High speed adaptation system for controlled electrical drives. *Automatica*, **11**, 119.

Csaki, K., L. Keviczky, J. Hetthessy, M. Hilger and J. Kolostori (1978). Simultaneous adaptive control of chemical composition, fineness and maximum quantity of ground materials at a closed circuit ball mill. *7th IFAC Congress*, Helsinki.

de Larminat, Ph. (1979). On overall stability of certain adaptive control systems. *Proc. 5th FAC Symp. on Identification and System Parameter Estimation*, Darmstadt.

Desoer, C. A. and M. Vidyasagar (1975). *Feedback Systems: Input–Output Properties*. Academic Press, New York.

Dumont, G. A. and P. R. Béelanger (1978). Self-tuning control of a titanium dioxide kiln. *IEEE Trans Aut. Control*, **AC-23**, 532.

Edmunds, J. M. (1976). Digital adaptive pole-shifting regulators. Ph.D. thesis, UMIST.

Egardt, B. (1979a). *Stability of Adaptive Controllers*. Lecture Notes in Control and Information Sciences, Vol. 20. Springer, Berlin.

Egardt, B. (1979b). Unification of some continuous-time adaptive control schemes. *IEEE Trans Aut. Control*, **AC-24**, 588.

Egardt, B. (1980a). Unification of some discrete time adaptive control schemes. *IEEE Trans Aut. Control*, **AC-25**, 693.

Egardt, B. (1980b). Stability analysis of discrete-time adaptive control schemes. *IEEE Trans Aut. Control*, **AC-25**, 710.

Egardt, B. (1980c). Stability analysis of continuous-time adaptive control systems. *Siam J. Control & Optimiz.*, **18**, 540.

Elliott, H. (1982). Direct adaptive pole placement with application to nonminimum phase systems. *IEEE Trans Aut. Control*, **AC-27**, 720.

Eykhoff, P. (Ed.) (1981). *Trends and Progress in System Identification*. Pergamon Press, Oxford.

Falconer, D. D. (1980). Adaptive filter theory and applications. In A. Bensoussan and J. L. Lions (Eds.). *Analysis and Optimization of Systems*. Springer, Berlin.

Feldbaum, A. A. (1960–1961). Dual control theory I–IV. *Aut. & Remote Control*, **21**, 874; **21**, 1033; **22**, 1; **22**, 109.

Feldbaum, A. A. (1965). *Optimal Control Systems*. Academic Press, New York.

Feuer, A. and A. S. Morse (1978). Adaptive control of single-input, single-output linear systems. *IEEE Trans Aut. Control*, **AC-23**, 557.

Fjeld, M. and R. G. Wilhelm, Jr (1981). Self-tuning regulators—the software way. *Control Engng*, Oct., 99.

Florentin, J. J. (1971). Optimal probing, adaptive control of a simple Bayesian system. *Int. J. Control*, **13**, 165.

Fortescue, T. R., L. S. Kershenbaum and B. E. Ydstie (1981). Implementation of self-tuning regulators with variable forgetting factors. *Automatica*, **17**, 831.

Fuchs, J. J. (1979). Commande adaptative directe des systemes linéaires discrets. These D.E. Université de Rennes.

Gawthrop, P. J. (1977). Some interpretations of the self-tuning controller. *IEE Proc.*, **124**, 889.

Gawthrop, P. J. (1980). On the stability and convergence of a self-tuning controller. *Int. J. Control*, **31**, 973.

Gilbart, J. W. and G. C. Winston (1974). Adaptive compensation for an optical tracking telescope. *Automatica*, **10**, 125.

Glattfelder, A. H., A. H. Huguenin and W. Schaufelberger (1980). Microcomputer based self-tuning and self-selecting controllers. *Automatica*, **16**, 1.

Goodwin, G. C., P. J. Ramadge and P. E. Caines (1980). Discrete time multivariable adaptive control. *IEEE Trans Aut. Control*, **AC-25**, 449.

Goodwin, G. C., P. J. Ramadge and P. E. Caines (1981). Discrete time stochastic adaptive control. *SIAM J. Control & Optimiz.*, to be published.

Gregory, P. C. (Ed.) (1959). *Proceedings of the Self Adaptive Flight Control Systems Symposium*. WADC Technical Report 59-49, Wright Air Development Centre, Wright-Patterson Air Force Base, Ohio.

Haber, R., J. Hetthessy, L. Keviczky, I. Vajk, A. Feher, N. Czeiner, Z. Csaszar and A. Turi (1979). Identification and adaptive control of a glass furnace by a portable process computer laboratory. *Preprints 5th IFAC Symposium on Identification and System Parameter Estimation*, Darmstadt.

Halme, A. (1980). A two level realization of a self-tuning regulator in a multi-microcomputer process control system. In A. Bensoussan and J. L. Lions (Eds.), *Analysis and Optimization of Systems*. Springer, Berlin.

Hang, C. C. and P. C. Parks (1973). Comparative studies of model-reference adaptive control systems. *IEEE Trans Aut. Control*, **AC-18**, 419.

Harris, T. J., J. F. MacGregor and J. P. Wright (1980). Self-tuning and adaptive controllers: an application to catalytic reactor control. *Technometrics*, **22**, 153.

Harris, C. J. and S. A. Billings (Eds.) (1981). *Self-tuning and Adaptive Control: Theory and Applications*. Peter Peregrinus, London.

Holst, J. (1979). Local convergence of some recursive stochastic algorithms. *Preprints 5th IFAC Symposium on Identification and System Parameter Estimation*, Darmstadt.

Horowitz, I. M. (1963). *Synthesis of Feedback Systems*. Academic Press, New York.

Huber, P. J. (1981). *Robust Statistics*. John Wiley, New York.

Hägglund, T. (1983). Recursive least squares identification with forgetting of old data. Report CODEN: LUTFD2/(TFRT-7254)/1-038/(1983), Department of Automatic Control, Lund Institute of Technology.

IEEE (1977). Mini-issue on NASA's advanced control law program for the F-8 DFBW aircraft. *IEEE Trans Aut. Control*, **AC-22**, 752.

Ionescu, T. and R. V. Monopoli (1977). Discrete model reference adaptive control with an augmented error signal. *Automatica*, **13**, 507.

Irving, E. (1979). Improving power network stability and unit stress with adaptive generator control. *Automatica*, **15**, 31.

Irving, E. and H. D. van Mien (1979). Discrete-time model reference multivariable adaptive control: Applications to electrical power plants. *Preprints 18th CDC*, Fort Lauderdale.

Isermann, R. (1980a). Parameter adaptive control algorithms—a tutorial. *Proc. 6th IFAC/IFIP Conference on Digital Computer Applications to Process Control.*

Isermann, R. (1980b). Digital control methods for power station plants based on identified models. *IFAC Symposium on Automatic Control in Power Generation, Distribution and Protection*. Pretoria.

Jacobs, O. L. R. (1961). A review of self-adjusting systems in automatic control. *J. Elect. & Control*, **10**, 311.

Jacobs, O. L. R. and J. W. Patchell (1972). Caution and probing in stochastic control. *Int. J. Control*, **16**, 189.

Jacobs, O. L. R., P. F. Hewkins and C. White (1980). On-line computer control of pH in an industrial process. *IEE Proc.*, **127D**, 161.

Jacobs, O. L. R. (1980). When is adaptive control useful? *Proc. IMA Conference on Control Theory*. Sheffield.

Jensen, L. and R. Hänsel (1974). Computer control of an enthalpy exchanger. Report TFRT-3081, Department of Automatic Control. Lund Institute of Technology.

Johnson, Jr. C. R., D. A. Lawrence and J. P. Lyons (1982). A flaw in the reduced-order behavior of a direct adaptive pole placer. *Proc. 21st IEEE CDC*, Orlando, Vol. 1, p. 271.

Källström, C. G., K. J. Åström, N. E. Thorell, J. Eriksson and L. Sten (1979). Adaptive autopilots for tankers. *Automatica*, **15**, 241.

Kaczmarz, S. (1937). Angenäherte Auflösung von Systemen linearer Gleichungen. *Bull. Int. Acad. Polon. Sci. Mat. Nat. Ser A*, 355.

Kalman, R. E. (1958). Design of a self-optimizing control system. *Trans ASME*, **80**, 468.

Keviczky, L., J. Hetthessy, M. Hilger and J. Kolostori (1978). Self-tuning adaptive control of cement raw material blending. *Automatica*, **14**, 525.

Krylov, A. N. and N. N. Bogoliubov (1937). *Introduction to Non-linear Mechanics*. (English translation), Princeton University Press, Princeton (1943).

Kunze, E. and M. Salaba (1978). Praktische Erprobung Eines Adaptiven Regelungsverfanrens an Einer Zementmahlanlage.

Preprints Fachtagung "Regelungstechnik in Zementwerken", Bielefeld.
Kurz, H., R. Isermann and R. Schumann (1980). Experimental comparison and application of various parameter-adaptive control algorithms. *Automatica*, **16**, 117.
Kushner, H. J. (1977). Convergence of recursive adaptive and identification procedures via weak convergence theory. *IEEE Trans Aut. Control*, **AC-22**, 921.
Kushner, H. J. and D. S. Clark (1978). *Stochastic Approximation Methods for Constrained and Unconstrained Systems.* Springer, New York.
Landau, I. D. (1974). A survey of model-reference adaptive techniques: theory and applications. *Automatica*, **10**, 353.
Landau, I. D. (1979). *Adaptive Control—The Model Reference Approach.* Marcel Dekker, New York.
Lawson, C. L. and R. J. Hanson (1974). *Solving Least Squares Problems.* Prentice-Hall, Englewood Cliffs.
Lieuson, H., A. J. Morris, Y. Nazer and R. K. Wood (1980). Experimental evaluation of self-tuning controllers applied to pilot plant units. In Unbehauen (Ed.), *Methods and Applications in Adaptive Control.* Springer, Berlin.
Ljung, L. (1977a). Analysis of recursive stochastic algorithms. *IEEE Trans Aut. Control*, **AC-22**, 551.
Ljung, L. (1977b). On positive real transfer functions and the convergence of some recursive schemes. *IEEE Trans Aut. Control*, **AC-22**, 539.
Ljung, L. and B. Wittenmark (1974). Analysis of a class of adaptive regulators. *Proc. IFAC Symposium on Stochastic Control Theory*, Budapest, p. 431.
Ljung, L. and T. Söderström (1983). *Theory and Practice of Recursive Identification*, MIT Press, Cambridge.
Mannerfelt, C. F. (1981). Robust Control Design with Simplified Models. Report: CODEN:LUTFD2/(TFRT-1021)/1-153/(1981). Department of Automatic Control, Lund Institute of Technology.
Mendel, J. M. and K. S. Fu (1970). *Adaptive, Learning and Pattern Recognition Systems.* Academic Press, New York.
Menga, G. and E. Mosca (1980). MUSMAR: multivariable adaptive regulators based on multistep cost functionals. In D. G. Lainiotis and N. S. Tzannes (Eds.), *Advances in Control.* Reidel, Dordrecht.
Mishkin, E. and L. Braun (1961). *Adaptive Control Systems.* McGraw-Hill, New York.
Monopoli, R. V. (1973). The Kalman–Yakubovich lemma in adaptive control system design. *IEEE Trans Aut. Control*, **AC-18**, 527.
Monopoli, R. V. (1974). Model reference adaptive control with an augmented error signal. *IEEE Trans Aut. Control*, **AC-19**, 474.
Morgan, A. P. and K. S. Narendra (1977). On the stability of non-autonomous differential equations. *SIAM J. Control & Optimiz.*, **15**, 163.
Morris, A. J., T. P. Fenton and Y. Nazer (1977). Application of self-tuning regulator to the control of chemical processes. In H. R. van Nauta Lemke and H. B. Verbruggen (Eds.) *Digital Computer Applications to Process Control.* Preprints 5th IFAC/IFIP International Conference, The Hague.
Morse, A. S. (1980). Global stability of parameter-adaptive control systems. *IEEE Trans Aut. Control*, **AC-25**, 433.
Narendra, K. S. and Y.-H. Lin (1979). Design of stable model reference adaptive controllers. *Proc. of the Workshop on Applications of Adaptive Control*, Yale University.
Narendra, K. S. and R. V. Monopoli (Eds.) (1980). *Applications of Adaptive Control.* Academic Press, New York.
Narendra, K. S. and L. S. Valavani (1979). Direct and indirect adaptive control. *Automatica*, **15**, 653.
Narendra, K. S., Y.-H. Lin and L. S. Valavani (1980). Stable adaptive controller design—part II: Proof of Stability. *IEEE Trans Aut. Control*, **AC-25**, 440.
Narendra, K. S. and B. B. Peterson (1981). Adaptive control in the presence of bounded disturbances. *Proc. 2nd Yale Workshop on Applications of Adaptive Control.*
Negoesco, S., B. Courtiol and C. Francon (1978). Application de la commande adaptive à la fabrication des guides d'ondes hèlicoidaux. *Automatisme*, **23**, 1.7
Niemi, A. J. (1981). Invariant control of variable flow processes. *Proc. 8th IFAC World Congress*, Kyoto.
Parks, P. C. (1966). Lyapunov redesign of model reference adaptive control systems. *IEEE Trans Aut. Control*, **AC-11**, 362.
Parks, P. C., W. Schaufelberger, C. Schmid and H. Unbehauen (1980). Applications of adaptive control systems. In H. Unbehauen (Ed.), *Methods and Applications in Adaptive Control*, Springer, Berlin.
Peterka, V. (1970). Adaptive digital regulation of noisy systems. *Proc. 2nd IFAC Symposium on Identification and Process Parameter Estimation*, Prague.
Peterka, V. (1975). A square-root filter for real-time multivariable regression. *Kybernetica* **11**, 53.
Peterka, V. (1982). Predictor based self-tuning control, *Proc. 6th IFAC Symposium on Identification and System Parameter Estimation*, Arlington.
Peterka, V. and K. J. Åström (1973). Control of multivariable systems with unknown but constant parameters. *Proc. 3rd IFAC Symposium on Identification and System Parameter Estimation*, The Hague.
Poljak, B. T. and Ya. Z. Tsypkin (1976). Robust Identification. *Preprints 4th IFAC Symposium on Identification and System Parameter Estimation*, Tbilisi.
Richalet, J., A. Rault, J. L. Testud and J. Papon (1978). Model predictive heuristic control: applications to industrial processes. *Automatica*, **14**, 413.
Rohrs, C. L., Valavani, M. Athans and G. Stein (1982). Stability problems of adaptive control algorithms in the presence of unmodelled dynamics. *Proc. 21st IEEE CDC*, Orlando, Vol. 1, p. 3.
Saridis, G. N., J. M. Mendel and Z. Z. Nikolic (1973). Report on definitions of self-organizing control processes and learning systems. *IEEE Control Systems Society Newsletter.*
Saridis, G. N. (1977). *Self-organizing Control of Stochastic Systems.* Marcel Dekker, New York.
Sastry, V. A. (1979). Self-tuning control of Kamyr digester chip level. *Pulp & paper*, **5**, 160.
Schumann, R. (1980). Digital parameter-adaptive control of an air conditioning plant. *5th IFAC/IFIP Conference on Digital Computer Applications*, Dusseldorf.
Schumann, R., K.-H. Lachmann and R. Isermann (1982). Towards applicability of parameter-adaptive control algorithms. *Proc. 8th IFAC World Congress*, Kyoto.
Seyfried, H. W. and D. Stöle (1975). Application of adaptive control in rolling mill area, especially for plate mills. *Proc. 6th IFAC Congress*, Boston.
Shinsky, F. G. (1974). Adaptive pH control monitors nonlinear process. *J. Control Engng*, **2**, 57.
Speth, W. (1969). Leicht realisierbares Verfahren zur schnellen Adaption von Reglern der Antriebstechnik. *Preprints 4th IFAC World Congress*, Warsaw.
Stein, G. (1980). Adaptive flight control—a pragmatic view. In K. S. Narendra and R. V. Monopoli (Eds.), Applications of Adaptive Control. Academic Press, New York.
Sternby, J. (1977). On consistency for the method of least-squares using martingale theory. *IEEE Trans Aut. Control*, **AC-22**, 346.
Sternby, J. (1980). Extremum control systems—an area for adaptive control. *Proc. JACC American Automation Council.*
Sternby, J. and H. Rootzen (1982). Martingale theory in Bayesian least squares estimation. *Preprints 6th IFAC Symposium on Identification and System Parameter Estimation*, Arlington.
Stromer, P. R. (1959). Adaptive or self-optimizing control systems—a bibliography. *IRE Trans*, **AC-7**, 65.
Taylor, L. W. and E. J. Adkins (1965). Adaptive control and the X-15. *Proc. Princeton University Conference on Aircraft Flying Qualities*, Princeton University.
Truxal, J. G. (1964). Theory of self-adjusting control. *Proc. 2nd IFAC World Congress.*
Tse, E. and Y. Bar-Shalom (1975). Generalized certainty equivalence and dual effect in stochastic control. *IEEE Trans Aut. Control*, **AC-20**, 817.
Tsypkin, Ya. Z. (1971). *Adaptation and Learning in Automatic Systems.* Academic Press. New York.
Tsypkin, Y. Z. (1973). *Foundations of the Theory of Learning Systems.* Academic Press, New York.
Unbehauen, H. (Ed.) (1980). *Methods and Applications in Adaptive Control.* Springer, Berlin.
Unbehauen, H. and C. Schmid (1975). Status and application of adaptive control systems. *Automatic Control Theory & Applic.*, **3**, 1.

van Amerongen, J. (1981). A model reference adaptive autopilot for ships—practical results. *Proceedings IFAC World Congress*, Kyoto, Japan.

van Amerongen, J. (1982). Adaptive steering of ships. A model-reference approach to improved manoeuvring and economical course keeping. Ph.D. thesis, Delft University of Technology.

Wellstead, P. E., J. M. Edmunds, D. Prager and P. Zanker (1979). Self-tuning pole/zero assignment regulators. *Int. J. Control*, **30**, 1.

Wellstead, P. E., D. Prager and P. Zanker (1979). Pole assignment self-tuning regulator. *IEE Proc.*, **126**, 781.

Westerlund, T., H. Toivonen and K-E. Nyman (1980). Stochastic modelling and self-tuning control of a continuous cement raw material mixing system. *Modelling, Identification & Control*, **1**, 17.

Westerlund, T. (1981). A digital quality control system for an industrial dry process rotary cement kiln. *IEEE Trans Aut. Control*, **AC-26**, 885.

Whitaker, H. P., J. Yamron and A. Kezer (1958). Design of model-reference adaptive control systems for aircraft. Report R-164, Instrumentation Laboratory, MIT, Cambridge.

Wittenmark, B. (1973). A self-tuning regulator. Thesis, TFRT-1003, Department of Automatic Control, Lund Institute of Technology.

Wittenmark, B. (1975). Stochastic adaptive control methods—a survey. *Int. J. Control*, **21**, 705.

Wouters, W. R. E. (1977). Adaptive pole placement for linear stochastic systems with unknown parameters. *Preprints IEEE Conference on Decision and Control*, New Orleans.

Young, P. (1981). A second generation adaptive autostabilization system for airborne vehicles. *Automatica*, **17**, 459.

Zanker, P. M. (1980). Application of self-tuning. Ph.D. thesis. Control System Centre, University of Manchester.

Parameter Adaptive Control Algorithms—A Tutorial*

ROLF ISERMANN†

An introduction based on theoretical investigations and practical experience provides a basis for understanding and applying several adaptive control techniques.

Key Words—Adaptive control; identification; closed-loop parameter estimation; model structure determination; digital control.

Abstract—An introduction is given to adaptive (self-tuning) control algorithms with recursive parameter estimation, which have obtained increasing attention in recent years. These algorithms result from combinations of recursive parameter estimation algorithms and easy to design control algorithms. Firstly a short review is given on proper recursive parameter estimation methods, including their application in a closed loop. This is followed by the design equations for various control algorithms and ways for d.c.-value estimation and for offset compensation. Various explicit and implicit combinations can be designed with different properties of the resulting adaptive control algorithms for both deterministic and stochastic disturbances. Their convergence properties are discussed. Simulation examples are presented and examples for the adaptive control of an air conditioner and a pH-process are shown. The introduction of a third feedback level for coordination and supervision is considered. Finally further problems are discussed.

1. INTRODUCTION

Many different proposals for adaptive control have been made in the past. However, their application, mostly based on analog realization, has not been very successful and unconvincing until the early 1970s. The development of cheaper and reliable digital computers has meant that the field of adaptive control has been reactivated. Adaptive control algorithms have obtained much attention in the last few years because good results have been given by some applications.

Recent surveys on certain classes of adaptive control systems have been given by Åström and co-workers (1977); Tsypkin (1978); Landau (1979); Parks and co-workers (1980). This tutorial introduces some adaptive control principles based on process parameter estimation techniques. Common basic schemes of adaptive control are considered in Section 2 of this paper. Then one major class, the so-called parameter adaptive control algorithms are discussed in detail.

Suitable parameter estimation algorithms and their conditions for closed-loop application are treated in Sections 3 and 4. After selecting proper control algorithms in Section 5, parameter estimators and controllers are combined to create parameter adaptive control algorithms via explicit and implicit schemes, Section 6. Some simulations illustrate in Section 7 the overall behaviour and indicate the choice of *a priori* factors. Two applications are shown for the adaptive control of an air-conditioning unit and a pH-process, Section 8. The same principle can be used for adaptive feedforward control, Section 9. Experience shows that the introduction of a third feedback level for coordination and supervision of the adaptive controller is necessary and has several advantages, Section 10. Finally some further problems are discussed in the last section, Section 11.

2. SOME PRINCIPLES OF ADAPTIVE CONTROL

A comparison of contributions to this topic shows that many different definitions and classifications on adaptive control and related methods do exist. Saridis (1977) and the survey contributions cited above summarize the most important ones. This section is used to describe some basic schemes of adaptive control with regard to their application. For simplicity just linear single input–single output processes are considered. To describe some different adaptive control schemes their behavior is described in a black-box manner. Then space is left free for different adaptive control principles and algorithms. An attempt is made to use descriptions which are rather widely accepted.

Conversely to fixed control systems, *adaptive control systems* adapt their behavior to the (changing) properties of controlled processes and their signals. Two basic schemes of controller adaptation can be distinguished as shown in Fig. 1. If the process behavior changes are indicated by measurable signals (mass flow, speed, etc.) and if it is known in advance how the controller has to be adapted in dependence on these signals *feedforward adaptation* (open-loop adaptation) can be applied (Fig. 1a). Then no feedback for the adaptation exists from internal closed-loop signals to the controller.

The other scheme can be called *feedback adaptation* (closed-loop adaptation) (Fig. 1b). In a first step, information on the process behavior (structure, parameters) is gained by measuring process input and output signals. This can be performed, for example, by process identification (measuring of u and y) or by determination of the closed-loop performance (measuring of e_w and u).

Based on this information the controller can be calculated and adapted. A second feedback results leading to a closed-loop action with the signal flow path: control-loop signals–adaptation algorithm–controller–control-loop signals. This scheme is applied if the process behavior changes cannot be observed by direct measurable signals. In the following only feedback adaptation is considered because feedforward adaptation generally can be applied straightforwardly, if applicable.

Adaptive controllers with feedback can be divided mainly into two groups. *Self-optimizing adaptive controllers* try to

*Received 27 January 1981; revised 28 January 1982. The original version of this paper was presented at the 6th IFAC/IFIP Conference on Digital Computer Application to Process Control which was held in Düsseldorf, F.R. Germany during October 1980. The published proceedings of this IFAC meeting may be ordered from Pergamon Press Ltd, Headington Hill Hall, Oxford OX3 0BW, U.K. This paper was recommended for publication in revised form by associate editor A. van Cauwenberghe.

†Institut für Regelungstechnik, Technische Hochschule Darmstadt, 6100 Darmstadt, Federal Republic of Germany.

Reprinted with permission from *Automatica*, vol. 18, pp. 513–528, Sept. 1982.
Copyright © 1982, International Federation of Automatic Control IFAC. Published by Pergamon Press Ltd.

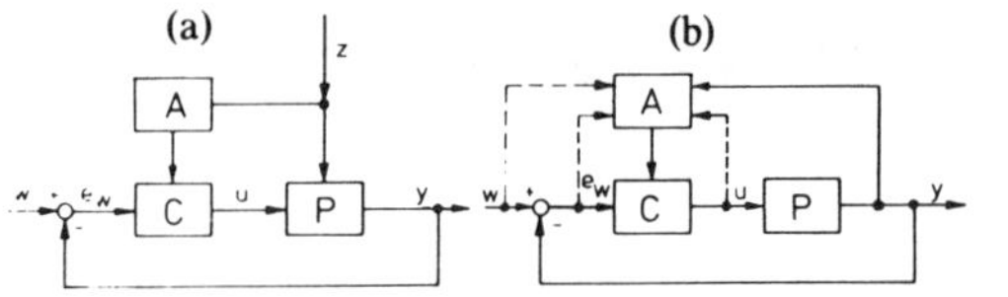

FIG. 1. Basic schemes of controller adaptation: (a) feedforward adaptation (open-loop adaptation). A: Adaptation algorithm (feedforward); (b) feedback adaptation (closed-loop adaptation). A: Adaptation algorithm (feedback).

reach an optimal control performance, subject to a given controller type and the obtainable information on the process and its signals (Fig. 2a).

Model reference adaptive controllers try to reach for a definite input signal, a control behavior close to a given reference model (Fig. 2b). They require a measurable external signal (e.g. the reference value), and they adapt to the model if this signal changes. If a fixed reference model is used, the closed loop approaches the *a priori* given reference behavior and not necessarily a possible optimal one. Their advantages are a quick adaptation for defined inputs (e.g. servo control systems) and a straightforward treatment of stability using nonlinear stability theories (Landau, 1979).

Self-optimizing adaptive controllers have the advantage of being able to adapt for any (especially unmeasurable) disturbances, if properly designed. Recent papers have shown that some equivalent properties exist between self-optimizing and model reference adaptive controllers, especially for the controller itself and the convergence analysis (Égardt, 1980; Landau, 1979; Narendra, 1980).

In the sequel *self-optimizing adaptive controllers* are considered which are *based on process identification*. This class of adaptive controllers can be classified according to:

process model;
process identification method;
information about the process;
criterion for controller design;
control algorithm.

Process model

The implementation on digital computers leads to discrete-time process models and the availability of proper controller design procedures recommends models in parametric form. Therefore *difference equations* are mostly used as process models. They enable a simple inclusion of deadtimes and also stochastic noise models.

Process identification methods

For the identification of parametric process models the use of parameter estimation methods is straightforward. *Recursive parameter estimation methods* are best suited for on-line identification in real time. However, special conditions for closed-loop identification have to be satisfied.

Information about the process

The information about the process then consists of the parameter estimates $\hat{\theta}$ and, if desired of their uncertainty $\Delta\hat{\Theta}$.

If the parameter estimates $\hat{\Theta}$ are assumed to be identical with the real process parameters, the resulting controllers are called *certainty equivalence controllers*. If, however, additionally the uncertainties $\Delta\hat{\Theta}$ of the parameter estimates are taken into account for the controller calculation, the

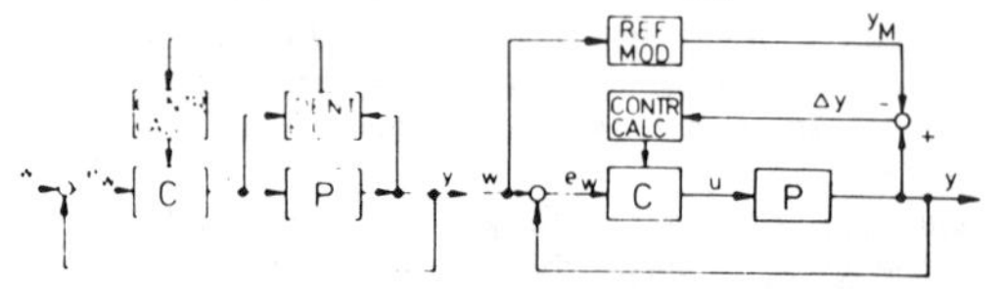

FIG. 2. Basic schemes of adaptive controllers with feedback adaptation: (a) self-optimizing adaptive controller (SOAC); (b) model reference adaptive controller (MRAC).

resulting controllers are called *cautious controllers* (Wittenmark, 1975).

Criterion for controller design

The performance of adaptive control systems which are based on process identication mainly depends on the quality of the identified process model and therefore on the process input signal. Hence, the input signal has to be determined so that as well as a good present control a good future process identification is achieved. The inclusion of both requirements in the controller design criterion leads to *dual controllers* (Feldbaum, 1965). *Nondual controllers* use only the present and past information about the process in their design criteria. Frequently used design criteria for nondual controllers are quadratic in $y(k)$ and $u(k)$ or are special design criteria as e.g. the principles of deadbeat, pole-zero cancellation or pole-assignment.

Control algorithms

The actual design of control algorithms is of course done before their implementation in the digital computer. Then the calculation of the controller parameters as a function of the process parameters remains. Control algorithms for adaptive control should satisfy:

closed-loop identifiability conditions;
small computational expense and storage requirement for the controller parameter calculation;
applicability for many classes of processes and signals.

Within the class of self-optimizing adaptive controllers based on process identification those types which have demonstrated most success in the last ten years, both in practice and theory, are characterized by *recursive parameter estimation*, the *nondual* and the *certainty equivalence principle*. The resulting algorithms will be called here *parameter adaptive control algorithms* (Fig. 3). This kind of algorithm is also called a *self-tuning regulator* (Åström and co-workers, 1977). A distinction between 'self-tuning' and 'adaptive' could be made by only applying the first term to constant process parameters. However, since there is no sharp distinction between both cases regarding their applicability the distinction is of less importance.

3. RECURSIVE PARAMETER ESTIMATION

It is assumed that a stable process is time invariant and linearizable so that it can be described by a linear *stochastic difference equation* with constant parameters

$$\begin{aligned} y(k) = &-a_1y(k-1) - \cdots - a_my(k-m) \\ &+ b_1u(k-d-1) + \cdots + b_mu(k-d-m) \\ &+ v(k) + d_1v(k-1) + \cdots d_mv(k-m) \end{aligned} \quad (1)$$

where $k = t/T_0 = 0, 1, 2, \ldots,$ is the discrete time, T_0 the sample time, d the discrete dead time, and

$$y(k) = Y(k) - Y_{00} \quad (2)$$

$$u(k) = U(k) - U_{00} \quad (3)$$

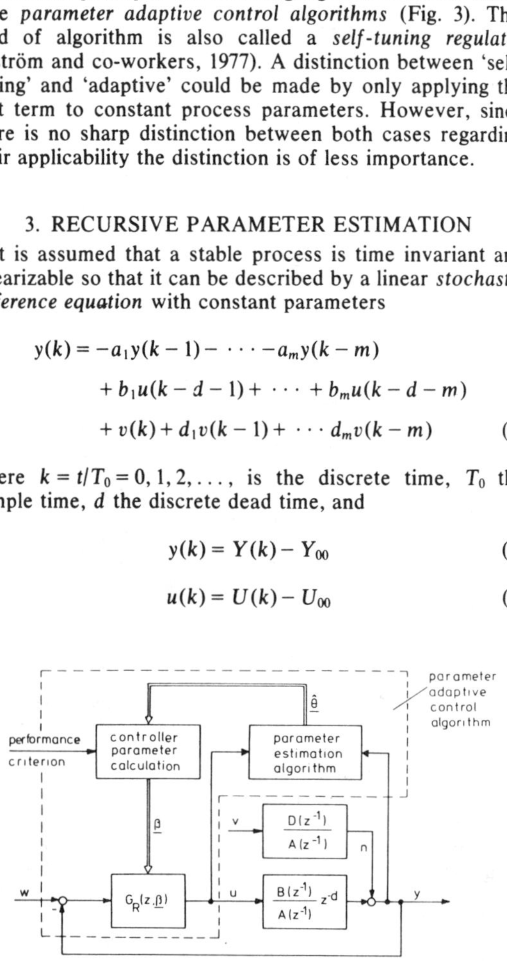

FIG. 3. Block diagram for adaptive controllers with recursive parameter estimation.

are the deviations of the measured process output signal $Y(k)$ and process input signal $U(k)$ from the d.c. values Y_{00} and U_{00}. The unmeasurable noise signal $v(k)$ is assumed to be statistically independent and stationary with

$$E\{v(k)\} = 0$$
$$E\{v(k)v(k+\tau)\} = \sigma_v^2\delta(\tau) \tag{4}$$

where σ_v^2 is the variance and $\delta(\tau)$ the Kronecker delta function.

The corresponding z-transfer function of (1) is

$$y(z) = \underbrace{\frac{B(z^{-1})}{A(z^{-1})}}_{G_P(z^{-1})} z^{-d} u(z) + \underbrace{\frac{D(z^{-1})}{A(z^{-1})}}_{G_v(z^{-1})} v(z) \tag{5}$$

with the polynomials

$$A(z^{-1}) = 1 + a_1 z^{-1} + \cdots a_m z^{-m}$$
$$B(z^{-1}) = b_1 z^{-1} + \cdots + b_m z^{-m} \tag{6}$$
$$D(z^{-1}) = 1 + d_1 z^{-1} + \cdots + d_m z^{-m}$$

where $z = e^{T_0 s}$ with $s = \delta + i\omega$ the Laplace variable. Equation (5) shows that the stochastic difference equation (1) consists of a *process model* $G_P(z^{-1})$ and a *noise model* $G_v(z^{-1})$. A certain specialization of this model is that the noise filter denominator is equal to the process model denominator. This simplifies the closed-loop parameter estimation and the controller design. The assumed stationarity of the noise implies that all zeros of the polynomial $z^m D(z^{-1})$ are inside the unit circle of the z-plane.

Equation (1) can be shortened by

$$y(k) = \boldsymbol{\psi}^T(k)\boldsymbol{\theta} + v(k) \tag{7}$$

with the data vector

$$\boldsymbol{\psi}^T(k) = [-y(k-1), \ldots, -y(k-m) \mid u(k-d-1), \ldots, u(k-d-m) \mid v(k-1), \ldots, v(k-m)] \tag{8}$$

and the parameter vector

$$\boldsymbol{\theta} = [a_1, \ldots, a_m \mid b_1, \ldots, b_m \mid d_1, \ldots, d_m]^T. \tag{9}$$

For the *estimation* of the unknown parameters of (6) the following model is assumed

$$y(z) = \frac{\hat{B}(z^{-1})}{\hat{A}(z^{-1})} z^{-d} u(z) + \frac{\hat{D}(z^{-1})}{\hat{A}(z^{-1})} e(z) \tag{10}$$

where the circumflex accent (ˆ) represents estimate. Here $e(z)$ corresponds to $e(k)$ the equation error or the one-step-ahead prediction error due to the unmeasurable noise $v(k)$, see (14).

For the recursive parameter estimation in an open loop several methods can be used, for example

recursive least squares (RLS);
recursive extended least squares (RELS);
recursive instrumental variables (RIV);
recursive maximum likelihood (RML);
stochastic approximation (STA).

These methods are described, for example, by Eykhoff (1974); Isermann (1974, 1981); Strejc (1980). The least-squares method for example is based on the minimization of the loss function

$$V = \sum_k e^2(k) \tag{11}$$

due to the unknown parameter vector $\hat{\theta}$. For normal distributed noise the same loss function holds for the maximum likelihood method. The various parameter estimation methods can be written in a unified form (Söderström, Ljung and Gustavsson, 1976)

$$\hat{\boldsymbol{\theta}}(k+1) = \hat{\boldsymbol{\theta}}(k) + \boldsymbol{\gamma}(k)e(k+1) \tag{12}$$

$$\boldsymbol{\gamma}(k) = \mu(k+1)\boldsymbol{P}(k)\boldsymbol{\varphi}(k+1) \tag{13}$$

$$e(k+1) = y(k+1) - \boldsymbol{\psi}^T(k+1)\hat{\boldsymbol{\theta}}(k). \tag{14}$$

The definitions of $\hat{\boldsymbol{\theta}}$, $\boldsymbol{\psi}^T$, $\boldsymbol{\varphi}$, and $\boldsymbol{P}$, which depend on the very parameter estimation method, are given in Table 1 for RLS, RELS and RML, which have shown to be best suited for parameter adaptive control (Kurz, Isermann and Schumann, 1980). (RIV yields biased parameter estimates in closed loop, STA converges too slowly and unreliably.)

For RML and RELS the assumed model is (10). However, RLS $\hat{D}(z^{-1}) = 1$, is a specialized case. The *convergence conditions* for the methods RLS, RELS and RML can be summarized as follows: to obtain unbiased estimates

$$E\{\hat{\boldsymbol{\theta}}(N)\} = \boldsymbol{\theta}_0 \tag{15}$$

TABLE 1. ELEMENTS OF THE UNIFIED RECURSIVE PARAMETER ESTIMATION ALGORITHM FOR RECURSIVE LEAST SQUARES (RLS), RECURSIVE EXTENDED LEAST SQUARES (RELS) AND RECURSIVE MAXIMUM LIKELIHOOD METHOD (RML)

Method	$\hat{\underline{\theta}}$	$\underline{\psi}^T(k+1)$	$\underline{\varphi}^T(k+1)$	$\mu(k+1)$	$\underline{P}(k+1)$
RLS	$[a_1,\ldots,a_m,$ $b_1,\ldots,b_m]^T$	$[-y(k),\ldots,-y(k-m+1),$ $u(k-d),\ldots,u(k-d-m+1)]$	$\underline{\psi}^T(k+1)$	$\dfrac{1}{\lambda(k+1)+\underline{\psi}^T(k+1)\underline{P}(k)\underline{\psi}(k+1)}$	$\dfrac{[\underline{I}-\underline{\gamma}(k)\underline{\psi}^T(k+1)]\underline{P}(k)}{\lambda(k+1)}$
RELS	$[a_1,\ldots,a_m,$ $b_1,\ldots,b_m,$ $d_1,\ldots,d_m]^T$	$[-y(k),\ldots,-y(k-m+1)$ $u(k-d),\ldots,u(k-d-m+1),$ $e(k),\ldots,e(k-m+1)]$	$\underline{\psi}^T(k+1)$	as RLS	as RLS
RML	as RELS	as RELS	$[-y'(k),\ldots,-y'(k-m+1),$ $u'(k-d),\ldots,u'(k-d-m+1)$ $e'(k),\ldots,e'(k-m+1)]$	$\dfrac{1}{\lambda(k+1)+\underline{\varphi}^T(k+1)\underline{P}(k)\underline{\varphi}(k+1)}$	$\dfrac{[\underline{I}-\underline{\gamma}(k)\underline{\varphi}^T(k+1)]\underline{P}(k)}{\lambda(k+1)}$

$\underline{I}$ ... identity matrix

$$y'(z) = \frac{1}{D(z^{-1})}y(z); \quad u'(z) = \frac{1}{D(z^{-1})}u(z); \quad e'(z) = \frac{1}{D(z^{-1})}e(z).$$

and consistent estimates in mean square

$$\lim_{N\to\infty} E\{\hat{\boldsymbol{\theta}}(N)\} = \boldsymbol{\theta}_0 \tag{16}$$

$$\lim_{N\to\infty} E\{[\hat{\boldsymbol{\theta}}(N) - \boldsymbol{\theta}_0][\hat{\boldsymbol{\theta}}(N) - \boldsymbol{\theta}_0]^{\mathrm{T}}\} = 0 \tag{17}$$

the following necessary conditions have to be satisfied:

(a) process order m and deadtime d are known;
(b) d.c.-values U_{00} and Y_{00} are known;
(c) the process input $u(k)$ is persistently exciting of order $n \geq m$ (Åström and Bohlin, 1966; Isermann, 1980);
(d) the error signal $e(k)$ is not correlated with $\boldsymbol{\psi}^{\mathrm{T}}(k)$, which means that $e(k)$ must be uncorrelated;
(e) $E\{e(k)\} = 0$.

Condition (d) is valid for the assumed basic model structures. The means that for RLS $D(z^{-1}) = 1$ is required, which leads to 'LS-model'

$$A(z^{-1})y(z) = B(z^{-1})z^{-d}u(z) + e(z) \tag{18}$$

whereas for RML and RELS the 'ML-model' or 'ARMAX-model' is to be used

$$A(z^{-1})y(z) = B(z^{-1})z^{-d}u(z) + D(z^{-1}) \cdot e(z). \tag{19}$$

The very special assumption on the noise filter $G_v(z^{-1}) = 1/A(z^{-1})$ for the RLS-method leads to biased estimates in practice.

If the process parameters are not constant, but slowly time-varying, the parameter estimation algorithms can be modified by a forgetting memory. The weighted least-squares method shows that this can be realized by minimizing weighted squares

$$V = \sum_{k=0}^{N} \varepsilon(k)e^2(k). \tag{20}$$

The choice of

$$\varepsilon(k) = \lambda^{N-k}, \quad 0 < \lambda < 1 \tag{21}$$

leads to an exponential forgetting memory (see Table 2). The forgetting factor λ is a constant, and it has to be chosen within $0.95 \leq \lambda \leq 0.995$ for most cases. Table 1 shows that λ appears in μ and $\boldsymbol{P}$. The slow convergence of RML and RELS in the starting phase can be accelerated by an increasing weighting

$$\lambda(k+1) = \lambda_0\lambda(k) + (1 - \lambda_0) \tag{22}$$

with $\lambda_0 < 1$, $\lambda(0) < 1$ (Söderström, Ljung and Gustavsson, 1976).

Combination of (22) and (21) results in

$$\lambda(k+1) = \lambda_0\lambda(k) + \lambda(1 - \lambda_0) \tag{23}$$

Example 1. *Parameter estimation with the recursive least-squares method for a first-order model*

The process model is

$$y(k) + a_1y(k-1) = b_1u(k-1) + v(k).$$

For the parameter estimation

$$y(k) = \boldsymbol{\psi}^{\mathrm{T}}(k)\hat{\boldsymbol{\theta}}(k) + e(k)$$

with

$$\boldsymbol{\psi}^{\mathrm{T}}(k) = [-y(k-1)u(k-1)]$$

$$\hat{\boldsymbol{\theta}}(k) = [\hat{a}_1(k)\hat{b}_1(k)]^{\mathrm{T}}.$$

The estimation algorithms can be programmed in this way (as suitable for adaptive control):

1. New measurements $y(k)$ and $u(k)$ are made at time k

2. $e(k) = y(k) - [-y(k-1)u(k-1)]\begin{bmatrix}\hat{a}_1(k-1)\\ \hat{b}_1(k-1)\end{bmatrix}.$

3. New parameter estimates

$$\begin{bmatrix}\hat{a}_1(k)\\ \hat{b}_1(k)\end{bmatrix} = \begin{bmatrix}\hat{a}_1(k-1)\\ \hat{b}_1(k-1)\end{bmatrix} + \underbrace{\begin{bmatrix}\gamma_1(k-1)\\ \gamma_2(k-1)\end{bmatrix}}_{\text{from step 7}} e(k)$$

4. Measurements $y(k)$ and $u(k)$ inserted in $\boldsymbol{\psi}^{\mathrm{T}}(k+1) = [-y(k)u(k)]$

5. $\underbrace{\boldsymbol{P}(k)}_{\substack{\text{from}\\ \text{step 8}}}\boldsymbol{\psi}(k+1) = \begin{bmatrix}p_{11}(k) & p_{12}(k)\\ p_{21}(k) & p_{22}(k)\end{bmatrix}\begin{bmatrix}-y(k)\\ u(k)\end{bmatrix}$

$$= \begin{bmatrix}-p_{11}(k)y(k) + p_{12}(k)u(k)\\ -p_{21}(k)y(k) + p_{22}(k)u(k)\end{bmatrix} = \begin{bmatrix}i_1\\ i_2\end{bmatrix}.$$

6. $\boldsymbol{\psi}^{\mathrm{T}}(k+1)\boldsymbol{P}(k)\boldsymbol{\psi}(k+1) = p_{11}(k)y^2(k) - (p_{12}(k) + - p_{21}(k))u(k)y(k) + p_{22}(k) \cdot u^2(k) = j.$

7. $\begin{bmatrix}\gamma_1(k)\\ \gamma_2(k)\end{bmatrix} = \dfrac{1}{j+\lambda}\begin{bmatrix}i_1\\ i_2\end{bmatrix}$

8. $\boldsymbol{P}(k+1) = \dfrac{1}{\lambda}\begin{bmatrix}1+\gamma_1(k)y(k) & -\gamma_1(k)u(k)\\ \gamma_2(k)y(k) & 1-\gamma_2(k)u(k)\end{bmatrix} \times \begin{bmatrix}p_{11}(k)p_{12}(k)\\ p_{21}(k)p_{22}(k)\end{bmatrix}.$

9. Replace $(k+1)$ by k and start again with step 1.

To start the recursive algorithms at time $k = 0$ one uses

$$\hat{\boldsymbol{\theta}}(0) = \begin{bmatrix}0\\ 0\end{bmatrix} \text{ and } \boldsymbol{P}(0) = \begin{bmatrix}\alpha & 0\\ 0 & \alpha\end{bmatrix}$$

where α is a very large number.

d.c.-Values

For the parameter estimation the variations $u(k)$ and $y(k)$ of the measured signals $U(k)$ and $Y(k)$ have to be used [see (2) and (3)]. Therefore in general the d.c.-values are required, which, however, are unknown. Following methods can be used to tackle this problem.

Method 1. *Differencing*

The easiest way to obtain the variations without knowing the d.c.-values is just to calculate the differences

$$U(k) - U(k-1) = u(k) - u(k-1) = \Delta u(k) \quad \text{or} \quad \Delta u(z) = u(z)[1 - z^{-1}] \tag{24}$$

$$Y(k) - Y(k-1) = y(k) - y(k-1) = \Delta y(k) \quad \text{or} \quad \Delta y(z) = y(z)[1 - z^{-1}] \tag{25}$$

at the process input and output. Instead of $u(z)$ and $y(z)$ the signals $\Delta u(z) = u(z)[1 - z^{-1}]$ and $\Delta y(z) = y(z)[1 - z^{-1}]$ are used for the parameter estimation. Because this special high

TABLE 2. WEIGHTING FACTORS $\varepsilon(k)$ DUE TO (21) FOR $N = 50$

k	1	10	20	30	40	47	48	49	50
$\lambda = 0.99$	$\varepsilon(k) = 0.61$	0.67	0.73	0.82	0.90	0.97	0.98	0.99	1
$\lambda = 0.95$	0.08	0.13	0.21	0.35	0.60	0.85	0.90	0.95	1

pass filtering is applied for both the process input and output, the process parameters can be estimated as in the case of measuring $u(k)$ and $y(k)$ without the filters.

If the d.c.-values are to be taken into account, other methods have to be used.

Method 2. *Averaging*

The d.c.-values can be estimated by simple averaging

$$\hat{Y}_{00} = \frac{1}{M}\sum_{k=1}^{M} Y(k). \quad (26)$$

Its recursive version is

$$\hat{Y}_{00}(k) = \hat{Y}_{00}(k-1) + \frac{1}{k}[Y(k) - \hat{Y}_{00}(k-1)]. \quad (27)$$

For slowly time varying d.c.-values recursive averaging with exponential forgetting leads to

$$\hat{Y}_{00}(k) = \lambda\hat{Y}_{00}(k-1) + (1-\lambda)Y(k) \quad (28)$$

with $\lambda < 1$. The same can be applied for U_{00}. By using (2) and (3) the variations $u(k)$ and $y(k)$ can be determined.

Method 3. *Estimation of a constant*

Inserting (2) and (3) into (1) results in

$$\begin{aligned} Y(k) = &-a_1Y(k-1) - \cdots - a_mY(k-m) \\ &+b_1U(k-d-1) + \cdots + b_mU(k-d-m) \\ &+v(k) + d_1v(k-1) + \cdots + d_mv(k-m) + C \quad (29) \end{aligned}$$

with

$$C = (1 + a_1 + \cdots + a_m)Y_{00} - (b_1 + \cdots + b_m)U_{00}. \quad (30)$$

Extending the parameter vector $\hat{\boldsymbol{\theta}}$ by C and the measurement vector $\boldsymbol{\psi}^T(k)$ by 1, the measured signals $Y(k)$ and $U(k)$ can be directly used for the parameter estimation algorithms and C can be estimated too. Then for one given d.c.-value the other can be calculated using (30) (Isermann, 1974). An alternative with quicker convergence is to first estimate the $\hat{a}_i$, $\hat{b}_i$, and $\hat{d}_i$ via (24) and (25) and then to estimate $\hat{C}$ separately via (29) and recursive least-squares method for this one parameter.

For closed-loop identification

$$Y_{00} = W(k) \quad (31)$$

can be used where $W(k)$ is the absolute value of the reference value to remove controller offsets and to calculate U_{00} (Kurz, Isermann and Schumann, 1980). See also Section 6.

4. PARAMETER ESTIMATION IN A CLOSED LOOP

The recursive parameter estimation methods described in the last section have been developed for identification in an open loop, where the input signal could be chosen freely with the exception that at least the condition of persistent excitation of order m has to be satisfied. However, if the process is driven in a closed loop together with a linear controller having a z-transfer function

$$G_R(z) = \frac{u(z)}{e_w(z)} = \frac{Q(z^{-1})}{P(z^{-1})} = \frac{q_0 + q_1z^{-1} + \cdots + q_\nu z^{-\nu}}{p_0 + p_1z^{-1} + \cdots + p_\mu z^{-\mu}} \quad (32)$$

with the control deviation

$$e_w(z) = w(z) - y(z) \quad (33)$$

where $w(z)$ corresponds to a variation $w(k)$ of the reference value W, and the only exciting input signal is the unmeasurable stochastic noise $v(k)$ (see Fig. 3), the process input $u(k)$ depends on the process output $y(k)$. Therefore special conditions for closed-loop parameter estimation have to be considered.

Several cases of closed-loop identification can be distinguished. For *indirect process identification* a model of the closed loop is identified. Then the process model can be determined, if the controller is known, and only the output signal $y(k)$ has to be measured. In the case of *direct process identification* the process is directly identified by measuring $u(k)$ and $y(k)$. The controller then need not be known. Additionally consideration must be given to whether or not the closed-loop identification is performed *without a perturbation signal* or *with a perturbation signal* (measurable or unmeasurable), in addition to the acting noise $v(k)$.

Conditions for closed-loop identification have been stated by Gustavsson, Ljung and Söderström (1977); Kurz and Isermann (1975). The description here follows Isermann (1981).

When considering self-optimizing control systems, direct process identification without a perturbation signal is primarily of interest. Therefore only this case will be treated next.

The task becomes obvious if a nonparametric identification method such as the correlation method is applied directly to $u(k)$ and $y(k)$ in the closed loop, then, because of

$$\left.\begin{aligned} \frac{u(z)}{v(z)} &= \frac{-G_R(z)G_v(z)}{1 + G_R(z)G_P(z)} \\ \frac{y(z)}{v(z)} &= \frac{G_v(z)}{1 + G_R(z)G_P(z)} \end{aligned}\right\} \quad (34)$$

and

$$\frac{y(z)}{u(z)} = \frac{y(z)/v(z)}{u(z)/v(z)} = -\frac{1}{G_R(z)} \quad (35)$$

the negative inverse controller transfer function would be identified, not the process. The reason is that the undisturbed process output $y_u(k) = y(k) - n(k)$ is not used, where $n(k)$ is a disturbance signal at the process output, $n(z) = v(z)D(z)/A(z)$ (see Fig. 3). If $y_u(k)$ is known, the process

$$\frac{y_u(z)}{u(z)} = \frac{y(z) - n(z)}{u(z)} = \frac{y(z)/v(z) - n(z)/v(z)}{u(z)/v(z)} = G_P(z) \quad (36)$$

could be identified by a nonparametric identification method. This shows that for direct closed-loop identification the knowledge of the noise filter $n(z)/v(z)$ is required. Therefore the process and noise model

$$A(z^{-1})y(z) = B(z^{-1})z^{-d}u(z) + D(z^{-1})v(z) \quad (37)$$

is used and parametric identification methods will now be considered. To discuss a *first identifiability condition* it must be remembered that (see Example 1) the (nonrecursive) least-squares method is based on a system consisting of equations

$$\begin{aligned} y(k) = \boldsymbol{\psi}^T(k)\hat{\boldsymbol{\theta}} + e(k) = [&-y(k-1)\cdots \\ &-y(k-m) \mid u(k-d-1)\cdots u(k-d-m)]\hat{\boldsymbol{\theta}} + e(k) \end{aligned} \quad (38)$$

for different k. Because of the feedback by the controller indicated in (32), the input and output signals are related by

$$\begin{aligned} u(k-d-1) = &-p_1u(k-d-2) - \cdots - p_\mu u(k-\mu-d-1) \\ &-q_0y(k-d-1) - \cdots - q_\nu y(k-\nu-d-1) \quad (39) \end{aligned}$$

if the reference signal $w(k) = 0$, i.e. $e_w(k) = -y(k)$. $u(k-d-1)$ in (38) therefore is linearly dependent on the other elements of $\boldsymbol{\psi}^T(k)$, if $\mu \le m-1$ and $\nu \le m-d-1$. Hence, no unique solution for the unknown process parameters does exist. Only if $\mu \ge m$ or $\nu \ge m-d$ does the linear dependence vanish.

After inserting the controller equation (32) into (34) it

follows that

$$\frac{y(z)}{v(z)} = \frac{D(z^{-1})P(z^{-1})}{A(z^{-1})P(z^{-1}) + B(z^{-1})z^{-d}Q(z^{-1})} = G_{id}(z)P(z^{-1}) \quad (40)$$

where $P(z^{-1})$ and $Q(z^{-1})$ are known and relatively prime. If $G_{id}(z^{-1})$ has p common poles and zeros, the first identifiability condition becomes (Gustavsson, Ljung and Söderström, 1977)

$$\max\{\mu;\ \nu + d\}, \qquad -p \geq m. \quad (41)$$

The *second identifiability condition* concerns the determination of the order of the process polynomials. The process equation is

$$Ay = Bu + Dv. \quad (42)$$

Inserting the controller equation leads to

$$Ay = B\left(-\frac{Q}{P}\right)y + Dv. \quad (43)$$

This equation is extended by an arbitrary polynomial $S(z^{-1})$

$$(A+S)y = B\left(-\frac{Q}{P}\right)y + Sy + Dv$$

$$(A+S)y = \left(B - S\frac{P}{Q}\right)\left(-\frac{Q}{P}y\right) + Dv$$

$$\underbrace{(A+S)Q}_{A^*}y = \underbrace{(BQ - SP)}_{B^*}\underbrace{\left(-\frac{Q}{P}y\right)}_{u} + \underbrace{DQ}_{D^*}v$$

which leads to

$$A^*y = B^*u + D^*v. \quad (44)$$

This shows that the process B/A and the noise filter D/A can be replaced by

$$\frac{B^*}{A^*} = \frac{BQ - SP}{AQ + SQ}; \qquad \frac{D^*}{A^*} = \frac{DQ}{AQ + SQ} \quad (45)$$

without changing the signals $u(k)$ and $y(k)$ for a given $v(k)$. Since $S(z^{-1})$ is arbitrary the order of A, B, and D cannot be determined uniquely based on measurements of $u(k)$ and $y(k)$. Therefore the order of the process and noise model must be known exactly (Bohlin, 1971).

The preceding considerations have shown, based on the basic equation system used for parameter estimation, that the process parameters can be estimated in a closed loop as in the case of an open loop by measuring the process input $u(k)$ and the output $y(k)$, if the identifiability conditions are satisfied. Additionally the specific convergence conditions of the parameter estimation methods least squares, extended least squares and maximum likelihood have to be satisfied. The relevant convergence condition is (cf. Section 3) that the equation error or prediction error

$$e(k) = y(k) - \psi^T(k)\hat{\theta}(k-1) \quad (46)$$

is independent of the elements of $\psi^T(k)$, which are

$$\text{RLS:} \quad \psi^T(k) = [-y(k-1) \cdots \mid u(k-d-1) \cdots] \quad (47)$$

$$\text{RML/RELS:} \quad \psi^T(k) = [-y(k-1) \cdots \mid u(k-d-1) \cdots \mid \hat{v}(k-1) \cdots]. \quad (48)$$

In the case of convergence $e(k) = v(k)$ can be assumed. Since $v(k)$ only influences $y(k)$, $y(k+1), \ldots$ and these elements do not appear in $\psi^T(k)$, the equation error $e(k)$ is truly independent of the elements $\psi^T(k)$. This holds also with a feedback to $u(k)$ by the controller. Therefore the convergence conditions are satisfied and parameter estimation methods which are based on the one-step-ahead prediction error can be applied in a closed loop as in an open loop.

Summary of closed-loop estimation

If a stochastic process can be described by equation (5), direct process identification in a closed loop, without use of an extra perturbation signal, can be performed by measuring the process input and output and applying a proper parameter estimation method as in an open loop, if the following conditions are satisfied:

Identifiability condition 1. The order of the linear controller, (32) has to be

$$\max\{\mu;\ \nu + d\}, \quad -p \geq m \quad (49)$$

where p is the number of common poles and zeros of

$$G_{id}(z) = \frac{D(z^{-1})}{A(z^{-1})P(z^{-1}) + z^{-d}B(z^{-1})Q(z^{-1})}. \quad (50)$$

Identifiability condition 2. The order m and deadtime d of the process equation (6) must be known *a priori*. If the identifiability condition 1 is not satisfied by the controller, identifiability can be reached by (either, or):

Identifiability condition 3. Applying two different switching controllers or timevarying controllers.

Identifiability condition 4. Applying an extra perturbation signal persistently exciting at least of order m to the closed loop but not between the measured signals $u(k)$ and $y(k)$.

5. CONTROL ALGORITHMS

As already discussed in Section 2 control algorithms for adaptive control should satisfy the closed-loop identifiability conditions and should have small computational expense and storage for parameter calculation. These properties are met especially by deadbeat and minimum variance controllers. Only a short description of these and other control algorithms is given here. For more details see, for example, Isermann (1981).

Deadbeat controller

The deadbeat controller is designed for finite settling time

$$e_w(k) = 0 \qquad \text{for } k \geq m + d$$

$$u(k) = \text{const.} \quad \text{for } k \geq m$$

[see (33)] after a step change of the reference variable $w(k)$. Its transfer function is (Kalman, 1954)

$$G_{DB1} = \frac{u(z)}{e_w(z)} = \frac{q_0 A(z^{-1})}{1 - q_0 B(z^{-1})z^{-d}} \quad (51)$$

$$q_0 = \left[\sum_{i=1}^{m} b_i\right]^{-1}. \quad (52)$$

Because the changes of the manipulated variable are too large in most cases when using this deadbeat controller, the settling time is increased by one unit, leading to (Isermann, 1981)

$$G_{DB2} = \frac{q_0 A(z^{-1})(1 - \alpha z^{-1})}{1 - q_0 B(z^{-1})z^{-d}(1 - \alpha z^{-1})} \quad (53)$$

with

$$\alpha = 1 - 1/q_0 \Sigma b_i. \quad (54)$$

Here q_0 can be chosen within

$$q_{0\,min} \leq q_0 \leq q_{0\,max}.$$

$q_{0\,max}$ follows from the original deadbeat controller, equation (52), and

$$q_{0\,min} = 1/(1-a_1)\sum b_i$$

is obtained by assuming $u(0) = u(1)$ for a unit step in $w(k)$, which leads to the smallest obtainable magnitudes of $u(0)$ and $u(1)$ in that case. Instead of choosing q_0 directly, a factor r' may be introduced

$$q_0 = q_{0\,min} + (1-r')(q_{0\,max} - q_{0\,min}) \tag{55}$$

yielding the original deadbeat controller for $r' = 0$ and the modified one for $0 < r' \leq 1$. The deadbeat controllers satisfy the identifiability condition 1 completely as $\nu = m$ and $\mu = m + d$ for DB1 and $\nu = m + 1$ and $\mu = m + d + 1$ for DB2 and $p = 0$. The expense for the parameter calculation is very small, because the controller parameters depend directly on the process parameters. However, because of the cancellation of process poles $A(z^{-1})\,z^m = 0$ the deadbeat controller may be applied only for asymptotically stable processes.

Minimum variance controllers

Minimum variance controllers are designed for stochastic disturbances $v(k)$ by minimizing a quadratic performance function

$$I(k+d+1) = E\{y^2(k+d+1) + ru^2(k)\}. \tag{56}$$

The variance of the predicted output $y(k+d+1)$ is minimized for $w(k) = 0$ taking into account the weighted variance of $u(k)$. [Since $b_0 = 0$ is assumed, $y(k+1)$ is the first signal influenced by $u(k)$, if $d = 0$, and $y(k+d+1)$ by $u(k)$, if $d \neq 0$.] The original minimum variance controller with $r = 0$ was proposed by Åström (1970). For $r \neq 0$, the variance of the process input can be reduced with an increase in output variance, Clarke and Hasting-James (1971). This (extended) minimum variance controller for the process model (5) is designed by

$$G_{MV3} = -\frac{L(z^{-1})}{zB(z^{-1})F(z^{-1}) + (r/b_1)D(z^{-1})} \tag{57}$$

(see Isermann, 1981), where the controller parameters follow from

$$D(z^{-1}) = A(z^{-1})F(z^{-1}) + z^{-(d+1)}L(z^{-1}) \tag{58}$$

with

$$\begin{aligned} F(z^{-1}) &= 1 + f_1 z^{-1} + \cdots + f_d z^{-z} \\ L(z^{-1}) &= l_0 + l_1 z^{-1} + \cdots + l_{m-1} z^{-(m-1)}. \end{aligned} \tag{59}$$

Equation (58) shows that the noise filter has been separated

$$\frac{n(z)}{v(z)} = \frac{D(z^{-1})}{A(z^{-1})} = F(z^{-1}) + \frac{L(z^{-1})}{A(z^{-1})}z^{-(d+1)} \tag{60}$$

where $F(z^{-1})$ contains that part which cannot be controlled by $u(k)$, because of the process dead time d. For $r = 0$ the minimum variance controller becomes (Åström, 1970)

$$G_{MV4} = -\frac{L(z^{-1})}{zB(z^{-1})F(z^{-1})}. \tag{61}$$

The controlled variable then results in

$$y(z) = F(z^{-1})v(z) \tag{62}$$

which is a moving average process of order d. Its variance is

$$E\{y^2(k)\} = [1 + f_1^2 + \cdots + f_d^2]\sigma_v^2. \tag{63}$$

The larger the dead time the larger the variance. For $d = 0$ one obtains $y(k) = v(k)$. The controlled variable then is a white noise process.

The order of the minimum variance controllers is $\nu = m - 1$ and $\mu = m + d - 1$ for MV4 and $\mu = \max\{m; m + d - 1\}$ for MV3. However, if the process and its process model agree exactly, $p = m$ common poles appear in $G_{id}(z)$, equation (40) and the identifiability condition 1 is only satisfied for $d \geq m + 1$. If there is no exact agreement, the identifiability condition is satisfied for any order, except for MV4 $d \geq 1$ is required. To circumvent the closed-loop identifiability problems with the minimum variance controllers, one or more process model parameters can be assumed as known constant values, see Section 6. As for $r = 0$ the controller cancels the process zeros, the process must have zeros within the unit circle (minimum phase behavior). The controller for $r \neq 0$ does not have such restrictions.

Because the minimum variance controllers have been designed for $E\{v(k)\} = 0$ and $w(k) = 0$ they have no integral action. Therefore offsets may occur for other disturbances, if this is not avoided by modifications (see the next section).

Parameter optimized PID-controllers

A control algorithm with transfer function

$$G_{PC} = \frac{Q(z^{-1})}{P(z^{-1})} = \frac{q_0 + q_1 z^{-1} + \cdots + q_\nu z^{-\nu}}{1 - z^{-1}} \tag{64}$$

which belongs to the class of parameter optimized controllers satisfies the identifiability condition 1 for $\nu \geq m - d$, if $p = 0$. Hence PID-controllers with $\nu = 2$ are only suitable for process orders $m \leq 2 + d$, i.e. for $d = 0$ or 1 or 2 the permitted process orders are $m \leq 2$ or 3 or 4. However, as small time lags can often by approximated by a dead time $d = 1$ or 2, the permitted model order range may be large enough for a variety of processes.

There are several ways in which PID-controllers for parameter adaptive control can be designed. A simple way is the *pole-assignment design*. In this case the characteristic equation

$$\begin{aligned} P(z^{-1})A(z^{-1}) + Q(z^{-1})B(z^{-1})z^{-d} \\ = 1 + \alpha_1 z^{-1} + \cdots + \alpha_{m+d+2} z^{-(m+d+2)} = 0 \end{aligned} \tag{65}$$

is used and the coefficients α_i are determined based on the assigned poles. Then the controller parameters q_0, q_1, q_2 can be directly calculated, see, for example, Isermann (1981). Examples for second order process models are given by Åström and Wittenmark (1980) and Wittenmark and Åström (1980). However, the design depends, of course, on the proper placement of at least three selected poles, which makes too much trouble in many practical cases. Another way is a design by *approximation of an easily designed control algorithm*, such as a deadbeat-control algorithm (Kurz, Isermann and Schumann, 1980).

State controllers

The derivation of the identifiability conditions equations (49) and (50) is based on input–output controllers. The results can therefore be only transferred to state controllers with state observers or state estimators, if the control algorithms can be put into an input–output representation. Generally the order of the input–output equations is high enough to satisfy the identifiability conditions. The state controller may be designed by a pole-assignment technique or by recursive solution of the matrix Riccati equation (Berry and Kaufman, 1975; Schumann, 1979b; Buchholt and Kümmel, 1979). To reduce the effort required in designing and updating the state reconstruction, an observable canonic state model may be used for a direct state reconstruction (Schumann, 1979b, 1981).

The identifiability conditions for these control algorithms have been discussed so far for stochastic disturbances $v(k)$ with $E\{v(k)\} = 0$. If stochastic or deterministic disturbances act somewhere on the loop, but not between the measurements of $u(k)$ and $y(k)$, these signals can be regarded as perturbations, for which the identifiability is satisfied for any controller, if they are persistently existing at least of order m.

Example 2. Controller parameter calculation
The process model is

$$y(k) + a_1 y(k-1) + a_2 y(k-2) = b_1 u(k-1) + b_2 u(k-2) + v(k) + d_1 v(k-1) + d_2 v(k-2).$$

Deadbeat controller $[v(k) = 0]$
DB1:

$$\begin{aligned} q_0 &= 1/(b_1 + b_2) & p_0 &= 1 \\ q_1 &= a_1 q_0 & p_1 &= -b_1 q_0 \\ q_2 &= a_2 q_0 & p_2 &= -b_2 q_0. \end{aligned}$$

DB2: $r' = 1$

$$\begin{aligned} q_0 &= q_{0\,\min} = 1/(1-a_1)(b_1+b_2) \\ q_1 &= q_0(a_1 - 1) + 1/(b_1 + b_2) \\ q_2 &= q_0(a_2 - a_1) + a_1/(b_1 + b_2) \\ q_3 &= -q_0 a_2 + a_2/(b_1 + b_2) \end{aligned}$$

$$\begin{aligned} p_0 &= 1 \\ p_1 &= -q_0 b_1 \\ p_2 &= -\{q_0(b_2 - b_1) + b_1/(b_1 + b_2)\} \\ p_3 &= b_2 - b_2/(b_1 + b_2). \end{aligned}$$

Minimum variance controller MV3
Equation (58) leads to

$$l_0 = d_1 - a_1, \qquad l_1 = d_2 - a_2$$

and (57) to

$$\begin{aligned} q_0 &= -l_0 = a_1 - d_1 & p_0 &= b_1 + \frac{r}{b_1} \\ q_0 &= -l_1 = a_2 - d_2 & p_1 &= b_2 + \frac{r}{b_1} d_1 \\ & & p_2 &= \frac{r}{b_1} d_2. \end{aligned}$$

6. PARAMETER ADAPTIVE CONTROL ALGORITHMS

Combinations

The previous sections have shown that various methods are available for

- recursive parameter estimation;
- d.c.-value estimation (U_{00}, Y_{00});
- control algorithms;
- compensation of offsets for control algorithms without integral action.

Parameter adaptive control algorithms result from proper combinations of these methods.

There are two different ways to organize the combinations of recursive parameter estimators and control algorithms. They are frequently called *explicit* and *implicit*. Both use the process model

$$y(k) = \boldsymbol{\psi}^{\mathrm{T}}(k)\boldsymbol{\theta}(k-1) \tag{66}$$

and the controller equation

$$u(k) = q_0 e_w(k) + q_1 e_w(k-1) + \cdots + q_\nu e_w(k-\nu) - p_1 u(k-1) - \cdots - p_\mu u(k-\mu) \tag{67}$$

now abbreviated as

$$u(k) = \rho^{\mathrm{T}}(k)\Xi(k-1) \tag{68}$$

with

$$\rho^{\mathrm{T}}(k) = [e_w(k) e_w(k-1) \cdots e_w(k-\nu) \mid -u(k-1) \cdots - u(k-\mu)] \tag{69}$$

$$\Xi^{\mathrm{T}} = [q_0 \quad q_1 \cdots q_\nu \mid p_1 \cdots p_\mu]. \tag{70}$$

For an *explicit combination* the process parameters are estimated explicitly

$$\hat{\boldsymbol{\theta}}(k) = \hat{\boldsymbol{\theta}}(k-1) + \boldsymbol{\gamma}(k-1)[y(k) - \boldsymbol{\psi}^{\mathrm{T}}(k)\hat{\boldsymbol{\theta}}(k-1)] \tag{71}$$

and stored as an intermediate result. Then the controller parameters are calculated

$$\Xi(k) = f[\hat{\boldsymbol{\theta}}(k)] \tag{72}$$

and the new process input $u(k+1)$ follows from (68).

In the case of an *implicit combination* the controller design equation (72) is introduced into the process model equation (66), so that

$$y(k) = \zeta^{\mathrm{T}}(k)\Xi(k-1) \tag{73}$$

results. Then the controller parameters are estimated directly via

$$\hat{\Xi}(k) = \hat{\Xi}(k-1) + \gamma_\Xi(k-1)[y(k) - \zeta^{\mathrm{T}}(k)\hat{\Xi}(k-1)]. \tag{74}$$

Therefore the process parameters $\boldsymbol{\theta}$ do not appear as intermediate results. Two examples will show both ways.

Example 3. An explicit RLS-DB1 adaptive controller
The process model used is

$$A(z^{-1})y(z) - B(z^{-1})z^{-d}u(z) = v(z)$$

and the process parameters $\hat{\boldsymbol{\theta}}^{\mathrm{T}} = [\hat{a}_1 \cdots \hat{a}_m \quad \hat{b}_1 \cdots \hat{b}_m]$ are estimated with the recursive least squares method, according to (71) or (12)–(14), see also Example 1. Then the controller parameters of the controller

$$G_{\mathrm{DB}}(z) = \frac{Q(z^{-1})}{P(z^{-1})} = \frac{q_0 A(z^{-1})}{1 - q_0 B(z^{-1})z^{-d}}$$

are calculated by

$$\Xi_{\mathrm{T}} = [q_0 \cdots q_m \mid p_{1+d} \cdots p_{m+d}] = \frac{1}{\sum_{i=1}^{m} \hat{b}_i}[1 \hat{a}_1 \cdots \hat{a}_m \mid \hat{b}_1 \cdots \hat{b}_m] = \frac{1}{\sum_{i=1}^{m} \hat{b}_i}\hat{\boldsymbol{\theta}}^{*\mathrm{T}}$$

(see Example 2) and inserted in the controller equation (67)

$$u(k) = q_0 e_w(k) + \cdots + q_m e_w(k-m) - p_{1+d}u(k-d-1) - \cdots - p_{m+d}u(k-d-m).$$

Example 4. An implicit RLS-DB1 adaptive controller
The controller design equation (51) can be written as

$$G_{\mathrm{DB1}}(z) = \frac{u(z)}{e_w(z)} = \frac{A(z^{-1})}{\frac{1}{q_0} - B(z^{-1})z^{-d}} = \frac{A(z^{-1})}{(1 - z^{-1})B^*(z^{-1})}$$

with

$$B^*(z^{-1}) = b_1^*(1 + z^{-1} + \cdots + z^{-d}) + b_2^* z^{-(1+d)} + \cdots + b_m^* z^{-(m+d-1)}$$

$$b_j^* = \sum_{i=j}^{m} b_i.$$

Hence

$$1/q_0 = \sum_{i=1}^{m} b_i = b_1^*$$

and

$$B(z^{-1})z^{-d} = b_1^* - (1 - z^{-1})B^*(z^{-1}).$$

This introduced into the process model as in Example 3 yields

$$A(z^{-1})y(z) + B^*(z^{-1})(1 - z^{-1})u(z) - b_1^*u(z) = v(z).$$

This process model now contains $2m$ controller parameters which can be directly estimated by applying RLS to the corresponding difference equation

$$y(k) = \zeta^T(k)\Xi(k-1) + e(k)$$

with

$$\zeta^T(k) = [-y(k-1)\ -y(k-2)\cdots -y(k-m) \mid u(k-d-1) -\Delta u(k-d-1)\cdots -\Delta u(k-m-d+1)]$$

$$\Xi^T(k-1) = [a_1\ a_2 \cdots a_m \mid b_1^*\ b_2^* \cdots b_m^*]$$

and

$$\Delta u(k) = u(k) - u(k-1)$$
$$\Delta u(k-1) = u(k-1) - u(k-2)$$
$$\vdots \qquad \vdots \qquad \vdots$$

With these estimated parameters the control algorithm becomes

$$u(k) = u(k-d-1) + \frac{1}{\hat{b}_1^*}[-\hat{b}_2^*\Delta u(k-d-1) - \cdots - \hat{b}_m^*\Delta u(k-d-m+1) + e_w(k) + \hat{a}_1 e_w(k-1) + \cdots + \hat{a}_m e_w(k-m)].$$

In this case no calculation time can be saved by the implicit adaptive controller in comparison to the explicit scheme of Example 3.

Peterka (1970) and Åström and Wittenmark (1973) proposed an implicit combination of RLS and MV4. This is shown in the next example.

Example 5. An implicit parameter adaptive controller RLS/MV4 'self-tuning regulator'

The process model equation (18)

$$A(z^{-1})y(z) - B(z^{-1})z^{-d}u(z) = v(z)$$

is used together with the minimum variance controller equation (61)

$$G_{MV4} = \frac{Q(z^{-1})}{P(z^{-1})} = -\frac{L(z^{-1})}{zB(z^{-1})F(z^{-1})}.$$

The controller design equation

$$1 = F(z^{-1})A(z^{-1}) + z^{-(d+1)}L(z^{-1})$$

is introduced into the process model, by first multiplying it with $F(z^{-1})$

$$FAy - BFz^{-d}u = Fv$$

yielding

$$[1 - z^{-(d+1)}L]y - BFz^{-d}u = Fv.$$

Introducing the controller equation results in

$$y = -Qz^{-(d+1)}y + Pz^{-(d+1)}u + Fv.$$

This modified model contains the controller parameters q_i and p_i according to (73) which now can be directly estimated by applying RLS to the corresponding difference equation

$$y(k) = -q_0 y(k-d-1) - \cdots - q_\nu y(k-d-m) + p_0[u(k-d-1) + \cdots + p_\mu u(k-m)] + \varepsilon(k-d-1)]$$

with $\nu = m-1$ and $\mu = m+d+1$.

Now $2m+d$ parameters must be estimated, instead of $2m$ of the original model. Since only $2m+d-1$ parameters can be estimated, because of the closed loop identifiability condition, one parameter must be assumed as known. If $p_0 = b_1$ is assumed known, it follows

$$y(k) = \zeta^T(k)\Xi(k-1) + p_0 u(k-d-1) + \varepsilon(k-d-1).$$

For the estimation of Ξ the RLS method is used. A nice property of this parameter-adaptive control algorithm is, that under certain conditions it also converges to the optimal minimum variance controller, if it is applied for an ARMAX-model, equation (19).

Another implicit combination of RLS with an extended minimum variance controller similar to MV3 was developed by Clarke and Gawthrop (1975). These authors have called their implicit schemes 'self-tuning regulators'.

Explicit parameter adaptive controllers, based on different combinations of RLS, RELS and RML with deadbeat, minimum variance, PID and linear pole assignment controllers have been developed and compared by Kurz, Isermann and Schumann (1980) and Kurz (1980). These investigations have shown that various combinations have good overall properties (see Table 3). In general DB2 and MV3 are to be prefered, because of the possibility to weight the process input by r or r'.

The explicit combination is evaluated to be the more general way, because it allows many combinations, makes modular programming easier, and allows direct access to the basic process parameters, e.g. for supervision purposes. The implicit combination requires the possibility of inserting the controller design equation into the process model equation, so that (73) results. At first glance some calculation time may be saved by circumventing the controller design equations. However, as the number of parameters to be estimated may increase it is not certain that considerable calculation time can be saved. An advantage of the implicit combinations is that the theories for the convergence of recursive parameter estimation methods can be directly applied.

Consideration of d.c.-values and offsets

Assuming that only stochastic disturbances with $E\{v(k)\} = 0$ act on the loop the d.c.-values U_{00}, Y_{00} can be estimated by simple averaging (see Section 3), before the adaptive control starts. Then also minimum variance controllers can be applied as shown in Section 5, since no offsets occur. However, if the disturbances do not have zero mean as in most real cases and also if reference variable changes $w(k)$ have to be taken into account, the d.c.-value estimation and the compensation for offsets must be considered for controllers without integral action, such as the minimum variance controllers and state controllers. The simplest way to remove the d.c.-value problem is to use the first order differences $\Delta u(k)$ and $\Delta y(k)$ for the parameter

TABLE 3. PARAMETER ADAPTIVE CONTROL ALGORITHMS WITH GOOD PROPERTIES

Parameter estimation \ Control algorithm	deterministic		stochastic	
	DB1	DB2	MV3	MV4
RLS	RLS/DB1	RLS/DB2	RLS/MV3 [1]	RLS/MV4 [1]
RELS	-	-	RELS/MV3	RELS/MV4
RML	-	-	RML/MV3	RML/MV4

[1] $\hat{D}(z^{-1}) = 1$ assumed.

estimation (see Section 3). Offsets can be avoided by adding a pole at $z_1 = 1$ to the estimated process model, and by designing a minimum variance controller for this extended model. However, this does not give the best control performance.

Another possibility is to replace $y(k)$ by $(y(k) - w(k))$ and $u(k)$ by $\Delta u(k) = u(k) - u(k-1)$ in both the parameter estimation and the control algorithm as proposed by Wittenmark (1973). But this leads to unnecessary changes of the parameter estimates after setpoint changes and therefore to a negative influence during the transient period.

Relatively good results have been obtained by the estimation of constant C, method 3 in Section 3. Using $Y_{00} = W(k)$, the right d.c.-value U_{00} is automatically calculated so that offsets do not appear (Kurz, Isermann and Schumann, 1980). Then controllers without integral action can be used directly.

In the case of a control algorithm with integral action, for example the deadbeat controller, the d.c.-values have only to be considered for the parameter estimation.

Convergence

Because the parameter adaptive control algorithms are nonlinear and time variant, a comprehensive analytical treatment of the transient behavior is difficult to perform. However, the problem can be divided into several different steps which are easier to solve.

First, *stability conditions* have to be considered for the case when the parameter estimates converge to certain values. It is assumed that the real process can be described exactly by the assumed process model. Then it is obvious that the calculated control algorithm together with the real process must be stable, if the process model converges to the real process behavior. Therefore a necessary condition is that the closed loop consisting of the control algorithm and the real process is stable with the exactly tuned fixed controller. Furthermore the closed loop should also be stable in the neighborhood of the exactly tuned controller. Thus a sufficient condition is that all parameter estimates required for designing the control algorithms converge to their true values.

These conditions can be satisfied by the choice of suitable combinations of parameter estimators and control algorithms, so that the closed-loop identifiability conditions are satisfied and consistent parameter estimates are obtained.

A description of the *convergence* may be divided in several phases

convergence at the beginning (short term behavior);
convergence far from the steady state;
convergence near the steady-state (asymptotic behavior).

Because the parameter estimates are the basis, the convergence conditions for *recursive parameter estimation* methods have first to be discussed. In recent years mainly two methods for convergence analysis have been developed, the so-called ODE-method (ordinary differential equation method), where a system of differential equations is associated with the recursive parameter estimation equations (Ljung, 1977a) and convergence analysis via Ljapunov-functions, for the deterministic case by de Larminat (1979) and for the stochastic case using the martingale-theory by Kumar and Moore (1979) and Solo (1979). A comparison of the results of both theories and their application to several recursive parameter estimation methods applied in open loop (Matko and Schumann, 1982) shows that a sufficient condition for the convergence of the parameter errors near the steady state

$$\lim_{k\to\infty} E\{\Delta\boldsymbol{\theta}^{\mathrm{T}}(k)\Delta\boldsymbol{\theta}(k)\} = \lim_{k\to\infty}\{\|\hat{\boldsymbol{\theta}}(k) - \boldsymbol{\theta}_0(k)\|^2\} = 0 \tag{75}$$

is that the matrix

$$\lim_{k\to\infty}\left[\frac{1}{k}\boldsymbol{P}^{-1}(k)\right] = \lim_{k\to\infty}\frac{1}{k}\boldsymbol{\Psi}^{\mathrm{T}}(k)\boldsymbol{\Psi}(k) \tag{76}$$

is positive definite.

This implies that the process input $u(k)$ has to be persistently exciting of order m. Taking into account the convergence conditions given in section 3 the RLS method then possesses local and global convergence. The RELS method converges locally if the noise filter $G_v(z) = 1/D(z^{-1})$ is positive real and converges globally, if $H(z) = 1/D(z^{-1}) - 1/2$ is positive real, see also Ljung (1977b). The last requirement means that the transfer function $H(z)$ must be stable and $\mathrm{Re}\{H(z)\} > 0$ or $\mathrm{Re}\{D(z^{-1})\} > 2$. Therefore the satisfaction of this condition depends on the noise filter parameters $d_1, d_2, \ldots, d_m$, which, however, are not known beforehand.

These methods for the convergence analysis of recursive parameter estimation methods can be directly transferred to the convergence analysis of *implicit parameter adaptive control algorithms*. Ljung (1977a) has shown that the implicit self-tuning regulator RLS-MV4 shows global convergence for nonminimum phase processes, if $H(z) = 1/D(z^{-1}) - 1/2$ is positive real. This positive real condition appears also in the analysis of the deterministic model reference adaptive controllers (MRAS) (Ljung and Landau, 1978; Narendra, 1980) so that there are similarities to the stochastic implicit self-tuning controller RLS-MV4 (Egardt, 1980).

The convergence of *explicit parameter adaptive controllers* can be investigated by analyzing the behavior of a time variant closed-loop system in state space form containing the estimated process model and the corresponding control algorithm (de Larminat, 1979).

It then can be shown for a variety of explicit parameter adaptive controllers that for large time the elements of the overall system matrix approach constant values, so that based on the resulting eigenvalues of the corresponding time invariant system stability can be proven as well for deterministic as for stochastic signals (Schumann, 1981). A necessary condition is that the parameter estimates converge to constant values for which the controller can stabilize the control loop. This is for example the case when (76) and the closed-loop identifiability conditions are satisfied.

So far the convergence of parameter adaptive control algorithms can be shown for the cases not far from the steady state, i.e. mainly for the asymptotic behavior. However, simulations and practical experience show that in many cases also convergence far from steady state can be obtained, if the basic stability conditions are fulfilled. For the convergence at the beginning, no general results are available as yet. However, proper actions can be taken to overcome this problem, see Section 10.

Considering the explicit parameter adaptive control algorithms given in Table 3, the following specific properties with regard to the steady-state convergence can be observed:

Stochastic disturbances. It is assumed that the process and the noise filter is described by

$$y(z) = \frac{B(z^{-1})}{A(z^{-1})}z^{-d}u(z) + \frac{D(z^{-1})}{A(z^{-1})}v(z) \tag{77}$$

where the orders m and d are exactly known and the forgetting factor is $\lambda = 1$ and $\mathrm{Re}\{D(z^{-1})\} < 2$.

The adaptive control algorithms based on combinations of RML or RELS with MV3 or MV4 ($d \geq 1$) then converge to the exactly adjusted control algorithms MV3 or MV4 despite of the fact that $p = m$ common poles arise in (50) and the identifiability condition for closed loop is violated for the exact controller. The reasons are first that the identifiability condition is satisfied for the adaptation phase. Secondly, if the adaptive controller approaches the exact controller, the parameter estimates become close to the exact ones and the identifiability condition would be violated, if the parameter estimation were to start at that time. However, the estimation is based on the common solution during the transient phase and the exactly tuned phase. Therefore the algorithms converge to the exact controllers, because this is the unique common solution.

The adaptive control algorithms RLS/MV3 and RLS/DB2 and RLS/DB1 do not converge to the exactly adjusted control algorithms, since the parameter estimates are biased because of $D(z^{-1}) \neq 1$. However, they show a reasonable control performance.

Reference variable changes $w(k) \neq 0$, $v(k) = 0$. It is assumed that the process is described by

$$y(z) = \frac{B(z^{-1})}{A(z^{-1})} z^{-d} u(z) \quad (78)$$

and the reference variable $w(k)$ is deterministic or stochastic persistently exciting of order $n \geq m$. Furthermore a proper d.c.-value estimation procedure (e.g. method 3) or just differencing (method 1) has to be applied. Then the parameter estimates $\hat{a}_i(k)$ and $\hat{b}_i(k)$ converge to their exact values a_i and b_i, if RLS, RELS, or RML is applied, because identifiability condition 4 is satisfied. In the case of RELS and RML the estimates $\hat{d}_i$ result in arbitrary values (missing excitation of the noise filter). Therefore the minimum variance control algorithms should only be applied with a fixed polynomial $D(z^{-1})$, for example $D(z^{-1}) = 1$. For $D(z^{-1}) = 1$ RLS can be used instead of RELS or RML. RLS/DB2 and RLS/DB1 converge to the exactly adjusted control algorithms. Analogously RLS/MV3 and RLS/MV4 converge to the exactly adjusted control algorithms for $D(z^{-1}) = 1$. In the following section some simulation results are presented.

7. SIMULATION RESULTS

A second-order process with continuous transfer function

$$G_P(s) = \frac{1}{(1 + 3.75s)(1 + 2.5s)}$$

was simulated on an analog computer and sampled with $T_0 = 2$ s, leading to the z-transfer function

$$G_P(z) = \frac{0.1387z^{-1} + 0.0889z^{-2}}{1 - 1.036z^{-1} + 0.2636z^{-2}} \quad (79)$$

including a zero-order hold. The adaptive control algorithms have been programmed on a process computer HP 21 MX-E. Figure 4 shows a run for stochastic disturbances with RML/MV3. During the adaptation phase the input shows increasing and then decreasing changes. A good control performance is achieved after a few samples, which is as good as with the exactly adjusted minimum variance algorithms.

In Fig. 5 one typical run is shown for stepwise reference value changes using RLS/DB1. This run also shows what happens if no external disturbances act on the loop. After closing the loop at $k = 0$ changes of the process input arise, leading to a rough model which is good enougn to stabilize the loop and for converging to a zero steady state. The first step change of the reference variable leads to an acceptable control behavior. The adaptive loop has now been excited

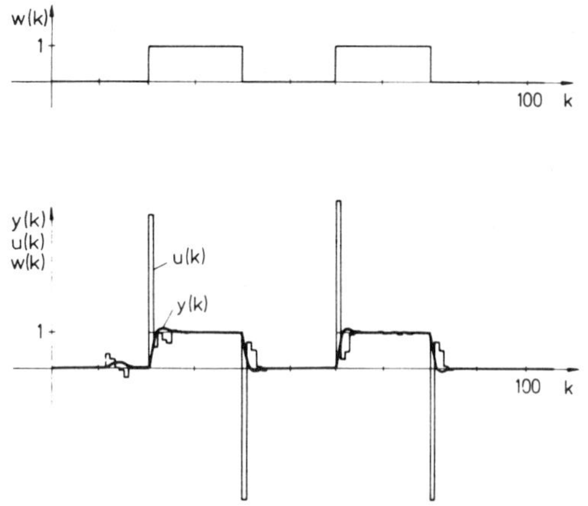

FIG. 5. Adaptive control for deterministic disturbances: (a) stepwise reference value changes; (b) adaptive control with RLS/DB1.

sufficiently, so that already for the second step change the control performance is almost the same as with the exactly adjusted deadbeat algorithm. Many other simulations have shown similar good results also for other combinations given in Table 3. The parameter adaptive algorithms do not only adapt to the process behavior, but also to the type of disturbances, Kurz and co-workers (1978, 1980). The quickest adaptation was achieved with RLS. Extra perturbation signals did not appear to improve the control performance in the case of stochastic disturbances $v(k)$.

Different types of processes

Simulations with other types of stable linear processes, as, for example, higher order processes, processes with non-minimum phase behavior and processes with integral behavior also indicated a quick convergence and good control performance for several types of adaptive algorithms (Kurz, 1980). It was even possible to control unstable processes with RLS/MV3.

Choice of a priori *factors*

To start the adaptive control algorithms the following factors have to be specified *a priori*: T_0, sampling time; $\hat{m}$, process model order; $\hat{d}$, process model dead time; λ, forgetting factor; r, process input weighting factor.

In general digital parameter adaptive control is not very sensitive to the choice of the *sample time* T_0. For proportional action processes good control can be obtained mostly within the range

$$\tfrac{1}{15}T_{95} \leq T_0 \leq \tfrac{1}{4}T_{95} \quad (80)$$

where T_{95} is the 95%-settling time of the process transient response. To avoid changes in the process input which are too large, the sample time should not be chosen too small for DB and MV controllers.

Simulations with a process of order $m_0 = 3$ have shown that the adaptive control was not sensitive to wrong *model orders* within the range

$$m_0 - 1 \leq \hat{m} \leq m_0 + 2. \quad (81)$$

Also other simulations have shown that the order need not be known exactly.

However, the adaptive control algorithms are sensitive to the choice of the *dead time* $\hat{d}$. If d is not known exactly or changes with time, the control is either poor or unstable, especially for combinations with minimum-variance controllers. But this can be overcome by including a dead time estimation, as shown by Kurz (1979).

The choice of the *forgetting factor* λ for the parameter estimation depends on the speed of process parameter changes, the model order and the kind of disturbances. For constant processes or very slowly time varying processes

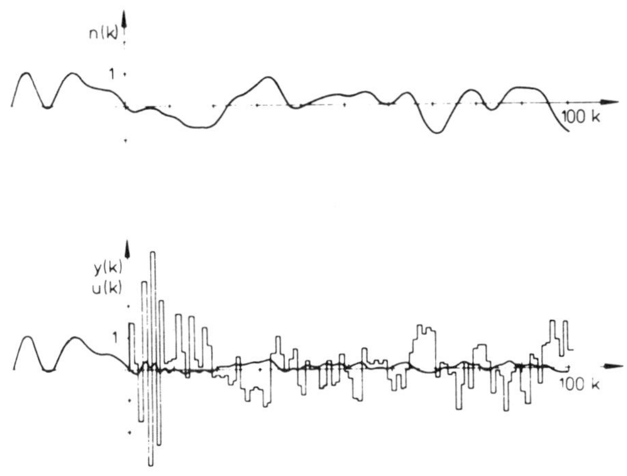

FIG. 4. Adaptive control for stochastic dsiturbances: (a) noise signal $n(k)$, generated by a correlated noise at the process input; (b) adaptive control with RML/MV3. $r = 0.01$.

$\lambda = 0.99$ is recommended. For slowly time varying processes and stochastic disturbances $0.95 \leq \lambda \leq 0.99$ and stepwise reference variable changes $0.85 \leq \lambda \leq 0.90$ are proper choices.

The influence of the *weighting factor r* on the manipulated variable can be estimated by looking at the first input $u(0)$ after a setpoint step $w(k) = 1(k)$. For the closed loop it is $u(0) = q_0 w(0) = g_0$. Therefore q_0 is a simple measure for the size of the process input. In the case of DB2 there is a linear relationship between q_0 and r' for $0 \leq r' \leq 1$, see equation (55). For MV3 it holds

$$q_0 = \frac{-l_0}{b_1 + \frac{r}{b_1}}. \tag{82}$$

In this case the choice of r is related to l_0 and b_1. $r = 0$ leads to MV4 with $q_0 = -l_0/b_1$. A reduction of $|q_0|$ by one half is obtained by choosing $r = b_1^2$. b_1 can be estimated from the process transient responses because for a step input u_0, $b_1 = y(1)/u_0$.

8. APPLICATIONS

Various examples have shown the applicability of parameter adaptive control algorithms to industrial and pilot processes. A summary of digital control with the implicit RLS/MV4 'self-tuning regulators' of the moisture content of a paper machine and the input of an ore crusher is given in Åström and co-workers (1977). The same type of self-tuning regulator has also been applied successfully for autopilots of tankers (Källström and co-workers, 1978), the level control of a glass furnace (Haber and co-workers, 1979) and the titan dioxide content in a kiln (Dumont and Bélanger, 1978). Clarke and Gawthrop (1980) report on the application of RLS/MV3 microprocessor based self-tuners to the control of the room temperature, pH-value and temperature of a batch chemical reactor. RLS/MV3 and RLS/DB algorithms have been programmed on a microcomputer and applied to an air heater (Bergmann, Radke and Isermann, 1978). Buchholt and Kümmel (1979) showed the application to a pH-neutralization process using a RLS/MV4 self-tuning regulator and a combination of RLS with an optimal state controller with better results for the latter.

In the following, two examples are shown for the adaptive control of an air heater and a pH-process.

Figure 6 represents the scheme of an air-conditioning unit.

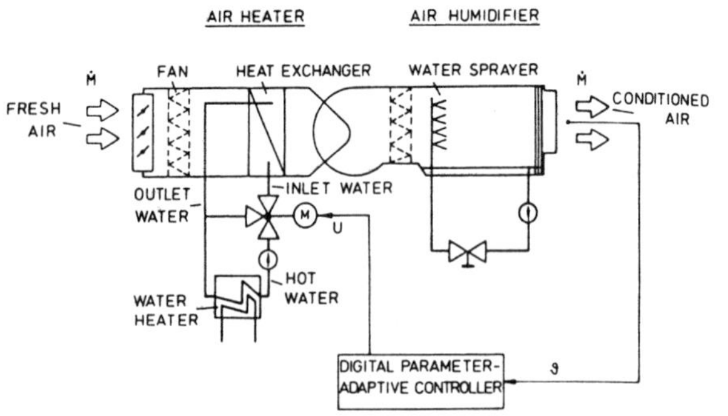

FIG. 6. Scheme of the air-conditioning unit.

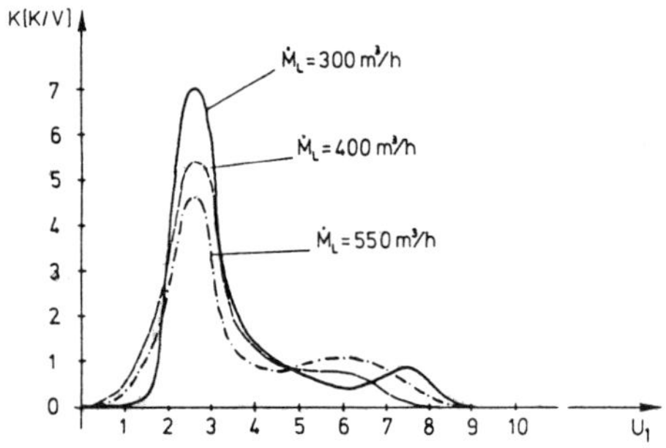

FIG. 7. Gain $K = \Delta\vartheta_L(\infty)/\Delta U(\infty)$ of the air-conditioning unit as function of the manipulated variable U for different air flow ratio M_L.

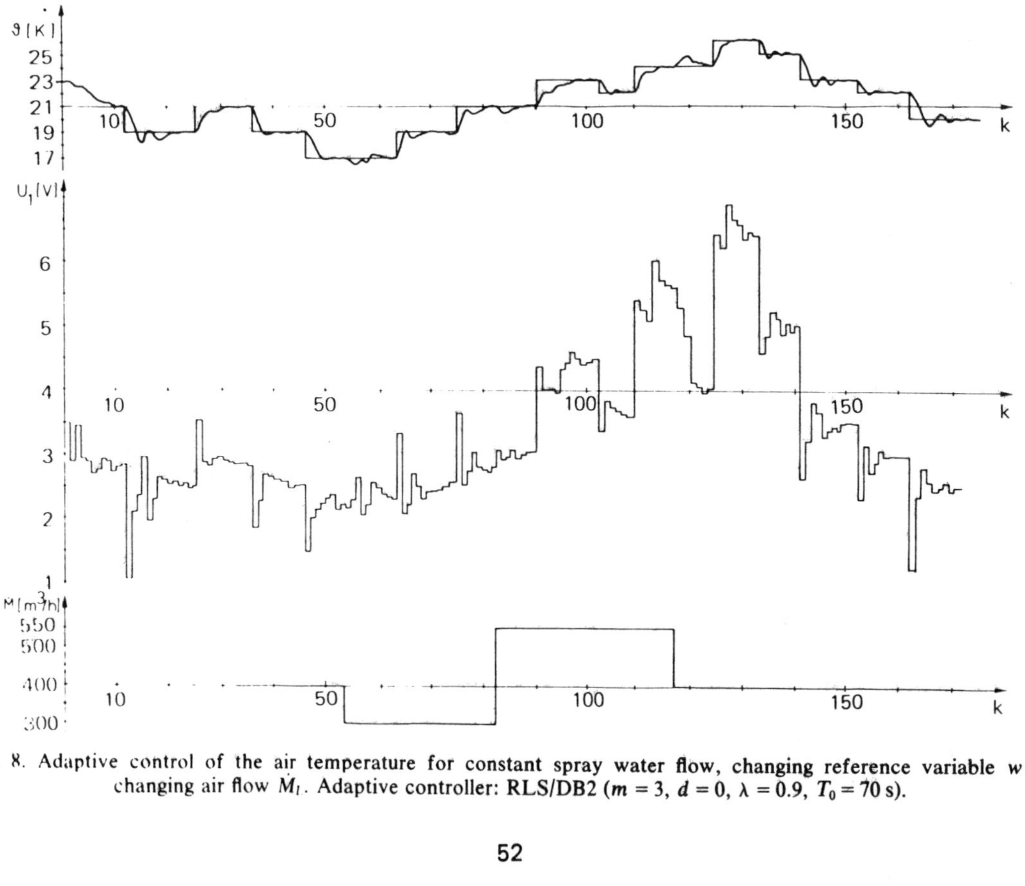

FIG. 8. Adaptive control of the air temperature for constant spray water flow, changing reference variable w and changing air flow $\dot{M}_l$. Adaptive controller: RLS/DB2 ($m = 3$, $d = 0$, $\lambda = 0.9$, $T_0 = 70$ s).

The variable gain K of the plant with position U of the split range valve for the warm water flow as input and the air temperature ϑ_L as output is shown in Fig. 7. For the adaptive control a microcomputer based on an Intel SBC 80/10 was used (Bergmann, Radke and Isermann, 1978). A quick adaptation to changing process gains (in the range of about 1:10) and changing dynamics can be observed, using the RLS/DB2 algorithm with d.c. estimation method 3 (Fig. 8). For more details see Bergmann and Schumann (1980).

pH-Value processes are known for their (at least static) nonlinear behavior. For the pilot process in Fig. 9 the gains in Fig. 10 have been measured. Also in this case a relatively good control performance could be obtained for varying reference variables, Fig. 11 using the RLS/MV3 algorithm with d.c. estimation method 3 and offset compensation by setting $Y_{00} = W(k)$. The change of the process gain was about 1:4 (Kurz, 1980).

In both cases the sampling time T_0, the process order m and deadtime d have been selected based on rough step response measurements and the rules given in the last section.

Based on these and other applications and based on many simulations the applicability of different adaptive control algorithms in dependence on the type of processes and signals is summarized in Table 4.

9. ADAPTIVE FEEDFORWARD CONTROL

So far parameter adaptive feedback control has been considered. The same principle can also be used for *adaptive feedforward control*. The output of the process is modelled by

$$y(z) = G_P(z)u(z) + G_v(z)v(z) \qquad (83)$$

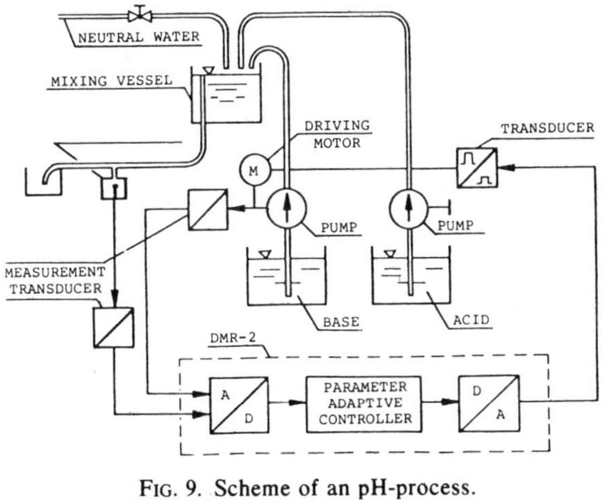

FIG. 9. Scheme of an pH-process.

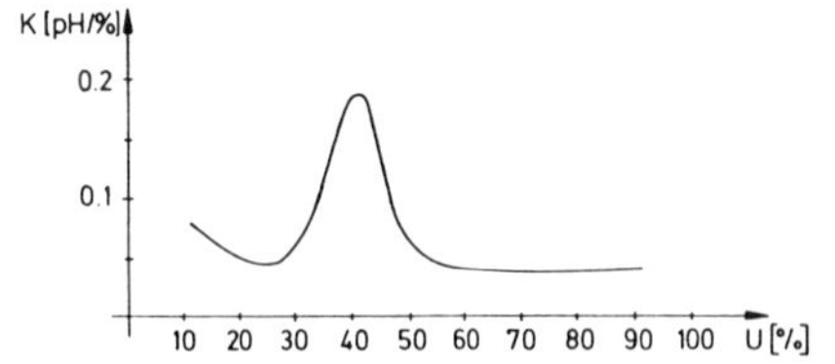

FIG. 10. Gain $K = \Delta pH(\infty)/\Delta U(\infty)$ of the pH-process as function of the manipulated variable U.

which can be approximated by

$$A(z^{-1})y(z) = B(z^{-1})z^{-d_p}u(z) + D(z^{-1})z^{-d_v}v(z) \qquad (84)$$

where $v(k)$ is a measurable disturbance signal. The parameters of this model are estimated using for example RLS or RELS or RML. Then different feedforward algorithms can be designed on the basis of pole-zero cancellation, the minimum variance and the deadbeat principle. The resulting adaptive algorithms are described by Schumann and Christ (1979) and Isermann (1981). They show very quick adaptation. It is well known that for many industrial processes the control performance can be considerably improved by feedforward control. However, no simple tuning rules exist as they do for PID controllers. Feedforward controllers have mostly just proportional, proportional–differential or lead–lag behavior and are often not well adjusted. Therefore adaptive feedforward control algorithms may give a good improvement in comparison to the present situation is practice where nonadaptive control is used.

10. SUPERVISION AND COORDINATION OF THE PARAMETER ADAPTIVE CONTROL ALGORITHMS

The previous discussion has shown that with regard to their application suitably chosen parameter-adaptive control algorithms converge rather quickly, if the following conditions are satisfied:

1. The linear process model and the noise model approximately correspond to the real process.
2. The process and noise parameters are constant.
3. A persistently exciting external signal of order $n \geq m$ acts on the closed loop.

Furthermore, some *a priori* factors have to be chosen properly in dependence on the process. Because in practice these conditions may be violated and suitable *a priori* factors may not be known, a *third feedback level* which coordinates and supervises the adaptive control algorithm, is proposed (Fig. 12). Then for example the following tasks can be carried out, Schumann and others (1981):

TABLE 4. APPLICABILITY OF DIFFERENT ADAPTIVE CONTROLLERS IN DEPENDENCE ON THE TYPE OF LINEAR PROCESSES AND SIGNALS

	TYPE OF PROCESSES			TYPE OF DISTURBANCES	
	asymptotic stable	integral behaviour	zeros outside unit circle	stochastic n(k)	deterministic w(k)
RLS/DB1 (2)	x	-	x	-	x
RLS/MV3	x	x	x	x	x
RLS/MV4	x	-	-	x	-
RML/MV3	x	x	-	x	x
RML/MV4	x	-	-	x	-

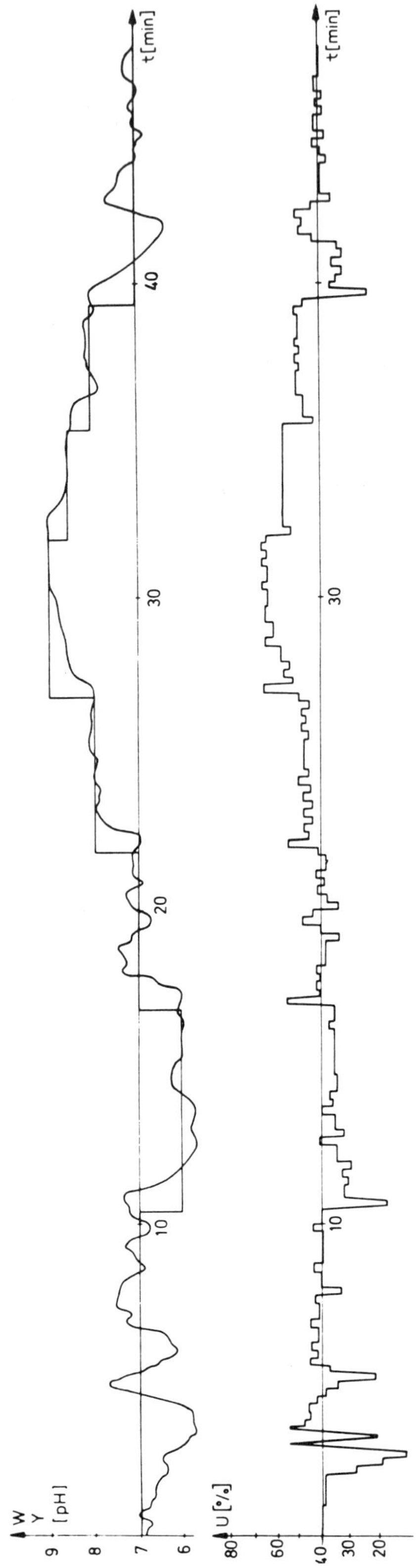

FIG. 11. Adaptive control of the pH value for changing reference variable w. Adaptive controller: RLS/MV3 ($m = 3$, $d = 2$, $\lambda = 0.88$, $r = 0.0008$, $T_0 = 15$ s).

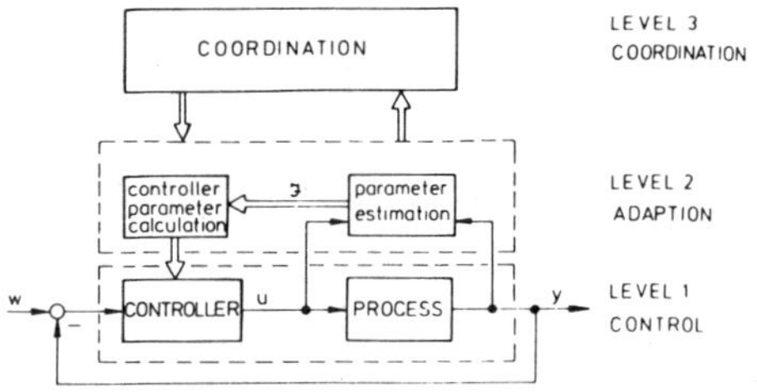

FIG. 12. Parameter-adaptive control with a third feedback level.

(a) on-line search for model order and deadtime;
(b) on-line search for the control sampling time;
(c) coordination of the first adaptation phase (e.g. starting with a defined probing signal);
(d) supervision of the parameter estimation;
(e) supervision of the control performance.

Whereas (a)–(c) are especially important for the commissioning phase, (d) and (e) should be used to guarantee a good control performance under any circumstances. For example, if after the adaptation phase the variations of the process input and output signals become rather small, the process parameters may diverge, especially, if $\lambda < 1$ is chosen, and the controller generates bursting process inputs, at least for a while. A simple remedy is the automatic switch-off for the parameter estimation, as long as the variance of process output signals does not exceed a certain value. An additional way to supervise the parameter estimation is to check the matrix $\boldsymbol{P}^{-1}(k)/k$, equation (76), which must remain positive definite. This means that $\det[\boldsymbol{P}^{-1}(k)/k] > 0$ or for large k $\det[\boldsymbol{P}^{-1}(k)]$ must be large. This can be tested for example by calculating the trace of $\boldsymbol{P}(k)$ which must stay small.

If $\mathrm{tr}\,\boldsymbol{P}(k)$ exceeds a certain value, the forgetting factor λ can be set to one, or a probing signal can be added, etc.

11. FURTHER PROBLEMS

The development of adaptive algorithms has been mainly focused on single input–single output processes. Open problems are for example the investigation of the behavior (stability, convergence, performance) for following cases: far from steady state, fast time varying processes, nonlinear processes. For fast processes with time constants in the range of some milliseconds, such as electrical drives, the computation time for the parameter-adaptive algorithm may be a problem. Then the calculations of the process and controller parameters may be distributed over several samples. Using the explicit parameter adaptive controllers, a smaller sampling time for parameter estimation than for the control algorithm can be used. In this way faster time varying processes may be controlled. These examples indicate that there are several ways to design the 'interface' between the parameter estimation and the control algorithm. This should also be investigated in the future.

Adaptive control of multivariable processes is a big field to be developed. First theoretical and experimental results have been published by, for example, Borisson (1976); Keviczky (1977); Schumann (1979b). Due to the manifold methods of multivariable identification and control many multivariable adaptive controllers can be designed (Schumann, 1982).

The concept of parameter-adaptive control can be directly transferred to certain nonlinear processes (Lachmann, 1982) if parameteric Volterra-models are used. It is to be expected that the control of nonlinear processes by this method may be improved considerably.

A special task is the operator–controller interface which is important for the introduction of adaptive controllers into practice. The adaptive controller may not be more complicated to operate than a conventional PID-controller. Therefore only very few *a priori* factors should be required for tuning.

Furthermore a good control performance must be guaranteed under all circumstances which may arise in practice. Therefore actions of a third feedback level must be further developed.

12. CONCLUSIONS

It has been shown how various parameter adaptive feedback and feedforward control algorithms can be designed by proper combinations of recursive parameter estimation algorithms and control algorithms.

The resulting algorithms show different properties, depending on the type of processes and their signals. A quick adaptation and a good control performance can be obtained for linearizable stable processes with arbitrary pole-zero distribution and dead times, if some *a priori* factors are chosen properly and the adaptive system is persistently excited by external signals. The discussed parameter adaptive feedback and feedforward control algorithms may be applied for

self-tuning during the implementation phase
for one operating point and then fixed controller parameters
for different operating points (load) and then feedforward adapting parameters;
adaptive control of slowly time variant processes.

Several applications have already shown that parameter adaptive control algorithms are an alternative to control processes with complex and slowly time varying behavior. However, additional research is required to guarantee a good control performance under any circumstances which may arise in practice.

REFERENCES

Åström, K. J. and T. Bohlin (1966). Numerical identification of linear dynamic systems from normal operating records. *IFAC Symposium on Theory of Self-adaptive Control Systems*, Teddington. Plenum Press, New York.

Åström, K. J. (1970). *Introduction to Control Theory.* Academic Press, New York.

Åström, K. J. and B. Wittenmark (1973). On self-tuning regulators. *Automatica*, **9**, 185.

Åström, K. J., U. Borisson, L. Ljung and B. Wittenmark (1977). Theoiy and applications of self-tuning regulators. *Automatica*, **13**, 457.

Åström, K. J. and B. Wittenmark (1980). Self-tuning controllers based on pole-zero placement. *IEEE Proc.*, **127**, 120.

Bergmann, S., F. Radke and R. Isermann (1978). Ein universeller digitaler Regler mit Mikrorechner. *Regelungstechnische Praxis*, **20**, 289 and 322.

Bergmann, S. and R. Schumann (1980). Digitale adaptive Regelung einer Lüftungsanlage. *Regelungstechnische Praxis*, **8**, 280.

Berry, P. and H. Kaufman (1975). Development of a digital adaptive optimal linear regulator flight controller. *6th IFAC Congress*, Boston.

Bohlin, T. (1971). On the problem of ambiguities in maximum likelihood identification. *Automatica*, **7**, 199.

Borisson, U. (1976). Self-tuning regulators for a class of multivariable systems. *4th IFAC Symposium on Identification and System Parameter Estimation*, Tiblisi. North-Holland, Amsterdam.

Buchholt, F. and M. Kümmel (1979). Self-tuning control of a pH-neutralization process. *Automatica*, **15**, 665.

Clarke, D. W. and R. Hasting-James (1971). Design of digital controller for randomly disturbed systems. *Proc. IEEE*, **118**.

Clarke, D. W. and P. J. Gawthrop (1980). Implementation and application of microprocessor-based self-tuners. *Automatica*, **16**, 117.

Dumont, G. A. and R. R. Bélanger (1978). Self-tuning control of a titanium dioxide kiln. *IEEE Trans Aut. Control*, **AC-23**, 532.

Egardt, B. (1980). Unification of some discrete-time adaptive control schemes. *IEEE Trans Aut. Control*, **AC-25**, 693.

Eykhoff, P. (1974). *System Identification.* John Wiley, London.

Feldbaum, A. A. (1965). *Optimal Control Systems.* Academic Press, New York.

Gustavsson, I., L. Ljung and T. Söderström (1977). Identification of processes in closed loop—identifiability and accuracy aspects. *Automatica*, **13**, 59.

Haber, R. and others (1979). Identification and adaptive control of a glass furnace by a portable process computer laboratory. *5th IFAC Symposium on Identification and System Parameter Estimation*, Darmstadt. Pergamon Press, Oxford.

Isermann, R. (1974). *Prozeßidentifikation.* Springer, Berlin.

Isermann, R. (1980). Practical aspects of process identification. *Automatica*, **16**, 575–587.

Isermann, R. (1981). *Digital Control Systems.* Springer, New York. (Translation of *Digitale Regelsysteme* (1977).)

Kalman, R. E. (1954). Discussion remark. *AIEE*, 236.

Källström, C. G., K. J. Aström, N. E. Thorell, J. Eriksson and L. Sten (1978). Adaptive autopilots for large tankers. *7th IFAC Congress, Helsinki, Proceed.* Pergamon Press, Oxford.

Keviczky, L. and J. Hetthéssy (1977). Self-tuning minimum variance control of MIMO discrete time systems. *Aut. Control Theory & Appl.*, **5**, 11.

Kumar, R. and J. B. Moore (1979). Convergence of adaptive minimum variance algorithms via weighting coefficient selection. Technical Report No. EE 7917, University of Newcastle, Australia.

Kurz, H. and R. Isermann (1975). Methods for on-line process identification in closed loop. *6th IFAC Congress, Boston. Proceed.* Pergamon Press, Oxford.

Kurz, H. (1979). Digital parameter adaptive control of processes with unknown constant or time varying dead time. *5th IFAC Symposium on Identification and Parameter Estimation, Darmstadt, Proceed.* Pergamon Press, Oxford.

Kurz, H. (1980). *Digitale adaptive Regelung auf der Grundlage rekursiver Parameterschätzung.* PDV-Bericht 188, Kernforschungszentrum Karlsruhe.

Kurz, H., R. Isermann and R. Schumann (1980). Experimental comparison and application of various parameter adaptive control algorithms. *Automatica*, **16**, 117. (Extended version of a paper in session 9B of the 7th IFAC Congress, Helsinki, 1978.)

Lachmann, K. H. (1982). Various parameter adaptive controllers for nonlinear processes. *6th IFAC Symposium on Identification*, Washington.

Landau, I. D. (1979): *Adaptive Control—The Model Reference Approach.* Marcel Dekker, New York.

de Larminat, Ph. (1979). On overall stability of certain adaptive control systems. *5th IFAC Symposium on Identification and System Parameter Estimation, Darmstadt, Proceed.* Pergamon Press, Oxford.

Ljung, L. and Wittenmark, B. (1974). Asymptotic properties of self-tuning regulators. Report 7404. Division of Automatic Control, Lund Institute of Technology, Sweden.

Ljung, L. (1977). On positive real transfer functions and the convergence of some recursions. *IEEE Trans Aut. Control*, **AC-22**, 539.

Ljung, L. (1977a). Analysis of recursive stochastic algorithms. *IEEE Trans Aut. Control*, **AC-22**, 551.

Ljung, L. (1977b). On positive real transfer functions and the convergence of some recursive schemes. *IEEE Trans Aut. Control*, **AC-22**, 539.

Ljung, L. and I. D. Landau (1978). Model reference adaptive systems and self-tuning regulators—some connections. *7th IFAC Congress Boston, Proceed.* Pergamon Press, Oxford.

Matko, D. and R. Schumann (1982). Comparative stochastic convergence analysis of seven recursive parameter estimation methods. *6th IFAC Symposium on Identification and System Parameter Estimation*, Washington.

Narendra, K. S. (1980). Stable discrete adaptive control. *IEEE Trans Aut. Control*, **AC-25**, 456.

Parks, P. C., W. Schaufelberger, Chr. Schmid and H. Unbehauen (1980). Applications of adaptive control systems. In methods and applications in adaptive control, Proc. Symposium, in Bochum, 1980. Springer Lecture Notes 24, Berlin.

Peterka, V. (1970). Adaptive digital regulation of noisy systems, *2nd IFAC Symposium on Identification and Process Parameter Estimation*, Prague. Pergamon Press, Oxford.

Saridis, G. (1977). *Self-organizing Control of Stochastic Systems*. Marcel Dekker, New York.

Schumann, R. (1979). Identification and adaptive control of multivariable stochastic linear systems. *5th IFAC Symposium on Identification and System Parameter Estimation, Darmstadt, Proceed.* Pergamon Press, Oxford.

Schumann, R. (1979b). Various multivariable computer control algorithms for parameter-adaptive control systems. *IFAC Symposium on Computer Aided Design of Control Systems. Zürich, Proceed.* Pergamon Press, London.

Schumann, R. and H. Christ (1979). Adaptive feedforward controllers for measurable disturbances. *Joint Automatic Control Conference*, Denver.

Schumann, R. (1982). Digitale parameter adaptive Mehrgrößen-Regelung—Ein Beitrag zu Entwurf und Analyse. Dissertation T.H. Darmstadt. PDV-Bericht 217, Kernforschungszentrum, Karlsruhe.

Schumann, R., K. H. Lachmann and R. Isermann (1981). Towards applicability of parameter-adaptive control algorithms. *8th IFAC Congress, Kyoto, Proceed.* Pergamon Press, Oxford.

Söderström, T., L. Ljung and I. Gustavsson (1976). A comparative study of recursive identification methods. Department of Automatic Control, Lund Institute of Technology, Report 7427.

Solo, V. (1979). The convergence of AML. *IEEE Trans. Aut. Control*, **AC-24**, 958.

Strejc, V. (1980). Least squares parameter estimation. *Automatica*, **16**, 535.

Tsypkin, Y. Z. (1978). Algorithms of optimization with *a priori* uncertainty (past, present, future). *6th IFAC Congress, Helsinki, Procced.* Pergamon Press, Oxford.

Wittenmark, B. (1973). A self-tuning regulator. Division of Automatic Control, Lund Institute of Technology, Sweden, Report 7311.

Wittenmark, B. (1975). Stochastic adaptive control methods; a survey. *Int. J. Control*, **21**, 705.

Wittenmark, B. and K. J. Åström (1980). Simple self-tuning controllers. In 'Methods and Applications in Adaptive Control', Proc. Symposium, in Bochum, 1980. Springer Lecture Notes 24, Berlin.

State-of-the-art and Prospects of Adaptive Systems*

A. A. VORONOV† and V. YU. RUTKOVSKY†

Works of Soviet authors in adaptive control describe the structure types of the basic loop and adaptation algorithms as well as possible methods for investigation and design analysis of adaptive systems including optimal adaptive systems.

Key Words—Adaptive systems; basic loop; adaptation algorithms; optimal adaptive control.

Abstract–The theory and applications of adaptive control have developed very intensively. There are two reasons for this: variable parameters of controlled plants change within a broad range, on the one hand, and accuracy requirements to process equipment, aircraft systems etc. increase, on the other. Moreover, the wide use of digital computers permits now even complex adaptation algorithms to be easily realized. The paper attempts to review methods of design analysis and synthesis of adaptive systems (AS). Extreme systems, systems with passive adaptation and identification problems have not been considered. Since contributions of Soviet authors has been largely overlooked by the writers of earlier reviews (Åström, 1983; Fujii, 1981; and others) we have decided to focus our attention mainly on the results obtained in the U.S.S.R.

1. INTRODUCTION

THE FIRST adaptive systems had probably appeared by the end of the World War II. The Germans utilized programmed alteration of gain in the FAU-2 and WASSERFALL missile control systems. Later gains in aircraft control systems were changed as a function of a dynamic head measured in flight by special sensors. This already has been an open-loop adaptation.

The term and the concept of adaptive control were introduced in the fifties when the complexity of aircraft led to the need for more efficient control systems for objects whose parameters may vary over a wide range.

A system is adaptive if it makes use of the information on external actions, dynamic characteristics of the plant, or its control system obtained in the course of operation, to change the structure or gains of the controller necessary to achieve the required properties of the closed-loop system (for example, the dynamics of a system should be independent of the variable parameters of the plant or should provide an example of some performance criterion under any feasible parameters of the plant, etc.). This definition is purely technical. There are mathematical definitions of an adaptive system too (Fomin *et al.*, 1981; Sragovich, 1981). Application of the principles of adaptation results in higher control accuracy under a considerable change of the plant's dynamic properties and allows optimization of is operation modes with varying coefficients, improves the system's reliability, typifies individual control systems and their units, reduces both technological requirements to manufacturing individual elements of the system and the times of its design and development.

The theory of adaptive systems rests upon the general theory of stability, theory of invariance, theory of optimal systems (analytical design) (Letov, 1960; Kalman, 1960; Krasovsky, 1963 and 1969), theory of stochastic control (Pugachev, 1962; Åström, 1970) method of stochastic approximation (Tzypkin, 1968) and theory of dual control (Feldbaum, 1965).

2. CLASSIFICATION OF ADAPTIVE SYSTEMS

Adaptive systems are classified into two classes (Åström, 1983): systems of direct adaptive control and indirect adaptive control. In the first class parameters are readjusted so that the output coordinate of the system coincides with that of the reference model. It is sometimes required that the transfer functions of the closed-loop system and the reference model are identical, too. The parameters are adjusted through the use of the value of mismatch between either the controlled coordinates of the system and the model, or their frequency characteristics, etc.

In indirect adaptive control the object is first identified and then the regulator coefficients are adjusted so that the closed-loop system features the given properties. The regulators in such systems are often referred to as self-tuning (Åström and Wittenmark, 1973; Peterka, 1970 etc.). Note that identification may be carried out on the basis of

* Received 28 November 1983; revised 14 April 1984. The original version of this paper was not presented at any IFAC meeting. This paper was recommended for publication in revised form by guest editor L. Ljung.

† Moscow, U.S.S.R.

Reprinted with permission from *Automatica*, vol. 20, pp. 547–557, Sept. 1984.
Copyright © 1984, International Federation of Automatic Control IFAC. Published by Pergamon Press Ltd.

direct adaptive control as well (Margolis and Leondes, 1961; Ashimov *et al.*, 1973). Most widely used however are the least squares technique (Åström, 1983; Ljung, 1977 and others), statistic approximation technique (Tzypkin, 1968), extended Kalman filter and maximum likelihood methods.

Systems with direct adaptive control may be divided into three classes (Petrov *et al.*, 1972):

(1) Systems with information on frequency characteristics (Smith, 1962; Krasovsky, 1963; Kukhtenko, 1970; Petrov *et al.*, 1972 and others).

(2) Systems with information on time characteristics (Solodovnikov, 1957; Braun, 1959 and others).

(3) Model reference adaptive systems (Whitaker *et al.*, 1958; Rutkovsky and Krutova, 1965 and others.

3. SYNTHESIS AND DESIGN OF ADAPTIVE SYSTEMS

Most widely used in the study of adaptive systems is the principle of decomposition. From physical considerations two subsystems can be isolated: a basic loop and a self-adjusting loop.

3.1. *Synthesis of the basic loop*

The major task in designing the basic loop is to find a structure of correcting devices, by tuning the parameters of which one may achieve matching of the closed-loop and reference model operators, i.e. a problem of adaptivity arises. The basic loop is fully adaptable by the output of the system with respect to external disturbances if for any vector of the object's variable parameters a single vector of adjusted parameters of correcting devices may be found such that the outputs of the system and the reference model are fully identical under the same initial conditions (Petrov *et al.*, 1976; Yadykin, 1981).

The basic loop synthesis may often rest upon the concept of the joint tuned object (JTO). At first a JTO is designed which incorporates the plant proper, sensors, actuators and correcting facilities with tuned coefficients. The law of adjusting these coefficients should be determined so that it promotes the description of the JTO by equations with constant coefficients (equations of the stationary model). Thus the problem of the plant variable coefficients matching is solved.

The control for the stationary JTO (Fig. 1) is then synthesized to provide the given dynamic characteristics of the closed-loop system. When optimal control is designed the problem is to be solved by the methods of the optimal system theory. Since JTO is stationary its solution is considerably simple.

Isolation of the JTO is a technique, which simplifies the synthesis of adaptive systems (AS). The JTO correcting elements may obviously be structurally combined with the correcting circuits generating the JTO-control.

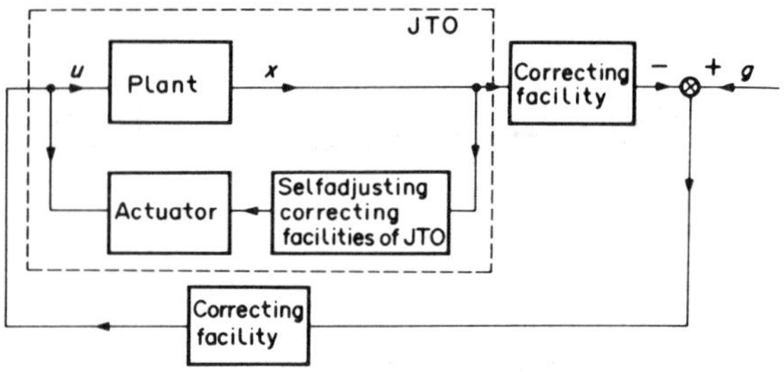

FIG. 1. Block diagram of a self-adjusting system.

Consider the synthesis of a JTO with the use of the invariance theory (Zemlyakov *et al.*, 1969; Pavlov, 1979 and Petrov *et al.*, 1980).

Let the object and model be described by equations

$$\dot{\mathbf{x}} = A(t)\mathbf{x} + \mathbf{D(t)}\boldsymbol{\mu} + C(t)\mathbf{f},$$
$$\dot{\mathbf{x}}_m = A_0\mathbf{x}_m + B_0\mathbf{u} + C_0\mathbf{f} \qquad (1)$$

where $\mathbf{x}$ is the state vector (n); $\boldsymbol{\mu}$ is the input coordinates vector of the actuators (m); $\mathbf{f}$ is the vector of disturbances (r); x_m is the state vector of the model (n); $\mathbf{u}$ is the control vector (m); $A(t)$, $D(t)$ and $C(t)$ are the variable matrices of dimensions $(n \times n)$, $(n \times m)$, $(n \times r)$, respectively; A_0, B_0, C_0 are stationary matrices $(n \times n)$, $(n \times m)$, $(n \times r)$.

To satisfy the condition $\mathbf{x} = \mathbf{x}_m$, the correcting JTO loops should be chosen as follows:

$$\boldsymbol{\mu} = D_0^+(D_{01}\mathbf{u} - \Delta K\mathbf{x} - \Delta N\boldsymbol{\mu} - \Delta R\mathbf{f}) \qquad (2)$$

where D_0^+ is pseudoinverse $(m \times n)$ matrix for D_0 and $D_0 = D(t) - \Delta D(t)$ is a constant matrix $(n \times m)$ whose coefficients correspond to the nominal operation mode of the plant; D_{01} is a constant $(n \times m)$ matrix and ΔK, ΔN and ΔC are matrices of adjusting coefficients of dimension $(n \times n)$, $(n \times m)$ and $(n \times r)$, respectively.

From the condition of parametric invariance and invariance with respect to $\mathbf{f}$ we obtain the following:

$$\begin{aligned} D_0D_0^+\Delta K &\equiv \Delta A(t), \\ D_0D_0^+\Delta N &\equiv \Delta D(t), \\ D_0D_0^+\Delta R &\equiv \Delta C(t) \text{ and} \\ D_0D_0^+D_{01} &= B_0 \end{aligned} \qquad (3)$$

where $\Delta A(t) = A(t) - A_0$, $\Delta C(t) = C(t) - C_0$.

The structure obtained (Fig. 2) may be considered as the limit structure and employed as a basis for designing a real system. To realize the system observing devices for $\mathbf{x}$ and $\mathbf{f}$ estimation should be added. Possible ways of simplifying this structure are considered by Petrov *et al.* (1976).

This structure gives us additive adaptation since the equation of a closed-loop system contains the

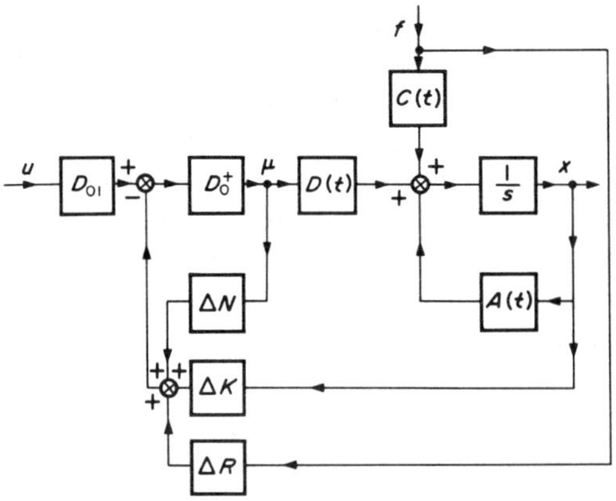

FIG. 2. Basic loop's structure of JTO.

sums of $D_0D_0^+\Delta K - \Delta A(t)$ etc. and according to (3) there occurs additive compensation of parametric disturbances.

Firsova (1981) suggests the principle of multiplicative adaptation according to which two terms are isolated in the denominator of the system transfer function: a multiplier with a constant coefficient system corresponding to the model transfer function, and a multiplier with adjustible coefficients, which becomes equal to the numerator due to adaptation and therefore cancels. It is evident that the multiplier in the numerator, in any combination of the parameters, should be asymptotically stable. The advantage of this method is the need to measure only one controlled coordinate (its derivatives are not required).

In his review paper Fujii (1981) presents a description of another scheme of the basic loop which also rests upon the principle of multiplicative adaptation.

Among other methods employed in the basic loop synthesis note the mode control widely used by Elyseev (1977, 1978) and the analytic method for the design of linearized systems described by Sokolov (1966).

3.2 *Synthesis of adaptation algorithms*

With a selected structure of the basic loop, the main problem in designing AS is synthesis of the adaptation algorithms.

These algorithms can be designed by different methods. The most used ones are the Lyapunov method (Zemlyakov, 1969; Zemlyakov and Rutkovsky, 1966 and 1967; Parks, 1966; Prokopov, 1974, and others), the method of hyperstability (Landau, 1969), the method of gradient (Whitaker, 1958; Donalson and Leondes, 1963; Evlanov, 1964; Kazakov, 1965, Fradkov, 1974, 1979 and others), the method of stochastic approximation (Tzypkin, 1968), recursive identification (Ljung, 1977; Peterka, 1970 and others) and other techniques.

Let us show the algorithms, obtained by the direct Lyapunov method (Petrov *et al*., 1980), for system (1), (2) and (3) when $D_0D_0^+ = E$ (E is a unity matrix).

Introduce the following notation:

$$\Delta A = \|\Delta a_{ij}\|, \quad \Delta D = \|\Delta d_{ij}\|, \quad \Delta C = \|\Delta C_{ik}\|,$$

$$\Delta K = \|\Delta k_{ij}\|, \quad \Delta N = \|\Delta n_{ij}\|, \quad \Delta R = \|\Delta r_{ik}\|,$$

$$Y = \Delta A - \Delta K, \quad Z = \Delta D - \Delta N, \quad S = \Delta C - \Delta R,$$

$$R_y = \|r_{yij}\|, \quad R_z = \|r_{zij}\|, \quad R_s = \|r_{sik}\|,$$

$$r_{yij} = \frac{d\Delta a_{ij}}{dt}, \quad r_{zij} = \frac{d\Delta d_{ij}}{dt}, \quad r_{sik} = \frac{d\Delta c_{ik}}{dt},$$

$$i,j = \overline{1,n}, \quad k = \overline{1,r}$$

Adaptation algorithms

$$\frac{d\Delta k_{ij}}{dt} = \kappa\sigma_i x_j, \quad = \text{const} > 0,$$

$$\frac{d\Delta n_{ij}}{dt} = \kappa\sigma_i \mu_j, \ i,j = \overline{1,n}, \tag{4}$$

$$\frac{d\Delta r_{ik}}{dt} = \kappa\sigma_i f_k, \quad k = \overline{1,r}$$

where σ_i are the components of vector $P\boldsymbol{\varepsilon}$, $\boldsymbol{\varepsilon} = \mathbf{x} - \mathbf{x}_m$, and $P = \|p_{ij}\|$ is the symmetric positive–definite matrix which provides for:

(1) Stability of zero solution

$$\boldsymbol{\varepsilon} = 0, \ Y = 0, \ Z = 0, \ S = 0 \tag{5}$$

and asymptotic stability along coordinate $\boldsymbol{\varepsilon}$ with $R_y = R_z = R_s = 0$.

(2) Asymptotic stability (8) on the whole, uniformly in t_0 and $Y(t_0)$, $Z(t_0)$, $S(t_0)$ with $R_y = R_z = R_s = 0$ and under the condition that the components of vectors $\mathbf{x}_m$, $\mathbf{f}$ and $\boldsymbol{\mu}$ make a linearly-independent uniformly nondisappearing system of functions (this condition will obviously be satisfied, for instance, when $\mathbf{u}$ and $\mathbf{f}$ contain a sufficient number of harmonics).

(3) Stability (5) under the condition that $\mathbf{x}_m$, $\mathbf{f}$ and $\boldsymbol{\mu}$ form the linearly-independent uniformly nondisappearing system of functions with $R_y, R_z, R_s \neq 0$.

It should be noted that under asymptotic stability (5) the problem of identification may be solved, i.e. algorithms (4) under these conditions may be used in systems of identification of either the plant or the closed-loop system. Condition $R_y, R_z, R_s \neq 0$ means that the plant coefficients are variable. Consequently (3) guarantees the system operability and with algorithms (4) at least at a low rate of plant parameter variation is possible.

The drawback of algorithms (4) obtained by many authors (Zemlyakov and Rutkovsky, 1966;

Parks, 1966; Aksyonov and Fomin, 1973; and others) is that the adaptation processes depend on control action $\mathbf{g}(t)$, disturbance $\mathbf{f}(t)$ and plant variable parameters. In a number of papers parametric feed is suggested to decrease the effects influencing the adaptation processes of the plant parameters (Petrov *et al.*, 1972; Yadykin, 1973, and others).

Putintsev and Yadykin (1980) suggested non-linear algorithms which, for the basic loop.

$$\dot{\mathbf{x}} = A(t)\mathbf{x} + B(t)\boldsymbol{\mu},\ \boldsymbol{\mu} = K\mathbf{e},$$
$$\mathbf{e} = \mathbf{u} - \mathbf{y},\ \mathbf{y} = \tilde{K}\mathbf{x}, \tag{6}$$
$$\dot{\mathbf{x}}_m = A_0\mathbf{x}_m + B_0\mathbf{u}$$

acquire the following form:

$$\frac{\tilde{k}^0_{ij}}{\alpha_{ij}}\cdot\frac{1}{\tilde{k}^3_{ij}}\frac{d\tilde{k}_{ij}}{dt} = \delta_{ij}\frac{\tilde{k}_{ij} - \tilde{k}^0_{ij}}{\tilde{k}^2_{ij}} + \sum_{\nu=1}^{n}\sum_{\lambda=1}^{n} \mathscr{E}_{0\nu i}p_{\nu\lambda}\boldsymbol{\varepsilon}_\lambda\mathbf{x}_j, \tag{7}$$

$$\frac{k^0_{ij}}{\alpha_{ij}}\cdot\frac{1}{k^3_{ij}}\frac{dk_{ij}}{dt} = -\beta_{ij}\frac{k_{ij} - k^0_{ij}}{k^2_{ij}} - \sum_{\sigma=1}^{n}\sum_{\nu=1}^{n}\sum_{\lambda=1}^{n} k^+_{\lambda i}\mathscr{E}_{0\nu\alpha}\boldsymbol{\varepsilon}_\sigma\mathbf{e}_j,\ i,j = \overline{1,m}.$$

Here $\mathbf{x}$, $\mathbf{x}_m$, $\boldsymbol{\mu}$, $\mathbf{e}$, $\mathbf{y}$ and $\mathbf{u}$ are vectors of dimension n, n, m, m, m and m respectively; $A(t)$, $B(t)$, A_0 and $B_0 = \|\mathscr{E}_{0ij}\|$ are plant parameter matrices and models of dimension $(n \times n)$ $(n \times m)$, $(n \times n)$ and $(n \times m)$; $K(t) = \|k_{ij}\|$ is the matrix of tuned coefficients of the regulator in the direct loop of the $(m \times m)$ dimension; $\tilde{K}(t) = \|\tilde{k}_{ij}\|$ is the matrix of tuned coefficients of the regulator in the $(n \times m)$ feedback loop; $\tilde{K}^0(t)$, $K^0(t)$ are the matrices of ideal tuning of the regulator coefficients with which the system is described by the equation of the model; $\boldsymbol{\varepsilon} = \mathbf{x} - \mathbf{x}_m$; $P = \|p_{ij}\|$ is a symmetric positive–definite $(n \times n)$ matrix; $K^+ = \|k^+_{ij}\|$ is a matrix pseudoinverse with respect to matrix $K(t)$; α_{ij}, δ_{ij} and β_{ij} are constant positive coefficients.

Algorithms (7) contain elements k^+_{ij} corresponding to parametric feedbacks and $\tilde{k}^0_{ij}$, k^0_{ij} which are computed from the equations

$$A_0 = A(t) - B_0\tilde{K}^0(t),\ B_0 = B(t)K^0(t). \tag{8}$$

To solve these equations matrices $A(t)$ and $B(t)$ should be known, i.e. the plant should be identified, and so using algorithms (7) make no sense. Therefore the self-tuning loops were assumed to be sufficiently fast-responding, and $\tilde{k}_{ij} = \tilde{k}^0_{ij}$, $k_{ij} = k^0_{ij}$. In this case the stability of system movement with respect to model movement cannot be guaranteed but some examples show that these simplified algorithms may ensure both stability and better adaptation as compared with algorithms (4).

In some papers (Kudva and Narendra, 1972: Fradkov, 1979) the terms proportional to Δk_{ij}, Δn_{ij} and Δr_{ik} are introduced into (4) to improve adaptation processes, i.e. following algorithms are obtained:

$$\frac{d\Delta k_{ij}}{dt} = \kappa_1\sigma_i\mathbf{x}_j + \kappa_2\Delta k_{ij},\ \kappa_1,\ \kappa_2 = \text{const}. \tag{9}$$

However the introduction of proportional terms leads to $\Delta k_{ij}(\infty) = 0$ $(\sigma_i(\infty) = 0)$ while in the ideal case one aims at getting $\Delta k_{ij} = \Delta a_{ij}$. Nevertheless, algorithms of the type (9) may give better results as compared to (4) when the plant parameters are rapidly changing.

The paper by Rutkovsky and Krutova (1965) provides the grounds for algorithms of the following form:

$$\Delta k_{ij} = \kappa_1\int_0^t \sigma_i x_j dt + \kappa_2\sigma_i. \tag{10}$$

These algorithms are not substantiated in a strict mathematical manner but simulation of specific systems shown that the σ_i-proportional terms improve the adaptation process performance.

With values of control action $\mathbf{g}(t)$ small in magnitude, i.e. with small $|\mathbf{u}(t)|$ when $\mathbf{f}(t)$ does not lend itself to measurement ($C_0 = 0$ in (1) and $\Delta R \equiv 0$), or with large $|\mathbf{f}(t)|$ one may observe infinite increase of Δk_{ij} and Δn_{ij} in time. This situation may be avoided via algorithm robustness suggested by some authors (Rutkovsky and Krutova, 1965; Aksyonov and Fomin, 1973 etc.) with the introduction of deadzones into the algorithms of adaptation, for example

$$\frac{d\Delta k_{ij}}{dt} = \kappa\phi(x_{mj})\sigma_i x_j \tag{11}$$

where

$$\phi(x_m) = \begin{cases} 1 \text{ with } \|x_m\| \geqslant \delta, \\ 0 \text{ with } \|x_m\| < \delta, \end{cases} \quad \delta = \text{const} > 0.$$

And finally, with $\kappa \to \infty$ algorithms (4) with due regard for the constraints imposed upon the multiplier output coordinates are reduced to the relay type (Petrov *et al.*, 1980)

$$\Delta k_{ij} = \overline{\Delta k}_{ij}\,\text{sign}\,(\sigma_i x_j),\ \overline{\Delta k}_{ij} = \text{const} > 0. \tag{12}$$

In this case a sliding mode occurs in the system with respect to the model's motion and AS becomes a general type system with variable structure (Emelyanov, 1967). However a sliding mode may take place only under limited parametric disturbances Δa_{ij} ($|\Delta a_{ij}| \leqslant \overline{\Delta k}_{ij}$) while with $|\Delta a_{ij}| > \overline{\Delta k}_{ij}$ it breaks loose. On the other hand choosing large $\overline{\Delta k}_{ij}$ will have a negative effect on the system's operation under noise and therefore it may be useful to use the integral and relay terms simultaneously, i.e. to design an algorithm of the type

$$\Delta k_{ij} = \kappa \int_0^t \sigma_i x_j \mathrm{d}t + \overline{\Delta k}_{ij} \operatorname{sign}(\sigma_i x_j). \quad (13)$$

AS operation with the use of algorithms (13) was treated in detail by Petrov *et al.* (1980).

Adaptation algorithms based on Lyapunov's method for discrete systems were suggested by Sebakhy (1976).

The structure of adaptation algorithms synthesized with the use of the theory of hyperstability coincides with (4). This is also true of those obtained by the gradient technique in which the regulator coefficients are adjusted on the condition of minimizing some quadratic function $J = [F(p)\varepsilon]^2$, where $F(p)$ is the polynomial of $p = \frac{\mathrm{d}}{\mathrm{d}t}$ and $\varepsilon = \mathbf{x} - \mathbf{x}_m$.

Minimization of J yields algorithms of the form

$$pk_{ij} = -2\gamma F(p)\varepsilon_i \cdot F(p)u_{ij}, \quad u_{ij} = \frac{\partial x}{\partial k_{ij}}, \gamma = \text{const} > 0 \quad (14)$$

Various methods developed within the framework of the theory of sensitivity are used to compute sensitivity functions $\mathbf{u}_{ij}$ including, in particular, the auxiliary operator method, the sensitivity points method, etc.

Adaptation algorithms close to gradient may be synthesized by means of the recurrent goal-oriented inequalities technique (Yakubovich, 1966; Yakubovich, 1968; Fomin *et al.*, 1981). In contrast to (14) this technique permits determination of the value γ. The recurrent goal-oriented inequalities technique employs finite-converging solution algorithms for infinite recurrent systems of inequalities. With these algorithms a solution is obtained within a finite number of stops.

Gradient algorithms (14) include gains γ. The problem of choosing the best, in a certain sense, adaptation algorithm was stated already in Tzypkin (1968) and it requires γ to be selected in a special manner. In the papers by Polyak and Tzypkin (1980) pseudogradient adaptation methods were suggested as providing optimal asymptotic convergence rate.

These algorithms were developed for identification problems by Tzypkin (1982). However they most likely may be applied for adaptation purposes as well.

3.3. *The problem of disturbance compensation*

Note here another problem associated not only with the adaptation algorithms but also with the entire structure of the basic loop. This implies the presence of disturbances **f**. It was previously assumed that **f** was measurable, however this is not always the case and it is well known that unmeasurable **f** results in loss of stability. Most vital therefore is the problem of indirect measuring of **f** in the form of estimates $\tilde{\mathbf{f}}$ by means of some observing devices. Two solutions are possible in this case. The basic loop may be designed as two-channel using the B. N. Petrov's principle, in which case $\tilde{\mathbf{f}}$ should not go into the model ($C_0 = 0$ in (1)), or the model may recognize $\tilde{\mathbf{f}}$ with a given operator.

In the first case $\tilde{\mathbf{f}}$ is compensated and self-tuning loops should only provide the compensation conditions under variable parameters of the plant. In the second, our goal is to achieve the given response on $\tilde{\mathbf{f}}$.

From the viewpoint of estimating the basic loop performance, the first case seems preferable. However in the second case when $\tilde{\mathbf{f}}$ is fed onto the model the self-tuning loops operate with a better accuracy. If for instance **g** (or **u** with $C_0 = 0$, which is the same) includes a lesser number of harmonics than is required for asymptotic stability (5), the disturbance broadens its spectrum and when fed into the model may compensate lacking harmonics, thus providing us with asymptotic stability.

3.4. *Analysis of adaptive systems*

Adaptive systems designed with the above techniques are nonlinear. To design specific systems both the basic loop and adaptation algorithms should be essentially simplified. Of great importance therefore are linearization techniques, methods of designing systems with the given properties using simplified algorithms and methods of investigating AS under random disturbances.

Linearized parametric AS models (Kukhtenko, 1970; Yadykin, 1973; Kosikov *et al.*, 1976) are widely used now for the design of self-adjusting loops for such systems. A parametric model is a stationary linear system whose input is the set of variable parameters of the plant and output, the tuned controller coefficients. The basic loop and nonlinear relationships in the adaptation algorithms are substituted here with an equivalent loop under certain conditions.

A structural diagram of a linear parametric model may be presented in a form shown in Fig. 3 where $\Delta\mathbf{a}$ and $\Delta\mathbf{k}$ are the vectors of variable parameters of the plant and re-adustable controller coefficients; **f** is the disturbance and $W^a(s)$, $W^k(s)$ and $W^f(s)$ are the

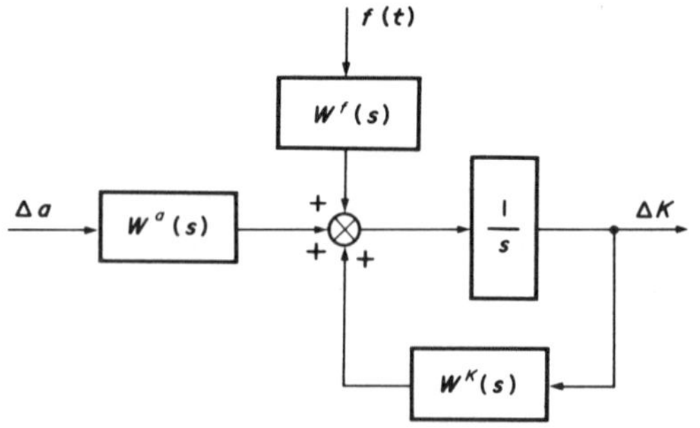

FIG. 3. Structural diagram of a linear parametric model.

transfer function matrices. The goal of all the above techniques is to find $W^a(s)$, $W^k(s)$ and $W^f(s)$.

The design and study of the self-adjusting loop dynamics with the use of linearized models may be carried out by all methods known for linear control systems. In particular one may apply frequency or analytical methods of synthesizing the correcting devices for linearized systems, find their transfer functions and introduce the appropriate corrections into the simplified adaptation algorithms.

Widely used in the basic loop design are the root locus and logarithmic frequency characteristics methods (Krutova, 1972).

4. OPTIMAL ADAPTIVE SYSTEMS

The requirement of optimality was not imposed on the systems considered above. At most, it was required that self-adjusting loops could tune the parameters of the regulator in an optimal way. Optimal systems however are now increasingly implemented. An optimal system with variable parameters of the plant should naturally be adaptive. One should note that the problem of AS design as stated by Solodovnikov (1957) is that of the optimal system synthesis and both Soviet and foreign authors have obtained significant results in this field.

4.1. *Optimal AS synthesized by the generalized performance criterion*

In Krasovsky *et al.* (1977), Bukov and Krasovsky (1974), Bukov (1982), and other papers, some universal optimal adaptive control algorithms were obtained resting upon the use of A. A. Krasovsky's method of analytical design (by the generalized performance criterion). A block-diagram of the designed system is shown in Fig. 4.

The control plant is described by the equation

$$\dot{\mathbf{x}} = F(\mathbf{x}, \mathbf{a}, \boldsymbol{\delta}, \mathbf{t}),\ \dot{\boldsymbol{\delta}} = \mathbf{u} \tag{15}$$

where $\mathbf{x}$ is the state vector; $\mathbf{a}$ the unknown parameters vector; $\boldsymbol{\delta}$ the controller coordinate vector and $\mathbf{u}$ the control. The minimized function (the generalized performance function) is

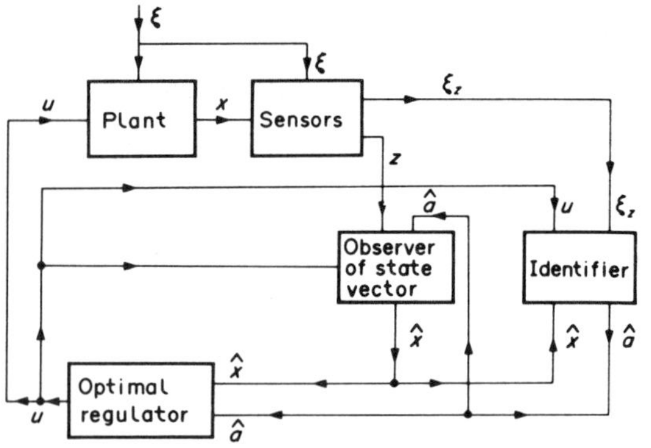

FIG. 4. Block diagram of optimal adaptive system by A. A. Krasovsky.

$$J = V_d(\mathbf{x}(t_f)) + \int_{t0}^{t_f} Q(\mathbf{x}(\tau)\mathrm{d}\tau + \frac{1}{2}\int_{t0}^{t_f} (\mathbf{u}^T K^{-1}\mathbf{u} + \mathbf{u}_{op}^T K^{-1}\mathbf{u}_{op})\mathrm{d}\tau. \tag{16}$$

Here t_0 and t_f are the initial and final time instants; Q and V_d, the specified positively defined functions and $\mathbf{u}_{op}$ the optimal control under the observations

$$\mathbf{z} = H(\mathbf{x}, \boldsymbol{\delta}, \mathbf{a}, t) + \xi_z \tag{17}$$

where ξ_z is the normal centered white noise with an intensity S_z. It is assumed that the problems of state and parameter estimation (identification) and optimal control design may be treated separately. It was proved that for a high degree of estimation this assumption is true. Estimation of state $\hat{\mathbf{x}}$ requires the use of a Kalman filter, identification is carried out with the help of either an adaptive model or a cycle Kalman filter or an algorithm with an 'empirical covariance matrix' or a 'time-saving' algorithm (Krasovsky, 1976; Krasovsky *et al.*, 1977). It is evident that other algorithms of identification may be used as well.

When designing algorithms of optimal control, operational algorithms or algorithms with a predicting model may be used. One of the control algorithms with a predicting model is due to Krasovsky *et al.* (1977).

The estimation of the current state of the plant is used to specify the initial conditions for the predicting model

$$\frac{\mathrm{d}}{\mathrm{d}\tau}\mathbf{x}_m = F(\mathbf{x}_m, \mathbf{a}_m, \boldsymbol{\delta}_m, \tau),\ \frac{\mathrm{d}}{\mathrm{d}\tau}\boldsymbol{\delta}_m = 0. \tag{18}$$

The predicting model is integrated within the optimization time interval. The results are used to set up initial conditions for the vectors

$$\mathbf{p}_x = \frac{\partial V_d^T(\tau_f)}{\partial \mathbf{x}_m(\tau_f)},\ \mathbf{p}_\delta = \frac{\partial V_d^T(\tau_f)}{\partial \boldsymbol{\delta}_m(\tau_f)}. \tag{19}$$

In reverse accelerated time the following system of equations is integrated:

$$\frac{d\mathbf{x}_m}{d\tau} = -F(\mathbf{x}_m, \mathbf{a}_m, \boldsymbol{\delta}_m, \tau), \quad \frac{d\boldsymbol{\delta}_m}{d\tau} = 0,$$
$$\frac{d\mathbf{p}_x}{d\tau} = \frac{\partial F^T(\tau)}{\partial \mathbf{x}_m(\tau)}\mathbf{p}_x - \frac{\partial Q^T(\tau)}{\partial \mathbf{x}_m(\tau)}, \qquad (20)$$
$$\frac{d\mathbf{p}_\delta}{d\tau} = \frac{\partial F^T(\tau)}{\partial \boldsymbol{\delta}_m(\tau)}\mathbf{p}_\delta - \frac{\partial Q^T(\tau)}{\partial \boldsymbol{\delta}_m(\tau)}$$

where all the functions are computed for the predicted motion. The value of the vector $\mathbf{p}_\delta$ is used to form the optimal control

$$\mathbf{u}_{op} = -K\mathbf{p}_\delta. \qquad (21)$$

The advantages of the above technique are in their applicability to adaptive control of essentially multidimensional plants with a large number of controllers, the possibility to realize multi-parameter adaptation when the plant's mathematical model includes a great number of parameters to be identified.

4.2. *Optimal AS synthesized on the principle of stochastic equivalency*

Another approach to the synthesis of optimal adaptive systems was suggested in Yadykin (1979) and Danilin and Yadykin (1982). These authors have obtained algorithms of optimal adaptive control of linear stochastic plants using the principle of stochastic equivalency and adaptive control with a reference model. A block diagram of such system is presented in Fig. 5.

The basic loop (BL) containing the plant, sensors, drives and correcting devices with tuned coefficients a and b is described by the equations of the form

$$\dot{\mathbf{x}} = A(t,a)\mathbf{x} + B(t,\mathscr{E})\mathbf{u} + D_{w_1}(t)w_1,$$
$$\mathbf{z} = C\mathbf{x} + D_{w_2}(t)w_2 \qquad (22)$$

where $\mathbf{x}$ is the state vector; $\mathbf{z}$ is the vector of output coordinates; $\mathbf{u}$ the control; A, B, D_{w_1} and D_{w_2} are the variable matrices of the plant; C is the unknown matrix and w_1 and w_2 the normal centred white noise with unity intensity.

To obtain estimates of the BL state use a linear Kalman filter

$$\dot{\hat{\mathbf{x}}} = A_0\hat{\mathbf{x}} + B_0\mathbf{u} + \tilde{K}(\mathbf{z} - C_0\hat{\mathbf{x}}),$$
$$\mathbf{y} = C_0\hat{\mathbf{x}}, \; C_0 = C \qquad (23)$$

yielding the estimate $\hat{\mathbf{x}}$ optimal in terms of minimal variance of the estimation error

$$J(\mathbf{x},t) = trM\{[\hat{\mathbf{x}}(t) - \mathbf{x}(t)][\hat{\mathbf{x}}(t) - \mathbf{x}(t)]^T\}. \qquad (24)$$

The basic idea of this approach is that the filter equation includes matrices of the BL model A_0 and B_0. This is admissible under the assumption that the adaptation unit (AU) provides small values of the errors $A(t,a) - A_0$ and $B(t,b) - B_0$. Therefore the solution of the Riccati matrix equation for the nonstationary plant equation is reduced to the solution of the corresponding equation of the BL stationary model which significantly simplifies the real time estimation procedure.

To adjust the BL parameters one may use a tuned reference model (TRM) of the BL and the Kalman filter which is generally nonstationary which is explained by the variable nature of the factor $\tilde{K}$ in the Kalman filter.

The plus of this approach is in its applicability to optimal adaptive stabilization adaptive programmed control and the possibility to generalize results for the case of nonlinear plant.

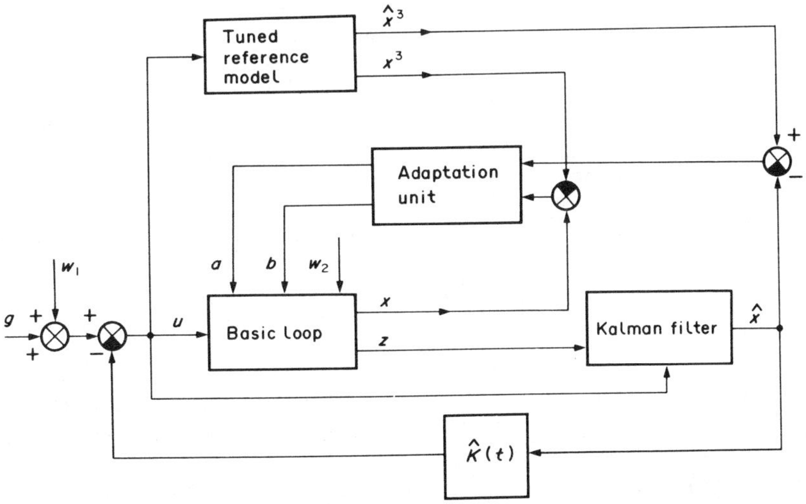

FIG. 5. Block diagram of optimal adaptive system by Yadykin.

4.3. *Optimal AS with a stochastic reference model*

The design of optimal adaptive systems with a stochastic reference model was suggested by Petrov (1978) and Petrov and Zubov (1981).

In the most general scheme the 'cautious' algorithms of adaptive control are synthesized (Wittenmark, 1975), i.e. the re-adjusted parameters are set not only by the plant parameters estimates, but also with due regard for the accuracy of these estimates (with regard for the covariance matrices $\mathbf{P}\mathbf{x}$, $\mathbf{P}\boldsymbol{\eta}$, $\mathbf{P}\mathbf{x}\boldsymbol{\eta}$, $\mathbf{x}$ is the state vector and $\boldsymbol{\eta}$ is the vector of random parameters of the plant). The estimates $\hat{\mathbf{x}}$, $\hat{\boldsymbol{\eta}}$, $\Delta\hat{\mathbf{x}}$ ($\Delta\mathbf{x} = \mathbf{x} - \mathbf{x}_m$) and the covariance matrices $\mathbf{P}\mathbf{x}$, $\mathbf{P}\boldsymbol{\eta}$, $\mathbf{P}\mathbf{x}\boldsymbol{\eta}$ may be obtained with the help of a nonlinear second-order filter in which one may conventionally isolate the filter $\mathbf{x}$, identifier $\boldsymbol{\eta}$, covariance computer and filter $\Delta\mathbf{x}$ (Fig. 6, w, v are random processes).

As a physically realizable unit, the reference model is excluded. It is substituted with the computer $\Delta\hat{\mathbf{x}}$ (filter of $\Delta\mathbf{x}$). However the model as such is tuned. Random in the model are not only the inputs and initial conditions but also the parameters which may vary with respect to $\boldsymbol{\eta}$ permitting the efficiency of the system to be increased due to adaptation of its characteristics to those of the plant.

The synthesis of the system rests upon the use of the dynamic programming technique and the theory of sensitivity. The minimized function is

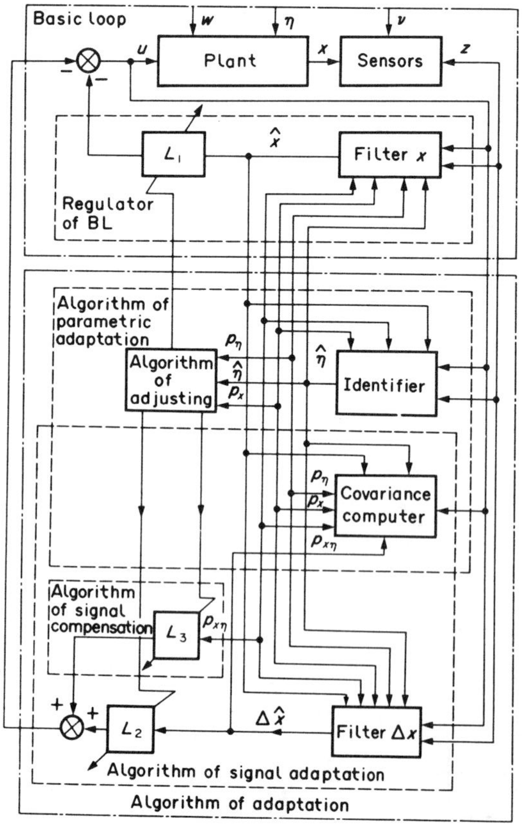

FIG. 6. Block diagram of an optimal adaptive system.

$$J = M\{\Delta\mathbf{x}^T(t_f)\lambda(\boldsymbol{\eta}(t_f))\Delta\mathbf{x}(t_f) + \int_0^{tf} [\Delta\mathbf{x}^T(t)Q_1(\boldsymbol{\eta}(t))\Delta\mathbf{x}(t) + \Delta\mathbf{u}^T(t)Q_2(\boldsymbol{\eta}(t))\Delta\mathbf{u}(t)]\mathrm{d}t \tag{25}$$

where $\Delta\mathbf{u} = \mathbf{u} - \mathbf{u}_m$; λ, Q_1, Q_2 are the weighting matrices and t_f is the final instant of the system operation time. Separation of the control unit into the basic loop regulator and adaptation algorithm was obtained as a result of the optimization problem solution rather than postulated in advance.

The block diagram includes parametric adaptation and signal adaptation. The parameters of matrices L_1, L_2 and L_3 are adjustable.

The system above features the advantages of both the system with stochastic models and that with identifiers: the processes of adaptation are of closed nature and the dynamic characteristics of the closed-loop system are optimized in accordance with the plant parameter variation.

4.4. *Optimal AS synthesized on the basis of computation of Hamiltonian*

Afanasyev and Danilina (1979, 1983) have suggested an idea of designing the optimal system adaptation algorithms based on computations of Hamiltonian.

The plant is described by the equations of the type (22) while matrix B is constant. The functional to be minimized is

$$J = \frac{1}{2}M\left\{\int_{t0}^{t} [\mathbf{x}^T(t)Q\mathbf{x}(t) + \mathbf{u}^T(t)R\mathbf{u}(t)]\,\mathrm{d}t\right\}. \tag{26}$$

The suggested approach may be employed in problems where is presentable in the form

$$J = J_1(\boldsymbol{\varepsilon}) + J_2(\hat{\mathbf{x}}, \mathbf{u}) + J_3(\hat{\mathbf{x}}, \boldsymbol{\varepsilon}) \tag{27}$$

where

$$\boldsymbol{\varepsilon} = \mathbf{x} - \hat{\mathbf{x}},$$

$$J_1(\boldsymbol{\varepsilon}) = \frac{1}{2}M\left\{\int_{t0}^{tf} \boldsymbol{\varepsilon}^T(t)Q\boldsymbol{\varepsilon}(t)\mathrm{d}t\right\},$$

$$J_2(\hat{\mathbf{x}}, \mathbf{u}) = \frac{1}{2}M\left\{\int_{t0}^{tf} [\hat{\mathbf{x}}^T(t)Q\hat{\mathbf{x}}(t) + \mathbf{u}^T(t)R\mathbf{u}(t)]\mathrm{d}t\right\}.$$

$$J_3(\hat{\mathbf{x}}, \boldsymbol{\varepsilon}) = M\left\{\int_{t0}^{tf} \hat{\mathbf{x}}^T(t)Q\boldsymbol{\varepsilon}(t)\mathrm{d}t\right\}.$$

It follows from the structure of the functional and the plant's equations that the basic loop should include an observer chosen from the condition of the minimum of $J_1(\boldsymbol{\varepsilon})$, a regulator to maintain the minimum $J_2(\hat{\mathbf{x}}, \mathbf{u})$, a plant and a measuring device. It must be noted that $J_3(\hat{\mathbf{x}}, \boldsymbol{\varepsilon})$ comes to zero under the minimum of $J_1(\boldsymbol{\varepsilon})$. The optimal control is

$$\mathbf{u} = -R^{-1}BS(t)\hat{\mathbf{x}}(t) \tag{28}$$

where $S(t)$ is the solution of the corresponding Riccati equation.

The block diagram of this system is shown in Fig. 7. The adaptation unit (AU) determines the adjustment algorithms for the parameters of plant α_p, observer α_{0_9} and regulator α_r. The algorithms for α_p and α_{0_9} are found from the modified Wiener–Hopf equation

$$\begin{bmatrix} \alpha_p \\ \alpha_{0_9} \end{bmatrix}^T = M\left\{[L(t,t_1)\mathbf{z}(t_1)]\,[\mathbf{z}(t) - C\hat{\mathbf{x}}(t)]\right\} \tag{29}$$

where the operator $L(t, t_1)$ is expressed through the sensitivity functions

$$\frac{\partial \mathbf{z}}{\partial \alpha_p}, \frac{\partial C\hat{\mathbf{x}}}{\partial \alpha_{0_9}}, \; t - t_1 = \mathrm{d} \neq 0.$$

The adjusted parameters of the regulator α_r (the elements of matrix $S(t)$ in (28) are found from the condition

$$\dot{S}(t) = \boldsymbol{\alpha}_r(t)\Delta H(\hat{\mathbf{x}}, \mathbf{u}) \tag{30}$$

where $\Delta H(\hat{\mathbf{x}}, \mathbf{u}) = H(\hat{\mathbf{x}}, \mathbf{u}) - H_{op}(\hat{\mathbf{x}}, \mathbf{u})$, $H(\hat{\mathbf{x}}, \mathbf{u})$ is the system's Hamiltonian and $H_{op}(\hat{\mathbf{x}}, \mathbf{u})$ is the Hamiltonian at the optimal trajectory.

The matrix $\alpha_r(t)$ is chosen so that (28) provide asymptotic system optimization by the regulator parameters:

$$\alpha_r(t) = -\frac{\partial \Delta H(\hat{\mathbf{x}}, \mathbf{u})}{\partial S(t)}. \tag{31}$$

Hence

$$\alpha_r(t) = \frac{1}{2} M\{\hat{\mathbf{x}}(t)\dot{\hat{\mathbf{x}}}^T(t_1) + \dot{\hat{\mathbf{x}}}(t_1)\hat{\mathbf{x}}^T(t)\} \tag{32}$$

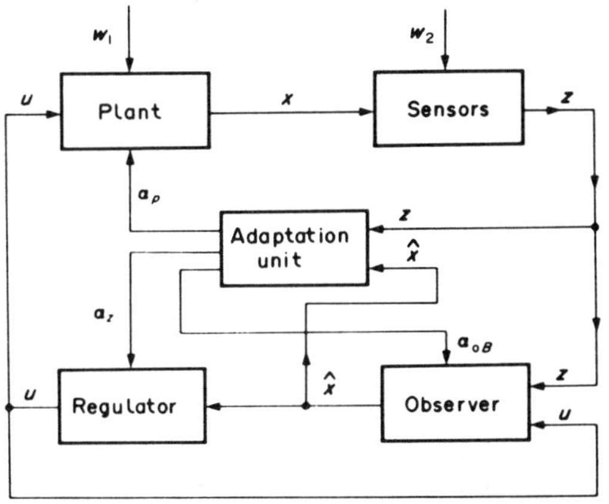

FIG. 7. Block diagram of an optimal adaptive system by Afanasyev and Danilina.

and within the interval $\mathrm{d} = t - t_1$ the parameters of the plant are assumed constant.

Thus the optimal adaptive control theory has been significantly developed and may be applied to the design of highly efficient control systems for most diverse plants.

5. CONCLUSION: TRENDS IN THE DEVELOPMENT AND IMPLEMENTATION OF AS

The overview of literature given above leads one to the conclusion that some interesting and significant results have been obtained in the theory of adaptive control. However this is a new direction of research in control theory which still requires much effort in its further development, design of new adaptation algorithms, new computational and investigation techniques and defining the terminology.

Most of the adaptation algorithms have been synthesized under the condition of convergence with $t \to \infty$, many algorithms are too complicated and require measurement of the complete state vector. Therefore researchers are facing the problem of designing new algorithms meeting the given requirements of the adaptation process performance, and the problem of meaningful simplification of overcomplicated algorithms.

In digital control systems it may be useful to apply information on the post coordinates of the system design. This information may to some extent substitute the actions by the derivatives in control and adaptation algorithms.

Most promising is the use of predicting devices which may essentially decrease the number of derivatives from controlled coordinates required for the adaptation algorithm convergence.

Very important for adaptive system realization are the observations of disturbances acting upon the plant since the model should be fed with all the actions effecting the system.

Application of adaptive systems is at present wide enough. As was noted above many processes in the most diverse industrial areas are characterized with varying parameters and their control by non-adaptive systems is impossible.

Besides their basic application—control of nonstationary plants with an accidental change of parameters—adaptive systems must find wide use in computer-aided design, in reconfiguration systems and in systems of automatic tuning for industrial controllers (Novosyolov *et al.*, 1975; Nishikawa *et al.*, 1981).

REFERENCES

Aksyonov, G. S. and V. N. Fomin (1973). On linear adaptive control systems. In *Methods of Computation*. Leningrad State Univ. Publ., No. 8, pp. 95–116 (in Russian).

Afanasyev, V. N. and A. M. Daniline (1979). Motion and tracking of a nonstationary plant along specified trajectory. *Aut. Rem. Control*, **40**, 1778–1783 (in Russian).

Afanasyev, V. N. and A. N. Daniline (1983). Stabilization problem for linear statistic non-stationary plant on a movable platform. *Izv. Tekh. Kibernetika*, No. 2, pp. 214–219 (in Russian).

Ashimov, A., Dh. Syzdykov and G. M. Tokhtabaev (1973). Searchless self-adjusting identification system. *Aut. Rem. Control*, **34**, 332–336.

Åström, K. J. (1970). *Introduction to Stochastic Control Theory*. Academic Press, New York.

Åström, K. J. and Wittenmark (1973). On self-tuning regulators. *Automatica*, **9**. 187–199.

Åström, K. J. (1983). Theory and applications of adaptive control—a survey. *Automatica*, **19**, 471–486.

Braun, L. (1959). On adaptive control systems. *IRE Trans*, **2**.

Bukov, V. N. and A. A. Krasovsky (1974). Operational optimal control algorithm. *Aut. Rem. Control*, **35**, 1541–1548.

Bukov, V. N. (1982). Static properties of dynamic systems that have controllers using a prediction model. *Aut. Rem. Control*, **43**, 1047–1053.

Danilin, A. B. and I. B. Yadykin (1982). Suboptimal control of stochastic dynamic plants. *Izv., Tekh. Kibernetika*, No. 3, pp. 194–204 (in Russian).

Donaldson, D. D. and C. T. Leondes (1963). A model referenced parameter tracking technique for adaptive control systems. I, II. *IEEE Trans Appl. Ind.*, No. 68, pp. 241–262.

Elyseev, V. D. (1977). Methods of design of multivariable self-adjusting control systems. *Aut. Rem. Control*, **38**, 506–512.

Elyseev, V. D. (1978). Modally invariant control systems. *Aut. Rem. Control*, **39**, 1598–1605.

Emeylanov, S. V. (1967). *Automatic Control Synthesis with Variable Structure*. Nauka, Moscow (in Russian).

Evlanov, L. G. (1964). Self-adjusting system employing gradient search by the auxiliary operator method. *Izv., Tekh. Kiberneitka*, No. 1, pp. 113–120. (in Russian).

Feldbaum, A. A. (1965). *Optimal Control Systems*. Academic Press, New York.

Firsova, Ye. M. (1981). Design of the basic loop structure for non-search self-adjustive systems in the absence of derivatives of the plant output variable. *Aut. Rem. Control*, **43**, 481–483.

Fomin, V. N., A. L. Fradkov and V. A. Yakubovich (1981). *Adaptive Control of Dynamic Plants*. Nauka, Moscow (in Russian).

Fradkov, A. L. (1974). Synthesis of adaptive system of stabilization of linear dynamic plants. *Aut. Rem. Control*, **35**, 1960–1966.

Fradkov, A. L. (1979). Speed-gradient scheme and its application in adaptive control. *Aut. Rem. Control*, **40**, 1333–1342.

Fujii, S. (1981). On the Trend of 'the researchers in adaptive control. *Syst. Control, Japan*, **25**, 715–726.

Kalman, R. (1960). Contributions to the theory of optimal control. *Bull. Soc. Mat. Mech.*, **5**.

Kazakov, I. E. (1965). Investigation of self-adjustment in systems with gradient search by the auxiliary operator method. In *Self-adjusting Systems. Proceedings of the 1st All-Union Conference on Theory and Practice of Self-adjusting System Design* (Moscow, 10–14 December 1963), pp. 23–33. Nauka, Moscow (in Russian).

Kosikov, V. S., I. N. Krutova and B. V. Pavlov (1976). Design of linear models of non-search self-adaptive systems. I, II. *Aut. Rem. Control*, **37**, 49–57; 225–237.

Krasovsky, A. A. (1963). *Dynamics of Continuous Self-adjusting Systems*. Fizmatgiz, Moscow (in Russian).

Krasovsky, A. A. (1969). *Analytical Design of Aircraft Control Loops*. Mashinostroyenie, Moscow (in Russian).

Krasovsky, A. A. (1976). Optimal algorithms in identification problem with an adaptive model. *Aut. Rem. Control*, **37**, 1851–1857.

Krasovsky, A. A. V. N. Bukov and V. S. Shendrik (1977). *Multi-purpose Algorithms for Optimal Control of Continuous Processes*. Nauka, Moscow (in Russian).

Krutova, I. N. (1972). Synthesis of searchless self-adjusting systems based on the root locus method. I, II. *Aut. Rem. Control*, **33**, 1641–1654; 1873–1981.

Kudva, P. and K. S. Narendra (1972). An identification procedure for linear multivariable systems. Tech. Report N-CT-48, Yale University, New Haven.

Kukhtenko, V. I. (1970). *Dynamics of Self-adjusting Systems with Frequency Characteristic Stabilization*. Mashinostroyenie, Moscow (in Russian).

Landau, I. D. (1969). Hyperstability criterion for model reference adaptive control systems. *IEEE Trans Aut. Control*, **AC-14**, 352–555.

Letov, A. M. (1960). Analytical design of control systems. I, II, III. *Avtomatika Telemekh.*, **21**, 436–411; 561–568; 661–665 (in Russian).

Ljung, L. (1977). On positive real transfer functions on the convergence of some recursive schemes. *IEEE Trans Aut. Control*, **AC-22**, 539–551.

Margolis, M. and S. T. Leondes (1961). On the theory of self-adjusting of control systems: a method of self-learning model. *Proceedings of the 1st Inter. IFAC Congress. Theory of Discrete, Optimal and Self-adjusting Systems* (Moscow, 27 June–7 July 1960). AN SSSR Publ., pp. 683–701 (in Russian).

Novosyolov, B. V., Yu. S. Gorokhov, A. A. Kobzev and A. J. Schitov (1975). *Automata-tuners for Follow-up Systems*. Energiya, Moscow (in Russian).

Nishikawa, Y., N. Sannomiya, T. Ohta, H. Tanaka and K. Tanaka (1981). A method for auto-tuning of PID-control parameters. *Preprints of the 8th Triennial World Congress*, Japan, 1981, Vol. VIII, pp. 65–70.

Parks, P. C. (1966). Lyapunov redesign of model reference adaptive control systems. *IEEE Trans Aut. Control*, **AC-11**, 362–367.

Pavlov, B. V. (1979). Synthesis of structure of basic loop of nonsearching self-adjusting systems. *Aut. Rem. Control*, **38**, 1790–1796.

Peterka, V. (1970). Adaptive digital regulation of noisy systems. *Proc. of the 2nd IFAC Symposium on Identification and Process Parameter Estimation*, Prague. Pergamon Press, Oxford.

Petrov, A. I. (1978). Stastical synthesis of terminal control adaptive systems with a reference model. *DAN SSSR*, **242**, 298–301.

Petrov, A. I. and A. G. Zubov (1981). On design of "cautions" adaptive regulators. *DAN SSSR*, **256**, 306–309 (in Russian).

Petrov, B. N., V. Yu. Rutkovsky, I. N. Krutova and S. D. Zemlyakov (1972). *Design Principles for Self-adjusting Control Systems*. Mashinostroyenie, Moscow (in Russian).

Petrov, B. N., V. Yu. Rutkovsky, S. D. Zemlyakov, I. N. Krutova and I. B. Yadykin (1976). Some problems of the non-search self-adjusting system theory. I, II. *Izv., Tekh. Kibernetika*, No. 2, pp. 154–163; No. 3, pp. 142–154 (in Russian).

Petrov, B. N., V. Yu. Rutkovsky and S. D. Zemlyakov (1980). *Adaptive Coordinate-Parametric Control of the Nonstationary Plants*. Nauka, Moscow (in Russian).

Polyak, B. T. and Ya. Z. Tzypkin (1980). Optimal pseudo-gradient adaptation algorithms. *Aut. Rem. Control*, **41**, 1101–1110.

Prokopov, B. N. (1974). On the design of model referenced adaptive systems by the direct Lyapunov method. *Izv., Tekh. Kibernetica*, No. 2, pp. 167–172 (in Russian).

Pugachev, V. S. (1962). *A Theory of Random Functions and its Application to Problems of Automatic Control*. Fizmatgiz, Moscow (in Russian).

Putintsev, V. A. and I. B. Yadykin (1980). Nonlinear adaptation algorithms for multivariable dynamic plants. *Avtomatika Telemekh.*, No. 6, pp. 85–95 (in Russian).

Rutkovsky, V. Yu. and I. N. Krutova (1965). Design principle and some theoretical problems for one class of self-adjusting model referenced systems. In *Self-adjusting Automatic Systems. Proc. of the 1st All-Union Conf. on Theory and Practice of Self-adjusting System Design* (10–14 December, Moscow, 1963). Nauka, Moscow, pp. 46–63 (in Russian).

Sebakhy, O. A. (1976). A discrete model referenced adaptive system design. *Int. J. Control*, **23**, 799–804.

Smith, K. C. (1962). Adaptive control through sinusoidal response. *IRE Trans*, **7**, 129–139.

Sokolov, N. I. (1966). *Analytical Design Method for Linearized Automatic Control Systems*. Mashinostroyenie, Moscow (in Russian).

Solodovnikov, V. V. (1957). Some design principles and theoretical problems of self-adjusting automatic control systems. *Proc. of the USSR Academy of Sciences Session on Scientific Problems of Automatic Control*. U.S.S.R. Academy of Sciences Publ., pp. 143–167 (in Russian).

Sragovich, V. G. (1981). *Adaptive Control.* Nauka, Moscow (in Russian).

Tzypkin, Ya. Z. (1968). *Adaptation and Learning in Automatic Systems.* Nauka, Moscow (in Russian).

Tzypkin, Ya. Z. (1982). Optimal algorithms of parameter estimates for identification. *Avtomatika Telemekh.*, No. 12, pp 9–23 (in Russian).

Whitaker, H. P., A. S. Jamrom and A. Kezer (1958). Design of model reference adaptive control systems for aircraft. Massachussets Technol. Instrum. Lab. Rept. Sept., R-164.

Wittenmark, B. (1975). Stochastic adaptive control methods: a survey. *Int. J. Control,* **21**, 705–730.

Yadykin, I. B. (1973). Frequency methods for the study of non-search systems dynamics. *Izv., Tekh. Kibernetika,* No. 5, pp. 185–194 (in Russian).

Yadykin, I. B. (1979). Optimal adaptive control designed on the principle of a non-search self-adjusting system with a learning reference model. *Avtomatika Telemekh.*, No. 2, pp. 65–79 (in Russian).

Yadykin, I. B. (1981). On the property of Controller Adaptivity in Adaptive Systems. DAN SSSR, **259**, 310–313 (in Russian).

Yakubovich, V. A. (1968). To the adaptive systems theory. DAN SSSR, **182**, 518–521 (in Russian).

Yakubovich, V. A. (1966). Recurrent finite-converging algorithms for the solution of inequalities systems. DAN SSSR, **6**, 1308–1311 (in Russian).

Zemlyakov, S. D. and V. Yu. Rutkovsky (1966). Synthesizing self-adjusting control systems with a standard model. *Aut. Rem. Control,* **27**, 407–414.

Zemlyakov, S. D. and V. Yu. Rutkovsky (1967). Generalized adaptation algorithms for a class of searchless self-adjusting systems. *Aut. Rem. Control,* **28**, 935–940.

Zemlyakov, S. D., B. V. Pavlov and V. Yu. Rutkovsky (1969). Structural synthesis of adaptive control systems. *Aut. Rem. Control,* **30**, 1233–1242.

Zemlyakov, S. D. (1969). Some problems of analysis and design of self-adjusting control systems. In *Self-Adjusting Control System Theory. Proceedings of the 2nd International IFAC Symposium on Self-adjusting Systems* (U.K., 14–17 September 1965). Nauka, Moscow, pp. 154–158 (in Russian).

Model Algorithmic Control (MAC); Basic Theoretical Properties*

RAMINE ROUHANI† and RAMAN K. MEHRA†

The time domain and the frequency domain analysis of model algorithmic control explains most of its heuristically observed properties, in particular with respect to stability, robustness, and nonlinear inputs.

Key Words—Industrial control; impulse response representation; stability; robustness; nonminimum phase systems.

Abstract—A mathematical framework for the analysis of model algorithmic control is developed and the operations of the main components of the control structure are described. The single input–single output case is treated in detail and results pertaining to the stability and the robustness of such systems are derived, both for deterministic and stochastic (colored output additive noise) environment. The case of the nonminimum phase plant is also considered. The robustness problem is discussed and performance 'measures' of robustness are proposed. The results derived in this paper are more generally applicable to optimizing type of control laws. In particular, the paper establishes a close link between the stability and closed loop properties of optimizing type and feedback type of control laws.

1. INTRODUCTION

THE INCREASING use of digital minicomputers has considerably helped the industrial implementation of algorithmic-adaptive types of control. Indeed, for this type of control to be practical and efficient, it is imperative for the controller to have access to fast digital computing facilities with fast memory access and substantial information storage capacity.

In recent years, model algorithmic control (MAC), or equivalently model predictive heuristic control (Richalet and co-workers, 1978), has been implemented on a number of industrial processes ranging from power plants to glass furnaces (Lecrique and co-workers, 1978; ADERSA/GERBIOS, 1978). The success of model algorithmic control schemes operating on complex multivariable processes is due, at least partially, to the particular representation of the processes that they use, namely impulse response representation from each input to each output. In fact, for most complex multivariable industrial processes, parametric models, such as state-space models based on physical laws, are difficult to obtain. It is known that parametric models can give results with large error if the order of the model does not agree with the order of the plant. Moreover in an industrial environment, perturbations affect the plant structure more often than the measurable variables. This requires a constant checking and updating of the model parameters. The impulse response representation is convenient, since in most industrial multivariable processes, the identification of impulse responses is relatively simple. With the current computational facilities, the memory requirement of impulse response representation, where each response has a typical dimension of 30–50, is not a very serious problem. The error due to truncations and approximations of impulse-responses can be looked at as a mismatch between the plant and its model. This mismatch is, partially, responsible for the discrepancy between the process output and its predicted value. The closed loop MAC, where this latter discrepancy enters the computation—and the adjustment—of the plant input, displays a particularly high degree of robustness against plant–model mismatch.

Theoretical investigation of MAC properties have been lacking in comparison of its ap-

*Received 29 June 1980; revised 17 February 1981; revised 13 October 1981; revised 22 January 1982. The original version of this paper was not presented at any IFAC meeting. This paper was recommended for publication in revised form by associate editor K. J. Åström.

†Scientific Systems, Inc., 54 Rindge Avenue Extension, Cambridge, MA 02140, U.S.A.

Reprinted with permission from *Automatica*, vol. 18, pp. 401–414, July 1982.

Copyright © 1982, International Federation of Automatic Control IFAC. Published by Pergamon Press Ltd.

plication to concrete processes. This paper intends to establish a mathematical basis and a conceptual framework for the analysis of MAC in order to study some of its important features such as stability, asymptotic behavior, and so on. Attention will mainly be focused on single input–single output systems. In order to gain some insight into the multivariable case, the single input–single output analysis is necessary and extremely helpful. In fact, much of the results pertaining to this latter analysis extend to multivariable systems.

The organization of this paper is as follows. Section 2 is a general overview of the major functions of MAC (prediction–optimization). Section 3 treats the case of a single input–single output system where the inputs are free of constraints, and introduces linear closed loop and open loop predictors. Section 4 is concerned with the stability analysis, both from a time domain and frequency domain viewpoint. Section 5 investigates the effect of an output additive noise and compares the performances of the control with respect to reference trajectories. Section 6 poses the robustness problem and its relation to stability. In Section 7 constraints on the input are introduced and the analysis of the constrained system is related to the analysis of the unconstrained one. Finally, Section 8 addresses the case of nonminimum phase systems.

2. GENERAL PHILOSOPHY

The model algorithmic control is conceptually similar to a model reference adaptive type of control with some important differences in practical implementation. It involves (i) dynamic models for system representation and prediction, (ii) a reference trajectory, and (iii) an optimality criterion leading to the optimal control.

2.1. *Representation and prediction*

The system is represented by its impulse responses, the identification of which can be done both on-line and off-line. However, in most cases, the off-line identification is accurate enough for the purpose of control and one can avoid the cost and complexity of an on-line identification procedure. This is due to the particular redundancy of the impulse response representation which allows a considerable enhancement of the robustness of the control scheme against identification errors and parameter perturbations (Mehra and co-workers, 1977; Mereau and co-workers, 1978).

The formal representation of the system is as follows:

$$y(t+1) = \mathbf{h}^T\mathbf{x}(t) = h_0x(t) + h_1x(t-1) + \cdots + h_Nx(t-N) \tag{1}$$

where $y(t+1)$ is the plant output at time $t+1$; $\mathbf{h}^T \in R^{N+1}$ denotes the plant impulse response; $x(t-j) \in \Omega \subset R$ for $j = 0, \ldots, N$. $x(t-j)$ is the input at time $(t-j)$ to the plant. Ω is the constraint set of the input.

The model (1) is also called the actual model to emphasize that $\mathbf{h}^T$ represents the actual process. But since such perfect knowledge of the plant impulse response is usually not possible, one has to use an approximation $\bar{\mathbf{h}}^T$ to $\mathbf{h}^T$. The model corresponding to this latter impulse response is then

$$y_M(t+1) = \mathbf{h}^T\mathbf{x}(t). \tag{2}$$

The above model, together with the past history of the plant output denoted by $Y(t) = \{y(\tau), \tau \le t\}$ is used to predict the future value of the output. Various prediction schemes are conceivable. In this paper we will limit ourselves to simple open loop and closed loop prediction.

2.2. *The reference trajectory*

The purpose of the control is to lead the output $y(t)$ along a desired, and generally smooth, path to an ultimate set point c. Such a path is called a reference trajectory. In the present paper the reference trajectory is of first order and is initiated on the output of the plant at time t_0

$$\begin{aligned} y_r(t_0+k) &= \alpha^k y_r(t_0) + (1-\alpha^k)c \\ k &= 1, \ldots, T, \quad |\alpha| < 1 \\ y_r(t_0) &= y(t_0). \end{aligned} \tag{3}$$

The reference trajectory can be chosen to be of higher order and the set point c can be made time variant. These latter extensions would not bring any conceptual difficulty in the solution of the control problem, although the actual computation of the inputs becomes more complex, but not intractable, as the order of the reference trajectory increases.

2.3. *The optimality criterion and the optimum control strategy*

The optimality criterion should reflect the previously mentioned purpose of following the reference path to the desired set point c. This can be done by defining the optimum control strategy as the one which minimizes over a certain horizon in the future, the deviation of the predicted outputs from the reference path. Formally, at each instant t_0, the optimum set of T future inputs $\{x^*(t_0), x^*(t_0+1), \ldots, x^*(t_0+T-1)\}$ are such that the predicted T outputs $\{y_P(t_0+1), \ldots, y_P(t_0+T)\}$ are as close as possible, in the sense of a weighted Euclidean norm, to the reference trajectory y_r. Therefore the function to minimize is

$$J_T = \sum_{k=1}^{T} (y_p(t+k) - y_r(t+k))^2 \omega_k \qquad (4)$$

where ω_k is a nonnegative weighting factor.

Note that at time t_0, the determination of T optimal inputs $x^*(t_0+k)$ $(k = 0, \ldots, T-1)$ is done by solving a static optimization problem

$$\begin{aligned} &\text{minimize } J_T(x(t_0), \ldots, x(t_0+T-1)) \\ &\quad \text{s.t.} \quad x(t_0+k) \in \Omega, \quad \text{for } k = 0, \ldots, T-1 \end{aligned} \qquad (5)$$

where Ω denotes the constraint set for the process inputs.

Various types of algorithms can be used to solve (5) and to determine the set $X^*(t_0) = \{x^*(t_0), \ldots, x^*(t_0+T-1)\}$ (Luenberger, 1973). In Richalet and co-workers (1978) a relatively simple projection type algorithm is proposed. The main advantage of such an algorithm is that it can be used both for control evaluation and identification of impulse responses, when the latter is done on-line. Recently a gradient type algorithm has been developed to solve (5) for systems with particularly fast modes (Scientific Systems, 1980). The interval $\{t_0, t_0+T\}$ is called the horizon of control evaluation, and sometimes the horizon of prediction since at time t_0 one has to predict a set of T outputs $y_P(t+k)$, $(k = 1, \ldots, T)$. Once the set $X^*(t_0)$ of the T optimum inputs is determined, it is possible to wait up to T periods before observing the process outputs $y(t)$, reinitializing $y_r(t)$, predicting y_Ps and computing the next set $X^*(t_0+T)$. This means that all the elements of the optimum control set $X^*(t_0)$ have been actually applied to the plant.

A more appealing strategy consists of applying the first few optimal inputs $x^*(t+k)$, $(k = 0, 1, \ldots, p$ with $p \ll T)$ before reinitializing y_r and computing the next T optimum inputs. In the limit, if the computing facilities allow, one would apply only the first optimum input $x^*(t)$ of the set $X^*(t)$, observe $y(t+1)$, initialize the reference trajectory y_r on the observed value $y(t+1)$, and solve the optimization problem (5) at time $t+1$. In this latter case the unused optimum inputs $x^*(t+1), \ldots, x^*(t+T-1)$ of $X^*(t)$ (computed at time t) will serve as starting values for the numerical algorithm determining $x^*(t+1)$, that is solving (5) at time $(t+1)$. Only the latter case will be considered in this paper. Figure 1 shows the optimum inputs and the corresponding predicted output over the horizon T. To visualize the overall control procedure corresponding to the last case, one must repeat the same illustration for $t+1, t+2, \ldots$ Figure 2 is a block representation of the whole control scheme.

3. UNCONSTRAINED INPUT

3.1. *Linear prediction*

So far the characterization of the prediction scheme has been kept general. Let us constrain the class of predictors to be linear. Notwithstanding the linearity of the plant, the above class of predictors is a practically convenient choice. Moreover, it simplifies substantially the determination of the optimum control sequence. Indeed, the linear character of $y_p(t)$ causes the function J_T to be quadratic in the inputs $x(t)$, and hence to become a candidate for fast quadratic optimization algorithms.

3.2. *One step prediction* (T = 1)

The assumption that the input vector $x(t)$ is free of constraints, i.e. $\Omega \equiv R$, results in a significant simplification in the optimum control determination. That is, the length T of the horizon of prediction does not affect the optimum value of the first input $x^*(t)$ to be applied. In other terms, the minimization of J_T and $J_{T'}$, with $T' \neq T$, will result in the same first input $x^*(t)$ of the optimum sequences of length T' and T. In particular, the first element of the sequence $\{x^*(t), \ldots, x^*(t+T-1)\}$ minimizing J_T is identical to the input minimizing J_1. This is an important simplification, since it reduces a T-variable minimization at each step to a one variable minimization. This property, which is based essentially on the principle of superposition of linear systems, can be easily established, and we leave proof of the following proposition to the reader.

Proposition 3.1. If the input to the plant is free of constraints, the minimization of J_T (4) and the minimization of J_1 at time t_0, result in the same first input $x^*(t_0)$. That is, the first element of the set $X^*(t_0)$ corresponding to the minimization of J_T is identical to the input minimizing J_1.

However, it is important to note that the T optimal inputs $x^*(t_0), \ldots, x^*(t_0+T-1)$ obtained by minimizing J_T are not identical to the input sequence obtained by T consecutive minimizations of J_1. Indeed, while minimizing the dis-

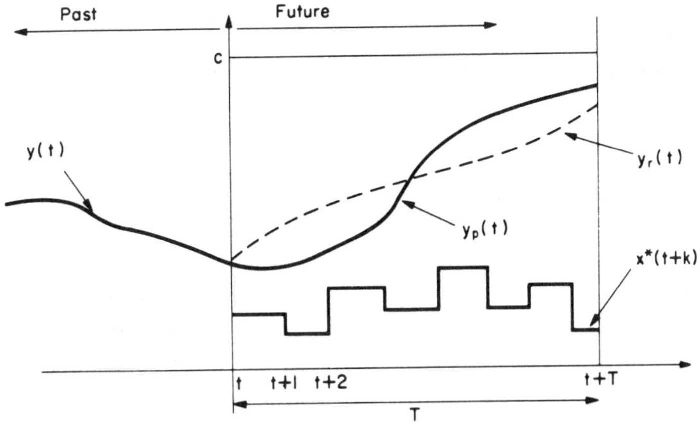

FIG. 1. Predicted output and the corresponding optimum input over a horizon T. Where $x^*(t)$, optimum input; $y_p(t)$, predicted output; $y_r(t)$, reference trajectory; and $y(t)$, process output.

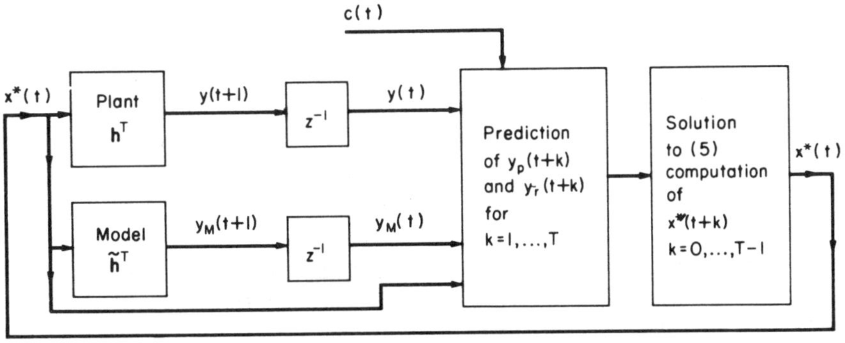

FIG. 2. $c(t)$, set point; $y(t)$, process (plant) output; $y_M(t)$ model output; $y_r(t)$, reference trajectory; $y_P(t)$, predicted output; $x^*(t)$, process (plant) input.

tance over a horizon of length T, the reference trajectory $y_r(t_0+j)$ is initialized on the process output $y(t_0)$ for the whole length of interval T; but in case of T consecutive minimizations of J_1 at time t_0, $t_0+1, \ldots, t_0+T-1$, the reference trajectory y_r is continuously initialized on the observed $y(t_0)$, $y(t_0+1), \ldots, y(t_0+T-1)$; that is

$$y_r(t_0+j) = \alpha y(t_0+j-1) + (1-\alpha)c, \quad \forall j = 1, \ldots, T.$$

3.3. *Open loop and closed loop prediction*

The simplest one step ($T=1$) open loop prediction scheme that one can imagine is

$$y_p(t+1) = y_M(t+1) = \mathbf{h}^T\mathbf{x}(t). \tag{6}$$

That is, the model of the plant is used to predict the output one step ahead. The main inconvenience of such a prediction is that, as one might expect, the output $y(t)$ of the controlled plant would not converge to the set point c (Richalet and co-workers, 1978). However it is easy to see that the bias is given by

$$\frac{\sum_{i=0}^{N} h_i - \sum_{i=0}^{N} \tilde{h}_i}{\sum_{i=0}^{N} \tilde{h}_i - \alpha \sum_{i=0}^{N} h_i} c \tag{7}$$

where, although the plant $\mathbf{h}$ is not exactly known, $\sum_{i=0}^{N} h_i$ can be estimated as the plant gain, or the response to a unit step function.

To alleviate the bias problem, one has to use a closed loop prediction scheme of the form

$$y_p(t+1) = y_M(t+1) + (y(t) - y_M(t)) \tag{8}$$

where $y(t) - y_M(t) = (\mathbf{h} - \tilde{\mathbf{h}})\mathbf{x}(t-1)$ represents a correction term, assuring the final convergence of the plant output to the set point. Indeed, the minimization of J_1 requires that

$$y_p(t+1) = y_r(t+1) = \alpha y(t) + (1-\alpha)c \quad (9)$$

or

$$(1-\alpha)[c - y(t)] = y_M(t+1) - y_M(t). \quad (10)$$

Now letting $t \to +\infty$, and assuming that the system is stable, one deduces

$$\lim_{t\to+\infty} y(t) = c.$$

4. STABILITY

The optimum sequence of inputs $x^*(t)$, in the sense of minimizing J_1, for a closed loop system, is generated by an auto-regressive equation, which results from (10), where $y_p(t+1)$ and $y_r(t+1)$ are expressed in terms of inputs

$$\begin{aligned} \tilde{h}_0 x^*(t) &= \sum_{j=0}^{N} \tilde{h}_j x^*(t-1-j) \\ &\quad - \sum_{j=1}^{N} \tilde{h}_j x^*(t-j) + (1-\alpha) \\ &\quad \left[c - \sum_{j=0}^{N} h_j x^*(t-1-j)\right] \\ y(t) &= \mathbf{h}^{\mathrm{T}} \mathbf{x}^*(t-1). \end{aligned} \quad (11)$$

Obviously, if the sequence $x^*(t)$ tends to an equilibrium value x^*, that is, if the corresponding autoregressive model is stable, then the output $y(t)$ tends to an equilibrium value $y(\infty)$ which equals c in the case of closed loop prediction but differs from c for open loop prediction. However, the converse is not true, that is, theoretically one may have a converging output $y(t)$, while the input $x^*(t)$ diverges. Intuitively it is clear that even though $|x^*(t)|$ might increase indefinitely, the linear function $\mathbf{h}^{\mathrm{T}}\mathbf{x}^*(t)$ may remain finite (at least in theory). This is, basically, what happens to nonminimum phase systems (Åström, 1970; Åström and Wittenmark, 1974).

The boundedness condition for the input sequence $x^*(t)$ is identical to the stability of the auto-regressive models, that is, the polynomial

$$(z-1)\sum_{i=0}^{N+1} \tilde{h}_i z^{-i} + (1-\alpha)\sum_{i=0}^{N+1} h_i z^{-i} = 0 \quad (12)$$

must have all its roots within the unit circle.

So far we mainly focused on the relationship between the output $y(t)$ and the optimum input sequence $x^*(t)$. Now let us turn our attention to the response of an optimally controlled system to the set point c, that is, the relationship between $y(t)$ and $c(t)$. In terms of the z-transforms, one deduces from (8)–(10), with some manipulation

$$\frac{X(z)}{C(z)} = \frac{(1-\alpha)}{z^{-1}(1-\alpha)H(z) + (1-z^{-1})\tilde{H}(z)} \quad (13)$$

$$\frac{Y(z)}{C(z)} = \frac{z^{-1}H(z)(1-\alpha)}{z^{-1}(1-\alpha)H(z) + (1-z^{-1})\tilde{H}(z)} \quad (14)$$

with $X(\cdot)$, $Y(\cdot)$, $H(\cdot)$, $\tilde{H}(\cdot)$ and $C(\cdot)$ denoting respectively the z-transforms of $x(t)$, $y(t)$, $\mathbf{h}$, $\tilde{\mathbf{h}}$ and $c(t)$, respectively (Fig. 3).

Let us first note that for a perfect identification of $\mathbf{h}$, that is, for $H(z) = \tilde{H}(z)$, the closed loop transfer function becomes

$$\frac{X(z)}{C(z)} = \frac{(1-\alpha)}{(1-\alpha z^{-1})\tilde{H}(z)} \quad (15)$$

$$\frac{Y(z)}{C(z)} = \frac{1-\alpha}{z-\alpha}. \quad (16)$$

Note that under such perfect identification the transfer function of $Y(z)$ with respect to $C(z)$ is of first order and identical to the reference trajectory. Now if the polynomial $H(z)$ has some of its roots outside the unit circle, then $X(z)$ corresponds to an increasing sequence $x(t)$. This latter case characterizes nonminimum phase systems where the cancellation of $z^{-1}H(z)$ in the original expression of $[Y(z)/C(z)]$, leading to (16), is not valid because

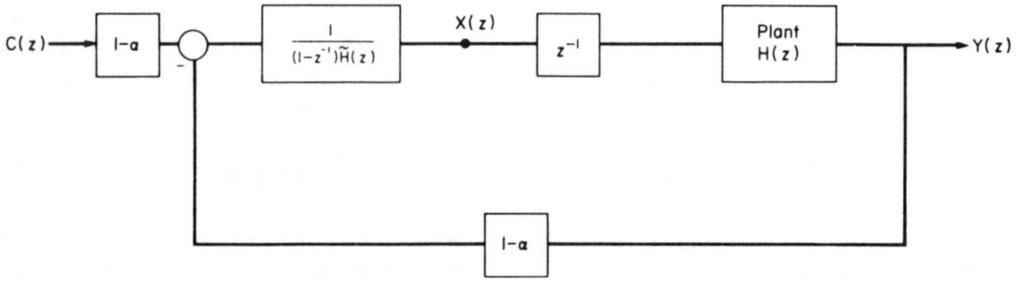

FIG. 3. Closed loop prediction.

$H(z)$ contains unstable zeros. This case will be studied in Section 8.

Let us come back to the case of imperfect identification $\tilde{H}(z) \neq H(z)$. From the transfer functions (13) and (14) it becomes clear that for the output $y(t)$ to be convergent and the optimum input sequence to be bounded, it is necessary and sufficient that the denominator of (13) has all its roots within the unit circle.

Note that in general the exact plant transfer function $H(z)$ is not known and therefore one cannot evaluate the exact expression of the above polynomial. However, if the identification of $H(z)$ is fairly good, then it is expected, by the continuity theorem (which states that the roots of a polynomial are continuous functions of its coefficients), that the roots of $z^{-1}(1-\alpha)H(z)+(1-z^{-1})\tilde{H}(z)$ be close to the roots of $(1-\alpha z^{-1})H(z)$. Therefore, the stability of the identified model $\tilde{H}(z)$ implies the stability of the system. But when the identification error becomes large so that the discrepancy between $\tilde{H}(z)$ and $H(z)$ becomes significant, then to determine the stability of the system one should have recourse to robustness analysis, which is discussed in Section 6 and involves determining the set of plant polynomials $H(z)$ for which the above characteristic polynomials have stable roots.

5. OUTPUT ADDITIVE NOISE

Assume that the uncertainties of the system are modeled as an additive zero mean noise on the plant output. That is

$$y(t+1) = \mathbf{h}^{\mathrm{T}}\mathbf{x}(t) + w(t). \tag{17}$$

The prediction model and the reference trajectory being as previously defined, the cost function to minimize over a horizon of T is now, instead of J_T

$$E[J_T] = E\left\{\sum_{j=1}^{T}(y_p(t+j) - y_r(t+j))^2\right\}$$

where E denotes the expectation operator.

Following the same line of argument as in Section 2, it is seen that when there is no constraint on inputs, the optimum input $x^*(t)$ is recursively determined by the equation

$$y(t) + \tilde{\mathbf{h}}^{\mathrm{T}}(\mathbf{x}^*(t) - \mathbf{x}^*(t-1)) = \alpha y(t) + (1-\alpha)c. \tag{18}$$

The difference with the deterministic case is that the present sequence of optimum control $x^*(t)$ is nondeterministic, which results from the noisy nature of the plant output $y(t)$. It is easy to verify that the mean value $\bar{y}(t)$ of the plant output and the mean $\bar{x}^*(t)$ of the optimum input sequence have a deterministic dynamic identical to the one governing the deterministic case as in Section 4. Thus, all the results of the previous section are valid for the calculation of the means $\bar{x}^*(t)$ and $\bar{y}(t)$. It remains to study the variance behavior, and in particular the variance of the controlled output $y(t)$. Let us start with the analysis of colored noise.

The colored nature of the additive noise is often due to the presence of not controllable disturbances of drift type at the input or at the output. It may also stem from the fact that the plant output $y(t)$ is observed through a filter, usually of low order.

For the sake of illustration, let us consider the simple case of a first-order Markov process noise

$$w(t+1) = \rho w(t) + e(t), \quad |\rho| < 1 \tag{19}$$

where $e(t)$ is the zero mean white noise with variance ν^2.

The stationary variance of the noise is then $\tau_w^2 = \nu^2/1-\rho^2$. Assuming that the identification of the plant impulse response is sufficiently good to allow the approximation $\tilde{\mathbf{h}} = \mathbf{h}$, from (18) one deduces

$$y(t) - w(t+1) + w(t) = \alpha y(t) + (1-\alpha)c. \tag{20}$$

From (19) and (20), with some rather lengthy but straightforward manipulations, the steady state output variance τ_c^2 is derived as

$$\tau_c^2 = \frac{2}{1+\alpha}\frac{1-\rho^2}{1-\alpha\rho}\sigma_w^2 = \frac{2}{1+\alpha}\frac{\nu^2}{(1-\rho\alpha)(1+\rho)}. \tag{21}$$

From the control point of view, it is interesting to analyze the effect of the parameter α, which characterizes the speed of the reference trajectory, on the above variance. The plots of Fig. 4 indicate that for a negatively correlated noise ($\rho < 0$), slowing down the reference trajectory (increasing α), decreases the variance τ_c^2. For positive ρ, three situations arise, depending on its value. For $\rho \leq \frac{1}{3}$, the variance still decreases with a slower reference trajectory. For $\frac{1}{3} < \rho < \frac{1}{2}$, the variance reaches its minimum for a reference trajectory corresponding to $\alpha = (1-\rho/2\rho)$, and moreover, it is always bounded by its value at $\alpha = 0$, i.e. $(2\nu^2/1+\rho)$. For $\frac{1}{2} < \rho < 1$, the situation is reversed in the sense that the variance is bounded by its value corresponding to the slowest trajectory

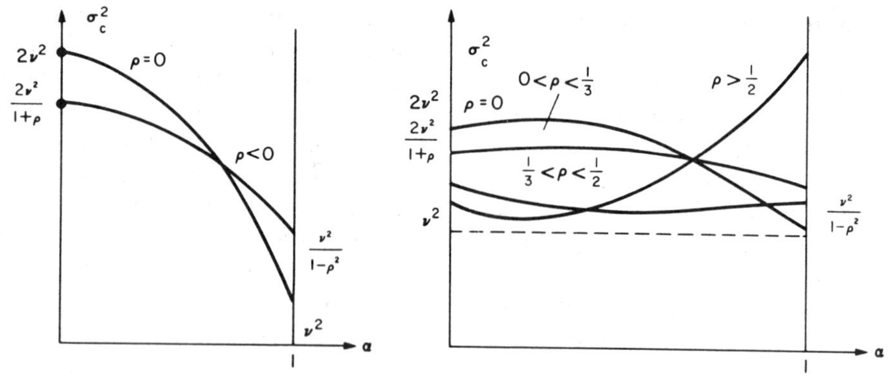

FIG. 4. Output closed loop variance (σ_c^2) as a function of α, for positively ($\rho > 0$) and negatively ($\rho < 0$) correlated noise.

($\alpha = 1$), that is by ($\nu^2/1 - \rho^2$). In these two latter cases, slowing down the reference trajectory does not necessarily lead to smaller output variance since the minimum is achieved for some $\alpha < 1$.

For $\rho = 0$, which corresponds to the case of white noise, the closed loop variance $\tau_c^2 = (2\nu^2/1 + \alpha)$ decreases by slowing down the reference trajectory (increasing α). It is interesting to note that when the prediction is open loop, as given by (6), then the output variance $\tau = (\nu^2/1 - \alpha^2)$ is an increasing function of α, so that a fast reference model results in a smaller output variance. Therefore under the assumption of white noise, when the plant output $y(t)$ is far from the set point c, one can operate in two consecutive phases. First use an open loop prediction with fast reference trajectory (small α) and then switch to a closed loop prediction control with slow reference trajectory. The latter is necessary, since as mentioned before, the open loop prediction yields a steady state bias. Moreover, in order to increase the speed of convergence of the output, one can use the estimated value of the open loop bias (as given in 7) for adjustment at the switching instant.

As the above analysis shows, in the presence of white noise, the selection of a fast reference trajectory increases the output variance. On the other hand, it is usually desirable to enhance the speed of the reference trajectory in order to shorten the convergence time of the output to the set point. With proper output filtering, one can reconcile, to a certain degree, these two conflicting objectives. In fact, through a filter, the white noise appears as colored in the output. The correlation factor ρ can be treated as the filter parameter whose proper adjustment—according to Fig. 4 and the related analysis—will allow the selection of a faster reference trajectory, while keeping the output variance within a certain desired range. This is probably why in almost all practical implementations of MAC, it has been found useful to perform an output filtering.

6. ROBUSTNESS

The control design is said to be robust if the plant output $y(t)$ converges to its ultimate desired value c, under a range of plant and behavior changes caused, for instance, by failure of some of its components or by large parameter deviations, etc.

With regard to the closed loop prediction control, the robustness problem is posed in terms of the stability of the input sequence $x(t)$ and the plant output $y(t)$. This is a consequence, as mentioned in Section 4, of the fact that the stability of the closed loop system guarantees the unbiased convergence of the output to the set point. Therefore, as long as the structural changes of the plant, which amount to altering the impulse response $\mathbf{h}^T$ from its nominal (identified) model $\hat{\mathbf{h}}^T$, are such that the characteristic polynomial (12) has all its roots within the unit circle, the control is robust.

Given the model $\hat{\mathbf{h}}$, it is of great interest to find a range θ for the $(N+1)$-dimensional vector $\mathbf{h}$ such that (12) is stable. Clearly, the delimitation of such a subspace of R^{N+1} requires complex search algorithms, and, often unrealistic amounts of computation. But, fortunately, in almost all practical problems, valuable additional information both on the physical properties of the plant and on the failure and perturbation history of its different components are available. The appropriate use of this information reduces substantially the size and dimension of the search space. For instance, if the robustness against identification errors or small perturbations is considered, then

$$h_i = \tilde{h}_i + \epsilon_i, \quad i = 0, \ldots, N$$

where $|\epsilon_i|$ are small with respect to $\tilde{h}_i$ and their upper bounds can be estimated from the particular identification scheme in operation. If the robustness against failure of some components—sensors or activators—is of concern, as in Ackerman (1979), then the impulse response $\mathbf{h}$ under such failure can be known (by simulation for example) and the stability of (12) tested.

So far the robustness has been considered in an absolute sense, that is, from the viewpoint of asymptotic convergence of the plant output to its desired value. Such a viewpoint does not provide an assessment of the relative robustness performance of different control schemes under various conditions. In particular, the effects of perturbations and component failures on the speed of convergence are not apparent at all, and clearly, this is very important for industrial applications.

In light of the above considerations, it becomes necessary to define some type of 'measure' for the performance of the control scheme with regard to robustness (Mehra and co-workers, 1979). Here we propose two possible performance indices. First consider

$$\mu = \frac{D(y(t), c\,|\mathbf{h} = \tilde{\mathbf{h}}, \alpha)}{\operatorname{Sup}_{h \in \theta} D(y(t), c\,|\mathbf{h}, \alpha)} \tag{22}$$

where D denotes the total weighted distance between the plant output $y(t)$ and the desired set point c

$$D = \sum_{t=0}^{\infty} (y(t) - c)^2 \omega_t, \quad \omega_t > 0.$$

A decreasing sequence of weight ω_t would enhance the importance of an early convergence to the set point. The numerator of the index is the distance under the assumption of perfect identification and for a given value of α. The denominator is the greatest possible distance while the controlled plant is still stable ($\mathbf{h} \in \theta$ implies robustness in the absolute sense). Both numerator and denominator depend on the reference trajectory through the value of α. Obviously the value of the performance index is within the range (0, 1), and the performance improves with increasing μ.

It is possible, and desirable, to define a performance index independent from the particular reference trajectory. Such an index can be used in comparison of MAC with other control schemes, such as LQ. One can define such an index by

$$\mu' = \frac{\operatorname{Inf}_{\alpha \in \Lambda} D(y(t), c, \alpha\,|h = \tilde{h})}{\operatorname{Sup}_{h \in \theta} \operatorname{Inf}_{\alpha \in \Lambda} D(y(t), c, \alpha\,|\mathbf{h})}$$

where $T\Delta \subset (0, 1)$ is the range of admissible reference trajectories. The main difference between this and the previous case is that in this case the controller is free to choose the best reference trajectory in the sense of minimizing the distance between the plant output and the desired set point c.

As a simple example, consider the case where the mismatch between $\mathbf{h}$ and $\tilde{\mathbf{h}}$ is a pure gain q ($\mathbf{h} = q\tilde{\mathbf{h}}$). Then the range θ is identical with the interval of possible variations of q

$$\theta = (q_1, q_2).$$

For the system to remain stable, it is necessary that

$$\theta = (q_1, q_2) \subset (0, 2/1 - \alpha)$$

(Mehra and co-workers, 1979). Then

$$\mu = \inf \frac{2 - (1 - \alpha)q}{1 + \alpha} = \frac{2 - (1 - \alpha)q_2}{1 + \alpha}, \quad q \in (q_1, q_2)$$

and

$$\mu' = \frac{\inf_{\alpha \in \Lambda} \frac{1}{1 + \alpha}}{\sup_{q \in \theta} \inf_{\alpha \in \Lambda} \left(\frac{1}{2 - (1 - \alpha)q} \right)} = \frac{2 - (1 - \alpha_2)q_2}{1 + \alpha_2}$$

$$\theta \equiv (q_1, q_2) \subset \left(0, \frac{2}{1 - \alpha}\right)$$

$$\Lambda \equiv (\alpha_1, \alpha_2) \subset (0, 1).$$

To see the behavior of μ with respect to α, consider

$$\frac{d\mu}{d\alpha} = \frac{2(q_2 - 1)}{(1 + \alpha)^2}.$$

It is seen that for $q_2 > 1$, that is if the maximum possible mismatch gain is larger than unity—which is usually the case—then slowing down the reference trajectory (decreasing α) improves the robustness of the system. Note also that the expression of μ' involves α_2, the maximum value of α, which corresponds to the slowest trajectory. Similarly μ' increases, the robustness improves, as α_2 becomes larger. It is interesting to note that these results on robustness enhancement via slowing down the reference trajectory have been confirmed in practical applications.

7. AMPLITUDE CONSTRAINTS ON INPUTS

In most applications, the input $x(t)$ is not free of constraints. In general there are both amplitude and rate constraints imposed by technical and cost considerations. Under such constraints, the results of Section 3 must be revised; in particular, the minimization of the distance J_T over a horizon of length T ($T > 1$) does not necessarily yield the same optimal input $x^*(t)$ as the minimization of J_1 ($T = 1$).

For the stability it can be shown that, under rather weak conditions, the stability of the system with inputs free of constraints implies the stability of the input constrained system. In fact, Praly (1979) has proved a more general result pertaining to nonlinearity in the recursive control equation determining $x(t)$. Let us describe briefly his result, a complete proof of which involves mathematical technicalities and can be found in Praly (1979). When there are no constraints, the recursive equation generating the inputs $x(t)$ is linear, of the type

$$x^*(t) = \frac{1}{h_0}\left[-\sum_{j=1}^{N} h_j x^*(t-j) + \alpha y(t-1) + (1-\alpha)c\right]. \qquad (23)$$

Assume that the above recursion is stable and denote its steady state solution by $\bar{x}^*$. Now consider a nonlinear time varying function $f_t(\cdot)$ and assume that instead of (23), the recursion governing the input sequence is

$$x^*(t) = f_t\left[\frac{1}{h_0}\left(-\sum_{j=1}^{N} h_j x^*(t-j) + \alpha y(t-1) + (1-\alpha)c\right)\right]. \qquad (24)$$

Then in order for the stability of (23) to imply the stability of (24) it is necessary that for each t

$$|f_t(\bar{x}^* + u) - \bar{x}^*| < \frac{1-\alpha}{r}|u| \qquad (25)$$

where r is the spectral radius of the companion matrix associated to the recursion (23).

In the case of constant amplitude bounds on the input $x^*(t)$, the functions $f_t(\cdot)$ are time independent and involve saturation type of nonlinearities

$$f(v) \overset{\Delta}{=} \begin{cases} M & \text{for } v \geq M \\ v & \text{for } m \leq v \leq M, \quad \text{with } \bar{x}^* \in [m, M] \\ m & \text{for } v \leq m. \end{cases}$$

It can be shown that, when the recursion (23) is stable, for a range or value of α, such functions verify the inequality (25). Therefore, the nonlinearities introduced by amplitude constraints do not make the system unstable.

The above result is important since it reduces the stability analysis of a MAC driven system with amplitude constraints to the one of the unconstrained system, which is relatively simple, thanks to its linearity. It should be noted that for a rather large class of nonlinear time varying functions, the inequality (25) is verified, and hence the stability analysis of a system with more general input constraints (such as time varying rate and amplitude constraints) can be related to the stability of the unconstrained system.

8. NONMINIMUM PHASE SYSTEM

Let us consider again the transfer function of Section 4

$$\frac{X(z)}{C(z)} = \frac{(1-\alpha)}{z^{-1}(1-\alpha)H(z) + (1-z^{-1})\tilde{H}(z)} \qquad (26)$$

$$\frac{Y(z)}{C(z)} = \frac{z^{-1}H(z)(1-\alpha)}{z^{-1}(1-\alpha)H(z) + (1-z^{-1})\tilde{H}(z)}. \qquad (27)$$

Assuming that $H(z)$ is known and has all its roots within the unit circle, then the best choice of $\tilde{H}(z)$ for the output to follow the reference trajectory is $\tilde{H}(z) = H(z)$. Cancellation of $H(z)$ in (27) will cause the output to follow exactly the reference model, and moreover the input sequence $x(t)$ remains bounded. But, if $H(z)$ has some of its roots outside the unit circle, then equating $\tilde{H}(z)$ with $H(z)$ will result in an unbounded input sequence $x(t)$. In fact, once $x(t)$ reaches its practically imposed bounds, then the cancellation of $H(z)$ in (27) is no longer perfect and the output starts to diverge from the reference trajectory (Fig. 5) (Åström, 1970; Peterka, 1972).

From the above analysis it is seen that unlike the minimum phase, where the natural choice of $\tilde{\mathbf{h}}$ is identical or close to $\mathbf{h}$, for nominimum phase case this choice does not lead to a stable and realizable control strategy. In fact, $\tilde{\mathbf{h}}$ should be chosen such that the characteristic polynomial of (26) has all its roots within the unit circle. This stabilizes the input while assurring an asymptotic convergence of the output to the set point. However, as one may expect, the output is no longer following the reference trajectory.

Obviously, there is an infinity of choices for the vector $\tilde{\mathbf{h}}$ resulting in a stable characteristic polynomial. Among all the admissible $\tilde{\mathbf{h}}$ one may choose those which satisfy some secondary criterion. A natural criterion is, for example, the minimization of the total distance between the

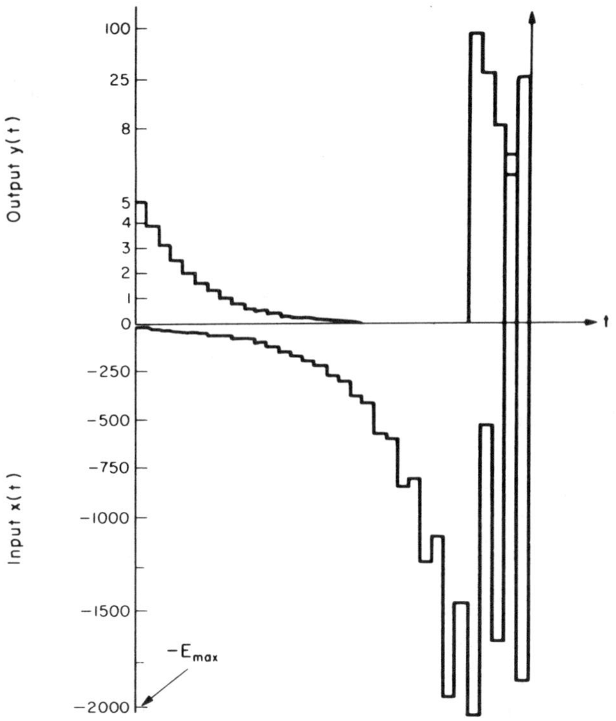

FIG. 5. Simulation of the nonminimum phase system $h^T = [1.0, -0.9, -1.26, 1.8060, -1.1092, 0.2856]$; $c = 0$, initial condition $y(0) = 5$, $\alpha = 0.8$ and the bound on the value of the amplitude of the input $E_{max} = 2000$.

process output $y(t)$ and the reference trajectory

$$\text{Min} \sum_{t=1}^{\infty} (y(t) - y_r(t))^2.$$

Other criterion may involve direct pole placements, or equivalently the design of the dynamics of the error $(y(t) - y_r(k))$ (Åström and Wittenmark, 1974). In the following subsection, we will solve the problem of determining the optimum model $\tilde{\mathbf{h}}$ in the sense of minimizing the Euclidean distance between the plant output and the reference trajectory.

8.1. *Least-square solution*

The problem can be stated as follows: Determine the $(N+1)$-dimensional vector $\tilde{\mathbf{h}}$ such that the characteristic equation $[(1-\alpha)H(z) + (z-1)\tilde{H}(z)]$ has all its roots within the unit circle and

$$J = \frac{1}{2} \sum_{j=1}^{\infty} (y(j) - y_r(j))^2 \tag{28}$$

is minimum.

To solve the above problem, as one might expect, we will use the jargon of linear quadratic control. But we have to formulate first the problem in a framework suitable for the linear quadratic approach.

For the sake of simplicity, and without loss of generality, we assume that the set point c is zero. Then the equation (11) which determines the input sequence can be written as

$$\begin{aligned} x^*(t) = \frac{1}{\tilde{h}_0} [(&\tilde{h}_0 - \tilde{h}_1 - (1-\alpha)h_0)x^*(t-1) \\ &+ \cdots (\tilde{h}_{N-1} - \tilde{h}_N - (1-\alpha)h_{N-1})x^*(t-N) \\ &+ (\tilde{h}_N - (1-\alpha)h_N)x^*(t-N-1)]. \end{aligned} \tag{29}$$

Or equivalently, defining $X(t)$ to be the $(N+2)$-*dimensional* vector of input sequence, a state equation is derived

$$X(t) = UX(t-1) + b_1[L^T \ 0]X(t-1) \tag{30}$$

with

$$X(t) = (x^*(t), x^*(t-1), \ldots, x^*(t-N-1))$$

$$U = \begin{bmatrix} 0 & 0 & 0 & 0 \\ 1 & 0 & 0 & 0 \\ 0 & 1 & 0 & 0 \\ 0 & 0 & 1 & 0 \end{bmatrix} \updownarrow (N+2), \quad b_1 = \begin{bmatrix} 1 \\ 0 \\ 0 \\ \cdot \\ \cdot \\ 0 \end{bmatrix} \updownarrow (N+2)$$

and

$$L^{\mathrm{T}} = \frac{1}{\tilde{h}_0}[\tilde{h}_0 - \tilde{h}_1 - (1-\alpha)h_0, \ldots,$$
$$\tilde{h}_{N-1} - \tilde{h}_N - (1-\alpha)h_{N-1}, \tilde{h}_N - (1-\alpha)h_N]$$
$$L \in R^{N+1}; \quad [L \;\; 0] \in R^{N+2}. \tag{31}$$

Next the expression of J in terms of the variable $x^*(t)$ is derived by writing $y(t)$ and $y_r(t)$ in terms of the vector $X(t)$

$$J = \frac{1}{2}\sum_{t=1}^{\infty}[X^{\mathrm{T}}(t)q \;\; q^{\mathrm{T}}X(t)] \tag{32}$$

with $q^{\mathrm{T}} = [h_0, \;\; h_1 - \alpha h_0, \ldots, h_N - \alpha h_{N-1}, -\alpha h_N]$. Within the above formulation, it becomes clear that the determination of the optimum $\tilde{\mathbf{h}}$ which minimize J is equivalent to the determination of a linear feedback gain $[L \;\; 0]$ of (30) minimizing J. At this point there is sufficient motivation to formulate the problem as follows

Find the optimum input $u^*(t)$ which minimizes

$$J = \lim_{k\to\infty}\frac{1}{2}[X^{\mathrm{T}}(k)qq^{\mathrm{T}}X(k) + \sum_{j=1}^{k-1}X^{\mathrm{T}}(j)qq^{\mathrm{T}}X(j)] \tag{33}$$

where the state equation is

$$X(t+1) = UX(t) + b_1u(t). \tag{34}$$

It is easy to check that the above system is controllable. The optimum control $u^*(t)$ is a linear function of the state vector: $u^*(t) = \hat{L}^{\mathrm{T}}X(t)$ (Sage and White, 1977; Aoki, 1976). It remains to show that $\hat{L}$ has its last element $\hat{L}_{N+2}$ equal to zero. The following lemma establishes the results.

Lemma: The optimal input sequence $u^*(t)$ for equation (17) does not depend on the last element of the state vector $X(t)$.

Proof: By linear quadratic control theory, we know that $u^*(t)$ has the following expression in terms of $X(t)$

$$u^*(t) = -[Q_2 + b_1^{\mathrm{T}}Pb_1]^{-1}b_1^{\mathrm{T}}PUX(t) = \hat{L}X(t)$$

where Q_2 is the cost matrix of the input $u(t)$ (here $Q_2 = 0$ since J depends only on the state vector, see the appendix for more detail) and P is the steady-state solution of a Riccati-type matrix equation. Then

$$b_1^{\mathrm{T}}Pb_1^{\mathrm{T}} = P_{11}$$
$$b_1^{\mathrm{T}}PU = [P_{11}, P_{12}, \ldots, P_{1(N+2)}]U$$
$$= [P_{12}, P_{13}, \ldots, P_{1(N+2)}, 0]$$

hence

$$\hat{L} = \frac{-1}{P_{11}}[P_{12}, P_{13}, \ldots, P_{1(N+2)}, 0].$$

With respect to the above solution, it should be mentioned that:

- The cost J (33) of the above formulated LQ problem is independent from the input $u(t)$. It is known that when the cost associated with the control sequence equates to zero, the existence of an optimum input is not guaranteed (Sage and White, 1977). However, for the particular problem that we are considering here, it can be demonstrated (see the appendix) that, for simple reasons relative to the structure of the state equation (34), an optimal input sequence exists, and depends linearly on the state vector $X(t)$.
- The determination of $\tilde{\mathbf{h}}$, i.e. the solution of Riccati equation, is done off line. Moreover the matrix Riccati equation and the recursions involved are very simple (see recursion A6 and Section 2 of the appendix). This is primarily due to the simplicity of U. Even for large values of N (of the order of 80), the optimum $\tilde{\mathbf{h}}$ can be computed without excessive cost.
- It has been suggested that the above solution is a time domain counterpart of the frequency domain solution proposed by Perterka (1972). Peterka has shown that the optimal closed loop system poles are at the zeros of the system transfers or at their reciprocal if outside the unit circle. However when the degree of $H(z)$ is large, which is usually the case when h represents a plant's impulse response, the determination of its zeros become intractable. The LQ solution proposed here replaces this computational burden with the Riccati recursion (A6) of the appendix which is much easier to resolve.

Let us recapitulate the results. It has been shown that the determination of the best vector $\mathbf{h}$ in the sense of minimizing the total Euclidean distance between the process output and the reference trajectory while keeping the control sequence $x^*(t)$ bounded, is equivalent to the determination of the gain matrix $\hat{L}$ of the linear quadratic control problem (33) and (34). The determination of $\hat{L}$ involves the solution of a Riccati equation of order $N+2$. Once $\hat{L}$ is known $\tilde{\mathbf{h}}$ can be computed uniquely from (31). It should be noted that $\tilde{\mathbf{h}}$ depends uniquely on $\mathbf{h}$ and α.

We end this section by applying the above considerations to the example in Fig. 5 (see Fig. 6).

9. CONCLUSION

A mathematical framework for the analysis of

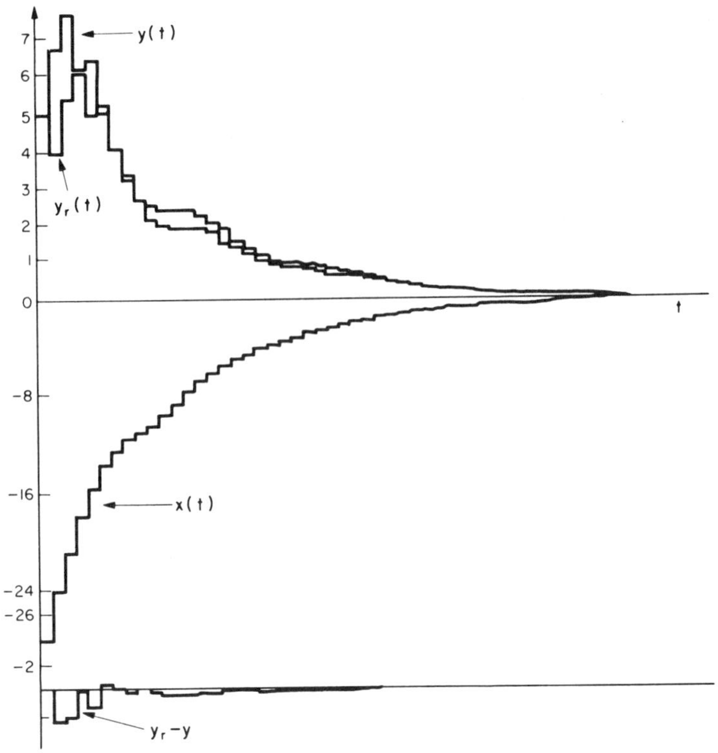

FIG. 6. LQ optimal control corresponding to $h = [1.0, -0.9, -1.26, 1.806, -1.1092, 0.2856]$ with $\alpha = 0.8$ and $c = 0$. The optimal $\tilde{h}$ is $[-0.5754, 0.0546, 0.0446, 0.1266, -0.2946, 0.0772]$.

model algorithmic control has been laid down. The main operations and components of MAC have been described.

A simplified version of MAC, the single input/single output case, has been analyzed using time domain and frequency domain techniques. The stability conditions has been derived and the robustness of the closed loop control has been investigated. In particular, the analytical relationship between the robustness and the components of the controller, i.e. the prediction model and the reference trajectory, has been established. The experimentally observed fact that slowing down the reference trajectory enhances the robustness of the control scheme, has been analytically explained and proved.

The case of nonminimum phase plants has been studied within the framework of MAC, and it has been shown that a controller can be designed to minimize the discrepancy between the plant output and the desired trajectory, while the input sequence remains bounded. Such design necessitates the solution of an off-line Riccati difference equation.

Much of the single input–single output results have natural extension to the multivariable cases. The study of the multivariable case has yielded partial results and will appear in a companion paper.

Acknowledgements—The work here was carried out in collaboration with J. Richalet, A. Rault and L. Praly of Adersa/Gerbios, France. The work was supported by contracts from DOE under ET-78-C-01-3394.

REFERENCES

Ackermann, J. E. (1980). Parameter space design of robust control systems. *T-AC*, pp. 1058–1071.

ADERSA/GERBIOS (1978). Report on glass furnace control.

Aoki, M. (1976). *Optimal Control and System Theory in Dynamic Economic Analysis.* North Holland, Amsterdam.

Åström, K. J. (1970). *Introduction to Stochastic Control Theory.* Academic Press, New York.

Åström, K. J. and B. Wittenmark (1974). Analysis of self-tuning regulators for non-minimum phase systems. *IFAC Symposium, Budapest.*

Lecrique, M., M. Tessier, A. Rault and J. Testud (1978). Multivariable control of a steam generator. Characteristics and results. *IFAC 7th Triennial World Congress*, p. 73.

Luenberger, D. G. (1973). *Introduction to Linear and Nonlinear Programming,* Addison Wesley, New York.

Mehra, R. K., W. C. Kessel, A. Rault and J. Richalet (1977). Model algorithmic control using IDCOM for the F-100 jet engine multivariable control design problem. *Int Forum of Alternatives for Multivariable Control.*

Mehra, R. K., R. Rouhani, A. Rault and J. G. Reid (1979). Model algorithmic control: theoretical results on robustness. *Proc. JACC*, pp. 387–392.

Mereau, P., D. Guillamme and R. K. Mehra (1978). Flight control application of MAC with IDCOM (identification and command). *Proc. IEEE Conf. on Decison and Control*, pp. 977–982.

Peterka, V. (1972). On steady state minimum variance control strategy. *Kybernetika*, **8**, 219.

Praly, L. (1979). General study of single input single output linear time invariant control laws. Application to an adapted model algorithmic control (AMAC). ADERSA/GERBIOS report.

Richalet, J., A. Rault, J. L. Testud and J. Papon (1978). Model predictive heuristic control: applications to industrial processes. *Automatica*, **14**, 413.
Sage, A. P. and C. C. White (1977). *Optimal System Control.* Prentice Hall, New York.
Scientific Systems (1980). Internal report, Basic research in digital stochastic model algorithmic control.

APPENDIX: DERIVATION OF THE RICCATI EQUATION

1. *Derivation of the Riccati equation for* $r = 0$

The optimum control problem of Section 8 is

$$X(k+1) = UX(k) + b_1 u(k) \tag{A1}$$

$$J(n) = \frac{1}{2} X^T(n)QX(n) + \frac{1}{2}\sum_{j=0}^{n-1} X^T(j)QX(j) \tag{A2}$$

where $Q = q\ q^T$ [see (32)]. Since dim $[b_1, Ub_1, \ldots, U^{N+1}b_1]$ is equal to $N+2$, the system (A1) is controllable.

The cost function $J(n)$ has the peculiarity of not depending explicitly on the input $u(j)$. But in order to be able to use the common discrete version of LQ, one must require that the cost associated with the input $u(j)$ be strictly positive (Sage and White, 1977; Aoki, 1978). That is, the cost function must be of the type

$$J_c(n) = \frac{1}{2} X^T(n)QX(n) + \frac{1}{2}\sum_{j=0}^{n-1}\{X^T(j)QX(j) + ru^2(j)\}, \quad \text{with } r > 0. \tag{A3}$$

In practice it is always possible to correct J with an additional term $r\sum_{j=1}^{n-1} u^2(j)$, where r is a small positive number, and obtain J_c. By minimizing J_c, one gets a suboptimal solution. Since this suboptimal solution depends on the value of r, one would expect that the smaller the value of r, the closer should be the suboptimal solution to the optimal one.

In LQ optimal control, the intuitive significance of requiring a strictly positive input weight in the cost function J_c, is to assure the boundedness of the input amplitude. But in this problem, the input $u(j)$ equates the first state component at time $(j+1)$: $X_1(j+1) = u(j)$. Therefore the positive weight q_{11} associated to $X_1(j+1)$ would prevent the value of $u(j)$ to become unbounded.

Motivated by the above consideration, we tackle the problem of finding the optimal input $u(n)$, if it exists, for the case $r = 0$. In order to do so, we derive the Riccati matrix for $r > 0$ and observing its continuous dependence on r, we let $r = 0$.

Following Sage and White (1977), the recursive Riccati equation is driven as

$$U^T P(k+1)\left[I + \frac{b_1 b_1^T}{r} P(k+1)\right]^{-1} u = P(k) - Q \tag{A4}$$

$$P(n) = Q.$$

Note that

$$\left[I + \frac{b_1 b_1^T}{r} P(k+1)\right]^{-1} = \frac{1}{r + P_{11}(k+1)} \begin{bmatrix} r & -P_{12}(k+1) & \cdots & -P_{1N}(k+1) \\ 0 & r + P_{11}(k+1) & & 0 \\ & & \ddots & \\ 0 & & & r + P_{11}(k+1) \end{bmatrix}$$

Carrying out the matrix multiplication of (A4) we deduce

$$U^T P(k+1) \begin{bmatrix} \frac{-P_{12}(k+1)}{r+P_{11}(k+1)} & \frac{-P_{13}(k+1)}{r+P_{11}(k+1)} & \cdots & \frac{-P_{1N}(k+1)}{r+P_{11}(k+1)} & 0 \\ 1 & 0 & \cdots & 0 & 0 \\ 0 & 1 & \cdots & 0 & 0 \\ \vdots & \vdots & \cdots & \vdots & \vdots \\ 0 & 0 & \cdots & 1 & 0 \\ 0 & 0 & \cdots & 0 & 1 \end{bmatrix} + Q = P(k) \tag{A5}$$

The dependence of the above Riccati equation on r is throughout $1/[r + P_{11}(k+1)]$ which is continuous for $r \geq 2$ *provided that* $P_{11}(k+1) \neq 0$. That is, if $P_{11}(k) \neq 0$ for all $k \in [1, n]$ and all $r \geq 0$, we can let $r \to 0$ in the above equation and deduce the corresponding gain matrix $P(k)$.

Let us show that the entry $P_{11}(k)$ never vanishes, i.e. $P_{11}(k) \neq 0$ for all $k \in [1, n]$. This results from the following induction

$P(n) = Q$ implies that $P_{11}(n) = q_{11} = h_0^2 > 0$ (see Section 8). From equation (A5) we have

$$P_{11}(k) = q_{11} + \frac{[P_{11}(k+1)P_{22}(k+1) - P_{12}^2(k+1)] + rP_{22}(k+1)}{r + P_{11}(k+1)}$$

$P(k+1)$ being a semi-definite matrix, it results that

$$P_{11}(k+1) \geq 0$$

$$P_{22}(k+1) \geq 0$$

$$P_{11}(k+1)P_{22}(k+1) - P_{12}^2(k+1) \geq 0$$

(all the minors of the nonnegative matrix $P(k+1)$ are nonnegative). Hence it follows that $P_{11}(k) \geq q_{11} > 0$ for all $r \geq 0$, and all $k \in [1, n]$.

Now we can let $r = 0$ in equation (A5) and the resulting equation (A6) will generate the sequence $P(k+1)$ corresponding to the case $r = 0$

$$P(k) = Q + U^T P(k+1) \begin{bmatrix} \frac{-P_{12}(k+1)}{P_{11}(k+1)} & \cdots & \frac{-P_{1N}(k+1)}{P_{11}(k+1)} & 0 \\ 1 & 0 \cdots & 0 & 0 \\ 0 & 1 \cdots & 0 & 0 \\ \vdots & \cdots & \vdots & \vdots \\ 0 & 1 \cdots & 1 & 0 \end{bmatrix} \tag{A6}$$

$$P(n) = Q.$$

2. *The optimum input* u(j)

The optimum input $u(j)$ is derived from $P(j+1)$ by

$$u(j) = L(j)X(j) = -[b_1^T P(j+1) b_1] \\ b_1^T P(j+1) UX(j)$$

$$u(j) = \frac{1}{P_{11}(j+1)}[P_{12}(j+1), \ldots, P_{N+1} \\ (j+1), 0]X(j).$$

Letting n go to infinity, the cost function J becomes

$$J = \frac{1}{2}\sum_{j=0}^{\infty} X^T(n) q \; q^T X(n).$$

Equation (A6) solved backward for large n, converges to 'steady-state' Riccati solution and the gain $\hat{L}(j)$ becomes time invariant

$$L = \frac{1}{P_{11}}[P_{12}, P_{13}, \ldots, P_{1N+1}, 0].$$

Since only the first column of the matrix P is of importance for the determination of L, we can write the Riccati recursive equation in terms of column vectors of P.

$$P = [p_1 \vdots \quad \vdots p_{N+2}] \text{ and } Q = [q_1 \vdots \quad q_{N+2}].$$

Then

$$\begin{cases} p_1(k) = U^T p_2(k+1) + q_1 - \dfrac{p_{12}(k+1)}{p_{11}(k+1)}, & U^T p_1(k+1) \\ p_j(k) = U^T p_{j+1}(k+1) + q_j - \dfrac{p_{1j+1}(k+1)}{p_{11}(k+1)}, & U^T p_1(k+1) \\ p_{N+2}(k) = q_{N+2}. & \end{cases}$$

A New Solution to Adaptive Control

JUAN M. MARTIN-SANCHEZ

Abstract—A new solution to adaptive control for a linear time-variant multivariable process with unknown parameters is presented. The proposed method requires knowledge of only input and output data and, consequently, no state estimation is necessary. The control signal is generated by a control block placed in series with the process. The control block behaves as the "adaptive" inverse of the process, and has as its input the desired output of the process. The whole control system is asymptotically hyperstable. The control system can solve the main problems encountered in process control: structural differences and parameter variations. In addition, it behaves satisfactorily under the influence of perturbations. The method is extremely simple to implement and quite general in scope. Various examples are included for illustration.

I. Introduction

During the last decade, a large amount of work has been done on the design of adaptive control systems. Of this, the work on model reference adaptive systems (MRAS) has received considerable attention [10]. More recently, a general method for the design of MRAS has been based on their equivalence to an autonomous feedback system, which verifies Popov's hyperstability criterion [2]. This technique has been extended to the synthesis of discrete parallel MRAS [3] and discrete series-parallel MRAS [4].

A new solution to the adaptive-control problem is presented here, which is based on Popov's hyperstability criterion as applied to discrete systems, but which is uniquely different from the MRAS formulation. The proposed solution was conceived with the following philosophy in mind.

Try to find an adaptive-control block, which will only use the information that it receives from the inputs and outputs of the plant and which will behave as the exact inverse of the plant. In this way, the control-block output will be equal to the process input and the process output will follow the control-block input, which will be the desired process output.

The main differences from classical MRAS are that the new system needs no reference model for its implementation and does not require *a priori* knowledge about the parameters of the plant. The system realizes an identification on real time of the parameters of the plant. This identification is equivalent to the one made by a gradient parameter estimation method [6]–[8]. A general and conceptual configuration of the adaptive-control system appears in Fig. 1, where two possible modes of system operation are shown.

i) Following path 1, a human or an automatic operator can directly control the plant and identification-block input. The comparison between plant and identification-block outputs generates an error, e, which is used by the adaptive mechanism to adjust the parameters of the identification block in order to have $e \to 0$ as $t \to \infty$. In this mode the system behaves as an identification system.

ii) Following path 2, the system behaves as a proper control system. The operator sends the desired plant output d (desired trajectory) as input to the control block, which generates the input to the plant and identification block. Given that the control and identification blocks are governed by the same equation, the input to the control block is equal to the output of the identification block. Consequently, this mode of operation can be represented, using a discrete formulation, by the general configuration shown in Fig. 2.

The plant is considered to be output-controllable and can be described by the input–output (I/O) equation

$$y(k) = \sum_{i=1}^{h} A_i(k)y(k-i) + \sum_{j=1}^{r} B_j(k)u(k-j) \qquad (1)$$

where $y(k)$ and $u(k-j)$ are the output and input vectors of the process at sampling moment k and $k-j$, respectively, and h and r are integers depending upon the structure of the plant. It is assumed that the input u and output y are vectors of equal dimension, $n \times 1$.[1] $A_i(k)$ and $B_j(k)$ are $(n \times n)$ real matrices with a finite number of bounded changes in their values as $k \to \infty$, and with the further assumption that $B_1(k)$ is a nonsingular matrix for all k.

Let $d(k+1)$ be a vector of the same dimension as $y(k)$. It will be referred to as the "desired output," because we may assign to it the value of the desired output for the process at moment $k+1$. The control block generates the control sequence $u(k)$ from the input $d(k+1)$, using an adaptive model. The parameters of this adaptive model are a function of the error, $e(k)(=y(k) - d(k))$. Furthermore, this is the same adaptive model which is used by the identification block. The characteristic that defines the control system is the following.

The system is asymptotically hyperstable. The error $e(k)$ will be a measure of the system's deviation from its equilibrium point (at which $e(k) = 0$). If for any reason, such as the action of perturbations or the variation of the parameters of the process, the system is displaced from its equilibrium ($e(k) \neq 0$), then the system takes appropriate control action in order to reach its equilibrium again; in other words, $e(k) \to 0$ at all times and, consequently, $y(k) \to d(k)$.

To implement the system, it is only necessary to provide to the adaptive model used by the control and identification blocks: 1) a certain prior structure, which need not necessarily be the same as the structure of the plant; and 2) a reasonable initial parameter value.

In order to simplify the theoretical analysis, we will at first only consider a noise-free time-invariant process defined by (2) and further assume that there are no structural differences between the process and the adaptive model. Later on in Sections XI–XIV we will consider other cases and the general process described by (1).

II. Statement of the Problem

Suppose the process is governed by the equation

$$y(k) = Ay(k-1) + Bu(k-1). \qquad (2)$$

Manuscript received March 15, 1976; revised April 20, 1976. This work was supported by the Juan March Foundation of Spain.

The author was with the Electronic Systems Laboratory, M.I.T., Cambridge, MA. He is now at Calle Alava n° 75, Barcelona-5, Spain.

[1] If the dimension of u is bigger than the dimension of y, supplementary conditions can be added to make them equal.

Reprinted from *Proc. IEEE*, vol. 64, pp. 1209–1218, Aug. 1976.

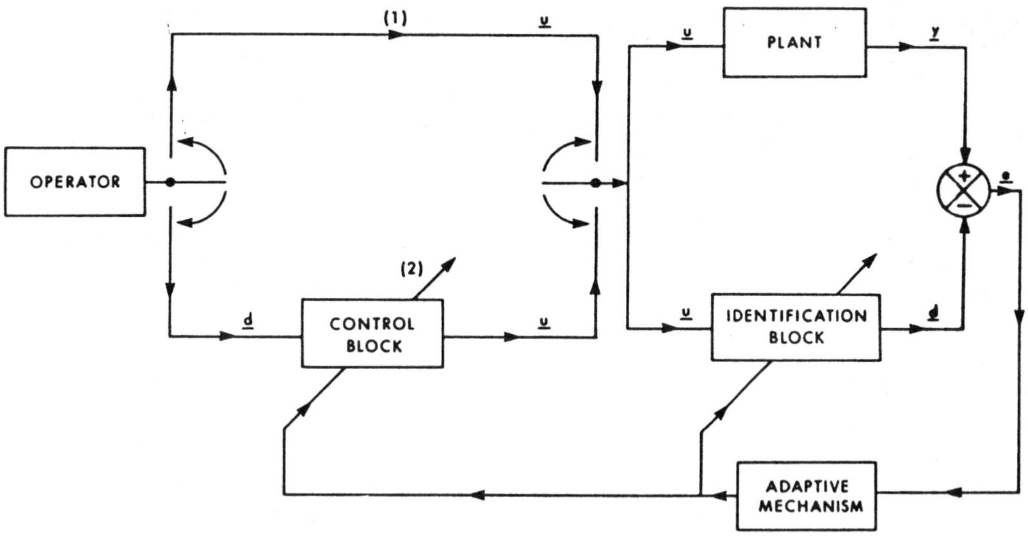

Fig. 1. General configuration of the adaptive control system.

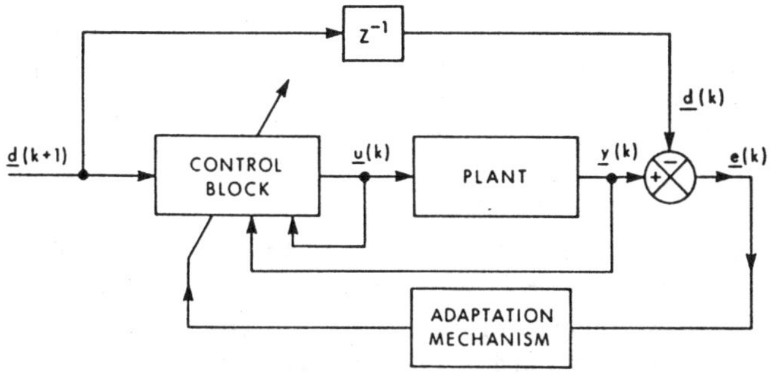

Fig. 2. General discrete structure of the adaptive control system in its control operation mode.

The identification block will be governed by

$$d(k) = \tilde{A}(k-1)y(k-1) + \tilde{B}(k-1)u(k-1) \tag{3a}$$

where $\tilde{A}(k-1)$ and $\tilde{B}(k-1)$ are estimates, at instant $k-1$, of the plant matrices A and B; $d(k)$ represents the predicted plant output, at instant k, from the information available at instant $k-1$.

The control block will be also governed by (3a), except that in this case the independent variable will be $d(k+1)$; i.e., equation (3a) can be rewritten as

$$u(k) = \tilde{B}(k)^{-1} d(k+1) - \tilde{B}(k)^{-1} A(k)y(k). \tag{3b}$$

In (3b), $d(k+1)$ represents the desired output for the process at the moment $k+1$. Consequently, the control-block output $u(k)$ as computed from (3b) ensures that the predicted plant output is equal to the desired output at any sampling moment.

The error will be defined by

$$e(k) = y(k) - d(k). \tag{4}$$

Let matrices $\tilde{A}(k)$ and $\tilde{B}(k)$ be generated by the following algorithms:

$$\tilde{A}(k) = \Delta\tilde{A}(k) + \tilde{A}(k-1). \tag{5}$$

$$\tilde{B}(k) = \Delta\tilde{B}(k) + \tilde{B}(k-1). \tag{6}$$

Using the matrix estimates $\tilde{A}(k)$ and $\tilde{B}(k)$, an estimate of the plant output at sampling-instance k, $g(k)$, can be obtained from the following equation:

$$g(k) = \tilde{A}(k)y(k-1) + \tilde{B}(k)u(k-1). \tag{7}$$

The error of this estimation will be defined by

$$s(k) = y(k) - g(k). \tag{8}$$

The problem to be solved is to find the adaptive mechanism that generates the incremental matrices $\Delta\tilde{A}(k)$ and $\Delta\tilde{B}(k)$ in order to verify that the whole control system has the desired properties of hyperstability, independently of the condition $(\tilde{A}(0), \tilde{B}(0), u(0), y(0))$ and for all possible desired outputs.

III. A Hyperstability Condition

Equation (8) can be considered as a linear transformation between (2) and (7), which establishes the equivalence between the whole control system and a nonlinear time-variant feedback autonomous system, such as that shown in Fig. 3, whose state can be defined by the vector $s(k)$.

A particular case of a Popovian theorem [2] extended to discrete systems [3] states that the equivalent feedback system mentioned above will be asymptotically hyperstable if the following condition is verified:

$$\sum_{k=0}^{k_1} s(k)' s(k) \leqslant \alpha_0^2, \qquad \alpha_0 = \text{constant } \forall k_1. \tag{9}$$

Therefore if the control system satisfies condition (9) it will be asymptotically hyperstable.

IV. Hyperstability of the Control System

The following theorem defines an adaptive mechanism that generates matrices $\tilde{A}(k)$ and $\tilde{B}(k)$ such that asymptotic hyperstability condition (9) is satisfied.

Theorem 1: The whole system defined by (2)–(8), and whose general configuration is shown in Fig. 1, will be asymptotically hyperstable if the matrices $\tilde{A}(k)$ and $\tilde{B}(k)$ are constructed as

$$\tilde{A}(k) = s(k)y(k-1)' + \tilde{A}(k-1) \tag{10}$$

$$\tilde{B}(k) = s(k)u(k-1)' + \tilde{B}(k-1). \tag{11}$$

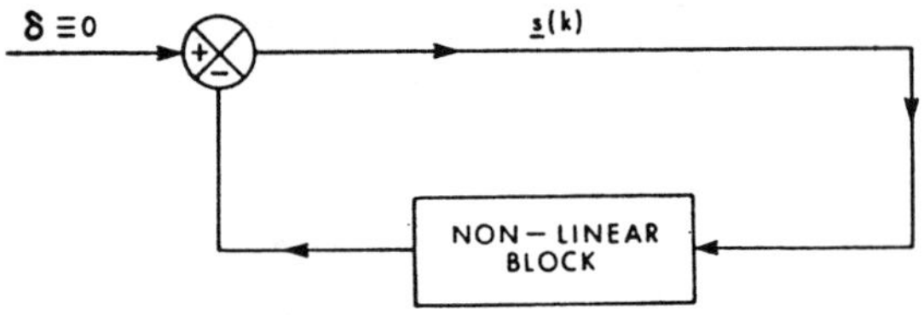

Fig. 3. Equivalent autonomous system.

Proof: From (2), (7), and (8),

$$s(k) = (A - \tilde{A}(k))y(k-1) + (B - \tilde{B}(k))u(k-1). \quad (12)$$

From (12), multiplying by $-s(k)'$,

$$-s(k)'s(k) = s(k)'(\tilde{A}(k) - A)y(k-1) + s(k)'(\tilde{B}(k) - B)u(k-1) \quad (13)$$

$$-\sum_{k=0}^{k_1} s(k)'s(k) = \sum_{k=0}^{k_1} s(k)'(\tilde{A}(k) - A)y(k-1) + s(k)'(\tilde{B}(k) - B)u(k-1). \quad (14)$$

From (14), condition (9) can be decomposed in the following two conditions:

$$\sum_{k=0}^{k_1} s(k)'(\tilde{A}(k) - A)y(k-1) \geqslant -\lambda_a^2 \ \forall k_1 \quad (15)$$

$$\sum_{k=0}^{k_1} s(k)'(\tilde{B}(k) - B)u(k-1) \geqslant -\lambda_b^2 \ \forall k_1 \quad (16)$$

where λ_a and λ_b are constants.

Conditions (15) and (16) can be written in scalar form:

$$\sum_{i=1}^{n} \sum_{j=1}^{n} \sum_{k=0}^{k_1} s_i(k)(\tilde{a}_{ij}(k) - a_{ij})\, y_j(k-1) \geqslant -\lambda_a^2 \ \forall k_1 \quad (17)$$

$$\sum_{p=1}^{n} \sum_{q=1}^{n} \sum_{k=0}^{k_1} s_p(k)(\tilde{b}_{pq}(k) - b_{pq})\, u_q(k-1) \geqslant -\lambda_b^2 \ \forall k_1. \quad (18)$$

These conditions will be satisfie.i if the parameters $\tilde{a}_{ij}(k)$ and $\tilde{b}_{pq}(k)$ are formed according to the algorithms

$$\tilde{a}_{ij}(k) = \sum_{h=0}^{k} s_i(h) y_j(h-1), \qquad i = 1, n \quad j = 1, n \quad (19)$$

$$\tilde{b}_{pq}(k) = \sum_{h=0}^{k} s_p(h) u_q(h-1), \qquad p = 1, n \quad q = 1, n. \quad (20)$$

Because, in this case, equations (17) and (18) correspond to a special case of the more general equation:

$$\sum_{k=0}^{k_1} x(k)\left(\sum_{h=0}^{k} x(h) + c\right) = \frac{1}{2}\left(\sum_{k=0}^{k_1} x(k) + c\right)^2 + \frac{1}{2}\sum_{k=0}^{k_1} x(k)^2 - \frac{1}{2}c^2 \geqslant -\lambda^2, \qquad c = \text{constant } \forall k_1, \quad \lambda = \text{constant} \quad (21)$$

equations (19) and (20) can be expressed in the form

$$\tilde{a}_{ij}(k) = s_i(k) y_j(k-1) + \tilde{a}_{ij}(k-1), \qquad i = 1, n \quad j = 1, n \quad (22)$$

$$\tilde{b}_{pq}(k) = s_p(k) u_q(k-1) + \tilde{b}_{pq}(k-1), \qquad p = 1, n \quad q = 1, n.^2 \quad (23)$$

Converting (22) and (23) to matrix form, we obtain (10) and (11). This proves the theorem.

In accord with Theorem 1, the equivalent feedback system is represented in Fig. 4.

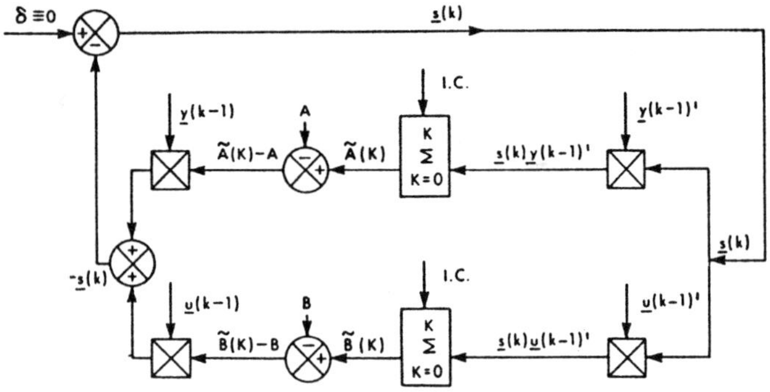

Fig. 4. Equivalent asymptotically hyperstable feedback autonomous system.

V. Relation Between the Error $e(k)$ and the State $s(k)$ of the Equivalent Feedback System

As a consequence of Theorem 1 the relation between $e(k)$ and $s(k)$ is established by the following: the state of the equivalent autonomous feedback system $s(k)$ is related to the error of the control system $e(k)$ by means of a variable gain $\alpha(k)$ defined by

$$\alpha(k) = 1/(1 + y(k-1)'y(k-1) + u(k-1)'u(k-1)). \quad (26)$$

Proof: From (10), (11), and (12),

$$s(k) = (A - \tilde{A}(k-1) - s(k)y(k-1)')y(k-1) + (B - \tilde{B}(k-1) - s(k)u(k-1)')u(k-1). \quad (27)$$

From (2), (3a), (4), and (27),

$$s(k) = e(k) - s(k)(y(k-1)'y(k-1) + u(k-1)'u(k-1)) \quad (28)$$

$$s(k) = e(k) \,/\, (1 + y(k-1)'y(k-1) + u(k-1)'u(k-1)). \quad (29)$$

From (29) and defining $\alpha(k)$ by (26),

$$s(k) = \alpha(k)e(k). \quad (30)$$

[2] Note that one more general expression for parameters $\tilde{a}_{ij}(k)$ and $\tilde{b}_{pq}(k)$ verifying the hyperstability conditions is

$$\tilde{a}_{ij}(k) = \beta_{aij}\, s_i(k) y_j(k-1) + \tilde{a}_{ij}(k-1) \quad (24a)$$

$$\tilde{b}_{pq}(k) = \beta_{bpq}\, s_p(k) u_q(k-1) + \tilde{b}_{pq}(k-1). \quad (25a)$$

β_{aij} and β_{bpq} being positive constant coefficients that can be chosen conveniently for each application.

Algorithms (10) and (11) can now be written in the form

$$\tilde{A}(k) = \alpha(k)e(k)y(k-1)' + \tilde{A}(k-1) \quad (31)$$

$$\tilde{B}(k) = \alpha(k)e(k)u(k-1)' + \tilde{B}(k-1). \quad (32)$$

VI. Convergence of Control Block Parameters

Convergence of the control-block parameters is deduced from the asymptotical hyperstability of the control system. This convergence can also be analyzed by the gradient parameter estimation technique and cast into the scalar equation error formulation.

The ith component of the state $s(k)$ of the equivalent feedback system of Fig. 4 is related to the corresponding component of the error vector $e(k)$ by

$$s_i(k) = \alpha(k)e_i(k) \quad (33)$$

where $\alpha(k)$ is defined by (26) and $e_i(k)$ by

$$e_i(k) = y_i(k) - d_i(k). \quad (34)$$

Let $\boldsymbol{\theta}_i$ and $\tilde{\boldsymbol{\theta}}_i(k-1)$ be

$$\theta_i = [a_{i1}, \cdots, a_{in}, b_{i1}, \cdots, b_{in}]' \quad (35)$$

$$\tilde{\theta}_i(k-1) = [\tilde{a}_{i1}(k-1), \cdots, \tilde{a}_{in}(k-1), \tilde{b}_{i1}(k-1), \cdots, \tilde{b}_{in}(k-1)]' \quad (36)$$

and let $x(k-1)$ be

$$x(k-1) = [y_1(k-1), \cdots, y_n(k-1), u_1(k-1), \cdots, u_n(k-1)]'. \quad (37)$$

From (2), (35), and (37),

$$y_i(k) = x(k-1)'\boldsymbol{\theta}_i. \quad (38)$$

From (3a), (36), and (37),

$$d_i(k) = x(k-1)'\tilde{\boldsymbol{\theta}}_i(k-1). \quad (39)$$

The algorithms previously found for the adaptive parameters of the control and identification blocks and related to the component i of the error vector can be written in the following form:

$$\tilde{\boldsymbol{\theta}}_i(k) = \tilde{\boldsymbol{\theta}}_i(k-1) + \alpha(k)e_i(k)x(k-1). \quad (40)$$

These algorithms are in accordance with a gradient parameter estimation method, which minimizes the following function:

$$j[\tilde{\boldsymbol{\theta}}_i(k)] = \tfrac{1}{2}e_i(k)^2. \quad (41)$$

From (40) and (34),

$$\tilde{\boldsymbol{\theta}}_i(k) = \tilde{\boldsymbol{\theta}}_i(k-1) + \alpha(k)x(k-1)(y_i(k) - d_i(k)). \quad (42)$$

From (42), (38), and (39),

$$\tilde{\boldsymbol{\theta}}_i(k) = \tilde{\boldsymbol{\theta}}_i(k-1) + \boldsymbol{\alpha}(k)x(k-1)x(k-1)'[\boldsymbol{\theta}_i - \tilde{\boldsymbol{\theta}}_i(k-1)]. \quad (43)$$

Let $\hat{\boldsymbol{\theta}}_i(k)$ be the identification error

$$\hat{\boldsymbol{\theta}}_i(k) = \boldsymbol{\theta}_i - \tilde{\boldsymbol{\theta}}_i(k). \quad (44)$$

The finite-difference equation for parameter identification error is obtained from (43) and (44):

$$\hat{\boldsymbol{\theta}}_i(k) = [I - \boldsymbol{\alpha}(k)x(k-1)x(k-1)']\hat{\boldsymbol{\theta}}_i(k-1). \quad (45)$$

Using (45) and the contraction mapping approach [7], we can prove that the square of the Euclidean norm of the identification error $\|\hat{\boldsymbol{\theta}}_i(k)\|^2$ will contract from iteration to iteration and lim $\|\hat{\boldsymbol{\theta}}_i(k)\|^2 \to 0$, except in the case of orthogonality between $x(k)$ and $\hat{\boldsymbol{\theta}}_i(k)$. Similarly, the stability theory approach [8] permits us to conclude that "almost always [unless $x(k)$ and $\hat{\boldsymbol{\theta}}_i(k)$ are orthogonal] $\tilde{\boldsymbol{\theta}}_i(k) \to \theta_i$ as $k \to \infty$, regardless of the initial estimate $\tilde{\boldsymbol{\theta}}_i(0)$." It is possible to prove this statement by considering the square of the Euclidean norm of the identification error as a scalar Lyapunov function of the identification error system.

VII. Orthogonality Between Parameter Identification Error and I/O Vector

The condition of orthogonality depends on the frequency content of $x(k)$, as can be seen in [9].

Orthogonality between $\hat{\boldsymbol{\theta}}_i(k-1)$ and $x(k-1)$ implies that

$$e_i(k) = \hat{\boldsymbol{\theta}}_i(k-1)'x(k-1) = 0 \quad (46)$$

and from (30),

$$s_i(k) = 0. \quad (47)$$

This means that the autonomous hyperstable feedback system has reached its equilibrium point. However, this is only a local equilibrium since it is dependent on the orthogonality condition considered above. The absolute equilibrium will be reached when $\hat{\boldsymbol{\theta}}_i(k) = 0$, i.e., when the left-hand side of the inequality (9) reaches its maximum value.

When $x(k)$ and $\hat{\boldsymbol{\theta}}_i(k)$ are orthogonal, the identification error system is uniformly stable in the large, but is not uniformly asymptotically stable in the large, meaning that $\tilde{\boldsymbol{\theta}}_i(k)$ may not converge to $\boldsymbol{\theta}_i$ as $k \to \infty$. This is the reason why, from the identification point of view, the above mentioned orthogonality has been considered undesirable. But it is important to emphasize that in the method presented here the orthogonality condition makes it possible to control the plant in the desired way without a complete identification of its parameters. Consequently, it is enough to get an "identification with a view to the control."

VIII. Relation Between Control and Stability

In Theorem 1 the hyperstability conditions were stated in terms of the state $s(k)$. However, the performance of the control system really depends on the error vector $e(k)$. Theorem 2 will establish the relation between $e(k)$ and the stability characteristic of the system. To prove this theorem we will make use of the following two assumptions.

Assumption 1: The desired output will be always bounded.

Assumption 2: $\tilde{B}(k)$ is nonsingular for all k. Note that this is a fair assumption considering a reasonable initial parameter value.

Theorem 2: The control system defined by equations (2)–(8), (10), and (11) is asymptotically hyperstable. The control law imposed by the control block, defined by (3b), makes the output of the process equal to the desired output when the control system is at the equilibrium point. Away from the equilibrium point, the process output approaches the desired output as the system approaches equilibrium, i.e., $e(k) \to 0$.

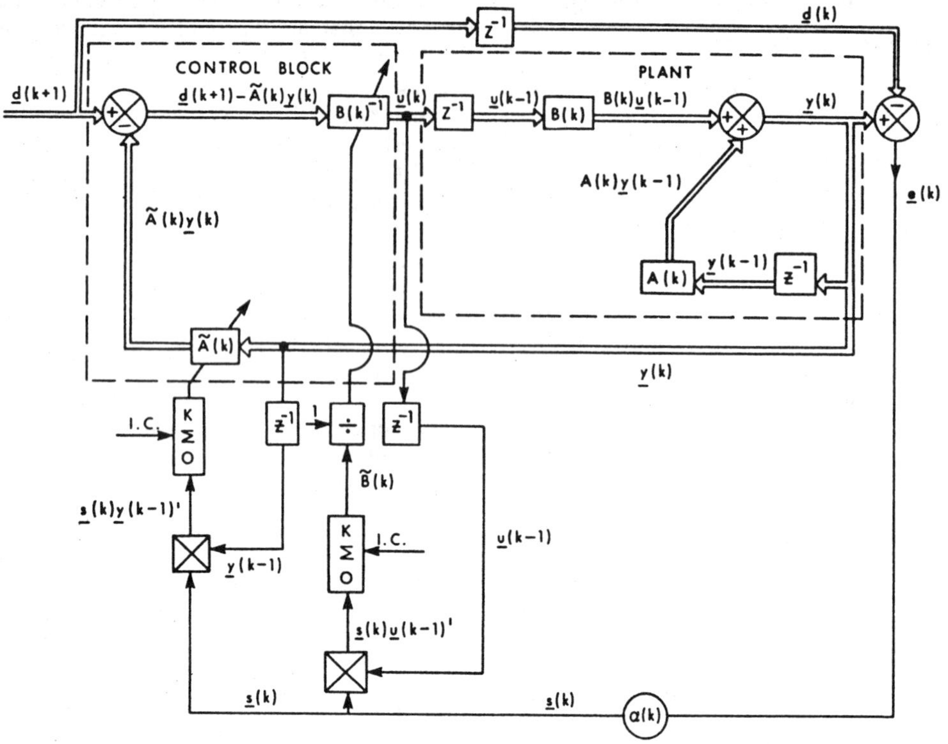

Fig. 5. Detailed flow diagram of the system in its control role: double line indicates control path, single line indicates adaptation mechanism.

Proof: From (2), (3a), and (4), the error vector can be written in the scalar form

$$e_i(k) = \sum_{j=1}^{n} (a_{ij} - \tilde{a}_{ij}(k-1))y_j(k-1)$$

$$+ \sum_{j=1}^{n} (b_{ij} - \tilde{b}_{ij}(k-1))u_j(k-1), \qquad i = 1, \cdots, n. \quad (48)$$

Similarly, from (12),

$$s_i(k-1) = \sum_{j=1}^{n} (a_{ij} - \tilde{a}_{ij}(k-1))y_j(k-2)$$

$$+ \sum_{j=1}^{n} (b_{ij} - \tilde{b}_{ij}(k-1))u_j(k-2), \qquad i = 1, \cdots, n. \quad (49)$$

In order to prove that $e_i(k) \to 0$ when $s_i(k-1) \to 0$, let us consider that $e_i(k)$ does not tend to zero. In this case, the previously analyzed parameter convergence states that $\tilde{a}_{ij}(k-1) \to a_{ij}$ and $\tilde{b}_{ij}(k-1) \to b_{ij}$ when $k \to \infty$. Consequently, from (48), at least one of the $y_j(k-1)$ or one of the $u_j(k-1)$ tends to infinite faster than the correspondent term $(a_{ij} - \tilde{a}_{ij}(k-1))$ or $(b_{ij} - \tilde{b}_{ij}(k-1))$ tends to zero.

From assumptions 1 and 2 and (2) and (3a), it can be deduced that the terms $y_j(k-1)$ and $u_j(k-1)$ $(j = 1, \cdots, n)$ cannot tend to infinite faster than the correspondent terms $y_j(k-2)$ and $u_j(k-2)$. Therefore, from (49), $s_i(k-1)$ will not tend to zero and the hyperstability condition will be violated, which is inconsistent with Theorem 1 previously proven. This proves Theorem 2.

IX. Passive or Identification Action

When the system follows path 1 in Fig. 1, the operation cycle to be carried out in real time at each instant k will be as follows:

1) application of $u(k)$;
2) measurement of $y(k)$;
3) calculation of $d(k)$ from (3a);
4) calculation of $e(k)$ by (4);
5) calculation of $s(k)$ by (26) and (30);
6) calculation of the value of the matrices $\tilde{A}(k)$ and $\tilde{B}(k)$ according to (10) and (11).

This computation can be performed on line or off line.

X. Active or Control Action

When the system behaves as a proper control system, in accordance with Fig. 2, the operation cycle to be carried out in real time at each instant k will be as follows:

1) measurement of $y(k)$;
2) calculation of $e(k)$ by (4);
3) calculation of $s(k)$ by (26) and (30);
4) calculation of $\tilde{A}(k)$ and $\tilde{B}(k)$ according to (10) and (11);
5) application of control $u(k)$, calculated from (3b), where $d(k+1)$ is the plant output desired at the moment $k+1$.

Fig. 5 shows a detailed scheme for implementing the control system. As a starting procedure for the control system, it is convenient that, initially, the system behaves in identification mode so that the control block obtains a reasonable set of parameters.

Once the control action is started, the system can take passive or identification action in parallel to the control action.

XI. Plant with Variable Parameters

Let the plant be governed by the equation

$$y(k) = A(k)y(k-1) + B(k)u(k-1) \tag{50}$$

$A(k)$ and $B(k)$ being matrices with a finite number of bounded changes in their values as $k \to \infty$. Similarly, as in Section II, let the control and identification blocks be governed by (3a). In this case, the control system will have the desired hyperstability properties if the matrices $\tilde{A}(k)$ and $\tilde{B}(k)$ verify the following conditions:

$$\sum_{k=0}^{k_1} s(k)'(\tilde{A}(k) - A(k))\,y(k-1) \geqslant -\lambda_a^2, \qquad \lambda_a = \text{constant} \tag{51}$$

$$\sum_{k=0}^{k_1} s(k)'(\tilde{B}(k) - B(k))\,u(k-1) \geqslant -\lambda_b^2, \qquad \lambda_b = \text{constant.} \tag{52}$$

Conditions (51) and (52) can be written in the scalar form

$$\sum_{i=1}^{n}\sum_{j=1}^{n}\sum_{k=0}^{k_1} s_i(k)(\tilde{a}_{ij}(k) - a_{ij}(k))\,y_j(k-1) \geqslant -\lambda^2{}_{aij} \tag{53}$$

$$\sum_{p=1}^{n}\sum_{q=1}^{n}\sum_{k=0}^{k_1} s_p(k)u_q(k-1)(\tilde{b}_{pq}(k) - b_{pq}(k)) \geqslant -\lambda_{bpq}^2. \tag{54}$$

If $\tilde{a}_{ij}(k)$ and $\tilde{b}_{pq}(k)$ are formed in accordance with algorithms (24) and (25), they will be convergent as $k \to \infty$. Using an argument similar to the one used in Section IV, Theorem 1, the hyperstability conditions (53), (54) will be satisfied, provided that

$$\sum_{k=0}^{k_1} x(k)\left(\sum_{h=0}^{k} x(h) + c(k)\right) = \frac{1}{2}\left(\sum_{k=0}^{k_1} x(k)\right)^2 + \frac{1}{2}\sum_{k=0}^{k_1} x(k)^2 + \sum_{k=0}^{k_1} c(k)x(k) \geqslant -\lambda_N^2. \tag{55}$$

To prove (55), we note that the first two terms on the right-hand side are positive. For the third term, using the fact that $c(k)$ will have only a finite number of bounded changes and that $\sum_{k=0}^{\infty} x(k)$ converges, it is easy to see that it is bounded from below. Denoting $-\lambda_N^2$ as the minimum between zero and that lower bound, we get (55).

Theorem 2 can be easily extended to this case, considering that the changes in the value of matrices $A(k)$ and $B(k)$ will not affect assumption 2.

XII. Generalizations

For the sake of simplicity, we have discussed systems governed by an equation such as (2). The method is easily extended to the general form described by (1), where the control and identification blocks will be governed by (56):

$$d(k) = \sum_{i=1}^{h} \tilde{A}_i(k-1)y(k-1) + \sum_{j=1}^{r} \tilde{B}_j(k-1)u(k-j). \tag{56}$$

For this general case, the desired hyperstability properties can be formally stated by the following theorem.

Theorem 3: When the process and the control and identification blocks are described by (1) and (56), respectively, the whole control system, whose configuration is shown in Fig. 1, will be asymptotically hyperstable if the components $\tilde{a}_{ipq}(k)$ $(i = 1, \cdots, h;\ p = 1, \cdots, n;\ q = 1, \cdots, n)$ and $\tilde{b}_{jpq}$ $(j = 1, \cdots, r;\ p = 1, \cdots, n;\ q = 1, \cdots, n)$ of the matrices $\tilde{A}_i(k)$ $(i = 1, \cdots, h)$ and $\tilde{B}_j(k)$ $(i = 1, \cdots, r)$, respectively, are constructed as

$$\tilde{a}_{ipq}(k) = \beta_{aipq}s_p(k)y_q(k-i) + \tilde{a}_{ipq}(k-1) \tag{24b}$$

$$\tilde{b}_{jpq}(k) = \beta_{bjp}s_p(k)u_q(k-j) + \tilde{b}_{jpq}(k-1) \tag{25b}$$

where $s_p(k)$, $y_q(k-i)$ and $u_q(k-j)$ are the components p, q of the vectors $s(k)$, $y(k-i)$ and $u(k-j)$, respectively. β_{aipq} and β_{bipq} are positive constant coefficients.

The proof of this theorem, taking into account Section XI, is exactly the same as the proof for Theorem 1. An analogous form of Theorem 2 also exists for this general case.

In order to compute the control $u(k)$ from (56), the matrix $\tilde{B}_1(k)$ would have to be inverted. There is, however, a different way of obtaining $u(k)$ from the parameter identification process without having to invert $\tilde{B}_1(k)$.

Equation (1) can also be written in the form

$$u(k) = C_1(k+1)\,y(k+1) + \sum_{i=2}^{h+1} C_i(k+1)\,y(k+2-i) + \sum_{j=1}^{r-1} D_j(k+1)\,u(k-j). \tag{57}$$

Similarly, the identification- and control-block equation (corresponding to (3a) in the particular case) will be replaced by

$$u(k) = \tilde{C}_1(k)\,d(k+1) + \sum_{i=2}^{h+1} \tilde{C}_i(k)\,y(k+2-i) + \sum_{j=1}^{r-1} \tilde{D}_j(k)u(k-j) \tag{58}$$

$$u_1(k) = \tilde{C}_1(k)\,y(k+1) + \sum_{i=2}^{h+1} \tilde{C}_i(k)\,y(k+2-i) + \sum_{j=1}^{r-1} \tilde{D}_j(k)u(k-j). \tag{59}$$

Equations (57) and (59) will be used to obtain the identification error

$$\epsilon(k) = u(k) - u_1(k). \tag{60}$$

This error will be used for the purpose of identifying matrices $\tilde{C}_i(k)$ and $\tilde{D}_j(k)$. But the control $u(k)$ will be computed from (58), where it appears explicitly as a function of the matrices $\tilde{C}_i(k)$ and $\tilde{D}_j(k)$ previously identified.[3]

[3] The hyperstability conditions can also be rigorously demonstrated for this implementation of the method. This implementation and the direct one requiring inversion of $\tilde{B}_1(k)$ have both been applied in practice successfully.

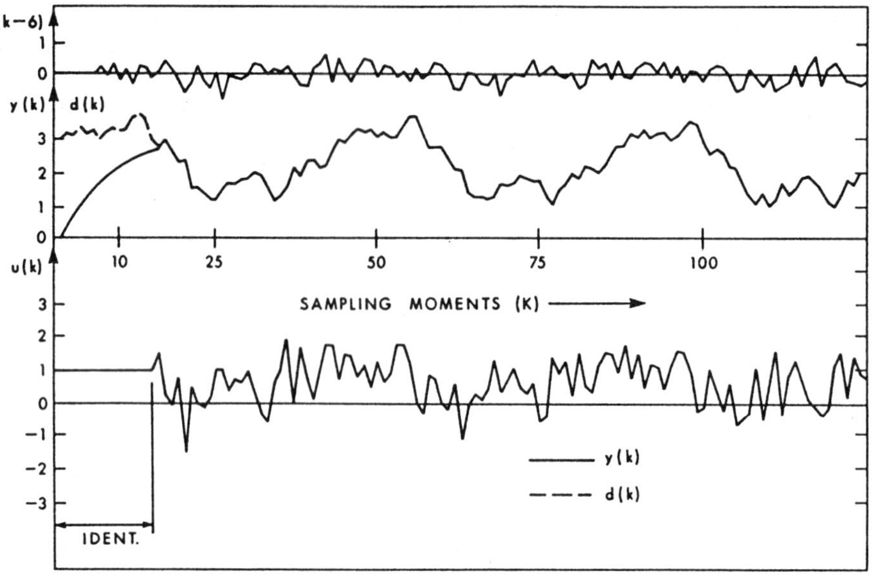

Fig. 6. Control in the absence of noise.

XIII. Structural Difference and Presence of Noise

The main characteristic of the identification, with a view to the control, is the following: when the control system is in the equilibrium point, the matrices $\tilde{A}_i(k)$ and $\tilde{B}_j(k)$ are equivalent representations of $A_i(k)$ and $B_j(k)$ with respect to the inputs and outputs, or equal to $A_i(k)$ and $B_j(k)$.

The equivalence considered above holds for the orthogonality between the identification error matrix and plant output and input vectors. When the frequency contents of the vector $x(k)$, whose components are the inputs and outputs of the plant, result in this orthogonality condition, the equivalence can be held even if there is a structural difference between the control-block matrices and the plant matrices. Therefore, the control system will be able to reach the equilibrium despite the structural differences.

The control will be separated from the equilibrium when the frequency contents of the desired output is such that the orthogonality conditions considered above do not exist. In spite of this, the control system always keeps the characteristic of asymptotic hyperstability and thus tends to equilibrium. In this case, a theoretical analysis can be made by considering the control system equivalent to an asymptotically hyperstable nonautonomous feedback system.

If the error should become intolerable, it will only be necessary to increase the order of the adaptive model, this being a simple matter for this control method.

The effects of the noise can be considered from the study of one equivalent hyperstable feedback nonautonomous system. The noise that acts upon the system displaces it from the equilibrium and consequently disturbs the control. However, given its property of hyperstability the system tends to return to equilibrium continually. From previous experience, despite the effect of noise measurements of the input and output, control has always been satisfactory. Of course the control performance depends on the levels of noise, and an adequate choice of the sampling time is necessary to minimize its perturbing effect.

XIV. Examples

Some examples will now be presented that serve to show the performance of the method on an ill-conditioned system that may be encountered in real applications. All examples have been carried out using digital simulation.

A. Control in the Absence of Noise

Consider the following noise-free monovariable process:

$$T_p(Z) = 0.4Z^{-1} - 0.3Z^{-2}/1-1.6Z^{-1} + 0.63Z^{-2}$$

The desired output, which we have imposed, is the integral of a Gaussian noise of zero mean value and standard deviation 0.2 starting from an initial value of 3. Fig. 6 shows the results obtained.

The first graph shows the Gaussian noise used for generating the desired output displaced six sample steps in advance.

The second graph represents the evolution of the process output and the desired output. As shown, both coincide practically from the beginning of the control action. In each of the examples considered below, a step input is introduced during the initial stage for the purpose of identification and, during this period, the represented desired output has no significance.

The third graph shows the action applied to the process. For the three graphs, the abscissas represent the number of sample periods.

B. Control in the Presence of Noise

For this case, the monovariable process considered is given by the transfer function

$$T_p(z) = 0.5Z^{-1} + 0.3Z^{-2}/1-1.2Z^{-1} + 0.35Z^{-2}.$$

The process is disturbed by noise added to its input and output. The desired output imposed is the same as the one considered in the preceding example. Fig. 7 shows the results.

The first graph shows the noise that is added to the process input, which is the same noise that is added to the output but displaced three sampling steps in advance. Both are Gaussian noises of zero mean value and standard deviation 0.2.

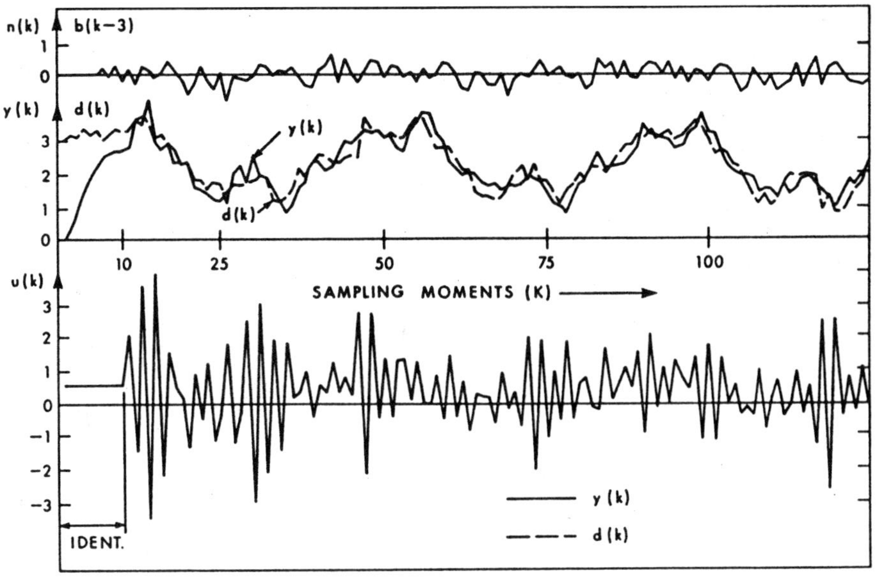

Fig. 7. Control in the presence of noise.

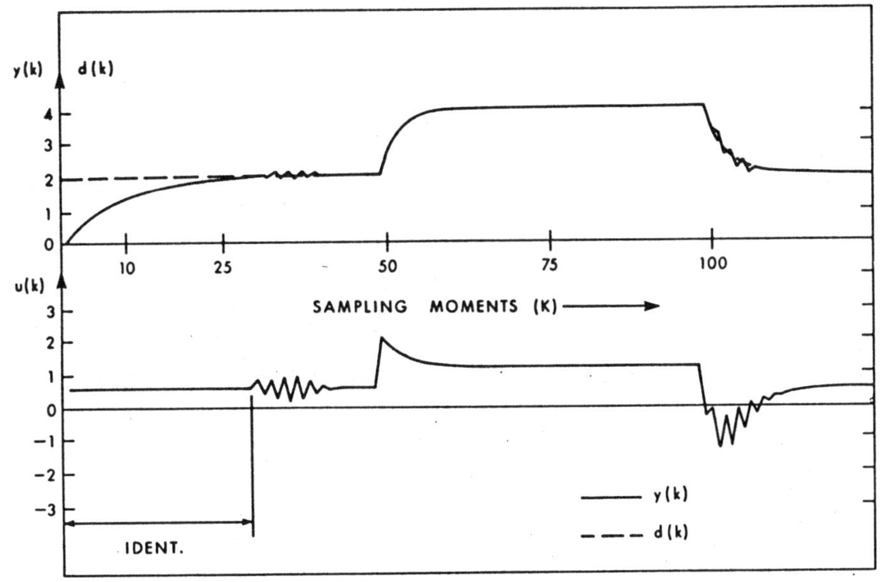

Fig. 8. Regulation system with change in the process' parameters.

The second graph shows the evolution of the process output and the desired output. The third graph shows the control action applied to the process.

C. Performance of the Regulation System with an Abrupt Change of the Process Parameters

The results obtained can be seen in Fig. 8. The first graph shows the desired output and the process output. The desired output changes from a value of 2 to 4 and later from 4 to 2. The dynamics of these changes can be described by a system whose transfer function is

$$T_m(Z) = 0.6Z^{-1}/1\text{-}0.7Z^{-1}$$

and whose time constant is three times the sample period. The process corresponds to the following transfer function up to sample step 75:

$$T_{p1}(Z) = 0.4Z^{-1} - 0.3Z^{-2}/1 - 1.6Z^{-1} + 0.63Z^{-2}.$$

Starting from step 75 the transfer function of the process is

$$T_{p2}(Z) = 0.4Z^{-1} - 0.3Z^{-2}/1 - 1.2Z^{-1} + 0.23Z^{-2}$$

which has much slower dynamics, because its most significant time constant is approximately 23 sample steps. However, the steady-state gains of the systems are the same. The second graph shows the process input.

D. Performance of the Regulation System with an Abrupt Change of the Process Parameters in the Presence of Noise.

This experiment is analogous to the one performed in example *C*, except that in this case there is a measurement noise over the process output. The results obtained are presented in Fig. 9.

The first graph shows the noise that acts on the process output, which is a zero-mean Gaussian noise with a standard deviation of 0.137.

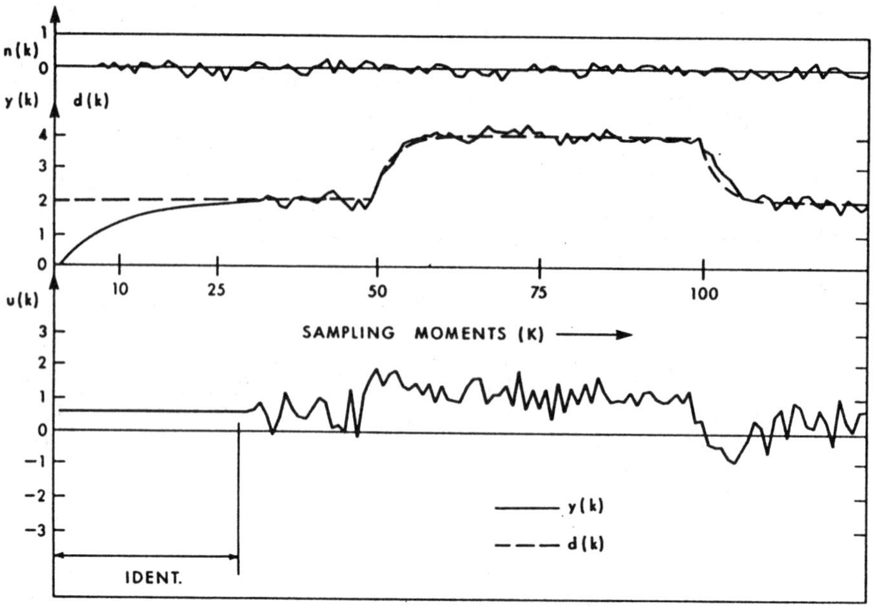

Fig. 9. Regulation system with change in the process' parameters in the presence of noise.

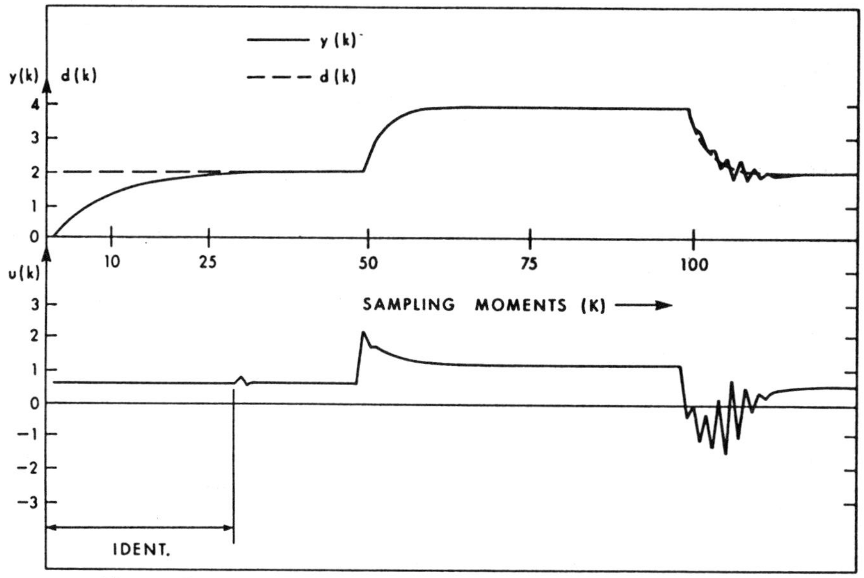

Fig. 10. Structural difference between plant and control block and change of plant parameters.

E. Performance with a Structural Difference and Variation of the Process Parameters.

In all previous examples, the control block had the same structure as the process. Example *C* is now repeated with a control block that has a first-order structure. The results are shown in Fig. 10.

F. Multivariable Control: Two inputs–one output

The output of the multivariable process is the sum of outputs of two monovariable systems, whose inputs are the inputs of the multivariable process to be considered.

The results obtained can be seen in Fig. 11. The desired output is the same as the one used in example *A* and, as can be seen, the process output follows the desired output right from the start of the control action.

To obtain only one solution, an additional condition on the inputs has been imposed, as can be seen in Fig. 11.

XV. Conclusions

Fig. 5 shows the similarity in structure between the adaptive control system presented here and the classical structure of the control system from the optimal control theory. The principal characteristics of the new adaptive control system are the following.

1) Knowledge of the plant's parameters or its variation with time is not required. For practical applications, no *a priori* data are required.

2) The system only uses the information given by plant inputs and outputs and does not require an estimation of the plant state variables.

3) The adaptive mechanism generates matrices $\tilde{A}(k)$ and $\tilde{B}(k)$ by means of simple arithmetic operations and the calculation of the control signal is also extremely simple.

4) When the control system is in its equilibrium point, the process output is equal to the desired output. In this way, the

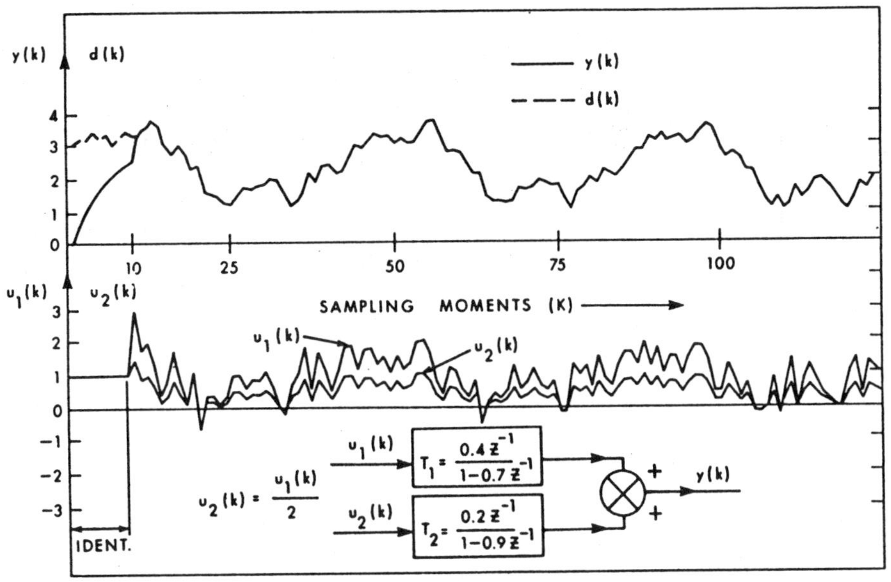

Fig. 11. Multivariable control.

main control problem becomes the design of the desired output, in accord with the limitations of the plant and the desired specifications.

5) We can always modify the desired output, in real time, making it possible to take care of severe perturbations that may act on the system.

The main advantages of this method over MRAS are:

a) no reference model is needed;

b) *a priori* knowledge of the plant parameters is not required;

c) it can work with structural differences;

d) it uses only I/O data, consequently no state estimation or adaptive observer is required.

The new system takes into account perturbations, disturbance inputs, and measurement noise. The rapidity of self-adaptation of the method and the orthogonality properties mentioned in Section VII enable it to solve the important problems of structural differences and variation of parameters. The examples that have been presented illustrate the potential and validity of the method.

In addition, the adaptive control system presented has the following characteristics in relation to its application.

Stability: Given its conception, the system, if implemented in a proper way, will maintain its stability in the presence of the perturbations (noise, structural changes, etc.) acting upon it.

Generality: It may be applied as readily to regulatory control as to servo control, to monovariable plants as to multivariable ones, and using digital or analog computation. At the same time, the system provides an "identification with a view to the control" of the plant in real time. This identification can be stopped if the error signal remains lower than a pre-established limit.

Additional work on this subject is needed which may yield interesting results:

1) introduction of stochastic processes;

2) analysis of specific nonlinearities such as time delays;

3) practical implementation problems;

4) comparative analysis of the various algorithms that can be used to verify the hyperstability conditions or the control block parameter convergence criterions;

5) a rigorous proof for the nonsingularity of $\tilde{B}_1(k)$, which would be convenient to have. (The experience gained with the method so far suggests that assumption 2, in Section VIII, is a fair one.)

With respect to the third area, the author is currently engaged in the implementation of a pitch-loop autopilot system for a supersonic research aircraft (NASA F-8 fly-by-wire test bed) based on the new concept. A future paper will present the results of that effort. Practical results obtained so far confirm the expected theoretical performance of the method.

Acknowledgment

The author is very grateful for the assistance rendered him by Prof. G. Ferraté of the Universidad Politécnica de Barcelona and Prof. M. Athans of the Massachusetts Institute of Technology.

References

[1] S. S. Codbole and C. F. Smith, "A new control approach using the inverse system," *IEEE Trans. Automat. Contr.*, pp. 698-720, Oct. 1972.

[2] V. M. Popov, "The solution of a new stability problem for controlled systems," *Automation and Remote Control*, vol. 24, pp. 1-23, Jan. 1963.

[3] I. D. Landau, "Synthesis of discrete model reference adaptive systems," *IEEE Trans. Automat. Contr.*, vol. AC-16, pp. 507-508, Oct. 1971.

[4] I. D. Landau and J. M. Martin-Sanchez," "Sistemas adaptivos discretos con modelo de referencia, analysis y sintesis a partir de la teoría de la hiperestabilidad," in *Congr. Automatic 72*, Barcelona, Spain, pp. 1183-1211, Oct. 1972.

[5] J. M. Martin-Sanchez, "Aportacion a los sistemas adaptativos con modelo de referencia a partir de la teoría de la hiperestabilidad," Tesis Doctoral, Universidad Politecnica de Barcelona (U.P.B.), Sept. 1974.

[6] J. M. Mendel, "Gradient, error-correction identification algorithms," *Inform. Sci.*, vol. 1, pp. 23-42, 1968.

[7] J. Nagumo and A. Nodd, "A learning method for system identification," *IEEE Trans. Automat. Contr.*, vol. AC-12, pp. 282-287, 1967.

[8] J. M. Mendel, *Discrete Techniques of Parameter Estimation: The Equation Error Formulation.* New York: Marcel Dekker, 1973.

[9] P. N. Lion, "Rapid identification of linear and nonlinear systems," *AIAA J.*, vol. 5, pp. 1835-1842, 1967.

[10] I. D. Landau, "A survey of model reference adaptive techniques (Theory and applications)," *Automatica*, vol. 10, pp. 353-379, 1974.

Optimality in Discrete Adaptive Systems

Ya. Z. Tsypkin
Institute of Control Sciences, Moscow, U.S.S.R.

Abstract. The problem of generating optimal adaptation algorithms in direct and indirect adaptive systems for control of a linear dynamic process with a specified level of a priori data on the noise is stated and solve.

The properties and specifics of these algorithms are determined. As particular cases all conventional, generally non-optimal, algorithms are obtained.

Keywords. Discrete adaptive systems, optimality, criterial optimality in a class, identification, prediction, recurrent algorithms.

INTRODUCTION

Adaptive control, or control of dynamic plants under uncertainty when no data are available on its parameters and accurate characteristics of the disturbances, is in the focus of attention of today's control theory.

In the context of extensive use of micro-computers, discrete adaptive control systems (DACS) are of special interest. DACS's design under uncertainty includes the determination of an optimal, in terms of a specified criterion, structure and using an adaptation algorithm. This algorithm which processes the observable quantities (the input and output quantities and the setpoint) changes the controller parameters so as to minimize the criterion [1, 2]. Until very recently choice of the best DACS adaptation algorithms was not discussed. In what follows this problem is stated and solved. The justification of the best optimization and identification algorithms is the subject of Refs [3-7]. These algorithms operate through argument optimization and are to find an optimal solution (argument) which minimizes the criterion (mean losses). The discussion will be devoted to criterion optimization the objective of which is to minimize the criterion. This minimum may be found on the set of optimal solutions [8, 9].

1 STATEMENT OF THE PROBLEM

Let us have a look at a dynamic plant which is described by the linear difference equation

$$Q(q)y(n)=qP_u(q)u_n+P_\xi(q)\xi(n), \quad (1)$$

where q is the delay operator; $q^m x(n)=x(n-m)$; $y(n)$ is the output quantity; $u(n)$ is the control signal; $\xi(n)$ is a disturbance signal, a stationary sequence of random quantities such that $M\{\xi(n)\}=0$; $M\{\xi(n)\xi(n-m)\}=0$ with $m\neq 0$ and $M\{\xi^2(n)\}=\sigma_\xi^2$. $Q(q)$, $P_u(q)$, and $P_\xi(q)$ are polynomials of q of N-th power, N_1-th, N_2-th respectively and such that $Q(0)=P_\xi(0)=1$ and $P_u(0)=\sigma_0$.

The dynamic plant (1) must be controlled so that with a stationary random setpoint $r(n)$ the control system would behave in a way most closest to that of a system described by the difference equation

$$G^0(q)y_0(n)=qH^0(q)r(n), \quad (2)$$

where $G^0(q)$ and $H^0(q)$ are specified N_3-th, and N_4-th power, respectively, polynomials of q.

The closeness of $y(n)$ and $y_0(n)$ is usually measured in terms of the quadratic criterion

$$J=M\{[G^0(q)e(n)]^2\}, \quad (3)$$

where $e(n)=y(n)-y_0(n)$ is the error.

Design of an optimal DACS proceeds in two stages. Stage 1 is design of the optimal DACS structure and stage 2 determining the optimal adaptation algorithm with the highest possible asymptotic criterial rate of convergence.

Stage 1 is performed in a well-known way with specified parameters of the plant (1); it determines an optimal plant structure [1-10]. For a stable and minimal phase plant the optimal control function which minimizes the criterion (3) is given by the equation

$$P_u(q)u(n)=P_\xi(q)G^0(q)y_0(n+1)-P(q)y(n) \quad (4)$$

or, by virtue of (2),

$$P_u(q)u(n)=P_\xi(q)H^0(q)r(n)-P(q)y(n), \quad (5)$$

where

$$P(q)=q^{-1}(G^0(q)P_\xi(q)-Q(q)). \quad (6)$$

With optimal control

$$G^0(q)e(n)=\xi(n);\ J_*=\inf J=M\{\xi^2(n)\}=\sigma_\xi^2.$$

Stage 2 is the specific subject of this paper.

2 TYPES OF ADAPTIVE SYSTEMS

All DACS's can be classified into direct and indirect. Indirect DACS's identify the plant from observations of its state and the estimates of the plant parameters are used in computing the controller parameters by the polynomial equation (6).

Reprint with permission from *Preprints of the 9th World Congress of IFAC*, 1984, pp. 29–32, vol. X.

In direct DACS's the observed states of a closed-loop system lead directly to estimates of the controller parameters. A similar classification is employed in [12, 13] and is different from the commonly used classification of DACS's into self-adjusting systems and adaptive systems with a reference model [11, 13].

Indirect DACS's are made optimal through selection of their optimal structure and optimal identification algorithms [5-7] and direct DACS's through optimal prediction or adaptation algorithms. The controller equation (5) can be represented as

$$u(n) = \frac{1}{b_0(n)}\left[\bar{r}(n)-\alpha^T(n)Z_0(n;\bar{r})\right] \quad (7)$$

The observation vector

$$Z_\nu(n;s)=(\nu u(n),u(n-1),\dots,u(n-N_1),$$

$$s(n-1+\nu),\dots,s(n-Np+\nu)-y(n+\nu),\dots,-y(n-N+\mu))^T$$

enters (7) with $\nu=0$, $p=4$, $s(n)=\bar{r}(n)$, and $\bar{r}(n)=H^0(q)r(n)$. The components of the parameter vector α are estimates of unknown parameters of the polynomials $P_u(q)$, $P_\xi(q)-1$, and $P(q)$. It is required to generate algorithms which ultimately change the controller parameters $\alpha(n)$ so as to minimize the criterion (3) as soon as possible.

3 ALGORITHMS AND THEIR PERFORMANCE

The performance of adaptive algorithms can be measured by an asymptotic error covariance matrix (AECM)

$$V = \lim_{n\to\infty} nM\{(\alpha(n)-\alpha^*)(\alpha(n)-\alpha^*)^T\}$$

or the asymptotic deflection

$$w = \lim_{n\to\infty} n[MJ^a(\alpha(n))-J^a_*]$$

where

$$J^a(\alpha(n))=M\{e_a^2(n)\};\quad J^a_*=\min J^a(\alpha(n))=M\{\xi^2(n)\}$$

and $e_a(n)$ is the generalized error. AECM minimization is equivalent to argument optimization [6], and minimization of asymptotic deflection, to criterion optimization [9].

For the quadratic criterion $J^a=M\{e_a^2\}$ the optimality condition

$$\nabla_\alpha J^a(\alpha)=2M\{e_a(n)v_a(n)\},\ \text{where}\ v_a(n)=\frac{de_a(n)}{d\alpha}$$

is the sensititivy function, can be generalized to

$$M\{\varphi(e_a(n))v_a(n)\}=0;\quad \varphi(-e)\equiv-\varphi(e). \quad (9)$$

This condition generates a criterion optimization algorithm

$$\alpha(n)=\alpha(n-1)-\Gamma(n)\varphi(e_a(n))v(n),$$

$$\Gamma(n)=[nM\{\varphi'(\xi)\}]^{-1}B^+;\quad B=M\{v_a(n)v_a^T(n)\}\ge 0, \quad (10)$$

where B is a non-negative definite matrix of rank $N_0=N_1+N_2+N_3$ and B^+ is a pseudoinverse matrix. The asymptotic deflection for the algorithm (10) is equal to

$$w=w(\varphi,p_0)=\tilde{N}\frac{M\{\varphi^2(\xi)\}}{[M\{\varphi'(\xi)\}]^2};\quad \tilde{N}=\text{rank}B\le N_0 \quad (11)$$

and depends on the noise distribution density and the nonlinear transformation $\varphi(e)$. Let us choose $\varphi(e)$ so as to minimize the asymptotic deflection. This optimal transformation $\varphi_0(e_a)=\arg\min w(\varphi,p_0)$ is known [6] to be equal to

$$\varphi_0'(e_a)=-(\ln p_0(\xi))'=-\frac{p_0'(\xi)}{p_0(\xi)}\Big|_{\xi=e_a} \quad (12)$$

Furthermore,

$$w_{min}=w(\varphi_0,p_0)=\tilde{N}I^{-1}(p_0), \quad (13)$$

where

$$I(p_0)=M\{[\varphi_0'(\xi)]^2\}=M\{[\frac{p_0'(\xi)}{p_0(\xi)}]^2\}$$

is the Fischer information. If the noise distribution density $p_0(\xi)$ is unknown and only the class P where it is a member is known, then the nonlinear transformation optimal in a class is

$$\varphi_*'(e_a)=-(\ln p_*(\xi))'_{\xi=e_a}=-\frac{p_*'(\xi)}{p_*(\xi)}\Big|_{\xi=e_a}; \quad (14)$$

where

$$p_*(\xi)=\arg\min_{p_0\in P} I(p_0). \quad (15)$$

The principle of optimality on a class holds

$$w(F_*,p_0)\le w(F_*,p_*)\le w(F,p_*). \quad (16)$$

Examples of optimal and optimal on a class nonlinear transformations have been given in Ref. [6] which was devoted to robust methods of estimating the parameters of regression plants. Now they also hold for autoregression and regression-autoregression plants.

For the class of distribution densities

$$P_1=\{p_0(\xi):p_0(0)=\frac{1}{2s}>0\}\quad \varphi_0(e_a)=\text{sign}\ e_a$$

is a relay transformation, for the class of distribution densities with a limited variable

$$P_2=\{p_0(\xi):\sigma^2=\int_{-\infty}^{\infty}\xi^2p_0(\xi)d\xi\le\sigma_1^2\}\quad \varphi_0(e_a)=e_a$$

is a linear transformation. For the class of near-normal distributions

$$P_3=\{p_0(\xi):p_0(\xi)=(1-\lambda)p_N(\xi)+\lambda p_1(\xi)\},$$

where

$$p_N(\xi)=N(0,\sigma^2)$$

is a normal distribution and $p_1(\xi)$ is any symmetrical distribution, $\varphi_0(e_a)=\text{sat}\ e_a$ is a transformation of the saturation type. Let us emphasize that the argument optimization involves more complicated transformations in these cases.

Algorithms, optimal in a class, are obtained from (10) with $\varphi(e_a)=\varphi_*(e_a)$. They are not physically implementable in that the matrix B and, consequently, B^+ are unknown.

4 PHYSICALLY IMPLEMENTABLE ALGORITHMS

To find physically implementable algorithms let us replace the inverse amplitication matrix

$$\Gamma_*^{-1}(n)=nM\{\varphi_*'(\xi)\}B$$

by a sample matrix

$$\hat{\Gamma}_*^{-1}(n)=\sum_{m=1}^{n}\varphi_*'(e_a(m))v_a(m)v_a^T(m),$$

i.e.

$$\hat{\Gamma}_*^{-1}(n)=\hat{\Gamma}_*^{-1}(n-1)+\varphi_*'(e_a(n))v_a(n)v_a^T(n), \qquad (17)$$

or, which is equivalent,

$$\hat{\Gamma}_*(n)=\hat{\Gamma}_*(n-1)-\hat{\Gamma}_*(n-1)v(n)[\varphi_*'^{-1}(e(n))+ v_a^T(n)\hat{\Gamma}_*(n-1)v_a(n)]\cdot v_a^T(n)\hat{\Gamma}_*(n-1). \qquad (18)$$

Implementable algorithms, optimal in a class, are represented in the form

$$\eta(n)=\eta(n-1)+\hat{\Gamma}_*(n)\varphi(e_a(n))v_a(n), \qquad (19)$$

$$v_a(n) = \frac{de_a(n)}{d\alpha},$$

$$\hat{\Gamma}_*^{-1}(n)=\hat{\Gamma}_*^{-1}(n-1)+\varphi'(e_a(n))v_a(n)v_a^T(n),$$

The initial values in these algorithms are $\eta(0)=\eta_1$, $v(m)=0$ with $m=0, -1, -2, \ldots, -N_2$ and $\hat{\Gamma}_*^{-1}(0) = \Gamma_1^{-1}$.

Specific representations of the algorithms (19) lead to adaptive algorithms for indirect and direct adaptive control systems.

5 CRITERIAL ALGORITHMS, OPTIMAL IN A CLASS, IN INDIRECT ADAPTIVE SYSTEMS

Adaptive algorithms in indirect DACS's are identification algorithms. Let Θ denote the plant parameter vector and

$$x(n)=(-y(n-1),,\ldots,-y(n-N),u(n-1),\ldots,u(n-N_1), \varepsilon(n-1),\ldots,\varepsilon(n-N_2))^T$$

the plant state observation vector where

$$\varepsilon(n)=y(n)-\hat{y}(n)$$

is the error and

$\hat{y}(n) = \Theta^T(n-1)x(n)$. Then following the replacement in (19) of $\alpha(n)$ by $\Theta(n)$ and computation of the sensitivity functions $v_a(n)=v_x(n)$ we have

$$\Theta(n)=\Theta(n-1)+\hat{\Gamma}_*(n)\varphi(y(n)-\Theta^T(n-1)x(n))v(n),$$

$$v_*(n)=(1-\hat{P}_\xi(q))v(n)+x(n), \qquad (20)$$

$$\hat{\Gamma}_*^{-1}(n)=\hat{\Gamma}_*^{-1}(n-1)+\varphi_*'(y(n)-\Theta^T(n-1)x(n))v_x(n)v_x^T(n).$$

From parameter estimates $\Theta(n)$ estimates of the polynomials

$$\hat{Q}(q),\ \hat{P}_u(q) \text{ and } \hat{P}_\xi(q)$$

are found and from the polynomial equation (6) $\hat{P}(q)$ and, consequently, the vector $\alpha(n)$ and the control function are found.

6 CRITERIAL ALGORITHMS, OPTIMAL IN A CLASS, IN DIRECT ADAPTIVE SYSTEMS

Direct DACS's can be implemented either by using a predictor of the dynamic plant output one cycle in advance or without it. The equation of an optimal predictor is obtained from that of an optimal controller (4) where $y_0(n)$ is replaced by $\hat{y}(n)$

$$P_\xi(q)G^0(q)\hat{y}(n)=P(q)y(n-1)+P_u(q)u(n-1). \qquad (21)$$

Denoting

$$\hat{\bar{y}}(n) = G^0(q)\hat{y}(n)$$

write (21) as

$$\hat{\bar{y}}(n)=\alpha^T Z_1(n-1;\hat{\bar{y}})$$

which is analogous to the equation of an adjustable model.

Adaptive prediction algorithms follow from (19) with

$$e_a(n)=G^0(q)y(n)-\hat{\bar{y}}(n)$$

and computation of the sensitivity functions $v_a(n)=v_z(n)$. They have the form

$$\alpha(n)=\alpha(n-1)-\hat{\Gamma}_*(n)\varphi_*(G^0(q)y(n)- -\alpha^T(n-1)Z_0(n;\hat{\bar{y}}))v_z(n);$$

$$v_z(n)=(1-\hat{P}_\xi(q))v_z(n)+Z_1(n-1;\hat{\bar{y}}); \qquad (22)$$

$$\hat{\Gamma}_*^{-1}(n)=\hat{\Gamma}_*^{-1}(n-1)+\varphi_*'(G^0(q)y(n)- -\alpha^T(n-1)Z_0(n;\hat{\bar{y}}))v_z(n)v_z^T(n).$$

The estimate $\alpha(n)$ defines the control equation (7).

Adaptive algorithms in direct adaptive systems without prediction are obtained from (19) with $e(n)=G^0(q)(y(n)-y_0(n))$ and computation of sensitivity functions $v_a(n)=v_z(n)$. They have the form

$$\alpha(n)=\alpha(n-1)-\hat{\Gamma}_*(n)\varphi_*(G^0(q)(y(n)-y_0(n))v_z(n),$$

$$v_z(n)=(1-P_\xi(q))v_z(n)-Z_0(n;y_0), \qquad (23)$$

$$\hat{\Gamma}_*^{-1}(n)=\hat{\Gamma}_*^{-1}(n-1)-\varphi'(G^0(q)(y(n)-y_0(n)))v_z(n)v_z^T(n),$$

the estimates $\alpha(n)$ determine the control equation (7). In the algorithm (20) $G^0(q)y_0(n)$ can be replaced by $H^0(q)r(n-1)=\bar{r}(n-1)$.

7 DISCUSSION

If the matrix $B=M\{v(n)v_z^T(n)\}$ is positive definite, i.e. $\tilde{N}=\text{rank}B=N_0$, then $B^+=B^{-1}$ and criterial optimal algorithms such as (20), (22) and (23) coincide with argument optimal algorithms when $\varphi_*(\cdot)$ is replaced by $\varphi_0(\cdot)$. As for algorithms, optimal in a class, (20), (22) and (23), argument optimal algorithms feature a more complicated form of nonlinear transformation $\varphi_*(e_a)$. The above algorithms can be simplified by eliminating sensitivity algorithms, or by assuming that

$$v(n)=x(n);\ \bar{v}_z(n)=Z_1(n;\hat{\bar{y}})$$

All identification, prediction, and adaptation algorithms, known at this point and employed in DACS's are simplified linear algorithms with various, often simplified amplification matrices. The matrix B is often assumed to be non-singular. The efficiency of these algorithms is significantly lower than that of criterial optimal algorithms. With noises of unlimited variance they become ineffectual. The implementable algorithms, optimal in a class, (20), (22), and (23) can be classified with accelerated algorithms which feature accelerated convergence with a finite number of observations. They "reach the asymptotics" faster. Comparing the adaptive identification algorithms (20) and prediction algorithms (22) shows that they are kindred. The utmost potential of all identification algorithms (20), prediction algorithms (22), and adaptation algorithms (23) are the same if there is no delay in the plant.

SUMMARY

Criterial optimal and optimal in a class adaptation algorithms have been determined. They include identification algorithms in direct adaptive systems and prediction and adaptation algorithms in indirect adaptive systems. Their utmost potential has been found. Their simplification is shown to lead, with some loss of effectiveness, to all nonoptimal adaptation algorithms available today.

REFERENCES

[1] Åström, K. J.. Introduction to Stochastic Control Theory. New York, Academic Press, 1970.

[2] Tsypkin, Ya. Z.. Adaptation and Learning in Automatic Systems. New York and London, Academic Press, 1971.

[3] Polyak, B. T. and Ya. Z. Tsypkin. Optimalnye pseudogradientnye algoritmy stokhasticheskoi optimizatsii. Doklady Akademii Nauk, 1980, t. 250, No 5, p. 1082.

[4] Tsypkin, Ya. Z. and A. S. Posnyak. Optimalnye poiskovye algoritmy stokhasticheskoi optimizatsii. Doklady Akademii Nauk, 1981, t. 260, No 3, p. 550.

[5] Tsypkin, Ya. Z. Sintez optimalnoi nastraivaemoi modeli v zadachakh identifikatsii. Avtomatika i telemekhanika, 1981, No 11, p. 62.

[6] Tsypkin, Ya. Z. Optimalnye kriterii kachestva v zadachakh identifikatsii. Avtomatika i telemekhanika, 1982, No 11, p. 5.

[7] Tsypkin, Ya. Z. Optimalnye algoritmy otsenivanija parametrov v zadachakh identifikatsii. Avtomatika i telemekhanika, 1982, No 12, p. 9.

[8] Tsypkin, Ya. Z. Optimizatsyja v uslovijakh neopredelennosti. Doklady Akademii Nauk, 1976, t. 228, No 6, p. 1306.

[9] Tsypkin, Ya. Z., A. S. Poznyak and A. M. Pesin. Poiskovye algoritmy kriterialnoi optimizatsii. Doklady Academii Nauk, 1983, t. 270, No 3, p. 565.

[10] Narendra, K. S. and L. S. Valavani. Direct and indirect adaptive control. Automatica, 1979, v. 15, No 6, p 653.

[11] Gawthrop, P. J. On the stability and convergence of a selftuning controller. International Journal of Control, 1980, v. 31, No 5, p. 973.

[12] Isermann, R. Parameter adaptive algorithms. A tutorial. Automatica, 1982, v. 18, No 5, p. 513.

[13] Landau, I. D. A survey of model reference adaptive techniques theory and applications. Automatica, 1974, v. 10, No 4, p. 353.

Global Stability of Adaptive Pole Placement Algorithms

HOWARD ELLIOTT, MEMBER, IEEE, ROBERTO CRISTI, MEMBER, IEEE, AND MANOHAR DAS, MEMBER, IEEE

Abstract—**This paper presents direct and indirect adaptive control schemes for assigning the closed-loop poles of a single-input, single-output system in both the continuous- and discrete-time cases. The resulting closed-loop system is shown to be globally stable when driven by an external reference signal consisting of a sum of sinusoids. In particular, persistent excitation of the potentially unbounded closed-loop input–output data, and hence convergence of a sequential least-squares identification algorithm is proved. The results are applicable to standard sequential least squares, and least squares with covariance reset.**

I. Introduction

ONE of the long standing issues to theorists studying adaptive control systems is global stability of indirect schemes, and schemes which are applicable to nonminimum phase systems. Parameter convergence, and hence global stability of such systems has been linked by a number of investigators, e.g., [1]–[4], [7], [8], to persistent excitation of the control signal entering the unknown process to be controlled. This signal is typically the sum of a reference signal and a nonlinear, time varying, adaptive feedback signal. The feedback signal can in principle cancel any excitation contained in the reference signal. This problem and the potential for unbounded growth of the control and output signals has made verification of persistant excitation difficult. This paper resolves this issue for an adaptive pole placement algorithm in the case where the external reference signal consists of a sum of w sinusoids; w being half the number of parameters estimated.

A recent report by Goodwin *et al.* [4] addressed this same issue for an indirect discrete adaptive pole placement controller. The authors proposed the use of block processing where a number of data samples are gathered while the process is operated in open loop. The unknown process parameters, and hence the control parameters are then obtained using a batch estimation procedure. The scheme discussed in this paper makes use of block processing. However, the analysis is done for the sequential least squares estimation procedure operating with closed-loop data obtained from the adaptive network. Block processing is used in the sense that N data samples are taken, and N iterations of the least-squares algorithm are performed between control parameter updates. In this case the identification data are obtained from a sequence of linear time-invariant systems, and linear system theory can be applied to the analysis.

Two recent reports by Anderson and Johnstone [7], [8] also address the discrete adaptive pole placement problem and make use of block processing in a manner similar to that discussed below. However, the philosophy is fundamentally different. Their goal is to first guarantee boundedness of the system input–output data and then show persistency of excitation. We show persistency of excitation without requiring bounded data. Boundedness of all signals then follows indirectly from proof of parameter convergence. Although the scheme in [8] allows for a more relaxed definition of persistency of excitation of the reference signal in comparison to the sinusoidal constraint, the control parameter update algorithm is much more complex. Furthermore, only indirect adaptive pole placement is considered and prior knowledge of a lower bound on the Sylvester resultant associated with the numerator and denominator polynomials in the plant transfer function is required. Hence, prior knowledge regarding system parameters is necessary. Although for the case of indirect control we require the same type of prior information, the proof given below for direct adaptive pole placement requires *no* prior information regarding plant parameters. One need only know the order of a minimal realization of the unknown process. The direct algorithm used is described in [5] and [11]. It has the property that the control parameter estimates remain bounded. This feature is the primary reason prior parameter information is not necessary.

The limitation of this paper to sinusoidal reference signals is severe. However, persistency of excitation of input–output data for pole placement adaptive controllers has been a long outstanding problem in the adaptive field. We thus feel that a solution, even for this restricted form of the reference, is an important contribution to theory in this field.

This paper first considers the case of direct adaptive control in detail. The indirect case is then briefly considered. Rather than repeat all details of the analysis, only points of difference with the direct case are highlighted.

Results are presented for both a continuous time (CT) hybrid implementation similar to that used in [9], and for a purely discrete time (DT) version. As was done in [11], the two formulations are presented together. This allows the similarities and differences to be highlighted and keeps the presentation compact. Although technicalities of the proof require a random sampling scheme in the hybrid case, this is *not* necessary for the DT formulation.

As is typically done in analysis of adaptive systems, the unknown plant is assumed to be time invariant. However, one step toward extension of such theory to the case of truly time varying plant models is use of an estimation scheme which discounts data a certain distance in the past. To this end, the analysis given below is carried out for both standard sequential least squares which does not have this property, and sequential least squares with covariance reset which does. Another advantage of the covariance reset algorithm is that it leads to exponential parameter convergence which has been tied by other authors, e.g., [16] to improved sensitivity to noise and modeling errors.

This paper is organized as follows. Section II discusses models and presents a fixed strategy for pole placement. Section III then modifies this strategy for direct adaptive implementation and presents a scheme for parameter estimation. Stability analysis for the direct case is presented in Section IV. Section V then presents an indirect formulation and summarizes modifications of Section IV which are necessary to complete a similar analysis of the

Manuscript received September 26, 1983; revised April 2, 1984. This paper is based on a prior submission of January 3, 1983. Paper recommended by Past Associate Editor, C. R. Johnson, Jr. This work was supported in part by the National Science Foundation under NSF Grant ECS-8214534 and in part by the Air Force Office of Scientific Research under Grant AFOSR-80-0155.

H. Elliott was with the Department of Electrical and Computer Engineering, University of Massachusetts, Amherst, MA 01003. He is now deceased.

R. Cristi was with the Department of Electrical and Computer Engineering, University of Massachusetts, Amherst, MA 01003. He is now with the University of Michigan, Dearborn, MI 48128.

M. Das was with the Department of Electrical and Computer Engineering, University of Massachusetts, Amherst, MA 01003. He is now with the School of Engineering and Computer Science, Oakland University, Rochester, MI 48063.

Reprinted from *IEEE Trans. Automat. Contr.*, vol. AC-30, pp. 348–356, Apr. 1985.

indirect controller. Finally, concluding remarks are given in Section VI.

II. Modeling Assumptions and Fixed Control Strategy

Let us consider the problem of controlling a linear, shift invariant (DT case) or time invariant (CT case) system characterized by either of the following models:

$$p(D)z(t)=u(t) \tag{1a}$$

$$y(t)=r(D)z(t) \tag{1b}$$

or

$$p(D)y(t)=r(D)u(t). \tag{2}$$

Here $u(t)$ is the system input, $y(t)$ is the system output, and $z(t)$ is an internal state variable. The polynomials $p(D)$ and $r(D)$ in the differential (CT case) or delay (DT case) operator D have the form:

CT Case:

$$p(D)=D^n+\sum_{i=0}^{n-1} P_iD^i \tag{3a}$$

$$r(D)=\sum_{i=0}^{n} r_iD^i \tag{4a}$$

DT Case:

$$p(D)=1+\sum_{i=1}^{n} p_{i-1}D^i \tag{3b}$$

$$r(D)=\sum_{i=1}^{n} r_{i-1}D^i. \tag{4b}$$

Although, in the adaptive control problem the parameters in $p(D)$ and $r(D)$ are assumed unknown, it will be assumed that n, the system order, is known, and that the model is minimal, i.e., $r(D)$ and $p(D)$ are relatively prime.

Consider the fixed control strategy

$$q(D)u(t)=h(D)y(t)+k(D)u(t)+q(D)v(t) \tag{5}$$

where $v(t)$ is an external reference signal. Using this control the closed-loop system becomes

$$[q(D)p(D)-h(D)r(D)-k(D)p(D)]z(t)=q(D)v(t) \tag{6a}$$

$$y(t)=r(D)z(t). \tag{6b}$$

Choosing $q(D)$ and $p^*(D)$ as stable polynomials of degree n of the following forms:

CT Case:

$$q(D)=D^n+\sum_{i=0}^{n-1} q_iD^i$$

$$p^*(D)=D^n+\sum_{i=0}^{n-1} p_i^*D^i$$

DT Case:

$$q(D)=1+\sum_{i=1}^{n} q_{i-1}D^i$$

$$p^*(D)=1+\sum_{i=1}^{n} p_{i-1}^*D^i$$

and defining $h(D)$ and $k(D)$ as the solutions to

$$h(D)r(D)+k(D)p(D)=q(D)[p(D)-p^*(D)] \tag{7}$$

the closed-loop system simplifies to

$$q(D)p^*(D)z(t)=q(D)v(t) \tag{8a}$$

$$y(t)=r(D)z(t). \tag{8b}$$

It might be pointed out here that in the DT case, although we assumed for the sake of simplicity that $q(D)$ and $p^*(D)$ are each of degree n, the actual technical requirement of the algorithm [11] is that

$$\deg\,[q(D)p^*(D)]=2n.$$

As is readily seen from (8), the closed-loop poles are the zeros of $q(D)p^*(D)$. The zeros of $q(D)$ are interpreted as uncontrollable observer poles, and those of $p^*(D)$ are the new assigned pole locations.

Let the polynomials $h(D)$ and $k(D)$ satisfying (7) take the following forms:

CT Case:

$$h(D)=\sum_{i=0}^{n-1} h_iD^i \tag{9a}$$

$$k(D)=\sum_{i=0}^{n-1} k_iD^i \tag{9b}$$

DT Case:

$$h(D)=\sum_{i=1}^{n} h_{i-1}D^i \tag{9c}$$

$$k(D)=\sum_{i=1}^{n} k_{i-1}D^i \tag{9d}$$

and define the desired control parameter vector as

$$\alpha^*=[h_0,\ h_1,\ \cdots,\ h_{n-1},\ k_0,\ \cdots,\ k_{n-1}]^T. \tag{10}$$

III. Adaptive Control and Parameter Estimation

Let $\{t_k\}_0^\infty$ be a sequence of time instants where $t_0 = 0$ and $t_k > t_{k-1}$ and define I_k as

$$I_k=\begin{cases}\text{time interval } [t_k,\ t_{k+1}) & \text{CT case}\\ \text{sets of times } \{t: t_k\le t<t_{k+1}\} & \text{DT case.}\end{cases}$$

Assuming that the control parameters are updated only at the times t_k, the adaptive version of (5) can be defined as

$$q(D)u(t)=h_k(D)y(t)+k_k(D)u((t)+q(D)v(t), \qquad t\in I_k \tag{11}$$

where $h_k(D)$ and $k_k(D)$ are estimates of $h(D)$ and $k(D)$ satisfying (7).

Writing for the CT case

$$h_k(D)=\sum_{i=0}^{n-1} h_{k_i}D^i \tag{12a}$$

$$k_k(D)=\sum_{i=0}^{n} k_{k_i}D^i \tag{12b}$$

and for the DT case

$$h_k(D)=\sum_{i=1}^{n} h_{k_{i-1}}D^i \tag{12c}$$

$$k_k(D)=\sum_{i=1}^{n} k_{k_{i-1}}D^i \tag{12d}$$

the estimate of the control parameter vector α^* can also be defined as

$$\alpha_k=[h_{k0},\ h_{k1},\ \cdots,\ h_{k,n-1},\ k_{k0},\ k_{k1},\ \cdots,\ k_{k,n-1}]^T. \tag{13}$$

Calculating α_k: To derive a direct scheme for estimating α_k, first define a filter polynomial $f(D)$ as follows. For the CT case $f(D)$ must be chosen as a monic Hurwitz polynomial of degree at least $3n-1$ to assure physical realizability of the regression vector used for estimation. For convenience we will assume it to be of degree exactly $3n-1$. In the DT case, it is convenient to simply choose $f(D)=1$.

Define $\bar{z}(t)$, $\bar{y}(t)$, $\bar{u}(t)$ by the relations

$$f(D)\bar{z}(t)=z(t) \tag{14a}$$

$$f(D)\bar{y}(t)=y(t) \tag{14b}$$

$$f(D)\bar{u}(t)=u(t) \tag{14c}$$

combining (14) and (1) one can show

$$p(D)\bar{z}(t)=\bar{u}(t)+\delta_1(t) \tag{15a}$$

$$\bar{y}(t)=r(D)\bar{z}(t)+\delta_2(t) \tag{15b}$$

where the signals $\delta_i(t)$ are linear combinations of decaying exponentials corresponding to modes of the filters and will be zero if all initial conditions are zero. In the DT case assuming $f(D)=1$, the signals $\delta_i(t)=0$.

Multiplying the diophantine design equation (7) by $\bar{z}(t)$ and using (15) yields

$$h(D)\bar{y}(t)+k(D)\bar{u}(t)+q(D)p^*(D)\bar{z}(t)=q(D)\bar{u}(t)+\delta_3(t). \tag{16}$$

In order to use (16) as a basis for estimating $h(D)$ and $k(D)$, the filtered internal state $\bar{z}(t)$ must be replaced by an estimate in terms of the available signals $\bar{y}(t)$ and $\bar{u}(t)$. This can be done using the technique discussed in [5] and briefly described below. Since $r(D)$ and $p(D)$ are prime, there also exist polynomials $b(D)$ and $c(D)$ such that

$$b(D)r(D)+c(D)p(D)=1. \tag{17}$$

Multiplying (17) by $\bar{z}(t)$ and using (15) yields

$$b(D)\bar{y}(t)+c(D)\bar{u}(t)=\bar{z}(t)+\delta_4(t). \tag{18}$$

Substituting (18) into (16) yields

$$h(D)\bar{y}(t)+k(D)\bar{u}(t)+b(D)q(D)p^*(D)\bar{y}(t)$$
$$+c(D)q(D)p^*(D)\bar{u}(t)=q(D)\bar{u}(t)+\delta_5(t). \tag{19}$$

Assuming $b(D)$ and $c(D)$ are of the following forms:

CT Case:

$$b(D)=\sum_{i=0}^{n-1} b_iD^i$$

$$c(D)=\sum_{i=0}^{n-1} c_iD^i$$

DT Case:

$$b(D)=\sum_{i=1}^{n} b_{i-1}D^i$$

$$c(D)=1+\sum_{i=1}^{n} c_{i-1}D^i$$

and defining the vectors

$$\bar{\phi}(t)=\begin{cases}[\bar{y},\ D\bar{y},\ \cdots,\ D^{n-1}\bar{y},\ \bar{u},\ D\bar{u},\ \cdots,\\ \quad D^{n-1}\bar{u},\ q(D)p^*(D)\bar{y},\ \cdots,\\ \quad D^{n-1}q(D)p^*(D)\bar{y}, q(D)p^*(D)\bar{u},\ \cdots,\\ \quad D^{n-1}q(D)p^*(D)\bar{u}]^T \quad \text{CT case}\\ [D\bar{y},\ D^2\bar{y},\ \cdots,\ D^n\bar{y},\ D\bar{u},\ D^2\bar{u},\ \cdots,\\ \quad D^n\bar{u},\ Dq(D)p^*(D)\bar{y},\ \cdots,\\ \quad D^nq(D)p^*(D)\bar{y},\ Dq(D)p^*(D)\bar{u},\ \cdots,\\ \quad D^nq(D)p^*(D)\bar{u}]^T \quad \text{DT case}\end{cases} \tag{20}$$

$$\eta^*=[b_0,\ b_1,\ \cdots,\ b_{n-1},\ c_0,\ c_1,\ \cdots,\ c_{n-1}]^T \tag{21}$$

$$\theta^*=[\alpha^{*T},\ \eta^{*T}] \tag{22}$$

(19) can be written as

$$\bar{\phi}(t)^T\theta^*=\begin{cases}q(D)\bar{u}(t)+\delta_5(t) & \text{CT case}\\ (q(D)-1)\bar{u}(t)+\delta_5(t). & \text{DT case.}\end{cases} \tag{23}$$

Equation (23) can be used to estimate the $4n$ vector θ^* of which the $2n$ control parameter vector α^* is a component. Notice that in the case if it is known that the unknown process has a strictly proper transfer function so that deg $(r(D))\le n-1$, then $q(D)$ can be chosen of degree $n-1$, $k(D)$ and $c(D)$ can be chosen of degree $n-2$, and $f(D)$ can be chosen of degree $3n-2$. This reduces the number of control parameters in α^* to $2n-1$ and the length of θ to $4n-2$. An analogous reduction in parameters is possible in the DT case if the system is known to have a time delay. Next a specific estimation scheme is given for generating an estimate θ_k of θ^*.

Estimating θ_k: Let us define

$$t_{k+1}-t_k=\begin{cases}N\tau_k & \text{CT Case}\\ N & \text{DT Case.}\end{cases} \tag{24}$$

The choices for the positive integer N and the real positive sequence $\{\tau_k\}_0^\infty$ will be given later. In addition, let

$$t_k^i=\begin{cases}t_k+i\tau_k, & 0\le i\le N \quad \text{CT Case}\\ t_k+i, & 0\le i\le N \quad \text{DT Case.}\end{cases} \tag{25}$$

Since (24) holds we can generate a sequence of parameter estimates θ_k^i, $0\le i\le N$, during the time interval I_k, which attempts to minimize the equation error sequence

$$\epsilon(t_k^i)=\begin{cases}\bar{\phi}(t_k^i)^T\theta_k^i-q(D)\bar{u}(t_k^i) & \text{CT case}\\ \bar{\phi}(t_k^i)^T\theta_k^i-\sum_{i=1}^{n} q_{i-1}D^i\bar{u}(t_k^i) & \text{DT case.}\end{cases} \tag{26}$$

Using the sequential least-squares procedure we obtain

$$\theta_k^i=\theta_k^{i-1}-\frac{P_k^{i-2}\bar{\phi}(t_k^{i-1})\epsilon(t_k^{i-1})}{1+\bar{\phi}(t_k^{i-1})^TP_k^{i-2}\bar{\phi}(t_k^{i-1})} \tag{27a}$$

$$P_k^i=P_k^{i-1}-\frac{P_k^{i-1}\bar{\phi}(t_k^i)\bar{\phi}(t_k^i)^TP_k^{i-1}}{1+\bar{\phi}(t_k^i)^TP_k^{i-1}\bar{\phi}(t_k^i)} \tag{27b}$$

$$i=1,\ 2,\ \cdots,\ N$$

$$\theta_k^0=\theta_{k-1}^N=\theta_k$$

$$\theta_{-1}^N=\theta_0 \text{ arbitrary}$$

$$P_k^{-1}=\begin{cases}\sigma_0 I & k\in k_s\\ P_{k-1}^{N-1} & \text{otherwise}\end{cases}$$

$$P_{-1}^{N-1}=\sigma_0 I$$

$$\sigma_0>0$$

$$k_s=\{k_1, k_2, \cdots\}.$$

Remark 1: The idea is to block process the data in the sense that the estimator is run N-times during each interval I_k. The final estimate is then defined as θ_{k+1} and used to generate the control vector α_{k+1}, which in turn is used for control during the interval I_{k+1}. What is critical to the analysis which follows is that the controller only be adjusted at the beginning of each interval.

Remark 2: Note that this algorithm is the standard sequential least squares when k_s is empty. If k_s is not empty, one obtains least squares with covariance reset. We will assume one of these two models is in use below. The superscript notation has only been introduced so that the block processing can be clearly defined. As a result all known properties of sequential least-squares are applicable. Those necessary for this paper are summarized below.

Lemma 1: Using the identification scheme given by (27) there exists a $c_1 < \infty$ such that

$$\|\theta_k^i\| \le \|\theta_k^{i-1}\| < c_1 \forall k, \ i. \tag{28}$$

Furthermore, if

$$\Phi_k\Phi_k^T \ge \epsilon_1 I \tag{29a}$$

infinitely often, where ϵ_1 is finite and positive, and where

$$\Phi_k=[\tilde{\phi}(t_k^0), \ \tilde{\phi}(t_k^1), \ \cdots \ \tilde{\phi}(t_k^{N-1})] \tag{29b}$$

then

$$\lim_{k\to\infty} \theta_k=\theta^*. \tag{30}$$

Proof: Equation (28) has been shown for the case where k_s is empty (standard sequential least squares) in [9] and [12], and for the case where k_s is not empty (covariance reset) in [13]. Since the manipulations are straightforward, for brevity, they will not be repeated here.

To show (30) for the case where k_s is empty (standard least squares) define the Lyapunov function

$$V_k^i=(\tilde{\theta}_k^i)^T(P_k^{i-1})^{-1}(\tilde{\theta}_k^i)$$

$$\tilde{\theta}_k^i=\theta_k^i-\theta^*.$$

Applying the matrix inversion lemma to (27b), we can then write

$$V_k^N=(\tilde{\theta}_k^N)^T(P_k^{N-1})^{-1}\tilde{\theta}_k^N$$

$$=(\tilde{\theta}_k^N)^T\left((P_{-1}^{N-1})^{-1}+\sum_{j=0}^{k}\sum_{l=0}^{N-1}\tilde{\phi}(t_j^l)\tilde{\phi}(t_j^l)^T\right)\tilde{\theta}_k^N$$

$$=(\tilde{\theta}_k^N)^T\left((P_{-1}^{N-1})^{-1}+\sum_{j=0}^{k}\Phi_j\Phi_j^T\right)\tilde{\theta}_k^N.$$

Using (29) one obtains

$$V_k^N \ge (\sigma_0-(\bar{k}+1)\epsilon_1)\|\tilde{\theta}_k^N\|$$

where $\bar{k} \le k$ but $\bar{k} \to \infty$ as $k \to \infty$. The proofs in [9] and [12] also show that V_k^i is nonincreasing, and hence must be bounded. Thus, $\|\tilde{\theta}_k^N\|$ must converge to zero as k increases which proves (30).

For the case of covariance reset define V_k^i as before, but observe that in this case, at the reset instances, k_i

$$V_{k_i}^0=\sigma_0\|\tilde{\theta}_{k_i}^0\|^2.$$

Assume $k_{i-1} \le k < k_i$, then using (28) and (29) we have

$$V_k^N \ge (\sigma_0+\epsilon_1)\|\tilde{\theta}_k^N\|$$
$$\ge(\sigma_0+\epsilon_1)\|\tilde{\theta}_{k_i}^0\| \quad \text{i.o.} \tag{31a}$$

But using the monotonicity of V_k^N [13]

$$V_k^N \le V_{k_{i-1}}^N$$
$$\le V_{k_{i-1}}^0$$
$$\le \sigma_0\|\tilde{\theta}_{k_{i-1}}^0\|^2. \tag{31b}$$

Combining (31a) and (31b) yields

$$\|\tilde{\theta}_{k_i}^0\|^2 \le \frac{\sigma_0}{\sigma_0+\epsilon_1}\|\tilde{\theta}_{k_{i-1}}^0\|^2 \quad \text{i.o.} \tag{31c}$$

which implies

$$\lim_{k\to\infty}\|\tilde{\theta}_{k_i}^0\|=0,$$

and hence established (30).

Remark 3: As will be clear in the next section, the qualifier "infinitely often" is only necessary for the hybrid analysis. For the discrete-time proof, to follow, it would have been sufficient to have replaced all occurrences of the phrase "infinitely often" with "for all k" in Lemma 1 and its proof.

Remark 4: Notice that one implication of the inequality (31c) is that if the reset instants occur at equally spaced intervals for all time, then convergence is exponential. It has been pointed out, e.g., [16] that algorithms with exponential convergence will be more robust with respect to noise and modeling errors.

IV. Global Stability Analysis

Since the conditions of Lemma 1 ensure that the controller parameters are bounded, to complete the stability analysis, one needs only ensure that Φ_k satisfies (29). This is summarized with the following corollary to Lemma 1.

Corollary 1: If $\Phi_k\Phi_k^T > \epsilon_1 I$ infinitely often, then all system signals remain bounded and the closed-loop system converges to that characterized by (8).

Remark 5: Although no formal proof of Corollary 1 is given here, analogous results have appeared many times in the literature, e.g., [1] and [2].

The remainder of this section will be devoted towards verifying that Φ_k satisfies (29) or equivalently that

$$\mu_k>\epsilon_1>0$$

infinitely often where μ_k is the minimum elgenvalue of $\Phi_k\Phi_k^T$. The analysis requires the following specific assumptions.

Assumption 1: The External Input $v(t)$: It will be assumed that

$$v(t)=\sum_{i=1}^{2n} B_i e^{j\omega_i t}+B_i^c e^{-j\omega_i t} \tag{32}$$

$$B_i \ne 0 \ \forall i$$

$$B_i^c = \text{ complex conjugate of } B_i$$

$$\omega_i \ne \omega_j \quad \text{if } i \ne j.$$

Assumption 2: The Sampling Period τ_k (for the CT Case Only): For the CT case it will be assumed that for the intervals I_k, the sampling periods $\{\tau_k\}$ are a sequence of independent random variables uniformly distributed over an interval $[T_1, T_2]$, $0 < T_1 < T_2 < \infty$. (Recall the sequence $\{\tau_k\}$ is not used for the DT formulation.)

Assumption 3: The Number of Samples N: It will be assumed that during each interval (CT case) or set of times (DT case) I_k the identifier uses N samples where N is given by

$$N \geq \begin{cases} 9n-1 & \text{CT case} \\ 6n & \text{DT case when } f(D)=1. \end{cases}$$

It might be pointed out that more than $2n$ sinewaves can be used. However, each additional sinewave will require the minimum value of N, the block size, to increase.

A Factorization of Φ_k: Our approach in analysis of Φ_k will be to factor it into the product of four matrices and then analyze the properties of each factor. To begin, define the $(4n \times 4n)$ matrix M^* as

$$M^* = \begin{bmatrix} M_1 \\ \hdashline M_2 \\ \hdashline M_3 \\ \hdashline M_4 \end{bmatrix} \tag{33}$$

where the $(n \times 4n)$ matrix M_1 is of the form

$$M_1 = \begin{bmatrix} r_0 & r_1 & r_2 & \cdots & r_{n-2} & r_{n-1} & r_n & 0 & 0 & \cdots & 0 \\ 0 & r_0 & r_1 & \cdots & r_{n-3} & r_{n-2} & r_{n-1} & r_n & 0 & \cdots & 0 \\ \cdot & & & & & & & & & & \\ \cdot & & & & & & & & & & \\ \cdot & & & & & & & & & & \\ 0 & 0 & 0 & \cdots & 0 & r_0 & r_1 & r_2 & & \cdots & r_n \end{bmatrix}$$

and the r_i are the coefficients of the polynomial $r(D)$. M_2, M_3, and M_4 are analogously defined with coefficients of $p(D)$, $q(D)p^*(D)r(D)$, and $q(D)q^*(D)p(D)$ replacing those of $r(D)$. Using the process equations (15) and the definition of M^* one can see that

$$\tilde{\phi}(t) = M^*\tilde{\Psi}(t) + \bar{\delta}(t) \tag{34a}$$

$$\tilde{\Psi}(t) = \begin{cases} [\tilde{z}, D\tilde{z}, \cdots, D^{4n-1}\tilde{z}]^T & \text{CT case} \\ [D\tilde{z}, D^2\tilde{z}, \cdots, D^{4n}\tilde{z}]^T & \text{DT case} \end{cases} \tag{34b}$$

where $\bar{\delta}(t)$ is a vector of decaying exponentials unless $f(D) = 1$ in which case $\bar{\delta}(t) = 0$.

This implies that

$$\Phi_k = M^*\Psi_k + \Delta_k \tag{35a}$$

where

$$\Psi_k = [\tilde{\Psi}(t_k^0), \tilde{\Psi}(t_k^1), \cdots, \tilde{\Psi}(t_k^{N-1})] \tag{35b}$$

and Δ_k is a matrix of exponentials or zero.

Using (1) and (11), it can be seen that during each I_k, $\tilde{z}(t)$ satisfies

$$[q(D)p(D) - h_k(D)r(D) - k_k(D)p(D)]f(D)\tilde{z}(t) = q(D)v(t). \quad t \in I_k \tag{36}$$

Defining $s(D)$ as

$$s(D) = \begin{cases} \prod_{i=1}^{2n} (D^2 + \omega_i^2) & \text{CT case} \\ \prod_{i=1}^{2n} (1 - 2D\cos\omega_i + D^2) & \text{DT case} \end{cases}$$

it is also clear that the signal $v(t)$ satisfies

$$s(D)v(t) = 0. \tag{37}$$

Multiplying (36) by $s(D)$ and using (37) yields

$$p_k(D)\tilde{z}(t) = 0, \qquad t \in I_k \tag{38a}$$

$$p_k(D) = s(D)[q(D)p(D) - h_k(D)r(D) - k_k(D)p(D)]f(D) \tag{38b}$$

$$\begin{aligned} &\deg(P_k(D)) = 9n-1 && \text{CT case} \\ &4n < \deg(p_k(D)) \leq 6n && \text{DT case.} \end{aligned} \tag{38c}$$

Solutions to (38) take the following form:

$$\tilde{z}(t) = \sum_{i=1}^{m}\sum_{j=1}^{q_i} A_{ij}(t-t_k)^{q_i-1}\rho_i^{(t-t_k)}, \qquad t \in I_k \tag{39a}$$

where in CT case s_i, $1 \leq i \leq m$ are roots of $p_k(D)$ of multiplicity q_i and $\rho_i = e^{s_i}$, $1 \leq i \leq m$.

In the DT case, $\rho_i = 1/\eta_i$, where η_i, $1 \leq i \leq m$ are the roots of $p_k(D)$ of multiplicity q_i.

Notice that m, q_i, s_i, ρ_i, A_{ij} can vary with k, however, to simplify notation, dependence on k has been suppressed.

Using (39) $\tilde{\Psi}(t)$ can be written as

$$\tilde{\Psi}(t) = Q_k\gamma(t-t_k) \tag{40a}$$

where

$$\gamma(t) = [\rho_1^t, t\rho_1^t, \cdots t^{q_1-1}\rho_1^t, \rho_2^t, \cdots, t^{q_m-1}\rho_m^t]^T. \tag{40b}$$

In the CT case the matrix Q_k is $(4n) \times (9n-1)$ and its entries are functions of s_i and A_{ij}. In the DT case the matrix Q_k is $(4n) \times L$, $L = \deg(p_k(D))$ and its entries are functions of ρ_i and A_{ij}. This matrix plays an important role in the analysis to follow and its structure is discussed below. Combining (35) and (40) yields

$$\Phi_k = M^*Q_k\Gamma_k + \Delta_k \tag{41}$$

where

$$\Gamma_k = \begin{cases} [\gamma(0), \gamma(\tau_k), \cdots, \gamma((N-1)\tau_k)] & \text{CT case} \\ [\gamma(0), \gamma(1), \cdots, \gamma(N-1)] & \text{DT case.} \end{cases} \tag{42}$$

As is commonly the case with linear system theory, multiple roots do not affect the results presented below, however, they do considerably complicate the analysis. Thus, for brevity we will restrict detailed analysis to the case where $\tilde{z}(t)$ takes the simpler distinct root form

$$\tilde{z}(t) = \sum_{i=1}^{M} A_i\rho_i^{(t-t_k)}, \qquad t \in I_k \tag{43a}$$

where

$$M = \begin{cases} 9n-1 & \text{CT case} \\ L,\ 4n < L \leq 6n & \text{DT case} \end{cases} \tag{43b}$$

modifications to the analysis for the case where $\tilde{z}(t)$ takes the more general form (30) are available in [10].

For (43), the coefficients A_i will have the structure

$$A_i = \begin{cases} \dfrac{g(s_i)}{d_i} & \text{CT case} \\ \dfrac{g(\rho_i)}{d_i} & \text{DT case} \end{cases} \tag{44}$$

where

$$d_i=\begin{cases}\prod\limits_{\substack{1\le j\le 9n-1\\ j\ne i}}(s_i-s_j) & \text{CT case}\\ \prod\limits_{\substack{1\le j\le L\\ j\ne i}}(\rho_i-\rho_j) & \text{DT case}\end{cases} \tag{45}$$

and $g(\cdot)$ is a polynomial whose coefficients are a function of the coefficients of $p_k(D)$, and the initial conditions $\tilde{z}(t_k)$, $D\tilde{z}(t_k)$, $D^2\tilde{z}(t_k)$, $\cdots$, at the beginning of the continuous or discrete interval I_k.

In this case then

$$Q_k=G_kD_k^{-1} \tag{46}$$

where in the CT case

$$G_k=\begin{bmatrix} g(s_1) & g(s_2) & \cdots & g(s_{9n-1})\\ s_1g(s_1) & s_2g(s_2) & \cdots & s_{9n-1}g(s_{9n-1})\\ \cdot & & & \\ \cdot & & & \\ \cdot & & & \\ s_1^{4n-1}g(s_1) & s_2^{4n-1}g(s_2) & \cdots & s_{9n-1}^{4n-1}g(s_{9n-1})\end{bmatrix} \tag{47a}$$

and in the DT case

$$G_k=\begin{bmatrix} \eta_1g(\rho_1) & \eta_2g(\rho_2) & \cdots & \eta_Lg(\rho_L)\\ \eta_1^2g(\rho_1) & \eta_2^2g(\rho_2) & \cdots & \eta_L^2g(\rho_L)\\ \cdot & & & \\ \cdot & & & \\ \cdot & & & \\ \eta_1^{4n}g(\rho_1) & \eta_2^{4n}g(\rho_2) & \cdots & \eta_L^{4n}g(\rho_L)\end{bmatrix}. \tag{47b}$$

In both cases

$$D_k=\text{ diag }(d_i), \qquad 1\le i\le M. \tag{48}$$

Also when (43) holds, Γ_k has the Vandermonde structure

$$\Gamma_k=\begin{bmatrix} 1 & \rho_1 & \rho_1^2 & \cdots\rho_1^{N-1}\\ 1 & \rho_2 & \rho_2^2 & \cdots\rho_2^{N-1}\\ \cdot & & & \\ \cdot & & & \\ \cdot & & & \\ 1 & \rho_M & \rho_M^2 & \cdots\rho_M^{N-1}\end{bmatrix}. \tag{49}$$

To summarize the results of the subsection, when $\tilde{z}(t)$ has the structure (43), Φ_k can be factored as

$$\begin{aligned}\Phi_k&=M^*\Psi_k+\Delta_k\\ &=M^*G_kD_k^{-1}\Gamma_k+\Delta_k\end{aligned} \tag{50}$$

where M^*, G_k, D_k, and Γ_k are given by (33), (47), (48), and (49), respectively.

Analysis of $\Phi_k\Phi_k^T$: Having derived (50), we can now relate condition (29) on Φ_k, Φ_k^T to M^*, G_k, D_k, and Γ_k by the following lemma and corollary.

Lemma 2: If a constant $\epsilon_0 > 0$ exists such that

$$(M^*\Psi_k)(M^*\Psi_k)^T>\epsilon_0 I \qquad \text{i.o.} \tag{51a}$$

then a positive constant ϵ_1 exists for which

$$\Phi_k\Phi_k^T>\epsilon_1 I \qquad \text{i.o.} \tag{51b}$$

Proof: Let x be an arbitrary vector such that $\|x\| = 1$. Using properties of norms

$$\begin{aligned}\|x^T\Phi_k\|&=\|x^TM^*\Psi_k+x^T\Delta_k\|\\ &\ge\|x^TM^*\Psi_k\|-\|x^T\Delta_k\|\\ &\ge\|x^TM^*\Psi_k\|-\|x\|\|\Delta_k\|.^1\end{aligned}$$

But by assumption

$$\|x^TM^*\Psi_k\|>\sqrt{\epsilon_0}\,\|x\| \qquad \text{i.o.}$$

Thus,

$$\|x^T\Phi_k\|\ge(\sqrt{\epsilon_o}-\|\Delta_k\|)\|x\| \qquad \text{i.o.}$$

Since $\|\Delta_k\|$ converges to zero as k increases, there exists a finite k^* such that for any $\epsilon_1 < \epsilon_0$ $\|\Delta_k\| < \sqrt{\epsilon_0} - \sqrt{\epsilon_1}$. Thus for $k > k^*$

$$\|x^T\Phi_k\|\ge\sqrt{\epsilon_1}\,\|x\| \qquad \text{i.o}$$

This implies (51b) and completes the proof.

Corollary 2: A sufficient condition for $\Phi_k\Phi_k^T > \epsilon_1 I$ i.o. where ϵ_1 is a positive constant is

$$(D_k^{-1}\Gamma_k)(D_k^{-1}\Gamma_k)^T\ge\epsilon_2 I \qquad \text{i.o.} \tag{52a}$$

$$G_kG_k^T\ge\epsilon_3 I \qquad \text{i.o.} \tag{52b}$$

$$M^*M^{*T}\ge\epsilon_4 I \qquad \text{i.o.} \tag{52c}$$

for positive constants ϵ_2, ϵ_3, ϵ_4.[2]

Proof: By the factorization in (50) it is easy to see that the above conditions imply $(M^*\Psi_k)(M^*\Psi_k)^T > \epsilon_0 I$ infinitely often with $\epsilon_0 = \epsilon_2\epsilon_3\epsilon_4$. Lemma 2 completes the proof.

Lemma 3: If τ_k is chosen according to Assumption 1 (for the CT case only) and N according to Assumption 2, then there exists an $\epsilon_2 > 0$ such that

$$(D_k^{-1}\Gamma_k)(D_k^{-1}\Gamma_k)^T > \begin{cases}\epsilon_2 I \text{ for all } k\ge0 & \text{DT case}\\ \epsilon_2 I \text{ with probability 1 for all } k\ge0 & \text{CT case.}\end{cases} \tag{53}$$

Proof: First observe that the boundedness of the control parameter vector α_k guarantees the coefficients of $p_k(D)$ to be bounded. This in turn guarantees the $s_i(\rho_i$ in the DT case), $1 \le i \le M$, to be bounded for all k. Hence, the diagonal entries of the diagonal matrix D_k^{-1} are bounded away from zero since each entry d_j^{-1} is the reciprocal of a polynomial in either the variables s_i or ρ_i. Thus, (53) will be satisfied provided the rows of Γ_k are independent and not converging toward dependence. By choosing N according to Assumption 3, Γ_k will have at least as many columns as rows. Furthermore, since the rows have a Vandermonde structure, they can be dependent or converging toward dependence if for a pair of indexes i and j, $i \ne j$, $\rho_i = \rho_j$ or $\lim_{k\to\infty} \rho_i = \rho_j$.

By assuming no multiple poles, $s_i \ne s_j$ in the CT case and $\rho_i \ne \rho_j$ in the DT case. However, in the CT case, since $\rho_i = e^{S_i\tau_k}$, $\rho_i = \rho_j$ if and only if

$$s_i=\alpha+j\beta_i \tag{54a}$$

$$s_j=\alpha+j\beta_j \tag{54b}$$

$$(\beta_i-\beta_j)\tau_k=2\pi l \qquad l=0,\ \pm1,\ \pm2,\ \cdots. \tag{54c}$$

The set of τ_k satisfying (54c) is a set of measure zero. Consequently, by choosing τ_k randomly as discussed under Assumption 2, $\rho_i \ne \rho_j$ with probability 1.

[1] $\|\Delta_k\|$ is the matrix norm induced by the Euclidean vector norm.
[2] In Corollary 2 and Lemmas 3 and 4 "T" denotes conjugate transpose.

The only other possibility is one or more s_i (ρ_i in the DT case) converging to some s_j(ρ_j in the DT case) as k increases which corresponds to $p_k(D)$ having multiple roots asymptotically. For this case one must consider the effects of s_i converging to s_j(ρ_i to ρ_j in DT case) on both D_k^{-1} and Γ_k simultaneously. A detailed analysis is given in the Appendix.

Next the matrix G_k is considered. The main result is given by Lemma 4.

Lemma 4: The matrix $G_kG_k^T$ satisfies

$$G_kG_k^T \geq \epsilon_3 I \ \forall_k \geq 0 \tag{55}$$

for some positive constant ϵ_3.

Proof: Without loss of generality we can assume the s_i to be ordered so that

$$s_{2i-1}=j\omega_i \ (\rho_{2i-1}=e^{j\omega_i} \quad \text{in the DT case})$$

$$s_{2i}=-j\omega_i \ (\rho_{2i}=e^{-j\omega_i} \quad \text{in the CT case}) \ i=1, \cdots, 2n. \tag{56}$$

Partition G_k as

$$G_k=[G_{1k} \vdots G_{2k}] \tag{57}$$

where G_{1k} contains the first $4n$ columns of G_k. Since

$$G_kG_k^T=G_{1k}G_{1k}^T+G_{2k}G_{2k}^T$$

it is sufficient to show that there exists an $\epsilon_5>0$ such that $G_{1k}G_{1k}^T \geq \epsilon_5 I$. Using (47), G_{1k} can be written as

$$G_{1k}=SE_k \tag{58a}$$

where for the CT case

$$S=\begin{bmatrix} 1 & 1 & \cdots & 1 \\ s_1 & s_2 & & s_{4n} \\ s_1 & \cdot & & \cdot \\ \cdot & \cdot & & \cdot \\ \cdot & \cdot & & \cdot \\ \cdot & & & \\ s_1^{4n-1} & s_2^{4n-1} & & s_{4n}^{4n-1} \end{bmatrix} \tag{58b}$$

$$E_k= \text{diag } (g(s_i)) \tag{58c}$$

and in the DT case

$$S=\begin{bmatrix} \eta_1 & \eta_2 & \cdots & \eta_{4n} \\ \eta_1^2 & \eta_2^2 & & \eta_{4n}^2 \\ \cdot & \cdot & & \\ \cdot & \cdot & & \\ \cdot & \cdot & & \\ \eta_1^{4n} & \eta_2^{4n} & & \eta_{4n}^{4n} \end{bmatrix} \tag{58d}$$

$$E_k= \text{diag } (g(\rho_i)). \tag{58e}$$

Clearly, $G_{1k}G_{1k}^T \geq \epsilon_5 I$, provided there exist, $\epsilon_6>0$ and $\epsilon_7>0$ such that $SS^T \geq \epsilon_6 I$, $E_kE_k^T \geq \epsilon_7 I$. Since the reference does not change between intervals, S is constant. Furthermore, since it has a Vandermonde structure, $SS^T \geq \epsilon_6 I$.

Finally, observe that because the system is driven by a nondiminishing signal containing $2n$ sinusoids, these sinusoids will always be observable in $\tilde{z}(t)$. This is most easily seen by solving (36) for $\tilde{z}(t)$ rather than (38) and observing that $q(D)$ is strictly Hurwitz by assumption. This guarantees that there exists an $\epsilon_7>0$ such that

$$g(s_i) \geq \epsilon_7 \quad 1 \leq i \leq 4n \qquad \text{CT case}$$

or

$$g(\rho_i) \geq \epsilon_7 \quad 1 \leq i \leq 4n \qquad \text{DT case}$$

which established $E_kE_k^T > \epsilon_7 I$, and hence (55). The following theorem then summarizes the results of this section.

Theorem 1: Let the unknown system to be controlled be given by (1), where $r(D)$ and $p(D)$ are prime and n, the system order, is known. Let the external driving signal be a sum of $2n$ sinusoids as discussed in Assumption 1. Let the control law (11) be implemented, and the control parameters estimated using (27). In the CT case, let the sampling period for each interval, I_k, be given by Assumption 2. Let the number of samples per interval be given by Assumption 3. Then the closed-loop system is globally stable. In particular, $u(t)$, $y(t)$, and $z(t)$ are bounded, and the closed-loop system converges to one which satisfies (8).

V. Indirect Adaptive Pole Placement

In this section, we briefly present a globally stable indirect adaptive pole placement scheme which is basically a reformulation of the ideas that were discussed in the preceding sections. Unless mentioned otherwise, the symbols used in the discussion to follow retain their old meanings.

Once again, consider the problem of controlling a linear shift invariant (DT case) or time invariant (CT case) system characterized by (2) and use the same fixed control strategy that was discussed in Section II. Thus, the closed-loop system is once again given by (8). However, in the adaptive formulation we estimate the system parameters, i.e., the coefficients of $r(D)$ and $p(D)$ directly using (2) and then use these estimates to solve the design equation (7) for $h_k(D)$ and $k_k(D)$, the estimates of $h(D)$ and $k(D)$ at $t=t_k$.

Once again block processing is necessary in the sense that the system identifier will be run N times during each interval I_k and the control parameter vector α_k as given by (13), will be indirectly updated only at instants t_k, $k=0, 1, 2, \cdots$.

Estimating θ_k: For identification of the system parameters, first of all, notice that the system equation (2) can be rewritten as follows:

$$D^n y(t)=\phi^T(t)\theta^* \qquad \text{CT case} \tag{59a}$$

$$y(t)=\phi^T(t)\theta^* \qquad \text{DT case} \tag{59b}$$

where

$$\phi(t)=\begin{cases} [y, Dy, \cdots, D^{n-1}y, u, Du, \cdots, D^{n-1}u]^T & \text{CT case} \\ [Dy, D^2y, \cdots, D^n y, Du, D^2u, \cdots, D^n u]^T & \text{DT case} \end{cases} \tag{60}$$

$$\theta^*=[-p_0, -p_1, \cdots, -p_{n-1}, r_0, r_1, \cdots, r_{n-1}]^T. \tag{61}$$

Defining once again $\tilde{z}(t)$, $\tilde{y}(t)$, $\tilde{u}(t)$ by (14) where $f(D)$, the filter polynomial, is now of degree $(2n-1)$ in the CT case and $f(D)=1$ in the DT case, one can show that

$$\tilde{\phi}^T(t)\theta^*=\begin{cases} D^n\tilde{y}(t)=\delta_7(t) & \text{CT case} \\ \tilde{y}(t)=\delta_7(t) & \text{DT case.} \end{cases} \tag{62}$$

Equation (62) can be used to find θ_k^i, an estimate of θ^* at $t=t_k^i$, by following the procedure described in Section III. The only difference is that the error sequence to be minimized at each step is now given by

$$\epsilon(t_k^i)=\begin{cases} \tilde{\phi}(t_k^i)^T\theta_k^i-D^n\tilde{y}(t_k^i) & \text{CT case} \\ \tilde{\phi}(t_k^i)^T\theta_k^i-\tilde{y}(t_k^i) & \text{DT case.} \end{cases} \tag{63}$$

Thus, Lemma (1) holds in this case as well.

Calculating α_k: In the CT case, it is well known, e.g., [15] that α^* satisfies an equation of the form

$$M^*\alpha^*=F^*$$

where M^* is the transpose of the eliminant matrix for the polynomials $r(D)$ and $p(D)$ [15], and f^* is a vector containing the coefficients of $q(D)[p(D) - p^*(D)]$. A similar equation results in the DT case. Thus, an estimate α_k of α^* can be obtained by using θ_k to generate estimates M_k and f_k of M^* and f^* and calculating

$$\alpha_k = M_k^{-1} f_k.$$

However, to guarantee the boundedness of α_k we assume, similarly to [8], a lower bound μ on $|\det(M_k)|$ and define

$$\alpha_k = \begin{cases} \alpha_{k-1} & \text{if } |\det(M_k)| < \mu \\ M_k^{-1} f_k & \text{otherwise.} \end{cases}$$

This guarantees α_k remains bounded even when the estimates of $r(D)$ and $p(D)$ have common roots or roots very close to each other. This corresponds to uncontrollability or near uncontrollability of the estimated model.

Stability Analysis: The stability analysis now proceeds similarly to that given in Section IV. In particular, Lemma 1 guarantees $\lim_{k\to\infty} \theta_k = \theta^*$, and hence $\lim_{k\to\infty} \alpha_k = \alpha^*$ provided $\Phi_k \Phi_k^T > \epsilon_1 I$ infinitely often. In this case

$$\Psi_k = M^* \Phi_k + \Delta_k$$

where M^* is the $(2n \times 2n)$ transposed eliminant matrix and

$$\Psi_k = [\bar{\psi}(t_k^0), \bar{\psi}(t_k^1), \cdots, \bar{\psi}(t_k^{N-1})]$$

$$\bar{\Psi}(t) = \begin{cases} [\bar{z}, D\bar{z}, \cdots, D^{2n-1}\bar{z}]^T & \text{CT case} \\ [D\bar{z}, \cdots, D^{2n}\bar{z}]^T & \text{DT case.} \end{cases}$$

The factorization and analysis of Ψ_k then proceeds analogously to that given in Section IV. However, one important modification is necessary. In particular, the reduction in dimensionality of θ^*, and hence Φ_k allows Assumption 1 and 3 to be modified so that $v(t)$ need only consist of n sinusoids and

$$N \geq \begin{cases} 5n-1 & \text{CT case} \\ 3n & \text{DT case.} \end{cases}$$

That is, the minimum block sizes are smaller.

VI. Concluding Remarks

This paper has presented a proof of global stability for direct and indirect adaptive control schemes which arbitrarily assign the closed-loop poles of a single-input, single-output system. The results are applicable to the control of either minimum or nonminimum phase systems. Both hybrid continuous-time and discrete-time formulations were presented. To the authors' knowledge, this is the first verification of the persistent excitation condition for the potentially unbounded data resulting from closed-loop adaptive control using a sequential estimation procedure. The main limitation of the result is that it is only applicable to the case of sinusoidal excitation. An important extension of this work would be to other persistently exciting input signals. In fact, in the discrete case any signal over a finite interval has a discrete Fourier transform (DFT). The inverse DFT then gives a representation of any signal over that interval as a sum of sinewaves. By considering the inverse DFT of the reference over each block of length N it is possible to extend this result to arbitrary discrete reference signals. The key to developing such a result is to show that the block length N can be chosen independently of the number of sinusoids in the reference. Thus, it is felt that future analysis along the lines given in this report can yield a rather general persistency of excitation condition in the discrete time case and alleviate the restrictions imposed in this report.

A key to the proof was block processing. The minimum length of a block ranged from $9n - 1$ in the direct CT case to $3n$ in the direct DT case. This value resulted from technicalities in the proof. More work needs to be done to see if smaller blocks still result in stable operation. This is important since long block lengths can cause a deterioration in performance. Similarly, the CT case random sampling was necessary because of technicalities in the proof. Hopefully, further analysis will show that periodic sampling is sufficient.

Appendix

For brevity, we will consider the CT case only to show that the rows of $D_k^{-1}\Gamma_k$ do not converge toward dependence if s_i converges to s_j for some index pair $i \neq j$. A specific case will be considered. In particular, it will be assumed that s_2 converges to s_1, but there exists an $a_1 > 0$ such that

$$|s_i - s_j| > a_1, \qquad 2 \leq i < j, \ k \geq 0. \tag{A1}$$

Theorem A: If (A1) holds, then there exists a constant $a_2 > 0$, and a $K > 0$ such that for any unity norm vector sequence X_k

$$X_k^T (D_k^{-1}\Gamma_k)(D_k^{-1}\Gamma_k) X_k > a_2 X_k^T X_k \tag{A2}$$

for all $k > K$.

Proof: Assume that (A2) does not hold. Since X_k is bounded for all k, then for a subsequence k_l,

$$\lim_{l\to\infty} X_{k_l}^T (D_{k_l}^{-1}\Gamma_{k_l})(D_{k_l}^{-1}\Gamma_{k_l})^T X_{k_l} = 0 \tag{A3}$$

or equivalently

$$\lim_{l\to\infty} (D_{k_l}^{-1}\Gamma_{k_l})^T X_{k_l} = \bar{0} \tag{A4}$$

where $\bar{0}$ is the zero vector.

Columns 3 through N or Γ_k have a Vandermond structure. Thus, since (A1) and the random choice for τ_k assure for some $a_3 > 0$

$$|\rho_i - \rho_j| > a_3 \qquad 2 \leq i < j, \ k \geq 0. \tag{A5}$$

Equation (A4) can only hold if x_1 of x_2 are bounded away from zero where x_i is defined as the ith entry of X_{k_l}. Let c_i denote the ith column of $(D_{k_l}^{-1}\Gamma_{k_l})^T$ and consider $x_1 c_1 + x_2 c_2$. Since it is assumed that $\lim_{k\to\infty} \rho_2 = \rho_1$, let $\rho_2 = \rho_1 + \delta$ where $\lim_{k\to\infty} \delta = 0$. We can then write

$$x_1 c_1 + x_2 c_2 = \frac{-x_1}{d_1'\delta} \begin{bmatrix} 1 \\ \rho_1 \\ \rho_1^2 \\ \cdot \\ \cdot \\ \cdot \\ \rho_1^{N-1} \end{bmatrix} + \frac{x_2}{d_2'\delta} \begin{bmatrix} 1 \\ \rho_1 e^{\delta\tau_k} \\ \rho_1^2 e^{2\delta\tau_k} \\ \cdot \\ \cdot \\ \cdot \\ \rho_1^{N-1} e^{(N-1)\delta\tau_k} \end{bmatrix} \tag{A6}$$

$$d_1' = \prod_{j=3}^{M} (s_1 - s_j) \tag{A7}$$

$$d_2' = \prod_{j=3}^{M} (s_1 + \delta - s_j) \tag{A8}$$

observe that if $|x_1 - x_2|$ is bounded away from zero the first entry in $x_1 c_1 + x_2 c_2$ will become unbounded as δ converges to zero. Since $\sum_{j=3}^{M} x_i c_i$ will be bounded, one must conclude that x_2 converges to x_1. Application of L'Hospital's rule as δ converges to zero then yields

$$\lim_{k\to\infty} x_1 c_1 + x_2 c_2 = \frac{x_1}{d_1'} \begin{bmatrix} 0 \\ \tau_k \rho_1 \\ 2\tau_k \rho_1^2 \\ \cdot \\ \cdot \\ \cdot \\ (N-1)\tau_k \rho_1^{N-1} \end{bmatrix} \tag{A9}$$

$$\triangleq x_1 c_1'.$$

Thus, defining

$$X'_{k_l}=[x_1,\ x_3,\ \cdots,\ x_N] \tag{A10}$$

$$(D'^{-1}_{k_l}\Gamma'_{k_l})^T=[c'_1,\ c_3,\ \cdots,\ c_N] \tag{A11}$$

$$D'_{k_l}=\ \text{diag}\ [d'_1,\ d_2,\ \cdots,\ d_M] \tag{A12}$$

condition (A4) simplifies to

$$\lim_{k\to\infty}(D'^{-1}_{k_l}\Gamma'_{k_l})^T X'_{k_l}=\bar{0}. \tag{A13}$$

However, since the diagonal entries of D'_{k_l} are nonzero and bounded and since Γ'_k will have a generalized Vandermonde structure similar to that discussed in [10], (A13) cannot hold provided the ρ_i satisfy (A5). Thus, by contradiction, we must conclude that (A2) is satisfied.

References

[1] H. Elliott and W. A. Wolovich, "Parameter adaptive identification and control," *IEEE Trans. Automat. Contr.*, vol. AC-24, Aug. 1979.

[2] G. Kreisselmeier, "Adaptive control via adaptive observation and asymptotic feedback matrix synthesis," *IEEE Trans. Automat. Contr.*, vol. AC-25, Aug. 1980.

[3] K. J. Astrom and B. Wittermark, "On self-tuning regulators," *Automatica*, vol. 9, 1973.

[4] G. C. Goodwin, E. K. Teoh, and B. C. McInnis, "Globally convergent adaptive controllers for linear systems having arbitrary zeros," Univ. Newcastle, New South Wales, Australia, Tech. Rep., May 1982.

[5] H. Elliott, "Direct adaptive pole placement with application to nonminimum phase systems," *IEEE Trans. Automat. Contr.*, vol. AC-27, June 1982.

[6] G. E. Shilov, *Linear Algebra.* Englewood Cliffs, NJ: Prentice-Hall, 1971.

[7] R. M. Johnstone and B. D. O. Anderson, "Global adaptive pole placement: Detailed analysis of a first order system," Dep. Syst. Eng., Australian Nat. Univ., Canberra, Australia, Tech. Rep.

[8] B. D. O. Anderson and R. M. Johnstone, "Global adaptive pole positioning," Dep. Syst. Eng., Australian Nat. Univ., Canberra, Australia, Tech. Rep.

[9] H. Elliott, "Hybrid adaptive control of continuous time systems," *IEEE Trans. Automat. Contr.*, vol. AC-27, no. 2, Apr. 1982.

[10] H. Elliott, R. Cristi, and M. Das, "Global stability of direct adaptive pole placement algorithms," Univ. Mass., Amherst, Tech. Rep. UMASS-ECE-No-82-1, Nov. 1982.

[11] H. Elliott, W. A. Wolovich, and M. Das, "Arbitrary adaptive pole placement for linear multivariable systems," *IEEE Trans. Automat. Contr.*, vol. AC-29, Mar. 1984.

[12] G. C. Goodwin, P. J. Ramadge, and P. E. Caines, "Discrete time multivariable adaptive control," *IEEE Trans. Automat. Contr.*, vol. AC-25, June 1980.

[13] G. C. Goodwin, H. Elliott, and E. K. Teoh, "Deterministic convergence of a self-tuning regulator with covariance resetting," *IEEE Proc. Inst. Elec. Eng.*, vol. 130, pt. D, no. 1, Jan. 1983.

[14] G. C. Goodwin and R. L. Payne, *Dynamic System Identification: Experiment Design and Data Analysis.* New York: Academic, 1977.

[15] W. A. Wolovich, *Linear Multivariable Systems.* New York: Springer-Verlag, 1974.

[16] B. D. O. Anderson and C. R. Johnson, "Exponential convergence of adaptive identification and control algorithms," *Automatica*, vol. 18, no. 1, 1982.

Part II
Model Reference Adaptive Approach

IN THE second part of this volume, we present the adaptive approaches based upon model reference adaptive control (MRAC or MRAS) techniques. These approaches have been very popular for the last two decades. These types of adaptive systems are supposed to operate satisfactorily in the presence of environmental and parameter uncertainties. Whitaker *et al.* were the first to formulate such an adaptive system in 1958 at the Massachusetts Institute of Technology. However, their design, called the M.I.T. rule, was based upon the minimization of certain performance index, and had a basic inherent stability problem.

P. C. Parks was the first to suggest a redesign of the MRAS in 1966 using the Liapunov stability method which is reported in the tenth paper. To some extent, using this new design procedure, he came over the basic stability problem. A number of interesting but simple examples are included. Readers are encouraged to study the basic design philosophy given in this paper, along with his other paper (twelfth paper).

In the eleventh paper, Nikiforuk, Gupta, and Choe report a new method for the design of model reference adaptive system for a class of unknown plants by introducing the concept of a "characteristic variable." The characteristic variable possesses all the information of the unmodeled plant dynamics in an implicit way. This characteristic variable is generated and used in an adaptive feedback mode to generate a suitable control signal. The Liapunov stability method is used to synthesize an adaptive controller which guarantees the stability of the overall system. The introduction of the characteristic variable which contains all the information implicitly regarding the plant uncertainties, and its unknown parameters, reduces the synthesis problem to one that involves a known linear and time-invariant lower-order plant. An extension of this approach is given in the thirtieth paper with applications to an aircraft gust alleviation control system. Although an interesting notion, the concept of characteristic variable needs further investigations especially for a general class of plants. This concept may find useful applications for industrial systems as shown in the thirtieth paper.

In the thirteenth paper, Monopoli introduces the concept of an augmented error signal in the design of an adaptive controller. Controllers for single-input single-output, nonlinear and nonautonomous continuous and discrete plants are reported in this paper. The concept of an augmented error signal is shown to be useful in designing a stable adaptive control system without the use of plant output derivatives. A fruitful area for further studies appears to be an extension of this concept to systems that involve a practical situation.

Landau and Lozano present a unification of discrete time explicit model reference adaptive control designs in the fourteenth paper. In this article, the authors give a general structure and design procedure for adaptive control design. The general structure proposed helps to independently specify the tracking and regulation specifications. Some of these design methods have been reported in the design of controllers for practical applications.

In the fifteenth paper, Kreisselmeier and Narendra report their findings on the design of stable model reference adaptive control in the presence of bounded disturbances. The class of plants considered is linear and time-invariant with bounded disturbances. This is one of many papers written by Kreisselmeier and Narendra either jointly or separately. This paper illustrates the mathematical complexities associated even in the design of linear time-invariant systems. Modern technological systems are much more complex, of course.

Lastly, Narendra, Khalifa, and Annaswamy present, in the sixteenth paper, several stable adaptive algorithms for the control of hybrid and discrete systems in which the control parameters are updated infrequently. The proposed algorithms seem to be robust, and they provide a common framework for the control of systems with two time-scales. The authors give various simulation studies for different types of adaptive algorithms proposed in this paper. The extention of the error models for a general class of systems, if it can be achieved, may find practical applications.

Seven papers included in this part of the volume merely represent a small sample of papers from a large amount of literature that exists in the areas of model reference adaptive techniques. Hopefully, this small sample along with the cited references will provide a direction to the readers for further works in the field. Some useful applications of MRAS are reported in Parts V, VI and VII of this volume.

Selected Bibliography

R. E. Bowles, "Self-adaptive control system," U.K. Patent I 242 286, Oct. 14, 1968; Publ. Sept. 17, 1971.

R. L. Butchart, "Design of model reference control systems using a Lyapunov synthesis technique," *Proc. IEE,* vol. 114, no. 9, pp. 1363–1364, 1967.

R. M. Casciano and H. K. Staffin, "Model reference adaptive control systems," *Amer. Inst. Chem. Eng. J.,* vol. 13, no. 3, pp. 485–491, 1967.

A. J. U. T. Cate and N. D. L. Verstoep, "Improvement of Lyapunov model reference adaptive control systems in a noisy environment," *Int. J. Contr.,* vol. 20, no. 6, pp. 977–966, 1974.

H. H. Choe and P. N. Nikiforuk, "Inherently stable feedback control of a class of unknown plants," *Automatica,* vol. 7, no. 5, pp. 607–625, 1971.

B. Colburn and J. S. Boland, III, "A unified approach to model-reference adaptive systems. Part II: Application of conventional design techniques," *IEEE Trans. Aerosp. Electron. Syst.,* pp. 514–523, 1978.

D. D. Donalson and C. T. Leondes, "A model reference parameter tracking technique for adaptive control systems—The principle of adaptation," *IEEE Trans. Appl. Ind.,* vol. 82, pp. 241–262, 1963.

R. M. Dressler, "An approach to model referenced adaptive control system," *IEEE Trans. Automat. Cont.*, vol. AC-12, no. 1, pp. 75–80, 1967.

L. Dugard, G. C. Goodwin, and C. De Souza, "Prior knowledge in model reference adaptive control of multi-input multi-output systems," presented at the 22nd IEEE Decision and Control Conf., 1983.

B. Egardt, "Unification of some discrete-time adaptive control systems," *IEEE Trans. Automat. Contr.*, vol. AC-24, no. 4, pp. 693–697, 1980.

——, "Stability analysis of discrete-time adaptive control systems," *IEEE Trans. Automat. Contr.*, vol. AC-25, no. 4, pp. 710–717, 1980.

H. Elliott and W. A. Wolovich, "Parameter adaptive identification and control," *IEEE Trans. Automat. Contr.*, vol. AC-24, no. 4, pp. 592–599, 1979.

F. Fortunato and H. Kaufman, "Model reference adaptive control of thyristor driven DC motor systems with resonant loading," in *Proc. 23rd IEEE Conf. Decision and Control,* pp. 375–380, 1984.

——, "Model reference adaptive control of thyristor driven DC motor systems with resonant loading," *Proc. 23rd IEEE Conf. Decision and Control;* "Optimal controllers based on input matching," *IEEE Trans. Automat. Control,* vol. AC-26, no. 6, pp. 1269–1273, 1984.

G. C. Goodwin, P. J. Ramadge, and P. E. Caines, "Discrete time multivariable adaptive control," *IEEE Trans. Automat. Contr.*, vol. AC-25, pp. 449–456, 1980.

G. C. Goodwin, R. Johnson, and K. S. Sin, "Global convergence for adaptive one step ahead optimal controllers based on input matching," *IEEE Trans. Automat. Contr.*, vol. AC-26, no. 6, pp. 1269–1273, 1981.

M. M. Gupta, P. N. Nikiforuk, and Y. Yamane, "Adaptive control for single-input single-output linear discrete systems with deadbeat convergence," in *Proc. 1979 JACC,* pp. 474–480, 1979.

M. M. Gupta, P. N. Nikiforuk, and H. H. Choe, "Model reference invariant control for a class of unknown multivariable plants," in *Int. Federation of Automat. Contr.*, 8th Triennial World Congress, Kyoto, Japan, pp. VII.114–VII.119, 1981.

M. M. Gupta and P. N. Nikiforuk, "On the multilevel optimization using on-line coordination," *Problems of Cont. and Inform. Theory,* Academy of Sci. of the USSR and Hungarian Academy of Sci., vol. 7, no. 4, pp. 237–262, 1978.

M. M. Gupta, N. Minamide, and P. N. Nikiforuk, "Design of an adaptive observer and its application to an adaptive pole placement controller," *Int. J. Contr.*, vol. 37, no. 2, pp. 349–366, 1983.

C. C. Hang and P. C. Parks, "On accelerated model-reference adaptation via Lyapunov and steepest descent design techniques," *IEEE Trans. Automat. Contr.*, vol. AC-17, no. 6, pp. 842–843, 1972.

——, "Comparative studies of model reference adaptive control systems," *IEEE Trans. Automat. Contr.*, vol. AC-18, no. 5, pp. 419–428, 1973.

C. C. Hang, "The design of model reference parameter estimation systems using hyperstability theories," in *Proc. 3rd IFAC Symp. Ident. Simulation and Parameter Ident.*, pp. 741–744, 1973.

——, "A new form of stable adaptive observer," *IEEE Trans. Automat. Contr.*, vol. AC-21, pp. 544–547, 1976.

T. Horrocks, "Investigations into model reference adaptive control systems," *Proc. IEE* (London), vol. 11, p. 1894, 1964.

P. A. Ioannou and P. V. Kokotovic, "Singular perturbations and robust redesign of adaptive control," presented at the 21st Conf. Decis. Control, 1982.

P. Ioannou, "Robust direct adaptive control," in *Proc. 23rd IEEE Conf. on Decision and Control,* 1984.

T. Ionescu and R. V. Monopoli, "Discrete model reference adaptive control with an augmented error signal," *Automatica,* vol. 13, no. 5, pp. 507–518, 1977.

D. J. G. James, "Stability analysis of a model reference adaptive control system with sinusoidal inputs," *Int. J. Contr.*, vol. 9, pp. 311–321, 1969.

——, "Stability of model-reference systems with random inputs," *Proc. Int. Symp. Stability of Stoch. Dyn. Syst.*, pp. 147–159, 1972.

C. R. Johnson, Jr. and G. C. Goodwin, "Robustness issues in adaptive control," presented at 21st Conf. Decision Control, 1982.

C. R. Johnson and R. R. Bitmead, "A robust, locally optimal model reference adaptive controller," in *Proc. 23rd IEEE Conf. on Decision and Control,* pp. 993–997, 1984.

R. M. Johnstone, D. G. Fisher, and S. L. Shah, "Hyperstable adaptive control—An indirect approach with explicit model identification," in *Proc. 1979 JACC,* pp. 495–499, 1979.

S. L. Kilmentor and B. J. Prakopov, "Synthesis of an asymptotically stable algorithm of adaptive system with reference mode by Lyapunov's direct method," *Automat. and Remote Contr.*, vol. 35, pt. 2, pp. 1625–1632, 1974.

M. Lal and R. Mehrotra, "Design of model reference adaptive control systems for non-linear plants," *Int. J. Cont.*, vol. 16, no. 5, pp. 993–966, 1972.

I. D. Landau, "A hyperstability criterion for model reference adaptive control systems," *IEEE Trans. Automat. Contr.*, vol. AC-14, no. 4, pp. 552–555, 1969.

——, "Synthesis of discrete model reference adaptive systems," *IEEE Trans. Automat. Contr.*, vol. AC-16, no. 5, pp. 507–508, 1971.

——, "Model reference adaptive systems. A survey (What is possible and why?)" *Trans. ASME J. Dyn. Syst. Measure. Contr.,* 94 Series G, pp. 119–132, 1972.

——, "A generalization of the hyperstability conditions for model reference adaptive systems," *IEEE Trans. Automat. Contr.*, vol. AC-17, no. 2, pp. 246–247, 1972.

——, "A survey of model reference adaptive techniques—Theory and applications," *Automatica,* vol. 10, no. 4, pp. 353–379, 1974.

——, "Unbiased recursive identification using model reference adaptive techniques," *IEEE Trans. Automat. Contr.*, vol. AC-21, no. 2, pp. 194–202, 1976.

I. D. Landau, *Adaptive Control—The Model Reference Approach.* New York: Marcel-Dekker, 1979.

R. E. Larson and W. S. Krecker, "Optimum adaptive control in an unknown environment," *IEEE Trans. Automat. Contr.*, vol. AC-13, no. 4, pp. 438–439, 1968.

N. T. Loan, "Some methods of synthesizing model reference adaptive control systems," *Eng. Cybern.*, vol. 9, pp. 386–394, 1971.

E. H. Lowe and J. R. Roland, "Improved signal synthesis techniques for model reference adaptive control systems," *IEEE Trans. Automat. Contr.*, vol. AC-19, no. 2, pp. 119–121, 1974.

J. C. Luxat and L. H. Less, "Suboptimal adaptive control of a class of nonlinear systems," *Int. J. Contr.*, vol. 17, pp. 965–975, 1973.

E. P. Maslov and L. M. Osovskii, "Adaptive control systems with models," *Automat. and Remote Contr.*, vol. 27, pp. 1116–1136, 1966.

R. V. Monopoli, "Lyapunov's method for adaptive control system design," *IEEE Trans. Automat. Contr.*, vol. AC-12, no. 3, pp. 334–335, 1967.

R. V. Monopoli, J. W. Gilbert, and C. F. Price, "Improved convergence and increased flexibility in the design of model reference adaptive control systems," in *Proc. 1970 IEEE Symp. Adapt. Proc.*, pp. IV.3.1–IV.3.10, 1970.

R. V. Monopoli, "Model reference adaptive control using only input and output signals," presented at the Proc. 1973 IEEE Conf. Dec. and Contr., 1973.

——, "The Kalman–Yakubovich lemma in adaptive control systems design," *IEEE Trans. Automat. Contr.*, vol. AC-18, no. 5, pp. 474–484, 1973.

——, "Model reference adaptive control with an augmented error signal," *IEEE Trans. Automat. Contr.*, vol. AC-19, pp. 474–482, 1974.

R. V. Monopoli and C. C. Hsing, "Parameter adaptive control of multivariable systems," *Int. J. Contr.*, vol. 22, no. 3, pp. 313–327, 1975.

R. V. Monopoli and V. N. Subbarao, "A new algorithm for model reference adaptive control with variable gains," *IEEE Trans. Automat. Contr.*, vol. AC-25, no. 6, pp. 1245–1248, 1980.

——, "A simplified algorithm for model reference adaptive control," *IEEE Trans. Automat. Contr.*, vol. AC-25, no. 3, pp. 579–581, 1980.

R. L. Morris and C. P. Neuman, "Model reference adaptive control with multiple sample between parameter adjustments," *IEEE Trans. Automat. Contr.*, vol. AC-26, no. 2, pp. 534–537, 1981.

K. S. Narendra, "Some problems in model reference adaptive control," in *Proc. 1972 Int. Conf. Cybern. and Soc.*, pp. 360–362, 1972.

K. S. Narendra and L. S. Valavani, "Stable adaptive observers and controllers," *Proc. IEEE,* vol. 64, pp. 1198–1208, 1976.

——, "Stable adaptive controller design—Direct control," *IEEE Trans. Automat. Contr.*, vol. AC-23, no. 4, pp. 570–583, 1978.

K. S. Narendra and Y. H. Lin, "Stable discrete adaptive control," *IEEE Trans. Automat. Contr.*, vol. AC-25, no. 3, pp. 456–461, 1980.

K. S. Narendra and L. S. Valavani, "A comparison of Lyapunov and hyperstability approaches to adaptive control of continuous systems," *IEEE Trans. Automat. Contr.*, vol. AC-25, pp. 243–247, 1980.

K. S. Narendra, Y. H. Lin, and L. S. Valavani, "Stable adaptive controller design, Part II: Proof of stability," *IEEE Trans. Automat. Contr.*, vol. AC-21, no. 3, pp. 440–449, 1980.

K. S. Narendra and A. M. Annaswamy, "A general approach to the stability analysis of adaptive systems," in *Proc. of the 23rd IEEE Conf. Decision and Control,* pp. 1298–1303, 1984.

P. N. Nikiforuk, M. M. Gupta, and H. H. Cho, "Control of unknown plants in reduced state space," *IEEE Trans. Automat. Contr.*, vol. AC-14, no. 5, pp. 489–496, 1969.

P. N. Nikiforuk, M. M. Gupta, and K. Kanai, "Stability analysis and design for aircraft gust alleviation control," *Automatica,* vol. 10, no. 5, pp. 494–506, 1974.

——, "A deterministic type controller for a class of uncertain plants with partial identification," in *Proc. 3rd IFAC Symp. on Ident. and Parameter Est.,* The Hague, Netherlands, pp. 555–565, 1973.

P. N. Nikiforuk, M. M. Gupta, K. Kanai, and L. Wagner, "Synthesis of two-level controller for a class of linear plants in an unknown environment," in *Proc. 3rd IFAC Symp. on Sens., Adapt., and Opt.,* pp. 431–439, 1973.

P. N. Nikiforuk, M. M. Gupta, and K. Tamura, "On the design of a signal synthesis model reference adaptive control systems for linear unknown plants," in *Proc. 4th IFAC Symp. Ident. and Syst. Parameter Est.,* Paper #20.5, 1976.

——, "Design of model reference adaptive control system for linear plants," *ASME J. Dyn. Syst., Meas., and Contr.,* Series G, no. 2, pp. 123–129, 1977.

P. N. Nikiforuk, H. Ohta, and M. M. Gupta, "Adaptive observer and identifier design for multi-input multi-output systems," in *Proc. IFAC Symp. in Multivariable Syst.,* pp. 189–196, 1977.

P. N. Nikiforuk, N. Minamide, and M. M. Gupta, "Design of a continuous-time adaptive regulator," in *Int. Federation of Automatic Control,* 8th Triennial World Congress, pp. VII.41–VII.46, 1981.

P. N. Nikiforuk, H. Ohta, and M. M. Gupta, "Adaptive identification and control of linear MIMO discrete systems in a noisy environment," in *Int. Federation of Automatic Control,* 8th Triennial World Congress, Kyoto, Japan, pp. VII.71–VII.76, 1981.

P. N. Nikiforu, N. Minamide, and M. M. Gupta, "Design of a continuous-time adaptive regulator," in *Int. Federation of Automatic Control,* 8th Triennial World Congress, Kyoto, Japan, pp. VII.41–VII.46, 1981.

D. Orlicki, L. Valavani, M. Athans, and G. Stein, "Robust adaptive control with variable dead zones," in *Proc. 23rd IEEE Conf. on Decision and Control,* 1984.

P. C. Parks, "Lyapunov redesign of model reference adaptive control systems," *IEEE Trans. Automat. Contr.,* vol. AC-11, no. 3, pp. 362–367, 1966.

——, "Stability problems of model reference and identification self-adaptive control system," in *Proc. IFAC Symp. on Proc. Ident.,* 1967.

J. S. Pazdera and H. F. Spence, "Comments on accelerated model-referenced adaptation via Lyapunov and steepest descent design techniques," *IEEE Trans. Automat. Contr.,* vol. AC-17, no. 6, p. 812, 1972.

A. E. Pearson, "An adaptive control algorithm for linear systems," *IEEE Trans. Automat. Contr.,* vol. AC-14, no. 5, pp. 497–503, 1969.

P. H. Phillipson, "Concerning Lyapunov redesign of model reference adaptive control systems," *IEEE Trans. Automat. Contr.,* vol. AC-12, no. 5, p. 625, 1967.

B. Porter and M. I. Tatnall, "Performance characteristics of multivariable model reference adaptive systems synthesized by Lyapunov's direct method," 1969.

——, "Stability analysis of a class of multivariable model reference adaptive systems having time varying process parameters," *Int. J. Contr.,* vol. 11, pp. 325–332, 1970.

F. D. Powell, "Predictive adaptive control," *IEEE Trans. Automat. Contr.,* vol. AC-14, no. 5, pp. 550–552, 1969.

B. R. Pradhan, "On the speed of response of a model-reference adaptive control system," *Int. J. Contr.,* vol. 14, no. 4, pp. 775–780, 1971.

L. Praly, "Robust model reference adaptive controllers, Part I: Stability analysis," in *Proc. 23rd IEEE Conf. on Decision and Control,* pp. 1009–1014, 1984.

B. Riedle and P. Kokotivic, "A stability-instability boundary for disturbance-free slow adaptation and unmodeled dynamics," in *Proc. 23rd IEEE Conf. on Decision and Control,* pp. 998–1001, 1984.

C. Rohrs, L. Valavani, M. Athans, and G. Stein, "Analytical verification of undesirable properties of direct model reference adaptive control algorithms," presented at the 20th IEEE Conf. Decis. Control, 1981.

——, "Robustness of adaptive control algorithms in the presence of unmodelled dynamics," presented at the 21st Conf. Decis. Control, 1982.

B. Shackcloth and R. L. Butchart, "Synthesis of model reference adaptive systems by Lyapunov's second method," *Theory of Self-Adaptive Control Systems,* P. H. Hammond, Ed. New York: Plenum Press, 1966, pp. 145–152.

B. Shackcloth, "Design of model reference control system using a Lyapunov synthesis technique," *Proc. IEE,* vol. 114, no. 2, pp. 299–230, 1968.

H. I. H. Shahein, M. A. R. Ghonaimy, and D. W. C. Shen, "Accelerated model reference adaptation via Lyapunov and steepest descent techniques," *IEEE Trans. Automat. Contr.,* vol. AC-17, no. 1, pp. 125–128, 1971.

——, "Accelerated model-reference adaptation via Lyapunov and steepest design techniques," *IEEE Trans. Automat. Contr.,* vol. AC-17, no. 1, pp. 125–128, 1972.

D. W. Sutherlin and J. S. Boland, "Model reference adaptive control system design technique," *ASME Trans. Series E, J. Dyn. Syst. Meas. and Contr.,* vol. 95, pp. 374–379, 1973.

B. H. Swanick, "A high speed deterministic adaptive controller," *Int. J. Contr.,* vol. 15, pp. 833–838, 1972.

U. Ten Cate, "Improved convergence of Lyapunov model-reference adaptive systems by a parameter misalignment function," *IEEE Trans. Automat. Contr.,* vol. AC-20, pp. 132–134, 1975.

C. J. Wenk and Y. Bar-Shalom, "A multiple model adaptive dual control algorithm for stochastic systems with unknown parameters," *IEEE Trans. Automat. Contr.,* vol. AC-25, no. 4, pp. 703–710, 1980.

A. J. White, "The analysis and design of model reference adaptive control systems," *Proc. IEE* (London), vol. 113, pp. 175–184, 1966.

C. A. Winsor and R. L. Roy, "Design of model reference adaptive control systems by Lyapunov's second method," *IEEE Trans. Automat. Contr.,* vol. AC-13, no. 2, p. 205, 1968.

Y. Yamamoto, M. M. Gupta, and P. N. Nikiforuk, "Some new adaptive algorithms for a class of model reference control systems," IX Triennial World Congress, International Federation of Automatic Control, vol. VII, pp. 93–98, 1984.

Liapunov Redesign of Model Reference Adaptive Control Systems

PATRICK C. PARKS

Abstract—The model reference adaptive control system has proved very popular on account of a ready-made, but heuristically based, rule for synthesizing the adaptive loops—the so-called "M.I.T. rule." A theoretical analysis of loops so designed is generally very difficult, but analyses of quite simple systems do show that instability is possible for certain system inputs.

An alternative synthesis based on Liapunov's second method is suggested here, and is applied to the redesign of adaptive loops considered by some other authors who have all used the M.I.T. rule. Derivatives of model-system error are sometimes required, but may be avoided in gain adjustment schemes if the system transfer function is "positive real," using a lemma due to Kalman.

This paper amplifies and extends the work of Butchart and Shackcloth reported at the IFAC (Teddington) Symposium, September, 1965.

Introduction

THE MODEL reference system has proved to be one of the most popular methods in the growing field of adaptive control, particularly for practical application to devices such as autopilots where rapid adaption is required. This popularity is undoubtedly due to a ready-made, but heuristically based, rule for synthesizing the adaptive loops due originally to Whitaker et al. [1] of the Massachusetts Institute of Technology. However, as will be shown, such adaptive schemes lead to unstable adaption for certain types of input signals passing into quite simple systems. This is not a satisfactory feature of such a synthesis, and casts doubts on the stability properties of more complex systems using the "M.I.T. rule."

An alternative synthesis based on Liapunov's second method [2] is suggested, and is here applied to a number of problems considered previously by other authors. Besides guaranteeing stability for all kinds of inputs, the Liapunov method allows high gains in the adaptive loops to be used, and, often, considerable simplification of such loops. The Liapunov method has, of course, been used to a limited extent for analysis of adaptive control loops, notably by Leondes and Donalson [3].

The present paper amplifies and extends the work at Southampton University of Butchart and Shackcloth [4] in particular; it is shown that the use of derivatives of error in the Liapunov synthesis may be avoided if the model transfer function is "positive real." This interesting result follows from a lemma used by Kalman in his treatment of the Luré problem [7].

Manuscript received December 14, 1965; revised April 29, 1966.
The author is with the Department of Aeronautics and Astronautics, University of Southampton, England. He was a Visiting National Science Foundation Fellow at Kansas State University, Manhattan, Kan., from September, 1965, to May, 1966.

Stability Analyses of Simple Examples of the M.I.T. Scheme

Consider the simple model reference adaptive control system of Fig. 1, where the problem is to find a suitable adaptive loop to adjust K_c so that K_cK_v eventually equals the model gain K. The M.I.T. rule, based on minimizing $\int e^2dt$, is that

$$K_c = -Be\left(\frac{\partial e}{\partial K_c}\right) \qquad (B,\ \text{constant} > 0)$$

where

$$\left(\frac{\partial e}{\partial K_c}\right) = \frac{-K_v r}{(1+Ts)}$$

and is found by differentiating partially the transfer function

$$e = \frac{(K - K_cK_v)r}{(1+Ts)}$$

with respect to K_c. The signal $(\partial e/\partial K_c)$ is usually generated by additional circuitry, but here the signal $-\theta_m$ is all that is required effectively, leading to the scheme shown in Fig. 2, where $\dot{K}_c = B'e\theta_m$.

The equations of Fig. 2 are

$$\left.\begin{aligned} T\dot{e} + e &= (K - K_vK_c)r(t)\\ T\dot{\theta}_m + \theta_m &= Kr(t)\\ \dot{K}_c &= B'e\theta_m \end{aligned}\right\}. \qquad (1)$$

The analysis of these equations even for simple inputs is quite difficult. For example, suppose that a step input in $r(t)$ of magnitude $+R$ is applied at time $t=0$, when θ_m, θ_S are zero and K_cK_v is at that time not equal to K. Subsequently, K_v remains constant, but K_c is adjusted according to (1). θ_m will be given by $\theta_m = KR(1-\exp(-t/T))$, and the equation for $e(t)$ is

$$T\ddot{e} + \dot{e} + KK_vR^2B^1e(1 - \exp(-t/T)) = 0. \qquad (2)$$

Now, the third coefficient tends to KK_vR^2B' as $t\to\infty$, and the equation $T\ddot{e}+\dot{e}+KK_vR^2B'e=0$ is asymptotically stable, and so, by an extension of the Dini-Hukuhara theorem, is (2), and thus as $t\to\infty$ $e\to 0$ and $K_c\to K/K_v$, which is what one would hope.

Reprinted from *IEEE Trans. Automat. Contr.*, vol. AC-11, pp. 362–367, July 1966.

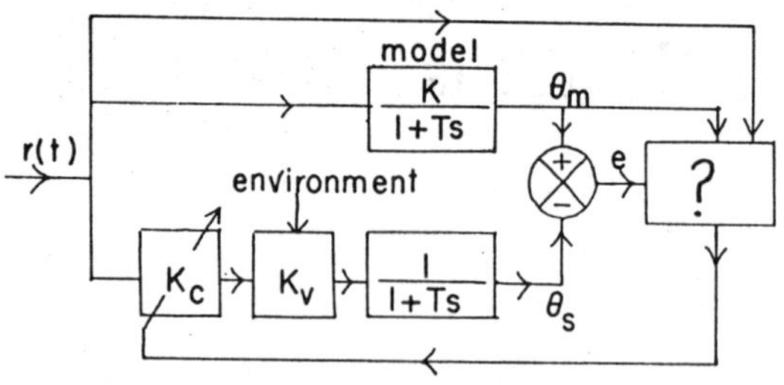

Fig. 1. Gain adaption problem.

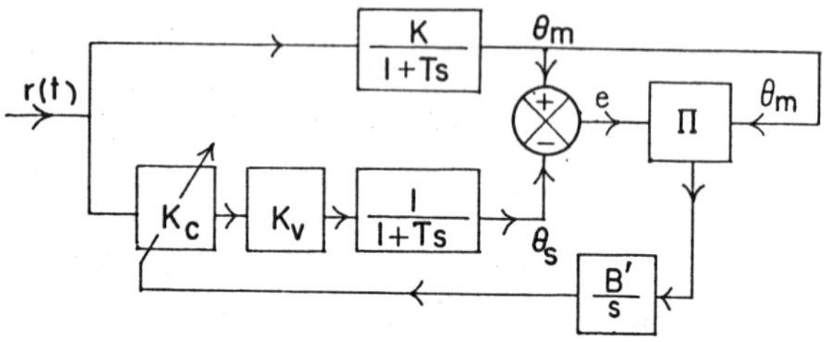

Fig. 2. First-order system—M.I.T. gain adaption.

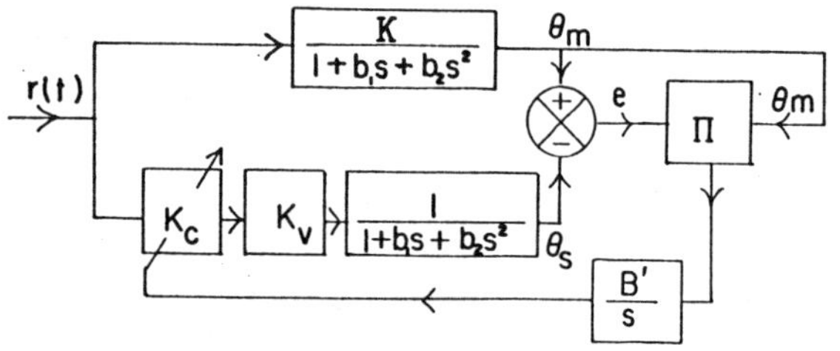

Fig. 3. Second-order system—M.I.T. gain adaption.

Now consider a sinusoidal $r(t)$ and that the adaption is switched in when $\theta_M(t)$ has reached its steady state, another sinusoid out of phase with $r(t)$. Equations (1) can now be written

$$\begin{bmatrix} \dot{x} \\ \dot{e} \end{bmatrix} = \begin{bmatrix} 0 & -K_vB'\theta_m(t) \\ \dfrac{r(t)}{T} & -\dfrac{1}{T} \end{bmatrix} \begin{bmatrix} x \\ e \end{bmatrix} \tag{3}$$

where $x=(K-K_vK_c)$, $\dot{x}=-K_v\dot{K}_c=-K_vB'e\theta_m(t)$. This is now a linear matrix equation of the form $\dot{\boldsymbol{x}}=\boldsymbol{A}(t)\boldsymbol{x}$ with $\boldsymbol{A}(t)$ periodic. Investigation of the stability properties of this matrix equation may be carried out, for example by dividing up the period τ of $r(t)$ into a large number N of equal subintervals, and forming the approximate transfer matrix over the period

$$\left\{I+\frac{\tau}{N}\boldsymbol{A}\left(\frac{N-1}{N}\tau\right)\right\}\left\{I+\frac{\tau}{N}\boldsymbol{A}\left(\frac{N-2}{N}\tau\right)\right\}\cdots \cdot\left\{I+\frac{\tau}{N}\boldsymbol{A}(0)\right\}.$$

This matrix (2×2) is then tested to see if its latent roots lie inside the unit circle, which is what is required for asymptotic stability of (3). A quick answer about stability clearly is not possible, and at least one published stability analysis [5] of adaption with sinusoidal $r(t)$ based on averaging time varying coefficients appears to be theoretically unsound.

However, if the model and system are of second order, then the following situation leads clearly to possible instability. The scheme is shown in Fig. 3.

The equations using the M.I.T. synthesis rule are

$$\left.\begin{aligned} b_2\ddot{e}+b_1\dot{e}+e &= (K-K_vK_c)r(t) \\ \dot{K}_c &= B'e\theta_m \\ b_2\ddot{\theta}_m+b_1\dot{\theta}_m+\theta_m &= Kr(t) \end{aligned}\right\}. \tag{4}$$

Suppose $r(t)=R$ constant, and that the adaptive loop is closed when θ_M and θ_S have taken up the steady-state values KR and $K_vK_c^0R$, K_c being initially constant at a value K_c^0. The equation for the error e is then

$$b_2\dddot{e}+b_1\ddot{e}+\dot{e} = -K_vRB'eKR$$

or

$$b_2\dddot{e}+b_1\ddot{e}+\dot{e}+KK_vB'R^2e = 0. \tag{5}$$

Clearly, instability is possible if the input R or the gain B' is large enough for

$$KK_vB'R^2 > \frac{b_1}{b_2}$$

on applying the Routh-Hurwitz criterion to (5).

From these examples it can be seen that the M.I.T. rule leads to difficult analysis problems for simple systems and inputs, as well as possible instability.

The Liapunov Scheme

Now reconsider the system of Fig. 1 and consider its "adaptive step response" [4]; that is, its behavior with $K_v(>0)$ constant, but K_vK_c initially different from K using a tentative Liapunov function

$$V = e^2+\lambda x^2 \qquad (\lambda \text{ constant} > 0)$$

where $x=K-K_vK_c$, $\dot{x}=-K_v\dot{K}_c$

$$\frac{dV}{dt} = 2e\dot{e}+2\lambda x\dot{x} = 2e\left(-\frac{e}{T}+\frac{xr(t)}{T}\right)+2\lambda x\dot{x}$$

Now make $\dot{x}$ equal to $-er(t)/\lambda T$ so that $dV/dt=-2e^2/T$. This gives rise to the new system of Fig. 4, where $B'=1/\lambda TK_v$.

This scheme is guaranteed to be asymptotically stable, provided, of course, that $r(t)\not\equiv 0$, and the difficult analysis of the M.I.T. synthesis is avoided. Under the same conditions, (2) now becomes

$$T\ddot{e}+\dot{e}+K_vR^2B'e = 0 \tag{6}$$

with the initial conditions $e=0$, $\dot{e}=(1/T)(K-K_vK_c^0)R$ at $t=0+$.

The stability of the adaptive step response would seem to be a basic requirement for any proposed adaptive loop for K_c if it is to adapt for a more general time-varying K_v.

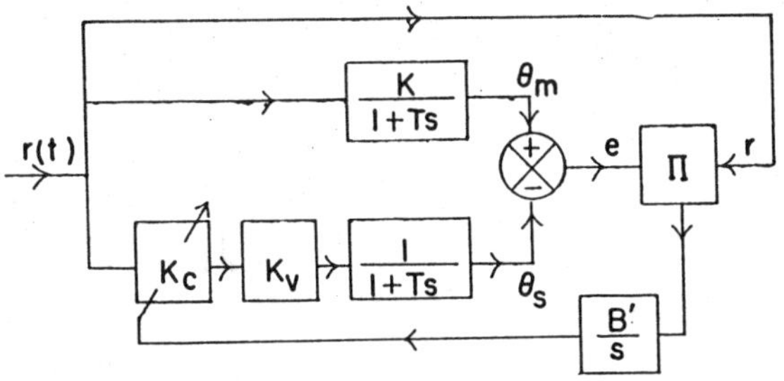

Fig. 4. Liapunov redesign of Fig. 2.

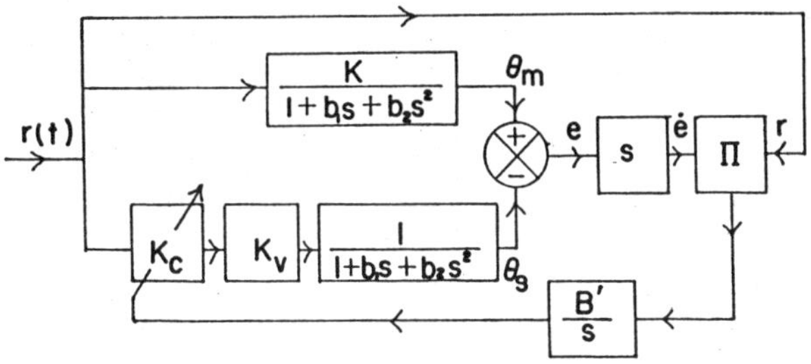

Fig. 5. Liapunov redesign of Fig. 3.

Now reconsider the second-order system and model of Fig. 3. The Liapunov function will be formed as a quadratic form Q in e and $\dot{e}$ for the second-order equation $b_2\ddot{e}+b_1\dot{e}+e=0$ and to this the form λx^2 will be added, as in the previous example. There are many possibilities for the quadratic form in e and $\dot{e}$; a very convenient choice which permits Q and $\dot{Q}$ to be written immediately is the Hermite matrix of Parks [6] for $b_2\ddot{e}+b_1\dot{e}+e=0$. That is,

$$Q = \frac{b_1}{b_2{}^2}e^2 + \frac{b_1}{b_2}\dot{e}^2, \qquad \dot{Q} = -2\left(\frac{b_1}{b_2}\right)^2 \dot{e}^2$$

thus,

$$V = \frac{b_1}{b_2{}^2}e^2 + \frac{b_1}{b_2}\dot{e}^2 + \lambda x^2 \qquad \text{where } x = K - K_vK_c$$

then,

$$\frac{dV}{dt} = -2\left(\frac{b_1}{b_2}\right)^2 \dot{e}^2 + 2\frac{b_1}{b_2}\dot{e}xr(t) + 2\lambda x\dot{x}$$

and then, choosing

$$\dot{x} = -K_v\dot{K}_c = \frac{-b_1\dot{e}r}{\lambda b_2}$$

or $\dot{K}_c = B'\dot{e}r$ where $B' = b_1/\lambda b_2 K_v$.

This scheme, shown in Fig. 5, employs the derivative of the error, and has been criticized on this ground [4]; however, the error rate might well be available directly from the comparison of a system rate gyro output, for example, with the corresponding model rate rather than by analog differentiation of the error signal. But, it is instructive to consider under what circumstances the use of derivatives of error can be avoided in the Liapunov scheme.

The Liapunov Scheme and Positive Real Transfer Functions

Now consider the more general arrangement of Fig. 6. The model transfer function $K(q(s)/p(s))$ is supposed to be such that the degree of $p(s)$ is n, and that of $q(s)$ is $n-1$ or less. The equations of the system may be written as

$$\dot{\mathbf{e}} = A\mathbf{e} + xr(t)\mathbf{c} \tag{7}$$

where A is the $n\times n$ companion matrix of $p(s)$, $x=K-K_vK_c$, $\mathbf{e}'=[e_1, e_2 \cdots e_n]$. If $p(s)=s^n+a_1s^{n-1}+a_2s^{n-2}+\cdots+a_{n-1}s+a_n$ then

$$A = \begin{bmatrix} 0 & 1 & 0 & \cdots & 0 \\ 0 & 0 & 1 & \cdots & 0 \\ \cdot & \cdot & \cdot & \cdot & \cdot \\ 0 & 0 & 0 & \cdots & 1 \\ -a_n & -a_{n-1} & \cdots & & -a_1 \end{bmatrix}.$$

Let $\mathbf{c}^1=[c_1, c_2 \cdots c_n]$ and $g(s)=b_1s^{n-1}+b_2s^{n-2}+\cdots+b_n$; $\mathbf{e}'=e_1,e_2 \cdots e_n$ involve derivatives of e, also xr and its derivatives, the c_i and b_i being linearly related. Now seek a Liapunov function

$$V = \mathbf{e}'P\mathbf{e} + \lambda x^2 \qquad (P \text{ symmetric})$$

with

$$\frac{dV}{dt} = \mathbf{e}'(PA + A'P)\mathbf{e} + 2\mathbf{e}'P\mathbf{c}xr(t) + 2\lambda x\dot{x}$$

and require that $P\mathbf{c}=[\mu, 0, 0, \cdots, 0]'$ so that $\dot{x}$ or $\dot{K}_c$ involves er only, and no derivatives of e. This is a special case of Kalman's lemma in his treatment of the Luré problem [7].

Kalman's Lemma

"Given a real number γ two real n vectors $\mathbf{g}$, $\mathbf{k}$ and a real $n\times n$ matrix $\mathbf{F}$. Let $\gamma\geq 0$, $\mathbf{F}$ stable and $\mathbf{F}$, $\mathbf{g}$ completely controllable. Then a real n vector $\mathbf{q}$ satisfying 1) $\mathbf{F}'\mathbf{P}+\mathbf{P}\mathbf{F}=-\mathbf{q}\mathbf{q}'$ and 2) $\mathbf{P}\mathbf{g}-\mathbf{k}=\sqrt{\gamma}\,\mathbf{q}$ exists if and only if

$$\tfrac{1}{2}\gamma + \text{Re}\,\{k'(i\omega I - F)^{-1}\mathbf{g}\} \geq 0$$

all real ω."

Here, $\gamma=0$, $k=[1, 0, 0 \cdots 0]'$, $\mathbf{g}=\mathbf{c}$, $\mathbf{F}=\mathbf{A}$. The transfer function $k'(sI-F)^{-1}\mathbf{g}$ is precisely that relating e to $(K-K_vK_c)r(t)$; that is, $q(s)/p(s)$. The Kalman lemma implies that $q(s)/p(s)$ must be "positive real" for a stable adaptive gain scheme using no error derivatives to be synthesized. Kalman [7] gives a procedure for finding $\mathbf{q}$, and hence $\mathbf{P}$. Here, an example is given for $n=3$, using the Hermite matrix [6]; Fig. 7 shows the scheme. Suitable values for b_1 and b_2 shall be sought for the adaptive scheme to be proven stable.

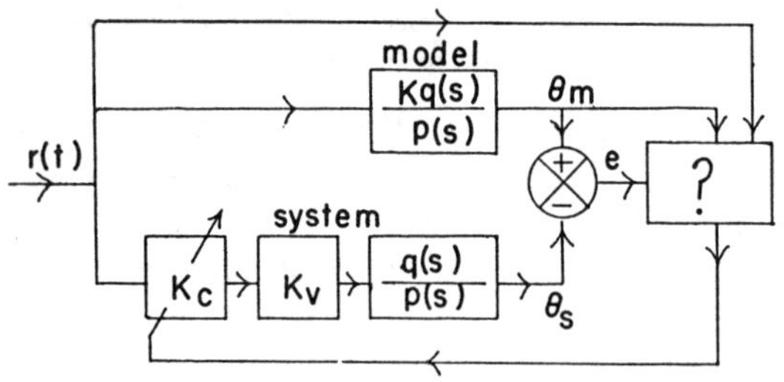

Fig. 6. Gain adaption with "numerator dynamics."

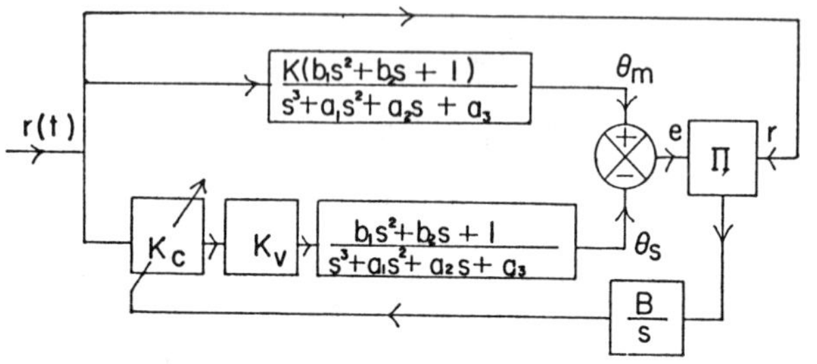

Fig. 7. Liapunov design with positive real transfer function.

The system equations are

$$\begin{bmatrix}\dot e_1\\ \dot e_2\\ \dot e_3\end{bmatrix} = \begin{bmatrix}0 & 1 & 0\\ 0 & 0 & 1\\ -a_3 & -a_2 & -a_1\end{bmatrix}\begin{bmatrix}e_1\\ e_2\\ e_3\end{bmatrix} + \begin{bmatrix}c_1\\ c_2\\ c_3\end{bmatrix}(K - K_vK_c)r(t) \tag{8}$$

where

$$e_1 = e$$

$$e_2 = \dot e - c_1(K - K_vK_c)r(t)$$

$$e_3 = \ddot e - c_1[(K - K_v\dot K_c)r(t)] - c_2[K - K_vK_c]r(t)$$

and

$$\dddot e + a_1\ddot e + a_2\dot e + a_3 e = [c_1D^2 + c_2D + a_1c_1D + a_1c_2 + c_3](K - K_vK_c)r(t)$$

so that

$$a_2c_1 + a_1c_2 + c_3 = 1, \qquad a_1c_1 + c_2 = b_2, \qquad c_1 = b_1 + a_2c_1$$

or

$$\begin{bmatrix}a_2 & a_1 & 1\\ a_1 & 1 & 0\\ 1 & 0 & 0\end{bmatrix}\begin{bmatrix}c_1\\ c_2\\ c_3\end{bmatrix} = \begin{bmatrix}1\\ b_2\\ b_1\end{bmatrix}$$

The Liapunov function

$$V = e'\begin{bmatrix}a_2a_3 & 0 & a_3\\ 0 & a_1a_2 - a_3 & 0\\ a_3 & 0 & a_1\end{bmatrix}e + \lambda x^2$$

where $x = (K - K_vK_c)$ yields

$$\frac{dV}{dt} = -2(a_1e_3 + a_3e_1)^2 + 2e'\begin{bmatrix}a_2a_3 & 0 & a_3\\ 0 & a_1a_2 - a_3 & 0\\ a_3 & 0 & a_1\end{bmatrix}\begin{bmatrix}c_1\\ c_2\\ c_3\end{bmatrix}xr + 2\lambda x\dot x.$$

It is required that

$$\begin{bmatrix}a_2a_3 & 0 & a_3\\ 0 & a_1a_2 - a_3 & 0\\ a_3 & 0 & a_1\end{bmatrix}\begin{bmatrix}c_1\\ c_2\\ c_3\end{bmatrix} = \begin{bmatrix}\mu\\ 0\\ 0\end{bmatrix}$$

and $\dot x = -(1/\lambda)\mu er$, or

$$c_1 = a_1\nu, \qquad c_2 = 0, \qquad c_3 = -a_3\nu,$$

for example, and hence

$$b_1 = a_1\nu, \quad b_2 = a_1^2\nu, \quad 1 = a_1a_2\nu - a_3\nu, \quad \nu = \frac{1}{a_1a_2 - a_3}.$$

Leading to a transfer function,

$$G(s) = \frac{q(s)}{p(s)} = \frac{\dfrac{1}{a_1a_2 - a_3}[a_1s^2 + a_1^2s + (a_1a_2 - a_3)]}{s^3 + a_1s^2 + a_2s + a_3}.$$

Notice that

$$\text{Sign Re } G(i\omega) = \text{sign }(a_3[a_1a_2 - a_3]) > 0$$

for all real ω, so that $G(s)$ is "positive real," and that the complete controllability of (8) may also be confirmed. The point, however, is that if $G(s)$ is positive real with a stable denominator, then $\dot K_c = B'er$ will work; it is not necessary to find the Kalman vector q or the matrix P; that is, to worry about finding the actual Liapunov function to prove the stability, and it is not necessary to use derivatives of e. The system of Fig. 4 is a special case, $1/1+Ts$ being a positive real function. The importance of positive real transfer functions in Liapunov stability theory has been emphasized by Brockett [8].

Some Liapunov Redesigns

Dymock, Meredith, Hall, and White [9] considered an autopilot system of which Fig. 8 is a slightly simplified version. This is based on the M.I.T. scheme, adjusting k according to the rule $\dot k \propto -e(\partial e/\partial k)$. The idea is to adjust k to allow for environmental changes in K, and to keep $K/k = 1$.

This problem will be examined from the Liapunov point of view. Supposing K constant, but $K/k \neq 1$, then

$$(D^2 + 2\zeta\omega_nD + \omega_n^2)\theta_m = \omega_n^2r(t)$$

$$(D^2 + 2\zeta\omega_nD)\frac{\theta_s}{K} = \omega_n^2\frac{r(t)}{k} - \omega_n^2\frac{\theta_s}{k}$$

and so

$$(D^2 + 2\zeta\omega_nD + \omega_n^2)e = \omega_n^2(r - \theta_s)\left(1 - \frac{K}{k}\right). \tag{9}$$

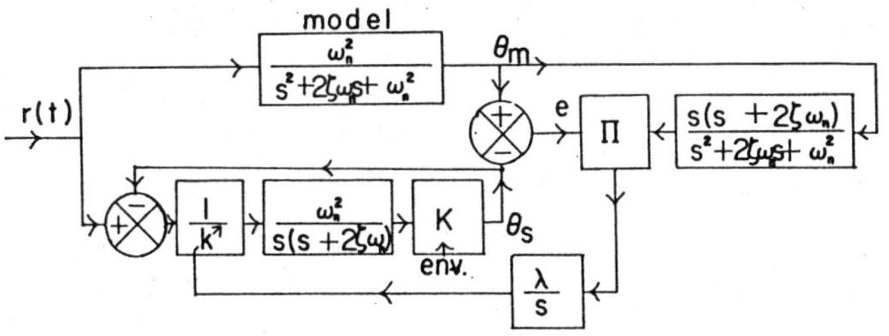

Fig. 8. Dymock et al.—M.I.T. rule.

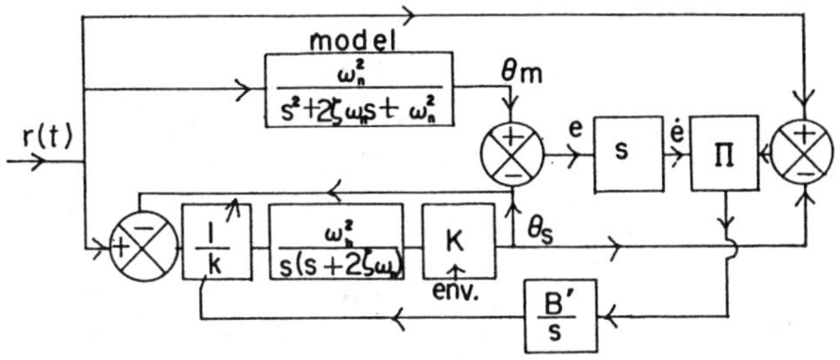

Fig. 9. Liapunov redesign of Fig. 8.

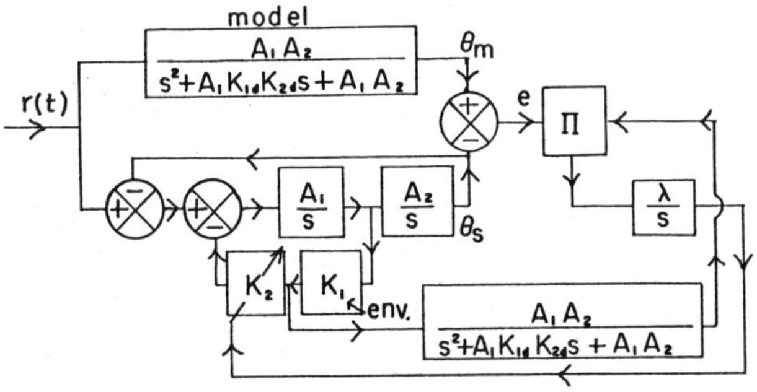

Fig. 10. Horrocks—M.I.T. rule.

Consider the Liapunov function

$$V = 2\zeta\omega_n^3e^2 + 2\zeta\omega_n\dot{e}^2 + \lambda\left(1 - \frac{K}{k}\right)^2$$

then,

$$\frac{dV}{dt} = 4\zeta\omega_n^3e\dot{e} + 4\zeta\omega_n\dot{e}\left(-\omega_n^2e - 2\zeta\omega_n\dot{e} + \omega_n^2[r - \theta_s]\left[1 - \frac{K}{k}\right]\right) - 2\lambda\left(1 - \frac{K}{k}\right)K\left(\frac{\dot{1}}{k}\right)$$
$$= -8\zeta^2\omega_n^2\dot{e}^2,$$

if

$$4\zeta\omega_n^3\dot{e}(r - \theta_s) = 2\lambda K\left(\frac{\dot{1}}{k}\right)$$

which suggests the simpler scheme of Fig. 9.

Horrocks [10] considers the system of Fig. 10, again based on the M.I.T. scheme. By tapping the signal from the inner feedback loop it may be noted that he is using the signal $K_1\theta_s/A_2$. Now consider a Liapunov synthesis making use of the differential equation for the error e

$$(D^2 + A_1K_{1d}K_{2d}D + A_1A_2)e = -A_1(K_{1d}K_{2d} - K_1K_2)\theta_s. \quad (10)$$

Putting $x = K_{1d}K_{2d} - K_1K_2$, consider the Liapunov function

$$V = A_1^2A_2K_{1d}K_{2d}e^2 + A_1K_{1d}K_{2d}\dot{e}^2 + \lambda x^2$$
$$\frac{dV}{dt} = 2A_1^2A_2K_{1d}K_{2d}e\dot{e} + 2A_1K_{1d}K_{2d}\dot{e}(-A_1K_{1d}K_{2d}\dot{e} - A_1A_2e - A_1x\dot{\theta}_s) + 2\lambda x\dot{x}$$
$$= -2A_1^2K_{1d}^2K_{2d}^2\dot{e}^2$$

if

$$2\lambda\dot{x} = 2A_1^2K_{1d}K_{2d}\dot{e}\dot{\theta}_s.$$

That is,

$$2\lambda K_1\dot{K}_2 = -2A_1^2K_{1d}K_{2d}\dot{e}\dot{\theta}_s$$

or

$$\dot{K}_2 = -B'\dot{e}\dot{\theta}_s$$

leading to the system of Fig. 11.

Fishwick and Davies [11] consider parameter adjustments as shown in Fig. 12, which is a special case of a more general scheme. The idea is to derive p_0 to p_0' and a_0 to A by suitable adaptive loops. Once again, their own scheme is effectively the M.I.T. one. The equations are

$$\left.\begin{aligned}(D + p_0')\theta_m &= Ar(t)\\ \theta_s &= a_0y\\ (D + p_0)y &= r(t)\end{aligned}\right\}.$$

Hence,

$$(D + p_0')e = (A - a_0)r(t) + (p_0' - p_0)(-\theta_s) - y\dot{a}_0. \quad (11)$$

Consider the Liapunov function

$$V = e^2 + \lambda_1(A - a_0)^2 + \lambda_2(p_0' - p_0)^2$$
$$\frac{dV}{dt} = -2\lambda_1(A - a_0)\dot{a}_0 - 2\lambda_2(p_0' - p_0)\dot{p}_0 + 2e(-p_0'e + (A - a_0)r(t) + (p_0' - p_0)(-\theta_s) - y\dot{a}_0).$$

If

$$\dot{p}_0 = -\frac{1}{\lambda_2}e\theta_s, \qquad \dot{a}_0 = \frac{1}{\lambda_1}er,$$

then

$$\frac{dV}{dt} = -2p_0e^2 - \frac{2}{\lambda_1}e^2yr.$$

Although λ_1 may be made large so that $(1/\lambda_1)\,|yr| < p_0'$, this is not entirely satisfactory. The difficulty arises on account of the physical position of the variable gain a_0 in Fig. 12. If the arrangement is changed to that of Fig. 13, the following equations result:

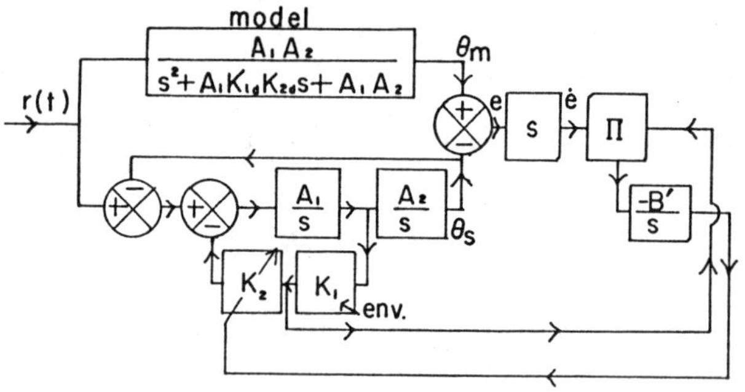

Fig. 11. Liapunov redesign of Fig. 10.

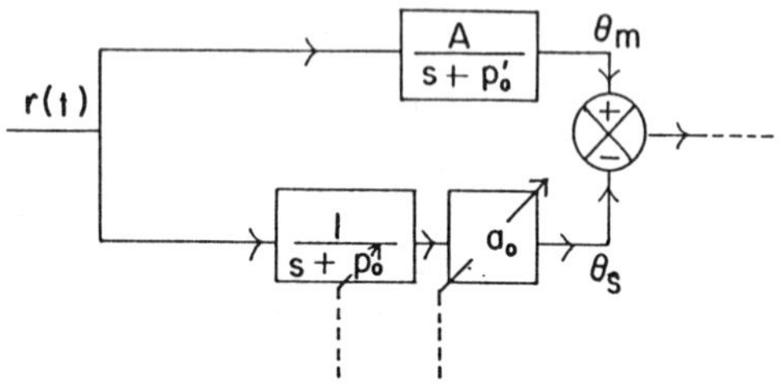

Fig. 12. Fishwick and Davies scheme.

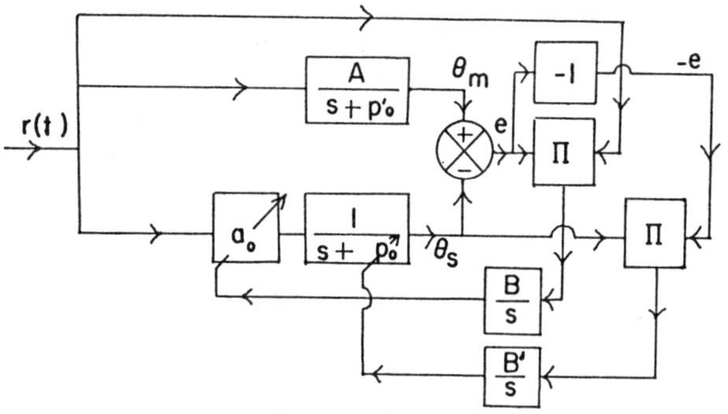

Fig. 13. Liapunov redesign of Fig. 12.

$$\left.\begin{aligned}(D + p_0')\theta_m &= Ar(t)\\ (D + p_0)\theta_s &= a_0 r(t)\end{aligned}\right\}$$

and so

$$(D + p_0')e = -(p_0' - p_0)\theta_s + (A - a_0)r(t). \quad (12)$$

Hence, the Liapunov function

$$V = e^2 + \lambda_1(A - a_0)^2 + \lambda_2(p_0' - p_0)^2$$

yields

$$\frac{dV}{dt} = -2\lambda_1(A - a_0)\dot{a}_0 - 2\lambda_2(p_0' - p_0)\dot{p}_0 + 2e[-p_0'e - (p_0' - p_0)\theta_s + (A - a_0)r(t)]$$

and so, if $\dot{p}_0 = -(1/\lambda_2)e\theta_s$, and $\dot{a}_0 = (1/\lambda_1)er$ are chosen as before,

$$\frac{dV}{dt} = -2p_0'e^2.$$

The adaptive circuits are shown in Fig. 13, the analysis agreeing with the two-parameter adjustment scheme of Butchart and Shackcloth [4]. This example emphasizes the importance of the order in which the blocks in the circuit diagrams are placed, a point also made by Dymock et al. [9].

Conclusions

The difficult problems of stability analysis of model adaptive control systems designed by the M.I.T. method, which seems to be, at present, in common use, may be avoided if such systems are synthesized or redesigned on the basis of Liapunov theory applied to the stability of the adaptive step response. Generally, this method uses a Liapunov function for the differential equation for the model-system error plus such terms as squares of parameter differences. A number of examples have been given where, it will be noticed, derivatives with respect to time of error are sometimes used for forming the adaptive loop signals. Use of these derivatives in the gain adjustment case, which is usually encountered, may be avoided if the model and system have positive real transfer functions. The model and system are of the same order, so that, when equilibrium is achieved, the model-system error and all parameter differences are zero. Clearly, the question of optimum adaptive performance based on suitable criteria, and the problem of added noise in the loops, are suitable subjects for further research. The choice of Liapunov function will influence such performance; the Hermite form used here is particularly convenient, as it and its derivative can be written down at once, but many other Liapunov functions are available after suitable computation.

However, it is believed that what is important is the provision of an alternative way of thought about designing model reference adaptive systems. Such thoughts, so far, have been dominated by the M.I.T. rule which must, in fact, be used with caution on account of possible instabilities.

References

[1] P. V. Osburn, H. P. Whitaker, and A. Kezer, "New developments in the design of adaptive control systems," Inst. of Aeronautical Sciences, paper 61–39.

[2] J. P. LaSalle and S. Lefschetz, *Stability by Liapunov's Direct Method with Applications.* New York: Academic, 1961.

[3] D. D. Donalson and C. T. Leondes, "A model reference parameter tracking technique for adaptive control systems—The principal of adaptation," *IEEE Trans. on Applications and Industry,* vol. 82, pp. 241–262, September 1963.

[4] R. L. Butchart and B. Shackcloth, "Synthesis of model reference adaptive control systems by Liapunov's second method," *Proc. 1965 IFAC Symp. on Adaptive Control* (Teddington, England) (Instrument Soc. America, 1966).

[5] A. J. White, "The analysis and design of model reference adaptive control systems," *Proc. IEE (London),* vol. 113, pp. 175–184, January 1966.

[6] P. C. Parks, "A new look at the Routh-Hurwitz problem using Liapunov's second method," *Bull. Polish Acad. Sci., Tech. Sci. Ser.,* vol. 12, pp. 19–21, June 1964.

[7] R. E. Kalman, "Liapunov functions for the problem of Luré in automatic control," *Proc. Nat'l Acad. Sci.,* vol. 49, pp. 201–205, February 1963.

[8] R. W. Brockett, *Qualitative Behavior of Time Varying and Nonlinear Feedback Systems.* New York: Wiley, 1966.

[9] A. J. Dymock, J. F. Meredith, A. Hall, and K. M. White, "Analysis of a type of model reference adaptive control system," *Proc. IEE (London),* vol. 112, pp. 743–753, April 1965.

[10] T. Horrocks, "Investigations into model reference adaptive control systems," *Proc. IEE (London),* vol. 111, p. 1894, November 1964.

[11] W. Fishwick and W. D. T. Davies, *Proc. 1965 UKAC Conf. on Automatic Control* (Nottingham, England) paper 20; *Proc. Inst. Mech. Engineers,* vol. 179, part 3H, 1966.

Control of Unknown Plants in Reduced State Space

PETER N. NIKIFORUK, MADAN M. GUPTA, MEMBER, IEEE, AND HO H. CHOE, STUDENT MEMBER, IEEE

Abstract—A method is proposed in this paper for the synthesis of an adaptive controller for a class of model reference systems in which the plant is not known exactly, but which is of the following type: single variable, time varying, either linear or nonlinear, of nth order, and capable of mth order input differentiation. The model is linear, stable, and of n'th order, where $(n-m) \leq n' \leq n$. The only knowledge of the plant that is required in this synthesis procedure is the form of the plant equation and the bounds of $b_m(t)$, the coefficient of the mth order plant input derivative. The synthesis procedure makes use of an unique function, called the characteristic variable, and Lyapunov type synthesis. The introduction of the characteristic variable reduces the synthesis problem to one that involves a known, linear time-invariant lower order plant. The control signal is generated by measuring the plant and model outputs, and their first $(n-m)$ derivative signals. This ensures that the norm of the $(n-m)$-dimensional error vector is ultimately bounded by ϵ, an arbitrarily small positive number provided $\xi(t)$, the characteristic variable, is bounded. Two nontrivial simulation examples are included.

I. Introduction

Recently, a number of papers [2]–[4] have discussed the control of plants within a class of model reference systems using a Lyapunov type synthesis. These techniques use a reference model that is of the same order as the plant, or one that is of lower order [3]. In order that a control signal may be generated using these techniques, all of the state variables must be measured and the form of the plant equations, the bounds within which the plant parameters may vary, and the form of the plant nonlinearities must be known. These techniques suffer from the following disadvantages: 1) for an nth order plant, derivative signals up to the $(n-1)$th order must be measured which may not be practically possible for higher order plants; 2) the bounds of the plant parameters and the form of the plant nonlinearities may not be known; and 3) it is usually more practical to use a reference model that is of lower order than the plant. Attempts at bypassing some of these disadvantages have already been reported [5]–[7]. In these studies methods of synthesizing a control signal from the plant output and its lower order derivatives have been considered for a restrictive class of linear plants. From a practical viewpoint, it is desirable to continue the development of such synthesis techniques, but to make them applicable to a more general class of plants. One such technique has been developed by the authors and is described in this paper.

In this technique a uniquely defined function called the characteristic variable is introduced. It is related implicitly to all of the plant parameters and nonlinearities through the available state variables. By introducing this characteristic variable an unknown, nonlinear, and time-varying plant of high order is replaced in the procedure by a known, linear, and time-invariant plant that is of lower order. A Lyapunov type synthesis technique is used to obtain a control signal which is expressed in terms of the characteristic variable. It is then shown that for an nth order time-varying plant, which may be either linear or nonlinear, a control signal can be synthesized with the aid of a suitably defined reference model which may be of lower order than the plant. This technique uses the plant and model outputs and their first $(n-m)$ derivative signals, where m is the order of the plant input differentiation. The only knowledge of the plant that is required is the form of the plant equation and the bounds of $b_m(t)$, the coefficient of the mth-order plant input derivative, if the boundedness of the characteristic variable is known a priori. Otherwise, the boundedness of this variable must be established by simulation.

Throughout this paper the term unknown plant is used according to this definition: this is a plant whose parameters and nonlinearities are not known exactly, but about which sufficient information is available to permit some meaningful simulations to be made.

Manuscript received November 18, 1968. This paper was presented at the 1969 Joint Automatic Control Conference, Boulder, Colo. This work was supported by the National Research Council of Canada under Grant A-5625 and the Defence Research Board of Canada under Grant 4003-02.

The authors are with the Division of Control Engineering, University of Saskatchewan, Saskatoon, Sask., Canada.

II. Problem Formulation

For purposes of discussion let I represent the set of nonnegative real numbers, $I = \{t | t \geq 0, t \in R\}$, and let t_0 be an arbitrary but fixed element of I. In addition, let $C^k(T)$ be the class of functions that is k times differentiable on T, where $T = \{t | t \geq t_0 \in I\}$.

Consider a class of single-input single-output dynamic systems, each of which is comprised of an unknown plant and a known stable reference model. The plant is defined by the following time-varying nonlinear nth order differential equation

$$\sum_{k=0}^{n} a_k(t)x^{(k)}(t) + f(t,x,\dot{x},\cdots,x^{(n-1)}) = \sum_{k=0}^{m} b_k(t)w^{(k)}(t) \quad (1)$$

where n and m are known positive integers such that $n > m$; $w(t)$ and $x(t)$ are respectively the scalar input and scalar output of the plant; $a_n(t) = 1$, $a_i(t)$, $i = 0,1,\cdots,(n-1)$, and $b_j(t)$, $j = 0,1,\cdots,m$, are unknown time-varying parameters; and f is an unknown time-varying nonlinear function of x, $\dot{x}$, $\cdots$, $x^{(n-1)}$. Furthermore, for this plant the following assumptions apply.

Assumption 1: For a given $w(t) \in C^m(T)$, (1) is defined for all $t \in T$ and enjoys the usual smoothness conditions so that no questions arise as to the existence and uniqueness of a solution on T for a given set of initial conditions.

Reprinted from *IEEE Trans. Automat. Contr.*, vol. AC-14, pp. 489–496, Oct. 1969.

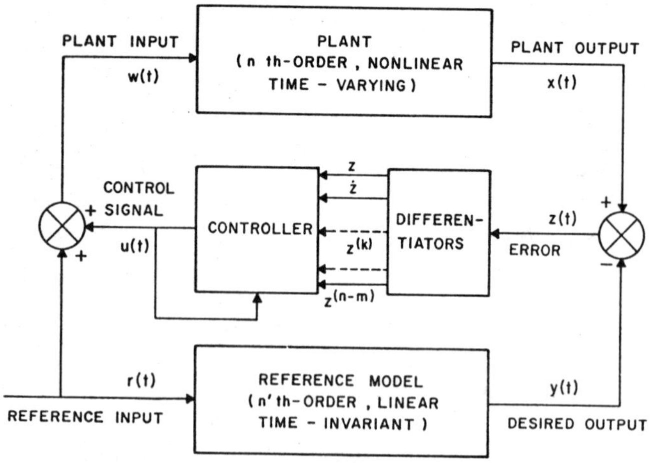

Fig. 1. Model reference control system.

Assumption 2: The time-varying parameter $b_m(t)$ is bounded with known bounds; that is, $0 < \beta_0 \leq b_n(t) \leq \beta_1 < \infty$, for all $t \in T$, where β_0 and β_1, the bounds, are known positive numbers.

The stable reference model is defined by the following time-invariant linear n'th order differential equation.

$$\sum_{j=0}^{n'} a_j^0 y^{(j)}(t) = \sum_{j=0}^{m'} b_j^0 r^{(j)}(t), \quad a_{n'}^0 = 1 \tag{2}$$

where $r(t) \in C^m(T)$ and $y(t)$ are respectively the scalar reference input and the scalar output of the model, and n' and m' are known positive integers such that $n' > m'$ and $n \geq n' \geq (n - m)$.

Let A_0 be a $n' \times n'$ constant stable matrix, formed by the coefficients of the model (2) and given by

$$A_0 = \begin{bmatrix} 0 & 1 & 0 & \cdots & 0 \\ 0 & 0 & 1 & & \\ \cdot & \cdot & \cdot & & \cdot \\ \cdot & \cdot & \cdot & & \cdot \\ \cdot & \cdot & \cdot & & \cdot \\ -a_0^0 & -a_1^0 & -a_2^0 & \cdots & -a_{n'-1}^0 \end{bmatrix} \tag{3}$$

A block diagram of such a system, together with a controller which is yet to be selected, is shown in Fig. 1. From this figure it is seen that the plant input is

$$w(t) = u(t) + r(t) \tag{4}$$

and the error between the plant output and the model output is

$$z(t) = x(t) - y(t). \tag{5}$$

The problem to be considered in this study is that of synthesizing a control signal $u(t) \in C^m(T)$ using only $z(t)$, $\dot{z}(t), \cdots, z^{(n-m)}(t)$ so as to ensure that $x(t)$ follows $y(t)$ as closely as possible.

Before tackling this problem it is worthwhile to emphasize the following points which have already been stated: 1) as far as the plant is concerned, only its general structures (1) and the two constants β_0 and β_1, the bounds of the coefficient $b_m(t)$, are known; 2) the only restrictive conditions which are imposed on the plant are those implied by Assumptions 1 and 2.

III. Controller Synthesis

Let a constant β and a time function $\theta(t)$ be defined as follows

$$\beta = (\beta_0 + \beta_1)/2 \tag{6}$$

$$\theta(t) = b_m(t) - \beta, \quad t \in T. \tag{7}$$

From Assumption 2, $|\theta(t)| < \beta$, for all $t \in T$. With the definition of β as in (6), the following lemma is stated.

Lemma 1

The right-hand side of (1) may be written as $b_m(t)u^m(t) + \zeta(t)$, where

$$\zeta(t) = b_m(t)r^{(m)}(t) + \sum_{k=0}^{m-1} b_k(t)w^{(k)}(t), \quad t \in T.$$

Let p be an integer such that $0 \leq p \leq m$ and $\boldsymbol{\alpha}$ be a row vector defined by

$$\boldsymbol{\alpha} = (\alpha_0 \; \alpha_1 \cdots \alpha_{n-p}) \tag{8}$$

where α_k, for $k = 0,1,\cdots,(n-p)$, is a real number with $\alpha_{n-p} = 1$. There then exists a uniquely defined function $\xi_{p,\alpha}(t) \in C^p(t)$, such that

$$\sum_{k=0}^{n-p} \alpha_k x^{(k)}(t) = \beta u^{(m-p)}(t) + \xi_{p,\alpha}(t) \tag{9}$$

is equivalent to (1). This can be proved as follows.

Proof: Using (4) and (6)–(8), it is found that (1) can be expressed in the form

$$\sum_{k=0}^{n-p} (a_{p+k}(t) - \alpha_k + \alpha_k)x^{(p+k)}(t) + \sum_{k=0}^{p-1} a_k(t)x^{(k)}(t) + f(t,x,\dot{x},\cdots,x^{(n-1)}) = \beta u^{(m)}(t) + \theta(t)u^m(t) + \zeta(t). \tag{10}$$

This equation can now be rewritten to give

$$\sum_{k=0}^{n-p} \alpha_k x^{(p+k)}(t) = \beta u^{(m)}(t) + \phi(t,x,\dot{x},\cdots,x^{(n-1)},f,u^m,\zeta) \tag{11}$$

where

$$\phi = \sum_{k=0}^{n-p} (\alpha_k - a_{p+k}(t))x^{(p+k)}(t) - \sum_{k=0}^{p-1} a_k(t)x^{(k)}(t) - f(t,x,\dot{x},\cdots,x^{(n-1)}) + \theta(t)u^{(m)}(t) + \zeta(t).$$

On the basis of Assumption 1 each term in (11) is integrable on T and p integrations of (11) from t_0 to t produce the result [8]

$$\sum_{k=0}^{n-p} \alpha_k x^{(k)}(t) = \beta u^{(m-n)}(t) + \phi_0(t) + \int_{t_0}^{t} \frac{(t-\lambda)^{p-1}}{(p-1)!} \phi(\lambda)\, d\lambda \tag{12}$$

where $\phi_0(t)$ represents all of the terms which arise from the initial values $x^{(k)}(t_0)$, $k = 0,1,\cdots,(n-1)$ and $u^{(j)}(t_0)$, $j = m-p, (m-p+1),\cdots,(m-1)$. The variables $x, \dot{x}\cdots$ are dropped from ϕ for convenience.

By defining

$$\xi_{p,\alpha}(t) = \phi_0(t) + \int_{t_0}^{t} \frac{(t-\lambda)^{p-1}}{(p-1)!}\phi(\lambda)\,d\lambda \tag{13}$$

it is found that (9) and (12) are the same. The interval of definition of $\xi_{p,\alpha}(t)$, its uniqueness property, and differentiability follow from the interval for which the solution of (1) is defined, the uniqueness of the solution of (1) stated in Assumption 1, and the differentiability of the right side of (13). By substituting (13) into (9), differentiating the resulting equation p times, and rearranging the terms so obtained, (1) is obtained. Lemma 1 is thus proved.

To summarize, it is seen that (1), which describes the time-varying nonlinear nth order unknown plant, can be replaced by (9), a $(n-p)$th order linear equation with known constant parameters. This is done by introducing a term $\xi_{p,\alpha}(t)$ which contains all of the time-varying and nonlinear dynamic characteristics of the plant. It is termed the characteristic variable of the plant and is associated with p and α.

In view of Lemma 1, the substitution of $(n-p) = n'$, the order of the model, and $\alpha = a^0 = (a_0^0\, a_1^0 \cdots a_{n'}^0)$ the model parameters into (9) yields

$$\sum_{k=0}^{n'} a_k^0 x^{(k)}(t) = \beta u^{(m-n+n')}(t) + \xi_{(n-n'),a^0}(t). \tag{14}$$

Using (14) and (2), the system error signal in (5) is found to satisfy

$$\sum_{k=0}^{n'} a_k^0 z^{(k)}(t) = \beta u^{(m-n+n')}(t) + \xi_{(n-n'),a^0}(t) - \sum_{k=0}^{m'} b_k^0 r^{(k)}(t). \tag{15}$$

Obviously, Lemma 1 is also applicable to (15). In particular, with $p = m-n+n' = s$ and $\alpha = (a_s^0\, a_{s+1}^0 \cdots a_{n'}^0)$ (15) is found to be equivalent to

$$\sum_{k=0}^{n-m} a_{s+k}^0\, z^{(k)}(t) = \beta u(t) + \xi(t) \tag{16}$$

where the subscripts of $\xi(t)$, the system characteristic variable, are dropped. In this equation, note that $\xi(t) \in C^m(T)$.

Let A_0^* be the $(n-m)\times(n-m)$ principal minor that is located in the lower right-hand corner of A_0; that is,

$$A_0^* = \begin{bmatrix} 0 & 1 & 0 & \cdots & 0 \\ 0 & 0 & 1 & & 0 \\ \cdot & \cdot & \cdot & & \cdot \\ \cdot & \cdot & \cdot & & \cdot \\ \cdot & \cdot & \cdot & & \cdot \\ -a_s^0 & -a_{s+1}^0 & -a_{s+2}^0 & \cdots & -a_{n'-1}^0 \end{bmatrix}. \tag{17}$$

Equation (16) can now be equivalently written in the vector differential equation form

$$\dot{z}(t) = A_0^* z(t) + (0\, 0 \cdots \beta u(t) + \xi(t))^T \tag{18}$$

where $z = (z_1 z_2 \cdots z_{n-m})^T$, $z_1 = z$ and $z_{i+1} = \dot{z}_i$, for $i = 1, 2, \cdots, (n-m-1)$.

As far as A_0^* is concerned, the following points should be noted. For $(n-m) \le 3$, it can be shown using the Hurwitz criterion [9] that a stable A_0 gives a stable A_0^*. If $(n-m) > 3$, some analysis may have to be carried out in order to establish such an A_0. However, it is always possible to select $n' = (n-m)$, from which it follows that $A_0^* = A_0$. Thus, in the discussions to follow it is assumed that a stable A_0 implies a stable A_0^*.

Lemma 2

Let the dimension of z in (18) be g, A_0^* be a stable matrix, β be a positive number, and $\xi(t)$ be continuous on T. If $|\xi(t)| \le M < \infty$, for all $t \in T$, where M is a positive number, then there exists a control signal of the form

$$u(t) = -(1/\beta)\xi(t)\tan^{-1}(K_1\xi(t))\tan^{-1}(K_2\gamma(z)) \tag{19}$$

which ensures that

$$\|z\| < \epsilon, \quad \text{for } t \to \infty, \quad \epsilon = \max\left(C_1/K_1, \sqrt{(C_2/K_2)}\right) \tag{20}$$

where K_1 and K_2 are positive numbers, $\gamma(z)$ is a linear combination of the components of z, $\|\cdot\|$ represents the Euclidean norm, and C_1 and C_2 are positive numbers. This can be proved as follows.

Proof: Let P and Q be $g \times g$ positive definite symmetric matrices such that

$$A_0^{*T}P + PA_0^* = -Q. \tag{21}$$

The existence of such P and Q is guaranteed by a theorem due to Lyapunov [1]. Let $V(z)$ be a positive definite quadratic form of z

$$V(z) = z^T p z. \tag{22}$$

Then $\dot{V}$, the time derivative of V along the solution of (18), is

$$\dot{V}(t,z) = -z^TQz + 2\gamma(z)(\beta u(t) + \xi(t)) \tag{23}$$

where

$$\gamma(z) = \sum_{k=1}^{g} p_{kg} z_k \tag{24}$$

and p_{ij} is an element of matrix P. The substitution of (19) into (23) yields

$$\dot{V}(t,z) = -z^TQz + 2(1 - \tan^{-1} \cdot (K_1\xi(t))\tan^{-1}(K_2\gamma(z)))\xi(t)\gamma(z). \tag{25}$$

It is thus seen that $\dot{V}(.,.)$ is a continuous mapping from $T \times R_g$ into R.

Now let the behavior of $\dot{V}$ on $T \times R_g$ be examined. For this purpose let $Q = I$, a $g \times g$ identity matrix. Then $z^TQz = \|z\|^2$. Also let $S = T \times R_g$. Define S_0, S_-, and S_+

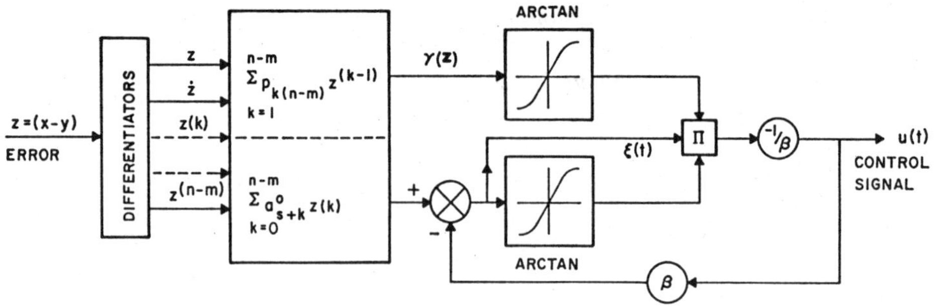

Fig. 2. Controller.

to be mutually disjoint subsets of S such that

$$S_0 = \{(t,z)|\xi(t)\gamma(z) = 0, \quad (t,z) \in S\}$$
$$S_- = \{(t,z)|\xi(t)\gamma(z) < 0, \quad (t,z) \in S\}$$
$$S_+ = \{(t,z)|\xi(t)\gamma(z) > 0, \quad (t,z) \in S\}.$$

Hence $S = S_0 \cup S_- \cup S_+$, and from this it is seen that

$$\dot{V}(t,z) < 0 \text{ on } S_0 \cup S_-, \quad \text{with } \dot{V} = 0, \quad \text{for } z = 0. \qquad (26)$$

Let $S_+ = S_1 \cup S_2 \cup S_3$, where

$$S_1 = \{(t,z)|\ |\tan^{-1}(K_1\xi(t))| \geq 1, \quad |\tan^{-1}(K_2\gamma(z))| \geq 1, \quad (t,z) \in S_+\}$$
$$S_2 = \{(t,z)|\ |\tan^{-1}(K_1\xi(t))| < 1, \quad (t,z) \in S_+\}$$
$$S_3 = \{(t,z)|\ |\tan^{-1}(K_2\gamma(z))| < 1, \quad (t,z) \in S_+\}.$$

If $(t,z) \in S_1$, then $\xi(t)\gamma(z) > 0$ and $\{1 - \tan^{-1}(K_1\xi(t)) \tan^{-1}(K_2\gamma(z))\} \leq 0$. Therefore,

$$\dot{V}(t,z) < 0 \text{ on } S_1. \qquad (27)$$

If $(t,z) \in S_2$, then $|\xi(t)| < \tan 1/K_1$ and

$$\dot{V}(t,z) < -\|z\|^2 + 2(1 + \pi/2) \tan 1/K_1 N\|z\| = -\|z\|^2 + \|z\|C_1/K_1 \qquad (28)$$

where $C_1 = 2(1 + \pi/2)N \tan 1$ and N is a positive number such that $|\gamma(z)| \leq N\|z\|$. Therefore,

$$\dot{V}(t,z) < 0 \text{ on } S_2, \quad \text{for } \|z\| \geq C_1/K_1. \qquad (29)$$

Also, if $(t,z) \in S_3$, then $|\gamma(z)| < \tan 1/K_2$ and

$$\dot{V}(t,z) < -\|z\|^2 + 2(1 + \pi/2)M \tan 1/K_2 = -\|z\|^2 + C_2/K_2 \qquad (30)$$

where $C_2 = 2(1 + \pi/2)M \tan 1$. Hence,

$$\dot{V}(t,z) < 0 \text{ on } S_3, \quad \text{for } \|z\| \geq \sqrt{(C_2/K_2)}. \qquad (31)$$

Therefore, from relations (26), (27), (29), and (31) it can be concluded that $\dot{V}(t,z) < 0$ on Ω', the complement of Ω relative to S, where

$$\Omega = \{(t,z)|\|z\| < \max(C_1/K_1, \sqrt{(C_1/K_2)}), \quad (t,z) \in S\}. \qquad (32)$$

It follows from the above discussion that $\|z\|$ is ultimately bounded by ϵ as is given in (20).

Comments: In order for Lemma 2 to be of any practical significance two points remain to be clarified: 1) the boundedness of the system characteristic variable $\xi(t)$ in (16), and 2) the arbitrary smallness of ϵ in (20). These two problems are closely related and an analytical treatment of them is highly complicated and not very obvious. However, the computer simulation studies for various systems have shown the boundedness of $\xi(t)$. Also, it has been observed that a selection of a large K_1 and K_2 does not affect M, the bounds of $\xi(t)$. Hence, ϵ can be made arbitrarily small by choosing K_1 and K_2 large enough. If the dependence of $M = M(K_1,K_2)$ on K_1 and K_2 is not obvious, then by making K_2 dependent on system variables [5] as $K_2 = K_{21} + K_{22}\xi^2$ where K_{21} and K_{22} are positive numbers, it can be shown with slight modification in the proof of Lemma 2 that ϵ can be made arbitrarily small by choosing K_1, K_{21}, and K_{22} large enough. In the simulation studies, some of which are shown in Section V, it has been observed that the controller with $K_{22} = 0$ is satisfactory for making ϵ sufficiently small in all cases that ξ is bounded.

Lemmas 1 and 2, and the preceding comments suggest a procedure, as given in the following section, for synthesizing an adaptive controller in relation to the formulation of the problem stated in Section II.

IV. Unified Procedure for Controller Synthesis

Consider a model reference system comprised of a plant and a model characterized respectively by (1) and (2). Lemma 1 and Lemma 2 with $g = n - m$ suggest a control signal $u(t)$ of the form of (19) rewritten below to achieve the control objective stated previously

$$u(t) = -(1/\beta)\xi(t) \tan^{-1}(K_1\xi(t)) \tan^{-1}(K_2\gamma(z)) \qquad (33)$$

where $\beta = (\beta_0 + \beta_1)/2$, K_1 and K_2 are positive numbers, and $\gamma(z)$ is defined as

$$\gamma(z) = \sum_{k=0}^{n-m} p_{k(n-m)}z_k \qquad (34)$$

where p_{ij} is an element of P, the solution matrix of (21) with dimension $(n - m)$. Recalling (16), the characteristic variable $\xi(t)$ can be synthesized as

$$\xi(t) = \sum_{k=0}^{n-m} a_{s+k} z^{(k)} - \beta u(t), \quad s = n' - (n - m). \quad (35)$$

Equations (33)–(35) then generate the required control signal if $\xi(t)$ is bounded, for it requires the measurement of only $(z, \dot{z}, \cdots, z^{(n-m)})$, $u(t) \in C^m(T)$ since $\xi(t) \in C^m(T)$ by Lemma 1, and from Lemma 2 and its comments $\|z\| < \epsilon$ for $t \to \infty$ where ϵ can be made arbitrarily small by choosing K_1 and K_2 sufficiently large. The implementation of this control scheme is shown in Fig. 2. It may be noted that the realization of the arctangent function is not essential. It may be replaced by a saturating amplifier without affecting the quality of the control.

It should now be noted that the control signal is not explicitly depending upon any of the plant parameters or plant nonlinearities except $b_m(t)$, the coefficient of the mth order plant input derivative, through its bounds β_0 and β_1. It is implicitly dependent upon the parameters and the nonlinearities through $\xi(t)$, which is expressed in terms of the available state variables. Hence, if a bounded ξ is a priori known, then the only knowledge of the plant that is required for the synthesis is the form of plant equation and the constants β_0 and β_1, the bounds of $b_m(t)$.

If $\xi(t)$ is not bounded, the controller with $K_2 = K_{21} + K_{22}\xi^2$ may still ensure an arbitrarily small ϵ. However, this may result in an instability of the control loop or an unbounded control input [5], [6] which is practically undesirable. An investigation of boundedness of ξ and the related stability problem for a plant as general as (1) may be best done only through simulation studies. However, the stability problems for some restricted class of systems have been analyzed and will be reported soon. The synthesis procedure for the adaptive controller is then as follows.

1) Select a stable time-invariant reference model of order n' such that $(n - m) \leq n' \leq n$. This model may have numerator dynamics in its transfer function.

2) Construct A_0^*, as defined by (17), which is a $(n - m) \times (n - m)$ principal minor, located in the lower right-hand corner of A_0, the model coefficient matrix as defined by (3).

3) Write (16) or (18) without the knowledge of $\xi(t)$.

4) Solve the $(n - m) \times (n - m)$-dimensional matrix (21) for P by choosing some suitable $(n - m) \times (n - m)$ positive definite symmetric matrix Q, and check for the positive definiteness of the matrix P.

5) The required control signal $u(t)$ is then given by (33)–(35).

V. Simulation Studies

To substantiate the preceding analysis, an extensive simulation study was carried out on a digital computer. Two examples of this study will now be described since they illustrate the more important aspects of this technique.

Example 1

Consider a system having a sixth order plant and sixth order model. The plant, for which $n = 6$ and $m = 4$, is unstable and is characterized by the following parameters and nonlinearities:

$$a_0(t) = 100 + K(t)\,(250 + 200 \cos \omega t),$$
$$a_1(t) = 200 - 100 \cos \omega t + K(t)\,(330 + 300 \sin \omega t),$$
$$a_2(t) = 260 + 130 \sin \omega t + K(t)\,(91 + 50 \cos \omega t),$$
$$a_3(t) = 162 + 81 \sin \omega t + K(t)\,(12 + 10 \sin \omega t),$$
$$a_4(t) = 61 + 30 \cos \omega t + K(t),$$
$$a_5(t) = 12 - 6 \cos \omega t,$$
$$a_6(t) = 1,$$
$$b_0(t) = K(t)\,(250 + 200 \cos \omega t),$$
$$b_1(t) = K(t)\,(330 + 300 \sin \omega t),$$
$$b_2(t) = K(t)\,(91 + 50 \cos \omega t),$$
$$b_3(t) = K(t)\,(12 + 10 \sin \omega t),$$
$$b_4(t) = K(t),$$
$$f = 500\dot{x}^2 - 300x^3; \quad K(t) = 3 + 0.4t, \quad \omega = \pi/2,$$
$$x^{(k)}(0) = 0, \quad \text{for } k = 0, 1, \cdots, 5.$$

For the reference model, $n' = n = 6$ and $m' = 0$. The following parameters also apply to this model:

$$a_0{}^0 = 2500, \quad a_1{}^0 = 3250, \quad a_2{}^0 = 2250, \quad a_3{}^0 = 880,$$
$$a_4{}^0 = 194, \quad a_5{}^0 = 22, \quad a_6{}^0 = 1, \quad b_0{}^0 = 2500,$$
$$y^{(k)}(0) = 0, \; k = 0, 1, \cdots, 5.$$

The input $r(t)$ is taken to be the unit step function.

Since $b_m(t) = b_4(t) = 3 + 0.4t$, (up to $t = 14$) $\beta_0 = 3$, $\beta_1 = 8.6$ and $\beta = 5.8$. Also since $n = 6$ and $m = 4$, $n - m = 2$ and

$$A_0^* = \begin{bmatrix} 0 & 1 \\ -194 & -22 \end{bmatrix}$$

a stable matrix. Equation (16) then becomes

$$\ddot{z} + 22\dot{z} + 194z = 5.8u(t) + \xi(t).$$

Let

$$Q = \begin{bmatrix} 399 & 0 \\ 0 & 42 \end{bmatrix}$$

then

$$P = \begin{bmatrix} 216 & 1 \\ 1 & 1 \end{bmatrix}.$$

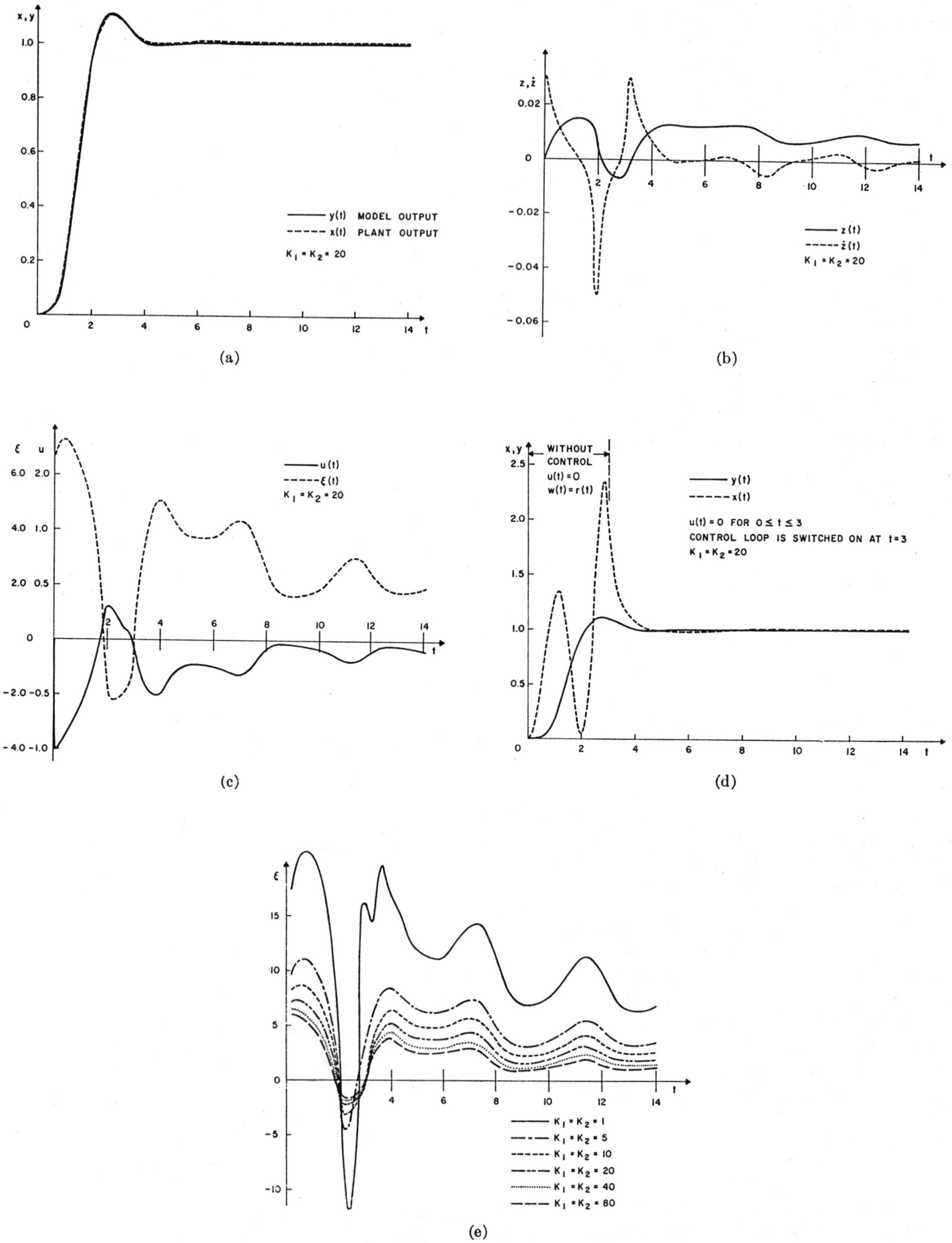

Fig. 3. Example 1. (a) Control system response. (b) Error $z(t)$ and error derivative $\dot{z}(t)$. (c) Control signal $u(t)$ and characteristic variable $\xi(t)$. (d) Response of plant with and without control $u(t)$. (e) Dependence of $\xi(t)$ on K_1 and K_2.

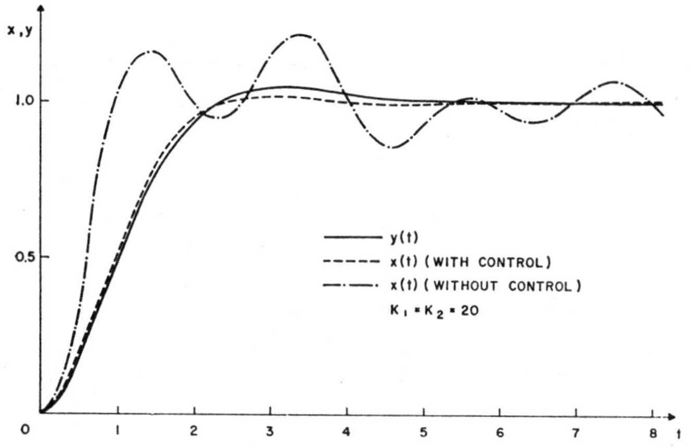

(a)

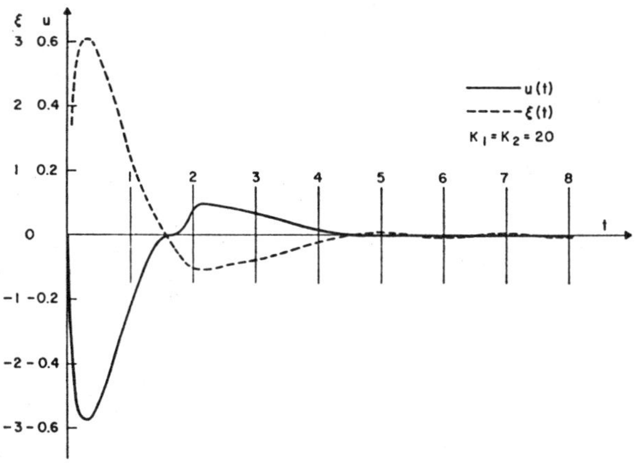

(b)

Fig. 4. Example 2. (a) Control system response. (b) Control signal $u(t)$ and characteristic variable $\xi(t)$.

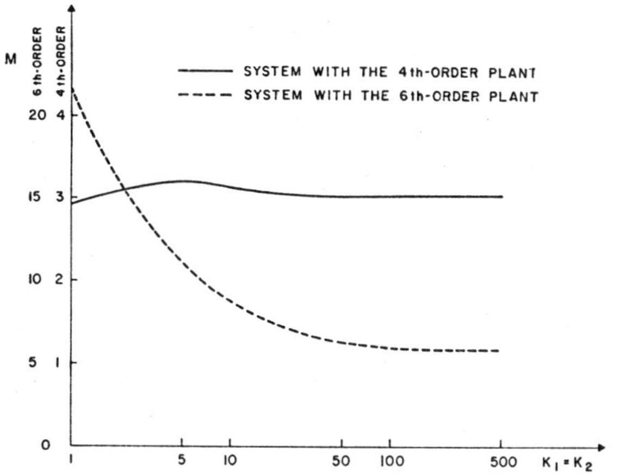

Fig. 5. Dependence of M, the bound of $\xi(t)$, on K_1 and K_2.

Note that q_{11} is made larger than q_{22} in Q in order to improve the convergence of the process [5]. The control signal $u(t)$ is then

$$u(t) = -(1/5.8)\xi(t) \tan^{-1}(K_1\xi(t)) \tan^{-1}(K_2\gamma(z))$$

where $\gamma(z) = z + \dot{z}$ and $\xi(t) = \ddot{z} + 22\dot{z} + 194z - 5.8u(t)$.

The resulting system response, $x(t)$ and $y(t)$, for $K_1 = K_2 = 20$ is shown in Fig. 3(a). Fig. 3(b) shows the error signal $z(t)$ and its derivative $\dot{z}(t)$ on an expanded scale. These figures show that the plant output follows the model output very closely. The control signal $u(t)$ and the characteristic variable $\xi(t)$ are shown in Fig. 3(c). Note that due to the unstable nature of the uncontrolled system, $u(t)$ cannot approach zero as t becomes large. Fig. 3(d) shows that the system without control ($u = 0$, $w = r$) is unstable. For this test the control loop was kept open until $t = 3$, at which time the control loop was suddenly closed. The dependence of $\xi(t)$ on K_1 and K_2 is shown for some values of K_1 and K_2 in Fig. 3(e), which suggests that $\xi(t)$ is bounded.

Example 2

In this example the plant is stable and of fourth order. The model is of second order. For the plant $n = 4$, $m = 2$, and

$$\begin{aligned}
a_0(t) &= 3K(t),\\
a_1(t) &= 6 - 2\cos\omega t + 4K(t),\\
a_2(t) &= 7 + 2.5\sin\omega t + K(t),\\
a_3(t) &= 3,\\
a_4(t) &= 1,\\
b_0(t) &= 3K(t),\\
b_1(t) &= 4K(t),\\
b_2(t) &= K(t) = 3 + 0.4t,\\
\omega &= \pi/2, \quad x^{(k)}(0) = 0, \quad \text{for } k = 0, 1, 2, 3.
\end{aligned}$$

For the model $a_0^0 = a_1^0 = b_0^0 = 2$, $a_2^0 = 1$, and $A_0^* = A_0$. The input $r(t)$ is again a step function. For the synthesis $\beta_0 = 3$, $\beta_1 = 7.4$ (up to $t = 11$), $\beta = 5.2$;

$$Q = \begin{bmatrix} 4 & 0 \\ 0 & 0.4 \end{bmatrix},$$

$$P = \begin{bmatrix} 3.2 & 1 \\ 1 & 0.6 \end{bmatrix},$$

and $K_1 = K_2 = 20$. The simulation results are shown in Fig. 4(a) and 4(b). Note that $u(t)$ and $\xi(t)$ in Fig. 4(b) approach zero as t becomes large.

In Fig. 5, $M = |\xi(t)|_{max}$ is plotted versus $K_1 = K_2$ up to unrealistically large values of K_1 and K_2 for each of the above examples. It is seen that M approaches a constant as K_1 and K_2 increases.

VI. Conclusions

A synthesis technique has been described in this paper for the design of an adaptive controller. This technique is based upon the existence and boundedness of the characteristic variable and Lyapunov type synthesis. It has been shown that this technique may be applied to a wide variety of plants in a class of model reference control systems. In the generation of the control signal no knowledge of the plant parameters and its nonlinearities is required. Only the form of the plant equation and the bounds of $b_m(t)$, the coefficient of the mth order plant input derivative, are needed. Also, this technique requires the measurements of only the output and its first $(n - m)$ derivatives. Simulation results have been given which show the boundedness of characteristic variable $\xi(t)$.

It appears that the proposed technique may be applicable to a wider class of control problems than has previously been possible. A proof of the boundedness of the characteristic variable is required. This has yet to be done, but in the meanwhile reliance has been placed upon the results of simulation studies. The effects of measurement noise and an imperfect controller on adaptation have not been discussed. Further work in this direction is in progress.

Acknowledgment

The author H. H. Choe wishes to acknowledge the Navy Department of the Republic of Korea for granting him a leave of absence for his graduate studies in Canada.

References

[1] R. E. Kalman and J. E. Bertram, "Control system analysis and design via the 'second method' of Lyapunov, pt. I: continuous-time systems," *Trans. ASME, J. Basic Engrg.*, pp. 371–393, June 1960.
[2] L. P. Grayson, "Design via Liapunov second method," Preprints, 1963 JACC (Minneapolis, Minn.), pp. 589–595.
[3] J. G. Hiza and C. C. Li, "Analytical synthesis of a class of model-reference time-varying control systems," *IEEE Trans. Applications and Industry*, vol. 82, pp. 356–362, November 1963.
[4] L. P. Grayson, "The status of synthesis using Lyapunov's method," *Automatica*, vol. 3, pp. 91–121, 1965.
[5] R. V. Monopoli, "Engineering aspects of control system design via the 'direct method' of Liapunov," NASA Contractor Rept. CR-564, December 1966.
[6] ——, "Control of linear plants with zeros and slowly varying parameters," *IEEE Trans. Automatic Control* (Short Papers), vol. AC-12, pp. 80–83, February 1967.
[7] R. V. Monopoli and C. R. Gitlitz, "Control of linear time varying plants with differential operators acting on the input," *Proc. 5th Ann. Allerton Conf. Circuit and System Theory*, 1967, pp. 608–613.
[8] F. B. Hildebrand, *Methods of Applied Mathmatics.* Englewood Cliffs, N.J.: Prentice-Hall, 1952.
[9] P. C. Parks, "A new proof of the Hurwitz stability criterion by the second method of Liapunov, with application to optimum transfer functions," Preprints, 1963 JACC (Minneapolis, Minn.), pp. 471–477.

Comparative Studies of Model Reference Adaptive Control Systems

CHANG-CHIEH HANG AND PATRICK C. PARKS

Abstract—**A brief but up-to-date survey on existing methods of designing a class of model reference adaptive control systems is given in this paper. A comparison of the merits of the various design rules is also made with particular attention to the M.I.T. rule and the Lyapunov synthesis technique. Subsequently a systematic performance comparison of the various designs, with deterministic as well as stochastic inputs, is presented using the computer simulation studies of two simple gain adjustment schemes. From the dimensionless characteristic graphs obtained the Lyapunov schemes are found to exhibit superior performance over other designs. These graphs also uncover the strange adaptive response of those designs based on gradient methods in that the performance index may increase or decrease with increasing system gain parameters.**

I. Introduction

THE model reference adaptive control (MRAC) technique has been a popular approach to the control of systems operating in the presence of parameter and environmental variations. In such a scheme, the desirable dynamic characteristics of the plant are specified in a reference model and the input signal or the controllable parameters of the plant are adjusted, continuously or discretely, so that its response will duplicate that of the model as closely as possible. The identification of the plant dynamic performance is not necessary and hence a fast adaptation can be achieved.

This paper is concerned with the continuous parameter adaptive schemes. Generally speaking, there are two approaches to the synthesis of this class of MRAC systems. One is based on the minimization of a performance index [1] and the other on a Lyapunov function [2]. Each of these approaches has its own merits and limitations, although many modifications have been suggested to improve them further. A direct contrast of the merits of these designs has been briefly mentioned in the literature [2], [8] but a rigorous comparison, especially that from a performance viewpoint, has not been reported. Hence a comparative study of the various design rules will be of great interest to the designers who have long been faced with the difficulty of selecting a suitable one for certain applications.

Since there are already some detailed accounts of the various design rules in the literature [17], [18], only a brief but up-to-date survey is conducted in this paper. Attention will then be concentrated on single-input single-output plant gain adjustment systems. Some of the more popular rules are critically analyzed to point out their relative merits with regards to the stability, realization, and adaptive response, which will also be supported by some simulation results. Subsequently a systematic performance comparison based on some well-known criteria is attempted through simulation studies. Deterministic as well as stochastic inputs are employed. Sensitivities of the performance to the input frequency bands are also examined. The interesting and useful performance characteristics are presented in the form of similitudes [20].

Manuscript received October 24, 1972. Paper recommended by G. N. Saridis, Chairman of the IEEE S-CS Adaptive and Learning Systems, Pattern Recognition Committee. The work of C. C. Hang was supported by the Royal Commission for the Exhibition of 1851 and the British Council.

The authors are with the Inter-Institute of Engineering Control, University of Warwick, Coventry, England.

II. A Brief Survey of the Design Rules

The MRAC system was first designed by the performance index minimization method proposed by Whitaker [1] of the M.I.T. Instrumentation Laboratory and has since then been referred as the M.I.T. design rule. The performance index is the integral squared of the response error. This rule has been very popular due to its simplicity in practical implementation, although it may require a large number of sensitivity filters for multiparameter adjustments. An improved design rule with respect to the speed of response has then been proposed by Donalson [3], who used a more general performance index than that of Whitaker, but additional filters and the measurement of the state vectors are required. The need of the sensitivity filters can be avoided by a gradient method developed later on by Dressler [4], or by an "accelerated gradient method" suggested by Price [5]. The latter is easier to implement and is capable of achieving faster adaptations compared with other gradient techniques. Another contribution to the simplification of the design comes from the application of sensitivity analysis by Kokotovic *et al.* [24], [25], resulting in a design similar to the M.I.T. rule. Here, with further approximation, only one sensitivity filter is required for multiparameter adjustments. For some other particular applications, Winsor [17] has also modified the M.I.T. rule to reduce the sensitivity of the response to the loop gain, at the expense of additional instrumentation. All the design rules mentioned are not, however, globally stable and hence the adaptive gain that governs the speed of response is limited. A good compromise between the stability and the speed of adaptation will have to be decided by laborious simulation studies. A recent contribution by Green [6] has extended the

Reprinted from *IEEE Trans. Automat. Contr.*, vol. AC-18, pp. 419–428, Oct. 1973.

work of Dressler to form a "stable maximum descent" method. Though the problem of global stability is solved, this adaptive rule is not attractive from a practical viewpoint because the first derivatives of the state vectors are often required for its implementation.

On the other hand, in the Lyapunov synthesis approach, the adaptive rule is obtained by selecting the design equations to satisfy conditions derived from Lyapunov's second method, so that the system stability is guaranteed for all inputs. Butchart and Shackcloth [7] first suggested the use of a quadratic Lyapunov function, which was employed later on by Parks [2] to redesign systems formerly designed by the M.I.T. rule. The use of a different Lyapunov function by Phillipson [8] and Gilbart *et al.* [9] has resulted in the introduction of feedforward loops that would improve the damping of the adaptive response. The main disadvantage of the Lyapunov method is that the entire state vector must be available for measurement, which is not often possible. Recent efforts in the application of the idea of positive real transfer function, notably that by Monopoli, have allowed one to eliminate or reduce the number of differentiators required for implementing the design rule [2], [10]. Among other possible solutions to avoid the use of derivative networks, Currie and Stear [11] have envisaged the use of a Kalman filter, which would also handle the measurement noise problem, while the use of state observers [12] to estimate the states of an unknown time-varying plant is still an open question. Another disadvantage is that the Lyapunov design rule may not be applicable to cases where the plant parameters cannot be directly adjusted. Such a case was mentioned by Winsor *et al.* [13] and a solution, though quite complex, was offered by Gilbart *et al.* [9].

All these methods may be extended to treat both the single-dimension and multidimension systems. For the latter case, another synthesis technique [14] based on the concept of hyperstability has been recently proposed. The resulting design is asymptotically hyperstable and the structure of the adaptive controller is simple. The drawback of this scheme is that a series of differentiating networks is often necessary. In the single-dimension case, it results in a design identical to the Lyapunov redesign due to Parks [2].

From this brief survey it is observed that many design rules for MRAC systems are available. So far as the single-dimension case is concerned, the M.I.T. rule has been the most popular although it has a severe stability problem. The Lyapunov synthesis, in avoiding the problem of stability, is difficult to realize in practice because the measurement of the entire state vector is not often possible. While the degree of these considerations would vary with the particular process to be controlled, another important factor, the performance, such as the speed of adaptation, has to be considered. It is the intention of this investigation to attempt to supplement some useful information regarding the relative performance of the various designs.

III. A Critical Comparison of the Design Rules

The following analysis is based on the aggregate of knowledge scattered in the literature. This information is reviewed here and studied by means of simulations. We shall first compare the M.I.T. rule [1] and the Lyapunov synthesis [2], [9] through the design of a gain adjustment loop of a linear system as shown in Fig. 1. Following this we shall examine design rules due to Dressler [4], Price [5], and Monopoli [10]. The block diagrams of these designs are shown in Fig. 2.

A. M.I.T. Rule and Lyapunov Synthesis

The notation used below is that shown in Figs. 1 and 2. The performance index used in the M.I.T. rule [1] is $\int e_1^2\,dt$ and the parameter adjustment law is

$$\dot{K}_c = Be_1 \frac{\partial \theta_p}{\partial K_c}. \tag{1}$$

In this case the sensitivity function $\frac{\partial \theta_p}{\partial K_c}$ is proportional to θ_m and hence the above equation becomes

$$\dot{K}_c = B'e_1\theta_m \tag{2}$$

where the constant B' is the adaptive gain.

The Lyapunov synthesis is based on the use of a Lyapunov function. The most successful form of this function V used to date is that proposed by Gilbart *et al.* [9]. This is briefly described in the Appendix, Section A.

$$V = \boldsymbol{e}^T P \boldsymbol{e} + \lambda(x + \gamma K_v m)^2 \tag{3}$$

where

$$m = B'\boldsymbol{e}^T P \boldsymbol{b} r \tag{4}$$

$$x = K - K_c K_v \tag{5}$$

and the time derivative of V is given by

$$\frac{dV}{dt} = -\boldsymbol{e}^T Q \boldsymbol{e} - 2\lambda\gamma K_v^2 m^2. \tag{6}$$

These result in the stable adjustment law

$$\dot{K}_c = m + \gamma \dot{m} \tag{7}$$

where γ is a proportional constant that is chosen to provide additional damping if required. Putting $\gamma = 0$ results in the design rule used by Parks [2].

Equations (2) and (7) will be compared in the following manner.

1) Stability: A stability analysis of (2) is very difficult. The doubt about possible instability has been demonstrated by Parks [2] for a second-order system with step inputs. Even for a first-order system with a sinusoidal input, James [15] has obtained a complicated stability domain in the parameter space. Hence, extensive simulations during the design stage are necessary to establish the region of stable operations. On the other hand, (7) is

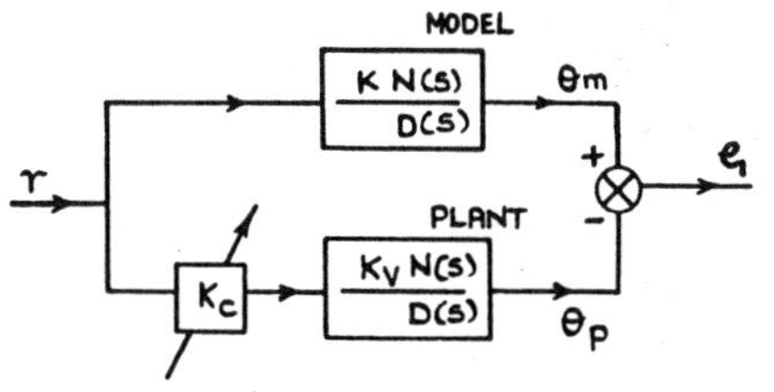

Fig. 1. A basic MRAC gain adjustment system.

assured to be stable for all inputs such that $\boldsymbol{e} \to 0$ asymptotically, with the assumption that K_v is slowly time varying. When this assumption is severely violated, a stability problem similar to that of (2) may exist. "Eventual stability," however, can be assured by using a theorem due to Lasalle and Rath if the time-varying function K_v belongs to a certain class [16].

2) Physical Realization: Equation (2) can be easily implemented and it is this distinct advantage that has made MRAC a popular adaptive control strategy. Equation (7) however, requires the estimation of the complete state vector, which is not often available and, hence, necessitates the use of differentiating networks causing a noise amplification problem.

3) Response: The speed of adaptation of both equations depends on the magnitudes of the adaptive gain B' and the input signal R. A large B' is always necessary to maintain a high speed of adaptation. However, as B' and R vary, the damping of the response will also vary. Root locus plots of these equations for a second-order system [8], [9] would show that when $B'R^2$ is large, the M.I.T. design will be underdamped while the Lyapunov design will be adequately damped with suitable values of γ.

B. Other Design Rules

We shall next examine the following rules.

1) Dressler [4]: The parameter adjustment law is

$$\dot{K}_c = B'e_1r. \tag{8}$$

The resulting controller is very simple and no sensitivity filter is required. The disadvantages are that the damping of the response suffers at larger loop gains and that the global stability is not guaranteed. Its stability problem is similar to that of the M.I.T. rule.

2) Price [5]: The parameter adjustment law is called the "accelerated gradient method."

$$\dot{K}_c = B'e_1r + \gamma_c \frac{d}{dt}(B'e_1r) \tag{9}$$

where γ_c is a constant. The controller is similar to that of Dressler except for the addition of the feedforward term. This term has the effect of improving the damping and the stability of the response. This stabilizing effect would be impaired as the order of the system increases, and generally the global stability cannot be guaranteed.

3) Monopoli [10]: This is based on a modification of the Lyapunov scheme. A differentiating block ($Z(s)$) is used to modify the plant transfer function such that $Z(s)N(s)/D(s)$ is positive real, and a Kalman–Meyer lemma [26], [27] is used to eliminate the error derivatives as required in (7). Hence,

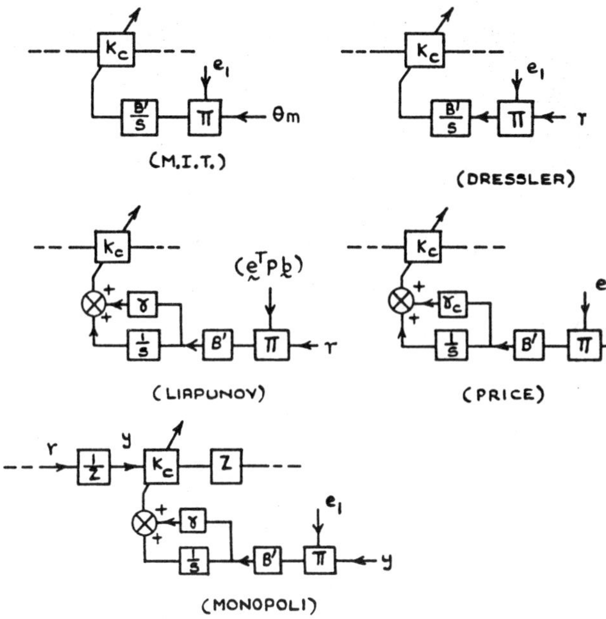

Fig. 2. Various designs of the adaptive loop.

$$\dot{K}_c = B'e_1y + \gamma \frac{d}{dt}(B'e_1y) \tag{10}$$

where y is the modified input signal to the plant and is obtained by passing the original signal through a filter $(1/Z(s))$. For an nth-order plant with m zeros, the order of $Z(s)$ is $(n - m - 1)$. Global stability of the adaptive loop will be guaranteed while the number of derivative network required is reduced to $(n - m - 1)$, or $(n - m - 2)$ if the extra damping loop is not in use. This technique can be easily extended to the case of a general time-varying gain [10] and also to the adjustment of other parameters [28].

C. Simulation Results

A simulation study of the adaptive response of these designs has shown that very often the Lyapunov designs could achieve excellent performance not attainable by other rules. As an example consider a second-order plant whose gain is to be adjusted. Referring to Figs. 1 and 2, the following values are assumed:

$$\frac{N(s)}{D(s)} = \frac{1}{1 + a_1s + a_2s^2}, \qquad a_1 = 2, \qquad a_2 = 1,$$

$$K = 1, \qquad K_v = 2, \qquad K_c(t_0) = 0.2.$$

From the Appendix, Section A, we obtain $\boldsymbol{e}^T P \boldsymbol{b} = e_1 + \dot{e}_1$. We shall choose $Z(s) = 2s + 2$ (as in [10]) and shall limit the values of γ and γ_c to, say, 50 percent because in actual use they may have to be quite small to reduce the effect of any noise at the plant output and the excessive transient overshoots due to large initial errors. Some of the typical adaptive responses of the various designs are depicted in Fig. 3 for step as well as sinusoidal inputs. The responses shown for the M.I.T., Dressler, and Price designs have been optimized with respect to the convergence time. The responses shown for the Lyapunov and Monopoli's schemes are, however, not optimized, i.e., they can still be further improved if required by increasing the adaptive gain. From this simulation study, the M.I.T. design is found to be unstable for $B' > 1$. Even when it is stable at lower values of B', the response is slow, the convergence time being well over five system time constants. The response of the Dressler scheme to a step input is similar to that of the M.I.T. scheme. However, for a sinusoidal input, the Dressler scheme shows a steady-state parameter error that is dependent on the loop gain as well as the input signal frequency. The design due to Price shows a better damping and stability, which improve as γ_c is increased. On the other hand, the Lyapunov design is always stable and the damping and convergence can be improved systematically by varying B' and γ. A convergence time of even less than one system time constant can be easily achieved. The design due to Monopoli, which does not require any differentiator in this case, exhibits quite a fast response. Although its damping would suffer at higher B', the system stability would always be maintained. These results support and substantiate the foregoing theoretical analysis.

It is appreciated that a further study on the relative performance of these rival designs is a complex problem. This is pursued in the next section.

IV. A Systematic Performance Comparison

Some commonly used performance criteria [19], which include the settling time (T_s), integral of squared error (ISE), integral of time absolute error (ITAE), and integral of time squared error (ITSE) will be employed to compare the response of the various designs against their system parameters. This will be studied experimentally through computer simulations of two gain adjustment schemes. The results will be presented in the form of similitudes by applying a dimensional analysis [20] to the system differential equations such that the quantities to be investigated are expressed in dimensionless groups. The dimensionless performance criterion is denoted by π_1 while the dimensionless system parameter is denoted by π_2. The performance characteristics are defined in this connection as the plots of π_1 against π_2.

A. First-Order Systems ($N(s)/D(s) = 1/(1 + sT)$)

In this case, the designs due to Dressler and Price are identical to the Lyapunov schemes. Also, the latter does not require any differentiators. Hence we only need to compare the M.I.T. and the Lyapunov designs.

1) Deterministic Inputs: Step and sinusoidal inputs are employed. From the dimensional analysis shown in the Appendix, Section B, the following are defined.

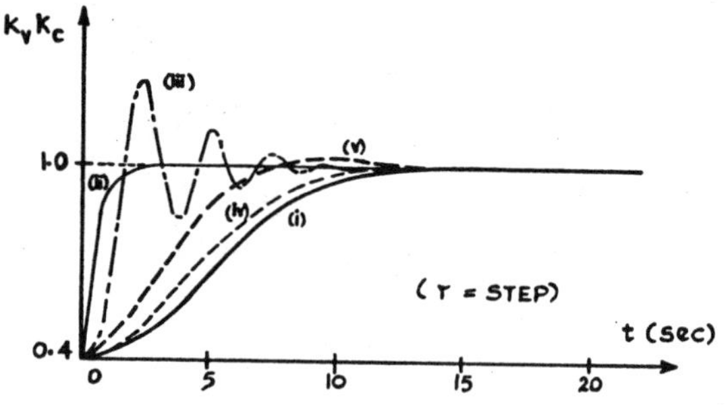

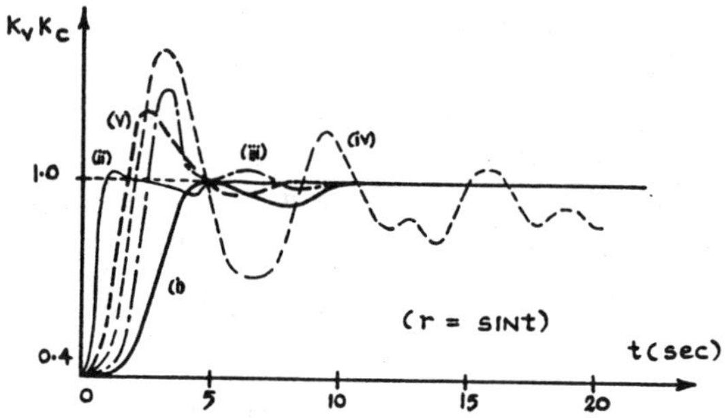

Fig. 3. Responses of the second-order systems: (i) M.I.T.; (ii) Lyapunov (γ = 0.1); (iii) Monopoli; (iv) Dressler; (v) Price (γ_c — 0.5).

$$\pi_2 = KK_vB'R^2T \quad \text{(M.I.T. design)}$$

$$= K_vB'R^2T \quad \text{(Lyapunov design)}$$

$$\pi_1 = T_s/T \quad \text{(5 percent } T_s \text{ criterion)}$$

$$= \frac{1}{K^2R^2T}\int e_1^2\,dt \quad \text{(ISE criterion)}$$

$$= \frac{1}{KRT^2}\int t|e_1|\,dt \quad \text{(ITAE criterion)}$$

$$= \frac{1}{K^2R^2T^2}\int te_1^2\,dt \quad \text{(ITSE criterion).}$$

The parameters that cannot be grouped into the above are fixed at frequency of sinusoidal input = 2.5 c/s, $K_c(t_0) = 0$, $\gamma = 0$ and 0.1.

The performance characteristics obtained are shown in Figs. 4 and 5. For step inputs, in which case the M.I.T. design is always stable, the T_s criterion shows a region where this design is unfavorable since π_1 may increase or decrease with an increment in π_2, whereas the same type of uncertainty does not appear in the Lyapunov design with $\gamma = 0.1$. For sinusoidal inputs, all the four characteristics for the M.I.T. design possess regions of uncertainty over a wide range of π_2. Furthermore, it has already been ensured that within the parameter ranges tested, that is $\pi_2 < 25$, this design is operated below the region of conditional stability as pointed out by James [15]. These findings suggest that an extensive simulation study would be necessary in order to determine a safe and economic value of π_2 to achieve any specific π_1 even though the system is

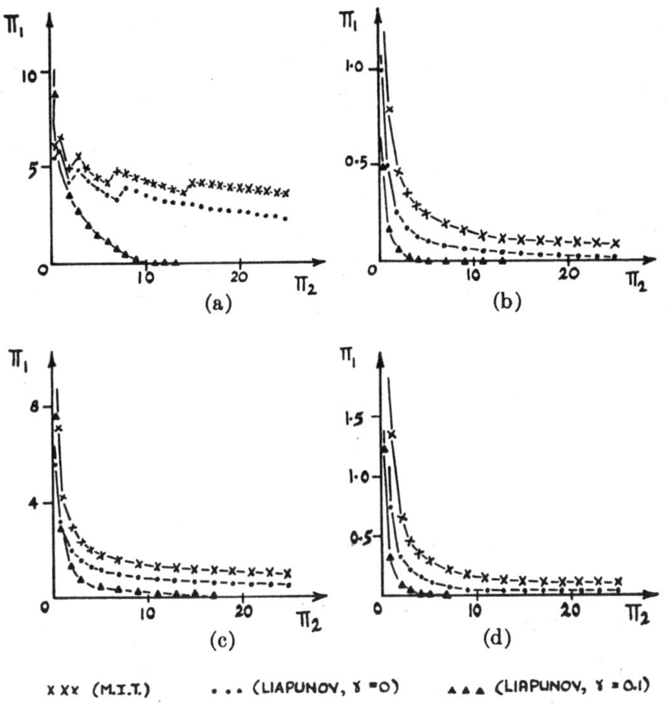

Fig. 4. Performance characteristics of first-order systems with step inputs. Criteria: (a) T_s; (b) ISE; (c) ITAE; (d) ITSE.

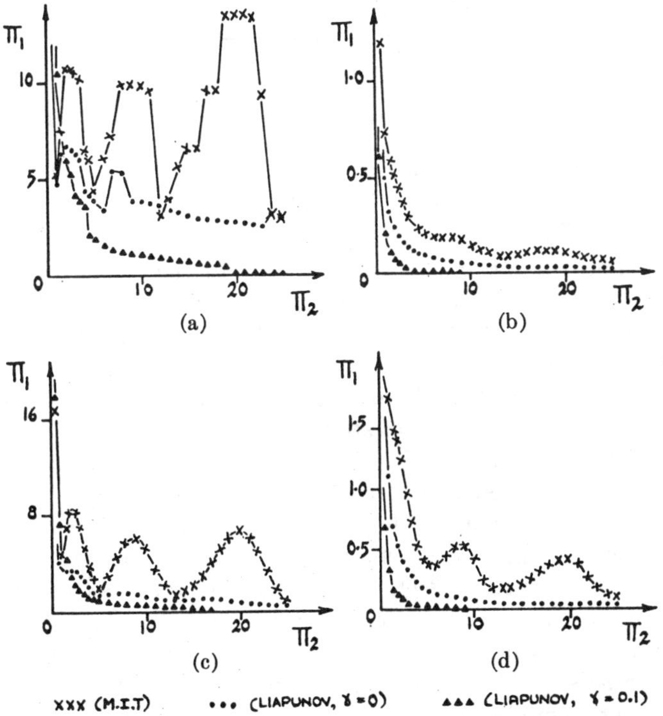

Fig. 5. Performance characteristics of first-order systems with sinusoidal inputs. Criteria: (a) T_s; (b) ISE; (c) ITAE; (d) ITSE.

operated in the stable region. On the other hand, the similitudes for the Lyapunov designs show a monotonic decrease of π_1 with increasing π_2. This is a desirable feature. In addition this design can achieve values of π_1 not attainable by the M.I.T. design. Examinations of the effect of changing the input signal frequency have also been conducted. The results, which are too long to show here, indicate that in the M.I.T. scheme the system performance is very sensitive to the change in frequency whereas in the Lyapunov scheme it is almost insensitive to the frequency especially at higher gains.

2) Stochastic Inputs: The above experiment is repeated with a band-limited Gaussian white-noise input. This stochastic signal is obtained by spacing a digitally generated, zero-mean, Gaussianly distributed sequence of pseudorandom numbers, by an interval of h s and with linear interpolations. The variance of the signal is denoted δ_N^2 and its power spectrum, which is approximately flat, possesses a cutoff frequency of $1/2h$ Hz. To reduce the complexity of this investigation, only the ISE criterion will be studied in detail.

The dimensionless quantities are

$$\pi_1 = \frac{1}{K^2\delta_N^2T}\,\mathrm{E}\left[\int e^2\,dt\right]$$

$$\pi_2 = KK_vB'\delta_N^2T \qquad \text{(M.I.T. design)}$$

$$= K_vB'\delta_N^2T \qquad \text{(Lyapunov design)}$$

where $E[\ ]$ denotes the expectation (i.e., ensemble average) operator. The fixed parameters are $h = 0.002$, $K_c(t_0) = 0.0$, $\gamma = 0$ or 0.1.

The results obtained are plotted in Fig. 6(a). The similitudes show that both the M.I.T. and Lyapunov designs exhibit the desirable characteristics that π_1 decreases monotonically with increasing π_2. The latter also achieves a much lower π_1, which cannot be reached by the former. Another important property that has been noted is that the variances about the expected values are different in each case. From the plot shown in Fig. 6(b), one observes that the variances in the M.I.T. design are very much larger than those in the Lyapunov scheme. This indicates that in the former scheme there may exist a considerable degree of uncertainty about its performance. This is confirmed by studying the ensemble members of the random process. One of these is shown here in Fig. 7. Also shown are ensembles of the corresponding results using the other two integral criteria. These similitudes reveal that the M.I.T. scheme possesses the undesirable property that π_1 may increase or decrease with increasing π_2 while that in the Lyapunov scheme shows an almost monotonic decrease.

In addition to the case just reported, other experiments have been carried out. The finding is that when the power spectrum of the input signal (proportional to $1/h$) is reduced, the performance of the M.I.T. design would deteriorate whereas that of the Lyapunov design would improve.

B. Second-Order System ($N(s)/D(s) = 1/(1 + a_1s + a_2s^2)$)

The five designs described in Section III will be examined here. It is noted that while the Lyapunov design requires one differentiator, that due to Monopoli does not need any.

1) Deterministic Inputs: The following dimensionless parameters are defined:

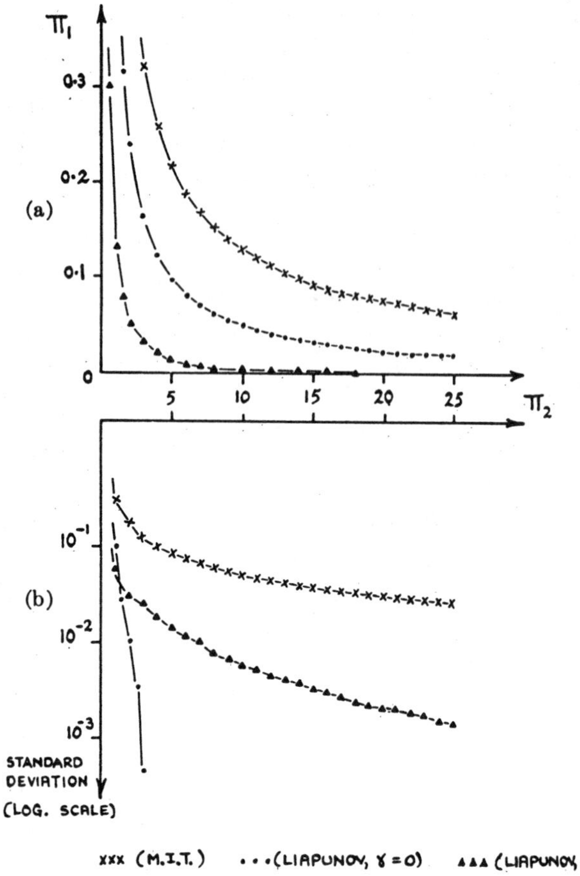

Fig. 6. Performance characteristics of first-order systems with stochastic inputs. Criterion: ISE.

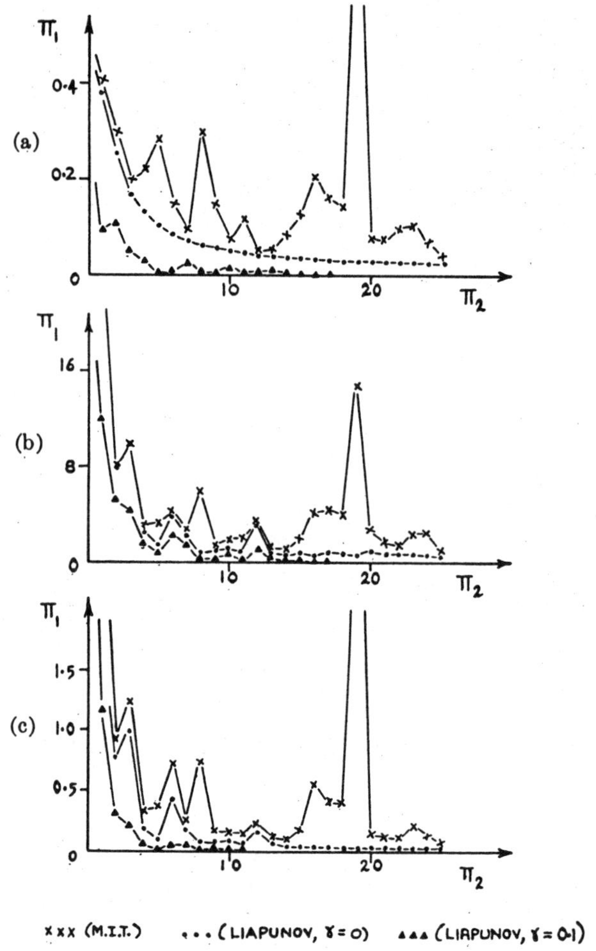

Fig. 7. A sample of the characteristics. Criteria: (a) ISE; (b) ITAE; (c) ITSE.

$$\pi_2 = KK_vB'R^2a_1 \quad \text{(M.I.T. design)}$$
$$= K_vB'R^2a_1 \quad \text{(others)}$$
$$\pi_1 = T_s/a_1 \quad \text{(2 percent } T_s \text{ criterion)}$$
$$= \frac{1}{K^2R^2a_1}\int e_1^2\,dt \quad \text{(ISE criterion)}$$
$$= \frac{1}{KRa_1^2}\int t|e_1|\,dt \quad \text{(ITAE criterion)}$$
$$= \frac{1}{K^2R^2a_1^2}\int te_1^2\,dt \quad \text{(ITSE criterion).}$$

Other fixed parameters are $a_2/a_1^2 = \frac{1}{4}$, $K_c(t_0) = 0.2$, $\gamma_c = 0.5$, $\gamma = 0$ and 0.1, frequency of sinusoidal input = 0.16 Hz.

The performance characteristics obtained are shown in Figs. 8 and 9. With step inputs, the M.I.T. and Dressler designs possess a minimum in π_1 as π_2 varies; for π_2 smaller or larger than this minimum value, π_1 increases sharply. Other designs show a monotonic reduction, especially at higher values of π_2. With sine inputs, both the M.I.T. and Dressler designs are again found to possess a minimum in π_1, and the latter is more critical than the former. The design by Price shows an unfavorable performance in that the uncertainty as discussed in the first-order systems occurs. The Lyapunov and the Monopoli designs, however, still maintain the desirable performance characteristics similar to that with step inputs.

The performance of these designs with different frequencies of the sinusoidal input signal has also been examined. The same range of π_2 is used. The general conclusion is that the Lyapunov and Monopoli designs are less sensitive to the signal frequency with regards to both the stability and the convergence rate. The M.I.T. system always possesses a minimum π_1 at some value of π_2 that increases with the frequency; at lower frequencies, more than one minimum point may be observed. The convergence rate decreases with increasing frequencies. The Dressler system is unstable at higher frequencies; at lower frequencies the system is stable for a small range of π_2 but this range may increase or decrease with decreasing frequencies. The design by Price improves at lower frequencies, in that the fluctuation in π_1 reduces, but deteriorates rapidly at frequencies higher than the resonant frequency of the plant and eventually becomes unstable.

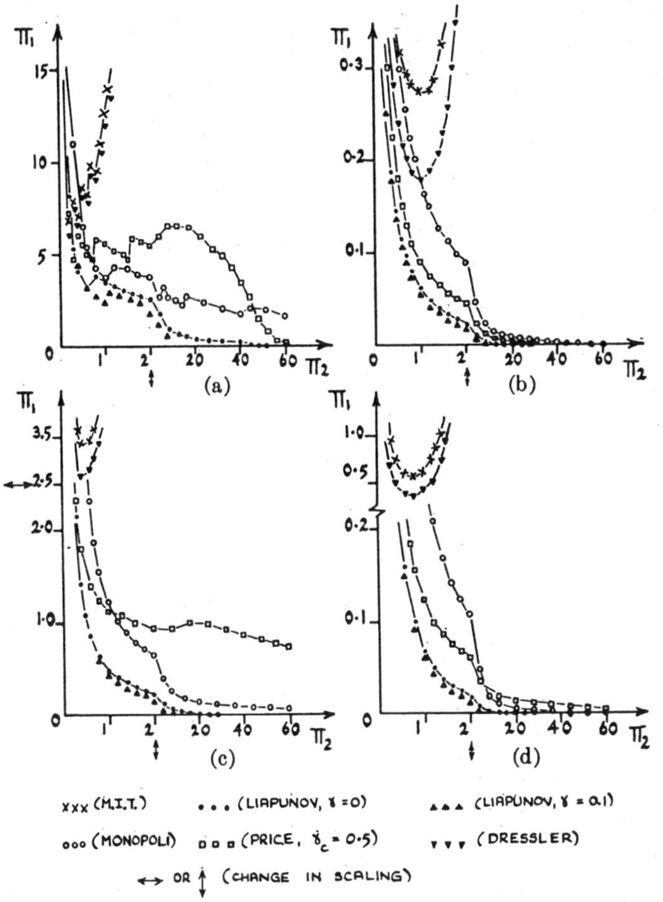

Fig. 8. Performance characteristics of second-order systems with step inputs. Criteria: (a) T_s; (b) ISE; (c) ITAE; (d) ITSE.

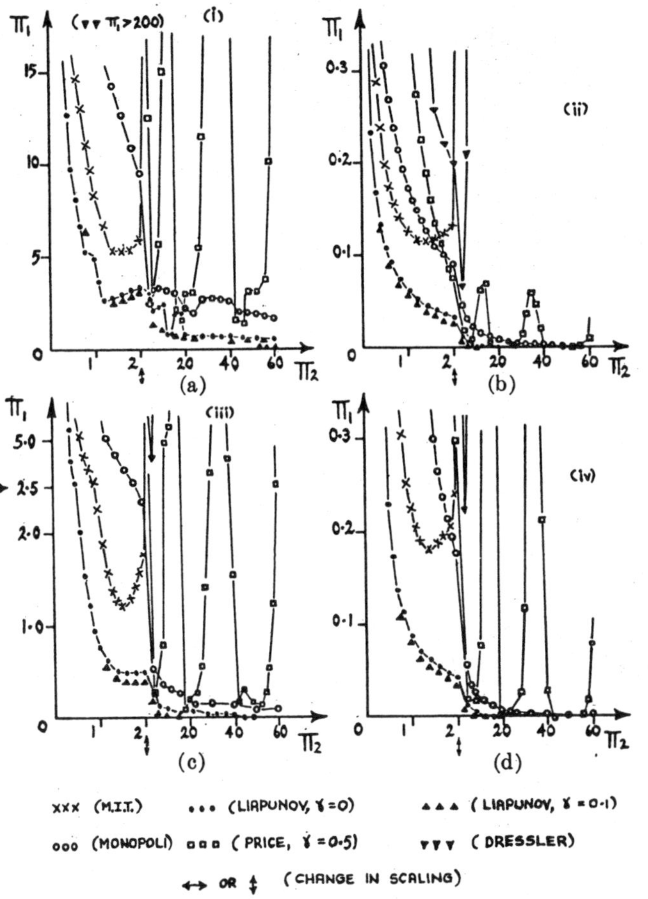

Fig. 9. Performance characteristics of second-order systems with sinusoidal inputs. Criteria: (a) T_s; (b) ISE; (c) ITAE; (d) ITSE.

2) Stochastic Inputs: The same experiment as in the first-order case is repeated. The main results with $h = 0.1$ are shown in Fig. 10(a) and (b). To summarize, the Lyapunov and Monopoli designs exhibit monotonic decrease of π_1 with increasing π_2 and the variances of π_1 are small; the other designs exhibit one or more minima in π_1 and the variances are also large indicating serious uncertainty as mentioned in the previous section. Different spectra of the input signal have also been used. The general observation is that the M.I.T. and Dressler designs exhibit worse performance when the bandwidth of the signal is reduced, while the other designs show improved performance.

V. Discussion and Conclusion

The extensive computer simulation study of the various MRAC designs reveal many interesting properties regarding the performance of the adaptive systems at different loop gains and under different input signals. These may be summarized as follows.

1) The designs that are not assured to be stable globally behave very differently when the gain parameter π_2 varies. They are also found to be sensitive to the frequency band of the input signal; one reason for this is that the total effective gain varies due to different attenuation by the system at different frequencies. The possible outcome of these two disadvantages is instability, poor damping, or poor convergence of the adaptation. It is unfortunate that in trying to compensate for the change in environment, the adaptive system may become sensitive to its own parameters.

2) The performance of those designs that are assured to be globally stable improves as the gain parameter π_2 increases. In addition, they can be made less sensitive to the input signal magnitude and frequency content by operating at larger values of π_2.

3) Among the three schemes based on gradient methods, the Dressler design exhibits the worst performance characteristics, especially when the input is sinusoidal or stochastic. The M.I.T. design is quite acceptable if the performance specification is not very strict. The design by Price performs better than the M.I.T. system with step or stochastic inputs but is inferior with sinusoidal inputs.

4) The two designs based on stability consideration may achieve low values of π_1 not attainable by other designs.

Between the two, the Lyapunov scheme is better as it requires a lower value of π_2 to satisfy the same performance specification.

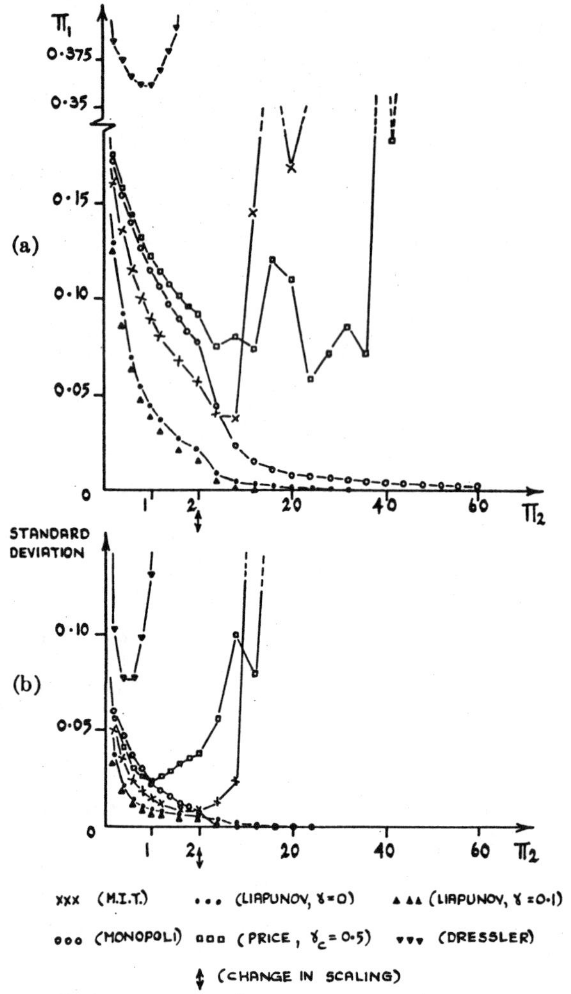

Fig. 10. Performance characteristics of second-order systems with stochastic inputs. Criterion: ISE.

Although the designs investigated in this paper are only low-order gain adjustment schemes, it seems possible that the general observations may hold for higher order systems and for the adjustment of other parameters. At least for those schemes based on the stability theory, the Lyapunov function always assures global stability and that the convergence rate reduces as the adaptive gain increases (the maximum value of $-\dot{v}/_v$ gives an estimate of the convergence rate, see [21]). This paper also assumes a noise-free system. The important problem of noise-biasing action is briefly discussed in the Appendix, Section C. On interchanging the roles of the plant and the model, the case studied becomes an identification system. Hence this investigation also reveals the shortcomings of those model reference identification schemes [22], [23] based on gradient methods.

We suggest that the globally stable design rules should be given prime considerations on account of their stability and performance properties. There are many cases where these design rules are not applicable and further research to relieve this situation is necessary. Meanwhile, the M.I.T. rule and other gradient methods that have been preferred by many designers due to the ease of their implementations should be used with great caution.

Appendix A

A. Lyapunov Design [9]

The state equations of the plant and model are

$$\text{plant: } \dot{\theta}_p = A\theta_p + bK_vK_cr \qquad \text{(A.1)}$$

$$\text{model: } \theta_m = A\theta_m + bKr. \qquad \text{(A.2)}$$

Define

$$e = \theta_m - \theta_p, \; X = K - K_vK_c.$$

We obtain

$$\dot{e} = Ae + bXr. \qquad \text{(A.3)}$$

Choose a V function

$$V = e^TPe + \lambda(X + \gamma K_vm)^2 \qquad \text{(A.4)}$$

where

$$m = B'e^TPbr. \qquad \text{(A.5)}$$

The time derivative of V is

$$\dot{V} = e^T(A^TP + PA)e + 2e^TPbXr + 2\lambda(X + \gamma K_vm)(\dot{X} + \gamma K_v\dot{m}). \qquad \text{(A.6)}$$

If we select the adaptive rule

$$\dot{X} + \gamma K_v\dot{m} = -K_vm \qquad \text{(A.7)}$$

then $\dot{V}$ becomes

$$\dot{V} = -e^TQe - 2\lambda\gamma K_v^2m^2. \qquad \text{(A.8)}$$

P and Q are positive definite symmetric matrices that satisfy the Lyapunov matrix equation

$$A^TP + PA = -Q. \qquad \text{(A.9)}$$

For example, if

$$A = \begin{bmatrix} 0 & 1 \\ -1 & -2 \end{bmatrix}, \qquad b = \begin{bmatrix} 0 \\ 1 \end{bmatrix}.$$

Let

$$Q = \begin{bmatrix} 2 & 0 \\ 0 & 2 \end{bmatrix}.$$

Solving (A.9), we obtain

$$P = \begin{bmatrix} 3 & 1 \\ 1 & 1 \end{bmatrix}.$$

Hence, $e^TPb = e_1 + e_2$.

B. Dimensional Analysis [20]

The equations describing the first order M.I.T. system are

$$T\dot{e} + e = (K - K_v K_c)r \quad \text{(A.10)}$$

$$\dot{K}_c = B'e\theta_m. \quad \text{(A.11)}$$

Define the following dimensionless variables:

$$\epsilon = e/(KR) \quad \text{(A.12)}$$

$$r_u = r/R \quad \text{(A.13)}$$

$$y_m = \theta_m/(KR) \quad \text{(A.14)}$$

$$X = (K - K_v K_c)/K \quad \text{(A.15)}$$

$$\tau = t/T. \quad \text{(A.16)}$$

Substituting these into (A.3) and (A.4), we obtain

$$\frac{d\epsilon}{d\tau} + \epsilon = X r_u \quad \text{(A.17)}$$

$$\frac{dX}{d\tau} = -(KK_v B' R^2 T)\epsilon\, y_m. \quad \text{(A.18)}$$

Hence the required dimensionless parameter π_2 is

$$\pi_2 = KK_v B' R^2 T. \quad \text{(A.19)}$$

Repeating the same process for the Lyapunov design with the same dimensionless variables, we obtain

$$\frac{d\epsilon}{d\tau} + \epsilon = X r_u \quad \text{(A.20)}$$

$$\frac{dX}{d\tau} = -(K_v B' R^2 T)\epsilon\, r_u. \quad \text{(A.21)}$$

Hence,

$$\pi_2 = K_v B' R^2 T. \quad \text{(A.22)}$$

The dimensionless performance indices are obtained by using the dimensionless error ϵ and the dimensionless time variable τ. For instance, to obtain π_1 using the ISE criterion, we have

$$\pi_1 = \int_0^\infty \epsilon^2 \, d\tau = \frac{1}{K^2 R^2 T}\int_0^\infty e^2 \, dt.$$

Likewise, dimensionless parameters and performance indices for other systems and for other inputs are derived.

C. Effects of Noise in Adaptive Control [31]

It can be readily shown that the noise, which is present in the plant states measurement or enters the plant as an additive disturbance, has a destabilizing effect on the adaptive systems considered so far [30]. In particular although the state error vector is bounded, the parameter error may become unbounded due to the noise biasing action. To see this the following equation, which represents a general form of parameter adjustment, is considered.

$$(\dot{\Delta p}) = f_1(\boldsymbol{e})\cdot f_2(\boldsymbol{\theta}_m, \boldsymbol{\theta}_p). \quad \text{(A.23)}$$

The parameter error Δp will approach zero when there is no noise. If noise is present, both $\boldsymbol{e}$ and $\boldsymbol{\theta}_p$ will have noise components. Hence, the expected value of Δp, which is now a cross correlation of two noise dependent functions, will contain a dc term. The situation becomes worse if the input is also not active enough, as Δp may not approach zero even without the noise.

Returning to the various designs considered in this paper, all except the Dressler scheme suffer from the noise biasing problem. The designs based on Lyapunov stability theory are still quite flexible in that the bound of the parameter error could be controlled using a modification due to Narendra *et al.* [29] even in the presence of both noise and insufficient frequencies. For the particular case of single plant gain adjustment, only the M.I.T. design has the noise biasing problem.

References

[1] P. V. Osburn, H. P. Whitaker, and A. Keezer, "New developments in the design of adaptive control systems," Inst. Aeronautical Sciences, Paper 61-39, 1961.
[2] P. C. Parks, "Lyapunov redesign of model reference adaptive control systems," *IEEE Trans. Automat. Contr.*, vol. AC-11, pp. 362–367, July 1966.
[3] D. D. Donalson and C. T. Leondes, "A model referenced parameter tracking technique for adaptive control systems," *IEEE Trans. Appl. Ind.*, pp. 241–262, Sept. 1963.
[4] R. M. Dressler, "An approach to model-reference adaptive control systems," *IEEE Trans. Automat. Contr.*, vol. AC-12, pp. 75–80, Feb. 1967.
[5] C. F. Price, "An accelerated gradient method for adaptive control," *Proc. 9th IEEE Symp. Adaptive Processes Decision and Control*, Dec. 1970, pp. IV.4.1–4.9.
[6] J. W. Green, "Adaptive control of multi-loop speed control systems with particular reference to the Ward-Leonard system," Ph.D. dissertation, Dep. Elec. Electron. Eng., Leeds Univ., Sept. 1969.
[7] R. L. Butchart and B. Shackcloth, "Synthesis of model reference adaptive control systems by Lyapunov's second method," *Proc. 1965 IFAC Symp. Adaptive Control* (Teddington, England), ISA, 1966, pp. 145–152.
[8] P. H. Phillipson, "Design methods for model reference adaptive systems," *Proc. Inst. Mechanical Engineers*, vol. 183, part I, pp. 695–706, 1968–1969.
[9] J. W. Gilbart, R. V. Monopoli, and C. F. Price, "Improved convergence and increased flexibility in the design of model reference adaptive control systems," *Proc. 9th IEEE Symp. Adaptive Processes Decision and Control*, Dec. 1970, pp. IV.3.1.–3.10.
[10] R. V. Monopoli, J. W. Gilbart, and W. D. Thayer, "Model reference adaptive control based on Lyapunov-like techniques," *Proc. 2nd IFAC Symp. System Sensitivity and Adaptivity* (Dubrovnik, Yugoslavia), Aug. 1968, pp. F.24–F.36.
[11] M. G. Currie and E. B. Stear, "State space structure of model reference adaptive control for a noisy system," *Proc. 2nd Asilomar Conf. Circuits and Systems*, 1968, pp. 401–405.
[12] D. G. Luenberger, "Observing the state of a linear system," *IEEE Trans. Mil. Electron.*, vol. MIL-8, pp. 74–80, Apr. 1964.
[13] C. A. Winsor and R. J. Roy, "Design of model reference adaptive control systems by Lyapunov's second method," *IEEE Trans. Automat. Contr.* (Corresp.), vol. AC-13, p. 204, Apr. 1968.
[14] I. D. Landau, "A hyperstability criterion for model reference adaptive control systems," *IEEE Trans. Automat. Contr.* (Short Papers), vol. AC-14, pp. 552–555, Oct. 1969.
[15] D. J. G. James, "Stability of a model reference control system," *AIAA J.*, vol. 9, pp. 950–952, May 1971.
[16] B. Porter and M. L. Tatnall, "Stability analysis of a class of multivariable model reference adaptive systems having time-varying process parameters," *Int. J. Contr.*, vol. 11, pp. 325–332, 1970.
[17] C. A. Winsor, "Model reference adaptive design," NASA-CR-98453, Nov. 1968.

[18] C. T. Leondes *et al.*, "An investigation study on model reference adaptive techniques as applied to altitude control system for launch vehicles," NASA-CR-102346, 1969.

[19] J. E. Gibson, *Nonlinear Automatic Control.* New York: McGraw-Hill, 1963, ch. 11.

[20] D. C. Ipson, *Units, Dimensions and Dimensionless Numbers.* New York: McGraw-Hill, 1960, ch. 9–11.

[21] R. E. Kalman and J. E. Bertram, "Control system analysis and design via the second method of Lyapunov," *Trans. ASME*, J. Basic Eng., pp. 371–393, June 1963.

[22] M. Margolis and C. T. Leondes, "On the theory of adaptive control systems; The learning model approach," *Proc. 1st IFAC Congr.* (Moscow), 1960, pp. 556–563.

[23] T. C. Hsia and V. Vimolvanich, "An on-line technique for system identification," *IEEE Trans. Automat. Contr.* (Short Papers), vol. AC-14, pp. 92–96, Feb. 1969.

[24] J. V. Medanić and P. V. Kokotovic, "Some problems in the development of adaptive systems using the sensitivity operator," *Proc. 1965 IFAC Symp. Adaptive Control* (Teddington, England), ISA, 1966, pp. 204–212.

[25] P. V. Kokotovic, J. V. Medanić, S. P. Bingulac, and M. I. Vušković, "Sensitivity method in the experimental design of adaptive control systems," *Proc. 3rd IFAC Congr.* (London), 1966, pp. 45B.1–B.12.

[26] R. E. Kalman, "Lyapunov functions for the problem of Luré in automatic control," *Proc. National Academy of Sciences* (U.S.A.), vol. 49, 1963, pp. 201–205.

[27] K. R. Meyer, "Lyapunov functions for the problem of Luré," *Proc. National Academy of Sciences* (U.S.A.), vol. 53, 1965, pp. 501–503.

[28] R. V. Monopoli, "The Kalman–Yacubovich lemma in adaptive control system design," to be published.

[29] K. S. Narendra and S. S. Tripathi, "The choice of adaptive parameters in model reference control systems," presented at the 5th Asilomar Conf., 1971.

[30] D. P. Lindorff, "Effects of incomplete adaptation and disturbance in adaptive control," in *1972 Joint Automat. Contr. Conf., Preprints.*

[31] C. C. Hang, Ph.D. dissertation Dep. Eng., Warwick Univ., Coventry, England, 1973, to be submitted.

Model Reference Adaptive Control with an Augmented Error Signal

RICHARD V. MONOPOLI, SENIOR MEMBER, IEEE

Abstract—**It is shown how globally stable model reference adaptive control systems may be designed when one has access to only the plant's input and output signals. Controllers for single input-single output, nonlinear, nonautonomous plants are developed based on Lyapunov's direct method and the Meyer-Kalman-Yacubovich lemma. Derivatives of the plant output are not required, but are replaced by filtered derivative signals. An augmented error signal replaces the error normally used, which is defined as the difference between the model and plant outputs. However, global stability is assured in the sense that the normally used error signal approaches zero asymptotically.**

Manuscript received August 31, 1973. Paper recommended by G. N. Saridis, Past Chairman and D. G. Lainiotis, Chairman of the IEEE S-CS Adaptive and Learning Systems, Pattern Recognition Committee. This work was supported in part by the National Aeronautics and Space Administration Grant NGL-22-010-018 and in part by the Office of Naval Research under Contract ONR-N0C014-68-A-0146-6A.

The author is with the Department of Electrical and Computer Engineering, University of Massachusetts, Amherst, Mass.

I. Introduction

Recent progress in applying Lyapunov designed model reference adaptive control systems to engineering problems has been encouraging [1]–[3]. The technique has been applied to the control of a satellite and star tracking system at the NASA Goddard Space Flight Center [1], a laboratory servo system at the University of Massachusetts [2], and adaptive autopilot for ships by Udink ten Cate [3]. These applications are based on considerable prior theoretical work [4]–[18]. In [19], Parks and Hang have demonstrated fairly conclusively that this method for adaptive control system design is the most promising suggested to date. It assures stability of the overall closed-loop adaptive system, and it avoids the stability analysis problem associated with designs based on gradient techniques [19]–[23].

Reprinted from *IEEE Trans. Automat. Contr.*, vol. AC-19, pp. 474–484, Oct. 1974.

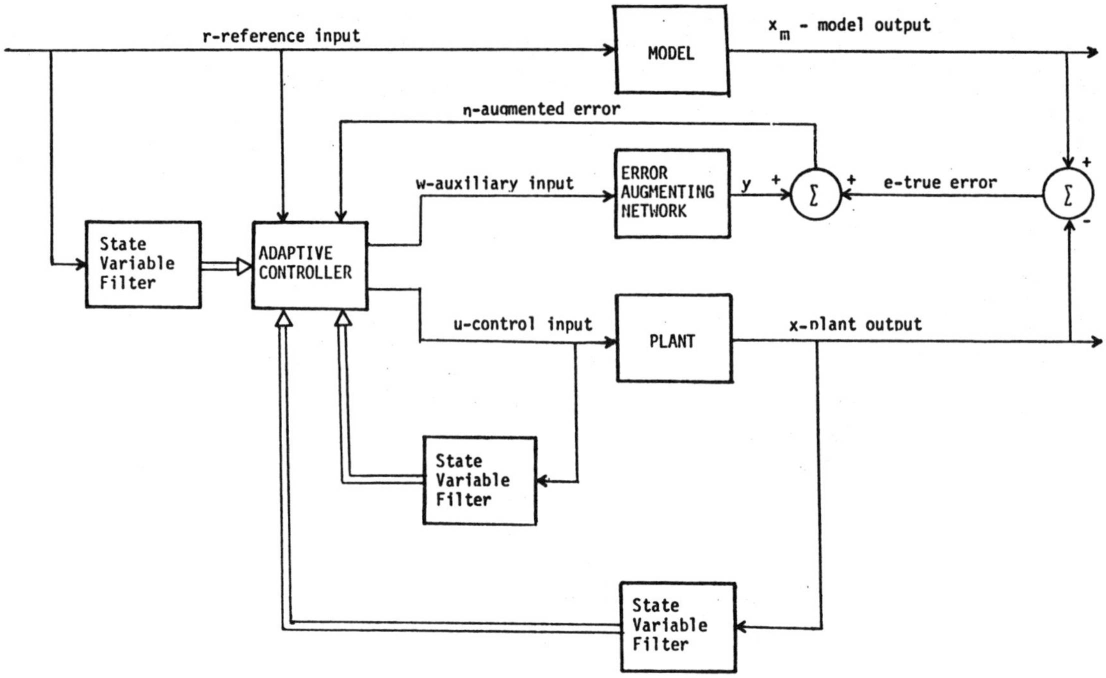

Fig. 1. General system layout.

A conceptually new approach to these designs is described in this paper. Instead of nulling the true error between the plant output and model output directly, an augmented error signal is introduced and the design assures that the augmented error approaches zero (see Fig. 1). However, the true error signal is driven to zero as well. Use of the augmented error signal allows a design which does not require derivatives of the plant output signal. This is an extremely attractive feature in an engineering sense, and was previously attainable for only a limited class of plants [4], [11], [16].

In these designs, feed-forward and feedback gains are adjusted to compensate for unknown plant parameters. This approach [7] is in contrast to that in which plant parameters are adjusted directly [4], [6]. Another distinguishing feature of these designs is that no explicit identification of plant parameters takes place. The desired objective is a stable adaptive control system, and whether or not adjustable parameters converge to unknown plant parameters is of secondary importance. Also, state estimation is not employed in this design, although filtered derivatives of the plant's input and output are used. These, in essence, give an indication of the plant state adequate for purposes of the design. The advantage of not using explicit parameter identification or state estimation is that the stability problem which arises in other implicit adaptive control schemes or when an adaptive loop is closed through an identifier alone or through an adaptive observer and identifier is avoided [23]–[26].

These designs have had the disadvantage that, generally speaking, $n-1$ derivatives of the output of an nth-order plant were required to implement the controller. In certain cases, some or all of the output derivatives can be eliminated. Parks [4] showed that all derivatives can be avoided if the plant has the gain as the only unknown parameter, and it has a positive real transfer function. He used the Meyer–Kalman–Yacubovich (MKY) lemma [27]–[29] to achieve this result. In [11] and [16], this lemma was coupled with the state variable filter concept, introduced in [30] and employed in [31], to extend the results of Parks to a much broader class of plants. A somewhat different approach to the problem of designing without all plant output derivatives was reported in [8]. However, the problem was not completely solved in any of the references [8], [11], or [16]. There was still the requirement that $n-m-2$ plant output derivatives be available for an nth-order plant with the mth-order derivative of the input appearing in the plant equations.

In this paper the augmented error signal concept, which was prompted by the work reported in [24], and the state variable filter concept combine to make it possible to put a large class of systems into the form where the Meyer–Kalman–Yacubovich (MKY) lemma may be used for design of the adaptive parameters as in [11] and [16]. Normally in these designs the error between the plant and model outputs (e in Fig. 1) is used in the parameter adjustment algorithm. Here the augmented error signal η is used. However, asymptotic convergence to zero of the true error between plant and model outputs is still assured.

Section II contains the problem statement and the design objectives and Section III gives a brief review of the application of the MKY lemma to these systems. The problem solution for a limited class of plants is discussed in Section IV, and for the general class in Section V. A computer simulation is included in Section VI, and the conclusions in Section VII.

II. Problem Statement and Design Objectives

A dynamic system (plant) is described by the nonlinear, nonautomonous, differential equation

$$D_p(p)\, x(t) = D_u(p)\, u(t) + cf(x,t) \tag{1}$$

where $u(t)$ and $x(t)$ are the plant input and output, respectively, $f(x,t)$ is a nonlinear time varying function of known form, p is the operator d/dt, $D_p(p) = p^n + a_1 p^{n-1} + \cdots a_n$, and $D_u(p) = b_0 p^m + b_1 p^{m-1} + \cdots b_m$. It is assumed that coefficients a_i, b_i, and c are unknown and constant or slowly varying. The term $cf(x,t)$ may be replaced by a sum of terms of the same type. Since each is handled the same way in the design, there is no loss of generality in carrying only the one term. It is also assumed that 1) the function f (after first being introduced, the arguments of functions and time variables will be omitted, with the exception of p operators) satisfies the continuity conditions necessary for solutions of (1) to exist and be unique, 2) all roots of $D_u(s)$ are in the open left half plane, 3) $b_0 \neq 0$, and 4) $m \leq n - 1$.

Assumption 2) is required to insure a bounded control input by not requiring the adaptive system to act to cancel nonminimum phase plant zeroes. For a similar reason, u is not included as an argument of f. In certain cases functions of the form $f(x,u,t)$ may be allowed. However, in general, it is not known what restrictions to impose on these functions to insure that the control input u is bounded. In the purely theoretical sense assumption 2) is not necessary. The system will adapt and follow the model but the control input will become unbounded—an undesirable practical result. Similarly $f(x,u,t)$ can be allowed theoretically provided it does not impose a magnitude constraint on u.

The design objective is to have the plant output follow the output of a model reference defined by the equation

$$D_m(p)\, x_m(t) = K_0\, r(t) + g(\mathbf{x}_m,r,t) \tag{2}$$

where $D_m(p) = p^n + a_{d1} p^{n-1} + \cdots a_{dn}$, x_m is the model output, r is the reference input $\mathbf{x}_m$ is the model state vector, and g is a nonlinear, time varying function with the smoothness properties required to insure the existence and uniqueness of solutions to (2). If it is considered desirable to have a model which differentiates the reference input, $r(t)$ may be replaced by $r(t) = D_r(p) r^1(t)$ where $D_r(p) = p^\alpha + R_1 p^{\alpha-1} + \cdots R_\alpha$. If $\alpha \leq m$, no derivatives of r^1 will be required in the design. However, if $\alpha > m$, they will be. Stability of the operator $D_m(p)$ is assumed, i.e., $D_m(s) = 0$ has roots in the open left half plane only. The design problem is to synthesize a parameter adaptive control system for (1) which will cause the error, $e(t) = x_m - x$, between plant and model outputs to approach zero. The distinguishing feature between this work and previous designs of this type is that derivatives of the plant output x are not to be used in the design for the obvious reason that in practice they may be too noisy.

A new concept, an augmented error signal, must be introduced to achieve the design objective. This signal used in conjunction with the state variable filter concept makes it possible to derive system differential equations in a form suitable for direct application of the MKY lemma, and as in other designs using this lemma [4], [11], [16], no derivatives of x are required.

An augmenting signal $y(t)$ is added to $e(t)$ to give the augmented error signal $\eta(t)$,

$$\eta(t) = e(t) + y(t). \tag{3}$$

The signal $y(t)$ is the output of the error augmenting filter defined by

$$D_m(p) y = D_w(p) w(t) \tag{4}$$

where $D_w(p) = p^{n-1} + c_1 p^{n-2} + \cdots c_n$, and w is an auxiliary system input to be determined along with the control input u as part of the design. Coefficients c_i in $D_w(p)$ must be chosen such that the transfer function $D_w(s)/D_m(s)$ is a positive real function of s in order to use the MKY lemma later in the development.

Subtracting (1) from (2) gives the differential equation for e which is

$$D_m(p)e = K_0 r + g - D_u(p)u - cf + D_\Delta(p)x \tag{5}$$

where $D_\Delta(p) = D_p(p) - D_m(p) = \Delta a_1 p^{n-1} + \cdots \Delta a_n$ and $\Delta a_i = a_i - a_{di}$ for $i = 1$ to n.

Next, (4) is added to (5) to obtain the differential equation for the augmented error signal η, which is

$$D_m(p)\eta = K_0 r + g - D_u(p)u - cf + D_\Delta(p)x + D_w(p)w. \tag{6}$$

Equation (6) is the starting equation for the synthesis procedure. The design problem is to generate u and w independent of derivatives of x and in such a way that $e \to 0$ as $t \to \infty$. In a sense, the augmenting signal y is a catalyst which plays a role only during the adaptive transient period and goes to zero once adaptation is completed. Its successful application in this problem leads one to speculate about the possibility of using the concept of an augmented error signal to insure convergence in other types of gradient algorithms.

III. The MKY Lemma in Adaptive Control System Design

A statement of the MKY lemma is given in Appendix I. The reason for its usefulness in designs of this type is explored in this section.

Consider the system

$$p\mathbf{e}(t) = A\mathbf{e}(t) + \mathbf{d}r(t) \tag{7}$$

with output $e = \mathbf{k}'\mathbf{e}$. Here A, $\mathbf{d}$, and $\mathbf{k}$ are the $n \times n$ matrix and n vectors referred to in the lemma. The lemma is useful in these designs when the inequality is satisfied with $\tau = 0$, i.e., when $\boldsymbol{k}'[sI - A]^{-1}\boldsymbol{d}$ is a positive real function of s. In this case $B\boldsymbol{d} = \boldsymbol{k}$ by condition (b) of the lemma. It is this fact which reduces the adaptive law to a function of e only, eliminating the need for any other components of $\boldsymbol{e}$. For example, as in the special case considered in [4], let the plant and model equations be (8) and (8a), respectively,

$$D_m(p)x = D_w(p)b_0u \tag{8}$$

$$D_m(p)x_m = D_w(p)K_0r \tag{8a}$$

where $D_w(s)/D_m(s)$ is positive real, and b_0 is an unknown constant gain. The equation for the error $e = x_m - x$ is

$$D_m(p)e = D_w(p)(K_0r - b_0u). \tag{9}$$

Since $m = n - 1$, no augmented error is needed in this case. Also, since the plant transfer function, excluding b_0, is positive real, no state variable filters are needed such as were used in [16]. Equation (9) may be put into the vector-matrix form of (7) as follows:

$$p\boldsymbol{e} = A\boldsymbol{e} + \boldsymbol{d}(K_0r - b_0u) \tag{10}$$

where $e = e_1$ (i.e., $k' = [1,0,\cdots,0]$), matrix A has zero elements everywhere except for ones above the diagonal and nonzero elements in the last row which is the row vector $[-a_{dn}, -a_{d(n-1)},\cdots, -a_{d1}]$. The elements of $\boldsymbol{d}$ are functions of coefficients a_{di} and c_i. If the control input u is taken to be $u = k_0(t)r$ where $k_0(t)$ is the adaptive (adjustable) gain then (10) becomes

$$p\boldsymbol{e} = A\boldsymbol{e} + \boldsymbol{d}\delta_0(t)r \tag{11}$$

where $\delta_0(t) = K_0 - b_0k_0(t)$.

Choose as a potential Lyapunov function for (11)

$$V(\boldsymbol{e},\delta_0) = \boldsymbol{e}'B\boldsymbol{e} + \lambda_0\delta_0^2 \tag{12}$$

where λ_0 is an arbitrary positive constant. The time derivative of (12) is

$$pV(t) = \boldsymbol{e}'(A'B + BA)\boldsymbol{e} + 2\boldsymbol{e}'B\boldsymbol{d}\delta_0r + 2\lambda_0\delta_0p\delta_0. \tag{13}$$

Since $\boldsymbol{k}'[sI - A]^{-1}\boldsymbol{d} = D_w(s)/D_m(s)$ is positive real, the second term on the right side of (13) becomes simply $2e\delta_0r$ by letting $\tau = 0$ and using condition (b) of the lemma. Condition (c) of the lemma insures that B is positive definite. Thus, if the adaptive law is taken to be

$$p\delta_0/b_0 = -pk_0 = -(1/\lambda_0b_0)er \tag{14}$$

then $pV = \boldsymbol{e}'[A'B + BA]\boldsymbol{e} < 0$ for $\boldsymbol{e} \neq 0$ by conditions (a), (c), and (d) of the lemma. It is seen that the very simple adaptive law given by (14) involving e and no other components of $\boldsymbol{e}$ guarantees that $e \to 0$ as $t \to \infty$.

In the extension of this method to the general problem, the objective will be to find appropriate system input signals u and w such that (6) may be put into the form

$$p\boldsymbol{\eta} = A\boldsymbol{\eta} + \boldsymbol{d}\left(\sum_{i=0}^{N} \delta_i\phi_i\right) \tag{15}$$

where A, $\boldsymbol{d}$, and $\boldsymbol{k}$ are as in (10), $\eta_1 = \eta = e + y$, $N + 1$ is the number of adjustable gains, each δ_i contains one or more unknown plant parameters along with one adjustable gain, and signals ϕ_i may be obtained without differentiating networks. The Lyapunov function to be used in conjunction with (15) is a generalization of (12), i.e.,

$$V(\boldsymbol{\eta},\boldsymbol{\delta}) = \boldsymbol{\eta}'B\boldsymbol{\eta} + \sum_{i=0}^{N} \lambda_i\delta_i^2 \tag{16}$$

where λ_i for $i = 0$ to N are arbitrary positive constants and $\boldsymbol{\delta}' = [\delta_0,\delta_1,\cdots,\delta_N]$. Again, if the adaptive algorithm is chosen to be

$$p\delta_i = -(1/\lambda_i)\eta\phi_i, \qquad \text{for } i = 0,1,2,\cdots,N \tag{17}$$

then $pV = \boldsymbol{\eta}'[A'B + BA]\boldsymbol{\eta} < 0$ for $\boldsymbol{\eta} \neq \boldsymbol{0}$ by the conditions of the lemma. Thus, $\eta \to 0$ as $t \to \infty$. As will be seen, the design also insures that $w \to 0$ as $t \to \infty$. Since $\eta \to 0$ and $w \to 0$ imply that $e \to 0$, the desired result is achieved by the design.

Use of the MKY lemma has been explored by considering the design for the special class of plants representable by the product of an unknown gain and a positive real function. The relationship of this example to the approach to be taken in the general case was discussed. In the next section, design for the class of plants which can be represented by the product of an unknown gain and a known, nonpositive real transfer function is considered.

IV. Problem Solution for a Limited Class of Plants

A. Class of Plants Considered and Problem Formulation

Details of the solution for the general class of plants given by (1) are deferred until Section V. In this section, the solution is developed for linear plants with an unknown gain only and with $m = 0$, i.e., it is assumed that in (6) $D_\Delta(p) = 0$, $D_u(p) = b_0$, and $c = 0$. For convenience it is also assumed that $g = 0$. This approach is taken in order to emphasize first the method of solution and not have it obscured by the details of the general case. Also, the adaptive plant gain case is a problem of interest in its own right. Since $m = 0$, the plant transfer function generally will not be positive real as it was in the example in Section III where $m = n - 1$. Therefore, in contrast to that example, an augmented error signal will be required in this case. In addition, state variable filters, which were not necessary in Section III, will be needed here.

The problem, then, is to have a plant given by $D_m(p)x = b_0u$, where b_0 is unknown, follow a model given by $D_m(p)x_m = K_0r$. The augmented error equation for this case is

$$D_m(p)\eta = K_0r - b_0u + D_w(p)w. \tag{18}$$

It is at this point that state variable filters must be introduced. In order to eventually put (18) into the form of (15), define filtered versions of u and r, called $z_0(t)$ and $z_1(t)$, respectively, by $D_w(p)z_0 = u$; $D_w(p)z_1 = r$. Then (18) may be written as

$$D_m(p)\eta = D_w(p)(-b_0z_0 + K_0z_1 + w). \tag{19}$$

Next, (19) is put into vector-matrix form [as in (10)]

$$p\mathbf{n} = A\mathbf{n} + \mathbf{d}(-b_0 z_0 + K_0 z_1 + w). \tag{20}$$

An invalid approach to the design from this point is to simply take $w = 0$ in (20), and let $z_0 = k_0(t)z_1$. Then a development like that from (10) to (14) will yield $pk_0(t) = (l/\lambda_0 b_0)\eta z_1$. The trouble with this approach, however, is that the operator $D_w(p)$ must be applied to z_0 to generate u and this requires $n - 1$ derivatives of z_0. This is not acceptable because these derivatives will involve derivatives of x. Therefore, a different approach must be taken. It is to choose z_0 and w in a way which allows (20) to be put in the form of (15), u and w to be generated without derivatives of x, and which causes $w \to 0$ as $\eta \to 0$. The last condition is needed to insure that $y \to 0$ as $\eta \to 0$ so that $e \to 0$ as required.

Two methods available for achieving these results are described next. The first pertains when no bounds are known for the unknown plant parameters (called Case I). The second, which offers certain advantages to be discussed, may be used when bounds are known for these parameters (called Case II).

B. Solution for Case I

In order to accomplish the result of putting (20) into the form of (15), an auxiliary adjustable gain, $k_1(t)$, must be associated with the signal w by defining $w = (1 + k_1)w_1$. Then if the linear combination $z_0 + w_1$ is taken to be

$$z_0 + w_1 = k_0 z_1 \tag{21}$$

and this expression is substituted into (20) the result is

$$p\mathbf{n} = A\mathbf{n} + \mathbf{d}(\delta_0(t)z_1 + \delta_1(t)w_1) \tag{22}$$

where $\delta_0(t) = -b_0 k_0(t) + K_0$ and $\delta_1(t) = k_1(t) + b_0 + 1$. Since (22) is of the form of (15), the adaptive algorithm is written immediately using (17)

$$pk_0 = (1/\lambda_0)\eta z_1; \; pk_1 = -(1/\lambda_1)\eta w_1. \tag{23}$$

Next, it is necessary to derive the input signals u and w from (12). This may be done by applying the operator $D_w(p)$ to both sides of (21) and making use of the identity (A.1) in Appendix II. This procedure yields

$$D_w(p)(z_0 + w_1) = u + D_w(p)w_1 = k_0 r + \sum_{i=0}^{n-2} D_{wi}(p)((pk_0)(p^{n-2-i}z_1)). \tag{24}$$

One way to choose u and w_1 from (24) to satisfy the design objectives is as follows:

$$u = k_0 r \tag{25a}$$

$$w_1 = D_w^{-1}(p) \sum_{i=0}^{n-2} D_{wi}(p)((pk_0)(p^{n-2-i}z_1)). \tag{25b}$$

No derivatives of x are required to generate either u or w_1 this way. Furthermore, $w_1 \to 0$ (hence, $w \to 0$) as $\eta \to 0$ since w_1 is the sum of the outputs of stable linear filters each of which has η as a factor (via the term pk_0) in its input. As mentioned, $\eta \to 0$ and $w \to 0$ imply that $e \to 0$.

It should be noted that if $n = 2$, $w_1 = 0$ may be chosen (no augmented error is required) and $u = k_0 r + D_w^{-1}(p)((pk_0)(z_1))$. This completes the design for Case I.

C. Solution for Case II

If a suitable bound is known for b_0, a different approach may be taken which does not require the adjustable gain k_1. However, it does require a nonlinear network for constraining the range of k_0 (which is defined slightly differently for this case). Constraining the range of adjustable parameters may be advantageous in maintaining stability in the presence of disturbances.

In this approach, the linear combination $z_0 + w$ is chosen to be

$$z_0 + w = -K_0 z_1 + k_0 z_0 \tag{26}$$

which, when substituted into (20), gives

$$p\mathbf{n} = A\mathbf{n} + \mathbf{d}\delta_0(t)z_0 \tag{27}$$

where $\delta_0(t) = k_0(t) - b_0 - 1$. Thus, if

$$pk_0 = -(1/\lambda_0)\eta z_0 \tag{28}$$

then $\eta \to 0$. Next, to generate u and w, the operator $D_w(p)$ is applied to both sides of (26), and (A.1) is used to obtain

$$u + D_w(p)w = -K_0 r + k_0(t)u + \sum_{i=0}^{n-2} D_{wi}(p)((pk_0)(p^{n-2-i}z_0)). \tag{29}$$

Appropriate choices for u and w from (29) are

$$u = -K_0 r + k_0 u \tag{30a}$$

$$w = D_w^{-1}(p) \sum_{i=0}^{n-2} D_{wi}(p)((pk_0)(p^{n-2-i}z_0)). \tag{30b}$$

Here again, derivatives of x are not required to generate either u or w. Also, for the same reasons given in the previous section, $w \to 0$ as $\eta \to 0$, and therefore $e \to 0$, the desired design objective.

One practical difficulty remains in the solution as developed so far. When (30a) is rewritten as

$$u = -(1/(1 - k_0))K_0 r \tag{31}$$

it is seen that an infinite gain is required when $k_0 = 1$ for any t. This difficulty may be overcome in cases where a suitable bound is known for b_0. Assume, for example, that $b_0 \geq \epsilon > 0$ and ϵ is known. The required steady-state value for k_0 is $k_0(\infty) = b_0 + 1 \geq \epsilon + 1$, and there is no need to have $k_0 = 1$ for any value of t. However, to avoid the situation $k_0 = 1$, the adaptive law in (28) must be modified in a way which still assures that $\eta \to 0$, $w \to 0$, and $e \to 0$. Such a modification is possible and is discussed below.

With ϵ known, the initial condition for k_0 may be chosen as $k_0(0) > 1 + \epsilon$. Then if (28) is modified as follows:

$$pk_0 = -(1/\lambda_0)(\eta z_0 - g_1(\eta z_0)g_2(k_0)) \tag{32}$$

where g_1 and g_2 are nonlinear functions to be determined, there is an additional term in pV which now becomes

$$pV = \mathbf{n}'(A'B + BA)\mathbf{n} + \delta_0 g_1 g_2. \qquad (33)$$

The term $\delta_0 g_1 g_2$ will be negative if δ_0 and $(g_1 g_2)$ are of opposite signs. Whenever $k_0 > 1 + \epsilon$ and $\eta z_0 < 0$ then $g_1 g_2$ may be set equal to zero since this will give $pk_0 > 0$ and there is no danger of approaching $k_0 = 1$. On the basis of this observation let g_1 and g_2 be partially specified as follows:

$$g_1(\eta z_0) = 0, \qquad \text{if } \eta z_0 < 0 \qquad (34a)$$

$$g_2(k_0) = 0, \qquad \text{if } k_0 \geq 1 + \epsilon. \qquad (34b)$$

Next, note that the sign of δ_0 is known if $k_0 < 1 + \epsilon$, i.e., $\delta_0 = k_0 - b_0 - 1 < 0$. Therefore, if $g_1 g_2 > 0$ when $k_0 < 1 + \epsilon$ the contribution of the term $\delta_0 g_1 g_2$ to pV will be negative. Accordingly, the remaining specifications on g_1 and g_2 are taken to be

$$g_1(\eta z_0) = \eta z_0, \qquad \text{if } \eta z_0 > 0 \qquad (35a)$$

$$g_2(k_0) = G_0 - G_0 k_0/(1 + \epsilon), \qquad \text{if } k_0 < 1 + \epsilon \qquad (35b)$$

where $G_0 = (1 + \epsilon)/(\epsilon - \epsilon_1)$ and $0 < \epsilon_1 < \epsilon$. This scheme causes pk_0 to become zero when $k_0 = 1 + \epsilon_1$. In fact it yields $k_0 \geq 1 + \epsilon_1$ for all t while insuring that $pV < 0$ for $\eta \neq 0$. Implementation is shown in Fig. 2.

In situations where upper and lower bounds are known for the uncertain parameters, the technique described above can be employed to keep the corresponding adjustable gains within known bounds. Constraining the magnitude of the adjustable gains in this way is helpful in determining the stability of a differential operator which arises in one approach to the general solution in Section V. Also, it provides a possible solution to the problem of unstable adaptive gains due to disturbances pointed out by Lindorff [17]. To illustrate this extension, let $\delta_i = k_i + a_i$ and $a_{im} \leq a_i \leq a_{iM}$ where a_{im} and a_{iM} are known. The required range for k_i is $-a_{iM} \leq k_i \leq -a_{im}$. Thus, if $k_i < -a_{iM}$ and $\eta\phi_i > 0$, then $pk_i = 0$ leaves the term $\delta_i \eta\phi_i < 0$ in pV. Similarly, if $k_i > -a_{im}$ and $\eta\phi_i < 0$, pk_i may be set equal to zero leaving $\delta_i \eta\phi_i$ in pV. Through the use of this type of logic, all of the adjustable gains may be kept within known bounds when the appropriate bounds for the plant parameters are known.

In this section, two approaches to the problem of adapting for an unknown plant gain have been given. These illustrate most of the steps needed to solve the general problem while avoiding unnecessary complications. In Section V, the solution for the general class of plants is considered. Again, Cases I and II are treated separately.

V. Problem Solution for the General Class of Plants

Basically the method of solution for the general case is the same as given in Section IV. However, there are differences in detail which are examined in this section. If $m = n - 1$, then the problem may be solved without resorting to the augmented error signal as shown in [16]. The differences in detail are concerned primarily with the terms $D_u(p)u$ and $D_\Delta(p)x$ which were taken simply as $b_0 u$ and 0, respectively, in Section IV. In general, higher order state variable filters than those used in Section IV will be required in order to avoid differentiating the output x. Also, signals other than u and r must undergo this filtering process. As in Section IV, Case I and Case II are considered separately in this section also.

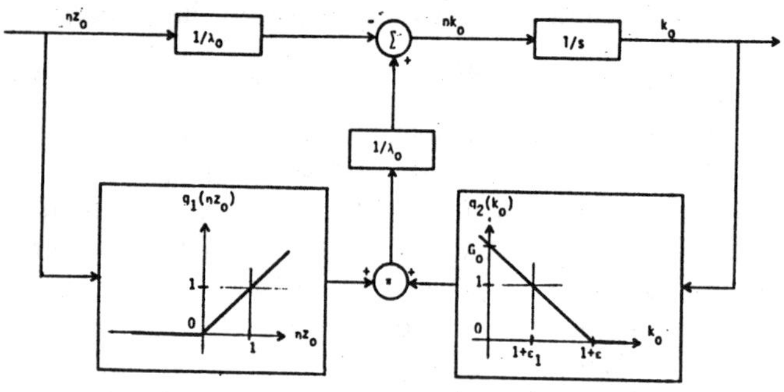

Fig. 2. Technique for bounding k_0 from below.

A. Solution for Case I

The augmented error equation (6) is written in the form

$$D_m(p)\eta = D_w(p)[-D_u(p)z_0 + K_0 z_1 + D_\Delta(p)z_2 - cz_3 + z_4 + w] \qquad (36)$$

where $z_2 = D_w^{-1}(p)x$, $z_3 = D_w^{-1}(p)f$, and $z_4 = D_w^{-1}(p)g$.

The additional filtering required is associated with the operator $D_f(p)$ where

$$D_f(p)x_f = (p^{n-m-1} + F_1 p^{n-m-2} + \cdots F_{n-m-1})x_f = x. \qquad (37)$$

Similarly $D_f(p)u_f = u$. It is assumed that $D_f(s)$ is a Hurwitz polynomial. Note that if $m = n - 1$, this filter is unnecessary.

Later in the development, it will be necessary to derive u and w_1 by applying the operator $D_f(p)$ to a certain expression in a manner analogous to the procedure followed in connection with (24). Derivatives of x will be generated by this step unless the $D_\Delta(p)z_2$ term in (36) is replaced by an alternate expression. In order to have the development here parallel that in Section IV, it is also necessary to replace the term $D_u(p)z_0$ in (36) by an alternate expression. It is shown in Appendix II that these terms may be replaced by the alternate expressions given by the right sides of (38) and (39).

$$D_u(p)z_0 = D_u(p)D_w^{-1}(p)u = b_0 D_f^{-1}(p)u + [A_0 p^{n-2} + \cdots A_{n-2}]D_f^{-1}(p)z_0 \qquad (38)$$

where constants A_0 to A_{n-2} are unknown as are the $3n - m - 2$ constants B_0 to B_{n-1}, C_0 to C_{n-2}, and D_0 to D_{n-m-2} in

$$D_\Delta(p)z_2 = \sum_{i=0}^{n-1} B_i p^{n-2-i} D_f^{-1}(p)z_2 + \sum_{i=0}^{n-2} C_i p^{n-2-i} D_f^{-1}(p)z_0 + \sum_{i=0}^{n-m-2} D_i p^{n-m-2-i} D_f^{-1}(p)cz_3. \qquad (39)$$

In (38) and (39) the idea is to replace a linear combination of signals with unknown coefficients with a linear combination of a different set of signals which also has unknown coefficients. It is important to note that the constants A_i, B_i, C_i, and D_i need not actually be found. All that is used in the development is the fact that they exist. Next (38) and (39) are substituted into (36) to yield

$$D_m(p)\eta = D_w(p)\left(-b_0u_f + w + \sum_{i=0}^{N-1} \beta_i\phi_i\right) \tag{40}$$

where $N = 3n - m + 1$, $\beta_0 = K_0$, $\phi_0 = z_1$; $\beta_1 = -c$, $\phi_1 = z_3$; $\beta_2 = 1$, $\phi_2 = z_4$; $\beta_{3+i} = A_i + C_i$, $\phi_{3+i} = p^{n-2-i}D_f^{-1}(p)z_0$ for $i = 0$ to $n - 2$; $\beta_{n+2+i} = B_i$, $\phi_{n+2+i} = p^{n-1-i}D_f^{-1}(p)z_2$ for $i = 0$ to $n - 1$; and $\beta_{2n+2+i} = cD_i$, $\phi_{2n+2+i} = p^{n-m-2-i}D_f^{-1}(p)z_3$ for $i = 0$ to $n - m - 2$. Again let $w = (1 + k_N(t))w_1$, and, proceeding as in (21), choose

$$u_f + \phi_N = \sum_{i=0}^{N-1} k_i(t)\phi_i \tag{41}$$

where $\phi_N = w_1$. When (41) and $w = (1 + k_N)\phi_N$ are substituted into (40), the vector matrix form (15) results with $\delta_i = -b_0k_i + \beta_i$ for $i = 0$ to $N - 1$, and $\delta_N = k_N + b_0 + 1$. The adaptive algorithm given by (17) assures that $\eta \to 0$.

The procedure used in deriving (25) may be employed to derive u and w_1 if $D_f(p)$ takes the place of $D_w(p)$ and the identity (A.2) is employed. Applying $D_f(p)$ to both sides of (41) yields the design equations

$$u = \sum_{i=0}^{N-1} k_iD_f(p)\phi_i \tag{42a}$$

$$w_1 = D_f^{-1}(p)\left[\sum_{i=0}^{N-1}\sum_{j=0}^{n-m-2} D_{fj}(p)((pk_i)(p^{n-m-2-i}\phi_i))\right]. \tag{42b}$$

It is noted that these inputs can be generated without derivatives of x and that they will cause $e \to 0$ as required. Implementation of (42) is shown in Figs. 3 and 4.

B. Solution for Case II

In certain cases it is possible to use an alternate method for avoiding derivatives of x. It consists of substituting derivatives of the model output for those of the plant output, and may lead to considerable simplification in the controller. Its success depends on knowledge of parameter bounds and assurance that adjustable parameters approach correct values, i.e., δ_i's $\to 0$. This method is described next.

If only (38) and not (39) is substituted into (36), then

$$D_m(p)\eta = D_w(p)\left[-b_0u_f + w + K_0z_1 - cz_3 + z_4 + \sum_{i=0}^{n-2} A_0p^{n-2-i}D_f^{-1}(p)z_0 + D_\Delta(p)z_2\right]. \tag{43}$$

Substituting an expression like (41) with appropriately defined ϕ_i's into (43) leads again to the vector-matrix equation (15) where now ϕ_i's and δ_i's are defined as follows:

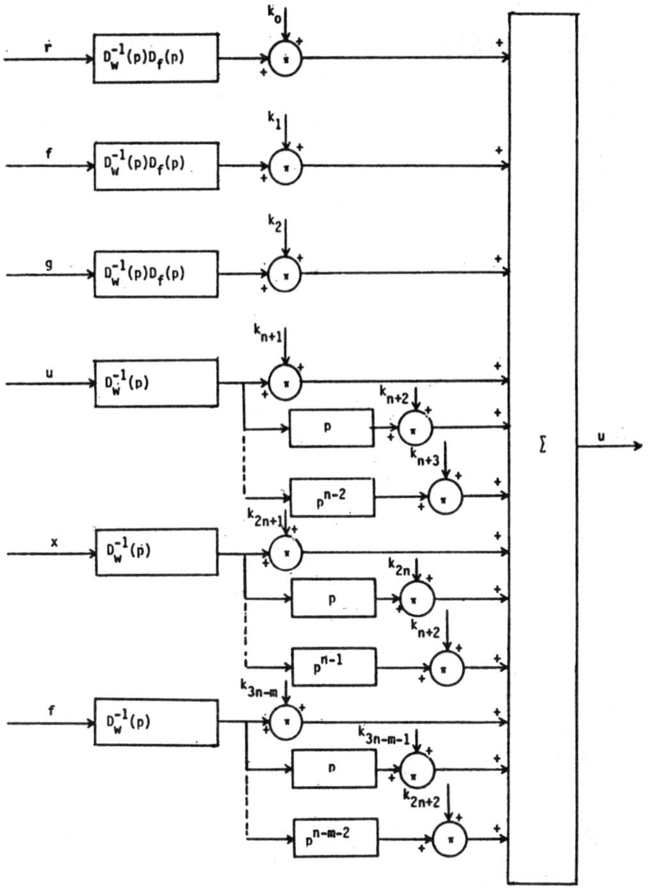

Fig. 3. Implementation for generating u.

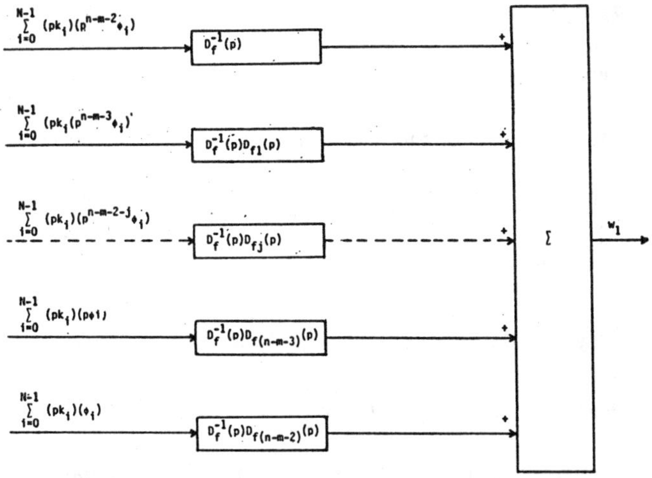

Fig. 4. Implementation for generating w_1.

$\phi_0 = z_1$, $\delta_0 = -b_0k_0 + K_0$; $\phi_1 = z_3$, $\delta_1 = -b_0k_1 - c$; $\phi_2 = z_4$, $\delta_2 = -b_0k_2 + 1$; $\phi_{3+i} = p^{n-2-i}D_f^{-1}(p)z_0$, and $\delta_{3+i} = -b_0k_{3+i} + A_i$, for $i = 0$ to $n - 2$; $\phi_{n+1+i} = p^{n-i}z_2$ and $\delta_{n+1+i} = -b_0k_{n+1+i} + \Delta a_i$, for $i = 1$ to n, $\phi_N = w_1$, $\delta_N = k_N + b_0 + 1$.

Adaptive laws are given by (17). It is noted that $N = 2n + 2$ as compared to $N = 3n - m + 1$ in the previous case.

If the procedure used in connection with (42) is followed to derive u and w_1, however, derivatives of x arise when $D_f(p)$ is applied to the $k_{n+2}\phi_{n+2}$ to $k_{2n}\phi_{2n}$ terms in the expression for $u_f + \phi_N$. Therefore, the procedure must be modified to avoid this. The modification required is to let $w_1 = w_2 + w_3$ and to rewrite (41) as

$$u_f + w_2 + w_3 = \sum_{i=0}^{n+1} k_i\phi_i + \sum_{i=n+2}^{2n+1} k_i p^{2n+1-i} z_5 - \sum_{i=n+2}^{2n+1} k_i p^{2n+1-i}(z_5 - z_2) \quad (44)$$

where $z_5 = D_w^{-1}(p)x_m$. Then, define w_3 by

$$w_3 = -\sum_{i=n+2}^{2n+1} k_i p^{2n+1-i} D_w^{-1}(p)e \quad (45)$$

which leaves

$$u_f + w_2 = \sum_{i=0}^{n+1} k_i\phi_i + \sum_{i=n+2}^{2n+1} k_i p^{2n+-1i} z_5. \quad (46)$$

Application of the $D_f(p)$ operator to (46) leads to equations like (42) for u and w_2. Now, instead of derivatives of x arising, they are replaced by derivatives of x_m which are available in the model.

In this process, the nature of the auxiliary input w is changed because of (45). It can no longer be argued that $w_1 \to 0$ as $\eta \to 0$ because of its w_3 component. It can still be argued that $w_2 \to 0$ as $\eta \to 0$ for the same reasons as before. In order to examine this problem, let $k_i(\infty)$ be the constant steady-state value of $k_i(t)$. Then, since $w_2(\infty) \to 0$,

$$w_1(\infty) = w_3(\infty) = -\sum_{i=n+2}^{2n+1} k_i(\infty) p^{2n+1-i} D_w^{-1}(p)e. \quad (47)$$

When this expression for w_1 is used in (4) it is seen that

$$D_m(p)y = D_w(p)(1 + k_N(\infty))\left(-\sum_{i=n+2}^{2n+1} k_i(\infty)p^{2n+1-i} \cdot D_w^{-1}(p)e\right) \quad (48)$$

Since $\eta = 0$ implies $e = -y$, (48) may be written as

$$\left[D_m(p) - (1 + k_N(\infty))\sum_{i=n+2}^{2n+1} k_i(\infty)p^{2n+1-i}\right]y = 0 \quad (49)$$

from which it is seen that stability in the sense that $y \to 0$ and $e \to 0$ as $\eta \to 0$ can only be assured if the operator in (49) is stable. In general, this is difficult to determine. However, when the adjustable gains can be bounded as discussed in Section IV, it may be possible to do so. When it is, this approach may offer the advantage of a reduction in the required number of adjustable parameters. For example, if a_1 in $D_p(p)$ is the only unknown coefficient and the approach using the substitution (39) is taken, then $3n - m - 2$ unknowns (coefficients B_i's, C_i's, and D_i's) must be introduced, and, of course, $3n - m - 2$ adjustable gains along with them. On the other hand, if this alternative approach can be shown to be stable, only one adjustable gain is needed for the unknown coefficient a_1.

Two approaches to the general problem have been considered in this section. One applies when no knowledge of parameter bounds is assumed and convergence of parameter errors cannot be assured. The second may be used if parameter bounds are known, parameter errors can be shown to converge, and operator (49) can be shown to be stable. These points are illustrated by example in the next section.

VI. Computer Simulation

A digital computer simulation was performed for a third-order plant described by

$$(p^3 + a_1p^2 + 8p + 6)x = b_0u \quad (50)$$

and a model given by

$$(p^3 + 5p^2 + 8p + 6)x_m = 6r \quad (51)$$

where a_1 and b_0 are the only unknown parameters. It is assumed that $b_0 > \epsilon > 0$ and ϵ is known, and that $a_1 > a_{1m} = 0.75$. Simulation values used were $b_0 = 1$ and $a_1 = 3$. Because these parameter bounds are known, it is possible to obtain a solution using a combination of Case II results from Sections IV and V. The advantages of this approach are 1) the $D_f^{-1}(p)$ filter is not required, 2) the adjustable gain k_N for w is not required, 3) only two adjustable gains are necessary (one for a_1 and one for b_0) whereas six would be needed following the Case I approach of Section V, i.e., utilizing (39). Advantage 1) is offset by the requirement for a $D_w^{-1}(p)$ type filter for x_m.

The $D_w(p)$ operator employed is $D_w(p) = p^2 + p + 1$ and it is noted that $(s^2 + s + 1)/(s^3 + 5s + 8s + 6)$ is a positive real function of s. The augmented error equation for this case is

$$(p^3 + 5p + 8p + 6)\eta = -b_0u + 6r + (a_1 - 5)p^2x + (p^2 + p + 1)w \quad (52)$$

which may be written as

$$(p^3 + 5p + 8p + 6)\eta = (p^2 + p + 1)[-b_0z_0 + 6z_1 + (a_1 - 5)p^2z_2 + w] \quad (53)$$

where $(p^2 + p + 1)z_0 = u$, $(p^2 + p + 1)z_1 = r$, and $(p^2 + p + 1)z_2 = x$.

Following a method like that used in connection with (44), $z_0 + w$ is taken to be

$$z_0 + w = b_0z_0 - 6z_1 - k_1p^2(z_3 - z_2) + k_1p^2z_3 \quad (54)$$

where $D_w(p)z_3 = x_m$.

When (54) is substituted into (52) the result is

$$(p^2 + 5p + 8p + 6)\eta = (p^2 + p + 1)(\delta_0z_0 + \delta_1p^2z_2) \quad (55)$$

where $\delta_0 = (k_0 - b_0 - 1)$ and $\delta_1 = (k_1 + a_1 - 5)$.

The adaptive laws may be written immediately as

$$p\delta_0 = pk_0 = -(1/\lambda_0)[\eta z_0 - g_1(\eta z_0)g_2(k_0)] \quad (56a)$$

$$p\delta_1 = pk_1 = -(1/\lambda_1)[\eta z_2 - g_3(\eta z_2)(g_4(k_1)] \quad (56b)$$

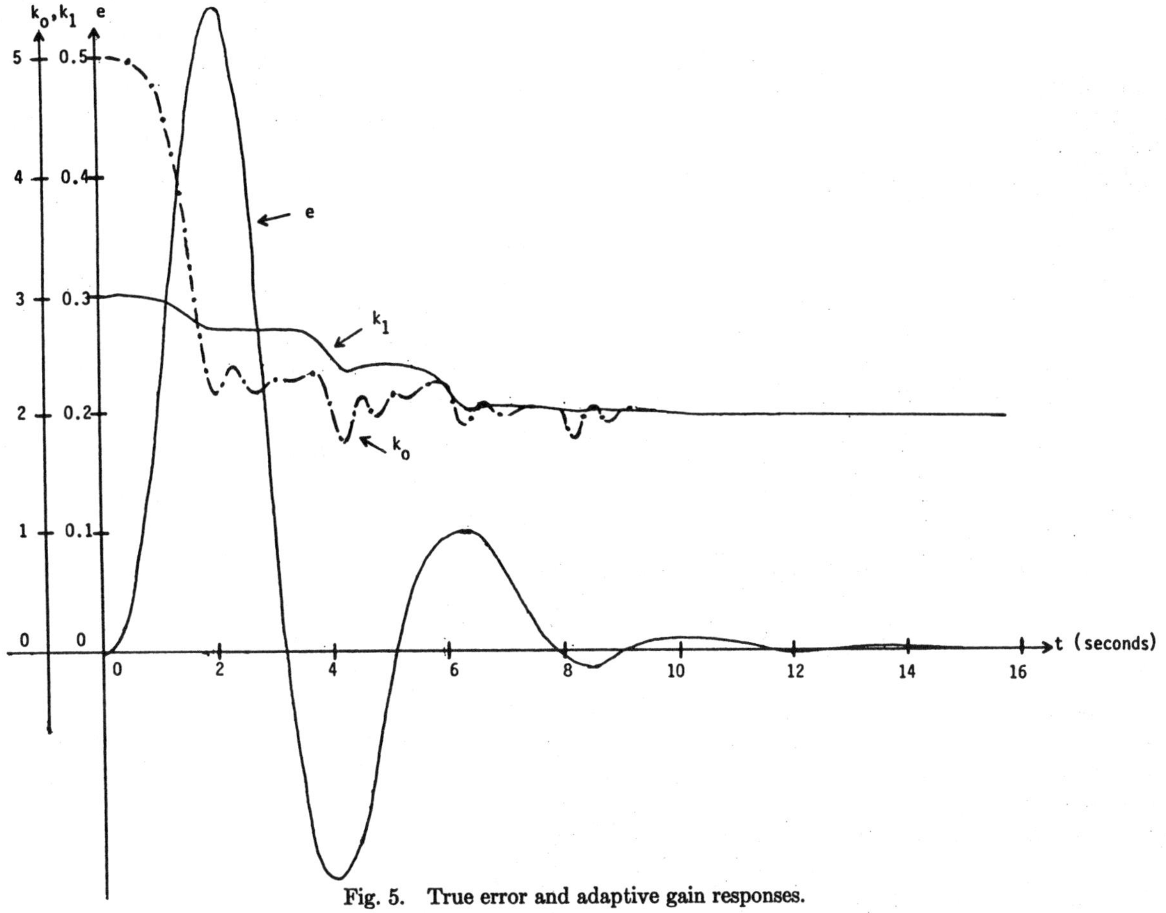

Fig. 5. True error and adaptive gain responses.

where $g_1 g_2$ and $g_3 g_4$ are functions for appropriately limiting k_0 and k_1.

Finally, the operator $D_w(p)$ is applied to both sides of (54) to give

$$u = k_0 u + k_1 p^2 x_m - 6r \tag{57a}$$

$$D_w(p) w_2 = \sum_{i=0}^{1} D_{wi}(p) [(pk_0)(p^{1-i} z_0) + (pk_1)(p^{3-i} z_3)] \tag{57b}$$

$$w_3 = -k_1 p^2 (z_3 - z_2) \tag{57c}$$

where, in this case, $w = w_1 = w_2 + w_3$ has been used.

Since there are only two adjustable parameters, an input r with one frequency is sufficient to assure that $\delta_0 \to 0$ and $\delta_1 \to 0$ [31]. Hence, with a square wave input, which was used, it is assured that $k_1(\infty) = 5 - a_1$. Therefore, the operator (49) becomes

$$p^3 + 5p^2 + 8p + 6 - k_1(\infty) p^2 = p^3 + a_1 p^2 + 8p + 6 \tag{58}$$

which is stable for $a_1 > 0.75$, the same condition as for stability of the plant. One would expect this restriction that the plant be stable since in this approach there is no direct feedback of x nor any of its derivatives.

Simulation results are shown in Fig. 5 where it is seen that k_0 and k_1 converge quite rapidly, after which excellent model following is achieved.

VII. Conclusions

The concept of an augmented error signal was introduced and shown to be useful in designing stable adaptive control systems without the use of plant output derivatives. Alternative designs are given for situations in which bounds on plant parameters are known and those in which they are not. In those cases where these bounds are known, it is shown that the adjustable gains may be bounded, thereby avoiding instability of these gains that might be caused by disturbances. The computer simulation demonstrates that the technique does work quite well, and that its implementation can be relatively simply in certain cases if the appropriate alternative design can be applied. Eliminating the requirement for plant output derivatives should make it possible for designs of this type to be employed in many more engineering problems. It is felt that the method merits further study. Fruitful areas for such study appear to be its extension to multivariable, time varying, and more general nonlinear systems, and also its behavior in the presence of noisy measurements and disturbances.

APPENDIX I

A statement of the MKY lemma as in [29, p .376] is given here following several definitions.

Definition 1: E^n is Euclidean n-space.

Definition 2: For A a real $n \times n$ matrix, $A(z) = zI - A$ is the characteristic matrix of A.

Definition 3: The subspaces of E^n generated by vectors $\boldsymbol{b}, A\boldsymbol{b}, \cdots$ are denoted by $[A,\boldsymbol{b}]$. The orthogonal complement of $[A,\boldsymbol{b}]$ in E^n is denoted by $[A,\boldsymbol{b}]^0$, and, in general,

$$[A,\boldsymbol{b}]^0 = \{x \in E^n : x'A^k\boldsymbol{b} = 0,\ k = 0,1,2,\cdots\}.$$

Lemma 2: Let A be a real $n \times n$ matrix all of whose characteristic roots have negative real parts; let τ be a real nonnegative number and let $\boldsymbol{d},\boldsymbol{k}$ be two real n-vectors. If

$$T(z) = \tau + 2\boldsymbol{k}'A(z)^{-1}\boldsymbol{d}$$

is a positive real function then there exist two $n \times n$ real symmetric matrices B,D and a real n-vector $\boldsymbol{q}$ such that

1) $A'B + BA = -\boldsymbol{q}\boldsymbol{q}' - D.$
2) $B\boldsymbol{d} - \boldsymbol{k} = \sqrt{\tau}\boldsymbol{q}.$
3) D is positive semidefinite and B is positive definite.
4) $\{x \in E^n : x'Dx = 0\} \cap [A',\boldsymbol{q}]^0 = \{0\}.$
5) $\boldsymbol{q} \notin [A,\boldsymbol{d}]^0.$
6) if $i\omega$, ω real, is a zero of $-\boldsymbol{q}'A(z)^{-1}\boldsymbol{d} + \sqrt{\tau}$, then it is a zero of $\boldsymbol{d}'A(-z)^{-1}DA(z)^{-1}\boldsymbol{d}$.

APPENDIX II

The identities used in deriving u, w, and w_1 in (24), (29), and (42) are given.

Let $D_w(p)$ be as given in (4) and $D_w(p)z_0 = u$.

Define polynomials in $p, D_{w0}(p) = 1$, $D_{w1}(p) = p + c_1$, $D_{w2}(p) = p^2 + c_1p + c_2, \cdots, D_{wi}(p) = p^i + c_1p^{i-1} + c_2p^{i-2} + \cdots, c_i, \cdots, D_{w(n-2)} = p^{n-2} + c_1p^{n-3} + \cdots, c_{n-2}$, where constants c_i are the coefficients of $D_w(p)$. Then the following identity is true:

$$D_w(p)(k_0(t)z_0) = k_0u + \sum_{i=0}^{n-2} D_{wi}(p)((pk_0)(p^{n-2-i}z_0)). \quad \text{(A.1)}$$

Similarly

$$D_f(p)(k_0(t)\phi_0) = k_0D_f(p)\phi_0 + \sum_{i=0}^{n-m-2} D_{fi}(p)((pk_0)(p^{n-m-2-i}\phi_0)) \quad \text{(A.2)}$$

where $D_f(p)$ is as in (37) and $D_{f0}(p) = 1$, $D_{f1}(p) = p + F_1, \cdots, D_{f(n-m-2)}(p) = p^{n-m-2} + F_1p^{n-m-3} + \cdots, F_{n-m-2}$.

APPENDIX III

It is to be shown that unknown constants A_0 to A_{n-2} exist which satisfy (38). Replace u on the right side of (38) by $D_w(p)z_0$. Then equate the right side to the left, cancel the factor z_0, and multiply by $D_f(p)$ leaving

$$D_f(p)D_u(p) = b_0D_w(p) + [A_0p^{n-2} + A_1p^{n-3} + \cdots, A_{n-2}]. \quad \text{(A.3)}$$

Both sides of (A.3) are polynomials in p of order $n - 1$. The coefficient of p^{n-1} in each is b_0. Equating the remaining coefficients of like powers of p on both sides of (A.3) leads to a set of $n - 1$ algebraic equations which may be solved for the $n - 1$ constants A_0 to A_{n-2} in terms of the unknown parameters of $D_u(p)$.

It is also true that $3n - m - 2$ unknown constants B_0 to B_{n-1}, C_0 to C_{n-2}, and D_0 to D_{n-m-2} exist which satisfy (39). To show this, multiply both sides of (39) by $D_f(p)D_w(p)$ to get

$$D_\Delta(p)D_f(p)x = [B_0p^{n-1} + \cdots B_{n-1}]x + [C_0p^{n-2} + \cdots C_{n-2}]D_u^{-1}(p)(D_p(p)x - cf) + [D_0p^{n-m-2} + \cdots D_{n-m-2}]cf. \quad \text{(A.4)}$$

The $n - 1$ unknown constants C_0 to C_{n-2} may be assumed to be those satisfying the set of $n - 1$ algebraic equations found by equating coefficients of like powers of p on both sides of (A.5).

$$[C_0p^{n-2} + C_1p^{n-3} + \cdots C_{n-2}] = [D_0p^{n-m-2} + D_1p^{n-m-3} + \cdots D_{n-m-2}]D_u(p). \quad \text{(A.5)}$$

This may be done since both sides of (A.5) are polynomials in p of order $n - 2$.

If (A.5) is substituted into (A.4) and the factor x is cancelled, the result is

$$D_\Delta(p)D_f(p) = [B_0p^{n-1} + B_{n-1}] + [D_0p^{n-m-2} + \cdots D_{n-m-2}]D_p(p). \quad \text{(A.6)}$$

Both sides of (A.6) are polynomials in p of order $2n - m - 2$. The $2n - m - 1$ algebraic equations derived from equating coefficients of like powers of p on both sides of (A.6) admit of a solution in the $2n - m - 1$ unknown constants B_0 to B_{n-1} and D_0 to D_{n-m-2}. Therefore, the right hand side of (39) is a valid alternative expression for the left hand side.

REFERENCES

[1] J. W. Gilbart, R. V. Monopoli, and G. C. Winston, "Combination open loop, closed loop, and adaptive compensation: Its application to an optical tracking system," in *Proc. 1972 Int. Conf. Cybernetics and Society*, Washington, D.C., Oct. 1972.

[2] E. F. Lizewski, "An experimental study of a hybrid adaptive control system," M.S. thesis, Univ. Mass., Amherst, 1973; also in *Proc. 1973 Allerton Conf. Circuits and Systems*.

[3] J. Van Amerongen and A. J. Udink ten Cate, "Adaptive autopilots for ships," in *Proc. 1973 IFAC/IFIP Symp. Ship Operation Automation*, Oslo, Norway.

[4] P. C. Parks, "Lyapunov redesign of model reference adaptive control systems," *IEEE Trans. Automat. Contr.*, vol. AC-11, pp. 362–367, July 1966.

[5] R. V. Monopoli, "Liapunov's method for adaptive control system design," *IEEE Trans. Automat. Contr.* (Corresp.), vol. AC-12, pp. 334–335, June 1967.

[6] C. A. Winsor and R. J. Roy, "Design of model reference adaptive control systems by Liapunov's second method," *IEEE Trans. Automat. Contr.* (Corresp.), vol. AC-13, p. 204, Apr. 1968.

[7] J. W. Gilbart and R. V. Monopoli, "Model reference adaptive control systems with feedback and prefilter adjustable gains," in *Proc. 4th Annu. Princeton Conf. Information Sciences and Systems*, Mar. 1970.

[8] J. W. Gilbart, R. V. Monopoli, and C. F. Price, "Improved

convergence and increased flexibility in the design of model reference adaptive control systems," in *Proc. 9th IEEE Symp. Adaptive Processes Decision and Control*, Austin, Tex., Dec. 1970.

[9] R. M. Dressler, "An approach to model-referenced adaptive control systems," *IEEE Trans. Automat. Contr.*, vol. AC-12, pp. 75–80, Feb. 1967.

[10] D. W. Sutherlin and J. S. Boland, III, "Design techniques for model reference adaptive control systems," in *Proc. 1972 Joint Automatic Control Conf.*, Stanford, Calif., Aug. 1972.

[11] R. V. Monopoli, "Adaptive control and identification via Liapunov's direct method," in *Proc. Allerton Conf.*, Univ. Ill., Urbana, Oct. 1972.

[12] G. Lüders and K. S. Narendra, "Lyapunov functions for quadratic differential equations with applications to adaptive control," *IEEE Trans. Automat. Contr.* (Corresp.), vol. AC-17, pp. 798–801, Dec. 1972.

[13] P. H. Phillipson, "Design methods for model-reference adaptive systems," in *Proc. Inst. Mech.*, vol. 183, part 1, 1968–1969, pp. 695–706.

[14] D. P. Lindorff, "Control of nonlinear multivariable systems," *IEEE Trans. Automat. Contr.*, vol. AC-12, pp. 506–515, Oct. 1967.

[15] B. Porter and M. L. Tatnall, "Performance characteristic of multivariable model-reference adaptive systems synthesized by Liapunov's direct method," *Int. J. Contr.*, vol. 10, no. 3, pp. 241–257, 1969.

[16] R. V. Monopoli, "The Kalman–Yacobovich lemma in adaptive control system design," *IEEE Trans. Automat. Contr.* (Tech. Notes and Corresp.), vol. AC-18, pp. 527–529, Oct. 1973.

[17] D. P. Lindorff, "Effects of incomplete adaptation and disturbance in adaptive control," in *Proc. 1972 Joint Automatic Control Conf.*, Stanford Univ., Stanford, Calif., Aug. 1972.

[18] J. W. Gilbart and R. V. Monopoli, "Parameter adaptive control of a class of time varying plants," in *Proc. 5th Annu. Princeton Conf. Information Sciences and Systems*, Princeton, N.J., Mar. 1971, pp. 404–408.

[19] C. C. Hang and P. C. Parks, "Comparative studies of model reference adaptive control systems," in *Proc. 1973 Joint Automatic Control Conf.*, Columbus, Ohio, June 1973, pp. 12–22.

[20] P. V. Osburn, H. P. Whitaker, and A. Kezer, "New developments in the design of adaptive control systems," Inst. Aeronautic Sciences, paper 61-39.

[21] C. F. Price and W. D. Koenigsberg, "Adaptive control and guidance for tactical missiles," The Analytic Sciences Corp., Reading, Mass., Rep. TR-170-1, 1960.

[22] C. F. Price, "An accelerated gradient method for adaptive control," in *Proc. 9th IEEE Symp. Adaptive Processes*, Univ. Tex., Austin, Dec. 1970.

[23] D. J. G. James, "Stability analysis of a model reference adaptive control system with sinusoidal inputs," *Int. J. Contr.*, vol. 9, no. 3, pp. 311–321, 1969.

[24] R. L. Carroll and D. P. Lindorff, "An adaptive observer for single-input–single-output linear systems," in *Proc. 1973 Joint Automatic Control Conf.*, Columbus, Ohio, June 1973, pp. 473–480; also *IEEE Trans. Automat. Contr.*, vol. AC-18, pp. 428–435, Oct. 1973.

[25] G. Luders and K. S. Narendra, "An adaptive observer and identifier for a multivariable linear system," *7th Annu. Princeton Conf. Information Sciences and Systems*, Mar. 1973, pp. 367–370.

[26] W. D. T. Davies, *System Identification for Self Adaptive Control*. New York: Wiley-Interscience, 1970.

[27] R. E. Kalman "Liapunov functions for the problem of Lure in automatic control," in *Proc. National Academy of Sciences*, vol. 49, pp. 201–205, Feb. 1963.

[28] S. Lefschetz, *Stability of Nonlinear Control Systems*. New York: Academic, 1965.

[29] K. R. Meyer, "On the existence of Liapunov functions for the problem of Lure," *J. SIAM Contr.*, ser. A., vol. 3, no. 3, pp. 373–383, 1966.

[30] R. A. Rucker, "Real time system identification in the presence of noise," in *Proc. Western Electron. Conv.*, paper 2.3, Aug. 1963.

[31] P. M. Lion, "Rapid identification of linear and nonlinear systems," *AIAA J.*, vol. 5, Oct. 1967.

Unification of Discrete Time Explicit Model Reference Adaptive Control Designs*

I. D. LANDAU† and R. LOZANO†

Evaluation of various designs for discrete time explicit model reference adaptive control (MRAC) based on a unified stability point of view, both in tracking and regulation, indicate that a 'constant trace' algorithm yields the best results.

Key Words—Adaptive control; discrete time systems; parameter estimation; regulator; stability; tracking systems.

Abstract—Various discrete time explicit MRAC designs based on stability considerations are presented from a unified point of view. A general structure and a design procedure are given, which contain as particular cases various possible designs. Some of these designs have been already given in the literature but some are new. The general structure considered removes some of the restrictions encountered in previous designs and solves the important problem of independent specification of tracking and regulation objectives. The proof of boundedness of the control variables is also given. Simulations are included in order to evaluate the various designs in a deterministic environment.

1. INTRODUCTION

THE FIRST design of a MRAC for discrete time SISO (single-input, single-output) systems in input output form was proposed by Ionescu and Monopoli (1977). Some of the basic characteristics of this design are

(1) It uses the augmented error concept in order to circumvent the problems related to the system delay and to the inherent one step delay existing in discrete time MRAS.

(2) It uses a 'parallel' type reference model.

(3) It uses an adaptation algorithm with constant adaptation gains.

(4) It uses filtered variables in the adaptation algorithm.

(5) It assumes that the sign of the leading coefficient of the plant numerator polynomial is known.

(6) The design is based on a stability point of view via the use of Lyapunov functions.

*Received 11 November 1979; revised 30 June 1980; revised 15 December 1980. The original version of this paper was not presented at any IFAC meeting. This paper was recommended for publication in revised form by Associate Editor Y. Bar-Shalom.

†Laboratoire d'Automatique de Grenoble, Institut National Polytechnique de Grenoble, B.P. 46, 38402 St Martin d'Heres, France.

Of course other designs can be based on a stability point of view which present different characteristics and are more appropriate for certain types of control problems. For example the parallel configuration considered by Ionescu and Monopoli (1977) is in fact designed for tracking purposes, but the properties of the design with respect to regulation have not been investigated. On the other hand, other configurations (series–parallel for example) can be considered which are more appropriate for regulation (Landau, 1979a; Irving and co-workers, 1979).

Other designs can also be considered which instead of augmented error and filtered variables, use a reference model with an *a priori* and an *a posteriori* output (Landau, 1979b) (this being an extension of the evolutive reference model introduced by Bénéjean (1977) and a linear compensator acting on the plant-model error (as used for recursive identification (Landau, 1976).

Extensive simulations have shown that algorithms with constant adaptation gains despite their simplicity and real time properties have two main disadvantages

(a) it is not clear how to choose the best values of the adaptation gains;

(b) the performance of the adaptive systems is often very sensitive with respect to the values of the adaptation gains.

Therefore it appears to be useful to develop adaptation algorithms with time varying adaptation gains inspired from algorithms used in recursive identification which eliminates the need to choose the adaptation gains and provides better performance. However the use of these algorithms makes the stability analysis more complicated. But the problem can be solved in a very direct way by using a theorem (Landau,

Reprinted with permission from *Automatica*, vol. 17, pp. 593–611, July 1981.
Copyright © 1981, International Federation of Automatic Control IFAC. Published by Pergamon Press Ltd.

1980) which is proved using some of the stability results given in Landau and Silveira (1979) and based on the use of the positivity concepts. The use of this theorem also drastically simplifies the derivation of the design given in Ionescu and Monopoli (1977). Finally the hypothesis that the sign of the numerator leading coefficient should be known can also be removed.

Our analysis of the various possible designs of discrete time explicit MRAC from a stability point of view leads to the following classification

MRAC for $\begin{cases} \text{tracking} \\ \text{regulation} \\ \text{tracking and regulation} \end{cases}$

MRAC with $\begin{cases} \text{parallel reference model} \\ \text{series–parallel reference model} \\ \text{parallel + series–parallel reference} \\ \text{model} \end{cases}$

MRAC using $\begin{cases} a\ priori\ \text{and}\ a\ posteriori \\ \text{reference model} \\ \text{augmented error} \end{cases}$

MRAC using $\begin{cases} \text{filtered variables} \\ \text{linear compensator acting on the} \\ \text{plant model error} \end{cases}$

MRAC using $\begin{cases} \text{constant adaptation gains} \\ \text{time varying adaptation gains.} \end{cases}$

In this paper, a unified design procedure including all the possibilities mentioned above is examined and a general structure for the explicit MRAC scheme models is given. In addition, the design procedure is focused towards solving the problem of independent specification of tracking and regulation objectives, which is not solved by any previous design given in the literature.

Some of the designs considered in this paper also appear in Egardt (1980a, b). However, the work of Egardt does not cover all the structures and their relations with tracking and regulation objectives, the basic adaptation algorithm is not as general and the design using a reference model with an *a priori* and *a posteriori* error is not considered.

An important observation to be made is that the various designs assure the global asymptotic stability either of the augmented error or of the error between the evolutive reference model and the plant. Through an additional analysis it is shown that these designs assure also the boundedness of the control applied to the plant and of the plant output.

This paper is organized as follows: in Section 2, the perfect model following control for tracking and regulation in the case of linear plants with stable zeroes and known parameters is discussed and the structure of the control law is defined. In Section 3, we present a stability theorem which is the basic tool for the design and leads to a clear understanding of need for introducing the augmented error concept or the *a priori* and *a posteriori* reference models, the presence of a positivity condition on a transfer function for convergence and its dependence on the type of the adaptation gain.

Section 4 discusses the general design procedure. Some important particular cases and their correspondence with various known algorithms are discussed in Section 5. In Section 6, the behaviour in regulation of the basic algorithms is discussed.

In Section 7, the evaluation by simulation of various designs is presented. These simulations emphasize in particular the following aspects

tracking and regulation behaviour of various structures;
comparison of adaptation performances with respect to the type of parametric adaptation algorithm used (with filtered variables or linear compensator, with constant or time varying adaptation gains).

2. PERFECT LINEAR MODEL FOLLOWING CONTROL FOR TRACKING AND REGULATION

Consider a SISO (single-input—single-output) discrete linear time-invariant plant described by

$$A(q^{-1})y(k) = q^{-d}B(q^{-1})u(k);$$

$$d > 0;\ y(0) \neq 0 \qquad (1)$$

where

$$A(q^{-1}) = 1 + a_1 q^{-1} + \ldots + a_{n_A} q^{-n_A} \qquad (2)$$

$$B(q^{-1}) = b_0 + b_1 q^{-1} + \ldots + b_{n_B} q^{-n_B};\ b_0 \neq 0 \quad (3)$$

$\{q^{-1}\}$ is the backward shift operator, $\{d\}$ represents the plant time delay, $\{u(k)\}$ and $\{y(k)\}$ are the plant input and output, respectively. We assume the zeroes of $B(z^{-1})$ are all in $|z| < 1$, therefore they can be cancelled without leading to an unbounded control input.

The objectives of the control which we are considering are the following:

(1) The control should be such that in tracking, the output of the process satisfies the equation

$$C_1(q^{-1})y(k) = q^{-d}D(q^{-1})u^M(k) \qquad (4)$$

where $C_1(q^{-1})$, $D(q^{-1})$ are polynomials in $\{q^{-1}\}$,

$C_1(q^{-1})$ is asymptotically stable and $u^M(k)$ is a bounded reference sequence.

(2) The control should be such that in regulation $[u^M(k)\equiv 0]$, an initial disturbance $[y(0)\neq 0]$ is eliminated with the dynamics defined by

$$C_2(q^{-1})y(k+d)=0 \quad k\geq 0 \tag{5}$$

where

$$C_2(q^{-1})=1+c_1^2q^{-1}+\ldots+c_{n_{c_2}}^2q^{-n_{c_2}} \tag{6}$$

is an asymptotically stable polynomial.

A solution is obtained by using an explicit reference model given by

$$C_1(q^{-1})y^M(k)=q^{-d}D(q^{-1})u^M(k) \tag{7}$$

$\{y^M(k)\}$ and $\{u^M(k)\}$ are the model output and input, respectively; they are both bounded.

The plant-model error is defined as

$$\varepsilon(k)=y(k)-y^M(k). \tag{8}$$

It is clear that the objectives of the control are accomplished if the following equation holds

$$C_2(q^{-1})\varepsilon(k+d)=0 \quad k>0. \tag{9}$$

Using the identity

$$C_2(q^{-1})=A(q^{-1})S(q^{-1})+q^{-d}R(q^{-1}) \tag{10}$$

where

$$S(q^{-1})=1+s_1q^{-1}+s_2q^{-2}+\cdots+s_{n_S}q^{-n_S} \tag{11}$$

$$R(q^{-1})=r_0+r_1q^{-1}+r_2q^{-2}+\ldots+r_{n_R}q^{-n_R} \tag{12}$$

which has a unique solution $S(q^{-1})$, $R(q^{-1})$ and $n_S=d-1$ and $n_R=\max(n_A-1,\ n_{C_2}-d)$ (see Appendix A). Equation (9) can be written

$$\begin{aligned}C_2(q^{-1})\varepsilon(k+d)&=B(q^{-1})S(q^{-1})u(k)+R(q^{-1})y(k)\\&\quad -C_2(q^{-1})y^M(k+d)\\&=b_0u(k)+p_0^{\mathrm{T}}\phi_0(k)\\&\quad -C_2(q^{-1})y^M(k+d)\\&=p^{\mathrm{T}}\phi(k)-C_2(q^{-1})y^M(k+d)\end{aligned} \tag{13}$$

where

$$\phi_0^{\mathrm{T}}(k)=[u(k-1),\ldots,u(k-d-n_B+1),\quad y(k),\ldots,y(k-n_R)] \tag{14}$$

$$p_0^{\mathrm{T}}=[b_0s_1+b_1,b_0s_2+b_1s_1+b_2,\ldots,\quad b_{n_B}s_{d-1},\ r_o,\ldots,r_{n_R}] \tag{15}$$

$$\phi^{\mathrm{T}}(k)=[u(k);\phi_0^{\mathrm{T}}(k)] \tag{16}$$

$$p^{\mathrm{T}}=[b_0;p_0^{\mathrm{T}}]. \tag{17}$$

Equating the right-hand side of equation (13) to zero, the control objective of (9) is achieved with the control input

$$u(k)=\frac{1}{b_0}[C_2(q^{-1})y^M(k+d)-R(q^{-1})y(k)\quad -B_s(q^{-1})u(k)] \tag{18}$$

where

$$B_s(q^{-1})=B(q^{-1})S(q^{-1})-b_0 \tag{19}$$

or equivalently

$$u(k)=\frac{1}{b_0}[C_2(q^{-1})y^M(k+d)-p_0^{\mathrm{T}}\phi_0(k)]. \tag{20}$$

The block diagram of the linear model following control scheme is given in Fig. 1.

Remarks

(1) Consider the following controller structure (Åström, Westberg and Wittenmark, 1978; Egardt 1980a).

$$B(q^{-1})S(q^{-1})u(k)=T(q^{-1})D(q^{-1})u^M(k)\quad -R(q^{-1})y(k) \tag{21}$$

where

$$T(q^{-1})=1+t_1q^{-1}+t_2q^{-2}+\cdots+t_{n_T}q^{-n_T}. \tag{22}$$

Supposing that $B(q^{-1})$ and $D(q^{-1})$ are coprime polynomials, this controller corresponds to the particular choice $C_2(q^{-1})=T(q^{-1})C_1(q^{-1})$ in the controller of equation (18).

(2) Consider that a disturbance is acting on the plant as follows:

$$A(q^{-1})y(k)=q^{-d}B(q^{-1})u(k)+w(k). \tag{23}$$

Suppose that we use the controller in equation (18), then using the identity (10), the expression for the plant-model error is

$$C_2(q^{-1})\varepsilon(k)=S(q^{-1})w(k). \tag{24}$$

When $w(k)=C_2(q^{-1})v(k)$ where $v(k)$ is a

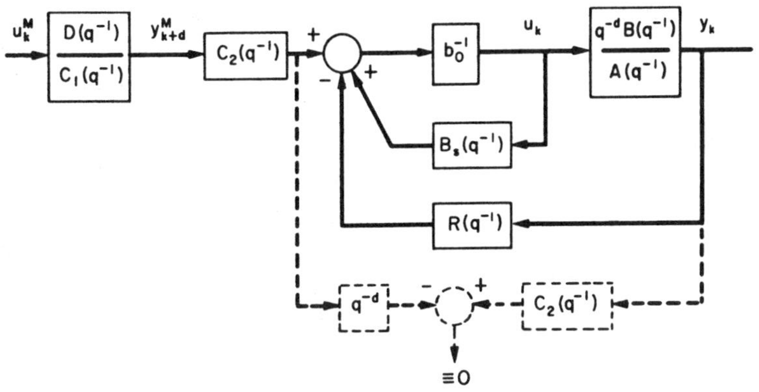

FIG. 1. Tracking and regulation control scheme for known plants. (The dashed line part of the scheme represents the control objectives.)

sequence of independent random variables $(0, \sigma)$, the controller of equation (13) with $y^M(k) \equiv 0$ corresponds to the minimum variance controller (Landau, 1979b; Åström and Wittenmark, 1973).

3. A STABILITY THEOREM USED FOR THE DESIGN OF MRAC

The design of MRAC which are natural extensions of the linear model following control configurations presented in Section 2 is done from a stability point of view, which is now a standard point of view (Ionescu and Monopoli, 1977; Landau, 1979a; Narendra, Lin and Valavani, 1980).

To design an MRAC the equivalent feedback representation (EFR) method will be considered. The MRAC to be designed will be represented by an equivalent feedback system and an appropriate adaptation mechanism will be chosen in order to insure the global asymptotic stability of the equivalent feedback system (Bénéjean, 1977; Landau and Courtiol, 1974; Landau, 1979).

To effectively derive the adaptation mechanism, the following theorem concerning the asymptotic stability of the equivalent feedback system representation will be used (Landau, 1980).

Theorem 3.1. Assume that the following adaptation algorithm for up-dating the parameter vector $\hat{p}(k)$ is used

$$\hat{p}(k) = \hat{p}(k-1) + F_k \phi(k-d) v_k \qquad (25)$$

where

$$F_{k+1}^{-1} = \lambda_1(k) F_k^{-1} + \lambda_2(k) \phi(k-d) \phi^T(k-d); \; F_0 > 0 \qquad (26)$$

with

$$0 < \lambda_1(k) \leqq 1; \quad 0 \leqq \lambda_2(k) < 2 \quad \forall k. \qquad (27)$$

Assume that the relation between $\phi(k-d)$ and v_k is given by

$$v_k = H(q^{-1})[p - \hat{p}(k)]^T \phi(k-d) \qquad (28)$$

where $\phi(k-d)$ is a bounded or unbounded vector sequence, $H(z^{-1})$ is a rational discrete transfer function normalized under the form

$$H(z^{-1}) = \frac{1 + h_1' z^{-1} + \cdots + h_\alpha' z^{-\alpha}}{1 + h_1 z^{-1} + \cdots + h_\beta z^{-\beta}} \qquad (29)$$

and p is a constant parameter vector. Then, if the transfer function

$$H'(z^{-1}) = H(z^{-1}) - \tfrac{1}{2}\lambda \qquad (30)$$

is strictly positive real where

$$2 > \lambda \geqq \max_{0 \leqq k < \infty} [\lambda_2(k)] \qquad (31)$$

one has for any v_0 and $\hat{p}(0)$ bounded

$$(1) \qquad \lim_{k \to \infty} v_k = 0. \qquad (32)$$

$$(2) \qquad \lim_{k \to \infty} x_k = 0. \qquad (33)$$

[Where x_k is the state vector of any of the state realizations of $H(z^{-1})$.]

$$(3) \quad \lim_{k \to \infty} \phi^T(k-d)\{[\hat{p}(k-1) - p] + F_k \phi(k-d) v_k\} = 0. \qquad (34)$$

$$(4) \quad \lim_{k \to \infty} [1 - \lambda_1(k)] \| [\hat{p}(k-1) - p] + F_k \phi(k-d) v_k \|_{F_k^{-1}}^2 = 0. \qquad (35)$$

(5) $\lim_{k\to\infty} \|\phi(k-d)v_k\|^2_{F_k} = \lim_{k\to\infty} \Delta\hat{p}(k)^{\mathrm{T}}$

$$\times F_k^{-1}\Delta\hat{p}(k)=0. \quad (36)$$

[Where $\Delta\hat{p}(k)=\hat{p}(k)-\hat{p}(k-1)$.]

(6) $[\hat{p}(k-1)-p]^{\mathrm{T}}F_k^{-1}[\hat{p}(k-1)-p]<M_1<\infty$

$$k>0. \quad (37)$$

(7) $\lim_{k\to\infty} [\hat{p}(k-1)-p]^{\mathrm{T}}F_k^{-1}[\hat{p}(k-1)-p]=\text{const.}$

(38)

If in addition $F_k^{-1}>\varepsilon F_0^{-1}$; $F_0>0$; $\varepsilon>0$, $k>0$ and F_k^{-1} is non-decreasing for $k \geqq k_0$. (39)

(8) $\lim_{k\to\infty} \Delta\hat{p}(k)=\lim_{k\to\infty} F_k\phi(k-d)v_k=0.$ (40)

(9) $\|\hat{p}(k)\| \leqq M_2<\infty \quad k\geqq 0.$ (41)

The proof of this theorem can be found in Landau (1980) [with the obvious change in notations: $\hat{p}(k)\to\hat{p}(k-1)$, $v_k\to\phi(k-d)$].

Remarks. (1) Equation (25) combined with (28) define a feedback system which has a linear block [defined by $H(z^{-1})$] and a time varying nonlinear block (Landau and Silveira, 1979; Landau, 1979a).

(2) The first result of this theorem will be used for the design ($\lim_{k\to\infty} v_k=0$). The others results will be used for proving the boundness of the input and output of the controlled plant.

Depending on the choices for $\lambda_1(k)$ and $\lambda_2(k)$ different types of adaptation algorithms are obtained as follows:

$\lambda_1(k)=1; \lambda_2(k)=0$, corresponds to a constant adaptation gain used for example in Ionescu and Monopoli (1977) and Bénéjean (1977).

$\lambda_1(k)=1$; $\lambda_2(k)=\lambda$; $0<\lambda<2$: corresponds to a time decreasing adaptation gain (Landau, 1976; Landau, 1979a; Irving, 1978).

$\lambda_1(k)=\lambda_1$; $\lambda_2(k)=\lambda_2$: corresponds to time varying adaptation gain, useful for the case of slowly time varying plants (usually with $0.95<\lambda_1<0.99$ (Wittenmark, 1974).

$\lambda_1(k)$, $\lambda_2(k)$ such that trace F_k=constant, corresponds to a real time adaptation algorithm for tracking time varying plants (Irving, 1979).

Remark. The condition (39) is automatically satisfied for all the particular algorithms mentioned above except the third, for this algorithm when $\lambda_2(k)=0$, and $\lambda_1(k)<1$ in order to satisfy (39), $\lim_{k\to\infty}\lambda_1(k)=1$ or more specially $\lambda_1(k)=1$ for $k>k_1$ where k_1 is determined by

$$\prod_{i=0}^{k_1} \lambda_1(i)=\varepsilon>0.$$

4. A GENERAL DESIGN PROCEDURE FOR DISCRETE TIME EXPLICIT MRAC

In this section, we first present a simple design procedure of a MRAC, that is an extension of the linear controller design given in Section 2 and that is applicable to minimum phase plants for which only the time delay $\{d\}$ and the upperbounds of the degrees of polynomials $A(q^{-1})$ and $B(q^{-1})$ denoted n_S and n_B are known. The design is made from a stability point of view using Theorem 3.1. An auxiliary error is added to the true plant-model error to obtain an equation of the form (28) that depends on $[p-\hat{p}(k)]$ which allows to use the Theorem 3.1.

For generality purposes, we will then introduce in the MRAC design a linear compensator acting on the augmented error and a filter acting on the various measured variables such that the positivity condition of equation (30) is still verified (Section 4.2). This generalization will then cover the various designs available in the literature and also present an interest in the case when MRAC operates in a stochastic environment (Dugard and Landau, 1980).

4.1. *A simple MRAC design*

When the plant parameters a_i and b_i are unknown, it is natural to replace the vector p_0 and the coefficient b_0 in equation (20) by the adjustable parameters $\hat{p}_0(k)$ and $\hat{b}_0(k)$ which will be updated by the adaptation mechanism. Therefore the control $u(k)$ in the adaptive case is given by

$$u(k)=\hat{b}_0^{-1}(k)[C_2(q^{-1})y^M(k+d)$$
$$-p_0^{\mathrm{T}}(k)\phi_0(k)] \quad (42)$$

or equivalently

$$\hat{p}^{\mathrm{T}}(k)\phi(k)=C_2(q^{-1})y^M(k+d) \quad (43)$$

with

$$\hat{p}^{\mathrm{T}}(k)=[\hat{b}_0(k);\hat{p}_0^{\mathrm{T}}(k)]. \quad (44)$$

Introducing equation (43) into (13), one has

$$C_2(q^{-1})\varepsilon(k+d)=[p-\hat{p}(k)]^{\mathrm{T}}\phi(k). \quad (45)$$

Let us now define the filtered plant-model error as

$$\varepsilon^f(k)=C_2(q^{-1})\varepsilon(k)=[p-\hat{p}(k-d)]^{\mathrm{T}}\phi(k-d). \quad (46)$$

The design objective is to find an adaptation mechanism ensuring that

$$\lim_{k\to\infty} \varepsilon(k)=0, \forall\varepsilon(0)\neq 0 \quad \hat{p}(0)\in\mathbb{R} \quad (47)$$

and

$$\|\phi(k)\| \leq M < \infty \quad \forall k.$$

This implies that the plant-model error converges to zero and that the input and output of the plant remain bounded. However, this problem cannot be solved directly because of the form of equation (46). Through the use of Theorem 3.1, one can solve in a first step the following problem: defining an *a posteriori* filtered plant-model error (called also the 'augmented' filtered plant-model error or the 'residual') satisfying the equation

$$\varepsilon^*(k) = [p - \hat{p}(k)]^{\mathrm{T}} \phi(k-d) \tag{48}$$

one can find an adaptation mechanism which will assure

$$\lim_{k \to \infty} \varepsilon^*(k) = 0. \tag{49}$$

Then, through an additional analysis, it will be shown that the objectives of equation (47) are also verified. In order to obtain the *a posteriori* filtered plant-model error, one should introduce an auxiliary error defined by

$$\bar{\varepsilon}(k) = [\hat{p}(k-d) - \hat{p}(k)]^{\mathrm{T}} \phi(k-d) \tag{50}$$

$$\varepsilon^*(k) = \varepsilon^f(k) + \bar{\varepsilon}(k). \tag{51}$$

Using Theorem 3.1, one concludes that if one uses the adaptation algorithm

$$\hat{p}(k) = \hat{p}(k-1) + F_k \phi(k-d) \varepsilon^*(k) \tag{52}$$

with

$$F_{k+1} = \frac{1}{\lambda_1(k)} \times \left[F_k - \frac{F_k \phi(k-d) \phi^{\mathrm{T}}(k-d) F_k}{\lambda_1(k)/\lambda_2(k) + \phi^{\mathrm{T}}(k-d) F_k \phi(k-d)} \right] \tag{53}$$

where

$$0 < \lambda_1(k) \leq 1; \quad 0 \leq \lambda_2(k) < 2, \quad F_0 > 0.$$

[Equation (53) is obtained from (26) by applying the matrix inversion lemma (Landau, 1979a).]

One has

$$\lim_{k \to \infty} \varepsilon^*(k) = 0. \tag{54}$$

Since the positivity condition (30) is verified [$H(z^{-1}) = 1$ in this case].

It is shown in Appendix B (an alternative proof is given in Appendix C) that this algorithm permits verification of the objective specified in equation (47).

To make the algorithm implementable an expression for $\varepsilon^*(k)$, which depends on the parameters estimated up to $k-1$, should be given.

From equations (51), (46), (50), (16), (43) and (52), one has

$$\begin{aligned}
\varepsilon^*(k) &= \varepsilon^f(k) + \bar{\varepsilon}(k) \\
&= C_2(q^{-1}) \varepsilon(k) + \hat{p}^{\mathrm{T}}(k-d) \phi(k-d) \\
&\quad - \hat{p}^{\mathrm{T}}(k) \phi(k-d) \\
&= C_2(q^{-1}) y(k) - \hat{p}^{\mathrm{T}}(k) \phi(k-d) \\
&= C_2(q^{-1}) y(k) - \hat{p}^{\mathrm{T}}(k-1) \phi(k-d) \\
&\quad - \phi^{\mathrm{T}}(k-d) F_k \phi(k-d) \varepsilon^*(k) \\
&= \frac{\hat{e}(k)}{1 + \phi^{\mathrm{T}}(k-d) F_k \phi(k-d)}
\end{aligned} \tag{55}$$

where

$$\hat{e}(k) = C_2(q^{-1}) y(k) - \hat{p}^{\mathrm{T}}(k-1) \phi(k-d). \tag{56}$$

Figure 2 shows the final adaptive control scheme corresponding to this design.

Remarks. (1) In order to avoid $\hat{b}_0(k) = 0$ for some k (it is known that $b_0 \neq 0$), one takes advantage of the fact that $\lambda_1(k-1)$ and $\lambda_2(k-1)$ can be varied as long as conditions of equation (27) are satisfied. If $|\hat{b}_0(k)| < \delta > 0$ for a certain k the up-dating equation for $\hat{p}(k)$ is recomputed for $\lambda'(k-1) = \lambda_1(k-1) + \Delta\lambda_1$ or for $\lambda'_2(k-1) = \lambda_2(k-1) + \Delta\lambda_2$ where $\Delta\lambda_1$ or $\Delta\lambda_2$ will be choosen such that $|\hat{b}_0(k)| \geq \delta$.

(2) We use in general the 'constant trace algorithm' i.e. we chose $\sigma = \lambda_1(k)/\lambda_2(k) = \text{constant}$ and then $\lambda_1(k)$ is adjusted at each step so that the trace of F_{k+1} is equal to that of F_k, (Irving, 1979), because this type of adaptation algorithm insures better performances than the other adaptation algorithms which can be obtained by particularizing $\lambda_1(k)$ and $\lambda_2(k)$ in equation (27) and because with this type of algorithm, one has

$$\lim_{k \to \infty} F_k \neq 0 \tag{57}$$

which is a required condition (as shown in Appendix B) for proving the convergence to zero

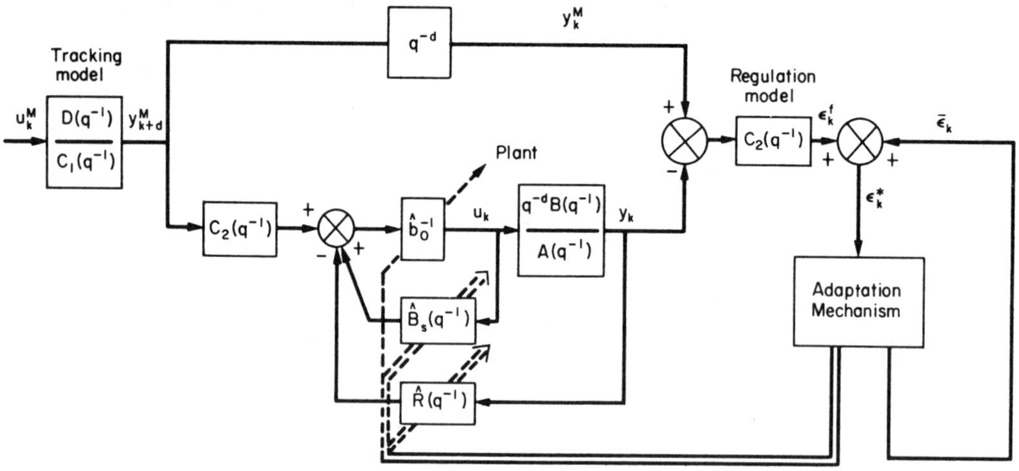

FIG. 2. Tracking and regulation adaptive control scheme in Section 4.1 $[H_1(q^{-1})=H_2(q^{-1})=1]$.

of $\varepsilon^f(k)$ and the boundedness of $\phi(k-d)$ for $d>1$.

(3) The same control algorithm can be obtained by using the concept of *a priori* and *a posteriori* reference model defined as follows (Landau, 1979; Bénéjean, 1977)

$$y_M^0(k)=-C_1^*(q^{-1})y_M(k)+q^{-d}D(q^{-1})u^M(k)$$

$$y_M(k)=y_M^0(k)+u_0(k) \tag{58}$$

where

$$C_1^*(q^{-1})=C_1(q^{-1})-1 \tag{59}$$

$\{y_M^0(k)\}$ is the *a priori* output $\{y_M(k)\}$ is the *a posteriori* output and $\{u_0(k)\}$ for this configuration is a transient signal equal to the auxiliary error $\bar{\varepsilon}(k)$ in equation (50) that tends asymptotically to zero (as the auxiliary error).

The *a priori* and *a posteriori* reference model is an extension of the *a priori* and *a posteriori* adjustable system for discrete time MRAS identifier (Landau, 1976; 1979a, b).

4.2. *General MRAC design*

In this section, we will give a general design procedure which will be particularized for various situations in Section 5. The design follows that of the previous section but in this case, a filter, which acts on the plant input and output, and a linear compensator, which acts on the plant-model error, will allow us

(a) to obtain enough flexibility in the design;
(b) to cover as particular cases various known designs;
(c) to satisfy a positive real condition for some particular schemes;
(d) to improve the performance of the adaptive control scheme in a stochastic environment (Dugard and Landau, 1980).

Let us define the following filtered variables

$$\left.\begin{aligned} L(q^{-1})y^f(k)&=y(k)\\ L(q^{-1})u^f(k)&=u(k)\\ L(q^{-1})y^{Mf}(k)&=y^M(k)\\ L(q^{-1})\phi_0^f(k)&=\phi_0(k)\end{aligned}\right\} \tag{60}$$

where

$$L(q^{-1})=1+l_1q^{-1}+\cdots+l_{n_L}q^{-nL} \tag{61}$$

is an asymptotically stable polynomial. The choice of n_L will be clarified after equation (78).

Using these filtered variables equation (13) can be rewritten as

$$C_2(q^{-1})\varepsilon(k+d)=L(q^{-1})[b_0u^f(k)+p_0^T\phi_0^f(k) -C_2(q^{-1})y^{Mf}(k+d)]. \tag{62}$$

It is clear from equation (62) that if in the case with known parameters, one uses the control input

$$u(k)=L(q^{-1})u^f(k) \tag{63}$$

with

$$u^f(k)=\frac{1}{b_0}[C_2(q^{-1})y^{Mf}(k+d)-p_0^T\phi_0^f(k)]. \tag{64}$$

The objective in equation (9) is achieved. However when the plant parameters are unknown they may be replaced by adjustable parameters. The control input in the adaptive case is given by (63) where $u^f(k)$ is given by

$$u^f(k)=\frac{1}{\hat{b}_0(k)}[C_2(q^{-1})y^{Mf}(k+d) - \hat{p}_0^{\mathrm{T}}(k)\phi_0^f(k)] \quad (65)$$

or equivalently

$$\hat{p}^{\mathrm{T}}(k)\phi^f(k)=C_2(q^{-1})y^{Mf}(k+d) \quad (66)$$

with

$$\phi^{f^{\mathrm{T}}}(k)=[u^f(k), \phi_0^{f^{\mathrm{T}}}(k)] \quad (67)$$

and

$$\hat{p}^{\mathrm{T}}(k)=[\hat{b}_0(k), \hat{p}_0^{\mathrm{T}}(k)]. \quad (68)$$

Introducing equation (66) into (62), one gets

$$\varepsilon_L^f(k)=C_2(q^{-1})\varepsilon(k) = L(q^{-1})\{[p-\hat{p}(k-d)]^{\mathrm{T}}\phi^f(k-d)\}. \quad (69)$$

For the same reason as in Section 4.1, let us define the auxiliary error in this case as

$$\bar{\varepsilon}_L(k)=L(q^{-1})\{[\hat{p}(k-d)-\hat{p}(k)]^{\mathrm{T}}\phi^f(k-d)\} \quad (70)$$

and the augmented error as

$$\varepsilon_L^*(k)=\varepsilon_L^f(k)+\bar{\varepsilon}_L(k). \quad (71)$$

Introducing equation (69) and (70) into (71)

$$\varepsilon_L^*(k)=L(q^{-1})\{[p-\hat{p}(k)]^{\mathrm{T}}\phi^f(k-d)\}. \quad (72)$$

Define the processed augmented error as

$$v(k)=\frac{H_1(q^{-1})}{H_2(q^{-1})}\varepsilon_L^*(k) \quad (73)$$

where $H_1(q^{-1})$ and $H_2(q^{-1})$ are monic polynomials in $\{q^{-1}\}$.

Then, $v(k)$ is given by

$$v(k)=\frac{H_1(q^{-1})L(q^{-1})}{H_2(q^{-1})}[p-\hat{p}(k)]^{\mathrm{T}}\phi^f(k-d) \quad (74)$$

and Theorem 3.1 can be straightforwardly applied. Consider the adaptation algorithm

$$\hat{p}(k)=\hat{p}(k-1)+F_k\phi^f(k-d)v(k) \quad (75)$$

with

$$F_{k+1}=\frac{1}{\lambda_1(k)} \times\left[F_k-\frac{F_k\phi^f(k-d)\phi^f(k-d)^{\mathrm{T}}F_k}{\lambda_1(k)/\lambda_2(k)+\phi^f(k-d)^{\mathrm{T}}F_k\phi^f(k-d)}\right] \quad (76)$$

in order to have

$$\lim_{k\to\infty} v(k)=0. \quad (77)$$

$H_1(q^{-1})$, $H_2(q^{-1})$ and $L(q^{-1})$ should be asymptotically stable polynomials satisfying

$$\frac{H_1(q^{-1})L(q^{-1})}{H_2(q^{-1})}-\frac{\lambda}{2}=\text{s.p.r.} \quad (78)$$

with λ as in equation (31). Thus $H_1(q^{-1})$, $H_2(q^{-1})$ and $L(q^{-1})$ are any finite dimensional asymptotically stable polynomials verifying the positivity condition (78). These polynomials play an important role when working in a stochastic environment (Dugard and Landau, 1980).

It is shown in Appendix B (an alternative proof is given in Appendix C), that if $C_2(q^{-1})$ is an asymptotically stable polynomial: $\lim_{k\to\infty}\varepsilon(k)=0$ and $\phi^f(k)$ is a bounded vector.

We will give (for implementation purposes) the expression of $v(k)$ which depends only on parameters estimated up to $(k-1)$

$$v(k)=\frac{C_2(q^{-1})y(k)-\hat{p}(k-1)^{\mathrm{T}}\phi^f(k-d)+\alpha(k)}{1+\phi^{f\mathrm{T}}(k-d)F_k\phi^f(k-d)} \quad (79)$$

where

$$\alpha(k)=[1-H_2(q^{-1})]v(k)+[H_1(q^{-1})-1]\varepsilon_L^*(q^{-1}) +[L(q^{-1})-1][(\hat{p}(k-d)-\hat{p}(k-1))^{\mathrm{T}} \times\phi^f(k-d)-\phi^{f\mathrm{T}}(k-d)F_k\phi^f(k-d)v(k)] \quad (80)$$

for $L(q^{-1})=H_1(q^{-1})=H_2(q^{-1})=1$, equation (79) is equal to the expression given in (55).

Remarks. (1) In the adaptive case, when using the filter $L(q^{-1})$, $u(k)$ is effectively computed using equation (63) where $u^f(k)$ is given by (65). It is clear that transiently $u(k)$ computed when $L(q^{-1})\neq 1$ *is different from* $u(k)$ when $L(q^{-1})=1$. Despite this the stability analysis shows that the control objectives are asymptotically achieved and simulation (see Section 7) have pointed out the differences during the adaptation process.

(2) As in Section 4.1, the same control algorithm can be obtained using the *a priori* and *a posteriori* reference model given in equations (58) and 59), where in this case the $u_0(k)$ is equal to the auxiliary error $\bar{\varepsilon}_L(k)$ in equation (70).

5. PARTICULAR CASES AND CONNECTIONS WITH VARIOUS EXPLICIT MRAC ALGORITHMS

In this section, we present some particular cases of the general control algorithm obtained in Section 4.2, and we establish the correspondence with various explicit MRAC algorithms, given in the literature.

5.1. *Independent tracking and regulation algorithm*

It is clear that if we choose $H_1(q^{-1})=H_2(q^{-1})=L(q^{-1})=1$, the positivity condition in equation (27) is automatically verified and the expression for $v(k)$ in (79) becomes equal to (55). In this case, the polynomials $C_1(q^{-1})$ and $C_2(q^{-1})$ can be chosen independently from each other.

5.2. *Series–parallel MRAC algorithm*

If in addition we consider $C_2(q^{-1})=C_1(q^{-1})$, $[H_1(q^{-1})=H_2(q^{-1})=L(q^{-1})=1]$, the expression for $v(k)$ [equation (79)] can be rewritten as follows [using equations (55), (48) and (7)].

$$v(k)=\frac{y(k)-y_s^M(k)+[\hat{p}(k-d)-\hat{p}(k-1)]^{\mathrm{T}}\phi(k-d)}{1+\phi^{\mathrm{T}}(k-d)F_k\phi(k-d)} \tag{81}$$

where $y_s^M(k)$ is the output of a series–parallel reference model given by

$$y_s^M(k)=-C_1^*y(k)+q^{-d}D(q^{-1})u^M(k) \tag{82}$$

[which can be used instead of the model given in equation (7)].

5.3. *Parallel MRAC algorithm with linear compensator*

We will call the parallel algorithm with a linear compensator the algorithm obtained with the selection $L(q^{-1})=1$, $C_1(q^{-1})=C_2(q^{-1})=H_2(q^{-1})$ and $H_1(q^{-1})$ such that the positivity condition [equation (78)] is verified. In this case, using equation (79), one has

$$v(k)=\frac{H_2(q^{-1})[y(k)-y^M(k)]+[\hat{p}(k-d)-\hat{p}(k-1)]^{\mathrm{T}}\phi(k-d)+\alpha'(k)}{1+\phi^{\mathrm{T}}(k-d)F_k\phi(k-d)} \tag{83}$$

with

$$\alpha'(k)=H_1^*(q^{-1})\varepsilon_L^*(k)-H_2^*(q^{-1})v(k) \tag{84}$$

where $y^M(k)$ is given by equation (7) $H_i^*(q^{-1})=H_i(q^{-1})-1$.

5.4. *Parallel MRAC algorithm with filter*

We will call the parallel algorithm with a filter the algorithm obtained with the choice $H_1(q^{-1})=1$, $C_1(q^{-1})=C_2(q^{-1})=H_2(q^{-1})$ and $L(q^{-1})$ such that the positivity condition [equation (78)] is verified. Equation (79) can be rewritten as follows:

$$v(k)=\frac{H_2(q^{-1})[y(k)-y^{Mf}(k)]+[\hat{p}(k-d)-\hat{p}(k-1)]^{\mathrm{T}}\phi^f(k-d)+\alpha''(k)}{1+\phi^{f\mathrm{T}}(k-d)F_k\phi^f(k-d)} \tag{85}$$

with

$$\alpha''(k)=L^*(q^{-1})\{[\hat{p}(k-d)-\hat{p}(k-1)]^{\mathrm{T}}\phi^f(k-d)\}-H_2^*(q^{-1})v(k) \tag{86}$$

where $y^{Mf}(k)$ is given by equations (7) and (60) and $L^*(q^{-1})=L(q^{-1})-1$.

This is a generalization of the algorithm given in Ionescu and Monopoli (1977) and Bénéjean (1977). Note also that the choice $C_2=TC_1$, $H_2=1$, where T is an observer, corresponds to the algorithm considered by Egardt (1980 a, b) with the notation $C_2=TA^M$, $C_1=A^M$, $L=P$, $H_1=Q$, $H_2=1$ (he has used a stochastic approximation algorithm).

5.5. *Generalization of the scheme* of Goodwin, Ramadge and Caines (1980)

If $C_2(q^{-1})=1$, the independent tracking and regulation algorithm (5.1) is a generalization of the adaptive control algorithm using recursive least squares given in Goodwin, Ramadge and Caines (1980) where $y^*(t+d)=y^M(t+d)$.

5.6. *Minimum variance self-tuning regulator* of Åström and Wittenmark (1978)

For regulation ($y^M(k)\equiv 0$], with the choice $C_2(q^{-1})=H_2(q^{-1})=H_1(q^{-1})=L(q^{-1})=1$ the algorithm is identical to the minimum variance self-tuning regulator [for $\lambda_1(k)=1$ and $\lambda_2(k)=\lambda=$const.). See also Dugard and Landau (1980).

6. BEHAVIOUR IN REGULATION $[y^M(k)\equiv 0]$

The regulation behaviour of the various MRAC proposed in the literature has not been widely examined in the past (Irving, 1979;

Landau, 1979a). It is, however, a very important problem, since in practice, regulation is the first task to be assured by a control system (disturbances are always present). Secondly, we have observed that various designs can have similar behaviour in tracking but very different behaviour in regulation.

In the regulation case, the control schemes of Sections 4 and 5 have different properties depending on the choice of $L(q^{-1})$, $H_2(q^{-1})$ and $H_2(q^{-1})$. The regulation situation corresponds to $y^M(k)\equiv 0$ and an initial output $y(0)\neq 0$. However the effect of an initial output disturbance is equivalent to an impulse

$$w(k)=\begin{cases} y(0) & \text{for } k=0 \\ 0 & k\neq 0 \end{cases}$$

acting on the plant

$$A(q^{-1})y(k)=q^{-d}B(q^{-1})u(k)+w(k). \qquad (87)$$

6.1. *The algorithm of Section* 4.1

Suppose that the controller parameters are already adjusted at the correct value (p) such that in tracking $y(k)-y^M(k)\equiv 0$. In the presence of $w(k)$ defined above, one has from equations (56) and (43)

$$\hat{e}(k)=C_2(q^{-1})y(k)-p^{\mathrm{T}}\phi(k-d)$$

$$=C_2(q^{-1})[y(k)-y^M(k)] \qquad (88)$$

and from equation (24)

$$\hat{e}(k)=S(q^{-1})w(k). \qquad (89)$$

Thus since we are working in regulation the measurement vector $\phi(k-d)$ is zero for $k\leqq d-1$ [see equation (24)] and since $\hat{e}(k)=0$ for $k\geqq d$ [see equations (89) and (16)] then

$$\Delta\hat{p}(k)=\hat{p}(k)-\hat{p}(k-1)$$

$$=\frac{F_k\phi(k-d)}{1+\phi^{\mathrm{T}}(k-d)F_k\phi(k-d)}\hat{e}(k)$$

$$=0 \quad \forall k>0 \qquad (90)$$

therefore, in regulation, the controller parameters remain at the values established for tracking (unique equilibrium point for tracking and regulation). Some examples illustrating this behaviour are given in Section 7.

6.2. *The algorithm of Section* 4.2

For $H_1(q^{-1})=H_2(q^{-1})=L(q^{-1})=1$, this algorithm corresponds to the previous one, however, for $H_1(q^{-1})\neq 1$, $L(q^{-1})\neq 1$, the properties of the algorithm in regulation are very different.

Consider for example $H_2(q^{-1})=C_2(q^{-1})=C_1(q^{-1})$, $L(q^{-1})=1$ and a polynomial $H_1(q^{-1})$ such that $[H_1(q^{-1})/H_2(q^{-1})]-\lambda/2$ be a strictly positive real transfer function as required by equation (78) (this corresponds to a parallel MRAC. In this case, for regulation, equation (73) becomes

$$v(k)=v^0(k)+\bar{v}(k)$$

$$=H_1(q^{-1})y(k)+\frac{H_1(q^{-1})}{C_2(q^{-1})}$$

$$\times[\hat{p}(k-d)-\hat{p}(k)]^{\mathrm{T}}\phi(k-d). \qquad (91)$$

Up to $k\leqq d-1$, $\Delta\hat{p}(k)=0$ because $\phi(k-d)=0$ but for $k\geqq d$, $\phi(k-d)\neq 0$, $v^0(k)=H_1(q^{-1})y(k)\neq 0$ and $\Delta\hat{p}(k)\neq 0$.

This implies that the parameters of the controller will vary from their values insuring $C_2(q^{-1})\,y(k)=0$, $k\geqq d$. In fact, they will tend to values insuring that $H_1(q^{-1})y(k)=0$ for $k\geqq d$. This is shown next by proving that if the controller parameters are adjusted such that $H_1(q^{-1})y(k)=0$ prior to the application of the disturbance, they will remain unchanged for $k\geqq d$. With this hypothesis, we will have $\Delta\hat{p}(k)=0$ for $k\leqq d-1$ because $\phi(k-d)=0$. For $k=d$, one has

$$v^0(d)=H_1(q^{-1})y(k)=0 \qquad (92)$$

$$v(k)=\bar{v}(k)=\frac{H_1(q^{-1})}{C_2(q^{-1})}[\hat{p}(0)-\hat{p}(d)]^{\mathrm{T}}\phi(0). \quad (93)$$

Equation (93) can be re-written as

$$C_2(q^{-1})\bar{v}(k)=H_1(q^{-1})[\hat{p}(0)-\hat{p}(k)]^{\mathrm{T}}\phi(0). \quad (94)$$

Since $\Delta\hat{p}(k)=0$ for $k\leqq d-1$, one has

$$\bar{v}(d-1),\ \bar{v}(d-2),\ldots,\bar{v}(0)=0$$

and

$$\hat{p}(d-2)=\ldots=\hat{p}(0).$$

Therefore, equation (94) becomes

$$\bar{v}(d)=[\hat{p}(d-1)-\hat{p}(d)]^{\mathrm{T}}\phi(0)$$

$$=-\phi^{\mathrm{T}}(0)F_d\phi(0)v(k) \qquad (95)$$

which leads to

$$[1+\phi^{\mathrm{T}}(0)F_d\phi(0)]\bar{v}(k)=[1+\phi^{\mathrm{T}}(0)F_d\phi(0)]$$

$$\times v(d)=0. \qquad (96)$$

One concludes therefore that

$$v(k)=0,\ \Delta\hat{p}(d)=0$$

and the controller parameters will remain unchanged also for $k \geqq d$. This behaviour is also illustrated in Section 7.

We conclude that this type of algorithm will have two equilibrium points in the parameter space and the disturbance will drift the parameters of the controllers from the values assuring the tracking objectives even if the plant parameters are constant. Therefore, to avoid this phenomenon, we need to choose $H_1=C_2$.

7. SIMULATION RESULTS

We will present next simulation results concerning some particular cases of the MRAC design in Section 4 which has been particularized in Section 5. The control algorithms 5.1–5.4 are studied for tracking and regulation. The discussion in Section 6 is illustrated by simulations and a comparison between the different parameter adaptation algorithms are also presented.

The reference model considered for simulation is characterized by the discrete transfer function

$$\frac{z^{-d}D(z^{-1})}{C_1(z^{-1})}=\frac{z^{-2}(0.28+0.22z^{-1})}{(1-0.5z^{-1})[1-(0.7+0.2j)z^{-1}]\times[1-(0.7-0.2j)z^{-1}]}. \tag{97}$$

The plant before a parameter change occurs is characterized by the discrete transfer function

$$\frac{z^{-d}B(z^{-1})}{A(z^{-1})}=\frac{z^{-2}(1+0.4z^{-1})}{(1-0.5z^{-1})[1-(0.8+0.3j)z^{-1}]\times[1-(0.8-0.3j)z^{-1}]}. \tag{98}$$

At time $k=t$, a change of the plant parameters is made. The plant is characterized by

$$\frac{z^{-d}B'(z^{-1})}{A'(z^{-1})}=\frac{z^{-2}(0.9+0.5z^{-1})}{(1-0.5z^{-1})[1-(0.9+0.5j)z^{-1}]\times[1-(0.9-0.5j)z^{-1}]}. \tag{99}$$

The notation is used in the figures is Y(k) is the plant output; YMR(k) is the reference model output.

In Figs 3–8, the solid lines represent the plant output when the plant parameters remain constant, which correspond also to the model reference output in tracking [equation (97)] and to the regulation model output of equation (5) in regulation, and the dashed lines represent the plant output when the plant parameters have been changed. The plant parameters are changed at $k=25$ in tracking [Figs 3(a), 4(a), 5–7(a), 8(a)] and at $k=0$, in regulation [Figs 3(b) and 4(b)]. In all the cases, the linear control is designed (Section 2) so that the error between the plant output [equation (98)] and the reference model output [equations (97)] vanishes with the dynamics of the C_2-polynomial. The control parameters resulting from this design are taken as the initial value of $\hat{p}(k)$ in the adaptive algorithm. Note that the change of plant parameters considered drastically alters the performance of the linear control (the closed loop is almost unstable).

The behaviour of the independent tracking and regulation algorithm (Section 5.1) is presented in Figs 3 and 4 for various choices of $C_2(q^{-1})$. In Figs 3(a) and (b), $C_2(q^{-1})=1$. In Figs 4(a) and (b), $C_2(q^{-1})=(1-0.4q^{-1})^3$. It can be seen that the C_2-polynomial avoids abrupt changes in the plant output. We use in all cases except for Figs 5 and 6 the constant trace gain matrix algorithm; ($t_rF_k=t_rF_0$; $F_0=\text{diag}\,[10]$, $\lambda_1(k)/\lambda_2(k)=1$).

Figure 5 illustrates the behaviour of a parallel MRAC using the filter $L(q^{-1})=C_1(q^{-1})$ (Section 5.4) instead of a compensator. In this case, we have used a constant matrix gain algorithm $F_k=\text{diag}\,[10]$. This curve can be compared with curve (b) in Fig. 6 which corresponds to the parallel MRAC using a linear compensator $H_1(q^{-1})=C_1(q^{-1})$ for which we have also used a 'constant matrix gain' algorithm. The control performances are slightly better in this last one.

The influence of the type of adaptation algorithm upon the performance of the MRAC is illustrated in Fig. 6 which shows the tracking regime for a 'series–parallel' MRAC (Section 5.2). Curve (a) corresponds to the 'constant trace' algorithm; curve (b) corresponds to the 'constant gain' algorithm ($\lambda_1=1;\lambda_2=0$); curve (c) corresponds to the 'decreasing gain' algorithm. ($\lambda_1=1;\lambda_2=1$). In all cases, $F_0=\text{diag}\,[10]$. The results are summarized in Table 1 where the criterion is defined as

$$J_N=\sum_{0}^{N}\varepsilon^2(k). \tag{100}$$

From Fig. 6 and the Table 1, one concludes that the 'constant trace' algorithm provides the best results.

The regulation behaviour discussed in Section 6.1 is illustrated in Figs. 7(a) and (b). The adaptive parameters reach almost their final

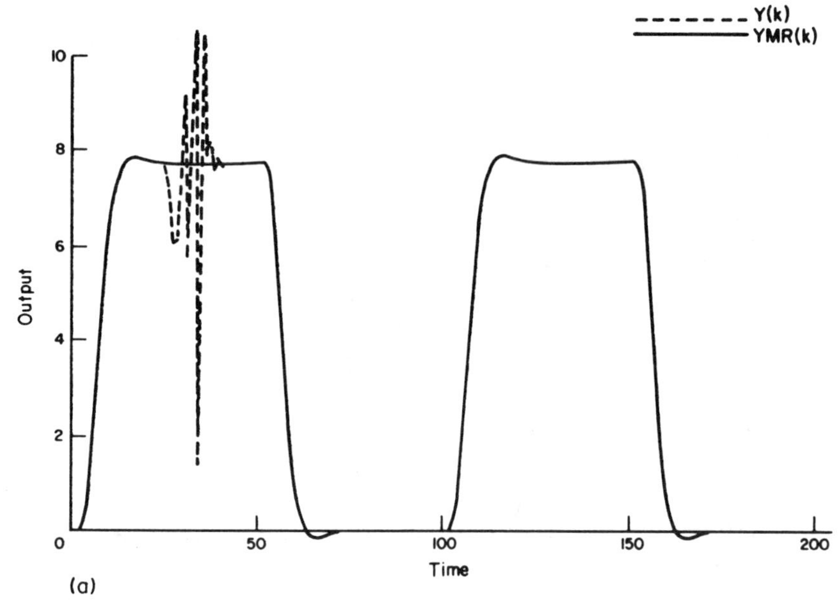

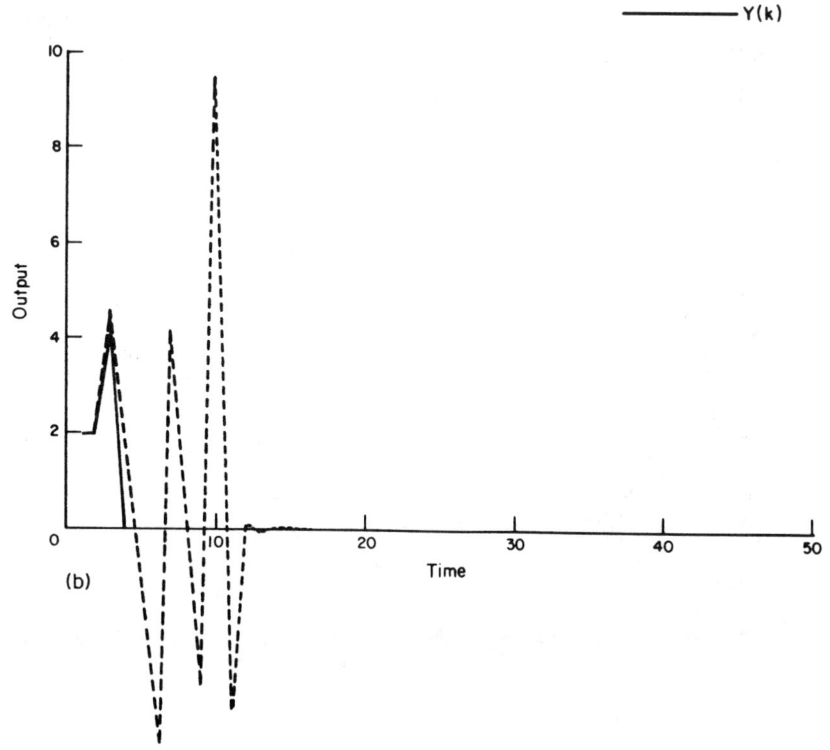

FIG. 3. Independent tracking and regulation control algorithm (Section 5.1) with $C_2(q^{-1}) = 1$. (a) Tracking; (b) regulation.

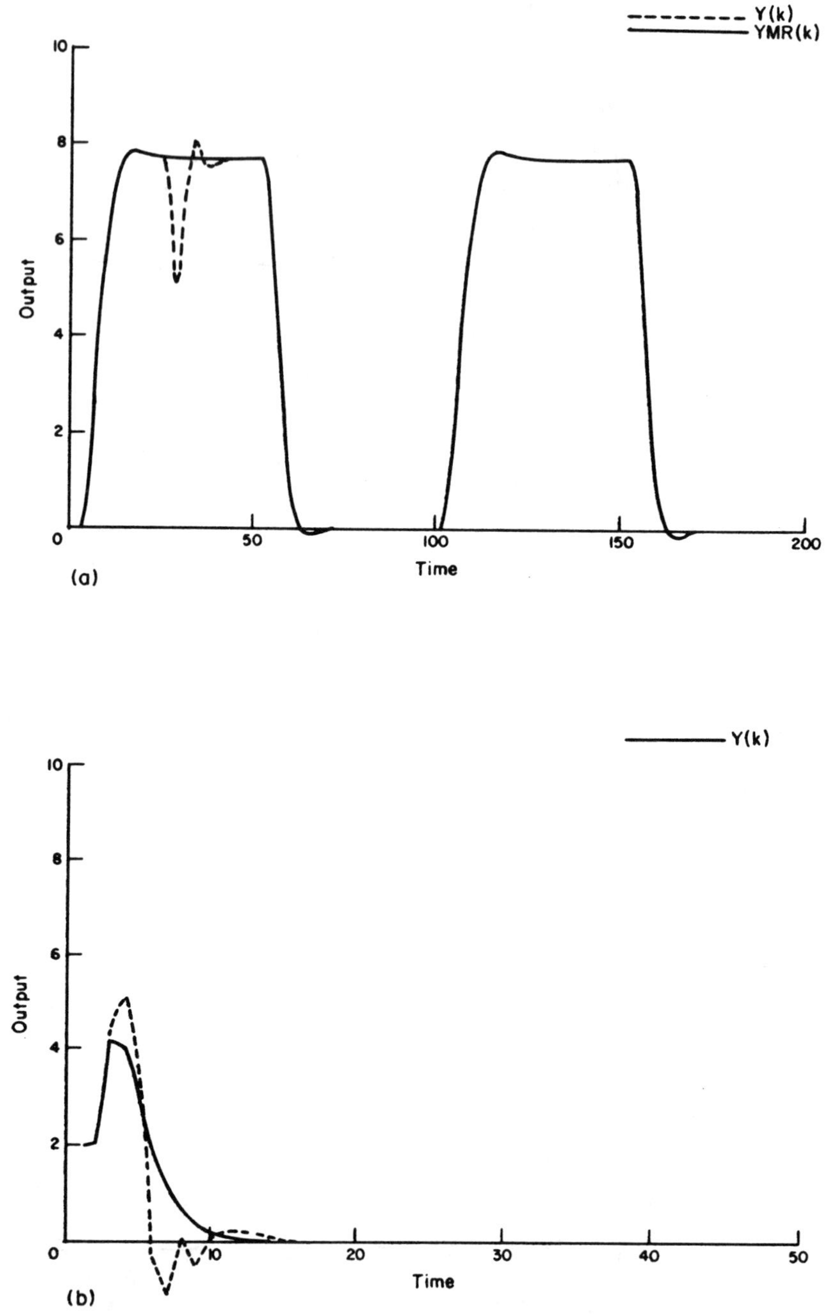

FIG. 4. Independent tracking and regulation control algorithm (Section 5.1) with $C_2(q^{-1}) = (1-0.4q^{-1})^3$. (a) Tracking; (b) regulation.

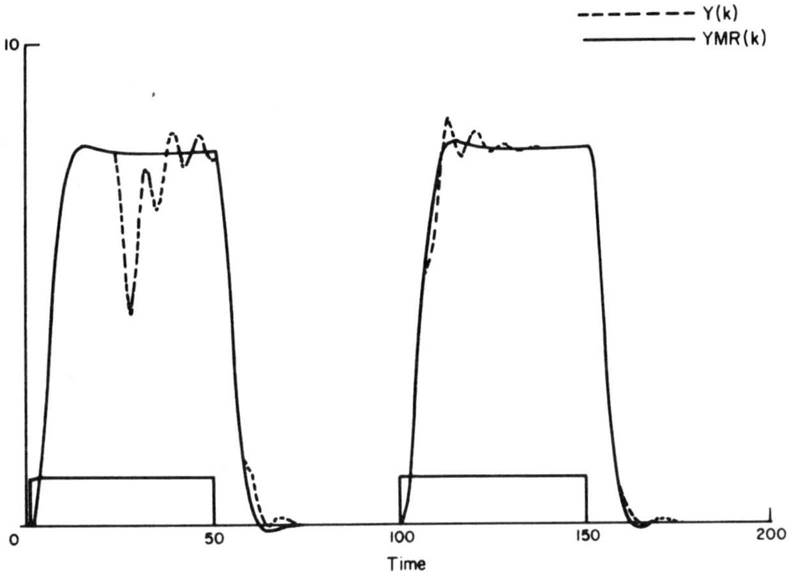

FIG. 5. Parallel MRAC algorithm with filter (Section 5.4) in tracking with $L(q^{-1}) = C_1(q^{-1})$.

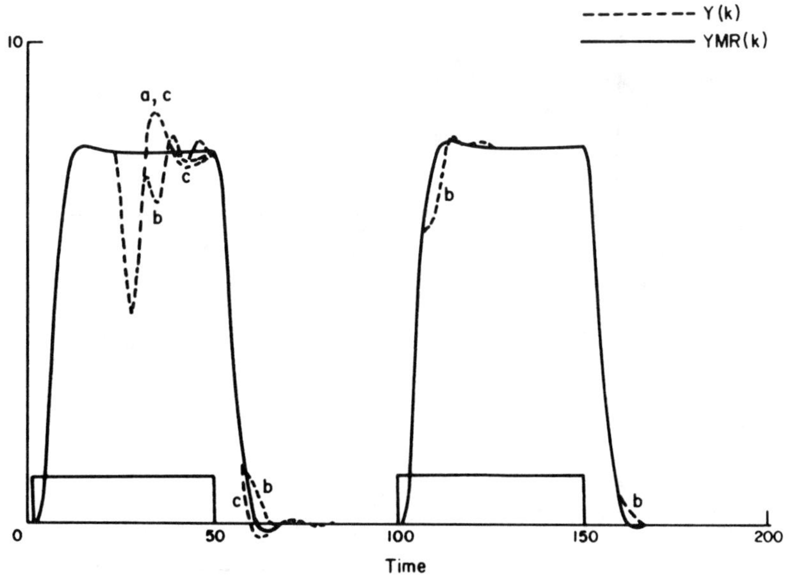

FIG. 6. Influence of the adaptation algorithm. (a) Constant trace algorithm; (b) constant gain algorithm ($F_0 = \text{diag}[10]$); (c) decreasing gain algorithm ($F_0 = \text{diag}[10]$, $\lambda_1 = \lambda_2 = 1$.

TABLE 1. INFLUENCE OF THE ADAPTATION ALGORITHM (F_0 =diag [10]).

Type of algorithm	J_{50}	J_{100}	J_{200}
$\lambda_1=1$ $\lambda_2=0$	42, 1	44, 5	50, 1
$\lambda_1=1$ $\lambda_2=0.1$	46, 3	47	47, 5
$\lambda_1=1$ $\lambda_2=1$	42, 7	44, 4	44, 9
$\lambda_1(k)=\lambda_2(k)$ $t_rF_k=t_rF_0$	42,3	42, 5	42, 5

values at step $k=50$ and after $k=200$, the control system operates in regulation [$y^M(k)=0$ for $k\geqq 200$].

An impulse perturbation acts on the plant at $k=300$ as in equation (87) with $w(300)=2$. Figure 7(a) shows the behaviour of the independent tracking and regulation MRAC (Section 5.1) with $C_2(q^{-1})=C_1(q^{-1})$ (which is also the series–parallel MRAC in Section 5.2) under such circumstances, and Fig. 7(b) shows some of the controller parameters that remain unchanged as expected.

In Fig. 8(a), we present the performances of the parallel MRAC with linear compensator (Section 5.3) to illustrate the discussion in Section 6.2. The linear compensator is in this case

$$H_1(z^{-1})=(1-0.74z^{-1})[1-(0.7+0.2j)z^{-1}]$$
$$\times[1-(0.7-0.2j)z^{-1}]. \qquad (101)$$

The evolution of some of the adaptive controller parameters is presented in Fig. 8(b). The adaptive parameter effectively change as proved in Section 6.2.

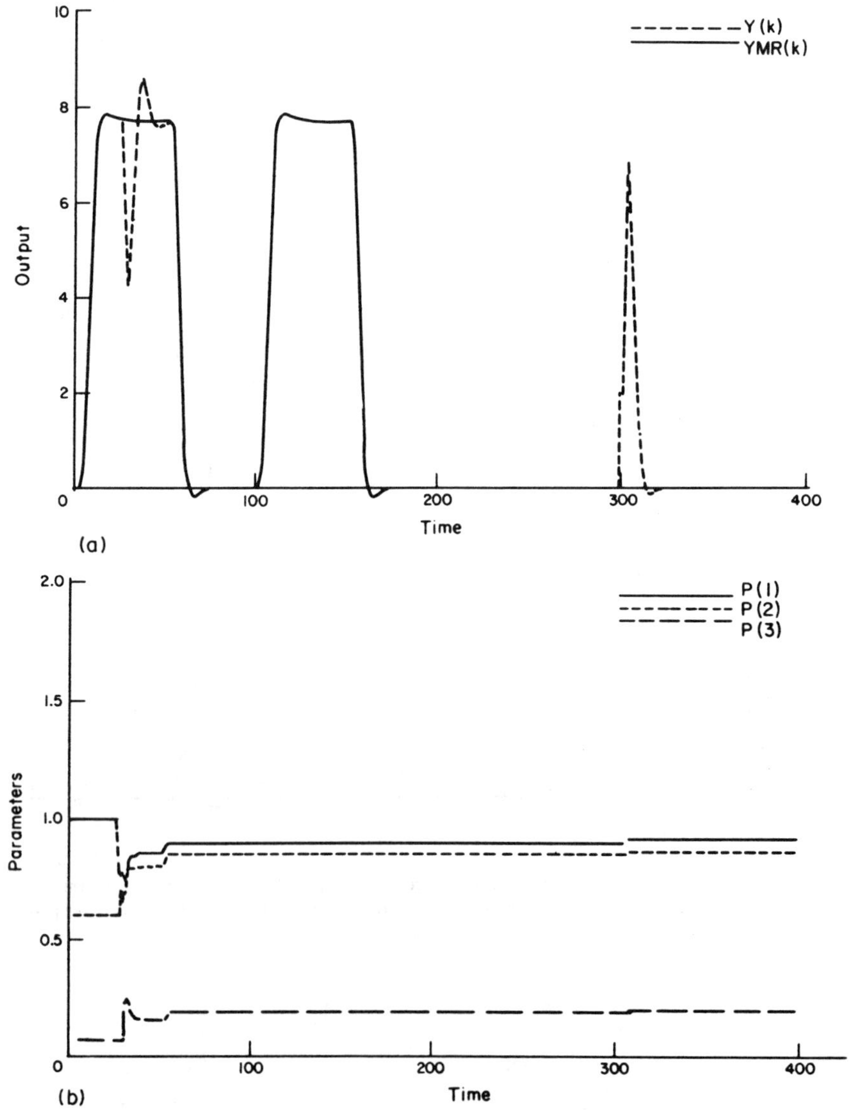

FIG. 7. The MRAC of Fig. 2 with $C_2(q^{-1})=C_1(q^{-1})$ when an impulse perturbation acts on the plant in regulation. (a) Plant and reference model outputs; (b) some of the controller parameters.

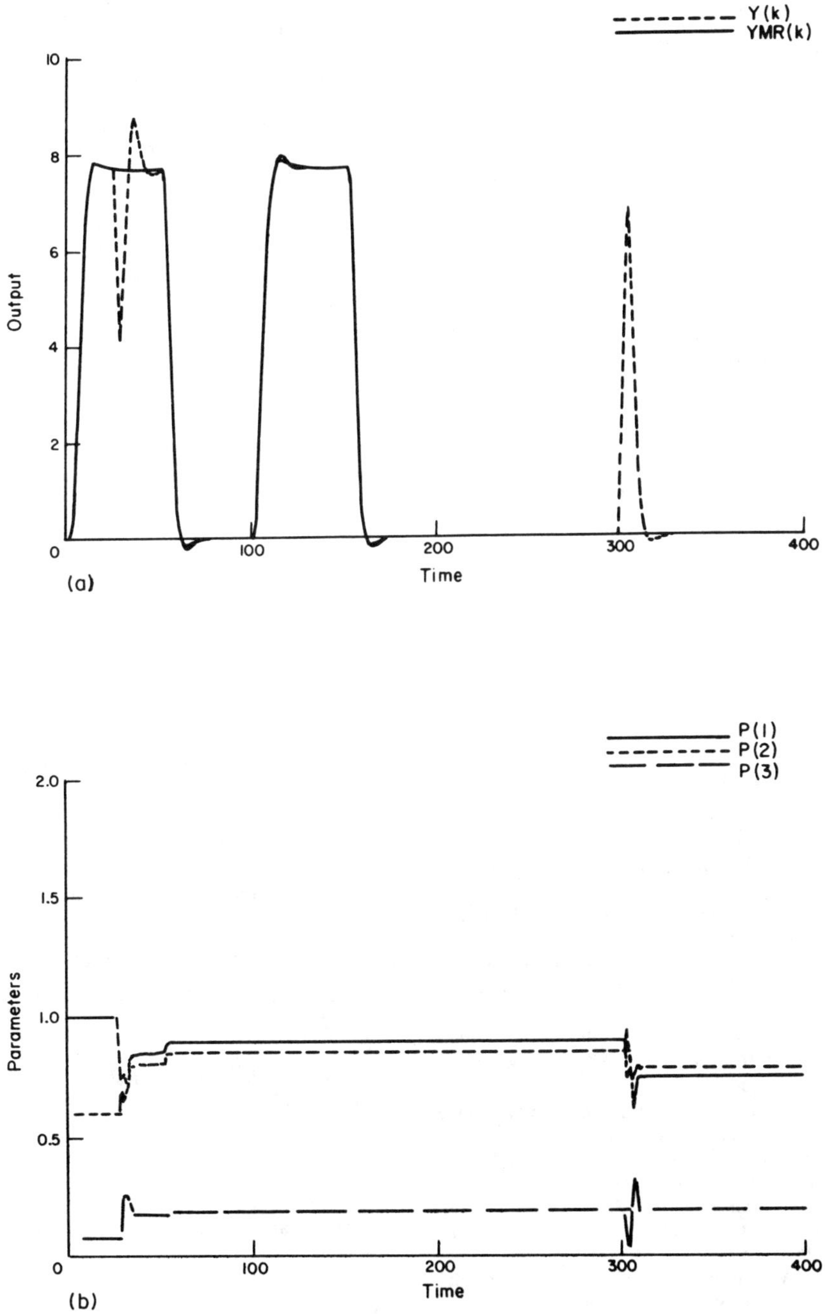

FIG. 8. Parallel MRAC with linear compensator $H_1(z^{-1})$ as in equation (101) when an impulse perturbation acts on the plant in regulation. (a) Plant and reference model outputs; (b) some of the controller parameters.

8. CONCLUSION

A general design for explicit MRAC using the stability point of view has been given. The corresponding scheme achieves both tracking and regulation objectives with a bounded control.

The tracking and regulation objectives can be specified independently, which is not the case for most of the previous schemes given in the literature. Various known MRAC can be obtained by the particularization of the general adaptive control algorithm in Section 4.2.

Simulation results have shown the importance of the C_2-polynomial, which specifies the regulation objectives, for smoothing the adaptation process. As shown by simulations, the performances of the MRAC depend on the type of adaptation algorithm. The best results are obtained when using a 'constant trace' algorithm followed by the 'time decreasing gain' and the 'constant gain' adaptation algorithms.

Similarities between MRAC and certain types of stochastic self-tuning regulators have been established. The analysis of these MRAC schemes in the presence of stochastic disturbance can be found in Dugard and Landau (1980).

REFERENCES

Åström, K. J., B. Westerberg and B. Wittenmark (1978). Self-tuning controllers based on pole-placement design. Dept. of Automatic Control, Lund Institute of Technology, Sweden (1978).

Åström, K. J. and B. Wittenmark (1978). On self-tuning regulators. *Automatica* **9**, 185.

Bénéjean, R. (1977). La commande adaptive à modèle de référence évolutif. Thèse de Docteur-Ingénieur, Grenoble, France.

Dugard, L. and I. D. Landau (1980). Stochastic model reference adaptive controllers. Presented at the 19th I.E.E.E. CDC, Albuquerque, U.S.A.

Egardt, B. (1980a). Unification of some discrete time adaptive control schemes. *IEEE Trans Aut. Control* **AC-25**, 693.

Egardt, B. (1980b). Stability analysis of discrete time adaptive control schemes. *IEEE Trans Aut. Control* **AC-25**, 710.

Goodwin, G. C., P. J. Ramadge and P. E. Caines (1980). Discrete time multivariable adaptive control. *IEEE Trans Aut. Control* **AC-25**, June 1980.

Ionescu, I. and R. V. Monopoli (1977). Discrete model reference adaptive control with an augmented error signal. *Automatica* **13**, 507.

Irving, E. (1978). Communication at the Summer School on Adaptive Control. Bréau sous Nappe, Juin 1978.

Irving, J. P. Barret, C. Charcossey and J. P. Monville (1979a). Improving power network stability and unit stress with adaptive generator control. *Automatica* **15**, 31.

Irving, E. (1979b). Private communication.

Landau, I. D. (1976). Unbiased recursive identification using model reference adaptive techniques. *IEEE Trans. Aut. Control* **AC-21**, 200.

Landau, I. D. (1979a). *Adaptive Control—The Model Reference Approach.* Dekker (1979).

Landau, I. D. (1979b). Adaptive controllers with explicit and implicit reference models and stochastic self-tuning regulators—equivalence and duality aspects. Proc. 17th IEEE Conf. on Decision and Control, pp. 1390–1395, San Diego.

Landau, I. D. (1980). An extension of a stability theorem applicable to adaptive control. *IEEE Trans. Aut. Control* **AC-25**.

Landau, I. D. and H. M. Silveira (1979). A stability theorem with application to adaptive control. *IEEE Trans Aut. Control* **AC-24.**

Landau, I. D. and B. Courtiol (1974). Design of multivariable adaptive model following control. *Automatica* **10**.

Landau, I. D. and R. Lozano L. (1979). On the design of the explicit model reference adaptive control for tracking and regulation. 18th IEEE CDC Conference, Fort Lauderdale.

Lozano L. R. and I. D. Landau (1980). Redesign of explicit and implicit discrete time model reference adaptive control schemes. *Int. J. Control* **33** (2).

Lozano L. R. (1981). Adaptive control with forgetting factor. *Proceedings IFAC/81*, Kyoto.

Narendra, K. S., Y. H. Lin and L. S. Valavani (1980). Stable adaptive controller design, Part II: Proof of stability. *IEEE Trans Aut. Control* **AC-25**, 440.

Narendra, K. S., L. S. Valavani (1978). Stable adaptive controller design-direct control. *IEEE Trans Aut. Control* **AC-23**, 570.

APPENDIX A: PROOF OF IDENTITY (10)

Let us rewrite equation (10) as a set of equations corresponding to the various powers of q^{-i}

$$\begin{aligned} 1 &= 1 \\ c_1^2 &= a_1 + s_1 \\ c_2^2 &= a_2 + a_1 s_1 + s_2 \\ &\vdots \\ c_{d-1}^2 &+ a_{d-1} + \cdots + s_{d-1} \\ c_d^2 &= a_d + \cdots + s_d + r_0. \\ &\vdots \end{aligned} \tag{A1}$$

The number of coefficients to be computed is $n_s + n_r + 1$. On the other hand, the number of equations in (A1) is $\max(n_a + n_s, n_r + d)$ and in order to have a unique solution, we must verify

$$n_s + n_r + 1 = \max(n_a + n_s, n_r + d) \geqq n_{C_2}. \tag{A2}$$

We can also see from equation (A1) that n_s must verify

$$n_s \geqq d - 1. \tag{A3}$$

Let us take the smallest number $n_s = d - 1$, then from equation (A2), we have

$$\begin{aligned} n_s &= d - 1 \\ n_r &= \max(n_a - 1, n_{C_2} - d). \end{aligned} \tag{A4}$$

Thus, taking into account equation (A4), (10) can be re-written

$$\begin{bmatrix} 1 & & & & & & & \\ a_1 & 1 & & & & & & \\ a_2 & a_1 & 1 & & \bigcirc & & & \\ \vdots & & & & & & & \\ a_{d-1} & a_{d-2} & \cdots & a_1 & 1 & & & \\ a_d & a_{d-1} & \cdots & & a_1 & 1 & & \\ a_{d+1} & \cdots & & & a_2 & 0 & 1 & \\ a_{d+2} & & & & a_3 & 0 & 0 & 1 \\ \vdots & & & & \vdots & \vdots & \vdots & & 1 \\ & & & & 0 & . & . & . & 0 & 1 \end{bmatrix} \begin{bmatrix} 1 \\ s_1 \\ s_2 \\ \vdots \\ s_{d-1} \\ r_0 \\ r_1 \\ r_2 \\ \vdots \\ r_{n_r} \end{bmatrix} = \begin{bmatrix} 1 \\ c_1^2 \\ c_2^2 \\ \vdots \\ c_{d-1}^2 \\ \vdots \\ \vdots \\ \vdots \\ \vdots \\ \vdots \end{bmatrix} \tag{A5}$$

and since the determinant of a triangular matrix is always different from zero, the solution exists and is unique.

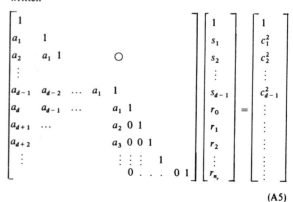

APPENDIX B: PROOF OF THE BOUNDEDNESS OF $\phi(k)$ AND CONVERGENCE TO ZERO OF $\varepsilon^f(k)$

For this proof, one uses Theorem 3.1 and the following lemmas

Lemma B.1. (Goodwin, Ramadge and Caines, 1980). The vector $\phi^f(k)$ given in equation (67) containing the filtered outputs and inputs of the plant in (1), which is minimum-phase, verifies

$$\|\phi^f(k-d)\| < C_1 + C_2 \max_{0 \leq l \leq k} |y^f(l)| \tag{A6}$$

$$0 \leq C_1 < \infty, \quad 0 < C_2 < \infty.$$

Lemma B.2. (Goodwin, Ramadge and Caines, 1980). If

$$\lim_{k\to\infty} \frac{\varepsilon_L^f(k)}{[1+\phi^{fT}(k-d)F_k\phi^f(k-d)]^{1/2}} = 0 \tag{A7}$$

where $\{\varepsilon^f(k)\}$ is a real scalar sequence, $\{\phi^f(k-d)\}$ is a real vector sequence and F_k is a sequence of positive definite real matrices, then subject to

$$(1) \qquad F_k^{-1} > \varepsilon F_0^{-1}, \varepsilon > 0, F_0 > 0 \quad \forall k > 1 \tag{A8}$$

$$(2) \qquad \|\phi^f(k-d)\| < C_3 + C_4 \max_{0 \leq l \leq k} |\varepsilon_L^f(k)| \tag{A9}$$

$$0 \leq C_3 < \infty; 0 < C_4 < \infty.$$

It follows that

$$\lim_{k\to\infty} \varepsilon^f(k) = 0$$

and

$$\|\phi^f(k-d)\| < C_5; 0 < C_5 < \infty \quad \forall k > 0. \tag{A10}$$

From Theorem 3.1, equation (32), one has

$$\lim_{k\to\infty} v(k) = 0 \tag{A11}$$

which implies that $\lim_{k\to\infty} \varepsilon_L^*(k) = 0$, given that $H_1(q^{-1})$ and $H_2(q^{-1})$ are asymptotically stable polynomials, see equation (73). Then, we can write

$$\lim_{k\to\infty} \frac{\varepsilon_L^*(k)}{[1+\phi^{fT}(k-d)F_k\phi^f(k-d)]^{1/2}} = 0 \tag{A12}$$

because

$$\frac{1}{[1+\phi^{fT}(k-d)F_k\phi(k-d)]^{1/2}} \leq 1. \tag{A13}$$

Consider now

$$\Omega(k) = \frac{\bar{\varepsilon}_L(k)}{[1+\phi^{fT}(k-d)F_k\phi^f_{k-d}]^{1/2}}$$

$$= \frac{L(q^{-1})\{[\hat{p}(k-d)-\hat{p}(k)]^T\phi^f(k-d)\}}{[1+\phi^{fT}(k-d)F_k\phi^f(k-d)]^{1/2}}. \tag{A14}$$

Using equation (75)

$$\Omega(k) = \frac{L(q^{-1})\left\{-\phi^{fT}(k-d)\sum_{i=0}^{d-1} F_{k-i}\phi^f(k-d-i)v(k-i)\right\}}{[1+\phi^{fT}(k-d)F_k\phi^f(k-d)]^{1/2}}$$

$$= -\sum_{j=0}^{n_L}\sum_{i=0}^{d-1} \times \frac{l_j\phi^{fT}(k-d-j)F_{k-i-j}\phi^f(k-d-i-j)v(k-i-j)}{[1+\phi^{fT}(k-d)F_k\phi^f(k-d)]^{1/2}} \tag{A15}$$

with $l_0 = 1$.

From Theorem 3.1 [with the additional requirement that $\lambda_1(k)$ and $\lambda_2(k)$ are such that $\lim F_k > \delta F_0$, $\delta > 0$] it results that $\hat{p}(k)$ is bounded, and from equations (60) and (1), one concludes that $u(k)$ and $y(k)$ cannot become unbounded for finite k [avoiding division by zero in equation (60) as indicated in Section 4]. Therefore, $\phi(k-d)$ can eventually become unbounded only asymptotically. Therefore the following relation is always true

$$\frac{\|\phi^f(k-d-j)\|}{[1+\phi^{fT}(k-d)F_k\phi^f(k-d)]^{1/2}}$$

$$\leq \frac{\|\phi^f(k-d-j)\|}{[1+\lambda \min(F_k)\|\phi^f(k-d)\|]^{1/2}} \leq C_6 < \infty \tag{A16}$$

and one concludes from equations (A15) and (40)

$$\lim_{k\to\infty} \Omega(k) = 0. \tag{A17}$$

From equations (A12), (A14), (A17) and (71), one obtains (A7) of the lemma (B.2), (A9) is obtained as follows from (8), we have

$$|\varepsilon(k)| \geq |y(k)| - |y^M(k)| \geq |y(k)| - C_7; 0 < C_7 < \infty \tag{A18}$$

since $y^M(k)$ is bounded.

From equation (46) and knowing that $C_2(q^{-1})$ is an asymptotically stable polynomial, we have

$$|\varepsilon(k)| \leq C_8 + C_9 \max_{0 \leq l \leq k} |\varepsilon^f(l)| \tag{A19}$$

$$0 \leq C_8 < \infty; 0 \leq C_9 < \infty.$$

From equation (60), and knowing that $L(q^{-1})$ is an asymptotically stable polynomial, we have

$$|y(k)| \geq C_{10} + C_{11} \max_{0 \leq l \leq k} |y^f(l)| \tag{A20}$$

$$0 \leq C_{10} < \infty; \; 0 \leq C_{11} < \infty.$$

Using equations (A18)–(A20) and (A6), we obtain (9). Thus from Lemma B.2, one concludes that $\lim \varepsilon_L^f(k) = 0$ and that $\phi^f(k-d)$ is bounded. And given that both the C_2-polynomial and the L-polynomial are asymptotically stable we can also say that $\lim_{k\to\infty} \varepsilon(k) = 0$ and that $\phi(k-d)$ is bounded.

APPENDIX C: ALTERNATIVE PROOF OF BOUNDEDNESS OF $\varepsilon^f(k)$

This analysis involves several steps. It will be shown first that $\phi^f(k)$ which contains the filtered plant input and output cannot be unbounded for finite k. Then, an equation for $\phi^f(k)$ will be derived and using also some of the results of Theorem 3.1 it will be shown by contradiction that $\phi^f(k)$ is bounded.

From equations (1) and (53), and neglecting the initial conditions in $L(q^{-1})$ which is an asymptotically stable polynomial, one can write

$$B(q^{-1})u^f(k-d) = A(q^{-1})y^f(k). \tag{A21}$$

Since the plant to be controlled is minimum-phase and the L-polynomial is asymptotically stable, if $y^f(k)$ is bounded this will imply that $u^f(k)$ is bounded. But equation (A21) allows to conclude also that even when $A(q^{-1})$ is unstable a bounded $y^f(k)$ can be obtained with a bounded control.

From the expression of $u^f(k)$ given by equation (64) it results that if $\hat{p}_0(k)$ and $\phi_0^f(k)$ are bounded $\{\phi^{fT}(k) = [u^f(k); \phi_0^{fT}(k)]$; $\hat{p}(k) = [\hat{b}_0(k), \hat{p}_0^T(k)]\}u^f(k)$ will be bounded if

$\hat{b}_0(k)\neq 0$ for all k. It also results from (1) and (64) that if $\hat{p}(k)$ is bounded for all k, $u^f(k)$ and $y^f(k)$ respectively cannot become unbounded for finite k and therefore $\phi(k)$ and $\phi_0(k)$ can become eventually unbounded but only asymptotically.

Therefore a necessary condition allowing to assure that $u^f(k)$ and $y^f(k)$ will be bounded is to avoid that $\hat{b}_0(k)=0$ (knowing that b_0 cannot be zero), by the technique discussed in Section 4.

To prove that $u^f(k)$ and $y^f(k)$ remains bounded, we will prove that $\phi^f(k)$ remains bounded and in order to do this, we will establish next a dynamic equation for $\phi^f(k)$. Note first from equations (14) and (60) that $\phi^f(k)$ can be expressed as

$$\phi^f(k)=\begin{bmatrix} 1 & 0 \\ q^{-1} & \vdots \\ q^{-m-d+1} & 0 \\ 0 & 1 \\ \vdots & q^{-1} \\ 0 & q^{-n+1} \end{bmatrix}\begin{bmatrix} u^f(k) \\ y^f(k) \end{bmatrix} \quad \text{(A22)}$$

on the other hand, $u^f(k)$ can be expressed in terms of $y^f(k)$ using equation (A21) and $y^f(k)$ can be expressed in terms of the parameter error vector as shown below.

From equations (69) and (60), one obtains

$$C_2(q^{-1})L(q^{-1})y^f(k)=C_2(q^{-1})L(q^{-1})y^{Mf}(k)+L(q^{-1})\{[\hat{p}(k-d)-p]^{\mathrm{T}}\phi^f(k-d)\} \quad \text{(A23)}$$

neglecting the initial conditions (A23) can also be written as

$$C_2(q^{-1})y^f(k)=\rho(k-d)+r^{\mathrm{T}}_{k-d+1}\phi^f(k-d) \quad \text{(A24)}$$

where

$$\rho(k-d)=C_2(q^{-1})y^M(k) \quad \text{(A25)}$$

$$r_{k-d+1}=\hat{p}(k-d)-p. \quad \text{(A26)}$$

Equations (A21), (A22) and (A24) allows to define a linear dynamic system (single input—multi outputs) with the input $\rho(k-d)+r^{\mathrm{T}}_{k-d+1}\phi^f(k-d)$ and output $\phi^f(k)$ which is characterized by a transfer vector $W(z)$ which is asymptotically stable

$$\phi^f(k)=W(q^{-1})[\rho(k-d)+r^{\mathrm{T}}_{k-d+1}\phi^f(k-d)] \quad \text{(A27)}$$

$$W^{\mathrm{T}}(q^{-1})=\left[\frac{A(q^{-1})q^d}{C_2(q^{-1})B(q^{-1})},\frac{A(q^{-1})q^{d-1}}{C_2(q^{-1})B(q^{-1})}\times\cdots\frac{A(q^{-1})q^{1-m}}{C_2(q^{-1})B(q^{-1})},\frac{1}{C_2(q^{-1})},\frac{q^{-1}}{C_2(q^{-1})}\cdots\frac{q^{-n+1}}{C_2(q^{-1})}\right]. \quad \text{(A28)}$$

Note that (A27) has the structure of a feedback system.

For the systems of the form of (A27) one has the following result (Narendra, Lin and Valavani, 1980)

Lemma C.1. Consider the system

$$\phi^f(k)=W(q^{-1})[\alpha(k-d)+\beta(k+1-d)^{\mathrm{T}}\phi^f(k-d)] \quad \text{(A29)}$$

where $W(z)$ in an asymptotically stable transfer vector and that $\|\beta_k\|<C_\beta<\infty$ and $\lim_{k\to\infty}\beta_k=0$ (or $\beta_k\in L^2$). Then

(1) $$\|\phi^f(k)\|<C<\infty \quad \forall k \text{ if } |\alpha(k)|<C_\alpha<\infty \quad \forall k \quad \text{(A30)}$$

(2) $$\lim_{k\to\infty}\phi^f(k)=0 \text{ if } \lim_{k\to\infty}\alpha(k)=0. \quad \text{(A31)}$$

One can now proceed to the proof of boundedness of $\phi^f(k)$ which follows the ideas of Narendra, Lin and Valavani (1980) and Narendra and Lin (1979)

C.1. *The case* $d=1$.

As shown previously $\phi^f(k)$ can be either bounded or asymptotically unbounded. We will show next that only the first possibility holds.

From equations (25) and (28) [with $\phi^f(k)$ instead of $\phi(k)$] and (A26) one has

$$r^{\mathrm{T}}_k\phi^f(k-1)=\frac{1}{H(q^{-1})}v_k-[F_k\phi^f(k-1)v_k]^{\mathrm{T}}\phi^f(k-1) \quad \text{(A32)}$$

and from equations (34) and (A32) one has

$$\lim_{k\to\infty}\phi^{f\mathrm{T}}(k-1)[r_k+F_k\phi^f(k-1)v_k]=\lim_{k\to\infty}\frac{1}{H(q^{-1})}v_k=0. \quad \text{(A33)}$$

Taking F_k such that it satisfies condition (39) one has from (40) and (41)

$$\lim_{k\to\infty}F_k\phi^f(k-1)v_k=\lim_{k\to\infty}\Delta\hat{p}(k)=0$$

$$\|\Delta\hat{p}(k)\|<M<\infty \quad \forall k. \quad \text{(A34)}$$

Using equation (A32) (A27) becomes

$$\phi^f(k)=W(q^{-1})\left[\rho(k-1)+\frac{1}{H(q^{-1})}v_k-\Delta\hat{p}(k)^{\mathrm{T}}\phi^f(k-1)\right]. \quad \text{(A35)}$$

Taking in account equations (33) and (34) and the fact that $\rho(k-1)$ is bounded one concludes applying Lemma C.1. to (35) that $\phi^f(k)$ will be bounded.

From the boundedness of $\phi^f(k)$, equations (40) and (70) one concludes also that the auxiliary error $\bar{\varepsilon}_L(k)$ goes to 0.

C.2. *The case* $d>1$

Equation (A33) becomes in this case

$$\lim_{k\to\infty}\phi^{f\mathrm{T}}(k-d)[r_k+F_k\phi^f(k-d)v_{(k)}]=\lim_{k\to\infty}\phi^{f\mathrm{T}}(k-d)\left[r_{k+1-d}+\sum_{i=0}^{d-1}F_{k-i}\phi^f(k-d-i)v_{(k-i)}\right]=0 \quad \text{(A36)}$$

and the proof is similar to the case $d=1$ by taking in account, that equation (A34) implies

$$\sum_{i=0}^{d-1}F_{k-i}\phi^f(k-d-i)v(k-i)=0. \quad \text{(A37)}$$

Note that this proof does not need the additional assumption $\lim_{k\to\infty}F_k>\delta F_0,\delta>0$, considered in the proof given in Appendix B.

Stable Model Reference Adaptive Control in the Presence of Bounded Disturbances

GERHARD KREISSELMEIER AND KUMPATI S. NARENDRA, FELLOW, IEEE

Abstract—The adaptive control of a linear time-invariant plant in the presence of bounded disturbances is considered. In addition to the usual assumptions made regarding the plant transfer function, it is also assumed that the high-frequency gain k_p of the plant and an upper bound on the magnitude of the controller parameters are known. Under these conditions the adaptive controller suggested assures the boundedness of all signals in the overall system.

I. Introduction

A MAJOR step in the development of adaptive systems theory was the establishment in recent years of the global stability of several equivalent adaptive schemes [1]–[3]. When the sign of the high-frequency gain k_p, the relative degree n^*, and the order n of the transfer function of a plant with zeros in the left-half plane are given, it was shown that these schemes can be used to adaptively control the given plant in a globally stable fashion. The latter implies that the parameters and signals of the plant and the controller are bounded, while the error between the plant output and the output of a reference model tends to zero. The question naturally arises as to how well these schemes perform in the presence of external disturbances. The adverse effects of measurement noise in model reference adaptive systems were first discussed in [4], [5] wherein practical methods were suggested to ensure system stability. In a recent report [6] it was shown that when bounded output disturbances are present with no modification of the adaptive laws, the parameter error vector $\phi(t)$ can grow without bound even though the state error vector between plant and model is bounded. This can be attributed to the fact that the overall nonlinear adaptive system is only uniformly stable, but not uniformly asymptotically stable. Hence it is evident that modifications in the basic adaptive schemes are necessary when bounded external disturbances are present.

The results presented in [6] reveal that when exact model matching is not possible due to the problem structure (e.g., as here, in the presence of an external disturbance) our notions regarding acceptable behavior of the overall system need to be revised. Evidently a relaxation in performance objective will afford greater freedom in the design of adaptive algorithms. The viability of a specific objective is generally contingent upon the availability of certain prior information, which, in turn, dictates the choice of the algorithm. From a practical standpoint, the minimal requirement of any such algorithm is that it assure the boundedness of all the signals in the adaptive loop. In general, the demonstration of this property is difficult and constitutes a principal theoretical problem. Once the theoretical conditions for boundedness of signals are established, the parameters of the overall system may be adjusted to optimize a specified performance criterion.

The adaptive algorithm suggested in [6] resulted from the observation that the adjustment of the parameter error vector [given by $\dot{\phi}(t)$] is known to be in the "right" direction only when the output error $e(t)$ is large. Hence, it was argued, if a bound on the disturbance is known and ϕ is adjusted only when the output error exceeds a computed threshold, the signals and the parameters of the system would be bounded. Such an adjustment corresponds to an adaptive law with a dead zone and in [6] it is shown that the foregoing reasoning is indeed correct. The presence of the dead zone, however, results in finite parameter and output errors even when no external disturbance is present.

In this paper an alternate approach is taken which retains the potential of obtaining zero output error $e(t)$ (and zero parameter error $\phi(t)$ if the input is rich) in the limit when no external disturbance is present. The structure of the controller used is identical to that used in [1] for the disturbance-free case. This assures the existence of a constant control parameter vector θ^* for which the transfer function of the plant, together with the controller, matches that of the model exactly. As part of the prior information it is assumed that θ^* has a norm less than a specified value $\|\theta^*\|_{\max}$ and that the disturbance is bounded, although no explicit value of a bound need be given. Since $\|\theta\|^*_{\max}$ is known, it is necessary to search, using the adaptive law for θ^*, only within a sphere S of radius $\|\theta^*\|_{\max}$ in parameter space; this, in turn, assures the boundedness of the parameter error vector $\phi(t)$. The adaptive law used to update $\theta(t)$ is identical to that in the disturbance-free case when θ lies in the interior of S and is modified when it reaches the boundary of S or lies outside it. It is demonstrated in this paper that such a scheme results in the boundedness of all signals in the adaptive system. It is worth noting that even as $t \to \infty$, the parameter vector does not converge to any

Manuscript received April 8, 1981; revised February 1, 1982. Paper recommended by B. W. Dickinson, Past Chairman of the Identification Committee. This work was supported in part by the Office of Naval Research under Contract N00014-76-C-0017.

G. Kreisselmeier was on leave at the Department of Engineering and Applied Science, Yale University, New Haven, CT 06520. He is with DFVLR-Institute für Dynamik der Flugsysteme, D-8031 Oberpfaffenhofen, West Germany.

K. S. Narendra is with the Center for Systems Science, Yale University, New Haven, CT 06520.

Reprinted from *IEEE Trans. Automat. Contr.*, vol. AC-27, pp. 1169–1175, Dec. 1982.

constant value as in [6]. However, if the external disturbance is removed $\lim_{t\to\infty}e(t)=0$ and $\theta(t)$ converges to θ^* (if the input is sufficiently rich).

Surprisingly enough, the difficulty in the proof of stability does not arise when the state vector grows rapidly (e.g., exponentially), but rather when it grows slowly with time. In addition, the state of an unstable system need not grow monotonically, so that the usual limiting arguments cannot be used directly. However, if $\xi(t)$ (where ξ is the state of the overall system) is assumed to grow without bound, it can be shown that $\xi(t)$ is large over an arbitrarily large interval of time $[t_1, t_2]$. In such a case the effect of a bounded disturbance is relatively small and hence arguments similar to those in the disturbance-free case apply. This, in turn, leads to a proof by contradiction of the boundedness of $\xi(t)$.

II. Structure of the Adaptive Control System

Let the plant to be controlled be represented by the equations

$$\dot{x}_p = A_p x_p + b_p u_p + d_p \nu_1$$

$$y_p = c_p^T x_p + \nu_2 \qquad (1)^1$$

where x_p, u_p, and y_p are the state, input, and output, respectively, and ν_1 and ν_2 are plant and output disturbances. The transfer function of the plant is given by

$$c_p^T(sI-A_p)^{-1}b_p = k_p\frac{N_p(s)}{D_p(s)} \triangleq W_p(s).$$

The following assumptions are made regarding $W_p(s)$ and the disturbances ν_1 and ν_2:

i) $N_p(s)$, $D_p(s)$ are monic polynomials of degrees m and n, respectively, and $\dim(x_p)=n$.

ii) m and n are known, but the coefficients of $N_p(s)$ and $D_p(s)$ are unknown.

iii) $N_p(s)$ is a strictly stable polynomial.

iv) k_p is known.

v) ν_1, ν_2 are piecewise continuous and uniformly bounded time functions defined for all $t \in R^+$.

A reference model is defined by the equations

$$\dot{x}_m = A_m x_m + b_m r$$

$$y_m = c_m^T x_m \qquad (2)$$

and its transfer function is defined as

$$c_m^T(sI-A_m)^{-1}b_m = k_m\frac{1}{D_m(s)} \triangleq \frac{k_m}{k_p}W_m(s).$$

It is assumed that

i) $D_m(s)$ is a monic, strictly stable polynomial;

ii) the degree of $D_m(s)$ is $n^* \triangleq n-m$, and $\dim(x_m)=n^*$; and

iii) r is a piecewise continuous, uniformly bounded reference signal.

The objective is to control the plant in such a fashion that the output error between the plant and the model, i.e., $e_1 \triangleq y_p - y_m$, as well as all the state variables, remain uniformly bounded. This objective applies both to the case of model following ($r \not\equiv 0$) as well as state regulation ($r\equiv 0$, $y_m \equiv 0$).

To meet the control objective, a controller described by the equations

$$\dot{v}_1 = Fv_1 + gu_p \qquad (3a)$$

$$\dot{v}_2 = Fv_2 + gy_p \qquad (3b)$$

$$u_p = v^T\theta + \frac{k_m}{k_p}r \qquad (4)$$

is set up, where $v^T \triangleq [v_1^T, v_2^T]$ and

i) $\dim(v_1)=\dim(v_2)=n$;

ii) F is an arbitrary, strictly stable matrix;

iii) (F, g) is a completely controllable pair;

and θ is a vector of controller parameters to be adapted.

The equations for the adaptive scheme are

$$\dot{\zeta}_1 = A_m\zeta_1 + b_m u_p \qquad (5a)$$

$$\dot{\zeta}_2 = A_m\zeta_2 + b_m y_p \qquad (5b)$$

$$\dot{\omega}_1 = F\omega_1 + gc_m^T\zeta_1 \qquad (6a)$$

$$\dot{\omega}_2 = F\omega_2 + gc_m^T\zeta_2 \qquad (6b)$$

$$\dot{\theta} = -\Gamma\frac{\omega\left[\omega^T\theta - c_m^T\zeta_1 + y_p\right]}{1+x^Tx} - \Gamma\theta f(\theta) \qquad (7)$$

$$f(\theta) = \begin{cases} (1-\|\theta\|/\|\theta^*\|_{\max})^2 & \text{if } \|\theta\| \geq \|\theta^*\|_{\max} \\ 0 & \text{elsewhere} \end{cases} \qquad (8)$$

where

$$x^T \triangleq [v^T, \zeta^T, \omega^T], \quad \zeta^T \triangleq [\zeta_1^T, \zeta_2^T], \quad \omega^T \triangleq [\omega_1^T, \omega_2^T]$$

and

$$S: \{\theta \| \|\theta\| \leq \|\theta^*\|_{\max}\}. \qquad (9)$$

i) $\Gamma = \Gamma^T > 0$ is an arbitrary gain matrix;

ii) $\|\theta^*\|_{\max}$ is a known upper bound on the norm of the (unknown) matching controller parameter vector θ^*.

Equations (3a), (3b), and (4) describe the plant feedback loop where $v^T\theta$ is the feedback signal. The structure of this loop is identical to that described in [1] for the disturbance-free case. Hence, it follows from [1] that a parameter vector θ^* exists such that when $\theta(t)\equiv\theta^*$ the transfer function of the plant, together with the controller, matches that of the model exactly. Equations (5a), (5b), (6a), and (6b) describe the auxiliary network needed to generate the augmented error used in the adaptive law and are essential for proving global stability in the disturbance-free case [7]. Equation (7) is the adaptive law for updating the parameter vector $\theta(t)$ and the right-hand side has two terms. The first one contains the augmented

[1]Throughout this paper, while representing a function of time, the argument "t" is omitted for the sake of simplicity of notation when no confusion can arise.

error $(\omega^T\theta - c_m^T\zeta_1 + y_p)$; the second nonlinear term, $-\Gamma\theta f(\theta)$, which is not needed in the disturbance free-case, is included to ensure that the variations of the parameter vector are confined primarily to the region S when a bounded disturbance is present.

The block diagram of the control loop, together with the disturbances ν_1 and ν_2, is shown in Fig. 1(a). If $\theta = \theta^* + \phi$, Fig. 1(a) can be transformed as shown in Fig. 1(b)–(d) to yield an error model with a bounded disturbance ν at the output. From Fig. 1(e) and (4)

$$y_p(t) - \nu(t) = W_m(s)\left[r'(t) - \phi^T(t)v(t)\right]^2$$
$$= W_m(s)\left[r'(t) + \theta^T(t)v(t) - \theta^{*T}v(t)\right]$$
$$= W_m(s)\left[u_p(t) - \theta^{*T}v(t)\right].$$

Since, by (5a),

$$c_m^T\zeta_1(t) = W_m(s)u_p(t),$$

we have

$$y_p(t) - \nu(t) = c_m^T\zeta_1(t) - W_m(s)\theta^{*T}v(t)$$

or

$$\theta^{*T}\omega(t) - c_m^T\zeta_1(t) + y_p(t) - \nu(t) \equiv 0 \tag{10}$$

where

$$W_m(s)\nu(t) = \omega(t).$$

Fig. 1(f)–(h) shows the above transformations in block diagram form.

Defining $\xi^T \triangleq [x_p^T, x^T]$ and using (10) in (7) the overall system may be written in the form

$$\dot{\xi} = A\xi + b\phi^T(t)v(t) + \mu(t) \tag{11}$$

$$\dot{\phi} = -\Gamma\frac{\omega(t)\left[\omega^T(t)\phi + \nu(t)\right]}{1 + x^T(t)x(t)} - \Gamma\theta f(\theta) \tag{12}$$

where A is a stable matrix, and $\mu(t)$ is a uniformly bounded input vector due to ν_1, ν_2, and r. The objective then, as stated earlier, is to show that the overall system (11), (12) is globally stable with the modified adaptive law (12).

III. Preliminary Analysis

The proof of global stability of the system described by (11) and (12) is given in Section IV. In this section we carry out a preliminary analysis to outline the method of proof used as well as to indicate the source of the difficulty when an external disturbance is present.

1) Adaptive Law: As in the disturbance-free case, we first show, using a quadratic function $V(\phi)$ and its time derivative $\dot{V}$ along a trajectory, that ϕ is bounded. In that case, this implies that $\phi^T\omega/(1 + x^Tx)^{1/2}$ belongs to the

[2] In the paper both differential equations and transfer functions are used in the arguments and, depending on the context, "s" is used as a differential operator or the Laplace transform variable.

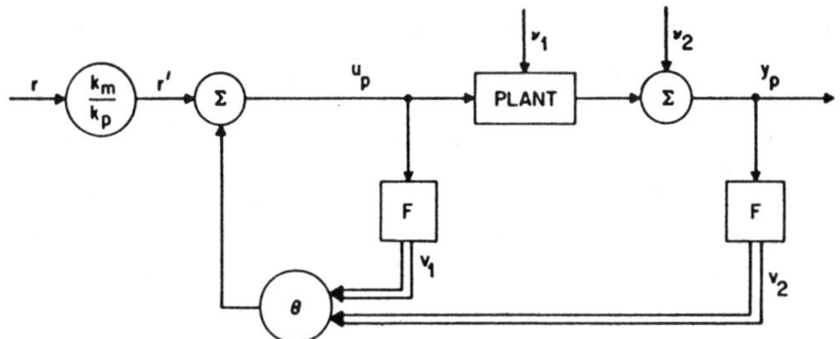

(a)

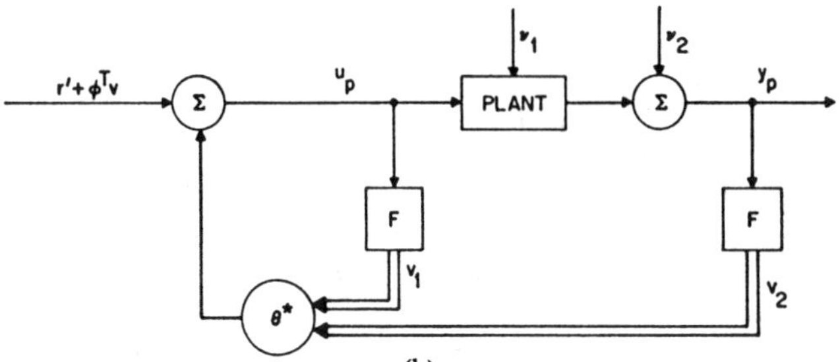

(b)

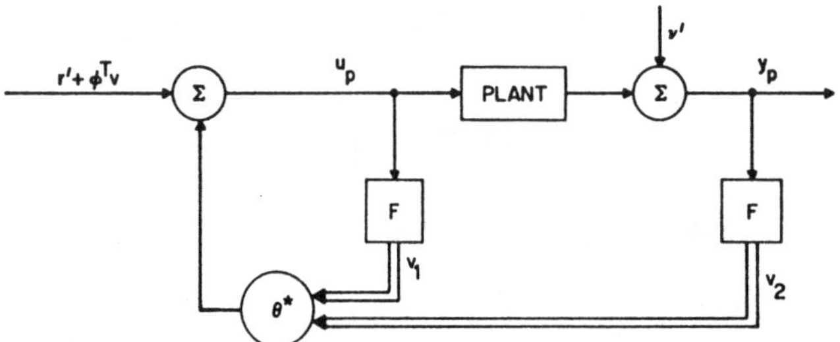

(c)

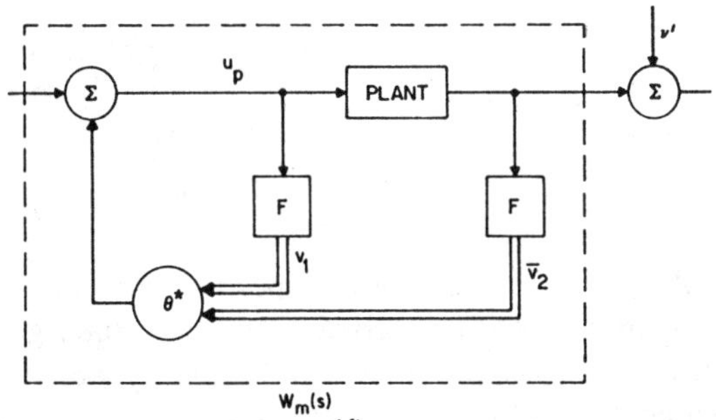

(d)

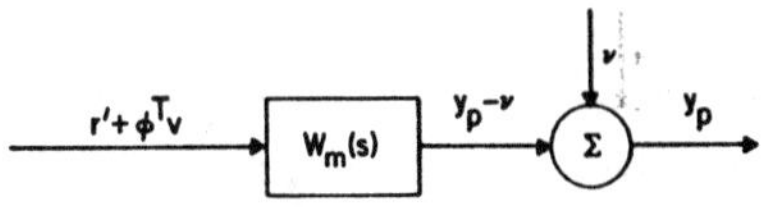

(e)

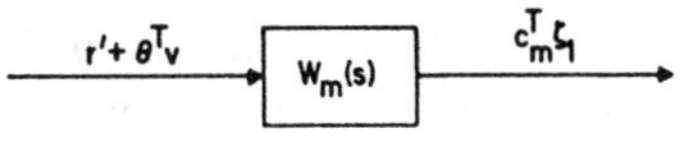

(f)

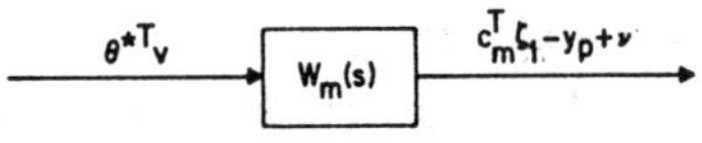

(g)

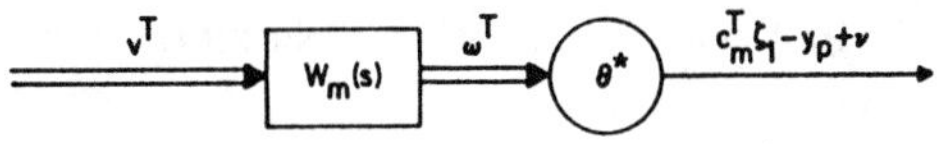

(h)

Fig. 1. Reformulation of the adaptation problem.

space $\mathcal{L}^2$ of square integrable functions or, equivalently, that $\dot{\phi} \in \mathcal{L}^2$, which is central to the proof of global stability [1]. The analysis in the present case, as shown below, becomes considerably more involved due to the presence of a disturbance, since it can no longer be deduced directly that $\phi^T\omega/(1+x^Tx)^{1/2} \in \mathcal{L}^2$ (which, in turn, implies that $\phi^T\omega/(1+x^Tx)^{1/2} \to 0$ [7]).

Let $V \triangleq \frac{1}{2}\phi^T\Gamma^{-1}\phi$ where Γ is a positive definite matrix. Evaluating its time derivative along the trajectory of (12) gives

$$\dot{V} = \frac{-\phi^T\omega[\phi^T\omega + \nu]}{1+x^Tx} - \phi^T\theta f(\theta). \tag{13}$$

From

$$\phi^T\theta = (\theta-\theta^*)^T\theta \geqslant \theta^T\theta - \|\theta^*\|\,\|\theta\| \tag{14}$$

and the definition of $f(\theta)$ it follows that $\phi^T\theta f(\theta) \geqslant 0$. Moreover, $\phi^T\theta f(\theta) \sim \|\phi\|^4$ for $\|\phi\| \gg \|\theta\|_{\max}$. Since $\omega/(1+x^Tx)^{1/2}$ is uniformly bounded, this implies that $\dot{V}$ is negative for all $V \geqslant V'$ and some $V' < \infty$. Therefore, V is uniformly bounded and hence $\phi(t)$ is also uniformly bounded. Together with (11) this implies that $\xi(t)$ cannot grow faster than an exponential.

Taking the integral of (13) we obtain

$$\begin{aligned}\int_{t_1}^{t_2}\left\{\frac{(\phi^T\omega)^2}{1+x^Tx} + \phi^T\theta f(\theta)\right\}d\tau &\leqslant V(t_1) - V(t_2) + \int_{t_1}^{t_2}\frac{|\phi^T\omega\nu|}{1+x^Tx}dt\\ &\leqslant \Delta_1 + \Delta_2\cdot\int_{t_1}^{t_2}\frac{1}{(1+x^Tx)^{1/2}}dt \end{aligned} \tag{15}$$

where Δ_1 and Δ_2 are appropriate positive constants.

In the absence of an external disturbance the right-hand side of (15) reduces to a constant. Setting $t_2 = \infty$ it can be concluded that $\phi^T\omega/(1+x^Tx)^{1/2} \to 0$ as $t \to \infty$ which, in turn, gives $\phi^Tv/(1+x^Tx)^{1/2} \to 0$ and $e \to 0$ as $t \to \infty$. The important point to note is the fact that the signal $\phi^Tv/(1+x^Tx)^{1/2}$ tends to zero can be concluded in a relatively straightforward fashion from (15) [7].

In the present situation, with disturbance present, such a conclusion cannot be drawn directly due to the additional integral term on the right-hand side. Even if $x(t) \to \infty$, this term may tend to ∞ as $t_2 \to \infty$. Hence, the conclusions drawn from (15) have to be modified suitably. The presence of the disturbance ν also results in nonzero values of all the signals in the system, hence limiting arguments (when $t_2 \to \infty$) of the type used in the disturbance-free case can no longer be used. In the proof of stability in Section IV the arguments are confined to a finite time interval and it is shown that $\phi^T\omega/(1+x^Tx)^{1/2}$ is small over most of this interval.

2) Control Loop: Since the parameter error vector $\phi(t)$ is bounded, it follows that the state of the system can grow at most exponentially. On the other hand, if the signal $\phi^Tv/(1+x^Tx)^{1/2}$ is made arbitrarily small, the state of the plant, together with the controller, will decay exponentially in the absence of the reference input and the bounded disturbances. In this section it is shown that a similar conclusion can be drawn even in the presence of bounded disturbances, provided the norm of the state of the system is sufficiently large. Assuming that over a finite interval $[t_1, t_2]$ the signal $\phi^Tv/(1+x^Tx)^{1/2}$ can take both small and large values, the state of the system can both grow and decay over that interval. The value of the state at time t_2 will then depend on the relative magnitudes of the intervals over which the two types of behavior occur. In Section IV it is shown that the period over which the state decays in the interval $[t_1, t_2]$ is substantially greater than the period over which it can grow and hence the state of the system will be bounded.

Let $W \triangleq \frac{1}{2}\xi^TP\xi$, where P is a positive definite matrix and satisfies the equation $A^TP + PA = -2I$. Evaluating $\dot{W}$ along the trajectory of (11) gives

$$\begin{aligned}\dot{W} &= -\xi^T\xi + \xi^TPb(\phi^Tv + \mu)\\ &\leqslant -\xi^T\xi\left[1 - \frac{|\xi^TPb|}{(\xi^T\xi)^{1/2}}\left(\frac{|\phi^Tv|}{(\xi^T\xi)^{1/2}} + \frac{\|\mu\|}{(\xi^T\xi)^{1/2}}\right)\right].\end{aligned} \tag{16}$$

Since μ is uniformly bounded there exist positive constants $\epsilon_1, a < \infty$ such that the conditions

$$|\xi^T\xi| \geqslant a^2 \tag{17a}$$

and

$$\phi^Tv/\sqrt{\xi^T\xi} \leqslant 2\epsilon_1 \tag{17b}$$

imply that the bracket term in (16) is greater than 1/2. Hence

$$\dot{W} \leqslant -\kappa W(t) \tag{18}$$

where $\kappa = 1/2\lambda_{\max}(P)$. Hence when conditions (17a) and (17b) are satisfied, W decreases exponentially with a rate at least equal to κ. On the other hand, if condition (17b) is not satisfied, the term within the brackets in (16) is still bounded and there exists a positive constant, $\lambda < \infty$, such that

$$\dot{W}(t) \leqslant \lambda W(t) \tag{19}$$

and W can at most grow exponentially with rate λ.

In what follows we make use of the fact that $v(t)$ is the internal state of a nonminimal representation of the plant, except for the bounded effect of the bounded disturbances. Therefore, positive constants c_1, c_2 exist such that $\|x_p\| \leqslant (1+c_1)\|v\| + c_2$. Hence

$$\|x\| \leqslant \|\xi\| \leqslant \|x_p\| + \|x\| \leqslant c_1\|x\| + c_2. \tag{20}$$

In view of inequality (20), $\|\xi\|$ and $\|x\|$, when they are large, are equivalent as far as the arguments in the following section are concerned.

IV. Stability of the Adaptive System

Based on the description of the plant and the controller and the assumptions imposed on them as listed in Section II, we can now state our main result as follows.

Theorem: The overall adaptive closed-loop system described by (11) and (12) is globally stable in the sense that for arbitrary bounded initial conditions and bounded signal $\mu(t)$ [or equivalently $\nu_1(t)$, $\nu_2(t)$, and $r(t)$], all states of the adaptive system remain uniformly bounded.

The basic idea of the proof of the theorem which was outlined in the previous section may be elaborated on as follows. The parameter error $\phi(t)$ has been made uniformly bounded at the outset by the introduction of the term $\Gamma\theta f(\theta)$ in the adaptive law (8) (see Section III). Therefore, only $x(t)$ [or equivalently $\xi(t)$] can grow without bound. If $x(t)$ grows without bound then, since it cannot grow faster than $\exp(\lambda t)$, it assumes values greater than a constant a over an interval of time of length a, where a is arbitrarily large. From (15) it follows that $(\phi^T\omega)^2/(1+x^Tx)$ and $\phi^T\theta f(\theta)$ become less than some ϵ and hence (see Appendix) that $|\phi^T v|/(1+x^Tx)^{1/2}$ becomes less than ϵ_1 most of the time within this interval. The latter implies [see (17b) and 18)] that $\|x(t)\|$ decreases at the rate $\exp(-\kappa t)$ most of the time, and can increase at most as $\exp(\lambda t)$ [see (19)] the rest of the time within the interval. Since the time interval over which $\|x(t)\|$ decreases is large compared to the interval on which it increases, it follows that $\|x(t)\|$ will assume a value less than a on the interval, which contradicts the original assumption that $\|x\| \geqslant a$ over the entire interval.

Proof: Let us assume that

$$\lim_{t\to\infty} \sup_{\tau \leqslant t} \|x(\tau)\| = \infty. \tag{21}$$

Then, since $\|x\|$ can grow at most exponentially, there exist monotonically increasing sequences $\{t_i\}$, $\{a_i\}$ with $\lim_{i\to\infty} t_i = \infty$, $\lim_{i\to\infty} a_i = \infty$ such that

$$\|x(t_i)\| = a_i \tag{22a}$$

$$\|x(t)\| \geqslant a_i \qquad \text{for all } t \in [t_i, t_i + a_i]. \tag{22b}$$

Using (22b) in (15) we obtain

$$\int_{t_i}^{t_i+a_i} \left\{ \frac{(\phi^T\omega)^2}{(1+x^Tx)} + \phi^T\theta f(\theta) \right\} dt \leqslant \Delta_1 + \Delta_2 \triangleq \Delta, \qquad (i \geqslant 0). \tag{23}$$

Since $\dot{\phi}(t)$ and the time derivative of $\omega/(1+x^Tx)^{1/2}$ are uniformly bounded (see Appendix), there exists a constant C, such that

$$\left| \frac{d}{dt} \frac{(\phi^T\omega)^2}{1+x^Tx} \right| \leqslant C \tag{24a}$$

and

$$\left| \frac{d}{dt} \{\phi^T\theta f(\theta)\} \right| \leqslant C. \tag{24b}$$

Let $\epsilon \in (0, C]$ be arbitrary and consider $i \geqslant i'$ such that $a_i \geqslant 2C\Delta/\epsilon^3$. Then the interval $T_i \triangleq [t_i, t_i + a_i]$ can be expressed as the union of $N = 2C\Delta/\epsilon^3$ disjoint subintervals, each being of length Δt_i greater than or equal to one.

If in any of these subintervals in T_i there exists an instant of time such that $(\phi^T\omega)^2/(1+x^Tx) + \phi^T\theta f(\theta) \geqslant \epsilon$, then (24) implies that this subinterval contributes to the integral on the left-hand side of (23) an amount of at least $\epsilon^2/2C$. Since the integral is bounded by Δ there can be at most $\Delta/(\epsilon^2/2C) = 2C\Delta/\epsilon^2$ of these subintervals. Let the set of all such subintervals be denoted by T_{i1}. The worst that can happen on T_{i1} is that W increases according to $\dot{W} = \lambda W$.

Let T_{i2} denote the set whose elements are the remaining $2C\Delta(1/\epsilon^3 - 1/\epsilon^2)$ subintervals so that $T_i = T_{i1} \cup T_{i2}$. We have $(\phi^T\omega)^2/(1+x^Tx) + \phi^T\theta f(\theta) < \epsilon$ when $t \in T_{i2}$. It is shown in the Appendix that this implies $|\phi^T v|/(1+x^Tx)^{1/2} \leqslant h(\epsilon)$, where $h(\cdot)$ is a continuous function with $\lim_{\epsilon\to 0} h(\epsilon) = 0$. Hence, if ϵ is sufficiently small, then we have $\dot{W} \leqslant -\kappa W$ on T_{i2}.

Therefore,

$$W(t_i + (j+1)\Delta t_i) \leqslant W(t_i + j\Delta t_i) \cdot e^{\lambda_j \Delta t_i}, \qquad (0 \leqslant j \leqslant N-1) \tag{25}$$

where $\lambda_j = -\kappa$ on T_{i1} and $\lambda_j = \lambda$ on T_{i2}. This implies

$$W(t_i + a_i) \leqslant W(t_i) \exp\left\{ \frac{2C\Delta}{\epsilon^3} [-\kappa(1-\epsilon) + \lambda\epsilon] \right\}. \tag{26}$$

Using $\xi^T = [x_p^T, x]$ and (20),(26) we finally obtain

$$\begin{aligned} \|x(t_i + a_i)\| &\leqslant \|\xi(t_i + a_i)\| \\ &\leqslant \|\xi(t_i)\| \cdot \left[\frac{\lambda_{\max}(P)}{\lambda_{\min}(P)} \right]^{1/2} \cdot \exp\left\{ \frac{C\Delta}{\epsilon^3} [-\kappa + \epsilon(\lambda+\kappa)] \right\} \\ &\leqslant a_i [c_1 + c_2 \epsilon^3] \cdot \left[\frac{\lambda_{\max}(P)}{\lambda_{\min}(P)} \right]^{1/2} \cdot \exp\left\{ \frac{C\Delta}{\epsilon^3} [-\kappa + \epsilon(\lambda+\kappa)] \right\}. \end{aligned} \tag{27}$$

Obviously, the right-hand side of (27) can be made less than a_i if ϵ is sufficiently small. But this is a contradiction to (22b). As a consequence, the assumption (21) was wrong, i.e., $x(t)$ is uniformly bounded, and so is $\xi(t)$ according to (20). □

V. Conclusions

The model reference adaptive control problem in the presence of bounded disturbances is considered in this paper. The principal result of the paper is that if an upper bound $\|\theta^*\|_{\max}$ on the norm of the unknown controller parameter vector θ^* is known and the adaptive law is suitably modified when $\|\theta\| \geqslant \|\theta^*\|_{\max}$, then in spite of the disturbances all parameters and signals in the adaptive loop remain bounded. In the absence of disturbances the modification is of no significance and the output error $e(t)$ between the plant and the model goes to zero as t goes to infinity. This implies that the scheme suggested retains the properties of earlier schemes and is, in addition, robust with respect to bounded disturbances, which makes it suitable for use in practical applications.

As mentioned in the Introduction, an alternate resolution of the problem is that if an upper bound on the magnitude of the disturbance is known, then a suitably designed dead zone in the adaptive law also guarantees the boundedness of all parameters and signals in the adaptive system [6]. In the latter approach the descent property of the parameter error vector $\phi(t)$ is always retained, whereas the output error $e(t)$ is only assured to lie within the deadzone as $t \to \infty$, even in the absence of disturbances. From this it is evident that the two approaches behave quite differently both in the presence of disturbances (in which $\phi(t)$ may not go to zero in the scheme suggested in this paper) and in the absence of disturbances. Therefore, the analysis of their relative behavior deserves further investigation. Combining the two approaches appears to be possible and also interesting for practical applications.

In the adaptive law, which is considered in this paper, the parameter adjustment is based on the instantaneously available information. Instead, all the information that has become available up to time t could be used in order to improve the speed of adaptation properties, as suggested in [7]. The adaptive law then becomes

$$\dot{\theta}(t) = -\Gamma \int_0^t \frac{\omega(\tau)\left[\omega^T(\tau)\theta(t) - c_m^T\zeta_1(\tau) + y_p(\tau)\right]}{1 + x^T(\tau)x(\tau)} \cdot e^{-q(t-\tau)}\,d\tau - \Gamma\theta f(\theta) \quad (28)$$

where $q > 0$ is an arbitrary constant. The equation for the parameter error then is given by

$$\dot{\phi}(t) = -\Gamma \int_0^t \frac{\omega(\tau)\left[\omega^T(\tau)\phi(t) + \nu(\tau)\right]}{1 + x^T(\tau)x(\tau)} \cdot e^{-q(t-\tau)}\,d\tau - \Gamma\theta f(\theta). \quad (29)$$

Since (12) and (29) are structurally the same, the stability arguments of this paper can also be applied to show the stability for the adaptive law (29).

It is assumed in this paper that the high-frequency gain k_p of the plant is known. This results in a relatively simple proof of stability. The basic idea of modifying the adaptive law based on knowledge of $\|\theta^*\|_{\max}$ can, of course, also be applied when k_p is unknown. A proof of stability may then be obtained along the same lines as in this paper, although the details may be more involved.

Appendix

i) We first verify that there exists a positive constant $C_1 < \infty$ such that

$$|\nu| \leqslant C_1 \quad (A1)$$

$$\|\phi\| \leqslant C_1 \quad (A2)$$

$$\|\dot{x}/(1 + x^T x)^{1/2}\| \leqslant C_1 \quad (A3)$$

$$\|\omega^{(i)}/(1 + x^T x)^{1/2}\| \leqslant C_1, \qquad (0 \leqslant i \leqslant n^* + 1) \quad (A4)$$

$$\left|\frac{d}{dt}\{\phi^T\omega^{(i)}/(1 + x^T x)^{1/2}\}\right| \leqslant C_1, \qquad (0 \leqslant i \leqslant n^*) \quad (A5)$$

where $\omega^{(i)}$ denotes the ith time derivative of ω.

(A1) holds by assumption. (A2) was shown in Section III. (A3) follows from (A2) and (11). (A4) is true since $\zeta^{(i-1)}$ is proportional to ζ for $i = 1, \cdots, n^*$ and proportional to ζ and u_p, y_p for $i = n^* + 1$. From $u_p = v^T\theta + r'$ it follows that $u_p/(1 + x^T x)^{1/2}$ is uniformly bounded because so are θ and r'. Since v is equal to the state of a nonminimal representation of the plant except for the bounded effect of the bounded disturbances ν_1, ν_2 it follows that $y_p/(1 + x^T x)^{1/2}$ is also uniformly bounded. This establishes (A4). (A5) follows from (A4) and (A2), (12).

ii) It is to be shown that

$$\|x(t)\| \geqslant 1/\epsilon^3 \quad (A6)$$

$$\frac{\left[\phi^T(t)\omega(t)\right]^2}{1 + x^T(t)x(t)} + \phi^T(t)\theta(t)f(\theta(t)) < \epsilon \qquad t \in [t_1, t_2] \quad (A7)$$

and $t_2 - t_1 \geqslant 1$ implies

$$|\phi^T(t)v(t)|/\left[1 + x^T(t)x(t)\right]^{1/2} \leqslant h(\epsilon) \qquad t \in [t_1, t_2] \quad (A8)$$

where $\{h: R^+ \to R^+\}$ is a continuous function and such that $\lim_{\epsilon \to 0} h(\epsilon) = 0$.

Since $\phi^T\theta f(\theta) \geqslant 0$, each term of the sum in (A7) is less than ϵ. Making use of the fact that $f(\theta) = 0$ for $\|\theta\| \leqslant \|\theta^*\|_{\max}$, we obtain

$$f(\theta) < \frac{\epsilon}{\phi^T\theta} = \frac{\epsilon}{\theta^T\theta - \theta^T\theta^*} \leqslant \frac{\epsilon}{\|\theta\|} \cdot \frac{1}{\|\theta^*\|_{\max} - \|\theta^*\|}. \quad (A9)$$

Equation (12), together with (A7),(A9), gives rise to

$$\|\dot{\phi}\| \leqslant \|\Gamma\| \cdot \left\{ \frac{\|\omega\|}{(1+x^T x)^{1/2}} \cdot \frac{|\phi^T \omega| + |\nu|}{(1+x^T x)^{1/2}} + \|\theta\| \cdot f(\theta) \right\}$$

$$\leqslant \|\Gamma\| \{ \sqrt{\epsilon} + \epsilon^3 C_1 + \epsilon / (\|\theta^*\|_{\max} - \|\theta^*\|) \}$$

$$\leqslant C_2 \sqrt{\epsilon}, \qquad (0 < \epsilon \leqslant 1). \qquad \text{(A1}$$

Now let

$$|\phi^T \omega^{(i)} / (1+x^T x)^{1/2}| \leqslant \epsilon_i, \qquad t \in [t_1, t_2] \qquad \text{(A11)}$$

hold for some $\epsilon_i > 0$ and $0 \leqslant i < n^*$. This is true for $i = 0$ and $\epsilon_0 = \sqrt{\epsilon}$ by the hypothesis (A7). For $i+1$ we obtain the following identity by partial integration:

$$\int_t^{t+\Delta t} \frac{\phi^T \omega^{(i+1)}}{(1+x^T x)^{1/2}} dt$$

$$= \left. \frac{\phi^T \omega^{(i)}}{(1+x^T x)^{1/2}} \right|_t^{t+\Delta t} - \int_t^{t+\Delta t} \frac{\dot{\phi}^T \omega^{(i)}}{(1+x^T x)^{1/2}} dt$$

$$+ \int_t^{t+\Delta t} \frac{\phi^T \omega^{(i)}}{(1+x^T x)^{1/2}} \cdot \frac{x^T \dot{x}}{1+x^T x} d\tau \qquad \text{(A12)}$$

where $[t, t+\Delta t] \subset [t_1, t_2]$. Defining $C \triangleq \max\{1, C_1, C_2\}$ it follows that

$$\left| \int_t^{t+\Delta t} \frac{\phi^T \omega^{(i+1)}}{(1+x^T x)^{1/2}} dt \right| \leqslant 2\epsilon_i + \Delta t \cdot (C^2 \sqrt{\epsilon} + \epsilon_i C).$$

$$\text{(A13)}$$

If the integrand on the left-hand side of (A13) is equal to ϵ_{i+1} at time $t' \in [t_1, t_2]$, then by choosing $\Delta t = \epsilon_{i+1}/C$ and t such that $t' \in [t, t+\Delta t]$, we obtain the result that the integral in (A13) is greater than or equal to $\Delta t \cdot \epsilon_{i+1}/2 = \epsilon_{i+1}^2/2C$. Inequality (A13) then can be rewritten as

$$\epsilon_{i+1}^2 - 2\epsilon_{i+1}(C^2 \sqrt{\epsilon} + \epsilon_i C) - 4\epsilon_i C \leqslant 0. \qquad \text{(A14)}$$

Solving (A14) for the maximum possible ϵ_{i+1}, we get

$$\epsilon_{i+1} = (C^2 \sqrt{\epsilon} + \epsilon_i C) + \{ (C^2 \sqrt{\epsilon} + \epsilon_i C)^2 + 4\epsilon_i C \}^{1/2}.$$

$$\text{(A15)}$$

In conclusion, (A11) holds for $\epsilon_0 = \sqrt{\epsilon}$ and $\epsilon_1, \cdots, \epsilon_{n_*}$ being defined recursively by (A15).

Since $v = \sum_{i=0}^{n^*} d_i \omega^{(i)}$ from (3), (5), and (6), it follows that

$$\left| \frac{\phi^T v}{(1+x^T x)^{1/2}} \right| = \left| \sum_{i=0}^{n^*} d_i \frac{\phi^T \omega^{(i)}}{(1+x^T x)^{1/2}} \right|$$

$$\leqslant \sum_{i=0}^{n^*} d_i \epsilon_i \triangleq h(\epsilon). \qquad \text{(A16)}$$

Hence $\lim_{\epsilon \to 0} h(\epsilon) = 0$, which was the result to be proven.

References

[1] K. S. Narendra, Y.-H. Lin, and L. S. Valavani, "Stable adaptive controller design, Part II: Proof of stability," *IEEE Trans. Automat. Contr.*, vol. AC-25, pp. 440–448, June 1980.

[2] A. S. Morse, "Global stability of parameter-adaptive control systems," *IEEE Trans. Automat. Contr.*, vol. AC-25, pp. 433–439, June 1980.

[3] G. C. Goodwin, P. J. Ramadge, and P. E. Caines, "Discrete time multivariable adaptive control," *IEEE Trans. Automat. Contr.*, vol. AC-25, pp. 449–456, June 1980.

[4] R. V. Monopoli and F. M. Kessler, Jr., "Measurement noise in model reference adaptive systems," in *Proc. JACC*, 1977, pp. 1599–1603.

[5] R. L. Gutmann, R. V. Monopoli, and R. Van Allen, "Adaptive control of a laser pointing and tracking system: A feasibility study," in *Applications of Adaptive Control*, K. S. Narendra and R. V. Monopoli, Eds. New York: Academic, 1980.

[6] B. B. Peterson and K. S. Narendra, "Bounded error adaptive control, Part I," S & IS Rep. 8005, Dec. 1980; and Part II, S & IS Rep. 8106, Apr. 1981; see also this issue, pp. 1161–1168.

[7] G. Kreisselmeier and D. Joos, "Stable model reference adaptive control with rapid adaptation," DFVLR-Institut für Dynamik der Flugsysteme, W. Germany, Tech. Rep., Feb. 1981.

Error Models for Stable Hybrid Adaptive Systems

KUMPATI S. NARENDRA, FELLOW, IEEE, IRAKY H. KHALIFA, AND ANURADHA M. ANNASWAMY

Abstract—**This paper presents several stable adaptive algorithms for the control of hybrid and discrete systems in which the control parameters are adjusted at rates slower than those at which the systems operate. Continuous algorithms of an integral type, recently suggested in the literature [5] are also shown to belong to this class. From a practical standpoint, the infrequent adjustment of the control parameters makes for more robust adaptive control while from a theoretical point of view, the algorithms are attractive since they provide a unified framework for the design of continuous, hybrid, and discrete adaptive systems. Simulation results are included to indicate the type of responses that can be expected using the different algorithms.**

INTRODUCTION

IN the past few years several continuous [1], [2] and discrete [3], [4] adaptive algorithms have been developed for the stable identification and control of linear time-invariant minimum phase plants with unknown parameters. In continuous adaptive systems, the plant operates in continuous time and the controller parameters are adjusted continuously. Similarly, in discrete adaptive systems, the plant is modeled in discrete time and the various signals as well as the control parameter vector are defined at discrete instants. Recent advances in microprocessors and related digital computer technology naturally favor the use of discrete systems in design. However, practical systems, for a variety of reasons, may contain both discrete and continuous elements. Such systems may be described as "hybrid" systems, where the term hybrid is used in a generic sense.

From a practical standpoint, continuous adaptive algorithms suffer from a lack of robustness since the control parameters are continuously adjusted, using an adaptive law involving pure integration. In such cases, it may be desirable to make infrequent adjustments of the control parameters at discrete instants, even as continuous signals are being processed in real time, making the systems hybrid. Even in purely discrete systems, the computational effort involved often necessitates the adjustment of the adaptive control parameters at rates significantly slower than the rate at which the system operates. Both hybrid as well as discrete adaptive systems of the latter type can be considered as systems with two time-scales.[1] Continuous algorithms of an integral type have recently been suggested in the literature [5] which can also be interpreted as belonging to this class.

This paper presents a unified approach to the design of hybrid, discrete, and continuous adaptive systems in which the control parameters are updated infrequently. Practically, the algorithms are attractive from robustness considerations, while from a theoretical viewpoint, they provide a common framework for control of systems with two time-scales.

Manuscript received May 2, 1983; revised February 21, 1984. This paper is based on a prior submission on August 25, 1982. Paper recommended by Past Associate Editor, G. C. Goodwin. This work was supported by the Office of Naval Research under Contract N00014-76-C-0017 and the National Science Foundation under Grant ECS-8300223.

The authors are with the Center for Systems Science and the Department of Electrical Engineering, Yale University, New Haven CT 06520.

[1] This term was generally used in the context of singular perturbation. All that is implied here is that the rate at which the control parameter is adjusted can be significantly smaller than the rate of the signals of the system.

Gawthrop [6] was the first to introduce the concept of hybrid self-tuning controllers which are partly realized in continuous time and partly in discrete time. In [7], Morris and Neuman considered a discrete-time model reference adaptive system wherein the parameter adjustment period is specified to be an integral multiple of the control period. Elliott [8] and Cristi and Monopoli [9] proposed methods of direct hybrid adaptive control of continuous time systems by sampling the relevant signals. The central question in both cases is the global stability of the overall system which can be best analyzed by considering the behavior of the corresponding error model. The main contribution of this paper is the detailed analysis of several error models using two distinct approaches. The fact that such error models can be studied independently provides considerable flexibility in the choice of an algorithm as well as the method of analysis most suited to a specific adaptive control problem. In particular, the insights obtained from the analysis of error models in this paper are found to be useful for the synthesis of stable hybrid and discrete adaptive controllers for general two time-scale systems.

In Section II, using a Lyapunov approach, different adaptive laws are developed for the various error models which assure the boundedness of the parameter error vector. Conditions under which the output and parameter error vectors tend to zero asymptotically, are also derived and are particularly useful in the identification problem. The main results of this section pertain to the behavior of the outputs of the various error models when their inputs grow in an unbounded fashion and find direct application in the adaptive control problem.

Applications, extensions, and modifications of the adaptive algorithms are treated briefly in Section III. The principal steps in the proof of global stability of a hybrid adaptive control system are first outlined in Section III-A. The same algorithms when suitably modified are also shown in Section III-B to be applicable to both discrete and continuous systems with two time-scales. Finally, in Section III-C it is shown that the use of time-varying adaptive gain matrices (as in the well-known recursive least-squares approach) in the algorithms does not affect the stability arguments of the earlier sections.

Section IV contains simulation results of different error models.

II. ERROR MODELS

The dynamical systems discussed in this section are continuous-time systems in which $t \in \mathcal{R}^+$, the set of all positive numbers. Let $u:\mathcal{R}^+ \to \mathcal{R}^m$ and $e_1:\mathcal{R}^+ \to \mathcal{R}^1$ be piecewise continuous functions referred to as the input and output functions of the error models, respectively. Let $\{t_k\}$ be a monotonically increasing unbounded sequence in $\mathcal{R}^+$ with $0 < T_{\min} \leq T_k \leq T_{\max} < \infty$ for $k \in N$, where $T_k \equiv t_{k+1} - t_k$. When $t_k = kT$ (or $T_k = T$) we shall call T the sampling period. Let $\phi:\mathcal{R}^+ \to \mathcal{R}^m$ be a piecewise constant function, referred to as the parameter error vector and assume values

$$\phi(t) = \phi_k \qquad t \in [t_k,\ t_{k+1})$$

where ϕ_k is a constant vector. The error models described in this section relate the output error e_1, the input vector u, and parameter error vector ϕ in terms of algebraic or differential equations.

Reprinted from *IEEE Trans. Automat. Contr.*, vol. AC-30, pp. 339–347, 1985.

The overall behavior of each error model is analyzed for the three specific cases: i) when $\|u(t)\|$ is uniformly bounded, ii) when $u(t)$ is persistently exciting[2] [10], and iii) when u is unbounded and $\epsilon\mathcal{L}_e^\infty$. Two different approaches are presented based on the manner in which ϕ_k is adjusted.

A. Approach 1

Error Model: The first error model is described by the equation

$$\phi_k^T u(t) = e_1(t) \qquad t \in [t_k, t_{k+1}),\ k \in N. \tag{1}$$

It is assumed that ϕ_0 (and hence ϕ_k) is unknown, the values $u(t)$ and $e_1(t)$ can be observed at every instant t and $\Delta\phi_k \triangleq \phi_{k+1} - \phi_k$ can be adjusted at $t = t_{k+1}$. The objective is to determine an adaptive law for choosing the sequence $\{\Delta\phi_k\}$ using all available input–output data so that

$$\lim_{t\to\infty} e_1(t) = 0.$$

Consider the Lyapunov function candidate

$$V(k) = 1/2\phi_k^T\phi_k.$$

Then

$$\Delta V(k) \triangleq V(k+1) - V(k) = \left[\phi_k + \frac{\Delta\phi_k}{2}\right]^T \Delta\phi_k. \tag{3}$$

Choosing the adaptive law

$$\Delta\phi_k = -\frac{1}{T_k}\int_{t_k}^{t_{k+1}} \frac{e_1(\tau)u(\tau)}{1+u^T(\tau)u(\tau)}\, d\tau \tag{4}$$

yields

$$\Delta V(k) = -\frac{1}{2}\phi_k^T[2I - R_k]R_k\phi_k \tag{5}$$

where R_k is the symmetric positive semidefinite matrix

$$R_k = \frac{1}{T_k}\int_{t_k}^{t_{k+1}} \frac{u(\tau)u^T(\tau)}{1+u^T(\tau)u(\tau)}\, d\tau \tag{6}$$

with all its eigenvalues within the unit circle. Since $[2I - R_k] > \beta I$ for some constant $\beta > 0$,

$$\Delta V(k) < -\frac{1}{2}\beta\phi_k^T R_k \phi_k \le 0. \tag{7}$$

Hence, $V(k)$ is a Lyapunov function and assures the boundedness of $\|\phi_k\|$ if $\|\phi_0\|$ is bounded. Further, since $\{\Delta V(k)\}$ is a nonnegative sequence with $\Sigma_{k=0}^\infty \Delta V(k) < \infty$, it follows that $\Delta V(k) \to 0$ as $k \to \infty$ or alternately from (7) $\phi_k^T R_k \phi_k \to 0$ as $k \to \infty$. This can also be expressed as

$$\frac{1}{T_k}\int_{t_k}^{t_{k+1}} \frac{e_1^2(\tau)}{1+u^T(\tau)u(\tau)}\, d\tau \to 0 \quad \text{as } k\to\infty. \tag{8}$$

Case i): If u is uniformly bounded in $\mathcal{R}^+$, it follows from (1) that e_1 is also uniformly bounded. Since $\Sigma_{k=0}^\infty \Delta V(k)$ is bounded, we have from (8) that $e_1 \in \mathcal{L}^2$. If $\dot{u}$ is also bounded, $\lim_{t\to\infty} e_1(t) = 0$. Hence, for a uniformly bounded u with a uniformly bounded derivative, e_1 tends to zero and $\Delta\phi_k \to 0$ as $k \to \infty$.

[2] In [10] the term sufficient richness" is used instead of "persistent excitation."

Case ii): If, in addition to being uniformly bounded, u is persistently exciting over an interval $T_{\min}$, the matrix R_k is uniformly positive definite for all $k \in N$. Then $\Delta V(k)$ in (7) is negative definite, and hence the parameter error vector tends to zero as $k \to \infty$.

Case iii): A more interesting case arises when u grows in an unbounded fashion which is relevant to the control problem. Since condition (8) is independent of the assumption of the boundedness of u, it follows that $e_1/(1 + u^Tu)^{1/2} = \rho \in \mathcal{L}^2$. Hence,

$$\phi_k^T u(t) = e_1(t) = \rho(t)(1+u^T(t)u(t))^{1/2}. \tag{9}$$

Equation (9) plays a central role in the proof of stability of the hybrid adaptive control problem and is discussed in Section III-A.

Error Model 2: The second error model is described by the error differential equation

$$\dot{e} = Ae + b\phi^T u \tag{10}$$

where $e(t)$, $b \in \mathcal{R}^n$, $A \in \mathcal{R}^{n\times n}$ and is stable, (A, b) is controllable, $\phi(t)$, $u(t) \in \mathcal{R}^m$ and $\phi(t) = \phi_k$, $t \in [t_k, t_{k+1})$, $k \in N$ where ϕ_k is a constant vector. In this case the parameter error vector is to be adjusted using the input u and the state error vector e. Since A is a stable matrix, a symmetric positive definite matrix $P = P^T > 0$ exists such that $A^TP + PA = -Q < 0$.

Using approach 1, error model (10) is first modified, as in the continuous case [1], using a feedback term $-\gamma u^T u b^T P e$ to the form

$$\dot{e} = Ae + b[\phi_k^T u - \gamma u^T u b^T P e]. \tag{11}$$

The corresponding adaptive law is the average gradient obtained in the continuous case over one period and is given by

$$\Delta\phi_k = -\frac{1}{T_{\max}}\int_{t_k}^{t_{k+1}} u(\tau)b^TPe(\tau)\, d\tau. \tag{12}$$

Defining the Lyapunov candidate as

$$V(k) = \frac{1}{T_{\max}} e^T(t_k)Pe(t_k) + \phi_k^T\phi_k, \tag{13}$$

$$\begin{aligned}\Delta V(k) = &-\frac{1}{T_{\max}}\int_{t_k}^{t_{k+1}} e^T(\tau)Qe(\tau)\, d\tau \\ &-\frac{2\gamma}{T_{\max}}\int_{t_k}^{t_{k+1}} [e^T(\tau)Pb]^2 u^T(\tau)u(\tau)\, d\tau \\ &+\frac{1}{T_{\max}^2}\left\|\int_{t_k}^{t_{k+1}} e^T(\tau)Pbu(\tau)\, d\tau\right\|^2.\end{aligned} \tag{14}$$

Since for any vector $x(t) \in \mathcal{R}^n$, $1/T \int_{t-T}^{t}\|x(\tau)\|^2\, d\tau \ge [\frac{1}{T}\|\int_{t-T}^{t} x(\tau)d\tau\|]^2$, the second term on the right-hand side of (14) dominates the third term for all $\gamma \ge T_{\max}/2T_{\min}$. Hence,

$$\Delta V(k) \le 0 \qquad \forall k \in N \text{ and } \gamma \ge \frac{T_{\max}}{2T_{\min}}. \tag{15}$$

This implies that $\|\phi_k\|$ and $\|e(t_k)\|$ are bounded if $\|\phi_0\|$ and $\|e(t_0)\|$ are bounded and $\Delta V(k) \to 0$ as $k \to \infty$, i.e.,

$$\int_{t_k}^{t_{k+1}} e^T(\tau)Qe(\tau)\, d\tau \to 0;$$

$$\int_{t_k}^{t_{k+1}} [e^T(\tau)Pb]^2 u^T(\tau)u(\tau)\, d\tau \to 0 \qquad \text{as } k\to\infty. \tag{16}$$

Again, since $\Sigma_{k=0}^\infty \Delta V(k)$ is bounded e, $[e^TPbu] \in \mathcal{L}^2$.

Case i): If u is uniformly bounded, it follows from (11) that $\dot{e}$ is also uniformly bounded. Since $e \in \mathcal{L}^2$ this results in $\lim_{t\to\infty} e(t) = 0$.

Case ii): As in the earlier case with error model 1, when $u(\cdot)$ is persistently exciting over a period $T_{\min}$, $\Delta V(k)$ in (14) is negative definite, and hence $\phi_k \to 0$ as $k \to \infty$.

Case iii): A somewhat more involved argument is needed when $u \in \mathcal{L}_e^\infty$ and is unbounded, to determine the relation between e and u corresponding to (9) in error model 1. The existence of the discrete Lyapunov function assures the boundedness of $e(t)$ at the discrete instants $t = t_k$. Also from (11), $\frac{d}{dt}[e^TPe] \leq 0$ for $\|e\| \geq c_1/\sqrt{2\gamma\lambda_{\min}}$ where $\|\phi\| \leq c_1$ and $\lambda_{\min}$ is the minimum eigenvalue of Q. Hence, $e(t)$ is uniformly bounded on $\mathcal{R}^+$.

Now, the same arguments as those used in the continuous case [1] can be used to demonstrate that $\|W(s)\phi^Tu\| = o\,[\sup_{t\geq\tau}\|u(\tau)\|]$ where $W(s) \equiv [sI - A]^{-1}b$. In (11), $[\phi^Tu - \gamma u^Tub^TPe]$ is the input to an exponentially stable system. Since $e^TPbu \in \mathcal{L}^2$ the component of e resulting from e^TPbu^Tu must grow at a slower rate than $\|u\|$, i.e., $o\,[\sup_{t\geq\tau}\|u(\tau)\|]$. If this is unbounded, then the response due to ϕ^Tu should also be $o\,[\sup_{t\geq\tau}\|u(\tau)\|]$, since we have already demonstrated that $e(t)$ is uniformly bounded for $t \in \mathcal{R}^+$.

Error Model 3: The third hybrid error model is merely a special case of the second error model but is important in view of its practical applications in adaptive control. It is described by the scalar differential equation

$$\dot{e}_1(t) = -\alpha e_1(t) + [\phi_k^Tu(t) - \gamma e_1(t)u^T(t)u(t)]. \tag{17}$$

The corresponding adaptive law can be specialized from (12) as:

$$\Delta\phi_k = -\frac{1}{T_{\max}}\int_{t_k}^{t_{k+1}} e_1(\tau)u(\tau)\,d\tau. \tag{18}$$

Using similar arguments as in error model 2, replacing e^TPbu by e_1u, the matrix A by $-\alpha$ and the vector b by 1, we obtain

$$|W(s)\phi_k^Tu(t)| = o\,[\sup_{t\geq\tau}\|u(\tau)\|] \tag{19}$$

where $W(s) = 1/(s+\alpha)$.

B. Approach 2

In this section, we present a second approach for adjusting $\Delta\phi_k$. By integrating the error equations over a finite interval, discrete error equations are derived for which discrete adaptive laws can be given by inspection. As in the first approach, all three error models are analyzed for three specific cases. However, in the third case where the input grows in an unbounded fashion, an additional assumption regarding the rate of growth of the input has to be made to derive the principal result that $e_1(t)$ grows at a slower rate than $\|u(t)\|$. As shown in Section III, the input to the error model derived in the adaptive control problem satisfies this assumption, and hence the results given here can be used to analyze hybrid adaptive systems.

Error Model 1: In this approach the error equation (1) is modified to the equivalent form

$$\phi_k^Te_1(t)u(t) = e_1^2(t) \qquad t \in [t_k, t_{k+1}),\ k \in N \tag{20}$$

by multiplying the two sides by $e_1(t)$. Integrating over the interval $[t_k, t_{k+1})$ yields:

$$\phi_k^T\int_{t_k}^{t_{k+1}} e_1(\tau)u(\tau)\,d\tau = \int_{t_k}^{t_{k+1}} e_1^2(\tau)\,d\tau$$

or equivalently the discrete error model

$$\phi_k^T\omega(k) = \epsilon(k) \tag{21}$$

where

$$\int_{t_k}^{t_{k+1}} e_1(\tau)u(\tau)\,d\tau \triangleq \omega(k) \text{ and } \int_{t_k}^{t_{k+1}} e_1^2(\tau)\,d\tau \triangleq \epsilon(k).$$

For the error model (21), the adaptive law for updating ϕ_k can be written by inspection [3] as

$$\Delta\phi_k = \frac{-\epsilon(k)\omega(k)}{1+\omega(k)^T\omega(k)}. \tag{22}$$

From well-known results in discrete adaptive control [3], [4] it also follows that i) $\|\phi_k\|$ is bounded if $\|\phi_0\|$ is bounded, ii) $\Delta\phi_k \to 0$ as $k \to \infty$, and iii) $|\epsilon(k)| = o[\{1+\omega(k)^T\omega(k)\}^{1/2}]$ or $\epsilon(k)$ grows more slowly than $\omega(k)$.

Case i): As in approach 1, if u is uniformly bounded in $\mathcal{R}^+$, e_1, and hence $\epsilon(k)$ and $\omega(k)$ are uniformly bounded for all $t \in \mathcal{R}^+$ and $k \in N$, respectively. If $\|\dot{u}\|$ is also bounded, $\lim_{t\to\infty} e_1(t) = 0$.

Case ii): If in addition to being uniformly bounded u is persistently exciting over an interval $T_{\min}$, $\int_{t_k}^{t_{k+1}} u(\tau)u^T(\tau)\,d\tau = P_k$ is unformly positive definite for all $k \in N$. If $V(k) = 1/2\phi_k^T\phi_k$, the adaptive law (22) results in

$$\Delta V(k) = -\frac{[\phi_k^TP_k\phi_k]^2}{1+\phi_k^TP_k^2\phi_k} < 0$$

for all $\phi_k \neq 0$ and $k \in N$. Hence, $\phi_k \to 0$ as $k \to \infty$.

Case iii): As mentioned earlier, when $u(t)$ grows in an unbounded fashion, the following further assumption on its rate of growth is made:

$$\|\dot{u}(t)\| \leq M_1 \sup_{t\geq\tau}\|u(\tau)\| + M_2 \qquad \text{for } M_1, M_2 \in \mathcal{R}^+. \tag{23}$$

This condition, as shown below, is sufficient to assure that $e_1(t) = o\,[\sup_{t\geq\tau}\|u(\tau)\|]$. From the adaptive law (22), it is known that

$$\lim_{k\to\infty}\frac{\epsilon^2(k)}{1+\omega^T(k)\omega(k)} = 0$$

or

$$\lim_{k\to\infty}\frac{\left[\int_{t_k}^{t_{k+1}} e_1^2(\tau)\,d\tau\right]^2}{1+\int_{t_k}^{t_{k+1}}\|e_1(\tau)u(\tau)\|^2\,d\tau} = 0. \tag{24}$$

If $e_1(t) \neq o\,[\sup_{t\geq\tau}\|u(\tau)\|]$, then there exists a sequence $\{t'_{n_i}\}$ and a constant $c_1 > 0$ such that

$$|e_1(t'_{n_i})| \geq c_1 \sup_{t'_{n_i}\geq\tau}\|u(\tau)\|, \qquad \forall i \in N.$$

Since, from error equation (1), $|\dot{e}_1(t)| \leq c_2 \sup_{t\geq\tau}\|u(\tau)\|$ for all $t \in \mathcal{R}^+$, it follows that constants c_3, Δ exist so that

$$|e_1(t)| \geq c_3 \sup_{t'_{n_i}\geq\tau}\|u(\tau)\|,\ t \in [t'_{n_i}, t'_{n_i}+\Delta),\ i \in N.$$

Assuming that

$$[t'_{n_i}, t'_{n_i}+\Delta) \subset [t_{n_i}, t_{n_i+1})\ i \in N,$$

$$\left[\int_{t_{n_i}}^{t_{n_i+1}} e_1^2(\tau)\,d\tau\right] \geq c_3^2\Delta \sup_{t'_{n_i}\geq\tau}\|u(\tau)\|^2.$$

Further, since $|e_1(t)| \leq c_4\|u(t)\|\ \forall\ t \in \mathcal{R}^+$ and u satisfies condition (23), it follows that

$$\left[\int_{t_{n_i}}^{t_{n_i+1}}\|e_1(\tau)u(\tau)\|^2\,d\tau\right]^{1/2} \leq c_5 \sup_{t'_{n_i}\geq\tau}\|u(\tau)\|^2 \tag{25}$$

for some constant $c_5 > 0$. Hence, the term on the left-hand side of (24) is greater than some constant c_6 on a subsequence $\{t_{n_i}\}$ which contradicts (24). Hence,

$$e_1(t) = o\,[\sup_{t\geq\tau}\|u(\tau)\|].$$

Error Model 2: In this case the error model is described by

$$\dot{e} = Ae + b\phi_k^T u \qquad t \in [t_k, t_{k+1}) \tag{26}$$

and does not contain the feedback term as in (11). Multiplying both sides of (26) by $e^T P$ and equating the integrals over an interval $[t_k, t_{k+1})$

$$e^T(t)Pe(t)\Big|_{t_k}^{t_{k+1}} + \int_{t_k}^{t_{k+1}} e^T(\tau)Qe(\tau)\, d\tau = 2\int_{t_k}^{t_{k+1}} e^T(\tau)Pbu^T(\tau)\phi_k\, d\tau \tag{27}$$

or

$$\phi_k^T w(k) = \epsilon(k). \tag{28}$$

where the left-hand side of (27) is $\epsilon(k)$ and $2\int_{t_k}^{t_{k+1}} e^T(\tau)Pbu(\tau)d\tau = w(k)$. Once again the adaptive law may be written by inspection as

$$\Delta\phi_k = \frac{-\epsilon(k)w(k)}{1 + w^T(k)w(k)} \tag{29}$$

and yields

$$|\epsilon(k)| = o[\{1 + w(k)^T w(k)\}^{1/2}]. \tag{30}$$

Case i): If u is uniformly bounded, from (26) we have $e(t)$ and $\dot{e}(t)$ uniformly bounded. Hence, $w(k)$ is uniformly bounded and $\epsilon(k) \to 0$ as $k \to \infty$. Defining $e^T(t)Pe(t)|_{t_k} = E_k$, since $\int_{t_k}^{t_{k+1}} e^T(\tau)Qe(\tau)d\tau \geq cE_k$ for some constant $c \in (0, 1)$, we have $E_{k+1} - (1 - c)E_k \to 0$ as $k \to \infty$ or $E_k \to 0$, and hence $\int_{t_k}^{t_{k+1}} e^T(\tau)Qe(\tau)d\tau \to 0$ as $k \to \infty$. Since $\dot{e}(t)$ is uniformly bounded, $\lim_{t\to\infty} e(t) = 0$.

Case ii): If u is persistently exciting over any interval of length $T_{\min}$, $\phi_k^T u(t) \to 0$ implies $\phi_k \to 0$ as $k \to \infty$.

Case iii): As in error model 1, when u satisfies the condition (23), we can prove by contradiction that $\|e(t)\| = o\,[\sup_{t\geq\tau} \|u(\tau)\|]$. From (30), we have

$$\lim_{k\to\infty} \frac{e^T(t)Pe(t)\Big|_{t_k}^{t_{k+1}} + \int_{t_k}^{t_{k+1}} e^T(\tau)Qe(\tau)\, d\tau}{\left[1 + 4\int_{t_k}^{t_{k+1}} \|e^T(\tau)Pbu(\tau)\|^2\, d\tau\right]^{1/2}} = 0. \tag{31}$$

While the proof follows along the same lines as in error model 1, some minor changes are warranted in view of the first term in the numerator. The following three cases are of interest.

i) If $\|e(t)\| \geq c_3 \sup_{t\geq\tau} \|u(\tau)\|$, $t \in [t'_{n_i}, t'_{n_i} + \Delta)$ as in error model 1 and in addition $e^T(t_k)Pe(t_k) = \beta_k\,[\sup_{t_k\geq\tau} \|u(\tau)\|]^2\ \forall\, k \in N$, where $\lim_{k\to\infty} \beta_k = 0$, the proof is identical to that given in case iii) in error model 1.

ii) The same result also follows when

$$e^T(t_{n_k})Pe(t_{n_k}) = \beta_k\,[\sup_{t_{n_k}\geq\tau} \|u(\tau)\|]^2$$

on any subsequence $\{t_{n_k}\}$.

iii) When $e^T(t_k)Pe(t_k) \geq c_1\,[\sup_{t_k\geq\tau}\|u(\tau)\|^2]$ for all $k \geq k_0$, k, $k_0 \in N$, where c_1 is some constant > 0, using (30) and the same arguments as in case i),

$$[E_{k+1} - (1 - c)E_k] \leq \gamma_k E_k$$

where $\lim_{k\to\infty} \gamma_k = 0$. This implies that $\lim_{k\to\infty} E_k = 0$ which contradicts the assumption that $E_k \geq c_1\,[\sup_{t_k\geq\tau}\|u(\tau)\|]^2$ when u grows in an unbounded fashion. Hence,

$$\|e(t)\| = o\,[\sup_{t\geq\tau} \|u(\tau)\|].$$

Error Model 3: Following along the same lines as in error model 2, the error differential equation is given by

$$\dot{e}_1(t) = -\alpha e_1(t) + \phi_k^T u(t) \tag{32}$$

if approach 2 is used. The adaptive law then becomes

$$\Delta\phi_k = \frac{-\epsilon(k)w(k)}{1 + w^T(k)w(k)} \tag{33}$$

where

$$\epsilon(k) \triangleq e_1^2(t)\,|_{t_k}^{t_{k+1}} + 2\alpha\int_{t_k}^{t_{k+1}} e_1^2(\tau)\, d\tau \tag{34}$$

and

$$w(k) \triangleq 2\int_{t_k}^{t_{k+1}} e_1(\tau)u(\tau)\, d\tau. \tag{35}$$

As before, we obtain

$$|W(s)\phi_k^T u(t)| = o\,[\sup_{t\geq\tau} \|u(\tau)\|] \tag{36}$$

where $W(s) = 1/(s + \alpha)$.

Comments:

1) The detailed analysis presented in this section is based on the conviction that efficient design methods for adaptive systems can arise only from a deeper understanding of the behavior of corresponding error models. For each model considered three specific cases have been discussed. The first two assume that the input u to the error model is uniformly bounded. The results are particularly relevant to the identification problem where the plant to be identified is assumed to be stable and the input to the plant is uniformly bounded. When the hybrid adaptive algorithms described in this section are used to identify such a plant, the output errors will tend to zero and the parameters will tend to the true values if the input is persistently exciting. The error model used and the specific algorithm chosen depend upon the parametrization of the plant as well as the application.

2) The main result of this section is relevant to the control problem as well as to identification problems where the vector u cannot be guaranteed to be uniformly bounded *a priori* (e.g., the parallel model). When any one of the adaptive laws given in this section is used, we obtain a relationship between the growth rates of the input and the output. While in approach 1, it is only assumed that $u \in \mathcal{L}_e^\infty$, in approach 2, we require the stronger assumption (23) that $\|\dot{u}(t)\| \leq M_1 \sup_{t\geq\tau} \|u(\tau)\| + M_2$ for $M_1, M_2 \in \mathcal{R}^+$. The principal results can be summarized as follows:

error model 1 : $\phi_k^T u = e_1 = \rho\|u\|\ \rho \in \mathcal{L}^2$ using approach 1

and $= o\ [\sup_{t\geq\tau} \|u(\tau)\|]$ using approach 2

error model 2: $\|W(s)\phi_k^T u\| = o\ [\sup_{t\geq\tau} \|u(\tau)\|]$

using approaches 1 and 2.

If the stronger assumption (23) is made even in approach 1, it can be shown that the same result $\phi_k^T u = o\,[\sup_{t\geq\tau} \|u(\tau)\|]$ obtained for all the other cases is valid even for error model 1. As shown in the next section, in the adaptive control problem, condition (23) is indeed satisfied by the relevant signals in the error models so that the analysis of this section can be directly applied.

3) The two approaches used to develop the adaptive algorithms in the three error models are conceptually different. In the first approach, the discrete Lyapunov function is a quadratic form in the parameter error vector (error model 1) or both parameter and output vectors (error models 2 and 3). The direction in which $\Delta\phi$ is adjusted is the average gradient of $e_1^2(t)$ with respect to ϕ over an interval $[t_k, t_{k+1})$. In contrast to this the second approach attempts to minimize the integral of $e_1^2(t)$ over an interval so that $\Delta\phi$ is the gradient of this performance index. Using the same Lyapunov function it can be shown that a convex combination of the two

adaptive laws also assures the boundedness of the parameter errors. While some qualitative statements are made in Section IV regarding the simulations of error models using both approaches, more work is needed to understand the conceptual differences between them and the manner in which they can be combined in time-varying and noisy situations.

4) The fact that the period T_k need not be a constant in the above error models is worth noting. This may have interesting implications in situations where a time varying period can be used to improve the response of the system.

III. Application, Extension, and Refinement

The concepts and techniques developed in Section II find wide application in adaptive systems where considerations of robustness demand a hybrid approach. The most obvious of such applications is the design of stable hybrid adaptive controllers. The proof of global stability of adaptive systems using such controllers is presented in Section III-A. The same proof can also be extended to discrete systems where data are collected at a faster rate than that at which the parameters are adjusted. This is briefly outlined in Section III-B. For the sake of completeness it is also shown in Section III-C that algorithms of an integral type [5] suggested for continuous systems can be considered as natural generalizations of the algorithms developed for hybrid and discrete systems. Finally, in Section III-C it is shown that the analytical results can also be extended to cases where time-varying gain matrices are used in the adaptive law, as for example, in the recursive least-squares approach.

A. Stable Hybrid Adaptive Control

A continuous-time plant P to be controlled is completely represented by the input–output pair $\{u(t), y_p(t)\}$ and can be modeled by a time-invariant system

$$\dot{x}_p = A_p x_p + b_p u$$
$$y_p = h_p^T x_p$$

where A_p is an $n \times n$ matrix, and h_p and b_p are n-vectors. The transfer function of the plant is $W_p(s)$ where

$$W_p(s) = h_p^T (sI - A_p)^{-1} b_p \equiv \frac{K_p Z_p(s)}{R_p(s)}$$

is rational with $W_p(s)$ strictly proper, $Z_p(s)$ a monic polynomial of degree $m(\leq n-1)$, $R_p(s)$ a monic polynomial of degree n, and K_p a constant gain parameter. It is further assumed that the sign of K_p, the order of the plant and the relative degree $n^* = n - m$ of the plant are known and that the monic polynomial $Z_p(s)$ is Hurwitz. A reference model M which represents the behavior desired from the plant has a uniformly bounded piecewise continuous input $r(\cdot)$, an output $y_m(\cdot)$, and a transfer function $W_M(s)$, where

$$W_M(s) \equiv \frac{K_M Z_M(s)}{R_M(s)}$$

where $Z_M(s)$ is a monic polynomial of degree $m < n$, $R_M(s)$ is a monic Hurwitz polynomial of degree n, and K_M is a constant. The aim of the adaptive control is to generate an input $u(\cdot)$ to the plant using a differentiator free controller so that $\lim_{t\to\infty} |e_1(t)| = \lim_{t\to\infty} |y_p(t) - y_m(t)| = 0$.

The solution to this problem for both continuous [1], [2] and discrete cases [3], [4] is well known. For the hybrid adaptive control problem we use an identical structure for the controller as in [1] but merely adjust the controller parameters at discrete instants. The controller structure may be described as follows:

$$\dot{v}^{(1)} = Fv^{(1)} + bu \qquad \dot{v}^{(2)} = Fv^{(2)} + by_p \tag{37}$$

$$u = c_0 \gamma + c^T v^{(1)} + d_0 y_p + d^T v^{(2)}$$

where $c^T = \{c_1, c_2, \cdots, c_{n-1}\}$, $d^T = \{d_1, d_2, \cdots, d_{n-1}\}$, $v^{(1)}$, $v^{(2)}: \mathcal{R}^+ \to \mathcal{R}^{n-1}$ and F is an $(n-1) \times (n-1)$ stable matrix.

Defining $\bar{\omega}^T(t) \triangleq [r(t), v^{(1)^T}(t), y_p(t), v^{(2)^T}(t)]$ and $\bar{\theta}^T(t) \triangleq [c_0(t), c^T(t), d_0(t), d^T(t)]$ the control input to the plant may be expressed as

$$u(t) = \bar{\theta}^T(t)\bar{\omega}(t). \tag{38}$$

We shall refer to $\bar{\omega}$ as the vector of sensitivity functions. It is well known that a unique constant vector $\bar{\theta}^*$ exists such that the transfer function of the plant together with a controller matches exactly the transfer function of the model when $\bar{\theta}(t) \equiv \bar{\theta}^*$. The objective of the adaptive control problem is to determine the adaptive laws for updating the parameter vector $\bar{\theta}(t)$ using all available data such that all signals in the system remain bounded while $e_1(t) \to 0$ as $t \to \infty$. In the continuous control problem $\bar{\theta}(t)$ is adjusted continuously. In the hybrid problem under consideration $\bar{\theta}(t)$ is updated at discrete instants $t_i (i \in N)$ and remains constant (i.e., $\bar{\theta}_i$) over each interval $[t_i, t_{i+1})$. It is desired to obtain a discrete adaptive law for adjusting $\bar{\theta}_i$, and hence $\bar{\phi}_i (\triangleq \bar{\theta}_i - \bar{\theta}^*)$ so that for any infinite unbounded sequence $\{t_i\}$ with $|t_i - t_{i+1}|$ bounded $\forall i \in N$, the overall system will be globally stable and $e_1(t) \to 0$ as $t \to \infty$.

Special cases exist for the control problem as for example when $W_M(s)$ is a strictly positive real transfer function. We consider below only the general case when $W_M(s)$ has a relative degree $n^* \geq 2$ but with the gain K_p known. In what follows we show that the adaptive law (4) results in global stability. Similar arguments can be used to show that all other adaptive algorithms in Section II when suitably applied also result in global stability. While due to space limitations, we cannot present the details of such arguments here, the remarks at the end of this section indicate their main features. In the following discussions it is assumed that $K_p = K_M = 1$.

Defining $\bar{\theta}^T(t) = [c_0, \theta(t)]$, $\bar{\omega}^T(t) = [r(t), \omega^T(t)]$ and $\bar{\phi}^T(t) = [0, \phi(t)]$ the input and the output to the plant can be expressed as

$$u(t) = r(t) + \theta^T(t)\omega(t) \tag{39}$$
$$y_p(t) = W_M(s)[r(t) + \phi^T(t)\omega(t)]$$

and the error equation as

$$e_1(t) = W_M(s)\phi^T(t)\omega(t). \tag{40}$$

To generate an adaptive law for adjusting $\theta(t)$, an auxiliary signal $y_a(\cdot)$ is added to $e_1(\cdot)$ where

$$y_a(t) = [\theta^T(t) W_M(s) I - W_M(s)\theta^T(t)]\omega(t) \tag{41}$$

so that

$$\phi^T(t)\xi(t) = e_1(t) + y_a(t) \triangleq \epsilon_1(t) \tag{42}$$

where $W_M(s) I\omega \triangleq \xi$. $\epsilon_1(t)$ is referred to as the augmented error.

Proof of Stability: Equation (42) relating the augmented error $\epsilon_1(t)$ to the parameter error $\phi(t)$ has the same form as the error model analyzed in Section II. Hence, the following adaptive law is used to update ϕ_k:

$$\Delta\phi_k = -\frac{1}{T_k}\int_{t_k}^{t_{k+1}} \frac{\epsilon_1(\tau)\xi(\tau)}{1 + \xi^T(\tau)\xi(\tau)} d\tau. \tag{43}$$

The following arguments show that the assumption that output of the plant $y_p(t)$ grows in an unbounded fashion results in a contradiction if the adaptive law (43) is used.

As shown in Section II, the adaptive law (43) assures

i) boundedness of the parameter error vector $\phi(t)$, and

ii) $\epsilon_1 = \rho\sqrt{1 + \xi^T\xi}$ where $\rho \in \mathcal{L}^2$.

Further, since $W_M(s) I\omega = \xi$ and $\lim_{k\to\infty} \Delta\phi_k = 0$ by the adaptive law, by Lemma 1 in the Appendix, we have

iii) $$W_M(s)\phi_k^T\omega(t) - \phi_k^T\xi(t) = o\ [\sup_{t\geq\tau} \|\omega(\tau)\|]. \tag{44}$$

Let $y_p(t)$ grow in an unbounded fashion. Since the control parameter vector $\theta(t)$ is uniformly bounded, it is known [1] that the signals $y_p(t)$, $\|v^{(2)}(t)\|$, $\|\omega(t)\|$, and $\|\xi(t)\|$ grow at the same rate, i.e.,

$$\sup_{t\geq\tau} |y_p(\tau)| \sim \sup_{t\geq\tau} \|v^{(2)}(\tau)\| \sim \sup_{t\geq\tau} \|\omega(\tau)\| \sim \sup_{t\geq\tau} \|\xi(\tau)\|. \tag{45}$$

From (42),

$$e_1(t)=\rho(t)\sqrt{1+\xi^T(t)\xi(t)}+o\,[\sup_{t\geq\tau}\|\omega(\tau)\|]. \tag{46}$$

Since $y_p(t) = y_m(t) + e_1(t)$ where $y_m(t)$ is uniformly bounded, from (37) it follows that $v^{(2)}(t) = o\,[\sup_{t\geq\tau}\|\omega(\tau)\|]$ which contradicts (45). Hence, $y_p(t)$ as well as all the signals in the feedback loop are uniformly bounded and $\lim_{t\to\infty} e_1(t) = 0$ and $\lim_{t\to\infty} \epsilon_1(t) = 0$.

Remarks:

1) The proof of stability given above assures that $\lim_{t\to\infty} e_1(t) = 0$ when the first approach is used with error model 1. If the second approach is used instead, the adaptive law would have the form

$$\Delta\phi_k=-\frac{\int_{t_k}^{t_{k+1}}\epsilon_1^2(\tau)\,d\tau\int_{t_k}^{t_{k+1}}\epsilon_1(\tau)\xi(\tau)\,d\tau}{1+\left\|\int_{t_k}^{t_{k+1}}\epsilon_1(\tau)\xi(\tau)\,d\tau\right\|^2}.$$

Since $W_M(s)I\omega = \xi$ where $W_M(s)$ is a transfer function with poles and zeros in the open left half plane and since $\|\dot{\omega}\| \leq M_1\|\omega\| + M_2$ for some M_1 and $M_2 \in \mathcal{R}^+$ in the adaptive feedback loop when the control parameter vector is bounded, it follows from [1] that $\|\dot{\xi}(t)\| \leq M_3 \sup_{t\geq\tau}\|\xi(\tau)\| + M_4$ for $M_3, M_4 \in \mathcal{R}^+$. Hence, the results derived using approach 2 apply to the error model. Once again, the proof of stability is established by contradictions by showing that $\epsilon_1(t) = o\,[\sup_{t\geq\tau}\|\omega(\tau)\|]$.

2) Similar results can be derived using error model 3 and either approach 1 or approach 2. In these cases, while the structure of the adaptive controller remains the same, the structure of the adaptive scheme used is different and is identical to that used in [1] for the continuous case.

B. Extension to Systems with Two Time-Scales

The hybrid error models described in Section II can be considered as systems which operate on two time-scales—a time scale associated with the continuous time functions and a second with the discrete parameters. Such situations also arise frequently in purely discrete systems where the inputs and outputs are observed at a certain rate but the control parameters are adjusted at a slower rate. It is also interesting to note that algorithms recently suggested for adjusting control parameters in continuous-time systems [5] can be interpreted as the continuous counterparts of such discrete-time systems operating on two time-scales. As shown in this section, the methods suggested in Section II can be considered to provide a unified approach to two time-scale problems in discrete, continuous, and hybrid systems.

Discrete-Time Models: As mentioned in the Introduction, recent advances in digital technology have made the implementation of discrete control algorithms in practical systems attractive. As shown in this section, the theory developed in Section II is applicable to the important class of discrete systems in which practical considerations require that control parameters be adjusted at a rate slower than that at which the information is collected.

The first error model corresponding to error model 1 of the hybrid case can be described by the equation

$$\phi_k^T u_l = e_l k,\quad l \in N,\ l \in [kT, (k+1)T-1] \tag{47}$$

where ϕ_k is a constant vector in the interval $[kT, (k+1)T-1]$ and u_l and e_l are the values of the input and output, respectively, at the time instant l. Using approach 1 it can be shown that if the adaptive law

$$\phi_{k+1}-\phi_k=\Delta\phi_k=-\frac{1}{T}\sum_{i=kT}^{(k+1)T-1}\frac{e_iu_i}{1+u_i^Tu_i}\equiv -R_k\phi_k \tag{48}$$

is used, $V(k) = 1/2\phi_k^T\phi_k$ is a Lyapunov function, $\|\phi_k\|$ is bounded if $\|\phi_0\|$ is bounded and

$$\Delta V(k)\equiv V(k+1)-V(k)=-\phi_k^T\left[I-\frac{R_k}{2}\right]R_k\phi_k\leq 0 \tag{49}$$

which yields

$$\lim_{i\to\infty}\frac{e_i}{(1+u_i^Tu_i)^{1/2}}=0 \qquad i\in N. \tag{50}$$

If the control parameters of a discrete adaptive system are to be adjusted using this error model, the augmented error is generated exactly as in Section III-A, using the discrete model transfer function $W_M(z)$. If $\phi_k^T\xi(l) = \epsilon_1(l)$ the adaptive law using approach 1 is given by (48) as

$$\Delta\phi_k=-\frac{1}{T}\sum_{i=kT}^{(k+1)T-1}\frac{\epsilon_1(i)\xi(i)}{1+\xi(i)^T\xi(i)}. \tag{51}$$

Similarly, if approach 2 is used,

$$\Delta\phi_k=-\frac{\sum_{i=kT}^{(k+1)T-1}\epsilon_1^2(i)\sum_{i=kT}^{(k+1)T-1}\epsilon_1(i)\xi(i)}{1+\left\|\sum_{i=kT}^{(k+1)T-1}\epsilon_1(i)\xi(i)\right\|^2}. \tag{52}$$

In both cases, the proof of stability follows along the same lines as in Section III-A.

Continuous-Time Models: In the continuous time error model

$$\phi^T(t)u(t)=e_1(t) \qquad t\in\mathcal{R}^+ \tag{53}$$

it is well known that the adaptive law

$$\dot{\phi}(t)=-\frac{e_1(t)u(t)}{1+u^T(t)u(t)} \tag{54}$$

results in a bounded parameter error vector. Recently, other continuous adaptive laws have been suggested [5] which utilize past input–output data in adjusting adaptive parameters. We shall refer to such adaptive algorithms as integral algorithms in contrast to the point algorithm (54). By a proper definition of the error model such algorithms can be shown to be generalizations of the hybrid and discrete algorithms developed in Sections II and III-B.

Let the unknown parameter error vector at time t be $\phi(t)$ and let the output $e_1:\mathcal{R}^+ \times \mathcal{R}^+ \to \mathcal{R}$ be defined by

$$\phi^T(t)u(\tau)=e_1(t,\tau) \qquad \tau\leq t,\ \tau,\ t\in\mathcal{R}^+. \tag{55}$$

The adaptive law

$$\dot{\phi}(t)=-\frac{1}{T}\int_{t-T}^{t}\frac{u(\tau)e_1(t,\tau)}{1+u^T(\tau)u(\tau)}\,d\tau \tag{56}$$

which uses input–output data over the interval $[t - T, t)$ is a generalization of (48) for the discrete case. However, unlike the discrete algorithm, (56) poses two major difficulties in implementation. The first involves the storage of the values of the function $u(\cdot)$ over the interval of integration. The second and significantly greater problem is caused by the fact that $e_1(t, \tau)$ cannot be measured directly for use in the adaptive law and cannot be computed from (55) since $\phi(t)$ is unknown.

The above problem can be circumvented by noting the error models of the type (55) arise in adaptive situations (as described in Section III-A) where a parameter vector $\theta(t)$ is adjusted and has to evolve to a desired but unknown constant vector θ^*, i.e., $\phi(t) = \theta(t) - \theta^*$.

Hence, the error model (55) becomes

$$[\theta(t)-\theta^*]^T u(\tau) = e_1(t, \tau) \tag{57}$$

or

$$\theta^T(t)u(\tau) - y_m(\tau) = e_1(t, \tau) \tag{58}$$

where $y_m(\tau)$ is the signal produced by the model and can be measured. Hence, in such cases the adaptive law can be implemented as

$$\dot{\phi}(t) = \dot{\theta}(t) = -\frac{1}{T}\int_{t-T}^{t} \frac{u(\tau)u^T(\tau)\theta(t) - u(\tau)y_m(\tau)}{1+u^T(\tau)u(\tau)}\, d\tau. \tag{59}$$

As mentioned earlier, the implementation of (59) is rendered difficult by the fact that the values of $u(\cdot)$ have to be stored over a window of length T. To overcome this problem the length of the interval T is increased to t so that the entire past data is used but a weighting factor $e^{-q(t-\tau)}$ is included to assure the convergence of the integral. Such an exponentially weighted adaptive algorithm has the form

$$\dot{\phi}(t) = \dot{\theta}(t) = -\int_{t_0}^{t} e^{-q(t-\tau)} \frac{u(\tau)u^T(\tau)\theta(t) - u(\tau)y_m(\tau)}{1+u^T(\tau)u(\tau)}\, d\tau$$

which can be conveniently realized by the matrix differential equations

$$\dot{\theta}(t) = -R(t)\theta(t) - r(t) \qquad \theta(t_0) = 0 \tag{60}$$

$$\dot{R}(t) = -qR(t) + \frac{u(t)u^T(t)}{1+u^T(t)u(t)} \qquad R(t_0) = 0$$

$$\dot{r}(t) = -qr(t) - \frac{u(t)y_m(t)}{1+u^T(t)u(t)} \qquad r(t_0) = 0.$$

The adaptive law (60) is precisely the one suggested in [5].

In conclusion, the approach developed in Section II is seen to unify discrete, continuous, and hybrid adaptive algorithms with two time-scales.

C. Adaptive Gain

In the discussions in the preceding sections, adaptive gains were not included in the adaptive laws to focus attention on the principal results. Experience with complex adaptive systems has, however, shown that the speed of convergence of the algorithms depends critically on the choice of the adaptive gains. In particular, a time-varying gain matrix obtained from least-squares considerations is found to be generally acceptable for most applications. In this section it is briefly shown that similar time-varying gain matrices can also be included in the hybrid adaptive laws. The details are included only for the first error model.

If the first error model is described by

$$\phi_k^T u(t) = e_1(t)$$

let

$$R_k \equiv \frac{1}{T_k}\int_{t_k}^{t_{k+1}} \frac{u(\tau)u^T(\tau)}{1+u^T(\tau)u(\tau)}\, d\tau.$$

The adaptive gain matrix Γ_k is defined by

$$\Gamma_{k+1}^{-1} = \Gamma_k^{-1} + R_k \qquad \Gamma_0 = I \tag{61}$$

and the adaptive law is given by

$$\Delta\phi_k = -\frac{\Gamma_k}{T_k}\int_{t_k}^{t_{k+1}} \frac{e_1(\tau)u(\tau)}{1+u^T(\tau)u(\tau)}\, d\tau. \tag{62}$$

For the system described by (1), (61), and (62) it can be shown that

$$V(k) = 1/2\phi_k^T\Gamma_k^{-1}\phi_k \tag{63}$$

is a Lyapunov function resulting in

$$\Delta V(k) = -\frac{1}{2T_k}\int_{t_k}^{t_{k+1}} \frac{e_1^2(\tau)}{1+u^T(\tau)u(\tau)}\, d\tau - \frac{1}{2T_k^2}\int_{t_k}^{t_{k+1}} \frac{e_1(\tau)u^T(\tau)}{1+u^T(\tau)u(\tau)}\, d\tau \times [I-\Gamma_k R_k]\Gamma_k \int_{t_k}^{t_{k+1}} \frac{e_1(\tau)u(\tau)}{1+u^T(\tau)u(\tau)}\, d\tau$$
$$\leq 0 \text{ if } I-\Gamma_k R_k > 0 \qquad \forall k \in N. \tag{64}$$

From (64) it follows that $\|\phi_k\|$ is bounded if $\|\phi_0\|$ is bounded, $\Delta\phi_k \to 0$ as $k \to \infty$ and

$$\int_{t_k}^{t_{k+1}} \frac{e_1^2(\tau)}{1+u^T(\tau)u(\tau)}\, d\tau \to 0 \text{ as } k \to \infty.$$

IV. Simulations

The error models described in Section II and their applications described in Section III have been simulated extensively on the digital computer. We include in this section three typical examples which compare the effectiveness of the two adaptive approaches proposed in Section II. In all cases the parameters are adjusted periodically with a period T, so that $t_k = kT (k \in N)$. The main interest in these simulations is on the effect of T on the speed and accuracy of the responses.

Example 1: The first hybrid error model, described by (1) was simulated when $u(t), \phi(t) \in \mathbb{R}^2$ and the input vector u is defined by

$$u_1(t) = \sin(0.75t) \quad u_2(t) = \sin(2.6t).$$

Fig. 1(a)–(d) shows the evolution of the output error $e_1(t)$ and the parameter error vector $\phi(t)$ when approaches 1 and 2 are used. In Fig. 1(a) and (b) $T = 0.5$ while $T = 5.0$ in Fig. 1(c) and (d). Approach 1 results in rapid convergence of $e_1(t)$ and $\phi(t)$ for $T = 0.5$. As seen from the simulation results the two adaptive algorithms lead to quite different responses of the overall system. Qualitatively speaking, the first approach is better when the sampling rate is high compared to the dominant frequencies of the input, while the second approach tends to get better as the sampling rate is reduced. Further, for a large sampling period, the procedure results in a single adjustment of the adaptive parameter vector after which it remains essentially a constant—a phenomenon observed in continuous time algorithms with a large adaptive gain. As mentioned earlier, more work is needed for a better understanding of the relationship between the sampling period and the effectiveness of the two approaches.

Example 2: Fig. 2(a)–(d) shows the evolution of $e_1(t)$ and $\phi(t)$ when the same experiments as in Example 1 were performed on error model 3. The basic features of the responses using the two approaches remain the same indicating that the approach rather than the specific error model chosen governs the behavior of the transient response.

Example 3: In this example, all the signals of interest are discrete, although input and output are defined for all $k \in N$ and the parameter error vector is adjusted periodically with period $T \in N$. In the second-order system simulated

$$u_1(k) = \sin(0.05k) \quad u_2(k) = \sin(0.25k).$$

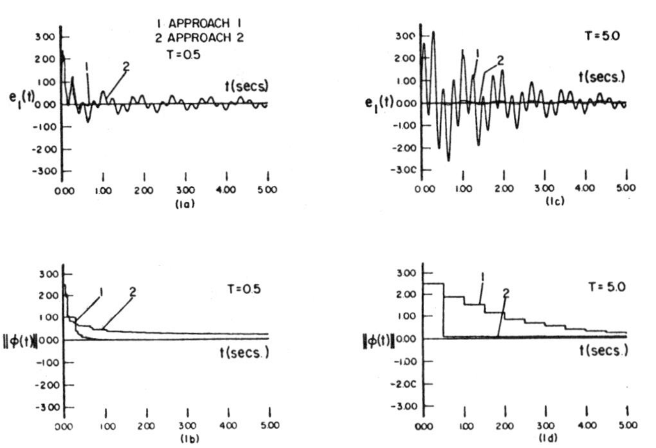

Fig. 1. Evolution of $e_1(t)$ and $\phi(t)$—error model 1.

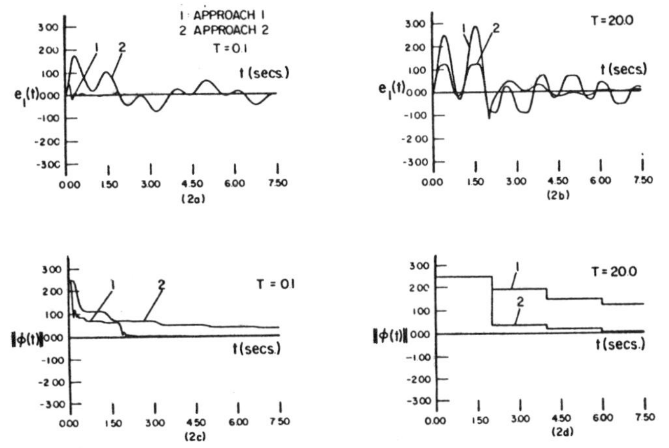

Fig. 2. Evolution of $e_1(t)$ and $\phi(t)$—error model 3.

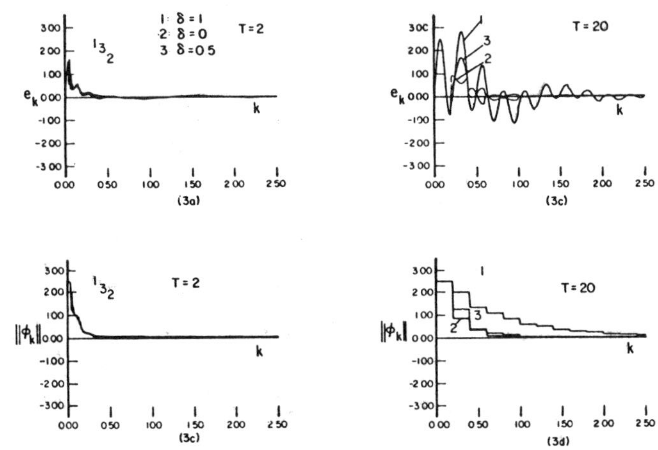

Fig. 3. Discrete error model with two time scales.

The adaptive law used in this case to adjust $\Delta\phi$ had the form

$$\Delta\phi_k = \delta\Delta\phi_1(k) + (1-\delta)\Delta\phi_2(k) \qquad 0 \le \delta \le 1$$

where $\Delta\phi_1(k)$ and $\Delta\phi_2(k)$ are the adaptive laws given by the two approaches. As might be expected $\delta \sim 1$ for small values of T and $\delta \sim 0$ for large values of T are found to be satisfactory as seen in Fig. 3(a)–(d).

Appendix I

Definition: Let $x(\cdot), y(\cdot) \in \mathcal{L}_e^\infty$. We denote $y(t) = o[x(t)]$ if there exists a function $\beta(\cdot)$ and $t_o \in \mathbb{R}^+$ such that $\beta(t) \to 0$ as $t \to \infty$ and $y(t) = \beta(t)x(t)$ for all $t \ge t_o$.

Definition: Let $\{x(k)\}$, $\{y(k)\}$ be two sequences in l_e^∞. If there exists a $k_o \in N$ and a sequence $\{\beta(k)\}$ such that $y(k) = \beta(k)x(k)$ for all $k \ge k_o$ and $\beta(k) \to 0$ as $k \to \infty$ then we denote $y(k) = o[x(k)]$.

Definition: If $x(\cdot)$ and $y(\cdot) \in \mathcal{L}_e^\infty$ and $|y(t)| \le M|x(t)|$ for all $t \ge t_o$, $t_o \in \mathbb{R}^+$ and some constant M, then we denote $y(t) = O[x(t)]$.

Definition: If $\{x(k)\}$ and $\{y(k)\}$ are two sequences in l_e^∞ and $|y(k)| \le Mx(k)\ \forall\, k \ge k_o$ for some constant M and $k_o \in N$, then we denote $y(k) = O[x(k)]$.

Definition: Let $x(\cdot)$ and $y(\cdot) \in \mathcal{L}_e^\infty$. If $y(t) = O[x(t)]$ and $x(t) = O[y(t)]$ then we say that $x(t)$ and $y(t)$ are equivalent and denote this by $x(t) \sim y(t)$. Similarly, if $\{x(k)\}$ and $\{y(k)\}$ are sequences and $y(k) = O[x(k)]$ and $x(k) = O[y(k)]$ we say that the two sequences are equivalent and denote it by $x(k) \sim y(k)$.

Definition: Let $x(\cdot), y(\cdot) \in \mathcal{L}_e^\infty$. We say that $x(t)$ and $y(t)$ grow at the same rate if $\sup_{t\ge\tau} |y(\tau)| \sim \sup_{t\ge\tau} |x(\tau)|$. Similarly two sequences $\{x(k)\}$ and $\{y(k)\}$ are said to grow at the same rate if $\sup_{k\ge\sigma} |x(\sigma)| \sim \sup_{k\ge\sigma} |y(\sigma)|$.

Lemma 1: Let $\omega(\cdot), \xi(\cdot): \mathbb{R}^+ \to \mathbb{R}^n$ be the input and output, respectively, of a transfer matrix $H(s)I$ where $H(s)$ is a rational transfer function and I is the $n \times n$ unit matrix. Let $H(s)$ have all its poles and zeros in the open left half plane. Further suppose that there is a vector $\phi(t) \in \mathbb{R}^n$ and $\|\phi\|$ is uniformly bounded and

$$\phi(t) = \phi_k \qquad t \in [t_k, t_{k+1}),\ k \in N$$

where ϕ_k is a constant vector and

$$\Delta\phi_k \triangleq \phi_{k+1} - \phi_k \to 0 \ \text{as}\ k \to \infty.$$

Then

$$[\phi^T(t)H(s)I - H(s)\phi^T(t)]\omega(t) = o\ [\sup_{t\ge\tau} \|\omega(\tau)\|]. \tag{A-1}$$

According to Lemma 1 if the input is $\omega(t)$, the outputs of the two systems $\phi^T(t)H(s)$ and $H(s)\phi^T(t)$ differ by $o\ [\sup_{t\ge\tau} \|\omega(\tau)\|]$ if $\Delta\phi_k \to 0$ as $k \to \infty$.

Proof: At time $t = t_{n+1}$

$$\phi^T(t)H(s)I\omega(t) = \left[\phi_0 + \sum_{i=0}^{n-1} \Delta\phi_i\right]^T H(s)I\omega(t). \tag{A-2}$$

If the impulse response of $H(s)$ is $h(t)$ where $|h(t)| \le \beta e^{-rt}$ for some positive constants β and r,

$$\begin{aligned} H(s)\phi^T(t)\omega(t)\Big|_{t=t_{n+1}} &= \phi_0^T H(s)I\omega(t) \\ &\quad + \sum_{i=0}^{n-1} \Delta\phi_i^T \int_{t_{i+1}}^{t_{n+1}} h(t_{n+1}-\tau)\omega(\tau)\, d\tau \\ &= \left[\phi_0 + \sum_{i=0}^{n-1} \Delta\phi_i\right]^T H(s)I\omega(t)\Big|_{t=t_{n+1}} \\ &\quad - \sum_{i=0}^{n-1} C_i^T \int_{t_i}^{t_{i+1}} h[t_{n+1}-\tau]\omega(\tau)\, d\tau \end{aligned} \tag{A-3}$$

where $C_o = \{\phi_n - \phi_0\}$ and $C_i = \Sigma_{j=i}^{n-1} \Delta\phi_j$. Since the vector ϕ is bounded $C_i (i = 0, \cdots, n)$ are bounded. Further since $\Delta\phi_n \to 0$, $C_{n-1} \to 0$ as $n \to \infty$. From (A-2) and (A-3) it follows that

$$\begin{aligned} [\phi^T(t)H(s)I - H(s)\phi^T(t)]\omega(t)\Big|_{t=t_{n+1}} &= \sum_{i=0}^{n-1} C_i^T \int_{t_i}^{t_{i+1}} h[t_{n+1}-\tau]\omega(\tau)\, d\tau \\ &= \sigma[t_{n+1}]. \end{aligned}$$

$$\begin{aligned} |\sigma[t_{n+1}]| &\le \sum_{i=0}^{n-1} |C_i|\beta e^{-rt_{n+1}} \int_{t_i}^{t_{i+1}} e^{+r\tau}\|\omega(\tau)\|\, d\tau \\ &\le \gamma_1\beta \sup_{t_{n+1}\ge\tau} \|\omega(\tau)\| \left\{\sum_{i=0}^{n-1} |C_i| e^{-rt_{n+1}}\right\} \end{aligned} \tag{A-4}$$

for some constant γ_1. Since $|C_n| \to 0$ as $n \to \infty$, the term in the brackets as $n \to \infty$ tends to zero. Hence,

$$|\sigma(t)| = o\ [\sup_{t \geq T} \|\omega(\tau)\|]$$

proving Lemma 1.

ACKNOWLEDGMENT

The authors would like to express their gratitude to two of the reviewers for their helpful suggestions.

REFERENCES

[1] K. S. Narendra, Y. H. Lin, and L. S. Valavani, "Stable adaptive controller design, Part II: Proof of stability," *IEEE Trans. Automat. Contr.*, vol. AC-25, pp. 440–448, June 1980.

[2] A. S. Morse, "Global stability of parameter-adaptive control systems," *IEEE Trans. Automat. Contr.*, vol. AC-25, pp. 433–439, June 1980.

[3] K. S. Narendra and Y. H. Lin, "Stable discrete adaptive control," *IEEE Trans. Automat. Contr.*, vol. AC-25, pp. 456–461, June 1980.

[4] G. C. Goodwin, P. J. Ramadge, and P. E. Caines, "Discrete time multivariable adaptive control," *IEEE Trans. Automat. Contr.*, vol. AC-25, pp. 449–456, June 1980.

[5] G. Kreisselmeier, "On adaptive state regulation," *IEEE Trans. Automat. Contr.*, vol. AC-27, pp. 3–17, Feb. 1982.

[6] P. J. Gawthrop, "Hybrid self-tuning control," *Proc. Inst. Elec. Eng.*, vol. 127, part D, no. 5, Sept. 1980.

[7] R. L. Morris and C. P. Neuman, "Model reference adaptive control with multiple samples between parameter adjustments," *IEEE Trans. Automat. Contr.*, vol. AC-26, pp. 534-537, Apr. 1981.

[8] H. Elliott, "Hybrid adaptive control of continuous time systems," *IEEE Trans. Automat. Contr.*, vol. AC-27, pp. 419–426, Apr. 1982.

[9] R. Cristi and R. V. Monopoli, "Model reference adaptive control: The hybrid approach," in *Proc. 1982 Automat. Contr. Conf.*, pp. 848–851.

[10] A. P. Morgan and K. S. Narendra, "On the uniform asymptotic stability of certain nonautonomous differential equations," *SIAM J. Contr. Optimiz.*, vol. 15, pp. 5–24, Jan. 1977.

[11] K. S. Narendra and I. H. Khalifa, "Stable hybrid adaptive control," Center Syst. Sci., Yale Univ., Tech. Rep. 8204.

Part III
Self-Tuning Regulators

SELF-TUNING regulators (STR) are an important class of self-adaptive systems with increasing theoretical and practical interests. The original concepts were conceived in 1958 by R. E. Kalman, and since the early 1970's a considerable amount of work has appeared on this approach. It now forms an important new branch of adaptive control systems in particular, and of control systems in general. This methodology is attracting the attention of an increasing number of researchers with theoretical and practical interests. The objective of the STR is to control systems with an unknown constant slowly varying parameters. Consequently, the theoretical interest is centered around stability, performance, and convergence of the recursive algorithms involved while the practical interest stems from its potential uses, both as a method for controlling time-varying and nonlinear plants over a wide range of operating points, and for dealing with batch problems over a wide range of operating points where the plant or materials may vary over successive batchs. Successful implementation, however, has been restricted primarily to applications in the pulp and paper, chemical and other resource based industries where the process dynamics are significantly slower than the STR algorithm. Also, some useful applications to adaptive autopilot for tankers is also reported. In this part of the volume, we report a small sample of representative papers which explore the theory of the STR. We will give some practical applications of the approach in Parts VI and VII of this volume.

A basic and important reference entitled "On-Self Tuning Regulators" authored by K. J. Astrom and B. Wittenmark is given in the seventeenth paper. This basic paper shows how the simplicity of adaptive algorithm in combination with its asymptotic properties can be useful for industrial process control. The authors demonstrate the feasibility of the approach by experiments on real processes in the paper and mining industries. No doubt, this paper introduces the basic notions which have enabled us to bring to fruition some of the theoretical results. This basic paper along with other recent papers will be of great interest to our readers for exploring this field in depth.

In the eighteenth paper, D. W. Clarke and P. J. Gawthrop present a summary of the work on STR's until 1979, and give some extension to the design of STR. The authors conclude that the STR has several potentials, but the problem of running a self-tuner for a long period of time on a strongly nonlinear process needs further investigations. It also discusses the technical problems involved with implementing self-tuners on microprocessors.

In the nineteenth paper, U. Borison gives a description of STR for a class of multivariable systems (minimum-phase system with number of inputs is equal to number of outputs). It uses a minimum variance strategy and a recursive parameter estimator in its design. The results of this paper also give insight into the case when several single-input single-output self-tuning regulators are operated in a cascade mode.

The pole-zero placement is an important concept that we have been using in the conventional control theory. Here we present two papers which discuss the pole-zero assignment for a class of self-tuning regulators. One of the earlier papers, ninth paper, also discusses the stability problem associated with such systems.

In the twentieth paper entitled "Self-Tuning Controllers Based on Pole-Zero Placement," Astrom and Wittenmark discuss some explicit and implicit schemes for the pole-zero placement design. The authors demonstrate usefullness of the algorithms by means of some simulation studies.

In a related paper (twenty-first paper) Prager and Wellstead give a theory and application of a multivariable self-tuning regulator where the control objective is the assignment of the closed-loop poles set to some prespecified locations. The algorithm described in the paper has a "self-tuning" property. The authors claim the more robustness of such tuners than the tuners that are based on optimal control synthesis method. This paper concludes with a discussion of the computer implementation of the method, together with applications to both simulated and real plants.

In the twenty-second paper, V. Peterka presents the design of predictor-based self-tuning controller. The paper deals with both servo and regulator problems. Unlike the standard Riccati equation, the method presented is numberically robust and, therefore, is suitable for real-time computation on microprocessor with reduced wordlength.

Finally, in a recent paper (the twenty-third paper), Wittenmark and Astrom discuss various practical issues in the implementation of self-tuning control. No doubt self-tuning controllers are being applied to many industrial processes, but still there are many practical issues that must be considered to get a properly working algorithm. The authors point out many practical problems that must be studied. This paper should help a newcomer to the field of adaptive control as there are a large number of methodologies and algorithms which may lead to confusion and unnecessary misunderstanding.

In summary, it is necessary to stress that, at the design stage, it is important to have as much *a priori* knowledge about the process as possible. Self-tuning regulators have been successful. It should be pointed out, however, that there is much to be done both theoretically and practically before adaptive and self-tuning controllers can be given in the hands of inexperienced users for routine applications. Microprocessor-based algorithmic controllers certainly, if designed properly, can help a practical user in this regard. Several applications of STR's are reported in the applications part of this volume.

Selected Bibliography

A. Y. Allidina and F. M. Hughes, "Generalized self-tuning controller with pole assignment," *Proc. IEE,* vol. 127, no. 1, pp. 13–18, 1980.

K. J. Astrom and B. Wittenmark, "On self-tuning regulators," *Automatica,* vol. 9, pp. 195–199, 1973.

——, "Analysis of a self-tuning regulator for non-minimum phase systems," *Automatica,* vol. 9, pp. 195–199, 1974.

K. J. Astrom, U. Borisson, L. Ljung, and B. Wittenmark, "Theory and application of self tuning regulators," *Automatica,* vol. 13, no. 5, pp. 457–476, 1977.

K. J. Astrom, B. Westerburg, and B. Wittenmark, "Self-tuning controllers based on pole-placement design," Lund Rep. LUTFD2/(TFRT-3/48)/1-052, Lund Institute of Technology, Lund, Sweden, 1978.

K. J. Astrom, "Self-tuning regulators-design principle and applications," in *Applications of Adaptive Control,* K. S. Narendra and R. V. Monopoli, Eds. New York: Academic Press, 1980.

K. J. Astrom and B. Wittenmark, "Self-tuning controllers based on pole-zero placement," *IEE Proc.,* vol. 127, pp. 120–130, 1980.

——, "Self-tuning controllers based on pole-zero placement," *IEE Proc.,* vol. 127, pt. D, no. 3, pp. 120–130, 1980.

U. Borisson, "Self-tuning regulators for a class of multivariable systems," *Automatica,* vol. 15, no. 2, pp. 209–217, 1979.

D. W. Clarke and J. P. Gawthrip, "Self tuning controller," *Proc. IEE,* vol. 122(a), pp. 929–934, 1975.

——, "Self-tuning control," *IEE Proc.,* vol. 126, pp. 633–640, 1979.

J. P. Clary and G. F. Franklin, "Self-tuning control with *a priori* plant knowledge," in *Proc. 23rd IEEE Conf. on Decision and Control,* pp. 369–374, 1984.

A. O. Cordero and D. Q. Mayne, "Deterministic convergence of a self-tuning regulator with variable forgetting factor," *Proc. IEE,* vol. 128, pt. D, no. 1, pp. 19–23, 1981.

R. M. C. De Keyzer and A. R. Van Cauwenberghe, "A self-tuning predictor as operator guide," IFAC Symp. Identif. Syst. Parameter Estim., 1979.

C. De Souza and G. C. Goodwin, "Robustness effects of sampling in mode reference and minimum variance control," presented at the 22nd CDC Conference, 1983.

E. P. Eremin, N. T. Loan, and G. S. Chkhartishvilli, "Synthesis of self-adjusting systems under random disturbances," *Automat. and Remote Contr.,* vol. 35, no. 7, pt. 1, pp. 1092–1096, 1973.

P. J. Gawthrop, "Some interpretations of the self-tuning controller," *Proc. IEEE,* vol. 124, pp. 889–894, 1977.

——, "On the stability and convergence of a self-tuning controller," *Int. J. Control,* vol. 31, no. 5, pp. 973–998, 1980.

P. Hagander and B. Wittenmark, "A self tuning filter for fixed-lag smoothing," *IEEE Trans. Inform. Theory,* vol. IT-23, no. 3, pp. 377–384, 1977.

L. Ljung and B. Wittenmark, "Asymptotic properties of self tuning regulators," Rep. 7404, Division of Automatic Control, Lund Institute of Technology, Lund, Sweden, 1974.

V. A. Sastry, D. E. Seborg, and R. K. Wood, "Self-tuning regulator applied to a binary distillation column," *Automatica,* vol. 13, no. 4, pp. 417–424, 1977.

B. L. Stevens, "STR for nonprime plant with some known dynamics," in *Proc. 23rd IEEE Conf. Decision and Control,* pp. 387–391, 1984.

P. E. Wellstead, D. Prager, and P. Zanker, "Pole assignment self tuning regulator," *Proc. IEE,* vol. 126, no. 8, pp. 781–787, 1979.

P. E. Wellstead and P. Zanker, "Servo self-tuners," *Int. J. Control,* vol. 30, pp. 27–36, 1979.

B. Wittenmark, "A self-tuning predictor," *IEEE Trans. Automat. Contr.,* vol. 19, pp. 848–851, 1974.

B. Wittenmark and K. J. Astrom, "Implementation aspects of adaptive controllers and their influence on robustness," 21st Conf. Decis. Control, 1982.

On Self Tuning Regulators*

Sur les Régulateurs Auto-Syntonisants

Über selbsteinstellende Regler

О самонастраивающихся регуляторах

K. J. ÅSTRÖM and B. WITTENMARK

Control laws obtained by combining a least squares parameter estimator and a minimum variance strategy based on the estimated parameters have asymptotically optimal performance.

Summary—The problem of controlling a system with constant but unknown parameters is considered. The analysis is restricted to discrete time single-input single-output systems. An algorithm obtained by combining a least squares estimator with a minimum variance regulator computed from the estimated model is analysed. The main results are two theorems which characterize the closed loop system obtained under the assumption that the parameter estimates converge. The first theorem states that certain covariances of the output and certain cross-covariances of the control variable and the output will vanish under weak assumptions on the system to be controlled. In the second theorem it is assumed that the system to be controlled is a general linear stochastic *n*th order system. It is shown that if the parameter estimates converge the control law obtained is in fact the minimum variance control law that could be computed if the parameters of the system were known. This is somewhat surprising since the least squares estimate is biased. Some practical implications of the results are discussed. In particular it is shown that the algorithm can be feasibly implemented on a small process computer.

1. INTRODUCTION

IT HAS been shown in several cases that linear stochastic control theory can be used successfully to design regulators for the steady state control of industrial processes. See Ref. [1]. To use this theory it is necessary to have mathematical models of the system dynamics and of the disturbances. In practice it is thus necessary to go through the steps of plant experiments, parameter estimation, computation of control strategies and implementation. This procedure can be quite time consuming in particular if the computations are made off-line. It might also be necessary to repeat the procedure if the system dynamics or the characteristics of the disturbances are changing as is often the case for industrial processes.

From a practical point of view it is thus meaningful to consider the control of systems with constant but unknown parameters. Optimal control problems for such systems can be formulated and solved using non-linear stochastic control theory. The solutions obtained are extremely impractical since even very simple problems will require computations far exceeding the capabilities of todays computers. For systems with constant but unknown parameters it thus seems reasonable to look for strategies that will converge to the optimal strategies that could be derived if the system characteristics were known. Such algorithms will be called *self-tuning* or *self-adjusting* strategies. The word adaptive is not used since adaptive, although never rigorously defined, usually implies that the characteristics of the process are changing. The problem to be discussed is thus simpler than the adaptive problem in the sense that the system to be controlled is assumed to have constant parameters.

The purpose of the paper is to analyse one class of self-adjusting regulators. The analysis is restricted to single-input single-output systems. It is assumed that the disturbances can be characterized as filtered white noise. The criterion considered is the minimization of the variance of the output. The algorithms analysed are those obtained on the basis of a separation of identification and control. To obtain a simple algorithm the identification is simply a least squares parameter estimator.

The main result is a characterization of the closed loop systems obtained when the algorithm is applied to a general class of systems. It is shown in Theorem 5.1 that if the parameter estimates converge the closed loop loop system obtained will be such that certain covariances of the inputs and the outputs of the closed loop system are zero. This is shown under weak assumptions on the system to be

* Received 2 March 1972; revised 12 September 1972. The original version of this paper was presented at the 5th IFAC Congress which was held in Paris, France during June 1972. It was recommended for publication in revised form by Associate Editor A. Sage.

Reprinted with permission from *Automatica*, vol. 9, pp. 185–199, Mar. 1973.

Copyright © 1973, International Federation of Automatic Control IFAC. Published by Pergamon Press Ltd.

controlled. Moreover if it is assumed that the system to be controlled is a sampled finite dimensional linear stochastic system with a time delay in the control signal it is demonstrated in Theorem 5.2 that, if the parameter estimates converge, the corresponding regulator will actually converge to the minimum variance regulator. This is true, in spite of the fact that the least squares estimate is biased.

The major assumptions are that the system is minimum phase, that the time delay is known and that a bound can be given to the order of the system. The first two assumptions can be removed at the price of a more complicated algorithm.

The paper is organized as follows: sections 2 and 3 provide background and a motivation. The algorithm is given in section 4. Control strategies for systems with known parameters are given in section 2. Least squares parameter estimation is briefly reviewed in section 3. Some aspects on the notion of identifiability is also given in section 3. The algorithm presented in section 4 is obtained simply by fitting the parameters of a least squares structure as was described in section 3 and computing the corresponding minimum variance control strategy as was described in section 2. The possible difficulty with non-identifiability due to the feedback is avoided by fixing one parameter.

The main result is given as two theorems in section 5. We have not yet been able to prove that the algorithm converges in general. In section 6 it is, however, shown that a modified version of the algorithm converges for a first order system. The convergence properties of the algorithm are further illustrated by the examples in section 7. Some practical aspects of the algorithm as well as some problems which remain to be solved are given in section 8. In particular it is shown that the algorithm is easily implemented on a minicomputer.

2. MINIMUM VARIANCE CONTROL

This section gives the minimum variance strategy for a system with known parameters. Consider a system described by

$$y(t)+a_1y(t-1)+\dots+a_ny(t-n)=b_1u(t-k-1)+\dots+b_nu(t-k-n)+\lambda[e(t)+c_1e(t-1)+\dots+c_ne(t-n)],\quad t=0,\pm 1,\pm 2,\dots \tag{2.1}$$

where u is the control variable, y is the output and $\{e(t), t=0, \pm 1, \pm 2, \dots\}$ is a sequence of independent normal (0, 1) random variables. If the forward shift operator q, defined by

$$qy(t)=y(t+1)$$

and the polynomials

$$A(z)=z^n+a_1z^{n-1}+\dots+a_n$$

$$B(z)=b_1z^{n-1}+\dots+b_n,\ b_1\neq 0$$

$$C(z)=z^n+c_1z^{n-1}+\dots+c_n$$

are introduced, the equation (2.1) describing the system can be written in the following compact form:

$$A(q)y(t)=B(q)u(t-k)+\lambda C(q)e(t). \tag{2.2}$$

It is well known that (2.1) or (2.2) is a canonical representation of a sampled finite dimensional single-input single-output dynamical system with time delays in the output and disturbances that are gaussian random processes with rational spectral densities.

The model (2.1) also admits a time delay τ in the system input which need not be a multiple of the sampling interval. The number k corresponds to the integral part of τ/h, where h is the sampling interval.

Let the criterion be

$$V_1=Ey^2(t) \tag{2.3}$$

or

$$V_2=E\frac{1}{N}\sum_1^N y^2(t). \tag{2.4}$$

The optimal strategy is then

$$u(t)=-\frac{q^kG(q)}{B(q)F(q)}y(t) \tag{2.5}$$

where F and G are polynomials

$$F(z)=z^k+f_1z^{k-1}+\dots+f_k \tag{2.6}$$

$$G(z)=g_0z^{n-1}+g_1z^{n-2}+\dots+g_{n-1} \tag{2.7}$$

determined from the identity

$$q^kC(q)=A(q)F(q)+G(q). \tag{2.8}$$

Proofs of these statements are given in [2]. The following conditions are necessary:

—The polynomial B has all zeroes inside the unit circle. Thus the system (2.1) is minimum phase.

—The polynomial C has all zeroes inside the unit circle.

These conditions are discussed at length in [1]. Let it suffice to mention here that if the system (2.1) is non-minimum phase the control strategy (2.5) will still be a minimum variance strategy. This strategy will, however, be so sensitive that the slightest variation in the parameters will create an unstable closed loop system. Suboptimal strategies which are less sensitive to parameter variations are also well known. This paper will, however, be limited to minimum phase systems.

3. PARAMETER ESTIMATION

For a system described by (2.1) it is thus straight forward to obtain the minimum variance regulator, if the parameters of the model are known. If the parameters are not known it might be a possibility to try to determine the parameters of (2.1) using some identification scheme and then use the control law (2.5) with the true parameters substituted by their estimates. A suitable identification algorithm is the maximum likelihood method which will give unbiased estimates of the coefficients of the A, B and C polynomials. The maximum likelihood estimates of the parameters of (2.1) are, however, strongly non-linear functions of the inputs and the outputs. Since finite dimensional sufficient statistics are not known it is not possible to compute the maximum likelihood estimate of the parameters of (2.1) recursively as the process develops. Simpler identification schemes are therefore considered.

The least squares structure

The problem of determining the parameters of the model (2.1) is significantly simplified if it is assumed that $c_i=0$ for $i=1, 2, \ldots, n$. The model is then given by

$$A(q)y(t)=B(q)u(t-k)+\lambda e(t+n). \qquad (3.1)$$

The parameters of this model can be determined simply by the least squares method [3]. The model (3.1) is therefore referred to as a *least squares model.*

The least squares estimate has several attractive properties. It can easily be evaluated recursively. The estimator can be modified to take different model structures, e.g. known parameters, into account. The least squares estimates will converge to the true parameters e.g. under the following conditions.

—The output $\{y(t)\}$ is actually generated from a model (3.1).
—The residuals $\{e(t)\}$ are independent.
—The input is persistently exciting, see Ref. [3].
—The input sequence $\{u(t)\}$ is independent of the disturbance sequence $\{e(t)\}$.

These conditions are important. If the residuals are correlated the least squares estimate will be biased. If the input sequence $\{u(t)\}$ depends on $\{e(t)\}$ it may not be possible to determine all parameters.

When the inputs are generated by a feedback they are correlated with the disturbances and it is not obvious that all the parameters of the model can be determined. Neither is it obvious that the input generated in this way is persistently exciting of sufficiently high order. A simple example illustrates the point.

Example 3.1

Consider the first order model

$$y(t)+ay(t-1)=bu(t-1)+e(t). \qquad (3.2)$$

Assume that a linear regulator with constant gain

$$u(t)=\alpha y(t) \qquad (3.3)$$

is used. If the parameters a and b are known the gain $\alpha=a/b$ would obviously correspond to a minimum variance regulator. If the parameters are not known the gain $\alpha=\hat{a}/\hat{b}$ where $\hat{a}$ and $\hat{b}$ are the least squares estimates of a and b could be attempted. The least squares parameter estimates are determined in such a way that the loss function

$$V(a, b)=\sum_1^N[y(t+1)+ay(t)-bu(t)]^2 \qquad (3.4)$$

is minimal with respect to a and b. If the feedback control (3.3) is used the inputs and outputs are linearly related through

$$u(t)-\alpha y(t)=0. \qquad (3.5)$$

Multiply (3.5) by $-\gamma$ and add to the expression within brackets in (3.4).

Hence

$$V(a, b)=\sum_1^N[y(t+1)+(a+\alpha\gamma)y(t)-(b+\gamma)u(t)]^2$$

$$=V(a+\alpha\gamma, b+\gamma).$$

The loss function will thus have the same value for all estimates a and b on a linear manifold. Thus the two parameters a and b of the model (3.2) are not identifiable when the feedback (3.3) is used.

The simple example shows that it is in general not possible to estimate all the parameters of the model (3.1) when the input is generated by a feedback. Notice, however, that all parameters can be estimated if the control law is changed. In the particular example it is possible to estimate both parameters of the model, if the control law (3.3) is replaced by

$$u(t)=\alpha y(t-1)$$

or

$$u(t)=\alpha_1 y(t)+\alpha_2 y(t-1)$$

or if a time varying gain is used.

4. THE ALGORITHM

In order to control a system with constant but unknown parameters the following procedure could be attempted. At each step of time determine the parameters of the system (3.1) using least squares estimation based on all previously observed inputs and outputs as was described in section 3. Then determine a control law by calculating the minimum variance strategy for the model obtained. To compute the control law the identity (2.8) must be resolved in each step. The problem of computing the minimum variance regulator is simplified if it is observed that by using the identity (2.8) the system (3.1) can be written as

$$y(t+k+1)+\alpha_1 y(t)+\ldots+\alpha_{\mathrm{m}}(t-m+1)$$
$$=\beta_0[u(t)+\beta_1 u(t-1)+\ldots+\beta_l u(t-l)]$$
$$+\varepsilon(t+k+1) \tag{4.1}$$

where $m=n$ and $l=n+k-1$. The coefficients α_i and β_i are computed from the parameters a_i and b_i in (3.1) using the identity (2.8). The disturbance $\varepsilon(t)$ is a moving average of order k of the driving noise $e(t)$.

The minimum variance strategy is then simply

$$u(t)=\frac{1}{\beta_0}[\alpha_1 y(t)+\ldots+\alpha_m y(t-m+1)]$$
$$-\beta_1 u(t-1)-\ldots-\beta_l u(t-l). \tag{4.2}$$

In order to obtain simple computation of the control strategy it could thus be attempted to use the model structure (4.1) which also admits least squares estimation. The trade-off for the simple calculation of the control law is that k more parameters have to be estimated.

As was shown in Example 3.1 all parameters of the model (4.1) can not necessarily be determined from input–output observations if the input is generated by a feedback (4.2) with constant parameters. In order to avoid a possible difficulty it is therefore assumed that the parameter β_0 is given. It will be shown in section 6 that the choice of β_0 is not crucial.

Summing up, the algorithm can be described as follows.

Step 1 *parameter estimation*

At the sampling interval t determine the parameters $\alpha_1,\ldots\alpha_m,\ \beta_1,\ldots,\beta_l$ of the model

$$y(t)+\alpha_1 y(t-k-1)+\ldots+\alpha_m y(t-k-m)$$
$$=\beta_0[u(t-k-1)+\beta_1 u(t-k-2)$$
$$+\ldots+\beta_l u(t-k-l-1)]+\varepsilon(t) \tag{4.1}$$

using least squares estimation based on all data available at time t, i.e.

$$\sum_{k=0}^{t}\varepsilon^2(k)$$

minimum. The parameter β_0 is assumed known.

Step 2 *control*

At each sampling interval determine the control variable from

$$u(t)=\frac{1}{\beta_0}[\alpha_1 y(t)+\ldots+\alpha_m y(t-m+1)]$$
$$-\beta_1 u(t-1)-\ldots-\beta_l u(t-l) \tag{4.2}$$

where the parameters α_i and β_i are those obtained in Step 1.

The control law (4.2) corresponds to

$$u(t)=\frac{\alpha_1+\alpha_2 q^{-1}+\ldots+\alpha_m q^{-m+1}}{\beta_0[1+\beta_1 q^{-1}+\ldots+\beta_l q^{-l}]}y(t)$$
$$=\frac{q^{l-m+1}\mathscr{A}(q)}{\mathscr{B}(q)}y(t). \tag{4.3}$$

Since the least squares estimate can be computed recursively the algorithm requires only moderate computations.

It should be emphasized that the algorithm is not optimal in the sense that it minimizes the criterion (2.3), or the criterion (2.4). It fails to minimize (2.3) because it is not taken into account that the parameter estimates are inaccurate and it fails to minimize (2.4) because it is not dual in FELDBAUM's sense [4]. These matters are discussed in [2]. It will however, be shown in section 5 that the algorithm has nice asymptotic properties.

The idea of obtaining algorithms by a combination of least squares identification and control is old. An early reference is KALMAN [5]. The particular algorithm used here is in essence the same as the one presented by PETERKA [6]. A similar algorithm where the uncertainties of the parameters are also considered is given in WIESLANDER–WITTENMARK [7].

5. MAIN RESULTS

The properties of the algorithm given in the previous section will now be analysed. We have

Theorem 5.1

Assume that the parameter estimates $\alpha_i(t)$, $i=1, \ldots, m$, $\beta_i(t)$, $i=1, \ldots l$ converge as $t \to \infty$ and that the closed loop system is such that the output is *ergodic* (in the second moments). Then the closed loop system has the properties

$$Ey(t+\tau)y(t)=r_y(\tau)=0 \qquad \tau=k+1, \ldots k+m \qquad (5.1)$$

$$Ey(t+\tau)u(t)=r_{yu}(\tau)=0 \qquad \tau=k+1, \ldots, k+l+1. \qquad (5.2)$$

Proof

The least squares estimates of the parameters $\alpha_1, \alpha_2, \ldots, \alpha_m, \beta_1, \beta_2, \ldots, \beta_l$ is given by the equations

$$\left[\begin{array}{cccc|ccc}
\Sigma y(t)^2 & \Sigma y(t)y(t-1) & \ldots & \Sigma y(t)y(t-m+1) & -\beta_0\Sigma y(t)u & \ldots & -\beta_0\Sigma y(t)u(t-l) \\
\Sigma y(t)y(t-1) & & & \Sigma y(t-1)y(t-m+1) & \vdots & & \vdots \\
\vdots & & & \vdots & & & \\
\Sigma y(t)y(t-m+1) & & & \Sigma y^2(t-m+1) & -\beta_0\Sigma y(t-m+1)u(t-1) & \ldots & -\beta_0\Sigma y(t-m+1)u(t-l) \\
\hline
-\beta_0\Sigma y(t)u(t-1) & & & & \beta_0^2\Sigma u^2(t-1) & \ldots & \beta_0^2\Sigma u(t-1)u(t-l) \\
\vdots & & & & & & \vdots \\
-\beta_0\Sigma y(t)u(t-l) & \ldots & & & & \ldots & \beta_0^2\Sigma u^2(t-l)
\end{array}\right].$$

$$\begin{bmatrix} \alpha_1 \\ \alpha_2 \\ \vdots \\ \alpha_m \\ \beta_1 \\ \vdots \\ \beta_l \end{bmatrix} = \begin{bmatrix} -\Sigma y(t+k+1)y(t)+\beta_0\Sigma u(t)y(t) \\ \Sigma y(t+k+1)y(t-1)+\beta_0\Sigma u(t)y(t-1) \\ \vdots \\ -\Sigma y(t+k+1)y(t-m+1)+\beta_0\Sigma u(t)y(t-m+1) \\ -\beta_0\Sigma y(t+k+1)u(t-1)-\beta_0^2\Sigma u(t)u(t-1) \\ \vdots \\ \beta_0\Sigma y(t+k+1)u(t-l)-\beta_0^2\Sigma u(t)u(t-l) \end{bmatrix} \qquad (5.3)$$

where the sums are taken over N_0 values. See Ref. [3].

Assume that the parameters converge. For sufficiently large N_0 the coefficients of control law (4.2) will then converge to constant values. Introduction of (4.2) into (5.3) gives

$$\begin{aligned}
&\Sigma y(t+k+1)y(t)=0 \\
&\Sigma y(t+k+1)y(t-1)=0 \\
&\quad\vdots \\
&\Sigma y(t+k+1)y(t-m+1)=0 \\
&\Sigma y(t+k+1)u(t-1)=0 \\
&\quad\vdots \\
&\Sigma y(t+k+1)u(t-l)=0.
\end{aligned}$$

Using the control law (4.2) it also follows that

$$\Sigma y(t+k+1)u(t)=0.$$

Under the ergodicity assumption the sums can furthermore be replaced by mathematical expectations and the theorem is proven.

Remark 1

Notice that the assumptions on the system to be controlled are very weak. In particular it is not necessary to assume that the system is governed by an equation like (2.1) or (3.1).

Remark 2

It is sufficient for ergodicity that the system to be controlled is governed by a difference equation of finite order, e.g. like (2.1), and that the closed loop system obtained by introducing the feedback law into (2.1) gives a stable closed loop system.

Remark 3

The self tuning algorithm can be compared with a PI-regulator. If the state variables of a deterministic system with a PI-regulator converge to steady state values, the control error must be zero irrespective of the properties of the system. Analogously theorem 5.1 implies that if the parameter estimates of the self tuning algorithm converge the covariances (5.1) and (5.2) are zero.

Remark 4

Theorem 5.1 holds even if the algorithm is modified in such a way that the parameter estimation (Step 1) is not done in every step.

If it is assumed that the system to be controlled is governed by an equation like (2.1) it is possible to show that the conditions (5.1) and (5.2) in essence imply that the self tuning regulator will converge to a minimum variance regulator. We have

Theorem 5.2

Let the system to be controlled be governed by the equation (2.1). Assume that the self tuning algorithm is used with $m=n$ and $l=n+k-1$. If the parameter estimates converge to values such that the corresponding polynomials $\mathscr{A}$ and $\mathscr{B}$ have no common factors, then the corresponding regulator (4.2) will converge to a minimum variance regulator.

Proof

Assume that the least squares parameter estimates converge. The regulator is then given by (4.3) i.e.

$$u(t)=\frac{q^{l-m+1}\mathscr{A}(q)}{\mathscr{B}(q)}y(t)=\frac{q^k\mathscr{A}(q)}{\mathscr{B}(q)}$$

where the coefficients of $\mathscr{A}$ and $\mathscr{B}$ are constant. Since the system to be controlled is governed by (2.1) the closed loop system becomes

$$[A(q)\mathscr{B}(q)-B(q)\mathscr{A}(q)]y(t)=\lambda C(q)\mathscr{B}(q)e(t). \qquad (5.4)$$

The closed loop system is of order $r=n+l$. Introduce the stochastic process $\{v(t)\}$ defined by

$$v(t)=\lambda\frac{q^lC(q)}{A(q)\mathscr{B}(q)-B(q)\mathscr{A}(q)}e(t). \qquad (5.5)$$

Then

$$y(t)=q^{-l}\mathscr{B}(q)v(t) \qquad (5.6)$$

and

$$u(t)=q^{-m+1}\mathscr{A}(q)v(t). \qquad (5.7)$$

Multiplying (5.6) and (5.7) by $y(t+\tau)$ and taking mathematical expectations gives

$$r_y(\tau)=r_{yv}(\tau)+\beta_1r_{yv}(\tau+1)\ldots+\beta_lr_{yv}(\tau+l) \qquad (5.8)$$

$$r_{yu}(\tau)=\alpha_1r_{yv}(\tau)+\alpha_2r_{yv}(\tau+1)$$
$$+\ldots+\alpha_mr_{yv}(\tau+m-1). \qquad (5.9)$$

Furthermore it follows from Theorem 5.1, equations (5.1) and (5.2), that the left member of (5.8) vanishes for $\tau=k+1,\ldots,k+m$ and that the left member of (5.9) vanishes for $\tau=k+1,\ldots,k+l+1$. We thus obtain the following equation for $r_{yv}(\tau)$.

$$\begin{bmatrix} 1 & \beta_1 & \cdots & \beta_l & 0 & \cdots & 0 \\ 0 & 1 & \beta_1 & & \cdot & & \cdot \\ \cdot & & \cdot & & \cdot & & \cdot \\ \cdot & & & \cdot & & \cdot & \cdot \\ \cdot & & & & \cdot & & 0 \\ 0 & \cdots & 0 & 1 & \beta_1 & & \beta_l \\ 0 & \alpha_1 & \cdots & \alpha_m & 0 & \cdots & 0 \\ \cdot & & \cdot & & \cdot & & \cdot \\ \cdot & & & \cdot & & \cdot & \cdot \\ \cdot & & & & \cdot & & \cdot \\ & & & & & & 0 \\ 0 & \cdots & & \alpha_1 & & \cdots & \alpha_m \end{bmatrix} \begin{bmatrix} r_{yv}(k+1) \\ r_{yv}(k+2) \\ \cdot \\ \cdot \\ \cdot \\ \\ \\ \\ \\ \\ \\ r_{yv}(k+l+m) \end{bmatrix} = \begin{bmatrix} 0 \\ 0 \\ \cdot \\ \cdot \\ \cdot \\ \\ \\ \\ \\ \\ \\ 0 \end{bmatrix} \qquad (5.10)$$

Since the polynomials $\mathscr{A}$ and $\mathscr{B}$ have no common factor it follows from an elementary result in the theory of equations [8, p. 145] that the $(l+m) \times (l+m)$-matrix of the left member of (5.10) is non-singular.

Hence

$$r_{yv}(\tau)=0 \qquad \tau=k+1,\ldots,k+l+m. \tag{5.11}$$

Since v is the output of a dynamical system of order $r=n+l=m+l$ driven by white noise it follows from (5.11) and the Yule–Walker equation that

$$r_{yv}(\tau)=0 \qquad \tau\geqslant k+1. \tag{5.12}$$

The equation (5.8) then implies

$$r_{y}(\tau)=0 \qquad \tau\geqslant k+1. \tag{5.13}$$

The output of the closed loop system is thus a moving average of white noise of order k. Denote this moving average by

$$y(t)=q^{-k}F(q)e(t) \tag{5.14}$$

where F is a polynomial of degree k. It follows from (5.4) and (5.14) that

$$q^kC=FA-\frac{BF\mathscr{A}}{\mathscr{B}}.$$

Since q^kC and FA are polynomials $BF\mathscr{A}/\mathscr{B}$ must also be a polynomial.

Hence

$$q^kC=FA+G \tag{5.15}$$

where

$$G=-\frac{BF\mathscr{A}}{\mathscr{B}} \tag{5.16}$$

is of degree $n-1$ and F of degree k. A comparison with (2.8) shows, however, that (5.15) is the identity which is used to derive the minimum variance strategy. It thus follows from (2.5) that the minimum variance strategy is given by

$$u(t)=-\frac{q^kG}{BF}y(t).$$

The equation (5.16) then implies that

$$-\frac{q^kG}{BF}=\frac{q^k\mathscr{A}}{\mathscr{B}} \tag{5.17}$$

and it has thus been shown that (4.2) is a minimum variance strategy.

Remark

The conditions $m=n$ and $l=n+k-1$ mean that there are precisely the number of parameters in the model that are required in order to obtain the correct regulator for the process (2.1). Theorem 5.2 still holds if there are more parameters in the regulator in the following cases.

(i) Theorem 5.2 still holds if $m=n$ and $l\geqslant n+k-1$. In this case the order of the system is $r=n+l$ and since $m=n$ the equation (5.10) implies (5.11)–(5.13) and the equation (5.16) is changed to

$$G=-\frac{q^{l-k-m+1}BF\mathscr{A}}{\mathscr{B}}. \tag{5.16$'$}$$

The rest of the proof remains unchanged.

(ii) If $m\geqslant n$ and $l=n+k-1$ the theorem will also hold. The closed loop system is of order $r\leq m+l$ but the equation (5.10) will still imply (5.11) and (5.16) is changed to (5.16$'$). The rest of the proof remains unchanged.

(iii) If $m>n$ and $l>n+k-1$ the theorem does not hold because $\mathscr{A}$ and $\mathscr{B}$ must have a common factor if the parameter estimates converge. It can, however, be shown that if the algorithm is modified in such a way that common factors of $\mathscr{A}$ and $\mathscr{B}$ are eliminated before the control signal is computed, Theorem 5.2 will still hold for the modified algorithm.

6. CONVERGENCE OF THE ALGORITHM

It would be highly desirable to have general results giving conditions for convergence of the parameter estimates. Since the system (2.1) with the regulator (4.2) and the least squares estimator is described by a set of nonlinear time dependent stochastic difference equations the problem of a general convergence proof is difficult. So far we have not been able to obtain a general result. It has, however, been verified by extensive numerical simulations that the algorithm does in fact converge in many cases. The numerical simulations as well as analysis of simple examples have given insight into some of the conditions that must be imposed in order to ensure that the algorithm will converge.

A significant simplification of the analysis is obtained if the algorithm is modified in such a way that the parameter estimates are kept constant over long periods of time. To be specific a simple example is considered.

Example 6.1

Let the system be described by

$$y(t)+ay(t-1)=bu(t-1)+e(t)+ce(t-1) \quad |c|<1. \tag{6.1}$$

Assume that the control law

$$u(t)=\alpha_n y(t) \qquad (6.2)$$

is used in the time interval $t_n < t < t_{n+1}$ where the parameter α_n is determined by fitting the least squares model

$$y(t+1)+\alpha y(t)=u(t)+\varepsilon(t+1) \qquad (6.3)$$

to the data $\{u(t), y(t), t=t_{n-1}, \ldots, t_n-1)\}$.

The least squares estimate is given by

$$\alpha_n=-\frac{\sum_{t=t_{n-1}}^{t_n-2} y(t)[y(t+1)-u(t)]}{\sum_{t=t_{n-1}}^{t_n-2} y^2(t)}$$

$$=\alpha_{n-1}-\frac{\sum_{t=t_{n-1}}^{t_n-2} y(t+1)y(t)}{\sum_{t=t_{n-1}}^{t_n-2} y^2(t)} \qquad (6.4)$$

where the last equality follows from (6.2). Assume that $t_n-t_{n-1}\to\infty$ and that

$$|a-b\alpha_{n-1}|<1 \qquad (6.5)$$

which means that the closed loop system used during the time interval $t_{n-1}<t<t_n$ is stable then

$$\alpha_n=\alpha_{n-1}-\frac{r_y(1)}{r_y(0)} \qquad (6.6)$$

where $r_y(\tau)$ is the covariance function of the stochastic process $\{y(t)\}$ defined by

$$y(t)+(a-b\alpha_{n-1})y(t-1)=e(t)+ce(t-1). \quad (6.7)$$

Straightforward algebraic manipulations now give

$$\alpha_n=\alpha_{n-1}-\frac{(c-a+b\alpha_{n-1})(1-ac+bc\alpha_{n-1})}{1+c^2-2ac+2bc\alpha_{n-1}}. \quad (6.8)$$

The problem is thus reduced to the analysis of the nonlinear differerence equation given by (6.8).

Introduce

$$x_n=\alpha_n-\frac{a-c}{b} \qquad (6.9)$$

the equation (6.8) then becomes

$$x_{n+1}=g(x_n)=(1-b)x_n+\frac{b^2cx_n^2}{1-c^2+2bcx_n}. \quad (6.10)$$

The point $x=0$ is a fixed point of the mapping g which corresponds to the optimal value of gain of the feedback loop, i.e. $\alpha=(a-c)/b$. The problem is thus to determine if this fixed point is stable. Since the closed loop system is assumed to be stable it is sufficient to consider

$$\frac{c-1}{b}<x<\frac{c+1}{b}. \qquad (6.11)$$

Three cases have to be investigated

1. $c=0$
2. $c>0$ or $c<0$, $0<b\leqslant 1$
3. $c>0$ or $c<0$, $1<b<2$.

For all cases $g(x)\approx(1-b)x$ if x is small. This implies that solutions close to 0 converge to $x=0$ if $0<b<2$.

Case 1

The equation (6.10) reduces to

$$x_{n+1}=(1-b)x_n$$

and the fixed point $x=0$ is stable if $|1-b|<1$ i.e. $0<b<2$.

Case 2

The pincipal behavior of $g(x)$ when $c>0$ and $0<b\leqslant 1$ is shown in Fig. 1. It is straightforward to verify that all initial values in the stability region $(c-1)/b<x<(c+1)/b$ will give solutions which converge to zero.

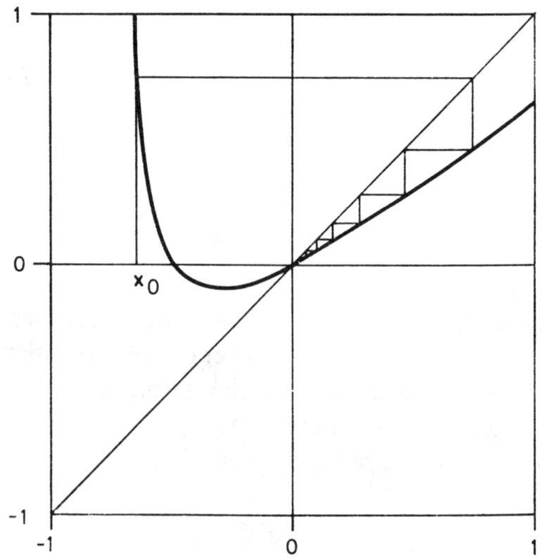

FIG. 1. Graph of the function g when $0<b\leqslant 1$. The figure is drawn with the parameter values $a=-0{\cdot}5$, $b=0{\cdot}5$ and $c=0{\cdot}7$.

Case 3

The function $g(x)$ for this case is shown in Fig. 2. It is not obvious that x will converge to zero, because there might exist a "limit cycle".

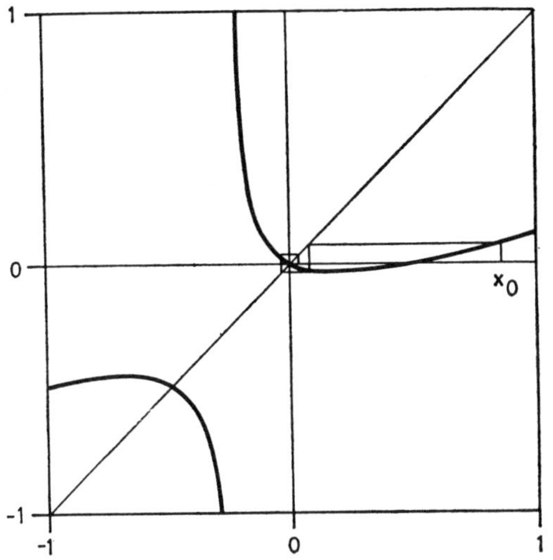

FIG. 2. Graph of the function g when $1<b<2$. The figure is drawn with the parameter values $a=-0{\cdot}5$, $b=1{\cdot}5$ and $c=0{\cdot}7$.

If $c>0$ and starting with $x_o>0$ then if $g(x_o)<0$ it can be shown that after two iterations the new value x_2 will satisfy $0<x_2<x_o$ i.e.

$$0<g(g(x))<x \qquad \text{if } x>0,\, g(x)<0.$$

If $(c-1)/b<x<0$ then $g(x)$ will be positive and can be taken as a new initial value for which the condition above will be satisfied. If $c<0$ then it can be shown that $g(g(x))>x$ if $x<0$ and $g(x)>0$.

Summary

From the analysis above we can conclude that $x=0$ is a stable fixed point if

$$-1<c<1$$

and

$$0<b<2. \tag{6.12}$$

The example shows that under the condition (6.12) the version of the self-tuning algorithm where the parameters of the control law are kept constant over long intervals will in fact converge. In the analysis above $\beta_o=1$ was chosen. If $\beta_o\neq 1$ then the condition (6.12) is replaced by

$$0<b/\beta_o<2 \tag{6.12'}$$

or

$$0.5b<\beta_o<\infty. \tag{6.13}$$

The condition (6.13) implies that it is necessary to pick the parameter β_o in a correct manner. The algorithm will always converge if β_o is greater than b. Under-estimation may be serious and the value $\beta_o<0{\cdot}5b$ gives an unstable algorithm.

The analysis presented in the simple example can be extended to give stability conditions for the modified algorithm in more complex cases. The analysis required is tedious.

7. SIMULATIONS

The results in section 5 are given under the assumption that the least squares estimator really converges, but yet we have not been able to give general conditions for convergence. But simulation of numerous examples have shown that the algorithm has nice convergence properties.

This section presents a number of simulated examples which illustrate the properties of the self-tuning algorithm.

Example 7.1

Let the system be

$$y(t)+ay(t-1)=bu(t-1)+e(t)+ce(t-1) \tag{7.1}$$

with $a=-0{\cdot}5$, $b=3$ and $c=0{\cdot}7$. The minimum variance regulator for the system is

$$u(t)=\frac{a-c}{b}y(t)=-0{\cdot}4y(t). \tag{7.2}$$

A regulator with this structure can be obtained by using the self-tuning algorithm based on the model

$$y(t+1)+\alpha y(t)=\beta_o u(t)+\varepsilon(t+1). \tag{7.3}$$

Figure 3 shows for the case $\beta_o=1$ how the parameter estimate converges to the value $\alpha=-0{\cdot}4$ which corresponds to the minimum variance strategy (7.2).

In Figure 4 is shown the expected variance of the output if the current value of α should be used for all future steps of time. Notice that the algorithm has practically adjusted over 50 steps.

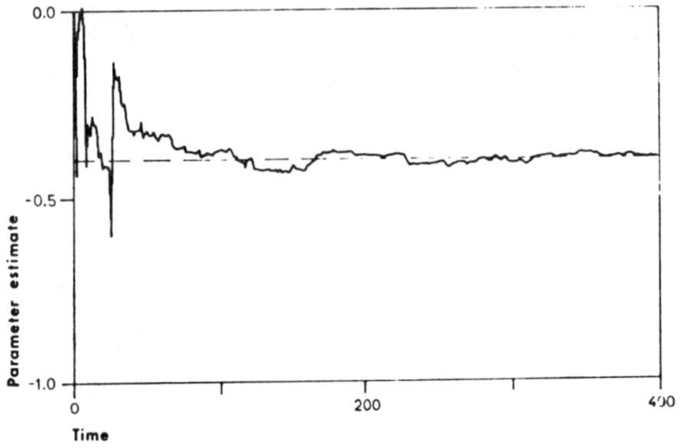

FIG. 3. Parameter estimate $\alpha(t)$ obtained when the self tuning algorithm based on the model (7.3) is applied to the system given by (7.1). The minimum variance regulator corresponds to $\alpha=-0{\cdot}4$ and is indicated by the dashed line.

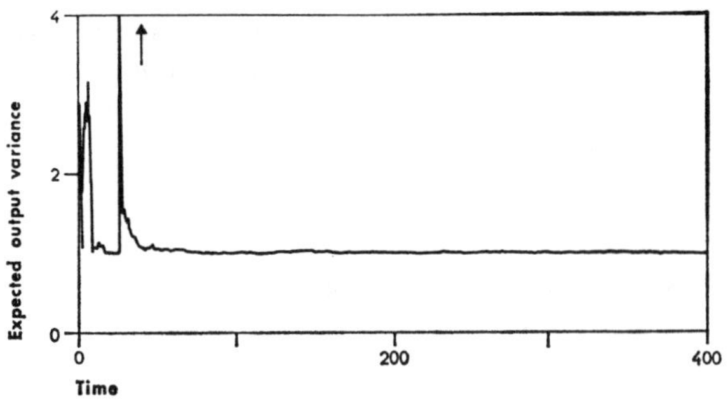

FIG. 4. Expected variance of the output of Example 7.1 if the control law obtained at time t is kept constant for all future times. Notice that the estimate at time $t=26$ would give an unstable system.

The analysis of Example 6.1 shows that, since $b>2$, and $\beta_o=1$ the modified self-tuning algorithm obtained when the parameters of the controller are kept constant over long intervals is unstable. The simulation in Example 7.1 shows that at least in the special case a conservative estimate of the convergence region is obtained by analysing the modified algorithm. If the value of b is increased further it has been shown that the algorithm is unstable. Unstable realizations have been found for $b=5$. In such cases it is of course easy to obtain a stable algorithm by increasing β_o. This requires, however, a knowledge of the magnitude of b.

The system of Example 7.1 is very simple. For instance, if no control is used the variance will still be reasonably small. The next example is more realistic in this aspect.

Example 7.2

Consider the system

$$y(t)-1.9y(t-1)+0{\cdot}9y(t-2)=u(t-2)$$
$$+u(t-3)+e(t)-0{\cdot}5e(t-1). \qquad (7.4)$$

If no control is used the variance of the output is infinite. Also notice that $B(z)=z-1$. The assumption that B has all zeroes inside the unit circle is thus violated. The minimum variance strategy for the system is

$$u(t)=-1{\cdot}76y(t)+1{\cdot}26y(t-1)-0{\cdot}4u(t-1)$$
$$+1{\cdot}4u(t-2). \qquad (7.5)$$

A regulator with this structure is obtained by using the self-tuning algorithm with the model

$$y(t+2)+\alpha_1y(t)+\alpha_2y(t-1)$$
$$=u(t)+\beta_1u(t-1)+\beta_2u(t-2)+\varepsilon(t+2). \qquad (7.6)$$

The convergence of the parameters is shown in Fig. 5. Figure 6 shows the accumulated losses when using the self-tuning algorithm and when using the optimal minimum variance regulator (7.5).

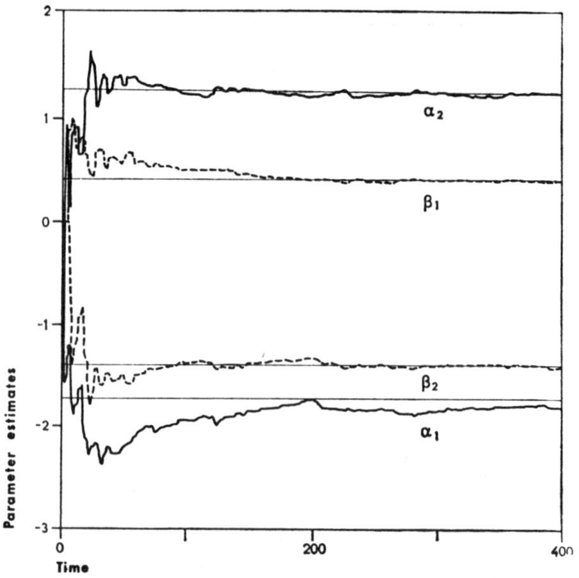

FIG. 5. Parameter estimates α_1, α_2, β_1 and β_2 obtained when the self tuning algorithm based on (7.6) is applied to the system given by (7.4). The thin lines indicate the parameter values of the minimum variance strategy.

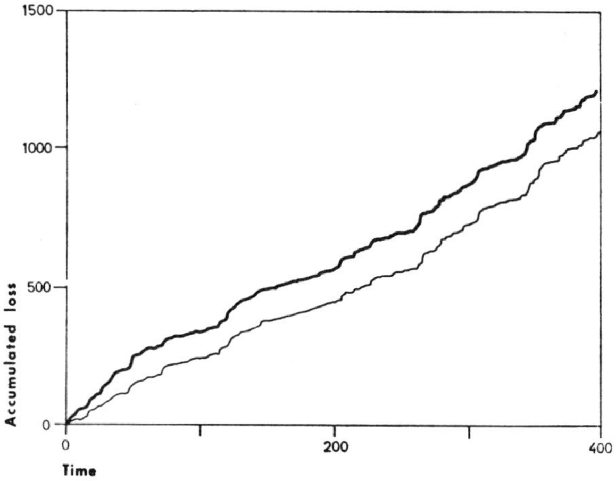

FIG. 6. Accumulated loss

$$\sum_{s=1}^{t} y^2(s)$$

for a simulation of the system (7.4) when using the self tuning algorithm (thick line) and when using the optimal minimum variance regulator (7.5) (thin line).

In both examples above, the models in the self-tuning algorithm have had enough parameters so it could converge to the optimal minimum variance regulator. The next example shows what happens when the regulator has not enough parameters.

Example 7.3

Consider the system

$$y(t)-1{\cdot}60y(t-1)+1{\cdot}61y(t-2)-0{\cdot}776y(t-3)$$
$$=1{\cdot}2u(t-1)-0{\cdot}95u(t-2)+0{\cdot}2u(t-3)+e(t)$$
$$+0{\cdot}1e(t-1)+0{\cdot}25e(t-2)+0{\cdot}87e(t-3). \qquad (7.7)$$

The polynomial $A(z)$ has two complex zeroes near the unit circle $(+0{\cdot}4\pm 0{\cdot}9i)$ and one real zero equal to 0·8.

If a self-tuning regulator is determined based on a model with $m=3$ and $l=2$ it will converge to the minimum variance regulator as expected. Figure 7(a) shows a small sample of the output together with the sample covariance of the output, $\hat{r}_y(\tau)$.

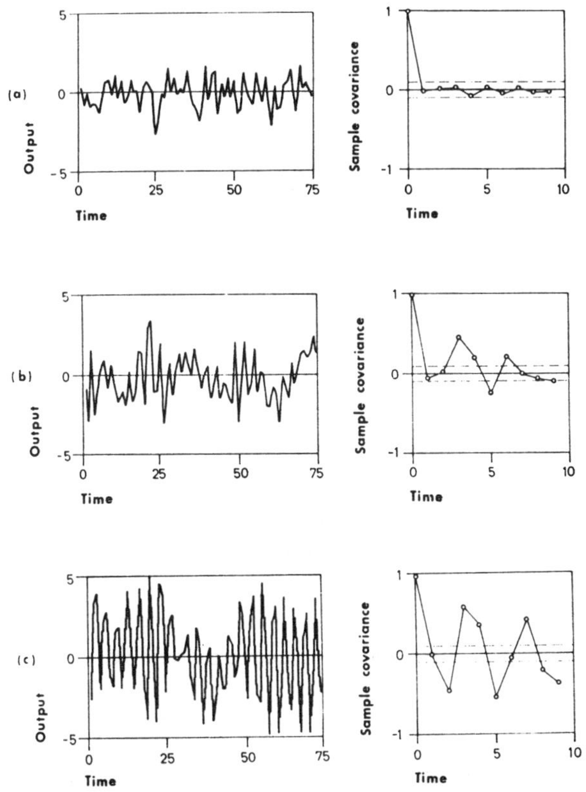

FIG. 7. Output of the system (7.7) and sample covariance of the output $\hat{r}_y(\tau)$ when controlling with self tuning regulator is having different number of parameters

(a) $m=3$, $l=2$ (b) $m=2$, $l=1$ (c) $m=1$, $l=0$.

The dashed lines show the 5 per cent confidence intervals for $\tau\neq 0$.

If the self-tuning algorithm instead is based on a model with $m=2$ and $l=1$ it is no longer possible to obtain the minimum variance regulator for the system since there are not parameters enough in the self-tuning regulator. Theorem 5.1 indicates, however, that if the self-tuning regulator converges, its parameters will be such that the covariances $r_y(1)$, $r_y(2)$, $r_{yu}(1)$ and $r_{yu}(2)$ are all zero. The simulation shows that the algorithm does in fact converge with $\beta_o=1{\cdot}0$. The covariance function of the output is shown in Fig. 7(b). It is seen that the sample covariances $\hat{r}_y(1)$ and $\hat{r}_y(2)$ are within the 5 per cent confidence interval while $\hat{r}_y(3)$ is not as would be expected from Theorem 5.1.

If a self-tuning algorithm is designed based on a model with $m=1$, $l=0$ then Theorem 5.1 indicates that $r_y(1)$ should vanish. Again the simulation shows that the algorithm does in fact converge and that the sample covariance $\hat{r}_y(1)$ does not differ significantly from zero. See Fig. 7(c).

When using regulators of lower order than the optimal minimum variance regulator, the parameters in the controller will not converge to values which for the given structure gives minimum variance of the output. In Table 1 is shown the variance of the output for the system above when using different regulators.

The loss when using the self-adjusting regulator is obtained through simulations. The optimal regulator is found by minimizing $r_y(0)$ with respect to the parameters in the controller.

TABLE 1

		Loss $\frac{1}{N}\sum_{t=1}^{N} Hy_2(t)$	
m	l	Self-adjusting	Optimal
3	2	1·0	1·0
2	1	2·5	1·9
1	0	4·8	3·4

The previous examples are all designed to illustrate various properties of the algorithm. The following example is a summary of a feasibility study which indicates the practicality of the algorithm for application to basis weight control of a paper machine.

Example 7.4

The applicability of minimum variance strategies to basis weight control on a paper machine was demonstrated in [9]. In this application the control loop is a feedback from a wet basis weight signal to thick stock flow. The models used in [9] were obtained by estimating the parameters of (2.1) using the maximum likelihood method. In one particular case the following model was obtained.

$$y(t)=\frac{b_1q^{-1}+b_2q^{-2}}{1+a_1q^{-1}+a_2q^{-2}}u(t-2)+v(t) \quad (7.8)$$

where the output y is basis weight in g/m^2 and the control variable is thick stock flow in g/m^2. The disturbance $\{v(t)\}$ was a drifting stochastic process which could be modelled as

$$v(t)=\lambda\frac{1+c_1q^{-1}+c_2q^{-2}}{(1+a_1q^{-1}+a_2q^{-2})(1-q^{-1})}e(t) \quad (7.9)$$

where $\{e(t)\}$ is white noise. The sampling interval was 36 sec and the numerical values of the parameters obtained through identification were as follows

$$a_1 = -1{\cdot}283$$

$$a_2 = 0{\cdot}495$$

$$b_1 = 2{\cdot}307$$

$$b_2 = -2{\cdot}025$$

$$c_1 = -1{\cdot}438$$

$$c_2 = 0{\cdot}550$$

$$\lambda = 0{\cdot}382.$$

To investigate the feasability of the self-tuning algorithm for basis weight control, the algorithm was simulated using the model (7.8) where the disturbance v was the actual realization obtained from measurements on the paper machine. The parameters of the regulator were chosen as $k=1$, $l=3$, $m=4$ and $\beta_o=2{\cdot}5$ and the initial estimates were set to zero. The algorithm is thus tuning 7 parameters.

The results of the simulation are shown in Figs. 8–10. Figure 8 compares the output obtained when using the self-tuning algorithm with the result obtained when using the minimum variance regulator computed from the process model (7.8) with the disturbance given by (7.9). The reference value was 70 g/m^2. In the worst case the self-tuning regulator gives a control error which is about 1 g/m^2 greater than the minimum variance regulator. This happens only at two sampling intervals.

After about 75 sampling intervals (45 min) the output of the system is very close to the output obtained with the minimum variance regulator.

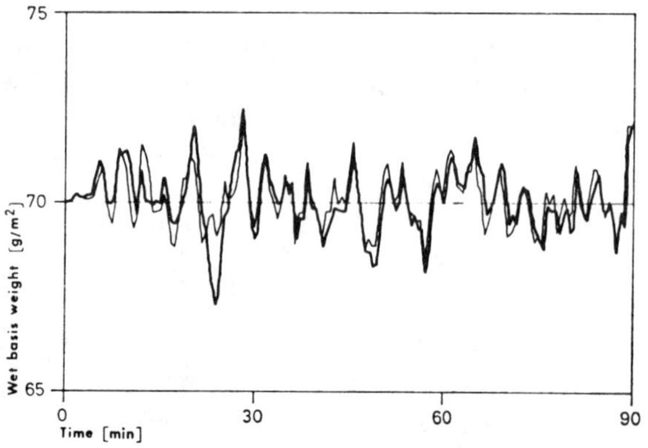

FIG. 8. Wet basis weight when using the self tuning regulator (thick line) and when using the minimum variance regulator based on maximum likelihood identification (thin line). The reference value for the controller was 70g/m².

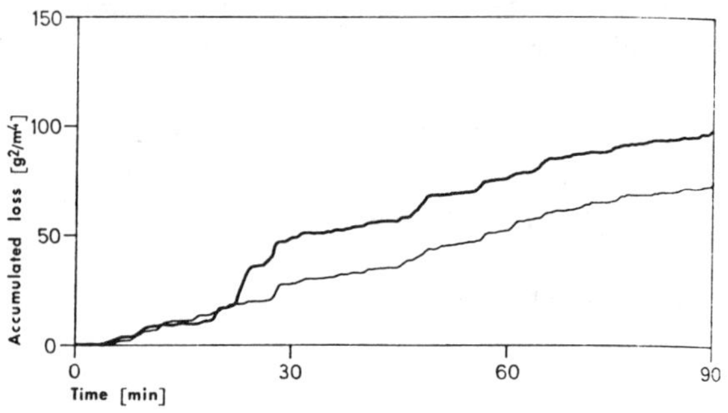

FIG. 9. Accumulated loss for Example 7.4 when using the self tuning regulator (thick line) and when using the minimum variance regulator (thin line).

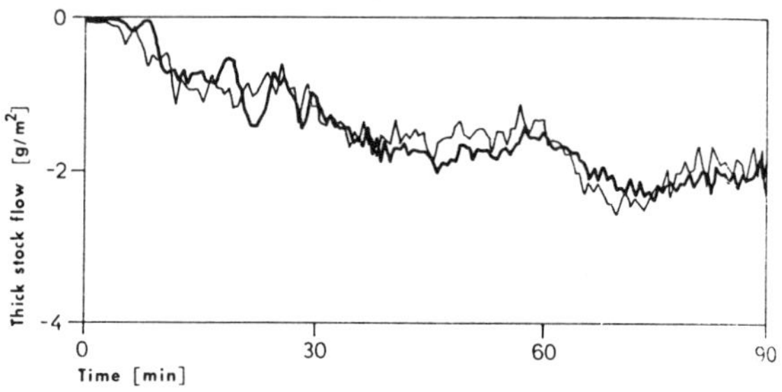

FIG. 10. The control signal in g/m² for Example 7.4 when using the self tuning regulator (thick line) and when using the minimum variance regulator (thin line).

Figure 9 compares the accumulated losses

$$V(t) = \sum_{n=0}^{t} y^2(n)$$

obtained with the minimum variance regulator and the self-tuning regulator. Notice that in the time interval (21, 24) minutes there is a rapid increase in the accumulated loss of the self-tuning regulator of about 17 units. The largest control error during this interval is 2·7 g/m^2 while the largest error of the minimum variance regulator is 1 g/m^2. The accumulated losses over the last hour is 60 units for the self-tuning regulator and 59 units for the minimum variance regulator.

The control signal generated by the self-tuning algorithm is compared with that of the minimum variance regulator in Fig. 10. There are differences in the generated control signals. The minimum variance regulator generates an output which has more rapid variations than the output of the self-tuning regulator.

The parameter estimates obtained have not converged in 100 sampling intervals. In spite of this the regulator obtained will have good performance as has just been illustrated. The example thus indicates that the self-tuning algorithm could be feasible as a basis weight regulator.

8. PRACTICAL ASPECTS

A few practical aspects on the algorithm given in section 4 are presented in this section which also covers some possible extensions of the results

A priori knowledge

The only parameters that must be known à *priori* are k, l, m and β_o. If the algorithm converges it is easy to find out if the à *priori* guesses of the parameters are correct simply by analyzing the sample covariance of the output. Compare Example 7.3. The parameter β_o should be an estimate of the corresponding parameter of the system to be controlled. The choice of β_o is not critical as was shown in the Examples 6.1 and 7.1. In the special cases studied in the examples an under-estimate led to a diverging algorithm while an over-estimate was safe.

Implementation on process computers

It is our belief that the self-tuning algorithm can be conveniently used in process control applications. There are many possibilities. The algorithm can be used as a tool to tune regulators when they are installed. It can be installed among the systems programs and cycled through different control loops repetitively to ensure that the regulators are always properly tuned. For critical loops where the parameters are changing it is also possible to use a dedicated version which allows slowly drifting parameters.

A general self-tuning algorithm requires about 40 FORTRAN statements. When compiled using the standard PDP 15/30 FORTRAN compiler the code consists of 450 memory locations. The number of memory locations required to store the data is $(l-1+m)^2+3(l-1+m)+2k+4$. Execution times on a typical process computer (PDP 15) without floating point hardware are given in the table below. The major part of the computing is to update the least squares estimate.

Number of parameters $l+m$	Execution time ms
1	5
3	16
5	34
8	69

Improved convergence rates

The results of this paper only shows that if the parameters converge the regulator obtained will tend to a minimum variance regulator. Nothing is said about convergence rates, which of course is of great interest from practical as well as theoretical points of view. There are in fact many algorithms that have the correct asymptotic properties. Apart from the algorithm given in section 4 we have the algorithm which minimizes (2.4). But that algorithm is impossible to use due to the computational requirements. It is of interest to investigate if other possible algorithms have better convergence rates than the algorithm of section 4. No complete answer to this problem is yet known. A few possibilities will be indicated. It could be attempted to take into account that the parameter estimates are uncertain. See Refs. [2, 7 and 10]. The least squares identifier can be improved upon by introducing exponential weighting of past data. This has in some cases shown to be advantageous in simulations. Algorithms of this type have in simulations been shown to handle slowly drifting parameters.

Another possibility is to assume that the parameters are Wiener processes, which also can be incorporated fairly easily [2, 7]. It has been verified by simulation that the region of convergence can be improved by introducing a bound on the control signal.

Feed forward

In many industrial applications the control can be improved considerable if feed forward is used. The self tuning regulators in this paper can include feed forward control by changing the process model (4.1) to

$$y(t+k+1)+\alpha_1 y(t)+\ldots+\alpha_m y(t-m+1)$$
$$=\beta_0[u(t)+\beta_1 u(t-1)+\ldots+\beta_l u(t-l)]+\gamma_1 s(t)$$
$$+\ldots+\gamma_p s(t-p+1)+\varepsilon(t+k+1) \qquad (8.1)$$

where $s(t)$ is a known disturbance.

The parameters α_i, β_i and γ_i can be identified as before and the control law (4.2) will be changed to

$$u(t)=\frac{1}{\beta_0}[\alpha_1 y(t)+\ldots+\alpha_m y(t-m+1)-\gamma_1 s(t)$$
$$-\ldots-\gamma_p s(t-p+1)]-\beta_1 u(t-1)$$
$$-\ldots-\beta_l u(t-l). \qquad (8.2)$$

Nonminium phase systems

Difficulties have been found by a straightforward application of the algorithm to nonminimum phase systems, i.e. systems where the polynomial B has zeroes outside the unit circle.

Several ways to get around the difficulty have been found. By using a model with $B(z)=\beta_o$ it has in many cases been possible to obtain stable algorithms at the sacrifice of variance.

It is well-known that the minimum variance regulators are extremely sensitive to parameter variations for nonminimum phase systems [1]. This is usually overcome by using suboptimal strategies which are less sensitive [1]. The same idea can be used for the self-tuning algorithms as well. The drawback is that the computations increase because the polynomials F and G of an identity similar to (2.8) must be determined at each step of the iteration. An alternative is to solve a Riccati-equation at each step.

Multivariable and nonlinear systems

It is possible to construct algorithms that are similar to the one described in section 4 for multivariable and nonlinear systems as long as a model structure which is linear in the parameters [3, p. 131] is chosen. For multivariable systems the structure given in equation (3.2) of Ref. [3] can thus be attempted. Analyses of the properties of the algorithm obtained when applied to a multivariable or a nonlinear system are not yet done.

9. CONCLUSIONS

The paper has been concerned with control of systems with constant but unknown parameters. The analysis has been limited to single-input single-output systems with disturbances in terms of filtered white noise. A control algorithm based on least squares parameter estimation and a minimum variance regulator computed from the estimated parameters has been analysed. Assuming that the parameter estimates converge the closed loop system has been analysed. A characterization of the closed loop system has been given under weak assumption on the system to be controlled. Under stronger assumptions on the system to be controlled it has been shown that the regulator obtained will actually converge to the minimum variance regulator if the estimates converge.

Since the closed loop system is characterized as a nonlinear stochastic system it is very difficult to give general conditions that guarantee that the estimates converge. The convergence has only been treated for simple examples and under further assumptions as in section 6. But simulations of numerous examples indicate that the algorithm has nice convergence properties.

The simplicity of the algrorithm in combination with its asymptotic properties indicate that it can be useful for industrial process control. The feasibility has also been demonstrated by experiments on real processes in the paper and mining industries.

10. REFERENCES

[1] K. J. ÅSTRÖM: *Introduction to Stochastic Control Theory*. Academic Press, New York (1970).

[2] K. J. ÅSTRÖM and B. WITTENMARK: Problems of identification and control. *J. Math. Analysis Applic.* **34,** 90–113 (1971).

[3] K. J. ÅSTRÖM and P. EYKHOFF: System identification—A survey. *Automatica* **7,** 123–162 (1971).

[4] A. A. FELDBAUM: Dual control theory I–IV. *Aut. Remote Control* **21,** 874–880, 1033–1039 (1961); **22,** 1–12, 109–121 (1962).

[5] R. E. KALMAN: Design of a self optimizing control System. *Trans. ASME* **80,** 468–478 (1958).

[6] V. PETERKA: Adaptive Digital Regulation of Noisy Systems, 2nd Prague IFAC Symposium on Identification and Process Parameter Estimation (1970).

[7] J. WIESLANDER and B. WITTENMARK: An approach to adaptive control using real time identification. *Automatica* **7,** 211–217 (1971).

[8] L. E. DICKSON: *First Course in the Theory of Equations*. Wiley, New York (1922).

[9] K. J. ÅSTRÖM: Computer control of a paper machine—An application of linear stochastic control theory. *IBM J. Res. Development* **11,** 389–405 (1967).

[10] J. B. FARISON *et al.*: Identification and control of linear discrete systems. *IEEE Trans. Aut. Control* **12,** 438–442 (1967).

Self-tuning control

D.W. Clarke, M.A., D.Phil., and P.J. Gawthrop, B.A., D.Phil.

Indexing terms: *Controllers, Control-system synthesis, Computerised control, Nonlinear control systems*

Abstract

Self tuning is an important new branch of control which is attracting increasing theoretical and practical interest. The objective of self tuning is to control systems with unknown constant or slowly varying parameters, so theoretical interest is concerned with the stability, performance and convergence of the recursive algorithms involved, while practical interest derives from its potential as a simple controller commissioning tool, both as a method for controlling time-varying or nonlinear plant over a range of operating points, and for dealing with batch problems where the plant or materials vary over successive batches. This paper summarises and expands previous work on the design of self-tuning controllers. It discusses the closed-loop properties of various classes of self tuner, convergence concepts and results, and some of the technical problems involved with implementing self tuners on small computers or microprocessors.

List of principal symbols

$A(z^{-1}), B(z^{-1}), C(z^{-1})$ = polynomials of order n corresponding to the system output, control input, and disturbance input, respectively; $a_0 = c_0 = 1$
$E\{\cdot\}$ = expectation operator
$E(z^{-1}), F(z^{-1}), G(z^{-1})$ = general polynomials in z^{-1}
k = system time delay (integer)
n = system order
$P(z^{-1}), Q(z^{-1}), R(z^{-1})$ = costing transfer functions acting on the system output, input and set point, respectively
t = time in sample instants (integer)
$u(t), w(t), y(t)$ = system input, set point and output, respectively, at time t
$v(t)$ = known disturbance measured at time t
$X(t)$ = vector containing measured data
z = forward-shift operator: $z^k u(t) = u(t+k)$
β = forgetting factor
ϵ = k-step-ahead prediction error of auxiliary output ϕ
λ = control weighting
$\phi(t)$ = auxiliary output at time t
θ = vector of parameters
$\xi(t)$ = member of an uncorrelated zero-mean random sequence

A polynomial of order r in the backward-shift operator z^{-1} is denoted by $A(z^{-1}) = a_0 + a_1 z^{-1} + \ldots + a_1 z^{-r}$. After definition, and where convenient the polynomial is written simply as A. The estimate of a parameter γ is denoted by $\hat{\gamma}$.

1 Introduction

Practical process control problems frequently result from plant characteristics which involve

(*a*) unknown parameters, for example in the commissioning of a control system on a new process
(*b*) appreciable dead time
(*c*) nonlinear behaviour, for example in complex chemical or biochemical reactions
(*d*) transfer functions that vary with time, because of changes in raw materials, plant throughput, or properties such as heat-transfer coefficients
(*e*) stochastic disturbances, such as variations in ambient temperature
(*f*) propagation of known disturbances along a chain of unit processes.

Although 3-term controllers are remarkably effective with many such processes, there are several cases where detuning is necessary to ensure stability over a wide range of conditions (giving generally mediocre control), where the initial tuning is a lengthy and difficult operation, or where the time variation or nonlinearity is so strong that a fixed-parameter control is completely inadequate. Some of these difficulties have been recognised for many years by control theoreticians, and a range of either *ad hoc* or theoretically abstruse solutions have been proposed. The disadvantage of *ad hoc* methods is that convergence and stability of the essentially nonlinear controllers is hard to predict or guarantee, whereas the theoretically respectable methods often require computing power far in excess of that of conventional controllers.

A relatively new technique, self-tuning control,[23,34] which has been proposed to solve the range of practical control problems outlined above, does not appear to suffer excessively from the disadvantages of previous adaptive methods. Stability and convergence proofs are available,[35,59–63] and the computing requirements, though rather greater than those of p.i.d. controllers, can be satisfied by current microprocessor technology.[20,31] More importantly as far as implementation is concerned, the algorithms are general purpose, so that a wide variety of processes are amenable to the method. Hence specific programs are not required for each case, and software costs (which are increasingly dominating overall costs) can be amortised over a range of possible applications. The self-tuning controller is a development and generalisation of the original self-tuning regulator of Åström and Wittenmark[4–11,89] (though Kalman[50] and Peterka[71] produced similar proposals). As its name implies, this important earlier theory is concerned principally with the regulation of the output of processes against disturbances without special regard to its transient response to set-point changes. However, it received a great deal of attention, and there have been several reported applications both on pilot scale and industrial plant. These cases include the control of an ore crusher,[14] a paper machine,[15] ship steering,[47–49] distillation[67,68,74] and absorption columns,[33] titanium-dioxide[27] and cement kilns.[51] In most cases, set-point changes are either rare, or dealt with by special adaptations to the original theory. The self-tuning controller specifically incorporates the following of set-point changes as part of the control problem, and solves it in a uniform way.

As self tuners are implemented digitally, an appropriate representation of the controlled process which models both the inherent time-delay and external disturbances in the single-input/single-output case is the difference equation

$$\sum_{i=0}^{n} a_i y(t-i) = \sum_{i=0}^{n} b_i u(t-i-k) + \sum_{i=0}^{n} c_i \xi(t-i); \tag{1}$$

where $u(t)$ and $y(t)$ are the control input and the measured output of the process at sample instant t, and ξ is an uncorrelated sequence of random variables with zero mean. The order of the system is n, and k ($\geqslant 1$) represents the time-delay in an integral number of sample intervals. The defintion of k implies that b_0 is nonzero, and both a_0 and c_0 are assumed to be equal to 1 without loss of generality. Using the backward shift operator z^{-1}, eqn. 1 is expressed in terms of polynomials as

$$A(z^{-1})y(t) = z^{-k}B(z^{-1})u(t) + C(z^{-1})\xi(t) \tag{2}$$

The roots of C may (for constant systems at least) be taken to be within the unit disc, and it is assumed that no root of C lies on the unit circle. No assumption is made concerning the roots of A or B; the plant may be open-loop unstable (root of A outside the unit disc), or nonminimum phase (root of B outside the unit disc), though care must be taken in the latter case.

Although a discrete-time model is used, the underlying process is continuous time, and the choice of sample interval is important and must be made with due regard to the natural time scales of the process response. Unlike the digital implementations of a 3-term controller, where a fast sample rate (at least up to the Nyquist frequency) produces a more 'accurate' representation, a self tuner works better with a longer sample interval, typically of 0·1 of the dominant plant time-constant. This is because, like other discrete-time controllers based on explicit process models, numerical problems can be

Paper 8325 C, first received 8th February and in revised form 6th March 1979

Dr. Clarke and Dr. Gawthrop are with the Department of Engineering Science, University of Oxford, Parks Road, Oxford OX1 3PJ, England

Reprinted with permission from *Proc. Inst. Elec. Eng.*, vol. 126, pp. 633–640, June 1979.
Copyright © 1979; The Institution of Electrical Engineers.

produced at fast sample rates due to the migration of all the poles of the model to the unit circle. Eqn. 1 should for completeness include an additional constant, as the model represents a local linearisation of a nonlinear process. However, a treatment of this case has appeared in previous papers,[20,23] so is omitted here for simplicity of exposition; it is nonetheless an important practical detail and will be discussed later.

The use of a self tuner involves a 2-stage process. The first consideration is to resolve what control performance is desirable if the process parameters were *exactly* known, bringing in engineering judgment and some prior knowledge of gross plant characteristics (time scales, drift, offsets etc.). The self tuner is then used to ensure this performance, despite the lack of detailed knowledge of model parameters and their variation with time. The significance of the self-tuning controller in this context is that it provides an extensive range of potential control objectives, each of which has the self-tuning property. This paper, therefore, develops a range of possible controllers for plant with known parameters, and shows that these controllers are both practically useful and provide the basis for self tuning.

2 Least-squares prediction and control

The self-tuning regulator of Åström and Wittenmark has as its target control objective the minimisation of the variance of the plant output $y(t)$. One reason for this choice is that for processes such as paper making, where a minimum quality such as paper thickness is specified, the set point may be taken nearer this minimum if fluctuations about the mean were reduced, hence saving material. It was shown that this minimum can be achieved by first predicting the future output $y(t+k)$, where k is the system dead time, and then choosing a current control to set this predicted value to zero. This would leave the effect of those unpredictable components of ξ that affect the plant over the dead time of the process as a remnant on the output.

The method suffers from the following drawbacks: (*a*) the control objective is only appropriate for a restricted class of processes, (*b*) the only modification the user can make is to vary the sample time, (*c*) excessive control inputs may be produced, (*d*) the closed loop is unstable for a process whose discrete-time model has a root of B outside the unit circle.

The self-tuning controller minimises instead the variance of an ***auxiliary*** output ϕ defined by

$$\phi(t) = P(z^{-1})y(t) + Q(z^{-1})u(t-k) - R(z^{-1})w(t-k) \qquad (3)$$

Here, w is the set point, and P, Q and R are *transfer-functions* in the backward shift operator z^{-1}, specified as

$$P(z^{-1}) = \frac{P_N(z^{-1})}{P_D(z^{-1})} \qquad \text{etc.} \qquad (4)$$

Fig. 1 depicts the plant model together with the auxiliary output. It is by the choice of P, Q and R that the user can obtain the wider range of control behaviour; special cases are discussed later.

At time $t-k$, the signals $Qu(t-k)$ and $Rw(t-k)$ are known, so the problem of predicting $\phi(t)$ simply reduces to that of the (least-squares) prediction of

$$\phi_y(t) = Py(t) \qquad (5)$$

given plant input/output data up to time $t-k$. This problem, of the prediction of one series from another, has a standard solution.[83,91]

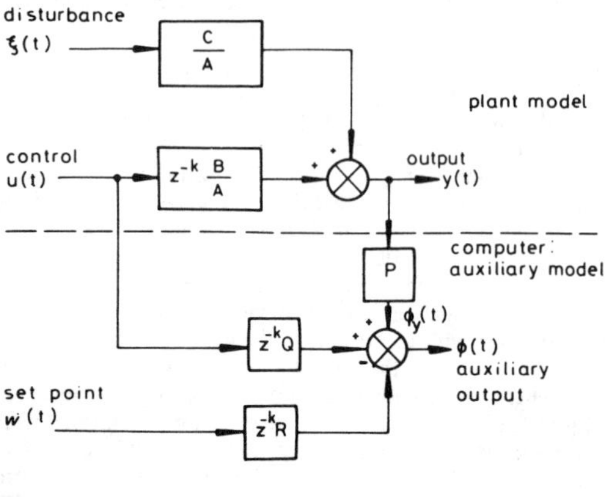

Fig. 1
Structure of the plant and the auxiliary model

$$(a)\text{ prediction:} \quad C\phi_y^*(t/t-k) = \frac{F}{P_D}y(t-k) + Gu(t-k) \qquad (6)$$

$$(b)\text{ remnant:} \quad \phi_y(t) = \phi_y^*(t/t-k) + E\xi(t) \qquad (7)$$

where the polynomials $E(z^{-1})$ and $F(z^{-1})$ are given by the identity

$$\frac{CP_N}{AP_D} = E(z^{-1}) + z^{-k}\frac{F(z^{-1})}{AP_D} \qquad (8)$$

The polynomial $G(z^{-1})$ is given by EB.

Eqn. 6 gives the reason for the restriction placed on C, i.e. that no root lies on the unit circle. In that case, an incorrect initial condition does not vanish for large t. If the roots are strictly within the unit circle, the effect of the error in the initial condition will decay asymptotically to zero at a rate depending on the positions of the roots. From eqns. 3 and 6 we obtain

$$\phi^*(t) = \phi_y^*(t) + Qu(t-k) - Rw(t-k) \qquad (9)$$

dropping the conditioning argument for convenience.

The control law is specified by choosing the control signal $u(t)$ to set the k-step-ahead prediction to zero, that is

$$\phi^*(t+k/t) = 0 \qquad (10)$$

This control will be in error initially due to the presumably unknown initial conditions of the least-squares prediction equations, but their effect will vanish. From eqns. 6 and 9, the feedback control that satisfies eqn. 10 is

$$u(t) = \frac{CRw(t) - Fy(t)/P_D}{EB + CQ} \qquad (11)$$

There are, however, other ways of development from eqn. 10. Eqn. 9 gives

$$u(t) = \frac{Rw(t) - \phi_y^*(t+k/t)}{Q} \qquad (12)$$

Define $u'(t)$ to be the control signal at time t which would set the prediction ϕ_y^* to equal $Rw(t)$, i.e. the signal required to cause exact model-following where $Q=0$, or minimum-variance control of the auxiliary output with respect to the modified set point $Rw(t)$. From eqn. 6 we obtain

$$\phi_y^*(t+k/t) - Rw(t) = \frac{F}{P_D C}y(t) + \frac{G}{C}u(t) - Rw(t)$$

$$= \frac{F}{P_D C}y(t) + \left[\frac{G}{C} - g_0\right]u(t) - Rw(t) + g_0 u(t)$$

$$= -g_0 u'(t) + g_0 u(t), \qquad \text{as } c_0 = 1$$

Hence, using eqn. 12

$$g_0 u'(t) = g_0 u(t) + Qu(t)$$

or

$$u(t) = \frac{1}{1 + (Q/g_0)}u'(t) \qquad (13)$$

This expression shows the effect of nonzero Q in reducing the amplitude of the control signal. In practice, Q plays a useful role in allowing control performance to be modified online without the necessity of changing the sample time. Note that Q appears only in eqn. 12 and not in eqns. 6 or 8; hence Q can be varied without changing any of the other controller parameters if eqn. 12 is used to realise $u(t)$ rather than eqn. 11.

It is often useful to analyse the continuous-time versions of discrete time control laws. The continuous-time analogues of the above control are summarised below.

The continuous-time system corresponding to eqn. 2 is

$$A'(s)y(s) = e^{-s\Delta}B'(s)u(s) + C'(s)\psi(s) \qquad (14)$$

where s is the Laplace operator, and Δ the pure delay. Deriving the sampled-data equivalent of eqn. 14 in the form of eqn. 2 is standard. The identity eqn. 6 is replaced by the realisability relation[34,36,83,91]

$$\frac{F'(s)}{A'P_D'} = \text{realisable part of}\left\{e^{s\Delta}\frac{C'}{A'}\frac{P_N'}{P_D'}\right\}$$

$$E'(s) = \frac{C'P_N' - e^{-s\Delta}F'}{A'P_D'} \qquad (15)$$

The primed quantities represent polynomials in s. Where appropriate the remaining discrete-time equations may be interpreted as continuous-time equations by substituting the corresponding primed polynomials.

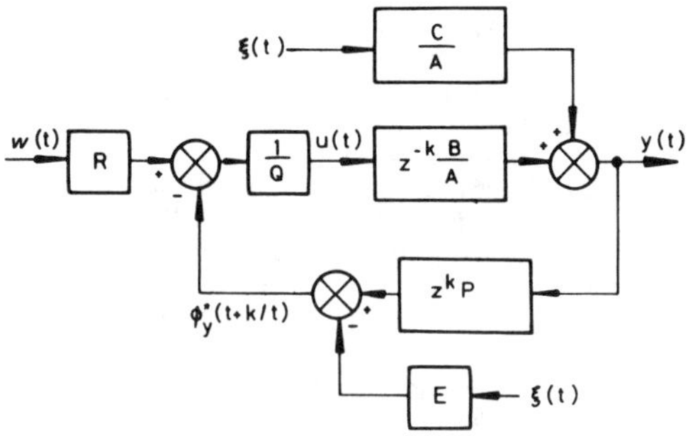

Fig. 2
Conceptual feedback system

3 Open and closed-loop properties

The controllers defined in Section 2 take the form of a feedback control law; when combined with the system eqn. 2 they may be analysed like any other feedback control law. In particular, the closed-loop transfer function may be derived so that the response of the controlled system to disturbances and set points can be determined. On the other hand, classical analysis using the loop transfer function leads to sensitivity results and stability margins.

The closed-loop transfer function may be derived by combining the explicit expression for the predictor (eqn. 6) with the control law (eqn. 9) and the system (eqn. 2). However, it is more direct to use the implicit expressions for the predictor (eqns. 5 and 7). Those simultaneous equations are displayed in block-diagram form in Fig. 2. It should be realised that the individual elements of this block diagram do not all represent causal systems; however, the relation between the system variables of interest y, u, ξ and w are, of course, causal.

Solving for the system output y:

$$y(t) = \frac{1}{PB+QA}\left[\frac{C}{A}\xi(t) + z^{-k}\frac{B}{QA}\{E\xi(t) + Rw(t)\}\right]$$

$$= \frac{EB+QC}{PB+QA}\xi(t) + z^{-k}\frac{B}{PB+QA}Rw(t) \qquad (16)$$

Solving for the control signal u:

$$u(t) = \frac{1}{1+\dfrac{PB}{QA}}\left[z^{-k}\frac{P}{Q}\frac{C}{A}\xi(t) + \frac{1}{Q}\{E\xi(t) + Rw(t)\}\right]$$

$$= -\frac{F}{PB+QA}\xi(t) + \frac{A}{PB+QA}w(t) \qquad (17)$$

where the identity eqn. 8 has been used in the last equality.

These two equations give the closed-loop properties of the system in terms of the system output y and control input u. It is of interest to note that the delay term z^{-k} does not appear in the denominator of eqns. 16 and 17; this is characteristic of predictive control. The implications of eqns. 16 and 17 are deferred to the following Sections; the loop properties are examined next. As noted in Section 1, most systems of interest are inherently continuous time. A rational method to analyse and synthesise discrete-time controllers is to consider the properties of the corresponding continuous-time controllers, and as a final design step to choose a suitable sample interval to specify the discrete-time control. From this point of view, it is natural to consider the open-loop transfer function $L'(s)$ obtained by combining the continuous-time control corresponding to eqn. 11 with the system of eqn. 14 as follows:

$$L'(s) = \exp(-s\Delta)\frac{B'F'}{A'P'_D(E'B' + C'Q')}$$

This expression leads to the familiar open-loop frequency response representations of Bode and Nyquist. Standard methods yield stability margins, and the effect of nonlinearities may be examined using standard methods such as the circle criterion or the describing function. For example, the model-reference control of Section 4 is found to be considerably less sensitive to nonlinearities than the simple minimum-variance control of Reference 3.

4 Model-reference control

The method of designing a control law to make the closed-loop system behave as a prespecified transfer function, or model, has been discussed both in the classical literature[43] and as a basis for adaptive control.[55,58] One version of the fixed parameter control laws of Section 2 can be interpreted as a model-reference control law,[35] and hence the corresponding self-tuning version of Section 7 may be viewed as a model-reference adaptive control law. The relation between self-tuning and model-reference adaptive control has been noted by Ljung and Landau,[64] and extensively examined by Egardt.[29,30] A feature of the method of this Section is that it explicitly includes the effect of stochastic disturbances.

The idea of specifying a closed-loop system response has some appeal. However, it is known to have certain disadvantages;[43] in particular, it is not always achievable with acceptable control effort. Once again, it is argued that good continuous-time design often leads to good behaviour of the corresponding discrete-time control.

If the transfer function Q (eqn. 3) is set to zero, eqn. 16 gives the closed-loop system as

$$y(t) = \frac{1}{P}E\xi(t) + z^{-k}\frac{R}{P}w(t) \qquad (18)$$

The system behaves as the model R/P with respect to the delayed set point $w(t-k)$. Moreover, the disturbance affects the system as the moving-average prediction error $E\xi$ passing through the model $1/P$; this is the additional model-reference property associated with these controllers.

The disadvantages of this method are revealed by the appropriate version of eqn. 17. Setting $Q = 0$

$$u(t) = -\frac{F}{PB}\xi(t) + \frac{A}{PB}w(t) \qquad (19)$$

The control signal u contains modes associated with the system numerator polynomial B; in particular, roots of this polynomial lying outside the unit circle lead to unstable control behaviour. The method is thus not appropriate to such systems.

Similar problems may arise through ignoring the underlying continuous system. It is known,[43] and may be deduced from the continuous-time analogue of eqn. 19, that continuous-time reference models with smaller relative order (number of poles − number of finite zeros) than the system B/A (eqn. 2) lead to unbounded control response to the disturbance ξ or to step changes in the set-point w. It follows that the chosen discrete-time model should correspond to a continuous model with allowable relative order. An illustrative example appears in Reference 35.

5 Detuned model-reference control

Intuitively, it seems that a reason for the unsatisfactory control sometimes associated with model-reference control is that control effort is not explicitly included in the criterion. With this in mind, this Section considers the effect of 'weighting control effort' by replacing $Q = 0$ by

$$Q = \lambda \qquad (20)$$

where λ is a scalar.

For particular systems, the effect of λ may be analysed using the frequency-response methods mentioned in Section 3. An alternative is the analysis of the closed-loop characteristic polynomial by standard root-locus techniques.[43]

Writing the transfer function P (eqn. 3) as the ratio of two polynomials as eqn. 4, eqn. 16 gives the closed-loop characteristic polynomial as

$$P_N B + \lambda P_D A \qquad (21)$$

Eqn. 21 is amenable to standard root-locus techniques with λ as the parameter.

The root-locus analysis of the continuous-time detuned model-follower leads to an approximate relation to the solutions of the standard linear-quadratic regulator problem[34,54,70] for systems with zeros of B (eqn. 2) in the left halfplane. For simplicity the regulator problem (zero set point) is considered.

Choose $P^1(s)$ to be the nth order Butterworth polynomial[54] with root radius ρ:

$$P' = \beta(p, n) \tag{22}$$

Also, choose n to be the relative order of the system B/A (order of A, order of B).

The corresponding control signal is given by eqn. 17 as

$$u = -\frac{F_0}{T_0}\xi \tag{23}$$

where

$$T_0 = \beta B' + \lambda A' \tag{24}$$

and F_0 is given by eqn. 15 as

$$\frac{F_0(s)}{A'} = \text{realisable part of}\left\{\exp(s\Delta)\frac{\beta C'}{A'}\right\} \tag{25}$$

The linear quadratic control signal minimising $E\{y^2 + \lambda_\infty u^2\}$ over realisable laws is[34, 54, 70]

$$u = -\frac{F'_\infty}{T'_\infty}\xi \tag{26}$$

where

$$T'_\infty(s)T'_\infty(-s) = B'(s)B'(-s) + \lambda_\infty A'(s)A'(-s) \tag{27}$$

and

$$\frac{F_\infty}{A} = \text{realisable part of}\left\{\exp(s\Delta)\frac{B'(-s)}{T'_\infty(-s)}\frac{C'(s)}{A'(s)}\right\} \tag{28}$$

It can be shown[54] that if all the roots of $B(s)$ are in the left-hand plane, then the closed-loop polynomial T_∞ is approximately given by

$$\beta\left(\frac{1}{\lambda_\infty}, n\right)B(s)$$

for λ_∞ sufficiently small. The form of the root-locus diagram for eqns. 24 and 27 give the degree of approximation involved. From eqn. 27

$$\frac{B(-a_i)}{T_\infty(-a_i)} = \frac{T_\infty(a_i)}{B(a_i)}; \quad \text{for} \quad A(a_i) = 0 \tag{30}$$

Hence from eqns. 25 and 28, T_∞ close to T_0 implies F_∞ close to F_0.

The conclusion is that for suitably chosen ρ and λ, the detuned model-reference regulator can approximate the linear-quadratic regulator for some λ_∞. The degree of correspondence may be determined from the appropriate root-locus diagrams.

6 Output predictor control

If in eqn. 3, the transfer functions P and R are chosen to be unity, eqn. 11 reveals that the control law is given by

$$u = \frac{1}{Q}[y^*(t+k/t) - w(t)] \tag{31}$$

This is analogous to classical unity-feedback control with the compensation $1/Q$ in the forward path. The difference is that in eqn. 31 the least-squares prediction of y, rather than y itself is used. The control may also be compared with Smith's method[76] where y^* is replaced by

$$y^s(t+k/t) = y(t) + (1 - z^{-k})\frac{B}{A}u(t) \tag{32}$$

The relation between the least-squares and Smith predictor is explored elsewhere.[35]

The closed-loop system is, from eqn. 16, given by

$$y = \frac{EB + QC}{B + QA}\xi + z^{-k}\frac{B}{PB + QA}Rw \tag{33}$$

If the system parameters (A, B) were known, it is possible (apart from controllability constraints) to choose Q to place the system poles as required. This can be done without affecting any parameter estimate when self tuning.

7 Self tuning

The control law of eqn. 12 involves the least-squares prediction of the quantity ϕ. This requires knowledge of the system and disturbance parameters: the polynomials A, B and C. Self-tuning methods are applicable to system where the structure is known but the parameters unknown. That is, an upper bound on the orders of A, B, C along with the exact value of the time delay k is required but the coefficients of the system are unknown. Certain algorithms, for example that of Wellstead,[80-82] do not require that the leading coefficient of B be nonzero, and thus effectively allow for unknown system time delay.

Self-tuning methods have been classified into the two broad catagories of implicit and explicit.[9] Implicit methods focus on generation of an approximation $\hat{\phi}^*$ to the prediction ϕ^*, by directly estimating the predictor parameters of eqn. 6. Analysis centres entirely on the error between ϕ^* and $\hat{\phi}^*$; the accuracy of parameter estimates is not of direct interest. On the other hand, explicit methods involve system parameter identification followed by some control law calculation. In this case, accuracy of the parameters and how it enters the control law calculation is important.

This distinction is not completely clear cut, and it is possible to combine the two approaches. In particular, the transfer function Q of eqn. 31 can be calculated as a function of estimated parameters (the explicit approach) and at the same time the control involves an estimate of y^* (the implicit approach). The emphasis in this paper is on implicit methods; explicit methods have been developed by Wellstead *et al.*[80-82] and by Åström *et al.*[9]

The self-tuning algorithm given here is written in a slightly different form than the form given in References 23 and 25. The approach is rather more rational, and moreover, constraints on control signal amplitude are easily included.

The predictor (eqn. 6) may be written as

$$\phi_y^*(t/t-k) = \frac{F}{P_D}y(t-k) + Gu(t-k) - \sum_{i=1}^{n} c_i\phi_y^*(t-i/t-k-i) \tag{34}$$

Define the vectors

$$X^T(t) = \left[u(t), \ldots; \frac{y(t)}{P_D}, \ldots; \phi_y^*(t+k-1/t-1), \ldots\right] \tag{35}$$

$$\theta^T = [g_0, \ldots; f_0, \ldots; -c_1, \ldots]$$

Then it is convenient to rewrite eqn. 34 as

$$\phi_y^*(t/t-k) = X^T(t-k)\theta \tag{36}$$

If $X(t-k)$ contained quantities known at time t, and the prediction error (eqn. 6) were white, then the estimation of θ according to eqn. 36 would be a suitable candidate for linear least squares.[32] However, ϕ_y^* is not known (unless either θ were known or ξ were zero) and, if $k > 1$, the prediction error is not white. The approach of self tuning (in common with other methods such as extended least squares[32] is to replace ϕ^* by a calculable approximation $\hat{\phi}_y^*$, and to use linear least squares as if $\hat{\phi}_y^* = \phi_y^*$ and the prediction error were white.

The recursive self-tuning algorithm is

$$\hat{X}^T(t) = \left[u(t), u(t-1), \ldots; \frac{y(t)}{P_D}, \frac{y(t-1)}{P_D}, \ldots; \hat{\phi}_y^*(t+k-1/t-1), \ldots\right] \tag{37}$$

$$\hat{\theta}(t) = \hat{\theta}(t-1) + \hat{S}^{-1}(t-k)\hat{X}(t-k)[\phi_y(t) - \hat{X}^T(t-k)\hat{\theta}(t-1)] \tag{38}$$

$$\hat{S}(t) = \beta\hat{S}(t-1) + \hat{X}(t)\hat{X}^T(t); \hat{S}(0) = S_0 > 0 \tag{39}$$

$$\hat{\phi}_y^*(t+k/t) = \hat{X}^T(t)\hat{\theta}(t) \tag{40}$$

Eqn. 38 requires eqn. 3 to generate ϕ_y. Eqn. 39 is usually updated in the form of the factorised inverse;[72, 78] β is the exponential forgetting factor.[32, 89]

Eqn. 40 may be rewritten in a useful alternative form. As in Section 2, define $\hat{u}'$ as the control which at time t satisfies

$$\hat{\phi}_y^*(t+k/t) = Rw(t) \tag{41}$$

Then, similarly to Section 2

$$\hat{\phi}_y^*(t+k/t) = Rw(t) + \hat{g}_0(u - \hat{u}') \tag{42}$$

Eqn. 42 may be used in place of eqn. 40; any control law, linear or nonlinear, may be used to generate u, and thus this version of the algorithm is divorced from the control law in use. It is, in fact, a self-tuning predictor[41] of the sequence ϕ_y.

The control law corresponding to eqn. 12 is

$$u(t) = -\frac{[\hat{\phi}^*(t+k/t) - Rw(t)]}{Q} \tag{43}$$

which may be written in the alternative form of eqn. 13 using eqn. 42. Substituting eqn. 43 into eqn. 40 gives

$$\hat{\phi}^*(t+k/t) = -Qu(t) + Rw(t) \tag{44}$$

This alternative equation leads to the previous formulation[23,35] by subtracting the right-hand side of eqn. 44 from both sides of eqn. 36 and making appropriate substitutions in eqns. 37–40. Eqn. 44 is, however, less general than eqns. 40 or 42 as it assumes eqn. 43 is always satisfied.

If $Q = R = 0$, that is the control law is such that

$$\hat{\phi}^*_y(t+k/t) = 0 \tag{45}$$

then the relevant portion of $\hat{X}$ in eqn. 37 is identically zero, and may be deleted from the calculation. This is the basis of the self-tuning regulator and the observation that C may be ignored.[11,89]

By applying suitable linear transformations to $\hat{X}$ (eqn. 35) and $\hat{\theta}$ (eqn. 36), different combinations of parameters may be estimated. This is examined by Holst.[41]

8 Stability and convergence

As mentioned in Section 7, analysis of implicit self-tuning algorithms centres on whether the estimated prediction $\hat{\phi}^*$ (eqn. 40) approaches the prediction ϕ^* (eqn. 36) as time increases; the parameter estimate $\hat{\theta}$ is not of direct interest.

Convergence analysis of such stochastic algorithms has given rise to some new theory.[30,53,59–63] The approach here is based on the idea of writing the adaptive algorithm as a feedback system.[56,57] This gives intuitive insight into the properties of the algorithm and is, moreover, amenable to stability analysis. As the scalar quantity $\hat{\phi}^*$ is of interest, input/output methods[85–88] are more appropriate than the hyperstability[73] methods used by Landau.[56,57] Further, these results may be combined with the martingale convergence theorem[18,26] to give a type of convergence with a probability of one.

The emphasis here is more on ideas and interpretation than on the technical details found elsewhere.[37,38] For simplicity, the unit delay case is considered.

To write the self tuner in the form of a feedback system, it is convenient to make the following definitions:

$$\tilde{\theta}(t) = \hat{\theta}(t) - \theta \tag{46}$$

$$\tilde{e}(t) = \hat{\phi}^*(t/t-1) - \phi^*(t/t-1) \tag{47}$$

$$v(t) = \hat{X}(t-1)\tilde{\theta}(t-1) \tag{48}$$

The estimator (eqn. 30) may then be written as the following time-varying system with scalar input $\xi(t) - \tilde{e}(t)$, state $\tilde{\theta}$, and output scalar $v(t)$ (eqn. 48):

$$\tilde{\theta}(t) = \tilde{\theta}(t-1) + S^{-1}(t-1)\hat{X}(t-1)[-\tilde{e}(t) + \xi(t)] \tag{49}$$

The predictor (eqn. 36) may be written in terms of the 'approximate' vector $\hat{X}$ as

$$\phi^*(t/t-1) = X^T(t-1)\theta \tag{50}$$

$$= \hat{X}^T(t-1)\theta - [\hat{X}^T(t-1) - X^T(t-1)]\,\theta$$

$$= \hat{X}^T(t-1)\theta - [\hat{\phi}^*(t/t-1) - \phi^*(t/t-1)]\,(C-1) \tag{52}$$

Hence the output error (eqn. 47) becomes

$$\tilde{e} = -\hat{X}^T(t-1)\hat{\theta}(t-1) - (C-1)\tilde{e} \tag{53}$$

i.e.

$$\tilde{e}(t) = -\frac{1}{C}v(t) \tag{54}$$

Eqn. 54 acts as a feedback system around the estimator system, eqns. 48 and 49. Hence the adaptive mechanism may be represented by the block diagram of Fig. 3.

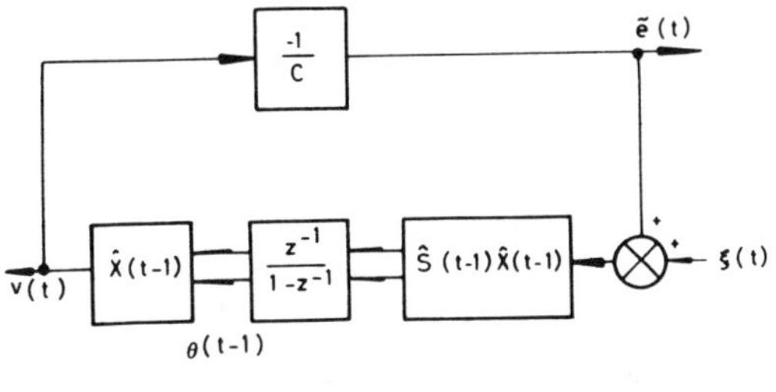

Fig. 3
Feedback structure for analysing convergence

The quantity of interest $\tilde{e}$ appears as the output of this feedback system; the white noise process ξ is the input to the feedback system. The behaviour of $\tilde{e}$ may thus be considered in two stages:

(*a*) the determination of the stability of the feedback system of Fig. 3
(*b*) the examination of the properties of ξ when applied to this system,

The stability of the feedback depends on two things:

(i) transfer function $1/C$
(ii) sequences $\hat{X}(t)$ and $\hat{S}(t)$.

The system polynomials A and B (eqn. 2) and the ϕ generating transfer functions P, Q and R (eqn. 3) do not explicitly appear in the equations; they merely enter indirectly through $\hat{X}$. Without further analysis, the important factors affecting stability are thus highlighted.

It may be shown[37,38] that, for the purposes of stability analysis, the properties of the sequences $\hat{X}$ and $\hat{S}$ are encapsulated in the scalar sequence

$$\sigma(t) = \hat{X}^T(t)\hat{S}^{-1}(t)\hat{X}(t) \tag{55}$$

It is readily shown[37,38] that

$$0 \leqslant \sigma(t) < 1 \tag{56}$$

The full stability criterion involves a combination of constraints on σ and C, but a useful special case is

$$\sigma(t) \rightarrow 0 \tag{57}$$

$$\frac{1}{C} - \frac{1}{2} \quad \text{is positive-real} \tag{58}$$

A positive-real system may be considered as a system that dissipates energy; a full discussion may be found in References 43, 73 and 85–88. Having a criterion in terms of an unknown quantity C seems of little use. However, it is to be expected, for if C were known it could be eliminated from the equations. Indeed, if an *a priori* estimate $\hat{C}$ is available, the self-tuning equations may be modified[37,38,63] to give the condition given in eqn. 58 in terms of $\hat{C}/C$ rather than $1/C$.

The condition given in eqn. 57 is nontrivial, and it seems that particular proofs for particular circumstances are required. For example, if the system (eqn. 2) is stable, the control input limited in the control algorithm and the disturbance uniformly bounded, then the output y is also uniformly bounded. If the forgetting factor β (eqn. 39) is unity and X 'sufficiently exciting' so that $\|S^{-1}\| \rightarrow$ zero then the condition given in eqn. 57 will hold. Analysis of more general cases is a subject of current research.

Having shown stability of the feedback system of Fig. 3, it can be shown[37,38] using the martingale convergence theorem[18,26] that, for unity forgetting factor and using some additional conditions, the error $\tilde{e}$ (eqn. 47) is constrained by

$$\lim_{N\rightarrow\infty} \frac{1}{N}\sum_{t=0}^{N} \tilde{e}^2(t) \rightarrow 0, \quad \text{w.p. 1} \tag{59}$$

This is slightly weaker than the statement that $\tilde{e} \rightarrow 0$.

It is of interest to obtain results if neither the condition given in eqn. 57 or eqn. 58 hold. The self-tuning regulator has been shown to have certain stabilisation properties.[65] Here a similar result is obtained based on the results of this Section. Consider a suboptimal predictor of the form of eqn. 34 designed as if $\xi' = C\xi$ were white. Then, redefining accordingly the quantities used in this Section, the corresponding feedback system is as Fig. 3 but with $C = 1$. Stability then follows with no extra constant on σ. The sequence ξ' is not white, and thus the convergence result (eqn. 59) does not apply. Instead, we obtain the weaker condition

$$\frac{1}{N}\sum^{N} \tilde{e}^2(t) < \frac{M}{N}\left(\sum^{N} \xi'^2(t) + M'\right) \tag{60}$$

where M and M' are positive constants. Eqn. 60 implies, for example, that the error term $\tilde{e}$ cannot increase exponentially if ξ' is uniformly bounded.

9 Feedforward and multivariable control

Some process disturbances, although unavoidable, are measurable. For example, the ambient temperature in central heating control, or the variations of feed flow or quality to a process, are disturbances likely to affect the controlled output. Feedforward is often advocated in such situations, such that the control signal is manipulated to counteract the disturbances before the process output is significantly affected. However, effective feedforward requires an accurate model of the process, so in practice only static feedforward is used, and even then the appropriate gains found only after a lengthy commissioning process. Self tuning on the other hand allows for dynamic feedforward parameters which are automatically adjusted to give the best possible disturbance rejection.[20,67] This is not possible with schemes using explicit identification if only simple least-squares estimation is used.

If $v(t)$ is the disturbance, a possible model expands eqn. 2 into

$$A(z^{-1})y(t) = z^{-k}B(z^{-1})u(t) + C'(z^{-1})\xi(t) + Z^{-k}D(z^{-1})v(t) \qquad (61)$$

where possible components of $v(t)$ which cannot be affected by the current control $u(t)$ are merged into $C'\xi(t)$. The prediction given by eqn. 6 becomes

$$C'\phi_y^*(t/t-k) = \frac{F'}{P_D}y(t-k) + G'u(t-k) + S(z^{-1})v(t-k) \qquad (62)$$

and the control law of eqn. 11 becomes

$$u(t) = \frac{C'Rw(t) - Sv(t) - F'y(t)/P_D}{E'B + C'Q} \qquad (63)$$

A straightforward extension of this theory allows for several feedforward terms and for delays in the feedforward measurements that are not equal to the system delay k.[25] Note that the usefulness of feedforward in this context depends on the relative dynamics by which $v(t)$ and $u(t)$ affect the plant output; if $v(t)$ acts much before $u(t)$ little would be gained from feedforward, but fortunately this is rarely the case. Self tuning of feedforward terms is a simple extension to the use of eqn. 37; the vector $X(t)$ is augmented by components $v(t)$, $v(t-1), \ldots,$ the vector $\hat{\theta}(t)$ is augmented by parameters $s_0, s_1, \ldots,$ and the new $\hat{\phi}_y^*$ is used in the control equation.

The self-tuning regulator[13] and controller[65] can be extended to the multivariable case, where the system model is specified as

$$A(z^{-1})y(t) = B(z^{-1})u(t-k) + C(z^{-1})e(t) \qquad (64)$$

Here u, y and e are p-vectors for a p-input/p-output system, and A, B and C are polynomial matrices. It transpires that the appropriate self tuners may be considered as a set of single input/single output self tuners each of which uses 'feed-forward' signals taken from the inputs and outputs of other loops.

10 Some practical implications

A detailed early report[20] showed that self tuning was practicable using an industry-standard 8-bit microprocessor, as the total computing time was found to be rather less than one second for typical numbers of parameters. The detailed programming, however, used a 'medium-level' language and lacked the flexibility required for a range of applications. Current work uses an interpretive language, Control Basic,[22] which allows programs to be modified easily on site, and a portable microcomputer system was constructed[31] and has been applied to the self tuning of industrial processes. (In one example, self tuning was operational within an hour of arrival at the site; papers describing this work are in preparation.) Although the use of an interpretive language considerably slows program execution, the main computational burden is in the recursive parameter estimator, and fortunately this is one aspect of self tuning which is used without modification in all applications, and hence can be optimised in a machine-language routine.

The recursive parameter estimation warrants particular numerical attention; the standard least-squares algorithm is such that the (inverse) covariance matrix estimate can become negative definite, and this causes rapid divergence of the parameter estimates and failure of the algorithm. Unfortunately, this phenomenon is seen in simulations only after many thousands of samples, and hence may not be detected if only short simulation runs are used. The problem is especially acute with low-precision number representations. There are two methods which are used to overcome this problem, both of which guarantee positive definiteness by factorising the matrix and by updating terms in the factors only. In the square-root method[20,68] the factors are $P(t) = S^{-1}(t) = R(t)R^T(t)$, R being an upper triangular matrix, whereas in the UD method[73] $P(t)$ is given by $P(t) = U(t)D(t)U^T(t)$, where D is a diagonal matrix and U is an upper triangular matrix with units along its diagonal. There is little to choose between the methods: both can be programmed simply, the UD method being slightly longer but having the individual variances directly in D and requiring no square-root extraction.

A related cause of difficulty arises when a self tuner operates for some time on a system which has little or no external disturbance. It is observed that the parameter covariances rapidly increase (so called 'blow up') so that when a disturbance does arrive the least-squares estimator gain has become so high that the estimates move rapidly, and the control exercised over the next few sample intervals is poor. This phenomenon is due to the use of a forgetting factor β which is less than one, for if no data arrive $[X(t) = 0,\ t > t_0]$, $P(t)$ behaves as $P(t_0)/\beta^t$. There are several proposed solutions to this problem:

(*a*) stop updating estimates if $|P(t)|$ exceeds some prescribed bound ($|P(0)|$, say)
(*b*) inject a suitable test signal if $|P(t)|$ exceeds the bound
(*c*) adjust the forgetting factor β according to some 'information measure', such that $\beta \to 1$ if there is little new information given by the data.[33]

Each of these methods has been successfully used.

As the system models assumed in self tuning are local linearisations to typically nonlinear responses (though certain types of parametric nonlinearity could be incorporated), the input/output signals are generally perturbations around nonzero mean levels $\bar{w}$, $\bar{u}$, $\bar{y}$. These mean levels are not usually related by the steady-state gain of the model in eqn. 1, so a constant term must be added for generality. This could be interpreted as $\xi(t)$ having nonzero mean. Moreover, a common requirement for the closed loop is that the steady-state error should be zero, i.e. $\bar{y} = \bar{w}$. If the plant contains no integrator, eqn. 12 shows that this is possible either if Q is a vanishingly small constant or if Q is of the form $(1 - z^{-1})Q^1$.

There are several approaches to the problem of constant offsets, though the most appropriate choice is a matter of current research. The simplest, and most commonly used to date, inserts an integrator into the loop after the self tuner which effectively computes increments in the control signal. The 'system' equation then becomes

$$(1 - z^{-1})Ay(t) = Bu(t-k) + (1 - z^{-1})C\xi(t) + (1 - z^{-1})d \qquad (65)$$

where d is the additive constant. As $(1 - z^{-1})$ is zero for constant d, it need not enter into the self tuning, and eqn. 3 shows that $\bar{y} = \bar{w}$ if $R(1) = P(1) = 1$, which is the usual case. The method suffers in that the integrator can detract from the closed-loop performance, but eqn. 7 shows that the disturbance polynomial $C^1 = (1 - z^{-1})C$ has a root on the unit circle. This can seriously affect the convergence of the self tuner. These difficulties can be alleviated to some extent by using a $P + I$ regulator in cascade with the self tuner instead.[67]

The second method estimates the constant directly, and uses $Q = \lambda(1 - z^{-1})$, or better, $Q = \lambda(1 - z^{-1})/(1 - \alpha z^{-1})$. The drawback with this method (which also pertains to the first method) is that the use of λ for obtaining stability for nonminimum phase systems does not hold for certain roots of B (i.e. those to the right of $z = 1$) as a closed-loop pole could approach the root of Q which lies on the unit circle from the right.

The third method is to add integrators into both the P and R transfer functions of the auxiliary model. Detailed analysis shows that, for a constant set point $w = w(t)$, the X vector is composed of $y(t) - w$ etc. (after filtering and integration) and $u(t)$ etc., and does not need to include a constant. Hence the offset can be eliminated without the convergence difficulties which can hamper the forward-path integrator, as the integration follows the measured output $y(t)$, and equally affects the disturbance term. The method, however, again fails for certain nonminimum phase systems.

The fourth method uses cascade control, in which a self tuner is placed in the inner loop and an integrator in the outer loop. The gain of the integrator is particularly easy to choose *a priori*, as the inner loop would have a transfer-function which is prescribed by the choice of P. Then Q could be taken as λ, and varied online to trade model-following performance against control fluctuations. This method can cope with (stable) nonminimum phase systems as the closed-loop poles depend on $PB + \lambda A$, which approaches λA for large enough λ(21, 25).

To commission a self tuner, a number of parameters need to be selected. For most of these the choice is not critical, and there are several relevant discussions in the literature.[7,19,20,35,74,89] For implicit methods, the system time delay k is an important parameter, and a value which is greater than the maximum expected is often acceptable. In some circumstances, for example in material-flow systems, the delay can be deduced from flow measurements. Again, the assumption of a model order at least equal to the system order is acceptable, and for model-reference control the order of P should be the same, as inferior results may be obtained if the desired closed-loop model is of lower order than that of the plant. The desired closed-loop model P^{-1} should be chosen with its continuous-time performance in mind.

A self tuner could be attached to a plant and set into operation immediately with zero initial choices of the feedback parameters; this is often done in simulations. However, it is more appropriate to use a 3-stage process:

(*a*) Close the loop with a conventional controller and simply monitor input/output data.

(*b*) Start least-squares estimation and observe the control signal the self tuner would impose. A variety of candidate self tuners could be inspected.

(*c*) Replace the conventional controller by the chosen self tuner.

In some critical plant, limits on the control signal have to be imposed; constraints that grow with time until attaining absolute upper and lower bounds on control can be useful in these cases.

11 Conclusions

This paper has summarised current developments of implicit self-tuning controllers. These controllers can be used to solve a wide variety of practical problems, and there is a growing amount in the literature on their applications. The self tuner can be arranged to adjust to a variety of servo and regulator characteristics, and feed-forward leading to multivariable control can readily be included. It is possible to derive convergence conditions using a variety of methods which are unavailable to explicit self tuners.

Although the applications studied so far have demonstrated the great potential of self tuning, a lot needs to be done before it can be applied in a routine fashion. Considerable progress has been made in simplifying and explaining the choice of initial parameters and structure, but the problem of running a self tuner for long periods of time on a strongly nonlinear process needs further exploration.

12 Acknowledgment

The authors would like to thank the UK Science Research Council for support into the implementation and application of microprocessor-based self tuning controllers.

13 References

1 ASHTON, R.P.: 'Simulation study of a *k*-step-ahead controller'. OUEL Report 1065/73, 1973

2 ASHTON, R.P.: 'Digital control of randomly disturbed systems'. D.Phil. thesis, Oxford University, 1974

3 ASTROM, K.J.: 'Introduction to stochastic control theory' (Academic Press, 1970)

4 ASTROM, K.J.: 'Stochastic control theory and some of its industrial applications'. Proceedings of the IFAC symposium on stochastic control, Budapest, 1974

5 ASTROM, K.J.: 'A self-tuning parameter estimator'. Lund Report 7419 (C), 1974

6 ASTROM, K.J.: 'Stochastic control problems'. Lund Report LUTFD2/(TFRT-3147)/1-068/(1977), 1977

7 ASTROM, K.J., BORRISON, U., LJUNG, L., and WITTENMARK, B.: 'Theory and applications of adaptive regulators based on recursive parameter estimation'. Lund Report 7507(C), 1975

8 ASTROM, K.J., BORRISON, U., LJUNG, L., and WITTENMARK, B.: 'Theory and applications of self-tuning regulators', *Automatica*, 1977, **13**, pp. 457–476

9 ASTROM, K.J., WESTERBERG, B., and WITTENMARK, B.: 'Self-tuning controllers based on pole-placement design'. Lund Report, LUTFD/(TFRI-3148)/1-052/(1978), 1978

10 ASTROM, K.J., and WITTENMARK, B.: 'Problems of identification and control', *J. Math. Anal. & Appl.*, 1971, **34**, pp. 90–113

11 ASTROM, K.J., and WITTENMARK, B.: 'On self-tuning regulators', *Automatica*, 1973, **9**, pp. 185–199

12 ATKINSON, R.S.: 'The application of self-tuning control'. M.Sc. thesis, Oxford University, 1978

13 BORRISON, U.: 'Self-tuning regulators: industrial applications and multivariable theory'. Lund Report 7513, 1975

14 BORRISON, U., and SYDING, R.: 'Self-tuning control of an ore-crusher'. Proceedings of the IFAC symposium on stochastic control, Budapest, 1974, pp. 491–497

15 BORRISON, U., and WITTENMARK, B.: 'Moisture control of paper machine: an application of a self-tuning regulator'. Lund Report 7337, 1973

16 BORRISON, U., and WITTENMARK, B.: 'An industrial application of a self-tuning regulator'. Lund Report 7310, 1973

17 CEGRELL, T., and HEDQVIST, T.: 'Successful adaptive control of paper machines'. Proceedings of the IFAC symposium on identification and system parameter estimation, The Hague/Delft, 1973, pp. 485–492

18 CHUNG, K.L.: 'A course in probability theory' (Academic Press, 1974, 2nd edn.)

19 CLARKE, D.W.: 'Implementation of self-tuning controllers'. Case study notes. SRC vacation school, Warwick University, 1978

20 CLARKE, D.W., COPE, S.N., and GAWTHROP, P.J.: 'Feasibility study of the application of microprocessors to self-tuning regulators'. OUEL Report 1137/75, 1975

21 CLARKE, D.W., DYER, D.A.J., HASTINGS-JAMES, R., ASHTON, R.P., and EMERY, J.B.: 'Identification and control of a pilot-scale boiling rig'. Proceedings of the IFAC symposium on identification and system parameter estimation, The Hague/Delft, 1973, pp. 355–366

22 CLARKE, D.W., and FROST, P.J.: 'A control Basic for microcomputers' *in* 'Trends in on-line computer control systems'. IEE Conf. Publ. 172, 1979, pp. 53–57

23 CLARKE, D.W., and GAWTHROP, P.J.: 'Self-tuning controller', *Proc. IEE*, 1975, **122**, (9), pp. 929–934

24 CLARKE, D.W., and HASTINGS-JAMES, R.: 'Design of digital controllers for randomly disturbed systems', *ibid.*, 1971, **118**, pp. 1503–1506

25 DE KAYSER, R., and VAN CAUWENBERGHE, A.R.: 'Simulation and self-tuning control in a nuclear power plant' *in* TROCH, I. (Ed.): 'Simulation of control systems' (North-Holland, 1978)

26 DOOB, J.L.: 'Stochastic processes' (Wiley, 1953)

27 DUMONT, G.A., and BELANGER, P.R.: 'Self-tuning control of a titanium dioxide kiln', *IEEE Trans.*, 1978, **AC-23**, pp. 532–538

28 EDMUNDS, J.M.: 'Digital adaptive pole-shifting regulators'. Ph.D. thesis, UMIST, 1976

29 EGARDT, B.: 'A unified approach to model-reference adaptive systems and self-tuning regulators'. Lund Report LUFTD2/(TFRT-7134)/1–67/(1978), 1978

30 EGARDT, B.: 'Stability of model-reference adaptive and self-tuning regulators'. Lund Report LUFD2/(TFRT-1017)/1-163/(1978), 1978

31 EVANS, S.E.: 'Application of microprocessors to adaptive control'. M.Sc. thesis, Brunel University, 1978

32 EYKHOFF, P.: 'System identification' (Wiley, 1974)

33 FORTESCUE, T.R., KERSHENBAUM, L.S., and YDSTIE, B.E.: 'Implementation of self-tuning regulators with variable forgetting factors', submitted for publication, 1979

34 GAWTHROP, P.J.: 'Frequency domain solution of the optimum steady-state regulator problem'. OUEL Report 1145/76, 1976

35 GAWTHROP, P.J.: 'Some interpretations of the self-tuning controller', *Proc. IEE*, 1977, **124**, (10), pp. 889–894

36 GAWTHROP, P.J.: 'Developments in optimal and self-tuning control theory'. OUEL Report 1239/78, 1978

37 GAWTHROP, P.J.: 'On the stability and convergence of self-tuning controllers'. Proceedings of the IMA conference on analysis and optimisation of stochastic systems, 1978

38 GAWTHROP, P.J.: 'On the stability and convergence of self-tuning algorithms'. OUEL Report 1259/78, 1978

39 GOODWIN, G.C., and RAMADGE, P.J.: 'Design of restricted complexity adaptive regulators'. Technical Report EE7804, University of New South Wales, Australia, 1978

40 HETTHESSY, J., and KEVICZKY, L.: 'Some innovations to the minimum variance control'. Proceedings of the IFAC symposium on stochastic control, Budapest, 1974, pp. 353–361

41 HOLST, J.: 'Adaptive prediction and recursive identification'. Lund Report, 1975

42 HORN, H.E.: 'Feasibility study of the application of self-tuning controllers to chemical batch reactors'. OUEL Report 1248/1978, 1978

43 HOROWITZ, I.M.: 'Synthesis of feedback systems'. (Academic Press, 1963)

44 JACOBS, O.L.R., and SARATCHANDRAN, P.: 'Comparisons of adaptive controllers'. OUEL Report 1202/77, 1977

45 JENSEN, L., and HANSEL, R.: 'Computer control of an enthalpy exchanger'. Lund Report 7417, 1974

46 JOSEPH, P., LEWIS, J., and TOU, J.: 'Plant identification in the presence of disturbances, and application to digital adaptive systems'. *Trans. Amer. Inst. Elect. Engrs.*, 1961, **AI-80**, pp. 18–24

47 KALLSTROM, C.G.: 'The Sea Scout experiments'. Lund Report 7407 (C), 1974

48 KALLSTROM, C.G., ASTROM, K.J., THOREL, N.E., ERIKSSON, J. STEN, L.: 'Adaptive autopilots for steering of large tankers'. Lund Report LUTFD2/(TFRT-3145)/1–66/(1977), 1977

49 KALLSTROM, C.G., and ASTROM, K.J.: 'Adaptive autopilots for large tankers'. Proceedings of the IFAC world congress, Helsinki, 1978

50 KALMAN, R.E.: 'Design of a self-optimising control system'. *Trans. ASME*, 1958, pp. 468–478

51 KEVICZKY, L., HETTHESSY, J., HILGER, M., and KOLOSTORI, J.: 'Self-tuning adaptive control of cement raw material blending', *Automatica*, 1978, **14**, pp. 525–532

52 KEVICZKY, L., and HABER, R.: 'Adaptive dual extremum control by Hammerstein model'. Proceedings of the IFAC symposium, Boston, 1977

53 KUSHNER, H.J.: 'Convergence of recursive adaptive and identification procedures via weak convergence theory', *IEEE Trans.*, 1977, **AC-22**, **6**, pp. 921–930

54 KWAKERNAAK, H. and SIVAN, R.: 'Linear optimal control systems' (Wiley, 1972)

55 LANDAU, I.D.: 'A survey of model-reference adaptive techniques: theory and applications', *Automatica*, 1974, **10**, pp. 353–379

56 LANDAU, I.D.: 'Unbiased recursive identification using model-reference adaptive techniques', *IEEE Trans.*, 1976, **AC-21**, pp. 194–202

57 LANDAU, I.D.: 'An addendum to unbiased recursive identification using model reference adaptive techniques' *ibid.*, 1978, **AC-23**, pp. 97–99

58 LINDORFF, D.P. and CARROLL, R.L.: 'Survey of adaptive control using Liapunov design', *Int. J. Control*, 1973, **18**, pp. 897–914

59 LJUNG, L.: 'Convergence of recursive stochastic algorithms'. Proceedings of the IFAC symposium on stochastic control', Budapest, 1974, pp. 215–223

60 LJUNG, L.: 'Convergence of recursive stochastic algorithms'. Automatic control division, Lund Institute of Technology, Report 7403, 1974

61 LJUNG, L.: 'Consistency of the least-squares identification method', *IEEE Trans.*, 1976, **AC-21**, pp. 779–781

62 LJUNG, L.: 'Analysis of recursive stochastic algorithms', *IEEE Trans.*, 1977, **AC-22**, pp. 551–575

63 LJUNG, L.: 'On positive real transfer, functions and the convergence of some recursive schemes', *ibid.*, 1977, **AC-22**, pp. 539–550

64 LJUNG, L., and LANDAU, I.D.: 'Model-reference adaptive systems and self-tuning regulators: some connections'. IFAC world congress, Helsinki, 1978

65 LJUNG, L., and WITTENMARK, B.: 'Asymptotic properties of self-tuning regulators'. Automatic control division, Lund Institute of Technology, Report 7404, 1974

66 MISHKIN, E., and BRAUN, L.: 'Adaptive control systems' (McGraw-Hill, 1961)
67 MORRIS, A.J., FENTON, T.P., and NAZER, Y.: 'Application of self-tuning regulators to the control of chemical processes'. IFAC Congress on digital computer applications to process control, Holland, 1977
68 MORRIS, A.J., and NAZER, Y.: 'Self-tuning controllers for single and multivariable systems. Part I: Theoretical evaluation of the control strategy; Part II: Evaluation of controller performance', submitted to *Automatica*, 1978
69 MORRIS, E.L., and ABAZA, B.A.: 'Adaptive digital control of a steam turbine'. *Proc. IEE*, 1976, **123**, pp. 549–553
70 NEWTON, G.C., GOULD, L.A., and KAISER, J.F.: 'Analytic design of linear feedback controls' (Wiley, 1957)
71 PETERKA, V.: 'Adaptive digital regulation of noisy systems'. Proceedings of the IFAC symposium on identification and process parameter estimation, Prague, 1970
72 PETERKA, V.: 'A square-root filter for real-time multivariable regression', *Kybernetika*, 1975, **11**, pp. 53–67
73 POPOV, V.M.: 'Hyperstability of control systems' (Springer-Verlag, 1973)
74 SASTRY, V.A., SEBORG, D.E., and WOOD, R.K.: 'A self-tuning regulator applied to a binary distillation column', *Automatica*, 1977, **13**, pp. 417–424
75 SCHWARTZ, S.C., and STEIGLITZ, K.: 'The identification and control of unknown linear discrete systems'. *Int. J. Control*, 1971, **14**, pp. 43–50
76 SMITH, O.J.M.: 'A controller to overcome dead-time', *Instrum. Soc. Am. J.*, 1959, **2**
77 STERNBY, J.: 'Topics in dual control'. Lund Report LUTFD2/(TFRT-1012)/1–135/(1977), 1977
78 THORNTON, C.L., and BIERMAN, G.J.: 'Filtering and error analysis via the UDU covariance factorisation'. *IEEE Trans.*, 1978, **AC-23**, 5, pp. 901–907
79 TURTLE, D.P., and PHILLIPSON, P.H.: 'Simultaneous identification and control', *Automatica*, 1971, **7**, pp. 445–453
80 WELLSTEAD, P.E.: 'On the self-tuning properties of pole/zero assignment regulators'. Control Systems Centre Report 402, UMIST, 1978
81 WELLSTEAD, P.E., ZANKER, P., and EDMUNDS, J.M.: 'Self-tuning pole-zero assignment regulators'. Control Systems Centre Report 404, UMIST, 1978
82 WELLSTEAD, P.E., and EDMUNDS, J.M.: 'On-line process identification and regulation' in 'Trends in on-line computer control systems'. IEE Conf. Publ. 127, 1975, pp. 230–237
83 WHITTLE, P.: 'Prediction and regulation by linear least-squares methods' (English University Press, 1963)
84 WIESLANDER, J., and WITTENMARK, B.: 'An approach to adaptive control using real-time identification', *Automatica*, 1971, **7**, pp. 211–217
85 WILLEMS, J.C.: 'Stability, instability, invertibility and causality', *SIAM J. Control*, 1969, **7**, pp. 645–671
86 WILLEMS, J.C.: 'The analysis of feedback systems' (MIT, 1971), Research Monograph 62
87 WILLEMS, J.C.: 'Dissipative dynamic systems. I: General theory; II: Linear systems with quadratic supply rates', *Arch. Ration. Mech. & Anal.*, 1972, **45**, pp. 321–395
88 WILLEMS, J.C.: 'Mechanisms for stability and instability in feedback systems', *Proc. IEEE*, 1976, **64**, pp. 24–35
89 WITTENMARK, B.: 'A self-tuning regulator'. Lund Report 7311, 1973
90 WOUTERS, W.R.E.: 'Parameter adaptive regulators control for stochastic SISO systems, theory and application'. Proceedings of the IFAC symposium on stochastic control', Budapest, 1974, pp. 287–296
91 YAGLOM, A.M.: 'An introduction to the theory of stationary random functions', (Dover, 1973), translated by R.A. Silverman

Self-Tuning Regulators for a Class of Multivariable Systems*

ULF BORISON†

Key Word Index–Adaptive control; multivariable control systems; parameter estimation; process control; self-adjusting systems; stochastic control.

Abstract—Control of a class of multivariable systems described by linear vector difference equations with constant but unknown parameters is discussed. A multivariable minimum variance strategy is first presented. This gives a generalization of the minimum variance strategy for single-input single-output systems. A multivariable self-tuning regulator based on the minimum variance strategy is then proposed. It uses a recursive least squares estimator and a linear controller obtained directly from the current estimates. The asymptotic properties of the algorithm are discussed. If the estimated parameters converge, the resulting controller will under certain conditions give the minimum variance strategy. The analysis also gives insight into the case when several single-input single-output self-tuning regulators are operating in cascade mode.

1. *Introduction*

MANY methods for the design of regulators for linear multivariable systems are based on the availability of a model of the process and its environment. However, in practical applications the process dynamics are often unknown, and some model building or identification technique must be used before the regulator can be designed. An algorithm will now be described, which simplifies this procedure for a class of multivariable systems.

The controlled process is assumed to be described by a linear vector difference equation including a moving average of white noise. The process is required to have as many inputs as outputs, to have a minimum phase property and to have an impulse response starting with a nonsingular matrix. For this type of multivariable system it will be shown how a generalization of the minimum variance strategy for single-input single-output systems given by Åström (1970) can be obtained.

The proposed self-tuning regulator uses a recursive parameter estimator, which is based on the least squares method, and a linear controller whose parameters are given by the current estimates. By choosing a special model structure for the identification, the parameters of the controller are obtained directly from the estimator without any further computations.

The idea of self-tuning control has been discussed by several authors, mainly for the control of single-input single-output systems. See for example Kalman (1958), Peterka (1970), Åström and Wittenmark (1973) and Clarke and Gawthrop (1975). Applications to multivariable systems with the linear controller computed from a Riccati equation have been reported by Peterka and Åström (1973).

*Received December 2 1977; revised August 8 1978. The original version of this paper was presented at the 4th IFAC Symposium on Identification and System Parameter Estimation which was held in Tbilisi Georgian Republic of the USSR during September 1976. The published Proceedings of this IFAC Meeting may be ordered from: North-Holland Publishing Company, P.O.B. 103 Amsterdam-West, Netherlands. This paper was recommended for publication in revised form by associate editor A. Van Cauwenberghe.

†Gränges Nyby AB, S-644 80 Torshälla, Sweden.

2. *Multivariable minimum variance control*

A minimum variance strategy for multivariable systems will now be discussed.

Process model. Let q^{-1} denote the backward shift operator and consider the process given by

$$A(q^{-1})y(t)=B(q^{-1})u(t-k-1)+C(q^{-1})e(t) \tag{2.1}$$

where k is a time delay, y the output vector, u the input vector and $\{e(t)\}$ a sequence of independent, equally distributed, random vectors with zero mean value and covariance $E[e(t)e^T(t)]=R$. The vectors y, u and e are all of dimension p. The polynomial matrices A, B and C are all of dimension $p\times p$, and they are given by

$$A(z)=I+A_1z+\ldots+A_nz^n$$

$$B(z)=B_0+B_1z+\ldots+B_{n-1}z^{n-1},\ B_0 \text{ nonsingular}$$

$$C(z)=I+C_1z+\ldots+C_nz^n$$

where $\det B(z)$ and $\det C(z)$ have all their zeros strictly outside the unit disc. The requirement on $\det C(z)$ can be considered as a weak condition. The requirement on $\det B(z)$ is not necessary for the derivation of the minimum variance strategy. However, the requirement is necessary in practical cases in order to get a stable closed loop system, see Borison (1975).

The assumption that the matrix B_0 is nonsingular implies that a nonsingular transformation of the following type can be introduced

$$\bar{u}(t)=B_0u(t). \tag{2.2}$$

Thus the process can be considered to have essentially one control variable for each loop ($\bar{u}_i(t)\,i=1,\ldots,p$), which influences the corresponding output before the other control variables. Furthermore, the outputs in the different loops will be influenced with the same delay.

Preliminaries. Introduce the identity

$$C(z)=A(z)F(z)+z^{k+1}G(z) \tag{2.3}$$

where

$$F(z)=I+F_1z+\ldots+F_kz^k$$

$$G(z)=G_0+G_1z+\ldots+G_{n-1}z^{n-1}$$

Since $A(0)$ is nonsingular, the polynomial matrices $F(z)$ and $G(z)$ are unique. Introduce also $\tilde{F}(z)$ and $\tilde{G}(z)$ given by

$$\tilde{F}(z)G(z)=\tilde{G}(z)F(z) \tag{2.4}$$

where $\det\tilde{F}(z)=\det F(z)$ and $\tilde{F}(0)=I$. The polynomial matrices $\tilde{F}(z)$ and $\tilde{G}(z)$ always exist but they are not unique, see e.g. Wolovich (1974).

Criterion. Let Q be a positively semidefinite matrix and consider the criterion

$$\min_{u(t)} E[y^T(t+k+1)Qy(t+k+1)] \tag{2.5}$$

Reprinted with permission from *Automatica*, vol. 15, pp. 209–215, Mar. 1979.
Copyright © 1979, International Federation of Automatic Control IFAC. Published by Pergamon Press Ltd.

where k is the time delay in the process model (2.1). The minimum is taken with respect to the admissible control strategies defined below.

Admissible control strategy. At time t the measurements $y(t)$, $y(t-1), \dots$ and the past control actions $u(t-1)$, $u(t-2), \dots$ are known. An admissible control strategy is such that $u(t)$ is a function of $y(t)$, $y(t-1), \dots$ and $u(t-1)$, $u(t-2), \dots$.

Solution. In Appendix 1 it is shown that the strategy

$$\tilde{G}(q^{-1})y(t)+\tilde{F}(q^{-1})B(q^{-1})u(t)=0 \tag{2.6}$$

minimizes the criterion (2.5) asymptotically. It will be called the minimum variance strategy. The asymptotic control error with this strategy is

$$y(t)=F(q^{-1})e(t).$$

When initial effects have settled, the minimum variance strategy (2.6) also minimizes the criterion

$$\min_{u(0),\dots,u(N-1)} E\frac{1}{N}\sum_{t=0}^{N-1} y^{\mathrm{T}}(t+k+1)Qy(t+k+1)$$

where the class of admissible control strategies is the same as above. This is shown in Appendix 2.

With the restrictions introduced on the class of systems, the obtained optimal strategy (2.6) can thus be described by a rational transfer function matrix, which does not depend on the weighting matrix Q of the loss function or on the covariance matrix R of the random vectors $\{e(t)\}$. These properties of the minimum variance strategy are fundamental for the multivariable self-tuning regulator described in the next section.

3. *A self-tuning minimum variance regulator*

Let the process to be controlled be given by (2.1). At each sampling interval the self-tuning algorithm performs a least squares identification based on a model given below. The obtained parameters are then used to compute the control signal.

Estimation. The algorithm estimates recursively the parameters of the model

$$y(t)+\mathscr{A}(q^{-1})y(t-k-1)=\mathscr{B}(q^{-1})u(t-k-1)+\varepsilon(t) \tag{3.1}$$

in such a way that the error $\varepsilon(t)$ is as small as possible in the sense of least squares. In (3.1) k is the time delay of the process model (2.1), and $\mathscr{A}(z)$ and $\mathscr{B}(z)$ are $p\times p$-dimensional polynomial matrices given by

$$\mathscr{A}(x)=\mathscr{A}_0+\mathscr{A}_1 z+\dots+\mathscr{A}_{n_{\mathscr{A}}}z_{n_{\mathscr{A}}}$$

$$\mathscr{B}(z)=\mathscr{B}_0+\mathscr{B}_1 z+\dots+\mathscr{B}_{n_{\mathscr{B}}}z_{n_{\mathscr{B}}} \tag{3.2}$$

Control. At each time t the control is computed from

$$\mathscr{B}(q^{-1})u(t)=\mathscr{A}(q^{-1})y(\text{[illegible]}$$

[illegible]ameters of the controller are [illegible] equal to the estimated parameters.

Introduce the [illegible] Θ given by

$$\Theta=[\theta_1\theta_2 \text{[illegible]} \theta_p]$$

$$=[\mathscr{A}_0\mathscr{A}_1\dots\mathscr{A}_{n_{\mathscr{A}}}\mathscr{B}_0\mathscr{B}_1\dots\mathscr{B}_{n_{\mathscr{B}}}]^{\mathrm{T}}. \tag{3.3}$$

In the algorithm the least squares estimation of Θ is done by estimating one vector θ_i at a time.

The least squares criterion implies that the vector θ_i at each step of time N is determined in such a way that

$$V_N(\theta_i)=\frac{1}{N}\sum_{t=1}^{N}\varepsilon_i^2(t), \quad i=1,\dots,p \tag{3.4}$$

is minimal. The following recursive equations can be used, see Åström and Eykhoff (1971).

$$\theta_i(t)=\theta_i(t-1)+K(t-1)[y_i(t)-\phi(t-k-1)\theta_i(t-1)]$$

$$K(t-1)$$
$$=P(t-1)\phi^{\mathrm{T}}(t-k-1)[1+\phi(t-k-1)P(t-1)\phi^{\mathrm{T}}(t-k-1)]^{-1}$$

$$P(t)$$
$$=P(t-1)-K(t-1)[1+\phi(t-k-1)P(t-1)\phi^{\mathrm{T}}(t-k-1)]K^{\mathrm{T}}(t-1)$$

where

$$\phi(t-k-1)=[-y^{\mathrm{T}}(t-k-1)\dots$$
$$-y^{\mathrm{T}}(t-k-1-n_{\mathscr{A}})u^{\mathrm{T}}(t-k-1)\dots$$
$$\dots u^{\mathrm{T}}(t-k-1-n_{\mathscr{B}})].$$

Thus it can be observed that the estimation is divided into p steps and that the initial values of $P(t)$ are assumed to be the same for these steps. The corresponding gain vectors $K(t-1)$ will then also be the same for all parameter vectors θ_i. In this way significant savings in the computations are obtained.

The matrix $\mathscr{B}_0$ in (3.2) can either be estimated as in (3.3) or be set equal to a constant nonsingular matrix. In the former case it must be required that the estimated matrix is nonsingular at each step of time. In the latter case the constant value of $\mathscr{B}_0$ must be chosen in such a way that it does not prevent the estimated parameters from converging, see Borison (1975).

Remark. By making a transformation of the type (2.2) so that $\mathscr{B}_0$ is a unit matrix, it follows that the algorithm can be interpreted as an interconnection of p single-input single-output self-tuning regulators. For example, the regulator in the first loop controls the output y_1 using the control variable u_1. The signals $y_2(t-i), \dots, y_p(t-i)$ and $u_2(t-1-i), \dots, u_p(t-1-i)$, $i\geqq 0$, can be considered as feed-forward signals.

4. *Analysis*

Convergence of a general algorithm of the type discussed here has been investigated in the work by Ljung (1974). There it has been shown that this type of algorithms does not always converge. However, the algorithm will stabilize a minimum phase system under weak conditions, even if the estimated parameters do not converge. See Ljung and Wittenmark (1974).

Considering the asymptotic properties of the multivariable self-tuning regulator, the minimum variance strategy is always a possible resulting strategy of the algorithm. This follows by applying Corollary 1 in Appendix 2 to the criterion function (3.4), using the model given by (3.1).

Conditions which guarantee that the algorithm gives the minimum variance strategy will now be discussed.

In Borison (1975) the following result is shown

$$\lim_{N\to\infty}\frac{1}{N}\sum_{t=1}^{N} y(t+\tau)y^{\mathrm{T}}(t)=0,$$
$$\tau=k+1,\dots,k+n_{\mathscr{A}}+1$$

$$\lim_{N\to\infty}\frac{1}{N}\sum_{t=1}^{N} y(t+\tau)u^{\mathrm{T}}(t)=0,$$
$$\tau=k+1,\dots,k+n_{\mathscr{B}}+1$$

It is assumed that the estimated parameters converge and that the sequences $\{u(t)\}$ and $\{y(t)\}$ are uniformly bounded in the sense of mean square. Using the result above it can be shown that the minimum variance strategy is obtained in the case when the disturbances are white noise ($C(z)=I$), if the number of estimated parameters is large enough. The same result can also be obtained for a system with a general C-polynomial matrix, if additional assumptions are made. These include assumptions on relatively prime polynomial matrices and requirements on Kronecker indices. See Borison (1975).

5. *Simulations*

Example 1. A simulation will now be shown, where the multivariable self-tuning regulator is applied to the control of a head-box of a paper machine. A schematic description is given in Fig. 1. In Borison (1975) the following head-box model is investigated.

$$y(t)+A_1 y(t-1)=B_0 u(t-1)+e(t)$$

where

$$A_1=\begin{bmatrix}-0.99101 & 8.80512\times 10^{-3}\\ -0.80610 & -0.77089\end{bmatrix}$$

$$B_0=\begin{bmatrix}0.89889 & -4.59328\times 10^{-3}\\ 19.390 & 0.88052\end{bmatrix}$$

$$E[e(t)e^{\mathrm{T}}(t)]=\begin{bmatrix}0.02 & 0.35\\ 0.35 & 7.6\end{bmatrix}$$

y_1 = stock level

y_2 = total pressure

u_1 = deviation in stock flow from steady state value

u_2 = deviation in air flow from steady state value.

The sampling interval is 1 second. The model is obtained from a continuous time version given in Åström (1972). The self-tuning algorithm estimates recursively the parameters of the following model

$$y(t)+\mathcal{A}_0 y(t-1)=\mathcal{B}_0 u(t-1)+\varepsilon(t).$$

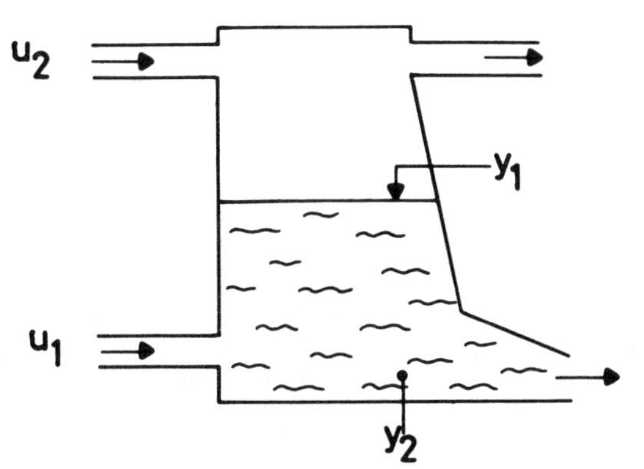

FIG. 1. Head-box of a paper machine.

At each step the control is obtained from

$$u(t)=\mathcal{B}_0^{-1}\mathcal{A}_0 y(t). \tag{5.1}$$

The estimated parameters are shown in Fig. 2. The regulator parameters given by (5.1) are shown in Fig. 3. It follows that the resulting regulator tends to the minimum variance regulator. There were no numerical problems with the inversion of the matrix $\mathcal{B}_0$. During the start-up the control was quite acceptable, and after some minutes it was almost optimal, although the estimated parameters had not yet converged.

Example 2. In this example a system with a time delay will be studied,

$$y(t)+A_1 y(t-1)=u(t-2)+e(t)+C_1 e(t-1)$$

where

$$A_1=\begin{bmatrix}-0.9 & 0.5\\ 0.5 & -0.2\end{bmatrix}\qquad C_1=\begin{bmatrix}-0.2 & -0.4\\ 0.2 & -0.8\end{bmatrix}$$

$$E[e(t)e^{\mathrm{T}}(t)]=R=\begin{bmatrix}0.1 & 0.1\\ 0.1 & 0.2\end{bmatrix}.$$

The zeros of $\det C(z)$ are $z_1=5/3$ and $z_2=5/2$, and they are outside the unit disc. Det $A(z)$ has the zeros $z_1\approx -16.6$ and $z_2\approx 0.86$. Since there is a zero inside the unit disc, the open loop system is unstable. The identity (2.3) gives

$$F_1=C_1-A_1=\begin{bmatrix}0.7 & -0.9\\ -0.3 & -0.6\end{bmatrix}$$

$$G_0=-A_1F_1=\begin{bmatrix}0.78 & -0.51\\ -0.41 & 0.33\end{bmatrix}.$$

The optimal control law given by (2.6) can in this case be written as

$$(I+\tilde{F}_1 q^{-1})y(t)=-\tilde{G}_0 y(t)$$

where

$$\tilde{G}_0=G_0$$

and

$$\tilde{F}_1=G_0F_1G_0^{-1}=\begin{bmatrix}1.4143 & 0.98571\\ -1.1857 & -1.3143\end{bmatrix}.$$

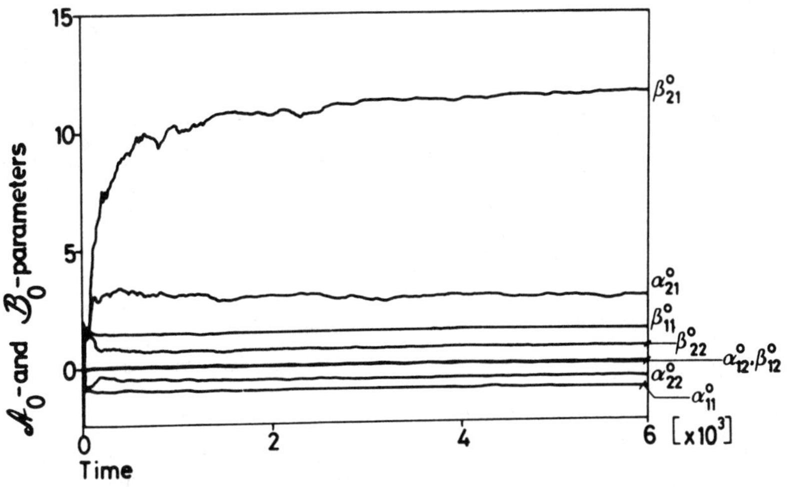

FIG. 2. Estimated parameters for the head-box.

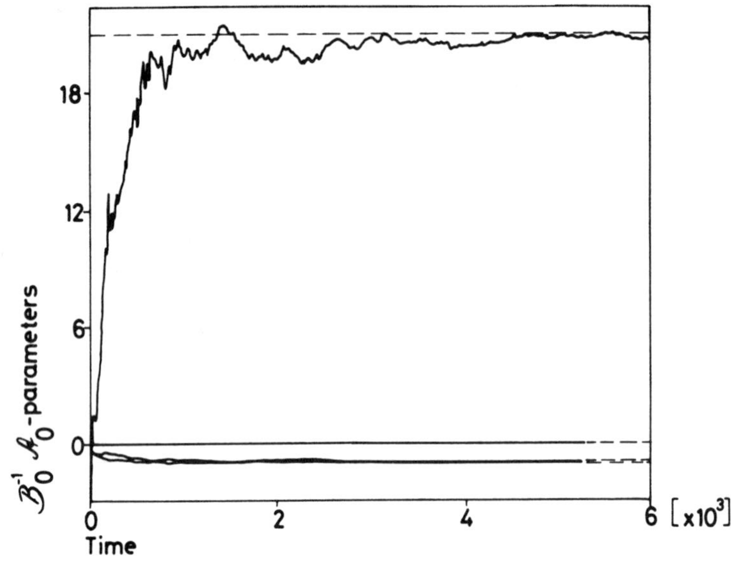

FIG. 3. Parameters of the matrix $\mathcal{B}_0^{-1}\mathcal{A}_0$ for the head-box. They have been computed from the estimated parameters shown in Fig. 2. The dashed lines show the optimal values.

With the model

$$y(t)+\mathcal{A}_0 y(t-2)=u(t-2)+\mathcal{B}_1 u(t-3)+\varepsilon(t)$$

and the regulator

$$(I+\mathcal{B}_1 q^{-1})u(t)=\mathcal{A}_0 y(t)$$

the optimal values are

$$\mathcal{A}_0=\begin{bmatrix}-0.78 & 0.51\\ 0.41 & -0.33\end{bmatrix}\quad \mathcal{B}_1=\begin{bmatrix}1.4143 & 0.98571\\ -1.1857 & -1.3143\end{bmatrix}$$

The result of a simulation is shown in Fig. 4. The estimated parameters approach the optimal values, but the $\mathcal{B}$-parameters converge slowly. The loss function

$$V=\sum_{t=1}^{N} y^{\mathrm{T}}(t)y(t)$$

is shown in Fig. 5, and it is almost optimal the whole time.

6. *Conclusions*

By restricting the class of systems to include only systems of minimum phase type that have the same number of inputs as outputs and an impulse response starting with a non-singular matrix, a minimum variance strategy for multivariable systems has been derived. It has then been shown that the simple self-tuning regulator for single-input single-output systems can be extended to the multivariable case with its simplicity retained. The results of the analysis also give insight into the case, when several single-input single-output self-tuning regulators are operated in cascade mode.

Acknowledgement—I wish to express my sincere gratitude to Professor K. J. Åström for his valuable support.

REFERENCES

Ashton, R. P. (1974). Digital control of randomly disturbed systems. PhD thesis, Balliol College, Oxford.

Åström, K. J. (1970). *Introduction to Stochastic Control Theory*, Academic Press, New York.

Åström, K. J. (1972). Lecture notes on paper machine control. *Head Box Flow Dynamics and Control*. Dept. of Automatic Control, Lund Institute of Technology, Lund, Sweden.

Åström, K. J. and P. Eykhoff (1971). System identification—a survey. *Automatica*, **7**, 123–162.

Åström, K. J. and B. Wittenmark (1973). On self-tuning regulators. *Automatica*, **9**, 185–199.

Borison, U. (1975). Self-tuning regulators—industrial application and multivariable theory. Report 7513, Dept of Automatic Control, Lund Institute of Technology, Lund, Sweden.

Clarke, D. W. and P. J. Gawthrop (1975). Self-tuning controller. *Proc. IEE*, **122**, 929–934.

Clarke, D. W. and R. Hastings-James (1971). Design of digital controllers for randomly disturbed systems. *Proc. IEE*, **118**, 1503–1506.

Kalman, R. E. (1958). Design of a self-optimizing control system. *Amer. Soc. Mech. Engineers Trans.*, **80**, 468–478.

Ljung, L. (1974). Convergence of recursive stochastic algorithms. Preprints of the IFAC Symposium on Stochastic Control, Budapest.

Ljung, L. and B. Wittenmark (1974). Analysis of a class of adaptive regulators. Preprints of the IFAC Symposium on Stochastic Control, Budapest.

Peterka, V. (1970). Adaptive digital regulation of noisy systems. *IFAC Symposium on Identification and Process Parameter Estimation*, Prague.

Peterka, V. and K. J. Åström (1973). Control of multivariable systems with unknown but constant parameters. *Preprints of the IFAC Congress on Systems Identification and Parameter Estimation*, the Hague.

Wolovich, W. A. (1974). *Linear Multivariable Systems*. Springer-Varlag, New York, p. 159.

Appendix 1: Derivation of the multivariable minimum variance strategy

Introduce the polynomial matrix $\tilde{C}(z)$ given by

$$\tilde{C}(z)=\tilde{F}(z)A(z)+z^{k+1}\tilde{G}(z). \tag{7.1}$$

Multiplying (2.3) by $\tilde{F}(z)$ from the left, (7.1) by $F(z)$ from the right and using (2.4), it follows that

$$\tilde{F}(z)C(z)=\tilde{C}(z)F(z). \tag{7.2}$$

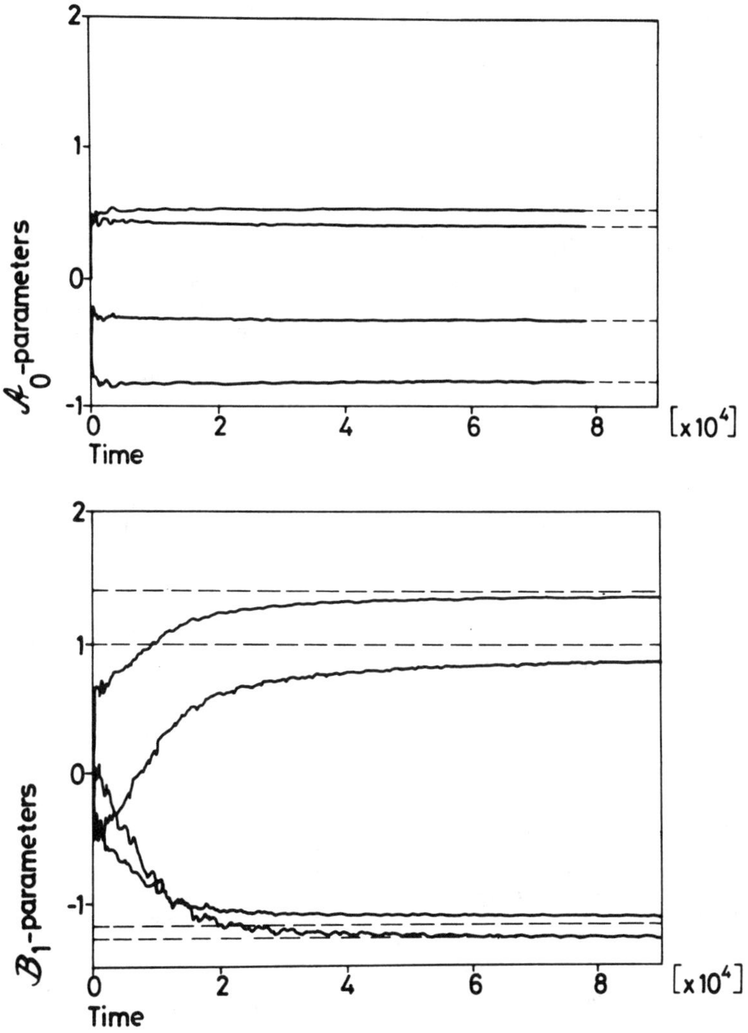

FIG. 4. Estimated parameters in Example 2. The dashed lines show the optimal values.

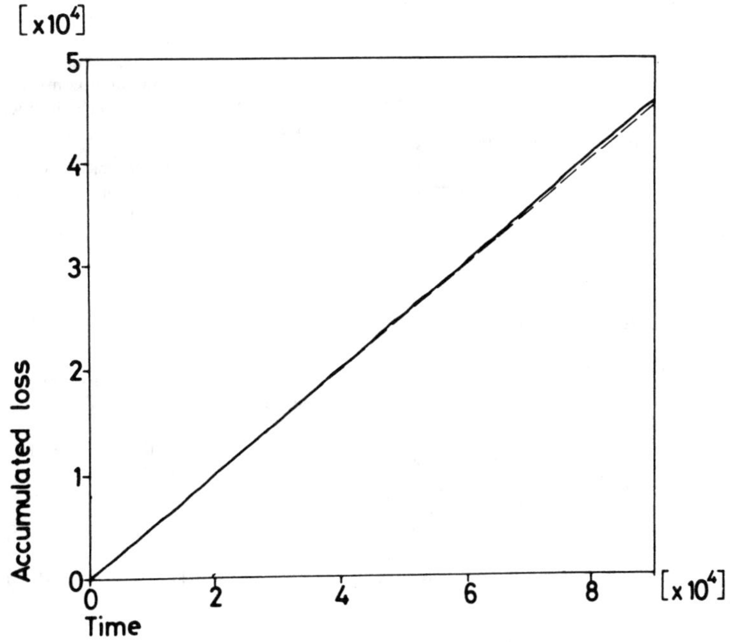

FIG. 5. Accumulated loss in Example 2. The dashed line shows the optimal value.

Since $\det \tilde{F}(z) = \det F(z)$ according to the definition in (2.4), it follows from (7.2) that

$$\det \tilde{C}(z) = \det C(z).$$

Since $\det C(z)$ has all its zeros strictly outside the unit disc, the same holds for $\det \tilde{C}(z)$.

Write the process model (2.1) as

$$A(q^{-1})y(t+k+1) = B(q^{-1})u(t) + C(q^{-1})e(t+k+1)$$

and multiply by $\tilde{F}(q^{-1})$ from the left. Then the following equation is obtained

$$\tilde{F}(q^{-1})A(q^{-1})y(t+k+1) = \tilde{F}(q^{-1})B(q^{-1})u(t) + \tilde{F}(q^{-1})C(q^{-1})e(t+k+1).$$

Introduction of (7.1) and (7.2) gives

$$\tilde{C}(q^{-1})[y(t+k+1) - F(q^{-1})e(t+k+1)] = \tilde{G}(q^{-1})y(t) + \tilde{F}(q^{-1})B(q^{-1})u(t). \quad (7.3)$$

Let $w(t)$ be defined by

$$w(t) = \tilde{G}(q^{-1})y(t) + \tilde{F}(q^{-1})B(q^{-1})u(t).$$

The equation (7.3) can then be written on the form

$$y(t+k+1) = F(q^{-1})e(t+k+1) + \sum_{s=t_0}^{t} M_{t-s}w(s) + I(t, t_0) \quad (7.4)$$

where M_{t-s} are $p \times p$ matrices and $I(t, t_0)$ is a p-*vector* function which depends on the initial conditions.

The random variables $e(t+k+1), e(t+k), \ldots, e(t+1)$ are independent of $y(t), y(t-1), \ldots$, and $u(t-1), u(t-2), \ldots$. If they are also independent of initial values, then

$$E[y^T(t+k+1)Qy(t+k+1)] \geqq E\{[F(q^{-1})e(t+k+1)]^T QF(q^{-1})e(t+k+1)\}. \quad (7.5)$$

It follows from (7.4) that equality will hold if $I(t, t_0) = 0$ and if $w(s) = 0$. When $w(s) = 0$, it follows that

$$\tilde{G}(q^{-1})y(s) + \tilde{F}(q^{-1})B(q^{-1})u(s) = 0. \quad (7.6)$$

The elements of $I(t, t_0)$ will tend to zero exponentially when $w(s) = 0$ since $\det \tilde{C}(z)$ is assumed to have all its zeros strictly outside the unit disc. Then equality will hold asymptotically in (7.5) for arbitrary initial conditions.

The equation (7.6) defines an admissible control strategy, since $\tilde{F}(0)$ and $B(0)$ are nonsingular. The strategy thus minimizes the criterion (2.5) asymptotically, and it will be called the minimum variance strategy.

The control error with the strategy (7.6) is asymptotically

$$y(t) = F(q^{-1})e(t).$$

Summing up, the following theorem is obtained

Theorem 1. Consider the process given by

$$A(q^{-1})y(t) = B(q^{-1})u(t-k-1) + C(q^{-1})e(t)$$

where $\{e(t)\}$ is a sequence of independent, equally distributed, random vectors which are independent of initial values and which have zero mean and covariance R. Let the polynomial matrix $B(z)$ have B_0 nonsingular and a determinant with all zeros strictly outside the unit disc. Let also $\det C(z)$ have all its zeros strictly outside the unit disc. The minimum variance strategy is then given by

$$\tilde{G}(q^{-1})y(t) + \tilde{F}(q^{-1})B(q^{-1})u(t) = 0$$

where $\tilde{F}(z)$ and $\tilde{G}(z)$ are given by

$$C(z) = A(z)F(z) + z^{k+1}G(z)$$
$$\tilde{F}(z)G(z) = \tilde{G}(z)F(z).$$

The control error with this strategy is asymptotically

$$y(t) = F(q^{-1})e(t).$$

Remark. It is not necessary to assume that the matrix B_0 is nonsingular. The necessary requirement is that equation (7.6) defines an admissible control strategy. The existence of an admissible control strategy is related to causality for systems described by linear vector difference equations. In the following example the matrix B_0 is singular, but the optimal strategy is still admissible.

$$y(t+1) + \begin{bmatrix} a_1 & a_2 \\ 0 & 0 \end{bmatrix} y(t) + \begin{bmatrix} a_3 & a_4 \\ a_5 & a_6 \end{bmatrix} y(t-1) = \begin{bmatrix} 1 & 0 \\ 0 & 0 \end{bmatrix} u(t) + \begin{bmatrix} b_1 & b_2 \\ 0 & 1 \end{bmatrix} u(t-1) + e(t+1).$$

The strategy minimizing the criterion (2.5) is

$$\begin{bmatrix} 1 & 0 \\ 0 & 0 \end{bmatrix} u(t) + \begin{bmatrix} b_1 & b_2 \\ 0 & 1 \end{bmatrix} u(t-1) = \begin{bmatrix} a_1 & a_2 \\ 0 & 0 \end{bmatrix} y(t) + \begin{bmatrix} a_3 & a_4 \\ a_5 & a_6 \end{bmatrix} y(t-1)$$

which gives an admissible control strategy. Except for special cases like the one above, it is usually more involved to determine a minimum variance strategy for a system which does not fulfil the conditions in Theorem 1. In a general case a Riccati equation must be solved. Modified criteria, which are simpler to minimize, can be introduced to get strategies that are more easily computed. This has been discussed by Ashton (1974). It will not be pursued further here.

Appendix 2: N-step control

The criterion (2.5) is sometimes called the one step control, while the criterion

$$\min_{u(0), \ldots, u(N-1)} E \frac{1}{N} \sum_{t=0}^{N-1} y^T(t+k+1)Qy(t+k+1) \quad (8.1)$$

where Q is positively semidefinite, is called the N-step control. Let the class of admissible control strategies be the same as before, see Section 2. The following result can now be obtained.

Theorem 2. Let the minimum variance strategy (2.6) be applied to the process model (2.1), and consider the system when initial effects have settled. The minimum variance strategy then minimizes the criterion (8.1), and the minimal loss is

$$\operatorname{tr} QR + \sum_{i=1}^{k} \operatorname{tr} F_i^T QF_i R.$$

Proof. At time $N-1$ the control signal $u(N-1)$ will influence only the last term of the loss function,

$$E[y^T(N+k)Qy(N+k)].$$

It follows from (7.5) that

$$E[y^T(N+k)Qy(N+k)] \geqq E\{[F(q^{-1})e(N+k)]^T QF(q^{-1})e(N+k)\} = \operatorname{tr} QR + \sum_{i=1}^{k} \operatorname{tr} F_i^T QF_i R$$

when any admissible control strategy is used. Observe that the minimal loss is constant and independent of previous controls.

At time $N-2$ the control signals $u(N-2)$ and $u(N-1)$ shall be chosen in such a way that the last two terms of the loss function are minimized. From (7.5) it follows that

$$E\{y^T(N+k)Qy(N+k)+y^T(N+k-1)Qy(N+k-1)\}$$

$$\geq \operatorname{tr} QR + \sum_{i=1}^{k} \operatorname{tr} F_i^T Q F_i R + E\{y^T(N+k-1)Qy(N+k-1)\}$$

$$\geq 2[\operatorname{tr} QR + \sum_{i=1}^{k} \operatorname{tr} F_i^T Q F_i R].$$

In the same way it follows that

$$E\frac{1}{N}\sum_{t=0}^{N-1} y^T(t+k+1)Qy(t+k+1) \geq \operatorname{tr} QR + \sum_{i=1}^{k} \operatorname{tr} F_i^T Q F_i R \quad (8.2)$$

with any admissible control strategy. But from Theorem 1 it follows that the minimum loss in (8.2) is obtained with the minimum variance strategy. Then the theorem is proved.

Corollary 1. The minimum variance strategy also minimizes the criterion

$$\min_{u(0),\ldots,u(N-1)} E\frac{1}{N}\sum_{t=0}^{N-1} y_i^2(t+k+1) \quad i=1,\ldots,p$$

for the system (2.1). The class of admissible control strategies is the same as before, see Section 2.

Proof. Let Q be a matrix of dimension $p \times p$ having all elements zero except the ith diagonal element, which is 1. Then the minimum variance strategy will minimize

$$E\frac{1}{N}\sum_{t=0}^{N-1} y_i^2(t+k+1).$$

This is valid for all i since the strategy is independent of Q.

Remark. A strategy minimizing the criterion

$$\min_{u(t)} E\{y^T(t+k+1)Q_1 y(t+k+1) + u^T(t)Q_2 u(t)\} \quad (8.3)$$

which includes a penalty on the control action, is calculated easily without solving a Riccati equation. It has been discussed by Clarke and Hastings-James (1971) in the single-input single-output case and by Ashton (1974) in the multivariable case. The variable Q_2 must be chosen in such a way that a stable closed loop system is obtained. The strategy minimizing (8.3) is not the same as the one minimizing

$$\min_{u(0),\ldots,u(N-1)} E\frac{1}{N}\sum_{t=0}^{N-1} \{y^T(t+k+1)Q_1 y(t+k+1) + u^T(t)Q_2 u(t)\}.$$

Self-tuning controllers based on pole-zero placement

Prof. K.J. Astrom and B. Wittenmark, Ph.D.

Abstract: The paper gives a review of pole-placement design for systems with known parameters, then focuses entirely on the servos. Adaptive pole-placement algorithms are discussed, leading to the formulation of *explicit* and *implicit* schemes, and simulation of the behaviour of the alorithms is given.

1 Introduction

The simple p.i. regulator is unquestionably the most common regulator in industry. In spite of this, there are cases where it is advantageous to use more complex control algorithms. More complex regulators have unfortunately often more adjustable parameters. It may thus be costly and time consuming to tune such regulators. Self-tuning control is one possibility to simplify the tuning.

The basic self-tuning regulator described in Reference 8 was designed for a situation where the control problem could be characterised as a minimum-variance-control problem. This means that the criterion is to minimise the variance of the output. The basic self-tuning regulator was designed based on a certainty-equivalence argument. The appropriate model of the process and its environment is thus estimated recursively. The control is determined as if the estimated model is equal to the true model. There are many problems which fit this problem formulation, and the basic self-tuning regulator has also been shown to work well in such cases. There are, however, also stochastic control problems where minimum-variance control is not appropriate. One case is a nonminimum-phase plant. Another case is when large control signals are required to achieve minimum variance. These cases can, however, be formulated as linear-quadratic-Gaussian (l.q.g.) control problems. A self-tuning regulator based on the l.q.g. design technique was described in Reference 9. Other versions are given in References 27 and 6. The self-tuning regulator based on the l.q.g. formulation has the drawback of being more complicated than the basic self-tuning regulator. The reason for this is that the design calculations which are done in each step involve the solution of a steady-state Riccati equation or, equivalently, a spectral factorisation. A simpler algorithm was proposed by Clarke and Gawthrop.[11] They proposed to use a l.q.g. formulation with a *one-step* criterion only. This simplifies the algorithm considerably. The algorithm can be made to work well in many cases but it is not foolproof. Further discussions of the algorithm are given in References 19 and 12.

There are many problems which do not fit the stochastic control formulation. Encouraged by the success of the self-tuning regulators for stochastic control problems, it is tempting to try a similar approach in other cases. Using the certainty equivalence argument, the design is straightforward: start with a design method for systems with known parameters, substitute the parameters of the known system model by estimates which are obtained recursively and recalculate the control parameters in each step. Self-tuning controllers of this type which are based on pole-placement design and least-squares estimation are discussed in this paper. The controllers obtained are useful in many situations. For instance, they can be used to tune control loops when the parameters or the controlled system are unknown or slowly time-varying. It is assumed that the main source of disturbances are changes in the reference value or, occasionally, large disturbances that have to be eliminated. The self-tuning regulator based on minimum variance control is not well suited for this case. The new self-tuning controllers can be used to solve the servo problem, and can thus be regarded as useful complements.

Self-tuning regulators based on pole-placement design have been discussed by several other authors. A digital adaptive pole-shifting algorithm was discussed in a dissertation by Edmunds.[13] This and similar algorithms are further discussed in References 33, 36, 34 and 15. In these works, the emphasis is, however, on the regulation problem and not on the servoproblem. The use of feed-forward is not discussed. Servo self-tuners have been discussed by Aström *et al.*[6, 7] and Wellstead and Zanker.[35] Self-tuning of p.i.d. controllers based on pole placement is discussed by Wittenmark.[32] Wouters[37] also proposes a stochastic pole-placement strategy. He also focuses on the stochastic regulation properties of the algorithm. The self-tuning controller proposed by Clarke[11, 12] can also be interpreted in a pole-placement framework (see Gawthrop[10]). Our paper differs from the previous treatments by focusing entirely on the servo problem. In our formulation, the links to a deterministic design procedure are also emphasised. This makes it possible to establish links to m.r.a.s. (see Egardt[14]). Another feature of this paper is that the notions of algorithms with implicit and explicit identification are introduced. Several of the algorithms proposed in this paper are also new.

The paper is organised as follows. Pole-placement design for systems with known parameters is reviewed in Section 2. The suitability at the pole-placement design as a basis for adaptive control is discussed in Section 3. It is shown that there are some difficulties which are inherent in the problem formulation. Adaptive pole-placement algorithms based on estimation of the parameters in an explicit process model are discussed in Section 4. This leads to the so called *explicit* schemes. In Section 5 it is shown that some simplification of the adaptive algorithms can be achieved by estimating parameters instead in a modified process model. This leads to the *implicit* schemes. Some simulations which illustrate the behaviour of adaptive algorithms based on the pole-placement design are given in Section 6.

Paper 698D, received 20th November 1979

The authors are with the Department of Automatic Control, Lund Institute of Technology, PO Box 725, S-220 07, Lund 7, Sweden

Reprinted with permission from *Proc. Inst. Elec. Eng.*, Part D, vol. 127, pp. 120–130, May 1980.
Copyright © 1980; The Institution of Electrical Engineers.

2 Pole-zero placement design

A brief review of the pole-zero-placement design method for systems with known parameters will now be given. This material is quite well known. See, for example the classic text on sampled data systems by Ragazzini and Franklin.[28] More recent discussions on design of digital control systems based on pole-placement design are found in References 1, 31 and 18. Owing to the algebraic nature of the problem there are strong similarities to the corresponding design procedure for continuous systems.[2] The discussion given here is limited to single-input systems.

2.1 Notation

The systems and regulators are described using a polynomial representation. The following notation is used:

$$A(q^{-1}) = a_0 + a_1 q^{-1} + \ldots + a_{n_a} q^{-n_a}, \qquad a_0 \neq 0$$

where q^{-1} is the backward shift operator. If $a_0 = 1$ the polynomial is said to be monic. The degree of a polynomial $A(q^{-1})$ is written either as $\deg A$ or as n_a. The argument of the polynomial is dropped if there is no ambiguity.

The input, output and command signals of the process are denoted $u(t)$, $y(t)$ and $u_c(t)$, respectively, and $v(t)$ is a disturbance. Z is a region well outside the unit disc. If the zeros of a polynomial belong to Z this implies that the corresponding modes are sufficiently well damped. This region is called the restricted stability region.

2.2 Formulation

Consider a process characterised by the rational operator

$$G(q^{-1}) = \frac{q^{-k}B(q^{-1})}{A(q^{-1})} \qquad (1)$$

It is assumed that A and B are coprime, that A is monic, and that the delay in the process is such that

$$k \geqslant 1 \qquad (2)$$

It is desired to find a controller such that the closed loop is stable and that the transfer function from the command input u_c to the output is given by

$$G_m(q^{-1}) = \frac{q^{-k}B_m(q^{-1})}{A_m(q^{-1})} \qquad (3)$$

where A_m and B_m are coprime and A_m is monic. The zeros of A_m are assumed to be inside Z.

For simplicity, it is assumed that the time delay in eqn. 3 is the same as that in eqn. 1. It is, however, sufficient to assume that the delay in eqn. 3 is at least as long as that in eqn. 1.

2.3 Design procedure

A general linear regulator can be described by

$$R(q^{-1})u(t) = T(q^{-1})u_c(t) - S(q^{-1})y(t) \qquad (4)$$

The closed-loop transfer function relating y to u_c is given by

$$\frac{q^{-k}TB}{AR + q^{-k}BS} = \frac{q^{-k}B_m}{A_m} \qquad (5)$$

where the right-hand side is the desired closed-loop transfer function G_m given by eqn. 3.

The design problem is thus equivalent to the algebraic problem of finding polynomials R, S and T such that eqn. 5 holds. It follows from eqn. 5 that factors of B which are not also factors of B_m must be factors of R. Since factors of B correspond to open-loop zeros it means that open-loop zeros which are not desired closed-loop zeros must be cancelled. Factor B as

$$B = B^+B^- \qquad (6)$$

where all the zeros of B^+ are in the restricted stability region Z and all zeros of B^- outside Z. This means that all zeros of B^+ correspond to well damped modes, and all zeros of B^- correspond to unstable or poorly damped modes.

A necessary condition for solvability of the servo problem is thus that the specifications are such that

$$B_m = B_{m_1}B^- \qquad (7)$$

Since $\deg A_m$ is normally less than $\deg(AR + q^{-k}BS)$ it is clear that there are factors in eqn. 5 which cancel. In state-space theory it can be shown that the regulator eqn. 4 corresponds to a combination of an observer and a state feedback.[2] It is natural to assume that the observer is designed in such a way that changes in command signals do not generate errors in the observer. This means that the factor which cancels in the right-hand side of eqn. 5 can be interpreted as the observer polynomial A_0. It is assumed that all zeros of A_0 are in the restricted stability region Z.

A block diagram of the closed-loop system is shown in Fig. 1. The regulator can be interpreted as being composed of a feedforward path with the transfer function T/R and a feedback path with the transfer function $-S/R$.

The design method can be described as follows.

Data
Given a mathematical model (eqn. 1) of the process characterised by the polynomials A and B, the desired response (eqn. 3) is characterised by the polynomials A_m and B_m and the desired observer polynomial A_0. Assume that the data satisfies the conditions of eqns. 2, 6 and 7, and that all zeros of A_0 are in Z.

Step 1
Solve the equation

$$AR_1 + q^{-k}B^-S = A_mA_0 \qquad (8)$$

with respect to R_1 and S.

Step 2
The regulator which gives the desired closed-loop response is given by eqn. 4, with

$$R = R_1B^+ \qquad (9)$$

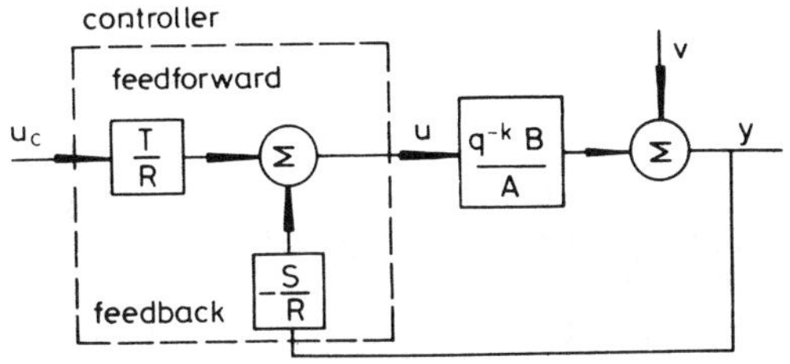

Fig. 1 *Block diagram of closed-loop system*

and

$$T = A_0 B_{m_1} \tag{10}$$

Eqn. 8 can always be solved because it was assumed that A and B were coprime. This implies, of course, that A and B^- are also coprime.

Eqn. 8 has infinitely many solutions. If R_1^0 and S^0 are solutions, then

$$R_1 = R_1^0 + B^- U$$

$$S = S^0 - AU$$

where U is an arbitrary polynomial, is also a solution. All solutions will give a closed-loop system with the desired closed-loop transfer function G_m. The different solutions will, however, give systems with different noise-rejection properties. The transfer function from the disturbance v acting on the process output to the output (see Fig. 1) is given by

$$\frac{AR}{AR + q^{-1k} BS} = \frac{AR}{A_m A_0 B^+}$$

The particular solutions used for the self-tuning regulators in this paper are such that the transfer functions S/R and T/R are causal, with no extra delay. The following are natural choices of solutions:

$$\deg S = \deg A - 1 \tag{11}$$

$$\deg R_1 = \deg A_m + \deg A_0 - \deg A$$

or

$$\deg S = \deg A_m + \deg A_0 = \deg B^- - k \tag{12}$$

$$\deg R_1 = \deg B^- + k - 1$$

The case of eqn. 11 corresponds to 'integral' compensation, and the case of eqn. 12 corresponds to 'derivative' compensation. There are, however, many other possibilities.

2.4 Interpretation as model following

The regulator of eqn. 4 can be interpreted as a model following. It follows from eqns. 8, 9 and 10 that

$$\frac{T}{R} = \frac{A_0 B_{m_1}}{B^+ R_1} = \frac{(AR_1 + q^{-k} B^- S) B_{m_1}}{A_m B^+ R_1}$$

$$= \frac{AB_{m_1}}{B^+ A_m} + \frac{q^{-k} SB^- B_{m_1}}{B^+ R_1 A_m} = \frac{A}{B} \cdot \frac{B_m}{A_m} + \frac{q^{-k} S}{R} \cdot \frac{B_m}{A_m}$$

The feedback law of eqn. 4 can thus be written as

$$u(t) = \frac{A}{B} y_c(t+k) + \frac{S}{R} [y_c(t) - y(t)] \tag{13}$$

where

$$y_c(t) = \frac{q^{-k} B_m}{A_m} u_c(t)$$

The signal y_c can be interpreted as the output obtained when the command signal u_c is applied to the model $q^{-k} B_m / A_m$. When the regulator (eqn. 4) is rewritten as eqn. 13 it is clear that it can be thought of as composed of two parts, one feedforward term $A/B\, y_c(t+k)$ and one feedback term $(S/R)(y_c(t) - y(t))$. The feedforward is a combination of the ideal model and an inverse of the process model. The feedback term is obtained by feeding the error $y_c - y$ through a dynamical system characterised by the operator S/R. The link between pole placement and model-following design is thus established. Notice, however, that the system $q^k A/B$ is not realisable, although the combination $AB_m/(BA_m)$ is.

2.5 Special cases

To perform the design, it is necessary to have procedures for decomposing a polynomial B into its factors B^+ and B^-, and for solving the linear polynomial eqn. 8. The decomposition is essentially a spectral factorisation problem. Eqn. 8 can be solved by using Gauss's elimination or by using Euclid's algorithm. In the adaptive algorithms, these calculations have to be repeated in each step of the iteration. It is then of interest to see if there are some special cases where the design calculations can be simplified. Two special cases, where the decomposition problem is avoided, are given below.

Example 1: (all process zeros cancelled)
Assume that all process zeros are cancelled and that no additional zeros are introduced. Further, assume that deg A_m = deg A and deg A_0 = deg $A - 1$. Eqn. 8 then reduces to

$$AR_1 + q^{-k} S = A_m A_0 \tag{14}$$

and the controller is then given by eqn. 4 with

$$R = R_1 B$$

$$B_m = B_{m_1} = K$$

$$T = A_0$$

where K is a constant. Notice that B appears as a factor of R which is the denominator of the regulator transfer function. Also, notice that the specifications are normally such that the desired closed-loop transfer function has unit gain at low frequencies. This means that the polynomials A_m and B_m should be normalised such that $B_m(1)/A_m(1) = 1$. Since A is monic the polynomials q^{-k} and A are always relatively prime. Eqn. 14 can thus always be solved. The solutions corresponding to

$$\deg S = \deg A - 1 \tag{15}$$

$$\deg R_1 = \deg A_m + \deg A_0 - \deg A$$

or

$$\deg S = \deg A_m + \deg A_0 - k \tag{16}$$

$$\deg R_1 = k - 1$$

are chosen. In the special case of example 1, the design calculations thus reduce to the solution of eqn. 14. Notice that eqn. 14 is easy to solve for the case of eqn 16. The coefficients of the polynomials R_1 and S can then be obtained one at a time.

Example 2: (no process zeros are cancelled)
Assume that the specifications are such that the desired closed-loop zeros are equal to the process zeros, i.e. $B_m = KB$, where K is a constant. The specifications are normally such that the low-frequency gain of the desired closed loop is unity. The constant factor of the polynomial B_m is then

chosen so that $B_m(1) = A_m(1)$. It is assumed that this normalisation is made. Eqn. 8 then gives

$$AR + q^{-k}BS = A_m A_0 \qquad (17)$$

The design procedure is thus again to choose the polynomials A_m and A_0. Eqn. 17 is then solved with respect to R and S, and the controller is then given by eqn. 4, where $T = KA_0$.

2.6 *Other alternatives*

There are many possible variations on the pole-placement design scheme presented in this Section. Franklin[18] has pointed out that the observer poles need not necessarily be cancelled precisely. Near-cancellations lead to an extension of the classical notion of dipole compensation. Similarly, Gawthrop[19] has pointed out that, in the case of stable but poorly damped process zeros, it is possible to cancel them, provided that the specified closed-loop transfer function has zeros close to the process zeros.

3 Self-tuning control

The basic idea when using the separation principle to design self-tuning regulators can be expressed as follows: start with a design procedure for systems with known parameters; when the parameters are not known they are estimated recursively and the regulator is redesigned in each step, using the estimated parameters instead of the true ones. This means that the certainty equivalence hypothesis is used to determine the controller. A block diagram of a general self-tuning regulator is shown in Fig. 2.

The pole-zero-placement design method was discussed in Section 2. The parameter estimation will be discussed in this Section. A general discussion of some properties of self-tuners based on pole-zero placement is also given in this Section. More details are given in the following Sections.

3.1 *Parameter estimation*

An overview of methods for recursive parameter estimation is given in Söderström *et al.*[29] There is, unfortunately, no recursive parameter estimator which is uniformly best. Least squares, which is one of the simplest recursive estimation schemes, will be used here. This procedure will give biased estimates if there are stochastic disturbances which are coloured noise. Since the discussion is focused on the servo problem, the major disturbances are, however, command inputs. This is also compatible with the pole-zero-placement design procedure, which is not suitable to handle trade-offs quantitatively between measurement noise and process noise.

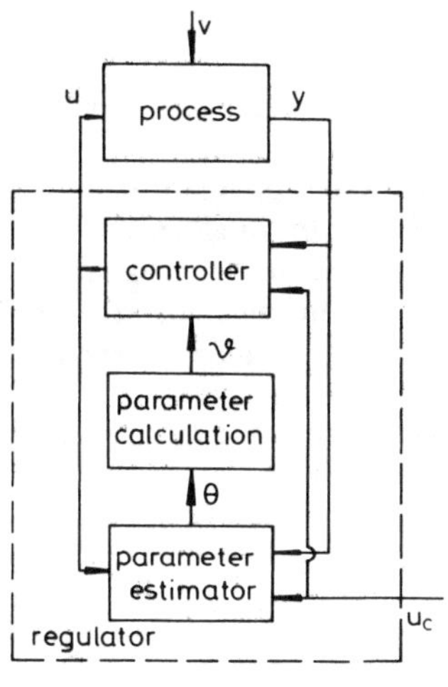

Fig. 2 *Schematic diagram of a self-tuning regulator*

3.2 *Recursive least squares*

Consider the process model

$$Ay(t) = Bu(t-k) \qquad (18)$$

which can be represented explicitly as

$$y(t) + a_1 y(t-1) + \ldots + a_{n_a} y(t-n_a) = b_0 u(t-k) + \ldots + b_{n_b} u(t-n_b-k)$$

Introduce a vector of parameter estimates

$$\theta = (\hat{a}_1 \ldots \hat{a}_{n_a} \hat{b}_0 \ldots \hat{b}_{n_b})^{\mathrm{T}} \qquad (19)$$

and a vector of regressors

$$\phi(t) = (-y(t-1) \ldots -y(t-n_a) \times u(t-k) \ldots u(t-n_b-k))^{\mathrm{T}} \qquad (20)$$

The recursive least-squares estimate is then given by

$$\theta(t+1) = \theta(t) + P(t+1)\phi(t+1)\epsilon(t+1) \qquad (21)$$

where

$$\epsilon(t+1) = y(t+1) - \phi(t+1)^{\mathrm{T}}\theta(t) \qquad (22)$$

and

$$P(t+1) = [P(t) - P(t)\phi(t) [1 + \phi^{\mathrm{T}}(t)P(t)\phi(t)]^{-1}\phi^{\mathrm{T}}(t)P(t)]/\lambda \qquad (23)$$

There are also other possibilities to perform the least-squares calculations. Square-root algorithms are useful if the problem is poorly conditioned. See, for example Peterka[26] and Bierman.[10] Fast algorithms can be used if computing time is critical.[24]

3.3 *Choice of λ and modifications of P-equation*

The factor λ in eqn. 23 is introduced to discount past data when performing the least squares. For the regulation problem, the estimator is excited by the process disturbances, which normally are reasonably uniform in time. It has been found empirically that a value of λ between 0·95 and 0·99 works well in such cases. For the servo problem, the major excitation comes from the changes in the command signal. Such changes may be irregular, and it has been found that there may be bursts in the process output if eqn. 23 is used with λ less than one. The presence of bursts can be understood intuitively as follows. The negative term in eqn. 23 represents the reduction in parameter uncertainty due to the last measurement. When there are no changes in the set point, the vector $P(t)\phi(t)$ will be zero. There will not be any changes in the parameter estimate and the negative term in the right-hand side of eqn. 23 will be zero. Eqn. 23 is then reduced to

$$P(t+1) = \frac{1}{\lambda}P(t)$$

and the matrix P will thus grow exponentially if $\lambda < 1$. If there are no changes for a long time, the matrix P may become very large. A change in the command signal may then lead to large changes in the parameter estimates and in the process output. The large values of the matrix P may also lead to numerical problems. Examples which illustrate this behaviour are found, for example, in Fortescue *et al.*[17] and Morris *et al.*[25]

There are many ways to eliminate bursts. Perturbation signals may be added to ensure that the process is properly excited. The estimation algorithm may be modified. One possibility is to stop the updating of the matrix $P(t)$ when the signal $P(t)\,\phi(t)$ or the prediction error is smaller than a given value. Another possibility is to substract a term like $\alpha P^2(t)$ from the right-hand side of eqn. 23 to ensure that the matrix $P(t)$ stays bounded. A third possibility is to choose the forgetting factor, so that a function of P like $\mathrm{tr}\,P$ is constant.

3.4 *Self tuners based on pole-zero placement*

Before presenting specific self-tuning algorithms, it will be discussed whether the pole-placement design procedures are suitable for design of adaptive regulators. There are several problems to be considered.

It follows from the discussion of the pole-zero assignment method in Section 2 that it is not possible to specify the closed-loop transfer function arbitrarily. First, it is assumed that the time delay in the model G_m (eqn. 3) is at least as long as that of the process of eqn. 1. Secondly, the specifications must be such that unstable or poorly damped process zeros must also be zeros of the desired closed-loop transfer function. This is formally expressed by the condition of eqn. 7. Notice that this does not mean that the poorly damped zeros must be known *a priori.* These zeros are estimated in the self tuner. Since the zeros cannot be cancelled, it means that the properties of the closed-loop system will change when the poorly damped process zeros change. In practice, it has been found that this is not of great importance if the poorly damped zeros correspond to frequencies which are higher than those of the dominating poles. Poorly damped process zeros within the servo bandwidth will, however, have a very noticeble influence on the system. Hence, it is not possible to ensure that properties like overshoot, bandwidth, static errors etc. remain invariant for the adaptive system.

It may be too restrictive to specify all closed-loop poles, at least for high-order systems. One possibility to avoid this difficulty for discrete-time systems is to specify only the dominant poles and require that the remaining poles are close to the origin. In practice, it is often satisfactory to choose A_m as

$$A_m(q^{-1}) = 1 - 2e^{-\zeta\omega h}\cos \omega h\sqrt{1-\zeta^2}\,q^{-2} + e^{-2\zeta\omega h}q^{-2} \qquad (24)$$

which corresponds to a second-order continuous-time with damping ζ and frequency ω sampled with period h. It is often easy to determine ζ and ω such that the system gets desired properties. The relative damping is often chosen in the interval 0·5–0·8. The resonance frequency ω is chosen based on the demands on the rise time and the solution time.

The pole-placement-design procedure requires that the observer poles are specified. The observer poles are not critical. Their choice should, however, reflect the characteristics of the disturbances. If an estimation procedure which gives the disturbance dynamics is used, e.g. in the form of a controlled a.r.m.a. model, it is natural to choose the observer polynomial proportional to the polynomial which characterises the moving average. In this paper, this is not done and the observer polynomial can thus be chosen arbitrarily.

4 Algorithms based on explicit identification

Some self-tuning algorithms based on the pole-zero placement design method will not be discussed. The algorithms are first presented. Their properties are then discussed briefly. Some practical aspects are then given.

4.1 *Algorithms*

A self-tuning controller can be obtained by implementing the system in Fig. 2 directly. The following algorithm is then obtained:

Algorithm E1: (basic explicit algorithm)

Data

The polynomials A_m and A_0, both with zeros in Z, and B_{m_1} are given.

Step 1

Estimate the parameters of the model

$$Ay(t) = Bu(t-k)$$

by least squares.

Step 2

Factor the polynomial B into B^+ and B^-.

Step 3

Solve the linear equation

$$AR_1 + q^{-k}B^-S = A_mA_0$$

with $\deg R_1$ and $\deg S$ chosen as in eqn. 11 or 12.

Step 4

Calculate the control signal from

$$Ru = Tu_c - Sy$$

with

$$R = R_1B^+$$

$$T = A_0B_{m_1}.$$

The steps 1, 2, 3 and 4 are repeated at each sampling period.

An algorithm of this type is called an algorithm based on *estimation of process parameters* or an algorithm with *explicit identification*, because the estimated parameters are the parameters of the process model in the standard form. In the terminology of model reference adaptive systems (m.r.a.s.), the corresponding algorithms are called indirect because the controller parameters are updated indirectly via estimation of the process model and design calculations.

Notice that the closed-loop transfer function obtained with this algorithm is

$$G = \frac{q^{-k}B_{m_1}B^-}{A_m}$$

where $\bar{B}^-$ is the polynomial which correspond to unstable or poorly damped process zeros. When these zeros change, the closed-loop response will also change.

One difficulty with the algorithm E1 is that the equation to be solved in step 3 is poorly conditioned for parameter values such that A and B almost have a common factor.

The factorisation in step 2 may be difficult and time-consuming. There are two special cases where the factorisation can be avoided. One case is when all process zeros are well damped. It is then reasonable to have a pole-placement design where all the process zeros are cancelled. Under this hypothesis, the pole-placement procedure can be simplified, as shown in example 1. The corresponding self-tuning pole-placement algorithm then becomes

Algorithm E2: (explicit algorithm with all process zeros cancelled)

Data
Given specifications in the form of the desired closed-loop poles and desired observer poles specified by the polynomials A_m and A_0 with zeros in Z. Further, B_m is a constant. The polynomial B_m is normalised so that $B_m(1)/A_m(1) = 1$. The polynomial A_0 is normalised arbitrarily.

Step 1
Estimate the parameters of the model

$$Ay(t) = Bu(t-k)$$

by least squares.

Step 2
Determine the polynomials R_1 and S such that

$$AR_1 + q^{-k}S = A_mA_0$$

with deg R_1 and deg S chosen as eqns. 15 or 16.

Step 3
Use the control law

$$BR_1u = Tu_c - Sy$$

where

$$T = A_0B_m.$$

The steps 1, 2 and 3 are repeated for each sampling period. As a safeguards, it should be tested that A and B do not have common factors and that B is a stable polynomial.

This algorithm cannot be expected to work well unless the corresponding design procedure for systems with known parameters work well. Since all process zeros are cancelled, the regulator will not be satisfactory for nonminimum-phase systems or for systems with zeros having poor damping. Such systems can, however, be handled using the design procedure in example 2. The corresponding self-tuning-control algorithm is given by

Algorithm E3: (explicit algorithm with no process zeros cancelled)

Data
Given specifications in the form of the desired closed-loop poles and the desired observer poles specified by the polynomials A_m and A_0 with zeros in Z. A_0 is normalised arbitrarily.

Step 1
Estimate the parameters of the model

$$Ay(t) = Bu(t-k)$$

by least squares.

Step 2
Introduce $B_m = KB$ and choose K such that $B_m(1) = A_m(1)$. Then determine the polynomials R and S such that

$$AR + q^{-k}BS = A_mA_0$$

Deg S and deg R are chosen as in eqns. 11 or 12, with deg $R = \deg R_1$, and deg $B = \deg B^-$.

Step 3
Use the control law

$$Ru = Tu_c - Sy$$

where

$$T = KA_0$$

The steps 1, 2 and 3 are repeated for each sampling period.

4.3 Remark

Possible common factors of A and B should be eliminated after the first step, to ensure that the equation in step 2 has a solution. Notice that the polynomial A_m cannot be normalised *a priori* because the normalisation requires knowledge of the polynomial B in the process model.

Notice that with algorithm E3 the properties of the closed-loop system will change even if A_m and A_0 are fixed because the closed-loop zeros will change if the process zeros change.

4.4 Properties

The properties of the closed-loop system obtained when the self-tuning regulator is applied to a given process will first be discussed. It is first assumed that the process to be controlled is described by the difference equation

$$A_sy(t) = B_su(t-k) \tag{25}$$

It is assumed that this equation is of the form of eqn. 18 and that deg $A = \deg A_s$ and deg $B = \deg B_s$. Eqns. 18 and 25 are then said to be compatible.

Using the notation of eqn. 20, eqn. 25 can also be written as

$$y(t) = \theta_s^T\phi(t) = \phi^T(t)\theta_s$$

where the components of θ_s are the coefficients of the polynomials.

The closed-loop system obtained with the algorithm E1 can then be described by eqn. 25 and the equations

$$\left\{\begin{aligned}
\theta(t+1) &= \theta(t) + P(t+1)\phi(t+1)\epsilon(t+1)\\
\epsilon(t+1) &= \phi^T(t+1)[\theta_s - \theta(t)]\\
P(t+1) &= [P(t) - P(t)\phi(t)\cdot\\
&\qquad [1 + \phi(t)^{-T}P(t)\phi(t)]^{-1}\phi^T(t)P(t)]/\lambda\\
Ru(t) &= Tu_c(t) - Sy(t)\\
R &= R_1B^+\\
T &= A_0B_{m_1}\\
AR_1 + q^{-k}B^-S &= A_mA_0
\end{aligned}\right. \tag{26}$$

The state of the closed-loop system can be chosen as θ, P the state of a representation of

$$\begin{cases} A_s y = q^{-k} B_s u \\ Ru = Tu_c - Sy \end{cases}$$

and possibly some additional delayed values of u and y, which are needed to represent the vector ϕ given by eqn. 20. To obtain a complete description, it is also necessary to specify the command signal. The equations describing the closed-loop system are nonlinear. Their global properties are not yet fully explored. A difficulty of the equations is that the mapping from the coefficients of the polynomials A, B to those of R and S is discontinuous at those points where A and B have common factors.

There are no proper stationary solutions to eqns. 25 and 26, in the sense that all state variables are constant unless the command signal is constant. There are, however, solutions such that the parameters estimates $\theta(t)$ assume constant values for arbitrary command signals. Assuming that the matrix $P(t)$ is positive definite for all t, it follows from eqn. 26 that $\theta(t)$ is constant if $\phi(t)\epsilon(t)$ is zero, i.e.

$$y(t-i)\epsilon(t) = 0, \qquad i = 1, 2, \ldots, n_a$$

$$u(t-i)\epsilon(t) = 0, \qquad i = k, k+1, \ldots, k+n_b$$

These equations imply that $\epsilon(t) = 0$. Assume, on the contrary, that $\epsilon(t) \neq 0$. Then

$$y(t-i) = 0, \qquad i = 1, 2, \ldots, n_a$$

$$u(t-i) = 0, \qquad i = k, k+2, \ldots, k+n_b+1$$

Eqn. 25 then implies that $y(t) = 0$. Since

$$\epsilon(t) = y(t) - a_1 y(t-1) - \ldots - a_n y(t-n_a) - b_0 u(t-k) - \ldots - b_m u(t-k-n_b)$$

We get $\epsilon(t) = 0$, which is a contradiction. When the parameters $\theta(t)$ are constant, it follows that

$$y = B_s v$$

$$u = A_s v$$

where

$$(A_s R + q^{-k} B_s) v = Tu_c$$

Hence,

$$\epsilon(t) = Ay - Bu = (AB_s - BA_s) v$$

Under modest requirements on u_c (e.g. piecewise deterministic with arbitrary generator[4]), it now follows that the condition $\epsilon(t) = 0$ implies that

$$AB_s = BA_s$$

The correct estimates are thus the only parameter values such that the estimates remain constant.

To investigate the local stability at the stationary solution $\theta(t) = \theta_s$, the equations are linearised. The linearised equation for $\theta(t)$ is decoupled from the rest of the equation. We get

$$\delta\theta(t+1) = [I - P_s(t+1)\phi_s(t+1)\phi_s^T(t+1)]\,\delta\theta(t) \tag{27}$$

where the subscript s indicates that the quantities have been evaluated at $\theta(t) = \theta_s$. Eqn. 27 is stable if

$$\sum_{t_0}^{t+r} \phi_s(k)\phi_s^k(k)$$

is positive definite. A proof of local stability for a similar algorithm is given by Goodwin and Sin.[21]

A more general model than eqn. 25 is

$$A_s y = B_s u + C_s e \tag{28}$$

where $\{e(t)\}$ is a sequence of independent random variables. If $C_s = 1$, then the parameter θ_s is a possible convergence point, which is locally stable. If $C_s \neq 1$, the parameter θ_s is not a possible convergence point because

$$E\phi(t)\epsilon(t) \neq 0$$

A pole-placement algorithm which has a self-tuning property for the process of eqn. 28 if the reference value is zero is described by Wellstead, Prager and Zanker.[34]

5 Algorithms based on implicit identification

The design calculations for the algorithms discussed in the previous Section may be time consuming. It is possible to obtain different algorithms where the design calculations are simplified considerably. The self-tuning regulator of Åström and Wittenmark[8] is a prototype for algorithms of this type. The basic idea is to rewrite the process model in such a way that the design step is trivial. For minimum-variance control, the process model can be rewritten so that the parameters of the minimum-variance regulator are the parameters of the rewritten model. By a proper choice of model structure, the regulator parameters are thus updated directly and the design calculations are thus eliminated. With reference to Fig. 2, it means that $v = 0$ and the block-marked design can be eliminated. Algorithms of this type are called algorithms based on implicit identification of a process model. In the terminology of m.r.a.s., the algorithms are also called direct methods because the parameters of the regulators are updated directly. Implicit algorithms and some of their properties will be discussed.

5.1 Algorithms

Consider a process described by

$$Ay = q^{-k} Bu \tag{29}$$

The regulator of eqn. 4 gives a closed-loop system with the transfer function

$$G = \frac{B^- B_{m_1}}{A_m} \tag{30}$$

Eqn. 8 gives

$$A_m A_0 y = AR_1 y + q^{-k} B^- Sy$$

Combination of this with eqn. 29 gives

$$A_m A_0 y = q^{-k} R_1 Bu + q^{-k} B^- Sy = q^{-k} B^- (Ru + Sy) \tag{31}$$

If the control signal is chosen such that

$$Ru = Tu_c - Sy$$

where $T = A_0 B_{m_1}$, then it follows from eqn. 31 that the closed-loop transfer function eqn. 30 is obtained. Notice that eqn. 31 can be regarded as a process model. The polynomials R and S of the feedback law appear directly in the model. The design problem is also trivial for the model of eqn. 31.

The following self-tuning algorithm is now obtained.

Algorithm I1: (basic implicit algorithm)

Data

The polynomials A_m and A_0, both with zeros in Z, and B_{m_1} are given.

Step 1

Estimate the parameters of the model

$$A_m A_0 y = q^{-k} B^-(Ru + Sy) \tag{32}$$

i.e. estimate B^-, R and S.

Step 2

Calculate the control signal from

$$Ru = Tu_c - Sy$$

where

$$T = A_0 B_{m_1}$$

The steps 1 and 2 are repeated at each sampling period.

Notice that the model of eqn. 32 is bilinear in the parameters. This means that the estimation problem is not trivial. For example, the parametrisation is not unique unless it is required that R and S do not have common factors. The polynomial B^- must also be such that it has all its zeros outside the stable region Z. A recursive estimation procedure for eqn. 32 is proposed by Åström.[5] Because of the difficulties of estimating the parameters of eqn. 32 it is of interest to consider special cases which lead to simpler calculations.

The special case when all process zeros were cancelled was discussed in example 1 for the case of known parameters. In that case, $B^- = 1$ and the self-tuning algorithm I1 is reduced to

Algorithm I2: (implicit algorithm with all process zeros cancelled)

Data

The polynomials A_m and A_0 with zeros in Z are given. Further

$$B_{m_1} = K = A_m(1)$$

Step 1

Estimate the parameters of the polynomials R and S in the model

$$A_m A_0 y = q^{-k}(Ru + Sy) \tag{33}$$

by least squares. The degrees of the polynomials S and R are chosen as

$$\deg S = \deg A_m + \deg A_0 - k \tag{34}$$

$$\deg R = \deg B + k - 1$$

or

$$\deg S = \deg A \tag{35}$$

$$\deg R = \deg A_m + \deg A_0 + \deg B - \deg A$$

Step 2

Compute the control signal from

$$Ru(t) = Tu_c(t) - Sy(t) \tag{36}$$

where

$$T = A_0 k$$

The steps 1 and 2 are repeated at each sampling interval.

This algorithm is identical to the self-tuning controller proposed by Clarke and Gawthrop.[11] The algorithm has also been explored by Kurz, Isermann and Schumann.[23]

A difficulty with algorithm I1 is that it may conceivably happen that the estimate of the leading coefficient r_0 of the polynomial R is zero. The feedback law (eqn. 36) is then no longer causal. There are various ways to overcome this difficulty. One possibility is to fix the value of the coefficient. Another possibility is to reparameterise the polynomial as

$$r_0[1 + r_1 q^{-1} + \ldots]$$

and use special techniques to estimate r_0. This is done in the m.r.a.s. systems.[14] Another possibility which is often used in practice is to increase the number k in the model.

5.2 Properties

It is assumed that the process to be controlled is described by the difference eqn. 28. The closed-loop system obtained when the implicit algorithms are applied to the process of eqn. 28 is governed by a set of nonlinear difference equations. These equations are similar to the ones obtained for the explicit algorithms. The equations obtained for the implicit algorithms are somewhat simpler because the regulator parameters are updated directly. There is no complete analysis for the general case. The special case of algorithm I2 is, however, reasonably well understood. The key results on stability are due to Egardt[14] and Goodwin *et al.*[20] A main result is that the closed-loop system is stable, and that the output of the system converges to the desired output. The assumptions required are that the time-delay k is known, that upper bounds on the degrees of the polynomials and that the system of eqn. 28 is minimum phase. The result is proven for the special case $C_s e(t) = 0$ and $k = 1$ in Goodwin *et al.*[20] Egardt[14] shows that the output is bounded even if there are disturbances $C_s e(t) \neq 0$ provided that the disturbance is bounded.

If $C_s = 1$ in eqn. 28, and if $B^- = 1$, it is easy to see that the process model can be written as

$$A_m A_0 y(t) = Ru(t-k) + Sy(t-k) + R_1 e(t)$$

(cf. eqn. 31).

Using the method of least squares, the estimates of R and S will be unbiased if the degrees are chosen as in eqn. 34. The degree of R_1 is then k and the regressors will be independent of $R_1 e(t)$.

5.3 Modifications

There are several modifications of the algorithms that are useful. When there are stochastic disturbances in the process it is shown in Åström[5] that it is advantageous to replace the model of eqn. 32 by

$$A_m y = q^{-k} B^-(R\bar{u} + S\bar{y}) \tag{37}$$

where

$$\bar{u} = \frac{1}{A_0} u \qquad (38)$$

$$\bar{y} = \frac{1}{A_0} y$$

Otherwise, the parameters will not converge to the correct values even if the observer polynomial is known. Similarly, it is sometimes useful to replace eqn. 33 by

$$A_m y = q^{-k}(R\bar{u} + S\bar{y}) \qquad (39)$$

where $\bar{u}$ and $\bar{y}$ are given by eqn. 38.

6 Simulations

Some of the properties of the algorithms are illustrated through simulations in this Section. The simulations have been done using the simulation program Simnon.[16] The special Simnon system for simulation of general adaptive controllers, described by Gustavsson[22], was used. More examples are found in References 7, 30 and 3.

6.1 Choice of parameters

There are several parameters which have to be selected in the algorithms. Unless otherwise stated, the following parameters have been used. Initial values of the parameters are chosen as zero except for $r_0 = 1$ in the implicit algorithm and $b_{n_b} = 1$ in the explicit algorithm. The initial value of the covariance matrix is chosen as one hundred times the unit matrix, and the forgetting factor is equal to one. Further $A_0(q^{-1}) = 1$ has been used in the simulations. The references signal was a square wave with amplitude ± 1 and a period of 100 samples.

Example 6.1

A continuous-time system with the transfer function

$$G(s) = \frac{0{\cdot}15\, e^{-0{\cdot}45\, s}}{s + 0{\cdot}15}$$

sampled with a sampling time of $T = 1$ gives the discrete-time system

$$y(t) = 0{\cdot}8607\, y(t-1) = 0{\cdot}0792\, u(t-1) + 0{\cdot}0601\, u(t-2) \qquad (40)$$

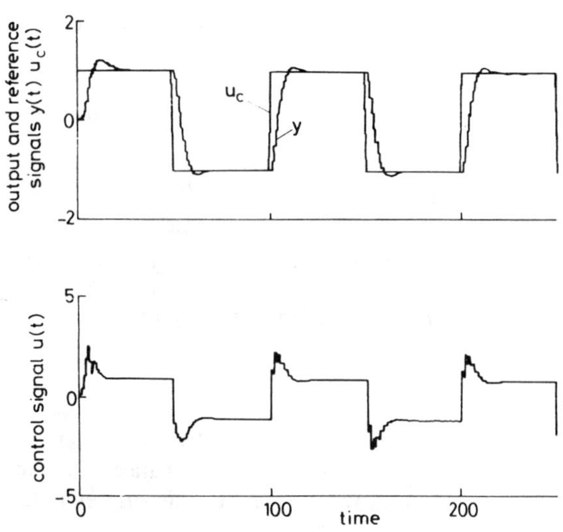

Fig. 3 *Output y, reference u_c and control u signals when process (eqn. 40) is controlled using the implicit algorithm I2*

Notice that the continuous-time system has a time delay which is not a multiple of the sampling time. The sampled modal has a zero $z = -0{\cdot}759$, which corresponds to a mode with damping $\zeta = 0{\cdot}087$. The solution time of the open-loop system is 20–25 s. The specifications for the closed-loop system have been chosen as a solution time of about 10 s and a damping of about $\zeta = 0{\cdot}7$. The desired characteristic equation has been chosen as

$$A_m(q^{-1}) = 1 - 1{\cdot}5\, q^{-1} + 0{\cdot}6\, q^{-2}$$

The behaviour of the implicit algorithm I2 with deg $R =$ deg $S = 1$ (i.e. four estimated parameters) and with the forgetting factor $\lambda = 0{\cdot}95$ is shown in Fig. 3. The behaviour of the closed-loop system is good already in the second transient. The parameters have converged at the fourth transient. The control signal has an oscillatory tendency because of the cancellation of the zero at $-0{\cdot}759$.

Fig. 4 shows the behaviour when the explicit algorithm E3 is used with deg $A =$ deg $B = 1$ (i.e. three estimated parameters). The behaviour of the two algorithms is, in this case, very much the same.

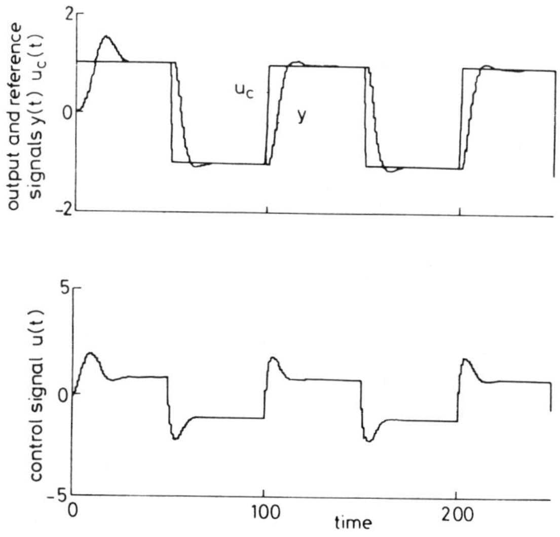

Fig. 4 *Output y, reference u_c and input u signals when process (eqn. 40) is controlled with the explicit algorithm E3*

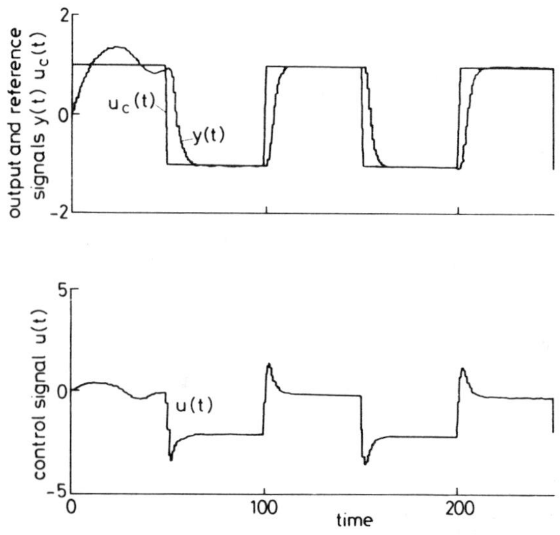

Fig. 5 *Result when the process (eqn. 40) is controlled with implicit algorithm I2 with the same parameters as in Fig. 3 but with a constant bias δ = 1 on the input to the process*

Assume that u in eqn. 40 is replaced by $u + \delta$, where δ is a constant bias. The adaptive controller does not know this bias, and only u and y are available for the controller. Figs. 5 and 6 show the behaviour of the closed-loop system when the implicit I2 and the explicit algorithms E3, respectively, are used with the same parameters as before. The implicit algorithm handles the bias by introducing an integrator in the controller. The R-polynomial after 250 steps is

$$R(q^{-1}) = 0{\cdot}116 - 0{\cdot}115\, q^{-1}$$

As seen from Fig. 5, the system will not converge to the desired closed-loop transfer function. By increasing the order of the R-polynomial, it is possible to get the same closed-loop performance as before. The identification in the explicit algorithm is disturbed by the bias term which explains the bad behaviour. It is, however, easy to also estimate the bias term and take it into consideration when computing the control signal (Clarke and Gawthrop[11]).

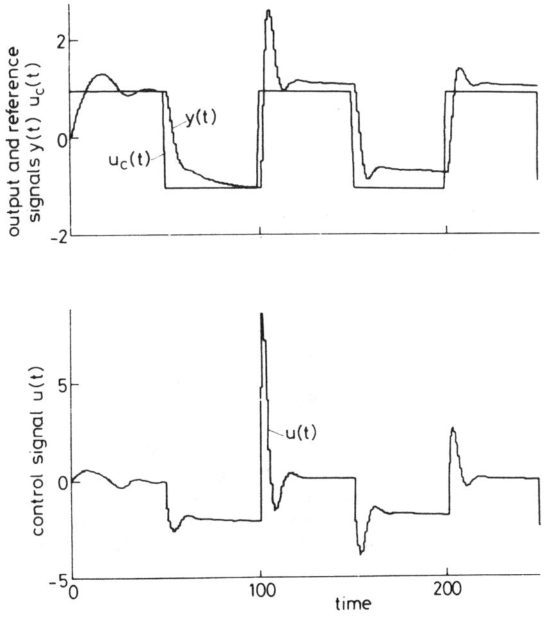

Fig. 6 *Same as in Fig. 4 but when there is a constant bias δ = 1 on the input*

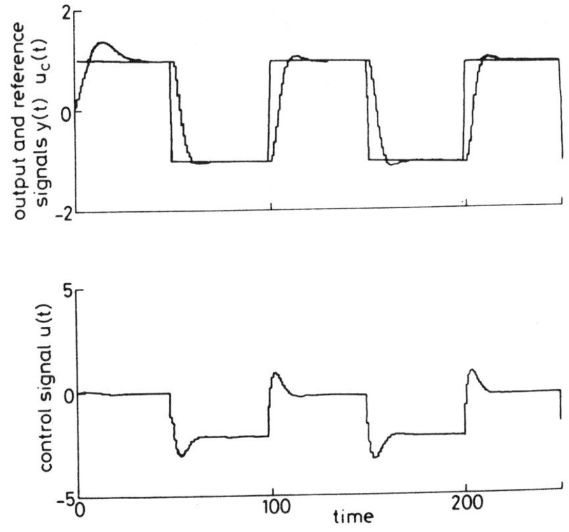

Fig. 7 *Result when using the explicit algorithms with compensation for the unknown bias δ = 1 on the input*

This is done in Fig. 7, and it is seen that it is possible to eliminate the bias.

Example 6.2

In this example, the adaptive regulator controls a time-continuous system. The system has the transfer function

$$G(s) = \frac{1}{(1+s)(1+0{\cdot}5\,s)(1+5\,s)} \tag{41}$$

Using a sampling interval of $T = 1$, we get a discrete-time model which is nonminimum phase. The zeros of the model are $z_1 = -1{\cdot}798$ and $z_2 = -0{\cdot}114$. This system was not possible to control with the implicit algorithm since this algorithm cancels all the zeros of the process. The computation of the control signal will then be unstable. The explicit algorithm could easily be used. Fig. 8 shows the behaviour when deg $A = 3$ and deg $B = 2$, and $A_m(q^{-1}) = 1 - q^{-1} + 0{\cdot}35\, q^{-2}$. Again, the behaviour of the closed-loop system is very good already in the second transient.

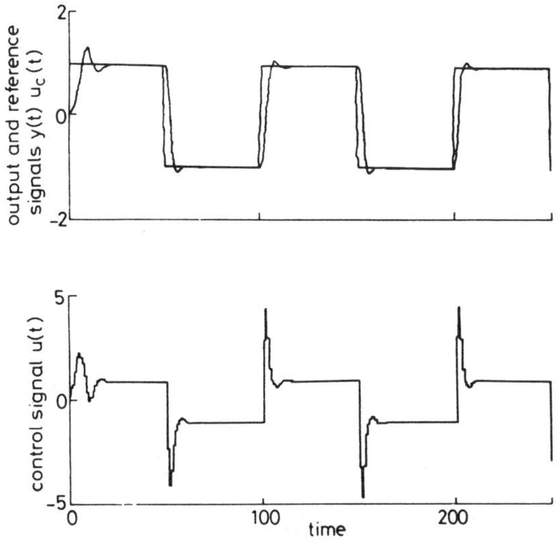

Fig. 8 *Results when explicit algorithm E3 is used to control the continuous-time system*

7 Acknowledgments

The work has been partially supported by the Swedish Board of Technology Development (STU) under contract no. 76-3804.

References

1 ANDERSSON, L.: 'DISCO - An educational microcomputer controller'. Proceedings of the IFAC Symposim on Trends in Automatic Control Education, Barcelona, Spain, 1977, pp. 32–45

2 ÅSTRÖM, K.J.: 'Reglerteori, Almqvist & Wiksell', Gebers

3 ÅSTRÖM, K.J.: 'Self-tuning control of a fixed bed chemical reactor system'. Dept. of Automatic Control, Lund Institute of Technology, Sweden, CODEN: LUTFD2/(TFRT-3151)/1-066/(1978)

4 ÅSTRÖM, K.J.: 'Piece-wise dete ninistic signals'. Dept. of Automatic Control, Lund Institu of Technology, Sweden, CODEN: LUTFD2/(TFRT-7171)/1-0J5/(1979)

5 ÅSTRÖM, K.J.: 'New implicit adaptive pole-zero placement algorithms for non-minimum phase systems'. Dept. of Automatic Control, Lund Institute of Technology, Sweden, CODEN: LUTFD2/(TFRT-7172)/1-021/(1979)

6 ÅSTRÖM, K.J., BORISSON, U., LJUNG, L., and WITTENMARK, B.: 'Theory and applications of self-tuning regulators', *Automatica*, 1977, **13**, pp. 457–476

7 ÅSTRÖM, K.J., WESTERBERG, B., and WITTENMARK, B.: 'Self-tuning controllers based on pole-placement design.' Dept. of Automatic Control, Lund Institute of Technology, Sweden, CODEN: LUTFD2/(TFRT-3148)/1-052/(1978)

8 ÅSTRÖM, K.J., and WITTENMARK, B.: On self-tuning regulators. *Automatica*, 1973, **9**, pp. 185–189

9 ÅSTRÖM, K.J., and WITTENMARK, B.: 'Analysis of a self-tuning regulator for non-minimum phase systems'. Presented at the IFAC Symposim on Stochastic Control, Budapest, Hungary, 1974

10 BIERMAN, G.J.: 'Factorization methods for discrete sequential estimation' (Academic Press, 1977)

11 CLARKE, D.W., and GAWTHROP, P.J.: 'Self-tuning controller', *Proc. IEE*, 1975, **122**, (9), pp. 929–934

12 CLARKE, D.W., and GAWTHROP, P.J.: 'Self-tuning control, *ibid.*, 1979, **126**, (6), pp. 633–640

13 EDMUNDS, J.M.: 'Digital adaptive pole shifting regulators'. Ph.D. dissertation, Control Systems Centre, Institute of Science and Technology, The University of Manchester, Manchester, England

14 EGARDT, B.: 'A unified approach to model reference adaptive systems and self-tuning regulators'. Dept. of Automatic Control, Lund Institute of Technology, Sweden, CODEN: LUTFD2/(TFRT-7134)/1-67/(1978)

15 ELLIOTT, H., and WOLOVICH, W.A.: 'Parameter adaptive identification and control'. *IEE Trans*, 1979, **AC-24**, pp. 592–599

16 ELMQVIST, H.: 'SIMNON - Users' manual'. TFRT-3091, Dept of Automatic Control, Lund Institute of Technology, Sweden, 1975

17 FORTESCUE, T.R., KERSHENBAUM, L.S., and YDSTIE, B.E.: 'Implementation of self-tuning regulators with variable forgetting factors'. Report, Dept. of Chemical Engineering and Chemical Technology, Imperial College, London, 1979

18 FRANKLIN, G.F.: Private communication, 1977

19 GAWTHROP, P.J.: Some interpretations of the self-tuning controller, *Proc. IEE*, 1977, **124**, pp. 889–894

20 GOODWIN, G.C., RAMAGE, P.J., and CAINES, P.E.: 'Discrete time multivariable adaptive control'. Report, Harvard University, Cambridge Mass.

21 GOODWIN, G.C., and SIN, K.S.: Adaptive control of non-minimum phase systems. Report EE 7918, The University of Newcastle, New South Wales, Australia, 1979

22 GUSTAVSSON, I.: 'Users' guide for a program package for simulation of self-tuning regulators. Dept. of Automatic Control, Lund Institute of Technology, Sweden, CODEN: LUTFD2/(TFRT-7149)/1-078/(1978)

23 KURZ, H., ISERMANN, and SCHUMANN, R.: 'Development, comparison and application of various parameter-adaptive digital control algorithms. Presented at the IFAC 7th World Congress, Helsinki, 1978

24 LEVINSON, N.: 'The Wiener RMS (root mean square) error in filter design and prediction'. *J. Math. Phys.*, 1947, **28**, pp. 261–278

25 MORRIS, A.J., FENTON, T.P., and NAZER, Y.: 'Application of self-tuning regulators to the control of chemical processes' *in* van Nauta Lemke, H.R., and Vervruggen, H.B. (eds.): 'Digital Computer Applications to Process Control'. Presented at the 5th IFAC/IFIP International Conference, The Hague, Netherlands, June 14th–17th 1977

26 PETERKA, V.: 'A square-root filter for real-time multivariable regression', *Kybernetica*, **11**, pp. 53–67

27 PETERKA, V., and ÅSTROM, K.J.: 'Control of multivariable systems with unknown but constant parameters'. Presented at the 3rd IFAC Symposium on Identification and System Parameter Estimation, The Hague, Netherlands, 1973

28 RAGAZZINI, J.R., and FRANKLIN, G.F.: 'Sampled-data control systems' (McGraw-Hill, New York, 1958)

29 SÖDERSTRÖM, T., LJUNG, L., and GUSTAVSSON, I.: 'A comparative study of recursive identification methods'. Report TFRT-3085, Dept. of Automatic Control, Lund Institute of Technology, Sweden, 1974

30 WESTERBERG, B.: 'Självinställande regulator baserad på polplacering. Dept. of Automatic Control, Lund Institute of Technology, Sweden, CODEN: LUTFD2/(TFRT-5198)/1-65/(1977)

31 WITTENMARK, B.: 'A design example of a sampled data system. Report TFRT-3130, Dept. of Automatic Control, Lund Institute of Technology, Sweden, 1976

32 WITTENMARK, B.: 'Self-tuning PID-controllers based on pole placement.' Dept. of Automatic Control, Lund Institute of Technology, Sweden, CODEN: LUTFD2/(TFRT-7179)/1-7179/1-037/(1979)

33 WELLSTEAD, P.E.: 'On the self-tuning properties of pole-zero assignment regulators'. Report 402, Control Systems Centre, The University of Manchester Institute of Science and Technology, Manchester, England, 1978

34 WELLSTEAD, P.E., PRAGER, D., and ZANKER, P.: 'Pole assignment selftuning regulator'. *Proc. IEE,* 1979, **126**, pp. 781–787

35 WELLSTEAD, P.E., and ZANKER, P.: 'Servo self-tuners'. *Int. J. Control,* 1979, **30**, pp. 27–36

36 WELLSTEAD, P.E., ZANKER, P., and EDMUNDS, J.M.: 'Self-tuning pole/zero assignment regulators'. Report 404, Control Systems Centre, The University of Manchester Institute of Science and Technology, Manchester, England, 1978

37 WOUTERS, W.R.E.: 'Adaptive pole, placement for linear stochastic systems with unknown parameters'. Proceedings of the IEEE Conference on Decision and Control, New Orleans, USA, 1977

Multivariable pole-assignment self-tuning regulators

D.L. Prager, B.Sc. (Eng.), M.Sc., and P.E. Wellstead, M.Sc., Ph.D.

Indexing terms: *Multivariable control systems, Closed-loop systems, Digital control, Control-system synthesis, Algorithms, Self-tuning regulators*

Abstract: The paper describes the theory and application of a multi-input/multi-output self-tuning regulator where the control objective is the assignment of the closed-loop pole set to prespecified locations. The key theoretical result is that the algorithm has the so-called 'self-tuning' property. From a practical viewpoint the pole-assignment concept leads to direct digital controllers which are more robust than self tuners based on optimal control synthesis methods.

1 Introduction

The seminal work of Astrom [1] and Peterka [2] on self-tuning digital control systems has given rise to a considerable literature. Indeed, it now seems clear that single-input/single-output systems are susceptible to self-tuning control [3] and pole-assignment regulation and control [4, 5, 6]. Furthermore, a number of practical applications have been recorded [7] and the connection between self-tuning and model reference adaptive control recognised and exploited [8].

The current effort on digital self-tuning/adaptive control is timely, coinciding as it does with the widespread use of microcomputers for industrial control. With this in mind, it is perhaps appropriate to recall previous contributions in the general area of what we might call 'combined identification and control'. In this connection the work of Young [9] and Hasting-James [10] is significant. By the same token it would be wrong not to make note of the dependence of self tuning on the design and synthesis procedures which are embodied in sampled data theory.

The work described here is addressed to the problem of self-tuning pole-assignment regulators for systems with more than one input and output. Previous work on multivariable self tuners has focused on minimum-variance design rules. In particular, Borisson [11] has extended the minimum-variance regulator to deal with MIMO systems in which there are an equal number of inputs and outputs. From a practical viewpoint, Keviczky, Hetthessy, Hilger and Kolostori [12] have implemented a variant of the minimum-variance law in which the self tuner also maintains a weighted average of the outputs over a finite interval as close as possible to a reference level. This variant of the minimum-variance self tuner has been applied to a cement material blending process.

Again related to minimum-variance regulation, the suggestion has been made [16] that a 'decoupling' controller could be cascaded with the plant, with a set of single-input/single-output self tuners operating on the decoupled subsystems. The decoupling controller would be designed using, for example, Rosenbrock's inverse Nyquist array method [13]. However, it is by no means clear how a *design* method based on an *inequality* could be put into the self-tuning format which by its very nature demands a *synthesis* rule based on identity.

Multiple minimum-variance self tuners suffer from similar disadvantages to their single-variable counterparts. In particular, the basic algorithm cannot control nonminimum phase systems. For such systems, a factorisation stage must be introduced during the control-law computation phase, and this increases the computational complexity. Minimum-variance self tuners are also sensitive to variations in system time delays [6].

This is a point of the utmost significance, since in discrete time systems, nonminimum-phase behaviour is the rule rather than the exception. In multivariable systems the further disadvantage exists in that minimum-variance design demands that all loops must have the same time delay. This requirement is not generally fulfilled by real systems.

The contribution of this paper is to show how these difficulties can be avoided by a multivariable version of the pole-assignment self tuner. To this end, the paper begins by reviewing the offline design rule for multivariable pole-assignment regulation of a stochastically disturbed system. The self-tuning version of the algorithm is then developed, with the important result that the technique has the 'self-tuning' property whereby the stochastic disturbance can be assumed to be an independent white-noise vector process.

It is understood that this does not constitute a proper convergence result for the algorithm, but does show that the desired controller is a possible convergence point for the algorithm.

The paper concludes with a discussion of the computer implementation of the method, together with applications to both real and simulated plant.

2 Offline controller design

Before discussing the self-tuning version, the offline design of the pole-assignment regulator to be employed is presented. The object is to design a regulator which assigns the closed-loop poles to values specified by the designer. It is assumed that the plant, which is both controllable and observable, may be modelled by the difference equation

$$[I + A(z^{-1})]\,y_t = z^{-k}B(z^{-1})u_t + [I + C(z^{-1})]\,e_t \qquad (1)$$

where u_t and y_t are p-vectors defining the measurable system input and output, respectively, and e_t is a p-vector representing a zero-mean white-noise process with covariance R. $A(z^{-1})$, $B(z^{-1})$ and $C(z^{-1})$ are polynomial matrices in the backward shift operator z^{-1}, and are of the form

$$X(z^{-1}) = X_1 z^{-1} + \ldots + X_{n_x} z^{-n_x} \qquad (2)$$

where X_i, $i = 1, 2, \ldots, n$ are $p \times p$ matrix coefficients. The component of the smallest system pure time delay that is an integer multiple of the sampling time is modelled by the term z^{-k}. Further time delays (integer or noninteger multiples of the sampling time) are absorbed in the $B(z^{-1})$ polynomial. Normally, however, $n_a = n_b = n$.

Paper 1126D, first received 23rd May and in revised form 3rd November 1980

The authors are with the Control Systems Centre, University of Manchester Institute of Science and Technology, PO Box 88, Sackville Street, Manchester M60 1QD, England

Reprinted with permission from *Proc. Inst. Elec. Eng.*, Part D, vol. 128, pp. 9-18, Jan. 1981.
Copyright © 1981; The Institution of Electrical Engineers

Introduce a control law of the form

$$u_t = G(z^{-1})(I + F(z^{-1}))^{-1} y_t \tag{3}$$

where

$$G(z^{-1}) = G_0 + G_1 z^{-1} + \ldots + G_{n_g} z^{-n_g}$$

and where $F(z^{-1})$ follows eqn. 2. The coefficient matrices G_i, $i = 0, 1, \ldots, n_g$ and F_i, $i = 1, 2, \ldots, n_f$ are of dimension $p \times p$.† Substituting eqn. 3 into eqn. 1 the closed-loop system becomes

$$y_t = [I + F(z^{-1})]\,[I + P(z^{-1})]^{-1}\,[I + C(z^{-1})]\, e_t \tag{4}$$

where

$$I + P(z^{-1}) = [I + A(z^{-1})]\,[I + F(z^{-1})] - z^{-k} B(z^{-1}) G(z^{-1}) \tag{5}$$

Now, if the coefficients of polynomials $F(z^{-1})$ and $G(z^{-1})$ are chosen so that

$$I + P(z^{-1}) = [I + C(z^{-1})]\,[I + T(z^{-1})] \tag{6}$$

where the order n_t of polynomial $T(z^{-1})$ is governed by

$$n_t \leqslant n_a + n_b + k - 1 - n_c \tag{7}$$

(where this condition is required for a solution of eqn. 5 to exist), then the closed-loop system becomes

$$y_t = [I + F(z^{-1})]\,[I + T(z^{-1})]^{-1} e_t \tag{8}$$

The poles of this system (assuming $I + F(z^{-1})$ and $I + T(z^{-1})$ are relatively prime) are given by $|I + T(z^{-1})|$ which is clearly open to the designer's choice. It is argued that in general $I + T(z^{-1})$ and $I + F(z^{-1})$ will be relatively prime, and the case when this is not true would be the exception rather than the rule.

The solution to eqns. 5 and 6 requires the solution of the set of simultaneous linear equations

$$(n_a + n_b + k - 1)p \left\{ \left[\underbrace{\begin{matrix} I & & \\ A_1 & \ddots & \\ \vdots & \ddots & I \\ A_{n_a} & & \vdots \\ & \ddots & \vdots \\ & & A_{n_a} \end{matrix}}_{(n_b + k - 1)p} \quad \underbrace{\begin{matrix} (pk)\left\{\begin{matrix} 0 \\ \vdots \\ 0 \end{matrix}\right. & \\ -B_1 & \\ \vdots & -B_1 \\ -B_{n_a} & \vdots \\ & \ddots \\ & -B_{n_b} \end{matrix}}_{n_a p} \right] \right. \times$$

$$\begin{bmatrix} F_1 \\ \vdots \\ F_{n_f} \\ G_0 \\ \vdots \\ G_{n_g} \end{bmatrix} = \begin{bmatrix} P_1 \\ \vdots \\ \vdots \\ \vdots \\ \vdots \\ P_{n_p} \end{bmatrix} - \begin{bmatrix} A_1 \\ \vdots \\ A_{n_a} \\ 0 \\ \vdots \\ 0 \end{bmatrix} \tag{9}$$

For the solution to exist, the matrix on the left-hand side must be nonsingular. This condition is met if, for example, the system is generic.

---

†N.B. Normally $n_g = n_a - 1$, $n_f = n_b + k - 1$

The control law may be implemented in two ways. The first method uses the fact that

$$[I + F(z^{-1})]^{-1} = [\det(I + F(z^{-1}))]^{-1} \operatorname{adj}(I + F(z^{-1})) \tag{10}$$

so that

$$\{\det [I + F(z^{-1})]\} I u_t = G(z^{-1}) \{\operatorname{adj} [I + F(z^{-1})]\} y_t \tag{11}$$

from which the control input u_t may be computed.

For the second method a further assumption on the form of the regulator (eqn. 3) must be introduced. Let

$$F^*(z) = z^{n_f}(I + F(z^{-1})) \tag{12}$$

$$G^*(z) = z^{n_g} G(z^{-1}) \tag{13}$$

Assumption: The regulator has $F^*(z)$ and $G^*(z)$ relatively right prime with all observability and controllability indices equal to n_f, and $n_a = n_b = n$.

It then follows [14, 17] that the regulator eqn. 3 may be written in the form

$$u_t = (I + \tilde{F}(z^{-1}))^{-1} \tilde{G}(z^{-1}) y_t \tag{14}$$

where

$$n_{\tilde{f}} = n_f \tag{15}$$

and

$$n_{\tilde{g}} = n_g \tag{16}$$

The assumption ensures that the matrix

$$n_f + n_g \left\{ \left[\underbrace{\begin{matrix} G_0^T & & \bigcirc \\ G_1^T & G_0^T & \\ \vdots & \ddots & G_0^T \\ G_{n_g}^T & & \vdots \\ & G_{n_g}^T & \vdots \\ \bigcirc & & G_{n_g} \end{matrix}}_{n_f} \quad \underbrace{\begin{matrix} -I & & \bigcirc \\ -F_1^T & -I & \\ \vdots & & \vdots \\ -F_{n_f}^T & & \vdots \\ & \ddots & \vdots \\ \bigcirc & & -F_{n_f}^T \end{matrix}}_{n_g} \right] \right. \tag{17}$$

required in the transformation from $[I + F(z^{-1}), G(z^{-1})]$ to $[I + \tilde{F}(z^{-1}), \tilde{G}(z^{-1})]$ has full rank and that eqns. 14 and 15 hold. It is generally not necessary for n_a to be equal to n_b so long as the transformation matrix has full rank.

From eqn. 13 the control law may be written as

$$u_t = -\tilde{F}(z^{-1}) u_t + \tilde{G}(z^{-1}) y_t \tag{18}$$

and is clearly very easily implemented.

The second method of implementing the control law may be preferred because it utilises routines already required in order to compute the pole shifting law parameters $F(z^{-1})$ and $G(z^{-1})$.

The pole shifting law has certain advantages over a minimum-variance law which carry through to the self-tuning version. These are:

(i) It can deal with nonminimum-phase systems without any difficulty.

(ii) Control excursions are often less violent.

(iii) Pure time delays may differ between loops.

(i) *Nonminimum-phase systems*

The term 'nonminimum phase' in this discrete-time application means that some zeros of $|B(z^{-1})|$ lie within the unit disc, i.e. $|z| > 1$. Since the minimum-variance law effectively attempts to cancel system zeros with controller poles, a highly sensitive closed-loop system will result if the system is nonminimum-phase, and in practice the system will become unstable. The pole-shifting law, however, is not sensitive to nonminimum-phase systems. This is a particularly important advantage because many systems which, in the s-domain may be minimum-phase, are nonminimum-phase in the discrete domain. Computational time delays in the control can also make the system appear nonminimum-phase [6].

(ii) *Differing loop pure time delays*

It is not uncommon in real applications for the system pure time delay in each loop to be different. In terms of the system model eqn. 1 this means that, for example, zeros may appear on the diagonal of the B_1 coefficient matrix leading to a singular B_1. With very few exceptions (see Reference 11, for example) the minimum-variance controller is unable to deal with such a situation. On the other hand, the pole-shifting control law will generally not be embarassed. It is only a requirement that a solution to the matrix polynomial eqn. 6 should exist. This, then, again reduces to the requirement that the transformation matrix on the left-hand side of eqn. 9 be nonsingular.

3 Self-tuning regulator

The offline design of the multivariable pole-shifting regulator (and, of course, also the minimum-variance regulator) presupposes a knowledge of the matrix polynomials $A(z^{-1})$, $B(z^{-1})$ and $C(z^{-1})$. These parameters may be obtained using estimation techniques, for example maximum-likelihood, but the problem is a nonlinear one and is tedious to solve (see, for example, Reference 15). Certainly, the online estimation of these parameters would be a formidable task. However, it will be shown here that a self-tuning property exists which considerably simplifies the online control problem.

An online model of the system is formed as follows:

$$y_t = -\alpha(z^{-1})y_t + \beta(z^{-1})u_t + \epsilon_t \qquad (19)$$

where

$$\alpha(z^{-1}) = \alpha_1 z^{-1} + \ldots + \alpha_{n_a} z^{-n_a} \qquad (20)$$

$$\beta(z^{-1}) = \beta_1 z^{-1} + \ldots + \beta_{n_b+k} z^{-n_b-k} \qquad (21)$$

and the $p \times p$ matrix coefficients α_i, β_i are evaluated using recursive least squares. Notice that the system pure time delay k enters the model by extending the order of the $\beta(z^{-1})$ polynomial. The control law is defined by eqn. 3, but unlike the offline design case, $F(z^{-1})$ and $G(z^{-1})$ are evaluated by solving the matrix polynomial identity

$$[I + \alpha(z^{-1})]\,[I + F(z^{-1})] - \beta(z^{-1})G(z^{-1}) = I + T(z^{-1}) \qquad (22)$$

where the order of $T(z^{-1})$, n_t, is governed by eqn. 7. As in the offline design, the determinant $|I + T(z^{-1})|$ specifies the closed-loop system poles. The self-tuning procedure is thus:

(i) At each iteration estimate the parameters of matrix polynomials $\alpha(z^{-1})$ and $\beta(z^{-1})$ in eqn. 19 using recursive least squares so that $\{\epsilon_t\}$ becomes the sequence of fitting errors.

(ii) Solve the pole-shifting eqn. 22. (This involves the solution of a set of linear simultaneous equations, analogous to eqn. 9.)

(iii) Solve for $\tilde{F}(z^{-1})$ and $\tilde{G}(z^{-1})$, where

$$[I + \tilde{F}(z^{-1})]\,G(z^{-1}) = \tilde{G}(z^{-1})[I + F(z^{-1})] \qquad (23)$$

and

$$n_{\tilde{f}} = n_f, \qquad n_{\tilde{g}} = n_g$$

so that the control law may be expressed in the form shown in eqn. 13. (This again is by solution of a set of linear simultaneous equations.)

(iv) Apply the control u_t calculated from the law derived in step (iii) and proceed to the next iteration (step (i)).

The self-tuning principle then states that, if the system converges, it converges to the closed-loop system that would have been obtained had the parameters $A(z^{-1})$, $B(z^{-1})$ and $C(z^{-1})$ been known *a priori* and the control law been derived from the offline design rule given in eqn. 6, i.e. the output converges to

$$y_t = [I + T(z^{-1})]\,[I + F(z^{-1})]^{-1} e_t \qquad (24)$$

Furthermore, the residual sequence $\{\epsilon_t\}$ will converge to the true system driving noise $\{e_t\}$. A proof of the self-tuning property is given in Appendix 9.1.

The self-tuning version retains the general advantages of a pole-shifting law over the minimum-variance law with respect to dealing with nonminimum-phase systems and systems in which the loop pure time delays differ. Furthermore, it is to be expected that the scheme will also be more tolerant of time-varying time delays than the minimum-variance regulator. This has already been shown for single-input/single-output systems (see Reference 6). As in the SISO version, the order of the $\beta(z^{-1})$ polynomial should be chosen to be $n_b + k_{max}$, where k_{max} is the maximum expected value of the time delay k. Variations in time delay are effectively treated in the same way as parameter variations.

4 Computer implementation of algorithm

Unfortunately the amount of computation required for the pole-shifting control is a good deal more than the minimum-variance law, yet the pole shifter is more generally applicable, and if a fast processor is used with floating-point hardware, computation times are not unduly excessive. For example, simulation carried out using a PDP-10 computer with $n_a = 2$, $n_b = 2$ and $p = 2$ (16 parameters) showed that computation required 58 ms per iteration.

It is worth noting that the estimation algorithm for the multivariable model (eqn. 19) does not involve much extra work over the single-output model. This is because if each output is regressed on the same set of data, the 'covariance' matrix required to compute the filter gain remains the same for the parameters in each of the regression equations describing the p outputs. The recursive-least-squares algorithm is given in Appendix 9.2.

A useful way to set up the initial estimation model is to use the least-squares parameter estimates for the system obtained from a short record of system input-output data. Simulation studies have shown that the choice of initial conditions can affect the controller parameter convergence rate. This is an area requiring further theoretical investigation.

The pole-shifting law requires the solution of $(n_f + n_g + 1)p$ simultaneous linear equations (eqn. 22) in order to find $F(z^{-1})$ and $G(z^{-1})$. A further $(n_f + n_g)p$ simultaneous equations must be solved to find $\tilde{F}(z^{-1})$ and $\tilde{G}(z^{-1})$ unless $n_f = 0$ $(n_b = 1)$ when

$$I + \tilde{F}(z^{-1}) = I; \qquad \tilde{G}(z^{-1}) = G(z^{-1})$$

or when $p = 1$, in which case the transformation is unnecessary.

The computational time delay has the effect of introducing a system of pure time delay. When the computational delay becomes significant in comparison with the sampling period, the $\beta(z^{-1})$ polynomial should be extended to take account of this.

The algorithm given here is intended to solve the regulator problem. Reference tracking is introduced most simply by incorporating a digital integrator in each loop and replacing the regulator equation by

$$u_t = -\tilde{F}(z^{-1})u_t + \tilde{G}(z^{-1})(y_t - y_{r_t})$$

where y_{r_t} is the reference input.

This method clearly will cause the controller to detune during a change in reference input, and is therefore not entirely satisfactory if frequent set point changes are expected. The self-tuning scheme described here is primarily a regulator, and the extension for set point following is a current topic of research.

5 Examples

The features of the pole-shifting algorithm are illustrated in two examples. The first is a simulation of a regulator problem in which the two system loops have differing time delays, and in addition the $B(z^{-1})$ has zeros within the unit disc. The second example is an application to a laboratory rig.

5.1 Simulation example

Consider the multivariable system

$$[I + A_1 z^{-1} + A_2 z^{-2}]y_t = [B_1 z^{-1} + B_2 z^{-2}]u_t + [I + C_1 z^{-1}]e_t \qquad (25)$$

where

$$A_1 = \begin{bmatrix} -1.4 & -0.2 \\ -0.1 & -0.9 \end{bmatrix} \qquad A_2 = \begin{bmatrix} 0.48 & 0.1 \\ 0 & 0.2 \end{bmatrix}$$

$$B_1 = \begin{bmatrix} 1 & 0 \\ 0 & 0 \end{bmatrix} \qquad B_2 = \begin{bmatrix} 1.5 & 1 \\ 0 & 1 \end{bmatrix}$$

$$C_1 = \begin{bmatrix} -0.5 & 0 \\ 0.1 & -0.3 \end{bmatrix}$$

and e_t is a white-noise vector sequence with zero mean and variance

$$R = \begin{bmatrix} 0.1 & 0 \\ 0 & 0.1 \end{bmatrix}$$

The closed-loop poles are to be placed at $z = 0.5$ and $z = 0.4$ so that a suitable choice of $I + T(z^{-1})$ is

$$I + \begin{bmatrix} -0.5 & 0 \\ 0 & -0.4 \end{bmatrix} z^{-1} \qquad (26)$$

Notice that B_1 is singular and that the system is non-minimum-phase.* Thus, a standard minimum-variance regulator could not be applied to the system.

The offline design of the pole-shifting regulator is achieved by solving eqn. 6 and invoking the transformation

$$(I + \tilde{F}_1 z^{-1})^{-1}[\tilde{G}_0 + \tilde{G}_1 z^{-1}] = [G_0 + G_1 z^{-1}] \times [I + F_1 z^{-1}]^{-1} \qquad (27)$$

The resulting control law is

$$u_t = -\tilde{F}_1 u_t + \tilde{G}_0 y_t + \tilde{G}_1 y_{t-1} \qquad (28)$$

where

$$\tilde{F}_1 = \begin{bmatrix} 0.213 & 0.203 \\ 0.155 & 0.286 \end{bmatrix}$$

$$\tilde{G}_0 = \begin{bmatrix} -0.10 & -0.048 \\ -0.16 & -0.115 \end{bmatrix}$$

$$\tilde{G}_1 = \begin{bmatrix} 0.0682 & 0.0265 \\ 0.0495 & 0.0469 \end{bmatrix}$$

The self-tuning version requires the system eqn. 25 to be modelled as

$$y_t = -\alpha_1 y_{t-1} - \alpha_2 y_{t-2} + \beta_1 u_{t-1} + \beta_2 u_{t-2} + \epsilon_t$$

where the coefficient matrices are estimated using recursive least squares, and the control law is found by solving eqn. 22 and eqn. 27 online. The convergence of the control parameters is shown in Fig. 1, where, after 3000 steps, the control matrices are

$$\hat{\tilde{F}}_1 = \begin{bmatrix} 0.275 & 0.213 \\ 0.139 & 0.292 \end{bmatrix}$$

$$\hat{\tilde{G}}_0 = \begin{bmatrix} -0.102 & -0.058 \\ -0.153 & -0.106 \end{bmatrix}$$

$$\hat{\tilde{G}}_1 = \begin{bmatrix} 0.0707 & 0.0382 \\ 0.0459 & 0.0332 \end{bmatrix} \qquad (29)$$

Although the parameters have not converged exactly, the proximity of the 'offline' and self-tuner designed laws is demonstrated in Fig. 2, where the residual sequence ϵ_t and system driving noise e_t have been superimposed. Note that in each of Figs. 2*a* and *b* two curves have been plotted, corresponding to components e_{1_t} and ϵ_{1_t}, and components e_{2_t} and ϵ_{2_t}. As predicted in the theory, the residual sequence has converged so closely to e_t that the traces are virtually indistinguishable.

*N.B. This also implies different time delays in the two input channels

5.2 *Control of hydraulic system [18]*

Fig. 3 shows a simple hydraulic system comprising two water tanks coupled by an interconnecting orifice. Each tank has an outlet tap and input water pump. The control objective is to independently control the level of water in each tank of this coupled multivariable system. The pumps saturated at a given flow rate, and this characteristic was reflected in the software by setting software saturation limits. A lower limit was also set so that a negative pump demand could not be selected. The self-tuner estimator is fed system input information after software saturation has been effected. The pump input voltage-flowrate characteristic was nonlinear, and the coupled tank system itself was also nonlinear.

Since the levels were to be controlled to a given set point, a digital integrator was incorporated in the loop, thus ensuring zero steady-state error. Furthermore, so as to avoid too much self tuner detuning due to the set-point input (which acts as a 'disturbance' to the self tuner) set-point inputs were rate limited. Rate limiting is a useful technique, but clearly should be small enough not to degrade system response unnecessarily.

The system gain and dynamics change with operating point, and thus a forgetting factor λ was introduced to allow the estimator to adapt more quickly to changes in the system characteristics as the set point changed. The system was started with empty tanks, and to aid initial tuning a low initial value of λ was chosen, as follows:

$$\lambda_{k+1} = \begin{cases} 0.99\lambda_k + 0.01 & \lambda_k \leqslant 0.99 \\ 0.99 & \text{otherwise} \end{cases}$$

$$\lambda_0 = 0.97$$

The estimation model was

$$[I + \alpha_1 z^{-1} + \alpha_2 z^{-2}] \begin{bmatrix} \Delta H_1 \\ \Delta H_2 \end{bmatrix} = \beta_1 z^{-1} \begin{bmatrix} \Delta Q_1 \\ \Delta Q_2 \end{bmatrix} + \epsilon_t$$

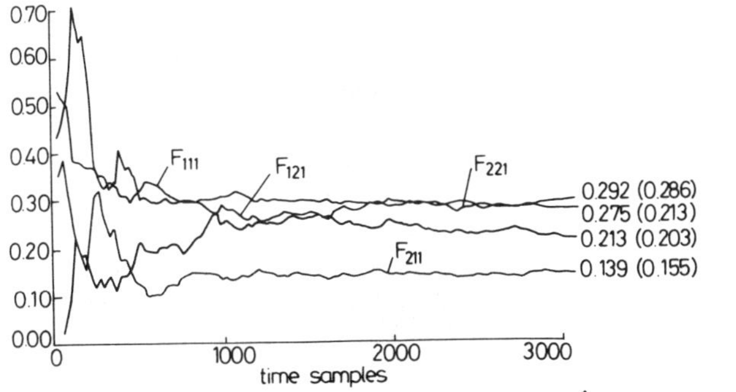

Fig. 1A *Time evolution of controller parameters* ($\hat{\tilde{F}}_1$)
Optimal values shown in brackets

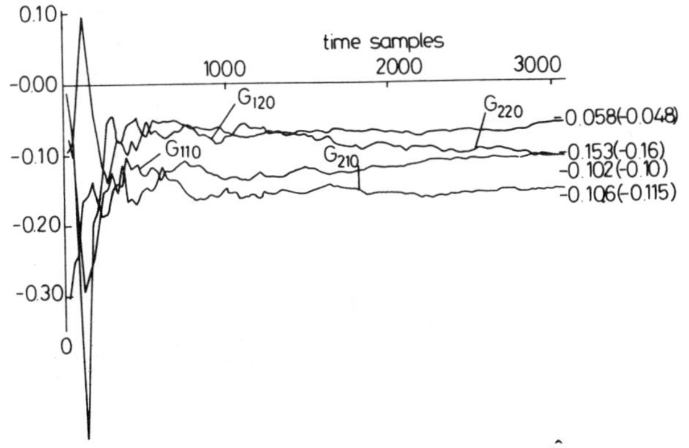

Fig. 1B *Time evolution of controller parameters* ($\hat{\tilde{G}}_0$)
Optimal values shown in brackets

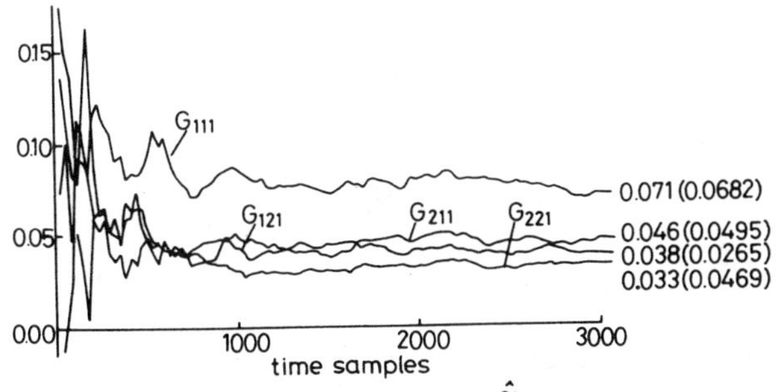

Fig. 1C *Time evolution of parameters* ($\hat{\tilde{G}}_1$)
Optimal values shown in brackets

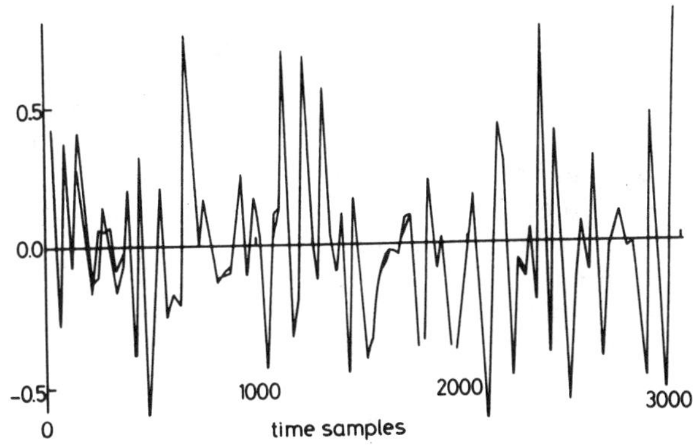

Fig. 2A *Driving noise* (e_1) *and residual sequence* (ϵ_1) *superimposed*

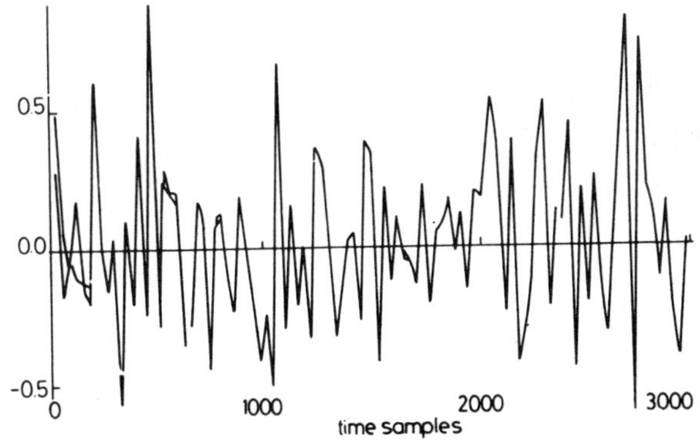

Fig. 2B *Driving noise* (e_2) *and residual sequence* (ϵ_2) *superimposed*

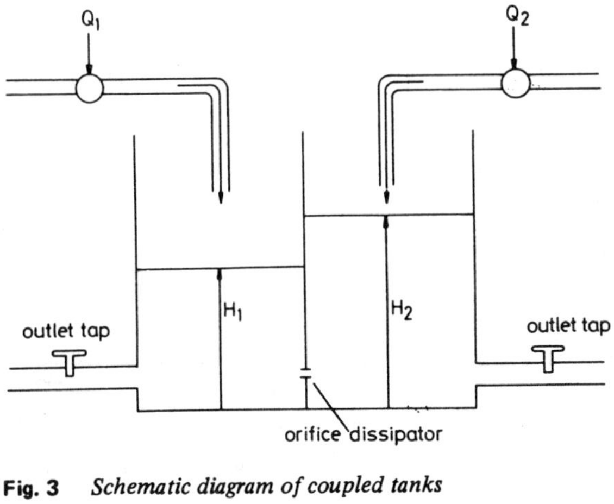

Fig. 3 *Schematic diagram of coupled tanks*

where

$\Delta H_i = H_i - H_{d_i}$
Q_i = flow rate control signal for pump i, V
ΔQ_i = incremental flowrate control signal
H_i = measured level of tank i, V
H_{d_i} = demanded level of tank i, V

Also,

$$I + T(z^{-1}) = I + \begin{bmatrix} -0.6 & 0 \\ 0 & -0.6 \end{bmatrix} z^{-1}$$

A sampling period of 5 s was chosen. Since $n_\beta = 1$, the transformation in eqn. 23 was not necessary, and total computation time for one iteration to update the control matrix coefficients

$$G_0 = \begin{bmatrix} g_{111} & g_{121} \\ g_{211} & g_{221} \end{bmatrix} \qquad G_1 = \begin{bmatrix} g_{112} & g_{122} \\ g_{212} & g_{222} \end{bmatrix}$$

of control law $\Delta Q_t = G_0 \Delta H_t + G_1 \Delta H_{t-1}$, and execute other necessary software, was 400 ms on a PDP-11/10 computer without floating-point arithmetic hardware. The computational time delay introduced was small enough in comparison with the sampling period to be neglected.

A typical set of results is shown in Figs. 4–6. Fig. 4 shows the demanded level and system output, and Fig. 5 the control demands to the pumps. The variation of the control parameters as the operating point (and thereby also the system characteristics) changed is well illustrated in Fig. 6. Although output 1 appears to rise very slowly after the set-point change at step 100, notice that the pump is in fact full on (saturated). Similarly, at step 200 when tank 1 is required to drain, the slow response is due to the limited draining capacity.

It is felt that these results are encouraging in that they demonstrate that the multivariable self-tuning pole-shifting controller is capable of good results when operating a real process.

5.3 *Fixed coefficient multivariable control*

The past decade has seen great progress in fixed coefficient design techniques for multivariable control systems. In particular, the work of Rosenbrock [13] and McFarlane [19] has led to a school of researchers whose interest is to reduce the interaction between servo-command channels. The work presented here does *not* address this problem. In particular, the multivariable self tuner given here is a regulator and will not in general lead to a decoupled or diagonally dominant controller in the sense of Rosenbrock [13]. Indeed, the closed-loop channels may experience strong interaction and hence permit crosscoupling of disturbances. Having said this, however, the following points are pertinent:

(*a*) The degree of crosscoupling is fixed by the polynomial matrix $F(z^{-1})$ (since this determines the closed-loop zeros) and the polynomial matrix $T(z^{-1})$. However, the latter can be selected (without loss of generality) to be diagonal, so it is predominantly the controller denominator polynomial matrix $F(z^{-1})$ which gives the interaction terms.

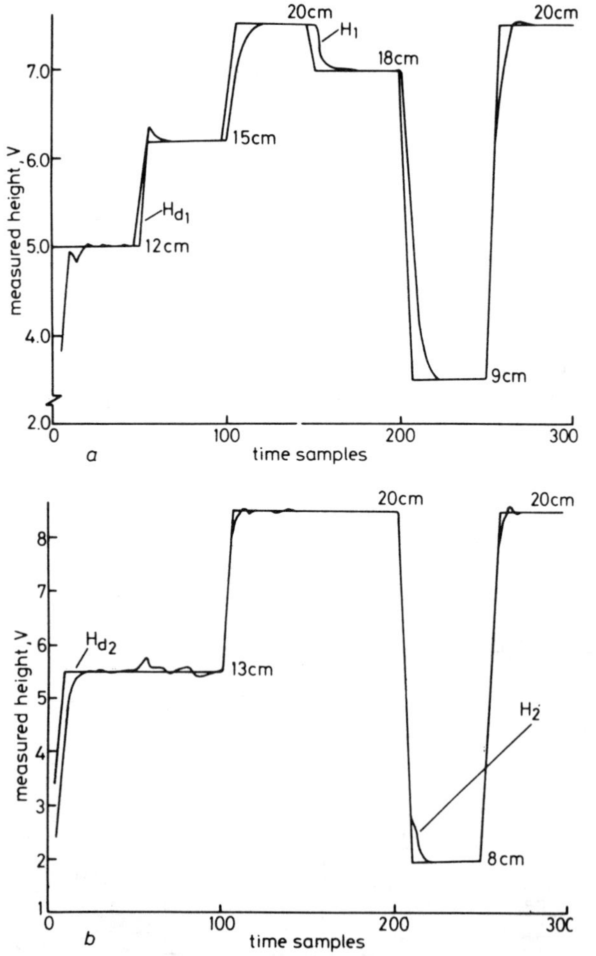

Fig. 4 *Demanded and measured height*

N.B. True height against measured output (in volts) characteristic is nonlinear

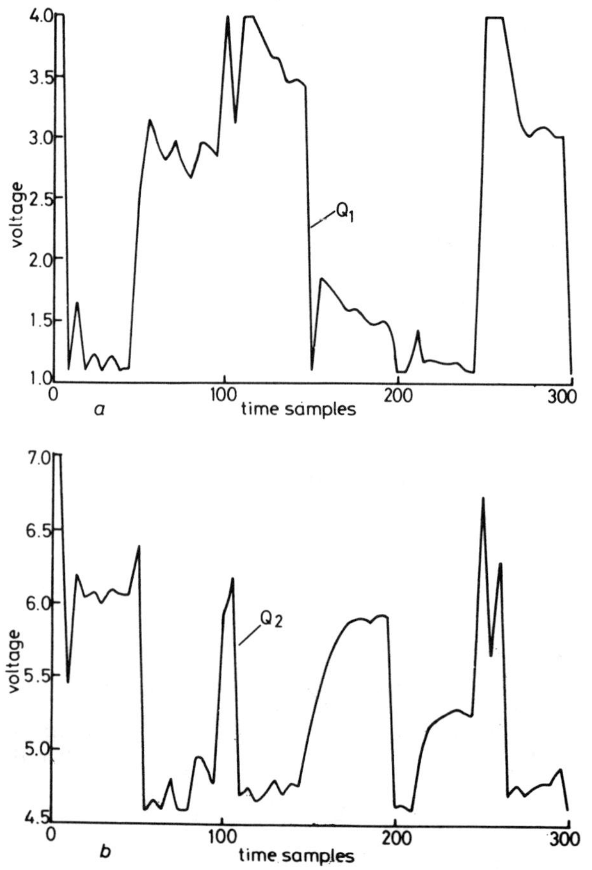

Fig. 5 *Control input*

(*b*) In the special case where the polynomial matrix $B(z^{-1})$ is first-order, the $I + F(z^{-1})$ polynomial matrix reduces to the unit matrix, so complete disturbance decoupling is achieved. Indeed, the coupled tanks (example 5.2) is an illustration of such a system, and Fig. 4 shows that negligible transient interaction occurs. In particular, and with reference to Fig. 4, observe H_2 at the transition of H_{d_1} from 12 cm to 15 cm, which occurs at time sample 50.

(*c*) In general, and as discussed in example 5.2, it is good practice to include integral action in all loops. In this case the loops are decoupled in the sense that their steady-state step response errors are independent and zero. For process regulation purposes this is often all that is required, although we readily admit that in the field of deterministic servo-command this is not so. To this end it is our intention to return to the question of self-tuning multivariable servo-systems in a subsequent paper.

However, as a concluding comment on the matter of interaction in self-tuning pole-assignment regulators, consider again the simulation example 5.1, and recall that the B_2 matrix is given by

$$B_2 = \begin{pmatrix} 1.5 & 1 \\ 0 & 1 \end{pmatrix}$$

In particular, note that the entries in column 2 are equal, indicating that the initial impact of a rapid change in the second input is of equal intensity on both outputs: in other words, 100% initial interaction, which by many standards is considered strong interdependence. As a result (and with no integral action) loop one would react significantly to set-point changes in channel two. However, as discussed previously, the addition of integral action ensures that this interaction is transient in nature. Fig. 7 reinforces this point by illustrating the two outputs of the closed-loop system with integral action in each loop. The system input is a step change in the set-point demand on channel two; note the clearly defined transient interaction which is subsequently taken out by the integral action.

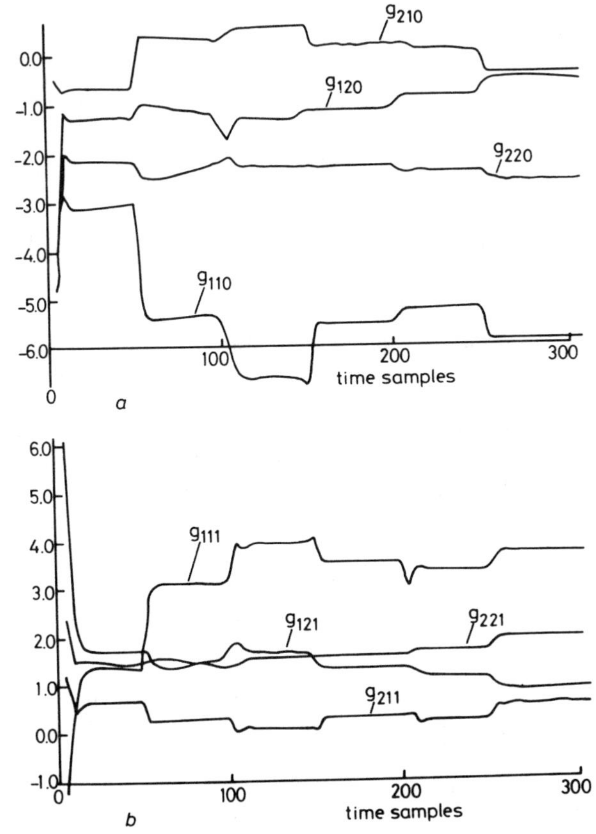

Fig. 6 *Time evolution of control parameters*

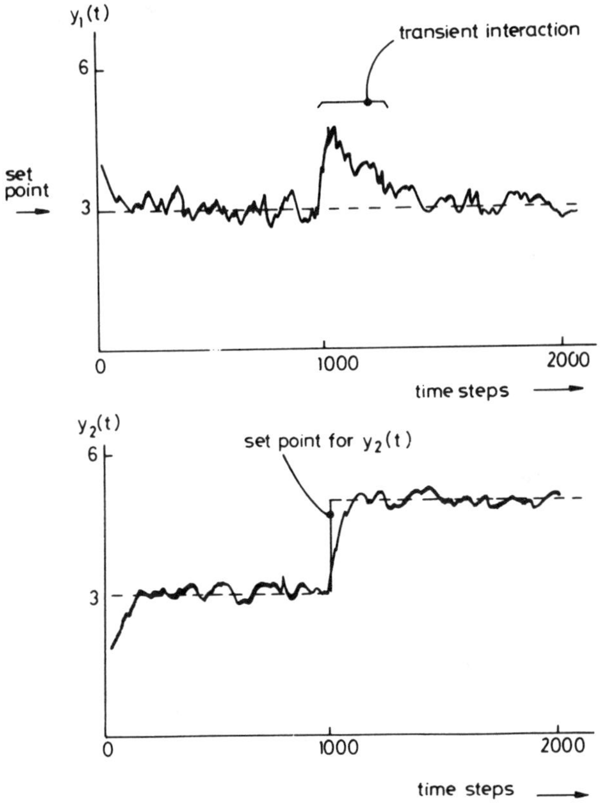

Fig. 7 *Illustrating transient interaction for simulation example 5.1*

6 Conclusion

This paper has described a new multivariable self-tuning controller based on a pole-assignment, rather than optimal-control approach. The controller exhibits certain advantages over the minimum-variance self tuner, the more significant of these being that it can deal with nonminimum-phase systems, and systems in which the various loops have differing pure time delays. In fact, just like its single-variable counterpart, its prime advantages may be summed up as robustness and generality. Its main disadvantage lies in the complexity of the computation required which restricts maximum allowable sampling frequency. This, however, will be remedied with the advent of even faster and more powerful microprocessors. By the same token, the robustness is won at the expense of optimality of the closed-loop configuration. However, this is not necessarily a disadvantage, since in many practical applications optimal control is not the central objective. In such situations a pole-assignment self tuner can provide a well-behaved closed-loop configuration with closely defined transient behaviour.

7 Acknowledgment

This work was supported by the Science Research Council under grant GR/A80358.

8 References

1 ASTROM, K.J., and WITTENMARK, B.: 'On self-tuning regulators', *Automatica*, 1973, **9**, pp. 185–199
2 PETERKA, V.: 'Adaptive digital regulation of noisy systems'. 2nd Prague IFAC symposium on identification and process parameter estimation, 1970
3 CLARKE, D.W., and GAWTHROP, P.J.: 'Self-tuning controller', *Proc. IEE*, 1975, **122**, (9), pp. 929–934
4 WELLSTEAD, P.E., EDMUNDS, J.M., PRAGER, D.L., and ZANKER, P.: 'Self-tuning pole/zero assignment regulators'. *Int. J. Control*, 1979, **30**, pp. 1–26
5 EDMUNDS, J.M. Ph.D. thesis, Control Systems Centre, UMIST, 1976
6 WELLSTEAD, P.E., PRAGER, D.L., and ZANKER, P.: 'Pole assignment self-tuning regulator', *Proc. IEE*, 1979, **126**, (8), pp. 781–789
7 ASTROM, K.J., BORISSON, U., LJUNG, L., and WITTENMARK, B.: 'Theory and applications of self-tuning regulators', *Automatica*, 1977, **13**, pp. 457–476
8 GAWTHROP, P.J.: 'Some interpretations of the self-tuning controller', *Proc. IEE*, 1977, **124**, (10), pp. 889–894
9 YOUNG, P.C., and YANCEY, C.B.: 'Second generation adaptive pitch autostabilization system for a missile or aircraft'. Technical note 404-109, Naval Weapons Centre, China Lake, Ca. 93555, USA, 1971
10 HASTINGS-JAMES, R. Ph.D. thesis, University of Cambridge, 1970
11 BORISSON, U.: 'Self-tuning regulators – industrial application and multivariable theory'. Lund report 7513, Department of Automatic Control, Lund Institute, Sweden, 1975
12 KEVICZKY, L., HETTHESSY, J., HILGER, M., and KOLOSTORI, J.: 'Self-tuning adaptive control of cement raw material blending', *Automatica*, 1978, **14**, pp. 525–532
13 ROSENBROCK, H.H.: 'Computer-aided control system design' (Academic Press, London, 1974)
14 WOLOVICH, W.A.: 'Linear multivariable systems' (Springer Verlag, New York, 1974)
15 PRAGER, D.L., and WELLSTEAD, P.E.: 'Interactive recursive maximum likelihood estimation'. 5th IFAC symposium on identification and system parameter estimation, Darmstadt, W. Germany, 1979
16 SINHA, A.K.: 'Minimum interaction minimum variance controller for multivariable systems', *IEEE Trans.*, 1977, **AC-22**, pp. 274–275
17 BORISSON, U.: 'Self-tuning regulators for a class of multivariable systems', *Automatica*, 1979, **15**, pp. 209–215
18 WELLSTEAD, P.E.: 'Scale models in control engineering'. Control Systems Centre report 482, UMIST 1980 (available on application to the Control Systems Centre Secretariat)
19 MacFARLANE, A.G.J.: 'Frequency response methods in control systems' (IEEE Press, 1979)
20 EYKHOFF, P.: 'System identification' (Wiley Interscience, London, 1974)

9 Appendixes

9.1 *Proof of self-tuning property*

Let the unknown system be modelled by

$$(I + \alpha(z^{-1}))y_t = \beta(z^{-1})u_t + \epsilon_t \tag{30}$$

where the matrix polynomials have the same form as in the main body of the paper and are estimated using a recursive least-squares algorithm minimising

$$\sum_{j=1}^{t} (\epsilon_j^i)^2, \qquad i = 1, 2, \ldots, p$$

where

$$\epsilon_t = (\epsilon_t^1 \ldots \epsilon_t^p)^T$$

Let

$$n_\alpha = n_a = n \tag{31}$$

$$n_\beta = n + k \tag{32}$$

and select the control input such that

$$u_t = G(z^{-1})[I + F(z^{-1})]^{-1} y_t \tag{33}$$

where

$$G(z^{-1}) = G_0 + G_1 z^{-1} + \ldots + G_{n_g} z^{-n_g} \tag{34}$$

$$I + F(z^{-1}) = I + F_1 z^{-1} + \ldots + F_{n_f} z^{-n_f} \tag{35}$$

$$n_g = n_\alpha - 1 \tag{36}$$

$$n_f = n_\beta - 1 \tag{37}$$

and $F(z^{-1})$ and $G(z^{-1})$ are obtained as solutions to the set of equations generated from

$$[I + \alpha(z^{-1})][I + F(z^{-1})] - \beta(z^{-1})G(z^{-1}) = I + T(z^{-1}) \tag{38}$$

Define also

$$I + L(z^{-1}) = [I + A(z^{-1})][I + F(z^{-1})] - z^{-k}B(z^{-1})G(z^{-1}) \tag{39}$$

Assume the following conditions are met:

C1: $n_t \leqslant n_a + n_b + k - 1 - n_c$.

C2: The offline regulator exists, and has $F^*(z)$ and $G^*(z)$ (defined in eqns. 11 and 12) relatively right prime with all observability and controllability indices equal to n_f.

C3: $[I + L(z^{-1})][I + T(z^{-1})]^{-1}$ may be represented as $[I + \tilde{T}(z^{-1})]^{-1}(I + \tilde{L}(z^{-1}))$, where $n_{\tilde{t}} \leqslant n_t$ and $n_{\tilde{l}} \leqslant n_l$.

C4: The model parameters converge and the output of the closed-loop system is ergodic.

Then the control converges to that obtained by solving the offline identity

$$[I + A(z^{-1})][I + F(z^{-1})] - z^{-k}B(z^{-1})G(z^{-1}) = [I + C(z^{-1})][I + T(z^{-1})] \tag{40}$$

and the residual ϵ_t equals the system driving noise e_t. Furthermore, the system output is

$$y_t = [I + F(z^{-1})][I + T(z^{-1})]^{-1} e_t \tag{41}$$

Notes on conditions

(*a*) Condition *C2* is required also for the minimum-variance self tuner and is met for a wide range of systems [11].

(*b*) Condition *C3* is required only for multivariable systems and is obviously always met by scalar systems (scalar polynomials commute). Let $I + L(z^{-1})$ and $I + T(z^{-1})$ be factored so that

$$[I + L(z^{-1})][I + T(z^{-1})]^{-1} = [I + L^*(z^{-1})][I + T^*(z^{-1})]^{-1}$$

where $I + L^*(z^{-1})$ and $I + T^*(z^{-1})$ are relatively right prime and have orders

$$n_{l*} \leqslant n_l$$

$$n_{t*} \leqslant n_t$$

Condition *C3* requires that a matrix

$$\underbrace{\begin{bmatrix} I & & \bigcirc \\ T_1^{*T} & \ddots & \\ \vdots & & I \\ T_{n_t}^{*T} & & \vdots \\ \bigcirc & \ddots & T_{n_t}^{*T} \end{bmatrix}}_{n_{l*}\text{ block columns}} \underbrace{\begin{bmatrix} -I & & \bigcirc \\ -L_1^{*T} & \ddots & \\ \vdots & & -I \\ -L_{n_{l*}}^{*T} & & \vdots \\ \bigcirc & \ddots & L_{n_{l*}}^{*T} \end{bmatrix}}_{n_{t*}\text{ block columns}}$$

is nonsingular. Now, $I + L(z^{-1})$ is a function of the control parameters, and it is argued that the stochastic element will ensure that the matrix is never singular. Note also that, when the parameters converge exactly,

$$[I + L(z^{-1})]\,[I + T(z^{-1})]^{-1} = I + C(z^{-1})$$

and the required relationship exists. (Put $I + L^*(z^{-1}) = I + C(z^{-1})$ and $T^*(z^{-1}) = \phi$.)

(*c*) In line with other self tuners, no general guarantee of convergence can as yet be given. Therefore the proof is based on an assumption of convergence and it is shown that the parameters then indeed converge to the correct values. Simulation studies indicate that asymptotic convergence is generally attained.

Proof: Assume the model parameters have converged. Then, from the properties of least-squares estimation, the following results hold at time t:

$$E(y_{t-i}\epsilon_t^T) = R_{y\epsilon}(i) = 0 \qquad i = 1, 2, \ldots, n_\alpha \quad (42)$$

$$E(u_{t-i}\epsilon_t^T) = R_{u\epsilon}(i) = 0 \qquad i = 1, 2, \ldots, n_\beta \quad (43)$$

where $E(\ )$ denotes the mathematical expectation operation.

Define an auxiliary p-vector ω_t such that

$$\omega_t = [I + F(z^{-1})]^{-1}\,y_t \quad (44)$$

Then, from eqns. 44 and 33,

$$y_t = [I + F(z^{-1})]\,\omega_t \quad (45)$$

$$u_t = G(z^{-1})\,\omega_t \quad (46)$$

and the following $(n_\alpha + n_\beta)p$ equations may be written, of which $(n_\alpha + n_\beta - 1)p$ are (by condition *C2*) independent:

$$\begin{matrix} n_\alpha\left\{ \right. \\ \\ n_\beta\left\{ \right. \end{matrix} \underbrace{\begin{bmatrix} I & F_1 & F_2 & \cdots & F_{nf} & & \bigcirc \\ \bigcirc & & I & \cdots & \cdots & \cdots & F_{nf} \\ G_0 & \cdots & \cdots & G_{ng} & & & \bigcirc \\ \bigcirc & & G_0 & \cdots & \cdots & \cdots & G_{ng} \end{bmatrix}}_{n_\alpha + n_\beta - 1} \times$$

$$\begin{bmatrix} R_{\omega\epsilon}(1) \\ \cdot \\ \cdot \\ \cdot \\ R_{\omega\epsilon}(n_\alpha + n_\beta - 1) \end{bmatrix} = \begin{bmatrix} R_{y}(1) \\ \vdots \\ R_y(n_\alpha) \\ R_{u\epsilon}(1) \\ \vdots \\ R_{u\epsilon}(n_\beta) \end{bmatrix} = \mathbf{0} \quad (47)$$

Thus,

$$R_{\omega\epsilon}(\tau) = 0 \qquad \tau = 1, 2, \ldots, m \quad (48)$$

where

$$m = n_\alpha + n_\beta - 1$$

By substituting the control law into the model eqn. 30, and noting eqns. 38 and 45, it is easy to establish that

$$[I + T(z^{-1})]\,\omega_t = \epsilon_t \quad (49)$$

Also, it can be shown quite easily by substituting the control law into the system equation and using eqns. 39 and 45 that

$$[I + L(z^{-1})]\,\omega_t = [I + C(z^{-1})]\,e_t \quad (50)$$

Together with condition *C3*, eqns. 49 and 50 enable the residual and true system noise e_t to be related thus:

$$[I + \tilde{L}(z^{-1})]\,\epsilon_t = [I + S(z^{-1})]\,e_t \quad (51)$$

where

$$I + S(z^{-1}) = [I + \tilde{T}(z^{-1})]\,[I + C(z^{-1})] \quad (52)$$

$$n_{\tilde{l}} \leqslant n_a + n_b + k - 1 = m \quad (53)$$

and

$$n_s \leqslant n_t + n_c \leqslant n_a + n_b + k - 1 = m \quad (54)$$

Transposing eqn. 51, premultiplying by ω_{t-m-1} and taking expectations yields

$$\begin{aligned} &R_{\omega\epsilon}(m+1) + R_{\omega\epsilon}(m)L_1^T + \ldots + \\ &\quad + R_{\omega\epsilon}(m + 1 - n_{\tilde{l}})L_{n\tilde{l}}^T = \\ &R_{\omega e}(m+1) + R_{\omega e}(m)S_1^T + \ldots + \\ &\quad + R_{\omega e}(m + 1 - n_s)S_{n_s}^T \end{aligned} \quad (55)$$

Now, it has been shown that the left-hand side of eqn. 55 reduces to $R_{\omega\epsilon}(m+1)$ (refer to eqn. 48) and since $n_s \leqslant m$ and e_t is uncorrelated with past data, the right-hand side vanishes. It follows that

$$R_{\omega\epsilon}(m+1) = 0 \quad (56)$$

and in a similar fashion it can be shown that

$$R_{\omega\epsilon}(\tau) = 0 \qquad \tau > 0 \quad (57)$$

Substituting for ω in the correlation eqn. 57, clearly

$$R_{\epsilon\epsilon}(\tau) = 0 \qquad \tau > 0 \quad (58)$$

i.e. ϵ_t is a white process; but, from eqn. 51, ϵ_t could be represented as

$$\epsilon_t = e_t + \sum_{i=1}^{\infty} \phi_i e_{t-i} \quad (59)$$

Thus, from eqn. 58, $\epsilon_t = e_t$ and the self-tuning property is proven.

9.2 *Recursive least squares*

The model equation may be written in the following way:

$$y_t^i = x_t^T \hat{\theta}_t^i + \epsilon_t^i \qquad i = 1, 2, \ldots, p$$

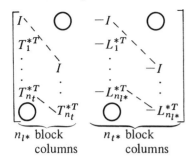

where

y_t^i is the ith component of the output vector y_t
ϵ_t^i is the ith component of the residual vector ϵ_t
x_t is the data vector

$$x_t = [y_{t-1}^1, \ldots, y_{t-n_a}^1, \ldots, y_{t-1}^p, \ldots, y_{t-n_a}^p, u_{t-1}^1, \ldots, u_{t-n_b}^1, \ldots, u_{t-1}^p, \ldots, u_{t-n_b}^p]^T$$

and θ_t^i is the associated parameter vector.

From standard recursive least-squares [20] the parameters are found as follows:

$$\hat{\theta}_{t+1}^i = \hat{\theta}_t^i + \gamma_t p_t x_{t+1} [y_{t+1}^i - x_{t+1}^T \hat{\theta}_t^i]$$

$$i = 1, 2, \ldots, p$$

where

$$\gamma_t = (1 + x_{t+1}^T p_t x_{t+1})^{-1}$$

$$p_{t+1} = p_t - \gamma_t p_t x_{t+1} x_{t+1}^T p_t$$

and p_0 is set to some initial diagonal matrix, i.e. $10I$.

Note that the parameter update equation only need be repeated p times. The other quantities, x_t, γ_t and p_t, are not dependent on the output component number.

Predictor-based Self-tuning Control*

V. PETERKA†

Heavy-duty conditions of microprocessor self-tuning control motivated the development of a new numerical method for LQ-optimum control design—a kind of Wiener approach revival.

Key Words—Control system synthesis; self-tuning control; adaptive control; self-adjusting systems; optimal control; direct digital control; stochastic control.

Abstract—Linear finite-memory output predictors updated in real-time appear to be a suitable internal representation of the system in a digital self-tuning controller. A new time-domain method of quadratic-optimum control synthesis for systems described by such predictors is presented. The synthesis covers both the servo problem and the regulation problem, including the program control (with preprogrammed command signal) and the feedforward from measurable external disturbances. Unlike the standard Riccati equation, the method leads to algorithms (or explicit formulae in low-order cases) which are numerically robust and therefore suitable for real-time computation using microprocessors with reduced wordlength.

1. INTRODUCTION

Most of the existing self-tuning control algorithms which appear practicable rest on the so-called certainty-equivalence hypothesis (Åström, 1980), i.e. on the enforced separation of identification and control. In this paper a similar simplification is employed using a linear finite-memory output predictor, updated in real-time, as an internal representation of the uncertain process to be controlled. In this sense the paper follows the line applied in Peterka (1970) and Clarke and Gawthrop (1975) but it differs substantially in the control strategy applied. Here the control strategy is designed to minimize, for the most recent predictor available, the expected value of a suitably chosen quadratic criterion embracing a given (finite or asymptotically infinite) control horizon. As in such a self-tuning controller the control law has to be repeatedly redesigned or updated, usually in each control step, it is required that the algorithm performing this task be numerically stable, free of singularities, uniform for all situations which might occur, and in addition as simple as possible. These requirements and the requirement to tune the controller also with respect to the preprogrammed sequence of the setpoints (program control) motivated the development of a new numerical method for LQ-optimum control synthesis which is the main topic of the paper.

The problem is stated in Section 2 where the control objectives are defined and the corresponding models and assumptions are introduced. The main results are summarized in Section 3. Their constructive proof and the optimization method itself are presented in Section 4. In the concluding Section 5 possible extensions are outlined and some practically important points are recalled and briefly discussed.

2. CONTROL OBJECTIVES AND MODELS

The scheme in Fig. 1 shows the situation considered throughout the paper. The process P, with single input u and single output y, is stochastic due to unmeasurable internal and/or external disturbances. The self-tuning controller STC may include the feedforward from the measurable external disturbance v if available. It is required that the controller be automatically tuned also with respect to possible changes of the setpoint (command signal) w which may be either *a priori* known (program control), or uncertain (servo), or it may be required that the process output follows the output of a reference model the input of which is *a priori* uncertain (model following). The extension

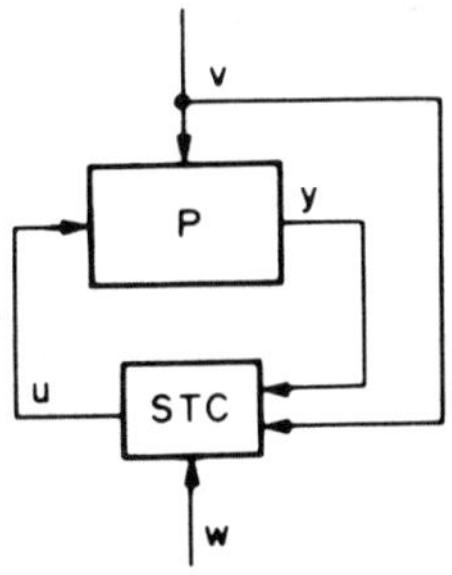

Fig. 1. Control loop considered.

* Received 11 June 1982; revised 23 May 1983. The original version of this paper was presented at the 6th IFAC Symposium on Identification and System Parameter Estimation which was held in Washington, D.C., U.S.A. during June 1982. The published proceedings of this IFAC meeting may be ordered from Pergamon Press Ltd, Headington Hill Hall, Oxford OX3 0BW, U.K. This paper was recommended for publication in revised form by associate editor P. Parks under the direction of editor B. D. O. Anderson.

† Institute of Information Theory and Automation, Czechoslovak Academy of Sciences, Prague 8, Pod vodárenskou věží 4, Czechoslovakia.

Reprinted with permission from *Automatica*, vol. 20, pp. 39–50, Jan. 1984.
Copyright © 1984, International Federation of Automatic Control IFAC. Published by Pergamon Press Ltd.

for the case of more than one measurable external disturbances is straightforward and simple, and therefore is not considered explicitly.

Time indexing

The chronology of signal samples is defined so that the samples $y_{(t-1)}$ and $v_{(t-1)}$ are available when $u_{(t)}$ is computed whereas $y_{(t)}$ and $v_{(t)}$ are not known at that moment. The availability of the future setpoints (command signal) will be discussed and defined separately.

Output predictors

Two kinds of finite-memory predictors are considered. The first one has the standard linear form

$$\hat{y}_{(t)} = -\sum_{i=1}^{\partial a} a_i y_{(t-1)} + \sum_{i=0}^{\partial b} b_i u_{(t-i)} + \sum_{i=1}^{\partial d} d_i v_{(t-i)} + k \quad (1)$$

where the integers ∂a, ∂b and ∂d define the memory of the predictor. It will be called the positional (P) predictor. The prediction error is introduced as

$$e_{(t)} = y_{(t)} - \hat{y}_{(t)} \quad (2)$$

and is available as soon as the true output $y_{(t)}$ has been observed.

In most cases of industrial process control the main task of a controller is to compensate stochastic disturbances which are inherently nonstationary like drifts, unpredictable load changes, etc. In such cases it is more appropriate to consider the following *incremental* (I) predictor

$$\hat{y}_{(t)} = y_{(t-1)} - \sum_{i=1}^{\partial a} a_i \Delta y_{(t-1)} + \sum_{i=0}^{\partial b} b_i \Delta u_{(t-i)} + \sum_{i=1}^{\partial d} d_i \Delta v_{(t-i)} \quad (3)$$

where $\Delta u_{(\tau)} = u_{(\tau)} - u_{(\tau-1)}$ and similarly for y and v. Note that in (3) the 'fixed point' in the input–output relation is the last observed output $y_{(t-1)}$ instead of the constant k in (1).

Process dead-time

The unknown time delay of the process is covered by the b-parameters in (1) or (3) which are estimated in real-time. For the sake of simplicity the *a priori* known delay, say κ, is not considered explicitly. It can be introduced in the following ways: (i) the corresponding number of leading b-parameters are not estimated and simply set to zero and (ii) the known time delay is neglected in the control-design step ($y_{(\tau)}$ is considered instead of $y_{(\tau+\kappa)}$) and the outputs which are not available but are required in the resulting control law are replaced by their predictions supplied by a predictor running in parallel. The latter appeared convenient in practice and is theoretically equivalent to the former if the prediction errors can be assumed uncorrelated.

Performance criterion

Let t_c be the current time index. The control law for $t > t_c$ is designed with the aim to minimize the criterion

$$J = \frac{1}{T} E\left\{ \sum_{t=t_c+1}^{t_c+T} [(y_{(t)} - w_{(t)})^2 + \omega(\Delta u_{(t)})^2] + \sum_{t=t_c+T-\partial b+1}^{t_c+T} \omega_s (\Delta u_{(t)})^2 \right\} \quad (4)$$

where $\omega \geq 0$, $\omega_s > 0$ and T is the control horizon. The case of $T \to \infty$ is of particular interest. Note that not the inputs $u_{(t)}$ but only their increments $\Delta u_{(t)}$ are penalized for both predictors. The reason is to ensure that the mean of the control error $(y_{(t)} - w_{(t)})$ be zero in steady state for any fixed setpoint. The choice of ω is a performance-oriented 'knob' which makes it possible for the user to damp the movements of the actuator if it is required by the operating conditions. The auxiliary term, penalizing only the last control actions, is introduced in order to guarantee the asymptotic stability of the inputs also for $\omega = 0$. This makes the difference between the nonstabilized minimum variance control (Åström, 1970) and stabilized minimum variance control (Peterka, 1972) of a nonminimum-phase process. Note that the auxiliary term $(0 < \omega_s < \infty)$ is negligible for $T \to \infty$ but only when the input is stable.

Separation of identification and control

According to the certainty-equivalence hypothesis the control law is determined in each control step for the most recent predictor available, but under the assumption that its parameters remain constant within the control horizon T considered. This means that the criterion (4) is minimized under the simplifying assumption that the predictor parameters a_i, b_i, d_i are known and equal to their estimates based on the data observed up to and including the current time index t_c. However, to be able to find the control strategy minimizing the criterion (4) in a well-defined manner some additional assumptions concerning the prediction errors have to be adopted. A standard way would be to assume that the prediction errors are mutually uncorrelated random variables with zero mean (discrete white noise). Instead of this the following assumptions, which lead to the same result but are somewhat more general and better suited for the conditions of self-tuning control, will be made.

A1: For $t_c < t \leq t_c + T$

$$Ee_{(t)} = 0$$

$$E\tilde{u}_{(t)}e_{(\tau)} = 0, \quad t < \tau \leq t_c + T \tag{5}$$

where $\tilde{u}_{(t)}$ means $u_{(t)}$ or $\Delta u_{(t)}$, depending on whether the *P*-predictor (1) or the *I*-predictor (3) is considered, and is understood as a function, which is to be determined, of the data available up to and including the time index $(t - 1)$.

A2: The covariances of the prediction errors do not depend on the control strategy in the set of strategies from which the optimal one is to be selected, otherwise they can be arbitrary but finite.

In this way, by simplification based on the certainty-equivalance hypothesis, the self-tuning controller is decomposed algorithmically into the identification part and control-law-design part. Only the latter is discussed in detail in this paper. The unknown parameters of the predictor (1) or (3) can be identified in real time using recursive least squares (possibly modified by exponential forgetting and other tricks). However, for numerical reasons it is strongly recommended to use a square-root filter (Peterka, 1975; Peterka and Halousková, 1976), or its square-root free and somewhat faster variant (Bierman, 1977), for this purpose. The main modelling problem is to choose the structure of the predictor so that the assumptions A1 and A2 could be fulfilled, at least asymptotically, with good approximation. It is known that the linear least-squares regression, when applied to ergodic time series, produces residuals which are orthogonal with respect to the variables considered in the regressor. This indicates that the number of delayed inputs incorporated into the predictor (the number of *b*-parameters) may be most important. It well accords with practical experience. Practical experience also shows that in most cases of industrial process control the *I*-predictors should be preferred.

Measurable external disturbance

If the self-tuning controller includes the feed-forward from the measurable external disturbance v and the predictor contains a linear combination of past values $\{v_{(t-i)}: i = 1, 2, \ldots, \partial d\}$ then the minimization of the criterion (4) requires the adoption of a suitable model for the future development of the external disturbance for $t > t_c$. As in most practical industrial cases the measurable external disturbances are nonstationary it is appropriate to design the self-tuning controller for a 'random walk'

$$v_{(t)} = v_{(t-1)} + e_{v(t)}$$

$$E[e_{v(t)}|v_{(t-1)}, v_{(t-2)}, \ldots] = Ee_{v(t)} = 0. \tag{6}$$

The variances of the uncorrelated increments

$$\Delta v_{(t)} = e_{v(t)} \tag{7}$$

are not required to be known and can be arbitrarily time-varying but finite. The 'random walk' (6) generalized in such way is a very realistic model for load changes at unpredictable time instants, drifts, etc. Note that such a model does not need to be identified. In order to cover other cases also, an autoregressive model will be considered in the general development of the control-law design procedure in the next section [see (13)].

Command signal

Because the criterion (4) depends on the command signal $w_{(t)}$ for $t > t_c$ it is necessary to introduce a suitable model for its future development and to define on what values of w an admissible control law can operate. The following typical situations will be considered.

(A) *Fixed setpoint.* If the process output y is measured as the deviation of the given setpoint, then

$$w_{(t)} = 0, \quad t > t_c$$

(B) *Program control.* The command signal is deterministically preprogrammed and the entire sequence of the future setpoints $\{w_{(t)}: t = t_c + 1, t_c + 2, \ldots, t_c + T\}$ is *a priori* known and available for the controller.

(C) *Positional servo.* The future course of the setpoints is *a priori* uncertain and is modelled as a random walk generalized in the above sense

$$w_{(t)} = w_{(t-1)} + e_{w(t)}. \tag{8}$$

It is assumed that $w_{(t)}$ (the desired value of $y_{(t)}$) is available when $u_{(t)}$ (preceding the output $y_{(t)}$) is computed.

(D) *Model following.* It is required that the process output follow the output of a reference model introduced by the relation

$$w_{(t)} + \sum_{i=1}^{\partial a_m} a_{m,i} w_{(t-i)} = \sum_{i=0}^{\partial b_m} b_{m,i} w_{c(t-i)} \tag{9}$$

where w_c is the input of the reference model (the overall command signal) modelled as a generalized random walk

$$w_{c(t)} = w_{c(t-1)} + e_{wc(t)} \tag{10}$$

and $a_{m,i}$, $b_{m,i}$ are the reference-model parameters chosen by the designer (usually so that the steady-state gain is equal to one). It is assumed that $w_{(t)}$ and $w_{c(t)}$ are available when $u_{(t)}$ is computed. There are two situations when it is appropriate to formulate the control problem as the model-following problem: (1) in case of unpredictable step-wise setpoint changes [which might be modelled by the generalized random walk (8)] the minimization of the quadratic criterion may lead to overshoots which may be undesirable. This disadvantage of the quadratic criterion can be, as a rule, removed by the reference model (9) of first order

$$w_{(t)} = \alpha w_{(t-1)} + (1-\alpha) w_{c(t)} \tag{11}$$

with just a single adjustable parameter α ($0 < \alpha < 1$). (2) If the overall command signal w_c is generated by a human operator (like in aircraft applications) it may be required to maintain the dynamic properties of the closed control loop as constant as possible and close to a nominal case prescribed by the reference model (9).

3. CONTROL SYNTHESIS

Notational conventions

In order to cover all cases discussed above, the following forms of the process models will be considered

$$y_{(t)} + \sum_{i=1}^{\partial \tilde{a}} \tilde{a}_i y_{(t-i)} = \sum_{i=0}^{\partial b} b_i \tilde{u}_{(t-i)} + \sum_{i=1}^{\partial d} d_i \tilde{v}_{(t-i)} + k + e_t \tag{12}$$

$$\tilde{v}_{(t)} + \sum_{i=1}^{\partial a_v} \tilde{a}_{v,i} \tilde{v}_{(t-i)} = e_{v(t)} \tag{13}$$

where for the P-predictor (1) $\tilde{u} = u$, $\tilde{v} = v$, $\partial \tilde{a} = \partial a$, $\tilde{a}_i = a_i$ and according to (6) $\partial \tilde{a}_v = 1$, $\tilde{a}_{v,1} = -1$; for the I-predictor (3) $\tilde{u} = \Delta u$, $\tilde{v} = \Delta v$, $\partial \tilde{a} = \partial a + 1$, $\tilde{a}_1 = a_1 - 1$, $\tilde{a}_{\partial a} = -a_{\partial a}$, $\tilde{a}_i = a_i - a_{i-1}$ ($1 < i < \partial a$), $k = 0$ and according to (7) $\partial \tilde{a}_v = 0$.

Operating with elements of square matrices the following indexing will appear convenient

$$M_{t,t-i} = M_{i(t)}$$

where i indicates the distance from the main diagonal to the left for the matrix element appearing in the row (t). Thus, for a *lower triangular* (LT) matrix it holds that $M_{i(t)} = 0$ for $i < 0$ and all t. We shall say that the matrix M is *strictly lower triangular* (SLT) if $M_{i(t)} = 0$ for $i \le 0$. The LT-matrix M for which it holds that $M_{i(t)} = 0$ for $i > \partial M$ will be called *lower band* (LB) matrix of *degree* ∂M. A LB-matrix with zero diagonal elements will be denoted as a *strictly lower band* (SLB) matrix. For a diagonal matrix the index i is superfluous and will be omitted. The matrix will be called *row-stationary* (RS) if it holds that $M_{i(t)} = M_i$ for all t considered.

It is also useful to introduce the matrix

$$\Theta = \begin{bmatrix} 1 & & & & \\ -1 & 1 & & \bigcirc & \\ & -1 & 1 & & \\ & & & 1 & \\ \bigcirc & & & -1 & 1 \end{bmatrix}. \tag{14}$$

General solution

The system of T equations obtained from (12) and (13) for $t = t_c + 1, t_c + 2, \ldots, t_c + T$ can be written in the following matrix form

$$\tilde{A} y = B\tilde{u} + D\tilde{v} + e + k^* \tag{15}$$

$$\tilde{A}_v \tilde{v} = e_v + k_v^* \tag{16}$$

where

$$y = \begin{bmatrix} y_{(t_c+1)} \\ y_{(t_c+2)} \\ \vdots \\ y_{(t_c+T)} \end{bmatrix}, \quad \tilde{u} = \begin{bmatrix} \tilde{u}_{(t_c+1)} \\ \tilde{u}_{(t_c+2)} \\ \vdots \\ \tilde{u}_{(t_c+T)} \end{bmatrix}$$

and similarly for $\tilde{v}$, e and e_v. $\tilde{A}$, B, D and $\tilde{A}_v$ are RS-LB-matrices of degrees $\partial \tilde{A} = \partial \tilde{a}$, $\partial B = \partial b$, $\partial D = \partial d$ and $\partial \tilde{A}_v = \partial \tilde{a}_v$, respectively. $\tilde{A}_{0(t)} = 1$, $\tilde{A}_{i(t)} = \tilde{a}_i$, $B_{i(t)} = b_i$, etc. in the appropriate range of the index i. The vectors k^* and k_v^* are defined as follows:

$$k^* = \begin{bmatrix} k + s_{1(t_c)} \\ \vdots \\ k + s_{n(t_c)} \\ k \\ \vdots \\ k \end{bmatrix}, \quad k_v^* = \begin{bmatrix} s_{v,1(t_c)} \\ \vdots \\ s_{v,m(t_c)} \\ 0 \\ \vdots \\ 0 \end{bmatrix} \tag{17}$$

where $n = \max(\partial \tilde{a}, \partial b, \partial d)$, $m = \partial \tilde{a}_v$

$$s_{j(t_c)} = \sum_{i=j}^{n} (-\tilde{a}_i y_{(t_c+j-i)} + b_i \tilde{u}_{(t_c+j-i)} + d_i \tilde{v}_{(t_c+j-i)}) \tag{18}$$

$$s_{v,j(t_c)} = -\sum_{i=j}^{m} \tilde{a}_{v,i} \tilde{v}_{(t_c+j-i)}. \tag{19}$$

The criterion (4) can be written in the following vector-matrix form

$$J = \frac{1}{T} E\{(y-w)'(y-w) + \tilde{u}'\tilde{\Theta}'\Omega\tilde{\Theta}\tilde{u}\} \tag{20}$$

where $(\cdot)'$ means transposition, Ω is a diagonal matrix

$$\Omega_{(t)} = \omega \quad \text{for } t_c < t \le t_c + T - \partial b$$

$$\Omega_{(t)} = \omega + \omega_s \quad \text{for } t_c + T - \partial b < t \le t_c + T$$

and

$$\tilde{\Theta} = \Theta \quad \text{for the } P\text{-predictor}$$

$$\tilde{\Theta} = I \quad \text{for the } I\text{-predictor.}$$

In case of the P-predictor the term

$$\omega(u_{(t_c)}^2 - 2u_{(t_c)} u_{(t_c)}) \tag{21}$$

is omitted in (20) because it is insignificant in the sequal.

To make the main results easier to survey, the control synthesis minimizing the criterion (20) will be presented first as Result 1. Its constructive proof, and the optimization method itself, will be given in the next Section 4.

Result 1. *Finite control horizon:* The control law minimizing the criterion (20) for the process models (12) and (13) has the form

$$\tilde{u}_{(t)} = -\Delta_{(t)}^{-1}\left[\sum_{i=1}^{\partial R} R_{i(t)}\tilde{u}_{(t-i)} + \sum_{i=1}^{\partial S} S_{i(t)}y_{(t-i)} + \sum_{i=1}^{\partial Q} Q_{i(t)}\tilde{v}_{(t-i)} + x_{k(t)} - x_{w(t)}\right] \tag{22}$$

and is obtained as the tth row, $t > t_c + \max(\partial\tilde{a}, \partial b, \partial d, \partial\tilde{a}_v)$, of the vector-matrix relation

$$\tilde{u} = -\Delta^{-1}[R\tilde{u} + Sy + Q\tilde{v} + x_k - x_w] \tag{23}$$

where R, S and Q are SLB-matrices and Δ is diagonal. The matrices can be determined in the following way. Compose the symmetric band matrix

$$B'B + \bar{A}'\Omega\bar{A} = M \tag{24}$$

where

$$\bar{A} = \tilde{\Theta}\tilde{A} = \Theta A \tag{25}$$

and perform its factorization

$$M = \Phi'\Delta\Phi \tag{26}$$

where Φ is a LB-matrix with unit diagonal elements. Determine the SLB-matrices S, H and the LB-matrices F, G in the matrix equalities

$$B' = \Phi'S + F'\tilde{A} \tag{27}$$

$$B'D = \Phi'H + G'\tilde{A}_v\tilde{A}. \tag{28}$$

The SLB-matrices R and Q are determined by the strictly lower parts of matrix equalities

$$\Phi'R \doteq F'B + \bar{A}'\Omega\Theta - \Phi'\Delta \tag{29}$$

$$\Phi'Q \doteq F'D - G'\tilde{A}_v \tag{30}$$

where $\doteq$ indicates that only the equality of SL – parts is significant.

Algorithms performing this task are described below. They operate recursively backwards with respect to natural time (for $t = t_c + T, t_c + T - 1, t_c + T - 2, \ldots$), can run in parallel, require only a fixed memory size (depending on degrees $\partial\tilde{a}$, ∂b, ∂d and $\partial\tilde{a}_v$, but not on T) and exhibit a high numerical stability.

The term $x_{k(t)}$ in the optimal control law (22) is zero for I-predictors where $k = 0$. For P-predictors with $k \neq 0$ this term is nonzero, in general, and can be calculated also recursively for decreasing t according to the 'upper-band' set of equations

$$\Phi'x_k = F'k^* + G'k_v^*. \tag{31}$$

The term $x_{w(t)}$, by which the command signal is incorporated into the control law (22), is obtained as follows:

(*A*) *Fixed setpoint.* The output $y_{(t)}$ is understood as the deviation from the given setpoint and

$$x_{w(t)} = 0.$$

(*B*) *Program control.* The term $x_{w(t)}$ is the output of the filter operating on the future course of the command signal w backwards with respect to natural time according to the relation

$$\Phi'x_w = B'w. \tag{32}$$

(*C*) *Positional servo.*

$$x_{w(t)} = \Gamma_{(t)}w_{(t)} \tag{33}$$

where $\Gamma_{(t)}$ is the element of the diagonal matrix Γ determined by the matrix equation

$$B' = \Phi'\Gamma + \Lambda'\Theta \tag{34}$$

where Λ is an auxiliary SLB-matrix. The matrix equation (34) is solved again recursively for decreasing t.

(*D*) *Model following.* The term $x_{w(t)}$ is obtained as the output of the filter operating on the overall command signal w_c in real-time in the following way

$$s_{m(t)} = -\sum_{i=1}^{\partial a_m} a_{m,i}s_{m(t-i)} + w_{c(t)}$$
$$x_{w(t)} = \sum_{i=0}^{\partial\Gamma_m} \Gamma_{m,i(t)}s_{m(t-i)}. \tag{35}$$

The coefficients $\Gamma_{m,i(t)}$ are obtained as the elements of the LB-matrix Γ_m defined by the matrix equation

$$B'B_m = \Phi'\Gamma_m + \Lambda_m'\Theta A_m \tag{36}$$

which is solved recursively for decreasing t. Λ_m is an auxiliary SLB-matrix, A_m and B_m are RS LB-matrices with the elements $A_{m,i(t)} = a_{m,i(t)}$, $B_{m,i(t)} = b_{m,i}$, respectively.

Remark 1. Recall that in the important case of an incremental predictor (3), when combined with the external disturbance modelled as a generalized random walk (7), $\tilde{A}_v = I$, $\tilde{\Theta} = I$. In this case only the first matrix terms on the right-hand sides of (29) and (30) have nonzero elements under the main diagonal. Omitting the insignificant terms the relations

$$\Phi'R \doteq FB, \qquad \Phi'Q \doteq F'D \tag{37}$$

are sufficient to determine R and Q. However, then neither the matrix G nor H are required and the decomposition (28) does not need to be performed.

Remark 2. The optimal control law (22) has been given only for $t > t_c + \max\ (\partial\tilde{a},\ \partial b,\ \partial d,\ \partial a_v)$. However, this does not restrict its generality because, for the known model parameters, the time index t_c can be shifted arbitrarily backwards. Nevertheless, it may be interesting to note that the first row in (23) gives the optimal control law operating on the state the components of which are defined, for $t = t_c$, by (18) and (19). For instance, for the case discussed in Remark 1 the first row in (23) gives

$$\Delta u_{(t_c+1)} = -\Delta^{-1}_{(t_c+1)}\,[x_{k(t_c+1)} + x_{w(t_c+1)}]$$

where, according to (31) and (17)

$$x_{k(t_c+1)} = \sum_{i=1}^{n} \Psi_{i-1(t_c+1)} s_{i(t_c)}$$

and $\Psi_{i(t_c+1)}$ are the elements in the first row of the matrix

$$\Psi' = (\Phi')^{-1} F'$$

Algorithms. The algorithm for matrix factorization (26) is well known (Bierman, 1977). As the 'upper-diagonal-lower' factorization is required the algorithm must start in the lower right-hand corner of the matrix equality (26) and runs backwards with respect to natural time (as in dynamic programming). An examination of the matrix decomposition (27) shows that also here the lower right-hand corner is the only place where the solution may start. The algorithm can be found by comparing the lower and upper parts of the matrix equality. Similarly, the algorithm to calculate the matrix R from the lower part of (29) can be derived. All the three algorithms can run in parallel and can be unified into a single Algorithm OCS given below. The algorithms for solving (28) and (30) are similar and therefore are omitted here. However, it is worthy to note that in the important special case, discussed in Remark 1, (28) is not required to be solved and $Q_{i(t)}$ can be calculated in the very same way as $R_{i(t)}$, just replacing b_i by d_i [see (37)].

Algorithm OCS (optimum control synthesis). In order to unify the algorithm let $n = \max\ (\partial b,\ \partial a + 1) = \partial\Phi$ and set zeros for parameter values where required. The algorithm operates backwards with respect to natural time with zero initial conditions for $t > t_c + T$ and for all calculated variables

$$\Delta_{(t)} = M_{0(t)} - \sum_{j=1}^{n} \Delta_{(t+j)}\Phi^2_{j(t+j)}$$

$$F_{0(t)} = b_0 - \sum_{j=1}^{n} (S_{j(t+j)}\Phi_{j(t+j)} + \tilde{a}_j F_{j(t+j)})$$

for $i = 1$ to n

$$\Phi_{i(t)} = \Delta^{-1}_{(t)}\left(M_{i(t)} - \sum_{j=1}^{n-i} \Phi_{j(t+j)}\Delta_{(t+j)}\Phi_{i+j(t+j)}\right)$$

$$F_{i(t)} = b_i - \sum_{j=1}^{n-i} (S_{j(t+j)}\Phi_{i+j(t+j)} + \tilde{a}_j F_{i+j(t+j)})$$

$$S_{i(t)} = -\tilde{a}_i F_{0(t)} - \sum_{j=1}^{n-i} (S_{i+j(t+j)}\Phi_{j(t+j)} + a_{i+j}F_{j(t+j)})$$

$$R_{i(t)} = b_i F_{0(t)} - \sum_{j=1}^{n-i} (R_{i+j(t+j)}\Phi_{j(t+j)} - b_{i+j}F_{j(t+j)})$$

$$R_{1(t)} := R_{1(t)} - \mu\omega$$

where $\mu = 1$ for the P-predictor and $\mu = 0$ for the I-predictor. Note that there is just a single division in the entire algorithm, namely by $\Delta_{(t)}$ in the factorization part. However, this number must be positive since the factorized matrix M is positive definite (except for the case when all b-parameters are zero and $\omega = 0$, i.e. when any control has no sense and the input is allowed to be arbitrary). The only potential numerical danger is the subtraction of positive numbers in the first row the result of which must be the positive number $\Delta_{(t)}$. This drawback can be removed by a modification of the algorithm developed by my colleague K. Šmuk. The idea of this modification is to transform the left-hand side of (24) directly into the right-hand side of (26) using an algorithm similar to the Bierman's UD-filter which guarantees the positive definiteness numerically without calculating the matrix M explicitly. Details will be reported elsewhere. Note also that the algorithm, when coded for some fixed $n_{\max}$, can solve all cases for $n \leq n_{\max}$.

Asymptotic solution

The solution for the infinite control horizon can be found, if it exists, by increasing the number of iterations in the Algorithm OCS until the stationarity is reached. However, in low-degree cases it may be more convenient to look for the stationary solution directly.

It is well known that the multiplication of RS–LB-matrices is equivalent to multiplication of polynomials. Hence, it is appropriate to look for the stationary solution in terms of polynomial equations. This can be done very simply by replacing the LB-matrices in the matrix relations (24) to (30) by polynomials introduced as follows: $\tilde{a} = \tilde{a}(\zeta) = 1 + \tilde{a}_1\zeta + \cdots + \tilde{a}_{\partial a}\zeta^{\partial a}$ in place of $\tilde{A}$, and in general

$$p = p(\zeta) = p_0 + p_1\zeta + \cdots + p_{\partial p}\zeta^{\partial p}$$

in place of a LB-matrix P. Note that $p(0) = p_0$,

$p(1) = p_0 + p_1 + \cdots + p_{\partial p}$ and that, according to (14), $\theta(\zeta) = 1 - \zeta$, and according to (25) for both P- and I-predictors

$$\bar{a} = \bar{a}(\zeta) = \tilde{\theta}(\zeta)\tilde{a}(\zeta) = (1 - \zeta)a(\zeta). \qquad (38)$$

It can be verified by inspection that, when considering the row-stationary zone in the matrix relations (24) to (30), the transpose P' must be replaced by the polynomial which can be called, with some abuse of language, reciprocal

$$p' = p(\zeta^{-1}) = p_0 + p_1\zeta^{-1} + \cdots + p_{\partial p}\zeta^{-\partial p}$$

Result 2. *Infinite control horizon:* If ζ is understood as the one-step-delay operator, e.g. $\zeta\tilde{u}_{(t)} = \tilde{u}_{(t-1)}$, then the stationary control law minimizing the criterion (4) for $T \to \infty$, if it exists, can be written in the form

$$\tilde{u}_{(t)} = -\delta^{-1}[r\tilde{u}_{(t)} + sy_{(t)} + q\tilde{v}_{(t)} + x_k - x_{w(t)}] \quad (39)$$

where the scalar δ and the polynomials r, s, q (with $r_0 = s_0 = q_0 = 0$) are determined by the polynomial equations which follow from RS-zones of matrix equations (24)–(30).

$$b'b + \bar{a}'\omega\bar{a} = m = \phi'\delta\phi \qquad (40)$$

$$b' = \phi's + f'\tilde{a} \qquad (41)$$

$$b'd = \phi'h + g'\tilde{a}_v\tilde{a} \qquad (42)$$

$$\phi'(\delta + r) \underline{+} f'b + \bar{a}'\omega\tilde{\theta} \qquad (43)$$

$$\phi'q \underline{+} f'd - g'\tilde{a}_v \qquad (44)$$

where $\underline{+}$ means that only the equality of terms with positive powers of ζ is considered. The polynomial ϕ in the factorization (40) has no roots inside the unit circle and is normalized so that $\phi_0 = 1$.

The constant term x_k in (39) is zero for the I-predictor while for the P-predictor

$$x_k = \frac{f(1)}{\phi(1)}k. \qquad (45)$$

The term $x_{w(t)}$ in the control law (39) is obtained as follows.

(*A*) *Fixed setpoint*. The output y means the deviation from the given setpoint and $x_{w(t)} = 0$.

(*B*) *Program control*. The term $x_{w(t)}$ is obtained as the output of the filter b/ϕ by which the future command signal is filtered backwards with respect to natural time

$$x_{w(t)} = \frac{b'}{\phi'}w_{(t)} \qquad (46)$$

(*C*) *Positional servo*. The term $x_{w(t)}$ depends only on the recent setpoint

$$x_{w(t)} = \frac{b(1)}{\phi(1)}w_{(t)} \qquad (47)$$

(*D*) *Model following*. The term $x_{w(t)}$ is generated by the filter operating on the overall command signal w_c in real time

$$x_{w(t)} = \frac{\gamma_m}{a_m}w_{c(t)}. \qquad (48)$$

The polynomial γ_m is determined by the polynomial equation

$$b'b_m = \phi'\gamma_m + \lambda_m'\theta a_m \qquad (49)$$

where $\lambda_m(0) = \lambda_{m,0} = 0$.

Remark 3. In the important special case discussed in Remark 1 $\tilde{a}_v = 1$, $\tilde{\theta} = 1$. Omitting the insignificant terms the relations (43) and (44) read

$$\phi'r \underline{+} f'b, \qquad \phi'q \underline{+} f'd.$$

However, then the (42) is superfluous and it is advantageous to proceed as follows. Perform $v = \max(\partial b, \partial d)$ steps of the polynomial division $f/\phi = \psi_0 + \psi_1\zeta + \cdots + \psi_v\zeta^v + \cdots$ to obtain the polynomial ψ, $(\partial\psi = v)$. The coefficients r_i and q_i of the control law are easily determined by comparing the terms with positive powers of ζ in the relations

$$r \underline{+} \psi'b, \qquad q \underline{+} \psi'd. \qquad (50)$$

The numbers ψ_i are, actually, the parameters of the stationary control law operating on the state introduced for $t = t_c$ by (18), see Remark 2.

Remark 4. If the order of the factorization (40) is not higher than 2 then the following simple formulae can be used to perform it explicitly

$$\lambda = m_0/2 - m_2 + \sqrt{[(m_0/2 + m_2)^2 - m_1^2]}$$

$$\delta = [\lambda + \sqrt{(\lambda^2 - 4m_2^2)}]/2$$

$$\phi_1 = m_1/(\delta + m_2), \qquad \phi_2 = m_2/\delta.$$

If the sampling period is chosen long enough compared to the significant part of the system step response then such low-order predictors appear satisfactory for a rather broad class of practical cases. However, then also the solution of (41)–(44) can be brought up to simple explicit formulae which can be well implemented on a microprocessor. For instance, a self-tuning controller based on the I-predictor with the structure $\partial a = 1$ ($\partial\tilde{a} = 2$), $\partial b = 2$, $\partial d = 2$, $\partial\tilde{a}_v = 0$ can well manage many practical situations if the number of samples within the transient part of the step response is, say, up to six (but often even more). An important feature of the procedure is that, unlike other methods of numerical synthesis, the only singularity which might be encountered is the unavoidable one discussed below.

Remark 5. Inspection of (40) and (41) shows that (41) does not have any solution if the polynomials $\tilde{a}$

and b have an unstable common factor. In this case the stationary solution does not exist. The algorithm OCS (its decomposition part) does not converge, but for a finite horizon T it still retains its correct physical meaning. However, the process is not stabilizable.

Remark 6. In case of model following with a reference model of first order (11) the polynomial γ_m in (48) is

$$\gamma_m = \gamma_{m,0} + \gamma_{m,1}\zeta.$$

Its coefficients can be easily determined using the formulae

$$\gamma_{m,0} = \frac{b(1)}{\phi(1)} - \alpha\frac{b(\alpha)}{\phi(\alpha)}, \quad \gamma_{m,1} = \alpha\left[\frac{b(\alpha)}{\phi(\alpha)} - \frac{b(1)}{\phi(1)}\right]$$

which follow from the polynomial equation (49) with $a_m(\zeta) = 1 - \alpha\zeta$, $b_m(\zeta) = b_{m,0} = 1 - \alpha$.

Remark 7. As in case of an I-predictor $\tilde{a}(1) = 0$, it follows from (41) that

$$\frac{b(1)}{\phi(1)} = s(1) = \sum_i s_i.$$

Then the control law (39) with $x_{w(t)}$ determined by (47) can be given the form

$$\Delta u_{(t)} = -\delta^{-1}\left[\sum_i r_i \Delta u_{(t-i)} + \sum_i s_i(y_{(t-i)} - w_{(t)}) + \sum_i q_i \Delta v_{(t-i)}\right] \tag{51}$$

operating only on the time increments of u and v and on the deviations of the past actual outputs from the present setpoint.

Remark 8. The self-tuning controller designed for program control must have a buffer where a finite number, say T_w, of the next setpoints are stored. To achieve the desired effect the memory of the buffer does not need to be large. As a rule, $T_w < 10$ is fully sufficient. Then it is natural to start the backward filtering (46) with the initial conditions

$$w_{(t+T_w+i)} = w_{(t+T_w)}$$
$$x_{w(t+T_w+i)} = \frac{b(1)}{\phi(1)} w_{(t+T_w)} \qquad \text{for } i > 0.$$

This is, actually, the precise solution for the case when the uncertainty of $w_{(t+k)}$ for $k > T_w$ is modelled as the generalized random walk (8). The interested reader, who will perform some simulations for himself, will appreciate the favourable and sometimes amazing effect of the term $x_{w(t)}$ calculated in this way. For instance, for $\omega = 0$ the sampled output of a deterministic system follows precisely any *a priori* given discrete command signal even when the system is of non-minimum phase. When the system input is damped by penalization of its increments, $\omega > 0$, then the controller naturally starts to follow a sudden change in the setpoint earlier than it really occurs in order to avoid large movements of the actuator. This makes the transition smoother and easier to realize.

4. OPTIMIZATION METHOD

In this section Result 1, from which the Result 2 directly follows, will be derived. The principle idea of the method is a suitable prearrangement of the criterion (20) into two nonnegative components the first one of which can be zeroed in one shot while the second one does not depend on the strategy applied and represents the attainable minimum of the criterion.

Recall that the matrices $\tilde{A}$ and $\tilde{A}_v$ in (15) and (16) are lower triangular with unit diagonal elements and therefore invertible. Moreover, the RS–LT-matrices commute

$$\begin{aligned} B\tilde{A} &= \tilde{A}B, & \tilde{A}^{-1}B &= B\tilde{A}^{-1} \\ DA &= \tilde{A}D, & \tilde{A}^{-1}D &= D\tilde{A}^{-1} \end{aligned} \tag{52}$$

which can be easily verified by inspection. Making use of these properties the vector of control errors $(y - w)$ and the criterion (20) can be expressed in the following way

$$\begin{aligned} y - w &= B\tilde{A}^{-1}\tilde{u} + z \\ z &= \tilde{A}^{-1}(e + k^*) + D\tilde{A}^{-1}A_v^{-1}(e_v + k_v^*) - w \end{aligned} \tag{53}$$

$$J = \frac{1}{T}E\{\tilde{A}^{-1}\tilde{u})'M(\tilde{A}^{-1}\tilde{u}) + 2(\tilde{A}^{-1}\tilde{u})'B'z + z'z\}$$

where M is the symmetric band matrix (24). This matrix is positive definite (except the very special and uninteresting case when simultaneously $B = 0$ and $\omega = 0$ and any control has no sense) and therefore can be uniquely factorized according to (26). Applying this factorization the rearrangement of the criterion can continue as follows:

$$J = \frac{1}{T}E\{\bar{u}'\Delta\bar{u} + 2\bar{u}'(\phi')^{-1}B'z + z'z\} \tag{54}$$

where

$$\bar{u} = \Phi\tilde{A}^{-1}\tilde{u}. \tag{55}$$

Note that $\bar{u}_{(t)}$ is a linear combination of $\{\tilde{u}_{(\tau)}; \tau \leq t\}$ but, due to the lower triangularity of $\Phi\tilde{A}^{-1}$, does not contain $\tilde{u}_{(\tau)}$ for $\tau > t$. This fact will appear very important. It is also the reason why the factorization (26) has been performed in the 'upper-diagonal-lower' form.

The cross-term in (54) can be written, using (53)

$$E\bar{u}'(\phi)^{-1}B'z = E\{\bar{u}'(\Phi')^{-1}B'\tilde{A}^{-1}(e + k^*) + \bar{u}'(\Phi')^{-1}B'D\tilde{A}^{-1}\tilde{A}_v^{-1}(e_v + k_v^*) - \bar{u}'(\Phi')^{-1}B'w\}. \quad (56)$$

Now the main trick of the method follows. Denote for a while

$$(\Phi')^{-1}B'A^{-1} = P \quad (57)$$

and decompose this $T \times T$-matrix into the strictly lower part and the upper part

$$P = P_+ + P_- \quad (58)$$

where the SLT-matrix P_+ has zeros on and above the main diagonal whereas P_- has zeros under the main diagonal. The point, and the reason for the decomposition, is that

$$E\bar{u}'P_-e = 0$$

whatever P_- might be. This is due to the fact that this expression contains the products $\tilde{u}_{(t)}e_{(\tau)}$ only for $\tau > t$ the mean values of which are zero according to the assumption A1, equation (5). Hence

$$E\bar{u}'Pe = E\bar{u}'P_+e.$$

It is more convenient to consider the decomposition (58) of (57) in the following equivalent form

$$(\Phi')^{-1}B'\tilde{A}^{-1} = S\tilde{A}^{-1} + (\Phi')^{-1}F' \quad (59)$$

$$B' = \Phi'S + F'\tilde{A} \quad (60)$$

where (60) coincides with (27). Note that the matrix relation (60) determines the SLB-matrix S and the LB-matrix F uniquely, including their orders. In this way the first bilinear form on the right-hand side of (56) can be expressed as follows:

$$E\bar{u}'(\Phi')^{-1}B'\tilde{A}^{-1}(e + k^*) = E\bar{u}'[S\tilde{A}^{-1}(e + k^*) + (\Phi')^{-1}F'k^*]. \quad (61)$$

In a very similar way also the second term on the right-hand side of (56) can be rearranged

$$(\Phi')^{-1}B'D(\tilde{A}_v\tilde{A})^{-1} = H(\tilde{A}_v\tilde{A})^{-1} + (\Phi')^{-1}G'. \quad (62)$$

This decomposition leads to the matrix equality (28) by which the SLB-matrix H and the LB-matrix G are determined. Recall that the sequence $\{v_{(t)}\}$ has been introduced as a disturbance which is external with respect to the control loop. This means that

$$E\bar{u}_{(t)}e_{v(\tau)} = 0, \quad \text{for } \tau \geq t$$

and in analogy with (61)

$$E\bar{u}'(\Phi')^{-1}B'D\tilde{A}^{-1}\tilde{A}_v^{-1}(e_v + k_v^*) = E\bar{u}'[H\tilde{A}^{-1}\tilde{A}_v^{-1}(e_v + k_v^*) + (\phi')^{-1}G'k_v^*]. \quad (63)$$

The suitable arrangement of the last term on the right-hand side of (56) depends on the way how the command signal w is generated and made available for the controller. To proceed generally at this stage express this term as follows:

$$E\bar{u}(\Phi')^{-1}B'w = E\bar{u}'x_w \quad (64)$$

and assume that $x_{w(t)}$ is available when $u_{(t)}$ is computed. The determination of the vector x_w for the four situations (A, B, C, D) will be discussed separately.

Making use of (61), (63) and (64) the bilinear term (56) of the quadratic criterion can be written as follows:

$$E\bar{u}'(\Phi')^{-1}B'z = E\bar{u}'x \quad (65)$$

where

$$x = S\tilde{A}^{-1}(e + k^*) + H\tilde{a}^{-1}\tilde{A}_v^{-1}(e_v + k_v^*) + x_k - x_w \quad (66)$$

and in coincidence with (31)

$$x_k = (\Phi')^{-1}(F'k^* + G'k_v^*). \quad (67)$$

Now, by completion of squares, the criterion (54) can be given the form

$$J = \frac{1}{T}E[\bar{u}'\Delta\bar{u} + 2\bar{u}'x + z'z]$$

$$= \frac{1}{T}E[(\bar{u} + \Delta^{-1}x)'\Delta(\bar{u} + \Delta^{-1}x) + z'z - x'\Delta^{-1}x]$$

$$= \frac{1}{T}E\left[\sum_{t=t_c+1}^{t_c+T} \Delta_{(t)}\left(\bar{u}_{(t)} + \Delta_{(t)}^{-1}x_{(t)}\right)^2\right] + J_0 \quad (68)$$

where

$$J_0 = \frac{1}{T}E[z'z - x'\Delta^{-1}x] \quad (69)$$

is the component of the criterion which cannot be influenced by any admissible control strategy. At the same time J_0 is the minimum of J which is attained by the control strategy satisfying the relation

$$\bar{u} + \Delta^{-1}x = 0. \quad (70)$$

Substitution for $\bar{u}$ and x from (55) and (66), respectively, gives

$$\Phi\tilde{A}^{-1}\tilde{u} + \Delta^{-1}[S\tilde{A}^{-1}(e + k^*) + H\tilde{A}^{-1}\tilde{A}_v^{-1}(e_v + k_v^*) + x_k - x_w] = 0. \quad (71)$$

Note that the control strategy fulfilling this relation is realizable as each $\tilde{u}_{(t)}$ is generated as a linear combination of the past inputs, of the past prediction errors $\{e_{(\tau)}, e_{v(\tau)}; t_c < \tau < t\}$ which are known at the moment, and of the initial state contained in k^* and k_v^* (17). However, this is not a very practical control law. Therefore express $\tilde{A}^{-1}$

$(e + k^*)$ and $\tilde{A}_v^{-1}$ $(e_v + k_v^*)$ from (15) and (16), respectively, and substitute into (71) to obtain

$$(\Phi - \Delta^{-1}SB)\tilde{A}^{-1}\tilde{u} + \Delta^{-1}[Sy + (H - SD)\tilde{A}^{-1}\tilde{v} + x_k - x_w] = 0$$

or equivalently

$$\tilde{u} + \Delta^{-1}[R\tilde{u} + Sy + Q\tilde{v} + x_k - x_w] = 0 \quad (72)$$

where

$$R = (\Delta\Phi - SB)\tilde{A}^{-1} - \Delta \quad (73)$$

$$Q = (H - SD)\tilde{A}^{-1}. \quad (74)$$

Note that the structure of the relation (72) coincides with (23) but from (73) and (74) it is not directly seen that the SLT-matrices R and S are also band limited. To show this calculate S from (60) and substitute into the right-hand sides of (73) and (74)

$$R = (\Phi')^{-1}[(\Phi'\Delta\Phi - B'B)\tilde{A}^{-1} + F'B] - \Delta$$

$$Q = (\Phi')^{-1}[(\Phi'H - B'D)\tilde{A}^{-1} + F'D].$$

From the matrix relations (24) and (26) and from (28) it is readily seen that

$$\Phi'\Delta\Phi - B'B = \tilde{A}'\Omega\tilde{A} = \tilde{A}'\Omega\tilde{\Theta}\tilde{A}$$

$$\Phi'H - B'D = -G'\tilde{A}_v\tilde{A}.$$

In this way it is finally obtained

$$R = (\Phi')^{-1}[F'B + \tilde{A}'\Omega\tilde{\Theta}] - \Delta \quad (75)$$

$$Q = (\Phi')^{-1}[F'D - G'\tilde{A}_v]. \quad (76)$$

Now it is already apparent that, in general, $\partial R = \max(\partial B, \partial\tilde{\Theta})$ and $\partial Q = \max(\partial D, \partial A_v)$. Note that the upper parts of the matrix equalities (75) and (76), including the main diagonals, are superfluous and must be identically zero. The matrix relations (29) and (30), which were to be proved, directly follow from (75) and (76).

It remains to derive the relations by which the vector x_w is determined.

(*A*) *Fixed setpoint*. In this trivial case $x_w = 0$ which does not need any comment.

(*B*) *Program control*. As the entire vector w is available for any time index t, $(t_c < t \leq t_c + T)$ it directly follows from (64) that

$$x_w = (\Phi')^{-1}B'w$$

which coincides with (32).

(*C*) *Positional servo*. The set of equations obtained from (8) for $t = t_c + 2, t_c + 3, \ldots, t_c + T$ can be written in the matrix-vector form

$$\Theta w = e_w \quad (77)$$

where, *pro forma*, $e_{w(t_c+1)} = w_{(t_c+1)}$. Substitution for w in (64) gives

$$E\bar{u}'x_w = E\bar{u}'(\Phi')^{-1}B'\Theta^{-1}e_w. \quad (78)$$

As w is by definition an external signal for the control loop it holds

$$E\bar{u}_{(t)}e_{w(\tau)} = 0, \quad \text{for } \tau > t. \quad (79)$$

It is important to realize that (79) does not hold for $\tau = t$. Thus, the same trick can be applied as before with the only difference that the matrix decomposition in (78)

$$(\phi')^{-1}B'\Theta^{-1} = \Gamma\Theta^{-1} + (\phi')^{-1}\Lambda' \quad (80)$$

$$B' = \phi'\Gamma + \Lambda'\Theta \quad (81)$$

must be performed so that the main diagonal of the upper part $(\phi')^{-1}\Lambda'$ be zero. Hence, Λ is a SLB-matrix. As $\partial\Theta = 1$ it can be seen from (81) that Γ is diagonal. Substitution of (80) into (78) gives

$$E\bar{u}'x = E\bar{u}'\Gamma\Theta^{-1}e_w = E\bar{u}'\Gamma w$$

which proves (33).

(*D*) *Model following*. As the reference model (9) must be chosen sufficiently stable it is possible and convenient to simplify the verification of the above given Results assuming the zero initial conditions for this model. With this simplification the system of T equations obtained from (9) for $t = t_c + 1, t_c + 2, \ldots, t_c + T$ can be written as follows:

$$A_m w = B_m w_c$$

where A_m and B_m are RS–LB-matrices with the elements $A_{m,i(t)} = a_{m,i}$ and $B_{m,i(t)} = b_{m,i}$, respectively. Similarly to (77) we also have

$$\Theta w_c = e_{wc}$$

Using these relations the bilinear term (64) can be expressed as follows:

$$E\bar{u}x_w = E\bar{u}'(\phi')^{-1}B'w$$

$$= E\bar{u}'(\phi')^{-1}B'B_m A_m^{-1}\Theta^{-1}e_{wc}$$

Proceeding in the very same way as in the case (C) the following matrix decomposition is applied

$$(\phi')^{-1}B'B_m A_m^{-1}\Theta^{-1} = \Gamma_m A_m^{-1}\Theta^{-1} + (\phi')^{-1}\Lambda_m' \quad (82)$$

where Λ_m is a SLB-matrix. This leads to the result

$$E\bar{u}'x_w = E\bar{u}'\Gamma_m A_m^{-1}\Theta^{-1}e_{wc} = E\bar{u}'\Gamma_m A_m^{-1}w_c$$

$$x_w = \Gamma_m A_m^{-1}w_c$$

or equivalently

$$x_w = \Gamma_m s_m, \qquad A_m s_m = w_c. \quad (83)$$

This verifies the real-time filter (35) while (82) is equivalent to the matrix relation (36) by which the LB-matrix Γ_m is determined.

5. CONCLUDING REMARKS

The matrix factorization (26) and the matrix decomposition (58) are the fundamental steps in the derivation of the presented method. Similar steps can be found also in the Wiener synthesis, namely spectrum factorization and partial fractioning, (see e.g. Newton, Gould and Kaiser, 1957 or Chang, 1961). In this sense the method can be viewed as a certain revival of the Wiener approach. However, here it is applied to finite time intervals and nonstationary situations.

Going through the derivation in Section 4 it can be seen that the feedforward paths from any number of measurable external disturbances can be synthetized in the very same way. Also the extension of the control-law synthesis for the case of preprogrammed loading, when one or more external variables v are *a priori* determined so that their future values $\{v_{(t+i)}, i > 0\}$ are available when $u_{(t)}$ is being determined, is simple and straightforward. Unfortunately, the extension for the multi-input multi-output case is not so easy. The difficulty is that the matrices $\tilde{A}$, B and D in (15) are no longer row-stationary but only block-row-stationary and the commutation (52) does not apply automatically. As this commutation is inherent for the method it must be performed using an additional algorithm. This makes the method less elegant and less suitable (compared to square-root dynamic programming) for multivariate cases.

It has been stated in Section 2 that in most cases of industrial process control the I-predictors should be preferred to P-predictors. There are two main arguments for this recommendation. First, the control law synthetized for an I-predictor generates the increments of the system input ($\tilde{u} = \Delta u$) so that the controller possesses an integral action. This accords with the well-tried feedback structure generally used in practice to remove the steady state bias and to compensate drifts and disturbances or load changes of relatively long duration which is, as a rule, the primary task of any regulator. The second but not less important argument in favor of I-predictors concerns the identification part of the self-tuning algorithm. Those who deal with self-tuning control systematically know that the plain application of the certainty-equivalence hypothesis in connection with exponential forgetting of old data may lead to the unfavourable phenomena called 'bursts' (Åström, 1979) and 'covariance wind-up' (Åström, 1980) if the closed control loop is not excited externally (for instance by setpoint changes) and if no special precautions are taken (Kárný, 1982). The main root of this difficulty is that the fixed linear control law, to which the selftuning algorithm eventually converges, creates a linear dependence on the data in the regressor used to update the estimates of the predictor parameters. An insight into this problem can be obtained from the simple Example 4.4 in Peterka (1981). The advantage of the I-predictor is that, unlike the P-predictor and other non-incremental models, the corresponding optimal control law does not create this linear dependence. For instance, the parameters of the I-predictor

$$\hat{y}_{(t)} = y_{(t-1)} + b_0 \Delta u_{(t)} + b_1 \Delta u_{(t-1)} + b_2 \Delta u_{(t-2)} - a_1 \Delta y_{(t-1)}$$

are estimated by recursive least squares applied to the observation equation

$$\Delta y_{(t)} = [b_0, b_1, b_2, -a_1] \rho_{(t)} + e_{(t)}$$

where $\rho_{(t)}$ is the regressor

$$\rho_{(t)} = [\Delta u_{(t)}, \Delta u_{(t-1)}, \Delta u_{(t-1)}, \Delta y_{(t-1)}]'.$$

The corresponding optimal structure of the control law, considered in the form (51) with the fixed setpoint $w_{(t)} = w$, can be written as follows:

$$[\delta, r_1, r_2, s_2] \rho_{(t)} = (s_1 + s_2)(y_{(t-1)} - w).$$

This clearly shows that any deviation of the output from the given setpoint perturbs the linear dependence within the regressor which helps to avoid the phenomena mentioned above.

The numerical and computational aspects were the main motive for the development of the presented method of control synthesis. To conclude the paper some additional comments will be given to this, for the self-tuning control ever so important, point.

As outlined in Remark (4) the polynomial equations (40)–(44) can be used to derive explicit formulae for the control synthesis in case of low order predictors. It may be interesting to compare, from the computational point of view, this procedure with the pole-placement technique used by Wellstead, Prager and Zanker (1979) and by Wittenmark and Åström (1980) for a similar purpose. In the polynomial equation (43) only the equalities of terms with positive powers of ζ are significant for the determination of the polynomial $r = r_1\zeta + r_2\zeta^2 + \cdots r_{\partial b}\zeta^{\partial b}$. However, the equation itself holds generally. The unknown polynomial f could be eliminated from the equations (43) and (41). If the former is multiplied by the polynomial $\tilde{a}$, the latter is multiplied by b, and the both equations are subtracted, then the following relation is obtained

$$\phi'[(\delta + r)\tilde{a} + sb] = b'b + \tilde{a}'\tilde{\Omega}'\omega\tilde{\Omega}\tilde{a}$$
$$= \phi'\delta\phi$$

where the second equality follows from (40). This means that for all ζ for which $\phi(\zeta^{-1}) \neq 0$ it holds

$$(\delta + r)\tilde{a} + sb = \delta\phi. \tag{84}$$

Since the left-hand side is the characteristic polynomial of the closed control loop, the relation can be understood as the placement of the poles to the positions prescribed by the polynomial ϕ which is optimal with respect to the quadratic criterion. A similar relation can be found in (Peterka, 1972) and might be used to determine the polynomials r and s directly. However, this is not a good way how to proceed. As pointed out in Wittenmark and Åström (1980) the relation (84) leads to formulae which are singular whenever the polynomials $\tilde{a}$ and b have a common factor (even when it is stable) and are numerically sensitive in case of an almost common factor. As discussed in Remark 4 the formulae obtainable via (41) and (43) do not have this defect and can be given the form which is universal for all situations when the stationary optimal control exists. Note also that (41) and (43) determine the calculated polynomials uniquely, including their orders, and do not allow the cancellation of stable common factors in $\tilde{a}$ and b which would lead to the nonoptimality of the controller in some transient situations.

In most practical cases it is possible to retain the model order sufficiently low (without violating the assumption A1, equation (5), significantly) by a proper choice of the sampling interval (and, if need be, by a suitable prefiltering of the controlled output). Simple robust self-tuning controllers obtainable in this way will be reported in a separate paper where also experimental results and further important details concerning the identification part will be given.

Acknowledgements—The author is indebted to his colleagues, namely Drs J. Böhm, A. Halousková and M. Kárný, for extensive cooperation both in discussing concepts and collecting practical experience. Also the impulses obtained as a feedback from industrial practice, namely from Drs A. Lízr and A. Lízrová (pulp and paper production), Dr J. Cendelín (rolling mills) and Ing. J. Fessl (power plants), are gratefully acknowledged.

REFERENCES

Åström, K. J. (1970). *Introduction to Stochastic Control Theory*. Academic Press.

Åström, K. J. (1979). Self-tuning regulators—design principles and applications. *Proceedings of Yale Workshop on Applications of Adaptive Control*. Academic Press.

Åström, K. J. (1980). Design principles for self-tuning regulators. *International Symposium on Methods and Applications in Adaptive Control*. Bochum, Springer.

Bierman, G. J. (1977). *Factorization Methods for Discrete Sequential Estimation*. Academic Press.

Chang, S. S. L. (1961). *Synthesis of Optimum Control Systems*. McGraw-Hill.

Clarke, D. W. and P. J. Gawthrop (1975). Self-tuning controller. *Proc. IEE*, **122**, 929.

Kárný, M. (1982). Identification with interruptions as an anti-bursting device, *Kybernetika*, **18**, 320.

Newton, G. C., L. A. Gould and J. F. Kaiser (1957). *Analytical Design of Linear Feedback Controls*. John Wiley.

Peterka, V. (1970). Adaptive digital regulation of noisy systems. *2nd IFAC Symposium on Identification and Process Parameter Estimation*. Paper 6.2, Academia, Prague.

Peterka, V. (1972). On steady-state minimum-variance control strategy. *Kybernetika*, **8**, 219.

Peterka, V. (1975). A square-root filter for real-time multivariate regression. *Kybernetika*, **11**, 53.

Peterka, V. and A. Halousková (1976). Effective algorithms for real-time multivariate regression. *Proceedings of 4th IFAC Symposium on Identification and System Parameter Estimation*, Tbilisi, Vol. 3, pp. 100–110.

Peterka, V. (1981). Bayesian approach to system identification. In P. Eykhoff (Ed.), *Trends and Progress in System Identification*, Chapter 8, Pergamon Press.

Wellstead, P. E., D. Prager, and P. Zanker (1979). Pole assignment self-tuning regulator. *Proc. IEE*, **126**, 781.

Wittenmark, B. and K. J. Åström (1980). Simple self-tuning controller. *International Symposium on Methods and Applications in Adaptive Control*. Bochum. Springer.

Practical Issues in the Implementation of Self-tuning Control*

BJÖRN WITTENMARK† and KARL JOHAN ÅSTRÖM†

Although self-tuning controllers are now being applied to many industrial processes, there are several practical issues that have to be considered in order to get a properly working algorithm.

Key Words—Adaptive control; control applications; process control; robustness; self-tuning regulators.

Abstract—Implementation aspects of self-tuning regulators are discussed in the paper. There is a large discrepancy between simulation or academic algorithms and practical algorithms. In the idealized environment of simulations it is easy to get different types of adaptive algorithms to perform well. In practice the situation is quite opposite. The adaptive or self-tuning controller must be able to handle nonlinearities, unmodelled dynamics and unmodelled disturbances over a wide range of operating conditions. Some aspects of how to implement self-tuning controllers are discussed in the paper. This includes robustness, signal conditioning, parameter tracking, estimator wind-up, reset action and start-up. Different ways to use the prior knowledge about the process are also discussed.

1. INTRODUCTION

A NON-SPECIALIST who tries to get an understanding of adaptive control is confronted with contradicting information such as:

- —Adaptive control works like a beauty in this particular practical application
- —Adaptive control is ridiculous. Look how the algorithm behaves in this simulation
- —You cannot possibly use adaptive control because there is no proper theory that guarantees stability and convergence of the algorithm.

Use of adaptive control is very tempting in many situations, especially if little is known about the process to be controlled. The sad fact of life is, however, that there are seldom any short cuts that really pay off. Nothing, not even an adaptive controller, can replace good engineering and physical insight. A newcomer to the field of adaptive control must find the situation somewhat confusing. There are large numbers of proposed algorithms. Some are quite different while others only differ in minor details. Are the differences important or do they just reflect personal preferences of the designer?

Many adaptive schemes can be represented with the block diagram in Fig. 1. The adaptive controller may be regarded as composed of two loops. An inner loop, which is the basic control loop for the system, and an outer loop which adjusts the parameters of the regulator in the inner loop by parameter estimation and control design. When considering an adaptive control scheme one must be sure that both loops work satisfactorily, both individually and together. The design method used must work well if the process is totally known. It should also be robust and insensitive to the underlying assumptions, for instance to the distribution of the noise, the character of the reference signal and to unmodelled high frequency process dynamics. This means that all physical knowledge about the process should be used to check if the design method is appropriate. Furthermore the *a priori* knowledge should be used to check the design specifications. For instance, that the desired bandwidth is not too high compared to allowed control signals. The choices of presampling filters and sampling interval are also influenced by the knowledge about the process.

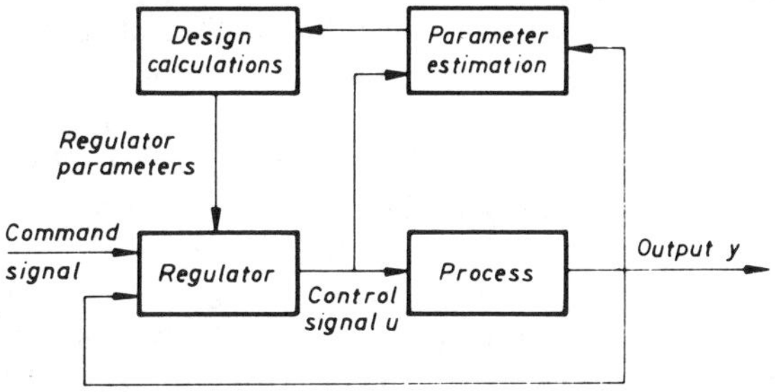

FIG. 1. Block diagram of a self-tuning regulator.

* Received 13 October 1983; revised 4 April 1984. The original version of this paper was not presented at any IFAC meeting. This paper was recommended for publication in revised form by guest editor L. Ljung.

† Department of Automatic Control, Lund Institute of Technology, Box 725, S-220 07 Lund, Sweden.

Reprinted with permission from *Automatica*, vol. 20, pp. 595–605, Sept. 1984.
Copyright © 1984, International Federation of Automatic Control IFAC. Published by Pergamon Press Ltd.

The estimation routine in the outer loop must be such that it can give good estimates in the intended application. Issues to be considered are for instance identifiability, the richness of the input signal, identification in closed loop and the possibility to follow changes in process parameters.

In summary the two loops must be based on good engineering practice. Each loop must work individually. The interaction between the loops is also important. This is, however, a very difficult problem to investigate. Interactions can normally be reduced by ensuring that the outer loop is much slower than the inner loop. This will unfortunately limit the adaptation rate.

Great progress has been made in the theory of adaptive control during the last years. Many theoretical problems such as stability and convergence have been solved for idealized cases. Shortcomings of adaptive algorithms have, however, also been pointed out. The limitations are often due to violation of the assumptions for the algorithms. These limitations must be circumvented in order to obtain robust algorithms that can be used in practice. An overview of theory and applications of adaptive control is given in Åström (1983a).

Adaptive control laws may be used in many different ways. One possibility is to make a control algorithm with automatic tuning. One example of such a controller is described in Åström and Hägglund (1984). The tuning may then be switched off when a satisfactory performance is obtained. The tuning can in this case be supervised by an operator and it is not necessary to have too much logic built into the algorithm. Another way to use an adaptive controller is to have the tuning switched on all the time in order to follow possible changes in the dynamic of the process. This is a much more difficult situation which requires a more robust algorithm.

Many feasibility studies and commercial installations indicate that adaptive control can be used successfully to control industrial processes, see for instance Åström (1983a). However, in all applications different types of 'safety-nets' and special tricks have to be used. The purpose of this paper, which is based on Wittenmark and Åström (1982), is to give some reflections on the state of the art and experiences from practical work on adaptive control. The paper is oganized in the following way. Section 2 discusses theory and practice. For instance what can be said theoretically and what are the practical limitations. Robustness issues are treated in Section 3. The idea of self-tuning algorithms is briefly reviewed in Section 4. Implementation aspects such as estimator wind-up, signal conditioning, and reset action is covered in Section 5. Section 6 gives some conclusions followed by a list of references.

2. THEORY AND PRACTICE

It is important to consider both theoretical and practical aspects when discussing adaptive control algorithms. The theory deals with idealized situations where all the conditions are under control. The theory thus gives the ultimate limit of what can be achieved and expected under idealized conditions. The practical situation is, however, such that there are all kinds of violations of the conditions of the theory.

Some important theoretical problems related to self-tuning control have been solved during the last years. Stability and convergence proofs for simple algorithms under idealized conditions are available. See for instance Egardt (1979, 1980), Goodwin *et al.* (1980), Morse (1980) and Narendra *et al.* (1980). There are, however, no results available for more realistic assumptions. One main criticism against adaptive control is the fact that the adaptive controllers, like most other controllers, are designed on models that are simpler than the real processes. The effect of unmodelled high frequency dynamics on some adaptive schemes is discussed for instance in Rohrs *et al.* (1981, 1982) and Gawthrop and Lim (1982). Other problems are due to process nonlinearities and actuator saturation. To avoid these kind of problems it is necessary to provide the practical algorithms with a safety-net or a supervisory level which can take care of unusual and undesired situations in a safe way. Much of the practical work is thus not done on a firm theoretical basis but consists of *ad hoc* solutions that often will depend on the considered application. It is often verified by extensive experimentation and simulation. Some of the main issues in adaptive control are:

(a) how to use prior information about the process;
(b) how to determine realistic specifications of the closed loop system;
(c) how to make robust estimation;
(d) unmodelled high frequency dynamics;
(e) signal conditioning;
(f) numerical problems;
(g) start-up and bumpless transfer;
(h) process and actuator nonlinearities.

Several of these points are not specific for adaptive control but are valid also for design of digital controllers in general.

3. ROBUSTNESS

Before going into the details of the implementation aspects of adaptive control it is relevant to discuss robustness issues in general. Many adaptive controllers are based on the certainty equivalence hypothesis. The regulators can then be regarded as a combination of a design procedure for known systems and a recursive estimation scheme (see Fig. 1). The true parameters of the process are replaced by the estimated parameters. All control design must be based on sound principles. For an adaptive controller it is necessary to have a robust design method and a robust estimation scheme in order to get a robust adaptive controller.

Robust control design

Robustness properties of the design of a controller can be discussed in terms of the loop gain of the system in cascade with the controller. A Bodeplot of a typical loop gain is shown in Fig. 2. It is common practice to make the design such that the loop gain is high below the cross-over frequency, ω_c. Further the loop gain should fall off rapidly above the cross-over frequency, see Horowitz (1963) and Doyle and Stein (1981). The high loop gain at low frequencies is obtained for instance by introducing integral action. The high loop gain at low frequencies will make it possible to eliminate low frequency disturbances. It will also make the system insensitive to the low frequency characteristics of the process model. The high frequency roll-off is necessary to eliminate the influence of high frequency disturbances or unmodelled high frequency dynamics. The high frequency roll-off is obtained by filtering the signals. For sampled data systems the sampling procedure will restrict the high frequency content in the sampled signal. To avoid aliasing it is necessary to provide the controller with an effective anti-aliasing filter. The choice of the sampling interval will then naturally be related to the closed loop behavior of the system. Aliasing and choice of sampling period are discussed, for instance, in Åström and Wittenmark (1984).

The discussion of the loop gain results in the conclusion that it is necessary to have a good process model for the frequencies around the cross-over frequency. It is possible to make a more quantitative statement for design procedures based on model following. Consider a system with the

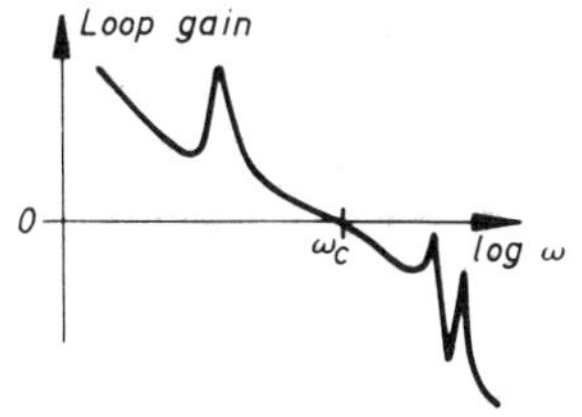

FIG. 2. Bodeplot of a typical loop gain.

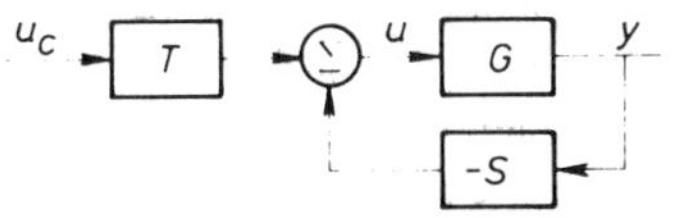

FIG. 3. Block diagram of process and regulator.

transfer function G_0. Assume that a design is based on the process model G. Select the feedforward and the feedback transfer functions such that the closed loop system from u_c to y is G_m (see Fig. 3). The closed loop system is stable if

$$|G_0 - G| < \left|\frac{G}{G_m}\right|\left|\frac{T}{S}\right| \tag{1}$$

on the imaginary axis and at infinity (or on the unit circle for a discrete time system). The statement is proven for discrete time systems in Åström (1980b). It also follows from the result of Doyle and Stein (1981). Other theorems of similar nature are given in Mannerfelt (1981) and Åström and Wittenmark (1984).

The left-hand side of (1) is the error in the transfer function of the model. The advantage of (1) is that the right-hand side depends only on known quantities on the used model, the desired model and the resulting controller. Using the inequality it is possible to investigate how the desired closed loop characteristics will influence the accuracy that is needed for the model. From (1) it also follows that it is necessary to have a good model for frequencies around the cross-over frequency. It also follows that requirements on model precision increases with increasing closed loop bandwidth.

The conclusion of this part is that a reliable design method should be used and that the design must be based on a model that is accurate at least for frequencies where the loop gain is around unity.

Robust estimation

The accuracy requirements for the design procedure lead to the question of how to obtain a good model. The key issues are good data and an appropriate model structure. The models used will invariably be simplified, e.g. linear and of low order. It is known from the theory of system identification that the estimates obtained in such a case will depend crucially on the properties of the input signal.

Above it has been demonstrated that robust control requires a model which is accurate around the cross-over frequency. To estimate an accurate reduced order model with this property it is essential that the input signal has sufficient energy content around the cross-over frequency and that it is so rich in frequency that it is persistently exciting; see Åström and Bohlin (1966). The conditions on persistent excitation are related to the complexity of the estimated model. This implies that the requirements on the input signal become more severe if the model order is increased. Since the input signal in an adaptive system is generated by feedback there is no guarantee that it will be persistently exciting. To guarantee a good model it is thus necessary to monitor the excitation and the energy of the input signal in the relevant frequency bands.

When the input signal generated by the feedback loop is not persistently exciting or when the signal energy is too low the estimated parameters will be poor. There are two ways to avoid this: by injecting extra perturbation signals or by switching off the adaptation when excitation is poor. The results by Egardt (1979) and Peterson and Narendra (1982) indicate that it is reasonable to estimate only when the absolute value of the useful input energy is above a certain level. Ways to give the signals a proper frequency content is to filter the signals or to introduce external perturbation signals. There are other safeguards of similar nature to make sure that the estimation is done only when the data is reasonable.

The difficulties with adaptive control reported by Rohrs *et al.* (1981, 1982) are due to high frequency reference signals and measurement noise in combination with unmodelled dynamics. The reported problems are analysed in Åström (1983b) and the difficulties can be eliminated if the precautions above are taken.

Example 3.1—*Rohrs' example.* A counterexample to adaptive control is presented in Rohrs *et al.* (1982). The algorithm is a continuous time model reference controller. The process is of third order and the dominating dynamic characteristic is a pole at -1. The other modes of the system represent dynamics with a natural frequency that is high compared to the dominating part and to the desired response of the system. The adaptive controller estimates a first order model, i.e. two parameters. The problems with the controller show up when the reference signal is constant and a low amplitude high frequency disturbance is added.

If the only excitation is a high frequency sinusoid the adaptive system will attempt to match the model to the system at this frequency. This will result in a low order model that will represent the dominating dynamics very poorly with expected disastrous results. (Fig. 4.) The remedy is either to switch off the adaptation when there is no excitation in the appropriate frequency band or to inject a relevant perturbation signal. One way to improve the algorithm is thus to filter the output of the process and to introduce an excitation signal of low frequency. Signal energy at low frequencies will make it possible to better estimate the dominating dynamics of the process. Figure 4 also shows the parameter estimates when the output is filtered with a low pass filter and when the excitation signal has the same amplitude as the high frequency disturbance. These precautions will prevent the parameter estimates from diverging. The estimates will now stay bounded and the closed loop system will be stable. □

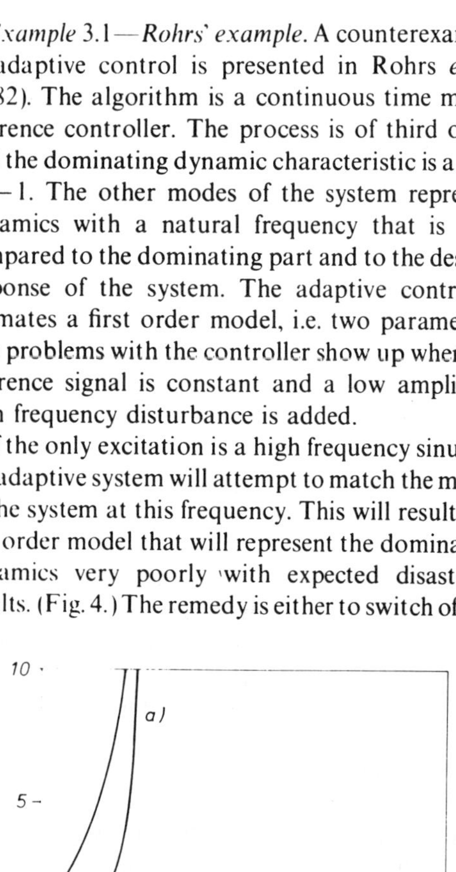

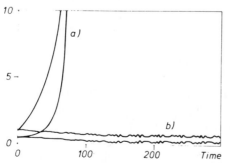

FIG. 4. Parameter estimates for the process in Example 3.1. (a) Rohrs' algorithm. (b) The modified algorithm.

Robust adaptive control

To obtain a robust adaptive control algorithm it is necessary to use both robust control and robust estimation. In the adaptive context there is also some new trade-offs to be made. Consider for instance the robustness properties obtained by having a high open loop gain at low frequencies. This may be obtained by having integral action in the control loop. It can also be obtained via adaptation. An adaptive controller with enough parameters will automatically introduce a high gain at those frequencies where there are low frequency disturbances. This will be discussed further in Section 5.

4. SELF-TUNING ALGORITHMS

Before discussing implementation aspects of self-tuning controllers it is necessary to be a bit more specific about the algorithms. A summary of self-tuning algorithms is given for instance in Clarke (1982) and in Åström (1983a). Assume that the process to be controlled can be described by the discrete time system

$$A^*(q^{-1})y(k) = B^*(q^{-1})u(k - d_0) + C^*(q^{-1})e(k) \tag{2}$$

where $y(k)$ is the output, $u(k)$ is the input and $e(k)$ is a white Gaussian noise disturbance. The time scale is normalized such that one sampling interval is one time unit. A^*, B^* and C^* are polynomials in the delay operator q^{-1}. The polynomials are defined as

$$A^*(q^{-1}) = 1 + a_1 q^{-1} + \cdots + a_n q^{-n}$$

etc. A general linear controller can be written in the form

$$R^*(q^{-1})u(k) = -S^*(q^{-1})y(k) + T^*(q^{-1})u_c(k) \tag{3}$$

where u_c is the command signal or reference value.

Self-tuning controllers can be divided into two categories, explicit and implicit algorithms; see Åström and Wittenmark (1980).

Explicit or indirect self-tuning algorithms

When using an explicit algorithm an explicit process model is estimated, i.e. the coefficients of the polynomials A^*, B^* and C^* in (2). An explicit algorithm can then be described by two steps. The first step is to estimate the polynomials A^*, B^* and C^* of the process model (2). In the second step a design method is used to determine the polynomials in the regulator (3) using the estimated parameters from the first step. The two steps are repeated at each sampling interval. The design procedure in the second step can be any good design method that is suitable for the problem on hand. A typical application of an explicit algorithm is for instance a pole placement algorithm described in Åström and Wittenmark (1980).

Implicit or direct self-tuning algorithms

In an implicit algorithm the parameters of the regulator are estimated directly. This can be made possible by a reparameterization of the process model. A typical example of an implicit algorithm is the minimum variance self-tuner in Åström and Wittenmark (1973).

One advantage with the implicit algorithms over the explicit algorithms is that the design computations are eliminated, since the controller parameters are estimated directly. The implicit algorithms usually have more parameters to estimate than the explicit algorithms, especially if there are long time delays in the process. Simulations and practical experiments indicate, however, that the implicit algorithms are more robust. The implicit algorithms usually have the disadvantage that all process zeros are cancelled. This implies that the implicit methods are intended only for processes with a stable inverse or minimum phase systems. Sampling of a continuous time system often gives a discrete time system with zeros on the negative real axis, inside or outside the unit circle. It is not good practice to cancel these zeros even if they are inside the unit circle. Cancellation of these zeros will give rise to 'ringing' in the control signal. Many implicit algorithms can, however, be used also if the system is nonminimum phase through a proper choice of parameters. An example is given in Example 5.5 below.

Feedforward

Feedforward control is a very effective way to reduce the influence of measurable disturbances. It requires, however, knowledge about the process dynamics. In adaptive controllers it is easy to include feedforward from different signals, see Åström and Wittenmark (1973). The estimation algorithm will automatically give the required dynamics. To be effective the models must also be reasonably accurate. Adaptation is thus almost a prerequisite for practical use of feedforward. Adaptive feedforward has been used in several of the applications of adaptive controllers discussed in Åström (1983b).

Real time estimation

Both explicit and implicit algorithms need a recursive estimation scheme. The most common is the method of least squares or its modifications. The least squares estimator is described by the equations

$$\boldsymbol{\theta}(k+1) = \boldsymbol{\theta}(k) + P(k+1)\boldsymbol{\phi}(k+1)\varepsilon(k+1) \quad (4)$$

$$P(k+1) = [P(k) - P(k)\boldsymbol{\phi}(k)R(k)\boldsymbol{\phi}^T(k)P(k)]/\lambda \quad (5)$$

$$R(k) = [\lambda + \boldsymbol{\phi}^T(k)P(k)\boldsymbol{\phi}(k)]^{-1} \quad (6)$$

where $\boldsymbol{\theta}$ is a vector consisting of the parameters to be estimated, $\boldsymbol{\phi}$ is a vector of delayed inputs and outputs (possibly filtered), and ε is the prediction error. P is proportional to the covariance matrix of the estimation error. Finally λ is an exponential forgetting factor. The forgetting factor is used to make it possible to follow time varying parameters. Further details of recursive estimation schemes are found in Ljung and Söderström (1983).

5. IMPLEMENTATION ASPECTS

The empirical knowledge that the authors have in the field of adaptive control stems mainly from using self-tuning regulators of different types, see Åström *et al.* (1977) and Källström *et al.* (1979). The estimation has mostly been done using the method of least squares. Different design methods have been used; for instance, minimum variance control, linear quadratic Gaussian control, and pole placement design based on complex and simplified models. Some of the experiences are discussed in the following.

Signal conditioning

In all digital control applications it is important to have a proper conditioning of the signals. Due to the aliasing problem connected with the sampling procedure it is necessary to eliminate all frequencies above the Nyquist frequency before sampling the signals. High frequencies may otherwise be interpreted as low frequencies and may introduce disturbances in the closed loop system. The filtering of the signals also has the effect that the system will be excited by frequencies where it is important to have good process models. Compare the example in Section 3.

Example 5.1—The effect of prefilter. A laboratory plant for concentration control has been used in

Åström and Zhao-Ying (1982) for experiments with a linear quadratic Gaussian self-tuner. Figure 5 shows the effect of prefiltering of the measurement. In Fig. 5a a typical sample of the continuous time process output is shown. In Figs 5b and 5c the sampled process output and the control signals are shown. (The curves in Figs 5a and 5b are not from the same experiment.) The sampling period is 15 s. At $t = 600$ an analog anti-aliasing filter is connected to the process output. The filter is a first order filter with a time constant of 75 s. The high frequency disturbance in the concentration measurement introduces a low frequency disturbance in the sampled process output and consequently also in the control signal. This disturbance is effectively eliminated by using a prefilter. □

Parameter tracking and estimator wind-up

The key property of an adaptive controller is the ability to track variations in the process dynamics and to do so it is necessary to discount old data. This will involve compromises: too fast discounting will make the estimates uncertain if the parameters are constant but too slow discounting will make it impossible to track rapid parameter variations.

Exponential forgetting is one way to discard old data. The algorithm described by (4)–(6) minimizes the loss function

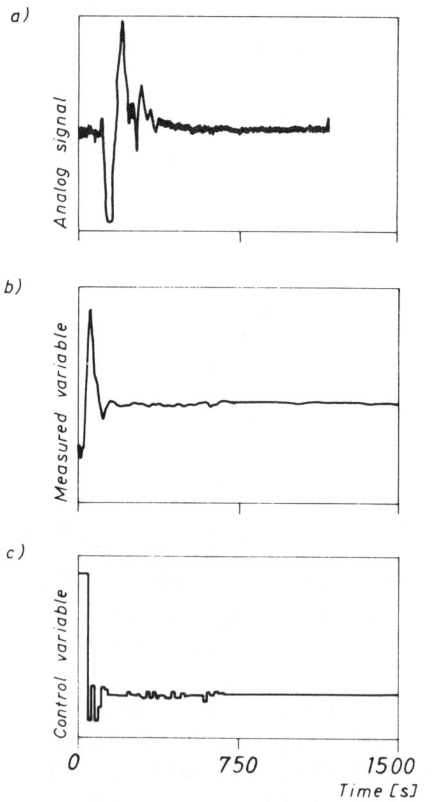

FIG. 5. The effect of a prefilter when controlling concentration in a laboratory process using a LQG self-tuner. (a) Analog process output. (b) Sampled process output. (c) Control signal (adapted from Åström and Zhao-Ying, 1982). Notice that curves (a) and (b) are not from the same experiment.

$$\sum_{k=0}^{N} \lambda^{N-k} \varepsilon(k)^2$$

where ε is the prediction error. With $\lambda = 1$ all data have the same weight. With $\lambda < 1$ more recent data are weighted more than old data. It is possible to generalize the method with exponential forgetting and have different forgetting factors for different parameters. Exponential forgetting works well only if the process is properly excited all the time. There are several problems with exponential forgetting when the excitation of the process changes. A typical situation is when the main source of excitation is changes in the set point. Then there may be long periods with no excitation, the estimator will then discount old information and the uncertainties of the parameters will grow. This may be called estimator wind-up (compare with integrator wind-up in conventional integral control). The problem can be understood from (5). The negative term on the right-hand side represents the reduction in uncertainty due to the last measurement. If there is no information in the last measurement then $P(k)\boldsymbol{\phi}(k)$ will be zero and (5) reduces to

$$P(k+1) = P(k)/\lambda.$$

$P(k)$ will thus grow exponentially until $\boldsymbol{\phi}$ changes direction if $\lambda < 1$. If there is no excitation for a long period of time then $P(k)$ may be very large. Since $P(k)$ also is the gain in (4) then there may be large changes in the estimated parameters when new information is coming into the system, for instance when the reference value changes. The estimator wind-up may then cause a burst in the output of the process.

There are several ways to avoid estimator wind-up. The main idea is to ensure that P stays bounded. This can be done, for instance, by ensuring that the trace of P is constant in each iteration; see Irving (1979). Another possibility is to adjust the forgetting factor automatically. Ways to do this are given in Fortescue *et al.* (1981) and Wellstead and Sanoff (1981). The automatic adjustment of λ in these references does not, however, guarantee that P stays bounded.

It has also been proposed to stop the updating of the parameters and the covariance matrix when $P\boldsymbol{\phi}$ or ε are sufficiently small, see Egardt (1979). Hägglund (1984) has proposed superior algorithms which only discount data in the directions where there are new information.

Example 5.2—*Estimator wind-up*. The problem with estimator wind-up is illustrated by a simple simulated example. Let the process to be controlled be described by

$$y(k) - 0.9y(k-1) = u(k-1).$$

It is desired that the pulse transfer operator from the reference signal to the output has a pole in 0.7 and that the gain is unity. This is achieved with the controller

$$u(k) = 0.3y(k) - 0.35y(k-1) + u_c(k) - 0.5u_c(k-1).$$

The process is controlled using an implicit pole placement algorithm where the parameters in the controller are estimated. Figure 6 shows the diagonal elements of the P-matrix (5) when different estimation schemes are used. The reference signal is a square wave with unit magnitude and period 100 up to time 300. After that the reference signal is constant. In Fig. 6a the estimation algorithm described by (4)–(6) is used with $\lambda = 0.99$. When the reference signal is constant and the output has settled there is no information in the measurements. The P-matrix will then start to increase. In Fig. 6b a new estimation routine described in Hägglund (1984) is used. The elements of the P-matrix will then settle on constant values and there is no estimator wind-up. □

There may be numerical problems in the updating of the parameters and the covariance matrix. This is especially true if the excitation of the process is poor. The square root method or the U–D factorization method, see Bierman (1977), are then good ways to organize the computations.

Start-up procedures

There are several ways in which a self-tuning algorithm can be initialized depending on the *a priori* information about the process.

One case is if the process has been controlled before with a conventional or an adaptive controller. The initial values should then be such that they correspond to the controller used before.

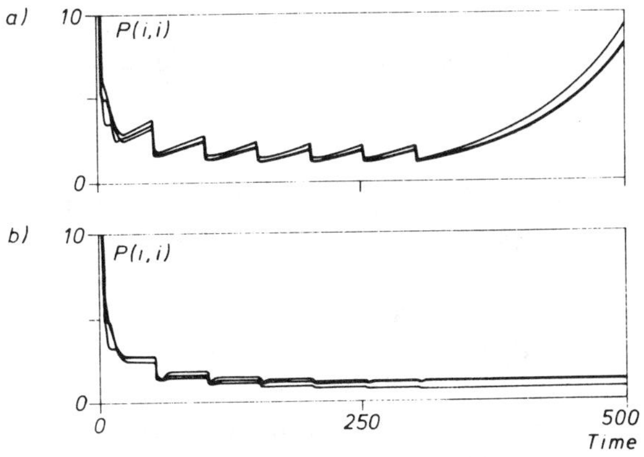

FIG. 6. The diagonal elements of the P-matrix when controlling the process in Example 5.2. (a) Constant exponential forgetting factor $\lambda = 0.99$. (b) Forgetting according to Hägglund (1984).

Another situation occurs if nothing is known about the process. The initial values of the parameters in the estimator can then be chosen to be zero or such that the initial controller is a proportional or integral controller with low gain. The inputs and outputs of the process should be scaled such that they are of the same magnitude. This will improve the numerical conditions in the estimation and the control parts of the algorithm. The initial value of the covariance matrix can be 1–100 times a unit matrix. These values are usually not crucial since the estimator will get reasonable values in a very short period of time. Our experience is that 10–50 samples is sufficient to get a very good controller. During the initial phase it can be advantageous to add a perturbation signal to speed up the convergence of the estimator. Auto-tuning discussed in Åström and Hägglund (1984) is a convenient way to initialize the algorithm because it generates a suitable input signal and safe initial values of the parameters.

Sometimes it is important to have the disturbances as small as possible due to the start-up of the self-tuning algorithm. There are then two precautions that can be taken. First the estimator can be used for some sampling periods before the self-tuning algorithm is allowed to put out any control actions. During that time a safe, simple controller should be used. It is also possible and desirable to limit the control signal. The allowable magnitude can be very small during the first period of time and can then be increased when better parameter estimates are obtained. This kind of soft start-up is for instance used in Asea's Novatune; see Bengtsson (1979). The drawback of having small input signals is that the excitation of the process will be poor and it will take a longer time to get good parameter estimates.

Reset action

It is important that a controller has the ability to eliminate steady state errors when the reference value is constant. The steady state error can be generated by many different mechanisms: calibration errors, nonlinearities, disturbances, etc. In a conventional controller, reset action is obtained by introducing an integrator in the controller. When using self-tuning controllers there are several ways to introduce reset action. Since there is no method that is uniformly best, different alternatives will be discussed. The introduction of integrators is discussed in Åström (1980a) and also in Allidina and Hughes (1982).

The simplest way to get reset action is to let the self-tuning regulator take care of the problem. Since it estimates a model of the process and the environment it can be expected that the self-tuner tries to model the offset and compensate for it. It is easy to check if a particular self-tuner has this ability by investigating possible stationary solutions. A typical example is given below.

Example 5.3—*Automatic reset action.* Consider the simple implicit self-tuning controller in Åström and Wittenmark (1973) which is based on least squares estimation and minimum variance control. The estimation is based on the model

$$y(k+d) = R^*(q^{-1})u(k) + S^*(q^{-1})y(k) \quad (7)$$

where d is an estimate of d_0 in (2) and the regulator is

$$u(k) = -\frac{S^*}{R^*}y(k). \quad (8)$$

The conditions for a stationary solution are that

$$\hat{r}_y(\tau) = \lim_{N\to\infty} \frac{1}{N}\sum_{k=1}^{N} y(k+\tau)y(k) = 0$$
$$\tau = d, \ldots, d + \deg S^* \quad (9)$$

$$\hat{r}_{yu}(\tau) = \lim_{N\to\infty} \frac{1}{N}\sum_{k=1}^{N} y(k+\tau)u(k) = 0$$
$$\tau = d, \ldots, d + \deg R^*. \quad (10)$$

These conditions are not satisfied unless the mean value of y is zero. When there is an offset the parameter estimates will get values such that $R^*(1) = 0$, i.e. there is an integrator in the controller. The convergence to the integrator may, however, be slow. It can be shown that other, both explicit and implicit, self-tuning regulators also can give reset automatically. □

A second way to introduce reset action is to estimate the bias in the process. A simple way to do this is to include a bias term, b, in the model (2). The model is then

$$A^*y(k) = B^*u(k-d_0) + C^*e(k) + b.$$

With this model it is easy to estimate b and to compensate for it. Such a scheme was proposed by Clarke and Gawthrop (1979). One drawback is that an extra parameter has to be estimated; furthermore it is necessary to have different forgetting factors on the bias estimate and the other estimates. Otherwise the convergence to a new level will be very slow. Finally if the bias is estimated in this way it is not possible to use the self-tuner as a tuner since there will be no reset when the estimation is switched off.

A third way to get reset action is to force an integrator into the regulator. That means that the controller has the form

$$R^*\nabla u(k) = -S^*y(k) + T^*u_c(k)$$

where

$$\nabla u(k) = (1-q^{-1})u(k) = u(k) - u(k-1).$$

This form has several advantages. First, there will always be an integrator even if the regulator parameters are not optimally tuned. Second, the high gain at low frequencies will increase the robustness of the system due to uncertainties in the process dynamics at low frequencies. This implies that the estimation can be concentrated at frequencies around the cross-over frequency. One drawback with this method is that the self-tuner will try to eliminate the integral action when it is not needed. This implies that the regulator will try to cancel a pole at the stability boundary.

Actuator nonlinearities

A controller, adaptive or not, usually contains several nonlinearities. For instance the magnitude of the control signal is limited. When using a self-tuning regulator it is especially important that the estimator is fed with the control signal that is sent out to the process. The estimator will otherwise get incorrect estimates, for instance, of the gain of the process.

Anti-reset wind-up

An integrator is an unstable system and it may happen that the integral can assume very large values if the control signal saturates when there is an error. This is called reset wind-up or integrator wind-up. Special precautions must be taken in order to avoid this. Ways to do this are discussed in Åström and Wittenmark (1984) for different controller structures.

Example 5.4—*Anti-reset wind-up controller.* Consider a regulator described by (3) where the regulator may contain unstable modes. One way to solve the reset wind-up problem is to rewrite (3) by adding $A_0^*(q^{-1})u(k)$ on both sides. This gives

$$A_0^*u(k) = T^*u_c - S^*y + (A_0^* - R^*)u.$$

A regulator with anti-reset wind-up compensation is then given by

$$\begin{cases} A_0^*v(k) = T^*u_c - S^*y + (A_0^* - R^*)u \\ u(k) = \mathrm{sat}[v(k)] \end{cases} \quad (11)$$

where sat[$\cdots$] is the saturation function. This regulator is equivalent to (3) when the control signal does not saturate. The polynomial A_0^* should be stable. It can be interpreted as the dynamics of the observer associated with the controller. A block diagram of (11) is shown in Fig. 7. A particular simple case is when $A_0^* = 1$, which corresponds to a dead beat observer. The controller is then

$$u(k) = \text{sat}[T^*u_c(k) - S^*y(k) + (1 - R^*)u(k)]. \quad \square$$

Tuning parameters

When adaptive control is mentioned the vision of the ideal black box easily appears, i.e. a system without any tuning knobs that would give a good closed loop performance no matter what it is connected to. Today we are far away from such a solution and it can also be questioned if such a solution is desired. It is at least necessary to tell the controller what we expect it to do. Usually it is also necessary to provide much more *a priori* information. Only for specialized applications may it be possible and desirable to have a regulator without any parameters to tune. Instead the important issue is what kind of parameters are to be tuned. One possibility is to introduce performance related knobs, i.e. knobs which relate to the performance of the closed loop system. Examples of performance related knobs are the bandwidth and the damping of the closed loop system or the gain and phase margins. Such parameters are easy to tune since they are directly related to the behaviour of the closed loop system.

There are two categories of parameters to be chosen in adaptive controllers, integers and reals. For self-tuning regulators the integer parameters are typically the order of the process model and possibly the time-delay in the process to be controlled. The real parameters are the performance related parameters and initial values for the estimation routine. The integer parameters are usually quite easy to determine. The performance can also be made insensitive to the initial values of the estimator. This was discussed above in connection with start-up procedures for adaptive algorithms. A discussion of the choice of the parameters in self-tuning regulators is found, for instance, in Wellstead and Zanker (1982).

One parameter that can influence greatly the behaviour of the algorithm is the sampling period. The choice of the sampling interval is not specific for adaptive controllers but is an important design parameter for all sampled data design methods. In general it is important that the sampling period be related to the desired performance of the closed loop system. There are several rules of thumb that can be given. One is to relate the sampling period, h, to the desired rise time, T_r, and define

$$N_r = T_r/h.$$

To get a good servo performance of the closed loop system, N_r should be chosen in the range 2–4. A similar rule of thumb is to relate the sampling period to the natural frequency, ω, of the control poles of the desired closed loop system. The sampling period can then be chosen such that $\omega \cdot h$ is in the range 0.25–1. The choice of the sampling period is also influenced by the character of the disturbances acting on the system. More details about the choice of the sampling period for different sampled data design methods are found in Åström and Wittenmark (1984). The choice of sampling period for minimum variance controllers is also discussed in MacGregor (1976) and Söderström and Lennartson (1981).

Example 5.5—*Choice of parameters.* In this example the choice of parameters are discussed for the self-tuning regulator in Åström and Wittenmark (1973). The controller is based on implicit estimation and minimum variance control. The controller parameters are estimated from the model (7) and the controller is given by (8). The tuning parameters are $\deg S^*$, $\deg R^*$, d, λ, $P(0)$, $\boldsymbol{\theta}(0)$ and the sampling period, h.

The estimator parameters λ, $P(0)$ and $\boldsymbol{\theta}(0)$ are not crucial and can often be given standard values.

The parameter d is the prediction horizon of the controller. For the optimal minimum variance controller d is the delay of the system in number of sampling periods, i.e. d is given by

$$d = d_0 = \text{int}[\tau/h] + 1$$

where τ is the time delay of the process. The self-tuning controller can be used with a longer time horizon. This will decrease the variance of the

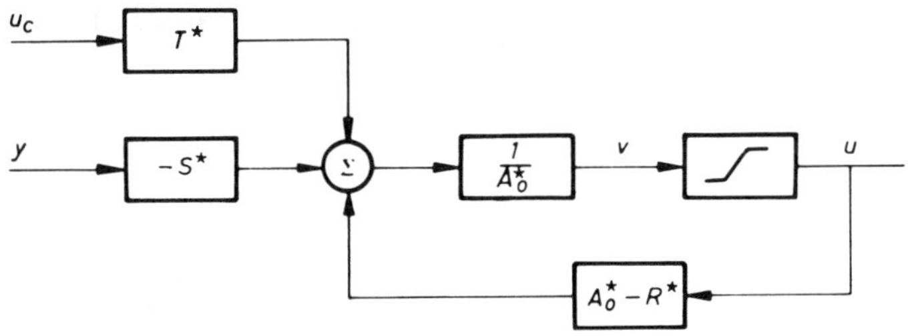

FIG. 7. Block diagram of (11) which avoids wind-up.

control signal at the expense of a larger variance of the output. It is important that dh not be shorter than τ while it is safe to have it longer than τ.

For the minimum variance controller the order is given by

$$\deg S^* = n - 1$$

$$\deg R^* = n + d_0$$

where n and d_0 are defined in (2). In practice it is often sufficient to choose $n = 1$–3. It is also possible to test if $\deg S^*$ and $\deg R^*$ are sufficiently large by monitoring $\hat{r}_y(\tau)$ and $\hat{r}_{yu}(\tau)$, see (9) and (10). If the system is controlled by an optimal minimum variance controller then $\hat{r}_y(\tau)$ and $\hat{r}_{yu}(\tau)$ are zero if $\tau \geq d - 1$.

The algorithm discussed here gives the optimal minimum variance controller only if the system to be controlled has a stable inverse. The algorithm will diverge if the process has zeros outside the unit circle. It is, however, possible to stabilize such systems by increasing the prediction horizon d. If $d = \max[\deg A^*, \deg B^* + d_0]$ then it can be shown that there exists a locally stable point for the parameter estimates. These parameter values correspond to a regulator such that the output is a moving average of order $n - 1$ and such that no process zeros are cancelled by the regulator. This regulator will thus not have any ringing in the control signal due to cancellation of process zeros. The algorithm will in this case converge to the suboptimal minimum variance controller for nonminimum phase systems given in Åström (1970). This new result is illustrated with a simulated example. Let the process be

$$y(k) + ay(k-1) = u(k-1) + bu(k-2) + e(k) + ce(k-1)$$

with $a = -1$, $b = 2$ and $c = -0.5$. Figure 8 shows the parameters estimates when the basic algorithm is used with $d = 2$, $\deg R^* = 1$ and $\deg S^* = 0$. The

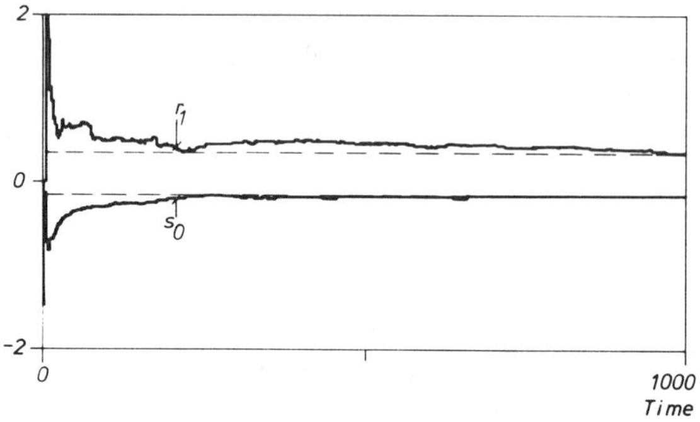

FIG. 8. Parameter estimates when controlling the process in Example 5.5 with the basic self-tuning algorithm with $d = 2$ and when there are two parameters in the controller. The dashed lines corresponds to the parameters of the suboptimal minimum variance controller.

dashed lines are the parameters to which the regulator should converge. The suboptimal minimum variance controller have an average loss of 1.11 per step while the stable optimal minimum variance controller gives a loss of 1.08 per step. The basic self-tuning regulator with increased prediction horizon will perform as well as the suboptimal minimum variance controller after a few steps of time. □

6. CONCLUSIONS

The experience from many feasibility studies and applications of self-tuning regulators indicate that they can be successfully used in many situations. The conclusions from the applications indicate that it is not quite straightforward to use self-tuning regulators. There are many precautions that must be taken, as when using conventional controllers. Some of these practical aspects of self-tuning regulators have been discussed.

To summarize, it is necessary to stress that it is important to have as much *a priori* knowledge about the process as possible. This knowledge is primarily used to choose design method and specifications. The parameters of the controllers are then estimated and tuned by the controller. The advantages compared to conventional control are that the controller can follow process variations, that more complex controllers can be used and that dead-time and feedforward compensations are easily handled. There is, however, much to be done both theoretically and practically before adaptive and self-tuning controllers can be applied routinely by inexperienced users.

Acknowledgements—The research in adaptive control at the Department of Automatic Control has for many years been supported by the Swedish Board of Technical Development (STU). This report was prepared under contract 82-3430. This support is gratefully acknowledged. The authors also want to thank colleagues and students who have shared their knowledge and experience. Special thanks are due to Tore Hägglund for the simulations given in Example 5.2.

REFERENCES

Allidina, A. Y. and F. M. Hughes (1982). Self-tuning controller with integral action. *Opt. Control Appl. Meth.*, **3**, 355–362.

Åström, K. J. (1970). *Introduction to Stochastic Control Theory*. Academic Press, New York.

Åström, K. J. (1980a). Design principles for self-tuning regulators. In H. Unbehauen (ed.), *Methods and Applications in Adaptive Control*. Springer-Verlag, Berlin.

Åström, K. J. (1980b). Robustness of a design method based on assignment of poles and zeros. *IEEE Trans Aut. Control*, **AC-25**, 588–591.

Åström, K. J. (1983a). Theory and applications of adaptive control—a survey. *Automatica*, **19**, 471–486.

Åström, K. J. (1983b). Analysis of Rohrs' counterexamples to adaptive control. *Proceedings of the 22nd Conference on Decision and Control*, San Antonio, Texas, U.S.A., pp. 982–987.

Åström, K. J. and T. Bohlin (1966). Numerical identification of linear dynamic systems from normal operating records. In P. N. Hammond (ed.), *Theory of Self-Adaptive Control Systems*. Plenum Press, New York.

Åström, K. J., U. Borisson, L. Ljung and B. Wittenmark (1977). Theory and applications of self-tuning regulators. *Automatica*, **13**, 457–476.

Åström, K. J. and T. Hägglund (1984). Automatic tuning of simple regulators with specifications on phase and amplitude margins. *Automatica*, **20**, 000–000.

Åström, K. J. and B. Wittenmark (1973). On self-tuning regulators. *Automatica*, **9**, 185–199.

Åström, K. J. and B. Wittenmark (1980). Self-tuning controllers based on pole-zero placement. *Proc. IEE*, **127**, 120–130.

Åström, K. J. and B. Wittenmark (1984). *Computer Controlled Systems—Theory and Design*. Prentice-Hall, Englewood Cliffs, New Jersey.

Åström, K. J. and Z. Zhao-Ying (1982). A linear quadratic gaussian self-tuner. *Proc. Workshop on Adaptive Control*, Florence, Italy.

Bengtsson, G. (1979). Industrial experience with a microcomputer based self-tuning regulator. *Proceedings of the 18th IEEE Conference on Decision and Control*, Fort Lauderdale, Florida.

Bierman, G. J. (1977). *Factorization Methods for Discrete Sequential Estimation*. Academic Press, New York.

Clarke, D. W. (1982). Model following and pole-placement self-tuners. *Opt. Control Appl. Meth.*, **3**, 323–335.

Clarke, D. W. and P. J. Gawthrop (1979). Implementation and application of microprocessor based self-tuners. *Preprints 5th IFAC Symposium on Identification and System Parameter Estimation*, Darmstadt, September 1979, pp. 197–208.

Doyle, J. C. and G. Stein (1981). Multivariable feedback design: concepts for a classical/modern synthesis. *IEEE Trans Aut. Control*, **AC-26**, 4–16.

Egardt, B. (1979). *Stability of Adaptive Controllers*. Lecture Notes in Control and Information Sciences, p. 20. Springer-Verlag, Berlin.

Egardt, B. (1980). Stability analysis of continuous-time adaptive control systems. *SIAM J. Control Optimiz.*, **18**, 540–557.

Fortescue, T. R., L. S. Kershenbaum and B. E. Ydstie (1981). Implementation of self-tuning regulators with variable forgetting factors. *Automatica*, **17**, 831–835.

Gawthrop, P. J. and K. W. Lim (1982). Robustness of self-tuning controllers. *Proc. IEE*, **129**, 21–29.

Goodwin, G. C., P. J. Ramadge and P. E. Caines (1980). Discrete multivariable adaptive control. *IEEE Trans Aut. Control*, **AC-25**, 449–456.

Horowitz, I. M. (1963). *Synthesis of Feedback Systems*. Academic Press, New York.

Hägglund, T. (1984). New estimation techniques for adaptive control. Ph.D. thesis, Report TFRT-1025, Department of Automatic Control, Lund Institute of Technology, Lund, Sweden.

Irving, E. (1979). Improving power network stability and unit stress with adaptive generator control. *Automatica*, **15**, 31–46.

Källström, C. G., K. J. Åström, N. E. Thorell, J. Eriksson and L. Sten (1979). Adaptive autopilots for tankers. *Automatica*, **15**, 241–254.

Ljung, L. and T. Söderström (1983). *Theory and Practice of Recursive Identification*. MIT Press, Cambridge, Mass.

MacGregor, J. F. (1976). Optimal choice of the sampling interval for discrete process control. *Technometrics*, **18**, 151–160.

Mannerfelt, C. F. (1981). Robust control design with simplified models. Ph.D. thesis, TFRT-1021, Department of Automatic Control, Lund Institute of Technology, Lund, Sweden.

Morse, A. S. (1980). Global stability of parameter adaptive control systems. *IEEE Trans Aut. Control*, **AC-25**, 433–439.

Narendra, K. S., Y. H. Lin and L. S. Valavani (1980). Stable adaptive control design. *IEEE Trans Aut. Control*, **AC-25**, 440–448.

Peterson, B. B. and K. S. Narendra (1982). Bounded error adaptive control. *IEEE Trans Aut. Control*, **AC-27**, 1161–1168.

Rohrs, C. E., L. Valavani, M. Athans and G. Stein (1981). Analytical verification of undesirable properties of direct model reference adaptive control algorithms. *Proceedings of the 20th IEEE Conference on Decision and Control*, San Diego, pp. 1271–1284.

Rohrs, C. E., L. Valavani, M. Athans and G. Stein (1982). Robustness of adaptive control algorithms in the presence of unmodelled dynamics. *Proceedings of the 21st IEEE Conference on Decision and Control*, Orlando, Florida, pp. 3–11.

Söderström, T. and B. Lennartson (1981). On linear optimal control with infrequent output sampling. In J. E. Marshall *et al.* (eds), *Third IMA Conference on Control Theory*. Academic Press, New York.

Wellstead, P. E. and S. P. Sanoff (1981). Extended self-tuning algorithm. *Int. J. Control*, **34**, 433–455.

Wellstead, P. E. and P. Zanker (1982). Techniques of self-tuning. *Opt. Control Appl. Meth.*, **3**, 305–322.

Wittenmark, B. and K. J. Åström (1982). Implementation aspects of adaptive controllers and their influence on robustness. *Proceedings of the 21st IEEE Conference on Decision and Control*, Orlando, Florida.

Part IV
Adaptive Control of Uncertain Plants

THE control of plants whose dynamics are changing and which are subjected to environmental and behavioral uncertainties is a problem of major theoretical and practical importance. In fact, theoretical and practical interest in the adaptive feedback control stems from such a class of problems. In this part of the volume, we discuss certain aspects of this class of problems which need to be explored further.

One of the earliest series of papers of historical importance in this field is due to A. A. Fel'dbaum. He published a series of four papers in 1960, and here we reproduce the theoretical basic paper entitled "Dual Control Theory I" in paper twenty-four. In this paper, the author compares some fundamental problems in communication theory and control theory, and then he formulates the problem of designing an optimum (in the statistical sense) closed-loop dual control systems. In this dual control problem, he considers the control signal as to help both in "learning process" and then to meet control objectives. Some of this philosophy has been employed in the other papers reported in this part of the volume.

In paper twenty-five E. Tse and Y. Bar-Shalom discuss an active adaptive control method for nonlinear stochastic systems. The authors use the term "active" in the sense that the adaptive controller utilizes, in addition to the available real-time information, the knowledge that future observations will be made, and regulate its adaptation (learning) accordingly. In this paper, several examples are given which illustrate the proposed actively adaptive control method. This method suffers from the computational complexity, however, which has to be carried out in real-time, and is much more suitable for problems where the sampling interval can be made reasonably long.

E. Tse and M. Athans examine the problems of controlling a linear discrete-time stochastic system with unknown, possible time-varying and random, gain parameters in paper twenty-six. The philosophy used is based on the use of open-loop feedback optimal control and a quadratic performance index. The simulation results are encouraging, however, this basic philosophy needs further work.

In paper twenty-seven entitled "Adaptive Control with Stochastic Approximation Algorithm: Geometry and Convergence," Becker, Kumar and Wei discuss new geometric properties possessed by the sequence of parameter estimate; these geometric properties give a valuable insight into the behavior of the stochastic approximation based algorithm as it is used in the minimum variance adaptive control. Also, the authors present certain probabilistic arguments which prove that if the system does not have a reduced order minimum variance controller, then the parameter estimate converge to a random multiple of the true parameter. This proposed stochastic approximation algorithm as presented in this paper seems to be very useful, however, it needs further exploration.

In paper twenty-eight, H. F. Chen and P. E. Caines describe a stochastic gradient algorithm of adaptive control having a strong consistency. The development of such algorithms for future adaptive type controller may prove to be very useful.

Lastly, in paper twenty-nine, Goodwin and his colleagues describe a stochastic adaptive control algorithm based on gradient estimation and minimum-variance control. The algorithm is shown to be globally convergent when the system parameters are exponentially convergent to values satisfying the conditions for a globally convergent time-invariant systems.

These six papers were selected and put arbitrarily under the category of "uncertain plants" though uncertainty has been treated in an explicit form in self-tuning regulators as well. However, some of the concepts briefly discussed in this section may be useful in developing future adaptive control algorithms for uncertain plants.

Selected Bibliography

J. Alster and P. R. Belanger, "A technique for dual adaptive control," *Automatica,* vol. 10, no. 6, pp. 627–634, 1974.

Y. Bar-Shalom and E. Tse, "Dual effort, certainty equivalence and separation in stochastic control," *IEEE Trans. Automat. Contr.,* vol. AC-19, pp. 494–500, 1974.

——, "Concepts and methods in stochastic control," in *Control and Dynamic Systems, Advances in Theory and Applications,* Vol. 12, C. T. Leondes, Eds. New York: Academic Press, 1976.

——, "Dual adaptive control and uncertainty effects in macroeconomic modelling," *Automatica,* vol. 16, no. 2, pp. 147–156, 1980.

P. E. Caines, "Stochastic adaptive control: Randomly varying parameters and continually disturbed controls," IFAC Congr., Kyoto, Japan, 1981.

H. F. Chen and P. E. Caines, "On the adaptive control of stochastic systems with random parameters," presented at Proc. 23rd IEEE Conf. on Decision and Control, 1984.

G. C. Goodwin, P. J. Ramadge, and P. E. Caines, "Stochastic adaptive control," *SIAM J. Cont. and Opt. Interactor Matrices,* IFAC Workshop on Adaptive Control, San Francisco, 1978.

——, "Stochastic adaptive control," Tech. Rep., Harvard University, Cambridge, MA, 1978.

G. C. Goodwin, K. S. Sin, and L. K. Saluja, "Stochastic adaptive control and prediction: The general delay-coloured noise case," *IEEE Trans. Automat. Contr.,* vol. AC-25, no. 5, pp. 946–950, 1980.

G. C. Goodwin and K. S. Sin, *Adaptive Filtering Prediction and Control.* Englewood Cliffs, NJ: Prentice Hall, 1984.

O. Gomart and P. E. Caines, "Robust adaptive control of time varying systems," in *Proc. 23rd IEEE Conf. on Decision and Control,* pp. 1021–1023, 1984.

M. M. Gupta, P. N. Nikiforuk, and J. M. Milne, "Jump type behavioral uncertainties in stochastic optimal control problems," in *Proc. 1971 JACC,* pp. XVI 3.1–3.4, 1971.

M. M. Gupta and P. N. Nikiforuk, "On the dynamic behavior of stochastic systems with discontinuities in parameters," *Automatica,* vol. 12, no. 1, pp. 97–101, 1976.

O. Hijab, "Entropy and dual control," in *Proc. 23rd IEEE Conf. Decision and Control,* pp. 45–50, 1984.

V. I. Ivanenko, O. A. Knokhel, and E. I. Shor, "Synthesis of an optimal adaptive control system by a Gaussian random process generator," *Cybern.,* vol. 4, no. 6, pp. 46–52, 1968.

J. Krause and G. Stein, "Adaptive control in the presence of unstructed dynamics," in *Proc. 23rd IEEE Conf. Decision and Control,* pp. 1019–1020, 1984.

T. L. Lai, "Some thoughts on stochastic adaptive control," in *Proc. 23rd Conf. Decision and Control,* pp. 51–56, 1984.

C. Mantredi, E. Mosca, and G. Zappa, "LQG adaptive control of armax plants by LS multistep predictors," in *Proc. 23rd IEEE Conf. Decision and Control,* pp. 683–684, 1984.

R. V. Monopoli and M. Troiani, "Discrete model reference adaptive systems with measurement noise," in *Proc. 7th IFAC Congr.,* pp. 1989–1994, 1978.

P. N. Nikiforuk, H. Ohta, and M. M. Gupta, "Adaptive identification and control of linear MIMO discrete systems in a noisy environment," in *Int. Federation of Automatic Control,* 8th Triennial World Congress, pp. VII.71–VII.76, 1981.

L. Praly, "Stochastic adaptive controllers with and without a positivity condition," in *Proc. 23rd IEEE Conf. Decision and Control,* pp. 58–63, 1984.

R. Rishel, "An exact formula for a linear quadratic adaptive stochastic optimal control law," in *Proc. 23rd IEEE Conf. Decision and Control,* pp. 64–68, 1984.

J. S. Riordon, "Dual control strategies for discrete state Markov processes I," *Int. J. Contr.,* vol. 6, no. 3, pp. 249–261, 1967.

——, "Dual control strategies for discrete state Markov processes II," *Int. J. Contr.,* vol. 6, no. 4, pp. 317–330, 1967.

G. N. Saridis, "Stochastic approximation methods in self-organizing control systems—A survey," *Int. Conf. Cybern., and Soc.,* pp. 442–430, 1972.

E. Tse and M. Athans, "Adaptive stochastic control for linear systems. II Asymptotic properties and simulation results," presented at Proc. 1970 IEEE Symp. Adapt. Processes, 1970.

B. Wittenmark, "Stochastic adaptive control methods—A survey," *Int. J. of Control,* vol. 21, pp. 705–730, 1975.

W. R. E. Wouters, "Adaptive pole placement for linear stochastic systems with unknown parameters," in *IEEE Conf. on Decision and Control,* pp. 159–166, 1968.

V. A. Yacubovich, "On a method of adaptive control under conditions of great uncertainty," presented at Proc. 1972 IFAC Congr., Paper 37.3, 1972.

DUAL-CONTROL THEORY. I

A. A. Fel'dbaum

Moscow

Translated from Avtomatika i Telemekhanika, Vol. 21, No. 9, pp. 1240-1249, September, 1960

Original article submitted March 23, 1960

Some fundamental problems in communication theory and control theory are compared. The problem of designing an optimum (in the statistical sense) closed-loop dual-control system, is formulated. Its solution, as well as examples and some generalizations, will be given in parts II, III, and IV.

Introduction

A general block diagram of signal transmission, as investigated in communication theory is shown in Fig. 1. The transmitted signal x^* proceeds from the transmitting device A to the communication channel H^*. The mixing of signal and interference (noise) h^* now takes place. The resultant signal y^* represents a mixture of the transmitted signal and the interference. The resultant signal proceeds to the input of receiver B. The optimum receiver problem consists in obtaining the signal $\underline{x}$ at its output, such that it is, in a specified sense, closest to the transmitted signal x^* or to some transformation of the signal x^*. The mathematical side of the problems related to such systems has been the subject of important investigations by A. N. Kolmogorov [1], N. Wiener [2], C. E. Shannon [3], and A. Wald [4]. This type of system was investigated in the works on communication theory by V. A. Kotel'nikov [5], D. Middleton, D. Van Meter [6], and others. The cited works differ in their various approaches to the problem, but are all basically concerned with the investigation of the scheme represented by the block diagram in Fig. 1. The results obtained in the above-cited works, and in particular the Kolmogorov-Wiener theory, have proved useful in formulating the statistical theory of automatic-control systems. This theory has been expounded in the books of V. S. Pugachev [7], J. H. Laning, Jr., and R. H. Battin [8], and others. The fullest consideration has been given to the theory of linear systems. If a system is linear, then whatever the closed-loop system, it is easy to obtain an open-loop system equivalent to it. That is why the automatic-control systems, which, as a rule, are closed-loop systems, enable one to use the scheme as in Fig. 1, provided that the system is linear. Some complications and difficulties arise when the interference does not appear at the input of the system, but at the input of the controlled object – the latter being inside the closed-loop network. This also creates difficulties which are not, however, of a fundamental nature. More serious difficulties arise due to the bounds to which the power of the system's signals are subjected. The problem becomes more involved when the controlled object is nonlinear or it is required that an optimum control system, which often proves to be nonlinear, is designed. It is not always possible to proceed from an open nonlinear system to an equivalent closed-loop one; furthermore, this is an extremely involved process. In such a case, the open-loop scheme depicted in Fig. 1 cannot be used in practice. A number of attempts have been made to reduce approximately nonlinear systems to equivalent linear ones (see, for example, the paper of I. E. Kazakov [9]). Such studies are of considerable practical value but do not, in their present state, provide a means of estimating how close the obtained approximation is to the true solution; neither do they enable one to synthesize the optimum system.

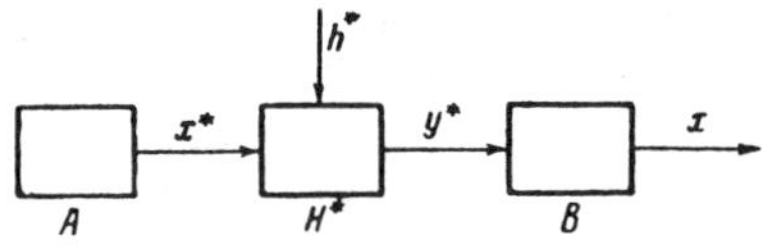

Fig. 1.

In order to be able to solve optimum problems of the control theory, a fundamentally different approach is required. Firstly, a different block diagram replacing that depicted in Fig. 1 is needed. Before selecting a common scheme to be used in automatic-control theory, it seems advisable to have a preliminary survey of certain basic concepts of the theory.

Figure 2, a shows the controlled object B with $\underline{x}$ as its output, $\underline{u}$ as the controller, and $\underline{z}$ as the disturbance (interference). When the system has several inputs and outputs, one can regard $\underline{x}$, $\underline{u}$, and $\underline{z}$ as vector quantities.

The output $\underline{x}$ depends on $\underline{u}$ and $\underline{z}$. This dependence can be described either by a linear or a nonlinear operator, and in a particular case of memoryless systems, only by a function. The interference $\underline{z}$ is generally a function

Reprinted with permission from *Optimal and Self-Optimizing Control*, The M.I.T. PRESS, Edited by Rufus Oldenburger, 1966, pp. 458–464.
Translated from *Avtomatika i Telemekhanika*, vol. 21, no. 9, pp. 1240–1249, Sept. 1960.
Copyright © 1966 by The Massachusetts Institute of Technology.

of time. Thus, since a change in the system's characteristics can be considered a particular result of interference (e.g., the parametric effect), then hereafter everything in the system's characteristics that changes with time will be attributed to interference. If, for example, x depends on u as in

$$x = [a_0 + f_0(t)]\, u^2 + [a_1 + f_1(t)]\, u + a_2 + f_2(t), \quad (1)$$

then the vector

$$z = \{z_0, z_1, z_2\},$$
$$z_0 = f_0(t), \; z_1 = f_1(t), \; z_2 = f_2(t) \quad (2)$$

gives the interference or disturbance, and the formula

$$x = [a_0 + z_0]\, u^2 + [a_1 + z_1]\, u + a_2 + z_2 \quad (3)$$

represents a particular operator.

If the object has memory and if, for example, its motion can be described by a differential equation of the nth order, then its state is considered as one of the characteristics of the object as described by the value of the vector x in the n-dimensional phase space.

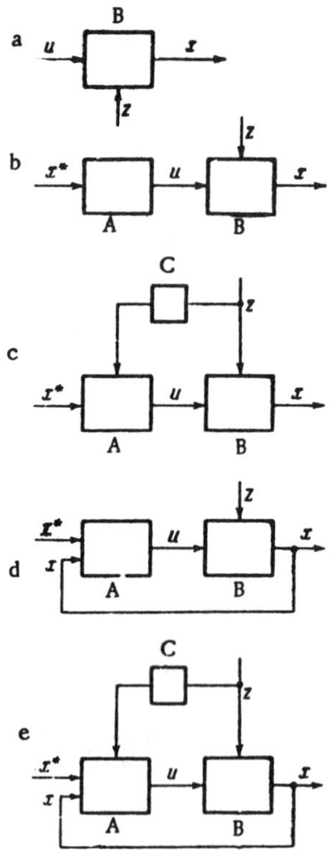

Fig. 2.

Complete information about a system thus consists of information about its operator, interference (noise), and the state of the system. The controlling device may be given and is considered as known. The open-loop systems are automatic-control systems of a simple type. A block diagram of an open-loop system is shown in Fig. 2, b and is of the same character as the one given in Fig. 1. The exciting quantity x* enters the input of the regulating member A, determining how the output quantity x should vary. The output u of the regulating device enters the input of the object B under control. The output x of the controlled object B does not proceed in this case to the regulating member A.

The required rule of change in x can only be implemented when full information about the controlled system is available, i.e., when its operator and state x are known, at least at the initial moment of time, as well as the interference z. The latter should be known a priori at all moments of time, including future moments. The required rule of change in x must be one of the admissible ones, such that it can be implemented for the given class of initial states of the system and for a class of controlling motions u staying within acceptable bounds.

The above conditions, and, in particular, the a priori full knowledge of the controlled system, cannot be satisfied in practice. This is why an accurate implementation of the required control rule cannot be obtained. Sometimes the interference z is not known a priori, but it is possible to measure it with a device C (see Fig. 2, c) and to introduce the outcome of the measurement into the controlling member A. One can then find in the latter the required controlling rule u. Such a scheme is also an open-loop one. But the scheme depicted in Fig. 2,c differs in some ways from the scheme with complete a priori information about the system, as now future magnitudes of the interference z remain unknown. Because of this, the exact implementation of the required rule of variation of the controlling quantity x is not always feasible.

When the state x of the system is not known, then, generally speaking, it is not possible to implement the required rule of the change in x. To be able to attain the required variation in x, or one near to it, a feedback network is needed to feed the output quantity x to the input of the controlling member A (see Fig. 2, d). Having compared x and x*, the controlling member generates the regulating action u, bringing x to its required value. The block diagram of Fig. 2, d is a closed-loop scheme, and is of the utmost importance in the automatic-control theory.

A closed-loop network offers far-reaching possibilities not available in an open-loop system. For example, it may be possible for a class of objects B of control to obtain a process x close to the one required even when the interference z remains unknown and incapable of measurement. Let, for example, the interference z, together with the controlling action u, be applied at the input of the controlled object B, the latter representing an inertia member. If the quantity $x^* = x$ is to be attained, the regulating member can be implemented in the form of an amplifier of high gain $k \gg 1$, the difference $x^* - x$ being sent to its input.

It is not difficult to see that the requirement $x^* = x$ will be satisfied approximately, whatever the continuously varying interference z, provided k is sufficiently high and the bounds of variability of u are such that the interference z can be compensated by u.

This principle of neutralizing the interference can be generalized and applied to cases considerably more involved, combining the system's accuracy with its stability. A detailed analysis of the applicability of this principle was carried out by M. V. Meerov in his monograph [10].

When the interference z can be measured, it is possible to implement a combined system (see Fig. 2, e) of measurement of the state of the controlled system x as well as of its interference. Such systems are of considerable practical value. We shall not, however, concern ourselves with them but shall limit the study to the "pure" type of closed-loop systems, considering them as being of primary importance.

The input quantity x^* may be previously unknown; usually, neither do we have any prior knowledge of the interference z. Consequently, these processes become random, and, in a favorable case, the a priori information is limited to our knowledge of their statistical characteristics. Such processes may be regarded as belonging to a class of curves $x^*(\lambda)$ and $z(\mu)$ where λ and μ are parameter vectors $(\lambda_1, \ldots, \lambda_q)$ and $(\mu_1, \ldots, \mu_m)$, respectively, with their probability distributions either known or unknown.

In communication channels connecting the blocks of a system, the errors of measurement or noise can be regarded as subsidiary random processes as well, with either known or unknown characteristics. Thus, the analysis of a control system and the synthesis of the regulating member can be regarded as problems of a statistical nature. The problem should be solved for an over-all block diagram in which all the above features of an automatic-control system are reflected. Such a block diagram is depicted in Fig. 3; it is the subject of the present paper as well as that of further papers in this series.

The input quantity x^* proceeds to the input of the controlling member A through channel H^* where it becomes mixed with noise h^*. Thus, the quantity y^* entering the input of A is generally not equal to the actual value of the input quantity x^*. There also exists a class of systems with the external input x^* altogether absent. Generally speaking, however, it cannot be neglected. A similar mixing takes place of the state x of controlled object B and noise h in channel H; quantity y entering A will not, as a rule, be equal to x. The regulating action u proceeds next from A to controlled object B having previously passed through channel G where it was mixed with noise g. The quantity v proceeding to the controlled object is not, as a rule, equal to u.

Dual Control

One cannot neutralize, in a general case, the interference z by a regulation u if the interference z is not known. Its direct measurement is not, as a matter of fact, often possible. In such a case, an open-loop system is useless. But the closed system in Fig. 3 shows how z can be indirectly determined by measuring the input and the output, in and out of object B, by studying its characteristics. The input of controlling member A enters both the input v and the output x of the object or, in any case, the quantities u and y related to v and x. The examination of the quantities u and y provides information on the characteristics of object B. It should be understood that this information is never complete, as the noises g and h render an exact measurement of B's characteristics impossible; if the actual form of the object's operator is not known either, a full determination of its characteristics would not be possible even in the absence of noise, unless the determination time is infinitely great. The lack of complete information on the disturbance z can assume the form of an a posteriori probability distribution of its parameters. Although the latter does not provide precise values of the parameters, it is more accurate than an a priori distribution, as the former reflects the real character of the interference.

If the random process can be measured directly, one is able eventually to specify its statistical characteristics more accurately. The method which provides such improvement with the aid of dynamic programming was discussed in examples by R. Bellman and R. Kalaba [11, 12] and also by M. Freimer [13]. One is able to find the characteristics of the process x^* more accurately in the open part of the block diagram in Fig. 3, or in a similar scheme in Fig. 1.

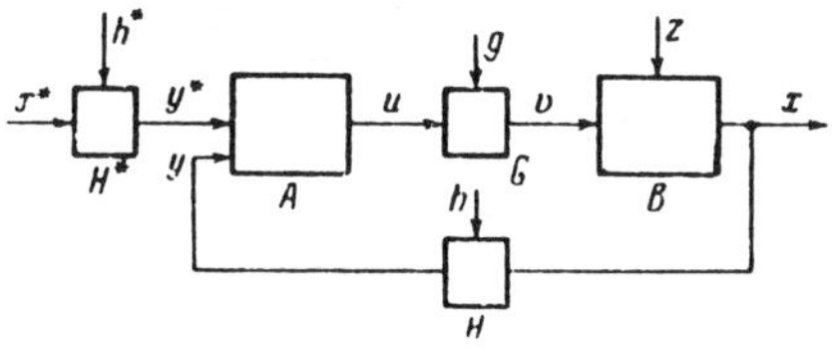

Fig. 3.

This formulation of the problem is characteristic for an open system. In a closed system its formulation becomes totally different. It is shown that some processes in the system of Fig. 3 may occur which have no counterpart in open-loop systems. Whereas open systems can only be studied by passive observations, the study may develop into an active one in closed systems. In order to improve the investigation one may vary the signals u (or v) which act on the controlled object B. The object is, as it were, "reconnoitered" by signals of an enquiring character whose purpose it is to promote a more rapid and more accurate study of the object's characteristics and of the methods of controlling it.

However, the controlling movements are necessary not only to study or to learn the characteristics of the object or the ways of controlling it, but also to implement the regulation, to direct the object to the required state. Thus, the controlling effects in the general block diagram in Fig. 3 must be twofold: they must, to a certain extent, be investigating as well as directing.

The control whose regulating effects are of this twofold character will in the sequel be called dual control; the papers in the present series will be devoted to tne theory of dual control.

Dual control is particularly useful and even indispensible in cases where the operator and the interference z in the object B are complex, and the object is thus distinguished either by its complexity or by the variability of its characteristics. Some typical examples of systems with dual control are to be found in automatic search systems, in particular, in automatic optimization systems (see, for example, [14 and 15]). In these systems, the investigating or "trial" part can usually be separated easily from thecontrolling or "operating" part of the signal, either by the difference in their frequency ranges or because they interweave in time. Such a separation, however, need not always take place; an effect can be twofold in character by virtue of being partly diagnostic and partly regulating.

Thus, in dual-control systems, there is a conflict between the two sides of the controlling process, the investigational and the directional. An efficient control can only be effected by a well-timed action on the object. A delayed action weakens the control process. But the control can only be effective when the properties of the object are sufficiently well known; one needs, however, more time to become familiar with them. A too "hasty" controlling member will carry out the operational movement without making proper use of the results of trial investigations performed on the object. A too "cautious" system will bide its time unnecessarily long and process the received information without directing the object to its required state at the right time. In each case, the control process may not prove the best one and may not even prove to be up to the mark. Our problem is to find out, one way or another, which combination of these two sides of the regulation would prove to be most suitable. The operations must be so selected as to maximize a criterion of the control's quality.

As shown above, the incomplete information about the object will be expressed by the presence of the probability distributions of potentially possible characteristics of the object. The regulating member compares, as it were, the various hypotheses on the object, with probability of its occurrence being attached to each hypothesis. These probabilities vary with time. There may be a control method such that the most probable hypothesis will always be selected and, therefore, assuming that it is valid, the optimum control method will be attained. Such a control system is not generally optimum in the absolute meaning of the word as the complete information on the object has not been utilized. The probability distribution of the different hypotheses extracted from the experiments is distorted as the probability 1 was ascribed to one of them and the probability 0 to others. A better control method will be one whereby the probabilities of all the hypotheses would be taken into account.

The probability distribution of hypotheses will vary with time, the higher probabilities concentrating more and more in the region of those hypotheses which approach the true characteristics of the object. The pace of concentration and, therefore, the success of the subsequent regulating movements, depends on the character of the preceding regulating movements, on how well they have "sounded" the object. Thus, two factors should be taken into account by the controlling member which decides the specific amount of regulating movement at any given moment of time:

(a) The loss occurring in the value of the quality criterion due to the fact that the outcome of the operation at a given moment, and at subsequent moments of time, will cause a deviation of the object either from the required state or from the best attainable one. The average value of this loss shall be called the action risk.

(b) The loss occurring in the value of the quality criterion due to the fact that the magnitude of the controlling action has not proved the best to obtain information on the characteristics of the object; in view of this, the subsequent actions will not be the best posssible ones either. The average value of this loss shall be called the investigation risk.

It will be shown that for a certain class of systems, the total risk will be equal to the sum of the action and investigation risks.

All systems of automatic search (see [14]) are characterized by trial actions. Dual control, therefore, is applicable to all systems of automatic search and, in particular, to automatic optimization systems. It can also be applied to other types of closed-loop systems which do not belong to the automatic search class at all. To illustrate the difference between the two types of dual-control systems, a few examples will be given.

Figure 4,a shows a system which operates as follows: the main regulating member A implements the control of object B, either in an open- or in a closed-loop network (the closed one is indicated by a dashed line). The

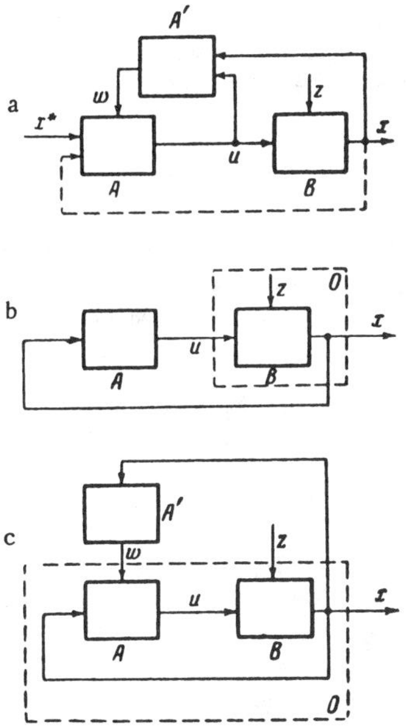

Fig. 4.

regulating movements u are of the investigational and action type simultaneously. The quantities u and x, from the input and the output of object B, respectively, enter an additional controlling member A'. The latter receives the characteristics of object B from the results of the investigations; subsequently, in accordance with an algorithm given in advance and fed into the device from outside, the parameters of controlling member A are so computed that its controlling action is optimal. Having the results of the computation, the additional regulating action w establishes the computed optimum parameters in the main controlling member A. This process may repeat itself periodically.

Such systems contain investigational movements; however, the automatic search is absent. The parameters of member A are established, not via automatic search, but from an algorithm given in advance, from a function of the determined characteristics of object B. There is no investigational component in the operation w. The channel w is not usually found in a closed network, as the change in the coefficients of A has no effect on the coefficients of B.

A block diagram of an automatic optimization system is presented in Fig. 4,b. Here the action u is dual in character, investigating object B as well as directing it to its optimum mode of action. The latter corresponds to an extremum of quantity Q dependent on x. The optimum mode is found by means of automatic search. The latter is conducted in such a way that the information received from the investigating action u and from the output x of the system is analyzed in controlling member A. This permits determination of the regulating part of the same action u on the same object B whose input and output were investigated, so that the same quantity x which was being investigated can be changed in the right direction.

The combination of the investigating and directing operations has, so far, not constituted the whole search, but only one of its distinct features. There is no search in the system depicted in Fig. 4,a; it takes place, however, in the system of automatic optimization shown in Fig. 4,b. But both systems are of the dual-control type.

In Fig. 4,b the controlled object is inside a dashed rectangle denoted by O. In this case, it does not differ from object B. The dual control, however, can also be applied to control the entire automatic system, considered as a complex object. For example, in Fig. 4,c the complex object O inside the dashed rectangle comprises the controlling member A and the object B of control. The auxiliary controlling member A' investigates the process x and, with the aid of the controlling process w, can vary the algorithm of regulation implemented by member A. The w processes are twofold in character. The investigation of changes in the algorithm of member A and their effect on the process x results in regulating processes w, bringing the algorithm of member A to such a form that the process x will either prove admissible, favorable, or optimum, depending for what purpose the system will be used. Here an automatic search takes place in the closed network of processes $w \to x \to u$.

Statement of the Synthesis Problem of an Optimum System of Dual Control

The problem of designing an optimum, in the specified meaning of the word, controlling member A, as shown in Fig. 3, is formulated below. It is advisable when formulating the problem to make use of certain concepts of the theory of games and of A. Wald's [4] theory of statistical decisions (see D. Blackwell and M. A. Girshick [16], and also Chow [17]). In solving the variational problem as stated later in the present series of papers, use is made of the concepts of R. Bellman's dynamic programming (see, for example, [18]). In the subsequent parts of this series, the mathematical exposition may appear somewhat cumbersome but is actually quite simple. The main contents of the papers deals with further development of the concepts of automatic control briefly described above.

Consider the scheme presented in Fig. 3. The following limitations of the statement of the problem are introduced.

1) A discrete-continuous system is investigated in which the time but not the level is quantized. All magnitudes occurring in the system are considered at discrete moments of time $t = 0, 1, 2, \ldots, n$ only. Any

magnitude at the sth moment of time will carry the index s. Thus, the considered quantities are x^*_s, x_s, y_s, v_s, g_s, etc.

Such limitation enables one to simplify the computation. Moreover, in many cases this actually occurs. The transition to the continuous time can in some cases be accomplished in an intuitive manner by making the time interval between the discrete values approach zero (see Part III). One meets with considerable difficulties in more fully examining the passage to the limit.

2) The time interval, or the number of cycles n within which the process is being investigated, is assumed to be a fixed constant. In certain cases no major difficulties arise when proceeding to the limit with $n \to \infty$. A wider generalization relating to a variable number n of cycles not known beforehand would be of interest, but will not be tackled in the present paper.

3) A Bayesian problem, in which a priori densities of random variables are given, is considered. Other formulations, for example, minimax, are also of considerable interest, but far more difficult to solve. This problem could also be formulated in relation to the concept of the so-called "inductive probability" (see, for example, the paper of L. S. Schwartz, B. Harris, and A. Hauptschein [19]).

We assume that h^*_s, h_s, g_s, are sequences of independent random variables with identical distribution densities $P(h^*_s)$, $P(h_s)$, $P(g_s)$. Further, let $z_s = z(s, \mu)$, and $x^*_s = x(s, \lambda)$ where μ and λ are random parameter vectors with coordinates μ_i and λ_i, respectively:

$$\mu = (\mu_1, \ldots, \mu_m), \qquad \lambda = (\lambda_1, \ldots, \lambda_q). \tag{4}$$

The a priori probability densities $P(\mu)$ and $P(\lambda^*)$ are given.

4) The object B is assumed to be memoryless; in other words, the values x_s of its output depend only on the values of the input quantities z_s and v_s at the same moment of time:

$$x_s = F_0(z_s, v_s). \tag{5}$$

The functions F_0 and z_s are assumed to be finite and single-valued, continuous and differentiable.

A generalization relating to objects with memory and with x_s depending on x_r, z_r, v_r $(r < s)$ will be given in Part IV. It should be pointed out that memoryless objects are of great practical value. Namely, if the input data (initial conditions or values of parameters) are given for a certain model, and one is able to carry out experiments using this model and also to register the results, then such an object becomes equivalent to a memoryless one.

5) A simple criterion W of quality is introduced. Let the partial loss function corresponding to the sth time moment be of the form

$$W_s = W(s, x_s, x^*_s). \tag{6}$$

Moreover, let the total loss function W for the total time be equal to the sum of partial loss functions (such a criterion shall be called a simple one):

$$W = \sum_{s=0}^{s=n} W(s, x^*_s, x_s). \tag{7}$$

The smaller the mathematical expectation of W, the better is the system. It shall be called optimum when its average risk R (i.e., the mathematical expectation M of the quantity W) is minimal. The amount of risk is given by the formula

$$R = M\{W\} = \tag{8}$$

$$= M\left\{\sum_{s=0}^{s=n} W(s, x^*_s, x_s)\right\} = \sum_{s=0}^{s=n} M\{W_s\} = \sum_{s=0}^{s=n} R_s.$$

Each $R_s = M\{W_s\}$ will be called a partial risk due to the sth cycle.

There may be many types of simple criteria, for example,

$$W_s = \alpha(s)\,[x_s - x^*_s]^2. \tag{9}$$

Criteria of practical importance need not always be simple, and generalizations relating to other criteria would therefore be of interest.

The formulation of the optimum strategy problem in terms of risks is not the only one in existence. There exist a number of studies in which closed systems are investigated from the point of view of the information theory (see, for example, R. L. Dobrushin's paper [20]). As the primary aim of a control system does not lie in transmitting information but in designing required processes, the formulation of the problem in the language of statistical decisions fits in better with the intrinsic nature of the problem.

6) All the quantities occurring in the sth cycle will be regarded as scalar. Our object, therefore, has only a single input v and a single output x. The exposition becomes more involved with generalizations relating to objects with several inputs and outputs (see Part IV).

7) We assume that the manner by which the signal and the noise are combined in H^*, H, or G blocks is

known and invariable, and that the blocks are memoryless. Thus,

$$v = v(u, g),$$
$$y^* = y^*(h^*, x^*), \quad (10)$$
$$y = y(h, x).$$

Therefore, the conditional probabilities $P(y^*|x^*)$ and $P(y|x)$ and $P(v|u)$ make sense.

8) We assume that the controlling member A generally possesses a memory and that, moreover, for the sake of generality, the algorithm of its action is a random one, i.e., the part A exhibits random strategy.

We introduce the vectors $(0 \leq s \leq n)$:

$$\mathbf{u}_s = (u_0, u_1, \ldots, u_s),$$
$$\mathbf{y}_s^* = (y_0^*, y_1^*, \ldots, y_s^*), \quad (11)$$
$$\mathbf{y}_s = (y_0, y_1, \ldots, y_s).$$

The controlling member can now be characterized by the probability densities

$$P_s(u_s) = \Gamma_s(u_s, \mathbf{u}_{s-1}, \mathbf{y}_s^*, \mathbf{y}_{s-1}) \quad (0 \leqslant s \leqslant n). \quad (12)$$

The problem consists in finding a sequence of functions F_s such that the average risk R (see [8]) becomes minimal.

LITERATURE CITED

1. A. N. Kolmogorov "Interpolation and extrapolation of stationary random sequences," Izvest. AN SSSR Ser. Matem. 5, 1 (1941).
2. N. Wiener, Extrapolation, Interpolation, and Smoothing of Stationary Time Series (J. Wiley and Sons, New York, 1949).
3. C. E. Shannon, "A mathematical theory of communication," Bell System Techn. J. 27, 3 (1948).
4. A. Wald, Statistical Decision Functions (J. Wiley and Sons, New York; Chapman and Hall, London, 1950).
5. V. A. Kotel'nikov, Theory of Potential Noise Stability [in Russian] (Gosénergoizdat, 1956).
6. D. Van Meter and D. Middleton, Modern Statistical Approaches to Reception in Communication Theory. Trans. IRE, IT-4 (Sept., 1954).
7. V. S. Pugachev, Theory of Random Functions and Its Applications to Automatic Control [in Russian] (Gostekhizdat, 1957).
8. J. H. Laning, Jr. and R. H. Battin, Random Processes in Automatic Control (McGraw-Hill, New York, 1956).
9. I. E. Kazakov, "An approximate statistical analysis of accuracy of essentially nonlinear systems," Avtomat. i Telemekh. 17, 5 (1956).*
10. M. V. Meerov, The Synthesis of Networks of Automatic Control Systems of High Accuracy [in Russian] (Fizmatgiz, 1959).
11. R. Bellman and R. Kalaba, On Communication Processes Involving Learning and Random Duration. IRE National Convention Record, Part 4 (1959).
12. R. Bellman and R. Kalaba, On Adaptive Control Processes. IRE National Convention Record, Part 4 (1959).
13. M. Freimar, A Dynamic Programming Approach to Adaptive Control Processes. IRE National Convention Record, Part 4 (1959).
14. A. A. Fel'dbaum, Computers in Automatic Systems [in Russian] (Fizmatgiz, 1959).
15. A. A. Fel'dbaum, Problems of statistical theory of automatic optimization, Proc. of the 1st. International Congress of Automatic Control (IFAC) [in Russian] (Moscow, 1960).
16. D. Blackwell and M. A. Girshick, Theory of Games and Statistical Decisions [Russian translation] (IL, 1959).
17. C. K. Chow, An Optimum Character Recognition System Using Decision Functions. IRE Trans. EC-6, No. 4 (1957).
18. R. Bellman, "Dynamic programming and stochastic control processes," Information and Control 1, 3 (Sept., 1958).
19. L. S. Schwartz, B. Harris, and A. Hauptschein, Information Rate from the Viewpoint of Inductive Probability. IRE National Convention Record, Part 4 (1959).
20. R. L. Dobrushin, "Transmission of information in channels with feedback," Teor. Ver. i ee Prim. 3, 4 (1958).

*See English translation.

Actively Adaptive Control for Nonlinear Stochastic Systems

EDISON TSE, MEMBER, IEEE, AND YAAKOV BAR-SHALOM, MEMBER, IEEE

Invited Paper

Abstract—An adaptive control method can be classified into two categories, actively adaptive or passively adaptive, according to how the available information is being utilized in the on-line calculation of the control. An actively adaptive controller utilizes, in addition to the available real-time information, the knowledge that future observations will be made, and regulates its adaptation (learning). This is done by anticipating how future estimation will be beneficial to the control objective. On the other hand, a passively adaptive controller, while utilizing the available real-time measurements, does not account for the fact that future observations will be made. Thus any learning in such a case will occur in an accidental manner. This paper summarizes a recent research effort in the development of an actively adaptive control method for nonlinear stochastic systems. A new and simpler set of equations for the original algorithm is given that provides further insight into the concepts of probing and caution in adaptive control. Several examples are chosen to illustrate the actively adaptive control method.

I. Introduction

THE STUDY of adaptive control is motivated by the desire to design a feedback control law which can regulate its characteristics when drift variations in the plant parameters occur. Methods of adaptive control have been applied to many different fields such as aerospace, electrical systems, satellite, and star-tracking telescopes, hydraulic servomechanisms, nuclear reactors, and many others. As one extends the range of applicability, new adaptive control formulations and new results are being obtained. In the fifties, development of adaptive control was based mainly on deterministic theory. The main tool was stability theory. The system was designed such that it was insensitive to parameter variations and noise disturbances. In the sixties, stochastic optimal control was developed within the framework of dynamic programming. This development, together with the advancement in the field of computers, provided a new approach to adaptive control that includes the statistical descriptions of noise and parameter variations.

We may classify an adaptive control method as either *actively adaptive* or *passively adaptive*. An important element of adaptive control is the learning of the drifting parameters. As the process unfolds, additional information becomes available which will provide learning for the purpose of control. This information may come about accidentally through past control actions or as a result of active probing, which itself is a possible control policy. Thus "learning" is present, whether it is "accidental" or "deliberate." Since more learning may improve overall control performance, the probing signal may indirectly help in controlling the stochastic system. On the other hand, excessive probing should not be allowed even though it may promote learning because it is "expensive" in the sense that it will, in general, increase the expected cost performance of the system. A "good" control law must then regulate its adaptation (learning) in an optimal manner. An adaptive control method is called passively adaptive if learning is not planned in the manner described above; and it is called actively adaptive if learning is planned and regulated for the ultimate objective of the control. In general, an actively adaptive control gives a better performance, but it is also more complicated than a passively adaptive control.

While the dynamic-programming method was formulated by Bellman in the fifties [6], it was Feldbaum who first introduced the concept of actively adaptive control in 1960 [14]. Feldbaum called it *dual-control theory* since he noted that the control when applied to a system can have two effects: in addition to its effect on the state, the present control might affect the future state uncertainty. The subject was studied under the framework of dynamic programming by Aoki [1], [2], Åström [3], [4], Florentin [13], Dreyfus [11], and Meier [18] via some simple two-stage examples. Adaptive control algorithms of the passive type are, among others, those of Saridis and Dao [22], Lainoitis *et al.* [16], [17], [27], [12], and Åström and Wittenmark [5].

A control algorithm that possesses the dual-control characteristics was developed by Tse, Bar-Shalom, and Meier in 1973 [23], [24]. They also called the method *actively adaptive* because it regulates learning for control purposes. The method has been applied to several classes of simulated example problems [24], [25], and the results are very promising. Through these studies, the concept of active learning also becomes more clear. In this paper, we shall describe the actively adaptive control method as developed by Tse *et al.* [23]. Instead of the original set of equations derived by Tse *et al.* [23], a new set of equations derived by Bar-Shalom and Tse [9] is given in Section II. These two sets of equations are equivalent (see, e.g., Kendrick and Kang [15]), but the set of equations given here allows us to gain deeper understanding on the interplay between learning and control in an adaptive system. In Sections III and IV, two classes of control problems are considered. For one class of problems there are unknown parameters in the system, whereas, in the other class, no unknown parameters are present but nonlinearities in the system allow enhancement of the estimation.

Aside from the algorithm described in this paper, there are two other actively adaptive control algorithms proposed in the econometric literature (MacRae [19] and Chow [10]). However, we shall not discuss these algorithms here since they are developed for a special class of econometric models.

Manuscript received November 12, 1975; revised February 10, 1976. This work was supported in part by the Air Force Office of Scientific Research under Contracts F44620-73-C-0028 and F44620-74-C-0026 and in part by NSF under Grant GS32271.

E. Tse is with the Department of Engineering-Economic Systems, Stanford University, Stanford, CA 94305.

Y. Bar-Shalom is with Systems Control, Inc., Palo Alto, CA 94304.

Reprinted from *Proc. IEEE*, vol. 64, pp. 1172–1181, Aug. 1976.

II. An Actively Adaptive Control Algorithm for Nonlinear Systems

Consider the system whose state, an n vector, evolves according to the equation

$$x(k+1) = f[k, x(k), u(k)] + v(k), \qquad k = 0, 1, \cdots, N-1 \tag{2.1}$$

and with observations (m vector)

$$y(k) = h[k, x(k)] + w(k), \qquad k = 1, \cdots, N \tag{2.2}$$

where $x(0)$ is the initial condition, a random variable with mean $\hat{x}(0|0)$ and covariance $\Sigma(0|0)$; $\{v(k)\}$ and $\{w(k)\}$ are the sequences of process and measurement noises, respectively, mutually independent, white, and with known statistics up to second order. For simplicity, we shall assume they are zero-mean. For the purpose of deriving the control algorithm, no assumptions about the distributions of these random variables are needed.

The cost function is taken as

$$C(N) = \psi[x(N)] + \sum_{k=0}^{N-1} L[x(k), k] + \phi[u(k), k] \tag{2.3}$$

and the performance index is

$$J(N) = E\{C(N)\}. \tag{2.4}$$

In the case of a linear system with unknown parameters, x is the "augmented" state, a stacked vector that includes the unknown parameters.

Rather than using the exact information state $\{Y^k, U^{k-1}\}$, the following approximate "wide-sense" information state is used

$$\mathcal{P}^k = \{\hat{x}(k|k), \Sigma(k|k)\} \tag{2.5}$$

i.e., the conditional mean and covariance of $x(k)$. The computation of $\mathcal{P}^k$ can be done by a number of approximate methods, e.g., extended Kalman filter, second-order filter, or nonlinear filter.

Assume now that the system is at time k and a closed-loop control [8] is to be computed using $\mathcal{P}^k$ and the present knowledge (statistical) about the future observations.

The cost-to-go for the last $N-k$ steps is

$$C(N-k) = \psi[x(N)] + \sum_{j=k}^{N-1} L[x(j), j] + \phi[u(j), j]. \tag{2.6}$$

The principle of optimality [6] with the information state (2.5) yields the following stochastic dynamic programming equation for the closed-loop-optimal expected cost-to-go at time k:

$$J^*(N-k) = \min_{u(k)} E\{L[x(k), k] + \phi[u(k), k] + J^*(N-k-1)|\mathcal{P}^k\}. \tag{2.7}$$

The main problem is to obtain an approximate expression for $J^*(N-k-1)$ while preserving its closed-loop feature, i.e., this expression should incorporate the "value" of the future observations. Note that $J^*(N-k-1)$ is obtained by the closed-loop minimization [7], [8] of $C(N-k-1)$. In order to find an explicit solution to this minimization, the cost-to-go C for the last $N-k-1$ steps is expanded about a nominal trajectory as follows. Let the nominal trajectory be

$$x_0(j+1) = f[j, x_0(j), u_0(j)], \qquad j = k+1, \cdots, N-1 \tag{2.8}$$

where $u_0(j)$, $j = k+1, \cdots, N-1$, is a sequence of nominal controls (to be discussed later) and the initial condition for this nominal trajectory is taken as

$$x_0(k+1) = \hat{x}[k+1|k; u(k)] \tag{2.9}$$

i.e., the predicted value of the state at $k+1$ given $\mathcal{P}^k$ and the control (yet to be found) $u(k)$. The expansion of the cost-to-go (2.6) with k replaced by $k+1$ is, with terms up to the second order,

$$C(N-k-1) = C_0(N-k-1) + \Delta C_0(N-k-1) \tag{2.10}$$

where

$$C_0(N-k-1) \triangleq \psi[x_0(N)] + \sum_{j=k+1}^{N-1} L[x_0(j), j] + \phi[u_0(j), j] \tag{2.11}$$

is the cost along the nominal and

$$\begin{aligned}\Delta C_0(N-k-1) \triangleq{}& \psi'_{0,x}\delta x(N) + \tfrac{1}{2}\,\delta x'(N)\psi_{0,xx}\delta x(N) \\ &+ \sum_{j=k+1}^{N-1} [L'_{0,x}(j)\delta x(j) \\ &+ \tfrac{1}{2}\,\delta x'(j)\, L_{0,xx}(j)\delta x(j) \\ &+ \phi'_{0,u}(j)\delta u(j) + \tfrac{1}{2}\,\delta u'(j)\,\phi_{0,uu}(j)\delta u(j)]\end{aligned} \tag{2.12}$$

is the variation of the cost about the nominal. The notations $L_{0,x}$ and $L_{0,xx}$ stand for the gradient and Hessian of L with respect to x evaluated along the nominal trajectory, and

$$\delta x(j) = x(j) - x_0(j) \tag{2.13a}$$

$$\delta u(j) = u(j) - u_0(j) \tag{2.13b}$$

are the perturbed state and control, respectively.

The approximation of the closed-loop-optimal expected cost-to-go for the last $N-k-1$ steps is done now as follows.

$$\begin{aligned}J^*(N-k-1) &= \min_{u(k+1)} E\{\cdots \min_{u(N-1)} E[C(N-k-1)|\mathcal{P}^{N-1}] \cdots |\mathcal{P}^{k+1}\} \\ &= J_0(N-k-1) + \Delta J_0^*(N-k-1)\end{aligned} \tag{2.14}$$

where

$$J_0(N-k-1) = C_0(N-k-1) \tag{2.15}$$

$$\Delta J_0^*(N-k-1) = \min_{\delta u(k+1)} E\{\cdots \min_{\delta u(N-1)} \cdot E[\Delta C_0(N-k-1)|\mathcal{P}^{N-1}] \cdots |\mathcal{P}^{k+1}\}. \tag{2.16}$$

Note that the closed-loop minimization of (2.16) is over a cost quadratic in $\delta x(i+1)$, $\delta u(i)$, $i = k+1, \cdots, N-1$, as can be seen from (2.12). Furthermore, from the definition of the nominal trajectory (2.8) and the dynamics of the system (2.1), the perturbations (2.13) obey the following dynamic equation (with terms up to the second order; $f^i_{0,xx}$ denotes the Hessian of the ith component of f, $i = 1, \cdots, n$):

$$\begin{aligned}\delta x(j+1) = f_{0,x}(j)\delta x(j) + f_{0,u}(j)\delta u(j) \\ + \sum_{i=1}^{n} e_i \left[\tfrac{1}{2}\, \delta x'(j) f^i_{0,xx}(j)\delta x(j)\right. \\ + \delta u'(j) f^i_{0,ux}(j) \delta x(j) \\ \left. + \tfrac{1}{2}\, \delta u'(j) f^i_{0,uu}(j)\delta u(j)\right] + v(j), \\ j = k+1, \cdots, N-1 \end{aligned} \tag{2.17}$$

with initial condition

$$\delta x(k+1) = x(k+1) - x_0(k+1). \tag{2.18}$$

Thus the problem defined by (2.16) consists of the minimization of the quadratic cost (2.12) for the quadratic system (2.17) and is very similar to the linear-quadratic problem. Up to terms of the second order, the solution of this problem can be assumed to be of the form

$$\begin{aligned}\Delta J_0^*(N-k-1) = g_0(k+1) + E\{p_0'(k+1)\delta x(k+1) \\ + \tfrac{1}{2}\, \delta x'(k+1) K_0(k+1) \delta x(k+1) | \mathcal{Y}^{k+1}\}.\end{aligned} \tag{2.19}$$

The proof by induction of the above is given in Bar-Shalom and Tse [9].

To emphasize the closed-loop property of ΔJ_0^*, i.e., the manner in which it is a function of the future uncertainties, it is rewritten as follows (see [9] for detailed derivations):

$$\begin{aligned}\Delta J_0^*(N-k-1) = \gamma_0(k+1) + E\{p_0'(k+1)\delta x(k+1) \\ + \tfrac{1}{2}\, \delta x'(k+1) K_0(k+1)\delta x(k+1) | \mathcal{Y}^{k+1}\} \\ + \tfrac{1}{2} \sum_{j=k+1}^{N-1} \mathrm{tr}\, [K_0(j+1) Q(j) \\ + \mathcal{A}_{0,xx}(j) \Sigma_0(j|j)]\end{aligned} \tag{2.20}$$

where $\Sigma_0(j|j)$ is the updated covariance of the state along the nominal trajectory.

The variables $\gamma_0(k+1)$, $p(k+1)$, and $K_0(k+1)$ are given by the following recursions:

$$\begin{aligned}\gamma_0(j) = \gamma_0(j+1) - \tfrac{1}{2}\, H'_{0,u}(j) \mathcal{H}^{-1}_{0,uu}(j) H_{0,u}(j), \\ j = N-1, \cdots, k+1 \\ \gamma_0(N) = 0\end{aligned} \tag{2.21}$$

$$\begin{aligned}p_0(j) = H_{0,x}(j) - \mathcal{H}'_{0,ux}(j)\mathcal{H}^{-1}_{0,uu}(j) H_{0,u}(j), \\ j = N-1, \cdots, k+1 \\ p_0(N) = \psi_{0,x}.\end{aligned} \tag{2.22}$$

$$\begin{aligned}K_0(j) = \mathcal{H}_{0,xx}(j) - \mathcal{A}_{0,xx}(j), \\ j = N-1, \cdots, k+1 \\ K_0(N) = \psi_{0,xx}\end{aligned} \tag{2.23}$$

with

$$H_0(j) \triangleq L_0(j) + \phi_0(j) + p_0'(j+1) f_0(j) \tag{2.24}$$

$$\mathcal{H}_{0,xx}(j) \triangleq H_{0,xx}(j) + f'_{0,x}(j) K_0(j+1) f_{0,x}(j) \tag{2.25}$$

$$\mathcal{H}_{0,ux}(j) \triangleq H_{0,ux}(j) + f'_{0,u}(j) K_0(j+1) f_{0,x}(j) \tag{2.26}$$

$$\mathcal{H}_{0,uu}(j) \triangleq H_{0,uu}(j) + f'_{0,u}(j) K_0(j+1) f_{0,u}(j) \tag{2.27}$$

$$\mathcal{A}_{0,xx}(j) \triangleq \mathcal{H}'_{0,ux}(j) \mathcal{H}^{-1}_{0,uu}(j) \mathcal{H}_{0,ux}(j). \tag{2.28}$$

Note that $\gamma_0(k+1)$, $p_0(k+1)$, and $K_0(k+1)$ are *not* stochastic and depend only on the nominal control sequence being chosen. The function $H_0(j)$ defined by (2.24) is the Hamiltonian for the corresponding deterministic optimal control problem. Therefore, if the nominal control is a local minimum for the deterministic problem, $H_{0,u}(j) = 0$ and thus $\gamma_0(j) = 0$; also $\mathcal{H}_{0,uu}(j)$ is positive definite for all $j > k+1$. The latter implies that the perturbation analysis is valid. In fact, a necessary and sufficient condition for the existence and uniqueness of the optimum perturbation solution is that $\mathcal{H}_{0,uu}(j)$ must be positive definite (regardless of whether the nominal control is a local minimum). If $\mathcal{H}_{0,uu}(j)$ is positive definite, then $\mathcal{A}_{0,xx}(j)$ defined by (2.28) must be nonnegative definite, with rank equal to the dimension of the control vector.

Combining (2.20) with (2.14), the stochastic dynamic programming equation (2.7) that will yield $u(k)$ becomes

$$\begin{aligned}J^*(N-k) = \min_{u(k)} \Big\{ E\{L[x(k), k] + \phi[u(k), k] \\ + C_0(N-k-1) + \gamma_0(k+1) + p_0'(k+1)\delta x(k+1) \\ + \tfrac{1}{2}\, \delta x'(k+1) K_0(k+1) \delta x(k+1) | \mathcal{Y}^k\} \\ + \tfrac{1}{2} \sum_{j=k+1}^{N-1} \mathrm{tr}\, [K_0(j+1) Q(j) \\ + \mathcal{A}_{0,xx}(j) \Sigma_0(j|j)] \Big\}.\end{aligned} \tag{2.29}$$

From (2.9) and (2.18), it follows that

$$E[\delta x(k+1) | \mathcal{Y}^k] = E[\tilde{x}(k+1|k) | \mathcal{Y}^k] = 0 \tag{2.30}$$

and

$$\begin{aligned}E[\delta x'(k+1) K_0(k+1) \delta x(k+1) | \mathcal{Y}^k] \\ = \mathrm{tr}\, [K_0(k+1) \Sigma(k+1|k)].\end{aligned} \tag{2.31}$$

Finally, dropping from (2.21) the first term which does not depend on $u(k)$ and using (2.22) and (2.23), the closed-loop control is obtained as

$$u^{\mathrm{CL}}(k) = \arg\min J^{\mathrm{CL}}(N-k) \tag{2.32}$$

$$J^{\mathrm{CL}}(N-k) \triangleq J_D(N-k) + J_C(N-k) + J_P(N-k) \tag{2.33}$$

where

$$J_D(N-k) \triangleq \phi[u(k), k] + C_0(N-k-1) + \gamma_0(k+1) \tag{2.34}$$

is the deterministic part of the cost and

$$\begin{aligned}J_C(N-k) \triangleq \tfrac{1}{2}\, \mathrm{tr}\, [K_0(k+1) \Sigma(k+1|k)] \\ + \tfrac{1}{2} \sum_{j=k+1}^{N-1} \mathrm{tr}\, [K_0(j+1) Q(j)]\end{aligned} \tag{2.35}$$

$$J_P(N-k) \triangleq \tfrac{1}{2} \sum_{j=k+1}^{N-1} \mathrm{tr}\, [\mathcal{A}_{0,xx}(j) \Sigma_0(j|j)] \tag{2.36}$$

are the stochastic terms in the cost.

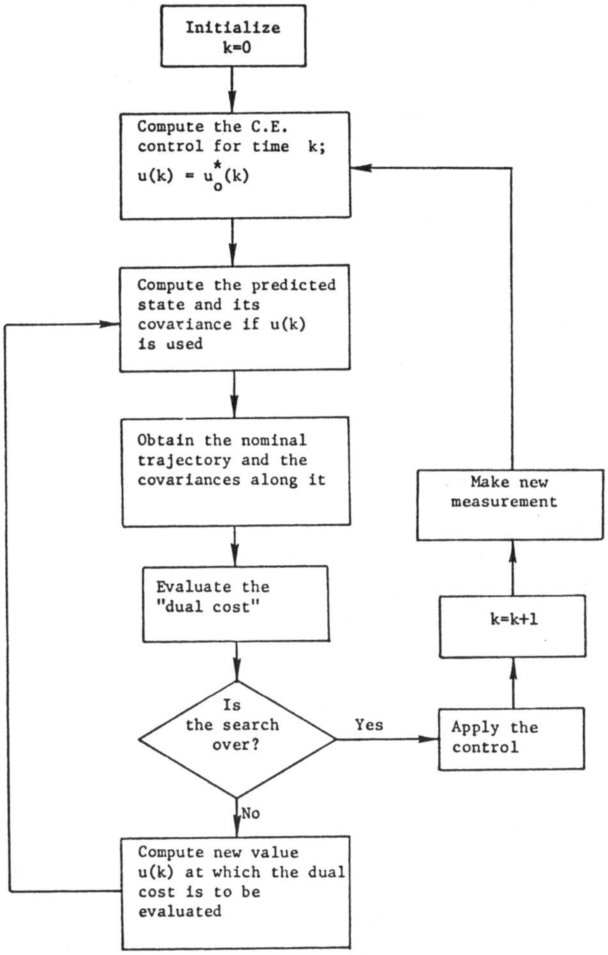

Fig. 1. Flow chart of the dual-control algorithm.

The first stochastic term, (2.35), reflects the effect of the uncertainty at time k and subsequent process noises on the cost. These uncertainties cannot be affected by $u(k)$ but their weightings do depend on it. The effect of these uncontrollable uncertainties on the cost should be minimized by the control; this term indicates the need for the control to be cautious and thus is called the *caution term*. The second stochastic term, (2.36), accounts for the effect of uncertainties when subsequent decisions will be made. As discussed above, if the perturbation problem has a solution, then the weighting of these future uncertainties is nonnegative ($\mathcal{A}_{0,xx}$ is positive semidefinite). If the control can reduce, by probing (experimentation), the future updated covariances, it can thus reduce the cost. The weighting matrix $\mathcal{A}_{0,xx}$ yields approximately the *value of future information* for the problem under consideration. Therefore this is called the probing term. That the rank of $\mathcal{A}_{0,xx}(j)$ is equal to the dimension of the control vector indicates that, along a certain nominal sequence, certain combinations of state uncertainties do not contribute (up to second order) to the total expected cost.

The benefit of probing is weighted by its cost and a compromise is chosen so as to minimize the sum of the deterministic, caution, and probing terms. The minimization of J^{CL} will also achieve a tradeoff between the present and future actions according to the information available at the time the corresponding decisions are made.

Preposterior analysis [21] can be seen as appearing explicitly in the decision on the present control which is to be done using the ("prior") estimate $\Sigma_0(j|j)$ of the future updated ("posterior") covariance $\Sigma(j|j)$ of the state.

To find the closed-loop control $u(k)$, the minimization of (2.33) is performed using a search procedure. At every k, to each control $u(k)$ for which (2.33) is evaluated during the search, there corresponds a predicted state (2.9) and to this predicted state a sequence of deterministic controls is attached that defines the nominal trajectory. The cost-to-go is then evaluated by expansion about this nominal and its variation (up to the second order) is minimized in a closed-loop fashion. This leads to (2.33), where the possible benefit from probing (active learning) as well as the need for caution appear explicitly. The only use of the nominals and perturbations is to make possible the evaluation of the cost-to-go optimized in a closed-loop manner. This procedure is repeated at each and every time that a new control is to be obtained. The algorithm will be referred to as actively adaptive or a dual-control algorithm in the following discussions.

A flow chart description of the algorithm is given in Fig. 1.

III. Control of Linear Systems with Unknown Parameters

In this section, two examples of controlling a third-order linear system with six unknown parameters will be presented to illustrate the algorithm developed above. The performance of this algorithm will be compared to those of the certainty equivalence (CE) control and the optimal control with the known parameters. The latter will serve as an unachievable lower bound. The cost is taken as quadratic in the state and control. The nominal trajectories used in the evaluation of the closed-loop cost-to-go were of the regulator type. In both examples, a second-order filter is used for real-time estimation. Details on this can be found in Tse *et al.* [26].

The examples illustrate the actively adaptive feature of the algorithm. In particular, we shall see how the actively adaptive controller anticipates the value of learning the system's unknown parameters according to the cost criterion.

The following third-order system is considered:

$$x(k+1) = A(\theta)\,x(k) + b(\theta)\,u(k) + v(k)$$
$$y(k) = [0 \;\; 0 \;\; 1]\; x(k) + w(k) \tag{3.1}$$

where

$$A(\theta) = \begin{bmatrix} 0 & 1 & 0 \\ 0 & 0 & 1 \\ \theta_1 & \theta_2 & \theta_3 \end{bmatrix} \qquad b(\theta) = \begin{bmatrix} \theta_4 \\ \theta_5 \\ \theta_6 \end{bmatrix} \tag{3.2}$$

and $\{\theta_i\}_{i=1}^{6}$ are unknown constant parameters with normal *a priori* statistics having mean and variance

$$\hat{\theta}(0|0) = [1, -0.6, 0.3, 0.1, 0.7, 1.5]' \tag{3.3}$$

$$\Sigma^{\theta\theta}(0|0) = \text{diag}\,(0.1, 0.1, 0.01, 0.01, 0.01, 0.1). \tag{3.4}$$

The true parameters are

$$\theta = [1.8 \; -1.01, 0.58, 0.3, 0.5, 1]'. \tag{3.5}$$

The initial state is assumed to be known:

$$\hat{x}(0|0) = x(0) = O. \tag{3.6}$$

Interception Type Example

In the first example, the objective is to bring the third component of the state to a desired value. This is expressed by the

TABLE I
SUMMARY OF RESULTS FOR THE INTERCEPTION EXAMPLE

CONTROL POLICY	OPTIMAL CONTROL WITH KNOWN PARAMETERS	C.E. CONTROL WITH UNKNOWN PARAMETERS	DUAL CONTROL WITH UNKNOWN PARAMETERS
AVERAGE COST	6	114	14
MAXIMUM COST IN A SAMPLE OF TWENTY RUNS	20	458	53
STANDARD DEVIATION OF THE COST	6	140	16
AVERAGE MISS DISTANCE SQUARED	12	225	22
WEIGHTED CUMULATIVE CONTROL ENERGY PRIOR TO FINAL STAGE	.1	1.4	3.2

cost

$$J = \frac{1}{2} E\left\{[x_3(N) - \rho]^2 + \sum_{i=0}^{N-1} \lambda u^2(i)\right\} \tag{3.7}$$

where ρ is some value and λ is chosen to be small. In this example, $\rho = 20$ and λ is chosen to be 10^{-3}. The noises $\{v_i(k)\}_{i=1}^{3}$ and $w(k+1)$ are assumed to be independent and are normally distributed with zero mean and unit variance. If we interpret x_3 as the position of an object, then this example corresponds to an "interception" problem; the guidance of an object to reach a certain point, without constraints on the velocity and acceleration of the object when it reaches that point. The difficulty lies in the fact that the *poles and zeros* of the system are both unknown. The initial condition (3.6) represents the fact that the system is initially at rest.

Twenty Monte Carlo runs were performed on the interception example, and average performances are summarized in Table I and Figs. 2–4. As shown in Table I, the dual-control algorithm's performance is an order of magnitude better than the CE control's performance. The second and third rows indicate that the dual-control performance is much more predictable than the CE control. Note that the dual control uses only about twice the energy of the CE control, at the same time achieving a dramatic improvement in the miss distance squared over the CE control. This indicates that the dual control does use control energy at appropriate times to improve learning and thus achieves satisfactorily the control objective.

As seen in Fig. 4, which shows the cumulative control energy (sum of the control values squared), the dual control invested considerable effort for learning at the beginning. The results of this can be seen mainly in θ_4, θ_5, and θ_6 (Fig. 3). On the other hand, the learning of θ_1, θ_2, and θ_3 (Fig. 2) was only slightly different than in the CE case. Nevertheless the dual control's performance is quite close to the unachievable lower bound given in the first column of Table I.

Soft Landing Example

In the second example, instead of bringing only the third component of the state to a desired value, the objective is to bring the final state to a certain point in the state space. This is expressed by

$$J = \frac{1}{2} E\left\{[x(N) - \boldsymbol{\rho}]' [x(N) - \boldsymbol{\rho}'] + \sum_{i=0}^{N-1} \lambda u^2(i)\right\} \tag{3.8}$$

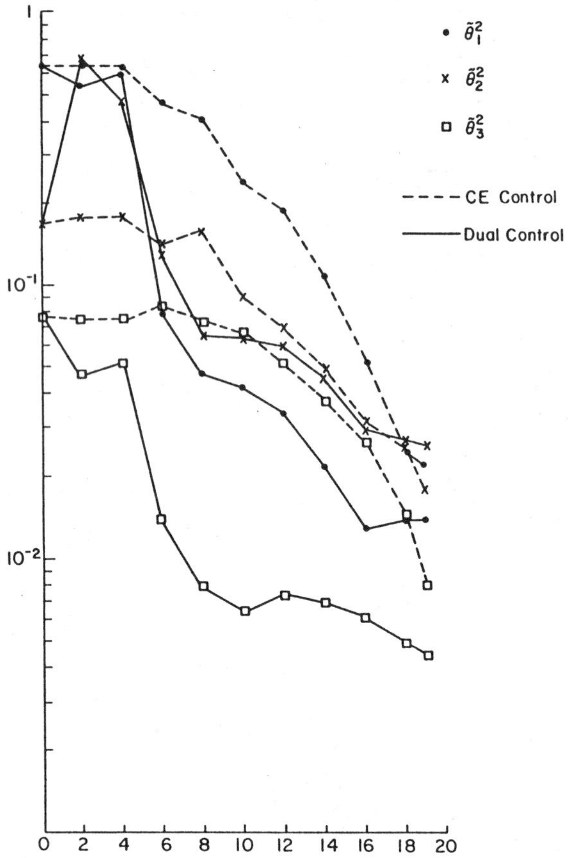

Fig. 2.

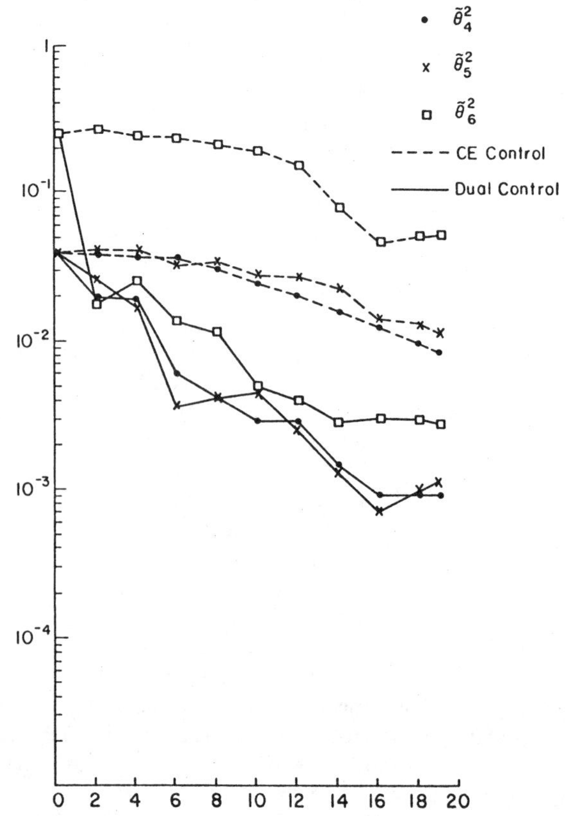

Fig. 3.

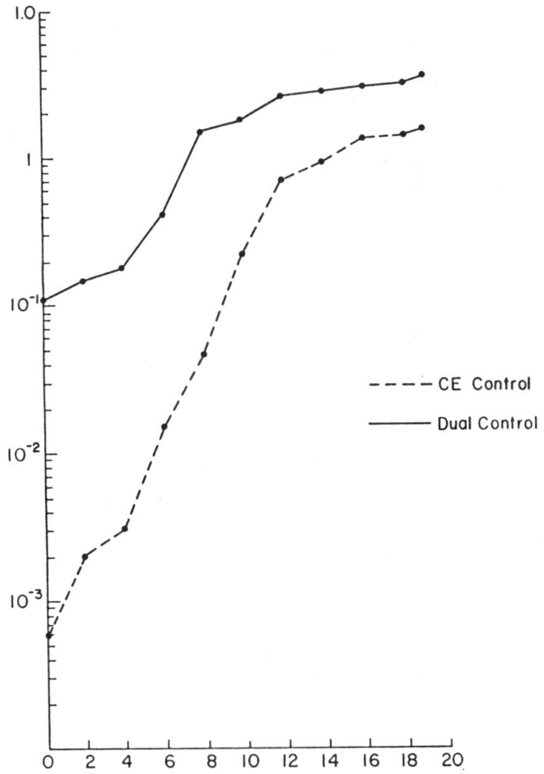

Fig. 4.

where $\boldsymbol{\rho}$ is a point in R^3 and λ is as before. This may be interpreted as a "soft landing" problem by selecting the terminal desired state to be

$$\boldsymbol{\rho} = \begin{bmatrix} 0 \\ 0 \\ 20 \end{bmatrix}. \tag{3.9}$$

Twenty Monte Carlo runs were carried out with known parameters for the CE control, the dual control, and the optimal control. The results are summarized in Table II and Figs. 5–7.

The soft landing is a "harder" problem than the interception problem because the aim now is to reach a point in the state space, while before the aim was to reach a surface. Therefore, it should be expected that the average cost is higher than in the previous example. This is seen to hold true as shown in Tables I and II for the dual and optimal controls with known parameters. However, for CE control, the opposite is true. This may look strange at first sight but a careful analysis of the simulation will offer an explanation for this.

Table II indicates the improvement of dual control over CE control, both in average performance and reliability. The terminal miss distance squared for the dual control is very close to the unachievable lower bound given by the optimal control with known parameters. To achieve this small miss distance, the dual control invests considerable effort for learning purposes. This can be seen in Fig. 7 where it is shown that a large amount of energy is invested at the initial time to promote future learning. As a result, the parameters are adequately learned and the dual control smoothly brings the system to a point close to $\boldsymbol{\rho}$.

Comparison of the Two Examples

To illustrate passive and active learning in stochastic control, the results of the two examples above will be compared. First,

TABLE II
SUMMARY OF RESULTS FOR THE SOFT-LANDING EXAMPLE

CONTROL POLICY	OPTIMAL CONTROL WITH KNOWN PARAMETERS	C.E. CONTROL WITH UNKNOWN PARAMETERS	DUAL CONTROL WITH UNKNOWN PARAMETERS
AVERAGE COST	15	104	28
MAXIMUM COST IN A SAMPLE OF TWENTY RUNS	35	445	62
STANDARD DEVIATION OF THE COST	9	114	11
AVERAGE MISS DISTANCE SQUARED	28	192	32
WEIGHTED CUMULATIVE CONTROL ENERGY PRIOR TO FINAL STAGE	1	7	12

consider the CE controls in the two cases. Note that the CE control energy used in the soft-landing example is much more than that used in the interception example. From Figs. 4 and 7, it can also be seen that, up to about $k = 12$, the CE controller uses about the same cumulative energy for the two examples. The fact that the ultimate aim is different has not yet become important enough to change the control strategy. As a consequence, the learning in both cases is about the same up to this time. In the first example, since the final destination is a surface, the controller can wait almost until the final time to apply a control to achieve the ultimate objective. Therefore, the CE control is still applying little energy after time $k = 12$. The knowledge of the parameters θ_4, θ_5, and θ_6 has only slightly improved. However, for the second example, since the final destination is a point in the state space, the control must work "harder" to achieve its objective (transferring from one point to another arbitrary point requires three time units in the deterministic case). Therefore, the control energy after time $k = 12$ increases very quickly for the second example. This results in a much better estimation of the gain parameters. Since the learning in the first example is much poorer than in the second example for the CE control, a higher cost is accrued in the first case than in the second. Note that, even though the second problem is "harder," a lower value is obtained for the cost. This is primarily because the "accidental" learning is enhanced by the difficulty of achieving the final aim.

For the dual control, quite a different control strategy at the beginning rather than at the end of the control interval can be noticed. The fact that a different end condition has to be fulfilled is anticipated by the control from the initial time, because it is of the closed-loop type. For the second example, the dual controller, realizing that the final mission is much more difficult to achieve, decides to invest more energy in the beginning, because learning is very important in this case to satisfactorily achieve the final objective. Note the "speed" of learning in the second example compared with the first example (see Figs. 2, 3, 5, 6). The dual control regulates its energy in learning: in the first example, where learning is less important, it does not insist on learning by applying large controls in the beginning; in the second example, the learning is much more important and thus more energy is utilized for the learning purpose. For both examples, the expected miss distances squared are comparable, thus, the increase in cost in the interception example is primarily due to the increase in cumulative input energy. This demonstrates the active-learning characteristic of the dual control.

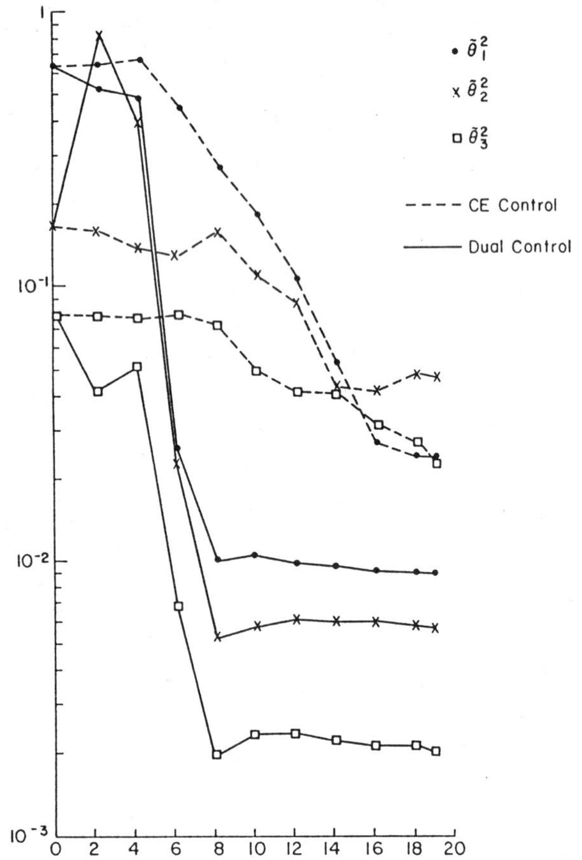

Fig. 5.

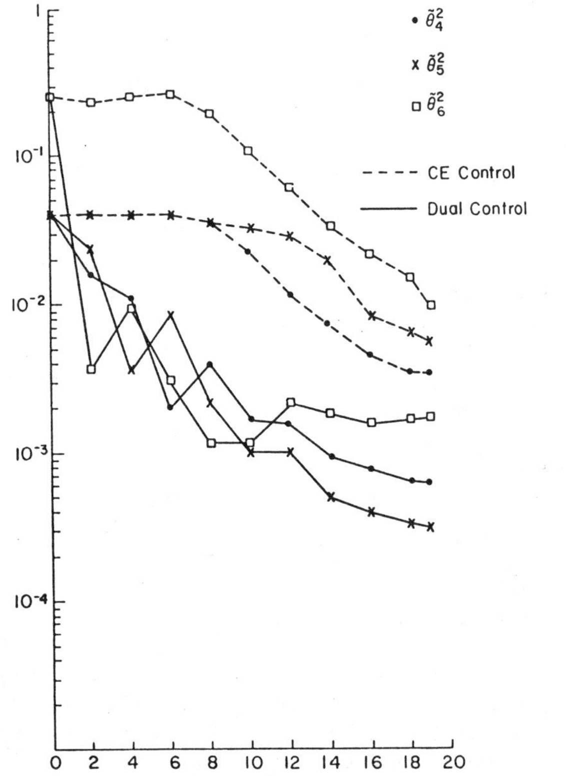

Fig. 6.

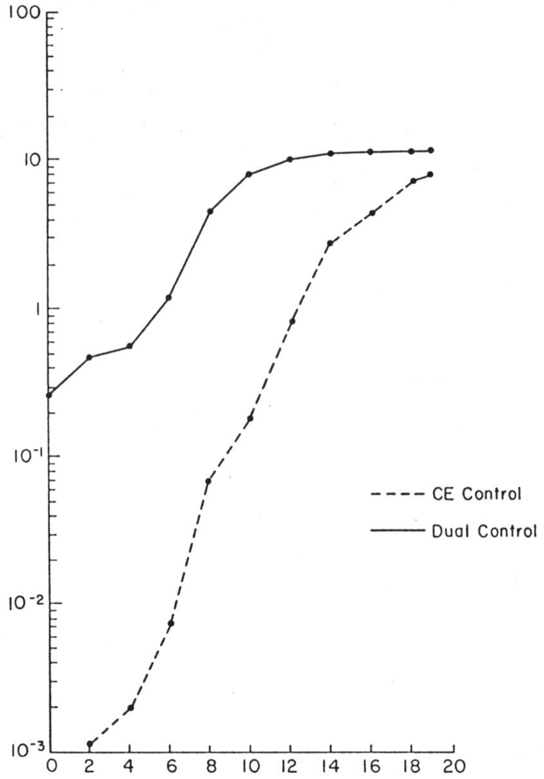

Fig. 7.

IV. Free End-Time Problem

In this section, we consider a problem where real-time shaping of the trajectory will significantly improve the estimation performance and thus improve the control performance. This reflects another important characteristic of the dual control: the capability of bringing the system state to a certain region which will improve state estimation.

Consider a missile which is moving in a plane with constant velocity. The only control is the lateral acceleration command whose magnitude is constrained. With a point-mass description, we have the dynamical equations given by

$$\ddot{x} = -\dot{y}\, u$$
$$\ddot{y} = \dot{x}\, u \tag{4.1}$$

where (x, y) is the position of the missile in the xy plane, and u is the lateral acceleration command.

Defining $x_1 = x$, $x_2 = \dot{x}$, $x_3 = y$, $x_4 = \dot{y}$, we have the state equation

$$\dot{x}(t) = f[x(t), u(t)] + v(t) \tag{4.2}$$

where

$$f[x(t), u(t)] = \begin{bmatrix} x_2(t) \\ -x_4(t)u(t) \\ x_4(t) \\ x_3(t)u(t) \end{bmatrix} \tag{4.3}$$

and $v(t)$ is the additive white Gaussian process noise with mean O and covariance $Q(t)\,\delta(t - \zeta)$. The process noise may account for the fact that the modelling is not exact.

We shall consider two cases separately. In the first case, we assume that at each sampling time t_k only an angle measurement is available. This would be the situation where the missile is only equipped with a passive sensor device. The observation equation is

$$y(t_k) = \tan^{-1}\frac{x_3(t_k)}{x_1(t_k)} + w(t_k), \qquad k = 0, 1, \cdots \tag{4.4}$$

where $w(t_k)$ is a white Gaussian noise with mean zero and covariance $r\delta_{kj}$. We further assume that $w(t_k)$ and $v(t_k)$ are independent. In the second case, we assume that at each sampling time measurements on range as well as angle are available. The measurement equations are

$$y_1(t_k) = \tan^{-1}\frac{x_3(t_k)}{x_1(t_k)} + w_1(t_k)$$

$$y_2(t_k) = \sqrt{x_1^2(t_k) + x_3^2(t_k)} + w_2(t_k) \tag{4.5}$$

where $w(t_k)$ is a white Gaussian noise with mean zero and covariance diag (r_1, r_2). Again $w(t_k)$ and $v(t_k)$ are independent. The sampling intervals are denoted as

$$t_{k+1} - t_k = \Delta_k. \tag{4.6}$$

One physical restriction is that the control $u(t)$ has magnitude bounded by some finite value, i.e.,

$$\mathcal{U}_k = \{u : |u| \leqslant a\}, \qquad k = 0, 1, \cdots. \tag{4.7}$$

This constraint will restrict the maneuverability of the missile; it will also restrict the enhancement of estimation that the dual control can achieve. It will be shown that despite this limitation the dual control can significantly improve the estimation.

The objective is to intercept a certain point in the plane. By appropriate translation, we can reformulate the problem as one in which the target is at the origin. The initial state of the missile is uncertain. Note that, because of the nonlinear observations, the covariance of the updated state will depend on the value of the state and thus depend on past controls, i.e., there is a dual effect in this problem.

The CE control used in this simulation was to align the heading of the missile with the line of sight of the target as quickly as possible. Using a second order expansion of (4.2), this control is obtained as

$$u^{CE}(t_j) = \begin{cases} a, & \text{if } u_j > a \\ -a, & \text{if } u_j < -a \\ \bar{u}_j, & \text{otherwise} \end{cases} \tag{4.8}$$

where, dropping for simplicity the arguments $(t_j|t_j)$,

$$\bar{u}_j = (\hat{x}_2\hat{x}_3 - \hat{x}_1\hat{x}_4)[(\hat{x}_1\hat{x}_2 + \hat{x}_3\hat{x}_4)\Delta_j + (\hat{x}_2^2 + \hat{x}_4^2)\Delta_j^2/2]^{-1}. \tag{4.9}$$

The stopping rule for this problem was chosen as follows. Let $\hat{x}(t_k|t_k)$ be the updated estimate of $x(t_k)$ at time t_k. If we apply a control value u within the interval $[t_k, t_k + \delta)$, where $\delta \leqslant \Delta_k$, then the predicted distance from the origin at $t_k + \delta$ can be approximated closely by the following second-order expansion of (4.2):

$$\rho_k(\delta, u) = \left[\hat{x}_3 + \hat{x}_4\delta + \hat{x}_2\frac{\delta^2}{2}u\right]^2 + \left[\hat{x}_1 + \hat{x}_2\delta - \hat{x}_4\frac{\delta^2}{2}u\right]^2. \tag{4.10}$$

At every time k, the time-to-go was obtained by minimizing the above expression with respect to δ for $u = u^{CE}(t_k)$ as given by (4.8) with $j = k$. If the resulting ϕ_k was less than Δ_k, then the process was to be stopped at $\tau = t_k + \phi_k$ after applying $u^{CE}(t_k)$. If ϕ_k exceeded Δ_k, then a nominal trajectory was computed, starting from $x^0(t_{k+1}) = \hat{x}[t_{k+1}|t_k; u(t_k)]$, using CE controls, as in (4.8), based on the nominal state. At every step the time-to-go was evaluated for the nominal, as in (4.10), to see whether the process should be terminated or continued.

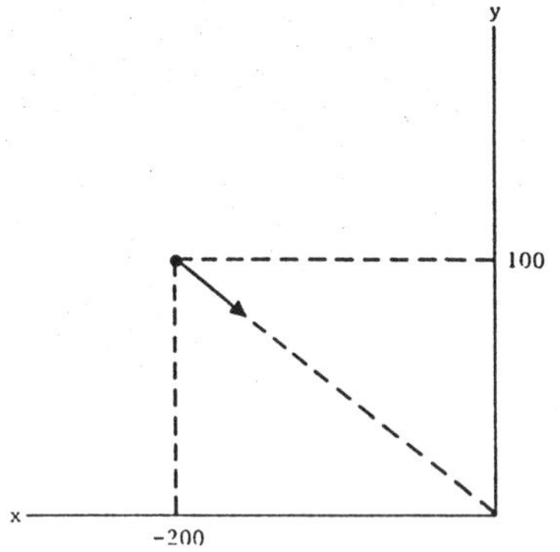

Fig. 8.

To complete the problem description, we have the cost performance given by

$$\psi[x(\tau)] = x_1^2(\tau) + x_3^2(\tau) \qquad \phi[u(\sigma)] = \lambda u^2 \qquad L[x(\sigma)] = 0 \tag{4.15}$$

with $\lambda = 10^{-2}$. Note that we do not attach any cost on the trajectory of the state other than its endpoint. This allows the control to make maneuvers so as to improve the estimation if it finds improvement necessary.

A modification of the control algorithm described in Section II that will enable us to handle a free-end time problem as described above was made by Tse and Bar-Shalom [25]. Simulation studies were performed for this example problem.

In the simulations, it was assumed that the initial position is given by $x_1(t_0) = 200$ and $x_3(t_0) = 100$. The initial velocity vector is given by $x_2(t_0) = 12$, $x_4(t_0) = -6$ (see also Fig. 8). Note that the initial condition of the missile is "pointing" towards the origin. The sampling periods were taken equal to unity, i.e., $\Delta_k = \Delta = 1$. The dual-control strategy is compared with the CE control. For state estimation, an adaptive filtering technique described in [26] was used in all cases. It was found that the use of a first- or second-order filter would not give reasonable estimation performance because of the high nonlinearity. Two cases were discussed.

Case 1. Angle sensor only on board (with a standard deviation of 0.5° or 1°).

Case 2. Angle sensor and range sensor on board (with standard deviations of 1° and 5°, respectively).

The initial estimate of the position is obtained by using one angle and one range measurement with the above-mentioned accuracies. The initial velocity uncertainty has a standard deviation of unity in both components. The maximum lateral acceleration was taken as $a = 9°/s^2$. The covariance of the process noise was $Q = \text{diag}\,(0, 10^{-2}, 0, 10^{-2})$.

Twenty Monte Carlo runs were performed for both cases and the results are summarized in Tables III and IV. We shall compare these results, which will give us more understanding about the importance of dual control. From Table III, we see that even though the small number of Monte Carlos do not allow

TABLE III
COMPARISON OF DUAL AND CE CONTROL FOR THE CASE WHERE ONLY ANGLE MEASUREMENT IS AVAILABLE (CASE 1)

		$\sqrt{R} = 1^\circ$		$\sqrt{R} = 1/2^\circ$	
		C.E.	Dual	C.E.	Dual
Terminal Miss Distance Squared	Mean	122	76	128	68
	Standard Deviation	106	56	106	47
Average Terminal Covariance		224	109	181	84

TABLE IV
COMPARISON OF DUAL AND CE CONTROL FOR THE CASE WHERE THERE ARE RANGE AND ANGLE MEASUREMENTS (CASE 2)

		C.E.	Dual
Terminal Miss Distance Squared	Mean	10.9	8.2
	Standard Deviation	10.2	8.1
Average Terminal Covariance		8.8	7.9

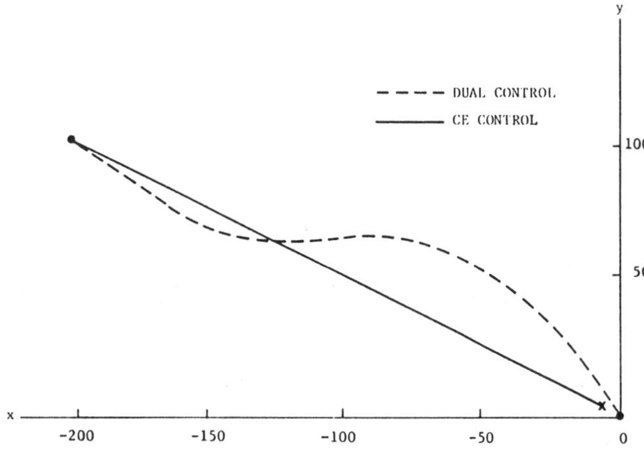

Fig. 9.

one to draw any conclusion with high confidence, the results clearly indicate that the use of dual control will improve the system performance. This is achieved because of the enhanced estimation due to the dual control. The effect of the control on the state uncertainty is seen in the final prediction error covariance, also shown in Tables III and IV.

Sample trajectories for CE and dual control for case 1 are shown in Fig. 9. It is seen that the dual-control trajectory will deviate from the nominal CE trajectory substantially in order to improve the estimation. Note that, since the measurement at each time is the line-of-sight angle, if the control aligns the line-of-sight with the heading as quickly as possible, this will prevent accumulation of further information about the distance to the target, and thus the distance uncertainty will be the same as the initial uncertainty, or even worse because of the process noise. By deviating and oscillating around the nominal CE trajectory, information about this distance can be extracted from the observations. However, since the degree of maneuverability is limited, the dual control tries to improve the estimation while making sure that it can redirect the missile to the origin in time. Therefore, improving the angle measuring device will yield larger benefits for the dual control's performance than for the CE control's performance (see Table III).

TABLE V
PERFORMANCE OF "WIGGLING" CE CONTROL

		Case 1 ($\sqrt{R} = 1^\circ$)	Case 2
Terminal Miss Distance Squared	Mean	79	8.6
	Standard Deviation	62	8.1
Terminal Covariance		160	8.0

Runs were also performed for the case where the initial condition of the missile is pointing in a horizontal direction towards the vertical axis. This would allow some accidental learning for the CE control. The results obtained are practically the same as in Table III and thus are not repeated here. These results indicate that the dual control did enhance estimation so as to improve the system performance.

In case 2, the control problem allows accumulation of further information about the range from the on-board range sensor. Nevertheless, the results in Table IV clearly indicate that the dual control yields an improvement over the CE control even though not as large as before. This happens because the dual control still has the capability of enhancing the estimation, as can be seen from the prediction error covariance.

Sample trajectories for CE and dual control are very similar to those for case 1, with slightly less maneuver for the dual control. However, if the range measurement error increases, dual control becomes more important (case 1 is the limiting case when the range error covariance goes to infinity).

In both cases, we note a "typical"shape of the dual-control trajectory. A control strategy which will give such trajectory shape has the following characteristics.

a) Maximum control magnitude in one direction (say +) for a certain period (say p_1 steps).
b) Maximum control magnitude in another direction (say −) for a certain period (say p_2 steps).
c) CE control applied for the rest of the period until terminated.

A simple control law can be designed with these characteristics. Consider a class of control laws as described above with tuning parameters (p_1, p_2). Simulations were performed so as to search for the pair (p_1, p_2) which will give optimum performance. It was found that in both cases, the optimum pair is

$$p_1 = 5, \quad p_2 = 6$$

and the corresponding performances are given in Table V. Comparing Table V with Tables III and IV, we see that such a simple strategy does give comparable performance with the dual control *while being as simple as the CE control.*

V. CONCLUSIONS

One distinguishing characteristic of the method described in the previous sections is that it explicitly takes into account that future measurements will be made and that these future measurements will provide additional information about the uncertain parameters which will enhance future control.

Therefore, learning is planned in accordance with the control objective.

In many classes of problems where the interplay between learning and control has an important role, the method is very effective. The two problems discussed in Sections III and IV are examples of these classes. By proper choice of the future nominals, it is believed that the algorithm will yield a control strategy which would be close to the optimum one. The method is also useful for the case of drifting parameters where the dynamic of the drift is unknown (Pekelman and Tse [20]).

The major disadvantage of the proposed method is its computational complexity which has to be carried out in real time. As an example, the time required for the computation of a dual-control algorithm is about seven times that of the CE control for the problem described in Section III. Thus the method may prove to be more useful in problems where the physical sampling interval is on the order of minutes or hours rather than those problems where the sampling interval is on the order of microseconds or seconds. An effort in trying to develop useful yet simple adaptive shochastic laws via the study of the dual-control method has also been attempted as shown in Section IV; however, much further work is to be done to take advantage of the control's actively adaptive feature in practical problems.

REFERENCES

[1] M. Aoki, "Optimal control of partially observable Markovian systems," *J. Franklin Institute*, pp. 367-386, Nov. 1965.

[2] ——, *Optimization of Stochastic Systems*. New York: Academic Press, 1967.

[3] K. J. Åström, *Introduction to Stochastic Control Theory*. New York: Academic Press, 1970.

[4] ——, "Optimal control of Markov processes with incomplete state information," *J. Math. Anal. & Appl.*, vol. 10, pp. 174-205, 1965.

[5] K. J. Åström and B. Wittenmark, "On self-tuning regulators," *Automatica*, vol. 9, pp. 185-199, 1973.

[6] R. Bellman, *Dynamic Programming*. Princeton, NJ: Princeton University Press, 1957.

[7] ——, *Adaptive Control Processes: A Guided Tour*. Princeton, NJ: Princeton University Press, 1961.

[8] Y. Bar-Shalom and E. Tse, "Dual effect, certainty equivalence and separation in stochastic control," *IEEE Trans. Automat. Contr.*, vol. AC-19, pp. 494-500, Oct. 1974.

[9] ——, "Concepts and methods in stochastic control," in *Control and Dynamic Systems: Advances in Theory and Applications*, C. T. Leondes, Ed. New York: Academic Press, 1975.

[10] G. Chow, *Analysis and Control of Dynamic Systems*. New York: Wiley, 1975.

[11] S. E. Dreyfus, *Dynamic Programming and the Calculus of Variations*. New York: Academic Press, 1965.

[12] J. G. Deshpande, T. N. Upadhyay, and D. G. Lainiotis, "Adaptive control of linear stochastic systems," *Automatica*, vol. 9, pp. 107-115, 1972.

[13] J. J. Florentin, "Optimal, probing, adaptive control of a simple Bayesian system," *J. Electr. & Control*, vol. 13, no. 2, pp. 165-177, 1962.

[14] A. A. Fel'dbaum, "Dual control theory I-IV," *Automation and Remote Control*, vol. 21, pp. 1240-1249, pp. 1453-1464, 1960; vol. 22, pp. 3-16, pp. 129-143, 1961; also *Optimal Control Systems*. New York: Academic Press, 1965.

[15] D. Kendrick and B. H. Kang, "An economist's guide to wide sense dual control," Project on Control in Economics, Dep. of Economics, Univ. of Texas, Austin, Report 75-1, 1975.

[16] D. G. Lainiotis, T. N. Upadhyay, and J. G. Deshpande, "Adaptive control of linear systems," in *Proc. 1971 IEEE Decision and Control Conf.*

[17] ——, "Optimal adaptive control of linear systems with unknown measurement subsystems," *J. Information Sciences*, vol. 6, pp. 217-233, 1973.

[18] L. Meier, "Combined optimal control and estimation," in *Proc. 3rd Allerton Conf. Systems and Circuits*, 1965.

[19] E. C. MacRae, "Linear decision with experimentation," *Annals of Economic and Social Measurement*, vol. 1, pp. 437-447, 1972.

[20] D. Pekelman and E. Tse, "Experimentation and control in advertisement: an adaptive control approach," in preparation.

[21] H. Raiffa and R. Schlaifer, *Applied Statistical Decision Theory*. Cambridge, MA: M.I.T. Press, 1972.

[22] G. N. Saridis and T. K. Dao, "A learning approach to the parameter-adaptive self-organizing control problem," *Automatica*, vol. 9, pp. 589-598, Sept. 1972.

[23] E. Tse, Y. Bar-Shalom, and L. Meier, "Wide-sense adaptive dual control of stochastic nonlinear systems," *IEEE Trans. Automat. Contr.*, vol. AC-18, pp. 98-108, Apr. 1973.

[24] E. Tse and Y. Bar-Shalom, "An actively adaptive control for discrete-time systems with random parameters," *IEEE Trans. Automat. Contr.*, vol. AC-18, pp. 109-117, Apr. 1973.

[25] ——, "Adaptive dual control for stochastic nonlinear systems with free end-time," *IEEE Trans. Automat. Contr.*, vol. AC-20, pp. 600-605, Oct. 1975.

[26] E. Tse, R. Dressler, and Y. Bar-Shalom, "Application of adaptive tuning of filters to exoatmospheric target tracking," in *Proc. 3rd Symp. Nonlinear Estimation Theory*, Univ. of California, San Diego, Sept. 1972.

[27] T. N. Upadhyay and D. G. Lainiotis, "Joint adaptive plant and measurement control of linear stochastic systems," *IEEE Trans. Automat. Contr.*, vol. AC-19, pp. 567-571, Oct. 1974.

Adaptive Stochastic Control for a Class of Linear Systems

EDISON TSE, MEMBER, IEEE, AND MICHAEL ATHANS, SENIOR MEMBER, IEEE

Abstract—**The problem considered in this paper deals with the control of linear discrete-time stochastic systems with unknown (possibly time-varying and random) gain parameters. The philosophy of control is based on the use of an open-loop feedback optimal (OLFO) control using a quadratic index of performance. It is shown that the OLFO system consists of 1) an identifier that estimates the system state variables and gain parameters and 2) a controller described by an "adaptive" gain and correction term. Several qualitative properties and asymptotic properties of the OLFO adaptive system are discussed. Simulation results dealing with the control of stable and unstable third-order plants are presented. The key quantitative result is the precise variation of the control system adaptive gains as a function of the future expected uncertainty of the parameters; thus, in this problem the ordinary "separation theorem" does not hold.**

I. Introduction

THE control of linear systems whose dynamics are not completely known is a problem of major theoretical and practical importance. The development of theoretical and design tools for adaptive control has been the subject of a multitude of investigations during the past fifteen years. Nonetheless, there is still a lack of a general method for formulating an "adaptive control problem." Often so-called adaptive systems utilized clever ways of using nonlinear feedback to compensate for parameter changes. Others used feedback to make the design insensitive to plant parameter variations.

There is much to be understood of the precise meaning of adaptive control. At least intuitively, one can set down certain qualitative properties of adaptive systems. Two such properties are the following.

1) The adaptive system should have some "learning" device which generates information about the unknown system variables. This can be thought of as an *estimation subsystem* which generates estimates of the plant state variables and plant parameters on the basis of perhaps noisy measurements.

2) The adaptive system should have some *controller* that generates the on-line inputs to the plant. It is reasonable to expect that the controller will have certain gains that operate upon the estimated state variables. Furthermore, it is reasonable to expect that these gains will depend also on the estimates of the plant parameters.

If one adopts this estimation–controller conceptual subdivision of an adaptive system, then one is interested in their interrelation. Clearly, the *estimation subsystem* will operate upon past data and past controls to generate current and future state and parameter estimates. It is to be expected that *control signals will affect the estimation accuracy since cleverly designed input waveforms can be used to excite only specific modes, isolate the effects of certain gain parameters, and regulate the signal-to-noise ratio at the sensor*. However, control signals that are good for estimation and identification may not be good for on-line control. This *duality* nature of the control was pointed out by Fel'dbaum [2].

The design of the gains in the *controller subsystem* should be based upon our estimate of the future plant responses. Thus, the control signals that we apply *now* must be such that the *future* plant response is in some sense optimal. This requires knowledge of the plant parameters. However, *well-designed adaptive systems should take into account not only the instantaneous parameter estimates but also the confidence that one can attach to these values.*

It is the opinion of the authors, that the latter point is of extreme importance in the design of adaptive control systems. To illustrate this, suppose that the estimate $\hat{\alpha}$ of a parameter α turns out to be the same at two different instants of time t_1 and t_2, i.e.,

$$\hat{\alpha}(t_1) = \hat{\alpha}(t_2), \qquad t_2 > t_1.$$

Let us also suppose that the system has "learned" something about α during the time interval $t \in [t_1, t_2]$. At the very least, this should be exhibited by a difference in the variance of the estimation error $\hat{\alpha}(t) - \alpha$. Hopefully,

$$\text{var}\,[\hat{\alpha}(t_1) - \alpha] > \text{var}\,[\hat{\alpha}(t_2) - \alpha].$$

If the identification of α is essential for future control (it may not be), then *other things being equal*, one would expect that the control to be applied at $t = t_1$ would be different from the control to be applied at $t = t_2$. Thus, the control (or the controller gains) must somehow depend on the *level of uncertainty* of the parameter estimation process. It is in this manner that the control can attempt to balance the often conflicting demands of the identification accuracy and system response. Because, if this were not the case, the knowledge about $\hat{\alpha}$ exhibited by the decrease in the variance of $(\hat{\alpha} - \alpha)$ would not have been utilized by the adaptive system, and hence it would not have acted any better in the presence of increased knowledge.

Manuscript received April 19, 1971; revised September 3, 1971. Paper recommended by J. M. Mendel, past Chairman of the IEEE S-CS Adaptive and Learning Systems, Pattern Recognition Committee. This research was carried out at the Decision and Control Sciences Group of the M.I.T. Electronic Systems Laboratory with support extended in part by the U.S. Air Force Office of Scientific Research under Grants AFOSR 69-1724 (Supplement A) and AFOSR 70-1941 and in part by NASA under Grant NGL-22-009-(124).

E. Tse is with Systems Control, Inc., Palo Alto, Calif. 94306.
M. Athans is with the Department of Electrical Engineering, Massachusetts Institute of Technology, Cambridge, Mass. 02139.

Reprinted from *IEEE Trans. Automat. Contr.*, vol. AC-17, pp. 38–51, Feb. 1972.

Such considerations indicate that the computation of adaptive controls would be more complex than that required in stochastic control in which the *separation theorem* [17] is enforced. The study presented in this paper illustrates this point vividly, in the sense that *the gains of the control system depend upon the parameter error covariance matrix.*

This paper deals only with a small class of problems in adaptive control where the gain vector of a discrete linear system is unknown. Essentially, the case that is treated is a generalization of the discrete-time version of controlling a linear system whose poles are known but whose zeros are not. Admittedly, such a situation is rare in actual practice; however, this research points out some of the complexities that will arise in controlling a linear system with unknown poles and zeros. Such problems have been considered by several people. It was noted that practical on-line computation of the optimal closed-loop control is not currently feasible due to computer limitations. Therefore, one must sacrifice true optimality in the hope of developing a suboptimal stochastic control algorithm which has a chance of being computationally feasible with current and projected computers. In the literature, different suboptimal solution methods were proposed. Farison *et al.* [3] enforced separation to obtain a suboptimal control scheme. Murphy [4] and Gorman and Zaborszky [5] proposed to approximate Bellman's equation. Bar-Shalom and Sivan [6], Dreyfus [9], Curry [18], Aoki [1], and Spang [16) used the open-loop feedback optimal approach. However, the latter approaches did not contain explicit quantitative results which indicate the precise way that the control gains are affected by parameter accuracy. Also, nontrivial simulation results that illustrate the qualitative and quantitative aspects of the problem are not available.

This paper is a condensed version of a two-part conference paper [15] and is based, in part, upon the first author's Ph.D. dissertation [13]. This paper stresses concepts and results. Detailed proofs and additional simulation results can be found in [13] and [15].

The structure of the paper is as follows. In Section II the problem formulation and the statistical assumptions are set forth. The open-loop feedback optimal (OLFO) approach is discussed. In Section III we present the identification equations that generate estimates of the plant state variables and the plant parameters as well as the associated error (auto- and cross-) covariance matrices; these are used to formulate in a precise manner the OLFO control problem. In Section IV the solution of the OLFO problem is presented. In Section V the solution is interpreted in a feedback sense. Section VI contains discussions of the qualitative properties of the derived adaptive system. The asymptotic behavior of the identification equation is considered and precise conditions under which the exact identification of the unknown parameters is obtained as the number of measurements tend toward infinity. In Section VII simulation results obtained for third-order systems (both stable and unstable) are presented and interpreted in the light of the general qualitative properties given in Section VI.

II. Problem Statement

Let us consider the discrete-time linear system[1]

$$S: \begin{cases} x(k+1) = A(k)x(k) + b(k)u(k) + \xi(k) \\ y(k) = C(k)x(k) + n(k) \end{cases} \tag{1}$$

where $x(k)$, $\xi(k) \in R^n$, $y(k)$, $n(k) \in R^m$, $A(k)$ is a *known* $n \times n$ matrix, $C(k)$ is a *known* $n \times m$ matrix, and $u(k)$ is a scalar control. *We assume that the "gain" vector $b(k)$ is unknown*, but we know that it satisfies the difference equation

$$b(k+1) = G(k)b(k) + \gamma(k) \tag{2}$$

where $G(k)$ is a *known* $n \times n$ matrix and $\gamma(k) \in R^n$. It is assumed that the vectors $[x(0), b(0), \xi(k), n(k), \gamma(k); k = 0, 1, \cdots]$ are mutually independent (white) Gaussian random vectors with known statistical laws:

$$\begin{aligned} &x(0) \sim g(x_0, \Sigma_{x0}) \qquad b(0) \sim g(b_0, \Sigma_{b0}) \\ &\xi(k) \sim g(0, R(k)) \qquad n(k) \sim g(0, Q(k)) \\ &\gamma(k) \sim g(0, N(k)); \qquad \text{with } \Sigma_{x0} > 0, \quad \Sigma_{b0} > 0, \quad R(k) \geq 0, \\ &\qquad\qquad Q(k) \geq 0, \; N(k) \geq 0. \end{aligned} \tag{3}$$

Thus, there is discrete white plant noise ($\xi(k)$), observation noise ($n(k)$), and "parameter noise" ($\gamma(k)$) that is used to model "unpredictable" effects in the state evolution, measurements, and parameter variations, respectively. The notation $y \sim g(\alpha, A)$ is used to denote that the random vector y is Gaussian, with known mean α and known covariance matrix A.

Our objective is to find a control sequence $[u(0), \cdots, u(N-1)]$ such that the cost functional

$$J(u) = \frac{1}{2} E\left\{x'(N)Fx(N) + \sum_{k=0}^{N-1} x'(k)W(k)x(k) + h(k)u^2(k)\right\} \tag{4}$$

is minimized subject to (1) and (2). The expectation is taken over all underlying random quantities. We shall assume that F and $W(k)$ are nonnegative-definite symmetric weighting matrices and that $h(k)$ is a positive scalar for each k (control weighting).

Depending on the kinds of *admissible* controls that we are allowed to choose, different formulations of the stochastic optimization problem are possible. In the most general setting, we may assume that the control is a *random function* of the observed data, i.e., $u(k)$ is a conditional probability measure on the control space. If the conditional probability measure is regular, then the control is said to be a *mixed* control law. If the conditional probability measure is singular (Radon measure), then the control is said to be a *pure* control law. Unfortunately, little can be done at this level of generality, where we consider both mixed and pure control laws.

In the next level of generality, we may confine ourselves

[1] Only the scalar control case is presented. The extension to the vector case is conceptually straightforward; however, the equations are extremely complex.

to consider only pure control laws to be admissible, i.e., the control at each instant is a *fixed* function of the observed data; in this case, the resulting control will be a random variable through its dependence on the random observed data. This type of restriction of admissible control leads to Bellman's equation [8] whose solution may only be approximated.

Finally, we may restrict ourselves to consider only deterministic open-loop controls to be admissible; this essentially means that we ignore the (zero-mean) random vectors and assume that the system will behave according to its average behavior. Of course, this may not lead to a good control system, especially whenever the covariances of the disturbances are large. The approach that we shall employ will be to recompute the open-loop optimal deterministic control after reevaluating the state of uncertainty of the system at each and every step (time). A control sequence which is optimal in this manner is called the *open-loop feedback optimal* control [9]. We remark that the OLFO approach to stochastic control problems was considered by Aoki [1], Bar-Shalom and Sivan [6], Spang [16], and Curry [18].

In this paper, we shall seek the OLFO control. We shall see that an OLFO control sequence is *adaptive* in nature.

III. Formulation of the OLFO Control Problem

The present time is indexed by k. Let us assume that the control sequence $U^*(0, k-1) \triangleq [u^*(0), u^*(1), \cdots, u^*(k-1)]$ has been applied to the system, and that the observation sequence $Y_{U^*(0,k-1)}(0, k) \triangleq [y_{U^*(0,k-1)}(i)]_{i=0}{}^k$ observed. We would like to find a "future" control sequence $U^*(k, N-1) \triangleq [u(k), \cdots, u(N-1)]$ so as to minimize the future cost (cost-to-go) conditioned on the total available information at the present time. Let us denote the σ-algebra generated by the observed data $Y_{U^*(0,k-1)})(0, k)$ as $\mathfrak{F}(k, U^*(0, k-1))$; the symbol $U^*(0, k-1)$ is used to denote that the data is really dependent on the past control history. Our aim now is to find the control sequence $U(k, N-1)$ such that the average cost-to-go

$$J(U(k, N-1); U^*(0, k-1), k) \triangleq \frac{1}{2} E\Big\{x'(N)Fx(N) + \sum_{j=k}^{N-1} x'(j)W(j)x(j) \,\Big|\, \cdot\mathfrak{F}(k, U^*(0, k-1))\Big\} + \frac{1}{2}\sum_{j=k}^{N-1} h(j)u^2(j) \tag{5}$$

is minimized subject to the constraints (1) and (2). The cost has the simple form (5) *because in OLFO problems the future control sequence $U(k, N-1)$ is assumed to be deterministic*. (If the future controls were assumed to depend on observed data, we could not take the last term of (5) outside the expectation operation.)

Define for $j \geq k$,

$$\hat{x}(j|k) \triangleq \hat{x}(j|k, U^*(0, k-1)) \triangleq E[x(j)|\mathfrak{F}(k, U^*(0, k-1))] \tag{6}$$

$$\hat{b}(j|k) \triangleq \hat{b}(j|k, U^*(0, k-1)) \triangleq E[b(j)|\mathfrak{F}(k, U^*(0, k-1))] \tag{7}$$

$$e_x(j|k) \triangleq e_x(j|k, U^*(0, k-1)) \triangleq \hat{x}(j|k, U^*(0, k-1)) - x(j) \tag{8}$$

$$e_b(j|k) \triangleq e_b(j|k, U^*(0, k-1)) \triangleq \hat{b}(j|k, U^*(0, k-1)) - b(j). \tag{9}$$

We note that $\hat{x}(j|k, U^*(0, k-1))$ is $\mathfrak{F}(k, U^*(0, k-1))$-measurable if $j \geq k$; so for $j \geq k$, we have

$$E\{x'(j)Mx(j)|\mathfrak{F}(k, U^*(0, k-1))\} = \hat{x}'(j|k)M\hat{x}(j|k) + E\{e_x'(j|k)Me_x(j|k)|\mathfrak{F}(k, U^*(0, k-1))\} \tag{10}$$

where M is an arbitrary $n \times n$ matrix. If we define the state-error moment matrix

$$\Sigma_x(j|k) \triangleq E\{e_x(j|k)e_x'(j|k)|\mathfrak{F}(k, U^*(0, k-1))\}, \tag{11}$$

then, using (10) and (11), the conditional cost (5) can be written as follows:

$$\begin{aligned} J(U(k, N-1); U^*(0, k-1), k) &= \frac{1}{2}\hat{x}'(N|k)F\hat{x}(N|k) + \frac{1}{2}\operatorname{tr}[F\Sigma_x(N|k)] \\ &+ \frac{1}{2}\sum_{j=k}^{N-1}\{\hat{x}'(j|k)W(j)\hat{x}(j|k) \\ &+ \operatorname{tr}[W(j)\Sigma_x(j|k)] + h(j)u^2(j)\}. \end{aligned} \tag{12}$$

We can now see that the cost-to-go (12) depends on the time evolution of $\hat{x}(j|k)$, $\Sigma_x(j|k)$, and $u(j)$ for $j = k, k+1, \cdots, N-1$. By assumption in the OLFO formulation the future controls $u(j)$ are treated as deterministic. It then follows that, if we get 1) a set of deterministic equations that describe the dynamics of $\hat{x}(j|k)$ and $\Sigma_x(j|k)$ for $j = k, k+1, \cdots, N-1$; and 2) a set of "initial conditions" at the present time (k), i.e., the values of $\hat{x}(k|k)$ and $\Sigma_x(k|k)$, then these, together with the deterministic cost-to-go (12), would define a *deterministic optimal control problem* whose solution would yield the optimal open-loop controls in the future.

Next we derive the dynamical equations for $\hat{x}(j|k)$ and $\Sigma_x(j|k)$.

Since all the noise sequences are assumed to be independent and white, it is easy to see from (1) that these equations are

$$\hat{x}(j+1|k) = A(j)\hat{x}(j|k) + \hat{b}(j|k)u(j) \tag{13}$$

$$\hat{b}(j+1|k) = G(j)\hat{b}(j|k) \tag{14}$$

$$\Sigma(j+1|k) = \tilde{A}(j, u(j))\Sigma(j|k)\tilde{A}'(j, u(j)) + \tilde{R}(j) \tag{15}$$

where we define

$$\Sigma(j|k) \triangleq E\left\{\begin{bmatrix} e_x(j|k) \\ \cdots \\ e_b(j|k)\end{bmatrix}[e_x'(j|k) \vdots e_b'(j|k)] \Big| \mathfrak{F}(k, U^*(0, k-1))\right\} \tag{16}$$

$$\tilde{A}(j, u(j)) \triangleq \begin{bmatrix} A(j) & \vdots & u(j)I_n \\ \cdots & & \cdots \\ 0 & \vdots & G(j)\end{bmatrix} \qquad \tilde{R}(j) \triangleq \begin{bmatrix} R(j) & \vdots & 0 \\ \cdots & & \cdots \\ 0 & \vdots & N(j)\end{bmatrix}. \tag{17}$$

The initial conditions at $j = k$ are $\hat{x}(k|k)$, $\hat{b}(k|k)$ and $\Sigma(k|k)$, which can be obtained by constructing a Kalman filter or an optimum observer–estimator [12] for the *linear augmented system*

$$\mathcal{S}: \left\{ \begin{array}{l} \left[\begin{array}{c} x(i+1) \\ \hline b(i+1) \end{array}\right] = \tilde{A}(i, u^*(i)) \left[\begin{array}{c} x(i) \\ \hline b(i) \end{array}\right] + \left[\begin{array}{c} \xi(i) \\ \hline \gamma(i) \end{array}\right] \\ y(i) = \tilde{C}(i) \left[\begin{array}{c} x(i) \\ \hline b(i) \end{array}\right] + n(i), \qquad i = 0, 1, \cdots, k-1 \end{array} \right\}. \tag{18}$$

(The detailed equations which describe the computation of $\hat{x}(k|k)$, $\hat{b}(k|k)$, and $\Sigma(k|k)$ are given in [15]. This is significant in that the computations involve: 1) one-step updating of a $2n$-vector difference equation; and 2) one-step updating of a $2n \times 2n$ matrix difference equation.)

We stress that if $U^*(0, k-1)$ is known, (18) is a linear system with *known* parameters. This is why optimal linear estimation theory can be applied.

We have now formulated the following *deterministic* control problem for the kth-step.

Statement of Open-Loop Control Problem ($k \leq j \leq N-1$)

Given

$$\hat{x}(j+1|k) = A(j)\hat{x}(j|k) + \hat{b}(j|k)u(j) \tag{19}$$

$$\hat{b}(j+1|k) = G(j)\hat{b}(j|k) \tag{20}$$

$$\Sigma(j+1|k) = \tilde{A}(j, u(j))\Sigma(j|k)\tilde{A}'(j, u(j)) + \tilde{R}(j) \tag{21}$$

with known initial conditions at $j = k$, $\hat{x}(k|k)$, $\hat{b}(k|k)$, $\Sigma(k|k)$, *we are to find a deterministic control sequence* $U(k, N-1)$ *such that it minimizes the cost-to-go*

$$J[U(k, N-1); U^*(0, k-1), k] = \frac{1}{2}\Big\{\hat{x}'(N|k)F\hat{x}(N|k) + \mathrm{tr}\,[\tilde{F}\Sigma(N|k)] + \sum_{j=k}^{N-1} \hat{x}'(j|k) \cdot W(j)\hat{x}(j|k) + \mathrm{tr}[\tilde{W}(j)\Sigma(j|k)] + h(j)u^2(j)\Big\} \tag{22}$$

subject to the constraints (19) *to* (21), *where the matrices* $\tilde{F}$ *and* $\tilde{W}(j)$ *are defined by*

$$\tilde{F} \triangleq \left[\begin{array}{c|c} F & 0 \\ \hline 0 & 0 \end{array}\right] \qquad \tilde{W}(j) \triangleq \left[\begin{array}{c|c} W(j) & 0 \\ \hline 0 & 0 \end{array}\right].$$

For the preceding deterministic control problem, we shall denote its optimal control sequence by $U^o(k, N-1) \triangleq \{u^o(j|k)\}_{j=k}^{N-1}$ *where the superscript o is used to denote optimal for the open-loop control problem; the symbol* $u^o(j|k)$ *is used to indicate that the control is open-loop optimal at time j conditioned on the observations up to the present time k.*

IV. Solution of the OLFO Control Problem

The deterministic formulation of the problem allows us to use the discrete matrix minimum principle [10] to derive the set of necessary conditions for optimality. Applying the minimum principle, we obtain a two-point boundary value problem where the state and costate are linearly related. Standard techniques are used to convert the two-point boundary value problem into an initial value problem; the problem is reduced to that of solving a nonlinear matrix Riccati difference equation. The detailed derivation can be found in [13] and [15]. The results are summarized below.

The optimal open-loop control sequence $U^o(k, N-1)$ is specified by

$$u^o(j|k) = -[\tilde{h}(j|k) + \tilde{b}^{o\prime}(j|k)\tilde{K}(j+1|k)\tilde{b}^o(j|k)]^{-1} \cdot \tilde{b}^{o\prime}(j|k)\tilde{K}(j+1|k)\Theta(j|k)\left[\begin{array}{c} \hat{x}^o(j|k) \\ \hline \delta^o(j|k) \end{array}\right] - \tilde{h}(j|k)d'(j+1)\left[\begin{array}{c} \hat{x}^o(j|k) \\ \hline \delta^o(j|k) \end{array}\right] \tag{23}$$

where $\tilde{K}(j|k)$, $j = k+1, \cdots, N-1$, satisfies the matrix difference equation

$$\tilde{K}(j|k) = \Theta'(j|k)[\tilde{K}(j+1|k) - \tilde{K}(j+1|k)\tilde{b}^o(j|k) \cdot (\tilde{h}(j|k) + \tilde{b}^{o\prime}(j|k)\tilde{K}(j+1|k)\tilde{b}^o(j|k))^{-1} \cdot \tilde{b}^o(j|k)\tilde{K}(j+1|k)]\Theta(j|k) + \tilde{D}(j|k)$$

$$\tilde{K}(N|k) = \begin{bmatrix} F & 0 \cdots 0 \\ 0 & 0 \cdots 0 \\ & \cdots, \\ 0 & 0 \quad 0 \end{bmatrix} \tag{24}$$

and, for $j = k, \cdots, N-1$,

$$\Theta(j|k) = \hat{A}(j) - \tilde{b}^o(j|k)\tilde{h}^{-1}(j|k)d'(j+1) \tag{25}$$

$$\hat{A}(j) \triangleq \begin{bmatrix} A(j) & 0 & \cdots & 0 \\ 0 & A(j)g_{11} & \cdots & A(j)g_{n1} \\ & & \cdots & \\ 0 & A(j)g_{1n} & \cdots & A(j)g_{nn} \end{bmatrix} \tag{26}$$

[2]

$$\tilde{D}(j|k) \triangleq \tilde{W}(j) - d(j+1)\tilde{h}^{-1}(j|k)d'(j+1) \tag{27}$$

$$d(j) \triangleq \begin{bmatrix} 0 \\ A(j-1)S(j)e_1 \\ \cdot \\ \cdot \\ \cdot \\ A(j-1)S(j)e_n \end{bmatrix} \in R^{n(n+1)}$$

$$\tilde{b}^o(j|k) \triangleq \begin{bmatrix} \hat{b}^o(j|k) \\ \Sigma_b{}^o(j|k)G'(j)e_1 \\ \cdot \\ \cdot \\ \cdot \\ \Sigma_b{}^o(j|k)G'(j)e_n \end{bmatrix} \in R^{n(n+1)} \tag{28}$$

$$\delta^o(j|k) \triangleq \begin{bmatrix} \Sigma_{xb}{}^o(j|k)e_1 \\ \cdot \\ \cdot \\ \cdot \\ \Sigma_{xb}{}^o(j|k)e_n \end{bmatrix} \tag{29}$$

[3]

[2] The g_{ij} are the elements of the G matrix [see (2)].
[3] The vectors $e_1, e_2, \cdots, e_n$ represent the natural basis in R^n.

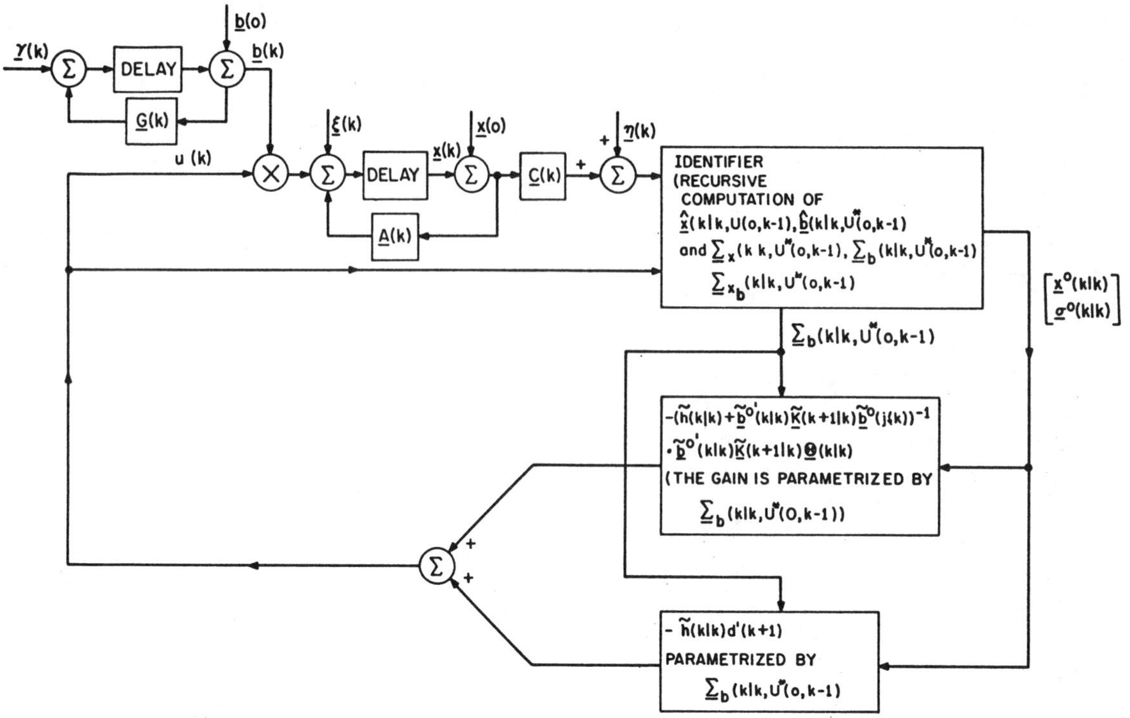

Fig. 1. The structure of the OLFO adaptive control system. The observation noise $\eta(k)$ may be nondegenerate, degenerate, or totally singular. The unknown parameter is $b(k)$; the system state vector is $x(k)$. The identifier generates the estimates of the state and of the unknown parameter vector, as well as of their covariance matrices, on the basis of past measurements and past controls.

$$\tilde{h}(j|k) \triangleq h(j) + \text{tr}\,[\Sigma_b^o(j|k)S(j+1)]. \tag{30}$$

The matrices $\Sigma_b^o(j|k)$, $S(j+1)$, $j = k, k+1, \cdots, N-1$, are given by

$$\Sigma_b^o(j+1|k) = G(j)\Sigma_b^o(j|k)G'(j) + N(j), \qquad j = k, \cdots, N-1$$

$$\Sigma_b^o(k|k) = \Sigma_b(k|k, U^*(0, k-1)) \tag{31}$$

and

$$S(j) = A'(j)S(j+1)A(j) + W(j); \qquad j = k, \cdots, N-1; S(N) = F, \tag{32}$$

and the vector $\hat{b}^o(j|k)$ satisfies

$$\hat{b}^o(j+1|k) = G(j)\hat{b}^o(j|k); \qquad j = k, \cdots, N-1,$$

$$\hat{b}^o(k|k) = \hat{b}(k|k, U^*(0, k-1)). \tag{33}$$

To find the OLFO control sequence, we have to solve the above open-loop control problem for $k = 0, 1, \cdots$. The OLFO control $\{u^*(k)\}_{k=0}^{N-1}$ is then given by

$$u^*(k) \triangleq u^o(k|k), \qquad k = 0, 1, \cdots, N-1, \tag{34}$$

where $u^o(k|k)$ is given by (23) to (33). The structure of the OLFO control system is illustrated in Fig. 1.

From (34) we see that if the optimal open-loop control $\{u^o(j|k)\}_{j=k}^{N-1}$ exists and is unique for all $k = 0, 1, \cdots, N-1$, we can conclude that the OLFO control $\{u^*(k)\}_{k=0}^{N-1}$ exists and is unique. If the solution of (24), i.e., the matrix $\tilde{K}(j|k)$, exists and is unique, then the control law given by (23) and (24) is the unique globally optimal open-loop control. Since $\tilde{D}(j|k)$ is an indefinite matrix, the solution of (37), $\tilde{K}(j|k)$ (if it exists), is not necessarily nonnegative definite; in fact it is always indefinite. Therefore, we cannot *a priori* conclude that $\tilde{h}(j|k) + \tilde{b}^{o\prime}(j|k)\tilde{K}(j+1|k)\tilde{b}^o(j|k)$ will always be nonzero, and thus deduce that $\tilde{K}(j|k)$ will remain bounded in finite time. It can be shown [13], [15] that if $G(j)$ satisfies

$$B \geq G(j)BG'(j) \text{ for all } B \geq 0; \qquad j = 0, 1, \cdots, N-1 \tag{35}$$

then $\tilde{h}(j|k) + \tilde{b}^{o\prime}(j|k)\tilde{K}(j+1|k)\tilde{b}^o(j|k) > 0$, $j = k, k+1, \cdots, N-1$; and so inductively $\{\tilde{K}(j|k)\}_{j=k}^{N}$ exists, is unique and bounded. Therefore, the OLFO control sequence $\{u^*(k)\}_{k=0}^{N-1}$ exists and is unique (almost surely) if condition (35) is satisfied.

V. Feedback Interpretation

Let us write

$$u^*(k|k) = \Bigg\{-(\tilde{h}(k|k) + \tilde{b}^{o\prime}(k|k)\tilde{K}(k+1|k)\tilde{b}^o(k|k))^{-1} \cdot \tilde{b}^{o\prime}(k|k)\tilde{K}(k+1|k)\Theta(k|k)\begin{bmatrix} I_n & \vdots & 0 \\ \cdots & & \cdots \\ 0 & \vdots & 0 \end{bmatrix}\Bigg\} \cdot \begin{bmatrix} \hat{x}^o(k|k) \\ \cdots \\ \hat{d}^o(k|k) \end{bmatrix}$$

$$- \Big\{(\tilde{h}(k|k) + \tilde{b}^{o\prime}(k|k)\tilde{K}(k+1|k)\tilde{b}^{o\prime}(k|k))^{-1}$$

$$\cdot \tilde{K}(k+1|k)\,\Theta(k|k)\begin{bmatrix} 0 & \vdots & 0 \\ \cdots & & \cdots \\ 0 & \vdots & I_n \end{bmatrix} + \tilde{h}(k|k)d'(k+1)\Bigg\}\begin{bmatrix} \hat{x}^o(k|k) \\ \cdots \\ \delta^o(k|k) \end{bmatrix}. \qquad (36)$$

We shall call the row vector ($1 \times n$)

$$\phi(k|k) \triangleq -(\tilde{h}(k|k) + \tilde{b}^{o\prime}(k|k)\tilde{K}(k+1|k)\tilde{b}^o(k|k))^{-1} \cdot b^{o\prime}(k|k)\tilde{K}(k+1|k)\Theta(k|k)\begin{bmatrix} I_n & \vdots & 0 \\ \cdots & & \cdots \\ 0 & \vdots & 0 \end{bmatrix} \qquad (37)$$

the *OLFO adaptive gain*, and the term

$$u^c(k|k) = -\Bigg\{(\tilde{h}(k|k) + \tilde{b}^{o\prime}(k|k)\tilde{K}(k+1|k)\tilde{b}^o(k|k))^{-1} \cdot \tilde{b}^{o\prime}(k|k)\tilde{K}(k+1|k)\Theta(k|k)\begin{bmatrix} 0 & \vdots & 0 \\ \cdots & & \cdots \\ 0 & \vdots & I_n \end{bmatrix} + \tilde{h}(k|k)d'(k+1)\Bigg\}\begin{bmatrix} \hat{x}^o(k|k) \\ \cdots \\ \delta^o(k|k) \end{bmatrix} \qquad (38)$$

the *control correction term*. Thus, the OLFO control (36) becomes

$$u^*(k|k) = \phi(k|k)\hat{x}^o(k|k) + u^c(k|k), \qquad (39)$$

which means that the current applied control $u^*(k|k)$ is obtained by multiplying the current estimate of the state $\hat{x}^o(k|k)$ by the *adaptive gain* (row vector) $\phi(k|k)$ and is corrected by the correction term $u^c(k|k)$.

Since the first n-components of the vector $d(\cdot)$ are zero, it follows that the current control correction term $u^c(k|k)$ is independent of the current state estimate $\hat{x}^o(k|k)$. From (24)–(28) we note that $\Sigma_b(k|k)$ *affects indirectly both the OLFO adaptive gain* $\phi(k|k)$ *and the control correction term* $u^c(k|k)$. *The cross-error covariance* $\Sigma_{xb}(k|k)$ *affects only the control correction term;* and if $\Sigma_{xb}(k|k)$ is zero, then from (29) and (38) we conclude that $u^c(k|k) = 0$.

In essence, the above equations show explicitly how the estimation accuracy of the b vector is expressed in terms of the estimation error covariance matrix $\Sigma_b(\cdot|k)$ and the cross covariance matrix $\Sigma_{xb}(\cdot|k)$. Hence, the parameter uncertainty does "modulate" in particular the adaptive gain $\phi(k|k)$. Note, however, that the uncertainty in the state vector [expressed by the state error covariance matrix $\Sigma_x(\cdot|k)$ does not affect $\phi(k|k)$ or $u^c(k|k)$. This is consistent with the results of the standard "separation theorem" in which the state uncertainty uncouples from the control gain determination.

To see this effect further, let us examine the case when the b vector is known. Then $\Sigma_b(k|k) = 0$ and $\Sigma_{xb}(k|k) = 0$. In this case there is no parameter uncertainty and one knows the structure (via the separation theorem) of the optimal stochastic system. To see this let us assume that $\Sigma_b(k|k) = 0$; then from (24)–(30), we have inductively that

$$\tilde{K}(k|k) = \begin{bmatrix} K(k) & \vdots & 0 \\ \cdots & & \cdots \\ 0 & \vdots & 0 \end{bmatrix} \qquad (40)$$

where $K(k)$ is the solution of a matrix Riccati difference equation, and from (37), the OLFO adaptive gain is

$$\phi(k|k) = -(h(k) + b'(k)K(k+1)b(k))^{-1} \cdot b'(k)K(k+1)A(k), \qquad (41)$$

which is the truly optimum gain demanded by the separation theorem. The assumption that $\Sigma_b(k|k) = 0$ also implies $\Sigma_{xb}(k|k) = 0$, and so the correction term $u^c(k|k)$ is zero and

$$u^*(k|k) = -(h(k) + b'(k)K(k+1)b(k))^{-1} \cdot b'(k)K(k+1)A(k)\hat{x}^o(k|k). \qquad (42)$$

Thus we see that *if for some* k, *the identification of* $b(k)$ *is assured to be exact, i.e., the level of confidence in the estimated gain parameters is very high, then the OLFO control will act optimally and use the obtained estimate of* $b(k)$ *as if it were the true gain vector.*

VI. Discussion

A. Interpretation of Equations

It is possible to examine the derived equations and interpret the results in a qualitative manner. In Section VII it will be shown that the simulation results agree with this qualitative discussion.

In essence, we are forcing some sort of "separation" in our formulation. The overall control problem is split into an identification and a deterministic control problem. *However, the effect of the identification error will be taken into account in the deterministic control problem.* Thus, this does not correspond to *pure* separation as it is in the case of stochastic control of linear system with known dynamics.

First let us clarify the role of the control. If $u(i) = 0$, then it is easy to show that

$$\hat{b}(i+1|i+1) = G(i)\hat{b}(i|i). \qquad (43)$$

From the above equation we see that *if no control is applied at time* i, *then the estimate of the parameter vector* $b(\cdot)$ *cannot be improved at the* $(i+1)$ *step.* Therefore, *a nonzero input is necessary to identify the gain parameter vector* $b(k)$. From (1) we see that if $u(i)$ is *very large*, then for the most part the value of $x(i+1)$ will be due to $b(k)u(k)$, and so *the observation* $y(k)$ *will contain a large amount of information about the gain parameter* $b(k)$. Therefore, we would expect that *large input magnitudes will be helpful in the identification of* $b(k)$. Thus a control sequence with a high energy would be useful for identification purposes. But large control energy will also give rise to a high cost. From the control point of view, we would like to use just enough control energy to regulate the state of the system. Thus, there is a conflict between identification and control, and a *reasonable control sequence should appropriately distribute its total energy to identify and/or control the system* S.

Let us consider (23) and (24). Comparing them with the Levis [11] results, it can be shown that $u^o(j|k)$ is also the optimal control for the problem of controlling the system $\tilde{S}_k$:

$$\tilde{S}_k: \tilde{x}(j+1|k) = \hat{A}(j)\tilde{x}(j|k) + \tilde{b}(j|k)u(j|k);$$

$$\tilde{x}(j+1|k) \triangleq \begin{bmatrix} \hat{x}(j+1|k) \\ \hline \sigma^o(j+1|k) \end{bmatrix} \quad (44)$$

with the cost criterion

$$J = \tilde{x}'(N|k)\tilde{F}\tilde{x}(N|k) + \sum_{i=k}^{N-1} \{\tilde{x}'(i|k)\tilde{W}(i|k)\tilde{x}(i|k) + \tilde{h}(j|k)u^2(j|k) + 2\tilde{x}'(i|k)d(i+1)u(i|k)\}. \quad (45)$$

Therefore, we can visualize $\tilde{h}(j|k)$ as being a *modified relative weighting on the control.* From (30) we note that $\tilde{h}(j|k)$ relates in a direct manner with $\Sigma_b{}^o(j|k)$. In a statistical sense, $\Sigma_b(k|k)$ reflects the level of uncertainty we have about the estimate of $b(k)$. *The modification on the relative weighting on the control is such that heavy weighting is put in the control if we have little confidence in the estimate of* $b(k)$; therefore, the control will be very "cautious" and control energy will not be wasted whenever the parameter estimation accuracy is bad.

The uncertainty $\Sigma_b{}^o(j|k)$ is also modified by the matrix $S(j+1)$ in the computation of $\tilde{h}(j|k)$ [see (30)]. The matrix $S(j)$ [see (32)] represents an effective change in the state weighting matrix $W(j)$. Roughly speaking, if we are controlling a stable system ($\|A(j)\| < 1$), then one can verify that

$$\|S(j)\| \approx \|A(j-1)\|^2\|W(j-1)\| + \|W_j\| \quad (46)$$

so that the effective state weighting provided by $S(j)$ is roughly of the same magnitude as the actual state weighting $W(j)$. On the other hand, if the system is unstable, $\|A(j)\| > 1$, then $\|S(j)\| \gg \|W(j)\|$, and hence, there is much more contribution to the value of $\tilde{h}(j|k)$. This means, that *if the system is stable, then at the initial stages, the control would tend to be larger than for an unstable system* (operating at the same level of uncertainty). This point is confirmed by the simulation results.

B. Asymptotic Behavior

The asymptotic behavior of the overall system can be studied by considering the behavior of $\Sigma_b(k|k)$, as $k \to \infty$. We shall now state a theorem concerning the asymptotic behavior of $\Sigma_b(k|k)$. The proof for this theorem can be found in [13] or [15].

Theorem 1: Let $\gamma(k) = 0$, $A(k)$, $G(k)$ be bounded and nonsingular and $G(k)$ satisfy (25), $k = 0, 1, \cdots$. If $\{(A(k), C(k))\}_{k=0}^{\infty}$ is uniformly completely observable of index ν, i.e., the matrix

$$[C'(k) \vdots \phi_A'(k,k)C'(k+1) \vdots \cdots \vdots \phi_A'(k+\nu-2,k) \cdot C'(k+\nu-1)]$$

has rank n for all $k = 0, 1, \cdots$, and $M < \mu(k) \leq \epsilon < 0$ for $k = 0, 1, \cdots$, then

$$\lim_{k\to\infty} \Sigma_b(k|k) = 0. \quad (47)$$

Since $\Sigma(k|k) \geq 0$, (47) also implies

$$\lim_{k\to\infty} \Sigma_{xb}(k|k) \to 0. \quad (48)$$

We remark that (47) holds when $G(k) = I$, $\gamma(k) = 0$, for all k, i.e., the unknown parameter vector b is constant.

Let us consider an observable system S, (1), in which the gain parameters are assumed to be unknown and satisfy

$$b(k+1) = G(k)b(k) \quad (49)$$

with $G(k)$ satisfying (35). Assume that we control the system S over an interval $N < \infty$ via the OLFO control. In the beginning, the modified weighting on the control is high, and thus in general the control magnitude will be low at the beginning. Thus, the trajectory of the overall control system would be pretty much the same as the input-free trajectory of the system S. *If the matrix* $A(k)$ *is exponentially stable, the true state of the system will evolve toward zero by using negligibly small control magnitudes (even zero). The result is that little effort of the input* $\{u^*(k)\}_{k=0}^{N-1}$ is spent for control and identification purposes. We would expect that the *estimated parameters will hardly converge to the true parameters* $b(k)$. *On the other hand if* $A(k)$ *is not exponentially stable, then the true state of the overall system will diverge. This diverging phenomenon will be noticed by the identifier, thus resulting in a high control magnitude because of (23).* Since little is initially known about the gain parameters, the high magnitude control will be utilized mainly for identification purposes. Therefore, the control will be kept bounded away from zero as long as exact identification of $b(k)$ has not been obtained. *Using Theorem 1, we predict that for unstable systems the estimated parameters of* $b(k)$ *will come close to the true gain parameters before the control magnitude goes to zero.* This is also borne out by the simulation results.

Analytical studies of the convergence rate of the OLFO system are not yet available. From the above discussion, we may predict roughly that the convergence rate for unstable system will be relatively fast and the convergence rate for stable system will be very slow.

Finally, we shall discuss some interesting implications of Theorem 1. Consider an observable system S, (1), with unknown gain parameters satisfying (49) and with $G(k)$ satisfying (35). Let $\phi_k(\hat{x}(k|k), \hat{b}(k|k), \Sigma_b(k|k))$ be any ad hoc control law which is "placed" after the identifier and with the following properties ($k \geq 0$).

Property 1:

$$\phi_k(\cdot,\cdot,\cdot): R^n \times R^n \times M_{nn} \to R.$$

Property 2:

$$\phi_k(x, b, \Sigma_b) \neq 0, \qquad x \in R^n,\ b \in R^n,\ \Sigma_b \in M_{nn},\quad x \neq 0,\ \Sigma_b \neq 0.$$

Property 3:

$$\phi_k(x, b, 0) = -(h(k) + b'(k)K(k+1)b(k))^{-1} \cdot b'(k)K(k+1)A(k)x, \qquad x \in R^n,\ b \in R^n.$$

From Property 2 we see that $\Sigma_b(k|k) \to 0$ as $k \to \infty$, and so from Property 3 the ad hoc control scheme will converge to the optimal control strategy when the full dynamics become known. This indicates that the ad hoc scheme $\phi_k(\hat{x}(k|k), \hat{b}(k|k), \Sigma_b(k|k), U(0, k-1))$ can provide reasonable results.

C. Control Over an Infinite Interval

Let us consider the problem of controlling the system S, which is time invariant with an unknown constant gain vector b, over an infinite interval, i.e., $N \to \infty$. To obtain a feasible solution we suggest the *window-shifting* approach. Assume that at all times we have N more steps to control; thus at all times we solve an open-loop control problem over an interval of N steps.

We note that in the OLFO approach we have to re-solve the open-loop control problem at every time k so as to adjust the control gains accordingly; in particular, we must compute $\tilde{K}(k|k)$ in a backward direction, starting from the terminal time N to the present time k for each k. If N is very large, this requires large computation times. From a computational standpoint then, one would like to "cut back" the terminal time. Conceptually, in trying to control over an infinite time period, the controller looks into all future effects caused by present action, and decides on the optimum decision for the next step. The window-shifting approach suggests that instead of looking at *all* future effects, the controller looks at only *near* future effects caused by present actions and arrives at suboptimal decisions; hence, this represents a "short term adaptive scheme."

Thus, from a conceptual and a computational point of view, such an approach may be desirable.

Assume that the time-invariant system S being controlled is observable and controllable. If b is known exactly and then if we consider control over an infinite time period, the optimal feedback gain is constant and is given by

$$\phi = -(h + b'Kb)^{-1}b'KA \tag{50}$$

where K is given by the steady-state solution of

$$K_{t+1} = A'(K_t - K_t b(h + b'K_t b)^{-1} b'K_t)A + W, \qquad K_0 = F. \tag{51}$$

Let N be the integer such that for $n \geq N$

$$\|K_n - K_{n-1}\| \leq \epsilon, \qquad \epsilon > 0. \tag{52}$$

Such an integer N can be found experimentally off line. Adjust the window width equal to N and apply the window-shifting approach. By Theorem 1, the estimate in b will converge asymptotically, and so when $b(k|k) \to b$, we have

$$\tilde{K}(k|k) \to \begin{bmatrix} K(k, N+k; F) & \vdots & 0 \\ \cdots & & \cdots \\ 0 & \vdots & 0 \end{bmatrix} \tag{53}$$

where $K(k, N+k; F)$ satisfies

$$\begin{aligned} K(k, N+k; F) = A'(&K(k+1, N+k; F) \\ &- K(k+1, N+k; F) \\ &\cdot b(h + b'K(k+1, N+k; F)b)^{-1} \\ &\cdot b'K(k+1, N+k; F))A + W, \\ K(N+k,& N+k; F) = F \end{aligned} \tag{54}$$

and the control converges to

$$u^*(k|k) \to -(h + b'K(k, N+k; F)b)^{-1} \cdot b'K(k, N+k; F)A\hat{x}^o(k|k). \tag{55}$$

(See discussion in Section V.) Comparing (51) and (54), we note that

$$K(k, N+k; F) = K_N \approx K. \tag{56}$$

Thus, asymptotically the time-varying adaptive system tends to be a time-invariant control system.

VII. Numerical Examples[4]

In the previous sections we have studied theoretically the adaptive control of a discrete-time linear system with an unknown gain vector. An adaptive system was derived using the OLFO approach and the asymptotic behavior of the control system was discussed. There are still some important questions which have not been treated theoretically. For example, rates of parameter convergence are, in general, of great interest, but this topic was not treated in detail. In this section we present simulation studies carried out for some specific third-order systems. The main purpose for these studies is to provide quantitative results about rates of convergence and to test the validity of the qualitative conclusions of Sections V and VI.

To enhance physical intuition, the discrete-time systems were obtained by sampling a continuous-time system. In this case, the uncertainty of the $b(k)$ vector is equivalent to uncertainty in: 1) the number of zeros; 2) the location of the zeros in the s-plane; and 3) the plant dc gain. It is assumed that the pole locations are known.

Let us consider a stochastic continuous time-invariant linear system described by

[4] Only a representative sample is given here; more extensive numerical results can be found in [13] and [15].

$$\dot{x}_f(t) = A_f x_f(t) + b_f u_f(t) + d_f \xi_f(t)$$
$$y_f(t) = c' x_f(t) + \eta_f(t)$$
$$x_f(0) \sim g(0, \Sigma_{x0}), \qquad b_f \sim g(0, \Sigma_{b0}) \tag{57}$$

where $\xi_f(t)$ is a scalar driving white Gaussian noise, $\eta_f(t)$ is the scalar observation white Gaussian noise. The statistical laws of $\xi_f(t)$ and $\eta_f(t)$ are assumed to be known

$$\int_{t_1}^{t_2} \xi_f(t)\, dt \sim g\left(0, \int_{t_1}^{t_2} r\, dt\right)$$
$$\int_{t_1}^{t_2} \eta_f(t)\, dt \sim g\left(0, \int_{t_1}^{t_2} q\, dt\right). \tag{58}$$

We can obtain a sampled-data version of the above system and apply the results of Section V. A computer program was designed which simulated the OLFO control for a third-order linear time-invariant system when the b vector is not known and also simulated the truly optimal control when the b vector is known. *It is important to stress that this comparison does not yield the difference between the OLFO design and the true closed-loop stochastic control system involving unknown parameters.* The comparison is to the optimal system response with *known* parameters. The relative deviations of the true responses will provide us the lost performance *from a control point of view* when parameter identification is necessary.

In the simulation results we present typical transient responses of the state variables, the changes in the estimates of some of the elements of the b vector as well as changes in adaptive gain. By examining these time responses one can deduce when the control is being used for identification, and when the identification is essentially complete.

In all the computer simulations, unless otherwise mentioned, we set the values (Δ represents the sampling time):

$$\Delta = 0.2\text{ s} \qquad q = 0.45 \qquad d_f = \begin{bmatrix} 1 \\ 2 \\ 3 \end{bmatrix} \qquad x_0 = \begin{bmatrix} 1 \\ 1 \\ 1 \end{bmatrix}$$
$$F = I_3 \qquad W = I_3 \qquad \Sigma_{b0} = \Sigma_{x0} = 4I_3 \qquad c' = [1 \quad 0 \quad 0]. \tag{59}$$

It is important to realize then that we deal with a third-order system. The only measurement is that of the output, every 0.2 s. This sampled-data measurement is corrupted by white noise whose variance is $g = (0.45)(0.2) = 0.09$ (or rms value 0.3). The plant could have none, one, or two zeros. We did not initially know how many zeros there were or their location(s). Hence, even though the poles are assumed known, the measurements are extremely meager since from the noisy measurement of one variable one must estimate six (three state variables and three parameters that define the number and location of zeros).

Example 1: Unstable System

It is assumed that the continuous-time system is described by

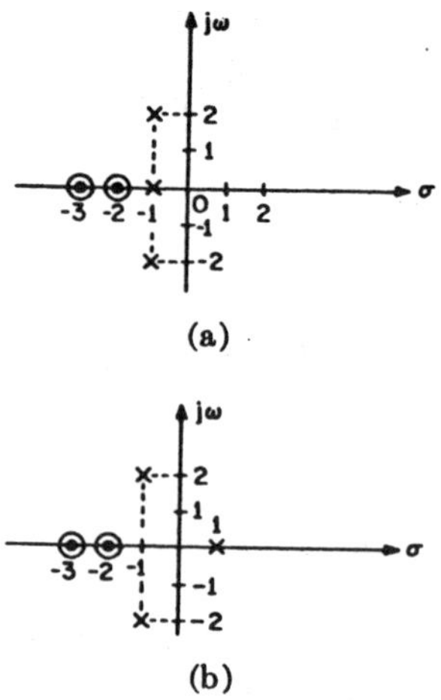

Fig. 2. (a) True pole-zero pattern for Example 1 (unstable system). (b) True pole-zero pattern for Example 2 (stable system). In both cases the initial guess was that the plant had no zeros.

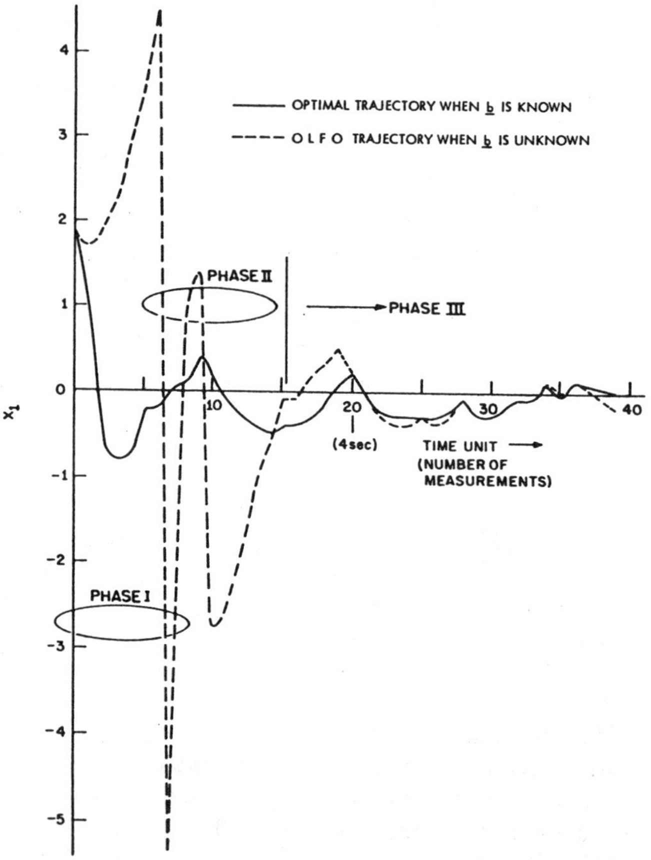

Fig. 3. Comparison of the response of the unstable system when the gain vector is known (optimal stochastic control) and when the gain vector is unknown (OLFO approach). The sample noises were identical in both cases. Similar behavior was observed in the other two (x_2 and x_3) state variables.

$$A = \begin{bmatrix} 0 & 1 & 0 \\ 0 & 0 & 1 \\ 5 & -3 & -1 \end{bmatrix} \qquad b_f = \begin{bmatrix} 1 \\ 2 \\ -3 \end{bmatrix} \qquad x_f(0) = \begin{bmatrix} 2 \\ -1 \\ 4 \end{bmatrix}. \tag{60}$$

Such a system has a true transfer function (see Fig. 2)

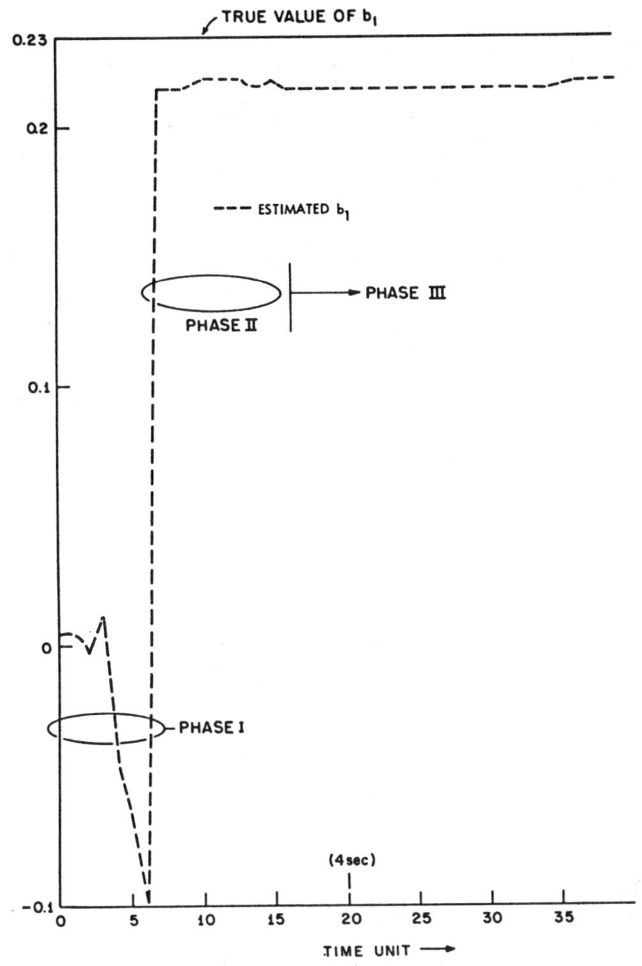

Fig. 4. Estimate of the first component of the b vector as a function of the number of measurements for the unstable system. Similar behavior was noted for the other two components of the unknown parameter vector.

$$H_1(s) = \frac{(s+3)(s+2)}{(s-1)(s^2+2s+5)}. \tag{61}$$

Note that it has an unstable pole at $s = 1$. Several runs have been made on the same system with 1) different noise samples; 2) different weightings on $h(h = 1, h = 10, h = 0.1)$; 3) different noise covariance ($r = 0.05$, $r = 0.45$); and 4) different initial guess on $\hat{b}_f(0|0)$. The plots for one particular run, $h = 1$, $r = 0.05$ and

$$\hat{b}_f(0|0) = \begin{bmatrix} 0 \\ 0 \\ -1 \end{bmatrix},$$

are shown in Figs. 3, 4, and 5. These plots compare the OLFO control system with the truly optimal control system.

Examination of Figs. 3, 4, and 5 brings into focus the nature of the adaptive system. From Fig. 3 we can see that during the first few measurements, phase I ($0 \leq k \leq 7$), the state vector diverges. Since this is an unstable system, this means that little control is being applied; this is confirmed by examining Fig. 5 in which the adaptive gain

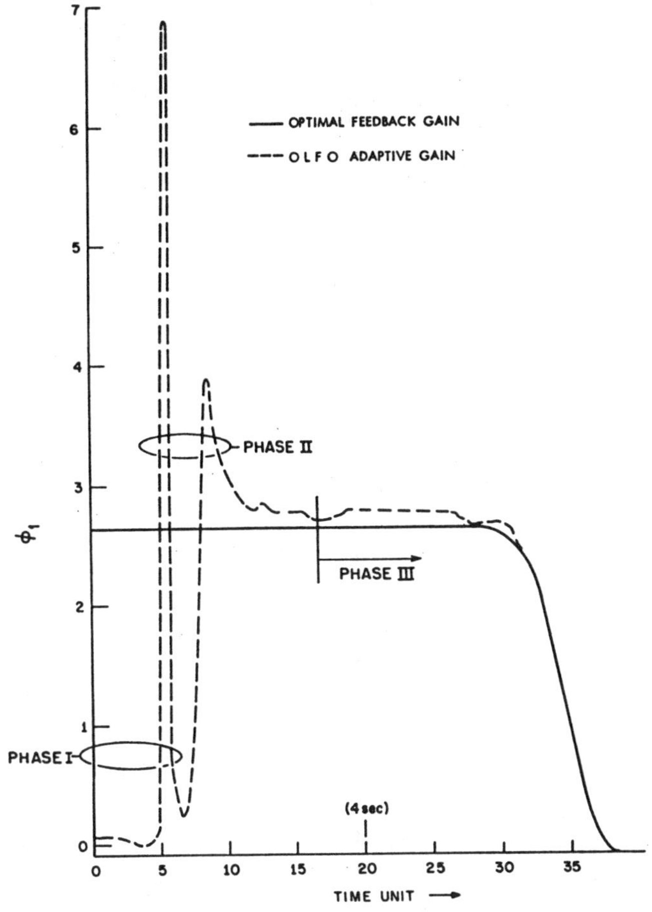

Fig. 5. Comparison between the first component of adaptive OLFO gain vector ϕ and the corresponding optimal feedback gain (when the parameter is known) for the unstable system. The other two components behaved similarly.

$\phi_1(k)$—the one that multiplies the estimate $\hat{x}_1(k)$—is almost zero. This verifies the conclusion reached in Sections V and VI that in unstable system with high initial parameter uncertainty the control is "cautious." Note that since little control is used, one cannot expect good identification; this is confirmed in Fig. 4, where the initial estimate of the parameter b_1 is far away from its true value.

Since the state of the system diverges, it eventually becomes large enough to cause larger controls although the adaptive gains are still small. Large controls in phase II become useful in identification. Within the next couple of measurements the parameter estimates approach their true value, as indicated in Fig. 4. This improvement in parameter accuracy results in larger adaptive gains as shown in Fig. 5 and larger control signals which cause the correction of state errors as shown in Fig. 3. After about 15 measurements, the identification is essentially complete and the OLFO adaptive system behaves almost exactly as the optimal one (see Fig. 3 as well as Fig. 5, which shows the convergence of the adaptive gain ϕ_1 to its optimal value).

Since this was a finite-time problem ($N = 40$) the estimate of b_1 never reaches the true value of b_1. However, such parameter inaccuracy is not essential from the control

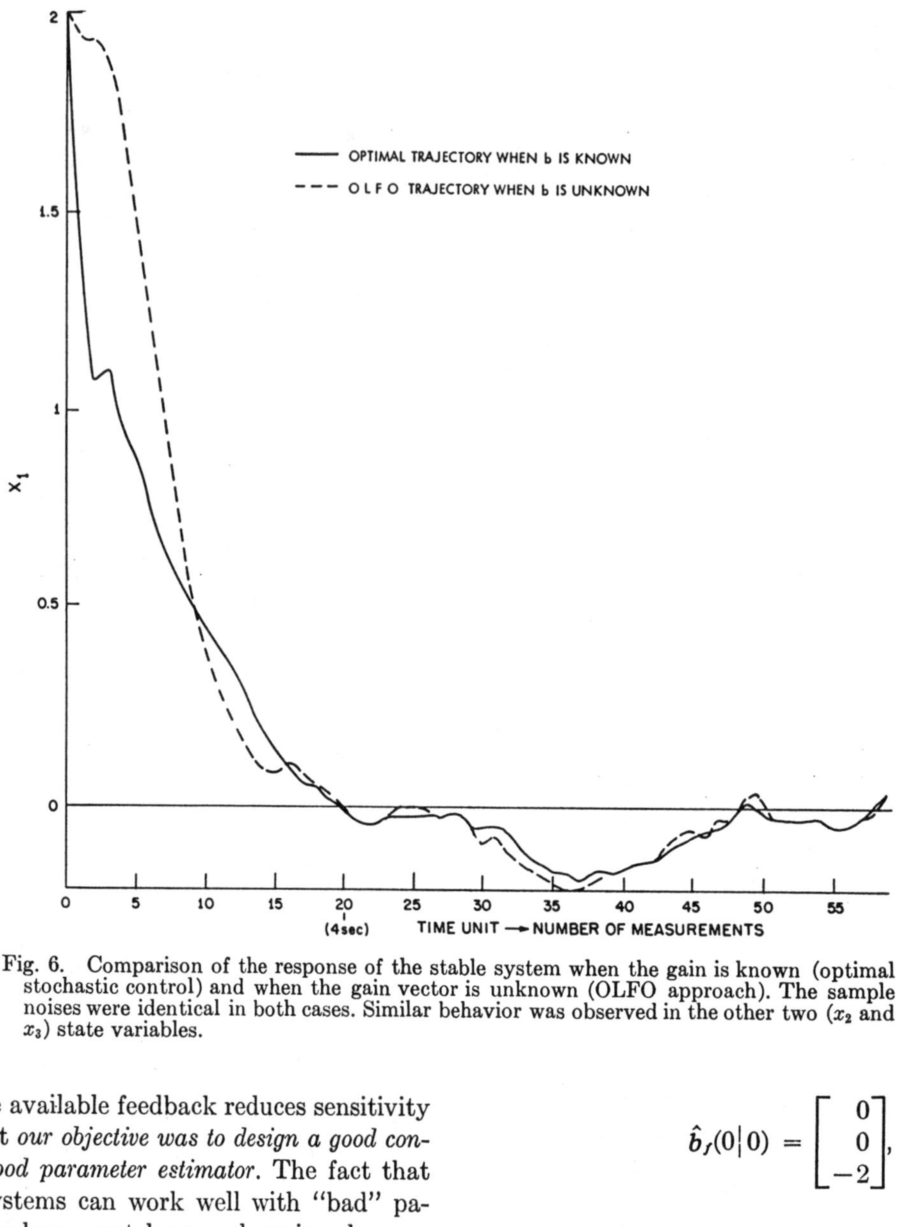

Fig. 6. Comparison of the response of the stable system when the gain is known (optimal stochastic control) and when the gain vector is unknown (OLFO approach). The sample noises were identical in both cases. Similar behavior was observed in the other two (x_2 and x_3) state variables.

viewpoint, since the available feedback reduces sensitivity anyway. Recall that *our objective was to design a good control system, not a good parameter estimator*. The fact that feedback control systems can work well with "bad" parameter estimates is borne out here and again when we examine stable systems.

Example 2: Stable System

It is assumed that

$$A_f = \begin{bmatrix} 0 & 1 & 0 \\ 0 & 0 & 1 \\ -5 & -7 & -3 \end{bmatrix} \quad b_f = \begin{bmatrix} 1 \\ 2 \\ -7 \end{bmatrix} \quad x_f(0) = \begin{bmatrix} 6 \\ -3 \\ 12 \end{bmatrix}. \tag{62}$$

The true transfer function for this stable system is (Fig. 2).

$$H_2(s) = \frac{(s+3)(s+2)}{(s+1)(s^2+2s+5)}. \tag{63}$$

Several runs have been made on the same system with 1) different noise samples, and 2) different initial guess on $\hat{b}_f(0|0)$. The plots for one particular run,

$$\hat{b}_f(0|0) = \begin{bmatrix} 0 \\ 0 \\ -2 \end{bmatrix},$$

are shown in Figs. 6, 7, and 8.

In contradistinction to the unstable case, Figs. 6, 7, and 8 do not present us with clear-cut phases of identification. It seems that combined identification and control are carried out together. From Fig. 6 we can see that the adaptive system behaves almost as the optimal system after about eight measurements. Large controls are used from the beginning (no phase I). From Fig. 7 we see that the estimate of b_1 was good with respect to magnitude, but it had opposite sign. *Hence, the adaptive system was not a good parameter estimator*. Similar behavior occurred in the adaptive gain ϕ_1 as shown in Fig. 8; the gain had the wrong sign. Nonetheless, the correct control polarity drove the system since the negative sign of the b_1 estimate and the negative sign of the adaptive gain resulted in a positive sign as far as the state evolution was concerned. Once more we stress that the OLFO philosophy does not reward the system for good identification; only for good control performance.

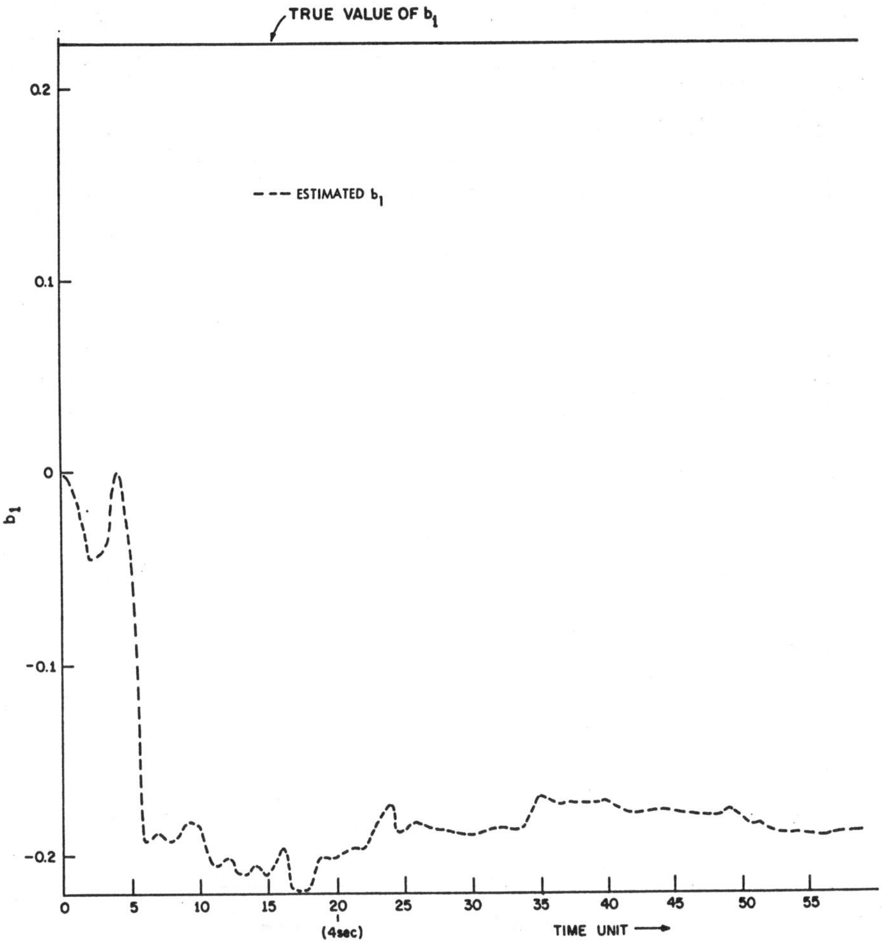

Fig. 7. Estimate of the first component of the b vector as a function of measurements for the stable system. Similar behavior was noted for the other two components of the unknown parameter vector.

In each set of experiments discussed above, the number of sample runs are not enough to enable us to draw specific statistical conclusions; yet the regularity in the sample runs enables us to draw some crude conclusions from the simulation presented here and in [13] and [15].

Conclusion 1: The rate of parameter convergence seems to be very dependent on the stability of the system. For unstable systems the convergence rate seems to be faster compared to that for stable systems.

Conclusion 2: It seems that large controls will help identification of the unknown gain parameters, and so convergence rate seems to relate directly to the magnitude of the control action.

Conclusion 3: For unstable systems, the rate of convergence seems to be fairly *independent* of the initial guess on the unknown gain, whereas for stable systems, the convergence rate may be quite *dependent* on the initial guess on the unknown gain.

Conclusion 4: For unstable systems the OLFO trajectory will depend on the initial guess in $\boldsymbol{b}_f$, but then for stable systems the OLFO trajectory will not vary drastically when we vary the initial guess in $\boldsymbol{b}_f$.

Conclusion 5: For the unstable system the OLFO trajectory seems to follow closely its input-free trajectory in the beginning until the diverging phenomenon tells the identifier to send back large controls for identification purposes. This causes some overshoots in the trajectory. The magnitude of the maximum overshoot seems to relate inversely with the values for the weighting constant h on the control. Nonzero control is applied until the identification of $\boldsymbol{b}$ is completed. This implies that identification of $\boldsymbol{b}$ is necessary for controlling an unstable system. For stable systems simultaneous identification and control seem to be carried out in the beginning. Since the system is stable, with little control energy, the state will go to zero. So after some time period when the state is near the origin, approximately zero control is applied, thus terminating the identification of $\boldsymbol{b}$. Even if the estimate of $\boldsymbol{b}$ does not converge to the true $\boldsymbol{b}$, the truly optimal trajectory and the OLFO trajectory are almost the same after the transient period. This indicates that for stable system, the identification of $\boldsymbol{b}$ is not absolutely necessary for control purposes.

Conclusion 6: Lastly, we would like to comment on the computational requirements of the proposed scheme. The above experiments were simulated using an IBM 360/64/40 system. It was found that the actual computation of the

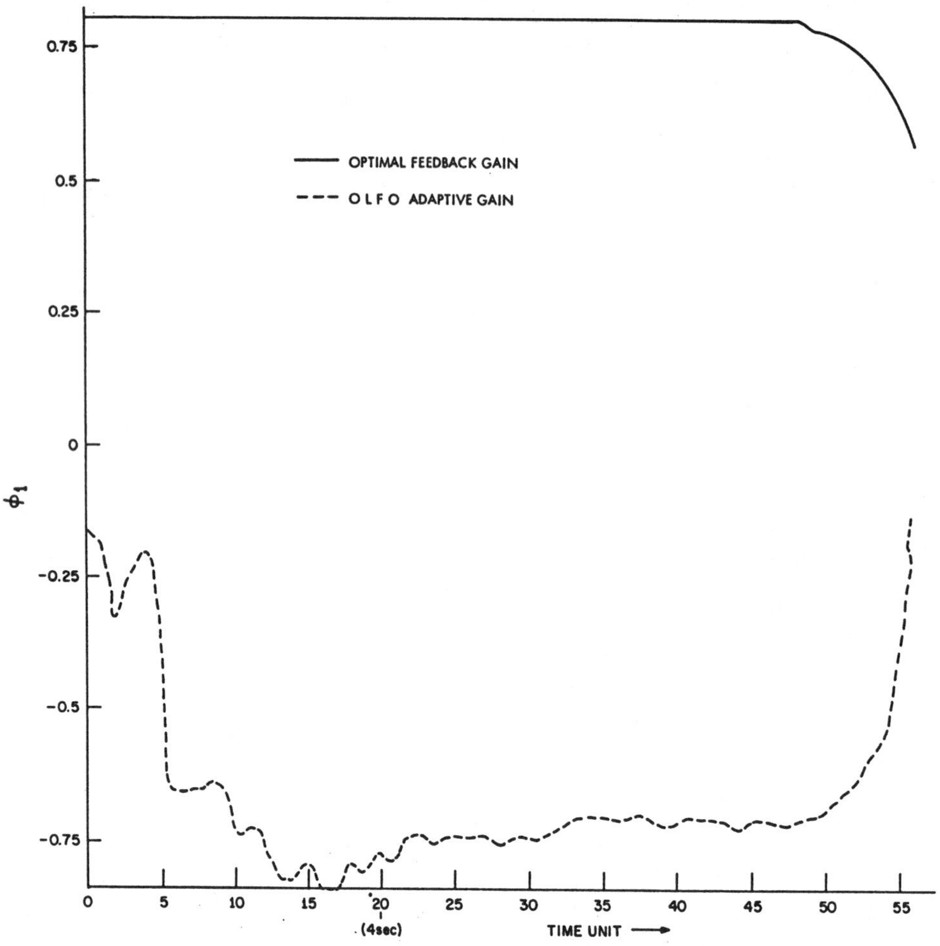

Fig. 8. Comparison between the first component of the adaptive OLFO gain vector ϕ and the corresponding optimal feedback gain (when the parameter vector is known) for the stable system. The other two components behaved similarly.

OLFO control sequence can be carried out almost in real time for $N = 40$. For higher order systems the computational time becomes formidable for on-line control; for this reason, the window-shifted approach is recommended.

VIII. Remarks

In this section we briefly summarize some items that pertain to the approach used, the results, and future research in this area.

Item 1: As mentioned in the main part of the paper, the OLFO approach has been used by many authors. The point of this paper is not to validate the OLFO approach, but rather to stress that its use leads to a design that enjoys adaptive properties and violates the ordinary separation theorem. As pointed out by one of the reviewers, the OLFO method makes the dual control problem *neutral* in Fel'dbaum's sense [2] and, therefore, forces some separation in the control action. In the case at hand, this separation is exhibited in the fact that the Kalman filter or observer–estimator can be designed independently of the control objectives. However, the controller design (as exhibited by the properties of the adaptive gain vector ϕ) depends strongly upon the state and parameter covariance matrices. Hence, there is one-way separation which is reminiscent of the control of *linear systems, with Gaussian statistics but nonquadratic criteria* (see Wonham [19] and Kramer [20]).

Item 2: The simulation results in this paper compared the OLFO design to the truly optimal feedback design when the gain vector $b(k)$ was known exactly. An intermediate design would be the one in which the separation theorem was arbitrarily enforced. By this we mean that the instantaneous estimate of the gain vector $\hat{b}(k|k)$ was used to compute the feedback gain from time k to the final time N. This would still require solving at each step a Riccati-type equation backward in time [since $\hat{b}(k|k) \neq \hat{b}(k+1|k+1)$]. However, the resulting algorithm would be much simpler to implement since no need for propagating the covariance matrices would arise. Such a comparison was not made; it is conjectured that the resultant system would be worse than that obtained by the OLFO design. Partial supporting evidence for this conjecture is provided by the recent paper by Saridis and Lobbia [21], although their parameter identification method was based on the technique of stochastic approximation. It should also be noted that this enforced separation was the topic of the paper by Farison *et al.* [3].

Item 3: The long-term stability of the OLFO scheme can be guaranteed if the conditions stated in Section VI-B hold. In this case, as $N \rightarrow \infty$, identification of the parameter vector is exact (provided $\gamma(k) = \mathbf{0}$); hence asymptotically the OLFO system will enjoy the same stability properties of the true stochastic optimal closed-loop system.

Item 4: The window-shifting approach (which was suggested by the works of Widrow and others) can be used to ease computational requirements (see Section VI-C). The sensitivity properties of such an approach remain, however, a subject for further research. Nonetheless, as discussed in Section VI-C, asymptotic stability should not be a problem.

Item 5: At the present time work is underway to extend the OLFO techniques to the case where some of the elements of the $\boldsymbol{A}(k)$ matrix in (21) are also unknown (R. Ku, M.S. thesis, M.I.T.). In such problems, additional complexities arise because the problem of simultaneous state and parameter estimation is a nonlinear one. Hence, the estimator cannot generate the exact conditional means and covariances. The approach taken is to approximate these by use of the extended Kalman filter and second-order filter [22]. The simulation results will compare the response of the resultant stochastic systems when a) the parameters are known, b) enforced separation is employed (for both the extended Kalman and second-order filter case), and c) OLFO design is employed (for both the extended Kalman and second-order filter case).

IX. Conclusions

A technique for adaptive control for a class of linear systems with unknown gain parameters has been presented. It is shown that the adaptive control law consists of 1) an identifier which generates the conditional mean estimates and the associated error covariance of the state and unknown gain, and 2) an OLFO adaptive feedback control law plus a correction term which are modified by the parameter-error covariance matrices. From the derived equations, some qualitative predictions can be made on the overall OLFO control system. Simulation results on some third-order systems have verified the qualitative predictions. Many more theoretical and simulation investigations are necessary to evaluate the use of this concept for the control of systems with more unknown parameters, e.g., poles.

References

[1] M. Aoki, *Optimization of Stochastic Systems* (Mathematics in Engineering, vol. 32). New York: Academic, 1967.

[2] A. A. Fel'dbaum, *Optimal Control Systems* (Mathematics in Science and Engineering, vol. 22). New York: Academic, 1965.

[3] J. B. Farison, R. E. Graham, and R. C. Shelton, Jr., "Identification and control of linear discrete systems," *IEEE Trans. Automat. Control* (Short Papers), vol. AC-12, pp. 438–442, Aug. 1967.

[4] W. J. Murphy, "Optimal stochastic control of discrete linear systems with unknown gain," *IEEE Trans. Automat. Contr.*, vol. AC-13, pp. 338–344, Aug. 1968.

[5] D. Gorman and J. Zaborszky, "Stochastic optimal control of continuous time systems with unknown gain," *IEEE Trans. Automat. Contr.*, vol. AC-13, pp. 630–638, Dec. 1968.

[6] Y. Bar-Shalom and R. Sivan, "On the optimal control of discrete-time linear systems with random parameters," *IEEE Trans. Automat. Contr.*, vol. AC-14, pp. 3–8, Feb. 1969.

[7] J. J. Florentin, "Optimal, probing, adaptive control of a simple Bayesian system," *J. Electron. Contr.*, vol. 11, 1962.

[8] R. Bellman, *Dynamic Programming.* Princeton, N.J.: Princeton University Press, 1957.

[9] S. E. Dreyfus, *Dynamic Programming and the Calculus of Variations* (Mathematics in Science and Engineering, vol. 21). New York: Academic Press, 1965.

[10] M. Athans, "The matrix minimum principle," *Inform. Contr.*, vol. 11, pp. 592–606, 1967.

[11] A. Levis, "On the optimal sampled-data control of linear processes," Ph.D. dissertation, Mass. Inst. Technol., Cambridge, 1968.

[12] E. Tse and M. Athans, "Optimal minimal-order observer estimators for discrete linear time-varying systems," *IEEE Trans. Automat. Contr.*, vol. AC-15, pp. 416–426, Aug. 1970.

[13] E. Tse, "On the optimal control of linear systems with incomplete information," Mass. Inst. Technol., Cambridge, Rep. ESL-R-412, Jan. 1970.

[14] D. L. Kleinman and M. Athans, "The discrete minimum principle with application to the linear regulator problem," Mass. Inst. Technol., Cambridge, Rep. ESL-R-260, Feb. 1966.

[15] E. Tse and M. Athans, "Adaptive stochastic control for linear systems—Part I: Solution methods; Part II: Asymptotic properties and simulation results," in *Proc. Decision and Control Conf.*, 1970.

[16] H. A. Spang, "Optimum control of an unknown linear plant using Bayesian estimation of the error," *IEEE Trans. Automat. Contr.* (Short Papers), vol. AC-10, pp. 80–83, Jan. 1965.

[17] K. J. Astrom, *Introduction to Stochastic Control Theory.* New York: Academic Press, 1970.

[18] R. Curry, *Estimation and Control with Quantized Measurements.* Cambridge, Mass.: M.I.T. Press, 1970.

[19] M. W. Wonham, "On the separation theorem of stochastic control," *SIAM J. Contr.*, vol. 6, 1968.

[20] L. S. Kramer, "On stochastic control and optimal measurements strategies," Ph.D. dissertation, Dep. Elec. Eng., Mass. Inst. Technol., Cambridge, 1971.

[21] G. N. Saridis and R. N. Lobbia, "Parameter identification and control of linear discrete time systems," in *1971 Joint Automatic Control Conf., Preprints*, pp. 724–730.

[22] M. Athans, R. P. Wishner, and A. Bertolini, "Suboptimal state estimation for continuous-time nonlinear systems from discrete noisy measurements," *IEEE Trans. Automat. Contr.*, vol. AC-13, pp. 504–514, Oct. 1968.

Adaptive Control with the Stochastic Approximation Algorithm: Geometry and Convergence

ARTHUR H. BECKER, JR., P. R. KUMAR, MEMBER, IEEE, AND CHING-ZONG WEI

Abstract—**New geometric properties possessed by the sequence of parameter estimates are exhibited, which yield valuable insight into the behavior of the stochastic approximation based algorithm as it is used in minimum variance adaptive control. In particular, these geometric properties, together with certain probabilistic arguments, prove that if the system does not have a reduced-order minimum variance controller, then the parameter estimates converge to a random multiple of the true parameter. An explicit expression for the limiting parameter estimate is also available. With strictly positive probability, the limiting parameter estimate is not the true parameter, and in some cases differs from the true parameter with probability one. If the system possesses reduced-order minimum variance controllers, then convergence to a minimum variance controller in a Cesaro sense is shown. The geometry of the limit set is described. Sufficient conditions are also given for some of these results to hold for parameter estimation schemes other than stochastic approximation.**

I. INTRODUCTION

WE consider the system

$$y(t+1)=\sum_{i=0}^{p}[a_iy(t-i)+b_iu(t-i)+c_iw(t-i)]+w(t+1) \tag{1}$$

where y, u, and w are the output, input, and noise, respectively. Suppose now that the parameters $(a_0, \cdots, a_p, b_0, \cdots, b_p, c_0, \cdots, c_p)$ are unknown and our goal is to minimize the variance of the output process. A well-known adaptive control algorithm for this purpose uses a recursive stochastic approximation (SA) scheme to obtain estimates of the parameters and then chooses a control input based on the estimated parameters. Specifically, the parameter estimates $\{\hat{\theta}(t)\}$ are given by the SA recursion

$$\hat{\theta}(t)=\hat{\theta}(t-1)+\frac{\mu\phi(t-1)}{r(t-1)}[y(t)-\phi^T(t-1)\hat{\theta}(t-1)] \tag{2}$$

where μ is a positive constant, and

$$r(t)=r(t-1)+\phi^T(t)\phi(t); \qquad r(0)=1 \tag{3}$$

$$\phi^T(t)=[y(t),\ y(t-1),\ \cdots,\ y(t-p),\ u(t),\ u(t-1),\ \cdots,\ u(t-p)]. \tag{4}$$

Manuscript received July 22, 1983; revised March 1, 1984. Paper recommended by Past Associate Editor, G. C. Goodwin. This work was supported by the U.S. Army Research Office under Contract DAAG-29-80-K0038 and by the National Science Foundation under Grants ECS-8304435 and MCS-8103448A01.

A. Becker is with the Department of Mathematics and Computer Science, University of Maryland-Baltimore County, Catonsville, MD 21228.

P. R. Kumar is with the Department of Electrical and Computer Engineering and the Coordinated Science Laboratory, University of Illinois, Urbana, IL 61801.

C.-Z. Wei is with the Department of Mathematics, University of Maryland, College Park, MD 20742.

Based on the parameter estimate $\hat{\theta}(t)$ available at time t, the control input $u(t)$ is chosen to satisfy

$$\phi^T(t)\hat{\theta}(t)=0 \tag{5}$$

or is more explicitly specified by

$$u(t)=\frac{-1}{\hat{\theta}_{p+2}(t)}[\hat{\theta}_1(t)y(t)+\cdots+\hat{\theta}_{p+1}(t)y(t-p)+\hat{\theta}_{p+3}(t)u(t-1)+\cdots+\hat{\theta}_{2p+2}(t)u(t-p)] \tag{6}$$

where $\hat{\theta}(t) := (\hat{\theta}_1(t), \hat{\theta}_2(t), \cdots, \hat{\theta}_{2p+2}(t))^T$.

Our goal in this paper is to analyze in detail the behavior of the adaptive control algorithm (2)–(5) when it is applied to system (1). First we expose some very basic and revealing geometric properties possessed by the sequence $\{\hat{\theta}(t)\}$ of parameter estimates. These geometric properties are then exploited together with probabilistic arguments to analyze the asymptotic behavior of the algorithm.

Our main results are as follows.

1) $\hat{\theta}(t) - \hat{\theta}(t-1)$ is orthogonal to $\hat{\theta}(t-1)$. This elementary observation is the basis of the geometric properties of the algorithm (Theorem 1).

2) If the system possesses no reduced-order minimum variance controllers, then $\lim \hat{\theta}(t) = k\theta^0$ where $\theta^0 := (a_0 + c_0, \cdots, a_p + c_p, b_0, \cdots, b_p)^T$ and k is a finite random constant. Thus, the parameter estimates converge to a random multiple of the true parameter (Theorem 21).

3) k, the random multiplier, is explicitly given by

$$k^2=\frac{\|\hat{\theta}(0)\|^2}{\|\theta^0\|^2}+\sum_{t=1}^{\infty}\frac{\mu^2y^2(t)\|\phi(t-1)\|^2}{\|\theta^0\|^2r^2(t-1)}.$$

Thus, the limit of $\{\hat{\theta}(t)\}$ can be explicitly written down to within sign ($\pm$) indefiniteness (Theorem 21).

4) $\mathcal{P}(\lim \hat{\theta}(t) \neq \theta^0) > 0$ always. If $\|\hat{\theta}(0)\| > \|\theta^0\|$, then $\mathcal{P}(\lim \hat{\theta}(t) \neq \theta^0) = 1$. Thus, in general, the parameter estimates will not converge to the true parameter. Further, if the norm of the initial estimate is chosen too large, then the limiting parameter estimate differs from the true parameter with probability one (Theorem 21).

5) If the system does possess reduced-order minimum variance controllers, then the parameter estimates converge to a hypersphere of dimension strictly less than that of the space [Theorem 19iii)]. Moreover,

$$\lim \frac{1}{N}\sum_{1}^{N} 1(\hat{\theta}(t) \in 0)=1 \text{ a.s.}$$

for every open set 0 containing a specified set D. D, the null space of a certain matrix, is a set of parameter values all of which produce minimum variance controllers for the true system. Thus,

Reprinted from *IEEE Trans. Automat. Contr.*, vol. AC-30, pp. 330–338, Apr. 1985.

the adaptive controller converges in a Cesaro-sense to a minimum variance controller [Theorem 19i), ii)].

6) Sufficient conditions for ii) and v) to be applicable to parameter estimation schemes other than SA are given (Theorem 22).

A. History and Discussion

The problem of adaptive control of linear systems has been the subject of a considerable amount of attention over the past decade. The pioneering work is that of Åström and Wittenmark [1] who consider a least-squares estimation scheme for the parameter estimates. Their surprising result was that if the parameter estimates converge to *any* limiting random variable, then the limiting control law is optimal. The next major step is due to Ljung [2], [3] who associates an ordinary differential equation with the sequence of parameter estimates. Ljung also demonstrates the central role played by a positive real condition on the noise color in guaranteeing convergence of this associated ordinary differential equation. The next breakthrough in the field is due to Goodwin, Ramadge, and Caines [4]. Utilizing some techniques introduced by Solo [5], Goodwin *et al.* were able to rigorously prove that the algorithm detailed above is an optimal adaptive control scheme. By this is meant that the sample path variance of the output converges to the minimum value that could be achieved if the system parameters were known. Since then, the techniques developed in [4] have found widespread use in establishing optimality of a variety of other adaptive control schemes, e.g., one based on a modified least-squares estimate in [6].

However, there appear to be no rigorous results available on the convergence of the parameter estimates themselves. Note that if "self-tuning" is the purpose of using an adaptive controller, then convergence of the parameter estimates is of central importance.

For any adaptive control scheme, the following questions are important.

i) Do the parameter estimates converge to the true parameter?

ii) Do the parameter estimates converge to *some* parameter which leads to an optimal control law (e.g., a multiple of the true parameter)?

iii) Does the adaptive control law converge to an optimal control law?

iv) Is the cost optimal?

Note the strict hierarchy of implications i) ⇒ ii) ⇒ iii ⇒ iv). i) of course is the most desirable property, but, as we show, it does not hold. If the system does not possess reduced-order minimum variance controllers, then we show that ii) holds, thus settling all the questions. On the other hand, if the system does possess reduced-order minimum variance controllers, then we show that the limit set of the parameter estimates is a hypersphere of dimension strictly less than that of the space. Further, we show that iii) holds in a Cesaro sense. Note that Goodwin *et al.* only prove iv).

It should be noted that overmodeling the system per se is not a problem. As we show by an example in Section VI, even if we over model the orders of the system, convergence of the parameters can occur.

The usual approach to convergence of parameter estimates is via the notion of "persistence of excitation." In *adaptive control*, however, such "persistency of excitation" conditions are not verifiable *apriori*. In particular, in this paper, persistency of excitation does *not* hold (i.e., it holds with probability zero) because of the linear relationship between the outputs and inputs induced by the convergence of the adaptive control law. One way, however, to ensure persistency of excitation, is to "inject" noise into the system; we refer the reader to Caines and Lafortune [12] and Chen and Caines [13] for details of this approach.

One of the reasons why stochastic adaptive control systems such as (1)-(6) are difficult to analyze with respect to their asymptotic properties is doubtless due to the highly nonlinear manner in which the variables $(y, \hat{\theta}, r, \phi, u)$ are coupled with each other. One of the purposes of theoretical analysis, as here, besides establishing results, is therefore also to develop methods to analyze such schemes. We note for example the successful use of methods proposed in [4], [6] in contexts other than which they were originally used. This, accordingly, is a goal of this paper also.

As potential possibilities in this regard we bring attention to the fact that the geometric properties developed in Section II are properties of the algorithm (2)-(5) itself and *not* the system, e.g., (1) to which it is applied. Thus, these results may, for example, find use in studying "robustness" properties, see Rohrs *et al.* [7]. As another possibility, we note that the probabilistic methods used in Section IV depend only on the assumptions that the parameter estimates are bounded, the first-order differences [i.e., $\hat{\theta}(t) - \hat{\theta}(t-1)$] converge to zero, and the optimality of the cost. Thus, these methods also apply to adaptive control schemes other than the one studied here; see Theorem 22.

It should be mentioned that the SA based adaptive control algorithm is not implemented in practice because of its poor convergence rate. However, it is easier to analyze than its least-squares counterpart, and so results have often first been proved for the SA scheme and then generalized to its least-squares like counterparts. Further work for the least-squares scheme is definitely needed, and this is an important area for future research. Our treatment in this paper is also limited by the fact that we only consider the "regulation" problem. The "tracking" problem is important in practice, and is thus another open issue. Lastly, note that we do not consider the case where the delay between the application of an input and its effect on the output is more than one time unit.

B. Assumptions

The following assumptions will be made about the system (1).

$$b_0 \neq 0, \qquad \theta^0 := (a_0+c_0, \cdots, a_p+c_p, b_0, \cdots, b_p)^T \tag{7.i}$$

θ^0 is called the "true parameter."

The polynomials

$$1+c_0q+\cdots+c_pq^{p+1} \text{ and } b_0+b_1q+\cdots+b_pq^p \tag{7.ii}$$

have all their zeros outside the closed unit disk.

The polynomial $[1-\mu/2+c_0q+\cdots+c_pq^{p+1}]$ is strictly positive real, i.e.,

$$\text{Re}\left[1-\frac{\mu}{2}+c_0e^{i\omega}+\cdots+c_pe^{i\omega(p+1)}\right]>0 \qquad \text{for all } \omega.$$

$\{w(-p), \cdots, w(0)\}$ and $\phi(0)\neq 0$ are fixed.

$$\{w(1), \cdots, w(t), \cdots\} \tag{8.i}$$

are scalar valued random variables on the probability space $(\Omega, \mathfrak{F}, \mathcal{P})$ whose distributions are all mutually absolutely continuous with respect to Lebesgue measure.

$$E(w(t)|\mathfrak{F}_{t-1})=0 \text{ for all } t\geq 1, \tag{8.ii}$$

where $\mathfrak{F}_t$ is the sub σ-algebra of $\mathfrak{F}$ generated by the random variables $\{w(1), \cdots, w(t)\}$. $\mathfrak{F}_0$ is the trivial σ-algebra.

$$E(w^2(t)|\mathfrak{F}_{t-1})=\sigma^2>0 \qquad \text{for all } t\geq 1. \tag{8.iii}$$

$$\sup_t E(|w(t)|^{2+\delta}|\mathfrak{F}_{t-1})<\infty \qquad \text{for some } \delta>0. \tag{8.iv}$$

We note that the assumptions (7), (8) are the same as those in [4] with the exception of (8.iv) which is slightly stronger. However, this leads to a slightly more relevant result on the optimality of the cost, and is therefore employed here.

Remark: The results of this paper are also valid when the

system is given by

$$y(t+1)=\sum_{i=0}^{p_1} a_i y(t-i)+\sum_{i=0}^{p_2} b_i u(t-i)+\sum_{i=0}^{p_3} c_i w(t-i)+w(t+1)$$

where for simplicity we have chosen $p = \max(p_1, p_2, p_3)$.

II. Geometric Properties of the Algorithm

We develop in this section the geometric properties possessed by the SA algorithm as used in adaptive control. We hasten to point out that these properties do *not* depend on the system (1), and thus they are a property of the algorithm itself. We also mention that it is the specific manner in which the control inputs are chosen in minimum variance adaptive control (5) that leads to these very useful properties. Thus, if control inputs were chosen in some other manner, these properties do not hold. Hence, we have the surprising fact that closing the loop as in (6), instead of complicating the analysis, actually simplifies it!

The basic property here is that the sequence $\{\hat\theta(t)\}$ of parameter estimates takes orthogonal jumps. In spite of its trivial proof, we elevate this result to the following.

Theorem 1:

$$\langle\hat\theta(t-1),\ \Delta\hat\theta(t)\rangle=0 \qquad \text{for all } t\geq 1$$

where

$$\Delta\hat\theta(t) := \hat\theta(t)-\hat\theta(t-1). \tag{9}$$

Proof: From (5) we have $\langle\hat\theta(t-1), \phi(t-1)\rangle = 0$. From (2) we obtain

$$\Delta\hat\theta(t)=\frac{\mu}{r(t-1)}\,(y(t)-\phi^T(t-1)\hat\theta(t-1))\phi(t-1)$$

$$=\frac{\mu y(t)}{r(t-1)}\,\phi(t-1). \tag{10}$$

Since $\mu y(t)/r(t-1)$ is a scalar, taking the inner product on both sides of (10) with $\hat\theta(t-1)$ and using (5) gives the result. □

Throughout the rest of this section, we study the evolution of $\{\hat\theta(t,\omega)\}$ for a fixed $\omega\in\Omega$, and so for simplicity of notation, the argument ω is omitted. Some consequences of Theorem 1 are the following.

Corollary 2:

$$\|\hat\theta(t)\|^2=\sum_{n=0}^{t}\|\Delta\hat\theta(n)\|^2 \text{ where } \Delta\hat\theta(0) := \hat\theta(0). \tag{11.i}$$

$$\|\hat\theta(t)\|\geq\|\hat\theta(t-1)\| \qquad \text{for all } t\geq 1. \tag{11.ii}$$

$$\text{If } \sup_t \|\hat\theta(t)\|<\infty, \text{ then } \{\|\hat\theta(t)\|\} \text{ converges.} \tag{11.iii}$$

$$\text{If } \sup_t \|\hat\theta(t)\|<\infty, \text{ then } \Delta\hat\theta(t)\to 0. \tag{11.iv}$$

$$\text{If } \{\|\hat\theta(t)-\theta^0\|\} \text{ is bounded, then } \sup_t \|\hat\theta(t)\|<\infty. \tag{11.v}$$

Proof: From Pythagoras' theorem and Theorem 1,

$$\|\hat\theta(t)\|^2=\|\hat\theta(t-1)\|^2+\|\Delta\hat\theta(t)\|^2 \qquad \text{for all } t\geq 1.$$

The rest follows. □

Let

$$\tilde\theta(t) := \hat\theta(t)-\theta^0.$$

Lemma 3: If

$$\{\|\tilde\theta(t)\|\} \text{ converges} \tag{12.i}$$

for some subsequence $\{t_n\}$ of the integers

$$\theta_i^0\hat\theta_{p+2}(t_n)-\theta_{p+2}^0\hat\theta_i(t_n)\to 0 \qquad \text{for every } i=1,2,\cdots,2p+2. \tag{12.ii}$$

Then $\lim\hat\theta(t) = k\theta^0$ for some constant k.

Proof: Note that if $\{\|\tilde\theta(t)\|\}$ converges, then it is bounded and so from (11.v) and (11.iii), $\{\|\hat\theta(t)\|\}$ converges. Since $\{\hat\theta(t)\}$ is bounded, we may as well assume that $\hat\theta(t_n)\to\theta^*$. If $\theta^*_{p+2}=0$, then from (12.ii) we see that $\theta^*=0$ and we are through by (11.ii). Suppose $\theta^*_{p+2}\neq 0$. Then $\hat\theta_i(t_n)\to[(\theta^*_{p+2})/(\theta^0_{p+2})\theta^0_i]$ for $i=1, 2, \cdots, 2p+2$. Hence, $\hat\theta(t_n)\to k\theta^0$ for some k. Now since $\{\|\tilde\theta(t)\|^2\}$ converges, and since $\tilde\theta(t)=\hat\theta(t)-\theta^0$, we see that $\{\|\hat\theta(t)\|^2-2\langle\theta^0,\hat\theta(t)\rangle+\|\theta^0\|^2\}$ converges, and since $\{\|\hat\theta(t)\|^2\}$ converges, we see that $\{\langle\theta^0,\hat\theta(t)\rangle\}$ converges. Hence, $\{\|\hat\theta(t)-k\theta^0\|^2\}=\{\|\hat\theta(t)\|^2-2k\langle\theta^0,\hat\theta(t)\rangle+k^2\|\theta^0\|^2\}$ converges. But its limit along the subsequence $\{t_n\}$ is zero, and so $\hat\theta(t)\to k\theta^0$. □

Remark: An explanation of Lemma 3 is useful. If both $\{\|\hat\theta(t)\|\}$ and $\{\|\tilde\theta(t)\|\}$ converge, then $\{\hat\theta(t)\}$ converges to the intersection of two spheres, one centered at the origin and the other centered at θ^0. This intersection is a hypersphere of dimension strictly less than the dimension of the space (here $2p+2$), and also has its center at some point on the line passing through θ^0 and the origin. Moreover, if the radius of this hypersphere is strictly positive, then this hypersphere does not intersect this line. However, (12.ii) says that $\{\hat\theta(t)\}$ does have a subsequence converging to the line. The only way for this to happen is for the hypersphere to "collapse" to a point—its center.

B. Special Situations

At this point we disgress to consider the special case where $\hat\theta(t)$ is a two-dimensional vector. This special case thus applies to systems of the form $y(t+1)=a_0y(t)+b_0u(t)+w(t+1)$; $y(t+1)=b_0u(t)+w(t+1)+c_0w(t)$ and even to $y(t+1)=b_0u(t)+b_1u(t-1)+w(t+1)$ (noting the Remark of Section I-B). In this special case, we see that (12.i) and Theorem 1 alone suffice to prove convergence of $\{\hat\theta(t)\}$. Thus, (12.ii) is unnecessary and the probabilistic arguments used in the sequel to establish it in the general case can be dispensed with.

Lemma 5: If $\{\hat\theta(t)\}$ is a sequence in $\mathbb{R}^2$ satisfying

i) $\langle\hat\theta(t-1),\ \hat\theta(t)-\hat\theta(t-1)\rangle=0$ for all $t\geq 1$

ii) $\{\|\hat\theta(t)-\theta^0\|^2\}$ converges for $\theta^0\neq 0$

then $\{\hat\theta(t)\}$ converges.

Proof: We provide a geometric proof, omitting some details. As above, since $\{\|\tilde\theta(t)\|^2\}$ converges, $\{\|\hat\theta(t)\|\}$ is bounded, and hence converges. Thus, $\{\hat\theta(t)\}$ converges to the intersection of two circles, one centered at $\theta^0\neq 0$ and the other centered at the origin. Since two circles in $\mathbb{R}^2$ can intersect in at most two points, suppose that $\{\hat\theta(t)\}$ has two limits with both, of course, being at the same distance from the origin. Let N_1 and N_2 be disjoint neighborhoods of the two limit points so small that $M\cap N_2$ is empty, where $M := \{\theta: \langle\theta-\bar\theta, \bar\theta\rangle = 0 \text{ for some } \bar\theta\in N_1\}$. (There exist such N_1 and N_2.) Since $\{\hat\theta(t)\}$ has only two limit points, we can assume that $\hat\theta(t)\in N_1\cup N_2$ for all $t\geq$ some T. However, once $\hat\theta(t)$ enters N_1, then $\hat\theta(t+1)\in M$ and so $\hat\theta(t+1)\notin N_2$. Thus, $\hat\theta(t)\in N_1$ for all $t\geq T$, a contradiction to the assumption that there are two limit points. □

In $\mathbb{R}^3$ (and higher dimensional spaces), the conditions of Lemma 5 are not sufficient to ensure convergence, as the following example shows.

Example 6: Let $\hat\theta(n)=(\rho_n\cos\alpha_n, \rho_n\sin\alpha_n, 1)$, $\theta^0=(0, 0, 1)$ where $\alpha_{n+1}=\alpha_n+1/n$ and $\rho_{n+1}=\rho_n\sec(1/n)$. It is easy to check that the conditions i) and ii) of Lemma 5 are true, but that $\{\hat\theta(n)\}$ does not converge.

III. Optimality of Cost and Consequences

In this section we obtain some consequences of the results of Goodwin *et al.* [4], which we find useful in our analysis.

Theorem (Goodwin, Ramadge, and Caines) [4]:

$$\lim \frac{1}{N}\sum_{1}^{N} E(y^2(t)|\mathfrak{F}_{t-1})=\sigma^2 \text{ a.s.} \tag{13.i}$$

$$\sum_{1}^{\infty} z^2(t)/r(t)<\infty \text{ a.s. where } z(t) := E(y(t+1)|\mathfrak{F}_t). \tag{13.ii}$$

$$\lim \frac{1}{N}\sum_{1}^{N} z^2(t)=0 \text{ a.s.} \tag{13.iii}$$

$$\lim\sup \frac{1}{N}\sum_{1}^{N} u^2(t)<\infty \text{ a.s.} \tag{13.iv}$$

Let $h(t) := \left(1-\frac{\mu+\rho}{2}\right)z(t)+\sum_{i=0}^{p} c_i z(t-i-1).$

For small enough $\rho > 0$,

$$S(t) := 2\mu\sum_{1}^{t} h(n-1)z(n-1)+L\geq 0 \text{ for some } L. \tag{13.v}$$

$$\left\{\|\tilde{\theta}(t)\|^2+\frac{S(t)}{r(t-1)}\right\} \text{ converges a.s.} \tag{13.vi}$$

Proof: All the above are proved in [4] if we can verify their assumption that

$$\lim\sup \frac{1}{N}\sum_{1}^{N} w^2(t)<\infty \text{ a.s.}$$

But by our assumptions (8.iii), (8.iv), from Minkowski's inequality, we see that $\sup E\{|w^2(t) - E(w^2(t)|\mathfrak{F}_{t-1})|^{1+\delta/2}|\mathfrak{F}_{t-1}\} < \infty$. From Chow's theorem [11, Theorem 3.3.1] it actually follows that

$$\lim \frac{1}{N}\sum_{1}^{N} w^2(t)=\sigma^2 \text{ a.s.} \tag{13.vii}$$ □

We now proceed to obtain some consequences which we shall use in our analysis. The first such consequence shows that the actual cost incurred is optimal. This is a slight strengthening of (13.i) and is also slightly more relevant with respect to demonstrating optimality of the adaptive control scheme.

Lemma 7:

$$\lim \frac{1}{N}\sum_{1}^{N} y^2(t)=\sigma^2 \text{ a.s.}$$

Proof: Noting that $y(t) = z(t-1) + w(t)$ and using the Cauchy-Schwarz inequality shows that for some $|\alpha_N| \leq 1$,

$$\begin{aligned}\frac{1}{N}\sum_{1}^{N} y^2(t)&=\frac{1}{N}\sum_{1}^{N}(z^2(t-1)+w^2(t))\\&\quad+2\alpha_N\left(\frac{1}{N}\sum_{1}^{N} z^2(t-1)\right)^{1/2}\left(\frac{1}{N}\sum_{1}^{N} w^2(t)\right)^{1/2}\\&=\left(\frac{1}{N}\sum_{1}^{N} z^2(t-1)\right)^{1/2}\left[\left(\frac{1}{N}\sum_{1}^{N} z^2(t-1)\right)^{1/2}\right.\\&\quad\left.+2\alpha_N\left(\frac{1}{N}\sum_{1}^{N} w^2(t)\right)^{1/2}\right]\\&\quad+\frac{1}{N}\sum_{1}^{N} w^2(t).\end{aligned}$$

Using (13.iii), (13.vii) and taking the limit, we obtain the result. □

Lemma 8:

$$r(t)\to\infty \text{ a.s.}$$

Proof: This follows from Lemma 7 since $r(t) \geq \Sigma_1^t y^2(n)$. □

Now we establish (12.i) of Lemma 3. As we have seen, this leads to some geometric consequences and together with (12.ii) guarantees convergence of $\{\hat{\theta}(t)\}$.

Lemma 9:

$$\{\|\tilde{\theta}(t)\|^2\} \text{ converges a.s.}$$

Proof: From (13.vi) it is clear that it suffices to establish that $S(N)/r(N-1) \to 0$ a.s. From the definitions of $S(N)$ and $h(N)$ in (13.v) this will follow if

$$\frac{1}{r(N)}\sum_{1}^{N} z(t-l)z(t)\to 0 \text{ a.s. for } 0\leq l\leq p+1.$$

The Cauchy-Schwarz inequality shows that for some $|\alpha_N| \leq 1$,

$$\frac{1}{r(N)}\sum_{1}^{N} z(t-l)z(t)=\alpha_N\left(\frac{1}{r(N)}\sum_{1}^{N} z^2(t-l)\right)^{1/2}\cdot\left(\frac{1}{r(N)}\sum_{1}^{N} z^2(t)\right)^{1/2}.$$

By using (13.ii), Lemma 8, and applying Kronecker's lemma [8], it is clear that the right side does converge to 0. □

The result of Lemma 9 is also mentioned in Caines and Lafortune [14] (see also [13]) along with some comments on the sequence of parameter estimates.

At this point it is already clear, by virtue of Theorem 1, Lemma 9, and (11.v), that $\{\hat{\theta}(t)\}$ converges to the intersection of two spheres, one centered at the origin and the other at θ^0. This will now be improved upon.

Before ending this section we note that the results of this section (appropriately generalized) are valid even for the "tracking" problem considered in [4], and are not just restricted to the "regulation" problem.

IV. Convergence of Parameter Estimates

We begin by mentioning that the results of this section do *not* depend on the fact that the parameter estimates are generated by an SA scheme such as (2). The results depend *only* on the system satisfying (1) and (8) and the facts that:

$$\lim \frac{1}{N}\sum_{1}^{N} z^2(t)=0 \text{ a.s.} \tag{14.i}$$

$$\lim \Delta\hat{\theta}(t)=0 \text{ a.s.} \tag{14.ii}$$

$$\{\hat{\theta}(t)\} \text{ is a bounded sequence a.s.} \tag{14.iii}$$

$$\phi^T(t)\hat{\theta}(t)=0 \text{ or, equivalently, (6) defines the control inputs.} \tag{14.iv}$$

$$\lim\sup \frac{1}{N}\sum_{1}^{N} u^2(t)<\infty \text{ a.s.} \tag{14.v}$$

$$\hat{\theta}_{p+2}(t)\neq 0 \quad \text{for all } t\geq 1 \text{ a.s.} \tag{14.vi}$$

Thus, if it can be demonstrated that the conditions of (14) are satisfied by some adaptive control algorithm, say one which uses a least-squares type procedure for obtaining parameter estimates, then the results of this section would hold for such an algorithm. In particular, of course, we have already demonstrated the validity of (14) for the SA scheme.

We first state a lemma on sequences, which we shall repeatedly use in the sequel.

Lemma 10: Let $\{\rho(t)\}$ be a real valued sequence satisfying

$$\lim \frac{1}{N}\sum_{1}^{N} \rho^2(t)=0.$$

Then for any real valued sequence $\{\xi(t)\}$,

$$\lim \frac{1}{N}\sum_{1}^{N} \xi^2(t)=0 \text{ if and only if } \lim \frac{1}{N}\sum_{1}^{N} (\rho(t)+\xi(t))^2=0.$$

Proof: This follows from the twin inequalities, $(\rho(t) + \xi(t))^2 \leq 2\rho^2(t) + 2\xi^2(t)$ and $\xi^2(t) \leq 2\rho^2(t) + 2(\rho(t) + \xi(t))^2$.□

Our approach uses the following result which is based on the local convergence theorem for martingales [9].

Lemma 11: Suppose $\{\eta(t)\}$, $\{\pi(t)\}$, and $\{s(t)\}$ are stochastic processes adapted to $\{\mathfrak{F}_t\}$ and satisfying

$$\{\eta(t)\} \text{ and } \{\pi(t)\} \text{ are bounded sequences a.s.} \tag{15.i}$$

$$\lim \frac{1}{N}\sum_{1}^{N} (\eta(t-1)y(t)+\pi(t-1)w(t)+s(t-1))^2=0 \text{ a.s.} \tag{15.ii}$$

Then

$$\lim \frac{1}{N}\sum_{1}^{N} (\eta(t)+\pi(t))^2=0 \text{ a.s.} \tag{16.i}$$

$$\lim \frac{1}{N}\sum_{1}^{N} s^2(t)=0 \text{ a.s.} \tag{16.ii}$$

Proof: Noting that $y(t) = z(t-1) + w(t)$, we can rewrite (15.ii) as

$$\lim \frac{1}{N}\sum_{1}^{N} (k(t-1)w(t)+q(t-1))^2=0 \text{ a.s.} \tag{17.i}$$

where $k(t) := \eta(t) + \pi(t)$ and $q(t) := \eta(t)z(t) + s(t)$. Expanding the quadratic gives

$$\frac{1}{N}\sum_{1}^{N} (k(t-1)w(t)+q(t-1))^2=\frac{1}{N}\sum_{1}^{N} k^2(t-1)w^2(t) +\frac{1}{N}\sum_{1}^{N} q^2(t-1) +\frac{2}{N}\sum_{1}^{N} k(t-1)q(t-1)w(t). \tag{17.ii}$$

From an extension of the local convergence theorem for martingales in Lai and Wei [10, Lemma 2.iii] we know that $\Sigma_1^N k(t-1)q(t-1)w(t)$ converges a.s. on the subset $\tilde{\Omega} := \{\omega: \Sigma_1^\infty k^2(t-1)q^2(t-1) < \infty\}$. Thus,

$$\frac{1}{N}\sum_{1}^{N} k(t-1)q(t-1)w(t)\to 0 \text{ a.s.}$$

on $\tilde{\Omega}$ and (17.i), (17.ii) give us

$$\frac{1}{N}\sum_{1}^{N} k^2(t-1)w^2(t)\to 0 \text{ and } \frac{1}{N}\sum_{1}^{N} q^2(t-1)\to 0 \text{ a.s. on } \tilde{\Omega}.$$

Now consider the complement $\tilde{\Omega}^c = \{\omega: \Sigma_1^\infty k^2(t-1)w^2(t) = \infty\}$. Here [10, Lemma 2.iii] shows that

$$\sum_{1}^{N} k(t-1)q(t-1)w(t)=o\left(\sum_{1}^{N} k^2(t-1)q^2(t-1)\right) \text{ a.s. on } \tilde{\Omega}^c.$$

Since $\{k(t)\}$ is a bounded sequence a.s., it follows that $\Sigma_1^N k(t-1)q(t-1)w(t) = o(\Sigma_1^N q^2(t-1))$ a.s. on $\tilde{\Omega}^c$. Hence, through (17.ii),

$$\frac{1}{N}\sum_{1}^{N} (k(t-1)w(t)+q(t-1))^2=\frac{1}{N}\sum_{1}^{N} k^2(t-1)w^2(t) +\frac{1}{N}\sum_{1}^{N} q^2(t-1)\left[1+\frac{o\left(\sum_{1}^{N} q^2(t-1)\right)}{\sum_{1}^{N} q^2(t-1)}\right]$$

a.s. on $\tilde{\Omega}^c$. However, since the left side converges to 0 a.s., we obtain that both

$$\frac{1}{N}\sum_{1}^{N} k^2(t-1)w^2(t)\to 0 \text{ and } \frac{1}{N}\sum_{1}^{N} q^2(t-1)\to 0 \text{ a.s. on } \tilde{\Omega}^c \text{ also.}$$

Hence, we obtain

$$\lim \frac{1}{N}\sum_{1}^{N} q^2(t-1)=0 \text{ a.s.} \tag{18}$$

and

$$\lim \frac{1}{N}\sum_{1}^{N} k^2(t-1)w^2(t)=0 \text{ a.s.} \tag{19}$$

Since $\{\eta(t)\}$ is a bounded sequence a.s., from (14.i) we obtain that

$$\lim \frac{1}{N}\sum_{1}^{N} \eta^2(t)z^2(t)=0 \text{ a.s.}$$

Using this, (18) and the definition of $q(t)$ shows from Lemma 10 that (16.ii) holds. Note also that from (8.iii), (8.iv) and since $k(t)$ is a bounded sequence a.s., we have

$$\sup_t E\{|k^2(t-1)w^2(t)-E(k^2(t-1)w^2(t)|\mathfrak{F}_{t-1})|^{1+\delta/2}|\mathfrak{F}_{t-1}) <\infty \text{ a.s.}$$

Hence, using Chow's theorem [11, Theorem 3.3.1], it follows that (16.i) also holds. □

To lay the framework for a repeated use of the above lemma we have the following result.

Lemma 12:

$$\lim \frac{1}{N}\sum_{1}^{N}\left(\sum_{i=0}^{p} (\alpha_i(t)y(t-i)+\gamma_i(t)w(t-i))+\sum_{i=1}^{p} \beta_i(t)u(t-i)\right)^2 =0 \text{ a.s.}$$

where

$$\alpha(t) := [a_0\hat{\theta}_{p+2}(t) - b_0\hat{\theta}_1(t),\ a_1\hat{\theta}_{p+2}(t) - b_0\hat{\theta}_2(t),\ \cdots,\ a_p\hat{\theta}_{p+2}(t) - b_0\hat{\theta}_{p+1}(t)]^T =: (\alpha_0(t),\ \cdots,\ \alpha_p(t))^T$$

$$\beta(t) := [b_1\hat{\theta}_{p+2}(t) - b_0\hat{\theta}_{p+3}(t),\ b_2\hat{\theta}_{p+2}(t) - b_0\hat{\theta}_{p+4}(t),\ \cdots,\ b_p\hat{\theta}_{p+2}(t) - b_0\hat{\theta}_{2p+2}(t)]^T =: (\beta_1(t),\ \cdots,\ \beta_p(t))^T$$

$$\gamma(t) := [c_0\hat{\theta}_{p+2}(t),\ c_1\hat{\theta}_{p+2}(t),\ \cdots,\ c_p\hat{\theta}_{p+2}(t)]^T =: (\gamma_0(t),\ \gamma_1(t),\ \cdots,\ \gamma_p(t))^T.$$

Proof: Note that $z(t)$ defined in (13.ii) is just

$$z(t) = \sum_{i=0}^{p} a_i y(t-i) + b_i u(t-i) + c_i w(t-i). \tag{20}$$

Substituting for $u(t)$ from (6) in (20), multiplying by $\hat{\theta}_{p+2}(t)$, and using the definitions of $\alpha(t)$, $\beta(t)$, and $\gamma(t)$ gives

$$\hat{\theta}_{p+2}(t)z(t) = \sum_{i=0}^{p} (\alpha_i(t)y(t-i) + \gamma_i(t)w(t-i)) + \sum_{i=1}^{p} \beta_i(t)u(t-i).$$

Since $\{\hat{\theta}_{p+2}(t)\}$ is a bounded sequence a.s. from (14.iii), the result follows from (14.i). □

We now show that in the conclusion of Lemma 12, we can replace the coefficients $\alpha_i(t)$, $\beta_i(t)$, and $\gamma_i(t)$ by $\alpha_i(t-l)$, $\beta_i(t-l)$, and $\gamma_i(t-l)$ for any $l \geq 1$. The purpose here is to make the coefficients "past-measurable" and so enable us to apply Lemma 11.

Lemma 13: For any $l > 1$

$$\lim \frac{1}{N}\sum_{1}^{N}\left[\sum_{i=0}^{p} (\alpha_i(t-l)y(t-i) + \gamma_i(t-l)w(t-i)) + \sum_{i=1}^{p} \beta_i(t-l)u(t-i)\right]^2 = 0 \text{ a.s.}$$

Proof: Consider the following situation. Let $\{x(t)\}$, $\{v(t)\}$, and $\{s(t)\}$ be real valued sequences satisfying

i) $\limsup \frac{1}{N}\sum_{1}^{N} v^2(t) < \infty$

ii) $\lim_t (x(t) - x(t-1)) = 0$

and

iii) $\lim \frac{1}{N}\sum_{1}^{N} (x(t)v(t) + s(t))^2 = 0.$

We now show that a consequence is

iv) $\lim \frac{1}{N}\sum_{1}^{N} (x(t-l)v(t) + s(t))^2 = 0 \quad$ for every $l \geq 1$.

From ii), for every $\epsilon > 0$, $|x(t) - x(t-l)|^2 < \epsilon$ for all $t >$ some T. So, for $N > T$,

$$\frac{1}{N}\sum_{1}^{N} [(x(t-l) - x(t))v(t)]^2 \leq \frac{1}{N}\sum_{1}^{T} [(x(t-l) - x(t))v(t)]^2 + \frac{\epsilon}{N}\sum_{T+1}^{N} v^2(t).$$

Taking the limit, and noting that $\epsilon > 0$ is arbitrary gives

$$\lim \frac{1}{N}\sum_{1}^{N} [(x(t-l) - x(t))v(t)]^2 = 0.$$

From this and iii), Lemma 10 shows the validity of iv). Now let $x(t)$ and $v(t)$ be any of $\alpha_i(t)\beta_i(t)$ or $\gamma_i(t)$ and $y(t-i)$, $u(t-i)$, or $w(t-i)$, respectively. In particular, suppose $x(t) = \alpha_j(t)$, $v(t) = y(t-j)$. Then let $s(t) = [\sum_{i=0}^{p} (\alpha_i(t)y(t-i) + \gamma_i(t)w(t-i)) + \sum_{i=1}^{p} \beta_i(t)u(t-i)] - x(t)v(t)$. By virtue of (14.ii), $x(t)$ satisfies ii); from (14.v), (13.vii), and Lemma 7, $v(t)$ satisfies i), and due to Lemma 12, iii) is satisfied. Thus, iv) holds and effectively shows that in Lemma 12, we can change $\alpha_j(t)$ to $\alpha_j(t-l)$. By repeated use of different choices for $x(t)$, $v(t)$, and $s(t)$, the result follows. □

Note now that it is our goal to investigate when (12.ii) holds, or equivalently, when $\alpha(t_n) + \gamma(t_n) \to 0$ and $\beta(t_n) \to 0$ along some subsequence. This, of course, will follow if

$$\frac{1}{N}\sum_{1}^{N} (\alpha_i(t) + \gamma_i(t))^2 \to 0 \text{ and } \frac{1}{N}\sum_{1}^{N} \beta_i^2(t) \to 0.$$

Now we see that Lemma 13 allows us to apply Lemma 11 to obtain such a conclusion at least as far as $\{(\alpha_0(t) + \gamma_0(t))\}$ is concerned. This is done in the following.

Lemma 14:

$$\lim \frac{1}{N}\sum_{1}^{N} (\alpha_0(t) + \gamma_0(t))^2 = 0 \text{ a.s.} \tag{21.i}$$

$$\lim \frac{1}{N}\sum_{1}^{N}\left(\sum_{i=1}^{p} \alpha_i(t-l)y(t-i) + \gamma_i(t-l)w(t-i) + \beta_i(t-l)u(t-i)\right)^2 = 0 \text{ for every } l > 1 \text{ a.s.} \tag{21.ii}$$

Proof: Identify $\eta(t-1)$ with $\alpha_0(t-l)$, $\pi(t-1)$ with $\gamma_0(t-l)$, and $s(t-1)$ with the inner sum in (21.ii). Now Lemma 11 gives us the required results since (15.i) is satisfied because of (14.iii), and (15.ii) because of Lemma 13. □

Note now that we want to obtain conclusions similar to (21.i) for each of $\{\alpha_i(t) + \gamma_i(t)\}$ and $\{\beta_i(t)\}$. To do this we intend to use Lemma 11 repeatedly on (21.ii). However, to use this procedure we need to deal with the terms in (21.ii) involving $u(t-l)$. The following result shows that we can substitute more desirable terms in place of terms involving $u(t-l)$.

Lemma 15:

$$\lim \frac{1}{N}\sum_{1}^{N}\left\{\sum_{i=1}^{p}\sum_{j=0}^{p} (b_j\alpha_i(t-l) - a_j\beta_i(t-l))y(t-i-j) + b_j\gamma_i(t-l) - c_j\beta_i(t-l))w(t-i-j)\right\}^2 = 0 \quad \text{for every } l \geq 1 \text{ a.s.} \tag{22}$$

Proof: Let

$$s(t) := \sum_{i=1}^{p} \alpha_i(t-l)y(t-i) + \gamma_i(t-l)w(t-i) + \beta_i(t-l)u(t-i).$$

Then Lemma 14 shows that

$$\frac{1}{N}\sum_{1}^{N} s^2(t) \to 0.$$

Hence, for every $j = 0, \cdots, p$,

$$\frac{1}{N}\sum_{1}^{N} s^2(t-j) \to 0$$

and so also

$$\frac{1}{N}\sum_{1}^{N} b_j^2 s^2(t-j) \to 0.$$

Lemma 10 now shows that

$$\frac{1}{N}\sum_{1}^{N}\left(\sum_{j=0}^{p} b_j s(t-j)\right)^2 \to 0,$$

which is equivalent to

$$\lim \frac{1}{N}\sum_{1}^{N}\left\{\sum_{j=0}^{p} b_j \sum_{i=1}^{p} \alpha_i(t-l-j)y(t-i-j)+\gamma_i(t-l-j)w(t-i-j)\right.$$
$$\left.+\beta_i(t-l-j)u(t-i-j)\right\}^2=0 \text{ a.s.}$$

The procedure in the proof of Lemma 13 shows, however, that in the above we can replace each of $\alpha_i(t-l-j)$, $\gamma_i(t-l-j)$, and $\beta_i(t-l-j)$ by $\alpha_i(t-l)$, $\gamma_i(t-l)$, and $\beta_i(t-l)$, respectively. This gives

$$\lim \frac{1}{N}\sum_{1}^{N}\left\{\sum_{j=0}^{p} b_j \sum_{i=1}^{p} \alpha_i(t-l)y(t-i-j)+\gamma_i(t-l)w(t-i-j)\right.$$
$$\left.+\beta_i(t-l)u(t-i-j)\right\}^2=0 \text{ a.s.} \tag{23}$$

Note now that since $\{\beta_i(t-l)\}$ is a bounded sequence a.s., by using (14.i) and Lemma 10, we obtain

$$\lim \frac{1}{N}\sum_{1}^{N}\left(\sum_{i=1}^{p} -\beta_i(t-l)z(t-i)\right)^2=0 \text{ a.s.} \tag{24}$$

Substituting now for $z(t-i)$ from (20) and, using Lemma 10 on (23) and (24), gives (22). □

Lemma 15 is now in a form where we can repeatedly use Lemma 11 to advantage. This yields the following.

Lemma 16:

$$\lim \frac{1}{N}\sum_{1}^{N}\left\{\sum_{j=0}^{k-1} b_j(\alpha_{k-j}(t)+\gamma_{k-j}(t))-(a_j+c_j)\beta_{k-j}(t)\right\}^2$$
$$=0 \text{ for } k=1, \cdots, p \text{ a.s} \tag{25.i}$$

$$\lim \frac{1}{N}\sum_{1}^{N}\left\{\sum_{j=k-p}^{p} b_j(\alpha_{k-j}(t)+\gamma_{k-j}(t))-(a_j+c_j)\beta_{k-j}(t)\right\}^2$$
$$=0 \text{ for } k=p+1, \cdots, 2p \text{ a.s} \tag{25.ii}$$

Proof: Let $q(t) := \sum_{i=1}^{p}\sum_{j=0}^{p}(b_j\alpha_i(t-l)-a_j\beta_i(t-l))y(t-i-j)+(b_j\gamma_i(t-l)-c_j\beta_i(t-l))w(t-i-j)$. By letting $k=i+j$ and interchanging the order of summation

$$q(t)=\sum_{k=1}^{p}\sum_{j=0}^{k-1}(b_j\alpha_{k-j}(t-l)-a_j\beta_{k-j}(t-l))y(t-k)+(b_j\gamma_{k-j}(t-l)$$
$$-c_j\beta_{k-j}(t-l))w(t-k)$$
$$+\sum_{k=p+1}^{2p}\sum_{j=k-p}^{p}(b_j\alpha_{k-j}(t-l)-a_j\beta_{k-j}(t-l))y(t-k)$$
$$+(b_j\gamma_{k-j}(t-l)$$
$$-c_j\beta_{k-j}(t-l))w(t-k).$$

We note from Lemma 15 that

$$\lim \frac{1}{N}\sum_{1}^{N} q^2(t)=0 \text{ a.s.}$$

Applying Lemma 11 once, we can conclude that (25) is true for $k=1$ and also we can eliminate terms involving $y(t-1)$ and $w(t-1)$ from $q(t)$. Again Lemma 11 is applied to show that (25) is valid for $k=2$, etc. In this way we apply Lemma 11 a total of $2p$ times each time showing that

$$\lim \frac{1}{N}\sum_{1}^{N}(\text{coefficient of } y(t-k) \text{ in } q(t)$$
$$+\text{ coefficient of } w(t-k) \text{ in } q(t))^2=0 \text{ a.s.}$$

thus achieving our result. □

Conclusions (21.i) and (25.i), (25.ii) are important and we analyze their implications in the next section.

V. Limit Set of Parameter Estimates

For any $\theta \in \mathbb{R}^{2p+2}$ define the polynomials

$$F(\theta, q) := \sum_{i=1}^{p} f_i(\theta)q^{i-1}$$

$$G(\theta, q) := \sum_{i=1}^{p} g_i(\theta)q^{i-1}$$

where

$$f_i(\theta) := (a_i+c_i)\theta_{p+2}-b_0\theta_{i+1}$$
$$g_i(\theta) := b_i\theta_{p+2}-b_0\theta_{p+2+i}$$

and

$$A(q) := \sum_{i=0}^{p} a_i q^i$$

$$B(q) := \sum_{i=0}^{p} b_i q^i$$

$$C(q) := \sum_{i=0}^{p} c_i q^i.$$

We now see the following interpretation of (21.i) and (25.i), (25.ii).

Lemma 17:

$$\lim \frac{1}{N}\sum_{1}^{N}[\text{coefficient of } q^{k-1} \text{ in } F(\hat{\theta}(t), q)B(q)$$
$$-G(\hat{\theta}(t), q)(C(q)+A(q))]^2=0 \text{ for } k=1, 2, \cdots, 2p \text{ a.s.} \tag{26.i}$$

$$\lim \frac{1}{N}\sum_{1}^{N}[(a_0+c_0)\hat{\theta}_{p+2}(t)-b_0\hat{\theta}_1(t)]^2=0 \text{ a.s.} \tag{26.ii}$$

Proof: Note that from the definitions of $\alpha(t)$, $\beta(t)$, and $\gamma(t)$ in Lemma 12, we have

$$f_i(\hat{\theta}(t))=\alpha_i(t)+\gamma_i(t)$$
$$g_i(\hat{\theta}(t))=\beta_i(t).$$

Now (25.i) gives the result in (26.i) for $k=1, 2, \cdots, p$ while

(25.ii) yields the result in (26.i) for $k = p + 1, \cdots, 2p$. Equation (26.ii) is just a restatement of (21.i) using the definition of $\alpha_0(t) + \gamma_0(t)$.

Remark: Suppose that the parameter estimate $\hat{\theta}(t)$ is fixed at θ for all t. Then by (5) the resulting control law is given by $\Sigma_{i=0}^{p} \theta_{i+1}y(t-i) + \Sigma_{i=0}^{p} \theta_{p+2+i}u(t-i) = 0$. Using transfer function notation where q is interpreted as the *backward* shift operator $qy(t) := y(t-1)$, we can rewrite this in the form

$$u(t) = H(\theta, q)y(t)$$

where

$$H(\theta, q) := \frac{-\sum_{i=0}^{p} \theta_{i+1}q^i}{\sum_{i=0}^{p} \theta_{p+2+i}q^i}.$$

Noting that for the system (1) the minimum variance control law is

$$u(t) = -\frac{A(q)+C(q)}{B(q)}y(t),$$

we make the following definition.

Definition 18: A parameter estimate $\theta \in \mathbb{R}^{2p+2}$ is said to produce a minimum variance control law if $H(\theta, q) = -[A(q) + C(q)]/B(q)$, after cancellation of common factors.

We now have the following result on the limit set of the parameter estimates.

Theorem 19: Let $D := \{\theta: F(\theta, q)B(q) = G(\theta, q)(A(q) + C(q))$ and $(a_0 + c_0)\theta_{p+2} = b_0\theta_1\}$. Then the following holds.

i) Every $\theta \in D$ produces a minimum variance control law.

ii) For every open set $0 \supseteq D$,

$$\lim \frac{1}{N}\sum_{1}^{N} 1(\hat{\theta}(t) \in 0) = 1 \text{ a.s.}$$

Here $1(\cdot)$ is the indicator function.

iii) $\{\hat{\theta}(t)\}$ converges a.s. to the intersection of two random spheres, one centered at θ^0 and the other centered at the origin.

Proof: Fix $\theta \in D$. Then since $b_0 \neq 0$, by using the defining relationships for D, we get

$$H(\theta, q) = -\frac{qF(\theta, q) - \theta_{p+2}(A(q)+C(q))}{qG(\theta, q) - \theta_{p+2}B(q)}.$$

Then i) follows from the relationship $FB = G(A + C)$. To show ii), let M be a compact (random) set containing the (a.s.) bounded sequence $\{\hat{\theta}(t)\}$. Let 0 be an open set containing D, and denote by 0^c its complement. Define

$$d(\theta) := \sum_{k=1}^{2p} [\text{coefficient of } q^{k-1} \text{ in } F(\theta, q)B(q) - G(\theta, q)(A(q)+C(q))]^2 + [(a_0+c_0)\theta_{p+2} - b_0\theta_1]^2.$$

Then $\theta \in D$ if and only if $d(\theta) = 0$. $d(\theta)$ is a continuous function on the compact set $0^c \cap M$, and hence attains a minimum which must be strictly positive since $D \cap (0^c \cap M)$ is empty, i.e., min

$$\lim \frac{1}{N}\sum_{1}^{N} d(\hat{\theta}(t)) = 0 \text{ a.s.,}$$

it follows that

$$\lim \frac{1}{N}\sum_{1}^{N} 1(\hat{\theta}(t) \in 0^c \cap M) = 0 \text{ a.s.}$$

However, since $\hat{\theta}(t) \in M$ for all t,

$$\lim \frac{1}{N}\sum_{1}^{N} 1(\hat{\theta}(t) \in 0^c) = 0 \text{ a.s.,}$$

giving the result. iii) follows from Lemma 9 and (11.iii). □

The above result shows that the parameter estimates converge in a Cesaro-sense to a set D of parameter values, where every element $\theta \in D$ produces a minimum variance control law. Note that D can be written as $\{\theta: L\theta = 0\}$ where L is a $(2p + 1) \times (2p + 2)$ matrix whose elements depend linearly on $\{a_i, b_i, c_i\}$.

To get stronger results we proceed as follows.

Remark: Note that if $FB = (A + C)G$ for some nontrivial polynomials F and G, then $u(t) = -F(q)/G(q)y(t)$ is a minimum variance controller. Further, if both F and G are of degree $\leq p - 1$, then we have a *reduced-order* minimum variance controller. This motivates the following.

Definition 20: If there are nonzero polynomials F and G of degree less than or equal to $p - 1$, satisfying $F(q)B(q) = G(q)(A(q) + C(q))$, then the system (1) will be said to possess reduced-order minimum variance controllers.

Note now that reduced-order minimum variance controllers occur when either i) $\deg (A + C) \leq p - 1$, $\deg B \leq p - 1$ or ii) $(A + C)$ and B have common factors. In what follows we exclude these cases.

Theorem 21: If the system (1) does not possess reduced-order minimum variance controllers, then

$$\lim \hat{\theta}(t) = k\theta^0 \text{ a.s. where } k \text{ is an a.s. finite scalar random variable.} \tag{27.i}$$

$$k^2 = \frac{\|\hat{\theta}(0)\|^2}{\|\theta^0\|^2} + \sum_{t=1}^{\infty} \frac{\mu^2 y^2(t)\|\phi(t-1)\|^2}{r^2(t-1)\|\theta^0\|^2}. \tag{27.ii}$$

$$\text{If } \|\hat{\theta}(0)\| > \|\theta^0\|, \text{ then } k^2 > 1, \text{ i.e., } \lim \hat{\theta}(t) \neq \theta^0 \text{ a.s.} \tag{27.iii}$$

$$\text{Prob } (k^2 > 1) > 0, \text{ i.e., Prob } (\lim \hat{\theta}(t) \neq \theta^0) > 0. \tag{27.iv}$$

Proof: Since there is no reduced-order minimum variance controller, $\deg F \leq p - 1$, $\deg G \leq p - 1$ and $FB = (A + C)G \Rightarrow F = 0, G = 0$. Hence,

$$D = \{\theta : F(\theta, q) = 0, \ G(\theta, q) = 0, \ (a_0 + c_0)\theta_{p+2} = b_0\theta_1\}.$$

From the definitions of F and G, this means

$$D = \{\theta : \theta_i^0\theta_{p+2} - \theta_{p+2}^0\theta_i \quad \text{for } i = 1, 2, \cdots, 2p+2\}.$$

Since $\{\hat{\theta}(t)\}$ is a bounded sequence a.s., it follows that there is a (random) subsequence $\{t_n\}$ along which $\lim \hat{\theta}(t_n) \in D$ a.s. This verifies (12.ii) and Lemma 3 now shows that (27.i) holds. To see (27.ii) note that from (11.i) we have $k^2\|\theta^0\|^2 = \lim \|\hat{\theta}(t)\|^2 = \|\hat{\theta}(0)\|^2 + \Sigma_1^{\infty} \|\Delta\hat{\theta}(t)\|^2$. Since

$$\Delta\hat{\theta}(t) = \frac{\mu y(t)\phi(t-1)}{r(t-1)},$$

(27.ii) follows. Equation (27.iii) is an easy consequence of (27.ii) since the summation is nonnegative. Also from (27.ii) we see that

$$k^2 \geq \frac{\|\hat{\theta}(0)\|^2}{\|\theta^0\|^2} + \frac{\mu^2 y^2(1)\|\phi(0)\|^2}{r^2(0)\|\theta^0\|^2}.$$

Given the initial conditions $\phi(0)$, $r(0)$, $\hat{\theta}(0)$, etc., there is a positive probability that the second term above is positive, in view of the mutual absolute continuity assumption (8.iii) and (8.i). (Note that this is the only place we use the assumption $\phi(0) \neq 0$ in (8.i).) □

It is clear that when the system possesses no reduced-order controllers, then the above theorem settles all issues.

It is important to note that the conclusions i) and ii) of Theorem 19 and i) of Theorem 21 are not really dependent on the fact that the parameter estimates are generated by an SA scheme, except through certain intermediate results. This we state as follows.

Theorem 22: If

i) $\{\hat{\theta}(t)\}$ is any $\{\mathfrak{F}_t\}$ adapted sequence

ii) the conditions of (14) are satisifed and the system satisfies (1) and (8); then i) and ii) of Theorem 19 are true. If, in addition,

iii) $\{\|\hat{\theta}(t)\|\}$ and $\{\|\tilde{\theta}(t)\|\}$ converge a.s.

iv) the system (1) has no reduced-order minimum variance controllers; then (27.i) of Theorem 21 is also true.

Thus, some of our methods and results may be of use for schemes other than SA.

VI. Concluding Remarks

We have shown convergence of parameter estimates to a random multiple of θ^0, if the true system has no reduced-order minimum variance controllers. Thus, overmodeling the system is not a problem per se. To illustrate this, consider the following example where there is some overmodeling of the system and yet no reduced-order controller exists.

Example:

$$y(t+1)=y(t)+12u(t)+8u(t-1)-u(t-2)-u(t-3)+w(t+1)$$

is the true system. However, if we model it as

$$y(t+1)=\theta_1^0y(t)+\theta_2^0y(t-1)+\theta_3^0y(t-2)+\theta_4^0y(t-3)+\theta_5^0u(t)$$
$$+\theta_6^0u(t-1)+\theta_7^0u(t-2)+\theta_8^0u(t-3)+w(t+1)$$

then the three parameters $(\theta_2^0, \theta_3^0, \theta_4^0)$ are redundant and we have overmodeled the system. However, Theorem 21 shows that

$$\hat{\theta}(t)\rightarrow k(1,\ 0,\ 0,\ 0,\ 12,\ 8,\ -1,\ -1)^T$$

where k is a random scalar. □

If the system does possess reduced-order controllers, then we have shown that the limit set of the parameter estimates is a hypersphere of dimension strictly less than that of the space. Further, we have shown convergence, in a Cesaro-sense, to the set D where every parameter in D yields a minimum variance control law. Consider, as an illustration the following example.

Example:

$$y(t+1)=2y(t)+y(t-1)+4u(t)+2u(t-1)+w(t+1).$$

Since the polynomials $A(q) + C(q) = 2 + q$ and $B(q) = 4 + 2q$ have a common factor, we actually have a reduced-order minimum variance controller

$$u(t)=-\frac{1}{2}y(t).$$

Consider $(1, 3, 2, 6)^T \in D$ (this is easily checked), and Theorem 19 shows that $\{\hat{\theta}(t)\}$ *may* converge to $(1, 3, 2, 6)^T$. This would lead us to believe that the system is of the form

$$y(t+1)=y(t)+3y(t-1)+2u(t)+6u(t-1)+w(t+1).$$

However, the resulting control law $u(t) = -1/2y(t) + -3/2y(t-1) - 3u(t-1)$ is a minimum variance control law. This is especially so if we cancel the common factor $(1 + 3q)$ from the representation $(1 + 3q)u(t) = -1/2(1 + 3q)y(t)$. □

Note that the geometric properties introduced here are properties of the SA scheme and do *not* depend on the nature of the system, which can even be nonlinear. Some investigations with regard to "robustness" may therefore be possible.

On the other hand the probabilistic arguments presented in Section IV and Theorem 19i) ii) and Theorem (21.i) apply to any algorithm as long as it is optimal, $\{\hat{\theta}(t)\}$ is bounded, and $\hat{\theta}(t) - \hat{\theta}(t-1) \rightarrow 0$, as Theorem 22 shows. Thus, some investigation of least-squares type schemes may be possible.

Acknowledgment

The authors would like to thank Prof. T. I. Seidman for several useful discussions.

References

[1] K. J. Åström and B. Wittenmark, "On self-tuning regulators," *Automatica*, vol. 9, pp. 185-199, 1973.

[2] L. Ljung, "Analysis of recursive stochastic algorithms," *IEEE Trans. Automat. Contr.*, vol. AC-22, pp. 551-575, 1977.

[3] L. Ljung, "On positive and real transfer functions and the convergence of some recursive schemes," *IEEE Trans. Automat. Contr.*, vol. AC-22, pp. 539-550, 1977.

[4] G. Goodwin, P. Ramadge, and P. Caines, "Discrete time stochastic adaptive control," *SIAM J. Contr. Optimiz.*, vol. 19, pp. 829-853, 1981.

[5] V. Solo, "The convergence of AML," *IEEE Trans. Automat. Contr.*, vol. AC-24, pp. 958-962, 1979.

[6] K. S. Sin and G. Goodwin, "Stochastic adaptive control using a modified least squares algorithm," *Automatica*, vol. 18, pp. 315-321, 1982.

[7] C. Rohrs, L. Valavani, M. Athans, and G. Stein, "Analytical verification of undesirable properties of direct model reference adaptive control algorithms," in *Proc. 20th IEEE Conf. Decision Contr.*, 1981, pp. 1272-1284.

[8] K. L. Chung, *A Course in Probability Theory*. New York: Academic, 1974.

[9] Y. S. Chow, "Local convergence of martingales and the law of large numbers," *Ann. Math. Stat.*, vol. 36, pp. 552-558, 1965.

[10] T. L. Lai and C. Z. Wei, "Least squares estimates in stochastic regression with applications to identification and control of dynamic systems," *Ann. Stat.*, vol. 10, pp. 154-166, 1982.

[11] W. F. Stout, *Almost Sure Convergence*. New York: Academic, 1974.

[12] P. E. Caines and S. Lafortune, "Adaptive optimization with recursive identification for stochastic linear systems," McGill Univ., Montreal, P.Q., Canada, 1981.

[13] H. F. Chen and P. E. Caines, "The strong consistency of the stochastic gradient algorithm of adaptive control," McGill Univ., Montreal, P.Q., Canada, 1983.

[14] P. E. Caines and S. Lafortune, "Adaptive control with recursive identification for stochastic linear systems," in *Proc. 6th IFAC Symp. Ident. Syst. Parameter Estimation*, Washington, DC, June 1982.

THE STRONG CONSISTENCY OF THE STOCHASTIC GRADIENT ALGORITHM OF ADAPTIVE CONTROL

H.F. Chen* and P.E. Caines†

*Institute of Systems Science, Academia Sinica, Beijing, China
and Dept. of Electrical Engineering, McGill University.
†Canadian Institute for Advanced Research and Dept. of Electrical
Engineering, McGill University, Montreal, P.Q., Canada

Abstract

By use of the technique of Chen [1983] sufficient conditions are established for the strong consistency of the stochastic gradient (SG) algorithm for MIMO ARMAX stochastic systems without monitoring. This result is then used in conjunction with the method of disturbed adaptive controls introduced in [Caines, 1981; Caines and Lafortune, 1984]; hence it is shown that the SG algorithm generates strongly consistent parameter estimates while it is operating as a part of the SG algorithm of adaptive control of Goodwin, Ramadge and Caines [1981].

1. Introduction

The stochastic gradient (SG) algorithm is probably the simplest method of parameter estimation for linear stochastic systems. It was used in [Goodwin, Ramadge and Caines, 1981] for the adaptive tracking control problem and later on in [Caines and Lafortune, 1984] for the adaptive tracking problem where disturbed controls were used for purposes of identification as explained below. In these papers the strong consistency of the SG algorithm was not established. In [Chen, 1981b] the strong consistency of the estimates generated by the quasi-least squares (QLS) method was proved for the case of a system subject to feedback control without monitoring.

In this paper, by using the technique given in [Chen, 1983], we first establish sufficient conditions for the strong consistency of the SG algorithm for MIMO stochastic systems without monitoring and then apply the method involving continually disturbed controls which was introduced in [Caines, 1981] and developed in [Lafortune, 1982] and [Caines and Lafortune, 1984]. Continually disturbing a system's control input provides a technique for ensuring that certain processes of regression vectors have the persistency of excitation property which is frequently required in recursive identification schemes (see e.g. [Moore, 1983]).

The ordinary differential equation method and the associated hypotheses used in this paper should be compared with those (involving monitoring) used to obtain the related results of Theorem 1 in [Ljung, 1979]. Concerning proof techniques, we also remark that the stochastic Lyapunov - or super martingale - method used in the basic first lemma below is fundamental to the consistency proofs of [Solo, 1979] and [Chen 1981a,b] and to all of the references of this paper concerned with stochastic adaptive control.

It will be shown that the SG algorithm used in [Lafortune, 1982] and [Caines and Lafortune, 1984] does in fact generate strongly consistent estimates. Hence, it is not necessary to introduce a second algorithm in order to generate consistent estimates, as in [Caines and Lafortune, 1984], where the Approximate Maximum Likelihood (AML) algorithm of [Solo, 1979] was used in addition to the SG algorithm.

We consider the MIMO system

$$y_n + A_1 y_{n-1} + \ldots + A_p y_{n-p} = B_1 u_{n-1} + \ldots + B_q u_{n-q} + w_n + C_1 w_{n-1} + \ldots + C_r w_{n-r} \quad (1)$$

where y_n, u_n and w_n are m-, ℓ- and m-dimensional respectively and $y_i=0$, $u_j=0$, $w_k=0$ for all $i < 0$, $j < 0$, $k < 0$.

Let F_n be a family of non-decreasing σ-algebras, assume that w_n and u_n are F_n-measurable and that

$$E(w_n/F_{n-1}) = 0, \quad E(\|w_n\|^2/F_{n-1}) \leqslant k_o r^{\varepsilon}_{b-1}, \quad 0 \leqslant \varepsilon < 1 \quad (2)$$

where k_o is a positive constant and r_n is defined below in (10).

A_i, B_j, C_k $i=1\ldots p$, $j=1\ldots q$, $k=1\ldots r$ are the unknown matrix coefficients to be estimated.

Let us write

$$A(z) = I + A_1 z + \ldots + A_p z^p \quad (3)$$

$$B(z) = B_1 + B_2 z + \ldots + B_q z^{q-1} \quad (4)$$

$$C(z) = I + C_1 z + \ldots + C_r z^r \quad (5)$$

where z denotes the unit backward shift operator.

We shall adopt the following notation:

$$\theta^\tau = [-A_1 \ldots -A_p \; B_1 \ldots B_q \; C_1 \ldots C_r] \quad m \times (mp+\ell q+mr) \quad (6)$$

$$\phi^\tau_n = [y^\tau_n y^\tau_{n-1} \ldots y^\tau_{n-p+1} \; u^\tau_n \ldots u^\tau_{n-q+1} \; y^\tau_n - \phi^\tau_{n-1}\theta_{n-1} \ldots y^\tau_{n-r+1} - \phi^\tau_{n-r}\theta_{n-r}] \quad (7)$$

$$\phi^{o\tau}_n = [y^\tau_n \; y^\tau_{n-1} \ldots y^\tau_{n-p+1} \; u^\tau_n \ldots u^\tau_{n-q+1} \; w^\tau_n \ldots w^\tau_{n-r+1}], \quad (8)$$

where θ_n is the estimate for θ given by the SG algorithm

$$\theta_{n+1} = \theta_n + \frac{\phi_n}{r_n}(y^\tau_{n+1} - \phi^\tau_n \theta_n), \quad (9)$$

$$r_n = 1 + \sum_{i=1}^{n} \|\phi_i\|^2, \quad r_o = 1 \quad (10)$$

with ϕ_{-1} and θ_0 deterministic and arbitrarily chosen.

The difference between the SG algorithm and the QLS algorithm lies in the fact that the residual term $y^\tau_n - \phi^\tau_{n-1}\theta_{n-1}$ in the SG algorithm is replaced by the term $y^\tau_n - \phi^\tau_{n-1}\theta_n$ in the QLS algorithm, in other words the a priori prediction error is replaced by the a posteriori prediction error.

Set

$$\tilde{\theta}_n = \theta - \theta_n \quad (11)$$

$$\xi_n = y_n - w_n - \theta^\tau_{n-1}\phi_{n-1}. \quad (12)$$

Then we have

$$C(z)(y^\tau_n - w^\tau_n - \theta^\tau_{n-1}\phi_{n-1}) = \{(y_n - C(z)w_n) + (C(z)-I)(y_n - \theta^\tau_{n-1}\phi_{n-1})\} - \theta^\tau_{n-1}\phi_{n-1}$$

Reprinted from *Proc. 23rd Conf. Decision and Control*, 1984, pp. 802–805, vol. 1.

$$= \theta^\tau\phi_{n-1} - \theta^\tau_{n-1}\phi_{n-1} = \tilde{\theta}^\tau_{n-1}\phi_{n-1},$$

hence

$$C(z)\xi_n = \tilde{\theta}^\tau_{n-1}\phi_{n-1} \qquad (13)$$

and

$$\tilde{\theta}_{n+1} = \tilde{\theta}_n - \frac{\phi_n}{r_n}(\xi^\tau_{n+1} + w^\tau_{n+1}) \qquad (14)$$

Using the now standard martingale convergence techniques (see e.g. [Chen, 1983] or [Goodwin, Ramadge, Caines, 1981]) we may establish the basic Lemma 1. For the system and algorithm (1) - (10), if $C(z) - \frac{1}{2} I$ is strictly positive real, then

$$\sum_{i=0}^{\infty} \frac{\|\xi_{i+1}\|^2}{r_i} < \infty, \text{ a.s.} \quad \sum_{i=0}^{\infty} \frac{\|\tilde{\theta}^\tau_i\phi_i\|^2}{r_i} < \infty, \text{ a.s.} \qquad (15)$$

and

$$\text{tr } \tilde{\theta}^\tau_n\tilde{\theta}_n \xrightarrow[n\to\infty]{} v < \infty \quad \text{on} \quad [\omega: \; r_n \xrightarrow[n\to\infty]{} \infty] \qquad (16)$$

□

Set

$$F = \begin{bmatrix} -C_1 & I & 0 & 0 \\ & 0 & & \\ & & & 0 \\ & & & I \\ -C_r & 0 & & 0 \end{bmatrix}, \; r > 0 \qquad (17)$$

$$F = [0]\}m, \quad F^o = I, \quad r = 0$$
$$\quad m$$

$$G = \begin{cases} [\underbrace{I \; 0 \, .. \, 0}_{mr}] & r > o \\ I & r = 0 \end{cases} \qquad (18)$$

It is easy to see that there exists a random mr-dimensional vector η_o depending on the initial values of $\{\xi_i\}$ such that

$$\xi_n = G\,\eta_n, \; \eta_{n+1} = F\,\eta_n + G^\tau\,\tilde{\theta}^\tau_n\,\phi_n, \qquad (19)$$

in other words, $\{\eta_n\}$ is the state process of a realization of the $\{\xi_n\}$ process. Clearly

$$\xi_{n+1} = G\sum_{i=0}^{n} F^{n-i}G^\tau\tilde{\theta}^\tau_i\phi_i + G\,F^{n+1}\eta_o. \qquad (20)$$

From this and (14) we see that

$$\tilde{\theta}_{n+1} = \tilde{\theta}_o - \sum_{j=0}^{n}\frac{\phi_j}{r_j}\,[\sum_{i=0}^{j}\phi^\tau_i\tilde{\theta}_i G F^{\tau(j-i)}G^\tau$$
$$+ \eta^\tau_o F^{\tau(j+i)}G^\tau + w^\tau_{j+1}] \qquad (21)$$

Now we list some conditions that we shall refer to later on.

a) Either r=0, or r>0 with $C(z) - \frac{1}{2} I$ strictly positive real and with the zeros of det C(z) lying outside the closed unit disk.

b) There exist random variables $\alpha > 0, \beta > 0$ and $T > 0$ such that

$$\sum_{i=m(t)}^{m(t+\alpha)} \frac{\phi_i\phi^\tau_i}{r_i} \geq \beta I, \forall\, t \geq T, \; \forall\, \omega \in [\omega: r_n \to \infty], \qquad (22)$$

where

$$m(t) = \max[n: t_n \leq t], \; t \geq 0, \; t_n = \sum_{i=0}^{n-1}\beta_i, \; t_o = 0, \qquad (23)$$

$$\beta_i = \frac{\|\phi_i\|^2}{r_i} \qquad (24)$$

(This condition first appeared [Chen, 1981].)

c^o) There exists a random variable γ, $0 < \gamma < \infty$ such that $\nu^{on}_{max}/\nu^{on}_{max} \leq \gamma$, $\forall\, n \geq 0$ where ν^{on}_{max} and ν^{on}_{min} are maximum and minimum eigenvalues of

$$N_n \triangleq \sum_{i=1}^{n}\phi^o_i\phi^{o\tau}_i + \frac{1}{d} I, \; d = mp+\ell+mr.$$

c) There exists a random variable γ, $0 < \gamma < \infty$ such that $\nu^n_{max}/\nu^n_{min} \leq \gamma$, $\forall\, n \geq 0$ where ν^n_{max} and ν^n_{min} are the maximum and minimum eigenvalues of

$$\sum_{i=1}^{n}\phi_i\phi^\tau_i + \frac{1}{d} I, \; d = mp+\ell q+mr.$$

d) B_1 is of full rank and zeros of $B^+_1 B(z)$ (B^+_1 denotes the psuedo-inverse of B_1) are outside the closed unit disk.

e) $B^+_1 A(z)$ and $B^+_1 B(z)$ are left coprime and $B^+_1 A_p$ and $B^+_1 B_q$ are of full rank.

The next two items specify alternative adaptive control laws:

f) The $\mathcal{F}_n$-measurable control u_n is selected such that

$$\theta^\tau_n\phi_n = y^*_{n+1} \qquad (25)$$

with v_n a $\mathcal{F}_n$-measurable a disturbance sequence.

(This is the so-called continually disturbed control law which was introduced in [Caines, 1981].)

Lemma 2. For the SG algorithm, under the conditions of Lemma 1, Conditions c) and c^o) are equivalent on $[\omega: r_n \to \infty]$ and each of them implies Condition b).

Proof. This fact is proved in [Chen, 1983] as Theorem 2, but ξ_i and ϕ_i in that paper should be understood to be defined by (7) and (12) above. □

Theorem 1. For the system and algorithm in (1) - (10), if Conditions a) and b) are satisfied then for almost all $\omega \in [\omega: r_n \to \infty]$

$$\theta_n \xrightarrow[n\to\infty]{} \theta.$$

Proof. Since the zeros of det C(z) lie outside the closed unit disk for r > 0 there exist constants $\rho \in (0,1)$ and $k_3 > 0$ such that for both the r=0 and the r>0 cases $\|F^i\| \leq k_3\rho^i$

Comparing (30) with (48) of [Chen, 1983] we find that the quantities $\frac{1}{r_j} I$ and $\tilde{\theta}^\tau_i\phi_i$ respectively of the present paper correspond to $\alpha_j R_j$ and $\tilde{\theta}^\tau_{i+1}\phi_i$ of that paper; in that paper $\|\alpha_j R_j\|$ is estimated by $\frac{k_7}{r_j}$ and instead of (16) above we have $\sum_{i=0}^{\infty}\frac{\|\tilde{\theta}^\tau_{i+1}\phi_i\|^2}{r_i} < \infty$.

With such a correspondence the analysis from [Chen, 1983] completely fits the present case for $\omega \in [\omega: r_n \to \infty]$. To be precise, we first introduce some notation by setting

$$G_{n+1,i} = -\frac{1}{\beta_i}\sum_{j=i}^{n}\frac{\theta_j}{r_j}\phi^\tau_i\tilde{\theta}_i\, G\, F^{\tau(j-i)}G^\tau, \; G_{n,n} = 0$$

and we also denote the last two terms in (21) by

J_{n+1} and H_{n+1} respectively. This yields

$$\tilde{\theta}_{n+1} = \tilde{\theta}_o + \sum_{i=0}^{n} \beta_i G_{n+1,i} + J_{n+1} + H_{n+1} . \qquad (27)$$

We now introduce two interpolating functions $X(t)$ and $\bar{X}(t)$ for any given matrix sequence $\{X_i\}$. These are given by:

(1) the linear interpolation

$$X(t_n) = X_n$$

$$X(t) = \frac{t_{n+1}-t}{\beta_n} X_n + \frac{t-t_n}{\beta_n} X_{n+1}, \quad t \in [t_n, t_{n+1}]$$

(2) the constant interpolation

$$\bar{X}(t) = X_n \ , \quad t \in [t_n, t_{n+1}) \ .$$

Let $G^o_{t,i}$ denote the linear interpolation of $\{G_{n,i}\}$ with the interpolating length $\{\beta_n\}$ for $t \geqslant t_i$ whenever i is fixed. When $t = t_k$, then $G^o_{t_k,i} = G_{k,i}$. Now for fixed t denote by $\bar{G}^o_{t,s}$ the constant interpolation for the sequence $\{G^o_{t,i}\}$ on $[0,t]$ with interpolating length $\{\beta_i\}$.

By use of these two interpolations the sum in (38) for $\tilde{\theta}_{n+1}$ turns into an integral of the function $\bar{G}^o_{t,s}$ when it appears in the formula for the interpolating function $\tilde{\theta}(t)$. To be specific, we have

$$\tilde{\theta}(t) = \tilde{\theta}_o + \int_0^t \bar{G}^o_{t,x} \, ds + J(t) + H(t) \qquad (28)$$

where $J(t)$ and $H(t)$ are the linear interpolating function for the sequences $\{J_n\}$, $\{H_n\}$.

Notice that since $\omega \in [\omega: r_n \to \infty]$ $\tilde{\theta}(t)$ is defined for all $t \geqslant 0$ and we notice that

$$\tilde{\theta}(t_n) = \tilde{\theta}_n . \qquad (29)$$

Define the family of matrix functions $\{\tilde{\theta}_n(t)\}$ by shifting the argument of $\tilde{\theta}(t)$ to the left as shown by

$$\tilde{\theta}_n(t) = \tilde{\theta}(t+n) \qquad t \geqslant 0 \ . \qquad (30)$$

We now invoke a major technical result of [Chen, 1983]; the proof of this lemma is valid in the present case subject only to the exchange of the symbols mentioned above.

Lemma 3. Under the conditions of Theorem 1, for any fixed $\omega \in [\omega: r_n \to \infty]$, $\{\tilde{\theta}_n(t)\}$ is uniformly bounded and equicontinuous. □

Hence, according to the Arzela-Ascoli Theorem, there exists a subsequence $\{\tilde{\theta}_{n_k}(t)\}$ of $\{\tilde{\theta}_n(t)\}$ and a continuous matrix function $\tilde{\theta}(t)$ which is the uniform limit of $\{\tilde{\theta}_{n_k}(t)\}$ over any finite interval.

It follows from the proof of the lemma that

$$\|\tilde{\theta}(t+\Delta) - \tilde{\theta}(t)\| = \lim_{k\to\infty} \|\tilde{\theta}_{n_k}(t+\Delta) - \tilde{\theta}_{n_k}(t)\|$$

$$\equiv \lim_{k\to\infty} \|\tilde{\theta}(t+\Delta+n_k) - \tilde{\theta}(t+n_k)\| = 0$$

and hence $\tilde{\theta}(t)$ is a constant matrix $\tilde{\theta}^o$.

Now we show that $\tilde{\theta}^o = o$; for a fixed $t \in [0,\infty]$ we have, by the Schwartz inequality, the definition of α and β in Condition b) above and the fact that $\beta_i \leqslant 1$, that

$$\lim_{k\to\infty} \| \sum_{i=m(t+n_k)}^{m(t+n_k+\alpha)} \frac{\phi_i \phi_i^T}{r_i} \tilde{\theta}_i \|$$

$$\leqslant \sqrt{2+\alpha} \lim_{k\to\infty} \left(\sum_{i=m(t+n_k)}^{m(t+n_k+\alpha)} \frac{\|\tilde{\theta}_i^T \phi_i\|^2}{r_i} \right)^{\frac{1}{2}} = 0 \qquad (31)$$

Notice that for all $i \in [0,1,\dots,m(t+n_k+\alpha)-m(t+n_k)] \triangleq S$

$$t < t_{m(t+n_k)+1} - n_k \leqslant t_{m(t+n_k)+1+i} - n_k$$

$$= t_{m(t+n_k)} + \sum_{j=m(t+n_k)}^{m(t+n_k)+i} \beta_j - n_k$$

$$\leqslant t_{m(t+n_k)} + \sum_{j=m(t+n_k)}^{m(t+n_k+\alpha)} \beta_j - n_k \leqslant t + \alpha + 2 ,$$

in other words, $t_{m(t+n_k)+i+1} - n_k \in [t, t+\ +2]$, $\forall\, i \in S$.

Hence

$\tilde{\theta}_{n_k}(t_{m(t+n_k)+i+1} - n_k) \to \tilde{\theta}^o$ as $k\to\infty$ and by (29), (30)

$$\tilde{\theta}_{m(t+n_k)+i+1} \xrightarrow[k \to \infty]{} \tilde{\theta}^o \qquad (32)$$

uniformly in $i \in S$. Consequently, we assert

$$\lim_{k\to\infty} \sum_{i=m(t+n_k)}^{m(t+n_k+\alpha)} \beta_i \| \tilde{\theta}_i - \tilde{\theta}^o \| = 0.$$

From (31), (32) it is easy to conclude that

$$\tilde{\theta}^{oT} \left[\lim_{k\to\infty} \sum_{i=m(t+n_k)}^{m(t+n_k+\alpha)} \frac{\phi_i \phi_i^T}{r_i} \right] \tilde{\theta}^o = 0 \qquad (33)$$

It is worth remarking that until this point all results have been obtained without involving Condition b).

Now by using b) it follows immediately from (33) that

$$\tilde{\theta}^o = 0 \text{ on } [\omega: r_n \to \infty]$$

and $\tilde{\theta}(t+n_k) \to 0$ uniformly in $t \in [a,b]$, where $[a,b]$ is any finite interval. Since $\tilde{\theta}(t_n) = \tilde{\theta}_n$ it is easy to see that for any fixed $\omega \in [\omega: r_n \to \infty]$ there exists a subsequence $\tilde{\theta}_{m_k} \to 0$ as $k \to \infty$. From here by Lemma 1 we conclude that $\lim_{n\to\infty} \operatorname{tr} \tilde{\theta}_n^T \tilde{\theta}_n = \lim_{k\to\infty} \operatorname{tr} \tilde{\theta}_{m_k}^T \tilde{\theta}_{m_k} = 0$ i.e.

$$\theta_n \xrightarrow[n\to\infty]{} \theta \qquad \forall\, \omega \in [\omega: r_n \to \infty] \ . \qquad □$$

Let $\{y_n^*\}$ be a bounded deterministic reference sequence.

Theorem 2. Assume that for (1) - (10) Conditions a) d) and f) are satisfied, and w satisfies

$$0 < \overline{\lim_{n\to\infty}} \frac{1}{n} \sum_{i=1}^{n} \|w_i\|^2 < \infty \quad \text{a.s.} \qquad (34)$$

Then $r_n \to \infty$ and

$$\overline{\lim_{n\to\infty}} \frac{1}{n} \sum_{i=1}^{n} \|u_i\|^2 < \infty, \quad \overline{\lim_{n\to\infty}} \frac{1}{n} \sum_{i=1}^{n} \|y_i\|^2 < \infty \qquad (35)$$

$$\lim_{n\to\infty} \| \frac{1}{n} \sum_{i=1}^{n} (y_i - y_i^*)(y_i - y_i^*)^T - \frac{1}{n} \sum_{i=1}^{n} w_i w_i^T \| = 0 \qquad (36)$$

Proof. Suppose $\lim_n r_n < \infty$, then $\phi_n \to 0$, $y_n \to 0$, $u_n \to 0$, and hence $w_i \to 0$. But, since $C(z)$ is asymptotically stable and (34) holds, this is seen to be an event of probability zero and so $r_n \to \infty$ a.s.

By d) there exist constants (which may depend upon ω) such that

$$\frac{1}{n}\sum_{i=1}^{n} \|u_i\|^2 \leq \frac{k_4}{n}\sum_{i=1}^{n} \|y_{i+1}\|^2 + k_5$$

Then using (12), (34) and f), it can be shown (as in e.b. [Goodwin, Ramadge, Caines, 1981]) that

$$\frac{1}{n}\sum_{i=1}^{n} \|y_{i+1}\|^2 \leq \frac{k_8}{n}\sum_{i=1}^{n} \|\xi_{i+1}\|^2 + k_0.$$

Consequently

$$\overline{\lim_{n\to\infty}} \frac{r_n}{n} < \infty \tag{37}$$

and hence via the Kronecker Lemma we have

$$\frac{1}{n}\sum_{i=1}^{n} \|\xi_{i+1}\|^2 \xrightarrow[n\to\infty]{} 0 \text{ a.s.} \tag{38}$$

But by (12) and Condition f)

$$(y_i - y_i^*)(y_i - y_i^*)^\tau - w_i w_i^\tau = \xi_i \xi_i^\tau - 2\xi_i w_i^\tau ,$$

and from this and (34) we obtain (36), while (35) follows from (37). □

Now we introduce the continually disturbed controls of Condition g).

Theorem 3. Let $\{w_i\}$ $\{v_i\}$ be two mutually independent i.i.d. sequences with $E\,v_i = E\,w_i = 0$, $E\,w_i w_i^\tau = R_1 > 0$, $E\,v_i v_i^\tau = R_2 > 0$, and let $\{y_i^*\}$ be a bounded deterministic reference sequence. For the system and algorithm (1) - (10) with $F_n \triangleq \sigma\{w_i, v_i, i \leq n\}$ if Conditions a), d), e), g) are satisfied, then $r_n \to \infty$

$$\overline{\lim_{n\to\infty}} \frac{1}{n}\sum_{i=1}^{n} \|u_i\|^2 < \infty , \quad \overline{\lim_{n\to\infty}} \frac{1}{n}\sum_{i=1}^{n} \|y_i\|^2 < \infty$$

$$\lim_{n\to\infty} \frac{1}{n}\sum_{i=1}^{n} (y_i - y_i^*)(y_i - y_i^*)^\tau = R_1 + R_2 \tag{39}$$

and

$$\theta_n \xrightarrow[n\to\infty]{} \theta \text{ a.s.}$$

Proof. (Sketch) From (12) and Condition g) we have

$$y_{n+1} = \xi_{n+1} + Y_{n+1}^* + (w_{n+1} + v_n). \tag{40}$$

By the strong law of large numbers we know that $\frac{1}{n}\sum_{i=1}^{n} w_i w_i^\tau \xrightarrow[n\to\infty]{} R_1$, and hence $r_n \xrightarrow[n\to\infty]{} \infty$ a.s. Recall that (37) and (38) still hold.

Consequently (39) follows immediately.

To complete the proof it is sufficient to show that Condition c°) holds for the regression vector ϕ_n^o in (8), since then, by Lemma 2, Condition b) is true and consequently Theorem 1 may be applied.

But this follows using the method of analysis contained in [Lafortune, 1982] and [Caines and Lafortune, 1984]. □

Concerning the sequence $\{y_n^*\}$, we observe that only the boundedness of the limits $\lim_{n\to\infty} \frac{1}{n}\sum_{i=1}^{n} y_{i-k}^* y_{i-\ell}^{*\tau}$, for all k,ℓ as assumed in the two papers cited above.

We also note that the i.i.d hypotheses on w and v in this theorem were only adopted for simplicity and that Theorem 3 holds, with random $R > 0$ and $\gamma < \infty$, if w and v are taken to be mutually uncorrelated ergodic martingale difference sequences as in [Lafortune, 1982] and [Caines and Lafortune, 1984].

Acknowledgements

The authors would like to acknowledge conversations with P.R. Kumar and C.Z. Wei who pointed out the possibility that the stochastic gradient adaptive control algorithm, with disturbed controls, generates strongly consistent parameter estimates.

References

Caines, P.E., Stochastic Adaptive Control: randomly varying parameters and continually disturbed controls, Control Science and Technology for The Progress of Society, (Ed. H. Akashi) Pergamon, 1981, 925-930.

Caines, P.E. and S. Lafortune, Adaptive Control with Recursive Identification for Stochastic Systems. IEEE Trans. Autom. Control, AC-29, No. 4, 1984, pp. 312-321. Presented at the Conference on Decision and Control, Orlando, Florida, December 1982.

Chen, H.F., Strong consistency of Recursive Identification under Correlated Noise, J. of Systems Science and Mathematical Science Vol. 1, No. 1, 1981, 34-52.

Chen, H.F., Recursive System Identification and Adaptive Control by use of the Modified Least Squares Algorithms, 1983, Research Report. Dept. of Electrical Eng., McGill University. Presented at the Conference on Decision and Control, San Antonio, Texas, December 1983 and to appear in SIAM J. Control and Optimization.

Chen, H.F., Quasi-lease squares identification and its strong consistency, Int. J. of Control, Vol. 34, No. 5, 1981, 921-936.

Goodwin, G.C., P.J. Ramadge and P.E. Caines, Discrete-time Stochastic Adaptive Control, SIAM J. Control and Optimization Vol. 19, No. 6, Nov. 1981, pp. 829-853. Corrigendum: Vol. 20, No. 6, Nov. 1982, p. 893.

Lafortune, S., Adaptive Control with Recursive Identification for Stochastic Linear Systems, M.Eng. Thesis, Dept. of Electrical Engineering, McGill University, Montreal, P.Q., March 1982.

Ljung, L., On positive real transfer functions and the convergence of some recursive schemes. IEEE Trans. Autom. Control AC-22, 1977, pp. 539-550.

Moore, J.B., Persistence of excitation in extended least squares. IEEE Trans. Autom. Control AC-28, No. 1, pp. 60-68.

Solo, V., The convergence of AML, IEEE Trans. Autom. Control AC-24, No. 6, 1979, pp. 958-962.

STOCHASTIC ADAPTIVE CONTROL FOR EXPONENTIALLY CONVERGENT TIME-VARYING SYSTEMS

Graham C. Goodwin and David J. Hill
Department of Electrical and Computer Engineering
University of Newcastle
New South Wales, 2308, Australia

Xie Xianya
Department of Computer Science
Shanghai University of Science and Technology
Shanghai, China

Abstract

This paper shows that the standard stochastic adaptive control algorithms for time-invariant systems have an inherent robustness property which renders them applicable, without modification, to time-varying systems whose parameters converge exponentially. One class of systems satisfying this requirement is those having non-steady-state Kalman Filter or innovations representations. This allows the usual assumption of a stationary ARMAX representation to be replaced by a more general state space model.

1. Introduction

A stochastic adaptive controller is an algorithm which combines on-line parameter estimation with on-line control to generate a control law applicable to systems having unknown parameters and random disturbances [1]. Control laws based on this philosophy have been studied for at least three decades [2], but it is only recently that rigorous convergence analyses have appeared. To gain insight into the operation of these algorithms, several special cases have been studied in detail. For example, the authors of [3] have examined the convergence properties of a particular scheme which combines a simple stochastic gradient parameter estimator with a minimum variance control law.

A number of interesting properties of these simple stochastic adaptive control laws have been established. For example, the tracking error is known to converge to zero in a sample mean square sense [3]. Also, if the desired output sequence is continuously disturbed then the parameters can be shown to converge to their true values [4]. In the case of regulation about a zero desired output, then it has been shown [5] that the parameter estimates converge to a fixed multiple of the true parameter values. Various extensions of the above results have also been studied. For example convergence results have been established [7], [8] for least squares based adaptive control algorithms.

The above papers deal with systems having constant parameters. However, in practice one is often confronted with systems whose parameters vary with time in some fashion. This has motivated several authors to investigate special classes of time-varying systems in an effort to gain insights into the convergence properties relevant to this case. For example, Caines [8] has analyzed the performance of the stochastic gradient algorithm of [3] applied to systems with (converging) martingale parameters. Further results for systems having random parameters are described in [9].

The current paper also deals with systems whose parameters are time-varying. Indeed, the work has much in common, at least philosophically, with the results in [8], [9]. However, here the parameter time variations are deterministic and thus a different method of analysis is necessary from that used in [8], [9].

Our analysis has three key steps: a proof that a system which is convergent toward a minimum phase system has an input which grows no faster than the output; a proof that a system which is exponentially convergent toward a Very Strictly Passive system is eventually Input Strictly Passive in a certain sense; and a martingale convergence proof along the lines of [3], but using a modified martingale result as first proposed in [8], [11] in a different context.

One application of the results developed here is to systems described by a state space model corresponding to a non-steady-state innovations representation. Subject to the assumption that the system has no uncontrollable modes (in the filtering sense [12]) on the unit circle, then it is known [13], that the parameters in the innovations model are time-varying and converge exponentially fast toward those of the steady-state optimal filter. Thus the results of this paper allow global convergence to be established for the standard adaptive control algorithms when applied to these systems. This represents a relaxation of the usual modelling assumption employed elsewhere in the literature (e.g. [3] to [9]) that the system is described by an ARMAX model or equivalently a steady-state Kalman Filter model. This particular robustness to modelling assumptions is often implicitly assumed in the literature and it is thus interesting for technical completeness to have a formal proof that the results go through in this case.

2. Preliminary Results

We verify two preliminary results which we will need in our subsequent proof. The first is straightforward and shows that if a system is convergent toward an asymptotically stable system then it is bounded-input bounded-output stable. The second result is less obvious and shows that a system which is exponentially convergent toward a Very Strictly Passive system is eventually Input Strictly Passive in a certain sense. For sake of brevity, proofs will be abbreviated or omitted throughout the paper. The details are presented in [19].

We consider the following time-varying system:

$$x(t+1) = A(t)x(t) + B(t)u(t) \tag{2.1}$$

$$y(t) = C(t)x(t) + D(t)u(t) \tag{2.2}$$

and we introduce the notation:

$$\Phi(n, n_0) = \prod_{k=n_0}^{n-1} A(k) \tag{2.3}$$

for the state transition matrix.

We will use $|\cdot|$ for the Euclidean norm and $\|\cdot\|$ for the ℓ_2 norm. (Similarly for the induced norms). The following result is well-known - see [14] for instance.

Reprinted from *Proc. 23rd Conf. Decision and Control*, 1984, pp. 39–44, vol. 2.

Lemma 2.1: Let A be asymptotically stable. Suppose $A(k) \to A$. Then there exists $\bar{n}$, $v > 0$, and $0 < \beta < 1$ such that

$$|\Phi(n, n_0)| \le v\beta^{n-n_0} \tag{2.4}$$

for all $n > n_0 \ge \bar{n}$.

Theorem 2.1: Consider the system (2.1), (2.2), then provided $A(k) \to A$ with A asymptotically stable:

(a) There exists $\bar{n}$, $0 < K_1 < \infty$, $0 \le K_2 < \infty$ independent of N such that

$$\sum_{t=n_0}^{N} |y(t)|^2 \le K_1 \sum_{t=n_0}^{N} |u(t)|^2 + K_2, \text{ for all } N \ge n_0 \ge \bar{n} \tag{2.5}$$

(b) There exists $0 \le m_3 \le \infty$, $0 \le m_4 < \infty$ which are independent of 't such that

$$|y_i(t)| \le m_3 + m_4 \max_{n_0 \le \tau \le N} |u(\tau)| \text{ for all } N \ge t \ge n_0 \tag{2.6}$$

In the sequel, we use the notation $\| x(t) \|_{n_0}^{N} \triangleq (\sum_{t=n_0}^{N} |x(t)|^2)^{\frac{1}{2}}$. If the sum is finite for all $N \ge n_0$, we say $x(t) \in \ell_{2e}(Z_+)$ where Z_+ is the set of integers $n_0, n_0+1, \ldots$.

Lemma 2.2: Consider a matrix A and sequence A(k) as in Lemma 2.1. Suppose $A(k) \to A$ exponentially. Then $\exists \bar{n}$, $\chi > 0$, and $0 < \eta < 1$ such that

$$|A^{n-n_0} - \prod_{k=n_0}^{n-1} A(k)| \le \chi\eta^{n-1} \tag{2.7}$$

for all $n > n_0 \ge \bar{n}$.

Consider the following special case of the model (2.1), (2.2)

$$x(t+1) = A(t)x(t) + Bu(t) \tag{2.8}$$

$$y(t) = Cx(t) \tag{2.9}$$

Let $A(t) \to A$ exponentially fast and define $\{y^*(t)\}$, $\{x^*(t)\}$ by

$$x^*(t+1) = Ax^*(t) + Bu^*(t) \tag{2.10}$$

$$y^*(t) = Cx^*(t) \tag{2.11}$$

with (A, B) controllable and A asymptotically stable.

The following result establishes that if (2.10), (2.11) is ISP then (2.8), (2.9) satisfies a property very close to ISP.

Theorem 2.2: Provided (2.10), (2.11) is Input Strictly Passive then there exists $\bar{n}_0$, β and $\delta > 0$ such that

$$\sum_{t=n_0}^{N} (y(t)u(t) - \delta u(t)^2) + \beta \ge 0 \tag{2.12}$$

for all $N \ge n \ge \bar{n}_0$ and for all $\{u(t)\} \in \ell_{2e}(Z_+)$. β, δ depend on u(t) for $t < n_0$.

Proof: Since (2.10), (2.11) is ISP, there exists $\delta^* > 0$ and $\beta^*(x^*(n_0))$ such that

$$\sum_{t=n_0}^{N} (y^*(t)u^*(t) - \delta^* u^*(t)^2) + \beta^*(x^*(n_0)) \ge 0 \tag{2.13}$$

for all $N \ge n_0$ and for all $u(t) \in \ell_{2e}(Z_+)$ [See [15], Appendix C]. From (2.19), we have

$$\sum_{t=n_0}^{N} (y^*(t) - y(t))u(t) + \sum_{t=n_0}^{N} y(t)u(t) \ge \delta^* \sum_{t=n_0}^{N} u(t)^2 - \beta^*(x^*(n_0))$$

Let

$$\alpha(n_0, N) \triangleq \sum_{t=n_0}^{N} (y^*(t) - y(t))u(t)$$

Then

$$\sum_{t=n_0}^{N} y(t)u(t) \ge \delta^* \sum_{t=n_0}^{N} u(t)^2 - \alpha(n_0, N) - \beta^*(x^*(n_0)) \tag{2.14}$$

Most of the proof involves working $\alpha(n_0, N)$ into something convenient. After somewhat lengthy calculations based on the Schwarz inequality and Lemma 2.2, we get

$$|\alpha(n_0, N)| \le (\varepsilon_1 + \varepsilon_2)(\| u(t) \|_{n_0}^{N})^2 + \varepsilon_1 |x(n_0)|^2$$

where ε_1, ε_2 can be made arbitrarily small by taking n_0 large enough.

Let $\varepsilon \triangleq \varepsilon_1 + \varepsilon_2$. Substituting into (2.14)

$$\sum_{t=n_0}^{N} y(t)u(t) \ge (\delta^* - \varepsilon)(\| u(t) \|_{n_0}^{N})^2 - \beta^*(x(n_0)) - \varepsilon_1 |x(n_0)|^2$$

Let

$$\delta \triangleq \delta^* - \varepsilon$$

$$\beta \triangleq \beta^*(x(n_0)) + \varepsilon_1 |x(n_0)|^2$$

By taking n_0 large enough, it can be guaranteed that $\delta > 0$.

∇∇∇

Remarks: 1. Since system (2.10), (2.11) is both Input Strictly Passive and asymptotically stable, it is in fact Very Strictly passive (see [15], Appendix C).

2. The system does not become Input Strictly Passive as usually defined because δ is dependent on $x(n_0)$.

3. The Adaptive Control Algorithm

We are concerned here with the adaptive control of a linear time-varying finite dimensional system admitting an autoregressive moving average representation of the form:

$$\begin{aligned} y(t) + a_1(t)y(t-1) + \ldots a_n(t)y(t-n) &= b_0(t)u(t-d) + \ldots b_m(t)u(t-d-m) \\ &+ \omega(t) + c_1(t)\omega(t-1) + \ldots c_\ell(t)\omega(t-\ell) \end{aligned} \tag{3.1}$$

We shall express (3.1) in compact notation as

$$A(t,q^{-1})y(t) = q^{-d}B(t,q^{-1})u(t) + C(t,q^{-1})\omega(t) \tag{3.2}$$

where q^{-1} represents the delay operator and $A(t,q^{-1}) = 1 + a_1(t)q^{-1} + \ldots + a_n(t)q^{-n}$; $B(t,q^{-1}) = b_0(t) + b_1(t)q^{-1} + \ldots b_m(t)q^{-m}$; $C(t,q^{-1}) = 1 + c_1(t)q^{-1} + \ldots c_\ell(t)q^{-\ell}$. The corresponding initial condition is $x_0 \triangleq \{y(0) \ldots y(1-k); u(1-d), \ldots, u(1-k); \omega(0), \ldots \omega(1-k)\}$ where $k = \max\{n, m+d, \ell\}$.

The process $\{x_0, \omega(1), \omega(2) \ldots\}$ is defined on the underlying probability space (Ω, F, P) and we define F_0 to be the σ-algebra generated by x_0.

We make the following assumptions on the process $\{\omega(t)\}$:

N.1 $E\{\omega(t)|F_{t-1}\} = 0$ a.s. $t \geq 1$ (3.3)

N.2 $E\{\omega(t)^2|F_{t-1}\} = \sigma_t^2 \leq \sigma^2 < \infty$ $t \geq 1$ (3.4)

N.3 $\limsup_{N\to\infty} \frac{1}{N} \sum_{t=1}^{N} \| \omega(t) \|^2 < \infty$ a.s. (3.5)

We wish to design an adaptive control law to cause $\{y(t)\}$ to track (in some sense) a given desired output sequence $\{y^*(t)\}$ and to ensure that $\{y(t)\}$, $\{u(t)\}$ remain bounded (in some sense). Reference [3] presents further background to this problem as well as giving a convergence analysis for a particular adaptive control algorithm for the case when $A(t,q^{-1})$, $B(t,q^{-1})$, $C(t,q^{-1})$ do not depend on t.

We make the following assumptions about the system (3.2):

S.1 d is known

S.2 Upper bounds for n, m and ℓ are known

S.3 $\theta(t) \to \theta_0$ exponentially fast, where

$$\theta(t)^T = (a_1(t),\ldots,a_n(t),\ b_0(t),\ldots,b_m(t),\ c_1(t),\ldots,c_\ell(t))$$

$$\theta_0 = (a_1,\ldots,a_n,\ b_0,\ldots,b_m,\ c_1,\ldots,c_\ell)$$

S.4 $B(z)$ and $C(z)$ have all zeros outside the closed unit circle where

$$B(z) = b_0 + b_1 z + \ldots + b_m z^m$$

$$C(z) = 1 + c_1 z + \ldots + c_\ell z^\ell$$

S.5 The system

$$C(q^{-1})z(t) = b(t)$$

is Input Strictly Passive.

For simplicity we shall treat the single input single output, unit delay $(d=1)$ case. However, natural extensions exist for the multi-input multi-output non-unit delay as explored for non-time-varying systems in [3], [15], etc.

The model (3.2) can be rearranged into the following predictor form:

$$C(t,q^{-1})[y(t) - \omega(t)] = \alpha(t,q^{-1})y(t-1) + \beta(t,q^{-1})u(t-1) \tag{3.6}$$

where

$$\alpha(t,q^{-1}) \triangleq [C(t,q^{-1}) - A(t,q^{-1})]q \tag{3.7}$$

$$\beta(t,q^{-1}) \triangleq B(t,q^{-1})q \tag{3.8}$$

The adaptive control algorithm which we propose to analyze is the following stochastic-gradient-minimum variance algorithm

A.1 $$\hat{\theta}(t) = \hat{\theta}(t-1) + \frac{\phi(t-1)}{r(t-2) + \phi(t-1)^T\phi(t-1)} \tag{3.9}$$

$\hat{\theta}(1)$ given such that $\hat{\theta}_{n+1}(1) \neq 0$

A.2 $$r(t-1) = r(t-2) + \phi(t-1)^T\hat{\phi}(t-1) \tag{3.10}$$

$r(0) > 0$ given

A.3 $$y^*(t) = \phi(t-1)^T\hat{\theta}(t-1) \tag{3.11}$$

A.4 $$\phi(t-1)^T = [y(t-1),\ldots,y(t-\bar{n}),\ u(t-1),\ \ldots,u(t-\bar{n}),\ -\bar{y}(t-1),\ldots,-\bar{y}(t-\bar{n})]$$

where $\bar{n}$ = upper bound on $\max(n, m+1, \ell)$

A.5 $$\bar{y}(t) = \phi(t-1)^T\hat{\theta}(t)$$

A.6 $$e(t) = y(t) - y^*(t)$$

This algorithm differs slightly to the time-invariant version in [3] by using a posteriori predictions. A discussion of the significance of this can be seen in [15].

The theoretical possibility of division by zero in A.5 can be avoided by a simple deterministic strategy as explained in detail in [17]. Alternatively, it can be argued inductively [3] that division by zero is a zero probability event. Since all the results in this paper are almost sure results, then division by zero can only affect the convergence on a set of measure zero. From either view point, the algorithm is well posed in the sense that all variables remain bounded in finite time (a.s.).

We then have the following global convergence result:

<u>Theorem 3.1</u>: Let assumptions N1 to N3 and S.1 to S.5 hold for the system (3.1). If the algorithm A.1 to A.6 is used, then with probability one, for any initial parameter estimate $\hat{\theta}(0)$

$$\limsup_{N\to\infty} \frac{1}{N} \sum_{t=1}^{N} y(t)^2 < \infty \tag{3.12}$$

$$\limsup_{N\to\infty} \frac{1}{N} \sum_{t=1}^{N} u(t)^2 < \infty \tag{3.13}$$

$$\lim_{N\to\infty} \frac{1}{N} \sum_{t=1}^{N} [E\{(y(t) - y^*(t))^2|F_{t-1}\} - \sigma_t^2] = 0 \tag{3.14}$$

where σ_t^2 is the minimum possible mean square control error at time t achievable with any causal feedback.

Proof: We shall present an outline proof only, highlighting the key departures from the usual proofs for time-invariant systems as in [3], [15].

Define

$$\eta(t) = y(t) - \bar{y}(t) \tag{3.16}$$

We then have the following preliminary properties of the algorithm:

P.1 $$\eta(t) = \frac{r(t-2)}{r(t-1)} e(t) \tag{3.17}$$

P.2 $$\lim_{N\to\infty} \sum_{t=1}^{N} \frac{\phi(t-1)^T\phi(t-1)}{r(t-1)r(t-2)} < \infty \tag{3.18}$$

P.3 $$C(t,q^{-1})z(t) = b(t) \tag{3.19}$$

where $$z(t) = \eta(t) - \omega(t) \tag{3.20}$$

$$b(t) = -\phi(t-1)^T\tilde{\theta}(t) \tag{3.21}$$

$$\tilde{\theta}(t) = \hat{\theta}(t) - \theta(t) \tag{3.22}$$

P.4 $E\{b(t)\omega(t)|F_{t-1}\} = -\frac{\phi(t-1)^T\phi(t-1)}{r(t-1)}\sigma_t^2$ (3.23)

The above properties are as in [3], [13].

Now subtracting $\theta(t)$ from both sides of (3.9) and using (3.10), (3.17) we have

$$\tilde{\theta}(t) = \tilde{\theta}(t-1) + \theta_e(t) + \frac{\phi(t-1)}{r(t-2)}\eta(t)$$

where

$$\theta_e(t) = \theta(t) - \theta(t-1)$$

$$\tilde{\theta}(t) - \frac{\theta(t-1)\eta(t)}{r(t-2)} = \tilde{\theta}(t-1) + \theta_e(t) \qquad (3.24)$$

Define

$$V(t) = \tilde{\theta}(t)^T\tilde{\theta}(t)$$

Then squaring both sides of (3.24) and using (3.21)

$$V(t) + \frac{2b(t)\eta(t)}{r(t-2)} + \frac{\phi(t-1)^T\phi(t-1)}{r(t-2)^2}\eta^2(t)$$
$$= V(t-1) + 2\tilde{\theta}(t-1)^T\theta_e(t) + \|\theta_e(t)\|^2$$

Hence using (3.20), (3.23)

$$E\{V(t)|F_{t-1}\} = V(t-1) - \frac{2}{r(t-2)}E\{b(t)z(t)|F_{t-1}\} + \frac{2\phi(t-1)^T\phi(t-1)}{r(t-2)r(t-1)}\sigma_t^2 - E\{\frac{\phi(t-1)^T\phi(t-1)}{r(t-2)^2}\eta(t)^2|F_{t-1}\} + \|\theta_e(t)\|^2 + 2\tilde{\theta}(t-1)^T\theta_e(t) \qquad (3.25)$$

From assumption S.3, there exists a G,λ with $0 < G < \infty$, $|\lambda| < 1$ such that

$$\|\theta_e(t)\|^2 \le G|\lambda|^t \qquad (3.26)$$

Also, from the Schwarz inequality:

$$2[\tilde{\theta}(t-1)^T\theta_e(t)] \le a(t)^2\|\tilde{\theta}(t-1)\|^2 + \frac{1}{a(t)^2}\|\theta_e(t)\|^2$$

Thus selecting $a(t)^2 = |\lambda|^{t/2}$ and using (3.26) we have

$$2[\tilde{\theta}(t-1)^T\theta_e(t)] \le |\lambda|^{t/2}\|\tilde{\theta}(t-1)\|^2 + G|\lambda|^{t/2}$$
$$= |\lambda|^{t/2}V(t-1) + G|\lambda|^{t/2} \qquad (3.27)$$

Substituting (3.27) into (3.25) gives

$$E\{V(t)|F_{t-1}\} \le V(t-1)[1 + |\lambda|^{t/2}] - \frac{2}{r(t-2)}E\{b(t)z(t)|F_{t-1}\} + \frac{2\phi(t-1)^T\phi(t-1)}{r(t-1)r(t-2)}\sigma_t^2 - E\{\frac{\phi(t-1)^T\phi(t-1)}{r(t-2)^2}\eta(t)^2|F_{t-1}\} + G[|\lambda|^{t/2} + |\lambda|^t] \qquad (3.28)$$

Define

$$S(t) \triangleq 2\sum_{j=n_0}^{t}[b(j)z(j) - \delta z(j)^2] + K \qquad (3.29)$$

$$X(t) \triangleq V(t) + \frac{S(t)}{r(t-2)} + 2\delta\sum_{j=n_0}^{t}\frac{z(j)^2}{r(j-2)} + \sum_{j=n_0}^{t}\frac{\phi(j-1)^T\phi(j-1)}{r(j-2)^2}\eta(j)^2 \qquad (3.30)$$

From Assumption S.5, S.3, Property P.3 and Theorem 2.2, we know that there exists a n_0, $\delta > 0$ and a K (depending on the conditions at n_0) such that $S(t) \ge 0$ for all $t \ge n$. Under these conditions $X(t) \ge 0$.

It is readily seen using (3.28), (3.30) that

$$E\{X(t)|F_{t-1}\} \le X(t-1)[1 + |\lambda|^{t/2}] + \frac{2\phi(t-1)^T\phi(t-1)}{r(t-2)r(t-1)}\sigma_t^2 + G[|\lambda|^{t/2} + |\lambda|^t] \quad \text{for } t \ge n_0$$

From property P.2, and Assumption N.2, we have that

$$\sum_{t=0}^{\infty}[\frac{\phi(t-1)^T\phi(t-1)}{r(t-2)r(t-1)}\sigma_t^2 + G[|\lambda|^{t/2} + |\lambda|^t]] < \infty$$

Thus we can apply the Martingale Convergence Theorem (Appendix A) to conclude

$$X(t) \to X < \infty \quad \text{a.s.} \qquad (3.31)$$

Using (3.30) we see that

$$\lim_{N\to\infty}\sum_{t=1}^{N}\frac{z(t)}{r(t-2)} < \infty \quad \text{a.s.} \qquad (3.32)$$

$$\lim_{N\to\infty}\sum_{t=1}^{N}\frac{\phi(t-1)^T\phi(t-1)}{r(t-2)^2}\eta(t)^2 < \infty \quad \text{a.s.} \qquad (3.33)$$

The lower summation limits in (3.32), (3.33) can be extended from n_0 to 1 because the algorithm ensures all variables remain bounded in finite time (a.s.).

A simple argument by contradiction can now be used to conclude (3.12) to (3.14) using equation (3.32) together with Assumption S.3, S.4 and Theorem 2.1. The steps are exactly as in [3] and as explained in general in [15]. ∇∇∇

4. Adaptive Control with General State Space Model

Consider a linear finite dimensional system described by the following time-invariant state space model:

$$x(t+1) = Fx(t) + Gu(t) + v_1(t) \qquad (4.1)$$

$$y(t) = Hx(t) + v_2(t) \qquad (4.2)$$

where $\{v_1(t)\}$, $\{v_2(t)\}$ are zero mean Gaussian white noise sequences satisfying:

$$E\{v_1(t)v_1(t)^T\} = Q = DD^T \ge 0 \qquad (4.3)$$

$$E\{v_2(t)v_2(t)^T\} = R \ge 0 \qquad (4.4)$$

The initial state $x(0)$ is also assumed to have a Gaussian distribution with mean $\bar{x}_0$ and convariance P_0. We make the following assumption

S.S.1: (H,F) is observable

S.S.2(a): (F,D) has no uncontrollable modes on the unit circle and $P_0 > 0$ or

(b): (F,D) is stabilizable and $P_0 \ge 0$.

Using standard Kalman Filtering ideas [12], the innovations model for (4.1), (4.2) is

$$\hat{x}(t+1) = F\hat{x}(t) + Gu(t) + K(t)\omega(t); \ \hat{x}(0) = \bar{x}_0 \qquad (4.5)$$

$$y(t) = H\hat{x}(t) + \omega(t) \qquad (4.6)$$

here K(t) is obtained from the solution of the following matrix Riccati equation:

$$\Sigma(t+1) = F\Sigma(t)F^T - F\Sigma(t)H^T(H\Sigma(t)H^T + R)^{-1}H\Sigma(t)F^T + Q \qquad (4.7)$$

$$\Sigma(0) = P_0 \qquad (4.8)$$

$$K(t) = F\Sigma(t)H^T(H\Sigma(t)H^T + R)^{-1} \qquad (4.9)$$

In view of Assumption S.S.1, we can assume that (H, F) are in observer canonical form, i.e. (4.5), (4.6) can be written as

$$\hat{x}(t+1) = \begin{bmatrix} -a_1 & 1 & & 0 \\ \vdots & & \ddots & \\ \vdots & & & 1 \\ -a_n & 0 & \dots & 0 \end{bmatrix} \hat{x}(t) + \begin{bmatrix} b_1 \\ \vdots \\ \vdots \\ b_n \end{bmatrix} u(t) + \begin{bmatrix} k_1(t) \\ \vdots \\ \vdots \\ k_n(t) \end{bmatrix} \omega(t) \quad (4.10)$$

$$y(t) = [\,1 \quad 0 \dots 0\,]\ \hat{x}(t) + \omega(t) \quad (4.11)$$

Using (4.11) in (4.10) gives the following time varying ARMAX model:

$$A(q^{-1})y(t) = B(q^{-1})u(t) + C(t, q^{-1})\omega(t) \quad (4.12)$$

where

$$A(q^{-1}) = 1 + a_1 q^{-1} + \dots + a_n q^{-n}$$

$$B(q^{-1}) = b_1 q^{-1} + \dots + b_n q^{-n}$$

$$C(t,q^{-1}) = 1 + (k_1(t-1) + a_1)q^{-1} + \dots (k_n(t-n) + a_n)q^{-n} \quad (4.13)$$

It is known [13] that Assumptions S.S.1, S.S.2 are sufficient to ensure:

$$K(t) \to \bar{K} \quad \text{exponentially fast} \quad (4.14)$$

and

$$C(q^{-1}) = 1 + (\bar{k} + a_1)q^{-1} + \dots (\bar{k}_n + a_n)q^{-n}$$

is asymptotically stable. (4.15)

We make the following additional assumptions:

S.S.3: $b_1 \neq 0$ (corresponding to $d=1$ in Section 3)

S.S.4: An upper bound for n is known

S.S.5: The system

$$C(q^{-1})z(t) = b(t)$$

is Input Strictly Passive

S.S.6: $B(z)$ has all zeros outside the unit circle and $b_1 \neq 0$ (the latter for simplicity only)

We then have the following elementary Corollary to Theorem 3.1:

Corollary 4.1: Let Assumptions S.S.1 to S.S.6 hold for the system (4.1), (4.2). If the algorithm A.1 to A.6 is used, then with probability one, for any initial parameter estimate $\hat{\theta}(0)$

$$\limsup_{N\to\infty} \frac{1}{N} \sum_{t=1}^{N} y(t)^2 < \infty \quad (4.16)$$

$$\limsup_{N\to\infty} \frac{1}{N} \sum_{t=1}^{N} u(t)^2 < \infty \quad (4.17)$$

$$\lim_{N\to\infty} \frac{1}{N} \sum_{t=1}^{N} E\{(y(t) - y^*(t))^2 | F_{t-1}\} = \sigma^2 \quad (4.18)$$

where

$$\sigma^2 = H\bar{\Sigma}H^T + R \quad (4.19)$$

and $\bar{\Sigma}$ is the steady state solution of (4.7).

Proof: The result is essentially immediate from Theorem 3.1 on noting that $\{\omega(t)\}$ is a Gaussian innovations sequence and therefore satisfies N.1 to N.3. Also, $\sigma_t^2 \to \sigma^2$ exponentially fast [13] and hence

$$\lim_{N\to\infty} \frac{1}{N} \sum_{t=1}^{N} \sigma_t^2 = \sigma^2 \quad (4.20)$$

∇∇∇

Remarks: 1. It should be pointed out that this result can be obtained directly from the corresponding result for the steady-state ARMAX model [3, 15]. Firstly, note that the covariance P_0 for the Kalman Filter (with $\bar{x}_0$) defines a Gaussian distribution for x_0. So use of the steady-state gain $\bar{K}$ corresponds to a particular choice of the initial condition distribution. Since the Martingale Convergence Theorem leads to a sample path convergence result then modification of the initial state distribution does not affect this result. The proof for this comes by noting that once a.s. convergence has been established with respect to one distribution, then it is also true for any other distribution which is absolutely continuous with respect to the original one [18]. Thus, having proved global convergence for the distribution corresponding to a steady-state Kalman Filter, it also holds for other distributions corresponding to the non-steady-state case. However, it should be realized that the original martingale properties will no longer apply.

2. If a non-steady-state initial condition distribution is used and one replaced (4.12) by the corresponding steady state ARMAX model, then the prediction error so defined will not satisfy N.1.

3. The result in Corollary 4.1 also applies to degenerate distributions, i.e. when the initial state is exactly known. In this case the argument in Remark 1 above cannot be used and the more complicated machinery of Section 3 is necessary to deal with this case.

5. Conclusions

This paper has analyzed a robustness property of the discrete time stochastic adaptive control algorithm based on gradient estimation and minimum variance control. The algorithm is shown to be globally convergent when the system parameters are exponentially convergent to values satisfying the conditions for a globally convergent time-invariant system. This result is applied to the special case where the time-variation is derived from a non-steady state Kalman Filter.

References

[1] Astrom, K.J., "Theory and applications of adaptive control - A survey", *Automatica,* Vol. 19, No. 5, pp. 471-487, September 1983.

[2] Draper, C.S. and Li, Y.T., "Principles of Optimalizing Control Systems and Application to Internal Combustion Engine", *ASME,* New York, 1951.

[3] Goodwin, G.C. Ramadge, P.J. and Caines, P.E., "Discrete time stochastic adaptive control", *SIAM J. Control and Optimization,* Vol. 19, No. 6, Nov. 1981, pp. 829-853.

[4] Chen, H.F. and Caines, P.E., "The strong consistency of the stochastic gradient algorithm of adaptive control", Technical Report, Department of Electrical Engineering, McGill University, July, 1983.

[5] Kumar, P.R., "Adaptive control with the stochastic approximation algorithm: Geometry and convergence", to appear, *IEEE Trans. on Auto. Control.*

[6] Sin, K.S. and Goodwin, G.C., "Stochastic adaptive control using a modified least squares algorithm", *Automatica,* Vol. 18, No. 3, 1983, po. 315-321.

[7] H.F. Chen, "Recursive system identification and adaptive control by use of the modified least squares algorithm", Research Report, Department of Electrical Engineering, McGill University, November, 1982.

[8] Caines, P.E., "Stochastic adaptive control: Randomly varying parameters and continuously disturbed controls", *IFAC Congress,* Kyoto, Japan, August, 1981.

[9] Chen, H.F. and Caines, P.E., "On the adaptive control of a class of systems with random parameters and disturbances", Research Report, Department of Electrical Engineering, McGill University, March, 1983.

[10] Hersh, M.A. and Zarrop, M.B., "Stochastic adaptive control of nonminimum phase systems", Control Systems Centre Technical Report, UMIST, 1981.

[11] M.A. Hersh, Ph.D. Thesis, Control Systems Centre, UMIST, U.K., 1983.

[12] Anderson, B.D.O. and Moore, J.B., Optimal Filtering, Prentice Hall, 1979.

[13] Chan, S.W., Goodwin, G.C. and K.S. Sin, "Convergence properties of the Riccati difference equation in optimal filtering of non-stabilizable systems", *IEEE Trans. on Auto. Control,* to appear, 1984.

[14] Fuchs, J.J., "On the good use of the spectral radius of a matrix", *IEEE Trans. on Auto. Control,* Vol. AC-27, No. 5, Oct. 1982, pp. 1134-1135.

[15] Goodwin, G.C. and K.S. Sin, Adaptive Filtering Prediction and Control, Prentice Hall, 1984.

[16] Neveu, J., Discrete Parameter Martingales, North Holland Pub. Co., Amsterdam, 1975.

[17] Goodwin, G.C., Ramadge, P.J. and Caines, P.E., "Discrete-time multivariable adaptive control", *IEEE Trans. on Auto. Control,* AC-25, June 1980, pp. 449-456.

[18] Loeve, M., Probability Theory, Springer-Verlag, New York, 1963.

[19] Goodwin, G.C., Hill, D.J. and Xianya, X., "Stochastic adaptive control for exponentially convergent time-varying systems", University of Newcastle, Technical Report EE8407, May, 1984.

Appendix

Modified Martingale Convergence Theorem

Let $\{X(t)\}$ be a sequence of non-negative random variables adapted to an increasing sequence of sub σ-algebras $\{F_t\}$. If

$$E\{X(t+1)|F_t\} \le (1+\gamma(t))X(t) - \alpha(t) + \beta(t) \quad \text{a.s.}$$

where $\alpha(t) \ge 0$, $\beta(t) \ge 0$ and $E\{X(0)\} < \infty$, $\sum_{j=1}^{\infty} |\gamma(t)| < \infty$, $\sum_{j=0}^{\infty} \beta(j) < \infty$ a.s. then $X(t)$ converges almost surely to a finite random variable and

$$\lim_{N\to\infty} \sum_{t=0}^{N} \alpha(t) < \infty \quad \text{a.s.}$$

Proof: See Neveu [16].

∇∇∇

The above theorem was first employed in a stochastic adaptive control proof in [10], [11]. Other applications are given in [15].

Part V
Applications to Aircraft Control Problems

IT should be recalled that the first type of adaptive controllers were designed for applications to aircraft control problems. In this part of the volume, we give some applications of various adaptive control methods described earlier to certain aircraft control problems.

In the first paper, paper thirty, P. N. Nikiforuk, M. M. Gupta, and K. Kanai examine the design of a gust alleviation control for high-speed aircraft operating over a wide range of environmental conditions. The control strategy uses a two-level adaptive control method designed using the Liapunov signal synthesis approach. The controller, however, does not require an explicit knowledge of the aircraft parameters or its environment. The design philosophy presented in this paper is an extension of paper eleven and seems to be applicable to both stochastic and nonlinear systems with bounded uncertain parameters. It is believed that more research on this design philosophy may result in a more effective controller with a greater saving in the design efforts.

D. McLean, in paper thirty-one, describes the use of active control technology to alleviate the effects on both the dynamics and structural responses of a flexible aircraft flying through atmospheric turbulence, while maintaining an acceptable degree of flight handling qualities. The author uses only output feedback since a linear state variable feedback in aircraft flight-control systems is usually impossible to implement. The paper considers some practical restricted situation, and presents two control schemes derived from optimal-control theory and the algebraic theory of model following systems. This paper is a good example of how the model following theory can be used in practice.

In paper thirty-two, P. N. Nikiforuk, H. Ohta, and M. M. Gupta describe a design of a two-level adaptive flight controller for a STOL aircraft with unknown dynamics. The approach appears to overcome some of the limitations that are inherent in the design of linear optimal and conventional adaptive controllers. It is assumed in the design that the state variables are not accessible (similar to what was stated in the previous paper), and are generated using an adaptive observer. Control at the first-level is provided by an adaptive optimal controller, while that at the second-level is provided by an error servo. Simulation studies carried out on pitch attitude control are encouraging. A translated version of this paper has also appeared in the *Russian Journal of Rocket Technology and Cosmonautics* (Moscow) in April 1979.

In paper thirty-three, U. Hartmann and V. Krebs describe a new digital adaptive control system for high performance aircraft. The approach presented uses a pole allocation principle and recursive least-square fading memory parameter estimation method to acquire a control which yields desired handling qualities.

In general, the design of command and stability systems, which can provide satisfactory handling qualities over the whole flight envelop of high-performance aircraft is a difficult task but, as shown in this small sample of papers, adaptive control is a very promising approach to achieve some of these goals. Algorithmic based adaptive controller may prove to be quite useful in the future aircraft control applications and must be a subject for future research.

Selected Bibliography

P. C. Gregory, Ed., Proceedings of the Self Adaptive Flight Control Systems Symposium, WADC Tech. Rep. 59-49, Wright-Patterson Air Force Base; "Design of a two-level adaptive flight control system," *J. Guidance and Contr.*, vol. 2, no. 1, pp. 79–85, 1959.

L. G. Hofmann, "Application of a new method in adaptive control," *J. Aircraft*, vol. 7, no. 1, pp. 32–38, 1970.

L. G. Holfmann and J. J. Best, "New methods in adaptive flight control," NASA Flight Research Center, Rep. CR-1152, 1968.

I. M. Horowitz, "Linear adaptive flight control for re-entry vehicles," *IEEE Trans. Automat. Contr.*, vol. AC-9, no. 1, pp. 90–97, 1964.

C. R. Johnson, Jr., "On adaptive model control of large flexible spacecraft," *J. Guidance and Contr.*, vol. 3, pp. 369–375, 1980.

K. Kanai, T. Degawa, P. N. Nikiforuk, and M. M. Gupta, "Synthesis of *C**—Model Reference adaptive flight controller," in *Int. Federation of Automatic Control,* 8th Triennial World Congress, pp. VII.126–VII.132, 1981.

I. D. Landua and B. Courtiol, "Adaptive model following systems for flight control and simulation," *J. Aircraft,* vol. 9, no. 9, pp. 668–674, 1972.

D. McLean, "Gust-alleviation control systems for aircraft," *Proc. IEEE,* Control and Science, 1979.

J. Mendel, "Invariant poles feedback control of flexible, highly variable spacecraft," *IEEE Trans. Automat. Contr.*, vol. AC-17, no. 6, pp. 814–821, 1972.

P. N. Nikiforuk, M. M. Gupta, and K. Kanai, "Stability analysis and design for aircraft gust alleviation control," *Automatica,* vol. 10, no. 5, pp. 494–506, 1974.

P. N. Nikiforuk, M. M. Gupta, K. Kanai, and H. Adachi, "On the design of model reference adaptive flight control systems," in *Proc. AIAA Guidance and Cont. Conf.*, Paper 75-1111, pp. 1–12, 1975.

P. N. Nikiforuk, H. Ohta, and M. M. Gupta, "A two-level adaptive controller for applications to flight control systems," in *Proc. 1977 AIAA Guidance and Contr. Conf.*, pp. 401–407, 1977.

——, "Design of a two-level adaptive flight control system," *J. Guidance and Control,* vol. 2, no. 1, pp. 79–85, 1979.

H. Ohta, M. M. Gupta, and P. N. Nikiforuk, "Design of a two-level adaptive flight control system," *Rocket Tech. and Cosmonautics,* vol. 629.7.015, (Moscow), pp. 166–174, 1979.

——, "Design of desirable handling qualities for aircraft lateral dynamics," *AIAA J. of Guidance and Contr.*, vol. 2, no. 1, pp. 31–39, 1979.

——, "Design of adaptive flight controllers for a STOL aircraft," presented at Int. Conf. on Policy Analysis and Information Systems, 1981.

——, "Design of adaptive flight controller for an aircraft," Workshop on Applications of Adaptive Systems Theory, Yale University, 1981.

H. Ohta, P. N. Nikiforuk, and M. M. Gupta, "Adaptive flight control of unstable helicopters during transition," IX Triennial World Congress, in *Int. Federation of Automatic Control,* vol. VIII, pp. 251–256, 1984.

A. P. Sage, J. T. Humphries, and R. E. Uhrig, "Model reference adaptive control of a nuclear rocket engine," *Proc. 1966 JACC,* 1966.

S. C. Shah, R. A. Walker, and H. A. Saberi, "Multivariable adaptive control algorithms and their mechamizations for aerospace applications," in *Proc. 23rd IEEE Conf. Decision and Control,* pp. 381–386, 1984.

O. H. Shuck, "Honeywell's history and philosophy in the adaptive control field," Wright Air Development Center, Wright-Patterson AFB, OH, Tech. Rep., 1959.

H. P. Whitaker, J. Yamron, and A. Kezer, "Design of model reference adaptive control systems for aircraft," Rep. R-164, Instrumentation Laboratory, Massachusetts Institute of Technology, Cambridge, MA, 1958.

P. Whitaker, "The MIT adaptive autopilot," in *Proc. Self-Adaptive Contr. Symp.,* Wright Air Dev. Center, vol. 113, pp. 175–184, 1959.

Stability Analysis and Design for Aircraft Gust Alleviation Control*†

Analyse de Stabilité et Conception pour le Contrôle d'Allégement de Bourrasque d'Avions

Stabilitätsanalyse und Entwurf für die Regelungzur Milderung des Einflusses von Windböen auf ein Flugzeug

Анализ устойчивости и расчет самолетных систем управления для компенсации ветровых нагрузок

P. N. NIKIFORUK, M. M. GUPTA and K. KANAI‡

Stability analysis indicate the need for an on-line, two-level, adaptive gust alleviation control system for a high speed aircraft operating in an unknown environment.

Summary—This paper is concerned with the stability analysis and the development of an on-line two-level gust alleviation control system for a high-speed aircraft operating in an unknown environment. As is well known, such an aircraft experiences dynamic instability in its lateral motion when subjected to the longitudinal and lateral gusts. This is due to the phenomenon of lateral–longitudinal aerodynamic cross-coupling.

In the first part of this paper the instability behaviour, in the presence of a longitudinal gust, of two typical high-speed aircraft is analyzed. From theoretical and simulation studies it is shown that for a sinusoidal gust, and for a given configuration, the instability region is a function of both the amplitude and frequency of the gust disturbance.

In the second part of the paper, an on-line gust alleviation control system is synthesized for such an aircraft operating over a wide range of unknown environmental conditions. This control system employs a two-level control structure. The controller, however, does not require an explicit knowledge of the aircraft parameters or the environment. Typical simulation results are presented which verify the theoretical predictions.

1. INTRODUCTION

1.1 *General*

THE design of a gust alleviation control system which can cope with the instabilities that occur in the lateral and longitudinal modes of modern high-speed aircraft is a complicated and difficult problem [1, 2]. The normal and usual approach in the design of such a system is to first establish a linearized model of the aircraft [1]. Once this is done, conventional control theory, either classical or modern, is then used for the design of the control system [1–3].

An inherent requirement [2, 3] of these conventional control techniques is that of identifying the plant parameters, and measuring the state variables. Not only are these difficult tasks, in most instances, but they become almost impossible in the case of on-line control for a high-speed aircraft operating over a wide range of unknown environments. For this reason on-line adaptive type control techniques which can preclude these rigid requirements are desirable [6].

The first part of this paper is concerned with the stability analysis of the lateral dynamic motion of an aircraft subjected to a sinusoidal type longitudinal gust. In this analysis the stability boundaries, shown in the amplitude–frequency plane of the sinusoidal gust, are first predicted theoretically and then verified using simulation studies for two types of aircraft configurations. This is then followed by the design of an on-line gust alleviation control system employing a two-level control structure [6]. It is shown that this design procedure does not require an explicit knowledge of the aircraft parameters or the gust acting upon it. Further, as will be shown later, it essentially bypasses the rigid requirements of identification and state variable measurements as required by the existing techniques.

* Received 27 December 1972, revised 5 September 1973, revised 7 March 1974. The original version of this paper was presented at the 5th IFAC Symposium on Automatic Control in Space, held at Genova, Italy, 4–8 June 1973. It was recommended for publication in revised form by Associate Editor P. Parks.

† This work is supported by the National Research Council of Canada under Grants A-5625 and A-1080, and the Defence Research Board of Canada under Grants 4003-02 and 9781-04.

‡ Systems and Adaptive Control Research Laboratory, Department of Mechanical Engineering, University of Saskatchewan, Saskatoon, Saskatchewan, Canada.

Reprinted with permission from *Automatica*, vol. 10, pp. 495–506, Sept. 1974.
Copyright © 1974, International Federation of Automatic Control IFAC. Published by Pergamon Press Ltd.

1.2. *Perturbation equation of lateral motion due to longitudinal gust*

In this section the stability of two high-speed aircraft subjected to a longitudinal gust is studied. For this purpose the usual aerodynamic assumptions that lead to the decoupling of the aircraft equations for small perturbations in the motion into lateral and longitudinal sets are made [1, 3]. It is also assumed that the various elements of the input disturbance vector which are associated with the different components of the gust are uncorrelated. On this basis the linearized small perturbation equation for the lateral motion of such an aircraft subjected to a longitudinal gust, $g(t)$, and lateral gust, $v_g(t)$, is derived in Appendix B, equation (b-1). The corresponding state evolution equation in a normalized form is given by

$$\dot{x}(t)=A(g(t), t)x(t)+Bu(t)+G(v_g(t), t) \qquad (1.1)$$

where the states $x(t)\underline{\Delta}[\beta(t), \phi(t), \hat{p}(t), \hat{r}(t)]'\varepsilon R^4$, and the control $u(t)\underline{\Delta}[\delta a(t), \delta r(t)]'\varepsilon R^2$.

It is known that the important stability derivatives of the matrix A are functions of the lift coefficient, C_l, where C_l may be written in the small perturbation form [1,5].

$$C_l(t)=C_{l0}[1+g(t)], \; |g(t)|\ll 1 \qquad (1.2)$$

where C_{l0} is the nominal value of the lift coefficient, and $g(t)$ is the longitudinal gust.

As shown in Appendix B, the lateral motion of the aircraft, equation (1.1), may equivalently be written as

$$\dot{x}(t)=\{A_0+A_g g(t)\}x(t)+Bu(t)+Gv_g(t) \qquad (1.3)$$

where A_g is an equivalent perturbation matrix on A due to gust $g(t)$, and matrix G represents the aircraft input matrix for the lateral gust, $v_g(t)$.

The state evolution equation of the normalized perturbation equation for the Dutch roll motion, which is derived in Appendix B, is written as

$$\dot{x}_d(t)=\{A_{d0}+A_{dg}g(t)\}x_d(t)+B_d u_d(t) + G_d v_g(tt) \qquad (1.4)$$

where the state $x_d(t)\underline{\Delta}[\beta(t), \hat{r}(t)]'\varepsilon R^2$ and the control $u_d(t)\underline{\Delta}\delta r(t)\varepsilon R^1$.

2. STABILITY ANALYSIS OF LATERAL DYNAMIC MOTION

There is no general method available for the study of the stability behaviour of systems of the ype considered in this paper. Recently, however, some results have been reported by ICHIKAWA and KANAI [7] for the stability analysis of higher-order linear systems with sinusoidally time-varying parameters. This method is based upon a successive approximations technique and yields a simple stability criterion. As explained in Appendix D. the method assumes a strong low-pass filtering effect in the system. For the system considered in this paper, the dynamic equations for the lateral motion possess, we believe, sufficient low-pass filtering characteristics to employ this successive approximation method of Ichikawa and Kanai.

For the sinusoidal gust disturbance $g(t)$ is

$$g(t)=\varepsilon\sin \nu t, \; \varepsilon\ll 1. \qquad (2.1)$$

The perturbed lift coefficient in (1.2) then becomes

$$C_l(t)=C_{l0}(1+\varepsilon\sin \nu t). \qquad (2.2)$$

The corresponding equations for the lateral and Dutch roll motions are given in (1.3) and (1.4). For input forcing terms equal to zero; that is, for $u(t)=v_g(t)=0$, these equations become respectively

$$\dot{x}(t)=(A_0+A_g\varepsilon\sin \nu t)x(t) \qquad (2.3)$$

$$\dot{x}_d(t)=(A_{d0}+A_{dg}\varepsilon\sin \nu t)x_d(t). \qquad (2.4)$$

In order to apply the stability boundary criterion of Appendix D to (2.3) and (2.4) these equations must be transformed into the phase canonical form. This transformation is given in Appendix B and the respective equations are

$$\dot{z}(t)=(F_0+F_g\varepsilon\sin \nu t)z(t) \qquad (2.5)$$

$$\dot{z}_d(t)=(F_{d0}+F_{dg}\varepsilon\sin \nu t)z_d(t) \qquad (2.6)$$

where F_0, F_g, F_{d0} and F_{dg} are defined in Appendix B.

Simulation studies

Two separate sets of stability derivatives were studied. The first corresponds to a hypothetical delta wing supersonic transport aircraft (#1) during low speed operating conditions, the second corresponds to a hypothetical subsonic interceptor #2). The parameters of these aircraft are tabulated in Appendix C.

Figures 1(a)–3(a) show the theoretically computed boundaries of the stability region in the ε–ν plane for $g(t)=\varepsilon \sin \nu t$. These theoretical stability boundaries were obtained employing only the first approximation (i.e. for $\omega=\nu/2$) in the stability method given in Appendix D. This first-order approximation gives sufficiently accurate results because of the low-pass characteristics of the aircraft dynamics.

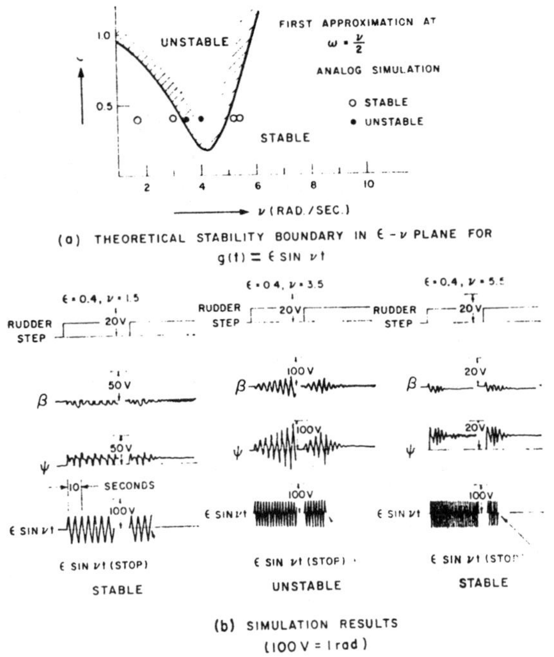

FIG. 1. Stability studies for lateral dynamic motion of supersonic aircraft No. 1.

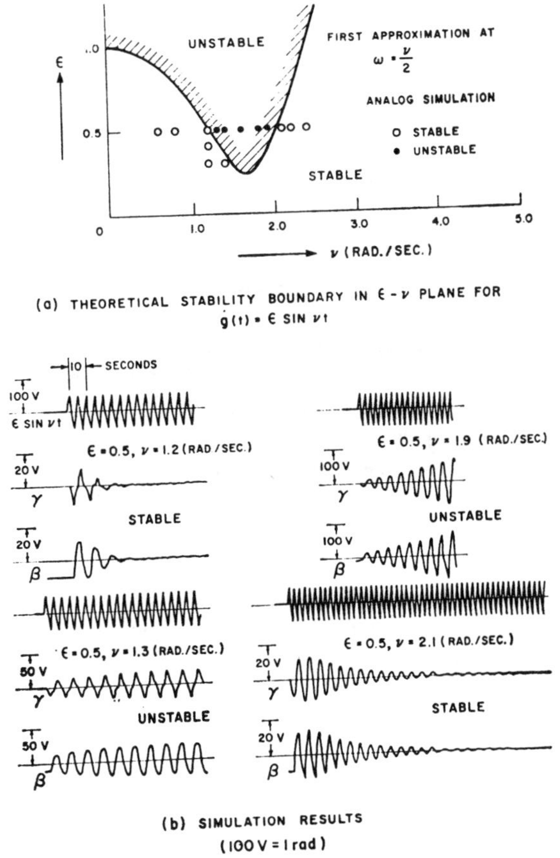

FIG. 2. Stability studies for Dutch roll motion of supersonic aircraft No. 1.

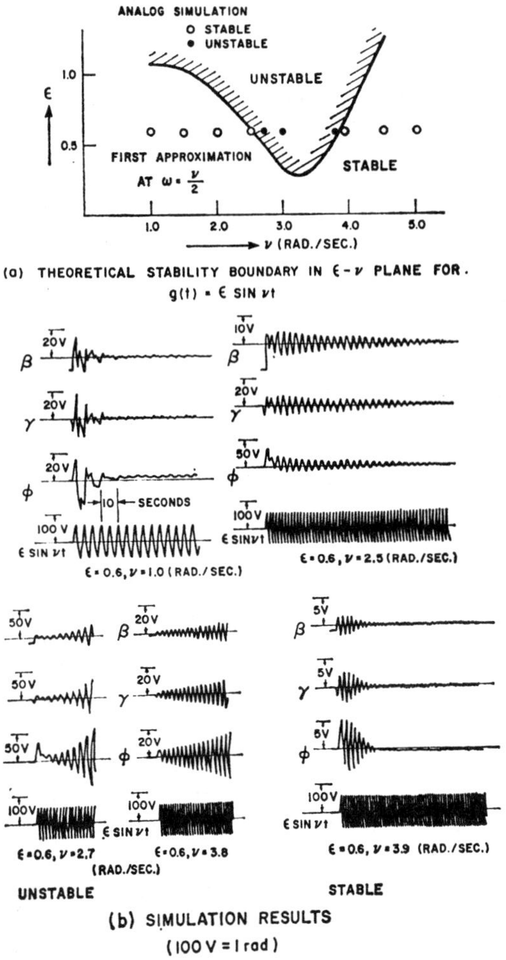

FIG. 3. Stability studies for Dutch roll motion of subsonic intercepter with sinusoidal gust and step action of rudder.

Figure 1(b) shows the dynamic response of aircraft (#1) in its lateral motion in $\beta(t)$, $r(t)$ and $\phi(t)$ for $\varepsilon=0.4$ and various values of the gust frequency ν. The theoretically predicted stability results agree with those obtained from the simulation studies. Similar results are shown in Fig. 2(b) for aircraft (#1) in its Dutch roll motion.

Figure 3(b) shows simulation results obtained for aircraft (#2) in its Dutch roll motion for a sinusoidal gust and with step action of the rudder. From this it is concluded that the Dutch roll motion of this aircraft is inherently stable in the absence of the gust but becomes unstable in the presence of the gust. Also, it can be seen that for this aircraft the stability conditions of the lateral dynamic motion are dominated by the Dutch roll motion.

Although the results for the stability studies were obtained assuming a sinusoidal gust, $g(t)=\varepsilon \sin \nu t$, the design of the gust alleviation control system, as given in the following section, is applicable to a wide class of random gust phenomenon.

3. DESIGN AND IMPLEMENTATION OF A GUST ALLEVIATION CONTROL SYSTEM

A controller synthesis technique for a class of unknown plants operating in an unknown environment which forces the plant to follow a given set of desired response characteristics, and which yields a Lagrange stability of the closed-loop system, is described in [6]. The controller employed has a two-level structure, a feedforward controller on its first-level and a conditional feedback controller on its second-level. The design procedure employed is the Liapunov signal synthesis technique. In this approach a mathematical model of the plant, based upon certain input/output operating conditions, is first formulated. The first-level controller is then synthesized for this known model so as to satisfy certain desired response characteristics. This is followed by the synthesis of the second-level controller which provides conditional feedback. It is a function of the measured error between the model output and the plant output, and the characteristic vector, $\lambda(t)$, which is characteristic to the unknown environments of the plant. This control procedure, as we mentioned earlier, does not require an explicit knowledge of the plant dynamics or its environment, and it thus bypasses the need for identifying the state variables as required by modern control theory.

In this section the design procedure for this controller is first described. Some typical simulation studies which were carried out are then presented using aircraft (#1) as an example. In particular, the lateral motion of this aircraft is described when it is subjected to random lateral and longitudinal gusts.

The general equation of this type of aircraft in a state evolution form is derived in Appendix B and is

$$\dot{x}(t)=[A_0+A_g g(t)]x(t)+Bu(t)+Gv_g(t). \qquad (3.1)$$

The corresponding numerical values are given in Appendix C. The response characteristics of interest of this aircraft are given by

$$y(t)=Cx(t) \qquad (3.2)$$

where

$$C=\begin{bmatrix} 1 & 0 & 0 & 0 \\ 0 & 0 & 0 & 1 \end{bmatrix}, \; y(t) \underline{\Delta} [\beta(t), \dot{r}(t)]'.$$

A general structural form of the gust alleviation control law may be written as

$$u(t)=f_u\{y(t), y_m(t), u_m(t), \lambda(t), r(t), t\} \qquad (3.3)$$

where $f_u\{v\}$ is, in general, a nonlinear time-varying function of its arguments and $u_m(t)\varepsilon R^{um}$, the first-level control, $r(t)\varepsilon R^r$ is a vector of pilot command inputs, $\lambda(t)\varepsilon R^m$ is a vector characteristic to the aircraft operating in an unknown environment. The design procedure for $u_m(t)$ and $u(t)$ will now be described.

The design of the first-level controller

For the design of the first-level controller a characteristic model representing the nominal Dutch roll motion will be assumed to be

$$\dot{x}_m(t)=A_m x_m(t)+B_m u_m(t) \qquad (3.4)$$

with desired response characteristics

$$y_m(t)=x_m(t) \qquad (3.5)$$

where, for aircraft #1

$$A_m=A_{do}=\begin{bmatrix} -0{\cdot}0057 & -1 \\ 0{\cdot}065 & -0{\cdot}057 \end{bmatrix}, \quad B_m=B_d=\begin{bmatrix} 0{\cdot}0015 \\ -0{\cdot}0034 \end{bmatrix}$$

and $u_m(t)$ is a linear feedback control law given by

$$u_m(t)=F_m x_m(t)+F_r\delta_r(t)$$

$$\delta_r(t)\underline{\Delta} r(t), \text{ the pilot command input.} \qquad (3.6)$$

Equation (3.4) represents the motion of the model in its Dutch roll motion, and it is assumed that the desired handling qualities of the controlled aircraft model (3.4) is given in terms of natural frequency ω_n and damping ratio ζ. Based upon the pilot iso-opinion curves [3] the following assignments* are assumed for simplicity

$$\zeta=0.2, \; \omega_n=1.0 \text{ rad/sec.} \qquad (3.7)$$

Using the canonical controller synthesis technique [9] and performing the appropriate numerical calculations.

$$F_m=[-264, -19], \; F_r=I. \qquad (3.8)$$

* In general, for the purpose of achieving certain design objectives which are defined for Dutch roll response as desirable handling quality, the desirable aircraft responses to lateral control inputs might be specified in the criteria for roll rate oscillations.

The design of the second-level controller

For this purpose, the aircraft tracking error, $e(t)$, is defined as

$$e(t)=y_m(t)-y(t),\ e(t)\varepsilon R^2. \tag{3.9}$$

The error differential equation is given by

$$\dot{e}(t)=A_m e(t)+B_m u_m(t)+\lambda(t)-D_0 u(t) \tag{3.10}$$

where,

$$\lambda(t)=A_m x(t)-C\{A_0+A_g g(t)\}x(t)-CGv_g(t)$$

and D_0 is an arbitrarily selected non-singular square matrix.

The characteristic vector $\lambda(t)$ is the function of the aircraft's unknown dynamics and its unmeasureable environment, but may be generated indirectly using (3.10).

Thus,

$$\lambda(t)=\dot{e}(t)-A_m e(t)-B_m u_m+D_0 u(t) \tag{3.11}$$

and D_0 is given by

$$D_0 \triangleq CB=\begin{bmatrix} {}^t0 & 0{\cdot}0015 \\ 0{\cdot}005 & -0{\cdot}0034 \end{bmatrix}.$$

Now, select a real positive definite scalar function $V(e, t)$ defined as

$$V(e, t)=e'(t)Pe(t) \tag{3.12}$$

where P is a real positive definite symmetric matrix

such that $(A_m'P+PA_m)\triangleq -Q<0$,

$$p=\begin{bmatrix} 2{\cdot}0 & 0{\cdot}1 \\ 0{\cdot}1 & 3{\cdot}0 \end{bmatrix}.$$

The bounded stability, in the sense of Lagrange, of (3.10) will be assured by the following theorem [6]:

Theorem 3.1

Let $V(e, t)$ defined in (3.12) be a positive definite scalar function with its time derivative given by

$$\dot{V}(t)=-e'(t)Qe(t)+2M(t),\ M(t)$$
$$=e'(t)P[D_m u_m(t)+\lambda(t)-D_0 u(t)],$$

and let $u(e, \lambda, t)=u(t)$ be the control policy given by

$$u(t)=D_0^{-1}B_m u_m(t)+D_0^{-1}u_c(t),$$
$$u_c(t)=G_1\lambda(t)\psi\{G_2 e'(t)p\lambda(t)\} \tag{3.13}$$

where $G_1 \geq G_2 > 0$ are scalar gain constants, and $\psi\{\cdot\}$ is a bounded asymmetric vector valued function. Then, for a stable A_m, $V(e, t)$ becomes a Liapunov function yielding asymptotic stability or a bounded stability in the sense of Lagrange of the output error differential equation (3.10). The proof of this theorem may be found in Ref. [6]. The gust alleviation control law implementation is shown in Fig. 4.

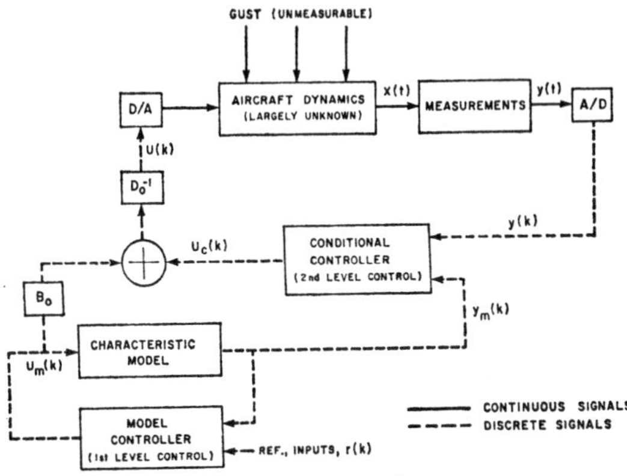

FIG. 4. Adaptive type digital computer control of a largely unknown process.

Simulation results

For this study simulation results were obtained on a hybrid computer. The lateral–dynamic motion of aircraft (#1) was simulated on the analog portion of this computer, while the characteristic model and the controller were simulated on the digital portion as shown in Fig. 5. The results of

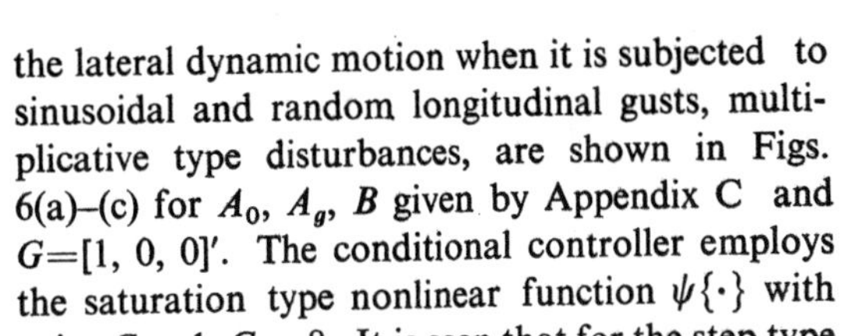

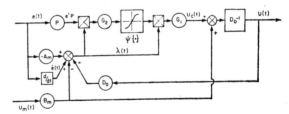

FIG. 5. Controller implementation.

the lateral dynamic motion when it is subjected to sinusoidal and random longitudinal gusts, multiplicative type disturbances, are shown in Figs. 6(a)–(c) for A_0, A_g, B given by Appendix C and $G=[1, 0, 0]'$. The conditional controller employs the saturation type nonlinear function $\psi\{\cdot\}$ with gains $G_1=1$, $G_2=8$. It is seen that for the step type command input, and with the aircraft operating in an unknown environment with known parameters, the desired response is followed with a very small

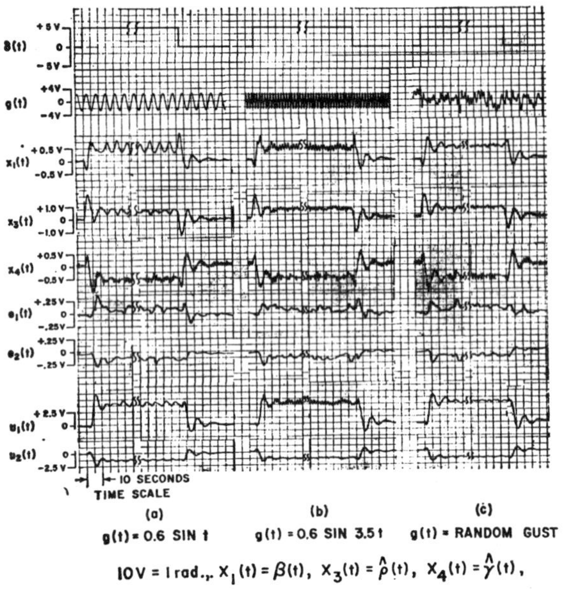

FIG. 6. Simulation studies of lateral dynamic motion of aircraft No. 1 subjected to longitudinal gust $g(t)$.

tracking error. Figure 7(a) shows the controlled behaviour of the dynamic motion, again with a step type command input for no gust disturbance, while Figure 7(b) shows the responses with

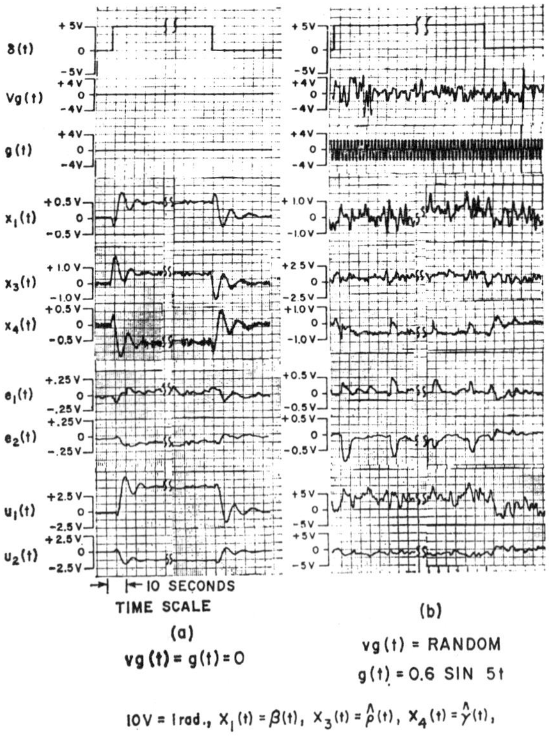

FIG. 7. Simulation studies of lateral dynamic motion of aircraft No. 1 subjected to longitudinal and lateral gust.

sinusoidal longitudinal gust and random lateral gusts. In the simulation the parameters of the aircraft were set arbitrarily, and it was found that the controller implementation is independent of these parameters or the gust characteristics.

4. CONCLUSIONS

In the first part of this paper the stability boundaries of two typical aircraft, subjected to a sinusoidal longitudinal gust, were derived using a successive approximation method. These stability boundaries were then verified using analog simulation. In the second part of the paper a gust alleviation control system using a two- level control structure was derived. The technique employs a Liapunov signal synthesis procedure and output variables which essentially bypasses the requirements for plant identification and state-variable measurements as required by modern control theory. This procedure also permits for relatively rapid design with inherent overall stability of the system, and the assurance of certain desired aircraft response characteristics of the aircraft. Further, it does not require a high-order dynamic controller as would be needed in the conventional model-reference adaptive techniques or in optimal control theory. The results presented in this paper are based upon some basic theoretical developments by the authors. However, it is believed that more research in the area may result in a more effective controller with a greater saving in the design efforts.

Acknowledgements—The authors would like to thank the reviewers for their interesting comments on this paper.

REFERENCES

[1] B. ETKIN: *Dynamics of Flight, Stability and Control.* Wiley, New York (1959).
[2] K. NAKAGAWA and Y. MUROTSU: Noninteracting control in automatic gust alleviation system for transport airplanes. *Jap. Assoc. Aut. Control Engrs* **9,** 18–25 (1965).
[3] B. ETKIN: *Dynamics of Atmospheric Flight,* Chapter 13. Wiley, New York (1972).
[4] J. H. BLAKELOCK: *Automatic Control of Aircraft and Missiles,* Chapter 3. Wiley, New York (1965).
[5] M. MASAK: On the lateral instability of aircraft due to parametric "excitation". *UTIAS* **TN-86** (1965).
[6] P. N. NIKIFORUK, M. M. GUPTA, K. KANAI and L. WAGNER: Synthesis of Two-level Controller for a Class of Linear Plants in an Unknown Environment. 3rd *IFAC Symposium on Sensitivity Adaptivity and Optimality,* pp. 431–439 Ischia, Italy (1973).
[7] K. ICHIKAWA and K. KANAI: Stability analysis of time-varying parameter system. *Trans. Jap. Soc. Mech. Engng (JSME)* **33,** 1063–1074 (1968).
[8] W. C. TRIPLETT: The Dynamic Response Characteristics of a 35° Swept-Wing Airplane as Determined from Flight Measurements. NASA TR-1950 (1955).
[9] P. N. NIKIFORUK, M. M. GUPTA and K. KANAI: Optimal control system synthesis with respect to desired response characteristics and available states as applied to VTOL aircraft. *Proc. JACC,* Stanford University, Session 9, pp. 16–18 (1972).

APPENDIX A

List of symbols

- s wing area
- b span
- m mass
- q_0 dynamic pressure, $q_0=\frac{1}{2}\rho U_0^2$
- I_x roll moment of inertia
- I_z yaw moment of inertia
- I_{xz} moment of inertia due to cross-coupling
- U_0 velocity of the centre of mass
- C_{ij} stability derivatives (i force or moment coefficient due to variable j) C_{lp}, C_{np}, C_{nr}, $C_{n\beta}$, $C_{y\phi}$, $C_{y\psi}$, $C_{y\beta}$, C_{lr}, $C_{l\beta}$, $C_{l\delta}$, $C_{l\delta r}$, $C_{y\delta a}$, $C_{y\delta r}$, $C_{n\delta a}$, $C_{n\delta r}$
- C_{l0} lift coefficient of the trimmed flight
- G_l, G_n, G_y input coefficient of additive gust
- ρ air density
- $\beta(t)$ angle of sideslip
- $\phi(t)$ roll Euler angle
- $\psi(t)$ yaw Euler angle
- $\hat{p}(t)$ non-dimensional rate of rotation about the x axis
- $\hat{r}$ non-dimensional rate of rotation about the z-axis
- $\delta r(t)$ rudder deflection
- $\delta a(t)$ aileron deflection
- $x(t)$ $\underline{\Delta}[\beta(t), \phi(t), \hat{p}(t), \hat{r}(t)]'$ state vector of lateral motion
- $z(t)$ $\underline{\Delta}Tx(t)$ phase canonical vector of lateral motion
- $u(t)$ $\underline{\Delta}[\delta a(t), \delta r(t)]'$ control vector for lateral motion
- A_0, A_g, B matrices of aerodynamic coefficients defined in (b-8)
- F_0, F_g phase canonical coefficient matrices defined in (b-17)
- $x_{d(t)}$ $\underline{\Delta}[\beta(t), \hat{r}(t)]$ phase canonical vector of Dutch roll motion
- $z_{d(t)}$ $\underline{\Delta}T_d x_d(t)$ phase canonical vector of Dutch roll motion
- $u_{d(t)}$ $\underline{\Delta}\delta r(t)$ control vector of Dutch roll motion
- A_{d0}, A_{dg}, B_d matrices of aerodynamic coefficients of Dutch roll motion defined in (b-11)
- F_{d0}, F_{dg} phase canonical coefficient matrices defined in (b-15)
- G input gust effectiveness matrix for lateral motion
- G_d input gust effectiveness matrix for Dutch roll motion
- T transformation matrix

APPENDIX B

Perturbation equation of lateral motion due to longitudinal gust disturbance

(i) *Lateral dynamic motion.* The linearized dynamic equation of aircraft motion subjected to the longitudinal gust $\alpha_g(t)\underline{\Delta}g(t)$ and lateral gust $v_g(t)$ may be written as [1, 4]

$$\begin{aligned}
&I_{x/sq_0b}\ddot{\phi}(t)-b_{/2U_0}C_{lp}\dot{\phi}(t)-I_{xz/sq_0b}\ddot{\psi}(t)\\
&\quad-b_{/2U_0}C_{lr}(g)\dot{\psi}(t)-C_{l\beta}(g)\beta(t)=C_{l\delta a}\delta_a(t)\\
&\qquad\qquad+C_{l\delta r}\delta_r(t)+G_l(v_g, t)\\
&-I_{xz/sq_0b}\ddot{\phi}(t)-b_{/2U_0}C_{np}(g)\dot{\phi}(t)+I_{z/sq_0b}\ddot{\psi}(t)\\
&\quad-b_{/2U_0}C_{nr}(g)\dot{\psi}(t)-C_{n\beta}(g)\beta(t)=C_{n\delta a}\delta_a(t)\\
&\qquad\qquad+C_{n\delta r}\delta_r(t)+G_n(v_g, t)\\
&-C_{y\phi}\phi(t)+\frac{mU_0}{sq_0}\dot{\psi}(t)-C_{y\psi}\psi(t)+\frac{mU_0}{sq_0}\dot{\beta}(t)\\
&\quad-C_{y\beta}\beta(t)=C_{y\delta a}\delta_a(t)+C_{y\delta r}\delta_r(t)\\
&\qquad\qquad+G_y(v_g, t). \quad \text{(b-1)}
\end{aligned}$$

By neglecting the small term of $C_{y\psi}$, the non-dimensional representation of (b-1) can be written as [3]

$$\begin{aligned}
&2\mu D\beta(t)=C_{y\phi}\phi(t)+C_{y\beta}\beta(t)-2\mu\hat{r}(t)+C_{y\delta a}\delta_a(t)\\
&\qquad+C_{y\delta r}\delta_r(t)+G_y(v_g, t)\\
&D\phi(t)=\hat{p}(t)\\
&i_F D\hat{p}(t)=\{i_c C_{lp}+i_E C_{np}(g)\}\hat{p}(t)\\
&\qquad+\{i_E C_{n\beta}(g)+i_c C_{l\beta}(g)\}\beta(t)\\
&\qquad+\{i_c C_{lr}(g)+i_E C_{nr}(g)\}\hat{r}(t)\\
&\qquad+(i_c C_{l\delta a}+i_E C_{n\delta a})\delta_a(t)\\
&\qquad+(i_c C_{l\delta a}+i_E C_{n\delta a})\delta_r(t)+i_c G_l(v_g, t)\\
&\qquad\qquad+i_E G_n(v_g, t)\\
&i_F D\hat{r}(t)=\{i_A C_{np}(g)+i_E C_{lp}\}\hat{p}(t)\\
&\qquad+\{i_A C_{n\beta}(g)+i_E C_{l\beta}(g)\}\beta(t)\\
&\qquad+\{i_A C_{nr}(g)+i_E C_{lr}(g)\}\hat{r}(t)\\
&\qquad+(i_E C_{l\delta a}+i_A C_{n\delta a})\delta_a(t)\\
&\qquad+(i_E C_{l\delta r}+i_A C_{n\delta r})\delta_r(t)+i_A G_n(v_g, t)\\
&\qquad\qquad+i_E G_l(v_g, t)
\end{aligned}$$

where

$$\begin{aligned}
&\mu=m/\rho sl,\ l=b/2,\ t^*=l/U_0,\ D=t^*\mathrm{d}/\mathrm{d}t,\\
&i_A=I_x/\rho sl^3,\ i_c=I_z/\rho sl^3,\ i_E=I_{xz}/\rho sl^3,\\
&\dot{\psi}=\hat{r},\ i_F=i_A i_c-i_E^2 \qquad \text{(b-2)}
\end{aligned}$$

Defining the state vector as

$$x(t)\underline{\Delta}[\beta(t), \phi(t), \hat{p}(t), \hat{r}(t)]'$$

and the control vector as

$$u(t)\underline{\Delta}[\delta a(t), \delta r(t)]',$$

a general form of the aircraft state evolution equation can equivalently be written as

$$\begin{aligned}
&\dot{x}(t)=A(g, t)x(t)+Bu(t)\\
&\qquad+G(v_g, t),\ x(t)\varepsilon R^4,\ u(t)\varepsilon R^2 \qquad \text{(b-3)}
\end{aligned}$$

where

$$A(g,t)=\begin{bmatrix} C_{y\beta}/2\mu & C_{y\phi}/2\mu & 0 & -1 \\ 0 & 0 & 1 & 0 \\ \dfrac{i_E C_{n\beta}(g)+i_c C_{l\beta}(g)}{i_F} & 0 & \dfrac{i_c C_{lp}+i_E C_{np}(g)}{i_F} & \dfrac{i_c C_{lr}(g)+i_E C_{nr}(g)}{i_F} \\ \dfrac{i_A C_{n\beta}(g)+i_E C_{l\beta}(g)}{i_F} & 0 & \dfrac{i_A C_{np}(g)+i_E C_{lp}}{i_F} & \dfrac{i_A C_{nr}(g)+i_E C_{lr}(g)}{i_F} \end{bmatrix}$$

$$B=\begin{bmatrix} C_{y\delta a}/2\mu & C_{y\delta r}/2\mu \\ 0 & 0 \\ \dfrac{i_c C_{l\delta a}+i_E C_{n\delta a}}{i_F} & \dfrac{i_c C_{l\delta r}+i_E C_{n\delta r}}{i_F} \\ \dfrac{i_E C_{l\delta a}+i_A C_{n\delta a}}{i_F} & \dfrac{i_E C_{l\delta r}+i_A C_{n\delta r}}{i_F} \end{bmatrix}$$

$$G(v_g, t)=[G_y, G_l, G_n]\, v_g(t)=Gv_g(t). \tag{b-4}$$

In (b-3), the matrix $A(g, t)$ embeds the multiplicative type of effect of the gust disturbance on the aircraft and $G(v_g, t)$ represents the additive type of effect of the gust.

In general, the longitudinal gust $g(t)$ perturbs the lift coefficient C_l and it has been shown [3] both theoretically and experimentally that many lateral stability derivatives may be represented as follows,

$$C_{l\beta}=a_\beta+b_\beta C_l(t),\ C_{lr}=a_l+b_l C_l(t),$$

$$C_{np}=a_p+b_p C_l(t),\ C_{n\beta}=a_n+b_n C_l^2(t)$$

$$C_{nr}=a_r+b_r C_l^2(t) \tag{b-5}$$

and

$$C_l(t)\simeq C_{l0}\{1+g(t)\},\ g(t)\ll 1 \tag{b-6}$$

where C_{l0} is the lift coefficient of the trimmed flight. Substituting (b-5) and (b-6) into (b-3), the state space equation of the aircraft becomes,

$$\dot{x}(t)=\{A_0+A_g g(t)\}x(t)+Bu(t)+Gv_g(t) \tag{b-7}$$

where, for small $g(t)$, $C_{l0}^2(t)\simeq C_{l0}^2\{1+2g(t)\}$, A_0 is the nominal matrix representing the aircraft dynamics, and A_g and G are respectively the equivalent effects of the multiplicative type and additive type gust disturbance. The elements of the matrix G define the forces and moments produced on the aircraft by the gust.

The nominal value of matrix A (i.e. for $g(t)=0$) is given by

$$A_0=\begin{bmatrix} C_{y\beta}/2\mu & C_{l0}/2\mu & 0 & -1 \\ 0 & 0 & 1 & 0 \\ \dfrac{i_E n_1+i_c\beta_1}{i_F} & 0 & \dfrac{i_c C_{lp}+i_E p_1}{i_F} & \dfrac{i_c r_1+i_E m_1}{i_F} \\ \dfrac{i_A n_1+i_E\beta_1}{i_F} & 0 & \dfrac{i_A p_1+i_E C_{lp}}{i_F} & \dfrac{i_A m_1+i_E r_1}{i_F} \end{bmatrix}$$

and A_g, the perturbation matrix due to gust, is

$$A_g=\begin{bmatrix} 0 & 0 & 0 & 0 \\ 0 & 0 & 0 & 0 \\ \dfrac{i_E n_2+i_c\beta_2}{i_F} & 0 & i_E p_2/i_F & \dfrac{i_c r_2+i_E m_2}{i_F} \\ \dfrac{i_A n_2+i_E\beta_2}{i_F} & 0 & i_A p_2/i_F & \dfrac{i_A m_2+i_E r_2}{i_F} \end{bmatrix} \quad \text{(b-8)}$$

where,

$\beta_1=a_\beta+b_\beta C_{l0}$, $\beta_2=b_\beta C_{l0}$, $r_1=a_l+b_l C_{l0}$, $r_2=b_l C_{l0}$,

$p_1=a_p+b_p C_{l0}$, $p_2=b_p C_{l0}$, $n_1=a_n+b_n C_{l0}^2$, $n_2=2b_n C_{l0}^2$

$m_1=a_r+b_r C_{l0}$, $m_2=2b_r C_{l0}^2$.

(ii) *Dutch roll motion.* Next, a state space model of the Dutch roll motion in the horizontal plane ("flat" yawing/side slipping motion) can be obtained approximately by putting $x_2(t)\underline{\Delta}\phi(t)=0$, $x_3(t)\underline{\Delta}\dot{\phi}(t)=0$ and dropping the rolling moment equation. By defining $x_d\underline{\Delta}[\beta(t),\ \dot{\psi}(t)]'\varepsilon R^2$ and $u_d(t)\underline{\Delta}\delta_r(t)\varepsilon R^1$, where the subscript d represents the motion in Dutch roll, one gets,

$$\dot{x}_d(t)=A_d(g,\ t)x_d(t)+B_d u_d(t)+G_d(v_g,\ t) \quad \text{(b-9)}$$

where

$$A_d(g,\ t)=A_{d0}+A_{dg}g(t),\ G_d(v_g,\ t)=G_d v_g(t) \quad \text{(b-10)}$$

$$A_{d0}=\begin{bmatrix} C_{y\beta/2\mu} & -1 \\ \dfrac{i_A n_1+i_E\beta_1}{i_F} & \dfrac{i_A m_1+i_E r_1}{i_F} \end{bmatrix}$$

$$A_{dg}=\begin{bmatrix} 0 & 0 \\ \dfrac{i_A n_2+i_E\beta_2}{i_F} & \dfrac{i_A m_2+i_E r_2}{i_F} \end{bmatrix}$$

$$B_d=\begin{bmatrix} C_{y\delta r/2\mu} \\ \dfrac{i_A C_{nr}+i_E C_{lr}}{i_F} \end{bmatrix}$$

$$G_d=\begin{bmatrix} G_y \\ G_n \end{bmatrix}. \quad \text{(b-11)}$$

(iii) *Phase canonical form representation.* In order to perform the stability analysis using the method of Appendix D, the phase canonical representation must be considered for the equation (b-7) and (b-9). For the case of (b-9), the following transformation is performed.

$$z_d(t)=T_d x_d(t),\ z_d(t)\underline{\Delta}[z_{d1}(t),\ \dot{z}_{d1}(t)=z_{d2}(t)]' \quad \text{(b-12)}$$

where, T_d is a nonsingular transformation matrix and $z_d(t)$ is a phase canonical vector.

Substituting (b-12) into (b-9) for $u_d(t)=G_d=0$ yield

$$\dot{z}_d=[F_{d0}+F_{dg}g(t)]z_d(t) \quad \text{(b-13)}$$

where,

$$F_{d0}=T_d A_{d0} T_d^{-1},\ F_{dg}=T_d A_{dg} T_d^{-1}. \quad \text{(b-14)}$$

After performing some calculations, the following canonical transformation is obtained.

$$F_{d0}=\begin{bmatrix} 0 & 1 \\ -a_7-a_5a_8 & a_5+a_8 \end{bmatrix},$$

$$F_{dg}=\begin{bmatrix} 0 & 0 \\ -b_6-a_5b_7 & b_7 \end{bmatrix},$$

$$T_d=T_{11}\begin{bmatrix} 1 & 0 \\ a_5 & -1 \end{bmatrix} \quad \text{(b-15)}$$

where T_{11} is an arbitrary non-zero element of the transformation matrix, T, and the elements a_5, – – – , b_7 are defined in (b-19).

Finally, since $\dot{z}_{d1}=z_{d2}$, for sinusoidal gust, $g(t)=\varepsilon\ \sin\nu t$, we have

$$\{D^2-(a_5+a_8)D+(a_7+a_5a_8)\}z_{d1}(t) =-\varepsilon\sin\nu t\{l_1D+l_0\}z_{d1}(t)$$

or

$$W_d^{-1}(D)z_{d1}(t)=-\varepsilon\ \sin\nu t\{l_1D+l_0\}z_{d1}(t) \quad \text{(b-16)}$$

where

$$W_d^{-1}(d)=D^2-(a_5+a_8)D+a_7+a_5a_8$$
$$l_0=(b_6+a_5b_7),\ l_1=-b_7.$$

For the case of (b-7), however, it is impossible to find analytically the transformation matrix of the form given in (b-12). However, one can approximately obtain the canonical transformation of (b-7) with the use of the assumption that the amplitude of the gust is small.

For $g(t)=\varepsilon \sin\nu t$, $\varepsilon \ll 1$, the phase canonical form of (b-7) is

$$\dot{z}(t)=\{F_0 \dagger F_g g(t)\}z(t) \qquad \text{(b-17)}$$

where F_0 and F_g are given by

$$F_0=\begin{bmatrix} 0 & 1 & 0 & 0 \\ 0 & 0 & 1 & 0 \\ 0 & 0 & 0 & 1 \\ -\alpha_0 & -\alpha_1 & -\alpha_2 & -\alpha_3 \end{bmatrix},$$

$$F_g=\begin{bmatrix} 0 & 0 & 0 & 0 \\ 0 & 0 & 0 & 0 \\ 0 & 0 & 0 & 0 \\ -\beta_0 & -\beta_1 & -\beta_2 & -\beta_3 \end{bmatrix}.$$

Since z is the phase vector, we write (b-17) equivalently for $g(t)=\varepsilon \sin\nu t$ as

$$W^{-1}(D)z_1(t)$$
$$=\varepsilon \sin\nu t[\beta_3D^3+\beta_2D^2+\beta_1D^1+\beta_o(t)]z_1(t) \qquad \text{(b-18)}$$

where $W^{-1}(D)=D^4+\alpha_3D^3+\alpha_2D^2+\alpha_1D+\alpha_o$ and the other coefficients (b-16) and (b-18) are

$$\alpha_o=a_2a_4a_8-a_3a_4a_7$$
$$\alpha_1=-\{a_5(a_1a_8-a_3a_6)+a_2a_4-a_2a_6+a_1a_7\}$$
$$\alpha_2=a_1a_8-a_3a_6+a_1a_5+a_5a_8+a_7$$
$$\alpha_3=-(a_1+a_5+a_8)$$
$$\beta_o=a_4(b_2a_8+a_2b_7-a_3a_6-a_7b_3)$$
$$\beta_1=-\{a_5(b_1a_8+a_1b_7-a_3b_5-b_3a_6)+a_4b_2 -a_6b_2-a_2b_5+b_1a_7+a_1b_6\}$$
$$\beta_2=b_1b_8+a_1a_7-a_3b_5-b_3a_6+b_1a_5+b_7a_5+b_6$$
$$\beta_3=-(b_1+b_7)$$

$$a_1=\frac{i_cC_{lp}+i_Ep_1}{i_F},\quad a_2=\frac{i_En_1+i_c\beta_1}{i_F},$$
$$a_3=\frac{i_cr_1+i_Em_1}{i_F},\quad a_4=C_{l0}/2\mu,\quad a_5=C_{y\beta}/2\mu,$$
$$a_6=\frac{i_Ap_1+i_EC_{lp}}{i_F},\quad a_7=\frac{i_An_1+i_E\beta_1}{i_F},$$
$$a_8=\frac{i_Am_1+i_Er_1}{i_F},$$
$$b_1=i_Ep_2/i_F,\quad b_2=\frac{i_En_2+i_c\beta_2}{i_F},$$
$$b_3=\frac{i_cr_2+i_Em_2}{i_F},\quad b_5=i_Ap_2/i_F,$$
$$b_6=\frac{i_An_2+i_E\beta_2}{i_F}. \qquad \text{(b-19)}$$

APPENDIX C

Numerical data

Quantity	Aircraft #1	Aircraft #2
$U_0 m/\sec$	76.25	148
I_x/sq_0b	0.0495	0.0079
I_z/sq_0b	−0.2845	0.025
I_{xz}/sq_0b	−0.0685	−0.0014
mU_0/sq_0	4.2	7.7
C_{l0}	0.5	0.514
C_{lp}	−0.1	−0.360
$C_{l\beta}$	$-0.488C_l$	$-0.199C_l$
C_{lr}	$0.24C_l$	$0.075+0.16C_l$
$C_{n\beta}$	$0{\cdot}8C^2$	$0{\cdot}14-0{\cdot}058C_l^2$
C_{np}	$-0.024C_l$	$0.064C_l$
C_{nr}	$-1{\cdot}2C_l^2$	$0{\cdot}68C_l^2$
$C_{y\beta}$	−0.16	−0.69
$C_{y\phi}$	C_{l0}	C_{l0}
$C_{y\psi}$	0	0

#1: Slender Supersonic Delta Transport Aircraft,
#2: Sub-sonic 35° Swept-Wing (Interceptor) Aircraft,

and the following data are considered for #1, where

$$C_{y\delta r}=0.04\quad C_{l\delta r}=0.005\quad C_{n\delta r}=-0.02$$
$$C_{l\delta a}=-0.015\quad C_{n\delta a}=0.02\quad C_{y\delta a}=0.$$

APPENDIX D

Stability analysis of lateral dynamic motion of aircraft subjected to sinusoidal gust disturbance

Reference [7] describes a method for the stability analysis of linear systems with sinusoidally varying parameters. The method is based upon the successive approximations technique and one may obtain the sufficiently accurate results by employing higher-order approximations in the stability analysis. If the system possesses sufficient low-pass characteristics, however, the first-order approximations will yield sufficiently accurate results. This reference considers many examples having different assignment conditions, however, in this appendix we will illustrate the method by a particular example of aircraft in its lateral dynamic motion. The motion in the phase-canonical form is described in (b-18), and since the aircraft possesses sufficient low pass characteristics, only first-order approximation will be employed.

The dynamic equation of lateral motion of the aircraft with zero input and subjected to gust disturbance on its parameters is described in (b-18) and is given by

$$W^{-1}(D)z_1(t) = \varepsilon \sin\nu t[\beta_3 D^3 + \beta_2 D^2 + \beta_1 D + \beta_o]z_1(t) \qquad \text{(d-1)}$$

where

$$W(D)^{-1} = D^4 + \alpha_3 D^3 + \alpha_2 D^2 + \alpha_1 D^1 + \alpha_o$$

The system is shown in Fig. 8.

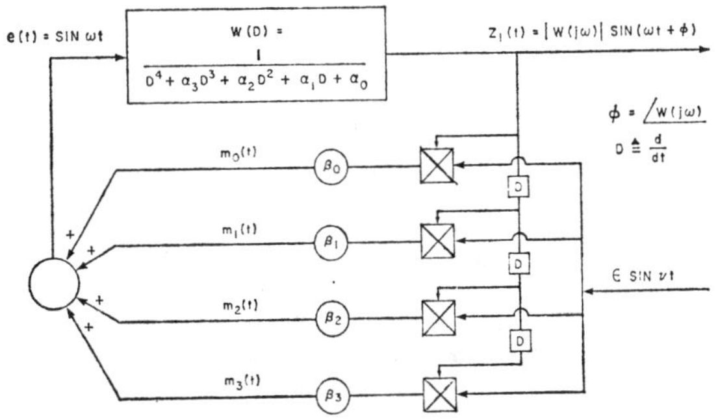

FIG. 8. Schematic diagram for the stability analysis of the aircraft in lateral motion subjected to sinusoidal gust, $g(t)=\varepsilon \sin\upsilon t$.

Now, assuming that the system is just on the stability boundary, one may consider an oscillatory signal around the loop with an amplitude which depends upon the initial conditions. As a first approximation, one may assume $e(t)$, the input to $W(D)$, to be a simple sinusoid given by

$$e(t) = \sin\omega t \qquad \text{(d-2)}$$

with $z_1(t) = |W(j\omega)| \sin(\omega t + \phi)$, $\phi = \underline{/W(j\omega)}$. Now since

$$e(t) = \sum_{i=0}^{3} \mathrm{m}_i(t), \mathrm{m}_i(t) = \varepsilon\beta\, |W(j\omega)|(\omega)^i \sin\left(\omega t + \phi + i\frac{\pi}{2}\right) \sin\nu t \qquad \text{(d-3)}$$

Comparing (d-2) and (d-3), one gets

$$\sin\omega t = \varepsilon \sum_{i=0}^{3} \beta_i |W(j\omega)|(\omega)^i \sin\left(\omega t + \phi + i\frac{\pi}{2}\right) \sin\nu t \qquad \text{(d-4)}$$

where the product $\sin(\omega t + \phi + i\pi/2)\sin\nu t$ on the right hand side of (d-4) produces two harmonics of frequencies $(\omega \pm \nu)$. Assuming that the aircraft possesses sufficient low-pass characteristics, one may safely assume that $\omega = \nu/2$ in (d-4). Thus (d-4) becomes

$$\sin\frac{\nu}{2}t = \frac{\varepsilon}{2}\sum_{i=0}^{3} \beta_i \left|W\left(j\frac{\nu}{2}\right)\right| \left(\frac{\nu}{2}\right)^i \cos\left(\nu t - \phi - i\frac{\pi}{2}\right). \qquad \text{(d-5)}$$

Equating the amplitudes on both sides of this equation, one gets

$$1 = \frac{\varepsilon^2}{4}\left[\left\{\beta_o - \beta_2\left(\frac{\nu}{2}\right)^2\right\}^2 + \left\{\beta_1\left(\frac{\nu}{2}\right) - \beta_3\left(\frac{\nu}{2}\right)^3\right\}^2\right] \left|W\left(j\frac{\nu}{2}\right)\right|^2. \qquad \text{(d-6)}$$

Equation (d-6) describes the stability boundary of the lateral motion of the aircraft in terms of the amplitude and frequency of the sinusoidal gust $g(t) = \varepsilon \sin\nu t$. The theoretical stability boundaries for two different types of aircraft have been computed, and they are given in Section 2.

Now, for the Dutch roll motion, defined in (b-16), the equation for stability boundary may be obtained similarly, and is given by

$$1 = \frac{\varepsilon^2}{4}\left[(b_6 + a_5 b_7)^2 + (b_7\nu/2)^2\right] |W_d(j^\nu/2)|^2. \qquad \text{(d-7)}$$

The corresponding simulation results are given in Section 2.

Gust-alleviation control systems for aircraft

D. McLean, Ph.D., F. Inst. M.C., C. Eng., M.I.E.E.

Indexing terms: *Aerospace control, Aircraft, Modelling, Model-reference adaptive control systems, Optimal control*

Abstract

The use of active control technology (a.c.t.) to alleviate the effects on both the dynamic and structural responses of an aircraft flying through atmospheric turbulence, while maintaining an acceptable degree of flight-handling qualities, is discussed.

Two control schemes, derived from optimal-control theory and the algebraic theory of model following, are presented, and their value as practical methods of achieving satisfactory ride comfort in a gusty environment is discussed. A possible weakness in current methods of mathematical modelling of flexible, or deformable, aircraft is indicated.

List of symbols

Scalar quantities

$a_{x_{cg}}, a_{y_{cg}}, a_{z_{cg}}$ = acceleration measured at the aircraft centre of gravity (c.g.) in the directions OX, OY, and OZ, respectively, m/s^2
$\bar{c}$ = chord of aircraft, m
g = acceleration due to gravity, m/s^2
m = aircraft mass, kg
p = change of roll rate, rad/s
q = change of pitch rate, rad/s
r = change of yaw rate, rad/s
u = change of forward speed, m/s
v = change of sideslip velocity, m/s
w = change of vertical velocity, m/s
x_{ac_i} = distance forward of aircraft c.g., m
C_D = coefficient of drag
C_{L_α} = coefficient of lift due to change in angle of attack
C_R = comfort rating
I_{rd} = ride-discomfort index
I_y = moment of inertia about axis OY, kgm^2
J = performance index
L_i, M_i, N_i, X_i, Y_i, Z_i = dimensional stability derivatives (subscript i is replaced by corresponding motion variable)
S = surface area of main lifting surface, m^2
U_0 = steady forward speed of aircraft, m/s
W = weight of aircraft, N
α = change in angle of attack, rad
β = change is sideslip angle, rad
γ_i = displacement of ith antisymmetrical bending mode, m
δ_i = control surface deflection, rad
θ = change in pitch altitude, rad
λ_i = displacement of ith symmetrical bending mode, m
ρ = air density, kg/m^2
ϕ = change in bank angle, rad
ψ = change in yaw angle, rad
$\eta_{i x_i}$ = bending-moment coefficient
$\mu_{i x_i}$ = bending-moment coefficient
$\Phi_{x_i, i}, \Phi_{x_i, i}$ = bending-displacement coefficients

Vector quantities

u, u_1 = longitudinal, lateral control vectors
w, v = longitudinal, lateral gust vectors
x, x_1 = longitudinal, lateral state vectors
y, y_1 = longitudinal, lateral output vectors
y_m = model vector

Matrices

$A, B, C, D, E, F, G, H, L, M, Q, R, S$ and T (dimensions are quoted in the text at appropriate points)

Subscripts

A = aileron
C_{hor} = horizontal canard
C_V = vertical canard
E = elevator
R = rudder
1, 5, 7, 8, 12 = symmetrical bending modes
1, 2, 3, 9, 10 = antisymmetrical bending modes
4·4, 21·9, 42 = body stations at which accelerations measured
10·8, 17·7 = body stations at which pitch and yaw rates were measured, respectively

1 Introduction

For a number of situations in aircraft operations there exists a problem of providing a satisfactory method of designing automatic flight-control systems (a.f.c.s.) for the specific purpose of alleviating the undesirable effects on both the structural and rigid-body motion responses of an aircraft when it encounters atmospheric turbulence. Of all the possible methods of solving these problems, those which use active control technology (a.c.t.), in conjunction with modern control theory, are considered[1-3] to be the most likely to provide acceptable solutions.

On an aircraft, a.c.t. means the use of an a.f.c.s. to drive simultaneously many control surfaces and auxiliary direct generators of force, or moment, to improve dynamically both the flight characteristics and the structural behaviour of that aircraft. Thus, it is the purpose of a.c.t. to provide, in conjunction with modern control theory and electronic technology, the potential to increase, at least cost, both the performance and the operational flexibility of an aircraft. Current design and mission requirements for military and commercial-transport aircraft are such that the resulting configurations of these vehicles have involved the use of thin lifting surfaces, long and slender fuselages, low mass-fraction structures, high-stress design levels and low dynamic load factors. In turn, these factors have resulted in aircraft that are structurally light, and, consequently, flexible. Such flexible aircraft can develop both large body displacements and accelerations, owing to structural deflection, in addition to the components due to the rigid-body motion of the aircraft. These structural deflections may

Paper 8118 C, first received 8th September 1977 and in revised form 9th February 1978

Dr. McLean is with the Department of Transport Technology, University of Technology, Loughborough, Leicestershire LE11 3TU, England

Reprinted with permission from *Proc. Inst. Elec. Eng.*, vol. 125, pp. 675–685, July 1978.
Copyright © 1978; The Institution of Electrical Engineers.

occur owing to either the aircraft flight manoeuvres, commanded by the pilot, or from the passage of the aircraft through a turbulent atmosphere. Aircraft motion of this kind may result in a reduction of the structural life of the airframe because of the high levels of stress and dynamic loads which result. By using controlled deflections of tail and wing control surfaces, it is possible to minimise the amplitude and/or the number of transient bending cycles to which the structure may be subjected in flight, and, in so doing, a measure of fatigue reduction may be effected. A ride-control system is derived to improve the ride comfort for the crew, or passengers, by reducing those objectionable vibrations due to the aircraft motion by means of controlled action of the appropriate surfaces and generators. For gust alleviation the system is designed specifically to reduce those dynamic loads which result from penetrating turbulence. In achieving gust alleviation some degree of ride-control enhancement and fatigue reduction may be achieved simultaneously. Thus, by using a.c.t., the following desirable features are expected to result:

(*a*) loads due to airframe flexibility will be reduced
(*b*) levels of acceleration at particular aircraft stations will be reduced
(*c*) the flying qualities of the aircraft will be improved.

If these features are achieved, then the following benefits may be expected:

(i) improved flight comfort
(ii) improved crew effectiveness
(iii) improved weapons delivery capability
(iv) increased fatigue life
(v) reduced gross weight
(vi) reduced fuel expenditure.

This paper deals particularly with the design of a gust-alleviation control system by using optimal control theory and model following so that the ride comfort, crew effectiveness and weapons delivery capability are all enhanced.

2 Gust alleviation

The air in which an aircraft flies is constantly in motion, and, as a result, those aerodynamic forces and moments acting on the aircraft fluctuate continuously about some equiibrium values. It is these variations that result in continuous changes in the aircraft's attitude so that the resultant accelerations are experienced by the crew, and passengers, if any, as unpleasant effects. To reduce the unpleasantness, the acceleration has to be reduced by cancelling the gust effects by other forces.

The general principle of gust alleviation is that particular sensors provide motion feedback signals to some controller to cause corresponding deflections of the control surfaces to create the aerodynamic forces and moments needed to cancel the accelerations induced when the aircraft penetrates the gust. A wide variety of methods to achieve this aim has been proposed almost since the inception of manned flight. The first reference to a gust alleviation system was in the USA patent awarded in 1914 to Sprater[4] for a 'stabilising device to counteract the disturbance and prevent it from having an injurious effect on the stability of the machine'. In the following year, the very first NACA report, by Hunsaker and Wilson,[5] reported work carried out on a study of the behaviour of aircraft in turbulence and how it could be ameliorated. Although several isolated and unrelated reports appear subsequently in the literature, it was not until 1937, with the appearance of the foundation papers by Von Karman[6] and Taylor,[7] which established suitable bases for the mathematical representation of turbulence, that a renewed interest in the problem was evident. A proposal for a gust-alleviation system was made by Hirsch, in 1938: his scheme was eventually flight tested in 1954 and was reported in 1957.[8] In 1949, the Bristol Brabazon was equipped with a gust-alleviation system so that the strength of the wing structure of the prototype aircraft could be reduced by 20% from the figure required, without the gust-alleviation system, to meet the airworthiness requirements. The Brabazon system, which was never tested in flight before the project was abandoned,[9] was an open-loop system that used symmetrical deflection of the ailerons in response to gust signals detected by a vane mounted on the aircraft nose. In the USA, gust-alleviation systems were tested in flight in 1950 by the Douglas Corporation using a DC-3 aircraft specially fitted with auxiliary flaps,[10] and in 1952 by NACA[11] using a specially modified C-54. The Royal Aircraft Establishment carried out some trials on a modified Avro Lancaster in 1953.[12] All these programmes, except that involving the Brabazon, were concerned with alleviating the effects of gusts on rigid-body motion only: in each case, however, the results were unsatisfactory. During the Lancaster trials it was found that the handling qualities of the aircraft were impaired because of a marked reduction of stability that arose from the very large adverse pitching moment that was created by the symmetrical aileron deflection required by the gust-alleviation system. The American systems were also unsatisfactory because, like the Brabazon system, they depended on a gust vane to detect entry to the turbulence field. Such vanes sense either changes of pressure or changes in direction of the relative wind. It was not, however, sufficiently well appreciated that a gust field has significant components normal to the plane of symmetry of the aircraft, nor that some secondary aerodynamic effects are also significant, Such secondary effects include downwash on the tailplane and the establishment of a pure time delay in the development of aerodynamic lift, because the wing enters the gust field first before the tail. Every vane system tried to provide, in effect, control correction before the gust acted on the lifting surfaces. In the patent application of Attwood, Cannon, Johnson and Andrew,[13] in 1955, it was proposed to sense the linear and angular accelerations of the aircraft, and to use auxiliary control surfaces to produce the control forces and moments needed to minimise such accelerations. The proposed system took into account the possibility of the deformation of the fuselage and wing by using as feedback signals the blended outputs from a pair of accelerometers and a pair of rate gyroscopes. All four sensors were so located that any unwanted signals due to the bending motion would be effectively cancelled. Nevertheless, the purpose of this gust-alleviation system was strictly to improve the ride comfort for the passenger. The use of gust alleviation to reduce structural loading effects due to gusts was first tried in the USAF programme involving the prototype bomber, the XB-70.[14,15]

In 1964 there occurred a dramatic event that intensified interest in gust-alleviation systems. While on a low-level mission over the western USA, a USAF bomber, a B-52H, encountered severe turbulence of estimated peak velocity 35 m/s. The intensity of the gust saturated the yaw channel of the standard stability-augmentation system (s.a.s.) of the aircraft so that the acceleration of the, then unaugmented, rigid-body dynamics was such that almost 80% of the vertical fin broke off in flight. As a result of this incident an extensive a.f.c.s. development programme, known as the load alleviation and mode suppression (l.a.m.s.) programme, was commenced in 1965 by the USAF and its contractors. The first results were reported in 1969[16] and were sufficiently good for the programme to be extended so that the gust-alleviation function could be incorporated into the ride-control system to improve the ride comfort for the crews of B-52s on low-level missions. This work stimulated the worldwide interest in the topic which is evident from the number of studies[18–24] reported to date.

3 Gust-alleviation control

The amplitude of the responses caused by gust-induced structural flexibility can be reduced if either the amount of energy transferred from the gust disturbance to the structural bending modes is reduced, or any energy absorbed from the gust disturbances by the structural modes is dissipated rapidly. Both methods should be employed simultaneously, if possible. It needs some countering moment, or force, to be generated by deflection of some surface to reduce the amount of energy transferred, but the method requires an accurate knowledge of the stability derivatives of the aircraft. These alter greatly, however, as changes occur in such quantities and properties as the aircraft mass, the distribution of that mass, the dynamic pressure, the aircraft speed and the nature of any atmospheric turbulence. Consequently, the dynamics of the aircraft and its environment are known too imperfectly to allow perfect cancellation of the forces and moments created by a gust. Once gust energy has been transferred, however, its dissipation in the structure can be effectively controlled by dynamically augmenting the stability of the elastic modes, although it is often difficult to achieve a sufficient increase in structural damping by such methods if the frequencies of two structural modes are close, because they are then usually closely coupled.

The two methods used in this paper counter the gust effects by using auxiliary controls to increase the damping of the structural modes and to reduce the accelerations to which those modes contribute. The first method is based on posing the problem as an example of a linear quadratic problem;[25] the second employs the algebraic technique that results from solving the implicit model-following problem.[26] The model equations and the weighting matrices are chosen so that they relate the problem to the achievement of specified measures of ride comfort. Obviously, gust-alleviation systems are dependent on how the structural flexibility of the aircraft affects its dynamics and also on the nature of the gust representation. These topics are dealt with first, before ride discomfort is discussed, and the two systems are presented.

3 Mathematical representation of the dynamics of a flexibile aircraft

3.1 Introduction

Using stability axes the linearised differential equations, which describe the small perturbation motion of a rigid aircraft about its equilibrium, straight and level, flight position, may be represented as two independent sets, provided that there is no crosscoupling between logitudinal and lateral motion. Thus:

longitudinal motion

$$\left.\begin{aligned} \dot{u} &= X_u u + X_w w - g\theta + X_{\delta_E}\delta_E + \sum_{i=1}^{3} X_{w_i}W_i \\ \dot{w} &= Z_u u + Z_w w + U_0 q + Z_{\delta_E}\delta_E + \sum_{i=1}^{3} Z_{w_i}w_i \\ \dot{q} &= M_u u + M_w w + M_{\dot{w}}\dot{w} + M_q q + M_{\delta_E}\delta_E + \sum_{i=1}^{3} M_{w_i}w_i \end{aligned}\right\}$$

$$\dot{\theta} = q \qquad (1)$$

lateral motion

$$\left.\begin{aligned} \dot{v} &= Y_v v + g\phi - r + Y_{\delta_A}\delta_A + Y_{\delta_R}\delta_R + \sum_{i=1}^{3} Y_{v_i}v_i \\ \dot{p} &= L'_v v + L'_p p + L'_p r + L'_{\delta_A}\delta_A + L'_{\delta_R}\delta_R + \sum_{i=1}^{3} L'_{v_i}v_i \\ \dot{r} &= N'_v v + N'_p p + N'_r r + N'_{\delta_A}\delta_A + N'_{\delta_R}\delta_R + \sum_{i=1}^{3} N'_{v_i}v_i \end{aligned}\right\}$$

$$\dot{\phi} = p$$

$$\dot{\psi} = r \qquad (2)$$

The meaning of the symbols and their corresponding units are given in the list of symbols.

If the aircraft problem does not involve a consideration of the long-period variation of forward speed, known as the phugoid motion, it is usual to employ, as a description of longitudinal motion, the two degrees of freedom representation:

$$\left.\begin{aligned} \dot{w} &= Z_w w + U_0 q + Z_{\delta_E}\delta_E + \sum_{i=1}^{3} Z_{w_i}w_i \\ \dot{q} &= (M_w + M_{\dot{w}}Z_w)w + (M_q + U_0 M_{\dot{w}})q + (M_{\delta_E} + M_{\dot{w}}Z_{\delta_E}) \\ &\quad \delta_E + \sum_{i=1}^{3} (M_{w_i} + M_{\dot{w}_i}Z_{w_i})w_i \end{aligned}\right\} \qquad (3)$$

Sometimes this 'short period' approximation is expressed in an alternative form by using, as a variable, the angle of attack α, instead of vertical velocity w, to which it is approximately related by

$$\alpha = w/U_0 \qquad (4)$$

When the aeroelastic effects associated with a flexible aircraft are period variation of forward speed, known as the phugoid motion, it the rigid-body equations by adding a set of generalised co-ordinates, associated with the normal modes, which have been calculated assuming that the structural behaviour is linear and that any structural displacement is small compared to the overall dimensions of the aircraft; e.g. if the ith bending mode is considered to be damped it may be represented as

$$A_l\ddot{q}_l + B_l\dot{q}_l + C_l q_l = Q_l \qquad (5)$$

where Q_l is a generalised force, A_l, B_l and C_l are coefficients of the lth generalised co-ordinate q_l and its associated rates. Set

$$\left.\begin{aligned} x_1 &= q_l \\ \text{and} \quad x_2 &= \dot{q}_l \end{aligned}\right\} \qquad (6)$$

then

$$\left.\begin{aligned} \dot{x}_1 &= x_2 \\ \dot{x}_2 &= \frac{-B_l}{A_l}x_2 - \frac{C_l}{A_l}x_1 + \frac{1}{A_l}Q_l \end{aligned}\right\} \qquad (7)$$

Thus, it is possible to represent the lth bending mode by the two 1st-order differential equations of eqn. 7. In this way it is possible to augment the rigid-body dynamic equations with pairs of 1st-order differential equations, for each bending mode being considered. The resultant state vector will be augmented to a degree that depends on the number of bending modes considered. It is convenient, therefore, to represent the dynamics of a flexible aircraft in terms of the state variable representation:

$$\dot{x} = Ax + Bu + Ew \qquad (8)$$

and

$$y = Cx + Du \qquad (9)$$

where x is the n-dimensional state vector, u is the n-dimensional control vector, y is the p-dimensional output vector and w is the k-dimensional gust vector. A, B, C, D and E are $n \times n$, $n \times m$, $p \times n$, $p \times m$, and $n \times k$ matrices, respectively, with elements in the real field, $\mathscr{R}$.

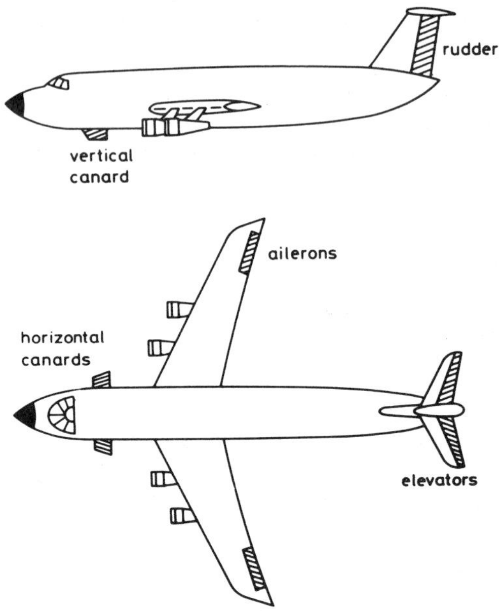

Fig. 1
Aircraft control surfaces

3.2 Mathematical model of the aircraft

3.2.1 State equations

The differential equations that represent the dynamics of the large aircraft used in this study are given in Appendix 8.1. It was assumed, in the longitudinal equations, that the rigid-body motion of the aircraft was adequately represented by the short-period approximation. Included in each independent set of equations were the dynamics associated with five structural bending modes only: in the longitudinal set these were the modes 1, 5, 7, 8 and 12, and in the lateral set they were 1, 2, 3, 9 and 10. It is conventional in the theory of structures to number modes in an ascending order as the frequency associated with the mode increases. Mode 1, therefore, is the mode with the lowest bending frequency. The terms associated with the effects of gust inputs acting at three different body stations were also included in each set of equations; hence, it appears that three different gusts w_1, w_2 and w_3 are acting, for example, in the longitudinal equations. It is merely a notational convenience, since w_2 and w_3 are related to w_1 and derived from it. The representation has been used to avoid dealing with the gust acting at the second body station, say, as a delayed gust input.

The control inputs employed for longitudinal motion were the deflections of elevator and horizontal canard; for lateral motion the three control inputs were the deflections of aileron, rudder and vertical canard. A canard is a small, independent control surface mounted on the fuselage forward of the wing (see Fig. 1).

If the equations in Appendix 8 are represented in state-variable form, they can be written as eqns. 8 and 9, where the state vector for longitudinal motion is defined as

$$x' = [q \;\; \lambda_1 \;\; \dot{\lambda}_1 \;\; \lambda_5 \;\; \dot{\lambda}_5 \;\; \lambda_7 \;\; \dot{\lambda}_7 \;\; \lambda_8 \;\; \dot{\lambda}_8 \;\; \lambda_{12} \;\; \dot{\lambda}_{12}] \qquad (10)$$

and the control vector as

$$u' = [\delta_E \quad \delta_{C_{hor}}] \qquad (11)$$

The gust vector is defined as

$$w' = [w_1 \quad w_2 \quad w_3] \qquad (12)$$

For lateral motion, the state and output equations are taken to be

$$\left.\begin{aligned} \dot{x}_1 &= Fx_1 + Gu_1 + Hv \\ y_1 &= Mx_1 + Lu_1 \end{aligned}\right\} \tag{13}$$

where the state vector is

$$x_1' = [v \; p \; r \; \phi \; \psi \; \gamma_1 \; \dot{\gamma}_1 \; \gamma_2 \; \dot{\gamma}_2 \; \gamma_3 \; \dot{\gamma}_3 \; \gamma_9 \; \dot{\gamma}_9 \; \gamma_{10} \; \dot{\gamma}_{10}]$$

The lateral control vector becomes

$$u_1' = [\delta_A \quad \delta_R \quad \delta_{C_V}] \tag{14}$$

and the side gust vector v_1 is defined as

$$v' = [v_1 \quad v_2 \quad v_3] \tag{15}$$

The matrices in the equations have the following dimensions:

A is of order 12×12
B is of order 12×2
E is of order 12×3
F is of order 15×15
G is of order 15×3
H is of order 15×3

The output matrices C, D, M and L are dealt with separately in Section 3.2.2. The values of matrices used in the study are given in Appendix 8.2, and the meaning of the variables used in the vectors is given in the list of symbols. The form of the state coefficient matrices A and F can be represented as

rigid-body terms	aeroelastic coupling terms
rigid-body coupling terms	structural-flexibility terms

from which it is evident that, in a deformable aircraft, there is a considerable degree of coupling between the rigid body and bending motions.

3.2.2 Output equations

If it is required to ensure correct measurement of the variables of the rigid-body motion, it is only necessary to position the appropriate motion sensors at the centre of gravity (c.g.) of the aircraft. However, in practice it is impossible to locate all the required sensors precisely at the c.g.; some correction is needed. A further adjustment is required when bending motion is also present. To illustrate this, consider measuring acceleration in a rigid aircraft with an accelerometer that can be located only at some distance x_A from the c.g. (x_A is taken to be positive when the accelerometer is located forward of the aircraft c.g.) The true normal acceleration a_{z_A} measured by the device is then

$$a_{z_A} = a_{z_{cg}} - x_A \dot{q} \tag{16}$$

where $\dot{q}$ is the pitching acceleration. The normal acceleration measured at the c.g. of the aircraft $a_{z_{cg}}$ is given by

$$a_{z_{cg}} = \dot{w} - U_0 q \tag{17}$$

However, for small angles and the axis system being used

$$\alpha = \frac{w}{U_0} \tag{18}$$

Thus

$$a_{z_A} = U_0(\dot{\alpha} - q) - x_A \dot{q} \tag{19}$$

If the measured normal acceleration is taken as the output y, it is easy to show, from eqns. 3 and 19, that

$$y \triangleq a_{z_A} = Cx + Du \tag{20}$$

where

$$C = [\{U_0 Z_\alpha - x_A \{(M_\alpha + M_{\dot{\alpha}} Z_\alpha)\}(M_q - M_{\dot{\alpha}} - 2U_0)] \tag{21}$$

and

$$D = [\{M_E + (U_0 + M_{\dot{\alpha}}) Z_{\delta_E}\}\{(U_0 + M_{\dot{\alpha}}) Z_{\delta_c} + M_{\delta_c}\}] \tag{22}$$

When bending effects are included in the aircraft dynamics, the accelerations due to the structural motion have to be added, so that the true normal acceleration becomes

$$a_{z_A} = U_0(\dot{\alpha} - q) - x_A \dot{q} + \Phi_{A,1}\ddot{\lambda}_1 + \Phi_{A,5}\ddot{\lambda}_5 + \Phi_{A,7}\ddot{\lambda}_7 + \Phi_{A,8}\ddot{\lambda}_8 + \Phi_{A,12}\ddot{\lambda}_{12} \tag{23}$$

Consequently, the matrices C and D must be altered to account for the augmenting of the state vector by the variables associated with the bending modes. Similarly, it can be shown that the lateral acceleration at some body station A, on the rigid aircraft, is given by

$$a_{y_A} = a_{y_{cg}} + x_A \dot{r} = (U_0 \dot{\beta} - g\phi + U_0 r) + x_A \dot{r} \tag{24}$$

For small angle theory

$$v \simeq U_0 \beta \tag{25}$$

Therefore

$$a_{y_A} = \dot{v} - g\phi + U_0 r + x_A \dot{r} \tag{26}$$

When lateral bending effects are included, the lateral acceleration becomes

$$y_1 = a_{y_A} = \dot{v} - g\phi + U_0 r + x_A \dot{r} + \Phi_{A,1}\ddot{\gamma}_1 + \Phi_{A,2}\ddot{\gamma}_2 + \Phi_{A,3}\ddot{\gamma}_3 + \Phi_{A,9}\ddot{\gamma}_9 + \Phi_{A,10}\ddot{\gamma}_{10} \tag{27}$$

The associated matrices M and L are given, with C and D, in Appendix 8.2. Similar spurious signals due to bending motion also affect the gyroscopes used as motion sensors. If a vertical gyroscope is used to measure the local inclination of either the fuselage or the wing, it means that, for longitudinal motion:

$$\theta_A = \theta + \Sigma_j \Phi_{A,j} \lambda_j \tag{28}$$

and, for lateral motion:

$$\Psi_A = \Psi + \sum_k \Phi_{A,k} \gamma_k \tag{29}$$

Rate gyroscopes, located at the same point A, will measure q_A and r_A, respectively, where

$$q_A = q + \sum_j \Phi_{A,j} \dot{\lambda}_j \tag{30}$$

$$r_A = r + \sum_k \Phi_{A,k} \dot{\gamma}_k \tag{31}$$

3.2.3 Gust and lift growth effects

It was intimated earlier in the paper that the gust inputs were treated as three separate gust inputs to avoid including pure time-delay effects in the equations of motion. Such time delays occur because of the nonstationary aerodynamic effects, e.g. lift at the horizontal tail is not generated instantaneously. Such effects are accounted for by the Wagner and Küssner functions. The Wagner function $w(t)$ defines the variation of the lift of an aerofoil with time for any unit-step change of angle of attack α. The Küssner function* $K(t)$ defines the variation of the lift of an aerofoil with time for any unit-step gust input. These functions are usually approximated; for this study the approximations used were:

$$w(t) = 0{\cdot}5 + 0{\cdot}165(1 - e^{-at}) + 0{\cdot}335(1 - e^{-bt}) \tag{32}$$

and

$$K(t) = 1{\cdot}0 - 0{\cdot}236\, e^{-0{\cdot}058\sigma} - 0{\cdot}513\, e^{-0{\cdot}364\sigma} - 0{\cdot}171\, e^{-2{\cdot}426\sigma} \tag{33}$$

where

$$a = 0{\cdot}0455 \times \frac{2U_0}{c} \tag{34}$$

$$b = 0{\cdot}3 \times \frac{2U_0}{\bar{c}} \tag{35}$$

and

$$\sigma = \frac{U_0}{2\bar{c}} t \tag{36}$$

U_0 has its usual meaning and $\bar{c}$ is the chord of the aircraft. Eqn. 33 is a common approximation known as the 'Jones' lift growth function.[26]

Figure 2 shows how the Wagner function is incorporated in the dynamics. For the aircraft, at the flight condition used in this study, the Jones' function is given by

$$K(t) = 1 - 0{\cdot}236 e^{-0{\cdot}8t} - 0{\cdot}513 e^{-10t} - 0{\cdot}171 e^{-33{\cdot}3t} \tag{37}$$

The associated Laplace transform of eqn. 37 is

$$K(s) = \frac{1}{s} - \frac{0{\cdot}236}{(s + 0{\cdot}8)} - \frac{0{\cdot}513}{(s + 10)} - \frac{0{\cdot}171}{(s + 33{\cdot}3)} \tag{38}$$

*Traditionally, the Küssner function is represented in aerodynamic literature by $\psi(t)$. To avoid confusion with the heading angle ψ, which is one of the state variables of the lateral motion, $K(t)$ has been used here.

This represents a transformed output obtained from applying the transform of unit-step gust $1/s$ to some transfer function $J(s)$, given by

$$J(s) = \frac{0{\cdot}189}{(s+0{\cdot}8)} + \frac{5{\cdot}139}{(s+10)} + \frac{5{\cdot}665}{(s+33{\cdot}3)} \tag{39}$$

The gust signal, say $W_g(s)$, is applied to this transfer function $J(s)$, and its output is the first gust input w_1 to the aircraft. The second gust w_2 is delayed by time U_0/x_1, where x_1 is the distance from the first body station, which encounters the gust first. The third input w_3 is delayed by time U_0/x_2. Rather than include such pure time delays in the equations of motion, a very simple approximation was adopted. The scheme is illustrated in the block diagram presented in Fig. 3. Although the Wagner function was not taken into explicit account in the equations of motion, it was incorporated in the digital simulation of the aircraft dynamics.

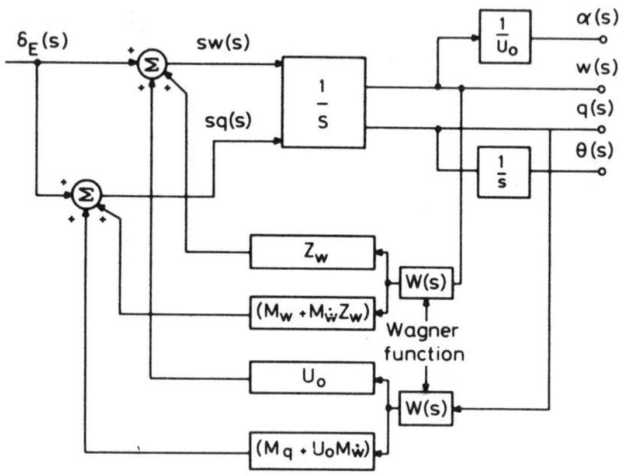

Fig. 2
Incorporation of Wagner lift function in aircraft dynamics

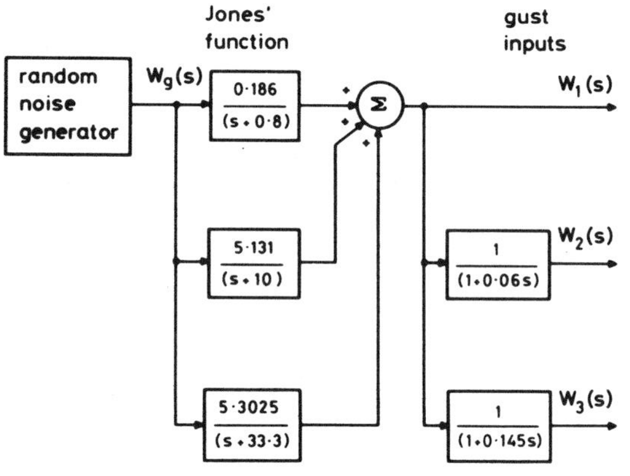

Fig. 3
Gust signal generation

4 Gust-alleviation control systems

4.1 Introduction

To devise a suitable gust alleviation control system by applying some appropriate theories of optimal control it is necessary that some quantitative measure of performance is available which reflects in its value the effectiveness of the control action on the ride quality of the aircraft. The final system must provide a suitable degree of gust alleviation while maintaining the aircraft handling qualities with least control action. A recently proposed index[23] is considered and it is shown that it can be expressed in a suitable form for use in a quadratic performance index which may then be minimised by using linear optimal control theory. The approach is related directly to meaningful flight characteristics and corresponds closely to the methods applied intuitively by investigators using conventional control theory.

4.2 Ride-quality criteria

Well defined criteria of ride comfort do not yet exist, although a great deal of research[23,27,28] is being pursued currently in the USA and the UK.[29] Two likely measures are the ride discomfort index (r.d.i.) and ride comfort rating.

4.2.1 Ride discomfort index

A r.d.i. for longitudinal motion has been proposed in the US military specification document MIL-F-9490D. This index is said to be proportional to the ratio of wing-lift slope to wing loading; i.e.

$$I_{rd} = \frac{kC_{L\alpha}}{W/S} = \frac{kSC_{L\alpha}}{W} \tag{40}$$

For longitudinal rigid-body motion, the dimensional stability derivative Z_w is given by

$$Z_w = \frac{-\rho S U_0}{2m}(C_{L_\alpha} + C_D) \tag{41}$$

For flight through turbulence it may be assumed that $C_{L_\alpha} \gg C_D$, thus

$$Z_w \simeq \frac{-\rho S U_0}{2m} C_{L_\alpha} \simeq \frac{-\rho U_0}{2k} g I_{rd} \tag{42}$$

Using the short-period approximation for the rigid-body dynamics:

$$\dot{w} = Z_w w + U_0 q + \sum_{j=1}^{p} Z_{\delta_j}\delta_j \tag{43}$$

where

$$j = 1, 2, \ldots p$$

The normal acceleration of the rigid-body aircraft is

$$a_z = \dot{w} - U_0 q$$

$$= Z_w w + \sum_{j=1}^{p} Z_{\delta_j}\delta_j$$

$$= \frac{-\rho U_0}{2k} g I_{rd} w + \sum_{j=1}^{p} Z_{\delta_j}\delta_j \tag{44}$$

Thus, if the normal acceleration and the control deflections are minimised by the optimal control system, I_{rd} is a minimum also. However, I_{rd} also influences the stability, and hence the flying qualities, of the aircraft. The static margin of an aircraft is defined as

$$\frac{C_{M_\alpha}}{C_{L_\alpha}} = \frac{-x_{ac}}{\bar{c}} \tag{45}$$

But

$$I_{rd} = -\frac{kS\bar{c}}{W}\frac{C_{M_\alpha}}{x_{ac}} = \frac{kS}{W}\frac{C_{M_\alpha}}{(\text{static margin})} \tag{46}$$

However

$$C_{M_\alpha} = \frac{2Iy}{\rho \bar{c} S U_0} M_w \tag{47}$$

Thus

$$I_{rd} = \frac{2kIy}{\rho U_0 W}\frac{M_w}{x_{ac}} \tag{48}$$

From eqn. 42, then

$$I_{rd} = -\frac{2k}{\rho U_0 g} Z_w \tag{49}$$

Thus, if the r.d.i. is minimised by a gust-alleviation control system, the static margin, and either Z_w or M_w, are also reduced. If C_{m_α} is small in the basic aircraft, then I_{rd} is low and the aircraft will have good gust behaviour: however, its handling qualities will be impaired unless Z_w or M_w are augmented by a control system.

By controlling normal acceleration a_z and control deflection δ_j, any value of I_{rd} can be obtained. From eqn. 20, any control which minimises $\boldsymbol{x}$ and $\boldsymbol{u}$ will also minimise I_{rd}, and hence a_z. When bending modes are present also, the associated displacements and rates, which are elements of the state vector, also influence the acceleration (see eqn. 23). Consequently, the linear quadratic optimal problem is a natural formulation for the gust-alleviation problem.

4.2.2 Ride-comfort rating

Ride comfort has been assessed[27,30] by conducting extensive surveys of opinions of airline passengers, or passengers in an airborne simulator, in the USA. The results of these surveys are very similar, although in one case the author employed a 7-point rating scale, in which a rating of 1 indicated a very comfortable ride and a rating of 7 indicated extremely uncomfortable, whereas in the other study a 5-point scale was used. Both studies indicated that passengers were twice as sensitive to lateral than to vertical acceleration. One result[27] indicated that the comfort rating C_R could be represented satisfactorily by

$$C_R = 1{\cdot}65 + 8{\cdot}32a_{x_{cg}} + 15{\cdot}1a_{y_{cg}} + 21{\cdot}5a_{z_{cg}} + 0{\cdot}183p - 1{\cdot}2q - 0{\cdot}238r \qquad (50)$$

In the other result[30] C_R was represented by

$$C_R = 2{\cdot}0 + 7{\cdot}6a_{y_{cg}} + 11{\cdot}9a_{z_{cg}} \qquad (51)$$

Obviously, comfort depends greatly on reducing both lateral and vertical accelerations, for if a_y and a_z are reduced, low values of C_R, and hence good comfort ratings, are obtained, and, from eqn. 44, the ride-discomfort index I_{rd} is reduced.

Earlier attempts at providing gust-alleviation systems depended on using suitable shaping networks obtained from applying conventional control theory, to reduce the r.m.s. accelerations at specified wing and body stations.

Table 1

EIGENVALUES OF UNCONTROLLED AIRCRAFT

λ_1 : $-0{\cdot}703 + j\ 2{\cdot}68$	λ_1 : $0{\cdot}0$
λ_2 : $-0{\cdot}703 - j\ 2{\cdot}68$	λ_2 : $-0{\cdot}0024$
λ_3 : $-2{\cdot}982 + j\ 6{\cdot}994$	λ_3 : $-2{\cdot}264$
λ_4 : $-2{\cdot}982 - j\ 6{\cdot}994$	λ_4 : $-0{\cdot}356 + j\ 1{\cdot}577$
λ_5 : $-0{\cdot}0455 - j\,15{\cdot}211$	λ_5 : $-0{\cdot}356 - j\ 1{\cdot}577$
λ_6 : $-0{\cdot}0455 - j\,15{\cdot}211$	λ_6 : $-0{\cdot}094 + j\ 9{\cdot}838$
λ_7 : $-1{\cdot}348 + j\,18{\cdot}651$	λ_7 : $-0{\cdot}943 - j\ 9{\cdot}838$
λ_8 : $-1{\cdot}348 - j\,18{\cdot}651$	λ_8 : $-0{\cdot}479 + j\,10{\cdot}952$
λ_9 : $-0{\cdot}229 + j\,19{\cdot}728$	λ_9 : $-0{\cdot}479 - j\,10{\cdot}952$
λ_{10} : $-0{\cdot}229 - j\,19{\cdot}728$	λ_{10} : $-2{\cdot}076 + j\,13{\cdot}626$
λ_{11} : $-0{\cdot}887 + j\,38{\cdot}279$	λ_{11} : $-2{\cdot}076 - j\,13{\cdot}626$
λ_{12} : $-0{\cdot}877 - j\,38{\cdot}279$	λ_{12} : $-0{\cdot}856 + j\,23{\cdot}068$
	λ_{13} : $-0{\cdot}856 - j\,23{\cdot}068$
	λ_{14} : $-1{\cdot}156 + j\,30{\cdot}466$
	λ_{15} : $-1{\cdot}156 - j\,30{\cdot}466$

4.3 Linear optimal control system

Since both normal and lateral accelerations are linear functions of both state and control vectors, it was decided to use in the gust-alleviation system the feedback control that results from a minimisation of the performance index J, where

$$J = \tfrac{1}{2}\int_0^{\infty} (x'Qx + u'Ru)dt \qquad (52)$$

and where Q is a symmetric, non-negative – definite state weighting matrix of order $n \times n$ and R is a symmetric positive-definite control weighting matrix of order $m \times m$, both for longitudinal and lateral motions.

Initially, Q was chosen to be a diagonal matrix and one in which the state variables corresponding to the rigid body motion were unity weighted, while those state variables associated with the flexible modes were weighted as 10·0. The control weighting matrix R was chosen in both cases to be I_m.

For the longitudinal motion, the resulting feedback control matrix was

Table 2

EIGENVALUES OF OPTIMALLY CONTROLLED AIRCRAFT

longitudinal motion	lateral motion
λ_1 : $-2{\cdot}908 + j\ 1{\cdot}919$	λ_1 : $-\ 0{\cdot}058$
λ_2 : $-2{\cdot}908 - j\ 1{\cdot}919$	λ_2 : $-\ 0{\cdot}3142$
λ_3 : $-2{\cdot}923 + j\ 6{\cdot}943$	λ_3 : $-\ 2{\cdot}0845$
λ_4 : $-2{\cdot}923 - j\ 6{\cdot}943$	λ_4 : $-\ 1{\cdot}204$
λ_5 : $-1{\cdot}144 + j\,15{\cdot}178$	λ_5 : $-\ 1{\cdot}304$
λ_6 : $-1{\cdot}144 - j\,15{\cdot}178$	λ_6 : $-14{\cdot}2124$
λ_7 : $-0{\cdot}228 + j\,19{\cdot}724$	λ_7 : $-75{\cdot}0254$
λ_8 : $-0{\cdot}228 - j\,19{\cdot}724$	λ_8 : $-\ 2{\cdot}898 + j\ 8{\cdot}214$
λ_9 : $-1{\cdot}652 + j\,20{\cdot}176$	λ_9 : $-\ 2{\cdot}898 - j\ 8{\cdot}214$
λ_{10} : $-1{\cdot}652 - j\,20{\cdot}176$	λ_{10} : $-\ 1{\cdot}703 + j\,11{\cdot}925$
λ_{11} : $-1{\cdot}696 + j\,38{\cdot}25$	λ_{11} : $-\ 1{\cdot}703 - j\,11{\cdot}925$
λ_{12} : $-1{\cdot}696 - j\,38{\cdot}25$	λ_{12} : $-\ 3{\cdot}166 + j\,23{\cdot}074$
	λ_{13} : $-\ 3{\cdot}166 - j\,23{\cdot}074$
	λ_{14} : $-\ 0{\cdot}865 + j\,30{\cdot}46$
	λ_{15} : $-\ 0{\cdot}865 - j\,30{\cdot}46$

$$S = \begin{bmatrix} 0{\cdot}094 & -0{\cdot}517 & 1{\cdot}873 & -0{\cdot}261 & -0{\cdot}476 & 1{\cdot}177 & 4{\cdot}537 & 0{\cdot}876 & -1{\cdot}196 \\ 0{\cdot}637 & -4{\cdot}627 & -1{\cdot}769 & & & & & & \\ 0{\cdot}021 & 0{\cdot}044 & -0{\cdot}07 & -0{\cdot}034 & -0{\cdot}112 & 0{\cdot}126 & 0{\cdot}048 & 0{\cdot}0546 & -0{\cdot}492 \\ 0{\cdot}353 & -0{\cdot}099 & -1{\cdot}198 & & & & & & \end{bmatrix} \qquad (53)$$

and, for lateral motion, the feedback matrix was

$$S = \begin{bmatrix} 0{\cdot}195 & 4{\cdot}83 & 7{\cdot}90 & 1{\cdot}61 & -0{\cdot}97 & 5{\cdot}27 & -1{\cdot}43 & 14{\cdot}37 & -0{\cdot}99 & -4{\cdot}43 \\ -2{\cdot}38 & -8{\cdot}46 & 0{\cdot}27 & -0{\cdot}85 & -0{\cdot}001 & & & & & \\ 0{\cdot}95 & 0{\cdot}24 & 19{\cdot}85 & 0{\cdot}78 & 0{\cdot}24 & 18{\cdot}84 & -0{\cdot}008 & -5{\cdot}67 & -0{\cdot}12 & -18{\cdot}62 \\ 0{\cdot}5 & -9{\cdot}36 & -0{\cdot}47 & 12{\cdot}67 & -0{\cdot}06 & & & & & \\ 0{\cdot}02 & -0{\cdot}56 & -1{\cdot}3 & -0{\cdot}13 & -0{\cdot}003 & -2{\cdot}49 & 0{\cdot}09 & -0{\cdot}2 & 0{\cdot}68 & 6{\cdot}29 \\ 0{\cdot}59 & 1{\cdot}25 & 0{\cdot}13 & -0{\cdot}47 & 0{\cdot}12 & & & & & \end{bmatrix} \qquad (54)$$

The eigenvalues of the uncontrolled aircraft dynamics are shown in Table 1 and the corresponding eigenvalues of the controlled aircraft are given in Table 2. Figs. 4 and 5 show the acceleration responses at specific stations of the uncontrolled aircraft. For the initial conditions shown, the acceleration responses at the same stations for the aircraft with feedback control are shown in Figs. 6 and 7. Comparison of Figs. 4 and 6, 5 and 7 indicates that the accelerations at the specific body stations have been reduced significantly. When the state weighting matrix was chosen to be $Q = C'C$ for longitudinal motion (or $Q = M'M$ for lateral motion), where the matrix $C(M)$ was the output matrix, then it was found that the control law depended very nearly on those state variables that generated the output, and not to any significant extent on the other state variables; e.g. for longitudinal motion, using the C matrix quoted in Appendix 8, the resulting matrix of feedback gains, for $G = I_2$, was determined to be

$$S = \begin{bmatrix} -0{\cdot}118\ -0{\cdot}498\ 0{\cdot}406\ 8{\cdot}97 \times 10^{-9}\ -4{\cdot}16 \times 10^{-10}\ -1{\cdot}2 \times 10^{-8}\ 3{\cdot}15 \times 10^{-10} \\ 1{\cdot}21 \times 10^{-8}\ -9{\cdot}4 \times 10^{-11}\ -1{\cdot}26 \times 10^{-8}\ -9 \times 10^{-12} \\ 0{\cdot}017\ 0{\cdot}038\ -0{\cdot}033\ 0{\cdot}0015\ -6{\cdot}73 \times 10^{-10}\ 3{\cdot}1 \times 10^{-11}\ 9{\cdot}3 \times 10^{-10}\ -2{\cdot}39 \times 10^{-11} \\ -8{\cdot}73 \times 10^{-10}\ 6{\cdot}65 \times 10^{-12}\ 9{\cdot}4 \times 10^{-10}\ 6{\cdot}29 \times 10^{-13} \end{bmatrix} \qquad (55)$$

The gains associated with the variables of bending motion are not significant (except only for the contribution of the 1st bending mode to the controlled deflection of the elevator and the canard), and, since the output vector is related directly to the r.d.i., minimisation of measures of this vector reduces acceleration levels and improves ride quality.

4.4 Algebraic model matching

From the theory of implicit model following,[26,32] it is possible to derive a state-feedback control law:

$$u = Sx \qquad (56)$$

which, when applied to the aircraft, will result in the output vector response being very close to that which is defined by an ideal, or model, system; namely

$$y_m = Ty_m \tag{57}$$

It is easy to show[26] that the feedback matrix S is given by

$$S = [CB]^+(TC - CA) \tag{58}$$

where $[CB]^+$ represents the generalised inverse of the rectangular matrix $[CB]$.

For longitudinal motion, in which the output vector was identical to the state vector, the model matrix T was chosen to be

$$T = diag\{-1{\cdot}0 \quad -2{\cdot}5 \quad 1-5{\cdot}0 \quad 1-8{\cdot}0 \quad 1-10{\cdot}0 \quad 1-15{\cdot}0 \quad 1-20{\cdot}0\} \tag{59}$$

so that the resultant controlled outputs, including the bending modes, might behave as rapid 1st-order modes. The resultant feedback matrix was

$$S = \begin{bmatrix} 0{\cdot}152 & 0{\cdot}055 & 5{\cdot}816 & -0{\cdot}131 & -29{\cdot}11 & 0{\cdot}79 & -57{\cdot}362 & 1{\cdot}016 & -23{\cdot}6 \\ 0{\cdot}751 & 356{\cdot}88 & -4{\cdot}441 & & & & & & \\ 24{\cdot}069 & 0{\cdot}834 & 59{\cdot}033 & 23{\cdot}43 & -396{\cdot}21 & 10{\cdot}761 & -751{\cdot}14 & 13{\cdot}272 & \\ -363{\cdot}77 & 1{\cdot}867 & 4809{\cdot}3 & -59{\cdot}85 & & & & & \end{bmatrix} \tag{60}$$

Evidently this feedback law depends very heavily on the bending motion rates and displacements; such dependence should be avoided if possible. One technique for achieving this is to restrict the output matrix C to be

$$C = [I_4 \quad : \quad 0] \tag{61}$$

so that only the rigid body motion and the first bending mode are controlled.

$$T = diag\{-10{\cdot}0 \quad -5{\cdot}0 \quad -1{\cdot}0 \quad -4{\cdot}0\} \tag{62}$$

The feedback matrix changed to

$$S = \begin{bmatrix} 9{\cdot}678 & 0{\cdot}95 & 66{\cdot}595 & 1{\cdot}67 & 0 & 0 & 0 & 0 \\ 0 & 0 & 0 & 0 & & & & \\ 152{\cdot}47 & 10{\cdot}348 & 879{\cdot}35 & 26{\cdot}44 & 0 & 0 & 0 & 0 \\ 0 & 0 & 0 & 0 & & & & \end{bmatrix} \tag{63}$$

In other words, only the variables of rigid body motion and the rate and displacement associated with the first bending mode were present in the feedback control law. Similar results were obtained when the output was restricted to rigid-body motion only. For lateral motion, very similar results were obtained, except that the rate and displacement associated with the 9th lateral bending mode were always present in the feedback. Some representative results from simulation studies are shown in Figs. 8 and 9.

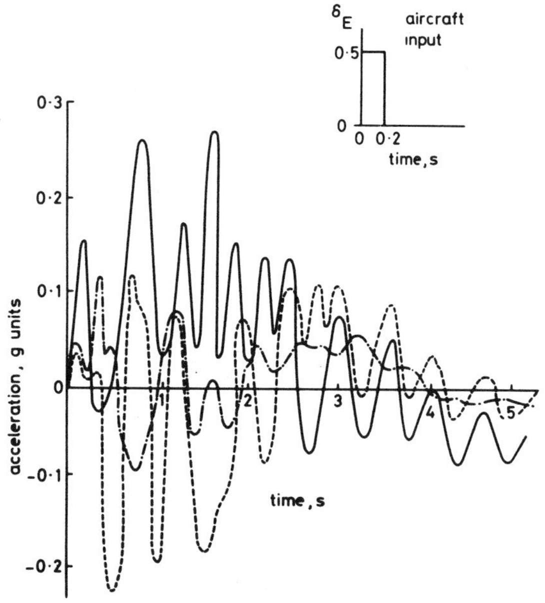

Fig. 4
Vertical acceleration of uncontrolled aircraft

——— *NZ* 42·0
– – – – *NZ* 21·9
–·–·– *NZ* 4·4

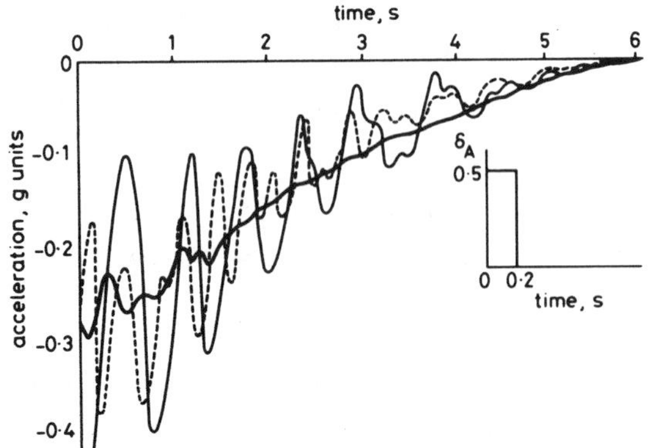

Fig. 5
Lateral acceleration of uncontrolled aircraft

——— *NY* 4·4
—·— *NY* 21·9
– – – – *NY* 42·0

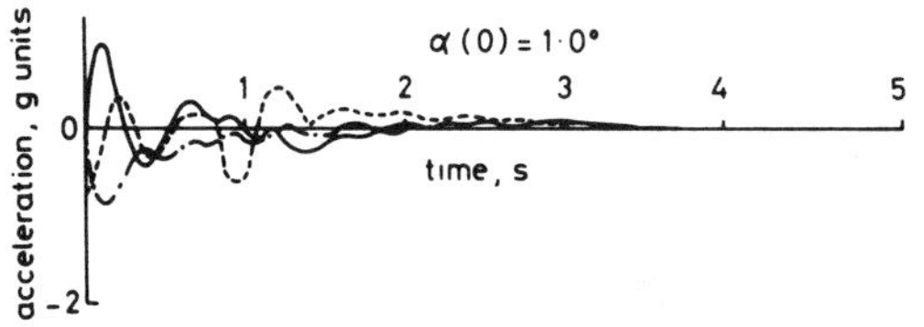

Fig. 6
Vertical acceleration: optimally-controlled aircraft

——— *NZ* 4·4
– – – – *NZ* 42·0
—·— *NZ* 21·9

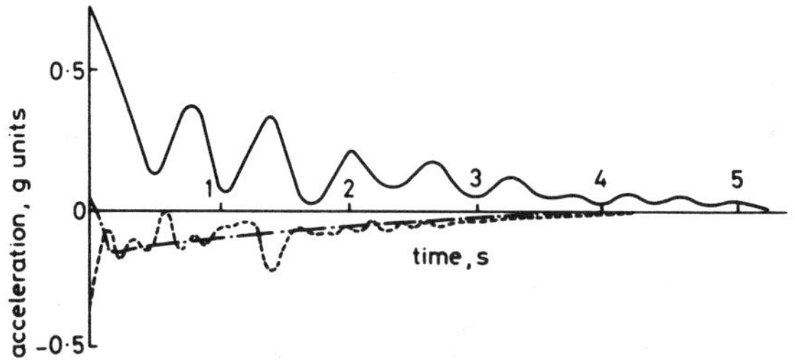

Fig. 7
Lateral acceleration: optimally-controlled aircraft

——— *NY* 42·0
– – – – *NY* 4·4
—·— *NY* 21·9

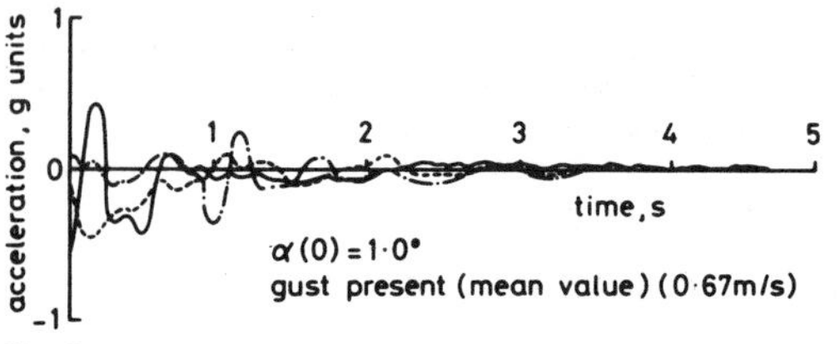

Fig. 8
Vertical acceleration: model-following control

——— *NZ* 4·4
——— *NZ* 21·9
—·— *NZ* 42·0

Table 3
EIGENVALUES OF MODEL-FOLLOWING CONTROLLED AIRCRAFT

full model		reduced model	
λ_1	$: -0{\cdot}059$	λ_1	$: -4{\cdot}866$
λ_2	$: -1{\cdot}832$	λ_2	$: -5{\cdot}1$
λ_3	$: -2{\cdot}5377$	λ_3	$: -1{\cdot}144 + j\ 1{\cdot}662$
λ_4	$: -9{\cdot}399$	λ_4	$: -1{\cdot}144 - j\ 1{\cdot}662$
λ_5	$: -3{\cdot}216 + j\ 8{\cdot}778$	λ_5	$: -0{\cdot}856 + j\,15{\cdot}192$
λ_6	$: -3{\cdot}216 - j\ 8{\cdot}778$	λ_6	$: -0{\cdot}856 - j\,15{\cdot}192$
λ_7	$: -1{\cdot}298 + j\,17{\cdot}174$	λ_7	$: -0{\cdot}213 + j\,19{\cdot}73$
λ_8	$: -1{\cdot}298 - j\,17{\cdot}174$	λ_8	$: -0{\cdot}213 - j\,19{\cdot}73$
λ_9	$: -0{\cdot}24 + j\,19{\cdot}682$	λ_9	$: -1{\cdot}361 + j\,20{\cdot}193$
λ_{10}	$: -0{\cdot}24 - j\,19{\cdot}682$	λ_{10}	$: -1{\cdot}361 - j\,20{\cdot}193$
λ_{11}	$: -2{\cdot}831 + j\,28{\cdot}024$	λ_{11}	$: -0{\cdot}877 + j\,38{\cdot}28$
λ_{12}	$: -2{\cdot}831 - j\,28{\cdot}024$	λ_{12}	$: -0{\cdot}877 - j\,38{\cdot}28$

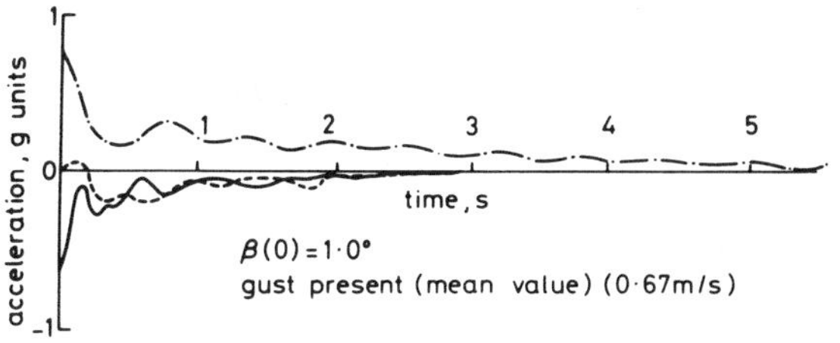

Fig. 9
Lateral acceleration: model-following control

——— *NY* 4·4
– – – – *NY* 21·9
—·— *NY* 42·0

The eigenvalues for both cases are presented in Table 3, and although the response of the aircraft, represented in Fig. 8, appears to be quite acceptable, it should be noted that some of the feedback gains are large; too large for practical synthesis since the associated rates of change required from the controlled deflections of the motivators are likely to be in excess of what can be produced by existing actuators, the dynamics of which were not taken into account here. Further experiments with the model dynamics will produce changes in the feedback control law until a suitable set is produced: in particular, the model dynamics of another aircraft with satisfactory gust-alleviation characteristics can be employed successfully.

5 Conclusions

This paper has presented an account of how gust-alleviation control systems for flexible aircraft may be considered to be candidates for linear optimal control, which, however, requires complete state-feedback for its implementation. Not every state variable of the mathematical model of dynamics of a deformable aircraft is either physically available for measurement, or can be measured accurately or conveniently. Only a few state variables are possible candidates as measurable outputs from the system: for this reason, linear state variable feedback control in aircraft flight-control systems is usually unsuccessful or impossible to implement. Attention must usually be restricted, therefore, to schemes that require only linear output feedback.

The very existence of bending modes caused by the deformation of the aircraft also leads to a difficult control problem: can a lightly damped (or an unstable) mode be allowed to exist, even if it is unobservable? For then, its displacement and associated rate would not be employed as feedback signals and its existence would not by definition affect the observable output response, although the control action might possibly excite the mode. The existence of an unstable bending mode will result in large levels of stress, and larger stress rates, at some particular structural stations. Consequently, such a mode could increase the fatigue of the structure at those stations, assuming that fatigue life depe. .s principally on cumulative levels of stress. Eventually, of course, the airframe structure would fail, due possibly to that very control action which has been regarded, from other technical considerations, as totally acceptable. This problem is made more difficult for the control engineer who must devise control laws from a given mathematical model of the motion of the aircraft, and this model contains only those bending modes considered by the structural engineer to be significant from his technical standpoint. It may be that the modes, which have been ignored, are more significant from the point of view of control: they may easily be excited by the control action. It should be remembered that, even if they do not appear in the mathematical model, these modes do exist physically. The difficulty occurs because the dynamic behaviour of the flexible aircraft is being described by ordinary differential equations, whereas it would probably be better represented by partial differential equations, since the flexible aircraft is behaving more as a distributed than as a lumped-parameter system.

It must be emphasised that a practically restricted situation has been considered: the gust penetration has been taken to occur in wings straight and level flight. This was done because the equations of motion that were available for use had no coupling between the longitudinal and lateral motion, and they had no nonlinear aerodynamic terms. However, the small amount of evidence available suggests that gust penetration more usually occurs when the aircraft is manoeuvring and induces severely coupled motion. To extend the work to such situations, a complete six degrees of freedom mathematical model, with flexibility terms, needs to be available. However, the greatest impediment to such further development lies in the difficulty of obtaining particular equations of motion, with the associated stability derivatives and bending mode coefficients, which can be freely disseminated.

6. Acknowledgment

This work was supported by a grant from the UK Science Research Council.

7 References

1 OSTGAARD, M.A., and SWORTZEL, F.R.: 'CCVs: active control technology creating new military aircraft design potential', *Astrophys. & Aero.*, 1977, **15**, pp. 42–57
2 BURRIS, P.M., and BENDER, M.A.: 'Aircraft load alleviation and mode stabilization (LAMS) – B52 system analysis, synthesis and design'. AFFDL-TR-68-161, WPAFB Dayton, Ohio 1969
3 HOFFMANN, L.G., and CLEMENT, W.F.: 'Vehicle design considerations for active control application to subsonic transport aircraft'. NASA CR-2408, Aug. 1974
4 SPRATER, A.: 'Stabilizing device for flying machines'. US Patent 1 119 234, 1914
5 HUNSAKER, J.C., and WILSON, E.B.: 'Report on behaviour of aeroplanes in gust turbulence'. NACA TM-1 (MIT), October 1915
6 VON KARMAN, T.: 'Fundamentals of the statistical theory of turbulence', *J. Aeropl. Sci.*, 1937, **4**, pp. 181–138
7 TAYLOR, G.I.: 'Statistical theory of turbulence, *ibid.*, 1937, **4**, pp. 311–315
8 HIRSCH, R.: 'Etudes et essais d'un absorbeur des rafales', *DOC Aeropl.* 1957, **42**, pp. 13–28
9 HARPUR, N.F.: 'Effect of active control systems on structural design criteria'. AGARD, 37th meeting of structures and materials panel, 1973
10 HAWK, J., CONNOR, R.J., and LEVY, C.: 'Dynamic analysis of the C-47 gust load alleviation system'. Douglas report SM 14456, Douglas Aircraft, Santa Monica 1952
11 KRAFT, C.C.: 'Initial results of a flight investigation of a gust alleviation system'. NASA TM-3612, 1956
12 ZBROZEK, J.K., SMITH, W., and WHITE, D.: 'Preliminary investigation on a gust alleviation investigation on a Lancaster aircraft', ARC R & M 2972, 1957
13 ATTWOOD, J.L., CANNON, R.H., JOHNSON, J.M., and ANDREW, G.M.: 'Gust alleviation system'. US Patent 2 985 409, 1961
14 SMITH, R.E. and LUM, E.L.S.: 'Linear optimal control theory and angular acceleration sensing applied to active structural bending control on the XB-70'. AFFDL-TR-66-68, WPAFB, Dayton, Ohio, 1966
15 WYKES, J.H., NARDI, L.V., and MORI, A.S.: 'XB-70 structural mode control system design and performance analyses'. NASA, CR-1577, July, 1970
16 DEMPSTER, J.B., and ARNOLD, J.I.: 'Flight test evaluation of an advanced SAS for the B.52 aircraft'. *J. Air.* 1969, **6**, pp. 343–348
17 STOCKDALE, C.R., and POYNEER, R.D.: 'Control configured vehicle – ride control system'. AFFDL-TR-73-83, WPAFB, Dayton, Ohio, 1973
18 VAN DIERENDONCK, A.J., STONE, C.R., and WARD, M.D.: 'Application of practical optimal control theory to the C-5A load improvement control system (LICS). AFFDL-TR-73-122, WPAFB, Dayton Ohio, 1973
19 HINSDALE, A.J.: 'Lateral ride quality of the B-1 aircraft subjected to a reduction of lateral static stability'. NASA, CR-148206, April 1976
20 HOBLIT, F.M.: 'Effect of yaw damper on lateral gust loads in design of the L-1011 transport'. AGARD, 37th meeting of structures and materials panel, 1973
21 GORDON, C.K., and DODSON, R.O.: 'STOL ride control feasibility study'. NASA, CR-2276, July, 1973
22 BRAINERD, C.H., and KOHLMAN, D.L.: 'A simulator evaluation of the use of spoilers on a light aircraft'. NASA, CR-2121, Oct. 1972
23 LANGE, R.H. et al.: 'Application of active controls technology to the NASA jetstar airplane'. NASA, CR-2561, June 1975
24 ERKELENS, L.J.J. and SCHURING, J.: 'Investigation on a passenger ride comfort improvement system with limited control surface actuator performance for a flexible aircraft'. Report NLR 75140U, National Aerospace Laboratory NLR, Netherlands, 1975
25 ATHANS, M.: 'The Role and Use of the Stochastic Linear-Quadratic-Gaussian Problem in Control System Design', *IEEE Trans.*, 1971, **AC-16**, pp. 529–552
26 ERZBERGER, H.: 'On the use of algebraic methods in the analysis and design of model-following control systems'. NASA, TN D-4663, July 1968
27 JONES, J.G.: 'Similarity theory of gust loads on aircraft'. RAE, TR 69171, Aug. 1969

28 SCHOONOVER, W.E.: 'Ride quality of terminal area manoeuvres'. NASA, TM-X-3295, Nov. 1975, pp. 387–408

29 WOLF, T.D., REZEK, J.W., and GEE, S.W.: 'Passenger ride quality response to an airborne simulator environment'. NASA, TM-X-3295, Nov. 1975. pp. 373–385

30 'Flying qualities requirements'. JAC 925, Joint Airworthiness Committee (UK) 73, April 1975

31 RICHARDS, L.G., KUHLTHAU, A.R., and JACOBSON, I.D.: 'Passenger ride quality determined from commercial airline flights'. NASA, TN-X-3295, Nov. 1975, pp. 409–436

32 MARKLAND, C.A.: 'Optimal model following control system synthesis techniques', *Proc. IEE*, 1970, **117**, (3), pp. 623–627

8 Appendixes

8.1.1 Equations of longitudinal motion

$$\dot{\alpha} = Z_\alpha \alpha - q + Z_{\delta_E}\delta_E + Z_{\delta_c}\delta_{c_{hor}} + Z_{\dot{\lambda}_1}\dot{\lambda}_1 + Z_{\lambda_1}\lambda_1 + \sum_{j=1}^{3} Z_{w_j} w_j$$

$$\dot{q} = M_\alpha \alpha + M_{\dot{\alpha}}\dot{\alpha} + M_q q + M_{\delta_E}\delta_E + M_{\delta_{c_{hor}}}\delta_{c_{hor}} + M_{\lambda_1}\lambda_1 + M_{\dot{\lambda}_1}\dot{\lambda}_1 + \sum_{j=1}^{3} M_{w_j} w_j$$

$$\ddot{\lambda}_1 = \eta_{1\alpha}\alpha + \eta_{1q} q + \eta_{1\delta_E}\delta_E + \eta_{1\delta_c}\delta_{c_{hor}} + (\eta_{1_{\dot{\lambda}_1}} - 2\zeta_1\omega_1)\dot{\lambda}_1 + (\eta_{1_{\eta_1}} - \omega_1^2)\lambda_1 + \sum_{j=1}^{3} \eta_{1_{w_j}} w_j$$

$$\ddot{\lambda}_5 = \eta_{5_\alpha}\alpha + \eta_{5_q} q + \eta_{5_{\delta_E}}\delta_E + \eta_{5_{\delta_c}}\delta_{c_{hor}} + (\eta_{5_{\dot{\lambda}_5}} - 2\zeta_5\omega_5)\dot{\lambda}_5 + (\eta_{5_{\lambda_5}} - \omega_5^2)\lambda_5 + \sum_{j=1}^{3} \eta_{5_{w_j}} w_j$$

$$\ddot{\lambda}_7 = \eta_{7_\alpha}\alpha + \eta_{7_q} q + \eta_{7_{\delta_E}}\delta_E + \eta_{7_{\delta_c}}\delta_{c_{hor}} + (\eta_{7_{\dot{\lambda}_7}} - 2\zeta_7\omega_7)\dot{\lambda}_7 + (\eta_{7_{\lambda_7}} - \omega_7^2)\lambda_7 + \eta_{7_{\dot{\lambda}_8}}\dot{\lambda}_8 + \eta_{7_{\lambda_8}}\lambda_8 + \sum_{j=1}^{3} \eta_{7_{w_j}} w_j$$

$$\ddot{\lambda}_8 = \eta_{8_\alpha}\alpha + \eta_{8_q} q + \eta_{8_{\delta_E}}\delta_E + \eta_{8_{\delta_c}}\delta_{c_{hor}} + (\eta_{8_{\dot{\lambda}_8}} - 2\zeta_8\omega_8)\dot{\lambda}_8 + (\eta_{8_{\lambda_8}} - \omega_8^2)\lambda_8 + \eta_{8_{\dot{\lambda}_7}}\dot{\lambda}_7 + \eta_{8_{\lambda_7}}\lambda_7 + \sum_{j=1}^{3} \eta_{8_{w_j}} w_j$$

$$\ddot{\lambda}_{12} = \eta_{12_\alpha}\alpha + \eta_{12_q} q + \eta_{12_{\delta_E}}\delta_E + \eta_{12_{\delta_c}}\delta_{c_{hor}} - 2\zeta_{12}\omega_{12}\dot{\lambda}_{12} - \omega_{12}^2\lambda_{12} + \sum_{j=1}^{3} \eta_{12_{w_j}} w_j \qquad (64)$$

8.1.2 Equations of lateral motion

$$\dot{v} = Y_v v + Y_p p + Y_r r + Y_\phi \phi + Y^*_{\delta_A}\delta_A + Y^*_{\delta_R}\delta_R + Y^*_{\delta_c}\delta_{c_V} + Y_{\gamma_9}\gamma_9 + Y_{\dot{\gamma}_9}\dot{\gamma}_9 + \sum_{j=1}^{3} Y_{v_j} v_j$$

$$\dot{p} = L'_v v + L'_{\dot{v}}\dot{v} + L'_p p + L'_r r + L'_{\delta_A}\delta_A + L'_{\delta_R}\delta_R + L'_{\delta_c}\delta_{c_v} + L'_{\gamma_9}\gamma_9 + L'_{\dot{\gamma}_9}\dot{\gamma}_9 + \sum_{j=1}^{3} L'_{v_j} v_j$$

$$\dot{r} = N'_v v + N'_{\dot{v}}\dot{v} + N'_p p + N'_r r + N'_{\delta_A}\delta_A + N'_{\delta_R}\delta_R + N'_{\delta_c}\delta_{c_V} + N'_{\gamma_9}\gamma_9 + N'_{\dot{\gamma}}\dot{\gamma}_9 + \sum_{j=1}^{3} N'_{v_j} v_j$$

$$\dot{\phi} = p$$

$$\dot{\psi} = r$$

$$\ddot{\gamma}_1 = \mu_{1_v} v + \mu_{1_{\dot{v}}}\dot{v} + \mu_{1_p} p + \mu_{1_r} r + \mu_{1_{\delta_A}}\delta_A + \mu_{1_{\delta_R}}\delta_R + \mu_{1_{\delta_c}}\delta_{c_V} - 2\zeta_A\omega_A\dot{\gamma}_1 - \omega_A^2\gamma_1 + \sum_{j=1}^{3} \mu_{1_{v_j}} v_j$$

$$\ddot{\gamma}_2 = \mu_{2_v} v + \mu_{2_{\dot{v}}}\dot{v} + \mu_{2_p} p + \mu_{2_r} r + \mu_{2_{\delta_A}}\delta_A + \mu_{2_{\delta_R}}\delta_R + \mu_{2_{\delta_c}}\delta_{c_V} - 2\zeta_B\omega_B\dot{\gamma}_2 - \omega_B^2\gamma_2 + \mu_{2_{\gamma_3}}\gamma_3 + \mu_{2_{\dot{\gamma}_3}}\dot{\gamma}_3 + \sum_{j=1}^{3} \mu_{2_{v_j}} v_j$$

$$\ddot{\gamma}_3 = \mu_{3_v} v + \mu_{3_{\dot{v}}}\dot{v} + \mu_{3_p} p + \mu_{3_r} r + \mu_{3_{\delta_A}}\delta_A + \mu_{3_{\delta_R}}\delta_R + \mu_{3_{\delta_c}}\delta_{c_V} - 2\zeta_C\omega_C\dot{\gamma}_3 - \omega_C^2\gamma_3 + \mu_{3_{\gamma_2}}\gamma_2 + \mu_{3_{\dot{\gamma}_2}} + \sum_{j=1}^{3} \mu_{3_{v_j}} v_j$$

$$\ddot{\gamma}_9 = \mu_{9_v} v + \mu_{9_{\dot{v}}}\dot{v} + \mu_{9_p} p + \mu_{9_r} r + \mu_{9_{\delta_A}}\delta_A + \mu_{9_{\delta_R}}\delta_R + \mu_{9_{\delta_c}}\delta_{c_V} - 2\zeta_D\omega_D\dot{\gamma}_9 - \omega_D^2\gamma_9 + \mu_{9_{\gamma_{10}}}\gamma_{10} + \mu_{9_{\dot{\gamma}_{10}}}\dot{\gamma}_{10} + \sum_{j=1}^{3} \mu_{9_{v_j}} v_j$$

$$\ddot{\gamma}_{10} = \mu_{10_v} v + \mu_{10_{\dot{v}}}\dot{v} + \mu_{10_p} p + \mu_{10_r} r + \mu_{10_{\delta_A}}\delta_A + \mu_{10_{\delta_R}}\delta_R + \mu_{10_{\delta_c}}\delta_{c_V} - 2\zeta_E\omega_E\dot{\gamma}_{10} - \omega_E^2\gamma_{10} + \mu_{10_{\gamma_9}}\gamma_9 + \mu_{10_{\dot{\gamma}_9}}\dot{\gamma}_9 + \sum_{j=1}^{3} \mu_{10_{v_j}} v_j \qquad (65)$$

8.1.3 Longitudinal acceleration and rate equations

$$a_{z_{4\cdot4}} = U_0\dot{\alpha} - U_0 q - x_{ac_{4\cdot4}}\dot{q} + \Phi_{4\cdot4,1}\ddot{\lambda}_1 + \Phi_{4\cdot4,5}\ddot{\lambda}_5 + \Phi_{4\cdot4,7}\ddot{\lambda}_7 + \Phi_{4\cdot4,8}\ddot{\lambda}_8 + \Phi_{4\cdot4,12}\ddot{\lambda}_{12}$$

$$a_{z_{21\cdot9}} = U_0\dot{\alpha} - U_0 q - x_{ac_{21\cdot9}}\dot{q} + \Phi_{21\cdot9,1}\ddot{\lambda}_1 + \Phi_{21\cdot9,5}\ddot{\lambda}_5 + \Phi_{21\cdot9,7}\ddot{\lambda}_7 + \Phi_{21\cdot9,8}\ddot{\lambda}_8 + \Phi_{21\cdot9,12}\ddot{\lambda}_{12}$$

$$a_{z_{42}} = U_0(\dot{\alpha} - q) - x_{ac_{42}}\dot{q} + \Phi_{42,1}\ddot{\lambda}_1 + \Phi_{42,5}\ddot{\lambda}_5 + \Phi_{42,7}\ddot{\lambda}_7 + \Phi_{42,8}\ddot{\lambda}_8 + \Phi_{42,12}\ddot{\lambda}_{12}$$

$$q_{10\cdot8} = q + \Phi_{10\cdot8,1}\dot{\lambda}_1 + \Phi_{10\cdot8,5}\dot{\lambda}_5 + \Phi_{10\cdot8,7}\dot{\lambda}_7 + \Phi_{10\cdot8,8}\dot{\lambda}_8 + \Phi_{10\cdot8,12}\dot{\lambda}_{12} \qquad (66)$$

8.1.4 Lateral accelerations and rate equations

$$a_{y_{4\cdot4}} = \dot{v} + x_{ac_{4\cdot4}}\dot{r} + U_0 r + \hat{\Phi}_{4\cdot4,1}\ddot{\gamma}_1 + \hat{\Phi}_{4.4,2}\ddot{\gamma}_2 + \hat{\Phi}_{4\cdot4,3}\ddot{\gamma}_3 + \hat{\Phi}_{4\cdot4,9}\ddot{\gamma}_9 + \hat{\Phi}_{4\cdot4,10}\ddot{\gamma}_{10}$$

$$a_{y_{21\cdot9}} = \dot{v} + x_{ac_{21\cdot9}}\dot{r} + U_0 r + \hat{\Phi}_{21\cdot9,1}\ddot{\gamma}_1 + \hat{\Phi}_{21\cdot9,2}\ddot{\gamma}_2 + \hat{\Phi}_{21\cdot9,3}\ddot{\gamma}_3 + \hat{\Phi}_{21\cdot9,9}\ddot{\gamma}_9 + \hat{\Phi}_{21\cdot9,10}\ddot{\gamma}_{10}$$

$$a_{y_{42}} = \dot{v} + x_{ac_{42}}\dot{r} + U_0 r + \hat{\Phi}_{42,1}\ddot{\gamma}_1 + \hat{\Phi}_{42,1}\ddot{\gamma}_2 + \hat{\Phi}_{42,3}\ddot{\gamma}_3 + \hat{\Phi}_{42,9}\ddot{\gamma}_9 + \hat{\Phi}_{42,10}\ddot{\gamma}_{10}$$

$$r_{17\cdot7} = r + \hat{\Phi}_{17\cdot7,1}\dot{\gamma}_1 + \hat{\Phi}_{17\cdot7,2}\dot{\gamma}_2 + \hat{\Phi}_{17\cdot7,3}\dot{\gamma}_3 + \hat{\Phi}_{17\cdot7,9}\dot{\gamma}_9 + \hat{\Phi}_{17\cdot7,10}\dot{\gamma}_{10} \qquad (67)$$

8.2 Aircraft matrices

8.2.1 Longitudinal motion

$$\dot{x} = Ax + Bu + Ew$$

$$A = \begin{bmatrix}
-1.6 & -1 & -1.1811 & -0.1181 & 0 & 0 & 0 & 0 & 0 & 0 & 0 & 0 \\
6.57 & -2.446 & -1.813 & 1.1805 & 0 & 0 & 0 & 0 & 0 & 0 & 0 & 0 \\
0 & 0 & 0 & 1 & 0 & 0 & 0 & 0 & 0 & 0 & 0 & 0 \\
-7.196 & -0.445 & -56.82 & -5.53 & 0 & 0 & 0 & 0 & 0 & 0 & 0 & 0 \\
0 & 0 & 0 & 0 & 0 & 1 & 0 & 0 & 0 & 0 & 0 & 0 \\
-1.349 & 0.2466 & 0 & 0 & -231.52 & -1.712 & 0 & 0 & 0 & 0 & 0 & 0 \\
0 & 0 & 0 & 0 & 0 & 0 & 0 & 1 & 0 & 0 & 0 & 0 \\
-2.093 & 0.242 & 0 & 0 & 0 & 0 & -408.86 & -2.679 & -10.71 & -0.518 & 0 & 0 \\
0 & 0 & 0 & 0 & 0 & 0 & 0 & 0 & 0 & 1 & 0 & 0 \\
0.3073 & 0.05588 & 0 & 0 & 0 & 0 & -1.24 & -0.176 & -390.1 & -0.474 & 0 & 0 \\
0 & 0 & 0 & 0 & 0 & 0 & 0 & 0 & 0 & 0 & 0 & 1 \\
-3.736 & 0.1331 & 0 & 0 & 0 & 0 & 0 & 0 & 0 & 0 & -1466.12 & -1.754
\end{bmatrix} \tag{68}$$

$$B' = \begin{bmatrix}
-0.07 & 3.726 & 0 & 0.572 & 0 & -0.465 & 0 & -0.582 & 0 & -0.112 & 0 & 0.929 \\
-0.006 & -0.28 & 0 & 0.019 & 0 & -0.054 & 0 & -0.0532 & 0 & -0.035 & 0 & 0.101
\end{bmatrix} \tag{69}$$

$$E' = \begin{bmatrix}
0.0042 & 0.06 & 0 & 0.0105 & 0 & 0.0065 & 0 & -0.0045 & 0 & -0.0021 & 0 & 0.015 \\
0.0037 & -0.0417 & 0 & 0.0393 & 0 & 0.0039 & 0 & 0.0101 & 0 & -0.0009 & 0 & 0.0033 \\
0.0012 & -0.056 & 0 & -0.0086 & 0 & 0.0059 & 0 & 0.0064 & 0 & 0.0012 & 0 & 0.0031
\end{bmatrix} \tag{70}$$

8.2.2 Lateral motion

$$\dot{x}_1 = Fx_1 + Gu_1 + Hv$$

$$F = \begin{bmatrix}
-0.18 & 0.12 & 171.0 & 9.8 & 0 & 0 & 0 & 0 & 0 & 0 & 0 & -103.6 & 5.5 & 0 & 0 \\
-0.015 & -2.26 & -0.47 & -0.008 & 0 & 0 & 0 & 0 & 0 & 0 & 0 & 34.41 & 1.79 & 0 & 0 \\
-0.015 & -0.032 & -0.54 & -0.007 & 0 & 0 & 0 & 0 & 0 & 0 & 0 & -15.57 & 0.015 & 0 & 0 \\
0 & 1 & 0 & 0 & 0 & 0 & 0 & 0 & 0 & 0 & 0 & 0 & 0 & 0 & 0 \\
0 & 0 & 1 & 0 & 0 & 0 & 0 & 0 & 0 & 0 & 0 & 0 & 0 & 0 & 0 \\
0 & 0 & 0 & 0 & 0 & 0 & 1 & 0 & 0 & 0 & 0 & 0 & 0 & 0 & 0 \\
-0.011 & -7.92 & -10.73 & -0.006 & 0 & -97.67 & -1.9 & 0 & 0 & 0 & 0 & 0.06 & -0.003 & 0 & 0 \\
0 & 0 & 0 & 0 & 0 & 0 & 0 & 0 & 1 & 0 & 0 & 0 & 0 & 0 & 0 \\
-0.003 & -32.3 & -1.63 & -0.001 & 0 & 0 & 0 & -151 & -3.57 & 150.1 & 3.73 & 0.015 & -0.001 & 0 & 0 \\
0 & 0 & 0 & 0 & 0 & 0 & 0 & 0 & 0 & 0 & 1 & 0 & 0 & 0 & 0 \\
0.004 & 0.17 & 0.11 & 0.002 & 0 & 0 & 0 & 8.71 & 0.6 & -160 & -1.54 & -0.02 & 0.001 & 0 & 0 \\
0 & 0 & 0 & 0 & 0 & 0 & 0 & 0 & 0 & 0 & 0 & 0 & 1 & 0 & 0 \\
0.003 & 0.11 & 0.14 & 0.001 & 0 & 0 & 0 & 0 & 0 & 0 & 0 & -532.6 & -1.72 & 3.25 & 0.36 \\
0 & 0 & 0 & 0 & 0 & 0 & 0 & 0 & 0 & 0 & 0 & 0 & 0 & 0 & 1 \\
-0.018 & -0.06 & -0.51 & -0.01 & 0 & 0 & 0 & 0 & 0 & 0 & 0 & 1.64 & 0.93 & -930.3 & -2.3
\end{bmatrix} \tag{71}$$

$$G' = \begin{bmatrix}
0.4 & -0.8 & 0.06 & 0 & 0 & 0 & -2.14 & 0 & -0.4 & 0 & -0.12 & 0 & +0.01 & 0 & -0.02 \\
-7.0 & -0.22 & -4.31 & 0 & 0 & 0 & +23.7 & 0 & +4.32 & 0 & -0.18 & 0 & -0.16 & 0 & +0.61 \\
-0.48 & 0.01 & 0.05 & 0 & 0 & 0 & -0.04 & 0 & +0.63 & 0 & +0.002 & 0 & +0.06 & 0 & +0.05
\end{bmatrix} \tag{72}$$

$$H' = \begin{bmatrix}
-0.03 & -0.006 & 0.002 & 0 & 0 & 0 & -0.0001 & 0 & 0.0005 & 0 & 0.0006 & 0 & -0.002 & 0 & -0.001 \\
-0.004 & -0.004 & 0.0002 & 0 & 0 & 0 & 0.0006 & 0 & -0.002 & 0 & 0.0005 & 0 & 0.0002 & 0 & 0 \\
-0.145 & -0.006 & -0.017 & 0 & 0 & 0 & -0.011 & 0 & -0.002 & 0 & 0.0001 & 0 & 0.004 & 0 & -0.02
\end{bmatrix} \tag{73}$$

8.2.3 Acceleration as output

(a) Longitudinal motion

$$y_A = C_A x + D_A u \qquad (74)$$

The matrices C_A and D_A depend on the location A at which the acceleration is measured. Three locations were used: 4·4 m, 21·9 m and 42·0 m from the tip of the nose of the aircraft. Consequently, there are three sets of matrices, which are:

$$C_{4\cdot4} = [-32{\cdot}65 \quad -28{\cdot}04 \quad 49{\cdot}93 \quad 2{\cdot}37 \quad -1405{\cdot}41 \quad -10{\cdot}39 \quad -1707{\cdot}61 \quad -13{\cdot}09 \quad -4472{\cdot}18 \quad -7{\cdot}53 \quad 4876{\cdot}78 \quad 5{\cdot}83] \qquad (75)$$

$$D_{4\cdot4} = [-12{\cdot}55 \quad -1{\cdot}02] \qquad (76)$$

$$C_{21\cdot9} = [-11{\cdot}69 \quad -35{\cdot}68 \quad 65{\cdot}02 \quad 6{\cdot}27 \quad 117{\cdot}33 \quad 0{\cdot}87 \quad 895{\cdot}25 \quad 5{\cdot}94 \quad 191{\cdot}7 \quad 1{\cdot}34 \quad 313{\cdot}96 \quad 0{\cdot}38] \qquad (77)$$

$$D_{21\cdot9} = [-1{\cdot}23 \quad 0{\cdot}00] \qquad (78)$$

$$C_{42\cdot0} = [3{\cdot}68 \quad -38{\cdot}04 \quad 153{\cdot}63 \quad 17{\cdot}68 \quad -739{\cdot}42 \quad -5{\cdot}47 \quad -821{\cdot}74 \quad -5{\cdot}66 \quad 672{\cdot}66 \quad -1{\cdot}83 \quad 1623{\cdot}60 \quad 1{\cdot}94] \qquad (79)$$

$$D_{42\cdot0} = [3{\cdot}69 \quad -1{\cdot}2] \qquad (80)$$

(b) Lateral motion

$$y_{1_A} = M_A x_1 + L_A u_1 \qquad (81)$$

Again, the matrices M_A and L_A depend on the location A at which the acceleration is measured. The same three locations as longitudinal motion were used. The corresponding sets of matrices are:

$$M^{4}_{4\cdot4} = [-0{\cdot}072 \quad -3{\cdot}566 \quad 34{\cdot}602 \quad 0{\cdot}975 \quad 0{\cdot}0 \quad 14{\cdot}939 \quad 0{\cdot}288 \quad -181{\cdot}765 \quad -4{\cdot}087 \quad 100{\cdot}485 \quad 3{\cdot}781 \quad -7530{\cdot}947 \quad -18{\cdot}539 \quad -3321{\cdot}345 \quad -3{\cdot}175] \qquad (82)$$

$$L_{4\cdot4} = [-1{\cdot}401 \quad -0{\cdot}939 \quad -1{\cdot}043] \qquad (83)$$

$$M_{21\cdot9} = [-0{\cdot}03 \quad 0{\cdot}045 \quad 34{\cdot}775 \quad 0{\cdot}991 \quad 0{\cdot}0 \quad 10{\cdot}956 \quad 0{\cdot}212 \quad 44{\cdot}345 \quad 1{\cdot}171 \quad -91{\cdot}84 \quad -1{\cdot}513 \quad 1158{\cdot}024 \quad 4{\cdot}776 \quad -462{\cdot}469 \quad -1{\cdot}923] \qquad (84)$$

$$L_{21\cdot9} = [0{\cdot}101 \quad -1{\cdot}208 \quad 0{\cdot}056] \qquad (85)$$

$$M_{42\cdot0} = [-0{\cdot}101 \quad 1{\cdot}418 \quad 32{\cdot}597 \quad 0{\cdot}0 \quad -808{\cdot}717 \quad -274{\cdot}573 \quad -6{\cdot}85 \quad 413{\cdot}086 \quad 8{\cdot}005 \quad -535{\cdot}333 \quad -2{\cdot}476 \quad -803{\cdot}553 \quad -1{\cdot}686] \qquad (86)$$

$$L_{42\cdot0} = [2{\cdot}031 \quad -6{\cdot}369 \quad -0{\cdot}177] \qquad (87)$$

Design of a Two-Level Adaptive Flight Control System

P.N. Nikiforuk,* H. Ohta,† and M.M. Gupta‡
University of Saskatchewan, Saskatoon, Sask., Canada

A design of a two-level adaptive flight controller for a STOL aircraft with unknown dynamics is described. The approach appears to overcome some of the limitations that are inherent in the design of linear optimal and conventional adaptive controllers. In particular, it is assumed that the state variables are not accessible. For estimating the state variables an adaptive observer is developed which has an exponential rate of convergence, and which simultaneously models the dynamics of the unknown plant. Control at the first level is provided by an updated optimal controller, while that at the second level is provided by an error servo. Some examples of simulation studies that were carried out for the pitch attitude control system for two different conditions are given.

Introduction

MUCH work has been done during the past few years on the design of controllers which can simultaneously satisfy the desired response characteristics of a system and cope with the modeling inaccuracies and environmental uncertainties. Typical examples of such works are found in the publications by Landau and Courtiol,[1] Horowitz et al.,[2] and Ohta and Sugiura.[3]

Of particular interest here is the work done on two-level controllers.[4,5] A common characteristic of these systems is that each controller operates in one of two modes and each plant is controlled by both feedback and feedforward. Basically, the first-level controller can produce the desired response characteristics if the plant is known and deterministic. If, however, the plant is unknown and nondeterministic, the second level compensates for modeling inaccuracies and uncertainties. More specifically, the system described by Preusche[4] consists of an optimal controller with model state feedback at the first level and an error servo at the second level. By assuming that the deviations between the dynamic behaviors of the plant and the model are small, the problems of state measurement and sensitivity, which are characteristic of optimal controllers, were avoided by Preusche. In the two-level adaptive controller reported by Nikiforuk et al.[5] a "characteristic vector," which is an implicit function of the unknown environment of the plant, was used. This adaptive controller was designed for multi-input multi-output unknown plants and employed a Liapunov-type signal synthesis procedure. This design procedure was subsequently applied to a gust alleviation control system.[6]

In this paper a two-level adaptive controller is developed for applications to aircraft-type systems – systems which experience substantial changes in their dynamic characteristics. Previous work in this area includes that done by Narendra and Tripathi,[7] who studied the identification and optimization of helicopter dynamics assuming the accessibility of the state variables. To avoid the use of state variables, an adaptive observer is used in this paper. Such observers, as has been shown by Carroll and Lindorff,[8] Kudva and Narendra,[9] Kreisselmeier,[10] and Nikiforuk et al.,[11] can be used in both single-input single-output and multi-input multi-output systems for providing estimates of the state variables and the parameters of an unknown plant using only input-output data. In the system to be described, an adaptive observer is used explicitly in the first level for the purpose of generating the signals needed for updating.

In the following paragraphs, first a description is given of the overall two-level adaptive control scheme that was adopted. This is followed by the development of an adaptive observer with an exponential rate of convergence. This observer, which is to be used for state estimation and plant indentification, is developed using an equiobservable canonical form previously described by the authors[12] and the least-squares method.[13] The optimal controller is then described. Finally, some simulation results are presented which were obtained from a study that was made of the controller's application to the pitch attitude control of a STOL.

Configuration of the Two-Level Adaptive Control Scheme

The purpose of the two-level adaptive control system that is proposed here is to control the outputs of a real but unknown plant so as to minimize a given performance criterion. This system, which is illustrated in Fig. 1, consists of four main blocks: 1) a real plant with unknown dynamics, 2) an adaptively identified model of the plant (the adaptive observer), 3) an updated optimal controller with model state variable feedback, and 4) an error servo. This system also contains two control switches. The first-level control becomes operative when switch 1 is closed and the second-level control becomes operative when switch 2 is closed. The estimation and the identification of the unknown plant are carried out by

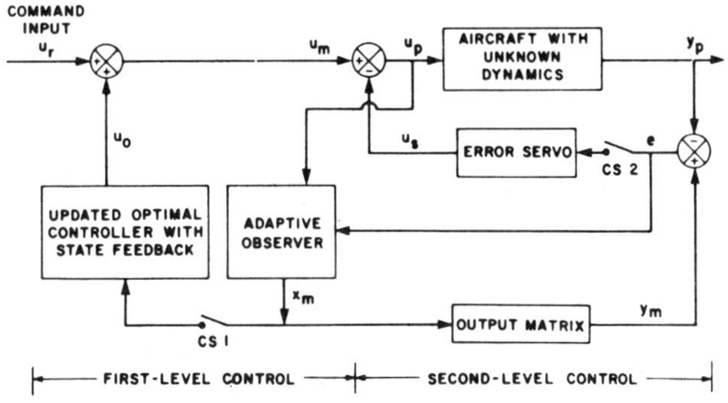

Fig. 1 Configuration of the two-level adaptive control system.

Presented as Paper 77-1092 at the AIAA Guidance and Control Conference, Hollywood, Fla. Aug. 8-10, 1977; submitted Sept. 12, 1977; revision received May 1, 1978. Copyright © American Institute of Aeronautics and Astronautics, Inc., 1977. All rights reserved.

Index categories: Spacecraft Dynamics and Control; Spacecraft Simulation.

*Professor and Dean, Systems and Adaptive Control Research Lab, College of Engineering.

†Research Associate, Systems and Adaptive Control Research Lab., College of Engineering; presently at Nagoya University, Japan. Member AIAA.

‡Professor, Systems and Adaptive Control Research Lab., College of Engineering.

Reprinted with permission from *J. Guidance and Control*, vol. 2, pp. 79–85, Jan–Feb 1979.
Copyright © 1979 by the American Institute of Aeronautics and Astronautics, Inc.

the adaptive observer regardless of the position of these switches.

This system operates as follows. All known properties (structure and parameters) of the plant and the command inputs are handled by the first-level control, which contains the adaptive observer and an updated optimal controller. These two elements are updated periodically, using the current estimate of the plant dynamics, and provide a first-level compensation. As a consequence, at the second level the error servo has to compensate only for the small deviations which exist between the model and the plant and which are due to the unidentified or unregarded properties of the plant.

Some of the advantages which are associated with this system are as follows:

1) The optimal controller uses estimated state variable feedback and generates an optimal policy that is based on the current knowledge of the unknown plant.

2) The adaptive observer considers only the essential dynamic properties of the plant. Thus, it may be of lower order and simpler than the plant itself.

3) A conventional controller may be used as an error servo system. Should the first-level control fail, the conventional control system may be used as a backup system.

Design of the Adaptive Observer

Consider a linear time-invariant dynamical plant described by the equation

$$\dot{x}(t) = Fx(t) + gu_p(t) \tag{1a}$$

$$y_p(t) = hx(t) \tag{1b}$$

where $x(t) \in R^n$ is a state vector and $u_p(t)$ is an input and $y_p(t)$ is an output. The unknown triple (F,g,h) is of appropriate size and is assumed to be completely controllable and observable. It is also assumed that the plant is free of disturbances and that the only accessible signals are $u_p(t)$ and $y_p(t)$, which are noise free.

A specific canonical form is selected for the design of an adaptive model of the plant. It is shown in the Appendix that if the plant described by Eq. (1) is completely observable, the pair (F,h) can be represented by

$$F = \Lambda + [f:\ldots:f] \qquad h = [1:\ldots:1] \tag{2}$$

where $\Lambda \in R^{n\times n}$ is a diagonal matrix with arbitrary but known constants and negative elements $-\lambda_i$ ($i = 1,\ldots,n$; and $\lambda_i \neq \lambda_j$ for $i \neq j$). The f and $g \in R^n$ are unknown vectors whose parameters are to be identified. Only the single-input single-output case is considered in this study. The multivariable case can be handled using the multivariable version of the canonical form discussed in Ref. 12.

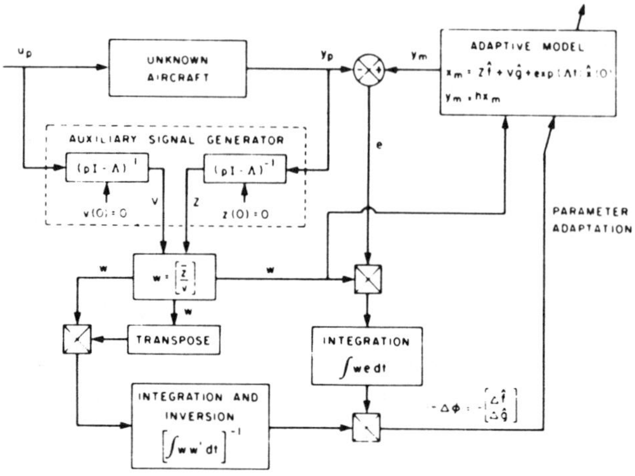

Fig. 2 Schematic diagram of the adaptive observer.

Let the control switches be in the off position in Fig. 1 (that is $u_p = u_m = u_r$). In addition, let the adaptive model of the plant be

$$\dot{x}_m(t) = \hat{F}x_m(t) + \hat{g}u_p(t) \tag{3a}$$

$$y_m(t) = hx_m(t) \tag{3b}$$

where $x_m(t) \in R^n$ and $\hat{F}$ has the same form as F in Eq. (2), and $\hat{f}$ and $\hat{g}$, the estimated values of f and g, which are estimated as follows.

Using the measurable signals $y_p(t)$ and $u_p(t)$, let the auxiliary signals, $Z(t)$ and $V(t) \in R^{n\times n}$, and $\bar{Z}(t)$ and $\bar{V}(t) \in R^n$, be defined as

$$Z(t) = [pI - \Lambda]^{-1} y_p(t) \tag{4a}$$

$$\bar{Z}(t) = Z(t)h' \tag{4b}$$

$$V(t) = [pI - \Lambda]^{-1} u_p(t) \tag{4c}$$

$$\bar{V}(t) = V(t)h' \tag{4d}$$

where $p \triangleq \mathrm{d}/\mathrm{d}t$ is the differential operator, and $[\cdot]'$ is a transpose. A filter whose eigenvalues are at the designer's discretion is introduced in Eq. (4). This filter, as illustrated in Fig. 2, is driven by either y_p or u_p of the plant, and generates the auxiliary signals $Z(t)$ and $V(t)$. These signals are available and are used for representing the plant dynamics as an algebraic function of them. Without loss of generality, the initial values of the filters, $Z(0)$ and $V(0)$, are set equal to zero. Then Eq. (1) can be expressed as

$$x(t) = Z(t)f + V(t)g + \exp[\Lambda t]\cdot x(0) \tag{5a}$$

$$y_p(t) = \bar{Z}'(t)f + \bar{V}'(t)g + h\exp[\Lambda t]\cdot x(0) \tag{5b}$$

Let the parameter vector to be identified, $\phi \in R^{2n}$, and the signal vector, $w(t) \in R^{2n}$, be defined as

$$\phi = \begin{bmatrix} f \\ g \end{bmatrix} \qquad w(t) = \begin{bmatrix} \bar{Z}(t) \\ \bar{V}(t) \end{bmatrix} \tag{6}$$

Hence, Eq. (5b) can be rewritten in the simple form

$$y_p(t) = w'(t)\phi + h\exp[\Lambda t]\cdot x(0) \tag{7}$$

Pre-multiplying Eq. (7) by $w(t)$ and taking integral over the time period T, the least-squares estimate of ϕ is given as[13]

$$\hat{\phi}(k+1) = \left[\int_{kT}^{(k+1)T} w(t)w'(t)\mathrm{d}t\right]^{-1} \times \left[\int_{kT}^{(k+1)T} w(t)\{y_p(t) - h\exp(\Lambda t)\cdot\hat{x}(0)\}\mathrm{d}t\right] \tag{8}$$

where

$$\hat{\phi}(k+1) \triangleq [\hat{f}'(k+1),\ \hat{g}'(k+1)]' \tag{9}$$

is the estimate of ϕ at $t = (k+1)T$, and is determined from measurements for $kT \leq t < (k+1)T$. $\hat{x}(0)$ is the initial estimate of the state at the designer's disposal. The inverse of $[\int_{kT}^{(k+1)T} w(t)w'(t)]$ exists if the input $u_p(t)$ is sufficiently rich in frequencies.[8] Substituting Eq. (7) into Eq. (8) gives

$$\hat{\phi}(k+1) = \phi\left[\int_{kT}^{(k+1)T} w(t)w'(t)\mathrm{d}t\right]^{-1} \times \left[\int_{kT}^{(k+1)T} w(t)\cdot h\exp(\Lambda t)\cdot(\hat{x}(0) - x(0))\mathrm{d}t\right] \tag{10}$$

On the other hand, $y_m(t)$ in Eq. (3) can be expressed in the same way as Eq. (7) using the estimated values $\hat{\phi}(\cdot)$ and $\hat{x}(0)$:

$$y_m(t) = w'(t)\hat{\phi}(k) + h\exp[\Lambda t]\cdot\hat{x}(0)$$

$$\text{for} \quad kT \leq t < (k+1)T \tag{11}$$

where, $x_m(0) = \hat{x}(0)$. Defining the output error between $y_p(t)$ and $y_m(t)$ as

$$\begin{aligned} e(t) &= y_m(t) - y_p(t) \\ &= w'(t)[\hat{\phi}(k) - \phi] + h\exp[\Lambda t]\cdot[\hat{x}(0) - x(0)] \end{aligned} \tag{12}$$

and substituting Eq. (12) into Eq. (10), the updating algorithm to be sought becomes

$$\hat{\phi}(k+1) = \hat{\phi}(k) - \left[\int_{kT}^{(k+1)T} w(t)w'(t)\mathrm{d}t\right]^{-1} \cdot \left[\int_{kT}^{(k+1)T} w(t)e(t)\mathrm{d}t\right] \tag{13}$$

As can be seen from Eq. (10), $\hat{\phi}(k+1)$ converges exponentially to ϕ as $k\rightarrow\infty(t\rightarrow\infty)$. In addition, $x_m(t)$, the estimated value of the state vector $x(t)$, described by

$$x_m(t) = Z(t)\hat{f}(k) + V(t)\hat{g}(k) + \exp[\Lambda t]\cdot\hat{x}(0)$$

$$\text{for} \quad kT \leq t < (k+1)T \tag{14}$$

also converges exponentially to $x(t)$ as $t\rightarrow\infty$. Hence, Eqs. (11, 13, and 14) define an adaptive observer and identifier whose schematic diagram is illustrated in Fig. 2.

It must be noted that this adaptive observer is similar in some respects to that described in Ref. 10, in that the observer dynamics are represented as an algebraic function of the filter states and the identified parameters, and the estimates of the state vector and unknown parameters converge to their true values exponentially. The identification algorithm of Eq. (13) does not require the selection of weighting parameters, which is unavoidable in the algorithms based on stability criteria.[8,9,11] Moreover, this observer does not require auxiliary signals for achieving global asymptotical convergence of the observation process. Such signals are characteristic of conventionally designed adaptive observers.[8,9] It is believed that the canonical form employed here simplifies the derivation of the observers discussed in Ref. 10.

Design of an Updated Optimal Controller

The design of the updated optimal controller is based here upon model Eq. (3) [or Eqs. (11) and (14)], which represents the current knowledge of the plant. Consider only the first-level control shown in Fig. 1 (that is, $u_p = u_m = u_0 + u_r$). The updated optimal controller is designed to produce an optimal input $u_m(t)$ so as to minimize a quadratic performance index J of the form

$$J = \frac{1}{2}\int_0^\infty [Qy_m^2(t) + Ru_m^2(t)]\mathrm{d}t \tag{15}$$

where Q and R are positive constants. The control law is given by

$$u_m(t) = -Lx_m(t) + u_r(t) \tag{16a}$$

$$L \triangleq R^{-1}\hat{g}'K \tag{16b}$$

where $n \times n$ positive definite symmetric matrix K satisfies the algebraic Riccati equation:

$$\hat{F}'K + K\hat{F} - K\hat{g}R^{-1}\hat{g}'K + h'Qh = 0 \tag{17}$$

According to Eqs. (16) and (17), the gain matrix L is updated every T seconds when new estimates of $\hat{F}$ and $\hat{g}$ are obtained. Since $(\hat{F},\hat{g})$ are not time-invariant, the same conditions as described in Ref. 7 are necessary for the derivation of the control law [Eq. (16)].

To solve Eq. (17) efficiently, the procedure suggested by Kleinman[14] is used. The iterative algorithm involves the following steps:

1) Selecting a matrix L_0 such that the matrix $A_0 = \hat{F} - \hat{g}L_0$ has eigenvalues with negative real parts.

2) Solving for V_k, the linear algebraic equation,

$$A_k'V_k + V_kA_k + h'Qh + L_k'RL_k = 0 \tag{18}$$

3) Computing

$$L_k = R^{-1}\hat{g}'V_{k-1} \tag{19a}$$

$$A_k = \hat{F} - \hat{g}L_k \tag{19b}$$

As $k\rightarrow\infty$ the foregoing procedure leads to the solution of Eq. (17).

Simulation Study

To evaluate the effectiveness of the system just described, a simulation study was carried out for the pitch attitude control of a STOL aircraft. The data that were used were taken from Ref. 15 and are those for the Dornier DO-28 D STOL "Sky Servant" aircraft.

The linearized longitudinal dynamics of the aircraft motion in the stability axis about the equilibrium point are given by the state equations

$$\dot{x}_p = A_px_p + b_pu_p \tag{20a}$$

$$y_p = c_px_p \tag{20b}$$

where

$$x_p = \begin{bmatrix} \theta \\ q \\ w \\ u \end{bmatrix} \begin{matrix} \text{(pitch attitude)} \\ \text{(pitch rate)} \\ \text{(vectical velocity)} \\ \text{(forward velocity)} \end{matrix}$$

$$u_p = \delta e \quad \text{(elevator deflection angle)}$$

$$y_p = \theta$$

Table 1 Definition of the flight condition[15]

No.	Flight conditions	Mass, kg	Altitude, m	Forward velocity, V_0, m/s	Flap position, deg	Glide path angle γ_0, deg	Dynamic pressure q_D, N/m^2
1	Cruise, fully loaded	3500	2000	77.8	0	0	3046
2	Cruise, without payload	2360	2000	77.8	0	0	3046
3	Descent	3400	0	33.4	20	− 5.4	683
4	Steep descent	3500	0	30.9	52	− 10.4	585

and the coefficient matrices in Eq. (20) are

$$A_p = \begin{bmatrix} 0 & 1 & 0 & 0 \\ -g\sin\gamma_0 M_{\dot{w}} & M_q + (V_0 + Z_q)M_{\dot{w}} & M_w + Z_w M_{\dot{w}} & M_u + Z_u M_{\dot{w}} + (y_F m/I_y)F_u \\ -g\sin\gamma_0 & V_0 + Z_q & Z_w & Z_u \\ -g\cos\gamma_0 & 0 & X_w & X_u + F_u \end{bmatrix} \quad (21a)$$

$$b_p' = [0 \quad M_{\delta e} + Z_{\delta e} M_{\dot{w}} \quad Z_{\delta e} \quad x_{\delta e}] \quad (21b)$$

$$c_p = [1 \quad 0 \quad 0 \quad 0] \quad (21c)$$

Simulation studies were conducted for the four flight conditions (FC) shown in Table 1. However, in this paper results are presented only for the two extremes, **FC 1** and **FC 4.** For purposes of information, the values of the coefficient matrices A_p and b_p are given in Table 2, from which it can be seen that there are large variations in the dynamics of the aircraft between these two flight conditions.

Study of the Identification Scheme

To illustrate the effectiveness of the identification and estimation procedure, digital simulations were first made using the short-period approximation of Eq. (21). The approximated dynamics consist of the three-dimensional system less the contribution of the forward velocity. The adaptive observer that is employed, Eqs. (11) and (14), is also of this order, which corresponds to the approximation of the plant. The control switches in Fig. 1 were set to be in the off position (that is $u_p = u_m = u_r$) for the identification and estimation purpose. The poles of the filter in Eq. (4), which was used for generating the auxiliary signals, were selected to be at

$$\lambda_1 = 4.0 \qquad \lambda_2 = 3.0 \qquad \lambda_3 = 0.5 \quad (22)$$

and the updating period T of the parameters was selected to be 1s. A rectangular pulse of period 2 s was used as a command input u_r.

Figures 3a and 3b show the time histories of the identified values of the unknown parameters for **FC** 1 and **FC** 4, where all of the initial conditions are zero. These figures illustrate the rapid and stable character of the adaptive observer. Figure 3a shows better responses than Fig. 3b because the poles, $-\lambda_i$, of the filter were selected for **FC** 1 and not **FC** 4. These figures also show that the output error e between y_m and y_p for each flight condition was nearly equal to zero after about 4 s. The state variables $x_m(t)$ of the adaptive observer were seen to converge to their true values $x(t)$ at almost the same rate. The updating period T also affects these responses. At the expense of overshoot, the speed of convergence of the parameters to their true values can be made faster by making T smaller. However, to avoid excessive overshoot, a value of $T=2$ was selected in the following simulation studies.

Table 2 Coefficient matrices for flight conditions 1 and 4

FC No.	A_p				b_p
1	0	1	0	0	0
	0	−4.625	−0.136	−0.0147	−39.488
	0	75.896	−1.643	−0.2520	−20.710
	−9.800	0	0.0739	−0.0490	0
4	0	1	0	0	0
	−0.0283	−2.233	−0.0643	−0.0214	−7569
	1.769	29.979	−0.735	−0.636	−3.977
	−9.639	0	0.205	−0.143	0

Simulation Studies of Short-Period Dynamics of STOL Aircraft

The effectiveness of the two-level adaptive controller was examined using the same approximate dynamics of the plant and adaptive observer and the same command input u_r as used in the above example. The updated optimal controller was synthesized using the weighting coefficients $Q=1$, $R=10$ in the performance index [Eq. (15)]. The matrix Riccati Eq. (17) was solved using the recursive algorithm given by Eqs. (18) and (19). For each updating period T, five iterations were carried out so as to calculate the feedback gain matrix L, based upon the current status of the adaptive observer.

The error servo consists of a feedback signal for the pitch attitude angle error between the plant and the adaptive observer. The numerical value of its gain, which was determined using modal control theory, was taken from Ref. 15 for the purpose of a conventional servo system, and is

$$K_\theta = 0.12 + 0.38\, q_{D1}/q_{Di} \quad (23)$$

where q_{Di} denotes the dynamic pressure of the ith flight condition.

Results of the simulation studies are shown in Figs. 4 and 5 for the flight conditions 1 and 4, respectively. Five time histories are illustrated for each flight condition: the output of the plant y_p, the output error $e = y_m - y_p$, the control input to the plant u_p, the estimated values of parameters $\hat{f}$ and $\hat{g}$, and the feedback gain L. Optimal values with a priori knowledge of the plant's dynamics are also shown for purposes of comparison. These simulation studies were performed by initially having switches CS 1 and CS 2 opened until the adaptive model yielded the first estimate of the plant at $t=2$s. The initial conditions of the plant and the adaptive observer [Eq. (14)] were set equal to zero for each flight condition, and the initial values of the unkown parameters were assumed to be

$$\hat{f}(0) = \begin{bmatrix} 5 \\ -5 \\ 0 \end{bmatrix} \qquad \hat{g}(0) = \begin{bmatrix} 10 \\ -10 \\ 2 \end{bmatrix}$$

As seen from Figs. 4a and 5a, the error e between the outputs of the plant and the adaptive observer is almost zero after the third adapatation at $t=6$ s. As shown by Figs. 4b and 5b, the proposed adaptive observer yields rapid estimations of the unknown parameters. The first-level and two-level adaptive controllers produce the same time histories of estimates values of unknown parameters and feedback gains. At $t=4$ s of the second estimation they are approximately equal to the optimum values. It is also to be noted that, owing to the additional control produced by the error servo at the second level, the output y_p of the two-level adaptive control scheme converges more rapidly to the optimal one than that of the first-level adaptive control scheme. Both of the

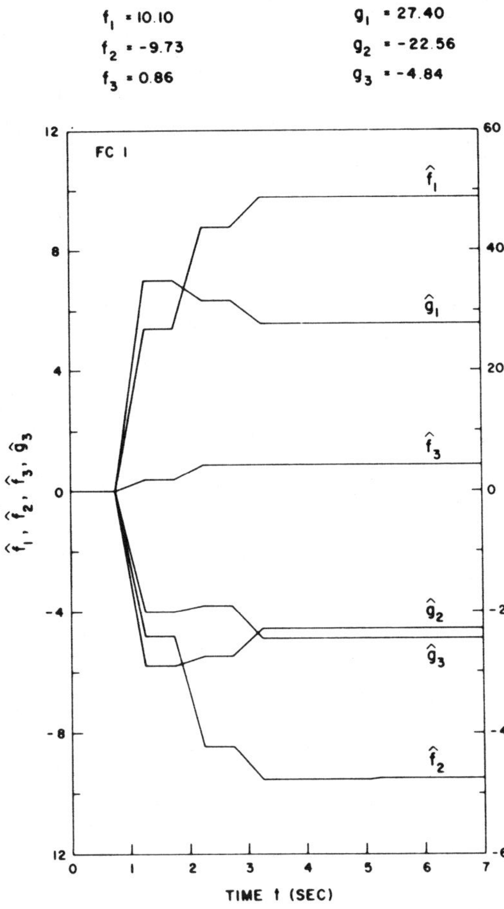

Fig. 3a Estimated values of the parameters for FC 1.

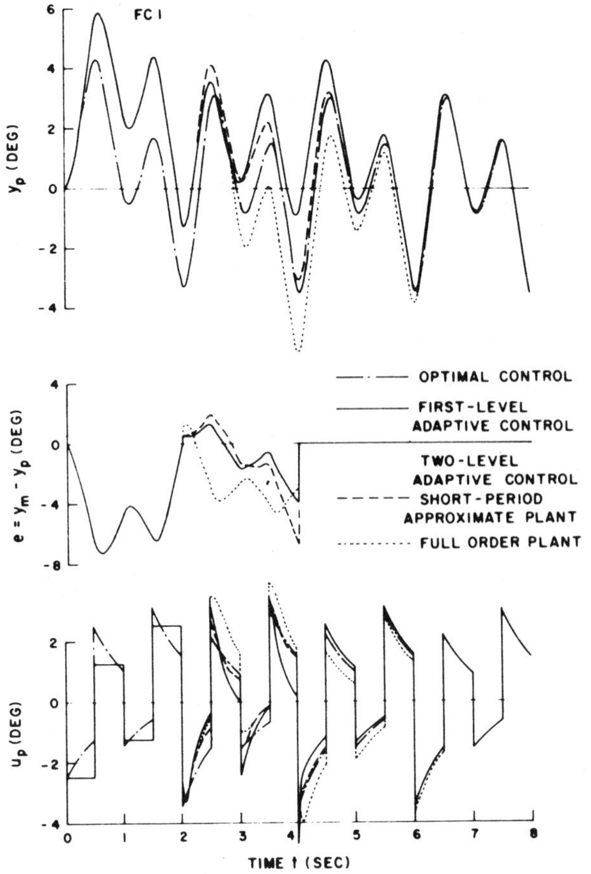

Fig. 4a Output y_p and input u_p of the plant and output error e for FC 1.

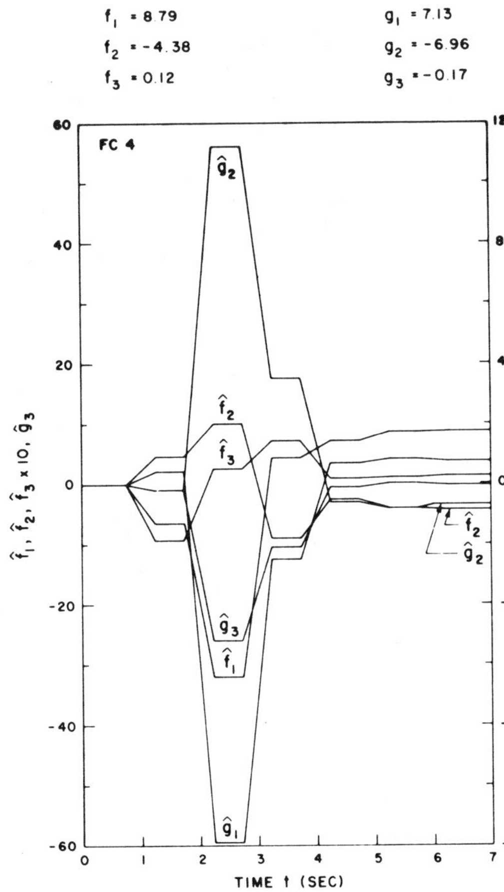

Fig. 3b Estimated values of the parameters for FC 4.

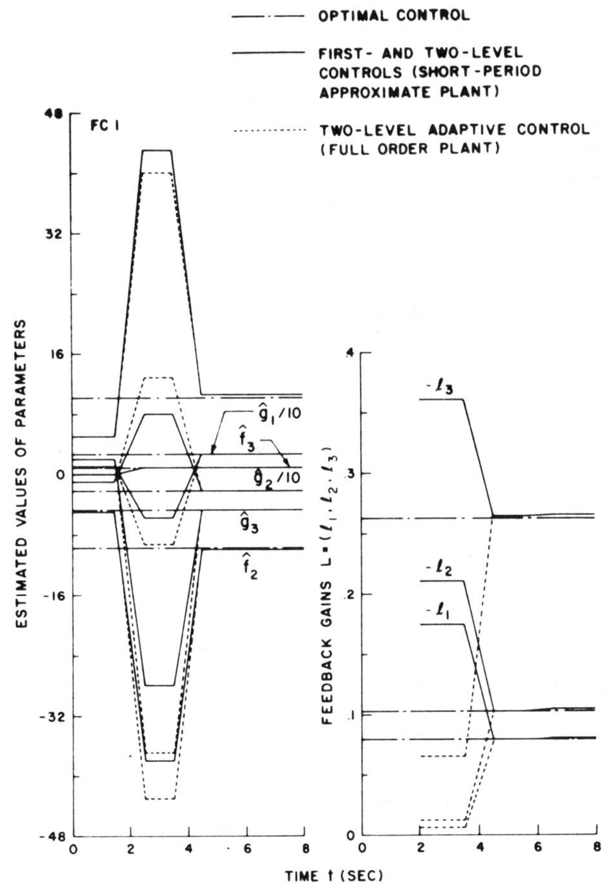

Fig. 4b Estimated values of parameters and feedback gain L for FC 1.

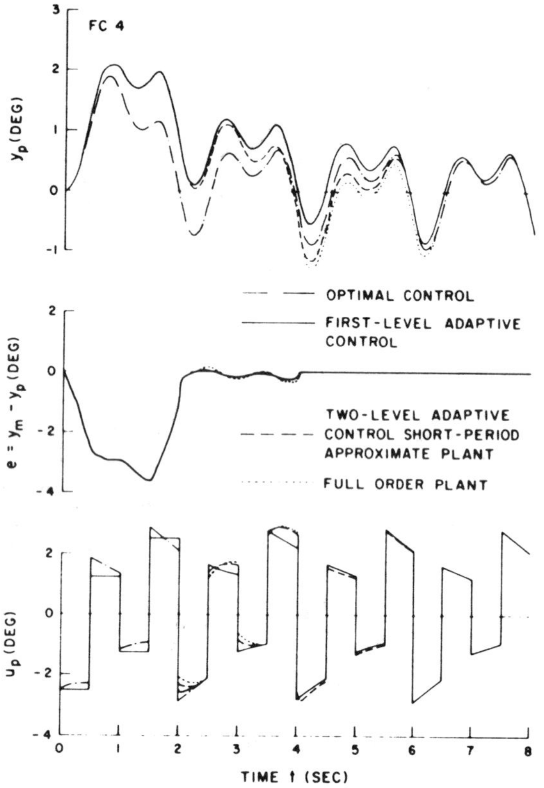

Fig. 5a The output y_p and input u_p of the plant and output error e for FC 4.

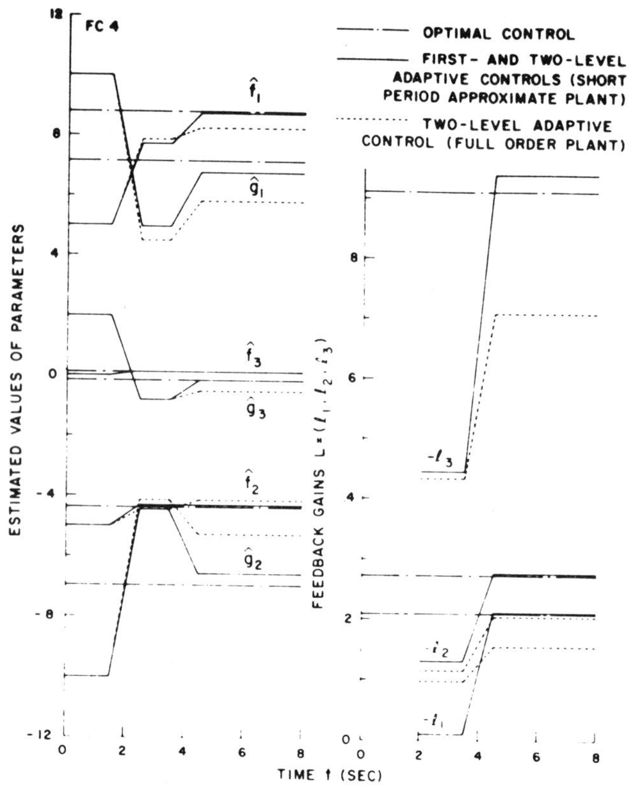

Fig. 5b Estimated values of the parameters and feedback gain L for FC 4.

schemes, however, produce almost optimal responses even when the adaptive observer does not yield a good estimate over the first estimation period.

Simulation Studies of Full-Order Dynamics of STOL Aircraft

In the next stage of the study, simulation studies were performed using the fourth-order dynamics of the STOL aircraft described in Eq. (21) and the previously described third-order adaptive observer; that is, the dimension of the observer is smaller than that of the plant. The numerical values of the parameters and initial conditions that were used here were the same as in the above example.

The time histories of the two-level adaptive control scheme using full-order dynamics of the plant are illustrated by the dotted lines in Fig. 4 for FC 1 and in Fig. 5 for FC 4. It is seen that the estimated values of the unknown parameters are almost the same as those obtained using the short-period dynamics. For FC 1 the output y_p, using the full-order dynamics, deviates from that using the short-period dynamics after the first adaptation. This is because the first estimates of $\hat{g}_1$ and $\hat{g}_2$ have greater overshoots, which lead to the poor estimates of the feedback gains. The output y_p converges, however, rapidly to the optimal value after the second adaptation. For FC 4, the output y_p, using full-order dynamics, also shows a good response.

Conclusion

In this paper, a two-level adaptive control scheme was developed for linear unknown plants. The unknown dynamics of the plant were identified and its state variables were estimated using an adaptive observer. Based upon this estimation, the feedback gains of an optimal controller were updated. This controller then provided a control signal at the first level. A conventional error servo provided the control at the second level. The adaptive observer was designed using a canonical form for modeling the unknown aircraft dynamics. The observer dynamics were represented as an algebraic function of certain filter states and have an exponential rate of convergence. The effectiveness of the proposed adaptive method was demonstrated using simulation studies of the pitch attitude control system of a STOL aircraft under two different flight conditions. Results of the simulation studies for short-period dynamics, as well as for full-order dynamics, show a rapid rate of convergence of the adaptive observer, and compare favorably with those obtained using an optimal controller with a priori knowledge of the dynamics of the plant.

Appendix: Derivation of the Canonical Form in Eq. (2)

Consider an n dimensional single output completely observable system (A, c) which is represented by the standard observable canonical form:

$$A = \left[\begin{array}{c|ccc} & 1 & \cdots & 0 \\ a & & \ddots & \\ & 0 & \cdots & 1 \\ & 0 & \cdots & 0 \end{array}\right] \qquad a = \begin{bmatrix} a_1 \\ a_2 \\ \vdots \\ a_n \end{bmatrix} \tag{A1}$$

$$c = [\,1 \,|\, 0 \cdots 0\,]$$

If a single-output system is completely observable, it can be transformed into the canonical form of Eq. (A1) by using the procedure given by Lüders and Narendra.[16]

The problem here is to find a nonsingular matrix T which transforms the pair (A,c) into the canonical form given by Eq. (2), that is,

$$F = TAT^{-1} \qquad h = cT^{-1} \tag{A2}$$

where

$$F = \Lambda + [f \vdots \ldots \vdots f] \qquad h = [1 \vdots \ldots \vdots 1]$$

$$\Lambda = \begin{bmatrix} -\lambda_1 & \cdots & 0 \\ \vdots & & \vdots \\ 0 & \cdots & -\lambda_n \end{bmatrix} \qquad f = \begin{bmatrix} f_1 \\ \vdots \\ f_n \end{bmatrix} \qquad (\lambda_i \neq \lambda_j \text{ for } i \neq j)$$

Let the following vectors be defined as:

$c[\Lambda]\epsilon R^{n+1}=$ a vector formed by the coefficients of the characteristic polynomial of $\prod_{k=1}^{n}(s+\lambda_k)$

$c(\Lambda/\lambda_i)\epsilon R^{n}=$ same as preceding, but the factor $(s+\lambda_i)$ $(i=1,...,n)$ is delected from the characteristic polynomial

A property of these vectors is

$$\begin{bmatrix} c(\Lambda/\lambda_i) \\ \hdashline 0 \end{bmatrix} + \lambda_i \begin{bmatrix} 0 \\ \hdashline c(\Lambda/\lambda_i) \end{bmatrix} = c(\Lambda) \qquad (i=1,...,n) \tag{A3}$$

Since the system matrices A and F must have the same characteristic polynomial, the two vectors a and f must satisfy the relation

$$\begin{bmatrix} 1 \\ \hdashline -a \end{bmatrix} = c(\Lambda) - \sum_{i=1}^{n} f_i \begin{bmatrix} 0 \\ \hdashline c(\Lambda/\lambda_i) \end{bmatrix} \tag{A4}$$

The transformation matrix T in Eq. (A2) is defined to be

$$T^{-1} = [c(\Lambda/\lambda_1) : c(\Lambda/\lambda_2) : ... : c(\Lambda/\lambda_n)] \tag{A5}$$

By carrying out the matrix multiplication $AT^{-1}=T^{-1}F$, and by using Eqs. (A3) and (A4), both sides of the equation can be shown to be identical. Further, $cT^{-1}=h$. Also, since the column vectors of T^{-1} are all linearly independent, because they are generated by the set of numers $\lambda_i (i=1,...,n)$ which are all distinct, T is a nonsingular matrix.

Acknowledgment

This work was supported by the National Research Council of Canada Grants A-5225 and A-1080.

References

[1] Landau, I.D. and Courtiol, B., "Adaptive Model Following Systems for Flight Control and Simulation," *Journal of Aircraft*, Vol. 9, Sept. 1972, pp. 668-674.

[2] Horowitz, I.M., Smay, J.W., and Shapiro, A., "A Synthesis Theory for the Externally Excited Adaptive Systems (EEAS)," *IEEE Transactions on Automatic Control*, Vol. AC-19, April 1974, pp. 101-107.

[3] Ohta, H. and Sugiura, I., "Optimal Landing Flare Control of Aircraft with Sensitivity Consideration," *Proceedings of the 3rd IFAC Symposium on Sensitivity, Adaptivity and Optimality*, Ischia, Italy, June 1973, pp. 251-259.

[4] Preusche, G., "A Two-Level Model Following Control System and Its Application to the Power Control of a Steam-Cooled Fast Reactor," *Automatica*, Vol. 8, March 1972, pp. 145-151.

[5] Nikiforuk, P.N., Gupta, M.M., Kanai, K., and Wagner, L., "Synthesis of Two-Level Controller for a Class of Linear Plants in an Unknown Environment," *Proceedings of the 3rd IFAC Symposium on Sensitivity, Adaptivity and Optimality*, Ischia, Italy, June 1973, pp. 431-439.

[6] Nikiforuk, P.N., Gupta, M.M., and Kanai, K., "Stability Analysis and Design for Aircraft Gust Alleviation Control," *Automatica*, Vol. 10, Sept. 1974, pp. 495-506.

[7] Narendra, K.S. and Tripathi, S.S., "Identification and Optimization of Aircraft Dynamics," *Journal of Aircraft*, Vol. 10, April 1973, pp. 193-199.

[8] Carroll, R.L. and Lindorff, D.P., "An Adaptive Observer for Single-Input Single-Output Linear Systems," *IEEE Transactions on Automatic Control*, Vol. AC-18, Oct. 1973, pp. 428-435.

[9] Kudva, P. and Narendra, K.S., "Synthesis of an Adaptive Observer Using Lyapunov's Direct Method, *International Journal of Control*, Vol. 18, Oct. 1973, pp. 1201-1210.

[10] Kreisselmeier, G., "Adaptive Observers with Experimental Rate of Convergence," *IEEE Transactions on Automatic Control*, Vol. AC-22, Feb. 1977, pp. 2-8.

[11] Nikiforuk, P.N., Ohta, H., and Gupta, M.M., "Adaptive Observer and Identifier Design for Multi-Input Multi-Output Systems," *Proceedings of the 4th IFAC Symposium on Multivariable Technological Systems*, Fredericton, Canada, July 1977, pp. 189-196.

[12] Nikiforuk, P.N., Ohta, H., and Gupta, M.M., "On the Development of Equiobservable Canonical Forms for Linear Multivariable Systems," *IEEE Transactions on Automatic Control*, (to be submitted), 78.

[13] Luenberger, D.G., *Optimization by Vector Space Method*, John Wiley, New York, 1969, Chap. 4.

[14] Kleinman, D.L., "On an Iterative Technique for Riccati Equation Computations," *IEEE Transactions on Automatic Control*, Vol. AC-13, Feb. 1968, pp. 114-115.

[15] Hartman, U., "Application of Model Control Theory to the Design of Digital Flight Control Systems," *AGARD-CP-137 on Advances in Control Systems*, Sept. 1973, pp. 5.1-5.21.

[16] Lüders, G. and Narendra, K.S., "An Adaptive Observer and Identifier for a Linear System," *IEEE Transactions on Automatic Control*, Vol. AC-18, Oct. 1973, pp. 496-499.

Command and Stability Systems for Aircraft: A New Digital Adaptive Approach*†

U. HARTMANN‡ and V. KREBS‡

Handling qualities requirements of high performance aircraft can be met by an adaptive control system using pole allocation principles and recursive least-squares fading memory parameter estimation, which is demonstrated by hybrid simulations and flight test data.

Key Words—Adaptive control; digital control; computer control; aerospace control; identification.

Abstract—Ensuring good handling qualities of aircraft in the whole flight envelope is a difficult and time consuming task, because aircraft parameters are subjected to drastic changes. Adaptive control is a very promising technique to solve these problems and to improve the performance deficiencies of conventional stability systems with preprogrammed control gains.

The approach used in this paper is based on digital control and on a special stabilization concept which requires only three dynamic parameters of the aircraft to meet standard handling qualities criteria. These three parameters cannot be measured directly. Therefore a recursive, weighted least squares algorithm is used, which delivers sufficient fast and accurate on-line parameter estimates. The adaptive control system has proven its very satisfactory performance in several digital and hybrid simulations including fast parameters variations and turbulence conditions. Real flight test data have been used to verify the performance of the identification process and recently flight tests of the complete adaptive system have been successfully carried out with the DO 28 D SKYSERVANT aircraft.

1. INTRODUCTION

THE DYNAMIC characteristics of aircraft are determined by a number of aerodynamic and configuration parameters like dynamic pressure, altitude, mass, wing-sweep-angle and flap position. The design of command- and stability systems which provide good handling qualities in the whole flight envelope of high performance aircraft is therefore a difficult task.

The programming of the controller gains is the approach generally used to solve this problem. The measurement of all parameters which influence the dynamics of the aircraft, however, would be very expensive especially in the case of triplex or quadruplex redundant systems. On the other hand, the programming of each controller gain as a function of several independent variables is not practicable with analog technology and rather difficult with digital technology. Therefore, in practical applications only a few of the most important influences are measured, e.g. dynamic pressure and altitude. This leads to certain compromises as far as the handling qualities are concerned.

Adaptive control is a very promising technique to overcome these difficulties and to establish good handling qualities for all flight conditions. The need for a complex sensor system in conventional systems is replaced by increased and more sophisticated computations. The efforts for the development of adaptive flight control systems have now been under way for more than 20 years (Gregory, 1959; AGARD Proceedings, 1970). However, in the past most of these developments failed because the problems connected with the practical realization and the pre- and inflight-testing turn out to be unsolvable when analog technology is used. The introduction of digital computers in the flight control field has now remarkably changed basic technical aspects, because the questions of realization and testing can be solved. Therefore, today increased activities in adaptive flight control can be observed. This is emphasized for instance by the NASA's advanced control law program for the F-8 digital fly-by-wire aircraft, which is discussed in a special issue of the IEEE Transactions on Automatic Control (Vol. AC-22, No. 5, October 1977).

The approach of this paper is based on three major steps. In the first step a design procedure for the stabilization of the longitudinal motion of aircraft is investigated which requires knowledge of only three dynamic parameters. This is essential, because these parameters cannot be measured directly and the difficulties of parameter

*Received 6 September 1978; revised 21 May 1979; revised 10 September 1979. The original version of this paper was presented at the 7th IFAC World Congress on A Link Between Science and Applications of Automatic Control, which was held in Helsinki, Finland during June 1978. The published Proceedings of this IFAC Meeting may be ordered from: Pergamon Press Limited, Headington Hill Hall, Oxford, OX3 0BW, England. This paper was recommended for publication in revised form by associate editor P. Dorato.

†This work was supported by the ministry of defense of the Federal Republic of Germany under grant T/R421/70002/72400.

‡Bodenseewerk Gerätetechnik GmbH, Postfach 11 20, D-7770 Überlingen, FRG.

Reprinted with permission from *Automatica*, vol. 16, pp. 135–146, Mar. 1980.
Copyright © 1980, International Federation of Automatic Control IFAC. Published by Pergamon Press Ltd.

identification increase significantly with the number of parameters to be identified.

The longitudinal motion was chosen, because it is the main control axis of aircraft; the parameter variations are strongly marked and the control problem is easy to understand. The design principle can be used for the lateral motion in a similar manner, but the problems are more involved as a result of the strong coupling of the roll and yaw axis of the aircraft. It should be noted that the design procedure is also of interest for the design of conventional stability systems because of its simplicity in adjusting the dynamic characteristics of the stabilization loop.

The second step defines the dynamic characteristics of the control loop in terms of damping, eigenfrequency and overshoot. These characteristics are derived from the generally used MIL-F-8785 B requirements (1969) and the so-called C^*-criterion (Tobie, Elliott and Malcolm, 1966).

The third step concerns the identification of the interesting aircraft parameters. For this purpose a recursive weighted least-squares estimation procedure is used. The identification process delivers sufficient accurate and fast estimates of the parameters.

2. DESIGN CONCEPT OF THE COMMAND AND STABILITY SYSTEM

2.1 *General approach*

The longitudinal motion of the aircraft can be described by a fourth order state equation (McRuer, Ashkenas and Graham, 1973)

$$\dot{\mathbf{x}} = \mathbf{A}\mathbf{x} + \mathbf{B}\mathbf{u} \qquad (1)$$

where the state vector

$$\mathbf{x} := [\theta, q, \alpha, u]^T \qquad (2)$$

contains the following variables

θ = pitch attitude, α = angle of attack,
q = pitch rate, u = forward velocity.

The input vector $\mathbf{u}$ generally consists of some control and disturbance variables, but in the case of longitudinal stabilization, to be discussed now, we are only interested in the elevator deflection η, so we have

$$\mathbf{u} = [\eta]. \qquad (3)$$

The elevator is deflected by an actuator which can be approximated by a first order lag with a time constant T of about 0.04 s. Due to this small time constant and the insensitive control law it was not necessary to take account of the actuator dynamics in the design procedure. The simulation results, which are shown in this paper, of course include the actuator dynamics in the mathematical model of the aircraft.

The characteristic polynomial of the state-space equation (1) shows in the general case two pairs of complex roots, a short period mode (angle of attack oscillation) and a long period mode (phugoid oscillation). The MIL-F-8785 requirements prescribe the values of the undamped natural frequency ω_{SP} and the damping ratio ζ_{SP} of the short period mode within certain boundaries shown in Figs. 1 and 2. During normal operation the values of ω_{SP} and ζ_{SP} should be placed in the shaded areas. The frequency ω_{SP} is associated with the so-called load factor sensitivity n/α, where n is the lift to weight ratio of the aircraft. For the long period mode only a minimum damping ratio is prescribed.

A very interesting design procedure can now consist in shifting the poles of the short period and the phugoid mode into appropriate areas of the complex plane. This problem has been treated in detail in Hartmann and Lonn (1977). The main results are:

(a) For all pole allocations of practical interest no feedback of the forward velocity u is required.

(b) A very advantageous pole configuration is obtained, if the short period mode is shifted to the desired values of frequency and damping ratio according to the MIL-F-8785 requirements and the phugoid poles are shifted into the zeros

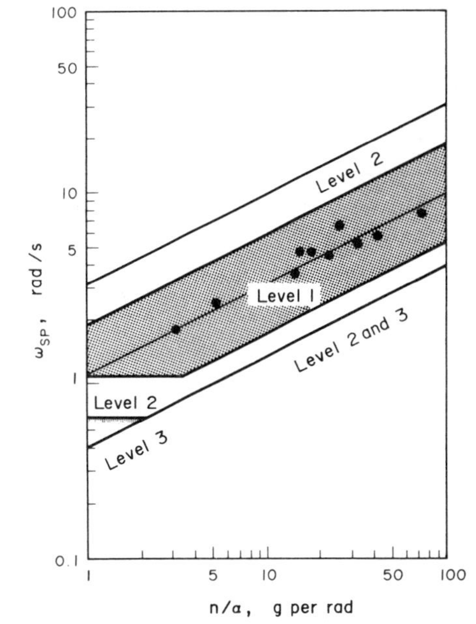

FIG. 1. Frequency requirements concerning the short period mode.

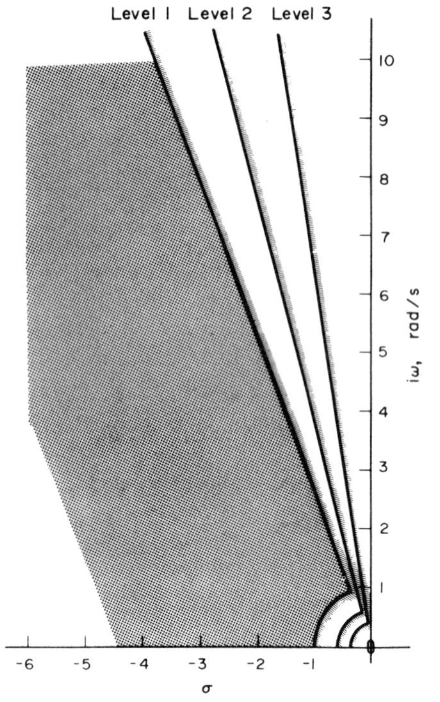

FIG. 2. Damping requirements concerning the short period mode.

of the transfer function $(\theta/\eta)(s)=F_\theta$. This is possible because these zeros are always placed in the left half of the complex plane.

The latter point will be shown in more detail. The transfer function $F_\theta(s)$ which can be easily derived from the state equation (1), has the following general form

$$\frac{\theta}{\eta}(s)=F_\theta(s)=\frac{A_\theta(s^2+B_\theta s+C_\theta)}{(s^2+2\zeta_{SP}\omega_{SP}s+\omega_{SP}^2)(s^2+2\zeta_P\omega_P s+\omega_P^2)}. \quad (4)$$

We are now looking for a state vector feedback, which shifts the poles of the phugoid mode so that

$$s^2+2\zeta_P\omega_P s+\omega_P^2=s^2+B_\theta s+C_\theta,$$

i.e. the poles of the degenerated phugoid mode s_3, s_4 correspond to the zeros of the numerator of $F_\theta(s)$. We consider now the natural mode of motion associated with the eigenvalues of the shifted phugoid, which are eigenvalues of the controlled aircraft. The residuals of these eigenvalues are zero in the transfer function $F_\theta(s)$, i.e. the phugoid mode is eliminated as far as the pitch angle is concerned.

Therefore, we have in the pitch control system considered here natural modes of motion which yield a change in angle of attack and forward speed, but no change in pitch attitude and pitch rate. To compute the gain factors of such a control system we make use of this condition.

The system matrix **A** and the input matrix **B** of the state equation (1) have the following general structure

$$\mathbf{A}=\begin{bmatrix}0 & 1 & 0 & 0\\ a_{21} & a_{22} & a_{23} & a_{24}\\ a_{31} & a_{32} & a_{33} & a_{34}\\ a_{41} & a_{42} & a_{43} & a_{44}\end{bmatrix},\quad \mathbf{B}=\begin{bmatrix}0\\ b_{21}\\ b_{31}\\ b_{41}\end{bmatrix} \quad (5)$$

and we assume a state feedback matrix

$$\mathbf{u}=\mathbf{K}\mathbf{x}$$

with

$$\mathbf{K}=[K_\theta K_q K_\alpha K_u]. \quad (6)$$

We are now looking for control laws which satisfy the conditions

$$\theta=0 \text{ and } q=0$$

for the natural motions considered here. For these motions the state equations are degenerating to

$$\begin{aligned}0&=a_{23}\alpha+a_{24}u+b_{21}\eta,\\ \dot{\alpha}&=a_{33}\alpha+a_{34}u+b_{31}\eta,\\ \dot{u}&=a_{43}\alpha+a_{44}u+b_{41}\eta\end{aligned} \quad (7)$$

and

$$\eta=K_\alpha\alpha+K_u u. \quad (8)$$

We obtain immediately from the first equation of (7)

$$K_\alpha=-\frac{a_{23}}{b_{21}}, K_u=-\frac{a_{24}}{b_{21}}. \quad (9)$$

It is now very easy to understand the physical meaning of the equations (9). Any angle of attack disturbance α causes a pitch angle acceleration $a_{23}\alpha$, but the control law (8) produces a counteracting elevator deflection, which compensates exactly the effect of the static longitudinal stability. The angle of attack disturbance does not cause a change of the pitch angle.

This feature can be very interesting for flight in turbulence conditions from a handling qualities viewpoint. It has been shown that pitch disturbances have a more adverse effect than heave disturbances (Franklin, 1972). From a flight mechanic's point of view we have now changed the usually static longitudinal stability of the aircraft into a condition of indifference. Note that the coefficient a_{21} is very small in comparison with a_{23}. To recover static longitudinal stability, we introduce an artificial stability dependent on the pitch angle deviation by the feedback gains K_θ and K_q.

Therefore we obtain with (6) and (9) the following control law

$$\eta = K_\theta \cdot \theta + K_q \cdot q - \frac{a_{23}}{b_{21}} \cdot \alpha - \frac{a_{24}}{b_{21}} \cdot u$$

which we introduce into the first two rows of the state equation (1). This yields

$$\dot{\theta} = q,$$

$$\dot{q} = (a_{21} + b_{21} K_\theta) \cdot \theta + (a_{22} + b_{21} K_q) \cdot q. \quad (10)$$

The characteristic equation of this short period mode is therefore

$$s^2 - (a_{22} + b_{21} K_q) s - (a_{21} + b_{21} K_\theta) = 0.$$

Comparing this with the desired characteristic equation

$$s^2 + 2\zeta_{SP} \omega_{SP} s + \omega_{SP}^2 = 0 \quad (11)$$

as prescribed by the handling qualities requirements we obtain

$$K_\theta = \frac{-\omega_{SP}^2 - a_{21}}{b_{21}}, K_q = \frac{-2\zeta_{SP}\omega_{SP} - a_{22}}{b_{21}}. \quad (12)$$

With this result we have reached the first goal. We have a direct link between the main requirements and the feedback gains of the control law. It can be shown in fact that the parameters a_{21} and a_{24} are negligible for aircraft configurations of today. Therefore we need for practical considerations only the three parameters a_{22}, a_{23} and b_{21}, which is very important from the identification viewpoint.

2.2 *Practical design aspects*

The control law (6) is based on the feedback of pitch angle, pitch rate and angle of attack. By reason of reliability and costs, the use of a vertical gyro (pitch angle) and an angle of attack sensor should be avoided in stability systems. It is easy to replace these variables. We use an integrated pitch rate signal instead of the pitch angle and a normal accelerometer instead of the angle of attack sensor. It is known that normal acceleration measured at an appropriate location

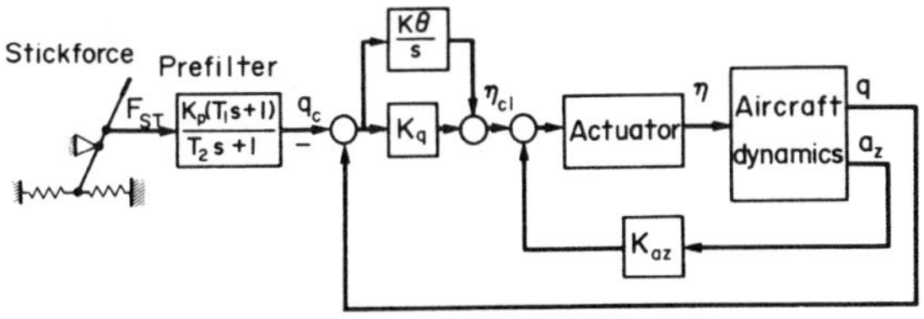

FIG. 3 Block-diagram of the command-and-stability system.

of the aircraft is a very good alternative for the angle of attack signal (McRuer, Ashkenas and Graham, 1973). Figure 3 shows a possible block diagram of the control system. The gain factor K_α is replaced by an equivalent gain K_{a_z}. Note that the relation between the normal acceleration and the angle of attack is approximately given by

$$a_z = Z_\alpha \cdot \alpha \quad (13)$$

where the normal force due to an elevator deflection is neglected.

Integration of the pitch rate signal is done in the forward loop to obtain adequate handling qualities by a rate command system. A lead-lag prefilter is used to shape the stick force inputs of the pilot. Additional requirements concerning the handling qualities such as stick force per g (load factor) and the C^*-criterion can thus be matched.

A short discussion of the block diagram Fig. 3 in terms of transfer functions shows the very transparent design concept. We use the essential part of the state equation (1)

$$\begin{bmatrix} \dot{q} \\ \dot{\alpha} \end{bmatrix} = \begin{bmatrix} a_{22} & a_{23} \\ a_{32} & a_{33} \end{bmatrix} \cdot \begin{bmatrix} q \\ \alpha \end{bmatrix} + \begin{bmatrix} b_{21} \\ b_{31} \end{bmatrix} \eta \quad (14)$$

and neglect the usually very fast actuator dynamics for this purpose. The first feedback loop (indifference loop) provides the indifferent equilibrium condition. We obtain the transfer function

$$\frac{q}{\eta_{c1}} = \frac{b_{21}}{s - a_{22}}. \quad (15)$$

The transfer function for the outer (pitch rate) loop is

$$\frac{q}{q_c} = \frac{1 + \frac{K_q}{K_\theta} s}{1 + \left(\frac{b_{21} K_q - a_{22}}{K_\theta b_{21}} \right) s + \frac{1}{K_\theta b_{21}} s^2}. \quad (16)$$

The natural frequency and the damping ratio of the closed loop can thus be easily determined by the gain factors K_θ and K_q.

For practical realizations we are interested in a discrete time version of the control system. Because the sampling frequency of the control system should be high in comparison with the typical time constants of the aircraft, we are looking for a discrete time approximation of the control law. This control law should be immediately suited for the automatic adaption. For this purpose we start with a discrete time version of the state equation (14) where we include the above mentioned relation between angle of attack and normal acceleration (13) and obtain

$$\begin{bmatrix} q(k+1) \\ a_z(k+1) \end{bmatrix} = \begin{bmatrix} \varphi_{11} & \varphi_{12} \\ \varphi_{21} & \varphi_{22} \end{bmatrix} \cdot \begin{bmatrix} q(k) \\ a_z(k) \end{bmatrix} + \begin{bmatrix} h_1 \\ h_2 \end{bmatrix} \eta(k). \tag{17}$$

The feedback law is put up in the form

$$\eta(k) = K_\theta^* \theta(k) + K_q^* q(k) + K_{a_z}^* a_z(k). \tag{18}$$

The condition for decoupling the pitch rate q from the normal acceleration can be used in the first equation of (17) with the feedback law (18)

$$\begin{aligned} q(k+1) &= (\varphi_{11} + h_1 K_q^*) q(k) - h_1 K_\theta^* \theta(k) \\ &= (\varphi_{12} + h_1 K_{a_z}^*) a_z(k) = 0. \end{aligned} \tag{19}$$

Thus the feedback of the normal acceleration is given by

$$K_{a_z}^* = -\frac{\varphi_{12}}{h_1}. \tag{20}$$

With the sampling period T we express

$$q(k+1) = q(k) + \dot{q}(k) T$$

and obtain as characteristic equation of (19)

$$Ts^2 + (1 - \varphi_{11} - h_{11} K_q^*) s - h_1 K_\theta^* = 0 \tag{21}$$

and by comparison with the desired characteristic equation (11)

$$K_\theta^* = -\frac{T\omega_{SP}^2}{h_1}, \; K_q^* = \frac{1 - \varphi_{11} - 2\zeta_{SP}\omega_{SP} T}{h_1}. \tag{22}$$

Therefore the identification process must deliver the three coefficients

$$\varphi_{11}, \varphi_{12}, \mathrm{h}_1$$

of the measurement equation

$$q(k+1) = \varphi_{11} q(k) + \varphi_{12} a_z(k) + h_1 \eta(k), \tag{23}$$

whereas the natural frequency ω_{SP} and the damping ratio ζ_{SP} are derived from the handling qualities requirements.

3. DEFINITION OF THE HANDLING QUALITIES REQUIREMENTS

As pointed out in Section 2.1, the natural frequency ω_{SP} is determined by the load factor sensitivity n/α. An evaluation shows that the load factor sensitivity can be well approximated by the following function of the dynamic pressure $q_c = p_t - p_s$ of the compressible flow

$$\frac{n}{\alpha} = g_1 \left(\frac{q_c}{10^3} \right)^{p_1}. \tag{24}$$

The natural eigenfrequency ω_{SP}, which depends on the load factor sensitivity n/α itself, can be expressed by a similar formula

$$\omega_{SP} = g_2 \left(\frac{n}{\alpha} \right)^{p_2}. \tag{25}$$

This leads to the final result

$$\omega_{SP} = g_3 \left(\frac{q_c}{10^3} \right)^{p_3} \tag{26}$$

with $g_3 = g_2 g_1^{p_2}$ and $p_3 = p_1 p_2$. An evaluation of the aerodynamic data of a F-4 ('PHANTOM') type fighter aircraft delivers the following values

$$g_1 = 1.29, \; p_1 = 0.79, \; g_2 = 1.05, \; p_2 = 0.49. \tag{27}$$

The results of equation (26) with the values of (27) are shown graphically as black dots for ten flight cases of the F-4 in Fig. 1. The requirements are met pretty well.

The required damping ratio of the aircraft does not depend on the flight case. Therefore we choose a constant damping ratio, which is significantly larger than the minimal damping ratio required, i.e. $\zeta_{SP} = 0.707$.

Another requirement concerning the response of the closed loop in the time domain is the C^*-criterion (Tobie, Elliott and Malcolm, 1966). This criterion uses the transient of a fictitious variable C^*, a blend of normal acceleration a_z, pitch rate q and pitch acceleration $\dot{q}$ according to the following definition

$$C^* := (-a_z + V_{c0} q + x_P \dot{q})/g.$$

The so-called cross-over velocity V_{c0} determines the velocity at which normal acceleration and pitch rate are equally sensed by the pilots ($V_{c0} \approx 120$ m/s). The distance between the center

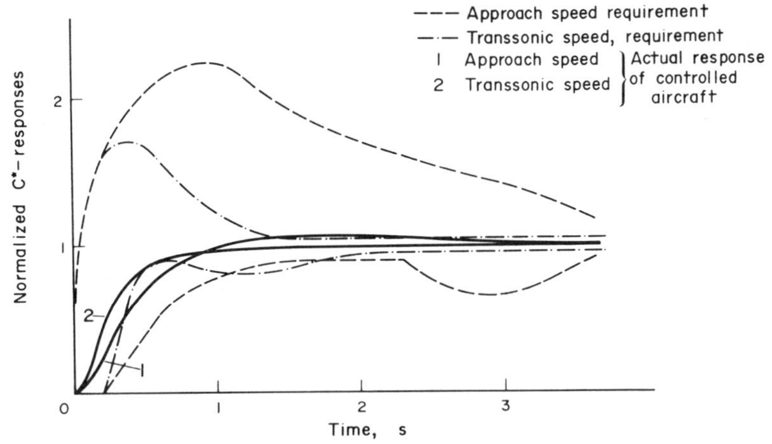

FIG. 4. Normalized C^*-responses of the controlled aircraft.

of gravity and the pilot's station is denoted by x_p. The time response required for the variable C^* as well as the actual C^*-response of the controlled aircraft is shown in Fig. 4 for two important cases.

To fulfil the requirements of the C^*-criterion, we have in the transsonic speed case a very fast normal acceleration demand. To obtain this fast acceleration response we use a pre-filter of lead-lag type as shown in the block diagram of Fig. 3.

4. IDENTIFICATION OF LONGITUDINAL AIRCRAFT DYNAMICS

In Section 2 it has been pointed out that realization of the adaptive stabilization concept requires knowledge of three coefficients φ_{11}, φ_{12} and h_1 of the discrete state-space equation (17). The estimation algorithm to be used should show the following properties:

- minimum variance unbiased estimates;
- recursive form since on-line identification is necessary;
- convergence independent of choice of the initial conditions of the estimator;
- identification of time variable parameters (due to different flight conditions);
- easy programming on a process computer.

We apply a least-squares approach in state-space representation with exponentially weighting of past data to ensure parameter tracking. Though this method gives biased estimates (Krebs and Thöm, 1974; Isermann, 1974) it has the advantage of a very simple structure. Moreover its efficiency has been proved already in wide-spread successful applications. Other algorithms which yield unbiased estimates like the generalized least squares method (Clarke, 1967), are too complex or give no practical improvement. This has been demonstrated for the instrumental variable method (Wong and Polak, 1967) applied to estimation of ship dynamics by Thöm (1976). Moreover, we remember that the mathematical model (17) is an approximation, hence our demands for accuracy of the estimates should not be emphasized too much.

The structure of the process with unknown parameters is given in state-space form

$$\mathbf{x}(k+1)=\boldsymbol{\phi}\mathbf{x}(k)+\mathbf{H}\mathbf{u}(k), \tag{28}$$

$$\mathbf{y}(k)=\mathbf{x}(k)+\mathbf{n}(k) \tag{29}$$

where $\mathbf{x}$ is the $n\times 1$ state vector, $\boldsymbol{\phi}$ is the $n\times n$ transition matrix, and $\mathbf{H}$ is the $n\times m$ input matrix. Measurements of the input $\mathbf{u}(k)$ ($m\times 1$ vector) and noisy measurements $\mathbf{y}(k)$ of the state $\mathbf{x}(k)$ are available at discrete time instants $k(k=0, 1, 2,\ldots)$. The model of the process is then

$$\mathbf{y}(k+1)=\hat{\boldsymbol{\phi}}(k)\mathbf{y}(k)+\hat{\mathbf{H}}(k)\mathbf{u}(k)+\mathbf{e}(k), \tag{30}$$

where '^' indicates that the matrices $\hat{\boldsymbol{\phi}}$ and $\hat{\mathbf{H}}$ contain parameter estimates as indicated in Fig. 5.

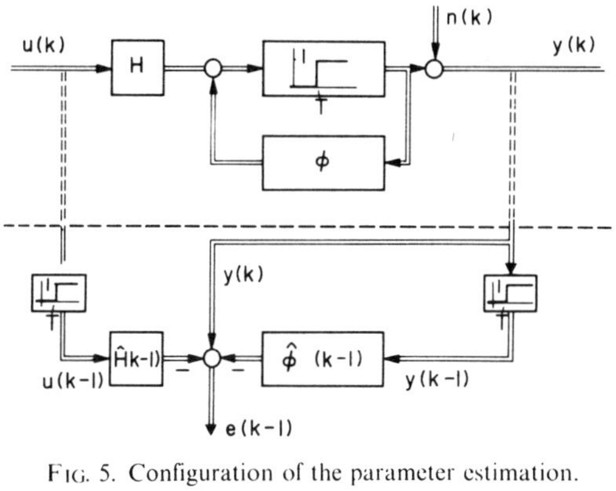

FIG. 5. Configuration of the parameter estimation.

The method of least squares means that we minimize the quadratic error

$$E := \mathbf{e}^T(k)\mathbf{e}(k) \rightarrow \text{Min} \tag{31}$$

for each k, where

$$\mathbf{e}(k) := \mathbf{y}(k+1) - \hat{\boldsymbol{\phi}}(k)\mathbf{y}(k) - \hat{\mathbf{H}}(k)\mathbf{u}(k). \tag{32}$$

This leads to

$$\frac{\partial E}{\partial \hat{\boldsymbol{\phi}}(k)} = -2\mathbf{e}(k)\mathbf{y}^T(k) \stackrel{!}{=} \mathbf{0} \tag{33a}$$

and

$$\frac{\partial E}{\partial \hat{\mathbf{H}}(k)} = -2\mathbf{e}(k)\mathbf{u}^T(k) \stackrel{!}{=} \mathbf{0}. \tag{33b}$$

Now we obtain the adjustment mechanism for the model parameters applying the gradient method, i.e. the rate of change in the parameter estimates is chosen proportional to the gradient of the cost functional E (see for the scalar case Isermann, 1971, p. 203):

$$\frac{\hat{\boldsymbol{\phi}}(k+1) - \hat{\boldsymbol{\phi}}(k)}{T} \approx \frac{\mathrm{d}\hat{\boldsymbol{\phi}}(k)}{\mathrm{d}t} \sim -\frac{\partial E}{\partial \hat{\boldsymbol{\phi}}(k)} \tag{34a}$$

$$\frac{\hat{\mathbf{H}}(k+1) - \hat{\mathbf{H}}(k)}{T} \approx \frac{\mathrm{d}\hat{\mathbf{H}}(k)}{\mathrm{d}t} \sim -\frac{\partial E}{\partial \hat{\mathbf{H}}(k)} \tag{34b}$$

Equations (33) and (34) yield the iterative estimation procedure

$$\hat{\mathbf{S}}(k) = \hat{\mathbf{S}}(k+1) + \boldsymbol{\Gamma}\mathbf{m}(k-1)[\mathbf{y}^T(k) - \mathbf{m}^T(k-1)\hat{\mathbf{S}}(k-1)] \tag{35}$$

with

$$\hat{\mathbf{S}}(k) := \begin{bmatrix} \hat{\boldsymbol{\phi}}^T(k) \\ \hat{\mathbf{H}}^T(k) \end{bmatrix}, \quad \mathbf{m}(k) := \begin{bmatrix} \mathbf{y}(k) \\ \mathbf{u}(k) \end{bmatrix}, \tag{36}$$

and the proportionality-matrix $\boldsymbol{\Gamma}$. Comparing (35) with the usual recursive least squares approach for single input–single output systems (Isermann, 1974, p. 71), we see that we may equate

$$\boldsymbol{\Gamma} := \mathbf{P}(k), \tag{37}$$

where $\mathbf{P}(k)$ is a matrix with decreasing norm evolving according to

$$\mathbf{P}(k) = \mathbf{P}(k-1) - \frac{\mathbf{P}(k-1)\mathbf{m}(k-1)\mathbf{m}^T(k-1)\mathbf{P}^T(k-1)}{\mathbf{m}^T(k-1)\mathbf{P}(k-1)\mathbf{m}(k-1)+1}. \tag{38}$$

Tracking of time-variable parameters is ensured if we provide an exponential weighting of past data in the estimation algorithm (35)–(38), i.e. new measurements influence the updating of the parameter estimates in a stronger way than older ones. This yields (see Eykhoff, 1974, p. 240)

$$\mathbf{P}(k) = f \cdot \left[\mathbf{P}(k-1) - \frac{\mathbf{P}(k-1)\mathbf{m}(k-1)\mathbf{m}^T(k-1)\mathbf{P}^T(k-1)}{\mathbf{m}^T(k-1)\mathbf{P}(k-1)\mathbf{m}(k-1)+\frac{1}{f}} \right] \tag{39}$$

with the weighting factor

$$f \geqq 1, \quad f \approx 1.$$

5. RESULTS

The properties of the digital adaptive stabilizing system have been tested by digital and hybrid simulations using the longitudinal dynamics of the McDonnell/Douglas F-4 'Phantom' aircraft. This aircraft is well suited for the adaptive stabilizing concept because its flight conditions span a wide envelope with drastic changes in the open-loop dynamics (Hartmann, 1975). The block-diagram of the system is given in Fig. 6.

As mentioned above the C^*-criterion requires the application of a pre-filter with positive phase characteristic. It should be carefully observed that the introduction of the dynamic pressure $p_t - p_s$ is not required for the adaption of the control laws. It is only used to match the desired handling qualities—especially the eigen-frequency of the closed loop—according to equation (26).

Extensive digital simulations of the adaptive system with complete fourth order longitudinal dynamic equations yielded good results (Hartmann, 1975) under fast parameter variations and in atmospheric disturbances which have been simulated using the Dryden form of a continuous random gust model. This gust model is defined and specified in the MIL-F-8785 requirement (1969). Furthermore the adaptive system has been implemented on a 16-bit airborne computer. Since this computer has a fixed-point arithmetic, the program should be as simple as possible. The results of hybrid simulations (the longitudinal aircraft dynamics was reproduced by an analogue computer) of the complete digital adaptive command-and-stability system are given in Fig. 7. The aircraft is disturbed by wind-gusts—Fig. 7h. A rapid change in the velocity from Mach 0.2 to Mach 1.0 and back to Mach 0.2—as indicated on Fig. 7a—yields a considerable change in the

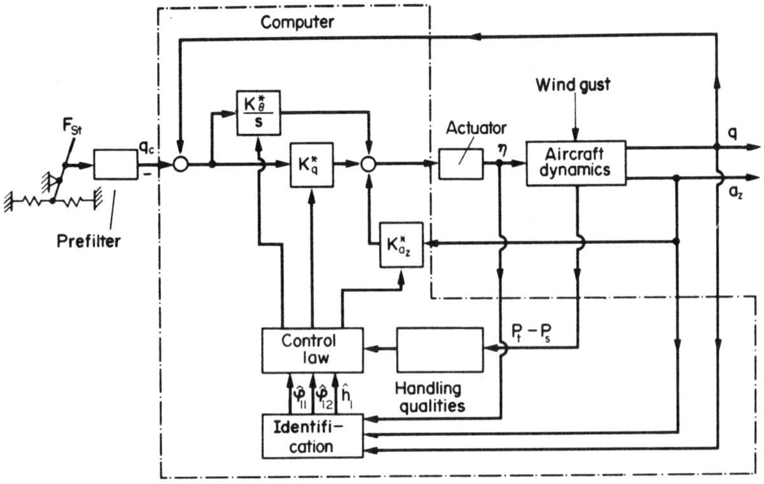

FIG. 6. Signal flow graph of the digital adaptive command-and-stability system.

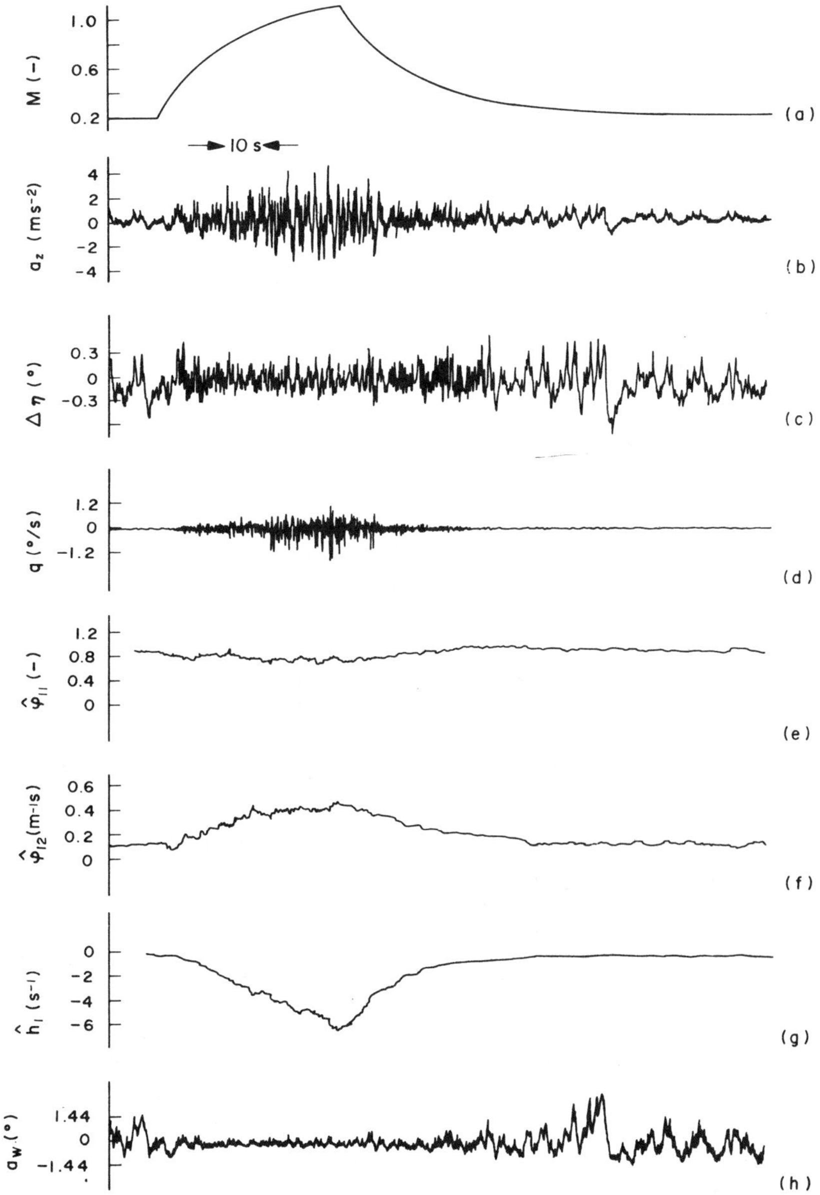

FIG. 7. Digital adaptive command-and-stability system for F-4 'PHANTOM'-hybrid simulation with exponential Mach number history.

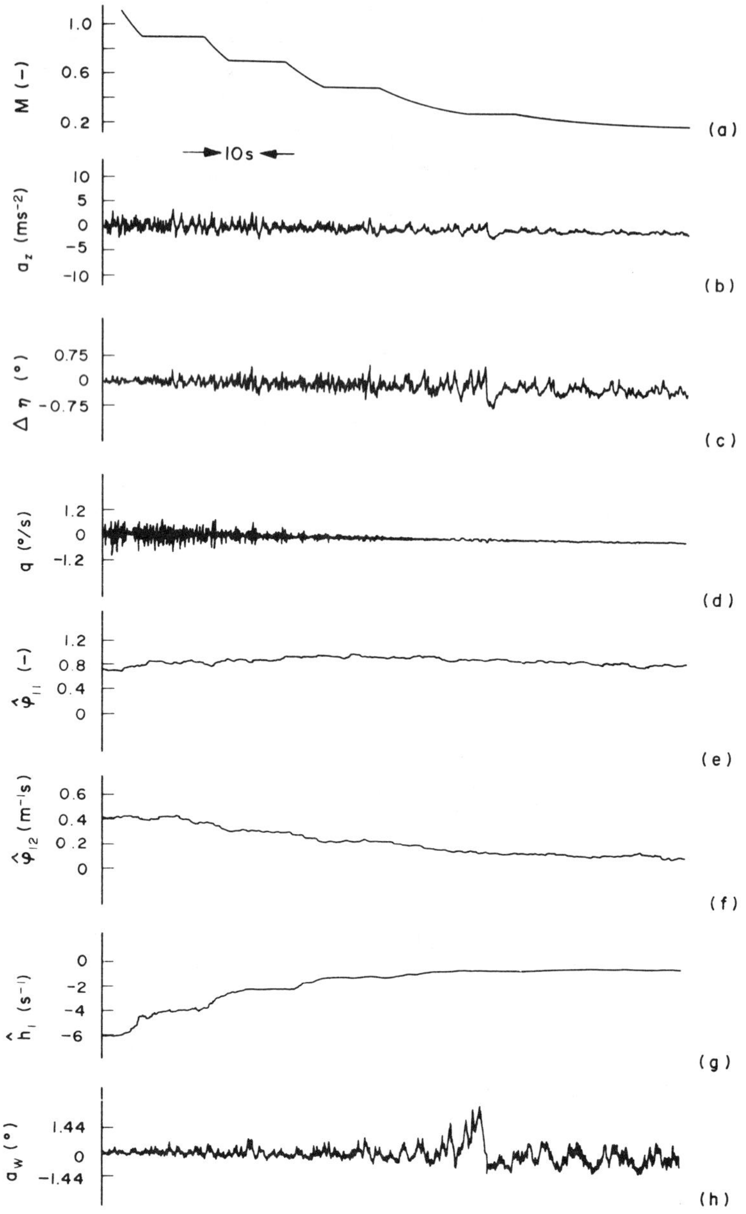

FIG. 8. Digital adaptive command-and-stability system for F-4 'PHANTOM'-hybrid simulation with stepwise Mach number history.

parameters and consequently a change in the estimates $\hat{\varphi}_{11}$, φ_{12} and $\hat{h}_1$ (Fig. 7e–7g). The rate of convergence of the estimation algorithm may be demonstrated even better by using another Mach number versus time function given in Fig. 8a. See the estimates in Fig. 8e–8g.

Finally the parameter estimation algorithm is applied on real flight test data. The input to the business jet HFB 320 'HANSA' is the deflection of the elevator, commanded by the pilot (Fig. 9a). The three parameter estimates as well as the estimate $\hat{b}_1$ of a bias-term is given in Fig. 9b, while Fig. 9c shows the reconstruction of the output q (rate of pitch) by

$$\hat{q}(k+1)=\hat{\varphi}_{11}q(k)+\hat{\varphi}_{12}a_z(k)+\hat{h}_1\eta(k)-\hat{b}_1(k)$$

in comparison to the measured output to verify the estimation as theoretical values of the parameters were not available. Since real flight test data have been used all adverse effects of the sensors (noise, nonlinearities) and the airframe (vibrations, structural bending modes) are included. The last picture (Fig. 9d) shows the quality of the recursive weighted least squares method developed here compared with the much

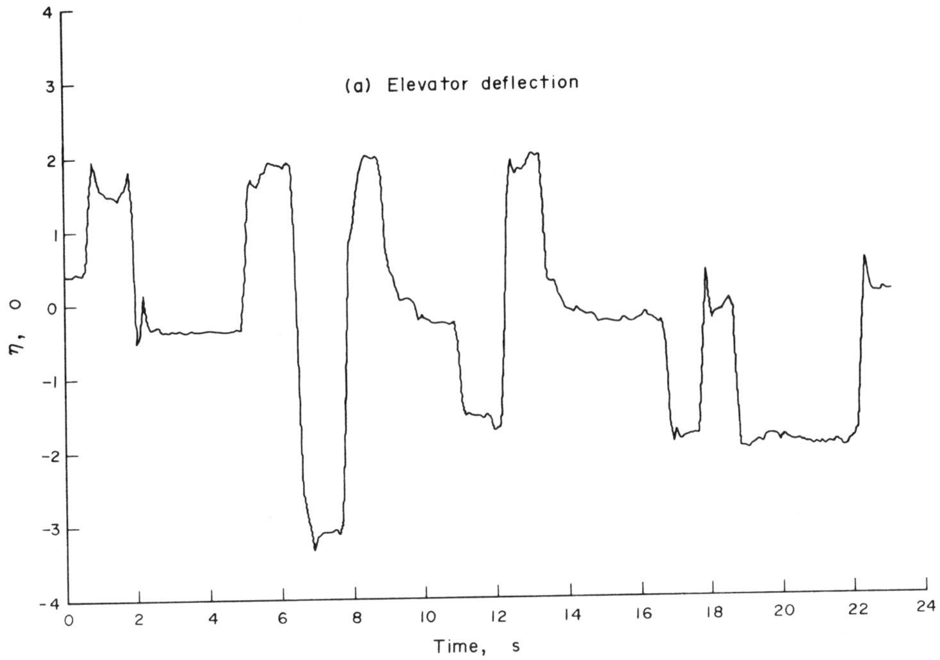

FIG. 9. Identification of dynamics of the HFB 320 'HANSA' Jet.

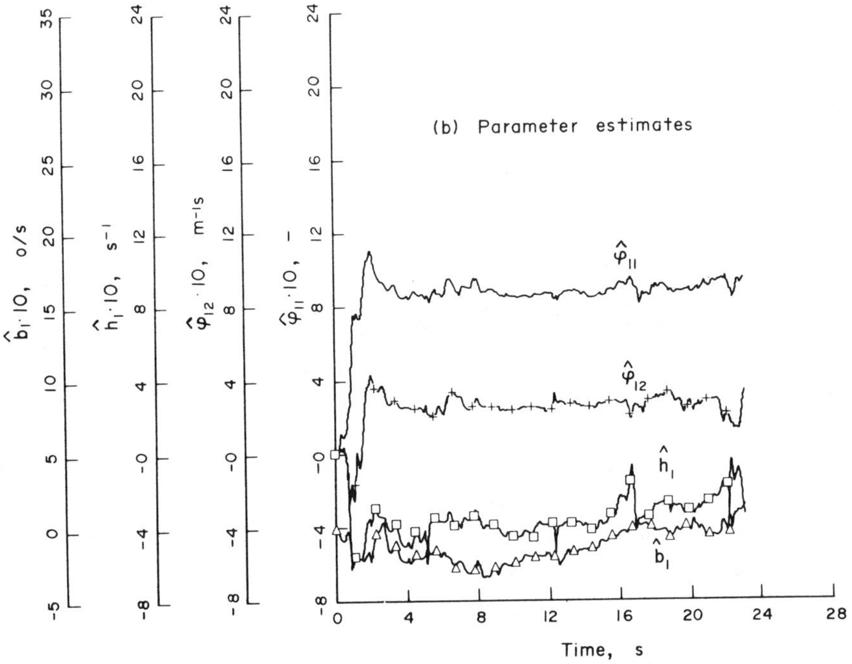

(c) Pitch rate q and estimate $\hat{q}$

(d) Absolute and relative errors

$$\text{Actual relative error (ARE)} := \frac{|q-\hat{q}|}{\Delta q} \cdot 100, \quad \Delta q = q_{max} - q_{min}$$

$$\text{Mean relative error (MRE)} := \frac{1}{N} \cdot \sum \frac{|q-q|}{\Delta q} \cdot 100$$

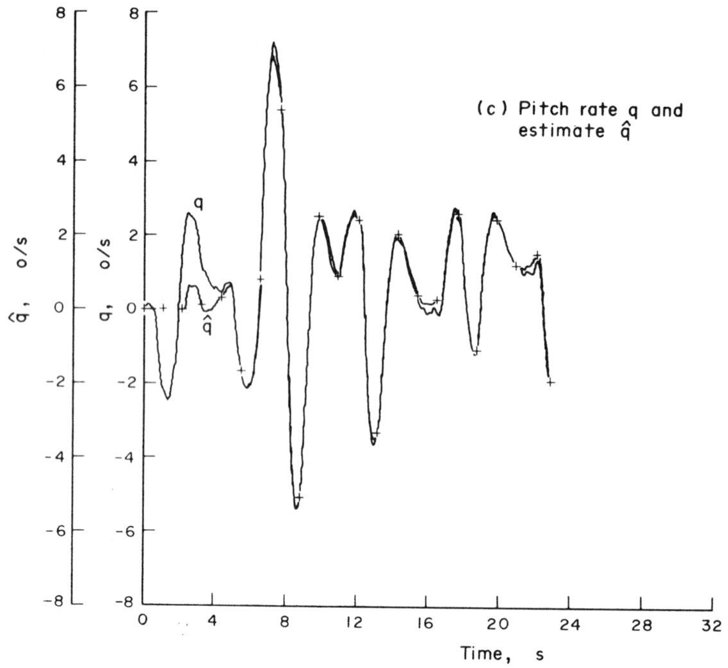

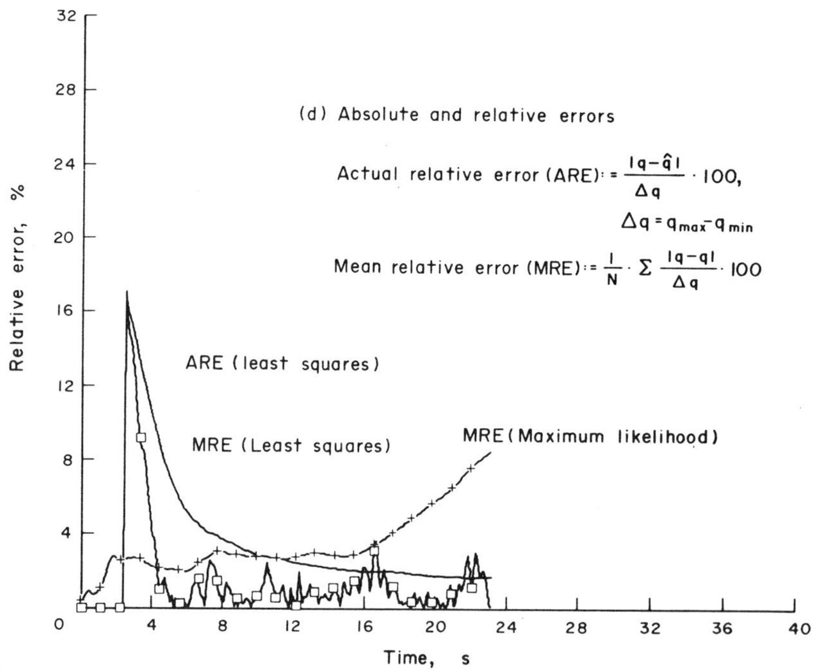

more sophisticated maximum likelihood estimation; the latter method used a string of data of only 12 s (Schulz, 1975) while the test covers 24 s. Thus the M.L.E. diverges slowly since no parameter tracking is possible. It should be mentioned however that it is possible to avoid the iterative nature of conventional maximum likelihood estimation by adopting a parallel approach to likelihood minimization. This method was utilized by Stein, Hartmann and Hendrick (1977), for tracking of time varying parameters in an adaptive flight control system.

Recently flight tests of the complete adaptive system have been carried out with the DO 28-D 'SKYSERVANT' of the Bodenseewerk. The flight tests confirmed the predicted performance of the system within the envelope of this aircraft (Krebs, 1977).

6. CONCLUSIONS

A new approach for the design of digital adaptive command and stability systems has been proposed, investigating first the longitudinal dynamics of the aircraft. The control law is obtained by state vector feedback using the requirements of

(i) decoupling the angle of attack and the pitch angle acceleration and

(ii) adjusting the roots of the characteristic equation to those values which are prescribed by the MIL-F-8785 B handling qualities.

The necessary gain factors contain unknown and variable plant parameters which are estimated by an on-line fading-memory least squares algorithm.

Hybrid simulations of the complete system with an airborne digital computer as well as flight test results of the complete system demonstrate the efficiency of the concept and challenge the operational application.

Acknowledgement—The authors are indebted to Mr. G. Engler and Mr. P. Wüst for the simulation results of Figs. 7 and 8.

REFERENCES

AGARD, NATO (1970). Advanced control system concepts. *AGARD Conference Proceedings*, No. 58.

Clarke, D. W. (1967). Generalized-least-squares estimation of the parameters of a dynamic model. *IFAC symp. Identification in Automatic Control Systems*, Prague, paper 3.17.

Eykhoff, P. (1974). *System Identification*. John Wiley, New York.

Franklin, J. A. (1972). Flight investigation of the influence of turbulence on longitudinal flying qualities. *J. Aircraft*, **9**, S. 273.

Gregory, P. C. (Ed.) (1959). Proceedings of the self adaptive flight control systems symposium. Wright-Patterson Air Force Base OH. WADC Techn. Rep. 59–49, ASTIA Doc. No. AD 209 389.

Hartmann, U. (1975). Flugmechanische Daten und Zustandsgleichungen von vier typischen Flugzeugen zur Untersuchung adaptiver Stabilisierungssysteme. Bodenseewerk Gerätetechnik GmbH, TB 000 D 950/75.

Hartmann, U. (1975) Untersuchung über die Auslegung digitaler adaptiver Stabilisierungssysteme. Bodenseewerk Gerätetechnik GmbH, TB 000 D 996/75.

Hartmann, U. and E. Lonn (1977). Anwendung der Polfestlegung beim Entwurf von Stabilisierungssystemen am Beispiel der Flugzeuglängsbewegung. *Zeitschr. Flugwissensch. Weltraumforsch.* **1**, 135.

Isermann, R. (1971). *Experimentelle Analyse der Dynamik von Regelungssystemen-Identifikation* I. Bibliographisches Institut, Mannheim.

Isermann, R. (1974). *Prozessidentifikation*, Springer, Berlin.

Krebs, V. and H. Thöm (1974). Parametererkennung nach der Methode der kleinsten Quadrate—Ein Überblick. *Regelungstech. Prozessdatenverarbeitung*, **22**, 1.

Krebs, V. (1977). Flugerprobung eines digitalen adaptiven und eines parameterunempfindlichen Regelungssystems. Bodenseewerk Gerätetechnik GmbH, TB 1145/77.

McRuer, D., I. Ashkenas and D. Graham (1973). *Aircraft Dynamics and Automatic Control*, Princeton University Press, Princeton.

Military Specification, Flying Qualities of Piloted Airplanes MIL-F-8785 B (ASG), 07. (Aug. 1969).

Schulz, G. (1975). Maximum-Likelihood-Identifizierung mittels Kalman-Filterung—kleinste Quadrate Schätzung. DFVLR Forschungsbericht 75–54.

Stein, G., G. L. Hartmann and R. C. Hendrick (1977). Adaptive control laws for F-8 flight test. *IEEE Trans. Auto. Control*, **AC-22**, 758.

Thöm, H. (1976). Modellbildung für das Kursverhalten von Schiffen. Dissertation D17, TH Darmstadt.

Tobie, H. N., E. M. Elliott and L. G. Malcolm (1966). A new Longitudinal Handling Qualities Criterion. National Aerospace Electronics Conference, Dayton, OH.

Wong, K. Y. and E. Polak (1967). Identification of linear discrete-time systems using the instrumental variable method. *IEEE Trans. Auto. Control*, **AC-12**, 707.

Part VI
Applications to Autopilots

THE design of autopilots for the steering of ships (e.g., large tankers) with variable characteristics is a major problem. This is especially true for supertankers where sudden variations in the depth of water may result in course instability. Due to their size, these ships are also manually difficult to handle. Adaptive autopilots which can be used to make accurate course changes in narrow coastal waters will be of great help. We will now discuss two papers which describe adaptive control solutions to these problems.

In the first paper, paper thirty-four, C. A. Kallstrom, K. J. Astrom, N. E. Thorell, J. Eriksson, and L. Sten report the design of two adaptive autopilots based on self-tuning regulator and Kalman filter for the steering of large tankers. The autopilots are based on velocity scheduling, a self-tuning regulator for steady-state course keeping, a high-gain turning regulator with variable structure and a Kalman filter. The authors report results from simulation and full-scale experiments with three different tankers. The autopilots were shown to work well under different load, speed, and weather conditions. The excellent performance of the turning regulator will guarantee smooth and precise turns, which implies a much safer handling of the ship. This is another excellent example which deals with a real-life situation. We hope that the reader's interest in the adaptive control problems will be stimulated by this and other related work being done by this group.

In paper thirty-five entitled "Adaptive Steering of Ships: A Model Reference Approach," J. Van Amerongen describes the application of model reference adaptive control to automatic steering of ships. Such adaptive control yields improved maneuvering, safer operation and saving in fuel. The model reference adaptive control is provided by means of direct adaptation, optimal control, and adaptive estimation of parameters. The paper gives the design details and describes a full-scale trial at sea. The author speculates that a new generation of autopilots based on modern control techniques will replace the present systems. The fast development of small and inexpensive microcomputers makes these autopilots practically realizable. The paper also gives a number of references of his own past work as well as those of other researchers; these references should be useful to our readers while exploring the area of design of autopilots.

SELECTED BIBLIOGRAPHY

C. G. Kallstrom, K. J. Astrom, N. E. Thorell, J. Eriksson, and L. Sten, "Adaptive autopilots for tankers," *Automatica,* vol. 15, no. 3, pp. 241–254, 1979.

J. Van Amerongen and Ten Cate Udink, "Model reference adaptive autopilots for ships," *Automatica,* vol. 11, pp. 441–449, 1975.

Adaptive Autopilots for Tankers*

C. G. KÄLLSTRÖM,† K. J. ÅSTRÖM,† N. E. THORELL,‡
J. ERIKSSON‡ and L. STEN‡

Simulations and full-scale experiments indicate that benefits can be obtained by using advanced autopilots based on self-tuning regulators and Kalman filters for the steering of large tankers.

Key Word Index Adaptive control; adaptive systems; computer-aided design; computer applications; computer control; Kalman filters; marine systems; optimal control; PID control; self-adjusting systems; ships.

Abstract—Two adaptive autopilots for ships are designed. The autopilots are based on velocity scheduling, a self-tuning regulator for steady state course keeping, a high gain turning regulator with variable structure and a Kalman filter. Methods for design of the autopilots are discussed. Results from simulations and full-scale experiments with three different tankers are presented. The autopilots are shown to work excellently under different load, speed, and weather conditions.

1. INTRODUCTION

THE REQUIREMENTS on ship steering are increasing for reasons of safety and economics. The autopilots commonly used today are based on simple PID-control. The measured heading signal is compared with the desired heading and the error is used as the input to the controller. The output of the controller is fed to the rudder servo. Special techniques such as limiting of the heading error or two mode operation, for example one mode for steady state course keeping and another for turning, are used to avoid saturation for large changes of the desired heading. Some autopilots are also using a rate gyro to obtain a good derivative feedback. An autopilot must be properly adjusted to give a good performance. Adjustments are required to compensate for wind, waves, currents, speed, trim, draught and water depth. The adjustments are tedious and time consuming. Fixed settings are therefore often used. It is a common experience that the autopilots do not work well in bad weather or when the speed is changed. It has also been the experience that autopilots may have difficulties when the ship is making large manoeuvres. The reason is partly that the autopilot is not properly tuned and partly that the PID-algorithm is too simple. The autopilot is therefore frequently switched off and manual steering is used in situations where automatic control is needed most.

Some of the disadvantages of conventional autopilots can be avoided by using an adaptive autopilot. Such a pilot can adjust its parameters to compensate for changes in the environment. Because the parameters of the controller are tuned automatically it is also possible to use control algorithms which are more complex than the ordinary PID-algorithm.

Adaptive control provides several benefits. There will be an induced drag due to steering, which can be described by a quadratic loss function (Norrbin, 1972). Adaptive control can minimize this loss. It is shown in Åström (1977) that the minimal loss depends on factors such as disturbance levels and load conditions. The minimal loss can easily change by a factor of 8 over typical operating conditions. An autopilot which guarantees a stable, well damped system for all operating conditions can be obtained simply by choosing sufficiently high gains in the feedback. Such a regulator will, however, generate excessive rudder motions and the induced drag will consequently be too high. It is shown in Åström (1977) that adaptation can typically reduce the steering loss by at least a factor of 2 compared to a fixed gain autopilot. Adaptive control also improves safety and makes the ship operation more convenient.

There are many ways to design an adaptive control system. Proposals to add adaptation heuristically to ordinary PID-autopilots are given in Oldenburg (1975) and Sugimoto and Kojima (1978). It has also been suggested to adjust the parameters of a PID-regulator automatically to minimize a loss function approximately describing the increase in drag due to steering (Schilling, 1976; Reid and Williams, 1978; Broome and Lambert, 1978; van Amerongen and van Nauta Lemke, 1978). Stochastic adaptive systems have also been proposed (Merlo and Tiano, 1975;

*Received 11 July 1977; revised 16 June 1978; revised 29 November 1978. The original version of this paper was presented at the 7th IFAC World Congress on a link between Science and Applications of Automatic Control which was held in Helsinki, Finland during June 1978. The published Proceedings of this IFAC Meeting may be ordered from: Pergamon Press Ltd., Headington Hill Hall, Oxford OX3 0BW, England. This paper was recommended for publication in revised form by associate editor I. Landau.

†Lund Institute of Technology, Department of Automatic Control, P.O. Box 725, S-220 07 Lund 7, Sweden.

‡Kockums Automation AB, Fack, S-201 10 Malmö, Sweden.

Reprinted with permission from *Automatica*, vol. 15, pp. 241–254, May 1979.
Copyright © 1979, International Federation of Automatic Control IFAC. Published by Pergamon Press Ltd.

Brink and co-workers, 1978). A stochastic controller with fixed gains is described in Ohtsu, Horigome and Kitagawa (1976) and Ohtsu and co-workers (1978). The model reference technique has been used in van Amerongen and Udink ten Cate (1975). This system worked well for manoeuvring but it is not well suited for course keeping because the disturbances are not taken into account explicitly in the model reference method. Adaptive filtering techniques have also been proposed in some autopilots (Ware, Fields and Bozzi, 1978; van Amerongen and van Nauta Lemke, 1978; Sugimoto and Kojima, 1978).

In this paper an attempt has been made to apply the self-tuning regulator (Åström and Wittenmark, 1973) for steady-state course keeping of ships. This approach was chosen because the self-tuning regulator is designed to handle disturbances and has the inherent property to tune parameters in response to changing characteristics of the disturbances for the purpose of minimizing a specific loss function. Two different adaptive autopilots have been investigated. The dependence of the ship velocity is handled by gain scheduling in both cases. The ship speed is thus measured or computed from the propeller rate of revolution and the feedback gains are changed accordingly. Adaptation is used to compensate for variations in the other factors. The simple system uses only heading measurements but the more complicated system may also use measurements of yaw rate, sway velocity, and rudder angle. The more complicated system has a Kalman filter. This uses the information from all sensors to obtain a reliable and smooth estimate of the heading, the yaw rate and the sway velocity. Both systems have a high gain regulator with variable structure for turning.

The paper is organized as follows. The different functions of the adaptive autopilots are described in Section 2. This section covers the velocity scheduling, the Kalman filter, the self-tuning regulator and the turning regulator. Simulations were used extensively. This is discussed in Section 3. Experiments with different tankers were a major activity. Over 130 tests were performed. They are summarized in Section 4. The major conclusions to be drawn are given in Section 5. It is found that adaptive ship steering has substantial advantages and that the chosen design concepts are sound.

2. AUTOPILOT FUNCTIONS

An autopilot has two main tasks: steady state course keeping and turning. Minimization of drag induced by the steering is the important factor in course keeping. Steering precision is the important factor when turning. It is therefore natural to have a dual mode operation. These two modes will be described below together with the basic autopilot functions. A brief description of the autopilots is given in Källström and co-workers (1978). A more detailed presentation can be found in Källström and co-workers (1977), where many additional references are also given.

Two autopilot structures

Several different adaptive regulators have been investigated. Particular attention has been given to autopilots that can be implemented economically. These considerations resulted in two main alternatives which differ mainly in the number of measurements used.

The structure of the simple adaptive autopilot KADPIL 1 (Kockums ADaptive autoPILot 1) is shown in Fig. 1. The simple autopilot uses measurements of forward speed, or propeller rate and heading.

A schematic diagram of the more advanced autopilot KADPIL 2 is shown in Fig. 2. It is similar to the simple autopilot but it also contains a Kalman filter. This implies that it may use more measurements than KADPIL 1, namely yaw rate, rudder angle, fore and aft sway velocities. The autopilot can function with the same measurements as KADPIL 1 but the additional measurements will improve the performance of the regulator. This design concept also offers interesting possibilities for sensor diagnosis. Because of the redundant measurements the advanced autopilot is less sensitive to instrument failures. The Kalman filter also gives smoothed values of heading, yaw rate, sway velocities, rudder angle, rudder bias and sensor biases. This information is useful not only for the autopilot functions but also for other functions such as navigation and collision warning.

Velocity scheduling

The influence of speed variations on ship steering dynamics is well known. To obtain a performance of the autopilot which is invariant to the ship velocity the parameters of the Kalman filter, the self-tuner and the turning regulator are changed as a function of the ship speed. This function is called velocity scheduling. It is included in both autopilots. The self-tuner can, of course, adapt to variations in velocity. The response to velocity variations by scheduling is, however, much quicker than that obtained by adaptation. The velocity variations are therefore compensated by gain scheduling both in the self-tuner and in the turning regulator.

The speed dependence of the ship steering dynamics can be approximately determined analytically. The general characteristics are that

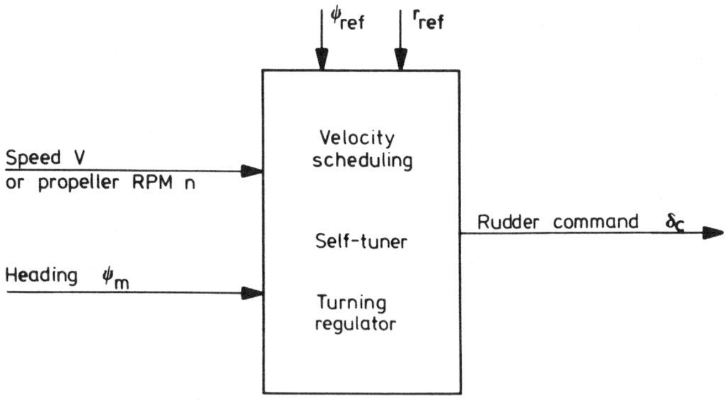

FIG. 1. Schematic diagram of the simple autopilot KADPIL 1.

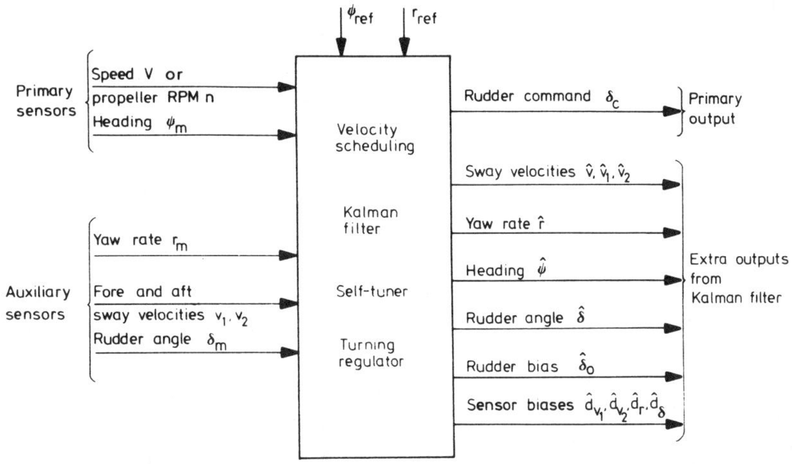

FIG. 2. Schematic diagram of the more advanced autopilot KADPIL 2.

the controller and filter parameters should be increased with decreasing speed. If true time invariance is desired the controller gains should be inversely proportional to the square of the speed, i.e. $G(V(t))=(V_0/V(t))^2$, where G is the gain schedule. The design speed V_0 is usually chosen equal to the service speed. If invariance in the path of the ship is the goal then the gains should be inversely proportional to the ship velocity, i.e. $G(V(t))=V_0/V(t)$. To avoid very large rudder motions at low speed the gains are limited. The signal from the speed log can be used for the scheduling. The propeller rate can also be used because in steady-state the ship velocity is a function of the propeller rate. A more detailed discussion of the velocity scheduling is given when discussing the other functions of the autopilots. Additional details are given in Källström and co-workers (1977).

Kalman filter

The purpose of the Kalman filter is to process all available measurements, heading angle, yaw rate, sway velocities and rudder angle, to obtain reliable estimates of heading, yaw rate and sway velocity. The Kalman filter provides several advantages. Some of the measured signals have to be filtered. When using a Kalman filter the filtering is done using a mathematical model of the ship. This results in a matched filter which is more efficient than an *ad hoc* low-pass filter. When the Kalman filter is used all measurements will contribute to the estimates.

By processing all the measurements, it is sometimes possible to detect sensor failures and to compensate for rudder and sensor biases. The resulting system is thus comparatively insensitive to sensor failures. It can be shown that the system is completely observable from the heading signal. This means that it is possible to obtain all estimates as long as the heading signal is available. The estimates will, however, not be as accurate as when all measurements are available. Kalman filtering thus gives a system with graceful degradation. With the Kalman filter it is also possible to separate the tasks of filtering and control. This means for example that it is not necessary to readjust the feedback gains when a sensor signal is lost.

The design of a Kalman filter is straightforward (Åström, 1970). It requires a mathematical model of the ship and its disturbances. The particular model used in this case is given in the Appendix. It includes a model of the ship steering dynamics and the steering engine. Biases in the measurements of sway velocities, yaw rate, and rudder angle are also included as well as models for disturbances in terms of stochastic processes. The parameter values of the mathematical model for a specific ship can be calculated approximately from ship construction data or estimated from tests with scale models. A system identification technique based on data from full-scale experiments is another method, which also gives models for the disturbances (Åström and co-workers, 1975; Åström and Källström, 1976; Byström and Källström, 1978).

The Kalman filter obtained has the following form

$$\hat{x}(t|t-1)=\Phi\hat{x}(t-1|t-1)+\Gamma u(t-1)$$
$$\hat{x}(t|t)=\hat{x}(t|t-1)+K\varepsilon(t)$$
$$\varepsilon(t)=y(t)-\theta\hat{x}(t|t-1)=y(t)-\hat{y}(t|t-1), \qquad (1)$$

where u is the rudder command and y a vector whose elements are the measured variables as defined in the Appendix. The quantity $\hat{x}(t-1|t-1)$ is the estimate of the state at time $t-1$ given all measurements up to and including time $t-1$ and $\hat{x}(t|t-1)$ is the prediction of the state one time unit ahead. Similarly $\hat{y}(t|t-1)$ is the prediction of the measurements at time t based on past measurements and $\varepsilon(t)$ is the prediction error. The prediction errors are weighted by the filter gain K, and used to update the state estimates. Analysis of the prediction errors can also reveal if the measurements are reasonable or if there are grounds for discarding a particular measurement.

Constant system matrices and gains were used in the implementation of the Kalman filter (see Appendix). To decrease the influence of speed variations the input, state and output variables were normalized using the length of the ship as the length unit and the time it takes to travel a ship length as the time unit. This is in fact one way to do the velocity scheduling. The dynamics of the ship and its environment will also change with trim, draught, wind and waves. It was, however, verified by extensive simulations that the quality of the estimates was only marginally improved by adapting the Kalman filter to the changing conditions, provided that the sampling period was chosen sufficiently short. A sampling period of 1 s was used.

Self-tuning regulator for steady state course keeping

There will be a retarding force mainly due to cross-coupling between yaw rate and sway velocity if the course is changing. The course deviations can be decreased by compensating rudder motions. The rudder motions will, however, also generate retarding forces. The objective of the course keeping regulator is to balance these effects so that the sum of the retarding forces is as small as possible and the requested course is maintained. It is intuitively clear that the best compromise will depend on many factors: water depth, trim, draught, ship speed, forces from wind, waves and currents. To get the best result it is therefore necessary to make readjustments when the operating conditions change.

It was shown in Koyama (1967) and Norrbin (1972) that the average increase in drag due to yawing and rudder motions can be approximately described by

$$\frac{\Delta R}{R}=k[\overline{\psi}^2+\lambda\overline{\delta}^2], \qquad (2)$$

where R is the drag and $\overline{\psi}^2$ and $\overline{\delta}^2$ denote the mean square of heading error and rudder angle amplitude, respectively. The parameters k and λ will depend on the ship and its operating conditions. In Norrbin (1972) the following numerical values are given for a typical tanker:

$$k=0.014\,\text{deg}^{-2}, \qquad \lambda=1/12.$$

It is thus natural to use the criterion

$$V=\frac{1}{T}\int_0^T[(\psi(t)-\psi_{\text{ref}})^2+\lambda\delta^2(t)]\,\mathrm{d}t \qquad (3)$$

as a basis for the design and evaluation of autopilots for steady-state course keeping.

A technique to design a self-tuning regulator which minimizes the loss function (3) for a linear system is discussed in Åström and Wittenmark (1974) and Åström and co-workers (1977). A block diagram of this regulator is shown in Fig. 3. The parameters of a model describing the steering dynamics of the ship and its environment are estimated at each sampling interval. The parameters of a control law which minimizes the criterion (3) for the estimated model are computed. The control signal is then determined from the control law. The controller obtained was fairly complex to implement because it requires spectral factorization or solution of steady state Riccati equations. It requires a code of over 2k words on a typical minicomputer with floating point arithmetic in hardware. Therefore an attempt was made to use a simpler version of a

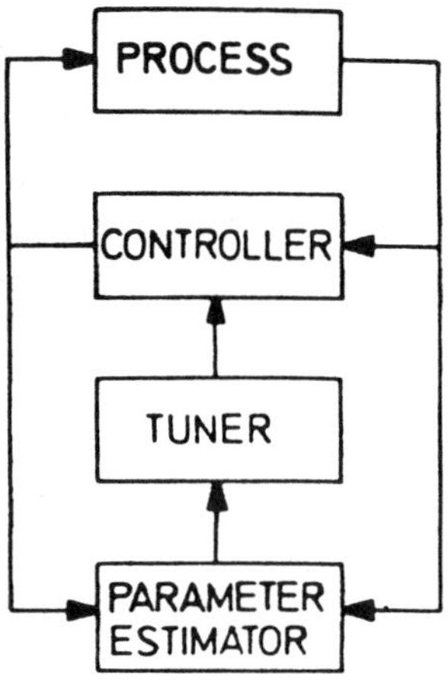

FIG. 3. Block diagram of the adaptive regulator.

self-tuner which can be implemented using a shorter code as discussed below.

If the control problem is sampled it can be shown that minimization of the loss function (3) is approximately equivalent to minimize the predicted value of the heading error, i.e.

$$V_1 = \frac{1}{N}\sum_{t=1}^{N} [\hat{\psi}(t+k+1|t) - \psi_{ref}]^2 \qquad (4)$$

where a longer prediction interval corresponds to a heavier weighting on the control variable. The self-tuning regulator which minimizes (4) is considerably easier to implement than the one that minimizes (3) as discussed in Åström and Wittenmark (1974). The code is about 5 times shorter. Simulations also showed that, at least for tankers, the simple self-tuner is almost as good as the more complex self-tuner which solves Riccati equations on line.

Adaptive regulators which minimize (4) have the same structure as shown in Fig. 3. In this particular case the implementation can, however, be considerably simplified because the calculations can be organized in such a way that the controller parameters are estimated directly (Åström and Wittenmark, 1973). At each step the parameters of the model

$$\begin{aligned} y(t) + a_1 y(t-k-1) + \ldots + a_n y(t-k-n) \\ = b_0[u(t-k-1) + b_1 u(t-k-2) + \ldots \\ + b_m u(t-k-m-1)] \\ + c_1 w_1(t-k-1) + c_2 w_2(t-k-1) \end{aligned} \qquad (5)$$

are estimated using least squares identification. The variables in (5) are defined by

$$y(t) = \hat{\psi}(t) - \psi_{ref}$$

$$u(t) = [\delta_c(t) - \delta_c(t-1)]/G(V(t)) = \nabla\delta_c(t)/G(V(t))$$

$$w_1(t) = V(t)[\hat{v}(t) - \hat{v}(t-1)]$$

$$w_2(t) = \hat{r}(t) - \hat{r}(t-1)$$

where the notations are introduced in Fig. 2. When the Kalman filter is used, the signals $\hat{\psi}$, $\hat{v}$ and $\hat{r}$ are taken from the filter otherwise the heading measurement is used directly, and the terms w_1 and w_2 are deleted. The parameter estimation is done using exponential forgetting to discount past data. It is advantageous to use the gain scheduling $G = (V_0/V)^2$ because this will make the parameters almost invariant with forward speed and the tuning is then simplified. The self-tuner then only has to adapt to changes in weather, sea and load conditions. Experiments have been performed with a self-tuner without gain scheduling to compensate for velocity variations. It was found that the self-tuner was capable also of adapting to speed changes. The parameter estimates were, of course, changed significantly when the speed was changed, and this is sometimes a drawback when the regulator is implemented. The scale factor b_0 and the parameters k, n, m, the sampling interval and the forgetting factor can be chosen from *a priori* knowledge of the ship and disturbance dynamics. The sampling interval and the parameter k are the crucial parameters.

The minimum variance regulator associated with (5) is

$$\begin{aligned} u(t) = \frac{1}{b_0}[a_1 y(t) + \ldots + a_n y(t-n+1)] - b_1 u(t-1) \\ - \ldots - b_m u(t-m) - \frac{1}{b_0}[c_1 w_1(t) + c_2 w_2(t)]. \end{aligned} \qquad (6)$$

An integrator is introduced in the regulator by the use of $\nabla\delta_c(t)$ instead of $\delta_c(t)$. This will assure that the requested heading ψ_{ref} is maintained.

A slight modification can be introduced by

$$\overline{\nabla\delta_c(t)} = \frac{b_0^2}{b_0^2 + qG^2(V)}\nabla\delta_c(t) \qquad (7)$$

which corresponds to the criterion

$$V_2 = [\hat{\psi}(t+k+1|t) - \psi_{ref}]^2 + q[\nabla\delta_c(t)]^2. \qquad (8)$$

The steering performance of a stable ship can be improved by using a non-zero value of q, but simulations and experiments have shown that $q = 0$ is the only acceptable value for unstable ships. Both stable and unstable tankers were investigated in the simulations and the practical experiments. The same ship could in fact be stable or unstable depending on trim and loading.

Turning regulator

The major concern when turning is to keep a tight control of the motion of the ship, possibly at the expense of rudder motions. This is true at least for tankers.

The steady state relation between turning rate and rudder angle for a typical tanker is shown in Fig. 4. It is seen from this figure that the relation

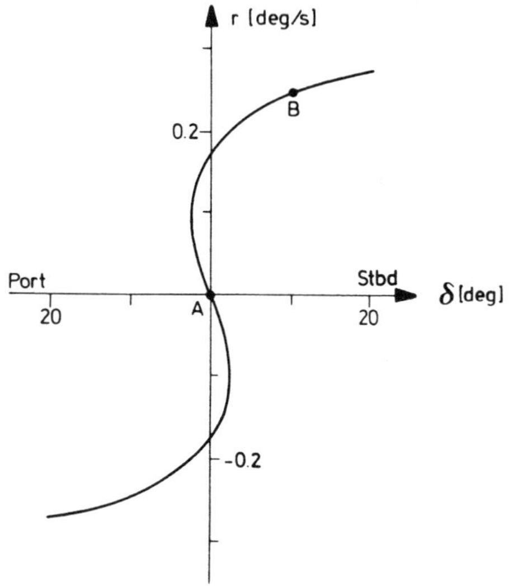

FIG. 4. Steady state relation between rudder angle δ and turning rate r for a 350,000 tdw tanker in an unstable loading condition.

is linear for small variations in turning rate. This is true for normal course keeping when the turning rate is typically below $0.05\,\mathrm{deg\,s^{-1}}$. It is also clear from Fig. 4 that the nonlinear characteristics must be taken into account in manoeuvres which involve high turning rates.

The normal course keeping regulator can handle small changes in heading. It can, however, not handle large manoeuvres because of the nonlinearities discussed above. A special turning regulator was therefore designed. Several different approaches were considered for this design. Both adaptive and non-adaptive schemes were explored. It was finally decided to use a high gain regulator. This decision was partly motivated by the results in Åström (1977) where it was shown that tankers could indeed be controlled by a regulator having fixed high gain. Such a regulator will control the motion of the ship well at the expense of increased steering drag. This property is in good harmony with the requirements on turning.

To explain the turning regulator let us first discuss how a typical turn is made. Let the ship initially be on a straight line path, i.e. at point A in Fig. 4. A turn is initiated by a rudder deflection, the yaw rate then increases until the desired yaw rate is obtained, e.g. at B in Fig. 4. The desired yaw rate is maintained until the ship has almost completed the turn. The yaw rate is then brought to zero in such a way that the ship points in the desired direction. It follows from Fig. 4 that the characteristics of the ship may change substantially when the turning is initiated and interrupted. Based on this discussion it seems natural to have a turning regulator with three different phases. These are described in detail below. A fourth phase is also included after the turn has been interrupted to make sure that the course keeping regulator starts with a good heading and a good estimate of the rudder bias.

Phase 1—*Initialization.* The turn is initialized using a proportional rate feedback. The following control law is used:

$$\delta_c(t) = G(V(t))\mathrm{sat}[k_1(\hat{r}(t) - r_{\mathrm{ref}}),\ \bar{c}_1|r_{\mathrm{ref}}|] + \bar{\delta}_c$$

where

$$\mathrm{sat}[x, y] = \begin{cases} x & \text{if } |x| \leq y \\ y\,\mathrm{sgn}(x) & \text{otherwise.} \end{cases}$$

The factor $G(V(t))$ is the velocity schedule, where $G = (V_0/V)^2$ is chosen to obtain time invariance. The saturation function is necessary to make sure that the rudder deflection is reasonable even if the commands are large.

Phase 2—*Control of steady state turning rate.* This phase is simply a rate controller designed to maintain a constant yaw rate. The control law used is

$$\delta_c(t) = G(V(t))[k_2(\hat{r}(t) - r_{\mathrm{ref}}) + k_3 T_s \sum_{n=0}^{t-1} (\hat{r}(n) - r_{\mathrm{ref}})] + \bar{\delta}_c$$

where T_s is the sampling interval.

Phase 3—*Stopping the turn.* It is essentially a dead beat control problem to stop the turn. The following control law is used:

$$\delta_c(t) = G(V(t))\mathrm{sat}[k_4(\hat{\psi}(t) - \psi_{\mathrm{ref}}) + k_5 \hat{r}(t),\ \bar{c}_3|r_{\mathrm{ref}}|].$$

This control law is a PD-algorithm. It is effective in stopping the turning. With reasonable values of the gains k_4 and k_5 it may still happen that the heading error is too large. A fourth phase which also includes an integrating term is therefore included.

Phase 4—Fine adjustment of heading. The control algorithm is

$$\delta_c(t) = G(V(t))[k_6(\hat{\psi}(t) - \psi_{\text{ref}}) + k_7\hat{r}(t) + k_8 T_s \sum_{n=0}^{t-1} (\hat{\psi}(n) - \psi_{\text{ref}})].$$

The rudder bias is updated as follows

$$\bar{\delta}_c(t+1) = \bar{\delta}_c(t) + \left(\frac{1-\gamma}{1+t} + \gamma\right)(\delta_c(t) - \bar{\delta}_c(t)) \quad (9)$$

where γ is a weighting factor.

The self-tuning regulator for steady state course keeping is used as a turning regulator if the requested heading change $\Delta\psi_{\text{ref}}$ is not larger than ψ_1. During a normal turn the turning regulator is switched sequentially from phase 1 to phase 4. The conditions for switching are given in the superdiagonal entries in Table 1. It is also possible to enter phase 4 immediately after the initiation. The condition for this is $\psi_1 < \Delta\psi_{\text{ref}} \leqq \psi_2$. Similarly phase 3 can be entered directly after phase 1. The condition is given in entry (1, 3) in Table 1. There are no ambiguities in the switching conditions since the entries (1, 3) and (2, 3) are the same. The turning mode is terminated and the self-tuning regulator for steady state course keeping is initiated when the time in phase 4 is equal to T_4. The turning regulator is a system with variable structure. Such regulators have been explored extensively in Emelyanov (1967).

The turning regulators used in the different autopilots differ only in the estimates that are used. In KADPIL 1 the value of $\hat{\psi}$ is taken as the heading measurement and the estimate $\hat{r}$ is obtained by numerical differentiation of the heading angle. In KADPIL 2 the estimates are instead taken from the Kalman filter. The reference values ψ_{ref} and r_{ref} can be changed at any time even during a turn.

It has been shown by a phase plane analysis that the turning regulator is stable when controlling a simple steering model of a ship. Simulations using a more sophisticated steering model and full-scale experiments also indicate good behaviour of the turning regulator although formal stability analysis has not been carried out for a more realistic model.

3. SIMULATIONS

Because full-scale experiments are costly and time consuming, simulations were used extensively. The feasibilities of the autopilots were first evaluated by simulations before the regulators were tested on real ships. The availability of the interactive simulation program SIMNON (Elmqvist, 1975) made it possible to carry out the simulations efficiently. A comprehensive summary of the simulations is given in Källström and co-workers (1977). The simulations were based on two fairly detailed nonlinear tanker models. The structure of the models used is described in Norrbin (1970). Identification experiments with full-scale ships were also performed in order to verify and improve upon the models (Åström and co-workers, 1975; Åström and Källström, 1976; Byström and Källström, 1978). The disturbances from wind, waves, and currents were modelled simply by adding a force and a moment to the equations of motion. The stochastic nature of wind turbulence was introduced by adding white noise realizations. Additive, white measurement noise was included in the simulation model as well as constant measurement biases.

The course keeping performances of the adaptive autopilots KADPIL 1 and KADPIL 2 were compared with the performance of a discrete, fixed gain PID-regulator

$$\delta_c(t) = k_1[\psi_m(t) - \psi_{\text{ref}}] + k_2 r_m(t) + k_3 \sum_{n=0}^{t-1} [\psi_m(n) - \psi_{\text{ref}}]. \quad (10)$$

The parameters k_1, k_2 and k_3 were always carefully tuned. The Kalman filter estimates $\hat{\psi}$ and $\hat{r}$ were sometimes used in (10) instead of the non-filtered measurements ψ_m and r_m. The loss function V (cf. (3)) was used to compare the steering quality of the different autopilots.

Simulations of a 255,000 tdw tanker

The course keeping performances of KADPIL 1 and a well-tuned PID-regulator, described by (10), were compared. The ship speed was 16 knots and the wind was 6–8 m s^{-1} in the simulations. In ballast conditions there was no significant difference between the regulators. The loss function (3) had the value $V = 0.4$ in both cases. At full draught the values of the loss function increased and it was found that the average drag when steering with KADPIL 1 was about 0.3% lower than when the PID-regulator was used.

The Kalman filter of KADPIL 2 was also tested under different load and weather conditions. The filter estimates obtained were very good. It was found that the Kalman filter was insensitive to changes of load and weather conditions. Comparisons between KADPIL 1 and KADPIL 2 showed that the steady state course keeping performance was significantly improved when the Kalman filter was used. The drag was decreased by 0.6% at full draught when the wind

TABLE 1. SWITCHING CONDITIONS BETWEEN THE DIFFERENT CONTROL MODES OF THE TURNING REGULATOR

	0	1	2	3	4
0		$\Delta\psi_{ref} > \psi_2$			$\psi_1 < \Delta\psi_{ref} \le \psi_2$
1			$r_{ref} \ge 0$ and $\hat{r}(t) - r_{ref} > -\varepsilon_1$ or $r_{ref} < 0$ and $\hat{r}(t) - r_{ref} < \varepsilon_1$ or (time in phase 1) $> T_1$	$r_{ref} \ge 0$ and $-\bar{c}_2\sqrt{\frac{V_0}{V(t)}}\ \hat{r}(t) < \hat{\psi}(t) - \psi_{ref}$ or $r_{ref} < 0$ and $-\bar{c}_2\sqrt{\frac{V_0}{V(t)}}\ \hat{r}(t) > \hat{\psi}(t) - \psi_{ref}$	
2				$r_{ref} \ge 0$ and $-\bar{c}_2\sqrt{\frac{V_0}{V(t)}}\ \hat{r}(t) < \hat{\psi}(t) - \psi_{ref}$ or $r_{ref} < 0$ and $-\bar{c}_2\sqrt{\frac{V_0}{V(t)}}\ \hat{r}(t) > \hat{\psi}(t) - \psi_{ref}$	
3					$\lvert\hat{r}(t)\rvert < \varepsilon_2$ or $r_{ref} \ge 0$ and $\hat{\psi}(t) - \psi_{ref} > -\varepsilon_3$ or $r_{ref} < 0$ and $\hat{\psi}(t) - \psi_{ref} < \varepsilon_3$ or (time in phase 3) $> T_3$
4	(time in phase 4) $> T_4$				

Steady state course keeping is denoted phase 0 and the requested heading change is denoted $\Delta\psi_{ref}$. The entry (0, 1) shows for example the condition to terminate phase 0 and to initiate phase 1.

speed was 6–8 $m\,s^{-1}$, but the decrease could be as much as 1.4% at wind speeds 17–20 $m\,s^{-1}$.

Simulations of a 355,000 tdw tanker

The different functions of KADPIL 1 and KADPIL 2 were tested at the speeds 4, 10 and 15.8 knots when the wind velocity was 11–14 $m\,s^{-1}$. The Kalman filter performed very well both in full load and ballast conditions. An example is shown in Fig. 5. Notice that the bias of the yaw rate measurements is estimated correctly after about 15 min and that the difference between the true yaw rate and the estimated yaw rate is very small, although the yaw rate measurements are very noisy.

The tuning rate of some of the parameters of the self-tuning regulator used in KADPIL 2 is illustrated in Fig. 6. It can be concluded that the parameters are roughly adjusted after about 10 min and that an acceptable course keeping is obtained in the same time. It is, of course, desirable to give the parameters good initial estimates instead of the values zero to avoid large, initial course errors. This has also been done in all simulations and experiments.

The average drag induced by the steering was also evaluated in the simulations. For a fully loaded ship at a speed of 15.8 knots the loss function (3) was $V=1.3$ for a well-tuned PID-regulator (10). This loss function was reduced to $V=1.0$ when the PID-regulator was provided with estimates from a Kalman filter. When the autopilot KADPIL 2 was used the loss function was further reduced to $V=0.7$. From (2) it thus follows that the drag is 0.4–0.8% lower with KADPIL 2 compared with a well-tuned PID-regulator.

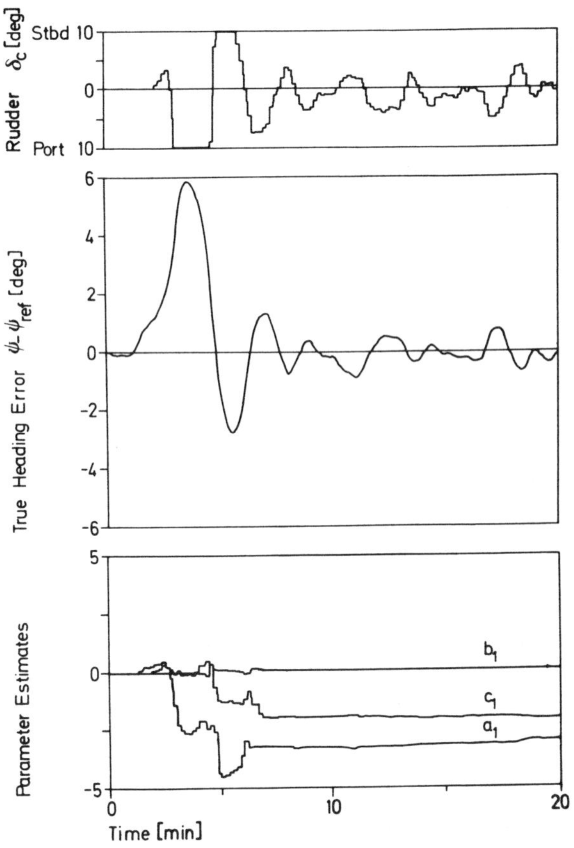

FIG. 6. Simulation of straight course keeping of a fully loaded 355,000 tdw tanker using KADPIL 2. The ship speed was 15.8 knots and the wind speed was 11–14 $m\,s^{-1}$. The initial values of all parameters of the self-tuning regulator were zero. Estimates of only three parameters are shown. The rudder deflection was limited to 10 deg.

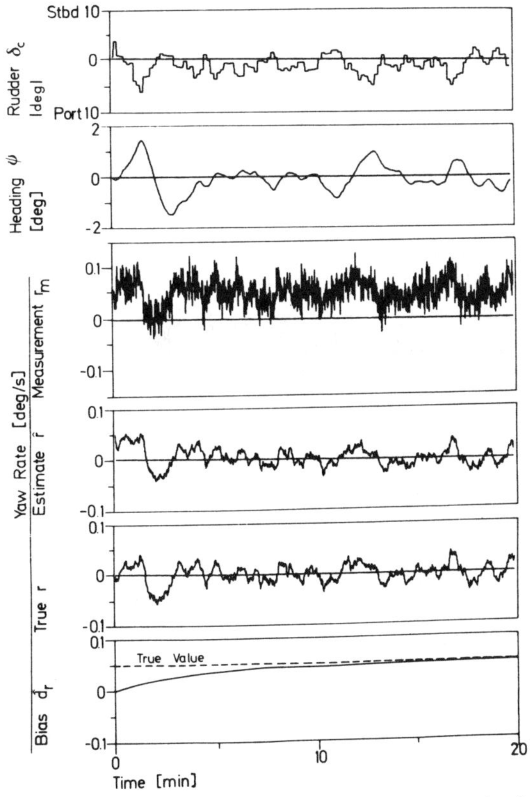

FIG. 5. Simulation of straight course keeping of a fully loaded 355,000 tdw tanker using KADPIL 2. The ship speed was 15.8 knots and the wind speed was 11–14 $m\,s^{-1}$. The value of the loss function V was 0.7.

The performance of the turning regulator of KADPIL 2 was excellent. Relatively large rudder motions were obtained from KADPIL 1 during turns, since no Kalman filter was used to provide smooth signals.

It has been possible to check the validity of some of the simulation results by comparisons with the full-scale experiments described in Section 4. A quite acceptable consistency was then obtained. However, the performance of a well-tuned PID-regulator seems to be slightly overestimated in the simulations. The improvements with the adaptive autopilots are thus greater in the experiments than in the simulations. A possible reason for this is that the model for the disturbances used in the simulations was too simple.

4. EXPERIMENTS

Experiments have been made with three different tankers: T/T *Sea Scout*, T/T *Sea Swift*, and T/T *Sea Stratus*. The ships were built for the Salén Shipping Companies in Stockholm by Kockums Shipyard in Malmö. The *Sea Scout* and the *Sea Swift* are sister ships of 255,000 tdw. The length between perpendiculars is 329 m and the beam is 52 m. The length of the 355,000 tdw tanker *Sea Stratus* is 350 m and the beam is 60 m. Each experimental voyage lasted for about two weeks. The tankers were equipped with an integrated navigation system, Kockums Bridge System. The experiments were carried out by linking the new adaptive autopilots with the integrated navigation system. The data gathered during the experiments were punched on paper tape. They were subsequently analysed by off-line methods. A large number of experiments of straight course keeping and turning were carried out with the ships under different load, speed, and weather conditions. Over 130 experiments were recorded but the testing was in fact done continuously during the trips, i.e. during about 6 weeks. The adaptive autopilot designed for the *Sea Swift* was used during at least one year after the experiments were finished in October 1974. A summary of the experiments is given in Källström and co-workers (1977).

The Sea Scout experiments

Preliminary experiments were performed with the *Sea Scout* in ballast condition. Early versions of KADPIL 1 and KADPIL 2 were tested. They were compared with a PID-autopilot. It was concluded that the proposed design scheme of the adaptive autopilots was sound.

The Sea Swift experiments

The adaptive autopilot KADPIL 1 was tested on the *Sea Swift* in full load condition. Different structures of the self-tuner for steady state course keeping were explored. The self-tuner was also compared with a well-tuned PID-regulator under different speed and weather conditions. The performances were compared using the loss function V defined by (3). The PID-regulator used a digitally filtered yaw rate. The results of some experiments are summarized in Table 2. Plots from four of the experiments are shown in Figs. 7 and 8. A reduced drag of 0.1–0.3% was achieved at full speed when KADPIL 1 was used instead

TABLE 2. LOSS FUNCTION V FROM EXPERIMENTS OF STRAIGHT COURSE KEEPING WITH THE *Sea Swift* IN FULL LOAD CONDITION

Ship speed [knots]	Wind speed [m/s]	KADPIL 1	PID using filtered yaw rate
18	4-8	0.3	0.4
17	4-8	0.5	0.7
5	17-24	3.7	5.6

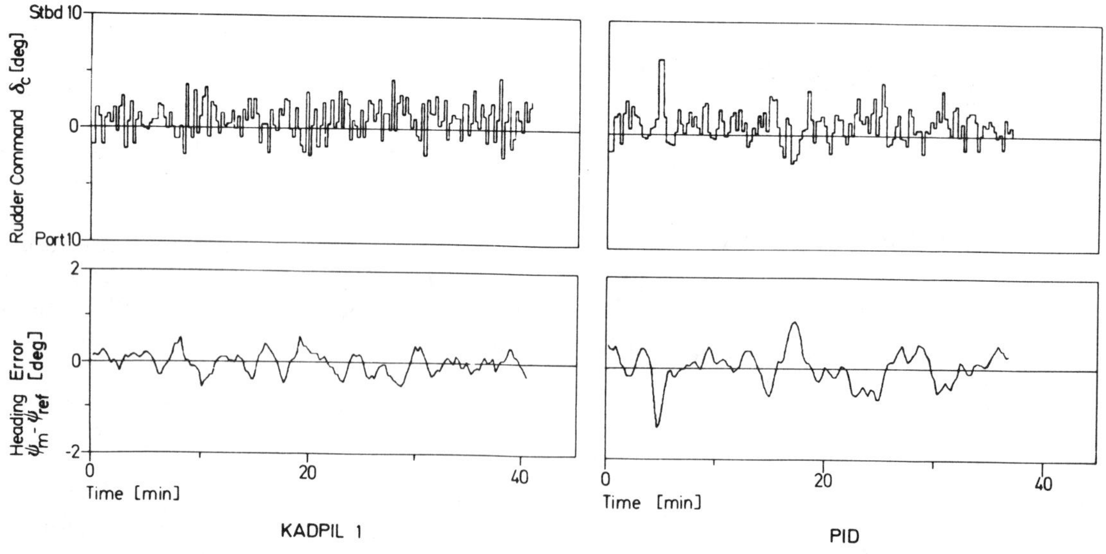

FIG. 7. Results from two experiments of straight course keeping with the *Sea Swift* in full load condition. The ship speed was 17 knots and the wind speed was 4–8 m s^{-1}. The autopilots KADPIL 1 and a well-tuned PID using filtered yaw rates were used. The values of the loss function V were 0.5 and 0.7 respectively. Under the conditions of the experiments the adaptive autopilot gives 0.3% less drag than the PID-autopilot.

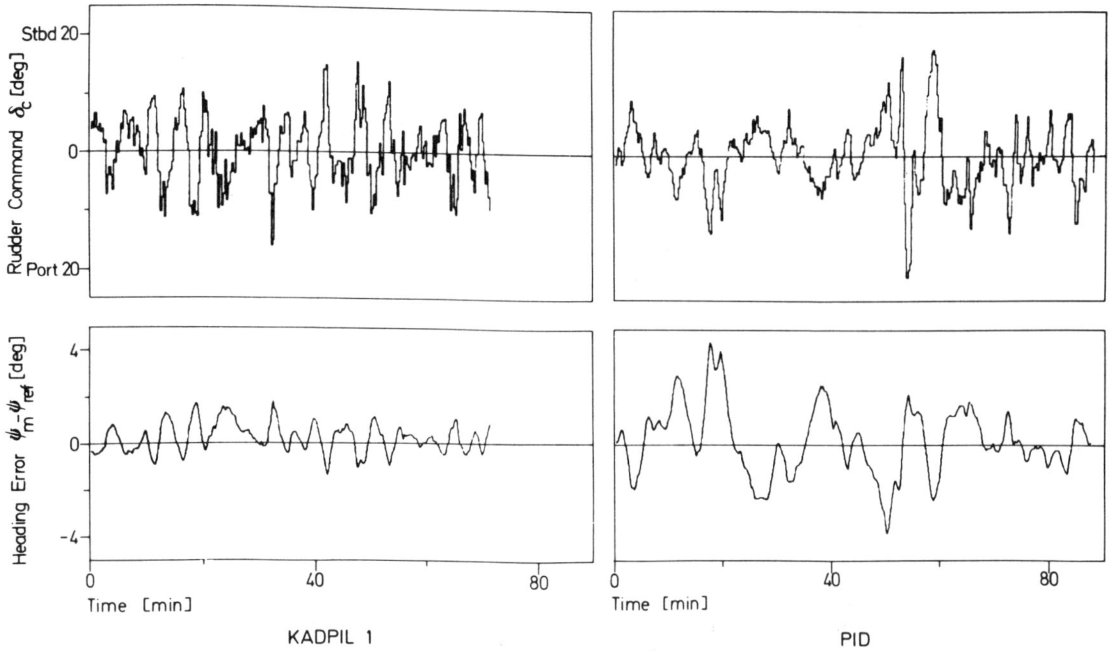

FIG. 8. Results from two experiments of straight course keeping with the *Sea Swift* in full load condition. The ship speed was 5 knots and the wind speed was 17–24 m s^{-1}. The autopilots KADPIL 1 and a well-tuned PID using filtered yaw rates were used. The values of the loss function V were 3.7 and 5.6 respectively. Under the conditions of the experiments the adaptive autopilot gives 2.7% less drag than the PID-autopilot.

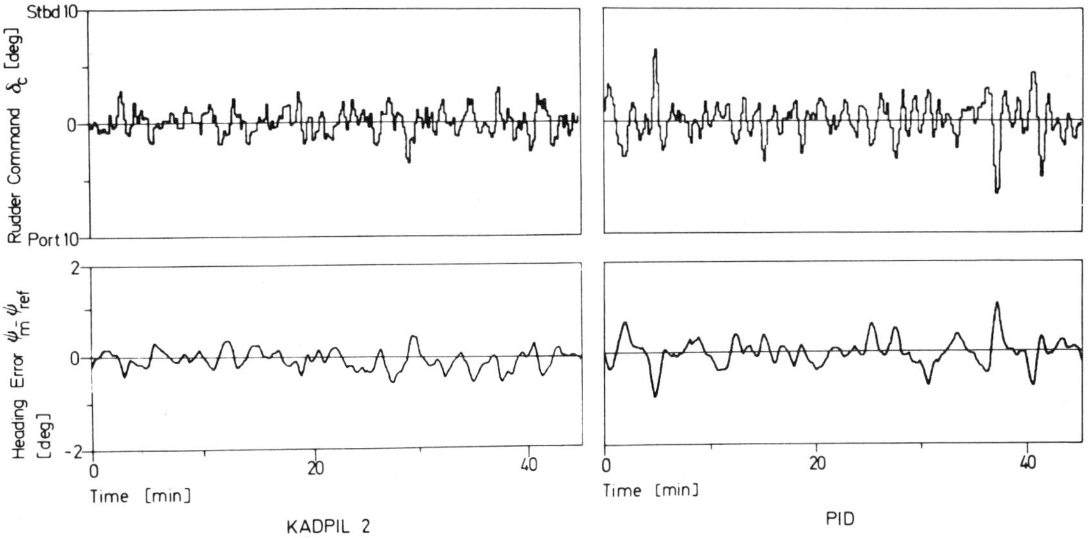

FIG. 9. Results from two experiments of straight course keeping with the *Sea Stratus* in ballast condition. The ship speed was 12 knots and the wind speed was about 11 m s^{-1}. The autopilots KADPIL 2 and a well-tuned PID using filtered headings and yaw rates from a Kalman filter were used. The values of the loss function V were 0.1 and 0.3 respectively. Under the conditions of the experiments the adaptive autopilot gives 0.3% less drag than the PID-autopilot.

of a well-tuned PID-regulator. The decrease in drag was as much as 2.7% at low speed and bad weather conditions. Figure 8 shows the excellent performance of KADPIL 1 in bad weather conditions. It was very difficult to adjust the parameters of the PID-regulator to obtain the performance shown in Fig. 8. According to the captain of the ship, a conventional autopilot would normally be switched off and replaced by manual control under these weather conditions.

The Sea Stratus experiments

The final tests of the adaptive autopilots KADPIL 1 and KADPIL 2 were carried out with the *Sea Stratus* in ballast condition. Results from experiments with KADPIL 2 and a well-tuned PID-autopilot under similar weather conditions are shown in Fig. 9. Based on all experiments it was concluded that the drag was reduced by 0–2.2% when the adaptive autopilots were used instead of well-tuned PID-regulators.

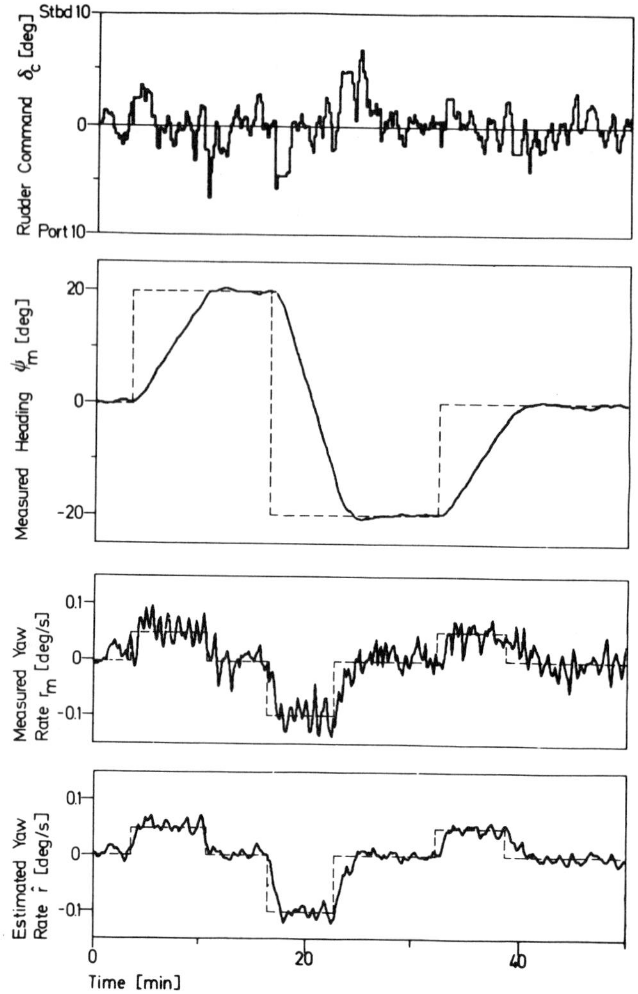

FIG. 10. Result from a turning experiment with the *Sea Stratus* in ballast condition. The ship speed was 13 knots and the wind speed was about $7\,\mathrm{ms}^{-1}$. The autopilot KADPIL 2 was used.

The reductions obtained varied with the weather conditions. They were larger in bad weather.

The turning regulator worked very well both for small heading adjustments and for large manoeuvres. An example is shown in Fig. 10. Notice that the requested yaw rate was $0.1\,\mathrm{deg\,s}^{-1}$ during the second turn and $0.05\,\mathrm{deg\,s}^{-1}$ in the other turns.

The Kalman filter performed well in all conditions. The experiments showed that it is worthwhile to include a Kalman filter in an ordinary autopilot. The quality of the estimates obtained from the Kalman filter is illustrated in Fig. 10.

5. CONCLUSIONS

The feasibility of using adaptive techniques for ship steering was explored. Several alternatives were analysed, simulated and tested. The simple adaptive autopilot uses measurements of forward speed, or propeller · rate and heading. It contains velocity scheduling, a self-tuning regulator for steady state course keeping and a turning regulator. The more advanced autopilot also contains a Kalman filter. This implies that more measurement signals may be used. It was concluded that it is worthwhile to include a Kalman filter in all autopilots for ship steering. Straightforward FORTRAN implementations on the computer PDP-15 require 2800 memory cells for KADPIL 1 and 4600 cells for KADPIL 2. Execution times for one step of the algorithms on the PDP-15 were typically 0.02 s for KADPIL 1 and 0.06 s for KADPIL 2. The execution time for one step of the Kalman filtering was 0.02 s. Typical sampling intervals are 1 s for the Kalman filter and 10–20 s for the autopilot. The computational requirements are thus reasonable. The adaptive autopilots were tested by extensive simulations and full-scale experiments. It was found that the adaptive autopilots are indeed feasible. The different functions of the autopilots all performed well under different load, speed, and weather conditions.

It has been attempted to estimate the reduced

drag by a quadratic loss function. The adaptive autopilots decrease the drag by 0–2% when compared with well-tuned PID-regulators. There were virtually no differences in performance under very good weather conditions provided that the PID-regulator used smoothed, filtered values of heading and yaw rate. A proper smoothing of the signals was found to be crucial when using a PID-regulator. The differences between the performances of adaptive and PID-autopilots are larger in bad weather. Experiments performed at low speed and bad sea conditions indicated that the drag was reduced by as much as 2.7% when an adaptive autopilot was used instead of a well-tuned PID-regulator. The more advanced autopilot KADPIL 2 also reduced the drag by approximately 0.5% compared with the simpler autopilot KADPIL 1.

Experiments in Schilling (1976) indicated that speed improvements of 1–2% were achieved on a tanker, when an optimally tuned PID-regulator was used instead of a PID-regulator tuned by the officer in charge. The gain will of course depend very much on operating conditions and on how well the PID-regulator was adjusted. Notice, however, that the adaptive autopilots discussed in this paper will give additional speed improvements over well-tuned PID-regulators. A reduced drag of 2% corresponds to a speed increase of approximately 1%. Much experimental work is naturally required before the gains due to improved steering can be established firmly. Based on the results of this paper and Schilling's experiments it is, however, not unreasonable to guess that an increased speed of the order of one or a few percent may be achieved in normal operation.

The economic benefits obtained for a 355,000 tdw tanker of Kockum's design, travelling between Europe and the Persian Gulf, are estimated in Källström and co-workers (1977) for two different economic situations. In a good economic situation, corresponding to World Scale 100, a speed increase of 1% corresponds to a net gain of $292,000 per year due to an increased transport capacity. When only part of the transport capacity of the ship is used, as in bad economic situations, the benefits of the adaptive control are obtained as a reduced fuel consumption. A decrease of the drag by 2% corresponds to a net gain of $24,000 per year, if it is assumed that the bunker oil price is $87 per ton. It is then also assumed that only three round trips between the Persian Gulf and Europe are made per year.

There are, however, not only economic benefits by using the adaptive autopilots. A conventional autopilot must be properly adjusted in order to obtain a good performance. Adjustments are required to compensate for wind, waves, currents, speed, trim, draught and water depth. No manual settings of the adaptive autopilots are necessary, since the regulator parameters are tuned automatically. This will greatly facilitate the management. The excellent performance of the turning regulator will guarantee smooth and precise turns, which implies a much safer handling of the ship.

It is our belief that the adaptive autopilots discussed easily can be implemented on many types of ships, although all experiments so far have been performed with tankers.

Acknowledgements—This work has been supported by the Swedish Board for Technical Development under contracts 73 41 87 and 74 41 27. The authors would also like to express their gratitude to the Salén Shipping Companies, Stockholm, for their willingness to allow experiments to be performed with their ships. We also thank Dr N. H. Norrbin at the Swedish State Shipbuilding Experimental Tank, Gothenburg, for many valuable discussions on ship steering dynamics. Several versions of the manuscript have been typed by Mrs E. Dagnegård. The figures were drawn by Miss B.-M. Carlsson.

REFERENCES

VAN AMERONGEN, J. and A. J. UDINK TEN CATE (1975). Model reference adaptive autopilots for ships. *Automatica* **11**, 441.

VAN AMERONGEN, J. and H. R. VAN NAUTA LEMKE (1978). Optimum steering of ships with an adaptive autopilot. *Proc. 5th Ship Control Systems Symp.*, Annapolis, Maryland, U.S.A.

ÅSTRÖM, K. J. (1970). *Introduction to Stochastic Control Theory*. Academic Press, New York.

ÅSTRÖM, K. J. and B. WITTENMARK (1973). On self tuning regulators. *Automatica* **9**, 185.

ÅSTRÖM, K. J. and B. WITTENMARK (1974). Analysis of a self-tuning regulator for nonminimum phase systems. *IFAC Symp. on Stochastic Control*, Budapest.

ÅSTRÖM, K. J., C. G. KÄLLSTRÖM, N. H. NORRBIN and L. BYSTRÖM (1975). The identification of linear ship steering dynamics using maximum likelihood parameter estimation. Publ. No. 75, Swedish State Shipbuilding Experimental Tank, Gothenburg, Sweden.

ÅSTRÖM, K. J. and C. G. KÄLLSTRÖM (1976). Identification of ship steering dynamics. *Automatica* **12**, 9.

ÅSTRÖM, K. J. (1977). Why use adaptive techniques for steering large tankers? Dept. of Automatic Control, Lund Institute of Technology, Lund, Sweden, CODEN: LUTFD2/(TFRT-3144)/1-35/(1977).

ÅSTRÖM, K. J., U. BORISSON, L. LJUNG and B. WITTENMARK (1977). Theory and applications of self-tuning regulators. *Automatica* **13**, 457.

BRINK, A. W., G. E. BAAS, A. TIANO and E. VOLTA (1978). Adaptive automatic course-keeping control of a supertanker and a containership—a simulation study. *Proc. 5th Ship Control Systems Symp.*, Annapolis, Maryland, U.S.A.

BROOME, D. R. and T. H. LAMBERT (1978). An optimising function for adaptive ships autopilots. *Proc. 5th Ship Control Systems Symp.*, Annapolis, Maryland, U.S.A.

BYSTRÖM, L. and C. G. KÄLLSTRÖM (1978). System identification of linear and non-linear ship steering dynamics. *Proc. 5th Ship Control Systems Symp.*, Annapolis, Maryland, U.S.A.

ELMQVIST, H. (1975). SIMNON—An interactive simulation program for nonlinear systems, user's manual. TFRT-3091, Dept. of Automatic Control, Lund Institute of Technology, Lund, Sweden.

EMELYANOV, S. V. (1967). *Automatic Control Systems of Variable Structure.* Nauka, Moscow.

KÄLLSTRÖM, C. G., K. J. ÅSTRÖM, N. E. THORELL, J. ERIKSSON and L. STEN (1977). Adaptive autopilots for steering of large tankers. Dept. of Automatic Control, Lund Institute of Technology, Lund, Sweden, CODEN: LUTFD2/(TFRT-3145)/1-66/(1977). Also available as MB34, Kockums Automation AB, Malmö, Sweden.

KÄLLSTRÖM, C. G., K. J. ÅSTRÖM, N. E. THORELL, J. ERIKSSON and L. STEN (1978). Adaptive autopilots for large tankers. *7th IFAC Triennial World Congress*, Helsinki.

KOYAMA, T. (1967). On the optimum automatic steering system of ships at sea. *J. Soc. Nav. Archit. Japan* **122,** 142.

MERLO, P. and A. TIANO (1975). Experiments about computer controlled ship steering. *Semana Internacional sobre la Automatica en la Marina*, Barcelona, Spain.

NORRBIN, N. H. (1970). Theory and observations on the use of a mathematical model for ship manoeuvring in deep and confined waters. *Proc. 8th Symp. on Naval Hydrodynamics*, Pasadena, California, U.S.A. Also available as Publ. No. 68, Swedish State Shipbuilding Experimental Tank, Gothenburg, Sweden.

NORRBIN, N. H. (1972). On the added resistance due to steering on a straight course. *13th International Towing Tank Conference*, Berlin/Hamburg.

OHTSU, K., M. HORIGOME and G. KITAGAWA (1976). On the prediction and stochastic control of ship's motion. *Proc. 2nd IFAC/IFIP Symp. on Ship Operation Automation*, Washington DC.

OHTSU, K., M. HORIGOME, M. HARA and G. KITAGAWA (1978). An advanced ship's autopilot system by a stochastic model. *Proc. 5th Ship Control Systems Symp.*, Annapolis, Maryland, U.S.A.

OLDENBURG, J. (1975). Experiment with a new adaptive autopilot intended for controlled turns as well as for straight course steering. *Proc. 4th Ship Control Systems Symp.*, The Hague.

REID, R. E. and V. E. WILLIAMS (1978). A new ship control design criterion for improving heavy weather steering. *Proc. 5th Ship Control Systems Symp.*, Annapolis, Maryland, U.S.A.

SCHILLING, A. C. (1976). Economics of autopilot steering using an IBM System/7 computer. *Proc. 2nd IFAC/IFIP Symp. on Ship Operation Automation*, Washington DC.

SUGIMOTO, A. and T. KOJIMA (1978). A new autopilot system with condition adaptivity. *Proc. 5th Ship Control Systems Symp.*, Annapolis, Maryland, U.S.A.

WARE, J. R., A. S. FIELDS and P. J. BOZZI (1978). Design procedures for a surface ship steering control system. *Proc. 5th Ship Control Systems Symp.*, Annapolis, Maryland, U.S.A.

APPENDIX: KALMAN FILTER MODEL

The design of the Kalman filter is based on the following standard model

$$\begin{cases} dx = Ax\,dt + Bu\,dt + dw \\ y(t_k) = Cx(t_k) + e(t_k), \quad k = 0, 1, 2, \ldots \end{cases} \qquad (11)$$

where

$u = \delta_c$ rudder command
$x_1 = v$ sway velocity at midship
$x_2 = r$ yaw rate
$x_3 = \psi$ heading angle
$x_4 = \delta - \delta_0$ deviation of rudder angle from bias
$x_5 = \delta_0$ rudder bias due to disturbances
$x_6 = d_{v_1}$ bias in fore sway velocity
$x_7 = d_{v_2}$ bias in aft sway velocity
$x_8 = d_r$ bias in yaw rate measurement
$x_9 = d_\delta$ bias in rudder angle measurement
$y_1 = \delta_m$ measured rudder angle
$y_2 = v_1$ measured fore sway velocity
$y_3 = v_2$ measured aft sway velocity
$y_4 = r_m$ measured yaw rate
$y_5 = \psi_m$ measured heading angle

and the disturbances are denoted w and e.

The matrices A, B and C are given by

$$A = \begin{bmatrix} a_{11}\frac{V}{L} & a_{12}V & 0 & b_{11}\frac{V^2}{L} & 0 & 0 & 0 & 0 & 0 \\ a_{21}\frac{V}{L^2} & a_{22}\frac{V}{L} & 0 & b_{21}\frac{V^2}{L^2} & 0 & 0 & 0 & 0 & 0 \\ 0 & 1 & 0 & 0 & 0 & 0 & 0 & 0 & 0 \\ 0 & 0 & 0 & -\frac{1}{T_r} & -\frac{1}{T_r} & 0 & 0 & 0 & 0 \\ 0 & 0 & 0 & 0 & 0 & 0 & 0 & 0 & 0 \\ 0 & 0 & 0 & 0 & 0 & 0 & 0 & 0 & 0 \\ 0 & 0 & 0 & 0 & 0 & 0 & 0 & 0 & 0 \\ 0 & 0 & 0 & 0 & 0 & 0 & 0 & 0 & 0 \\ 0 & 0 & 0 & 0 & 0 & 0 & 0 & 0 & 0 \end{bmatrix}$$

$$B^T = \begin{bmatrix} 0 & 0 & 0 & \frac{1}{T_r} & 0 & 0 & 0 & 0 & 0 \end{bmatrix}$$

$$C = \begin{bmatrix} 0 & 0 & 0 & 1 & 1 & 0 & 0 & 0 & 1 \\ 1 & L_1 & 0 & 0 & 0 & 1 & 0 & 0 & 0 \\ 1 & -L_2 & 0 & 0 & 0 & 0 & 1 & 0 & 0 \\ 0 & 1 & 0 & 0 & 0 & 0 & 0 & 1 & 0 \\ 0 & 0 & 1 & 0 & 0 & 0 & 0 & 0 & 0 \end{bmatrix}$$

where

V = ship speed
T_r = time constant of steering engine
L = ship length
L_1 = distance from midship to fore doppler log
L_2 = distance from midship to aft doppler log

and a_{11}, a_{12}, a_{21}, a_{22}, b_{11}, b_{21} are parameters which describe the steering dynamics of the specific ship. The parameter values, the covariance of the process noise w, and the measurement covariance are mainly based on results of system identification applied to full-scale experiments (Åström and co-workers, 1975; Åström and Källström, 1976; Byström and Källström, 1978).

The model (11) is first normalized using the length of the ship as the length unit and the time it takes to travel a ship length as the time unit. The model is then transformed to standard discrete time form and the constant filter gain K is computed by iterating the discrete time Riccati equation until stationarity is achieved (Åström, 1970). All matrices Φ, Γ, θ and K of (1) are thus determined in this procedure.

Adaptive Steering of Ships—A Model Reference Approach*

J. VAN AMERONGEN†

Model reference adaptive control, applied to automatic steering of ships, provides by means of direct adaptation, optimal control and adaptive state estimation, improved manoeuvring, and better fuel economy.

Key Words—Adaptive control; digital computer applications; optimal control; optimal filtering; parameter estimation; parameter optimization; ship control.

Abstract—This paper describes the application of model reference adaptive control (MRAS) to automatic steering of ships. The main advantages in this case are the simplified controller adjustment which yields safer operation and the decreased fuel cost. After discussion of the mathematical models of process and disturbances, criteria for optimal steering are defined. Algorithms are given for direct adaptation of the controller gains, applicable after setpoint changes, as well as for identification and adaptive state estimation, to be used when the input is constant. Solutions for applying MRAS to a certain class of nonlinear systems are dealt with. Full-scale trials at sea and tests with a scale model in a towing tank are described. It is shown that the autopilot designed indeed has the desired properties. Fuel savings up to 5 % in comparison to conventional PID control are demonstrated. These savings are mainly possible because of the adaptive state estimator.

1. INTRODUCTION

AUTOMATIC steering of ships was introduced many years ago (Minorsky, 1922; Sperry, 1922); with developing technology, the hardware of autopilots changed from purely mechanical devices to electronic systems, but the controller concept itself has hardly changed. However, it may be expected that, in the near future, a new generation of autopilots, based on modern control techniques, will replace the present systems. The fast development of small and inexpensive microcomputers makes these autopilots practically realizable.

In principle, a conventional autopilot is nothing more than a PID controller extended with a limiter to limit its output signal (the desired rudder angle) and a dead band and a filter to smooth the controller output. Two major disadvantages of this type of controller are: (1) it is difficult to adjust manually because the operator, the watch officer, has many other tasks and lacks the insight into control theory; his adjustment will seldom be optimal and (2) the optimal adjustment varies and is not known by the user. Changing circumstances require manual re-adjustment of a series of settings of the autopilot. This holds not only for variations in the parameters of the process but also when due to a varying traffic situation the required performance changes.

Because of the changing environment it is not possible to simply design an optimal controller. Various operating conditions require different controller structures (course changing and course keeping) and varying traffic situations demand other definitions of optimal performance. The first problem to be solved is thus how the operator can be provided with the means which enable him to simply adjust the autopilot according to his actual demands, without the necessity of setting all the conventional settings. These demands must then be translated into a performance index to be minimized by an optimal control system. The second problem is that the optimal performance has to be maintained when the process characteristics change due to changing forward speed, load condition, water depth, etc. This requires adaptive control.

Since 1973, the continuously rising fuel prices have made the savings from applying more sophisticated control algorithms obvious. Recently, several proposals for adaptive autopilots have been published (Van Amerongen and Udink ten Cate, 1975; Van Amerongen and Van Nauta Lemke, 1978, 1979; Van Amerongen, 1982; Kallström and co-workers, 1979; Ohtsu, Horigome and Kitagawa, 1979; Reid and Williams, 1978; Kojima and

* Received 1 June 1981; revised 20 July 1982; revised 19 April 1983. The original version of this paper was presented at the 8th IFAC World Congress on Control Science and Technology for the Progress of Society which was held in Kyoto, Japan, during August 1981. The published proceedings of this IFAC meeting may be ordered from Pergamon Press Ltd, Headington Hill Hall, Oxford OX3 0BW, U.K. This paper was recommended for publication in revised form by associate editor R. Isermann under the direction of editor H. Austin Spang, III.

† Control Laboratory, Electrical Engineering Department, Delft University of Technology, 2600 GA Delft, The Netherlands.

Reprinted with permission from *Automatica*, vol. 20, pp. 3–14, Jan. 1984.
Copyright © 1984, International Federation of Automatic Control IFAC. Published by Pergamon Press Ltd.

Sugimoto, 1978; Herther and co-workers, 1980). Several techniques are used to achieve the automatic adjustment of the controller parameters. The autopilot described in this paper is based on the theory of model reference adaptive systems (MRAS). In general, the theory based on stability methods requires that the process and reference model be linear. The solutions given in this paper to deal with certain classes of nonlinearities are also applicable to other systems where saturation effects in actuators dominate the response. In general, continuous time techniques will be used to design the adaptive controller. Because of the relatively high sampling rate which can be chosen, the algorithms can easily be digitally implemented as well.

The paper is organized as follows: Section 2 describes the mathematical models of the process and the disturbances. Section 3 discusses criteria for optimal steering. Section 4 gives the algorithms used for direct adaptation of the controller parameters. Indirect parameter adjustment by means of on-line identification and controller optimization, as well as adaptive state estimation, are discussed in Section 5. In Section 6 results of a series of full-scale trials as well as tests with a scale model in a towing tank are described. Section 7 summarizes the conclusions.

2. MATHEMATICAL MODELS

In the literature several mathematical models describing the steering dynamics of a ship are given. The most simple one is the first-order model of Nomoto (Nomoto and co-workers, 1957) which describes the transfer between the rudder angle δ and the rate of turn $\dot{\psi}$

$$\tau\ddot{\psi} + \dot{\psi} = K\delta \qquad (1)$$

where $\dot{\psi} = d\psi/dt$ and ψ is the ship's heading.

The influence of the forward speed of the ship can be added in this model by the relations (Van Amerongen, 1982)

$$K = K^*(U/L) \qquad (2)$$

and

$$\tau = \tau^*(L/U) \qquad (3)$$

where K^* and τ^* are dimensionless constants (with respect to speed variations) in the order of magnitude of 0.5–2.0; U is the ship's speed and L the length.

This very simple model, which is a simplification of the nonlinear model of Van Leeuwen (1970), does not describe, for instance, the nonlinear static relation between δ and $\dot{\psi}$, but for the purpose of designing an adaptive controller it is suitable.

The rudder is actuated by means of the hydraulic steering machine which has nonlinear dynamics. Both its output, the rudder angle, and the rudder speed are limited. Common values are

$$\delta_{max} = 35° \qquad (4)$$

and

$$\dot{\delta}_{max} = 2\text{–}7\,\deg\,s^{-1}. \qquad (5)$$

Compared with the limited rudder speed other time constants of the steering machine may be disregarded for the controller design.

Disturbances which play a role in ship steering are wind, waves and current. When only the ship's heading is controlled (no track control) a stationary current may be neglected. Wind causes a stochastic disturbance, with non-zero mean acting upon the hull. With respect to control of the heading only the moment caused by the wind plays a role. It can be added to the model of equation (1) by adding the moment of the wind to the moment excited by the rudder. This modifies (1) into

$$\tau\ddot{\psi} + \dot{\psi} = K(\delta + K_w) \qquad (6)$$

where K_w represents the influence of the wind.

The moments caused by the waves may be described by one of the standard spectra available in the literature (for instance, the Bretschneider spectrum), recommended by the 12*th International Towing Tank Conference.* Some typical spectra of fully developed seas for various wind speeds, V_w are given in Fig. 1.

The frequencies of the ship motions caused by this wave spectrum depend also on the angle between the direction of the waves and the heading of the ship and on the speed of the ship. Typical values for the peak frequency are 0.05–0.2 Hz. In the following the moments caused by the waves will be referred to as (high frequency) noise, added to the desired movements caused by the rudder.

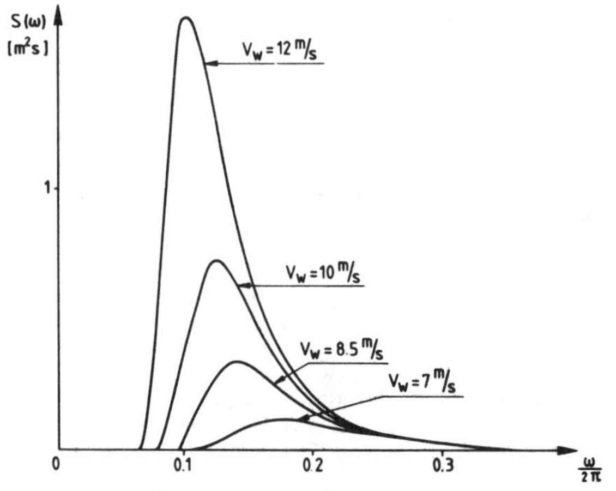

FIG. 1. Example of wave spectra.

3. STEERING CRITERIA

To be able to design an optimal controller a performance index has to be defined. Factors which play a role in this particular problem are: (1) economy (fuel cost); (2) safety (related to accuracy and manoeuvrability); and (3) user preferences. Maximum economy and safety cannot be realized without taking into account what the user subjectively considers to be good steering because he is ultimately responsible for the ship. Information about the user's ideas was obtained from an inquiry held among officers of the Royal Netherlands Navy and the Dutch Merchant Navy (Van Amerongen and Prins, 1980). It appears necessary to distinguish between two steering modes: course changing and course keeping.

Course changing

The inquiry indicates that during course changing the 'optimal' performance can be most easily defined as a step response in the time domain which has the form of the response given in Fig. 2. Three phases may be distinguished: (1) start of the turn; (2) stationary turning; and (3) end of the turn.

The turn should have a start which clearly shows to other ships the intention of the manoeuvre. The stationary phase of the turn is determined either by limiting the rudder angle, by controlling the rate of turn or by controlling the turning radius. Conventional autopilots have only the rudder limiter. Rate control or radius control will be preferable in most cases. It follows from (1) and (2) that by limiting the rudder angle a kind of radius control is implicitly achieved. Finally, the turn should stop without overshoot of the heading.

From the user's point of view there is no need for controller adjustment for the phases 1 and 3. Only the stationary phase should be adjustable (in terms of slow and fast turning), depending on the traffic situation, etc. All the conventional settings should be automatically adjusted, when varying process dynamics necessitate this. The only setting chosen to be provided to the operator for course changing is the stationary rate of turn.

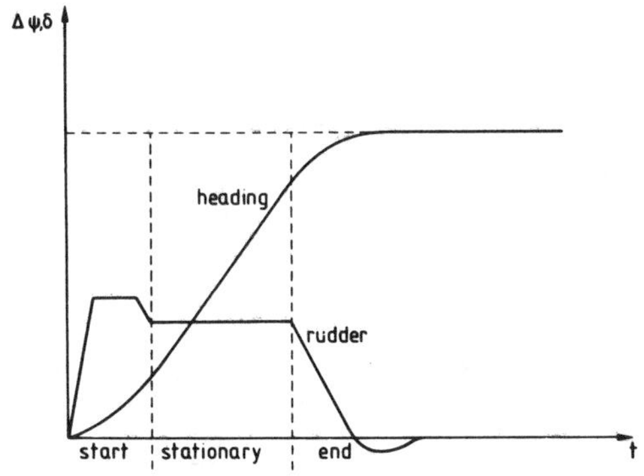

FIG. 2. Course-changing manoeuvre.

Course keeping

Optimization of the course-keeping controller is a more difficult problem. In confined waters with dense traffic the controller has to be above all accurate. This can be realized by selecting high controller gains. However, these gains are limited by the dynamics of the system.

On the ocean minimization of fuel cost will be the main goal. Assuming constant cruising speed this is realized by minimizing the extra drag due to steering. In other words, the loss of speed due to steering actions and the 'loss of speed' due to course errors (the elongation of the distance to be sailed) has to be minimized. There is no direct relation between fuel consumption and controller settings. Because of the difficult and inaccurate measurements involved, this problem cannot be solved by applying experimental optimization methods aimed at directly optimizing the fuel consumption. Attempts to define a more simple performance index have led to the criterion (Motora and Koyama, 1968; Norrbin, 1972)

$$J = \frac{1}{T}\int_0^T (\varepsilon^2 + \lambda\delta^2)\,\mathrm{d}t \tag{7}$$

where ε is the heading error, λ is a weighting factor, and δ is the rudder angle.

When the steering machine dynamics are neglected a state-feedback controller for the process of (1) is described by

$$\delta = K_p\varepsilon - K_d\dot{\psi} + K_i. \tag{8}$$

The factor K_i has been added to compensate for the slowly varying moment of the wind, K_w. When the sum of K_i and K_w is supposed to be zero the optimal feedback gains can be straightforwardly computed using standard LQ theory (see also Fig. 3)

$$K_p = 1/\sqrt{\lambda} \tag{9}$$

$$K_d = \frac{1}{K}\left\{\sqrt{\left(1 + 2\frac{K\tau}{\sqrt{\lambda}}\right)} - 1\right\}. \tag{10}$$

The integrating action K_i, which should compensate for the slowly varying disturbance K_w, can be

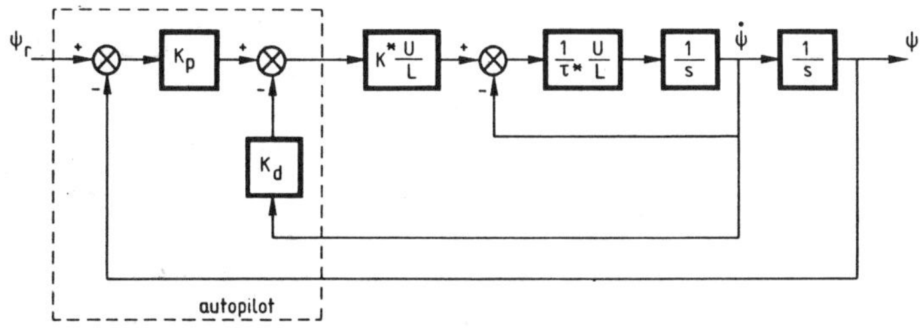

FIG. 3. Course-keeping control system.

separately computed by simply taking the mean of the rudder angle necessary for zero rate of turn

$$K_i = \frac{1}{T}\int_0^T \delta\,dt \qquad (11)$$

where T is in the order of 100 s.

In the literature there is no consensus about the value of λ. Values suggested range from 0.1 to 10. Van Amerongen and Van Nauta Lemke (1980) suggest the following extremes

$$\lambda = 0.1 \begin{cases} \text{for accurate steering} \\ \text{for large ships} \\ \text{for small ships in a calm sea} \end{cases}$$

and

$$\lambda = 4 \quad \text{for small ships in high sea states.}$$

This choice is based, among other things, on observations made during full-scale trials, towing-tank experiments and the previously mentioned inquiry.

It has also been shown that optimization of criterion (7) does not suffice to reach maximum economy. The frequencies of the ship motions caused by waves are so high that it makes no sense to try to compensate them by rudder movements. The latter will only cause extra loss of speed and, especially when the level of the 'noise' caused by the waves is high, the rudder movements will enlarge the motions of the ship, rather than reduce them: the steering machine introduces a considerable phase lag for large and fast rudder movements.

The presently commonly applied dead band is not the right solution. It is essential that a low-pass filter be designed to remove the high-frequency noise from the rudder signal. Because the noise frequencies are not too far from the bandwidth of the system, the filter must be carefully designed in order to avoid introducing stability problems. For optimal performance the amount of filtering should also be adaptive with respect to the level of the noise.

For optimal course-keeping performance it is thus essential to design a noise-reduction filter and to optimize criterion (7). The only setting chosen to be provided to the operator is the choice between maximum accuracy and maximum economy. This setting influences the value of λ and the amount of filtering. For maximum accuracy λ is small and there is a minimum amount of filtering. For maximum economy λ is large and, if necessary, the maximum amount of filtering is permitted. Corresponding to the foregoing λ is also influenced by the level of the noise.

4. COURSE-CHANGING CONTROLLER

The knowledge of the previous sections about mathematical models and steering criteria enables us to design the controller. During course changing the optimal performance has been defined as a step response with constant slope (rate of turn control). A suitable structure for realizing such a response is given in Fig. 4 (De Keizer, 1976). The heading-control system itself is preceded by a series model which modifies the heading reference, generating the type of response in Fig. 2. The slope of this response, the rate of turn, can be adjusted by the user.

The control algorithm of the actual autopilot is given by (8). The series model is shown in the block diagram of Fig. 5. In this figure ψ_m is the 'heading' of the series model, ψ_r', is the modified input signal for the course control loop, and ψ_r'' and f will be defined later. At this stage f is set to one.

The time constant τ_m is chosen approximately 2–3 times smaller than the dominating time constant of the ship at cruising speed [(1)]. This choice results from the following consideration: a reasonable course controller will have a rate-feedback gain K_d, which makes the time constant of the ship 2–3 times smaller than was the case for the open system. By choosing a similar time constant for the series model this guarantees that the process can follow the model. K_{pm} follows from the desired damping ratio of the system after the rate-of-turn limiter is neglected

$$K_{pm} = 1/(4z^2\tau_m).$$

When the actual heading control system is sufficiently tight the desired response will be realized. If the controller gains were not limited for reasons of stability, mainly due to the limited rudder speed, this could be achieved by selecting high controller gains. In practice, the controller gains should be carefully tuned to their maximum allowable values.

If the process parameters were constant and known this would not be a problem but when the circumstances change, this requires adjustment of the controller gains. When the influences of external variations on the process parameters are known this adjustment can be obtained by scheduling the controller gains. For example, to compensate for variations in the ship's speed, (2) and (3) may be used. However, this is only possible for a limited number of variables whose influence on the controller gains is well established. Other parameter variations can be dealt with by applying a second, parallel, reference model, according to Fig. 6.

A straightforward design of MRAS based on stability theory requires that process and parallel

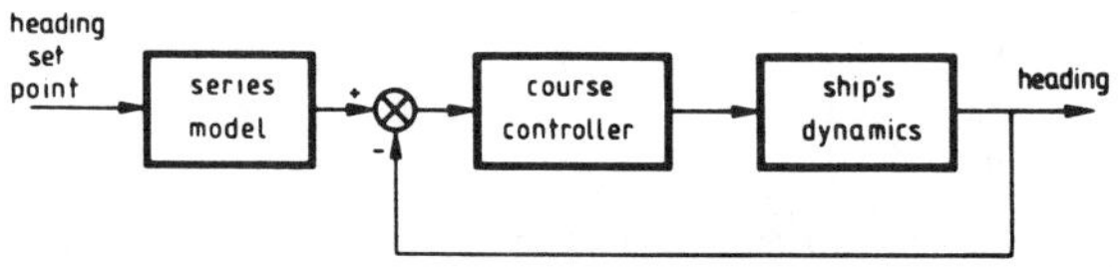

FIG. 4. Course-changing controller.

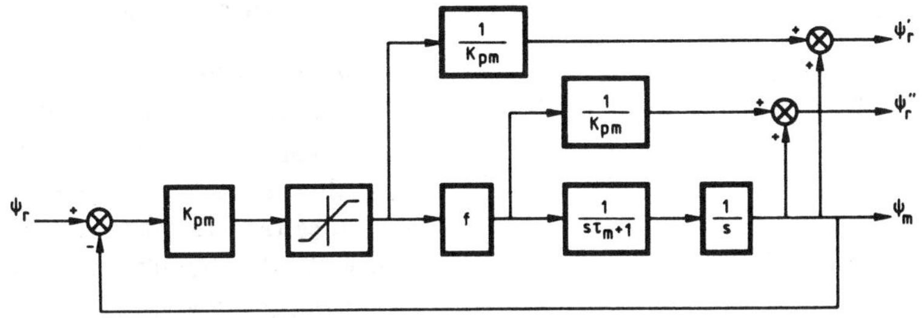

FIG. 5. The series model.

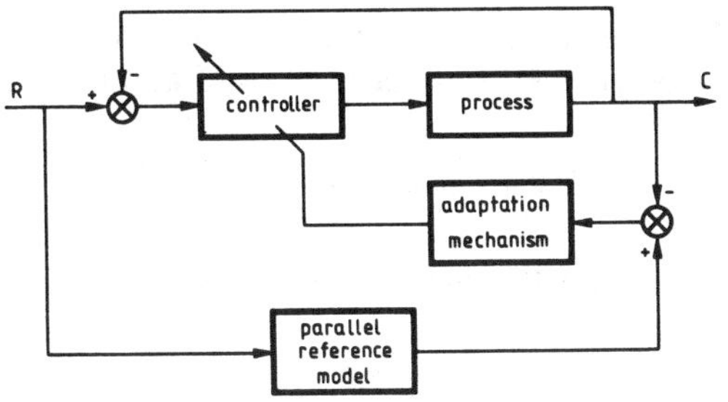

FIG. 6. Parallel MRAS.

reference model be linear and of the same order and structure. It has been shown by Van Amerongen, Nieuwenhuis and Udink ten Cate (1975) and by Johnson (1982) that for nonlinear ship's dynamics a stable MRAS can also be designed. For practical application, however, the proposed algorithms are too sensitive for small structural differences between process and reference model.

Without special precautions it is impossible to meet these requirements of linearity. The ship's dynamics themselves are nonlinear, but it has been shown (Van Amerongen, Nieuwenhuis and Udink ten Cate, 1975) that this is not a serious problem as long as simple adaptive algorithms are used. The major problems are introduced by the rudder limit (either as a controller parameter or as the absolute maximum of the rudder angle) and by the limited rudder speed. Because both nonlinearities are well known or easily measurable this problem can be circumvented by some minor modifications of the series model. When the series model only generates signals which saturate neither the rudder limiter nor the rudder-speed limiter, the influence of the steering machine has been removed from the inner control loop. This can be achieved as follows.

When the rudder limiter is known, the ratio between the maximum rudder angle δ_{max} and the desired rudder angle δ can be computed. When this ratio is smaller than 1 the factor f in Fig. 5, which was earlier disregarded by setting it to the value 1, is replaced by

$$f = \frac{\delta_{max}}{|\delta_r|} \tag{12}$$

with $f \leqslant 1$. Similar measures can be taken to introduce the effects of the limited rudder speed into the series model. Suppose that the actual rudder angle is δ and the desired rudder angle is δ_r. The time needed to move the rudder from δ to δ_r is then approximately

$$\tau_\delta = \frac{|\delta \; - \delta|}{\dot{\delta}_{max}} \tag{13}$$

where $\dot{\delta}_{max}$ is the maximum rudder speed. When the earlier defined factor f of the series model is extended by a first-order transfer function

$$H_\delta = \frac{1}{s\tau_\delta + 1} \tag{14}$$

with variable time constant τ_δ according to (13), the influence of the limited rudder speed is added to the series model as well.

The result of these measures is that there is in fact no more saturation of the steering machine; the process is thus linearized. This means that a linear parallel reference model of the same order and structure as (1) may be chosen. For the closed loop this yields the transfer function

$$\frac{\psi_m}{\psi_r''} = \frac{K_{pm}/\tau_m}{s^2 + s/\tau_m + K_{pm}/\tau_m} \tag{15}$$

where ψ_m is the 'heading' of the parallel reference model K_{pm} and τ_m are chosen similarly to the series model. Note that ψ_r'' is used as input signal.

Because the process is now linearized, the design of the adaptive controller is straightforward (Van Amerongen and Van Nauta Lemke, 1978, 1979; Landau, 1974, 1979). This yields the adjustment laws for the controller gains

$$\mathrm{d}K_p/\mathrm{d}t = \beta(p_{12}e + p_{22}\dot{e})\varepsilon \tag{16}$$

$$\mathrm{d}K_d/\mathrm{d}t = -\alpha(p_{12}e + p_{22}\dot{e})\dot{\psi} \tag{17}$$

$$\mathrm{d}K_i/\mathrm{d}t = \gamma(p_{12}e + p_{22}\dot{e}) \tag{18}$$

where e is defined as

$$e = \psi_m - \psi \tag{19}$$

and

$$\dot{e} = \dot{\psi}_m - \dot{\psi} \tag{20}$$

α, β, and γ are 'arbitrary' positive constants and p_{12} and p_{22} are elements of the matrix P. P can be solved from

$$A_m^T P + P A_m = -Q \tag{21}$$

where Q is an arbitrary positive definite matrix and A_m is the system matrix of the reference model according to (15), the states of the reference model being defined as $x_1 = \psi_m$ and $x_2 = \dot{\psi}_m$. The stability of the overall system can be proved, for instance with Liapunov's stability theory.

By computing K_i during course changing in an adaptive manner, according to (18), it is not necessary to stop the integration during course changing, which is common practice in conventional autopilots.

Because of the noise being present both on the signals, $\dot{\psi}$, ε and e, $\dot{e}$, the adjustment laws according to (16) and (17) require that measures be taken to prevent the controller gains from drifting away (Van Amerongen and Van Nauta Lemke, 1978, 1979). In the present design the concept of decreasing adaptive gains has been applied and the adaptation is totally switched off a certain period of time after a setpoint change. Decreasing adaptive gains are obtained by dividing the adaptive gains α and β by $(1 + T)$, where T is the time after the last setpoint change.

5. COURSE-KEEPING CONTROLLER

Parameter estimation

It has been shown in Section 2 that optimal course keeping can be achieved by optimizing criterion (7) and filtering noisy signals. The optimization procedure requires the parameters K and τ to be known. When scheduling of the gains does not suffice in view of the influence of changing load conditions, water depth, etc., an additional on-line identification procedure is required. For this purpose MRAS can be applied as well. This leads to the structure according to Fig. 7.

A simple first-order adjustable model is placed parallel with the transfer between δ and $\dot{\psi}$

$$\tau_m \ddot{\psi}_m + \dot{\psi}_m = K_m(\delta - K_{i,m}) \tag{22}$$

where $K_{i,m}$ is the rudder off-set. Defining

$$e = \dot{\psi}_m - \dot{\psi} \tag{23}$$

yields the simple adjustment laws

$$d(K_m/\tau_m)/dt = -\beta e(\delta - K_{i,m}) \tag{24}$$

$$d(1/\tau_m)/dt = \alpha e \dot{\psi}_m \tag{25}$$

$$d(K_{i,m})/dt = -\gamma e. \tag{26}$$

In this case there are no problems with nonlinearities and biasing due to noise. Stability can again be proved by applying the theory of Liapunov. Instead of (26) equation (11) could be used to compute $K_{i,m}$, but it appears that best results are obtained with (26).

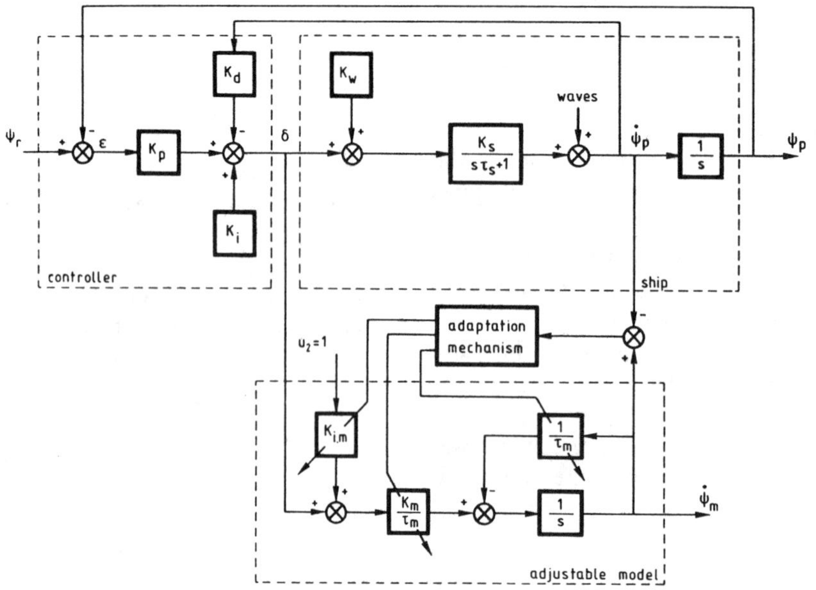

FIG. 7. Basic structure for parameter identification.

State estimation

Besides estimates of the process parameters the adjustable model also produces an estimate of the actual rate-of-turn signal. When the actual rate of turn signal is corrupted by 'noise', due to the influence of the waves, the estimate will be much smoother. The filtering problem is thus solved simultaneously. However, this filter is not the best possible one. When the level of the noise is low, it is not necessary to rely on the output of the adjustable model alone. The prediction may be updated, based upon the measurements. In order not to influence the identification process a second adjustable model is introduced whose parameters are adjusted simultaneously with the first model. The output of the second model is updated every sampling interval with the latest measurements. The weighting between prediction and measurements is determined by the relation between the low-frequency components of the error signal, which should not be filtered, and the high-frequency components which should be suppressed (Van Amerongen, 1982). This is not a straightforward Kalman filter because it does not distinguish between system noise and observation noise but between the low-frequency and high-frequency components of the system noise. Observation noise is supposed to be totally absent. The adaptive filter gains are computed on-line as follows: define

$$e = \hat{\dot{\psi}} - \dot{\psi} \tag{27}$$

where $\hat{\dot{\psi}}$ is the output of the second adjustable model. By means of a low-pass filter, e is split into a low-frequency and a high-frequency component

$$e_{\mathrm{lf}} = \frac{1}{s\tau_{\mathrm{f}} + 1} e \tag{28}$$

$$e_{\mathrm{hf}} = e - e_{\mathrm{lf}}. \tag{29}$$

Averaging yields the mean variances σ_{lf}^2 and σ_{hf}^2 of the low-frequency and high-frequency components of e. A gain factor ξ is now computed

$$\xi = \frac{\sigma_{\mathrm{lf}}^2}{\sigma_{\mathrm{lf}}^2 + \sigma_{\mathrm{hf}}^2} \tag{30}$$

which is used to update the predictions. In discrete form this yields

$$\hat{\dot{\psi}}(k + 1/k + 1) = \hat{\dot{\psi}}(k + 1/k) + \xi[\dot{\psi}(k + 1) - \hat{\dot{\psi}}(k + 1/k)]T/\tau_{\mathrm{m}} \tag{31}$$

where $\hat{\dot{\psi}}(k + 1/k)$ is the output of the adjustable model; $\hat{\dot{\psi}}(k + 1/k + 1)$ is this output, updated with the measured value $\dot{\psi}(k + 1)$ at $t = (k + 1)T$; T is the sampling interval; and τ_{m} is the time constant of the adjustable model.

The upper limit of ξ is 1, and the lower limit is influenced by the desired course-keeping accuracy. In a similar way estimates of the ship's heading can be obtained

$$\hat{\psi}(k + 1/k) = \hat{\psi}(k/k) + \hat{\dot{\psi}}(k + 1/k + 1)T \tag{32}$$

$$\hat{\psi}(k + 1/k + 1) = \hat{\psi}(k + 1/k) + \xi[\psi(k + 1) - \hat{\psi}(k + 1/k)]T. \tag{33}$$

6. PRACTICAL RESULTS

Most of the algorithms given above were continuous time algorithms, which can easily be combined with the proposed gain scheduling. However, practical realization is more robust and simple by using a digital computer. Because the sampling time of the computer can be chosen small (0.25 s), compared with the system's bandwidth the continuous-time algorithms can be used unmodified in the digital computer (sampling frequency >25 rad/s, closed-loop bandwidth <0.1 rad/s). The algorithms of the previous sections have been implemented in a digital computer in order to test the system under real-life conditions. A DECLAB 11/03 system with 28K words of memory, dual floppy disk and appropriate interfaces is used for control as well as data logging (Onkenhout, 1979). Up to 16 variables can be monitored on a graphical display unit and are stored on floppy disk for analysis afterwards. Full-scale tests on three different ships have been carried out, as well as experiments with a scale model in a towing tank. During these trials it had to be demonstrated whether the control algorithms which were derived for a simplified mathematical model also gave a satisfactory performance in practice.

In general, the adaptive autopilot, further referred to as ASA (from the Dutch: Adaptieve Stuur Automaat), showed the behaviour that might be expected from hybrid simulation experiments, which were carried out before the sea trials at the laboratory (Van Amerongen, 1982). Some modifications were necessary, however, mainly because the disturbances at sea differed from the disturbance models used in the simulation, leading to a low course-keeping accuracy. The values of λ suggested before result from these experiences. It was also necessary to use narrower bounds for the filter gains than previously expected.

Some typical results are given in the Figs 8–13. Figure 9 shows the course-changing performance under ASA control of H.Nl.M.S. *Tydeman*, the oceanographic survey vessel of the Royal Netherlands Navy (Fig. 8). The length of this ship is about 100 m. A standard series of course alterations is automatically carried out. In Fig. 9 the desired rate

Fig. 8. H.Nl.M.S. *Tydeman* (photograph courtesy of Royal Netherlands Navy).

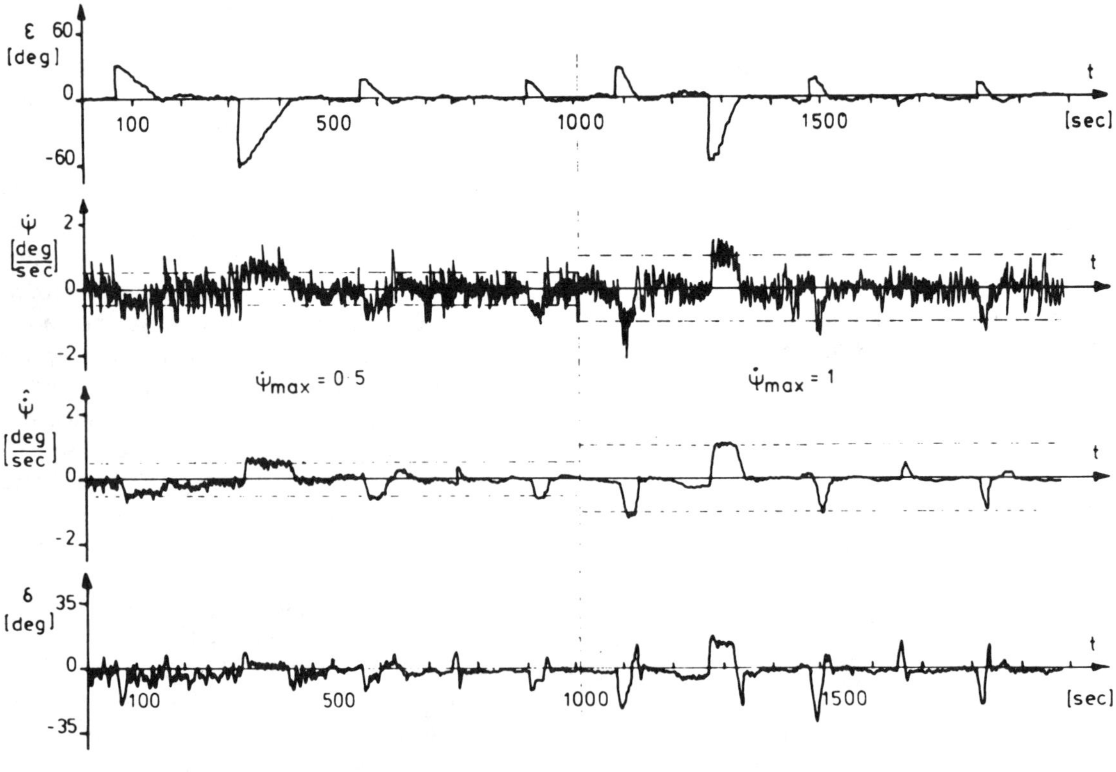

Fig. 9. Course-changing performance of ASA.

of turn was set to 0.5 and 1.0 deg/s. The performance of the adaptive state estimator can also be judged from this figure.

A comparison between the course-keeping performance of ASA, a conventional autopilot and an experienced helmsman is shown in Fig. 10. It can clearly be seen that both autopilots are superior to the helmsman and that ASA performs better than the conventional autopilot. The rudder is most smooth when steering with ASA.

The performance criterion (crit) which is shown in Fig. 10 is the criterion (7) with $\lambda = 10$. Criterion (7) is only an approximation of the real criterion: minimum fuel consumption. Attempts have been made to measure the latter in a more direct way.

Because the instrumentation necessary to measure the propeller thrust was not available, only the mean speed with a constant number of revolutions could be measured. During 8 h a fixed heading was sailed and after every hour control was switched from ASA to the conventional autopilot and back. During these trials, with sea state 4 (corresponding with a wave height of about 3 m), the mean speed during control by ASA was about 0.5% higher than with conventional control. Similar experiments in another area, with a sea state 3 (wave height approximately 2 m) indicate an increased speed between 0.3 and 1.5% in favor of ASA, depending on the direction of the waves. In terms of fuel consumption the savings are even bigger if the increased performance is used to decrease the propeller thrust and to maintain a constant speed.

The increased speeds were mainly obtained due to the smoother rudder movements. This will also lead to less wear and tear of the steering equipment. Figure 11 gives an illustration of the frequency spectra of both autopilots. The spectra in this figure were obtained after fast fourier transformation of the heading, rate-of-turn and rudder signals. For ASA the estimated rate-of-turn signal is also transformed. Because this signal is very smooth, the rudder spectrum also contains fewer high frequencies during control by ASA.

Because the performance measurements are difficult at full scale, it was decided to carry out additional experiments in a towing tank (Van Amerongen and co-workers, 1980). By measuring not only the ship's mean speed, but also the propeller torque and its number of revolutions, a more accurate measure of the fuel consumption was obtained. A model of a ship of 180 m length was used and tested in different sea states, generated by wave generators. Controllers with high gains and controllers with low gains, both with and without the adaptive filter, were tested. The improved performance, due to filtering in terms of increased speed, was between 0.3 and 1.5% for the high-gain controllers and between 1.5 and 5.6% for the low-gain controllers. During the experiments only pure head seas and pure following seas could be generated. Additional experiments are necessary to investigate the performance with other wave angles and to compare the high-gain and low-gain controllers.

During experiments with H.Nl.M.S. *Poolster*, a

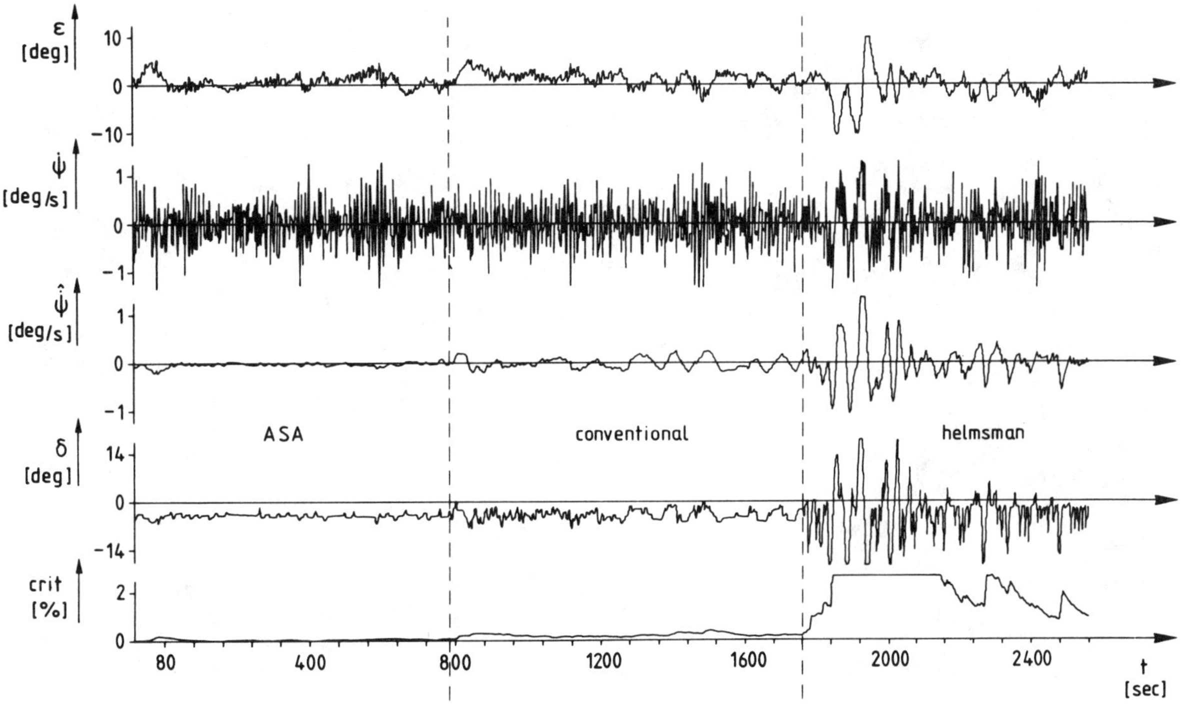

FIG. 10. Course-keeping performances.

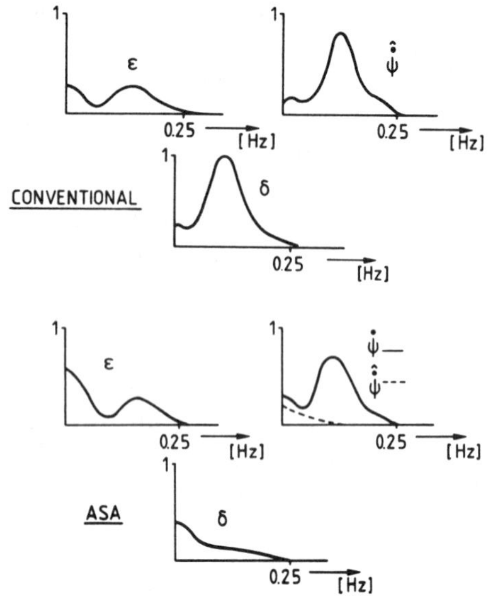

FIG. 11. Fast fourier spectra.

supply vessel of the Royal Netherlands Navy with a length of approximately 170 m, the necessity of adaptive control was clearly illustrated once again. Figure 13 shows the steering performance during a replenishment at sea (RAS) operation. During such a manoeuvre two or more ships are sailing very close together in order to transport goods between the different ships (Fig. 12).

This performance was not recorded during a special experiment, but it shows the normal use by the crew of the recently installed conventional autopilot. The conventional autopilot, which was adjusted for normal course keeping, was not re-adjusted before the RAS operation, where, due to the interaction of the other ships, a much tighter control is required. Because of the too-low accuracy the heading error became too large and the helmsman was ordered to take over the control. Although the heading error thereafter remained within the safety limits, the helmsman's control can also be qualified as poor. After defining 'RAS-settings' the conventional autopilot performed much better, as shown in Fig. 14. Control by ASA, easily adjusted for accurate steering, was a further improvement.

7. CONCLUSIONS

It has been shown that adaptive control enables the design of an autopilot with the following features: easier adjustment, course changing with predictable manoeuvres, and improved fuel economy. Full-scale experiments confirm the simulation results and demonstrate the practical usefulness of the autopilot designed.

The advanced filter algorithm, combining MRAS with the ideas of Kalman filtering, provides the major

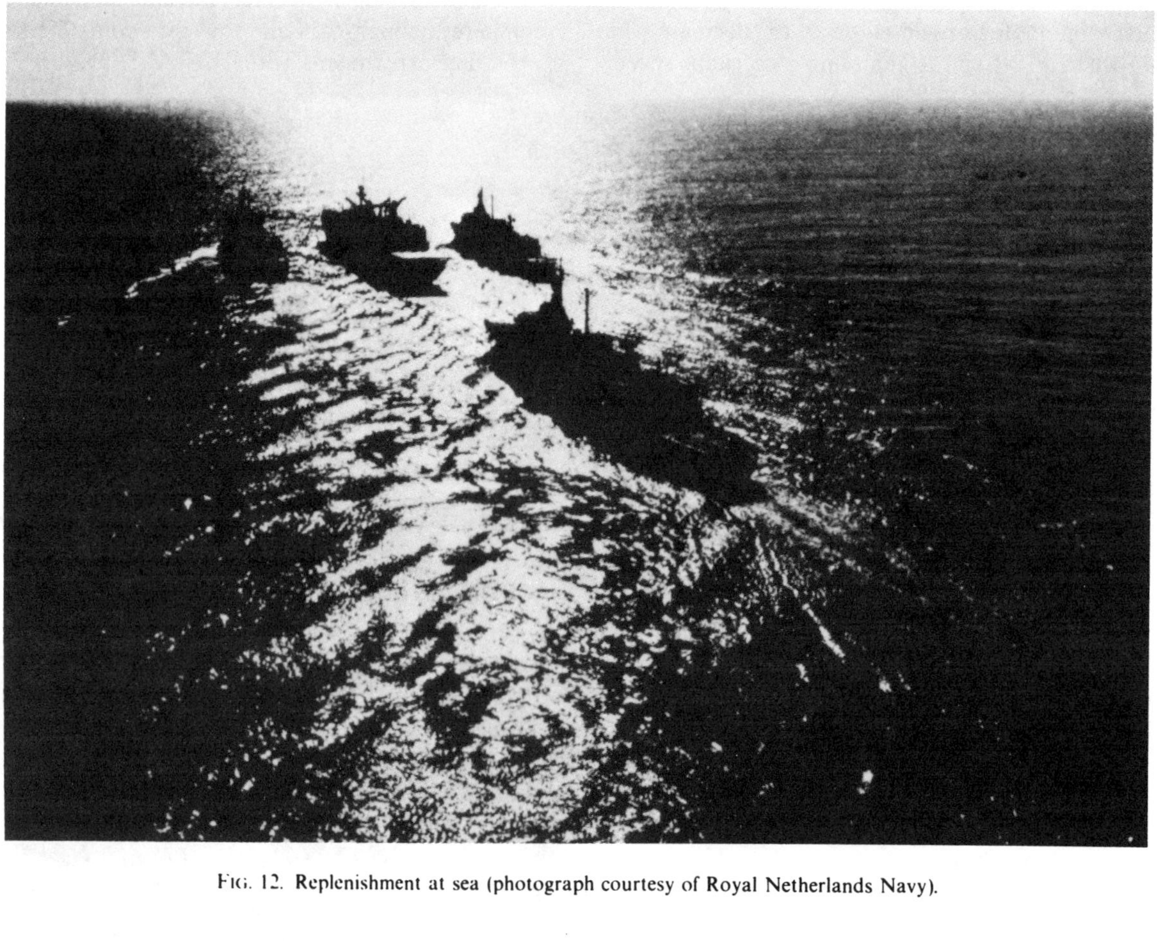

FIG. 12. Replenishment at sea (photograph courtesy of Royal Netherlands Navy).

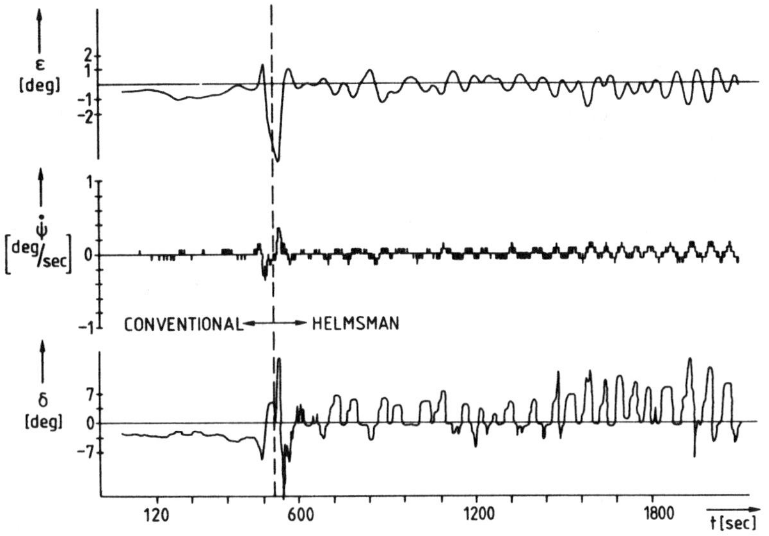

FIG. 13. Steering during replenishment at sea operation with a conventional autopilot and by the helmsman.

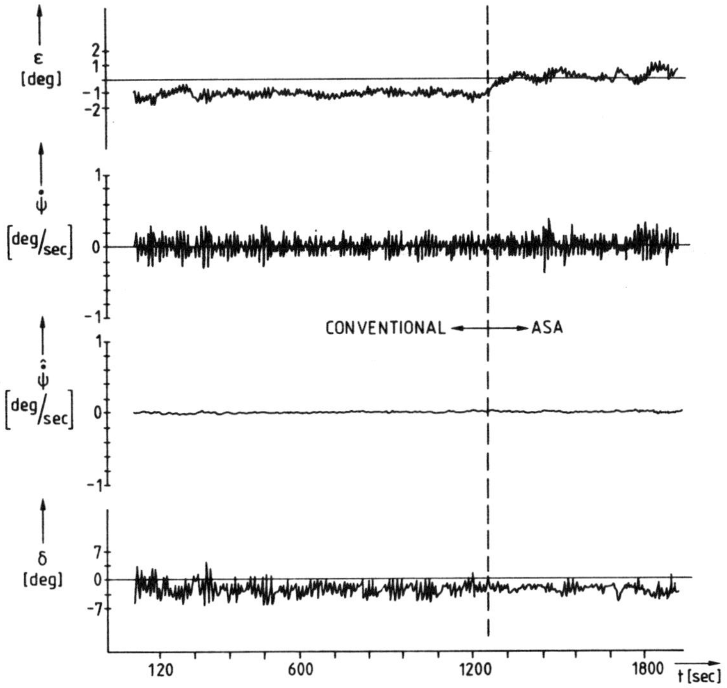

FIG. 14. Comparison of a conventional autopilot and ASA during a replenishment at sea operation.

contribution to improved fuel economy. During full-scale trials the speed increase has been shown to be 0.5–1.5 %. When the reduced drag is used to reduce the thrust, this leads to fuel savings between 1 and 3 %. During model tests, where the ship's speed could be corrected for variations in the thrust, increased speeds up to 5 % were demonstrated. The improved fuel economy is of growing importance because of the drastic increase of fuel prices since 1973.

In general, the users were enthusiastic about the features of the adaptive autopilot. In various cases the controlled rate-of-turn steering was purposefully used. During the experiments it was demonstrated several times that the users lack the insight and time to optimally adjust a conventional autopilot. All these features became practically realizable due to the availability of small and inexpensive digital hardware. Therefore, it may be expected that in the near future an increasing number of adaptive autopilots will be available on the market.

Acknowledgements—The support of Prof. ir. H. R. van Nauta Lemke throughout the research described in this paper is gratefully acknowledged. The measurements carried out with the adaptive autopilot were possible due to the cooperation of the Royal Netherlands Navy and the Netherlands Ship Model Basin. These measurements were supported by Observator BV, the future manufacturer of the adaptive autopilot.

REFERENCES

Amerongen, J. van and A. J. Udink ten Cate (1975). Model reference adaptive autopilots for ships. *Automatica*, **11**, 441.

Amerongen, J. van, H. C. Nieuwenhuis and A. J. Udink ten Cate (1975). Gradient based model reference adaptive autopilots for ships. *Proceedings 6th IFAC World Congress*, Boston/Cambridge, MA.

Amerongen, J. van and H. R. van Nauta Lemke (1978). Optimum steering of ships with an adaptive autopilot. *Proceedings 5th Ship Control Systems Symposium*, Annapolis, MD.

Amerongen, J. van and H. R. van Naute Lemke (1979). Experiences with a digital model reference adaptive autopilot. *International Symposium on Ship Operation Automation (ISSOA)*, Tokyo.

Amerongen, J. van and J. Prins (1980). Criteria for steering of ships. *Marineblad*, 58 (in Dutch).

Amerongen, J. van and H. R. van Nauta Lemke (1980). Criteria for optimum steering of ships. *Symposium on Ship Steering Automatic Control*, Genoa.

Amerongen, J. van, W. F. de Goeij, J. M. Moraal, J. W. Ort and A. Postuma (1980). Measuring the steering performance of ships during full-scale trials and model tests. *Symposium on Ship Steering Automatic Control*, Genoa.

Amerongen, J. van (1982). Adaptive steering of ships—a model reference approach to improved manoeuvring and economical course keeping. Ph.D. thesis, Delft University of Technology.

Herther, J. C., F. E. Warnock, K. W. Howard and W. VanVelde (1980). Digipilot—a self-adjusting digital autopilot for better manoeuvring and improved course and track keeping. *Symposium on Ship Steering Automatic Control*, Genoa.

Johnson, H. M. (1982). Adaptive control of non-linear self-oscillating systems using MRAS technique. *Proceedings* 1982 *American Control Conference*, Arlington, VA.

Kallström, C. G., K. J. Åström, N. E. Thorell, J. Eriksson and L. Sten (1979). Adaptive autopilots for tankers. *Automatica*, **15**, 241.

Keizer, C. de (1976). Adjusting the P, I and D control actions in an autopilot with the aid of model reference adaptation. Report from the Control Laboratory, Electrical Engineering Department, Delft University of Technology (in Dutch).

Kojima, T. and A. Sugimoto (1978). A new autopilot system with condition adaptivity. *Proceedings 5th Ship Control Systems Symposium*, Annapolis, MD.

Landau, I. D. (1974). A survey of model reference adaptive techniques—theory and applications. *Automatica*, **10**, 353.

Landau, I. D. (1979). *Adaptive Control—The Model Reference Approach*. Marcel Dekker, New York.

Minorsky, N. (1922). *Directional Stability of Automatically Steered Bodies*. American Society of Naval Engineers.

Motora, S. and T. Koyama (1968). Some aspects of automatic steering of ships. *Japan Shipbuilding & Marine Engineering*, July.

Nomoto, K., T. Taguchi, K. Honda and S. Hirano (1957). On the steering qualities of ships. *Int. Shipbuilding Prog.*, **4**.

Norrbin, N. H. (1972). On the added-resistance due to steering on a straight course. 13*th ITTC*, Berlin, Hamburg.

Ohtsu, K. M., Horigome and G. Kitagawa (1979). A new ship's autopilot design through a stochastic model. *Automatica*, **15**, 255.

Onkenhout, H. J. (1979). A microcomputer-based data-logging system. Master's thesis, Control Laboratory, Electrical Engineering Department, Delft University of Technology (in Dutch).

Pierson, W. J. and L. Moskovitz (1964). A proposed spectral form for fully developed wind seas based on similarity theory of S. A. Kitaigorodskii. *J. Geophys. Res.*, **69**.

Reid, R. E. and V. E. Williams (1978). A new ship control design criterion for improved heavy weather steering. *Proceedings 5th Ship Control Systems Symposium*, Annapolis, MD.

Sperry, E. (1922). Automatic steering. *Trans SNAME*.

Part VII
Applications to Process Control Robotics and Other Fields

THE satisfactory control of complex industrial processes operating under varying environmental conditions continues to present a challenge to control systems scientists working in the area of stochastic and adaptive control. In this last part of the volume, we present nine representative papers dealing with problems in the area of process control, robotics, biomedical, and power systems control.

In paper thirty-six, T. Cegrell and T. Hedqvist describe the application of adaptive control to paper machines so as to minimize the variance of moisture and basis weight. The control strategy requires very little *a priori* knowledge about the process dynamics, which is an added advantage and contrary to many adaptive and nonadaptive control algorithms presently available. The authors report that the algorithm has been applied to several real-time computer control systems for paper machines. The algorithm can be implemented easily on a minicomputer. The authors report a large number of tests that were conducted in order to evaluate the effectiveness of the adaptive controller, and the results were so successful that the adaptive algorithm was permanently installed in the system. This adaptive controller has been reported to reduce not only the variance in moisture and basis weight but also the losses due to quality changes and mill setups. Hopefully, this philosophy in the design of adaptive policy can be extended to many other types of processes. It is worth mentioning that T. Cegrell and his group at ASEA worked on this type of regulator as early as 1972, and hold patents on these regulators in a number of countries.

An adaptive self-tuning regulator based on a linear controller, where the parameters are updated by on-line parameter estimation, is reported by U. Borisson and R. Syding in paper thirty-seven to control successfully an ore crusher, a real industrial process. A process computer at the University of Lund was connected to the ore crushing plant at a distance of 1800 km. The self-tuning regulator was able to converge to a minimum-variance strategy in a truly adaptive manner with continually changing process variables.

G. A. Dumont and P. R. Belanger present in paper thirty-eight, an application of a self-tuning regulator to the control of a titanium dioxide kiln. The regulator is used in a feedback loop around a kiln already controlled by a discrete regulator designed on the minimum-variance principle. In this paper the authors also demonstrate the practical applicability of the STR. As well, they explore some novel features of the problem. The asymptotic study with a plant model of the order greater than that of the regulator is believed to be a new feature. The STR is capable of controlling the process through a grade change, though this feature has not been explored theoretically to any great extent but appears to be very promising for industrial practice.

The paper thirty-nine entitled "Self-Tuning Adaptive Control of Cement Raw Material Blending" by L. Kevczky, J. Hetthessy, M. Hilger, and J. Kolostori describes a new modified self-tuning minimum variance regulator algorithm for multivariable systems and applied to the control of cement raw material blending. The algorithm has some nice properties and advantages over the classical DPC system. The algorithm can be implemented on a microcomputer.

The paper forty, entitled "Adaptive Identification and Control Algorithms for Nonlinear Bacterial Growth Systems" by D. Dochain and G. Bastin, describes self-tuning type controllers for nonlinear bacterial growth processes. The process is time-varying and nonlinear. For this process, two different control problems are considered: substrate concentration control and production rate control. The authors present some interesting simulation results and the applications can be found in many industrial processes like waste treatment in sugar industry.

The following two papers deal with the applications on adaptive technique to robotics.

In paper forty-one, S. Nicosia and P. Tomei describe an application of model reference adaptive control algorithms to industrial robots (multifunction manipulators). Robots are, in general, nonlinear and time-varying, and it is difficult to apply the conventional control theory to such a situation. The authors present a decoupling algorithm which has some robust properties, and the parameter variations have little influence on the performance of the control. The paper reports some simulation studies.

In paper forty-two, R. B. McGhee and G. I. Iswandhi report their work on adaptive locomotion of a multilegged robot over rough terrain. This paper, unlike other papers, treats a control problem of locomotion inherent into man and animals. They present an adaptive walking machine. The paper includes a complet problem formalization, a heuristic algorithm for solution of the problems posed, and a preliminary evaluation of the proposed algorithm in terms of computer simulation study.

In paper forty-three, B.C. McInnis, Z. W. Guo, P. C. Lu, and J. C. Wang present a computer-based adaptive control system for left ventricular bypass assisted devices consisting of air driven diaphragm pumps. The system design includes an adaptive control algorithm which is a self-tuning PID-Controller based on pole placement. The authors demonstrate the performance of the system by *in vitro* experiments in a mock circulatory system. With increased atrial pressure, the system responds with increase in stroke volume. Following major changes in the circulatory system, the control algorithm returns itself and restores the system to the desired state. This is a useful application of the adaptive methods in the stabilization of blood pressure of a patient.

In the last paper, paper forty-four our colleagues from Hungary, Vajak, Vajta, Keviczky, Haber, Hetthessy, and Kovacs, give their experiences with an actually used adaptive load-frequency regulator designed using a self-tuning predictor. This paper deals with the modeling of the Hungarian power station and the interconnected power systems. The adaptive control strategy performs the following objectives:

- it eliminates the effects of the load functions to the tie-line power;
- it guarantees the scheduled value of the exported/imported energy; and
- it satisfies the control requirement with minimum cost.

This paper is another excellent example of real-life application of the adaptive controller. The authors claim improved dynamic performance, more efficient use of control action under a wide range of conditions. This regulator has been used in the Hungarian power system without any problem since June 1981. Hopefully, the experience given in this paper will be of some use to our readers.

This small sample of application papers present what has been done during the recent past in the area of applications of adaptive methods to industrial processes, robotics, biomedical, and power systems control. There are still many unexplored avenues, both in problems as well as adaptive approaches, that need to be explored. The design of a robust adaptive controller for industrial processes without any explicit identification or parameter and state-variable estimation, but with some learning capabilities, is still an open problem. Many new technological problems such as the application of robotics to manufacturing industries are emerging and some of these would need the help of adaptive technology, a useful and interesting area for further research and development. A marriage between adaptive concepts and notion of approximate reasoning and expert system may be very useful; this is still an unexplored area and readers are encouraged to give some serious consideration to these notions in their future research.

Selected Bibliography

R. S. Baheti, "Rapid tuning of a fuel gas saturator control system using Nyquist and Bode arrays," in *Proc. of the 23rd Conf. Decision and Control,* pp. 394–399, 1984.

K. C. Cheok, N. K. Loh, and H. D. McGee, "Optimal suspension design with microcomputerized parameter optimizing damper," in *Proc. of the 23rd IEEE Conf. Decision and Control,* pp. 400–403, 1984.

G. A. Dumont, "Self-tuning control of a chip refiner motor load," *Automatica,* vol. 18, no. 13, pp. 307–314, 1982.

G. A. Dumont and P. R. Belanger, "Control of titanium dioxide kiln," *IEEE Trans. Automat. Contr.,* vol. AC-23, no. 4, pp. 532–538, 1978.

——, "Self tuning control of a titanium dioxide kiln," *IEEE Trans. Automat. Contr.,* vol. AC-23, no. 4, pp. 532–538, 1978.

G. C. Goodwin, B. McInnis, and R. S. Long, "Adaptive control algorithms for waste water treatment and *p*H neutralization, Tech. Rep. EE8112, Department of Electrical and Computer Engineering, University of Newcastle, Newcastle, New South Wales, Australia, 1981.

L. Keviczky, J. Hettessy, M. Hilger, and J. Kolostori, "Self-tuning adaptive control of cement raw material blending," *Automatica,* vol. 14, no. 6, pp. 525–532, 1978.

——, "Self-tuning adaptive control of cement raw material blending," *Automatica,* vol. 14, no. 6, pp. 525–532, 1978.

K. Lau, H. Kaufman, V. Serna, and R. Roy, "Evaluation of three adaptive control procedures for multiple drug infusion," in *Proc. 23rd Conf. Decision and Control,* pp. 391–393, 1984.

R. L. Moore and F. Schweppe, "Model identification for adaptive control at nuclear plants," *Automatica,* vol. 9, no. 3, pp. 309–318, 1973.

N. V. Nadezhdina, A. F. Lariehava, and A. I. Potapov, "Adaptive control systems for batch production," *Mach. and Tool,* vol. 44, no. 10, pp. 12–13, 1973.

B. Pagurek, J. S. Riorden, and S. Mohmoud, "Adaptive control of the human glucose-regulatory system," *Medical and Biological Eng.* (GB.), vol. 10, no. 6, pp. 752–761, 1972.

G. Preusche, "A two-level model following control system and its application to the power control of a steam-cooled fast reactor," *Automatica,* vol. 8, pp. 145–151, 1972.

J. B. Priban and W. F. Fischam, "Self-adaptive control and the respiratory system," *Nature,* vol. 208, pp. 339–343, 1965.

F. Shinskey, "Adaptive *p*H controller monitors nonlinear process," *Control Engineering,* vol. 21, pp. 57–59, 1974.

——, "A self-adjusting system for effluent *p*H control," Spring Joint Conf. ISA, St. Louis, MO, 1978.

R. J. Teagle and K. A. McKenzie, "Hill-climbing adaptive control system with a view to power factor maximization," *Trans. S. Aft. Elec. Eng.,* vol. 64, pt. 9, pp. 174–175, 1973.

R. L. Thomas, E. J. Harter, and R. E. Goodson, "Adaptive computer control of a glass coating process," *Automatic,* vol. 8, no. 5, pp. 555–562; "Model reference adaptive autopilots for ships," *Automatica,* vol. 11, pp. 441–449, 1972.

E. G. Vogel and T. F. Edgar, "Application of an adaptive pole-zero placement controller to chemical processes with variable dead time," Amer. Control Conf., Washington, D.C., 1982.

C. R. Wasaff, "Model reference adaptive control theory with applications to manual control systems," *Diss. Abst.,* vol. 28, no. 7, 1968.

B. Widrow, "Adaptive model control applied to real-time blood-pressure regulation," Pattern Recognition and Machine Learning, Proc. Japan-U.S. Seminar on the Learning Processing, in *Control Systems,* K. S. Fu, Ed. New York: Plenum Press, 1971, pp. 310–324.

L. S. Wisler and E. R. Rang, "Oscillations in an adaptive aircraft control system," *Trans. ASME, J. Basic Eng.,* Ser. G, vol. 96, no. 1, pp. 100–101, 1974.

Successful Adaptive Control of Paper Machines*

Contrôle Adaptif avec Succès de Machines à Papier

Erfolgreiche adaptive Regelung von Papiermaschinen

Успешное адаптивное управление бумагоделательными машинами

TORSTEN CEGRELL† and TORBJÖRN HEDQVIST†

A self-adjusting regulator with useful properties based on an easily implemented technique may be successfully applied to industrial processes.

Summary—The papermaking process can be viewed as a time-varying stochastic system. The purpose of the control is to minimize the variance of moisture and basis weight of the manufactured paper. In the framework of stochastic theory and recently published papers on adaptive control an algorithm has been developed to cover the above-mentioned purpose. The strategy which hardly requires any *a priori* knowledge about the process is, in fact, a self-adjusting regulator. The central part of the algorithm is a compact identification scheme which gives directly the optimal control action. It is notable that contrary to normal identification and control methods the process dynamics never need to be calculated. The algorithm has been applied to several real-time computer control systems for paper machines. Up to now medium-scale computers have been used, but as the control program is powerful and short, some hundred words in length, it is in fact sufficient to perform the control with a mini-computer. Before the implementation of the algorithm for the above processes, the control system was built around digitalized PI-controllers, the parameters of which were carefully chosen and constant. A great number of comparative tests have been made in order to evaluate the efficiency of the adaptive regulator. The results were so successful that the algorithm has been permanently installed in the systems. The algorithm has not only decreased the output variance but also reduced the losses at quality changes and mill set-ups. Finally, it should be mentioned that the general nature of the algorithm permits application to many other types of processes.

1. INTRODUCTION

DURING the 1960's the linear stochastic control theory made considerable progress. One of the most important results was the solution of the minimum variance control problem for constant systems. This result and the development of effective identification methods [3], e.g. the maximum-likelihood method, made it possible to construct optimal control laws for linear stochastic systems. However, the practical use of these results has been limited. The main reason is that the identification phase in practical applications has to be repeated over and over again as industrial processes are never constant. In one word these methods are often too difficult to use.

In very recent years, however, new experiences in the theory of linear stochastic control have indicated that it may be possible to drop the separate identification phase and also that the parameters in a specified control law may be estimated within each of the control steps. These theories of self-adjusting control laws are still rather primitive and a number of problems remain to be solved. When working in this field one has to be aware of the fact that an algorithm to be adopted for practical use must be very simple. One such algorithm is given by Peterka [9]. During tests it turned out that this algorithm only worked satisfactorily sometimes. By a slight modification of the algorithm it has, however, been possible to extend the range of successful operation. This approach is discussed in Section 2, and the object there has been to give some practical aspects and rules of thumb when applying the regulator, which are not only simple to use but also easily accepted by practical working control engineers.

The same algorithm can also be derived in a more traditional analytical way. This has been done by Åström–Wittenmark [5]. A couple of other algorithms for self-adjusting regulators have also been published, e.g. refs. [4, 7, 10 and 12]. Most of these are however complicated and therefore not suitable for practical applications.

The main part of this paper concerns the behaviour of the algorithm, described in Section 2, as applied to a paper-machine control system, and it is outlined in the following way. The choice of a suitable control law is discussed in Section 3 and in Section 4 the paper-machine application is described. In Sections 5 and 6, finally, some results and general aspects of the applications are given.

* Received 24 July 1973; revised 28 January 1974. Revised 25 June 1974. The original version of this paper was presented at the 3rd IFAC Symposium on Identification and Parameter Estimation which was held in The Hague/Delft, The Netherlands, June 1973. It was recommended for publication in revised form by associate editor K. J. Åström.

† Development and Design Department, ASEA LME Automation AB Västerås, Sweden.

Reprinted with permission from *Automatica*, vol. 11, pp. 53–59, Jan. 1975.
Copyright © 1975, International Federation of Automatic Control IFAC. Published by Pergamon Press Ltd.

2. THE SELF-ADJUSTING REGULATOR

Consider a system described by the difference equation

$$y(t)+a_1 y(t-1)+\ldots+a_n y(t-n)$$
$$=b_0 u(t-k)+b_1 u(t-k-1)+\ldots+b_n u(t-k-n)$$
$$+\lambda\{e(t)+c_1 e(t-1)+\ldots+c_n e(t-n)\},$$
$$t=0,\ \pm 1,\ \pm 2,\ldots,\quad k>0, \qquad (2.1)$$

where u is the control variable, y is the output and $\{e(t), t\}$ is a sequence of independent normal (0, 1) random variables. Letting z^{-1} denote the backward shift operator the equation (2.1) can be written in a more compact form

$$A(z^{-1})\,y(t)=B(z^{-1})\,u(t-k)+\lambda C(z^{-1})\,e(t), \qquad (2.2)$$

where A, B and C are the polynomials

$$A(z^{-1})=1+a_1 z^{-1}+\ldots+a_n z^{-n},$$
$$B(z^{-1})=b_0+b_1 z^{-1}+\ldots+b_n z^{-n},\quad b_0 \neq 0,$$
$$C(z^{-1})=1+c_1 z^{-1}+\ldots+c_n z^{-n}.$$

The polynomial C can be written as

$$C(z^{-1}) \equiv A(z^{-1})\,F(z^{-1})+z^{-k}\,G(z^{-1}). \qquad (2.3)$$

It is well known, [1] and [8], that under certain circumstances the minimum variance control law can be written as

$$u(t)=-\frac{G(z^{-1})}{B(z^{-1})\,F(z^{-1})}y(t)=-\frac{A(z^{-1})}{B(z^{-1})}$$
$$\times\frac{G(z^{-1})}{C(z^{-1})-z^{-k}\,G(z^{-1})}y(t). \qquad (2.4)$$

Equation (2.4) is derived from the fact that under certain circumstances [1], the optimal control law is equivalent to the best k-step predictor. This justifies the introduction of the predictive model (2.5) for the system description

$$y(t)=b_0 u(t-k)+\theta\phi(t-k)+w(t), \qquad (2.5)$$

where

$$\theta^{\mathrm{T}}=\begin{bmatrix}\theta_1\\ \theta_2\\ \vdots\\ \theta_m\end{bmatrix},\quad \phi(t)=\begin{bmatrix}y(t)\\ y(t-1)\\ \vdots\\ y(t-r+1)\\ u(t-1)\\ u(t-2)\\ \vdots\\ u(t-s)\end{bmatrix}$$

with m, r and s of sufficient order and $m=r+s$. $w(t)$ is a random variable constituting the disturbance in the system.

Equation (2.5) is for this specific approach easier to use than (2.1). Although it does not, in general, describe the nature of the system completely, it represents the system sufficiently well for the purpose in view.

The basic idea of this self-adjusting regulator algorithm is to estimate b_0 and θ and use the estimates $\hat{b}$ and $\hat{\theta}$ for the control

$$u(t)=-(\hat{\theta}/\hat{b})\,\phi(t). \qquad (2.6)$$

As the estimation will be carried out in a closed-loop system some difficulties concerning identifiability may occur. To avoid this it is assumed that a 'once-for-all' estimate, $\tilde{b}$ (a reasonable guess of the parameter b_0), is made. The parameter vector ξ in the model

$$y(t+k)=\tilde{b}u(t)+\xi\phi(t)+v(t+k), \qquad (2.7)$$

where $v(t)$ is the error between the system and the model output, can then be estimated using a recursive least-squares method [2]. The estimate in each sample $\hat{\xi}_t$ is principally given by

$$\hat{\xi}_t=\left\{\sum_{i=1}^{t}[y(i)-\tilde{b}u(i-k)]\,\phi^{\mathrm{T}}(i-k)\right\}$$
$$\times\left[\sum_{i=1}^{t}\phi(i-k)\,\phi^{\mathrm{T}}(i-k)\right]^{-1} \qquad (2.8)$$

and the control law will thus be

$$u(t)=-(\hat{\xi}_t/\tilde{b})\,\phi(t). \qquad (2.9)$$

From (2.7) it is clear that, if the algorithm defined by (2.8) and (2.9) converges, the resulting model will simply degenerate to $y(t)=v(t)$. For a general system $v(t)$ is a function of all the system parameters in (2.1). In ref. [5] it has, however, been shown that convergence always implies that $y(t+k)$ will be uncorrelated to $u(t)$ and $\phi(t)$. Consequently, the disturbances $v(t+k)$ will contain no information correlated to $\phi(t)$, and it is intuitively clear that convergence implies minimum variance control, if the structure of $\phi(t)$ corresponds to the minimum variance structure. This fact is also proved in ref. [5].

A comparison of the algorithm defined by (2.8) and (2.9) to the algorithm given by Peterka [9] shows that Peterka assumes that $\tilde{b}\equiv +1$. Since the choice of $\tilde{b}$ influences the convergence properties this assumption will unfortunately limit the range of convergence of the algorithm, which will be discussed below.

2.0 *Some properties*

The algorithm, defined by (2.8) and (2.9), has turned out to be very useful. A number of simulations have, for example, shown that the choice of $\tilde{b}$ is not critical within certain limits.

A combination of (2.5), (2.8) and (2.9) gives

$$\hat{\xi}_t = \left\{ \sum_{i=1}^{t} [\theta + \hat{\xi}_{i-k}(1 - b_0/\tilde{b})] \phi(i-k) \phi^{\mathrm{T}}(i-k) \right\}$$
$$\times \left[\sum_{i=1}^{t} \phi(i-k) \phi^{\mathrm{T}}(i-k) \right]^{-1} + \left[\sum_{i=1}^{t} w(i) \phi^{\mathrm{T}}(i-k) \right]$$
$$\times \left[\sum_{i=1}^{t} \phi(i-k) \phi^{\mathrm{T}}(i-k) \right]^{-1}. \tag{2.10}$$

It is obvious that $\hat{\xi}_t$ is a stochastic process difficult to analyse. To simplify the analysis it is therefore assumed that in (2.5) $w(t)$ is uncorrelated to $u(t-k)$ and $\phi(t-k)$. This assumption is equivalent to having $C(z^{-1}) = 1$ in (2.1).

Consider (2.10) and use, during infinite time, an estimate $\hat{\xi}_N$ for the control law, i.e.

$$u(t) = -(\hat{\xi}_N/\tilde{b}) \phi(t). \tag{2.11}$$

If the closed-loop system is stable the estimate $\hat{\xi}_t$ will converge to

$$E\{\hat{\xi}_t\} = \theta + \hat{\xi}_N(1 - b_0/\tilde{b}). \tag{2.12}$$

Now use $E\{\hat{\xi}_t\}/\tilde{b} = \hat{\xi}_{N+1}/\tilde{b}$ for the control and repeat the estimation. The sequence of estimates can be described as

$$\hat{\xi}_{N+1} = \theta + (1 - b_0/\tilde{b}) \hat{\xi}_N. \tag{2.13}$$

or

$$\hat{\xi}_N = \frac{1}{1 - (1 - b_0/\tilde{b}) z^{-1}} \theta. \tag{2.14}$$

From the arguments above the following remarks can now be stated.

(1) The time series (2.13) will converge if and only if

$$|1 - b_0/\tilde{b}| < 1$$

or

$$b_0/\tilde{b} \in (0, 2).$$

(2) If it converges it will obviously converge to the optimal value, since

$$\hat{\xi}_\infty = (\tilde{b}/b_0) \theta,$$

which is equivalent to the minimum variance control law, as in the limit

$$u(t) = -(\hat{\xi}_\infty/\tilde{b}) \phi(t) = -(\theta/b_0) \phi(t).$$

(3) The rate of convergence depends on the choice of $\tilde{b}$, which can be seen from (2.14). Naturally $\tilde{b} = b_0$ will give the fastest convergence.

Although the analysis above is carried out for a simplified system, remark (1) is interesting and has turned out to be very useful in practice. It also explains why the algorithm by Peterka only works when b_0 is positive.

In the above it has been assumed that, besides an estimate $\tilde{b}$, the time delay parameter k is known. In many applications this is no disadvantage since it is often constant and well known. In other cases, however, k is not known *a priori* or is time dependent. In this case it is sometimes possible to modify the algorithm to produce an estimate of k as well.

In the discussions above it has further been assumed that the order of the system (2.5), i.e. the optimal structure of the ϕ-vector, is known. In practice, of course, this assumption is not valid. However, the regulator has turned out to work very well even if a wrong structure is postulated. In Section 3 the choice of structure is further discussed.

2.1 *An extension*

The system description (2.5) can be generalized to complicated systems and structures. As (2.5) is a predictor it is quite possible to allow other elements than y and u in the ϕ-vector. Every observation which is correlated to the output $y(t)$ and included in the ϕ-vector will decrease the expected variance of the prediction error. Hence the control law (2.6) in a general case is not a pure feedback regulator but a combined feedback and feedforward controller. This very nice property may be used for two important purposes.

(i) To compensate for coupling in multivariable systems.

(ii) To compensate for deterministic disturbances, i.e. observable but not by the system input $u(t)$ controllable variables, which in any way influence the system output $y(t)$.

The convergence properties will of course be complicated, but simulations and practical results have shown that this approach is very useful. This extension is further discussed in refs. [5] and [6].

2.2. *A practical aspect*

Up to now only time invariant systems have been taken into consideration. In reality, however, most processes are more or less time-varying. It is therefore important to study algorithms which are suitable even in such processes. The regulator will then be given by (compare 2.6))

$$u(t) = -\frac{1}{\tilde{b}} \hat{\theta}(t) \phi(t). \tag{2.15}$$

To adapt the unknown parameter vector $\xi(t)$ to different process properties, the real time least-squares method must be extended with some kind of forgetting property (refs. [9] and [11]). By introducing

$$\left. \begin{aligned} a_1(t) &= \lambda a_1(t-1) + [y(t) - \tilde{b}u(t-k)] \phi^{\mathrm{T}}(t-k), \\ B_1(t) &= \lambda B_1(t-1) + \phi(t-k) \phi^{\mathrm{T}}(t-k), \end{aligned} \right\} \tag{2.16}$$

the estimate of $\xi(t)$ will then be

$$\hat{\xi}_t = a_1(t) [B_1(t)]^{-1}, \tag{2.17}$$

where λ is the forgetting constant.

However, the vector a_1 and the matrix B_1 can also be considered as outputs from a first-order filter $1/(1-\lambda z^{-1})$. Thus it is possible to treat the 'forgetting problem' (the adaptive behaviour of the regulator) as a general filtering problem. If the filter function is given by $F(z^{-1})$, the estimate will be

$$\hat{\xi}_t = \{F(z^{-1})[y(t)-\tilde{b}u(t-k)]\phi^{\mathrm{T}}(t-k)\} \times [F(z^{-1})\phi(t-k)\phi^{\mathrm{T}}(t-k)]^{-1}. \quad (2.18)$$

By manipulating with different filters $F(z^{-1})$ it might be possible to alter the properties of the self-adjusting regulator in a wide range. Another advantage of viewing the problem in this way is that the number of necessary arithmetic operations will be small. This is of special importance when the control algorithm is applied in a mini-computer.

In the sequel only first-order filters will be considered. The number of various numerical operations for this special case are plotted in Fig. 1.

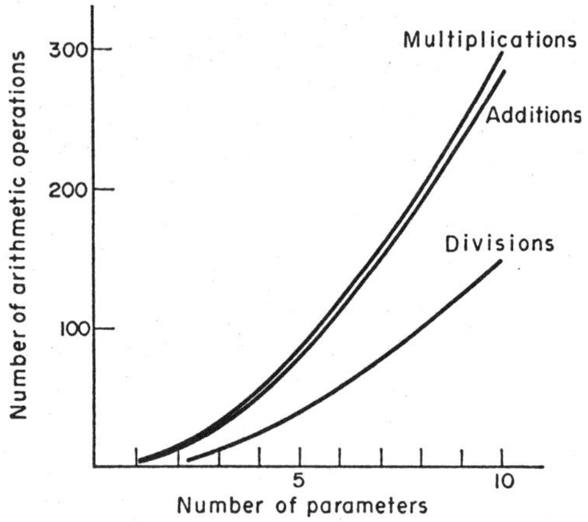

FIG. 1. Number of different arithmetic operations vs number of parameters to adjust in each control action computation.

3. CONTROL LAW STRUCTURE

When applying the self-adjusting regulator to an industrial process it is of great importance to choose a suitable structure for the controller.

This is equivalent to choosing r and s for the ϕ-vector in equation (2.5). From equation (2.4) it may be seen that the minimum variance control law consists of two parts. The first part is an inverse to the process dynamics with the time delay excluded and the second part can be regarded as a pseudo-inverse to the coloured noise. Thus if the process dynamics is of order n_1 and the noise of order n_2 then the denominator is of order n_1+n_2-1+k and the numerator of order n_1+n_2-1, where k is the pure time delay. When n_1 and n_2 are to be calculated common factors of the A- and B-polynomials respectively, the A- and C-polynomials will be disregarded. It is important to note that the factor k in the denominator provides a memory for the control activities during the period of time delay.

The dimension of the ϕ-vector will therefore be chosen to maximum $2n+k-1$, where $n = n_1+n_2$ is the order of the total system. In practice n is seldom greater than 3 and in most cases $n=2$ should suffice. A sampling rate that gives $k=3$ is normal for processes with time delays. This means that generally six parameters should be needed. In practical applications it is doubtful if anything can be gained by choosing more parameters. This is very important since the calculation time strongly depends on the dimension of the ϕ-vector as indicated in Fig. 1.

4. APPLICATION TO A PAPER-MACHINE CONTROL SYSTEM

A schematic picture of a paper machine is shown in Fig. 2. The object of control is to operate the thick stock flow valve before the headbox and also the steam pressure in the drier section in such a way that the basis weight and the moisture deviations of the manufactured paper from predetermined set points are minimized.

As shown in Fig. 3 the process is a two-input–two-output coupled system. The disturbances $w_1(t)$ and $w_2(t)$ are correlated. The time delays vary with the machine speed but are well known.

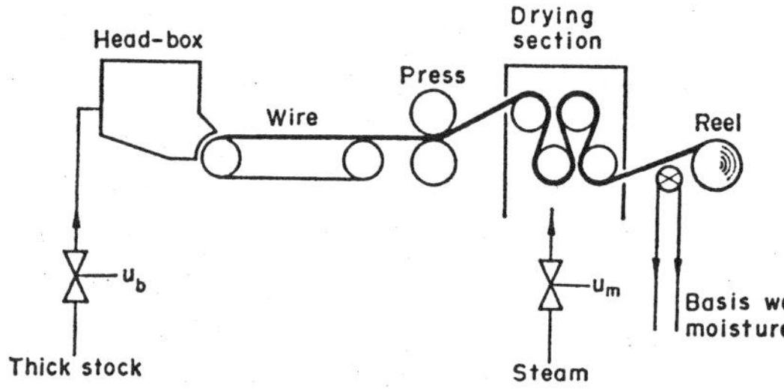

FIG. 2. A paper machine.

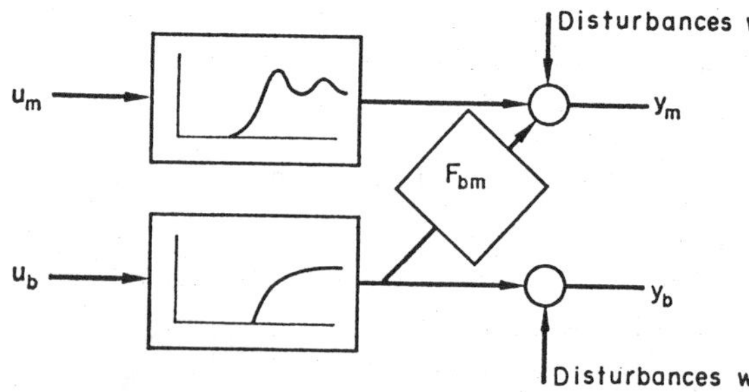

FIG. 3. Block diagram of a paper machine.

When testing the self-adjusting regulator several computer-controlled paper-machine systems were accessible. As the systems were alike it was easy to design an addition to the standard program package consisting of a small interface routine and a subroutine which was the self-adjusting regulator. The programs were written in FORTRAN. In Figs. 4(a) and (b) it is shown how this part of the program package had to be modified.

The original control algorithms were digitalized PI-algorithms

$$u(t) = u(t-1)+k_1 y(t)+k_2 y(t-1). \quad (4.1)$$

For analysing purposes a program was written, which restored every input–output sample and calculated the moving average, variance of the output $y(t)$ and its covariance function.

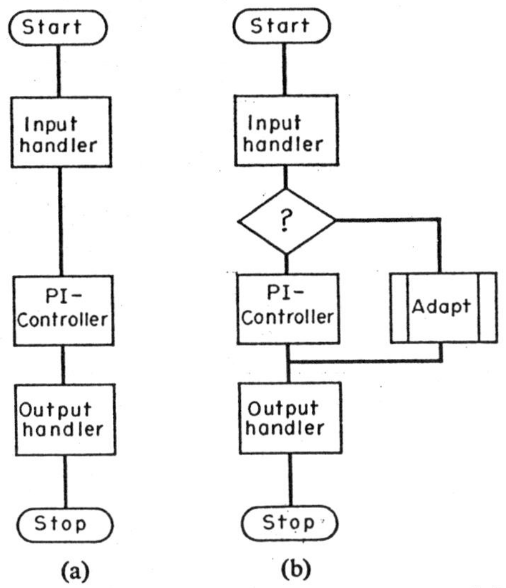

FIG. 4(a) and (b). Flow charts for control loops. "Adapt" is the name of the self-adjusting regulator subroutine.

5. RESULTS

The first practical tests took place in March 1972. Since then the self-adjusting regulator has been successfully applied to several other processes and also to three other paper machines. The results presented below originate from an installation which has now been in continuous operation for several months. As an example the behaviour of the moisture control loop will now be discussed. The dynamics of this loop is the most complicated and therefore the most interesting. The ϕ-vector dimension was chosen to be 6 and the time delay $k=4$. Two positions in the ϕ-vector are reserved for basis weight control actions since the decoupling of the system is being carried out in the regulator as well.

The ϕ-vector was organized as follows

$$\phi(t) = \begin{bmatrix} y_m(t) \\ y_m(t-1) \\ u_m(t-1) \\ u_m(t-2) \\ u_b(t) \\ u_b(t-1) \end{bmatrix}$$

where index m stands for moisture and index b for basis weight.

In Fig. 5 the six regulator parameters, the output $y(t)$ and the moving variance

$$\frac{1}{t}\sum_{1}^{t} y^2(t)$$

are plotted during 250 samples which is equivalent to 5 h. At time $t=1$ the regulator was started with the start values $\hat{\xi} = (0{\cdot}5, -0{\cdot}325, 0, 0, 0, 0)$ which were the parameters of the original PI-controller. As can be seen, the convergence is very smooth and resembles the step response of a first-order system.

The example shown originates from a start-up after a 12 hr production shutdown. During a shutdown of this length the normally wet parts of the paper machine will dry, which leads to blocking of pipes, screens and valves. For this reason a start-up is always very noisy and often paper breaks occur. In the example shown it was possible to record a number of state variables inside the paper machine as different flows and pressures. From these records it has been found that the start-up really was a noisy one. It is very probable that it was due to the regulator that no paper break occurred. As an example the largest disturbances marked in the figure will be examined.

Disturbance (1) was caused by a set point change of the basis weight. During the changing period the basis weight control was non-active. Disturbance (2) was caused by a large increase in the suction box pressure of a magnitude which without a powerful control law would have led to a paper break. It is notable that during normal operating conditions the algorithm needs only about 20 samples to find an estimate close to the optimal.

A very important question to answer, from a practical point of view, is that of profitability. Is the self-adjusting regulator significantly better than a conventional one? When answering, it is necessary to separate two different operating modes, namely the steady state and noisy state. This separation can be done since in practice the disturbances normally appear in bursts, often due to faults in the production process.

During a steady state the difference between a well-tuned PI-controller and the self-adjusting regulator is small. Problems with conventional controllers in a steady state will arise when, as in paper-making, different qualities are produced at the same unit. When quality is changed the process dynamics will change and the controllers should therefore be retuned. As retuning is rather complicated this will seldom be done. This leads to poor control result. With a self-adjusting regulator this problem will never occur, and the control result will always be optimal or near optimal.

During a noisy state large control actions may be necessary and since the conventional

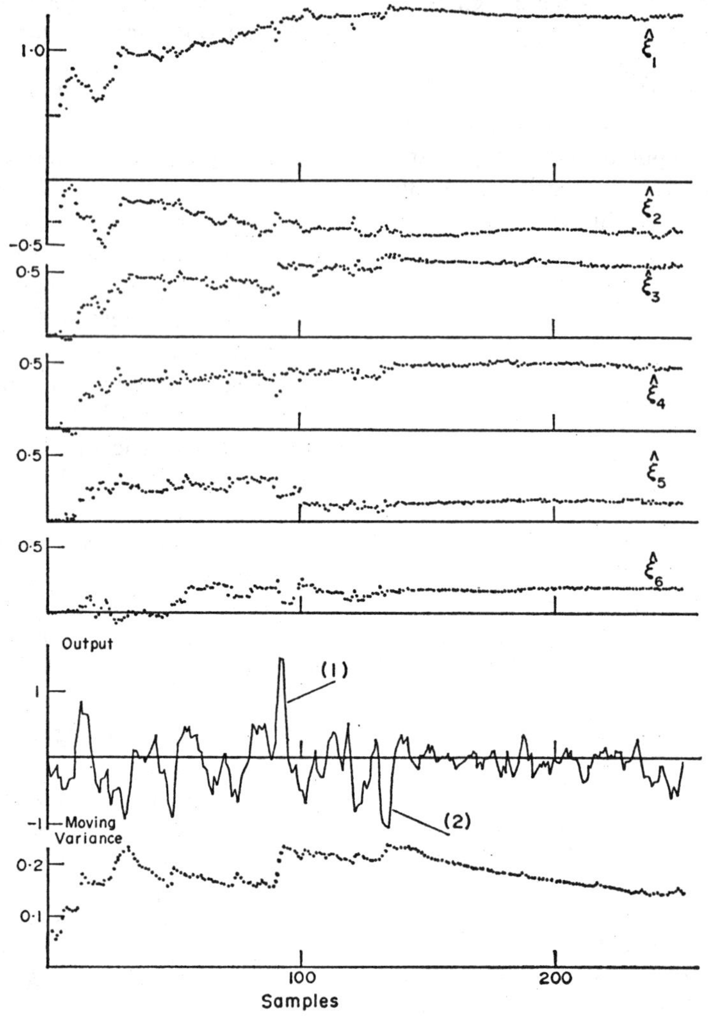

FIG. 5. Registration of parameter estimates and process variables.

PI-controller has no dead-time compensation terms it will give a poor control. The self-adjusting regulator, however, may have any number of parameters, all of which are well tuned. Hence the self-adjusting regulator can provide a very powerful control during bad conditions. In Fig. 6 a covariance function recorded during the above described start-up is shown. It is well known that in minimum variance control the function will vanish after k steps. As can be seen, the regulator was well tuned. The print-outs in the figure are from the above-mentioned analysing program. The upper line shows the time of day and the magnitudes of the five parameters in the basis weight loop. The second line shows the corresponding sample number, variance, average, last output and input. The third line shows the values of the covariance function, which is plotted below.

Another question of great importance is what will happen if the regulator is applied to processes where the assumptions made in the theory are not valid at all. Experience up to now does not show

ADB 2126 22 0·657 −0·272 0·471 0·521 0·472

B 268 0·308 −0·106 0·92 0

1·000 0·631 0·397 0·209 0·020 −0·060 −0·020

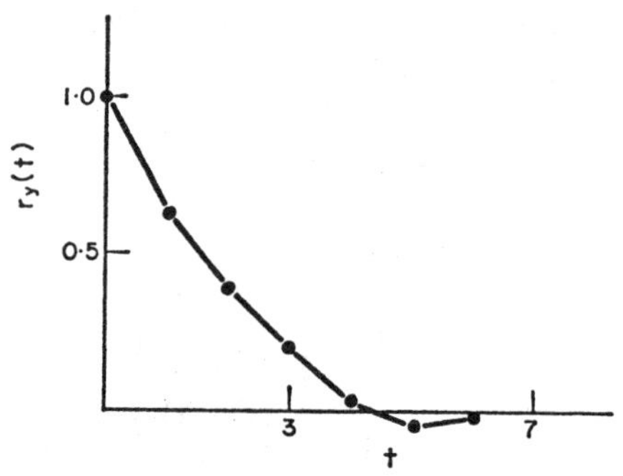

FIG. 6. An example from the analysing program print-out.

that anything serious will happen. In Fig. 7 the moisture error is shown along with the input and the parameter values of a self-adjusting regulator. During this recording the acceptance band for the input was very small. This led to a saturation phenomenon when a large disturbance occurred and the regulator tried to act powerfully. Because

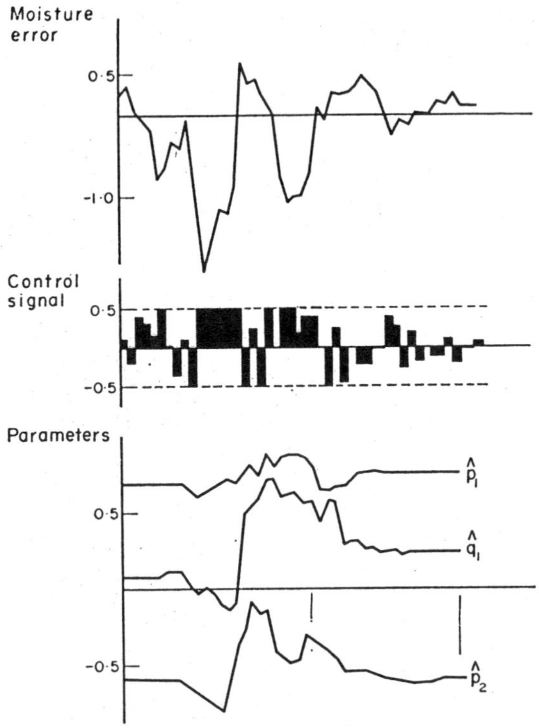

FIG. 7. A non-linearity phenomenon.

of the limit the calculated inputs were cut. This is equivalent to a non-linearity in the system. The self-adjusting regulator reacted immediately and during a couple of samples the regulator was trying to find a new linear model for the process. After the disturbance had been compensated the process returned to its linear state and the regulator in a few samples found the same model as before. A similar phenomenon can also be caused by too small a filter factor.

As mentioned before, the adaptive behaviour of the regulator is dependent on a learning parameter (λ). This fact can be used to shorten the time of transient behaviour during start-up, and, during one practical test, the regulator found its parameters close to the optimal within 12 samples when a suitable learning parameter was chosen.

6. CONCLUSIONS

The self-adjusting regulator described in Section 2 has turned out to be useful in practical applications. It has been possible to design powerful control laws due to the fact that regulators with many parameters can be tuned in an easy way. It has also been shown that the regulator may be used to control time-varying and multi-variable systems. Deterministic disturbances can be used to improve the control. Besides these important properties the control algorithm is very compact.

Acknowledgements—The authors would like to express their gratitude to Aktiebolaget Statens Skogsindustrier and to Svenska Cellulosa Aktiebolaget for their help during the test periods.

REFERENCES

[1] K. J. ÅSTRÖM: *Introduction to Stochastic Control Theory*. Academic Press, New York (1970).
[2] K. J. ÅSTRÖM: *Lectures on the Identification Problem—The Least Squares Method.* Report 6806, September 1968, Lund Institute of Technology.
[3] K. J. ÅSTRÖM and P. EYKHOFF: System identification—a survey. *Automatica* **7**, 123–162 (1971).
[4] K. J. ÅSTRÖM and B. WITTENMARK: Problems of identification and control. *J. Math. Anal. Appl.* **34**, 90–113 (1971).
[5] K. J. ÅSTRÖM and B. WITTENMARK: On self-tuning regulators. *Automatica*, **9**, 185–199 (1973).
[6] T. CEGRELL and T. HEDQVIST: Successful Adaptive Control of Paper Machines. *Proc. of the 3rd IFAC Symp., The Hague*, Paper PV-2 (1973).
[7] J. G. DESHPANDE, T. N. WPADHYAY and D. G. LAINIOTIS: Adaptive control of linear stochastic systems. *Automatica* **9**, 107–115 (1973).
[8] V. PETERKA: *On Steady-state Minimum Variance Control Strategy.* Institution Report, Institute of Information Theory and Automation, Prague, Czechoslovakia (1972).
[9] V. PETERKA: Adaptive Digital Regulation of Noisy Systems. 2nd Prague IFAC Symp. on Identification and Process Parameter Estimation (1970).
[10] G. N. SARIDIS and T. K. DAO: A learning approach to the parameter-adaptive self-organizing control problem. *Automatica* **8**, 589–597 (1972).
[11] J. WIESLANDER: *Real Time Identification—Part I.* Report 6908, Lund Institute of Technology, November 1969.
[12] J. WIESLANDER and B. WITTENMARK: An approach to adaptive control using real time identification. *Automatica* **7**, 211–217 (1971).

Self-Tuning Control of an Ore Crusher*

ULF BORISON† and ROLF SYDING‡

An adaptive regulator based on a linear controller, with parameters updated by on-line parameter estimation, successfully controls a real industrial process with tele-processing over a distance of 1800 *km.*

Summary—An industrial application is described where a self-tuning regulator is used for control of an ore crusher. There are difficult control problems associated with this process due to significant variations in operating conditions and long time delays in the system. The design of a self-tuning regulator for the crusher is discussed in detail. During this case study a temporary digital control loop was set up. A process computer at the University of Lund was connected to the ore crushing plant at 1800 km distance. When a self-tuning regulator is applied to a process with constant but unknown parameters, it will under certain conditions converge to a minimum variance strategy. The parameters of the ore crusher can hardly be considered as constant. However, it has been demonstrated practically that the use of exponential forgetting in the parameter estimation performed in the algorithm will give a very good regulator for the crusher.

1. INTRODUCTION

THE METHOD of designing regulators for industrial processes by first performing identification experiments to get a model of the process and its disturbances and then to determine an optimal controller from the obtained model may be difficult for ore crushers, since these are operating in a changing environment. It may be necessary to make repeated identification experiments to model the process properly under different operating conditions. Therefore, a self-tuning regulator capable of adapting to changing operating conditions was used for the ore crusher described here. Previous experiments on paper machines, Borisson and Wittenmark [4] and Cegrell and Hedqvist [5], have shown that this type of adaptive regulator has many attractive properties.

Self-tuning regulators are based on an assumption of separation of estimation of process parameters and computation of the control signal. The algorithm used in the application described here includes a least squares estimator and a linear controller, whose parameters are determined from the current estimates. As self-tuning regulators do not introduce any control action in order to reduce the uncertainty of the estimated parameters, they are not dual in the sense of Feldbaum. The idea of self-tuning control has been discussed several times, see e.g. Kalman [6], Peterka [8], Åström and Wittenmark [3] and Wittenmark [9]. Some different versions of self-tuning regulators have been described by Åström *et al.* [1]. In existing theoretical analysis of the algorithms it is assumed that the controlled system has constant parameters. As this assumption is questionable for the ore crusher, the facility of exponential forgetting was included in the parameter estimation. In this way a regulator with capability of adapting to process disturbances was obtained.

The process to be controlled is considered to be a single-input, single-output system with constant but unknown parameters

$$\begin{aligned} y(t)+a_1 y(t-1)+\ldots+a_n y(t-n) &= b_1 u(t-k-1)+b_2 u(t-k-2) \\ &\quad +\ldots+b_n u(t-k-n)+e(t)+c_1 e(t-1) \\ &\quad +\ldots+c_n e(t-n), \end{aligned} \tag{1.1}$$

where $y(t)$ is the output, $u(t)$ is the control variable and $e(t)$ is a sequence of independent, equally distributed random variables with zero mean. The algorithm uses a least squares model for recursive identification of the process (1.1). As the process disturbances in (1.1) are described by coloured noise, the least squares estimates will be biased.

* Received 10 March 1975; revised 7 July 1975. The original version of this paper was presented at the IFAC Symposium on Stochastic Control which was held in Budapest, Hungary, during September 1974. The published Proceedings of this IFAC meeting may be ordered from: 4th International Conference on Digital Computer Applications to Process Control, IFAC/IFIP Conference, Gloriastrasse 35, CH-8006 Zurich. The paper was recommended for publication in revised form by associate editor H. A. Spang III.

† Lund Institute of Technology, Department of Automatic Control, P.O. Box 725, S-220 07 Lund, Sweden.

‡ LKAB, Development Department Fack, S-981 01 Kiruna, Sweden.

Reprinted with permission from *Automatica*, vol. 12, pp. 1–7, Jan. 1976.
Copyright © 1976, International Federation of Automatic Control IFAC. Published by Pergamon Press Ltd.

However, it turns out that the bias will always influence the estimated parameters in such a way that the resulting strategy gives the optimal minimum variance regulator for the process (1.1), provided that the estimated parameters have converged and that the model order is sufficiently high. This was shown theoretically by Åström and Wittenmark [3].

The self-tuning algorithms have nice asymptotic properties, although they do not always converge. The parameter estimator is described by a non-linear, time-varying, stochastic difference equation, which makes the analysis difficult. The convergence properties have been discussed by Ljung and Wittenmark [7]. It has been shown that a deterministic, time-invariant, ordinary differential equation can be associated with the difference equation of the parameter estimator. Stability of the ordinary differential equation implies convergence with probability one for the algorithm. Using this result it is possible to construct examples where the self-tuning regulator does not converge.

Besides the experience gained from the practical use of self-tuning regulators this study has given valuable knowledge of tele-processing for remote control experiments. It gives an example of a possible means for universities to use real industrial processes for research in the field of automatic control.

2. CONTROL PROBLEM

This investigation was made as a joint project by the company LKAB, Kiruna, Sweden, and the Department of Automatic Control, University of Lund. In the crushing plant the ore is crushed to lumps with a maximum dimension of 25 mm, Fig. 1. A crushing line includes a crusher driven by an electrical motor and 2 screens. The ore enters the crushing line on an electro-mechanical feeder. A conveyor belt takes it to the first screen, where lumps which already have the desired dimension are separated. The rest of the ore proceeds to the crusher. Some part of the ore leaving the crusher consists of lumps with a dimension greater than 25 mm. These lumps will pass over the screen and will then be recycled to the crusher.

There is a special slip coupling limiting the moment of the crusher motor. If the load becomes too high, the crushing line is stopped automatically by a thermal overload protection. The purpose of the control is to keep the production on a level as high as possible without overloading. To maintain a high and constant power output it is necessary to control the input of ore to compensate for variations in crushability and lump size, as well as for changes in the crusher depending on wear of the jackets. The input of ore is controlled by varying the current to the electro-mechanical feeder.

The crushability of the ore depends mainly on the waste rock content. A large amount of waste rock in the ore will give a high load on the crusher. The variation in the size of the lumps may be considerable. The amount of ore that is separated at

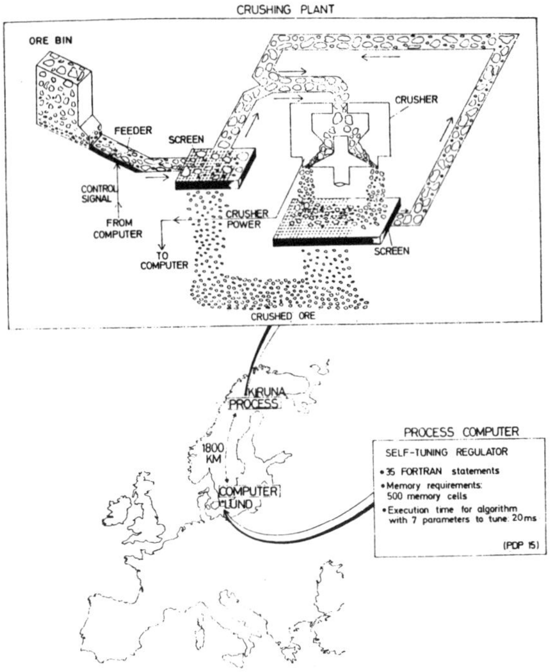

FIG. 1. A process computer at the University of Lund was connected to the crushing plant in Kiruna at 1800 km distance in a direct digital control loop.

the first screen may vary between 25 and 75 per cent of the total ore flow. To maintain a high efficiency the gap width of the crusher is adjusted every second day, and the jackets are changed every third month.

The transportation lags in the system, 40–50 sec between feeder and crusher and 70–80 sec in the recycle loop, are quite long compared with the time constants of the feeder and crusher, which are about 12 and 20 sec, respectively, if approximations to first-order systems are made. The long transportation lags contribute to the difficulties of the control.

Figure 2 shows a step response of the crusher under good operating conditions, when no regulator is controlling the power. The reference value of the feeder was increased with a step. The recycle loop can sometimes give rise to control problems. Figure 3 shows a step response when the ore is more difficult to crush. The successive accumulation of ore in the recycle loop increases rapidly the power output and the crusher is then overloaded.

The crusher is generally controlled by an analog PI regulator. This regulator cannot avoid great fluctuations in the crusher power. Therefore a rather low set point must be used to reduce the

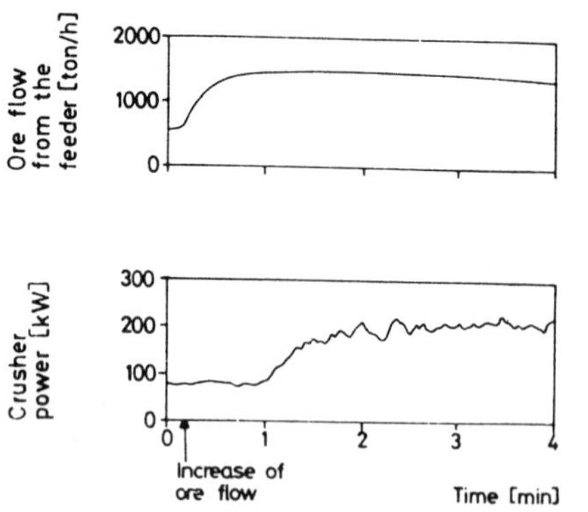

FIG. 2. Step response experiment under good operating conditions.

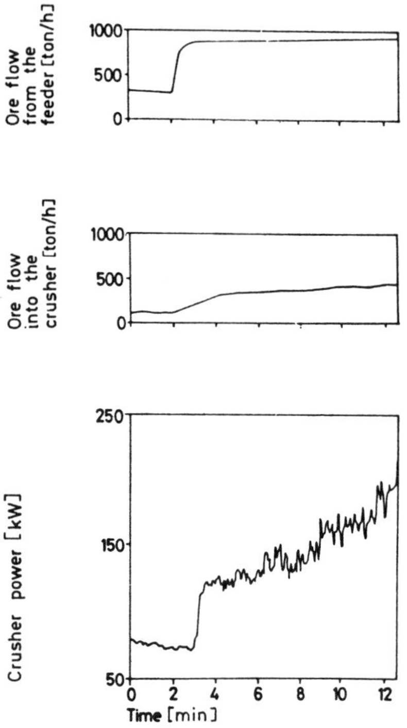

FIG. 3. Step response experiment under difficult operating conditions.

number of stops due to overload. As it may take about 1 hr before full production is achieved after a standstill, it is desirable to avoid unintended stops.

3. REGULATOR DESIGN

When using self-tuning algorithms some parameters describing the complexity of the regulator and the initial values for the start-up must be chosen. The choice of these parameters will be discussed for the ore crusher.

At each step of time the algorithm first estimates the parameters of the model

$$y(t+k+1)+\alpha_1 y(t)+\ldots+\alpha_m y(t-m+1) = \beta_0[u(t)+\beta_1 u(t-1)+\ldots+\beta_l u(t-l)] + \varepsilon(t+k+1) \quad (3.1)$$

with a least squares method. The parameter β_0 is assumed to be known. The control is determined from

$$u(t) = (1/\beta_0)[\alpha_1 y(t)+\alpha_2 y(t-1)+\ldots+\alpha_m y(t-m+1)] - \beta_1 u(t-1)-\beta_2 u(t-2)-\ldots-\beta_l u(t-l)$$

and the computation is based on the latest available parameter estimates. Introducing the vectors

$$\phi(t) = [-y(t)-y(t-1)\ldots-y(t-m+1)\,\beta_0 u(t-1)\;\beta_0 u(t-2)\ldots\beta_0 u(t-l)]$$

and

$$\theta = [\alpha_1\;\alpha_2\;\ldots\;\alpha_m\;\beta_1\;\beta_2\;\ldots\;\beta_l]^{\mathrm{T}}$$

the model (3.1) can be written as

$$y(t) = \beta_0 u(t-k-1)+\phi(t-k-1)\,\theta+\varepsilon(t).$$

The recursive least squares estimates, see e.g. Åström and Eykhoff [2], are given by

$$\theta(t+1) = \theta(t)+K(t)\,[y(t)-\beta_0 u(t-k-1) - \phi(t-k-1)\,\theta(t)],$$
$$K(t) = P(t)\,\phi(t-k-1)^{\mathrm{T}}[1+\phi(t-k-1) \times P(t)\,\phi(t-k-1)^{\mathrm{T}}]^{-1},$$
$$P(t+1) = P(t)-K(t)\,[1+\phi(t-k-1) \times P(t)\,\phi(t-k-1)^{\mathrm{T}}]\,K(t)^{\mathrm{T}}. \quad (3.2)$$

$P(t)$ is a normalized covariance matrix of the parameter estimates.

Sampling interval and number of time delays

A sampling interval of about 10–20 sec is reasonable with regard to the time constants of the system. The transportation lag between feeder and crusher is 40–50 sec. The parameter k of the model (3.1) corresponds to this time delay expressed in a number of sampling intervals. A value of k that is a little overestimated is better than an underestimated value, because in the latter case the control signals are often large. Preliminary experiments indicated that a sampling interval of 20 sec should be sufficient, and it was found empirically that the algorithm worked better with $k = 3$ than $k = 2$.

Number of α-parameters (m) and β-parameters (l)

The order of the controlled system (1.1) is denoted by n. The smallest values of m and l that theoretically give the minimum variance strategy are $m = n$ and $l = n+k-1$. Even if the feeder and the crusher are approximated by first-order systems, the recycle loop will increase the total order of the system due to the significant time delay within it. In practice, however, it has appeared that the choice of the number of parameters in the regulator is not crucial for the control result. Incremental control signals were computed by the regulator. By practical experiments it was found that $m = 4$ and $l = 3$ was a good choice.

Scale factor

β_0 corresponds to an estimation of b_1 in the system (1.1). The magnitude of β_0 thus depends on the magnitudes of the output and control signal of the process. However, the self-tuning regulator does not critically depend on a specific value of β_0. In this case, for example, a value between 1 and 100 gave good results.

Start-up

The algorithm was generally started in a rough but simple way with zero initial values of the α- and β-parameters and with a covariance matrix $P(0) = 10 \times I$, where I is a unit matrix. The magnitude of the control signal was limited during the start-up to avoid large control signals due to bad parameter estimates. After about 20 sampling intervals the estimates were often good enough to give a satisfactory control. Sometimes when parameters from earlier experiments were available, these were used as initial values. Then the elements of the covariance matrix were given smaller initial values.

Exponential forgetting

As was mentioned in Section 2, variations in crushability and size of the ore lumps, as well as changes in the crusher depending on wear, make the control difficult. By making use of exponential forgetting in the loss function used in the least squares estimation it is possible to obtain an algorithm that can adapt to variations in the process. Instead of minimizing

$$\sum_{i=0}^{t} [\varepsilon(i)]^2,$$

where $\varepsilon(i)$ are the residuals, the following loss function is considered

$$\sum_{i=0}^{t} \lambda^{t-i}[\varepsilon(i)]^2, \quad \lambda \leqslant 1. \tag{3.3}$$

Equation (3.2) for the covariance matrix of the parameter estimates is then changed to

$$P(t+1) = (1/\lambda)[P(t) - K(t)[1 + \phi(t-k-1) \times P(t)\phi(t-k-1)^{\mathrm{T}}]K(t)^{\mathrm{T}}].$$

For the ore crusher it was found empirically that $\lambda = 0{\cdot}99$ was feasible. The loss function (3.3) is then dominated by values of $\varepsilon(i)$ from roughly the past hour of controlled operation. During the start-up λ often had a value of about 0·95 to make the algorithm converge faster. It was then increased to 0·99.

4. IMPLEMENTATION

During the feasibility study of the self-tuning regulator a process computer at the Department of Automatic Control in Lund was used. There was no computer available at the crushing plant. The data were transmitted by a public telephone line and low speed modems. The distance between the computer and process is 1800 km. The direct digital control loop is sketched in Fig. 1.

Data entered and left the computer via an ordinary teletype interface. The process interface includes digital to analog conversion of the control signal. The self-tuning regulator can be implemented with about 35 FORTRAN statements. With the configuration used for the ore crusher the algorithm required about 500 memory cells on a PDP-15 computer. The computation time at each sampling interval is about 20 msec with floating point hardware.

5. CONTROL RESULTS

Many experiments were made to study the feasibility of self-tuning control compared with PI control. The ore crusher was in normal operation during the test period.

Two examples will now be discussed in detail. The control law was

$$\begin{aligned}\nabla u(t) = (1/\beta_0)[\alpha_1 y(t) + \alpha_2 y(t-1) + \alpha_3 y(t-2) \\ + \alpha_4 y(t-3)] - \beta_1 \nabla u(t-1) \\ - \beta_2 \nabla u(t-2) - \beta_3 \nabla u(t-3),\end{aligned}$$

where $u(t)$ is the input, the current to the electromechanical feeder which determines the ore flow [mA], $y(t)$ is the output (crusher power [kW]), $\nabla u(t) = u(t) - u(t-1)$ and $\beta_0 = 50$, a constant.

The α- and β-parameters were tuned by the algorithm. The number of time delays (k) discussed in Section 3 was 3. The base of the exponential weighting (λ) in (3.3) was 0·99.

Example 1 (*January* 9, 1973)

Figure 4 shows some registrations of crusher power, control signal and ore flow originating from an experiment with self-tuning control. The set point of the crusher power is 200 kW. The estimated standard deviation is 19·7 kW, which is a good result compared with conventional PI control for the same set point. About 30 min after the start of the registration in Fig. 4, the crusher power is becoming high. However, the self-tuning regulator decreases the ore flow quickly enough to avoid overloading.

The regulator parameters are shown in Fig. 5. In the beginning there are only small variations. Later on, when the crusher power becomes high, the parameters change more. The quick response of the regulator when the power increased depends

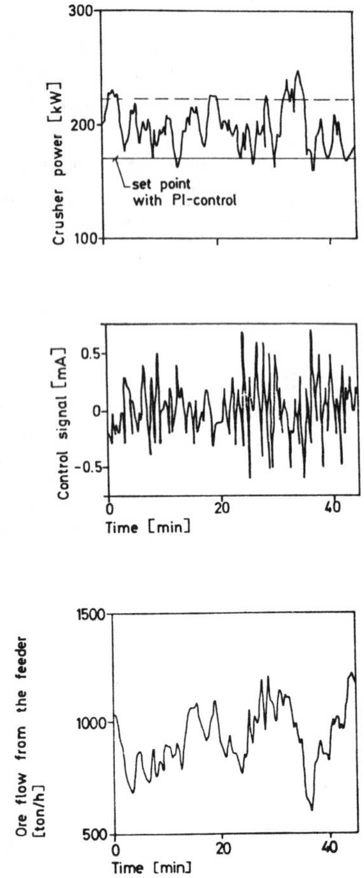

FIG. 4. Self-tuning control (example 1).

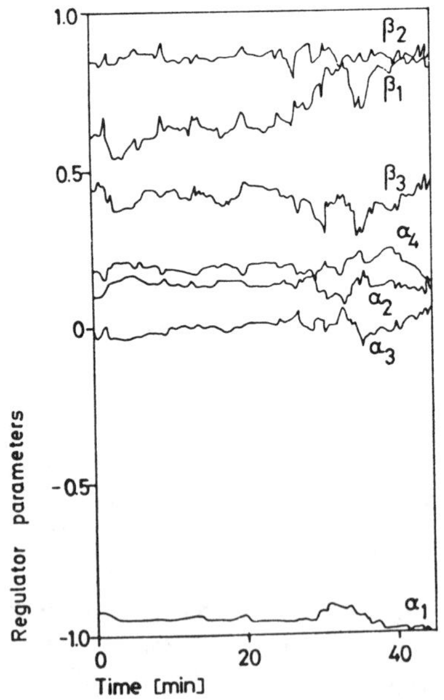

FIG. 5. Regulator parameters (example 1).

on a combination of larger control errors and adaption of the regulator parameters. In 4 min the ore flow is decreased by about 40 per cent. This action goes beyond the capacity of the PI regulator. If the parameter estimates converge the auto-covariance $r_y(\tau)$ and the cross-covariance $r_{yu}(\tau)$ have the properties

$$r_y(\tau) = 0, \quad \tau \geqslant k+1,$$

$$r_{yu}(\tau) = 0, \quad \tau \geqslant k+1$$

according to theoretical results [3]. Figure 6 shows the estimated covariances $\hat{r}_y(\tau)$ and $\hat{r}_{yu}(\tau)$ from

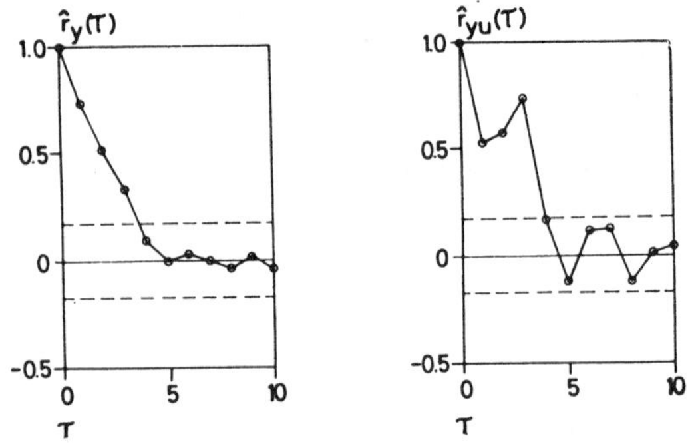

FIG. 6. Estimated covariance functions (example 1). They have been normalized. The covariance functions are almost zero for $\tau \geqslant 4$, which is in accordance with theoretical results.

example 1, where k is 3. The dashed lines indicate a 95 per cent confidence interval, where the covariances are regarded as zero. It follows that the covariance functions are almost zero for $\tau \geqslant 4$, which is in accordance with the theoretical results.

Example 2 (*February* 1, 1973)

In the crushing line there is an ore bin before the feeder, Fig. 1, where the ore is stored up before it is sent to the crusher. This example will show what may happen when the bin is filled. Many of the

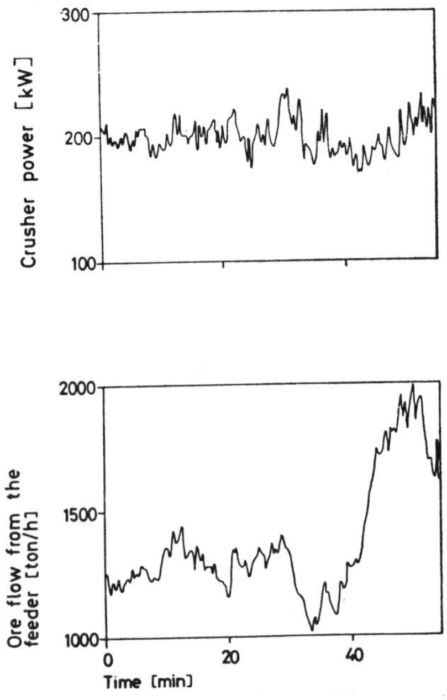

FIG. 7. Self-tuning control (example 2). After about 40 min the size of the ore lumps is becoming smaller, and the self-tuning regulator increases the ore flow quickly to keep the crusher power on a constant level.

ore lumps coming from the mine are already small. They tend to gather at the bottom of the bin. When they come to the first screen, they are separated and do not pass the crusher. In this situation the regulator has to increase the ore flow to maintain a constant power output.

In Fig. 7 the crusher power and the ore flow are shown. About 40 min after the start of the registration the bin is filled. It is seen how the self-tuning regulator then immediately increases the ore flow to keep the crusher power on the same level as before. With the standard PI regulator the same fast response was not achieved in situations like this. The estimated standard deviation of the power is 13·3 kW in this example. As the type of variation in the ore described here occurs several times a day, it is desirable to have a regulator that can do well in a situation like this.

6. EVALUATION

The crusher is usually controlled by an analog PI regulator with constant parameters which have been determined by thorough empirical investigations. However, this regulator cannot always maintain good control. In normal operation it must be used with a set point of about 170 kW, while the maximum capacity of the crusher is around 220 kW. In spite of the low set point the PI regulator does not manage to keep down the number of stops due to overlaod. During the test period it was established that the self-tuning regulator gives an estimated standard deviation of the crusher power that is lower than with analog PI control. This improvement can be utilized to increase the production capacity. The experiments showed that an average crusher power of 200 kW can be obtained. Figure 8 shows frequency histograms of the crusher power in 2 cases representative for the PI regulator and the self-tuning regulator, respectively, under similar operating conditions.

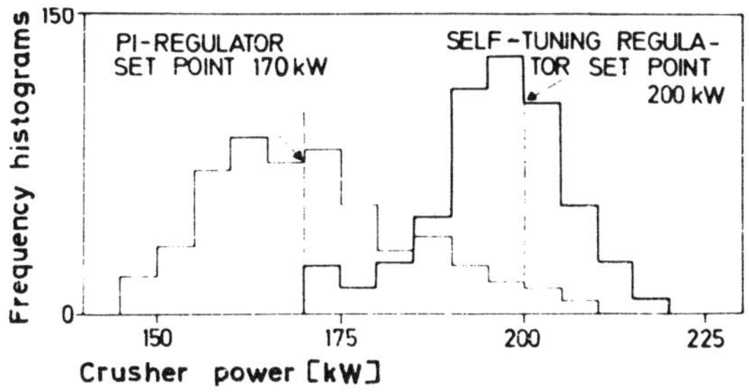

FIG. 8. Frequency histograms of the crusher power when the process is controlled by a PI regulator and a self-tuning regulator respectively.

As the autocovariance function of the crusher power tends to zero quickly with self-tuning control, the power reaches the upper limit only during short time intervals. The thermal overload protection tolerates this to a certain extent. With PI control the autocorrelation is greater and the overload protection is more easily released. The autocovariance function in Fig. 9, which results from the PI regulator, is typical.

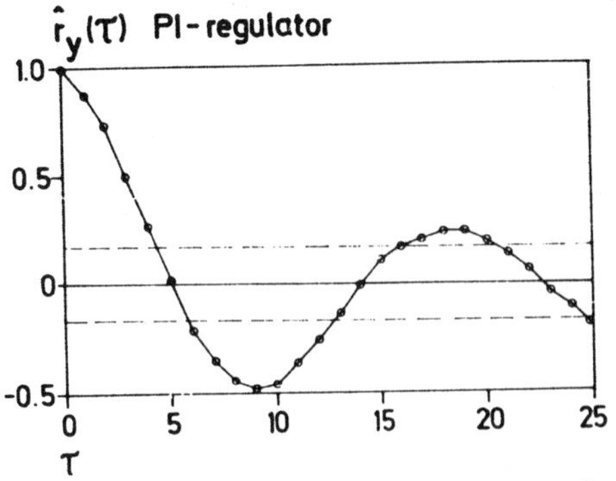

FIG. 9. The estimated autocovariance function from an experiment with the PI regulator that generally was used for the crusher. The autocovariance function has been normalized. It follows that the autocovariance function tends to zero very slowly with PI control.

The yearly production in the plant, which includes six crushing lines, is about 17 million tons. It is increasing at about 5 per cent per year. As the cost for a new crushing line is around 3 million dollars, any increase of the capacity within the existing lines is of importance. Figure 10 shows a

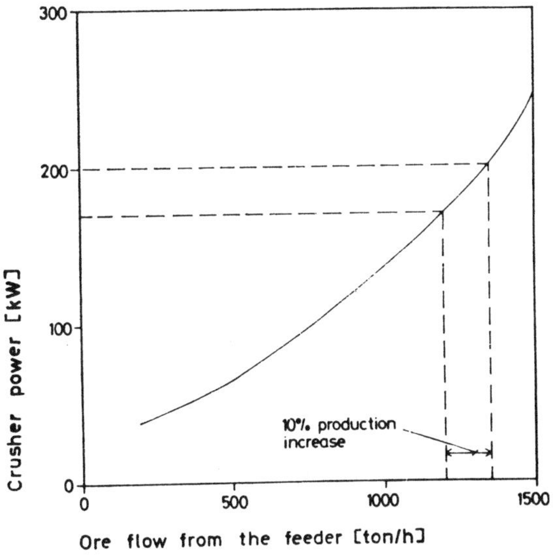

FIG. 10. The crusher power as a function of the ore flow. The self-tuning regulator makes it possible to increase the crusher power from 170 to 200 kW, and the production of ore will then be about 10 per cent higher.

diagram where the crusher power is plotted vs the ore flow in one crushing line. With the power output increased from 170 to 200 kW, the production of ore can be at least 10 per cent higher in that line. This gives economical reasons to support an installation of a digital control system to implement the self-tuning regulator.

7. CONCLUSIONS

The theoretical analysis of the self-tuning regulator shows that the algorithm will under certain conditions converge to a minimum variance regulator when applied to a plant with constant but unknown parameters. Intuitively it can also be expected that the self-tuning regulator will work for systems with slowly varying parameters as well, provided the estimation phase of the algorithm is suitably modified. This case study verifies that the self-tuning regulator will actually work well in a truly adaptive environment with continually changing process variables, when exponential forgetting is included in the parameter estimation.

Acknowledgements—We are very grateful to Professor K. J. Åström for his encouraging support during this work. We would also like to express our gratitude to the LKAB Company for providing the opportunities to realize this project. Special thanks are due to Mr. L. Andersson and Mr. R. Braun, at the Control Laboratory in Lund, who constructed the process interface and assisted kindly with all hardware equipments. Our appreciation also goes to Mr. S. B. Rönnkvist, LKAB, who participated in several of the experiments.

REFERENCES

[1] K. J. Åström, U. Borisson, L. Ljung and B. Wittenmark: Theory and applications of adaptive regulators based on recursive parameter estimation. *Triennal IFAC World Congress*, Boston (1975).

[2] K. J. Åström and P. Eykhoff: System identification—a survey. *Automatica* **7**, 123–162 (1971).

[3] K. J. Åström and B. Wittenmark: On self-tuning regulators. *Automatica* **9**, 185–199 (1973).

[4] U. Borisson and B. Wittenmark: An industrial application of a self-tuning regulator. *4th IFAC Conference on Digital Computer Applications to Process Control*, Zürich (1974).

[5] T. Cegrell and T. Hedqvist: Successful adaptive control of paper machines. *Automatica* **11**, 53–59 (1975).

[6] R. E. Kalman: Design of a self-optimizing control system. *Am. Soc. Mech. Engr. Trans.* **80**, 468–478 (1958).

[7] L. Ljung and B. Wittenmark: Analysis of a class of adaptive regulators. *IFAC Symposium on Stochastic Control*, Budapest (1974).

[8] V. Peterka: Adaptive digital control of noisy systems. *IFAC Symposium on Identification and Process Parameter Estimation*, Prague (1970).

[9] B. Wittenmark: A self-tuning regulator. Report 7311 Department of Automatic Control, Lund Institute of Technology (1973).

Self-Tuning Control of a Titanium Dioxide Kiln

GUY A. DUMONT, MEMBER, IEEE, AND PIERRE R. BÉLANGER, MEMBER, IEEE

Abstract—**This paper presents an application of the self-tuning regulator to the control of a TiO_2 kiln. The regulator is used in a feedback loop around a kiln already controlled by a discrete regulator designed on minimum-variance principles.**

The self-tuning regulator has only two parameters. A study of the asymptotic properties shows that the self-tuning regulator preserves optimality if the minimum-variance regulator is already optimal. The equilibrium points of the regulator are established for delays of one and two steps. Experimental results showed an increase in "in-spec" time over previous regulators. Grade-change results are also given, leading to marked improvement over previous practice.

I. Introduction

Titanium dioxide (TiO_2) is a whitening pigment used in the manufacture of paper, paints, and various other products where whiteness or opacity are desirable. The most important unit in the production of TiO_2 is the kiln, where the naturally abundant crystalline form anatase is transformed to the desired rutile form. The rutile percentage in the product should lie between 96.5 percent and 99.5 percent for good pigment quality and grindability.

The results of a previous study [1] show that good control can be obtained only by direct composition measurement. This is done by X-ray diffraction analysis of a sample taken from the process outlet. This can be done, at most, every half-hour. The same study also establishes a minimum-variance control law, using the fuel rate as a manipulated variable.

It was found experimentally that the model-based control law outperformed the previous control scheme, raising the in-spec percentage (percent of the time when the product is within specifications) from 64 percent of the total time to 78 percent on one particular kiln. Similar results were achieved on a second kiln, but only by tuning the control law parameter. Furthermore, a correlation analysis of the closed-loop output showed that the process was not, in fact, optimally tuned.

These considerations led us to consider the self-tuning regulator [2], [3] as a solution. Due to limitation in the available memory space, a problem formulation with a small number of parameters was required. This was achieved by using as a "plant" the closed-loop combination of kiln and fixed regulator. This leads to a self-tuning scheme where the tuned model has lower order than the model representing the plant. This idea was suggested in [4], but the present work appears to be the first application.

The self-tuning regulator was tried experimentally on a kiln, and the percent of time within specifications was raised to 85 percent. Finally, we attempted to use the self-tuning regulator to control the process through a grade of change. Under normal plant practice, a 10–12 hour period of off-spec product is expected: the self-tuning regulator cut this to two hours resulting in the best grade change ever experienced in that particular plant.

II. Self-Tuning Regulator

The theory of the self-tuning regulator is covered in [2], [3], and its value as a practical tool for process control has been established by application to a paper machine [6], an ore crusher [7], and a few other units. We review the theory briefly here, principally to put down definitions and symbols.

A single-input–single-output plant is represented by

$$y(t)=\beta_0 u(t-k-1)+\phi^T(t-k-1)\theta+\epsilon(t) \qquad (1)$$

where

$$\phi^T(t) \triangleq [-y(t)-y(t-1)\dots-y(t-n-k)\beta_0 u(t) \cdot\beta_0 u(t-1)\dots\beta_0 u(t-n-k+1)]$$

$$\phi^T \triangleq [\alpha_1 \quad \alpha_2\dots\alpha_n \quad \beta_i\dots\beta_{n+k-1}]$$

and where $\epsilon(t)$ is a moving average of order k. The parameter k is the delay, over and above the normal one-stage delay.

The minimum-variance control law (assuming minimum-phase dynamics) is

$$u(t)=-\frac{1}{\beta_0}\phi^T(t)\hat{\theta}(t). \qquad (2)$$

The parameter β_0 is assumed known; this does not restrict the control law, because the parameter of the estimate $\hat{\theta}(t)$ may be scaled up or down. However, the choice of β_0 does affect the convergence properties of the algorithm.

Manuscript received July 25, 1977; revised April 11, 1978. Paper recommended by H. A. Spang, III, Chairman of the Applications, Systems Evaluation, and Components Committee.

G. A. Dumont was with the Department of Electrical Engineering, McGill University, Montreal, P.Q., Canada. He is now with Tioxide S. A., Calais, France.

P. R. Bélanger is with the Department of Electrical Engineering, McGill University, Montreal, P.Q., Canada.

Reprinted from *IEEE Trans. Automat. Contr.*, vol. AC-23, pp. 532–538, Aug. 1978.

The estimate $\hat{\boldsymbol{\theta}}$ is generated by recursive least squares, as if $\epsilon(t)$ were white, by the equations

$$\boldsymbol{K}(t)=P(t)\phi(t-k-1)/\left[1+\phi^T(t-k-1)\cdot P(t)\phi(t-k-1)\right] \quad (3)$$

$$P(t+1)=\frac{1}{\lambda}\left\{P(t)-\boldsymbol{K}(t)\left[1+\phi^T(t-k-1)P(t)\phi(t-k-1)\right]^{-1}\boldsymbol{K}^T(t)\right\} \quad (4)$$

$$\hat{\boldsymbol{\theta}}(t+1)=\hat{\boldsymbol{\theta}}(t)+\boldsymbol{K}(t)\left[y(t)-\beta_0 u(t-k-1)-\phi^T(t-k-1)\hat{\boldsymbol{\theta}}(t)\right]. \quad (5)$$

The scalar parameter λ, $0<\lambda\leqslant 1$, is the discount factor. Its introduction implies the minimization of the time-weighted least squares cost.

$$J=\sum_{i=0}^{t}\lambda^{t-i}\epsilon^2(i). \quad (6)$$

The practical importance of λ is that $\lambda=1$ makes $\boldsymbol{K}(t)\to 0$, thus rendering the recursive estimator unable to track parameter changes. A choice of $\lambda<1$ weighs only more recent data, keeps the adaptive loop open, and allows for tracking of variable parameters.

The asymptotic properties of the self-tuning regulator may be studied using equations developed in [5], [8] by Ljung *et al.* Define $u(t,\boldsymbol{\theta})$ as the control obtained by replacing $\hat{\boldsymbol{\theta}}(t)$ in (2) by the fixed vector $\boldsymbol{\theta}$: i.e., $u(t,\boldsymbol{\theta})$ is a fixed linear control law. From this there results an output process $y(t,\boldsymbol{\theta})$, and a process $\epsilon(t,\boldsymbol{\theta})$ obtained in an obvious manner from (1). Now define

$$f(\boldsymbol{\theta}) \triangleq E\left[\phi(t,\boldsymbol{\theta})\epsilon(t,\boldsymbol{\theta})\right] \quad (7)$$

$$G(\boldsymbol{\theta}) \triangleq E\left[\phi(t,\boldsymbol{\theta})\phi^T(t,\boldsymbol{\theta})\right] \quad (8)$$

where $\phi(t,\boldsymbol{\theta})$ is simply $\phi(t)$ with the $y(t)$ and $u(t)$ processes replaced by the $y(t,\boldsymbol{\theta})$ and $u(t,\boldsymbol{\theta})$ processes.

It is shown [5] that the estimates of the self-tuning algorithm behave asymptotically, in probability, as the solution of

$$\dot{\boldsymbol{\theta}}=Rf(\boldsymbol{\theta}) \quad (9)$$

$$\dot{R}^{-1}=-R^{-1}+G(\boldsymbol{\theta}) \quad (10)$$

where R is a positive-definite matrix.

The equilibrium points $\boldsymbol{\theta}^*$ of (9), (10) satisfy

$$f(\boldsymbol{\theta}^*)=\boldsymbol{0}. \quad (11)$$

The vector $\boldsymbol{\theta}$ behaves in the neighborhood of an equilibrium point as the solution of the linear differential system

$$\Delta\dot{\boldsymbol{\theta}}=G^{-1}(\boldsymbol{\theta}^*)\left.\frac{\partial f(\boldsymbol{\theta})}{\partial\boldsymbol{\theta}}\right|_{\boldsymbol{\theta}=\boldsymbol{\theta}^*}\cdot\Delta\boldsymbol{\theta} \quad (12)$$

The functions f and G are obtained by computing correlations, at fixed $\boldsymbol{\theta}$, between the outputs and inputs. The plant model must be known, of course, to generate

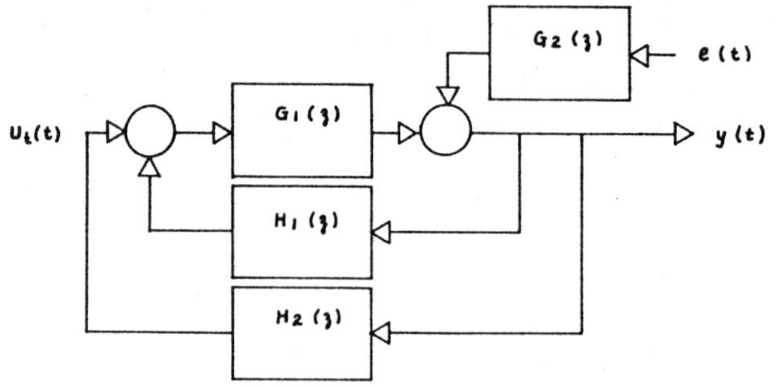

Fig. 1. Linearized model of kilns.

$y(t,\boldsymbol{\theta})$ from $u(t,\boldsymbol{\theta})$. Equation (11) yields the equilibrium points (there may be any number of them, including zero), and the eigenvalues of (12) establish the stability or instability of the point.

III. Self-Tuning for Kiln

The modeling of the kiln is discussed elsewhere (1). For present purposes, the kiln may be represented by the two transfer functions $G_1(z)$ and $G_2(z)$ appearing in Fig. 1, with

$$G_1(z)=\frac{264}{2670}\frac{z^{-1}}{0.67}\frac{z(1-0.33^{1-d})-0.33+0.33^{1-d}}{z-0.33} \quad (13)$$

$$G_2(z)=\frac{z^3-1.336z^2+0.344z-0.004}{(z-0.33)(z-0.85)(z-1)} \quad (14)$$

where d, $0\leqslant d\leqslant 1$, is the process delay as a fraction of the sampling period. The output is actually the reading of the diffraction apparatus, in counts; the number 2670 is a calibrating gain, carried explicitly here because it may vary according to the calibration. Note that $G_1(z)$ reduces to

$$G_1(z)=\frac{264}{2670}\frac{1}{z-0.33}$$

for $d=0$ (no delay) and to

$$G_1(z)=\frac{264}{2670}\frac{1}{z(z-0.33)}$$

for $d=1$ (delay of 1 period).

The minimum-variance law, designed for $d=0$, is given by

$$H_1(z)=\frac{-8.51z^2+11.25z-2.79}{2670(z^2-1.85z+0.85)} \quad (15)$$

Note that this compensator has a pole at $z=1$, i.e., contains a discrete integration.

The delay is introduced because of the manual operations involved in sampling and analysis: the total delay was estimated as possibly amounting to one sampling period of time.

The standard self-tuning method would proceed at this point to the design of a 5-parameter self-tuning controller of the same structure as the minimum-variance controller.

Another route was chosen, consisting of placing the self-tuning controller around the *controlled* plant. Three reasons prompted this decision. The first was that the minimum-variance controller can serve as a backup should the self-tuning control begin to diverge; the second was that the available computer memory could not accommodate a 5-parameter self-tuning control; third, since the minimum-variance regulator contains an integration, the problem of steady-state offset disappears.

In order to determine the form of the regulator, we note that, for $d=0$, the linear system with $u_t(t)$ as input (see Fig. 1) and $y(t)$ as output is described by

$$y(t)=0.33y(t-1) \\ +264[u_t(t-1)-0.848u_t(t-2)]/2670 \\ +e(t)-0.33e(t-1). \quad (16)$$

Based on (16), and allowing for the introduction of delay, the self-tuning model was chosen as

$$y(t+k+1)=-\alpha_1 y(t)+\beta_0[u_t(t)+\beta_1 u_t(t-1)] \\ +\epsilon(t+k+1) \quad (17)$$

which contains only two adjustable parameters.

The control law, for $|\hat{\beta}_1|<1$, is

$$u_t(t)=\frac{\hat{\alpha}_1}{\beta_0}y(t)-\hat{\beta}_1 u_t(t-1). \quad (18)$$

For $|\hat{\beta}_1|>1$, the suboptimal law used is [8]

$$u_t(t)=\frac{\hat{\alpha}_1}{\beta_0}\frac{y(t)-\hat{\beta}_1 y(t-1)}{1+\dfrac{\hat{\beta}_1^2}{\hat{\alpha}_1^2}}+\frac{\hat{\beta}_1^2}{1+\dfrac{\hat{\beta}_1^2}{\hat{\alpha}_1^2}}u(t-2). \quad (19)$$

Finally, the functions of (7) and (8) are

$$f\theta=\begin{bmatrix} -r_{yy}(k+1) \\ \beta_0 r_{yu}(k+2)\end{bmatrix}, \quad G(\theta)=\begin{bmatrix} r_{yy}(0) & -r_{yu}(1) \\ -r_{yu}(1) & r_{uu}(0)\end{bmatrix} \quad (20)$$

where $r_{yu}(i)=E[y(t)u_t(t-i)]$ with compatible definitions for r_{yy} and r_{uu}.

IV. Asymptotic Analysis

Simulations were carried out, under nominal minimum-variance control, to explore the effect of plant parameter variations. It was found that the closed-loop performance was not much affected by changes in the plant pole ($z=0.33$) and the disturbance pole ($z=0.848$). On the other hand, a substantial decrease in performance resulted from changes in the process gain K_s (nominal value$=264$) and the process delay. It was decided, therefore, to study the asymptotic behavior of the self-tuning regulator for variations in the process gain, for process delays d of one and two steps and self-tuning model delays k of one and two steps. Note that for given K_s, d and k, the correlation functions in (20) depend only on $\hat{\alpha}_1$ and $\hat{\beta}_1$ because these two parameters determine the control law. For given K_s, d and k, we may try to solve for the values of $\hat{\alpha}_1$ and $\hat{\beta}_1$ that yield $f(\theta)=0$, i.e., $r_{yy}(k+1)=r_{yu}(k+2)=0$. There may exist no solution, in which case we conclude that the self-tuning regulator does not have an equilibrium point. There may exist one or several solutions, in which case we should use (12) to weed out the unstable equilibrium points from the stable ones. If the exercise is repeated at fixed k and d for different values of K_s, we can, assuming that equilibrium points exist, generate loci of such points in the $\hat{\alpha}_1-\hat{\beta}_1$ plane, with K_s as a parameter.

One set of equilibrium points is readily available. If the plant is the nominal plant, the minimum-variance law is such that $f(\theta)=0$ without self-tuning. Considering (18), it is clear that, if $\hat{\alpha}_1=0$ and $|\hat{\beta}_1|<1$, the self-tuning input goes asymptotically to zero. This is also true in the case $|\hat{\beta}|>1$, using the suboptimal law (19). Therefore, the locus $\hat{\alpha}_1=0$ is an equilibrium locus, for the nominal plant.

To see that this is a stable locus, we compute (12) as

$$G^{-1}(\theta^*)\frac{\partial f(\theta)}{\partial\theta}\bigg|_{\theta^*}=\frac{1}{r_{yy}(0)}\begin{bmatrix} -\partial r_{yy}(1)/\partial\hat{\alpha}_1 & 0 \\ \beta_0\partial r_{yu}(2)/\partial\hat{\alpha}_1 & 0\end{bmatrix}. \quad (21)$$

It turns out that $\partial r_{yy}(1)/\partial\hat{\alpha}_1\hat{\alpha}_{1=0}>0$ for all $\hat{\beta}_1$, so that, for any starting point near the line $\hat{\alpha}_1=0$, $\hat{\alpha}_1\to 0$ and $\hat{\beta}_1$ remains constant. It is reassuring to see that the self-tuning algorithm preserves optimality.

For values of K_s, d and k other than the nominal, it is not possible to obtain analytical expressions for the correlation functions. Rather, these are evaluated numerically from a state representation combining realizations for the expressions of (13), (14), (15), and (16).

Given a locus point for a value of K_s, i.e., an $\hat{\alpha}_1$ and $\hat{\beta}_1$ such that $r_{yy}(k+1)=r_{yu}(k+2)=0$, we generate a point for a neighboring value of K_s by searching for a $\Delta\hat{\alpha}_1$ and a $\Delta\hat{\beta}_1$ such that $f(\theta)$ is once again nulled. The expression in (12) is evaluated, the Jacobian being done by finite differencing, and the eigenvalues of the 2×2 matrix in (12) are evaluated.

Three loci were generated; no equilibrium points could be found for the case $d=1$, $k=0$. The loci are shown in Fig. 2. Figs. 3, 4, and 5 show the behavior of the real parts of the eigenvalues in (12) as K_s changes. In the cases $k=d=0$ and $k=1$, $d=0$, the eigenvalues were always real, and therefore Figs. 3 and 4 show two values for each K_s over the stable K_s region. Fig. 5, corresponding to $k=d=1$, exhibits two real eigenvalues, followed by a range of K_s where the eigenvalues are complex, followed by another region with two real eigenvalues. Table I indicates the stability regions in all four cases.

By contrast, the minimum-variance regulator (designed for $d=0$) has stability region of 0–550 for $d=0$ and 0–310 for $d=1$.

It is also possible to generate loci by fixing the process gain and varying d. This line of work was explored, but not pursued.

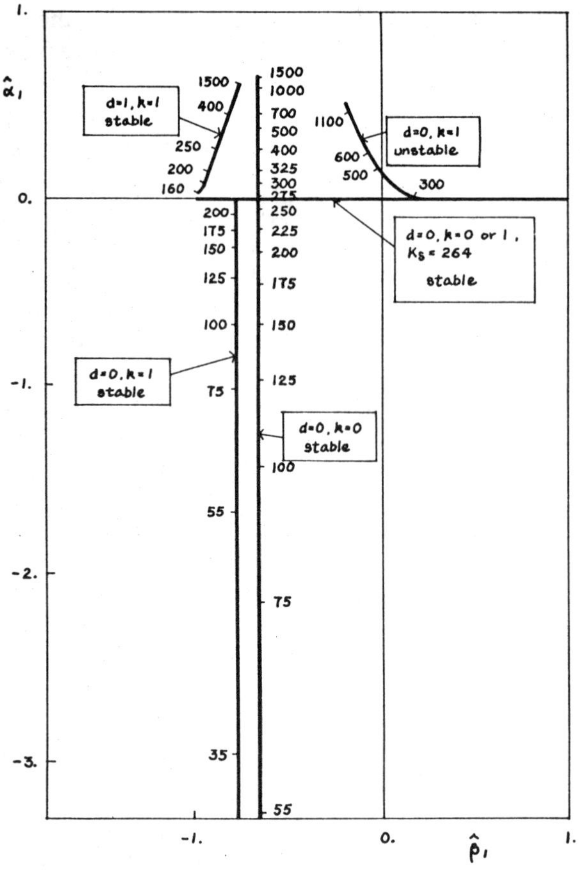

Fig. 2. Equilibrium loci of Ljung equations.

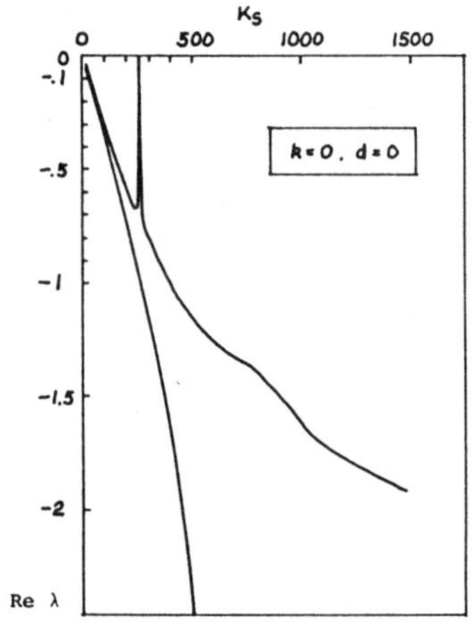

Fig. 3. Real parts of self-tuning stability matrix; $k=d=0$.

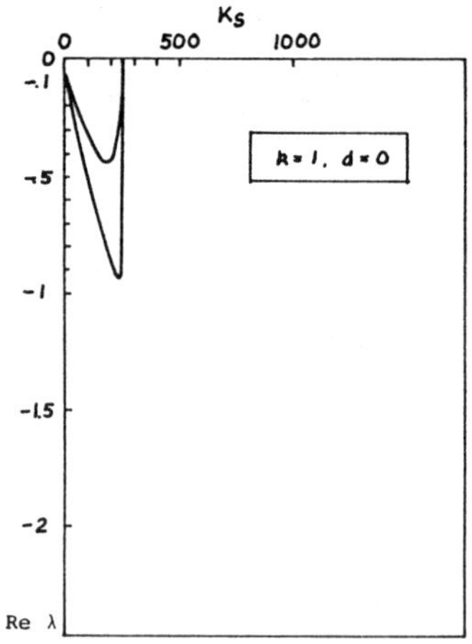

Fig. 4. Real parts of self-tuning stability matrix; $k=1$, $d=0$.

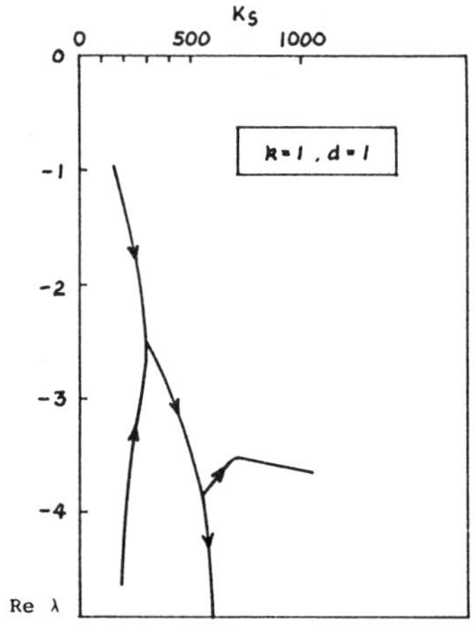

Fig. 5. Real parts of self-tuning stability matrix; $k=d=1$.

TABLE I

Process Delay d	Self-Tuning Delay k	Stability Region for Process Gain K_s
0	0	0 to 1550
0	1	0 to 264
1	0	no equilibrium points
1	1	160 to 1000

Fig. 6 illustrates the output variance obtained with the self-tuning controller (STC) and without, as K_s varies. The "optimal system" label refers to optimal regulators which use perfect knowledge of the plant. It is seen that the self-tuning scheme considerably reduces the sensitivity of the system.

V. Simulation

It is possible to study the average behavior of the self-tuning estimates by solving (9) and (10). However, in view of the virtual impossibility of obtaining closed-form expressions for $f(\theta)$, it was deemed easier simply to average a number of simulated runs. The system was simulated for $d=k=0$ and $K_s=500$, and the self-tuning regulator was allowed to run 40 times, each time with the same initial values of $\hat{\alpha}_1$ and $\hat{\beta}_1$. Fig. 7 shows the average over the 40 runs of the estimator $\hat{\alpha}_1$ and $\hat{\beta}_1$. To obtain Fig. 8, we computed at each time point of every run the stationary output variance that would result from using the current estimates; Fig. 8 is an average of those variance values over the 40 runs. For practical purposes, it appears that convergence takes about ten sampling periods, or five hours; that is considered satisfactory.

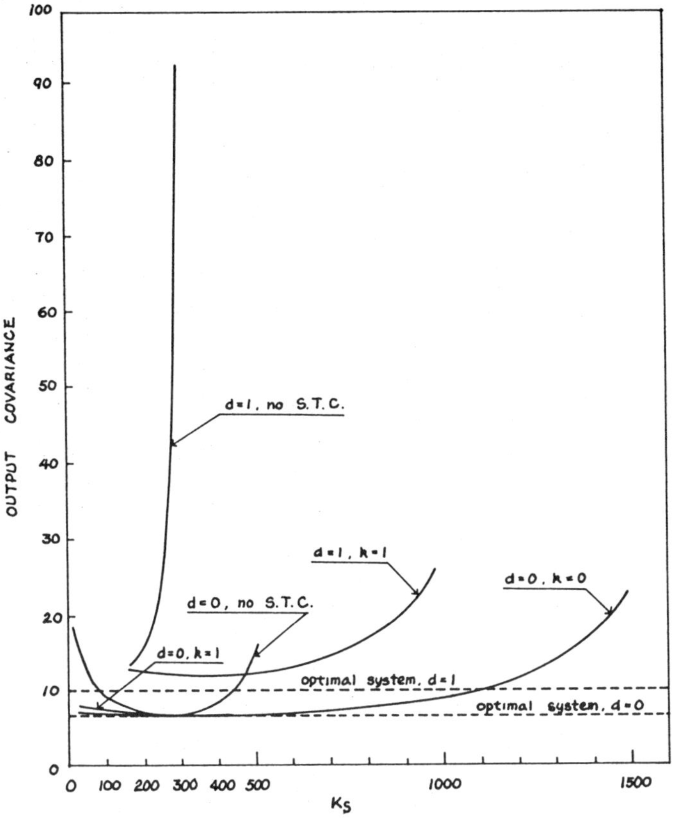

Fig. 6. Performance of self-tuning regulator.

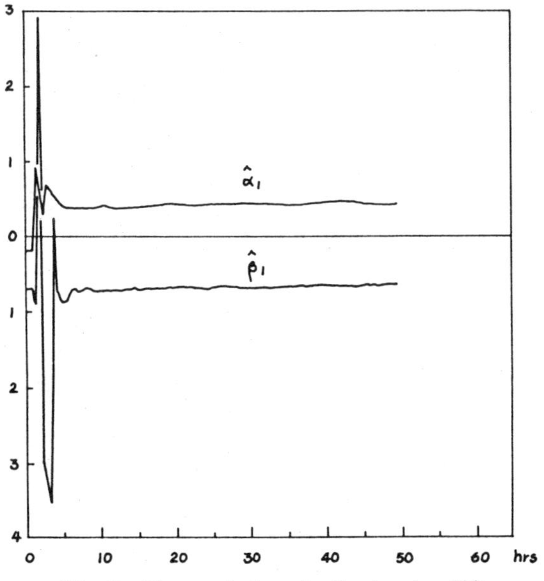

Fig. 7. Time evolution of estimates; $k_s = 500$.

VI. Experimental Results

The algorithm was implemented on an IBM7 computer. Memory required for the self-tuning regulator plus the minimum-variance control was 450 words. Limits of about ± 3 percent of normal steady-state fuel rate were placed on $u_t(t)$, the self-tuning input.

The kiln used for the experiment is a production unit located in Calais, France (kiln C in [1]).

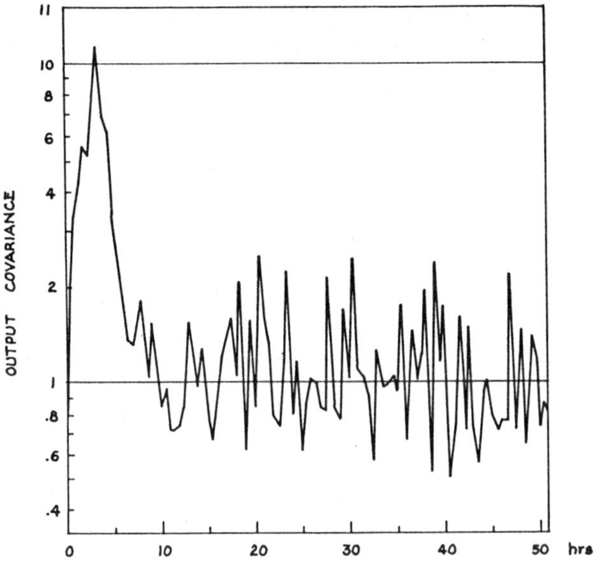

Fig. 8. Time evolution of output variance; $k_s = 500$.

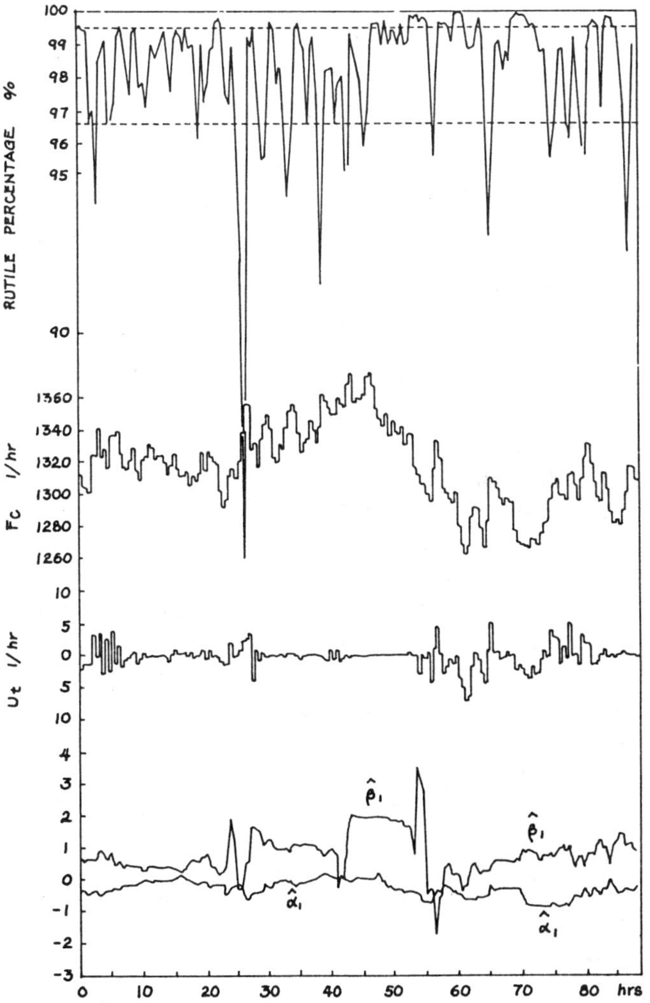

Fig. 9. Self-tuning control; experimental results.

A few experiments were performed with $\lambda = 1$, and the speed of convergence was as expected. It was found, however, that the parameters were subject to significant variations; accordingly, a discount factor, $\lambda = 0.90$, was finally selected. Fig. 9 shows records from a typical run,

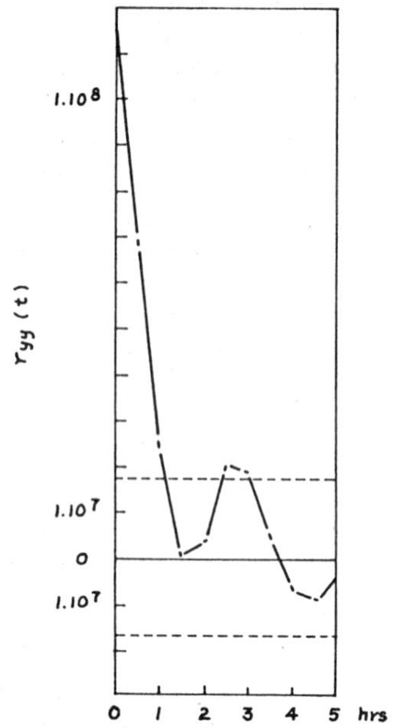

Fig. 10. Output autocorrelation function.

with $\beta_0 = 1600$. The large spikes in rutile percentage are believed to result from chemical changes affecting the reaction kinetics; the self-tuning regulator appears to respond by rapid changes in $\hat{\beta}_1$. Note that $|\hat{\beta}_1|$ is often greater than one. Fig. 10 shows the output correlation function; the dashed lines correspond to the 95 percent confidence interval for values theoretically equal to zero. The correlation function values for a shift of two sampling periods or more would be zero if the process were optimally controlled. The "in-spec" fraction of time achieved by self-tuning control was about 85 percent.

Given the success of those experiments, it was decided to attempt to use the self-tuning regulator to control the process through a grade change, i.e., through the transient period between the production of two different qualities of pigment. According to current practice, the new (chemically different) feed is introduced at $t=0$. The period from $t=18$ h to $t=24$ h has a mixed discharge, with the new grade being about completely achieved at $t=24$. From $t=10$ to $t=18$, one frequently observes an overcalcination period, followed until $t=24$ by a large rutile drop. Another five to six hours are usually required to enter the specification limits for the new product.

Because of the expected need for tracking changing parameters, the discount factor was lowered to $\lambda = 0.70$ at $t=0$. The immediate result was to cause instability in the algorithm; this was easily remedied by doubling β_0 to 3200. The new set point corresponding to the new grade was inserted at $t=21$. At $t=28$, steady-state having been achieved, the discount factor was increased to 0.95 and the sampling time increased to one hour. (The new grade is known to be much easier to control; the slower sampling rate is quite adequate and the parameters do not change as much as with the more difficult earlier grade.) Fig. 11 shows the history of this particular grade change. The old product, grade A, was held within specification up to $t=21$. After an "off-spec" period, inevitable in any case because of the mixed discharge, the kiln output was

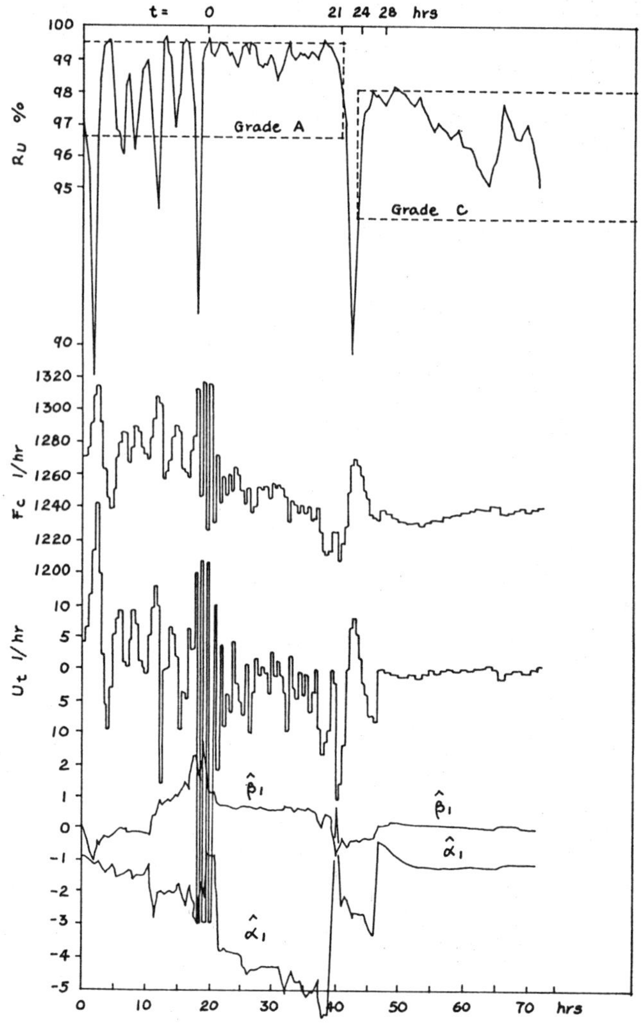

Fig. 11. Self-tuning control of a grade change.

within specifications and producing the new grade C at $t=24$. This grade change was by far the most successful performed in the plant.

VII. Conclusions

In addition to demonstrating once more the practical applicability of the self-tuning regulator, this paper has explored some novel features of the problem. The idea of using the self-tuning regulator to tune a plant already under fixed feedback control was useful, and allowed a reduction in the number of tuned parameters. The asymptotic study, with a plant model of order greater than that of the regulator, is believed to be new.

Finally, the self-tuning regulator proved capable of controlling the process through a grade change. This latter feature has not been explored theoretically to any great extent, but appears to be very promising in industrial practice.

Acknowledgment

The authors gratefully acknowledge the collaboration of Tioxide S.A., Calais, France.

REFERENCES

[1] G. Dumont and P. Bélanger, "Control of titanium dioxide kilns," this issue, pp. 521–531.

[2] V. Peterka, "Adaptive digital regulation of noisy systems," IFAC Symp. on Identification and Process Parameter Estimation, Prague, Czechoslovakia, 1970.

[3] K. Åström and B. Wittenmark, "On self-tuning regulators," *Automatica*, vol. 9, pp. 185–199, 1973.

[4] B. Wittenmark, "Self-tuning regulators," Lund Inst. of Technology, Div. of Automatic Control, Rep. 7311, Apr. 1973.

[5] L. Ljung and B. Wittenmark, "Asymptotic properties of self-tuning regulators," Rep. 7404, Lund Inst. of Technology, Div. of Automatic Control, 1974.

[6] T. Cegrell and T. Hedqvist, "Successful adaptive control of paper machines," *Automatica*, vol. 11, pp. 53–59, 1975.

[7] V. Borisson and R. Syding, "Self-tuning control of an ore crusher," *Automatica*, vol. 12, pp. 1–7, 1976.

[8] T. Soderström, L. Ljung, and I. Gustavsson, "A comparative study of recursive identification methods," Dept. of Automatic Control, Lund Inst. of Technology, Rep. 7427, 1974.

[9] K. Åström, *Introduction to Stochastic Control Theory*. New York: Academic, 1970.

Self-Tuning Adaptive Control of Cement Raw Material Blending*

LÁSZLÓ KEVICZKY,† JENÖ HETTHÉSSY,‡
MIKLÓS HILGER§ and JÁNOS KOLOSTORI||

A multivariable self-tuning regulator, and its modified version with a finite time criterion, provides accurate, efficient composition control of a cement material blending process.

Key Word Index—Cement industry; multivariable system; stochastic control; adaptive control.

Summary—A new, modified self-tuning (ST) minimum variance (MV) regulator algorithm developed for multiple input multiple output (MIMO) systems is presented with required average for finite time (RAFT). This control strategy has been developed for the self-tuning optimal control of cement raw material blending. After a brief survey of the technological background the algorithm of MIMO-ST-MV-RAFT control strategy is derived. A self-tuning blending control system is described with real-time experimental results illustrating the efficiency of the new multivariable regulator.

1. INTRODUCTION

THE QUICK pace of the social progress and the large-scale extension of the volume of industrial and social investments are responsible for the substantial growth in the building industry. This also applies to the silicate industry including the cement production. The size of cement production and consumption has been used for some time to characterize the state of development of a country. Besides the increase of the production, the reduction of costs and the improvement of the quality are also important objectives. The solutions may be the modernization of the technological methods and equipment as well as the increase of their efficiency on the one hand, and the introduction of the computer process control to a greater extent on the other. The automation of the cement plants was begun in the sixties and now there are centralized control rooms in almost all of the newly built factories which very often have a process computer.

The computer control is mostly applied for the blending control. The control of the chemical composition of the ground mix of the raw materials and its homogenization before feeding it into the kiln is a very essential problem in cement manufacture.

The aim of the blending control system is to produce a full silo of kiln feed at the desired chemical composition with minimum variation through the silo filling. The control problem arises from the fact that the chemical compositions of the various raw materials vary from time to time and they are not measured. Furthermore, each of the raw materials contains varying amounts of the constituent oxides. Only annual or monthly average values of raw material compositions are available, and it is the fluctuation of the actual raw material compositions about these long term average compositions which introduces disturbances to be eliminated by computer control.

So far some simple and more difficult closed control systems with multiple loops have been developed for direct digital control of the raw material blending[1–5]. The plant is multivariable and a coupled one because the feeder tanks do not contain chemically homogeneous raw materials. The time delays in the system are considerable in spite of X-ray fluorescence technique applied in recent years for analysing the chemical composition of raw meal. The optimum adjustment of the parameters of the regulators, which are usually designed in cascade loops with prediction, is a very difficult problem. Instead of a comprehensive matrix controller regulating the interactive system, noninteractive multiple PI regulators are generally used. If the factory is supported by quarries in which the components of the raw materials and their physical character-

*Received February 17, 1977; revised September 19, 1977; revised April 27, 1978. The original version of this paper was presented at the IFAC/IFIP Symposium on Software for Computer Control (SOCOCO) which was held in Tallinn, Estonian Republic USSR during May 1976. Information about the published Proceedings of this IFAC Meeting may be obtained from the IFAC Secretariat: Mrs. S. Saari, % EKONO, P.O.B. 27. SF-00131 Helsinki 13, Finland. This paper was recommended for publication in revised form by associate editor K. J. Åström.

†Department of Automation, Technical University of Budapest, H-1111, Hungary, Budapest, Goldmann György tér 3.

‡Institute for Electric Power Research, H-1051, Hungary, Budapest, Zrinyi u 1.

§Institute for Research and Planning in Silicate Industry, H-1300, Hungary, Budapest, Bécsi u 126.

||Cement and Lime Works, H-2601, Hungary, Vác.

Reprinted with permission from *Automatica*, vol. 14, pp. 525–532, Nov. 1978.
Copyright © 1978, International Federation of Automatic Control IFAC. Published by Pergamon Press Ltd.

istics (e.g. grindability) change considerably, it is necessary to readjust the regulator parameters. The best solution for this problem would be a multivariable regulator that adjusts itself to perform optimally while requiring little *a priori* information about the system. For this reason on the basis of the available results of stochastic control theory we have extended the algorithm of self-tuning regulators for single output systems to multiple output systems and have developed a modified version considering the special semi-batch operation of cement silos.

In Hungary the first cement factory with an on-line blending control by classical DDC (set-point) control algorithm has been built using an IBM SYSTEM-7. A good opportunity to test the above mentioned strategies was provided by this well equipped modern factory. The development of the software of self-tuning (or adaptive) on-line computer control of raw material blending has been made according to the following research project steps:

(1) Elaboration of a simulation language of very high level for testing new control and identification algorithms[6, 7].

(2) Development of self-tuning (ST) minimum variance (MV) regulators for multiple input multiple output (MIMO) systems[7, 8] considering a required average for finite time (RAFT) criterion, too.

(3) Testing the DDC of raw material blending by MIMO-ST-MV-RAFT algorithms using simulation with the model of the plant[9].

(4) Coding the algorithm into a programmable desk-top calculator (microprocessor) and testing it in real-time[10].

(5) Making the software of the on-line self-tuning blending control ready for the process computer.

This paper was written after point four, after having had many successful trials with the new methods and working on the design of new system software.

2. MINIMUM VARIANCE SELF-TUNING CONTROL OF MIMO SYSTEMS (MIMO-ST-MV)

Let us consider the multivariable extension of the Åström-model [8, 11–13]:

$$\mathbf{A}(z^{-1})\mathbf{y}(t)=\mathbf{B}(z^{-1})\mathbf{u}(t-d) + \mathbf{C}(z^{-1})\mathbf{\Lambda}\mathbf{e}(t) \quad (t=0,1,2,\ldots) \qquad (1)$$

where $\mathbf{y}$ is a q-dimensional output vector, $\mathbf{u}$ is a q-dimensional control vector, $\mathbf{A}$, $\mathbf{B}$ and $\mathbf{C}$ are polynomial matrices of the backward shift operator z^{-1}:

$$\mathbf{A}(z^{-1})=\mathbf{I}_q+\sum_{j=1}^{n} z^{-j}\mathbf{A}_j$$

$$\mathbf{B}(z^{-1})=\sum_{j=0}^{m} z^{-j}\mathbf{B}_j$$

$$\mathbf{C}(z^{-1})=\mathbf{I}_q+\sum_{j=1}^{k} z^{-j}\mathbf{C}_j,$$

further d is the time delay of the process to be controlled, $\mathbf{e}$ is a sequence of normally distributed independent vector variables with zero mean value and covariance matrix $\mathbf{I}_q$, $\mathbf{\Lambda}$ is the gain matrix of $\mathbf{e}$.

The aim of the minimum variance control is to minimize the loss function

$$V_1=\min_{\mathbf{u}(t)} E\{[\mathbf{y}(t+d)-\mathbf{y}_r]^{\mathrm{T}}[\mathbf{y}(t+d)-\mathbf{y}_r]\}, \qquad (2)$$

where $\mathbf{y}_r$ is the required reference value of the output, $E\{\ldots\}$ stands for mathematical expectation, and T denotes transposition.

If the system parameters are not known, following the self-tuning principle[14] it can be shown[8, 13] that the parameters of the minimum variance regulator can be estimated from the prediction equation

$$\mathbf{y}(t+d)=\mathbf{P}\mathbf{x}(t)+\boldsymbol{\varepsilon}(t+d), \qquad (3)$$

where $\boldsymbol{\varepsilon}$ is an independent residual,

$$\mathbf{x}^{\mathrm{T}}(t)=[\mathbf{u}^{\mathrm{T}}(t),\mathbf{u}^{\mathrm{T}}(t-1),\ldots,\mathbf{u}^{\mathrm{T}}(t-m-d+1),\mathbf{y}^{\mathrm{T}}(t),\mathbf{y}^{\mathrm{T}}(t-1),\ldots,\mathbf{y}^{\mathrm{T}}(t-n+1)]$$

is an $nq+mq+dq$ dimensional vector of observations, and further

$$\mathbf{P}=[\mathbf{P}_0\mathbf{P}_1\ldots\mathbf{P}_{m+d-1}\mathbf{P}_{m+d}\ldots\mathbf{P}_{n+m+d-1}]$$

is a $q(nq+mq+dq)$ dimensional parameter matrix of the regulator. From (3) the recursive least squares estimation of the regulator parameters is given by

$$\hat{\mathbf{P}}_t=\hat{\mathbf{P}}_{t-1}+[\mathbf{y}(t)-\hat{\mathbf{P}}_{t-1}\mathbf{x}(t-d)]\times\mathbf{x}^{\mathrm{T}}(t-d)\mathbf{R}_t, \qquad (4)$$

where

$$\mathbf{R}_t = \frac{1}{w^2} \times \left\{ \mathbf{R}_{t-1} - \frac{[\mathbf{R}_{t-1}\mathbf{x}(t-d)][\mathbf{R}_{t-1}\mathbf{x}(t-d)]^{\mathrm{T}}}{w^2 + \mathbf{x}^{\mathrm{T}}(t-d)\mathbf{R}_{t-1}\mathbf{x}(t-d)} \right\} \quad (5)$$

is the optimal weighting matrix with a forgetting factor $w \leq 1$.

Having obtained $\hat{\mathbf{P}}_t$, the d step minimum variance prediction for $\mathbf{y}(t+d)$ is given by

$$\hat{\mathbf{y}}(t+d \mid t) = \hat{\mathbf{P}}_t \mathbf{x}(t). \quad (6)$$

Combining (6) with

$$\hat{\mathbf{y}}(t+d \mid t) = \mathbf{y}_r \quad (7)$$

the optimal control strategy minimizing the loss function V_1 is

$$\mathbf{u}^0(t) = \hat{\mathbf{P}}_{0t}^{-1}[\mathbf{y}_r - \hat{\tilde{\mathbf{P}}}_t \tilde{\mathbf{x}}(t)], \quad (8)$$

where

$$\hat{\mathbf{P}}_t = [\hat{\mathbf{P}}_{0t}, \hat{\tilde{\mathbf{P}}}_t]$$

and

$$\mathbf{x}^{\mathrm{T}}(t) = [\mathbf{u}^{\mathrm{T}}(t), \tilde{\mathbf{x}}^{\mathrm{T}}(t)].$$

It is easy to see from (8) that uncertainties in the parameter estimation are not taken into consideration.

There are some remarks in connection with the MIMO-ST-MV algorithm given by (4), (5) and (8).

In case of $\mathbf{y}_r = \mathbf{0}$, one of the values

$$\mathbf{P}_i (i = 0, 1, \ldots, n+m+d-1)$$

usually $\mathbf{P}_0$, can be fixed[12] and the optimal control is

$$\mathbf{u}^0(t) = -\hat{\tilde{\mathbf{P}}}_t \tilde{\mathbf{x}}(t) \quad \text{if} \quad \mathbf{P}_0 = \mathbf{I}_q.$$

Denoting the jth row of $\hat{\tilde{\mathbf{P}}}_t$ by $\hat{\tilde{\mathbf{p}}}_{jt}^{\mathrm{T}} (j = 1, 2, \ldots, q)$, the jth control signal at t is

$$\mathbf{u}_j^0(t) = -\hat{\tilde{\mathbf{p}}}_{jt}^{\mathrm{T}} \tilde{\mathbf{x}}(t)$$

which can also be considered as a SISO-ST-MV algorithm having a feedforward control constructed from the observations of the components $l = 1, \ldots, j-1, j+1, \ldots, q$. [12]. Simulation results show that the control algorithm given by (4), (5) and (8) ensures nice convergence properties and has high performance [8, 10, 13].

3. MODIFIED MINIMUM VARIANCE SELF-TUNING STRATEGY

Due to the technological requirements in order to achieve a good control of cement raw material blending, it is necessary, not only to minimize the variance of the output signal, but it is very desirable to keep its weighted average for finite N

$$\mathbf{y}_a(N) = \frac{\sum_{i=1}^{N} h_i \mathbf{y}(i)}{\sum_{i=1}^{N} h_i} \quad (9)$$

as close to the reference value $\mathbf{y}_r$ as possible. In (9) h_i stands for the quantity of the ground raw material during the ith interval. To satisfy this demand a self-tuning strategy using a time varying reference value has been developed. As it will be shown, this time varying reference value $(\mathbf{y}_{rv})$ differs only slightly from $\mathbf{y}_r$ and it is to compensate the deviations of

$$\mathbf{y}_a(K) = \sum_{i=1}^{K} h_i \mathbf{y}(i) \Big/ \sum_{i=1}^{K} h_i \quad (K = 1, 2, \ldots, N)$$

from the required $\mathbf{y}_r$ for the whole period $(K = 1, 2, \ldots, N)$. First consider the case of $d = 1$. At the Kth instant

$$\mathbf{y}(K), \mathbf{y}(K-1), \ldots, \mathbf{y}_a(K), \mathbf{y}_a(K-1), \ldots$$

and $\mathbf{u}(K-1)$, $\mathbf{u}(K-2), \ldots$ are known and the task is to determine $\mathbf{u}(K)$. Having a one-step-ahead control policy $\mathbf{u}(K)$ controls only $\hat{\mathbf{y}}(K+1 \mid K)$, but to predict $\hat{\mathbf{y}}_a(N \mid K)$ we have no other choice than to compute with $\hat{\mathbf{y}}(K+1 \mid K)$ and h_k for $i = K+1$, $K+2, \ldots, N$. This latter means that h_i over the remaining intervals $(i = K+1, K+2, \ldots, N)$ is assumed to be the same as h_K. Thus we obtain a 'useful' prediction for $\mathbf{y}_a(N)$ as

$$\hat{\mathbf{y}}_a(N \mid K) = \frac{\mathbf{y}_a(K) \sum_{i=1}^{K} h_i + (N-K) h_K \hat{\mathbf{y}}(K+1 \mid K)}{\sum_{i=1}^{K} h_i + (N-K) h_K}, \quad (10)$$

and to ensure

$$\hat{\mathbf{y}}_a(N \mid K) = \mathbf{y}_r$$

the following value is required for $\hat{\mathbf{y}}(K+1 \mid K)$

$$\hat{\mathbf{y}}(K+1 \mid K) = \frac{\mathbf{y}_r \left[\sum_{i=1}^{K} h_i + (N-K) h_K \right]}{(N-K) h_K} - \frac{\mathbf{y}_a(K) \sum_{i=1}^{K} h_i}{(N-K) h_K}, \quad (11)$$

where the right side can be considered as $\mathbf{y}_{rv}(K+1)$, that is a time varying reference value.

For the sake of simplicity let us have

$$h_1=h_2=\ldots=h_N \tag{12}$$

then

$$\hat{\mathbf{y}}(K+1\,|\,K)=\frac{N}{N-K}\mathbf{y}_r-\frac{K}{N-K}\mathbf{y}_a(K)=\mathbf{y}_{rv}(K+1)$$

or

$$\mathbf{y}_{rv}(K+1)=\mathbf{y}_r-\frac{K}{N-K}[\mathbf{y}_a(K)-\mathbf{y}_r]$$

is given for $\mathbf{y}_{rv}(K+1)$. From this latter form it clearly follows that the term modifying $\mathbf{y}_r$ will have extreme importance only if $K>N-K$ and/or $\mathbf{y}_a(K)$ considerably differs from $\mathbf{y}_r$.

In this way the modified self-tuning algorithm will be based on (4), (5) as well as

$$\mathbf{u}(t)=\hat{\mathbf{P}}_{0t}^{-1}[\mathbf{y}_{rm}-\hat{\mathbf{P}}_t\tilde{\mathbf{x}}(t)], \tag{13}$$

where

$\mathbf{y}_{rm}=\mathbf{y}_{rv}(t+1)$ for $t=1,2,\ldots,N-1$;

and

$$\mathbf{y}_{rm}=\mathbf{y}_r \quad \text{for} \quad t=N.$$

This means that t runs periodically ($t=1, 2,\ldots, N, 1, 2,\ldots, N, 1, 2,\ldots$) and N is determined by the time necessary to fill the silo. If $d>1$ then at the Kth instant $\hat{\mathbf{y}}_a(K+d-1\,|\,K)$ is determined as follows, supposing that (12) still holds:

$$\hat{\mathbf{y}}_a(K+d-1\,|\,K)=\frac{K\mathbf{y}_a(K)+\sum_{i=K+1}^{K+d-1}\mathbf{y}_{rv}(i)}{K+d-1},$$

thus (10) turns into

$$\hat{\mathbf{y}}_a(N\,|\,K)=\frac{(K+d-1)\hat{\mathbf{y}}_a(K+d-1\,|\,K)+(N+1-K-d)\hat{\mathbf{y}}(K+d\,|\,K)}{N}$$

and we obtain

$$\mathbf{y}_{rv}(K+d)=\frac{N}{N-K-d+1}\mathbf{y}_r-\frac{K+d-1}{N-K-d+1}\hat{\mathbf{y}}_a(K+d-1\,|\,K).$$

The control law is given by (13), where

$$\mathbf{y}_{rm}=\mathbf{y}_{rv}(t+d) \quad \text{for } t=1,2,\ldots,N-d;$$

and

$$\mathbf{y}_{rm}=\mathbf{y}_r \quad \text{for } t=N-d+1,\ldots,N.$$

In summary, a modified multivariable minimum variance self-tuning regulator has been developed to ensure a required average for finite time (MIMO-ST-MV-RAFT). However the control law by (8) implies that $\mathbf{y}_r$=const., simulation results and practical applications show that the MIMO-ST-MV-RAFT algorithm using time varying reference value decreases the loss

$$E\{[\mathbf{y}_a(N)-\mathbf{y}_r]^{\mathrm{T}}[\mathbf{y}_a(N)-\mathbf{y}_r]\}$$

significantly. This feature is very advantageous from technological point of view.

Now some simulation results will be presented to illustrate the difference between the MIMO-ST-MV and MIMO-ST-MV-RAFT strategies.

Example. Consider the following second order system with $d=1$ and $q=2$ (the values come for [14] by the generalization of its SISO example):

$$n=2 \quad \mathbf{A}_1=\begin{bmatrix}-1.5 & 0.3\\ 0.2 & -1.5\end{bmatrix} \quad \mathbf{A}_2=\begin{bmatrix}0.54 & -0.1\\ 0.1 & 0.56\end{bmatrix}$$

$$m=1 \quad \mathbf{B}_0=\begin{bmatrix}2.0 & -0.3\\ 0.1 & 1.0\end{bmatrix} \quad \mathbf{B}_1=\begin{bmatrix}-1.8 & 0.2\\ -0.2 & -0.5\end{bmatrix}$$

$$k=2 \quad \mathbf{C}_1=\begin{bmatrix}0.2 & 0.1\\ -0.1 & 0.2\end{bmatrix} \quad \mathbf{C}_2=\begin{bmatrix}-0.48 & -0.2\\ 0.2 & -0.24\end{bmatrix}$$

$$\mathbf{\Lambda}=\begin{bmatrix}\lambda_1 & 0.0\\ 0.0 & \lambda_2\end{bmatrix} \quad \begin{matrix}\lambda_1=1.0\\ \lambda_2=1.0\end{matrix} \quad \mathbf{y}_r=\begin{bmatrix}1.0\\ 1.0\end{bmatrix}$$

To compare the MIMO-ST-MV and MIMO-ST-MV-RAFT algorithms, simulation results for this system will be presented with $N=30$. Figures 1 and 2 show 10 periods each for y_1 and y_2. Notice that y_i has been obtained by using MIMO-ST-MV strategy, while y_i^* by using MIMO-ST-MV-RAFT strategy. Below the values y_i and y_i^*, their averages

$$y_a(t)=\frac{1}{t}\sum_{i=1}^{t}y(i)$$

$$(t=1,2,\ldots,N,1,\ldots,N,1,\ldots)$$

$$y_a^*(t)=\frac{1}{t}\sum_{i=1}^{t}y^*(i)$$

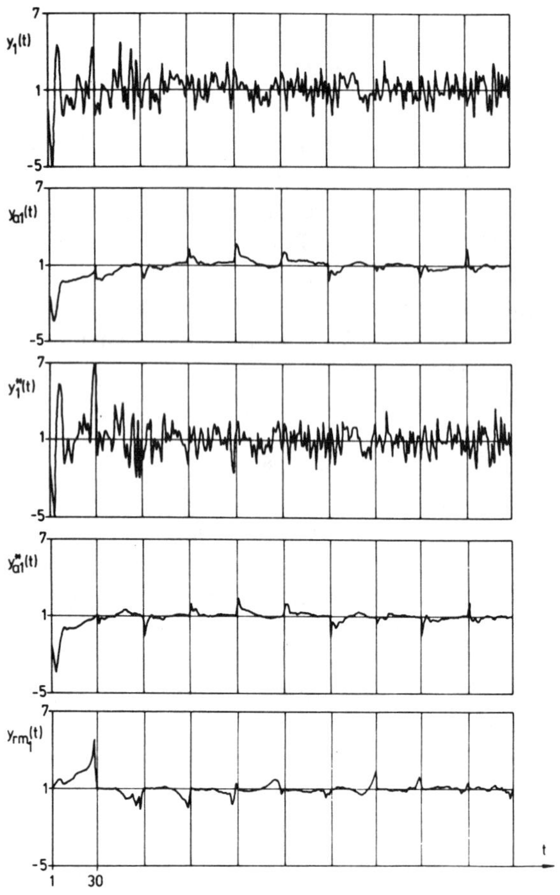

FIG. 1. Output y_1 and its average using MIMO-ST-MV and MIMO-ST-MV-RAFT (*) control law. y_{rm_1} represents the modified reference value computed by MIMO-ST-MV-RAFT strategy.

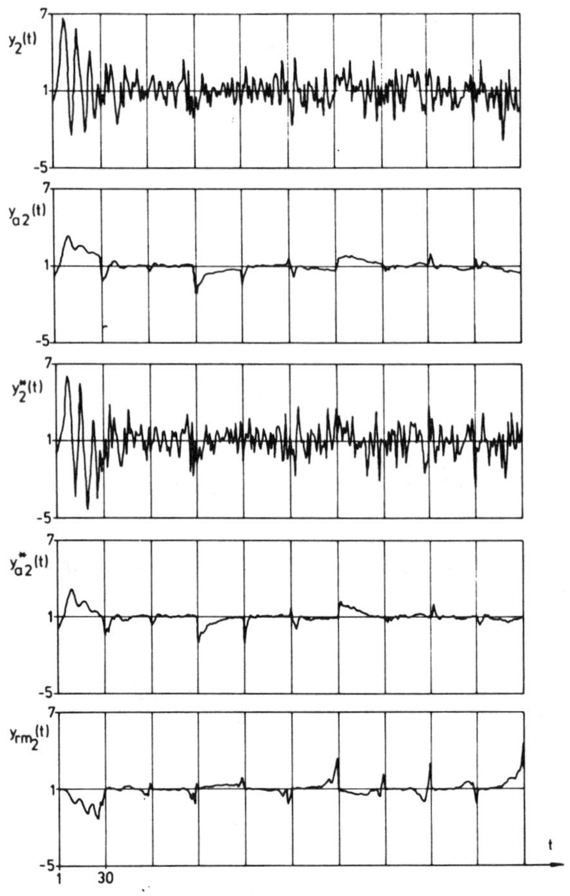

FIG. 2. Output y_2 and its average using MIMO-ST-MV and MIMO-ST-MV-RAFT (*) control law. y_{rm_2} represents the modified value computed by MIMO-ST-MV-RAFT strategy.

can be seen. Finally y_{rm_i} represents the modified reference value used by the MIMO-ST-MV-RAFT policy. It is easy to see that the method intends to ensure the proper average for N samples by properly changing y_{rm_i}. To show the difference between the two algorithms

$$y_a(iN) \quad (i=1,2,\ldots,10)$$

are separately drawn in Fig. 3. Their performances can be characterized numerically by the losses

$$v_{a_j}=\sum_{i=1}^{10}[y_{a_j}(iN)-y_{r_j}]^2 \quad (j=1,2).$$

In this example these values are

$$v_{a_1}=0.4750 \quad v^*_{a_1}=0.0181$$
$$v_{a_2}=1.2702 \quad v^*_{a_2}=0.0170$$

Notice that for $d=2$ these values are changed into

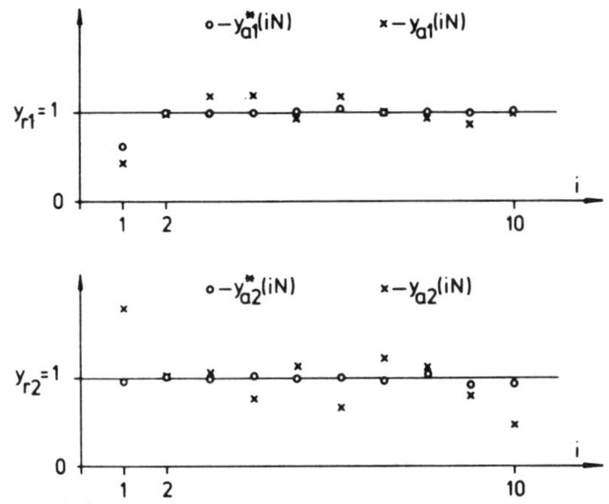

FIG. 3. Averages for N samples in 10 periods using MIMO-ST-MV and MIMO-ST-MV-RAFT (*) algorithm.

$$v_{a_1}=3.9497 \quad v^*_{a_1}=0.1589$$
$$v_{a_2}=5.3525 \quad v^*_{a_2}=0.4164$$

4. COMPUTER CONTROL OF RAW MATERIAL BLENDING

The simple technological scheme of the blending control is shown in Fig. 4. In the feeder tanks

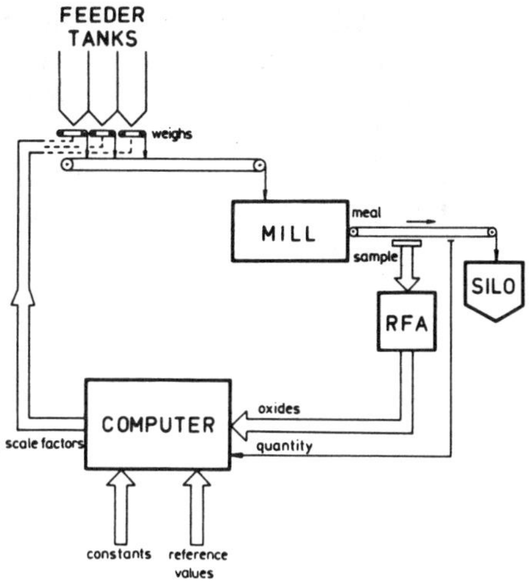

FIG. 4. Simple technological scheme of the blending control.

there are raw materials of different composition. The weigh feeders are controlled by computer. The rubble is fed into the mill by a conveyor belt. Afterwards the ground mix of raw materials (rawmeal) is placed into the silo. Before the silo the meal is sampled and the sample is fed into an X-ray Fluorescence Analyser (RFA). This equipment ensures the analysed oxide values for the computer. The primary purpose of the blending control is to reduce the feed composition disturbances to the kiln and improve the quality control.Every full silo content is homogenized to decrease the composition variations around the average values. The oxide compositions obtained by the X-ray analyser are provided for the computer which controls all this automatic sampling, conveying, preparation and analysing process and computes the new set points (scale factors) for the weigh feeders via the so-called moduli values. From the four most important oxides(C – CaO: S – SiO_2: A – Al_2O_3; F – Fe_2O_3) the following three moduli are computed by the computer in the factory where the experiments have been made:

(1) Lime standard (ML):

$$\mathrm{ML} = \frac{1000\,\mathrm{C}}{2.8\mathrm{S} + 1.1\mathrm{A} + 0.8\mathrm{F}}$$

(2) Aluminium modulus (MA):

$$\mathrm{MA} = \frac{\mathrm{A}}{\mathrm{F}}$$

(3) Silica modulus (MS):

$$\mathrm{MS} = \frac{\mathrm{S}}{\mathrm{A} + \mathrm{F}}$$

In the factory only two moduli can be controlled, since there are only three raw materials (lime stone, iron ore, clay); selecting values for the first two equations results in the value for equation (3) being determined.

Let us consider the output vector of the blending system for a possible process control, the vector of controlled variables being

$$\mathbf{y}(t) = [\mathrm{ML}(t), \mathrm{MA}(t)]^{\mathrm{T}} \tag{14}$$

formed from the analysed chemical composition of the rawmeal before the silo. Because the h_i silo feed during the ith interval is known, the integrated (or weighted average) silo composition $\mathbf{y}_a(t)$ can be computed by

$$\mathbf{y}_a(t) = \sum_{i=1}^{t} h_i \mathbf{y}(t) \Big/ \sum_{i=1}^{t} h_i; \qquad t = 1, 2, \ldots, N$$

where $\sum_{i=1}^{t} h_i$ is the actual content of the silo. The general objective is to control the average composition $\mathbf{y}_a(t)$ of a full silo according to the chemist's reference values and control the instantaneous mill output composition $\mathbf{y}(t)$ as closely as possible to the reference vector, but still allowing correction for the unavoidable past deviations. This combination of silo and mill composition control provides the desired kiln feed composition and minimizes blending (homogenizing) requirements. Thus at the end of silo feeding the average $\mathbf{y}_a(t)$ composition must reach the given reference value $\mathbf{y}_r$ at the Nth interval and the variation of $\mathbf{y}(t)$ around the $\mathbf{y}_r$ has to be minimized. One can recognize that the developed MIMO-ST-MV-RAFT regulator meets these requirements.

The adaptive control can be performed in two different ways. At the first approach on the basis of the measured moduli values the MIMO-ST-MV-RAFT regulator computes the so-called desired fictitious module values to be adjusted as optimal $\mathbf{u}^0(t)$ and the scale factors for the weigh feeders, as direct interventions, are obtained from the fictitious module values on the basis of an average composition matrix by solving an equation system[1]. This solution best fits the control systems previously used because actually only the PI regulator is replaced by a self-tuning one.

Since the system is overdetermined with respect to the module input/output, two moduli can be controlled on the basis of two measured moduli values. (It has to be mentioned that in principle it is possible to control all of three moduli and in this case a modified version of the ST algorithm is used with the restriction that

$$u_1(t) + u_2(t) + u_3(t) = 1.)$$

We have chosen the version, as a final solution,

which actually uses the self-tuning operation and does not need the average composition matrix, and the algorithm adjusts directly the scale factors of the weigh feeders on the basis of the moduli values computed from the measured oxides of the mill output and the quantity of the ground rawmeal fed to the silo.

The control strategy is illustrated in Fig. 5. Thus the vector of the controlled variables is according to (14), i.e. two dimensional, while the output vector of the MIMO-ST-MV-RAFT regulator is

$$\mathbf{u}(t)=[u_1(t), u_2(t)]^T.$$

Here u_1, u_2 mean the scale factors of lime and pyrites weigh feeders in percentage, respectively, while for the clay weigh

$$u_3(t)=1-u_1(t)-u_2(t)$$

holds. (Since there was no case when $0 \leqq u_i(t) \leqq 1$ would not be fulfilled we did not deal with the usual further optimization problem[3].) Thus, the applied control strategy corresponds to (13), where the modified reference value is computed from the integrated silo moduli.

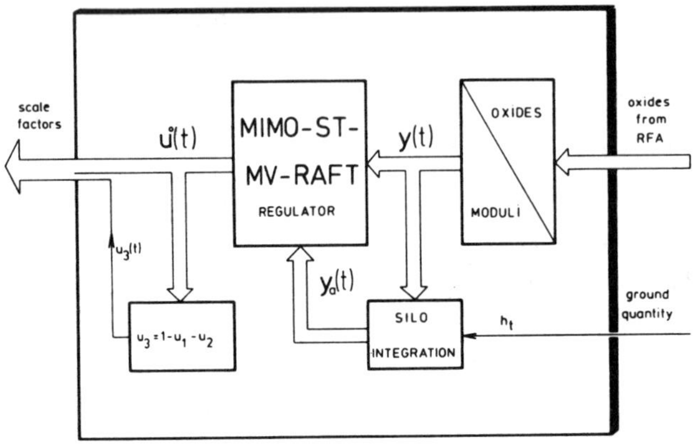

FIG. 5. Control strategy used in the real time experiment.

5. A REAL TIME EXPERIMENT

The self-tuning optimal blending control system was tested using the IBM SYSTEM-7 computer with original on-line connection cut and a desk-top calculator (microprocessor) inserted. Since the sampling time was half an hour the available time enabled a man to be in the control loop. The IBM SYSTEM-7 controlled the RFA, computed the moduli and integrated values, but the set points for the scale factors of weigh feeders were computed by the calculator.

A Hungarian made desk-top calculator EMG-666 was used, whose main facilities: microprocessor; 64kbit ROM (wired in); 8kbyte MOS-RAM (8000 program steps or 1008 data); 1 msec average operation time; cassette tape back memory.

In the real-time experiment the following structure of MIMO-ST-MV-RAFT regulator was chosen

$$d=2; n=2; m=1$$

with the initial values

$$\hat{\mathbf{P}}_0=0 \quad \hat{\mathbf{P}}_{0,0}=5\mathbf{I}_q \quad \mathbf{R}_0=100\,\mathbf{I}$$

and further a forgetting factor of $w=0.98$ was used.

Figure 6 shows the results of the real-time experiment using

$$\mathbf{y}_r=[\mathrm{ML}_r, \mathrm{MA}_r]^T=[94.0, 1.5]^T$$

as reference values. The figure shows the feeding of 6 silos. The first three were fed by the conventional DDC, the last three by the MIMO-ST-MV-RAFT regulator. The efficiency of the regulator is easy to see. The following average loss per step values were obtained with the conventional (c) and MIMO-ST-MV-RAFT (*) control:

$$v^c_{\mathrm{ML}}=26.88 \qquad v^*_{\mathrm{ML}}=11.59$$

$$v^c_{\mathrm{MA}}=9.8310^{-2} \qquad v^*_{\mathrm{MA}}=4.5210^{-2}$$

6. CONCLUSIONS

On the basis of the available results of stochastic control theory via some extensions and modifications an adaptive control strategy has been developed for the optimum control of raw material blending in a cement factory. The algorithm has very nice properties and advantages comparing favourably with the classical DDC systems generally applied for the same problem

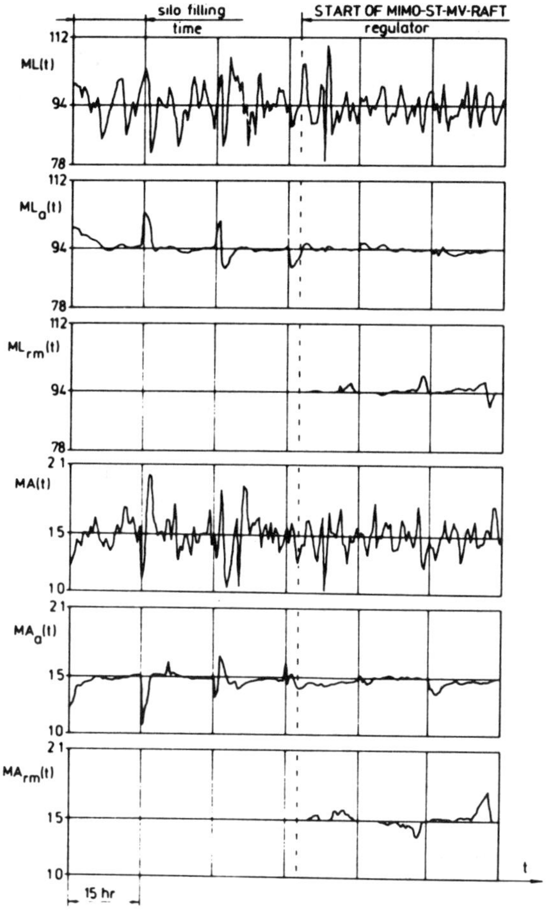

FIG. 6. Results from the real time experiment. The first three silos were fed by the conventional DDC, while the last three by the MIMO-ST-MV-RAFT regulator.

of chemical composition control. Our approach for demonstrating the goodness and efficiency of the new method with the application of a microcomputer has been proved to a good solution to make industrial companies apply the new results of control theory in reasonable and economic form.

REFERENCES

[1] H. HAMMER: Computer controlled rawmeal production in cement industry. *Regelungstechnik und Prozessdatenverarbeitung* **5**, 190–198 (1972).

[2] H. HOENIG: Component control in a cement plant using a process computer. *Zement-Kalk-Gips* **1**, 31–36 (1972).

[3] A. LUNDAN and O. MATTILA: A system for the control of the homogenization of the cement raw mill, Internal Report, Outokumpu Oy, Institute of Physics, Tapiola, Finland (1974).

[4] J. TEUTENBERG: Prozessautomation einer zementfabrik in argentinien. *Zement-Kalk-Gips* **4**, 141–151 (1971).

[5] S. C. K. YOUNG: Raw material blending—a multivariable control problem. *4th UKAC Control Convention on Multivariable Control*, Manchester (1971).

[6] L. KEVICZKY, I. VAJK and Cs. BÁNYÁSZ: On application of PROCAL in simulation of identification and control algorithms. *SIMULATION '75*, Zürich (1975).

[7] I. VAJK, L. KEVICZKY and Cs. BÁNYÁSZ: PROCAL in the service of control engineering. *AICA Symposium on Simulation Languages for Dynamic Systems*, London (1975).

[8] L. KEVICZKY and J. HETTHÉSSY: Minimum variance control of multivariable linear discrete time systems (in Hungarian). *Elektrotechnika* **5**, 170–175 (1976).

[9] M. HILGER, R. HABER and L. KEVICZKY: DDC of raw material blending, simulation investigations. *United Kingdom Simulation Conference*, Bowness-on-Windermere (1975).

[10] L. KEVICZKY, J. HETTHÉSSY, M. HILGER and J. KOLOSTORI: Self-tuning computer control of cement and raw material blending. *1st IFAC/IFIP Symposium on Software for Computer Control*, Tallin (1976).

[11] K. J. ÅSTRÖM: *Introduction to Stochastic Control Theory*. Academic Press, New York (1970).

[12] U. BORISSON: Self-tuning regulators—Industrial applications and multivariable theory. Report 7513 Division of Automatic Control, Lund Institute of Technology (1975).

[13] L. KEVICZKY and J. HETTHÉSSY: Self-tuning minimum variance control of MIMO discrete time systems. *Aut. Control*, **5** (1), January 1977.

[14] V. PETERKA: Adaptive digital control of noisy systems. *IFAC Symposium on Identification and Process Parameter Estimation*, Prague (1970).

[15] K. J. ÅSTRÖM and B. WITTENMARK: On self-tuning regulators. *Automatica* **9**, 185–199 (1973).

[16] J. HETTHÉSSY and L. KEVICZKY: Extension of the minimum variance control for the case of observable noise. *4th ICEE on Control and System Theory*, Shiraz (1974).

[17] J. HETTHÉSSY and L. KEVICZKY: Some innovations to the minimum variance control. *IFAC Symposium on Stochastic Control*, Budapest (1974).

[18] L. KEVICZKY, I. VAJK and Cs. BÁNYÁSZ: *PCSP—an effective simulator for modern Process Control Algorithms*. SCSC, Houston (1974).

[19] V. PETERKA and K. J. ÅSTRÖM: Control of multivariable systems with unknown but constant parameters. *IFAC Symp. on Identification and System Parameter Estimation*, The Hague/Delft (1973).

[21] A New Adaptive Method for Optimal Control of Closed Circuit Ball Grinding Mills. Hungarian Patent (1974).

[22] A New Method for Adaptive DDC of Raw Material Blending in a Cement Plant. Hungarian Patent (1975).

Adaptive Identification and Control Algorithms for Nonlinear Bacterial Growth Systems*

D. DOCHAIN† and G. BASTIN†

Simple self-tuning type controllers for nonlinear bacterial growth processes can be effective and their stability can be proven under mild conditions.

Key Words—Adaptive control, fermentation processes, nonlinear systems, parameter estimation.

Abstract—This paper suggests how nonlinear adaptive control of nonlinear bacterial growth systems could be performed. The process is described by a time-varying nonlinear model obtained from material balance equations. Two different control problems are considered: substrate concentration control and production rate control. For each of these cases, an adaptive minimum variance control algorithm is proposed and its effectiveness is shown by simulation experiments. A theoretical proof of convergence of the substrate control algorithm is given. A further advantage of the nonlinear approach of this paper is that the identified parameters (namely the growth rate and a yield coefficient) have a clear physical meaning and can give, in real time, a useful information on the state of the biomass.

1. INTRODUCTION

A COMMONLY used approach for the adaptive control of nonlinear systems is to consider them as time-varying linear systems and to use black-box linear approximate models to implement the control law. This approach has been used by the authors in previous works on the control of fermentation processes (Bastin and coworkers, 1983a, b).

But, since the underlying process is nonlinear, improved control can be expected by exploiting the nonlinear structure of the model. Such an idea is pursued in the present paper: we suggest how nonlinear adaptive control of nonlinear bacterial growth systems can be implemented. A similar idea has recently been used for the dissolved oxygen adaptive control in waste water treatment (Ko, McInnis and Goodwin, 1982), but under a somewhat different form than in the present paper.

The process is described by a nonlinear state space representation obtained from usual material balance equations (Section 2). However, this representation does not require any specific analytical description of the bacterial growth rate.

The system is then approximated by a discrete-time time-varying model which is linear in the parameters and in the control input though globally nonlinear. The time-varying parameters in this model (namely the growth rate and a yield coefficient) have a clear physical meaning and are identified in real time with a standard RLS algorithm (Section 3).

The parameter estimation algorithm is combined with minimum-variance and Clarke–Gawthrop controllers to obtain adaptive controllers in two different cases: substrate concentration control (Section 4) and production rate control (Section 5). The effectiveness of the parameter estimation algorithm and the adaptive control algorithms is demonstrated by simulation experiments. Furthermore a theoretical proof of the convergence of the substrate control is given in the Appendix.

Parameter estimation and nonlinear control of microbial growth systems have been, in the last decade, the object of growing interest. Among many others, we may mention the papers by D'Ans, Kokotovic and Gottlieb (1971), Aborhey and Williamson (1978), Holmberg and Ranta (1982) and a large number of papers (and references) contained in the proceedings of the first IFAC Workshop on Modelling and Control of Biotechnical Processes (Halme, 1983), especially the contributions of Marsili-Libelli (1983) and Stephanopoulos and Ka-Yiu San (1983). However, we believe that the algorithms proposed in this paper have some original features that we can summarize as follows:

(a) In our approach, the parameter estimation and the process control are performed *simultaneously*.

* Received 20 October 1983; revised 20 March 1984. The original version of this paper was presented at the IFAC Workshop on Adaptive Systems in Control which was held in San Francisco, California, U.S.A. during June 1983. The published proceedings of this IFAC meeting may be ordered from Pergamon Press Ltd, Headington Hill Hall, Oxford OX3 0BW, U.K. This paper was recommended for publication in revised form by associate editor P. Parks under the direction of editor B. Anderson.

† Laboratoire d'Automatique, de Dynamique et d'Analyse des Systèmes, University of Louvain, Bâtiment Maxwell, B-1348 Louvain-la-Neuve, Belgium.

Reprinted with permission from *Automatica*, vol. 20, pp. 621–634, Sept. 1984.
Copyright © 1984, International Federation of Automatic Control IFAC. Published by Pergamon Press Ltd.

(b) The specific growth rate is not modelled by an analytical function of the state but is considered as a time-varying unknown parameter estimated in real time by a simple least-squares algorithm.
(c) The control is performed by a very simple self-tuning scheme which contrasts with more sophisticated approaches followed elsewhere like, e.g. nonlinear optimal control (D'Ans, Kokotovic and Gottlieb, 1971), nonlinear state feedback with Riemanian geometric model (Takamatsu, Shioya and Kurome, 1983) or adaptive multimodel control (Cheruy, Panzarella and Denat, 1983).
(d) Global convergence of the substrate control algorithm is established under mild conditions.

2. DESCRIPTION OF THE SYSTEM

We consider the usual state–space representation of bacterial growth systems by mass-balance equations

$$\begin{aligned} \dot{X} &= [\mu(X,S) - U]\,X \\ \dot{S} &= -k_1\mu(X,S)\,X + U(V-S) \\ Y &= k_2\mu(X,S)\,X \end{aligned} \tag{1}$$

with *state variables:* X biomass concentration
S substrate concentration
inputs: U dilution rate (i.e. influent flow rate)
V influent substrate concentration
outputs: S substrate concentration
Y production rate of the reaction product
parameters: $\mu(X,S)$ growth rate
k_1 and k_2 yield coefficients.

We could think of adopting an analytical expression for the bacterial growth rate $\mu(X,S)$; the most popular expression is certainly the Monod law

$$\mu(X,S) = \frac{\mu^* S}{K_M + S} \quad \text{(Monod)} \tag{2}$$

but many other expressions have been suggested, like

$$\mu(X,S) = \begin{cases} \dfrac{\mu^*}{K_b} S & S \leqslant K_b \\ \mu^* & S \geqslant K_b \end{cases} \quad \text{(Blackman)} \tag{3}$$

$$\mu(X,S) = \frac{\mu^* S}{K_c X + S} \quad \text{(Contois)} \tag{4}$$

$$\mu(X,S) = \frac{\mu^* K_0 S}{1 + K_1 S + K_2 S^2} \quad \text{(Haldane).} \tag{5}$$

In these expressions, μ^* is the maximum growth rate.

The choice of an appropriate model for $\mu(X,S)$ is far from being an easy task and is the matter of continuing research (e.g. Roques and co-workers, 1982). Spriet (1982) lists no less than nine different models for $\mu(X,S)$ which have been proposed in the literature without even mentioning those which involve inhibitions (like the Haldane law (5)) or a pH-dependence (e.g. Vandenberg and coworkers, 1976).

Furthermore, it is well known that important identifiability difficulties occur when estimating the parameters (μ^* and K_m or K_b or K_c...) from real-life data (e.g. Holmberg and Ranta, 1982; Bastin and coworkers, 1983b, Holmberg, 1983).

Therefore we prefer to 'short-circuit' the problem of this choice and to identify the time-varying growth rate $\mu(X,S)$ in real-time by an adaptive algorithm.

Throughout this paper, we shall assume that:
(a) the dilution rate U is the *control* input;
(b) the influent substrate concentration V is an external *measurable* disturbance input;
(c) the substrate concentration S and the production rate Y are *measurable* outputs.

A typical example: the anaerobic digestion process

The state–space representation (1) is suited to describe the methanization stage in an anaerobic digestion process. The anaerobic digestion can be used, for instance, for the treatment of wastes in sugar industries: U is the influent acetic acid concentration (i.e. the input pollution level), S is the output pollution level and Y is a methane gas flow rate. V and S are observed through BOD measurements. The main interest of such a water treatment plant is obviously to yield methane gas which can be used as an auxiliary energy supply. Further details on the anaerobic digestion process can be found in Antunes and Installé (1981), Van den Heuvel and Zoetmeyer (1982), and Bastin and coworkers (1983a, b).

3. ADAPTIVE PARAMETER ESTIMATION

Using a first-order Euler approximation for $\dot{X}$ and $\dot{S}$, with a sampling period T, the following discrete-time equations are derived from the system equations (1)

$$\begin{aligned} X_{t+1} &= X_t + T\mu_t X_t - TU_t X_t + \tilde{v}_t \\ S_{t+1} &= S_t - Tk_1\mu_t X_t + TU_t(V_t - S_t) + \omega_t \\ Y_t &= k_2\mu_t X_t. \end{aligned} \tag{6}$$

In these equations, the subscript t is a discrete-time index ($t = 0,1,2,\ldots$) and the growth rate μ_t is a compact notation for $\mu_t = \mu(X_t, S_t)$.

We make the approximation

$$Y_{t+1} - Y_t = k_2\mu_t(X_{t+1} - X_t) + \varepsilon_t. \quad (7)$$

Then, substituting for X_t and X_{t+1} from (7) into (6), we have

$$Y_{t+1} = Y_t + \mu_t T Y_t - T U_t Y_t + v_t \quad (8)$$

$$S_{t+1} = S_t + kTY_t + TU_t(V_t - S_t) + \omega_t \quad (9)$$

with

$$v_t = \varepsilon_t + k_2\mu_t\tilde{v}_t$$
$$k = -\frac{k_1}{k_2}.$$

Equations (8) and (9) constitute the *basic discrete-time model* for the derivations of the parameter estimation and adaptive control algorithms. In this model, v_t and ω_t represent errors due to noise, discretization and approximation (7).

Since the basic model is linear in the parameters μ_t and k, recursive least-squares estimates can be readily obtained

$$\hat{\mu}_{t+1} = \hat{\mu}_t + TP_tY_t(Y_{t+1} - Y_t + TU_tV_t - \hat{\mu}_tTY_t) \quad (10)$$

$$\hat{k}_{t+1} = \hat{k}_t + TP_tY_t(S_{t+1} - S_t - TU_t(V_t - S_t) - \hat{k}_tTY_t) \quad (11)$$

$$P_t = \frac{P_{t-1}}{\lambda}\left(1 - \frac{T^2Y^2{}_tP_{t-1}}{\lambda + T^2Y^2{}_tP_{t-1}}\right)$$

with $P_0 \gg 0$ and $0 < \lambda \leqslant 1$.

λ is a forgetting factor to allow the tracking of the time-varying growth rate μ_t. This forgetting factor is also used for the estimation of the yield coefficient k to allow for variations 'due to unobservable physiological or genetic events' (Holmberg and Ranta, 1982). Notice that the estimation of both parameters is decoupled but with a common gain P_t.

In addition to these parameter estimates, the biomass concentration X can be estimated in real-time by writing $\hat{X}_t = Y_t/k_2\hat{\mu}_t$.

Simulation results

Simulation experiments have been performed using state equations (1) as the 'true' bacterial growth system, with a Monod growth rate (2). The following parameters and initial state values were used:

$$\mu^* = 0.4 \quad K_m = 0.4 \quad k = -0.3636 \quad X_0 = 0.069 \quad S_0 = 0.13.$$

The initial values of both estimated parameters $\hat{\mu}_t$ and $\hat{k}_t$ were set to zero. These values will be used for all the simulation experiments throughout the paper. Figure 1 shows the estimates $\hat{\mu}_t$ and $\hat{k}_t$ computed by the algorithm equations (10) and (11) with $P_0 = 10^6\text{I}$ and white noise input signals U_1 and V_t.

The same experiment is shown in Fig. 2, except that a jump is applied on the maximum growth rate ($\mu^* = 0.4 \rightarrow 0.45$) at time $t = 240$.

We observe a fast convergence, without bias, of the parameter estimate $\hat{k}_t$ and a slower convergence of $\hat{\mu}_t$ to the 'true' time-varying growth rate μ_t.

4. SUBSTRATE CONTROL

We consider the problem of regulating the substrate concentration S_t at a prescribed level S^* despite the disturbance input V_t, by acting on the dilution rate U_t.

In the anaerobic digestion example mentioned above, this is a depollution control problem with V_t and S_t the input and output pollution levels respectively.

A discrete-time minimum variance adaptive controller is adopted. At each sampling time, the control input U_t is computed by setting a one-step ahead prediction of the substrate concentration equal to the prescribed level

$$\hat{S}_{t+1} = S^*. \quad (12)$$

From the basic model equation (9), it is natural to define $\hat{S}_{t+1}$ as follows:

$$\hat{S}_{t+1} = S_t + \hat{k}_tTY_t + TU_t(V_t - S_t) \quad (13)$$

here $\hat{k}_t$ is updated by the parameter estimation algorithm (11).

A nonlinear control law is readily obtained from (12), since $\hat{S}_{t+1}$ is linear in U_t; in practice, the control action U_t is obviously constrained by the operating conditions. Therefore, the adaptive control algorithm is as follows:

$$\bar{U}_t = \frac{S^* - S_t - \hat{k}_tTY_t}{T(V_t - S_t)}$$
$$U_t = 0 \text{ if } \bar{U}_t < 0 \quad (14)$$
$$\bar{U}_t = U_{\max} \text{ if } \bar{U}_t > U_{\max}$$
$$U_t = \bar{U}_t \text{ otherwise.}$$

A block diagram of the closed-loop system is presented in Fig. 3. We note that a feedforward compensation of the measurable perturbation V_t is included.

Simulation results

Successful simulation experiments have been carried out, using the continuous-time state equations (1) as the 'true' system with a Monod growth rate (2) and $U_{\max} = 0.39$.

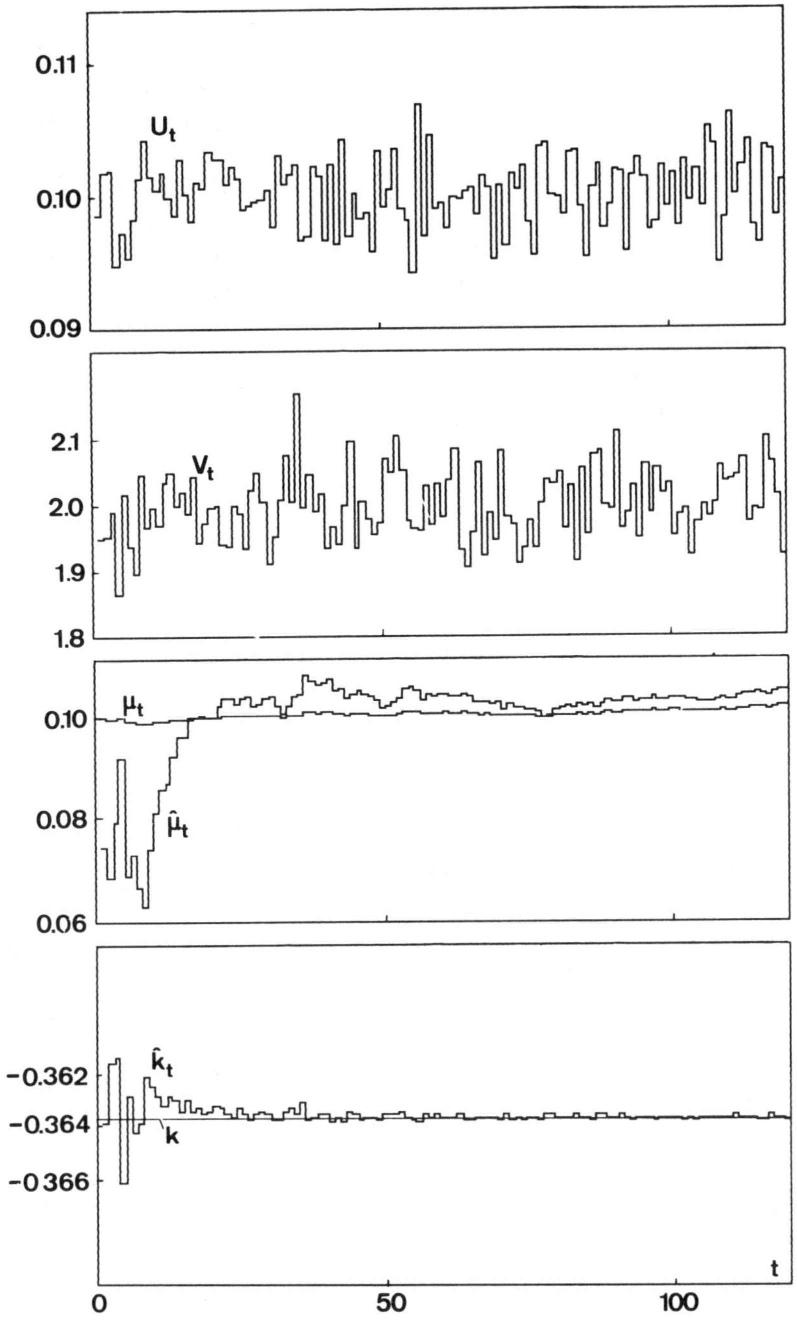

FIG. 1. Parameter estimation with white noise inputs.

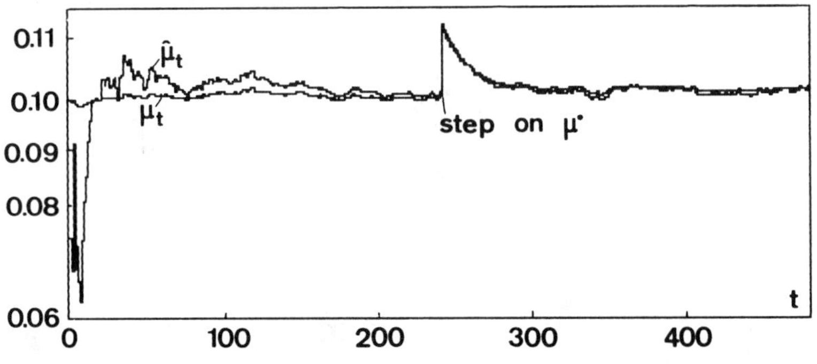

FIG. 2. Estimation of $\hat{\mu}_t$ with a step on μ^* at $t = 240$ s.

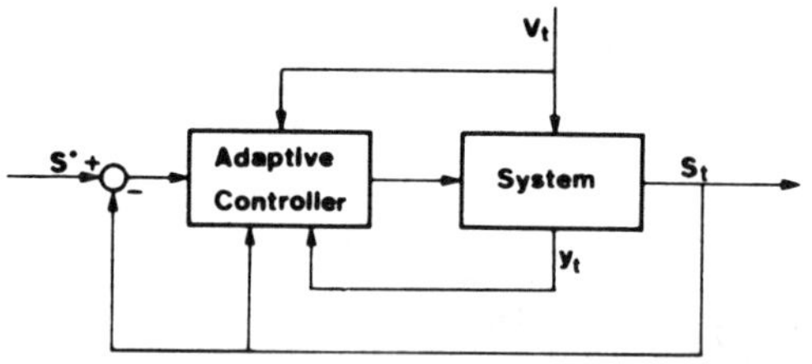

FIG. 3. Block diagram of the substrate concentration control.

Figure 4 shows the substrate concentration S_t, the control input U_t and the parameter estimates $\hat{\mu}_t$ and $\hat{k}_t$ in the case of a square-wave set point with a period of 96 sampling times and a constant perturbation $V_t = 2$. We observe that the controlled output S_t converges much faster than the parameter estimate $\hat{\mu}_t$, but this is not surprising since $\hat{\mu}_t$ is not actually used by the control algorithm.

Figure 5 shows the substrate concentration S_t, the control input U_t and the parameter estimates $\hat{\mu}_t$ and $\hat{k}_t$ in the case of a square-wave perturbation V_t and an additive white noise on the auxiliary output Y. Evidently, we observe a bias (due to the noise) in the parameter estimates but this is not important for the convergence of controlled output S_t.

Figure 6 shows the substrate concentration S_t and the control input U_t in the case of a 10% square-wave variation of the maximum growth rate.

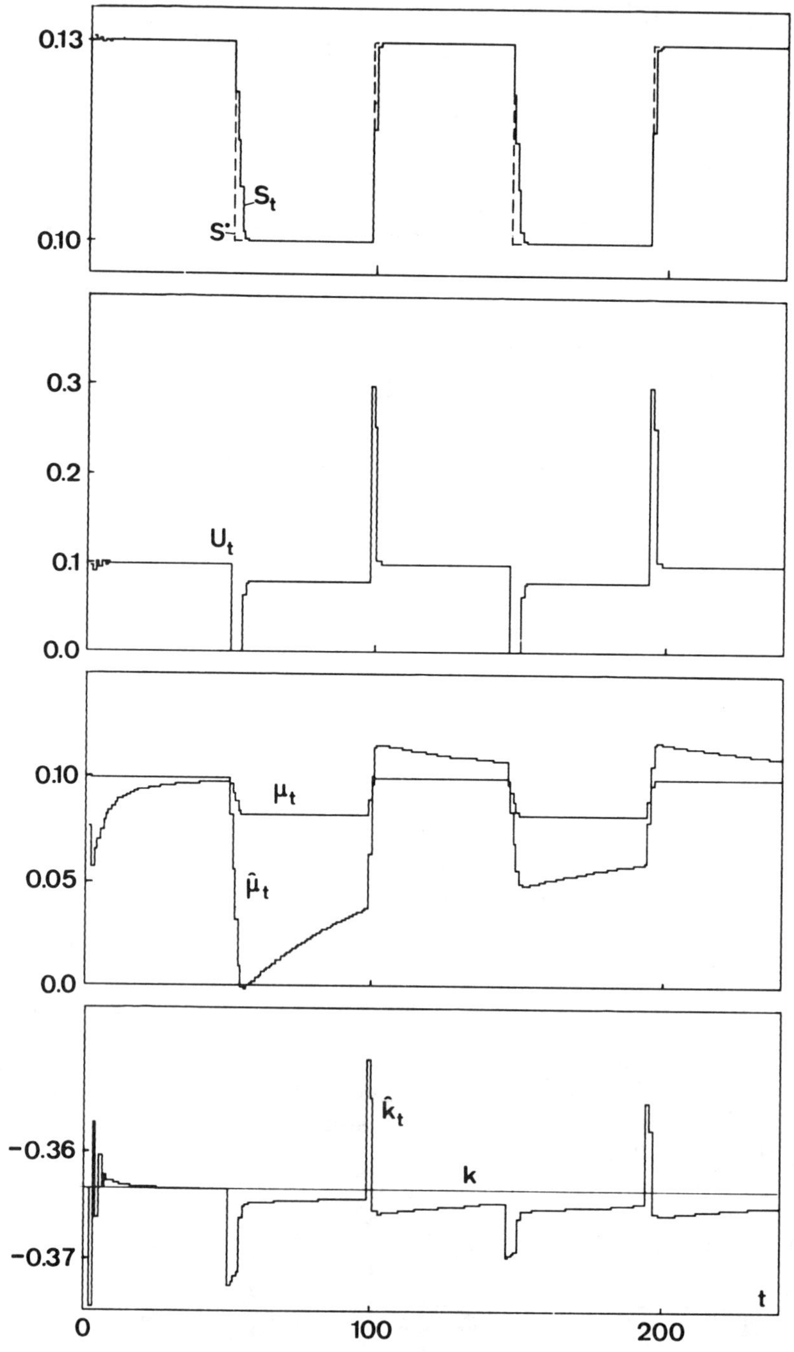

FIG. 4. Substrate concentration control with a square-wave set point.

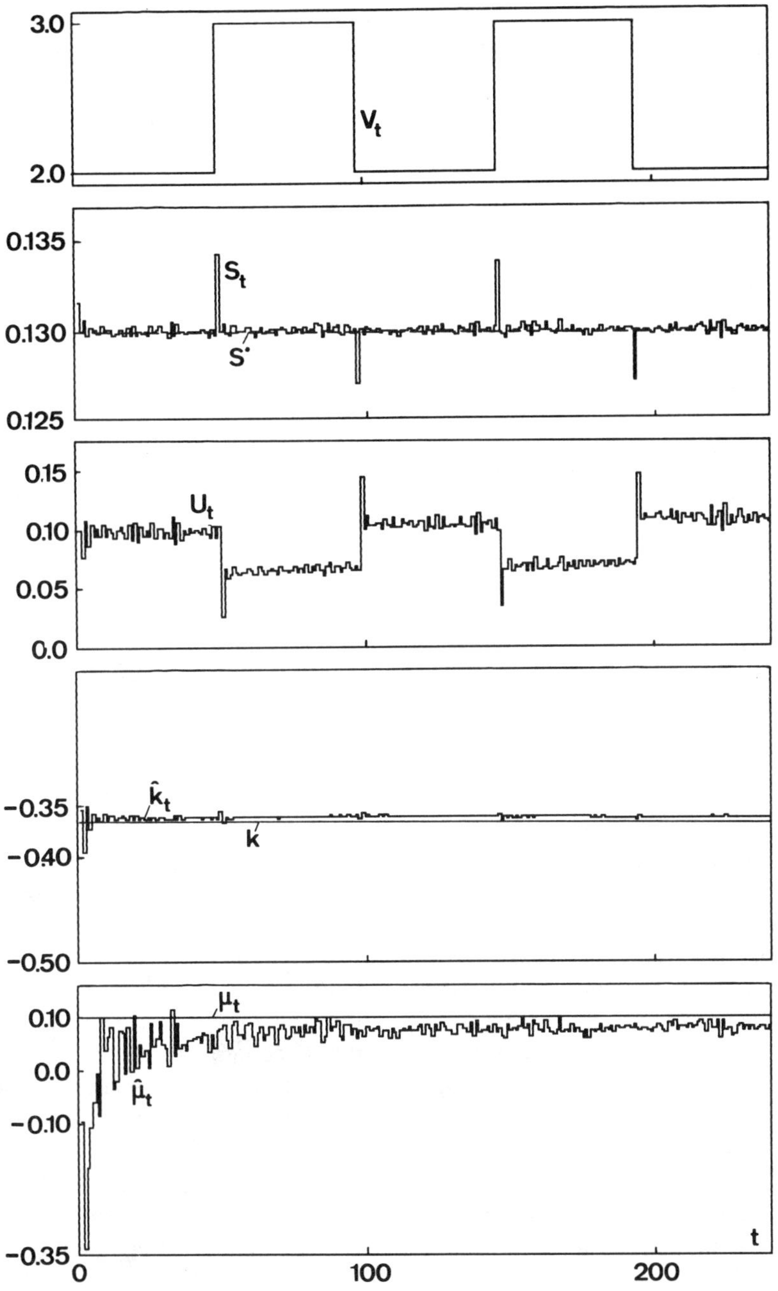

FIG. 5. Substrate concentration control with a square-wave input perturbation.

5. PRODUCTION RATE CONTROL

In order to facilitate the understanding of the later discussions, we refer here to the particular case of the anaerobic fermentation process described in Section 2 but, obviously, the results can also apply to other processes with the same structure.

The anaerobic digestion can be viewed as an energy conversion process. An amount of 'organic' energy is available in the influent under the form of the input organic load V_t. This energy is converted into methane gas Y_t by the anaerobic digestion. Obviously, the output energy Y_t cannot, in the mean, be larger than the available input energy. When the aim of the plant is not depollution but energy production (as in industrial farms), the control objective is to continuously adapt the output production Y_t to the available input load V_t. Therefore, the desired gas production Y_t^* is defined as follows:

$$Y_t^* = \beta V_t - \beta_0 \quad \beta > 0, \quad \beta_0 > 0. \qquad (15)$$

The coefficients β and β_0 have to be selected carefully by the user since if, by lack of knowledge, β is chosen too large or β_0 too small (i.e. if we require from the fermentor more methane gas than it can actually provide) then the process can be driven by the controller to a wash-out steady-state (Antunes

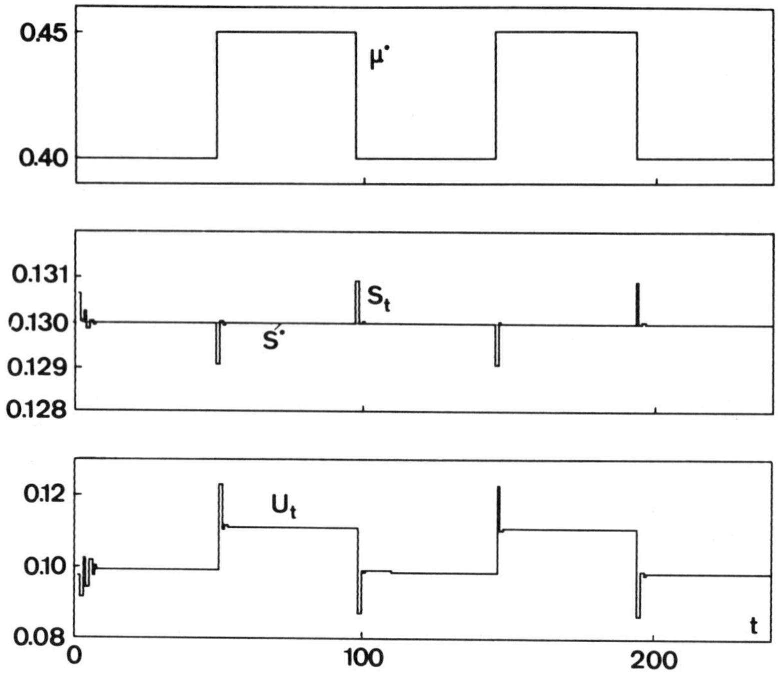

FIG. 6. Substrate concentration control with a square-wave perturbation on μ^*.

and Installé, 1981), i.e. to a state where the bacterial life has completely disappeared and where the reactor is definitely stopped.

In this section, we shall first demonstrate that a minimum variance adaptive control law using the basic model equation (8) may diverge. We shall then describe what kind of modification we bring to the model in order to improve the control algorithm.

Divergence of the minimum variance controller

As for the substrate concentration control, we first try to use a minimum variance control law derived from the basic model equation (8)

$$\begin{aligned} \bar{U}_t &= -\frac{Y^*_{t+1} - Y_t(1 + T\hat{\mu}_t)}{TY_t} \\ U_t &= 0 \text{ if } \bar{U}_t < 0 \\ U_t &= U_{\max} \text{ if } \bar{U}_t > U_{\max} \\ U_t &= \bar{U}_t \text{ otherwise.} \end{aligned} \tag{16}$$

Consider the case when $Y^*_{t+1} > Y_t\ (1 + T\hat{\mu}_t)$. Then $\bar{U}_t < 0$, i.e. U_t is set to zero.

If U_t is kept equal to zero, Y_t, possibly after a transient increasing period, will decrease and tend to zero (gas can no longer be produced if the influent has disappeared!). So, if the transient on Y_t is not important enough, U_t remains at the zero value, and Y_t tends to zero.

Figure 7 illustrates this feature: at time $t = 48$, the desired output level Y^*_{t+1} is set to a value 15% larger than the steady-state value of Y_t.

Modification of the basic discrete-time model

In order to improve the control algorithm, we introduce the following modifications of the basic model equations.

First, we consider the following approximate relation between μ and S:

$$\mu(X, S) = b(X, S).S \tag{17}$$

i.e. the parameter b is estimated, instead of μ, with a recursive least-square algorithm.

One may consider this approximation as a loss of generality with respect to the previous case where μ is left independent of any analytical expression and estimated as a parameter of the system. But this is plainly justified by the fact that all the proposed bacterial growth laws are compatible with (17).

Rewrite the expression of Y_t, from (6)

$$Y_t = k_2 b_t S_t X_t.$$

We modify the approximation (7) by the following one:

$$\begin{aligned} Y_{t+1} - Y_t = k_2 b_t [S_t(X_{t+1} - X_t) \\ + X_t(S_{t+1} - S_t)] + \epsilon_t \end{aligned} \tag{18}$$

i.e. the variation $\Delta Y_t = Y_{t+1} - Y_t$ is now dependent on both the variations ΔX_t in the bacterial

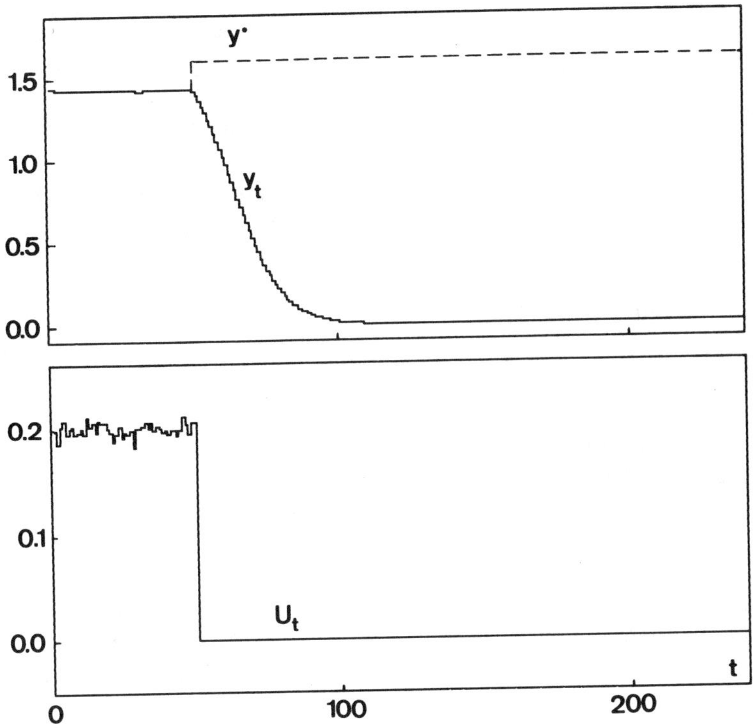

FIG. 7. Production rate control: divergence of a simple MV controller.

concentration and ΔS_t in the substrate concentration. Equation (8) becomes

$$Y_{t+1} = Y_t + b_t T S_t Y_t + kT\frac{Y_t^2}{S_t} + TU_tY_t\left(\frac{V_t}{S_t} - 2\right) + v_t \tag{19}$$

with $v_t = \varepsilon_t + k_2 b_t S_t \tilde{v}_t + k_2 b_t X_t \omega_t$.

Since (19) is linear in the parameter b_t, recursive least-squares estimates can be obtained

$$\hat{b}_{t+1} = \hat{b}_t + TS_tY_tP'_t\left(Y_{t+1} - Y_t - \hat{k}_t T\frac{Y_t^2}{S_t} + TU_tY_t\left(\frac{V_t}{S_t} - 2\right) - \hat{b}_t TS_tY_t\right) \tag{20}$$

$$P'_t = \frac{P'_{t-1}}{\lambda}\left(1 - \frac{T^2Y_t^2S_t^2P'_{t-1}}{\lambda + T^2Y_t^2S_t^2P'_{t-1}}\right). \tag{21}$$

In these expressions the value of $\hat{k}_t$ is assumed to be estimated by the recursive least-squares equation (11).

Notice that parameters $\hat{k}_t$ and $\hat{b}_t$ are estimated 'in cascade'. This allows us to decouple the estimation of both parameters, and to keep a very simple scalar identification algorithm.

Figure 8 shows the same experiment as Fig. 1, but for the estimation of $\hat{b}_t$.

New minimum variance control algorithm

As above, we choose a discrete-time minimum variance adaptive controller. Using (19), the control input U_t is given by

$$\bar{U}_t = \frac{Y^*_{t+1} - Y_t - T\hat{k}_tY_t^2/S_t - T\hat{b}_tS_tY_t}{TY_t(V_t/S_t - 2)} \tag{22}$$

$$\begin{aligned} U_t &= 0 && \text{if } \bar{U}_t < 0 \\ U_t &= U_{\max} && \text{if } \bar{U}_t > U_{\max} \\ U_t &= \bar{U}_t && \text{otherwise.} \end{aligned}$$

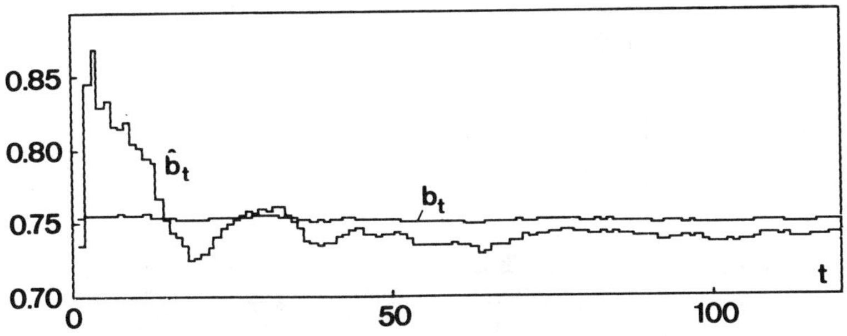

FIG. 8. RLS estimation of b_t.

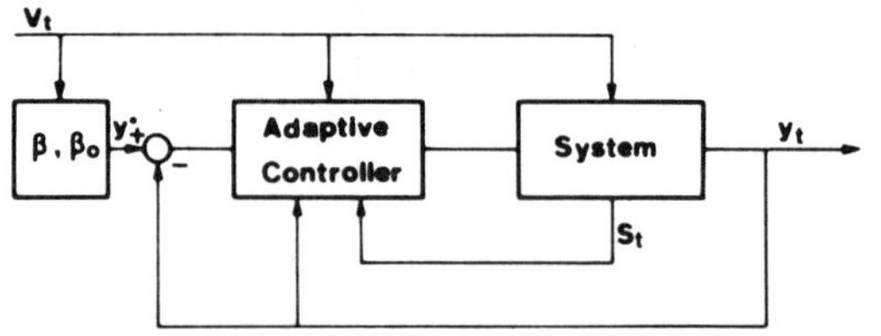

FIG. 9. Block diagram of the production rate control.

A block diagram of the closed-loop system is presented in Fig. 9.

Simulation results

The minimum variance adaptive controller, as written in (22), is more effective than the previous one (16). As a comparison, Fig. 10 shows the same experiment as Fig. 7, but with the control law (22).

In Fig. 11 steps of the influent substrate concentration V_t (external measurable perturbation), i.e. of the desired output level Y_t^* (see (15)), are applied to the system.

The control algorithm converges very quickly, although the convergence of the parameter $\hat{b}_t$ to its 'true' time-varying value is much slower.

Clarke–Gawthrop controller

It is evident, from (22), that the sign of $\bar{U}_t$ depends on the sign of $(V_t/S_t - 2)$. When the substrate concentration S_t reaches values close to 0.5 V_t, the minimum variance controller may appear not to be able to reach the desired set point. If S_t is larger than 0.5 V_t, $\bar{U}_t$ becomes negative, i.e. $U_t = 0$. As a consequence, S_t decreases. When S_t becomes smaller than 0.5 V_t, $\bar{U}_t$ is set to a positive value. If $(V_t/S_t - 2)$ is close to zero, $\bar{U}_t$ most likely reaches large values, larger than U_{max}, and S_t increases again so that $(V_t/S_t - 2)$ becomes negative, and so on.

In such a case, U_t is oscillating between 0 and U_{max}, leading to the oscillation of the system, and the control does not converge. A typical illustration is given in Fig. 12.

In order to solve these convergence problems, we introduce a Clarke–Gawthrop (1979) control law using a weight $Q(1 - z^{-1})$ in the performance index (Belanger, 1983). The control input is then computed so as to minimize the following criterion:

$$J = (\hat{Y}_{t+1} - Y_{t+1}^*)^2 + Q^2(U_t - U_{t-1})^2. \quad (23)$$

Using (19), we have

$$\bar{U}_t = \frac{Q^2}{Q^2 + T^2Y_t^2(V_t/S_t - 2)^2}U_{t-1} + \frac{TY_t(V_t/S_t - 2)}{Q^2 + T^2Y_t^2(V_t/S_t - 2)^2} \left[Y_{t+1}^* - Y_t - \hat{k}_tT\frac{Y_t^2}{S_t} - \hat{b}_tTS_tY_t\right]. \quad (24)$$

Figure 13 shows the improvement obtained by using this Clarke–Gawthrop controller. It is interesting to note that, as above, the convergence of the controlled output Y_t (Fig. 13) is much faster than the convergence of the parameter estimates (Fig. 14).

6. CONCLUSIONS

Simple adaptive controllers for a class of biotechnical systems have been proposed. Their

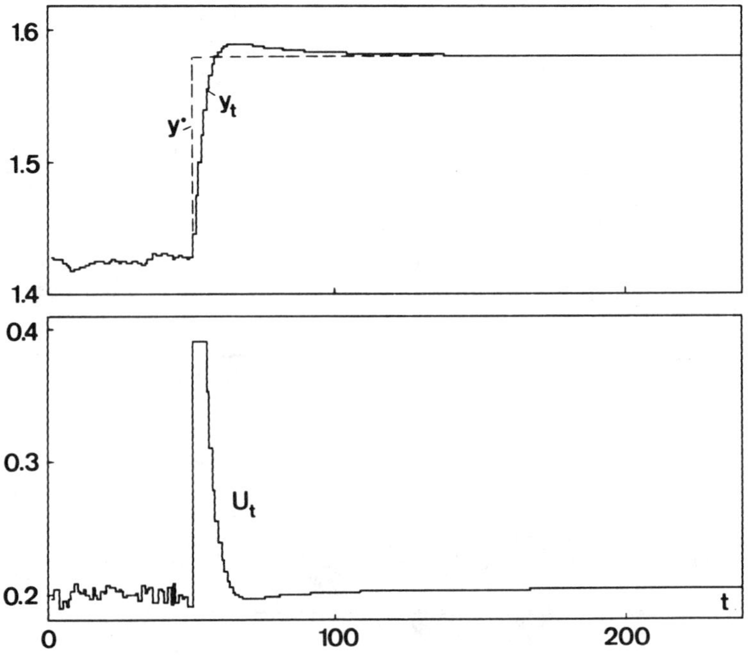

FIG. 10. Production rate control: convergence of the modified MV controller.

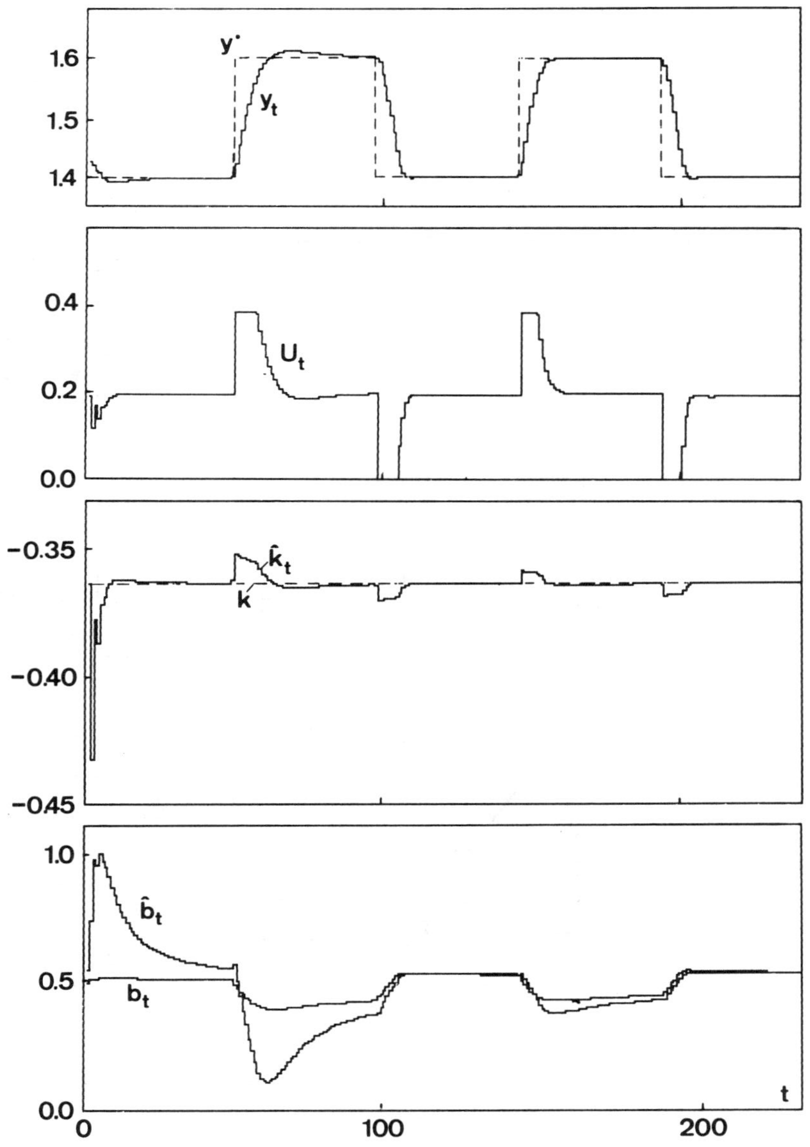

FIG. 11. Production rate control with a square-wave set point.

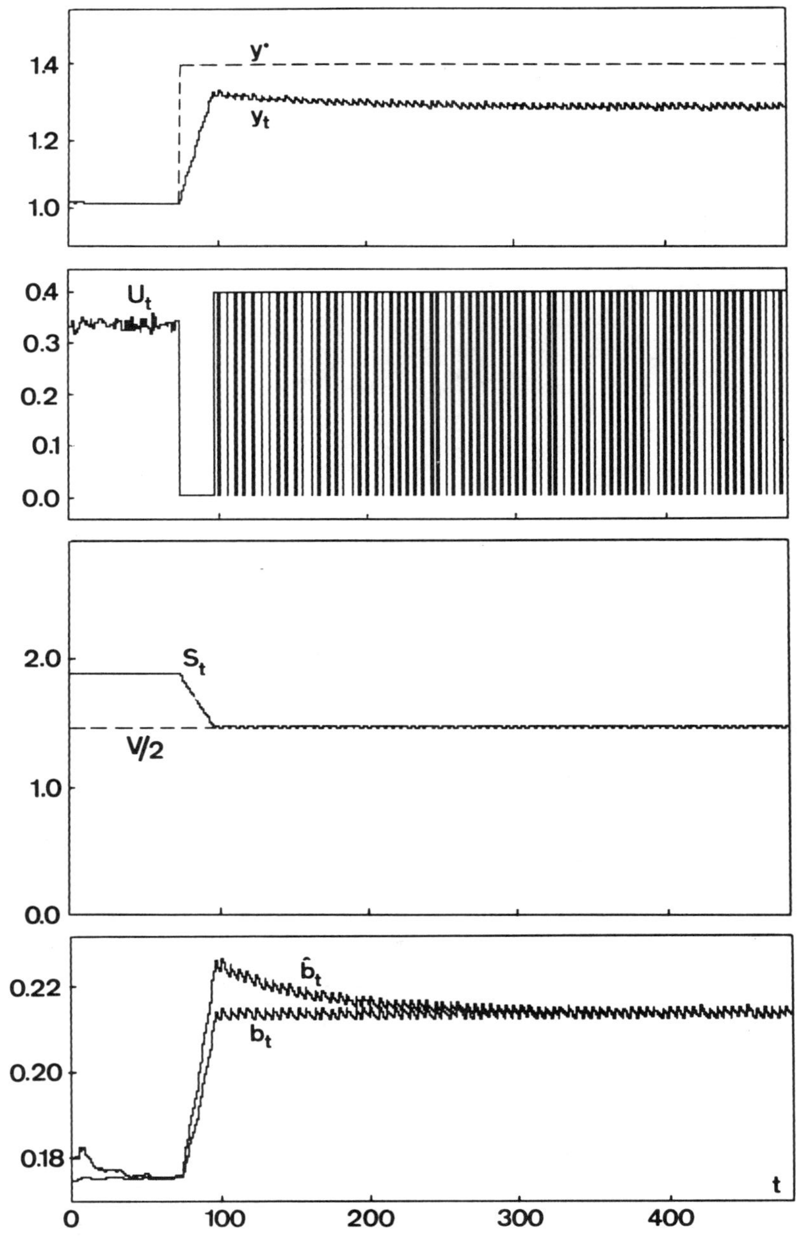

FIG. 12. Oscillation of the modified MV controller.

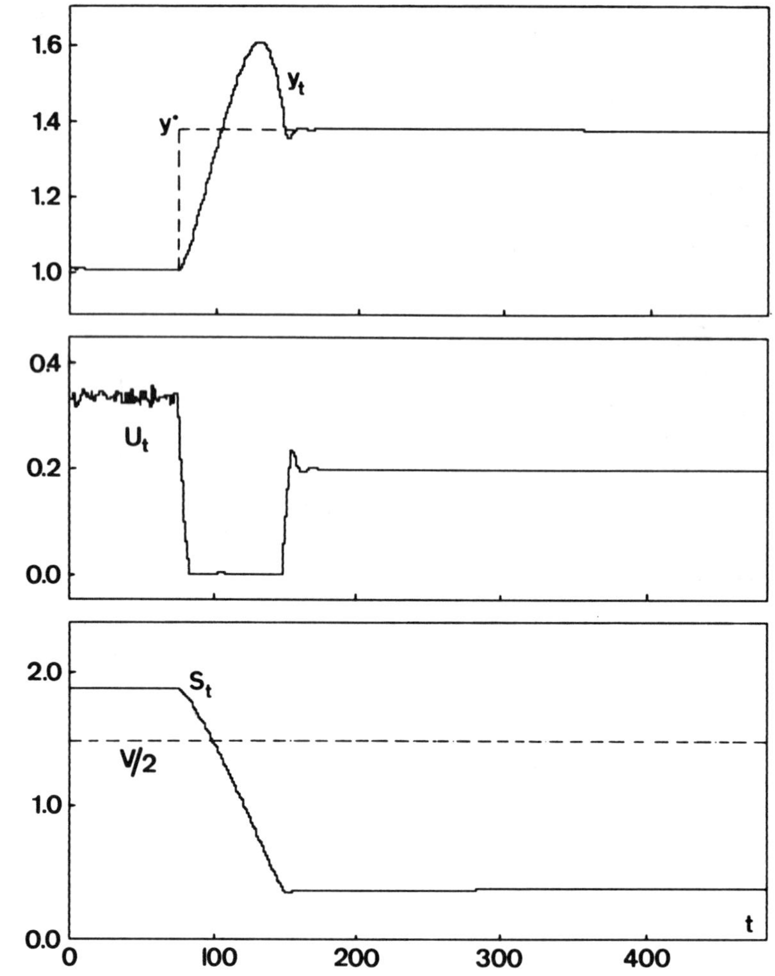

FIG. 13. Production rate control with a CG controller.

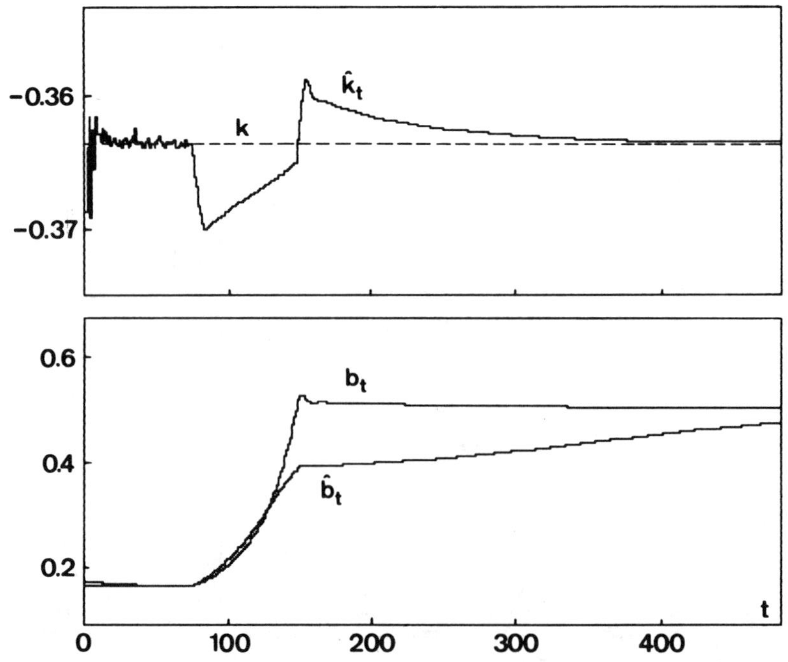

FIG. 14. Production rate control: evolution of the parameter estimates.

effectiveness has been demonstrated by simulation experiments.

A theoretical proof of the convergence of substrate concentration control algorithm is given in the Appendix. In the case of production rate control, the convergence of the algorithm has not been discussed and is obviously much more difficult to establish since the algorithm involves two cascaded steps together with the estimation of a truly time-varying parameter (μ_t).

In addition to the control itself, a further advantage of the nonlinear approach of this paper is that the identified parameters correspond clearly to physical parameters (namely growth rate and yield coefficient); therefore they can provide useful information, in real-time, on the state of the biomass.

Although the model (1) is well suited to industrial applications like waste treatment in sugar industries, in many other applications the model (1) is only the last stage of a complex multistage reaction: a typical situation is a five-state twelve parameter model (e.g. Bastin and coworkers, 1983b) describing a sequence of three reactions (solubilization, acidification, methanization). This is a further reason to explore the possibility of simple control schemes for the different stages of such high-order highly nonlinear systems.

Acknowledgements—The authors wish to thank M. I. Installé from the Laboratory of Automatic Control (University of Louvain), H. Naveau, E. J. Nyns and D. Poncelet from the Laboratory of Biotechnology (University of Louvain), A. Cheruy and L. Dugard from the Laboratory of Automatic Control (ENSIEG, Grenoble) for fruitful discussions about this work.

REFERENCES

Aborhey, S. and D. Williamson (1978). State and parameter estimation of microbial growth processes. *Automatica*, **14**, 493–498.

Antunes, S. and M. Installé (1981). The use of phase-plane analysis in the modelling and the control of a biomethanization process. *Proceedings of the 8th IFAC World Congress*, Kyoto, Japan, Vol. XXII, pp. 165–170.

Bastin, G., D. Dochain, M. Haest, M. Installé and P. Opdenacker (1983a). Modelling and adaptive control of a continuous anaerobic fermentation process. In *Modelling and Control of Biotechnical Processes*, A. Halme (ed). Pergamon Press, Oxford, pp. 299–306.

Bastin, G., D. Dochain, M. Haest, M. Installé and P. Opdenacker (1983b). Identification and adaptive control of a biomethanization process. In *Modelling and Data Analysis in Biotechnology and Medical Engineering*, G. C. Vansteenkiste and P. C. Young (eds). North-Holland, New York, pp. 271–282.

Belanger, P. R. (1983). On type I systems and the Clarke–Gawthrop regulator. *Automatica*, **19**, 91–94.

Cheruy, A., L. Panzarella and J. P. Denat (1983). Multimodel simulation and adaptive stochastic control of an activated sludge process. In *Modelling and Control of Biotechnical Processes*, A. Halme (ed). Pergamon Press, Oxford, pp. 127–183.

Clarke, D. W. and P. J. Gawthrop (1979). Self-tuning control. *Proc. IEEE*, **126**, 633–640.

D'Ans, G., P. V. Kokotovic and D. Gottlieb (1971). A nonlinear regulator problem for a model of biological waste treatment. *IEEE Trans. Aut. Control*, **AC-16**, 341–347.

Goodwin, G. C., B. McInnis and R. S. Long (1982). Adaptive control algorithms for waste water treatment and pH neutralization. *Opt. Control Appl. Meth.* **3**, 443–459.

Goodwin, G. C. and K. S. Sin (1983). *Adaptive Filtering Prediction and Control*. Prentice-Hall, Englewood Cliffs, NJ.

Halme, A. (1983). Modelling and control of biotechnical processes. *Proceedings of the 1st IFAC Workshop*, Helsinki, Finland, 17–19 August 1982. Pergamon Press, Oxford.

Holmberg, A. (1983). On the accuracy of estimating in the parameters of models containing Michaelis–Menten type nonlinearities. In *Modelling and Data Analysis in Biotechnology and Medical Engineering*, G. C. Vansteenkiste and P. C. Young (eds). North-Holland, New York, pp. 199–208.

Holmberg, A. and J. Ranta (1982). Procedures for parameter and state estimation of microbial growth process models. *Automatica*, **13**, 181–193.

Ko, K. Y., B. C. McInnis and G. C. Goodwin (1982). Adaptive control and identification of the dissolved oxygen process. *Automatica*, **18**, 727–730.

Marsili-Libelli, S. (1983). On-line estimation of bioactivities in activated sludge processes. In *Modelling and Control of Biotechnical Processes*, A. Halme (ed). Pergamon Press, Oxford, pp. 121–126.

Roques, H., S. Yve, S. Saipanich and B. Capdeville (1982). Is Monod's approach adaquate for the modelisation of purification processes using biological treatment? *Water Resources*, **16**, 839–847.

Spriet, J. A. (1982). Modelling of the growth of micro-organisms: a critical appraisal. In *Environmental Systems Analysis and Management*, Rinaldi (ed). North-Holland, New York, pp. 451–456.

Stephanopoulos, G. and Ka-Yiu San (1983). On-line estimation of time-varying parameters: application to biochemical reactors. *Modelling and Control of Biotechnical Processes*, A. Halme (ed). Pergamon Press, Oxford, pp. 195–199.

Takamatsu T., S. Shioya and H. Kurome (1983). Dynamics and control of a mixed culture in an activated sludge process. *Modelling and Control of Biotechnical Processes*, A. Halme (ed). Pergamon Press, Oxford, pp. 103–109.

Vandenberg, L., G. B., Patel, D. S. Clark and C. P. Lentz (1976). Factors affecting rate of methane formation from acetic acid by enriched methanogenic cultures. *Can. J. Microbiol.*, **22**, 1312–1319.

Van den Heuvel, J. C. and R. J. Zoetmeyer (1982). Stability of the methane reactor: a simple model including substrate inhibition and cell recycle. *Process Biochemistry*. May-June. 14–19.

APPENDIX: CONVERGENCE ANALYSIS OF THE SUBSTRATE CONCENTRATION CONTROL ALGORITHM

In this Appendix, we present a proof of the convergence of the substrate concentration control under a set of reasonable assumptions.

The demonstration has some similarities with that proposed by Goodwin, McInnis and Long (1982) in the case of dissolved oxygen control for waste water treatment.

It is organized in three steps:

(a) BIBO stability of the bacterial growth system,
(b) convergence of the parameter estimation algorithm,
(c) convergence of the adaptive control algorithm.

BIBO stability of the continuous-time bacterial growth system

Let us rewrite, for convenience, the state–space description of the bacterial growth system

$$\dot{X} = [\mu(X, S) - U]X \tag{A1a}$$

$$\dot{S} = -k_1\mu(X, S)X + U(V - S) \tag{A1b}$$

$$Y = k_2\mu(X, S)X. \tag{A1c}$$

In this section we prove the BIBO stability of this system (in accordance with the physical situation) under the following assumptions.

Assumptions.

(H1a) $0 \leqslant \mu(X, S) \leqslant \mu^*$
(H1b) $\mu(X, 0) = 0$.

(H2a) $0 \leqslant U \quad 0 \leqslant V \leqslant V_{max}$
(H2b) $0 \leqslant S(0) \quad 0 \leqslant X(0) \quad k_1 X(0) + S(0) \leqslant V_{max}$.
Notice that all the growth rate models of Section 2 fulfill assumption (H1).

Lemma 1. Under assumptions (H1) and (H2)

$$\text{(i)}\ 0 \leqslant S \leqslant V_{max} \quad \text{(ii)}\ 0 \leqslant X \leqslant \frac{V_{max}}{k_1} \quad \text{(iii)}\ 0 \leqslant Y \leqslant \frac{k_2}{k_1}\mu^* V_{max}\ \forall t > 0. \tag{A2}$$

Proof.

(1) $X \geqslant 0$ and $Y \geqslant 0$; straightforward by using (A1a), (A1c) and (H2b).
(2) For $S = 0$, we have $\dot{S} \geqslant 0$, using (H2a), (A1b) and (H1b). The conclusion $S \geqslant 0$ for all S follows from (H2b).
(3) For $S = V_{max}$, we have, using (A1b), (A1c) and (H2a),

$$\dot{S} = -\frac{k_1}{k_2} Y + U(V - V_{max}) \leqslant 0 \text{ since } Y \geqslant 0. \tag{A3}$$

The conclusion $S \leqslant V_{max}$ for all S follows from (H2b).
(4) Define the auxiliary variable $Z = k_1 X + S$.
Then, the following equation is readily derived from (A1a) and (A1b):

$$\dot{Z} = U(V - Z). \tag{A4}$$

For $Z = V_{max}$, we have $\dot{Z} \leqslant 0$. The conclusion $Z \leqslant V_{max}$ for all Z follows. Since $S \geqslant 0$, clearly we have $X \leqslant V_{max}/k_1$ for all X and it becomes obvious that, by using (A1c) and (H1a),

$$Y \leqslant \frac{k_2}{k_1}\mu^* V_{max}. \tag{A5}$$

Q.E.D.

It should be emphasized that from Lemma 1, the outputs S and Y and the state X are bounded without imposing any upper bound on the input U.

Convergence of the parameter estimation algorithm

We consider now the convergence of the estimation algorithm for the parameter $\hat{k}_t$ presented in Section 3.

The basic idea is due to Goodwin and Sin (1983) and can be roughly summarized as follows: if the noise term ω_t in (9) is bounded, then the convergence of the parameter estimate $\hat{k}_t$ can be guaranteed by involving, in the algorithm, a switching function to hold the parameter estimate constant wherever the prediction error becomes smaller than a prespecified bound. The algorithm (11) is considered without the forgetting factor ($\lambda = 1$) and modified as follows:

$$\hat{k}_{t+1} = \hat{k}_t + \sigma_t T P_t Y_t (S_{t+1} - \hat{S}_{t+1}) \tag{A6}$$

$$P_t = P_{t-1}\left(1 - \frac{\sigma_t T^2 Y_t^2 P_{t-1}}{1 + \sigma_t T^2 Y_t^2 P_{t-1}}\right) \quad P_0 > 0 \tag{A7}$$

with $\hat{S}_{t+1}$ defined by (13).

Assumptions.

$$\text{(H3)}\ \sup|\omega_t| \leqslant \Delta \tag{A8}$$

$$\text{(H4)}\ \sigma_t = 1 \text{ if } \frac{(S_{t+1} - \hat{S}_{t+1})^2}{1 + T^2 Y_t^2 P_{t-1}} > \Delta^2 \tag{A9}$$

$\sigma_t = 0$ otherwise.

Lemma 2. For the algorithm (A6) and (A7), subject to assumptions (H1)–(H4), then

$$\lim_{t\to\infty} \sup|S_t - \hat{S}_t| \leqslant C\Delta \tag{A10}$$

with C a positive constant independent of Δ.

Proof.

$$\text{Let } \tilde{k}_t = \hat{k}_t + \frac{k_1}{k_2}. \tag{A11}$$

Then, the following expression is readily derived from (A6) and (13):

$$\frac{\tilde{k}_{t+1}}{P_t} = \frac{\tilde{k}_t}{P_{t-1}} + \sigma_t T Y_t \omega_t. \tag{A12}$$

Then, by assumption (H4),

$$\frac{\tilde{k}_{t+1}^2}{P_t} - \frac{\tilde{k}_t^2}{P_{t-1}} \leqslant \left[\Delta^2 - \frac{(S_{t+1} - \hat{S}_{t+1})^2}{1 + \sigma_t T^2 Y_t^2 P_{t-1}}\right]. \tag{A13}$$

Then, $\dfrac{\tilde{k}_t^2}{P_{t-1}}$ is a nonincreasing function, bounded below by zero (since $P_{t-1} > 0$) and

$$\lim_{t\to\infty} \sigma_t\left[\Delta^2 - \frac{(S_{t+1} - \hat{S}_{t+1})^2}{1 + \sigma_t T^2 Y_t^2 P_{t-1}}\right] = 0. \tag{A14}$$

$$\text{Hence } \lim_{t\to\infty} \sup\left[\frac{(S_{t+1} - \hat{S}_{t+1})^2}{1 + \sigma_t T^2 Y_t^2 P_{t-1}}\right] \leqslant \Delta^2. \tag{A15}$$

Now, from (A7), the sequence P_t converges and we define

$$P_\infty = \lim_{t\to\infty} P_t \geqslant 0. \tag{A16}$$

Hence, in view of Lemma 1,

$$\lim_{t\to\infty} \sigma_t Y_t^2 P_{t-1} \leqslant T^2 \left(\frac{k_2}{k_1}\mu^* V_{max}\right)^2 P_\infty = C^2 - 1 \tag{A17}$$

and, from (A15),

$$\lim_{t\to\infty} \sup|S_{t+1} - \hat{S}_{t+1}| \leqslant C\Delta. \tag{A18}$$

Convergence of the adaptive control algorithm

We consider now the adaptive control algorithm (14). We have the following convergence result.

Theorem.

If (i) $V_{min} \leqslant V \leqslant V_{max}$
(ii) $S^* < V_{min}$
(iii) the parameter estimation algorithm (A6) and (A7) and the adaptive control algorithm (14) are used with

$$U_{max} \geqslant \frac{\mu^* V_{max}}{V_{min} - S^*} \tag{A19}$$

(iv) Assumptions (H1), (H2b), (H3) and (H4) hold

$$\text{then } \lim_{t\to\infty} \sup|S_t - S^*| \leqslant C\Delta \tag{A20}$$

with C the same constant as in Lemma 2.

Proof. From Lemma 2, for each $\eta > 0$, there exists $t_0 > 0$ such that

$$\hat{S}_t - C\Delta - \eta \leqslant S_t \leqslant \hat{S}_t + C\Delta + \eta \text{ for all } t \geqslant 0. \quad \text{(A21)}$$

Define the interval

$$I = [S^* - C\Delta - \eta,\ S^* + C\Delta + \eta]. \quad \text{(A22)}$$

(1) If the control algorithm gives $0 < U_{t_0} < U_{max}$ then $\hat{S}_{t_0+1} = S^*$ and hence $S_{t_0+1} \in I$.

(2) If the control algorithm gives $U_t = 0$ for $t = t_0 + k$ for $k = 0,1,2,3,\ldots$ then by definition of $\overline{U}_t$, we have $\hat{S}_{t_0+k} \geqslant S^*$ for $k = 1,2,3,\ldots$ and hence $S_{t_0+k} \geqslant S^* - C\Delta - \eta$.

But, if $U_t = 0$, S_t decreases and tends asymptotically to the steady-state $S = 0$, and there exists k' such that $S_{t_0+k'} \leqslant S^* + C_3\Delta + \eta$. The conclusion $S_{t_0+k'} \in I$ follows.

If the sequence $U_{t_0+k'} = 0$ terminates at time $t' > t_0$ so that $S_{t'} \in I$ and $0 < U_{t'} < U_{max}$, then $S_{t'+1} \in I$ as in (1) above.

(3) If the control algorithm gives $U_t = U_{max}$ for $t = t_0 + k$, $k = 0,1,2,3,\ldots$ then S_t increases since $\dot{S} \geqslant 0$ by definition of U_{max} and we can prove similarly that $S_{t_0+k'} \in I$ or $S_{t'+1} \in I$.

(4) So far, we have shown that there exists some $t_1 > t_0$ such that $S_{t_1} \in I$. Now it is easy to show that, if $S_{t_1} \in I$, then $S_t \in I$ for all $t \geqslant t_1$, by using the arguments of (1)–(3) above.

Thus we have

$$|S_t - S^*| \leqslant C\Delta + \eta \text{ for all } t \geqslant t_1. \quad \text{(A23)}$$

Since $\eta > 0$ may be chosen arbitrarily small, it follows from the definition of I:

$$\limsup_{t\to\infty} |S_t - S^*| \leqslant C\Delta. \quad \text{(A24)}$$

Q.E.D.

Comments.

(1) The controller achieves a zero steady-state error even with a varying disturbance V, since the algorithm includes feedforward compensation.

(2) In order to prove the convergence, the switching function σ_t has been included in the control algorithm and the following assumption has been stated

$$U_{max} \geqslant \frac{\mu^* V_{max}}{V_{min} - S^*}.$$

It is worth noting that these precautions were omitted in the simulation results presented above, since U_{max} was arbitrarily fixed at 0.39 and the switching function σ_t was not used in practice. These are necessary to prove the theoretical results but appear to be usually inoperative in the simulation experiments.

Model Reference Adaptive Control Algorithms for Industrial Robots*

S. NICOSIA† and P. TOMEI†

Key Words—Adaptive control; manipulation; model reference adaptive control; multivariable control systems; non-linear control systems; robots; state feedback; time-varying systems.

Abstract—In this paper some problems concerning the control of multifunctional manipulators (industrial robots) with high speed continuous movements are investigated. Although deterministic approaches to the control of robots, whose model are highly interconnected and non-linear, are known alternative approaches based on the Model Reference Adaptive System (MRAS) method of control are possible and useful. In the paper it is proved that a generalized MRAS control assures the convergence to a suitable reference model for a class of processes: the manipulator is shown to belong to such a class. The paper is completed by some applications evaluated by simulation.

1. INTRODUCTION

IN THE last years the use of adaptive control systems has been widely developed. Among these systems the so called Model Reference Adaptive System (MRAS) seems to be of a great relevance for the control of processes with variable parameters or with an unknown part.

A lot of work on this subject has been done especially with reference to continuous time linear systems; this may not be restrictive at all because of the possibility of considering non-linearities as model variations included in the unknown part of the process.

For a comprehensive exposition of MRAS applied to linear control systems the reader is referred to Landau (1979) and to references reported therein.

However, when the non-linearities are essential and the linear approximation cannot be used, complete non-linear models have to be considered.

The authors, who are engaged in a research programme on the applications of modern theories to the control of industrial robots whose models are highly non-linear, after some applications of the deterministic decoupling approach to the control (Nicosia *et al.*, 1981), are trying to apply MRAS methods to robotic processes. Some work has been done already along this line.

Dubowsky and Desforges (1979) derive an adaptive control algorithm which minimizes a quadratic function of the error defined as the difference between the desired state vector and the robotic process state vector. The model is assumed to be linear and decoupled; the control action has to be preceded by a 'learning period'. Tests have been performed referring to the movement of just a joint.

Arimoto and Takegaki (1981) apply adaptive control algorithms based on local parameter optimization to a robot described by a linear time-varying model derived from a linearization around the desired trajectory. The method assures just the stability of the error system. Simulation tests, referred to a non-linear complete robot model, give errors with a maximum of 1 cm when the trajectory is a straight line with a maximum speed of 0.1 m/s.

Tomizuka *et al.* (1982) refer to manipulators described by simplified dynamical equations obtained considering constant matrices in the model of the continuous case and just the linear term in the discrete one. The results are not easily comparable because step-responses are reported as examples.

Balestrino *et al.* (1983) consider a complete non-linear model and derive a control law based on the unit vector adaptation with a discontinuous process forcing signal similar to a pulse amplitude modulator. Simulation tests, limited to a continuous time control, are referred to a straight line trajectory with a speed of 0.75 m/s and exhibit a very good quality with a maximum error of 0.3 mm.

Nicolò and Katende (1983) use an original approach due to Miroshnik *et al.* (1982) where a separability property of Newtonian systems allows the construction of an adaptive control scheme with a partially linear control. A complete non-linear model is used to derive the control algorithm. Tests based on simulation and referred to a straight line with a maximum speed of 0.75 m/s, exhibit a maximum error of 0.35 mm when the control is continuous and of 3.5 mm when the control is discrete with a sampling time of 10 ms.

* Received 7 July 1983; revised 27 March 1984. The original version of this paper was not presented at any IFAC meeting. This paper was recommended for publication in revised form by associate editor P. Parks under the direction of editor B. Anderson.

† Department of Electrical Engineering, Seconda Università di Roma, Tor Vergata, Via Raimondo, 00173 Roma, Italy.

Reprinted with permission from *Automatica*, vol. 20, pp. 635–644, Sept. 1984.
Copyright © 1984, International Federation of Automatic Control IFAC. Published by Pergamon Press Ltd.

Referring to a time-varying non-linear model including all second order systems that can be derived from a generalized application of Lagrangian dynamics, we use the unit vector approach with a corrective term and without sign alternance.

2. PROCESS MODEL

We restrict ourselves to the class of models which are described by the following equations

$$\dot{y}(t) = A_p(y,t,\alpha)y(t) + B_p(y,t,\alpha)R_p u(t) + f(y,t,\alpha) \qquad (1)$$

where: $y \in \mathbb{R}^n$; $u \in \mathbb{R}^m$; $\alpha \in \Omega \subseteq \mathbb{R}^i$ $t, t_0 \in \mathbb{R}$, $t \geq t_0$; $\mathbb{R}^n$ is the state space, $\mathbb{R}^m$ the control space, $\mathbb{R}^i$ the parameter space, R_p a given non-singular $(m \times m)$ matrix and the vector field f and the matrices A_p and B_p have the following structure:

$$A_p(y,t,\alpha) \equiv \left[\begin{array}{c|c} 0 & I_{n-m} \\ \hline A_{p_1}(y,t,\alpha) & A_{p_2}(y,t,\alpha) \\ {[m \times m]} & \end{array}\right];$$

$$B_p(y,t,\alpha) \equiv \left[\begin{array}{c} 0 \\ {[(n-m) \times m]} \\ \hline B_{p_1}(y,t,\alpha) \\ {[m \times m]} \end{array}\right];$$

$$f(y,t,\alpha) \equiv \left[\begin{array}{c} 0 \\ \hline B_{p_1}(y,t,\alpha)\,[L_1(y,t,\alpha)Q(y,t) + L_2(y,t,\alpha)] \end{array}\right] = \left[\begin{array}{c} 0 \\ {[(n-m) \times 1]} \\ \hline f_1(y,t,\alpha) \\ {[m \times 1]} \end{array}\right]. \qquad (1a)$$

$Q(y,t)$ is known; $B_{p_1}(y,t,\alpha)$ is symmetric, positive definitive and bounded; $A_{p_1}(y,t,\alpha)$, $A_{p_2}(y,t,\alpha)$, $L_1(y,t,\alpha)$ and $L_2(y,t,\alpha)$ are bounded.

The assumptions in (1) and (1a) put a limit to the class of non-linear time-varying processes that can be handled,. However one can show that any 'natural' physical system, i.e. system in which the kinetic energy is a positive quadratic function of the state vector and the potential energy depends on the position (Arnold, 1978), falls in this class, provided that there is a control input for each nontrivial line in (1). This class embraces physical processes such as manipulator systems, power electrical networks (Marino and Nicosia, 1983), and many other processes including linear processes with known bounds on the parameters.

As far as manipulators are concerned, dynamical equations can be derived from well known results of the classical mechanics. For instance, using the Lagrangian approach, the equations are

$$\mu(t) = \frac{\mathrm{d}}{\mathrm{d}t}\left(\frac{\mathrm{d}L}{\mathrm{d}\dot{q}}\right) - \frac{\partial L}{\partial q} \qquad (2)$$

where: μ is the m dimensional vector field of the generalized forces to be supplied by the joint actuators; q and $\dot{q}$ are respectively the m dimensional vector fields of the relative rotations (or translations) between links and their time derivatives (velocities) and L is the Lagrange function expressed in terms of the kinetic energy $T(q,\dot{q})$ and of the potential energy $P(q)$.

Thus

$$L = T(q,\dot{q}) - P(q). \qquad (3)$$

The kinetic energy $T(q,\dot{q})$ is given by

$$T(q,\dot{q}) = \tfrac{1}{2}\dot{q}'B(q)\dot{q} \qquad (4)$$

where prime denotes transpose and $B(q)$, the inertia matrix, is symmetric and positive definite.

Eventually the robot motion is described by the matrix second order equation

$$\ddot{q}(t) = B^{-1}(q)[-f(q,\dot{q}) + r(q) + \mu(t)] \qquad (5)$$

where: $r(q)$ is the vector field of the generalized gravity forces given by

$$r(q) = -\frac{\partial P(q)}{\partial q}; \qquad (6)$$

$f(q,\dot{q})$ is the vector field of the centrifugal and Coriolis forces whose generic component is

$$f_i(q,\dot{q}) = \dot{q}'F_i(q)\dot{q}. \qquad (7)$$

$F_i(q)$ and $B(q)$ are related by the equation

$$F_i(q) = \frac{1}{2}\left[\frac{\partial b_i}{\partial q} - \frac{\partial B(q)}{\partial q_i} + \left(\frac{\partial b_i}{\partial q}\right)'\right] \qquad (8)$$

where b_i denotes the ith column of $B(q)$; $f(q,\dot{q})$ can be put in the form

$$f(q,\dot{q}) = C(q)h(\dot{q}) \qquad (9)$$

where $h(\dot{q}) \in \mathbb{R}^v$ $(v = m(m+1)/2)$ is

$$h(\dot{q}) \equiv \begin{bmatrix} \dot{q}_1 & \dot{q}_1 \\ \dot{q}_1 & \dot{q}_2 \\ \vdots & \\ \dot{q}_{m-1} & \dot{q}_m \\ \dot{q}_m & \dot{q}_m \end{bmatrix}$$

and the elements of $C(q)$ depend on the elements of $F_i(q)$.

With the above assumptions setting $G(q) = B^{-1}(q)$: (5) reduces to

$$\ddot{q}(t) = -G(q)[C(q)h(\dot{q}) - r(q)] + G(q)\mu(t). \tag{10}$$

Matrices $G(q)$, $C(q)$ and vector $r(q)$ depend on the physical parameters of the robot and its payload, which we arrange in vector α. Due to the characteristics of (4)–(9) the previous matrices and vectors are bounded for any allowed value of α and for any q; consequently, (10) can be considered as a special case of (1), when

$$y \equiv \begin{bmatrix} q \\ \dot{q} \end{bmatrix} \quad u \equiv \mu, \quad A_{p_1} \equiv 0, \quad A_{p_2} \equiv 0,$$

$$B_{p_1} \equiv G, \quad L_1 \equiv -C, \quad Q \equiv h, \quad L_2 \equiv r, \quad R_p \equiv I.$$

3. CONTROL SYSTEM STRUCTURE AND ALGORITHMS

Among the different approaches to MRAS the Adaptive Model Following Control (AMFC) has been adopted (Landau, 1979). A general scheme of AMFC is reported in Fig. 1.

The aim of the control is to nullify the differences between the behavior of the controlled process and the behavior of a reference model via a suitable adaptation mechanism.

A linear model that assures the fulfillment of suitable input–output specifications is usually chosen as the reference model. Since it has to be structurally similar to the process model, with reference to (1), we may adopt the following one

$$\dot{x}(t) = A_M x(t) + B_M u_M(t) \tag{11}$$

in which $x \in \mathbb{R}^n$; $u_M \in \mathbb{R}^p$;

$$A_M \equiv \left[\begin{array}{c|c} 0 & I_{n-m} \\ \hline A_{M_1} & A_{M_2} \\ [m \times m] & \end{array}\right];$$

$$B_M \equiv \left[\begin{array}{c} 0 \\ [(n-m) \times p] \\ \hline B_{M_1} \\ [m \times p] \end{array}\right].$$

A_M and B_M are constant matrices, and A_M is a strict Hurwitz matrix. Equations (1) and (11) may or may not fulfil the Erzaberger (1968) conditions of perfect model matching which are somewhat restrictive (Marino and Nicosia, 1984).

A block diagram representing the actual control structure is shown in Fig. 2 in which $e \equiv x - y \in \mathbb{R}^n$; $v = Je \in \mathbb{R}^n$; J is a suitable $[n \times n]$ constant matrix; $K_u(v, t)$ is a $[m \times p]$ matrix; $d(v, t) \in \mathbb{R}^m$ is a vector; T is a $[p \times n]$ constant matrix such that

$$T = \left[\begin{array}{c} I_n \\ \hline 0 \\ [(p-n) \times n] \end{array}\right] \qquad p \geq n$$

$$T = \left[\begin{array}{c|c} I_p & \begin{array}{c} 0 \\ [p \times (n-p)] \end{array} \end{array}\right] \qquad p \leq n.$$

In order to simplify the exposition the vector v is partitioned as follows

$$v = \begin{bmatrix} v_1 \\ v_2 \end{bmatrix} \qquad v_1 \in \mathbb{R}^{n-m}, \; v_2 \in \mathbb{R}^m$$

and the matrix $\bar{A}$ is defined as

$$\bar{A} = (A_M - A_p) + B_M T = \left[\begin{array}{c} 0 \\ [(n-m) \times n] \\ \hline \bar{A}_1 \\ [m \times n] \end{array}\right]. \tag{12}$$

In order to determine the control law it is customary to use the hyperstability approach (Popov, 1973) so that the asymptotic stability of the error system is assumed. The following definitions and theorems can be usefully applied.

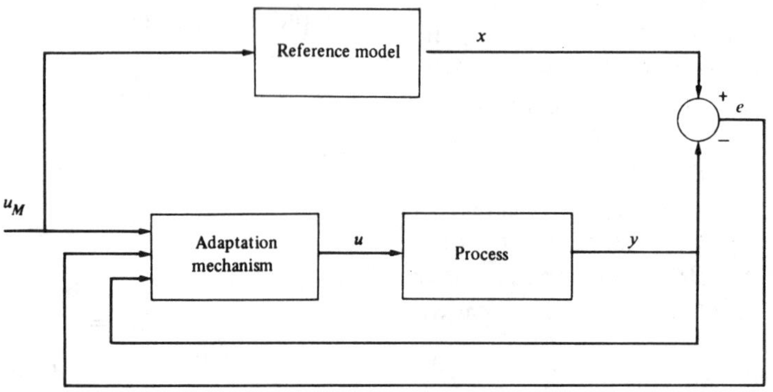

FIG. 1. General scheme of an Adaptive Model Following Control System (AMFC).

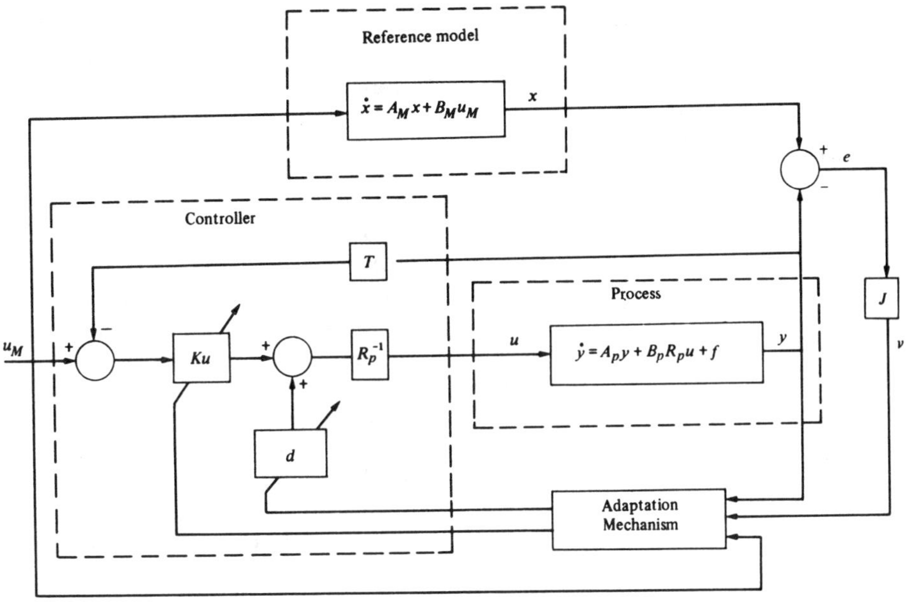

FIG. 2. Block diagram of the control structure.

Definition 1. Given a $[n \times m]$ matrix B whose transpose is B', a scalar $\|B\|$ is accepted as norm of B if

$$\|B\| = f(B'B) \quad \text{is a norm} \tag{13}$$

$$\|B\| = \|B'\| \tag{14}$$

and for $m = 1$

$$f(B'B) = \sqrt{\sum_{i=1}^{n} b_i^2}. \tag{15}$$

Notice that the function $f(\cdot)$ is any function for which statement (13) is true; generally $f(A)$, with A a generic square matrix, is not a norm.

The following

$$\|B\| = \sqrt{\operatorname{tr} B'B} = \sqrt{\sum_{i=1}^{n} \sum_{j=1}^{m} b_{ij}^2};$$

$$\|B\| = \sqrt{\text{max eigenvalue of } B'B}$$

are examples of norms satisfying conditions (13)–(15).

The subsequent results hold whatever norm satisfying conditions (13)–(15) are used.

Definition 2. Given a $[m \times n]$ matrix $F(\gamma)$ depending on a vector $\gamma \in \Gamma \subseteq R^h$ $M_f[F(\gamma)]$ and $M_n[F(\gamma)]$ are defined as follows

$$M_f[F(\gamma)] = \max_{\gamma \in \Gamma} f[F(\gamma)]$$

$$M_n[F(\gamma)] = \max_{\gamma \in \Gamma} \|F(\gamma)\| = M_f[F'(\gamma)F(\gamma)]$$

where $f(\cdot)$ is the same function used in Definition 1.

Theorem 1. Referring to the block diagram in fig. 2, let P_u be a $(p \times p)$ symmetric positive definitive constant matrix, $K_u(v, t)$ and $d(v, t)$ be defined as

$$K_u(v, t) = \frac{v_2}{\|v_2\|} \frac{(u_M - Ty)'}{\|u_M - Ty\|} P_u \tag{16}$$

$$d(v, t) = \frac{v_2}{\|v_2\|} (k_{d1} \|Q(y, t)\| + k_{d2} \|y\| + k_{d3}) \tag{17}$$

then $\lim_{t \to \infty} e(t) = 0$ for any initial condition $e(t_0)$, for any $\alpha \in \Omega$ (assumed constant during the adaptation process) and for any piece-wise continuous input vector functions u_M, if

$$\frac{1}{f^2(P_u^{-1})} \geq M_f^2(B_{p1}^{-1}) M_n(B_{M1}), \tag{18}$$

$$k_{d1} \geq M_f^2(B_{p1}^{-1}) M_n(B_{p1} L_1), \tag{19}$$

$$k_{d2} \geq M_f^2(B_{p1}^{-1}) M_n(\bar{A}_1), \tag{20}$$

$$k_{d3} \geq M_f^2(B_{p1}^{-1}) M_n(B_{p1} L_2), \tag{21}$$

and matrix J is solution of equation

$$J A_M + A_M' J = -R \tag{22}$$

where R is an arbitrary symmetric positive definite matrix.

Control laws of the type (17) are called unit vector control laws in which the vector $v_2/\|v_2\|$ defines the direction of the control action.

If the symmetric positive definite matrix B_{p1} can be assumed to be a positive definite kernel (see Definition A.4), the following theorem holds.

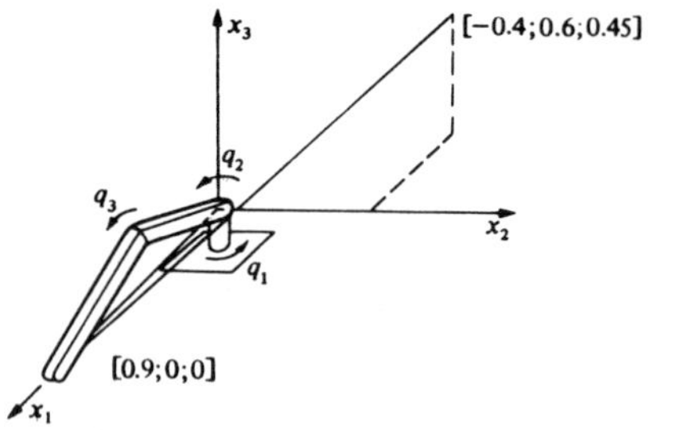

FIG. 3. Robot and trajectory considered in the case study.

Theorem 2. Let $K_u(v, t)$ be defined as in Theorem 1, and let $d(v, t)$ be defined as

$$d(v,t) = \frac{v_2(t)}{\|v_2(t)\|}(k_{d1}\|Q(y,t)\| + k_{d2}\|y\| + k_{d3}) + k_i \int_{t_0}^{t} v_2(\tau)\,d\tau$$

then $\lim_{t\to\infty} e(t) = 0$ for any initial condition $e(t_0)$, for any $\alpha \in \Omega$ (assumed constant during the adaptation process) and for any piecewise continuous input vector functions u_M if k_i is a non-negative constant and if conditions (18–22) of Theorem 1 hold.

The proofs of theorems are reported in Appendix 2.

4. APPLICATION TO THE CONTROL OF ROBOTS

A rigid joint robot is modeled by (10) which belongs to the class of systems (1).

The generalized forces $\mu(t)$ are actually applied to the manipulator by an appropriate actuator system. We refer to a set of d.c. motors with armature control whose equations are

$$\mu(t) = S\xi(t) - N\dot{q}(t) \qquad (23)$$

in which S and N are $(m \times m)$ constant diagonal matrices and ξ is the m-dimensional vector of applied voltages.

Using (10) and (23) together

$$\ddot{q}(t) = -G(q)\,[C(q)h(\dot{q}) + N\dot{q}(t) - r(q)] + G(q)S\xi(t). \qquad (24)$$

With the following assumptions (24) may be also considered as a particular case of (1), when

$$y \equiv \begin{bmatrix} q \\ \dot{q} \end{bmatrix}, \quad u \equiv \xi, \quad A_{p1} \equiv 0, \quad A_{p2} \equiv 0,$$

$$B_{p1} \equiv G, L_1 \equiv [-C \;\vdots\; -N], \qquad Q \equiv \begin{bmatrix} h \\ \dot{q} \end{bmatrix},$$

$$L_2 \equiv r, \quad R_p \equiv S. \qquad (25)$$

Results of Theorems 1 and 2 may be applied following a simple procedure.

If, as usual, a decoupled reference model is adopted A_{M1} and A_{M2} are diagonal matrices. Besides R may be chosen to be diagonal of type

$$R = \begin{bmatrix} R_1 & 0 \\ 0 & R_4 \end{bmatrix}$$

where R_1 and R_4 are two arbitrary $(m \times m)$ diagonal matrices. Then (22) can be easily solved.

In fact from (22), particularized for $n = 2m$, we have

$$J = \begin{bmatrix} J_1 & J_2 \\ J_2 & J_4 \end{bmatrix}$$

where J_1, J_2 and J_4 are diagonal matrices given by

$$J_1 = \tfrac{1}{2}(I - A_{M2}A_{M1}^{-1})R_1 - \tfrac{1}{2}A_{M1}A_{M2}^{-1}R_4,$$

$$J_2 = \tfrac{1}{2}A_{M1}^{-1}R_1,$$

$$J_4 = \tfrac{1}{2}A_{M2}^{-1}(R_4 - A_{M1}^{-1}R_1).$$

Without loss of generality it may be assumed $P_u = k_u I$, k_u being a positive constant. The values k_{d1}, k_{d2}, k_{d3}, k_i (if Theorem 2 is used), and k_u are computed with reference to the upperbounds of $\|G\|$, $\|C\|$, $\|h\|$, and $\|r\|$, that are supposed known. Since the hyperstability conditions invoked in the proofs of theorems are just sufficient conditions, the computed constants may be considered as a first approximation of more suitable values that should be obtained by interactive simulation. These values are indicated as 'a good choice' in the outlined cases. It should be noted that also the initial values may be good for control, but too high with respect to usually accepted control gains.

In the application the three degrees of freedom (DOF) robot of Fig. 3 has been considered. It corresponds to an actual robot whose links are 0.5 m long; the dynamical model of this robot, obtained neglecting elasticity and friction at joints, is reported in Appendix 3, together with numerical values of parameters when the load mass varies.

The tests were referred to a straight line trajectory (see Fig. 3) and to a trapezoidal velocity law with a maximum velocity of 0.75 m/s and maximum acceleration of 0.75 m/s^2.

Control laws obtained in Theorem 1 namely without integral action give satisfactory results only when the greatest velocity is of the order of 0.3 m/s: with larger velocities the torques applied to the axes of joints are not compatible with the size of the motors generally used for robots.

Various tests were performed with the control laws derived in Theorem 2. The more interesting ones are reported.

Case 1

Decoupled second order reference model. The model used is the same as in (11) with these assumptions

$$A_{M1} = -a_1 I_3 \qquad u_M(t) = q_d(t)$$
$$A_{M2} = -a_2 I_3$$
$$B_{M1} = a_1 I_3$$

in which $q_d(t)$ is the vector of the desired trajectory. With this choice the reference model is constituted by three decoupled identical second order systems.

The control law derived from Theorem 2 is the following (see Fig. 2):

$$u(t) = s^{-1}\left[\frac{v_2}{\|v_2\|}\left(k_u\|q_d - q\| + k_{d1}\left\|\begin{bmatrix} h \\ \dot{q} \end{bmatrix}\right\| + k_{d2}\left\|\begin{bmatrix} q \\ \dot{q} \end{bmatrix}\right\| + k_{d3}\right) + k_i \int_0^t v_2 \, dt\right]. \qquad (26)$$

With $a_1 = 10^6$ and $a_2 = 1400\,s^{-1}$, a good choice of parameters of control is

$$k_u = 1200; \quad k_{d1} = 1; \quad k_{d2} = 10; \quad k_{d3} = 0;$$
$$k_i = 2000; \quad v_2 = (j_2 I_3 \; \vdots \; j_4 I_3)e$$

with $j_2 = 10.0$ and $j_4 = 0.5$.

The results of simulation tests, with a load mass of 5 kg, are shown in Figs 4 and 5 where EPS is the absolute error; ELONG is the absolute error along the trajectory; ETRAS is the absolute error on the perpendicular to the trajectory; EMR is the relative error between the output of the model and the output of the robot; μ_1, μ_2 and μ_3 are the torques applied to each joint. Errors are in millimeters, couples in Newton meters.

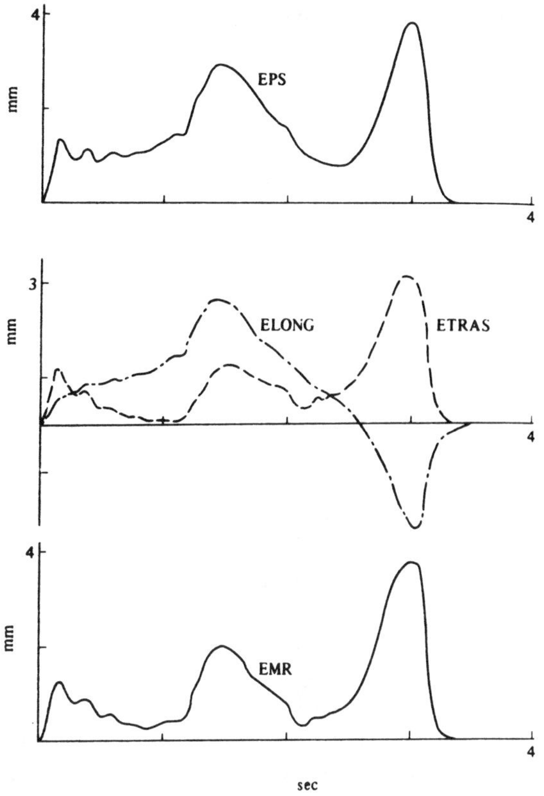

FIG. 4. Errors in the case of continuous control with a second order model ($m_l = 5$ kg).

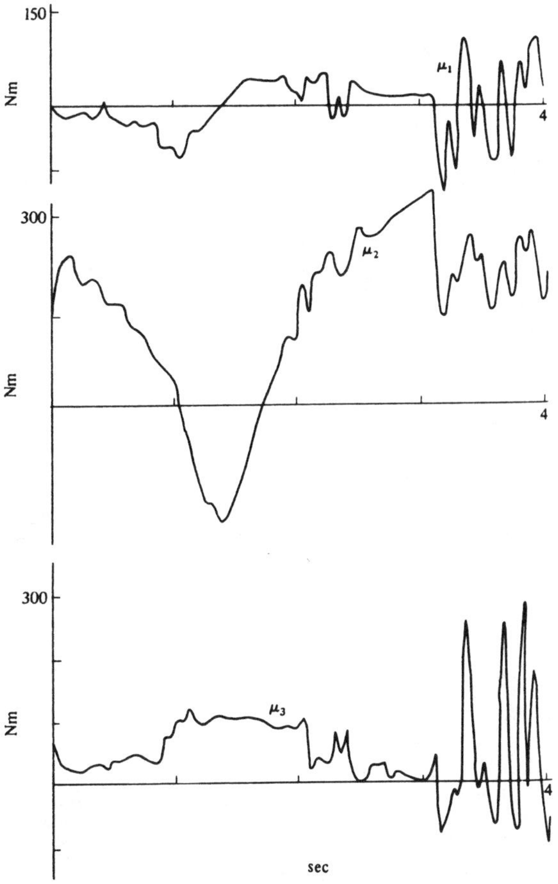

FIG. 5. Torques in the case of continuous control with a second order model ($m_l = 5$ kg).

Case 2

Decoupled second order reference model with augmented input vector. The model used is the following

$$A_{M1} = A_{M2} = -I_3 \qquad u_M(t) = \begin{bmatrix} q_d(t) \\ \dot{q}_d(t) \\ \ddot{q}_d(t) \end{bmatrix}.$$
$$B_{M1} = [I_3 \; \vdots \; I_3 \; \vdots \; I_3]$$

Notice that in this case

$$x(t) = \begin{bmatrix} q_d \\ \dot{q}_d \end{bmatrix}.$$

Namely the model exhibits the desired motion law of the robot, as the state vector.

The control law derived in Theorem 2 is the following

$$u(t) = S^{-1}\left[\frac{v_2}{\|v_2\|}\left(k_u\left\|\begin{bmatrix} q_d - q \\ \dot{q}_d - \dot{q} \\ \ddot{q}_d \end{bmatrix}\right\| + k_{d1}\left\|\begin{bmatrix} h \\ \dot{q} \end{bmatrix}\right\| + k_{d2}\left\|\begin{bmatrix} q \\ \dot{q} \end{bmatrix}\right\| + k_{d3}\right) + k_i\int_0^t v_2\,\mathrm{d}t\right]. \qquad (27)$$

A good choice of parameters is

$$k_u = 2900, \quad k_{d1} = 1.0, \quad k_{d2} = 0, \quad k_{d3} = 0,$$
$$k_i = 3000, \quad v_2 = (j_2 I_3 \,\vdots\, j_4 I_3)e$$

with $j_2 = 10.0$ and $j_4 = 0.5$.

In Figs 6 and 7 the results of simulation where the payload is 5 kg (nominal value) are reported. No significant variations have been observed when the payload is 0 or 10 kg.

If the control law is implemented on a suitable computer by using a rectangular approximation to the integral, the performance quality decreases substantially depending on the sampling time T. The shapes of the errors and of the applied torques remain the same whereas the maximum values are reported in Table 1.

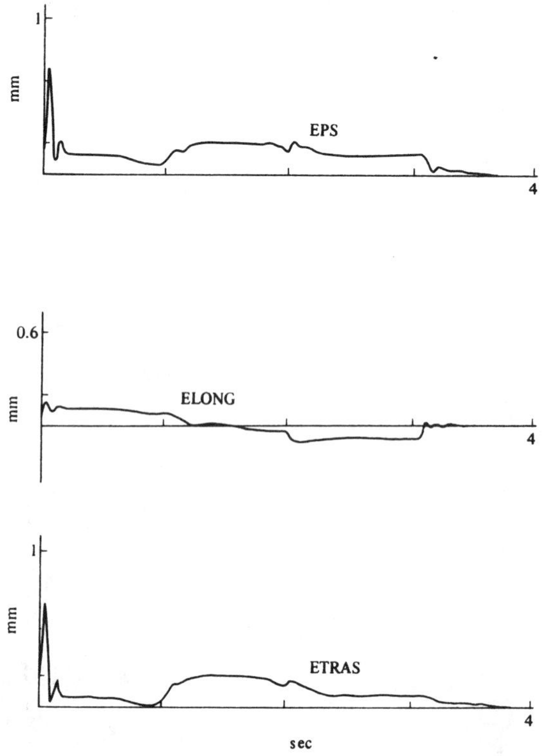

FIG. 6. Errors in the case of continuous control with an augmented model ($m_l = 5$ kg).

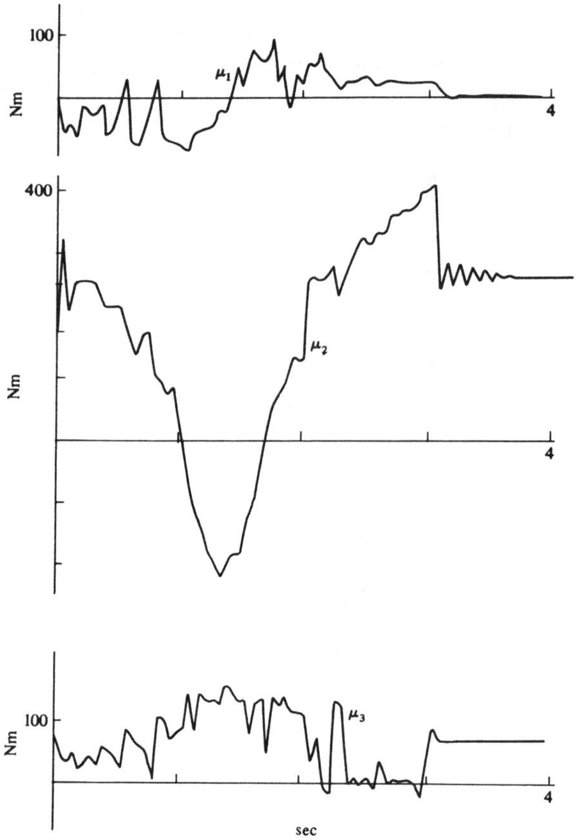

FIG. 7. Torques in the case of continuous control with an augmented model ($m_l = 5$ kg).

A complete set of responses is reported in Nicosia and Tomei (1983).

5. CONCLUSIONS

A first approach to the control of robots using a unit vector adaptive control law has been presented.

The results can be compared with the other ones obtained considering complete non-linear models (Nicolò and Katende, 1983; Balestrino *et al.*, 1983). These authors used as application the same robot and the same test (a continuous path control on a straight line). Comparisons may also be made with results obtained applying decoupling algorithms (Nicosia *et al.*, 1981).

Generally speaking, due to the robustness of the decoupling algorithms, parameter variations have little influence on the performances of the controls proposed by all these approaches. The use of

TABLE 1. MAXIMUM ERRORS AND TORQUES IN THE CASE OF DISCRETE-TIME CONTROL

T	10 ms	15 ms	20 ms
EPS_{max}	3.0 mm	5.3 mm	6.0 mm
$ELONG_{max}$	0.8 mm	1.8 mm	1.9 mm
$ETRAS_{max}$	2.9 mm	5.0 mm	5.5 mm
$\mu_{1\,max}$	70 Nm	75 Nm	80 Nm
$\mu_{2\,max}$	410 Nm	430 Nm	400 Nm
$\mu_{3\,max}$	170 Nm	130 Nm	200 Nm

adaptive algorithms is mainly justified by the significant simplifications on the implementation of the control structures.

Our control algorithm (27) is the simplest and gives results slightly less precise than those obtained via the decoupling algorithm which, on the other hand, is more complex.

Further studies on the use of MRAS with models of robots with elasticity at joints are in our programmes.

Acknowledgements—The authors are deeply grateful to Professor F. Nicolò for many helpful discussions and to A. Ficola for his support in developing simulation programs and tests. This work was supported by the Progetto Finalizzato Informatica of C.N.R. under a research contract.

REFERENCES

Arimoto, S. and M. Takegaki (1981). An adaptive method for trajectory control of manipulators. IFAC 8th Triennial World Congress, Kyoto, August.

Arnold, V. I. (1978). *Mathematical Methods of Classical Mechanics.* Springer-Verlag, New York.

Balestrino, A., G. De Maria and L. Sciavicco (1983). An adaptive model following control system for robotic manipulators. *Trans. ASME, J. Dynam. Syst. Measure. Control,* **105**, 143–151.

Cesareo, G., F. Nicolò and S. Nicosia (1981). Dynamical models of industrial Robots. First IASTED International Symposium on Applied Modeling and Simulation, Lyon, September.

Dubowsky, S. and D. T. Desforges (1979). Robotic manipulator control systems with invariant dynamic characteristics. *Proceedings of the 5th World Congress on Machines and Mechanism (IFIOMM),* Montreal, July.

Erzberger, H. (1968). Analysis and design of model following systems by state space techniques. *Proc. JACC,* 572–581.

Landau, Y. D. (1979). *Adaptive Control: the Model Reference Approach.* Marcel Dekker, New York.

Marino, R. and S. Nicosia (1983). Hamiltonian type Lyapunov functions. *IEEE Trans Aut. Control,* **AC-28**, 1055–1057.

Marino, R. and S. Nicosia (1984). Linear-model-following control and feedback-equivalence to linear controllable systems. *Int. J. Control,* **39**, 473–485.

Miroshnik, I., F. Nicolò and S. Nicosia (1982). Robust model reference adaptive control. *Proceedings of Workshop on Adaptive Control,* Florence, Italy, September.

Nicolò, F. and J. Katende (1983). A robust MRAC for industrial Robots. 2nd IASTED International Symposium on Robotics and Automation, Lugano.

Nicosia, S., F. Nicolò and D. Lentini (1981). Dynamical control of industrial robots with elastic and dissipative joints. IFAC VIII Triennal Congress, Kyoto, August.

Nicosia, S. and P. Tomei (1983). Model reference adaptive control algorithms applied to industrial robots. Report 07.83, Dept. Informatica e Sistemistica, University of Rome.

Popov, V. M. (1973). *Hyperstability of Control Systems.* Springer-Verlag, Berlin.

Tomizuka, M., R. Horowitz and Y. D. Landau (1982). On the use of model reference adaptive control techniques for mechanical manipulators. 2nd IASTED Symposium on identification, control and robotics, Davos, March.

APPENDIX 1

Hyperstability

In this appendix some theoretical results on hyperstability theory will be presented. For a more detailed exposition refer to Popov (1973) or Landau (1979).

Consider the system in Fig. A1, the forward path is described by the equations

$$\dot{x} = Ax + Bu \tag{A1}$$

$$v = Cx \tag{A2}$$

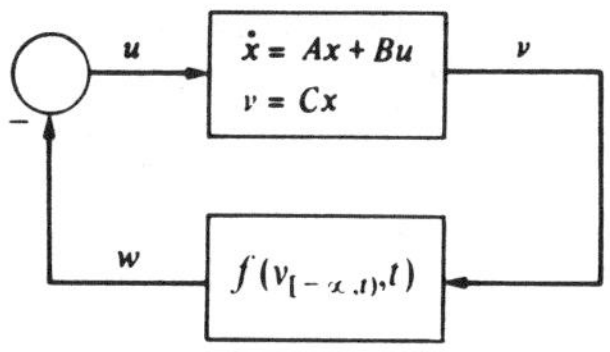

FIG. A1. Non-linear feedback multivariable system.

where $x \in \mathbb{R}^n$ is the state vector, $u \in \mathbb{R}^m$ and $v \in \mathbb{R}^m$ are respectively the input vector and the output vector of the forward path, A, B and C are constant matrices of suitable dimensions, the pairs (A, B) and (C, A) are respectively completely controllable and completely observable.

The backward path is described by the non-linear equation

$$w = -u = f(v_{[-\infty,t)}, t) \tag{A3}$$

Where f is a suitable functional depending on all values of v on the interval $[-\infty, t)$.

There exists a constant γ_0 such that the following inequality holds

$$\eta(t_0, t_1) = \int_{t_0}^{t_1} v'w\,dt \geq -\gamma_0^2 \quad \forall t_1 \geq t_0. \tag{A4}$$

The system described by (A1)–(A3) is said to be asymptotically hyperstable if it is globally asymptotically stable for all non-linear feedback blocks satisfying (A4).

On the hyperstability the following general results must be recalled that are useful for the proofs of the theorems.

Theorem A1. The necessary and sufficient condition for the feedback system described by (A1)–(A3) to be asymptotically hyperstable is that the forward path transfer matrix

$$W(s) = C(sI - A)^{-1}B \tag{A5}$$

be a strictly positive real matrix.

Definition A2. A square matrix $W(s)$ is said to be a strictly positive real transfer matrix if:

(i) all elements of $W(s)$ do not have poles in $R_e(s) \geq 0$;

(ii) matrix $W(j\omega) + W'(-j\omega)$ is positive definite for all real values of ω.

Theorem A3. Matrix $W(s)$, given by (A5), is strictly positive real if there exist two $(n \times n)$ symmetric positive definite matrices P and R such that

$$PA + A'P = -R \tag{A6}$$

$$B'P = C. \tag{A7}$$

Definition A4. The square matrix kernel $K(t, \tau)$ is called positive definite if for each interval $[t_0, t_1]$ and all the vector functions $f(t)$ that are piecewise continuous in $[t_0, t_1]$ the following inequality holds:

$$\eta(t_0, t_1) = \int_{t_0}^{t_1} f'(t) \int_{t_0}^{t} K(t, \tau) f(\tau)\,d\tau\,dt \geq 0. \tag{A8}$$

Notice that if K is a constant symmetric positive definite matrix (A8) is verified.

APPENDIX 2

Proof of the theorems

Prior to prove the theorems we recall some properties that have been proved by Nicosia and Tomei, 1983.

Property B1. Let $B \in \mathbb{R}^n$, $\gamma \in \Gamma \subseteq \mathbb{R}^h$, $\delta \in \mathbb{R}^n$ and $\eta \in H \subseteq \mathbb{R}^j$ be vectors, let $\phi(\gamma)$ be a symmetric, positive definite $[n \times n]$ matrix bounded for any γ, and $\psi(\eta)$ be a symmetric positive definite

$[n \times n]$ matrix for any η, let P_k be a constant symmetric positive definite $[n \times n]$ matrix and let $K(\beta,\delta)$ be a $[n \times n]$ matrix defined as

$$K(\beta,\delta) = \frac{\beta}{\|\beta\|}\frac{\delta'}{\|\delta\|}P_k$$

if

$$\frac{1}{f^2(P_k^{-1})} \geq M_n[\psi(\eta)]M_f^2[\phi^{-1}(\gamma)]$$

then

$$\beta'[\phi(\gamma)K(\beta,\delta) + \psi(\eta)]\delta \geq 0 \quad \forall \gamma \in \Gamma, \forall \eta \in H.$$

Property B2. Let β, γ, $\phi(\gamma)$ and η be defined as above, let $\rho \in \mathbb{R}^r$ be a vector, $M(\rho)$ be a $[m \times n]$ matrix, $N(\eta)$ be a $[n \times m]$ matrix bounded for any η and $\lambda(\rho)$ be defined as

$$\lambda(\rho) = \frac{\beta}{\|\beta\|}\|M(\rho)\|k_\lambda$$

if

$$k_\lambda \geq M_n[\phi(\gamma)N(\eta)]M_f^2[\phi^{-1}(\gamma)]$$

then

$$\beta'\phi(\gamma)[\lambda(\rho) + N(\eta)M(\rho)] \geq 0 \quad \forall \gamma \in \Gamma, \forall \eta \in H, \forall \rho \in \mathbb{R}^r.$$

Proof of Theorem 1. The proof will be divided into three parts. First the whole control system will be reduced to the scheme of Fig. A1; second, the inequality (A4) will be verified, and third, the conditions of Theorem A1 will be verified.

(i) The system of Fig. 2 is described by the equations

$$\dot{x}(t) = A_M x(t) + B_M u_M(t) \quad \text{(A9)}$$

$$\dot{y}(t) = A_p(y,t,\alpha)y(t) + B_p(y,t,\alpha)R_p u(t) + f(y,t,\alpha) \quad \text{(A10)}$$

$$u(t) = R_p^{-1}[K_u(v,t)(u_M(t) - Ty(t)) + d(v,t)] \quad \text{(A11)}$$

$$v(t) = J[x(t) - y(t)] = Je(t). \quad \text{(A12)}$$

From (A9)–(A11) and (12), we obtain

$$\dot{e} = A_M e + (B_M - B_p K_u)(u_M - Ty) + \bar{A}y - B_p d - f. \quad \text{(A13)}$$

(A12) and (A13) are equivalent to

$$\dot{e} = A_M e + I w_1 \quad \text{(A14)}$$

$$v = J \cdot e \quad \text{(A15)}$$

$$w = -w_1 = (B_p K_u - B_M)(u_M - Ty) + B_p d + f - \bar{A}y. \quad \text{(A16)}$$

(A14)–(A16) lead to the equivalent representation given in Fig. 1 (*cf.* Appendix 1). Moreover the pair (A_M, I) is completely controllable; since A_M is a strict Hurvitz matrix (and consequently from (22) J is a symmetric positive definite matrix) the pair (J, A_M) is completely observable.

(ii) The inequality (A4) of the Appendix 1 referred to the system of Fig. 2 is:

$$\int_{t_0}^{t_1} v'w\,dt = \int_{t_0}^{t_1} v'[(B_pK_u - B_M)(u_M - Ty) + B_pd + f - \bar{A}y]\,dt \geq -\gamma_0^2 \quad \forall t_1 \geq t_0. \quad \text{(A17)}$$

(A17) is true if the following inequalities hold

$$\int_{t_0}^{t_1} v'(B_pK_u - B_M)(u_M - Ty)\,dt = \int_{t_0}^{t_1} v_2'(B_{p1}K_u - B_{M1})(u_M - Ty)\,dt \geq -\gamma_{01}^2 \quad \forall t_1 \geq t_0 \quad \text{(A18)}$$

$$\int_{t_0}^{t_1} v'(B_pd + f - \bar{A}y)\,dt = \int_{t_0}^{t_1} v_2'(B_{p1}d + f_1 - \bar{A}_1 y)\,dt \geq -\gamma_{02}^2 \quad \forall t_1 \geq t_0 \quad \text{(A19)}$$

γ_{01} and γ_{02} are constant.

On the basis of property B1, and of hypotheses of this theorem, (A18) is true. (A19), taking into account (1A) and (17), becomes

$$\int_{t_0}^{t_1}\left[v_2'B_{p1}\left(\frac{v_2}{\|v_2\|}\|Q\|k_{d1} + L_1Q\right) + v_2'B_{p1}\left(\frac{v_2}{\|v_2\|}\|y\|k_{d2} - B_{p1}^{-1}\bar{A}_1y\right) + v_2'B_{p1}\left(\frac{v_2}{\|v_2\|}k_{d3} + L_2\right)\right]dt \geq -\gamma_{02}^2 \quad \forall t_1 \geq t_0. \quad \text{(A20)}$$

On the basis of hypotheses of this theorem and of property B2, applied to each of the three terms in the integrating function, (A20) is true.

(iii) In order for the forward path to fulfil the hypotheses of Theorem A1 it is sufficient to verify the existence of symmetric positive definite matrices P and R which solve the equations (see Theorem A3):

$$PA_M + A_M'P = -R \quad \text{(A21)}$$

$$P = J. \quad \text{(A22)}$$

A_M being a strict sense Hurwitz matrix (22) assures that (A21) and (A22) are verified and on the basis of Theorem A1 the theorem is proved.

Proof of theorem 2. The proof is exactly the same of the Theorem 1 provided that

$$\int_{t_0}^{t_1} v'(t)B_p(y,t,\alpha)k_i\int_{t_0}^{t} v_2(\tau)\,d\tau\,dt \geq -\gamma_{03}^2. \quad \text{(A23)}$$

(A23) is equivalent to

$$k_i\int_{t_0}^{t_1} v_2'(t)\int_{t_0}^{t} B_{p1}(y,t,\alpha)v_2(\tau)\,d\tau\,dt \geq -\gamma_{03}^2 \quad \text{(A24)}$$

which, k_i being a non negative constant and $B_{p1}(y,t,\alpha)$ a positive definite kernel, surely holds.

APPENDIX 3

Model of the robot

Following the notation of (10), the non-zero elements of matrices $B(q)$ and $C(q)$ and of vector $r(q)$ referred to the three DOF robot studied (see Fig. 3) are

$$B_{11} = A_1 + A_2\cos^2 q_2 + A_3\cos^2(q_2 + q_3) + A_4\cos q_2\cos(q_2 + q_3)$$

$$B_{22} = A_5 + A_4\cos q_3$$

$$B_{23} = A_6 + A_7\cos q_3$$

$$B_{32} = B_{23}$$

$$B_{33} = A_8$$

$$C_{12} = -A_2\sin 2q_2 - A_3\sin 2(q_2 + q_3) - A_4\sin(q_3 + 2q_2)$$

$$C_{13} = -A_3\sin 2(q_2 + q_3) - A_4\cos q_2\sin(q_2 + q_3) \quad \text{(A25)}$$

$$C_{21} = -C_{12}/2$$

$$C_{25} = -A_4\sin q_3$$

$$C_{26} = -C_{25}/2$$

$$C_{31} = -C_{13}/2$$

$$C_{34} = -C_{25}/2$$

$$r_2 = B_1\cos q_2 + B_2\cos(q_2 + q_3)$$

$$r_3 = B_2\cos(q_2 + q_3)$$

The equations (A25) were derived with code DYMIR described in Cesareo, Nicolò and Nicosia (1981).

In Table A1 the numerical values of coefficients in (A25) for three different values of load mass m_l are given.

The manipulator is actuated by d.c. motors. Referring to (23) numerical values are the following:

$$S_{11} = 29.59 \quad S_{22} = 25.42 \quad S_{33} = 29.59$$
$$N_{11} = 814.36 \quad N_{22} = 1163.32 \quad N_{33} = 814.36.$$

All values are in SI units.

TABLE A1. NUMERICAL VALUES OF COEFFICIENTS IN (A25)

	m_l		
	0	5	10
A_1	23.3803	23.3803	23.3803
A_2	9.2063	10.4563	11.7063
A_3	2.4515	3.7015	4.9515
A_4	5.4	7.9	10.4
A_5	82.399	84.899	87.399
A_6	2.6274	3.8774	5.1274
A_7	2.7	3.95	5.2
A_8	25.7778	27.0278	28.2778
B_1	−189.1708	−213.6748	−238.1788
B_2	−52.9286	−77.4326	−101.9366

Adaptive Locomotion of a Multilegged Robot over Rough Terrain

ROBERT B. McGHEE, MEMBER, IEEE, AND GEOFFREY I. ISWANDHI

Abstract—**Although the off-road mobility characteristics of wheeled or tracked vehicles are generally recognized as being inferior to those of man and cursorial animals, the complexity of the joint-coordination control problem has thus far frustrated attempts to achieve improved vehicular terrain adaptability through the application of legged locomotion concepts. Nevertheless, the evident superiority of biological systems in this regard has motivated a number of theoretical studies over the past decade which have now reached a state of maturity sufficient to permit the construction of experimental computer-controlled *adaptive walking machines*. At least two such vehicles are known to have recently demonstrated legged locomotion over smooth hard-surfaced terrain. This paper is concerned with an extension of the present theory of limb coordination for such machines to the case in which the terrain includes regions not suitable for weight-bearing and which must consequently be avoided by the control computer in deciding when and where to successively place the feet of the vehicle. The paper includes a complete problem formalization, a heuristic algorithm for solution of the problem thus posed, and a preliminary evaluation of the proposed algorithm in terms of a computer simulation study.**

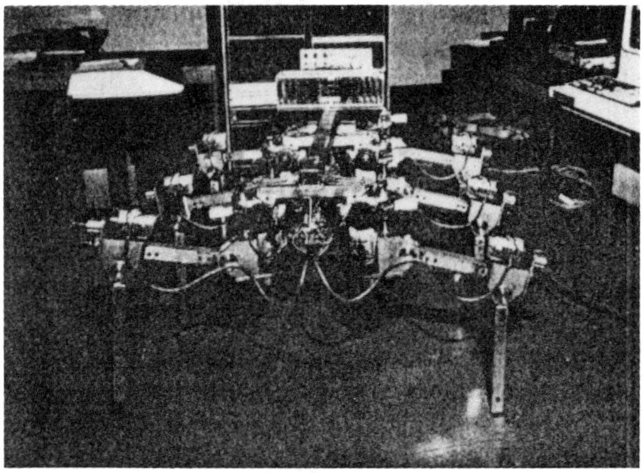

Fig. 1. Front view of OSU hexapod vehicle showing control computer in background. Vehicle length = 1.3 m, width = 1.4 m, total weight exclusive of cables = 103 kg.

INTRODUCTION

UP TO THE present time, nearly all vehicles for off-road locomotion have made use of systems of wheels or tracks for support and propulsion. This is in striking contrast to the locomotion of man and cursorial animals in which articulated systems of levers, individually powered and flexibly coordinated, are used to achieve this function. Vehicle designers have been aware for some time that the principles involved in natural "legged" locomotion systems result in superior mobility characteristics due to their inherently greater adaptability to terrain irregularities and to the fundamentally different nature of the interaction of such systems with the supporting terrain in comparison to wheeled or tracked vehicles [1]. Unfortunately, until recently, the complexity of the joint coordination control problem has frustrated attempts to apply these principles to obtain vehicles with off-road mobility characteristics comparable to those of living systems. While a theory for such *adaptive walking machines* has gradually evolved over the past decade [2], the attainment of the necessary joint coordination function by an on-board computer was not feasible prior to the introduction of microprocessors [3]. In response to the emergence of such devices, at least two experimental hexapod vehicles have been constructed for the purpose of supporting research on sensors and on computer hardware and software organization for automatic limb motion coordination [4]–[7]. Fig. 1 illustrates one of these machines.

Detailed consideration of the problem of automatic joint motion coordination reveals a natural hierarchical structure in which the highest levels are concerned with goal attainment, the intermediate levels involve motion planning, and the lower levels involve motion execution [2], [3], [8]. This paper is concerned with one of the more critical intermediate level problems, namely, the determination of an optimal schedule for the lifting and placing of the legs of a vehicle relative to the supporting terrain. In what follows, a summary of previously available results concerning this problem is presented together with a new problem formalization and a heuristic solution which has shown promising behavior in simulation studies.

Manuscript received March 29, 1978; revised November 20, 1978. This work was supported by the National Science Foundation under Grant ENG74-21664.

R. B. McGhee is with the Department of Electrical Engineering, Ohio State University, Columbus, OH 43210.

G. I. Iswandhi was with the Department of Electrical Engineering, Ohio State University, Columbus, OH. He is now with the Mead Paper Company, Chillicothe, OH 45601.

PROBLEM STATEMENT AND SUMMARY OF PREVIOUS WORK

A considerable amount of prior work has been devoted to the leg sequencing problem for the special case of straight-line constant-speed locomotion over level terrain [9]–[12]. Central to this work has been the notion of a finite characterization of leg states in which each leg of a machine or animal is idealized to a two-state device, namely, either the state of being on the ground (1-state) or in the air (0-state) [13], [14]. This concept leads to the following definitions.

Reprinted from *IEEE Trans. Syst., Man, Cybernetics*, vol. SMC-9, pp. 176–182, Apr. 1979.

Definition 1: The *support state* of a K-legged locomotion system is a binary row vector $y(t)$, such that at any time t, $y_i(t) = 1$ if leg i is in contact with the supporting surface and $y_i(t) = 0$ otherwise.

In general, then, the problem to be addressed in this paper is the determination of a sequence of support states for a given locomotion system over a particular terrain which is optimal with respect to some specified criterion. A periodic solution to this problem is usually referred to as a *gait* [11], [12], [14]–[16].[1]

In all but one prior study [17], the criterion used for support state sequence optimization has been one which relates to the degree of static stability of the system under consideration. A precise specification of this criterion, called the *longitudinal stability margin*, is provided by the following three definitions.

Definition 2: The *support pattern* associated with a given support state is the convex hull (minimum area convex polygon) of the point set in a horizontal plane which contains the vertical projections of the feet of all supporting legs [16].

Evidently, support patterns involve geometrical as well as temporal aspects of locomotion. For periodic gaits and straight-line locomotion, both of these attributes can be incorporated in a vector of initial foot positions and leg phasing relationships called a "kinematic gait formula" [15], [16]. Given this or any other complete parametric description of the kinematics of gait, the following definition formalizes the intuitive concept of static stability.

Definition 3: Consider any support state sequence with kinematics specified by a suitable parameter vector $p(t)$. Let $q(t)$ be the location of the vertical projection of the vehicle center of gravity onto any horizontal plane. Then the support pattern determined by $p(t)$ is *statically stable at time t* if and only if $q(t)$ is contained in its interior.

Finally, based upon the above concepts, two measures of the *degree* of static stability associated with a given support state sequence are provided by the following.

Definition 4: For a particular support state sequence and associated kinematic parameter vector $p(t)$, suppose that at time t the corresponding support pattern is statically stable. Then the *support state longitudinal stability margin* $s(t)$ is defined as the shortest distance from $q(t)$ to the front or rear boundary of the support pattern as measured in the direction of travel. If the support state sequence is periodic, then $\sigma(K)$, the *gait longitudinal stability margin* [16], is defined as

$$\sigma(K) = \min_{0 \le t \le T} s(t)$$

where T is the gait period and K is a kinematic gait formula. Fig. 2 illustrates all of the above definitions for an optimally stable quadruped gait [16].[2]

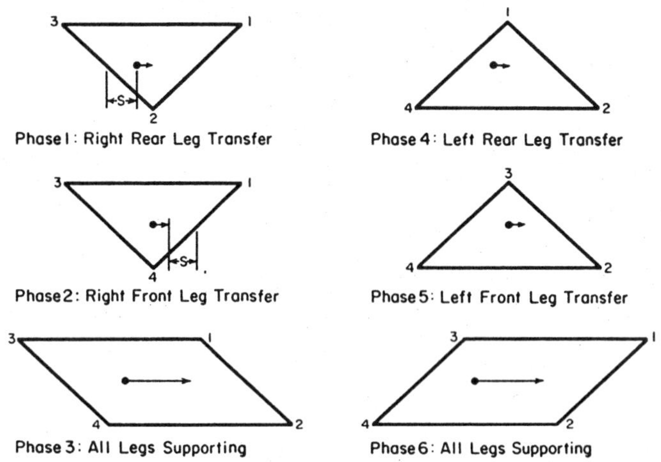

Fig. 2. Support patterns for successive phases of typical optimal quadruped crawl gait illustrating static stability. Arrows indicate total motion of vertical projection of center of gravity during each phase. Distance S is longitudinal stability margin for this gait.

For periodic support state sequences, it is possible, at least in principle, to examine the space of all kinematic gait formulas to seek a family of solutions which is optimal with respect to the longitudinal stability margin criterion. In order to obtain a meaningful answer to this problem, it has been found necessary to assume that all legs of the machine or animal operate with the same *duty factor* β, where this quantity is defined as the ratio of support time to cycle time for any leg [14]. If it is further assumed that the legs of the system are evenly spaced in right-left pairs along a longitudinal motion axis, and that each foot contacts the ground at a single point, then the optimal gaits are known for each allowable value of β for the cases of four, six, and eight legs [10]–[12], [16], [18]. Except for some anomalies for small values of β for eight-legged locomotion [12], these optimal solutions are all members of the family known as *wave gaits* [19], in which each gait is characterized by a forward wave of stepping actions on each side of the body with a half-cycle phase shift between the two members of any right-left pair. This family is used by natural quadrupeds for very low speed locomotion, by hexapods (insects) in all speed ranges, and is commonly employed by terrestrial octopods [19], [20]. The gait shown in Fig. 2 provides an example of a wave gait for a quadruped system.

More recently, Kugushev and Jaroshevskij [17] have suggested that the mathematical approach used in the study of periodic support state sequences for straight-line locomotion can be extended to include a more general case in which nonperiodic solutions known as "free gaits" may be expected. Specifically, in [17], a partial problem formalization is presented in which a trajectory is specified in advance for the motion of the center of gravity of a legged system over a given terrain containing certain regions which are unsuitable for support. This unsuitability could be due to excessive slope, soft soil conditions, holes or rocks, etc. The remainder of this paper is devoted to a completion of the formalization of this problem together with a description of one heuristic algorithm for its solution. A summary of results obtained from a computer simulation study employing the proposed algorithm is also included.

[1] Fig. 2 illustrates a typical quadruped gait. For this gait, the successive support states, beginning with Phase 1 and extending through Phase 6, are (1110), (1011), (1111), (1101), (0111), and (1111).

[2] Of course overall stability of motion is possible in gaits containing statically unstable phases, and in fact examples are easily found in man and higher animals. However, due to the relatively early stage of development of legged vehicles, such modes of locomotion are not treated in this paper.

THE FREE GAIT PROBLEM

A formal statement of the free gait problem in a computationally tractable form requires a number of further definitions as follows.

Definition 5: The terrain to be traversed is divided in square *cells*. Each cell is either *permitted* (denoted with 0) or *forbidden* (denoted with 1).

Definition 6: The *motion trace* [17] of a vehicle consists of the desired trajectory for the vehicle center of gravity. It is composed of successive circular arcs specified by either a human operator or by some type of automatic system. The vehicle longitudinal axis is assumed to be tangent to the motion trace at the vehicle center of gravity.

Since the motion trace is assumed to be specified in advance, the terrain around it should not contain obstacles so large that they cannot be walked over by the vehicle.[3] If such obstacles should be found along the motion trace, upon encountering them, the vehicle must stop. It is then the human operator or control computer's responsibility to define a new motion trace. While future research may produce algorithms for climbing over large three-dimensional obstacles, this case is not treated in this paper.

The two computer-controlled legged vehicles known to the authors both employ a main drive axis (azimuth axis) for each leg which is parallel to the vertical axis of the vehicle. This arrangement, which can be seen in Fig. 1, motivates the next two definitions.

Definition 7: Each vehicle leg has a *reachable area* in the form of a sector of an annulus. This area is specified by four parameters: minimum angle ψ_{min}, maximum angle ψ_{max}, minimum radius r_{min}, and maximum radius r_{max}. Reachable areas move as the body moves.

Definition 8: A cell is *reachable* by a leg at a given time if the center of that cell is covered by the current reachable area of that leg. Feet are always placed in the center of reachable cells.

Overlapping reachable areas raise interference problems. That is, a given leg may exclude regions of the reachable area of legs adjacent to it. Calculation of the excluded areas is not easy and is dependent on where the legs are at a given time. One way of dealing with this problem is to avoid it altogether by eliminating *a priori* all overlapping reachable areas so that each leg has a distinct region that can be accessed only by it and not by any other leg. This will be done in the present work, although it is recognized that in the future the leg interference problem should be dealt with in a more general way to insure that the full capability of a given vehicle can be utilized during free gait generation.

To provide a concrete example of some of the above ideas, the geometry of the machine shown in Fig. 1 will be assumed.

[3] While the *global* problem of choosing a good motion trace (usually referred to as the "navigation" problem) is of itself of considerable interest, this paper is concerned only with the *local* problem of foot placement (which arises only in connection with legged locomotion) and assumes that the navigation problem is solved either by human intelligence or by some higher level automatic system. The interested reader is referred to [8], [21], and [22] for some approaches to the difficult problem of robot navigation.

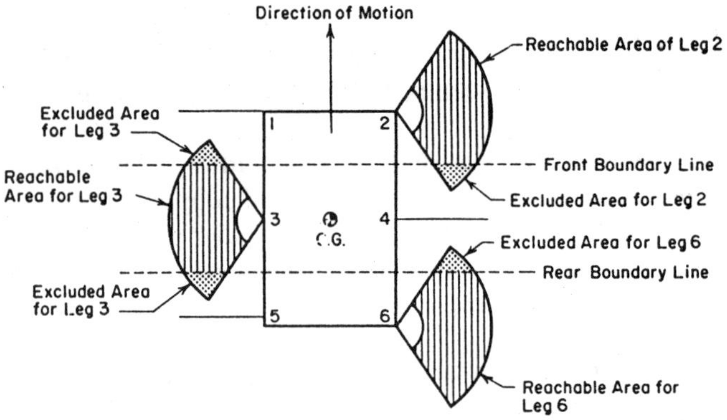

Fig. 3. Schematic top view of vehicle showing leg numbering and boundary lines for foot placing. Legs 1, 4, and 5 are shown in their rest configuration, while reachable and excluded areas are shown for legs 2, 3, and 6.

To solve the interference problem, two *boundary lines* perpendicular to the longitudinal axis of the vehicle can be specified. For the purposes of this paper, the front boundary line has been arbitrarily placed one-quarter of the body length in front of the vehicle center of gravity. The rear boundary line is similarly placed one-quarter body length behind the vehicle center. The front legs are constrained to be in front of the front boundary line, while the middle legs can only be between the two boundary lines, and the rear legs must be behind the rear boundary line. This arrangement is illustrated in Fig. 3. These constraints eliminate the interference problem entirely and motivate the following two definitions.

Definition 9: The *excluded area* for the front legs is that part of their reachable areas behind the front boundary line. The excluded area for the middle legs is the portion of their reachable areas in front of the front boundary line and behind the rear boundary line. The excluded area for the rear legs is the reachable area in front of the rear boundary line.

Excluded areas for the example system considered in this paper are also shown in Fig. 3.

Definition 10: A terrain cell i is a *foothold* [17] F_{ij} for leg j if and only if:

a) cell i is a permitted cell,
b) the center of cell i lies in a reachable area of leg j, and
c) the center of cell i is not in the excluded area of leg j.

For the present example, the form of the excluded areas is such that at any given time each cell can be a foothold of at most one leg. Moreover, each leg has a set of footholds of which not more than one can be in use at any given time. This situation suggests the next definition.

Definition 11: A foothold F_{ij} becomes a *support point* S_j for leg j whenever the foot of leg j is placed in its center.

The set of footholds for leg j is thus seen to be the set of all potential support points for that leg.

At this point, it is necessary to introduce some measure of the relative desirability of footholds to permit realization of an algorithm for selection of support points associated with successive support state sequences. The next two definitions provide one such measure.

Definition 12: The *existence segment of a foothold* is the largest segment of the motion trace such that when the projection of the vehicle center of gravity is on this segment, cell i is a foothold F_{ij} for leg j [17].

Informally, the existence segment of a foothold is that portion of the motion trace over which permitted cell i is a potential support point for leg j.

In determining the existence segment, two cases must be distinguished, namely, the *preview* case in which the motion trace is known for some time into future and the *nonpreview* case in which the motion trace is given only up to present time. In the latter circumstance, computation of the existence segment can be accomplished by assuming that the current vehicle turning radius will be maintained for an indefinite time into the future. In either case, a quantitative indication of the usefulness of a particular foothold is given by the following definition.

Definition 13: The *kinematic margin* for foothold F_{ij} is the arc length along the motion trace from the current vertical projection of the center of gravity of the vehicle to the forward end of the existence segment for F_{ij}.

The kinematic margin of a foothold evidently relates to how long a foot can be on a given cell before the leg reaches its kinematic limit. Clearly this notion also applies to each support point S_j, since every support point is also a foothold. For a particular circular arc of the motion trace, relative to the vehicle, the terrain moves in a circular arc concentric with the motion trace segment. A foothold F_{ij} will thus trace a circular arc in the opposite direction to the vehicle. This observation proves to be useful in calculating both the existence segment and the kinematic margin for a given foothold.

It is possible that even though the kinematic margins of all support points are greater than zero, a vehicle may nevertheless be forced to change its support state in order to maintain static stability. This observation leads to the following definitions.

Definition 14: The *stability segment* of a given support pattern is the largest segment of the motion trace such that when the projection of the vehicle center of gravity is on this trace, the vehicle is statically stable.

Combining several of the above ideas permits the following definition of a concept central to the solution of the free gait problem.

Definition 15: The *existence segment of a support pattern* is a maximal segment of the motion trace such that [17]:

a) the kinematic margins of all support points are greater than zero, and
b) the locomotion system is statically stable.

Evidently, movement of a legged locomotive system along a prescribed motion trace in a statically stable manner is possible if and only if there exists a sequence of support patterns and corresponding support states such that the *existence segments of successive support patterns overlap* [17]. The next section of this paper describes a heuristic algorithm for finding such sequences when they exist.

An Algorithm for Free Gait Generation

As stated above, the problem of free gait generation can be viewed as one of finding a sequence of support points such that there is an overlap of the existence segment of each support state with that of the support state which proceeded it. Evidently, in general, such sequences are not unique. Thus determination of a specific sequence requires either imposition of additional constraints or introduction of some type of global optimization criterion. These considerations motivate the following two definitions.

Definition 16: A support state sequence is *feasible* for motion along a specified motion trace over a given terrain if and only if there exists an associated sequence of support points such that the existence segments of successive support patterns overlap.

Definition 17: An *optimal support point sequence* for motion along a specified motion trace over a given terrain is one such that:

a) the corresponding support state sequence is feasible, and
b) some specified criterion function is maximized (minimized).

The *optimal support state sequence* is the sequence determined by the optimal support point sequence.

It is easy to imagine many different criteria for optimizing support point sequences. For example, the minimal stability margin over the whole motion trace could be maximized. Minimizing the maximum load placed on any leg might be important in traversing soft soils. If energy can be related to kinematics, then minimizing the total energy to traverse the entire motion trace might be important. The choice of a specific criterion clearly depends upon the characteristics of a given vehicle and the nature of its mission. Very likely, in any realistic situation, some combination of all of these factors and still other criteria would represent a more suitable basis for optimization.

In many circumstances, the combinatorial complexity of the problem of determining optimal support state sequences may be so great as to prohibit its computational solution in an acceptable period of time. While discrete dynamic programming [23] could perhaps be used to find the optimal sequence if the total number of terrain cells is not too great, a different approach is taken here. Specifically, as in [17], a suboptimal sequence will be sought by making use of an algorithm which determines the support pattern sequence only one stage forward rather than over the whole trajectory. While such an approach might be expected to give rather poor results, preliminary computer simulation studies have shown that this is not the case, providing that the motion trace has been appropriately selected. That is, if the motion trace is chosen so that a large number of feasible support state sequences exist, then the algorithm to be presented is typically capable of finding one such sequence. The departure of such sequences from optimality is yet to be determined.

Fig. 4 is a flowchart for the free gait algorithm utilized in this research. This algorithm is similar to that described in

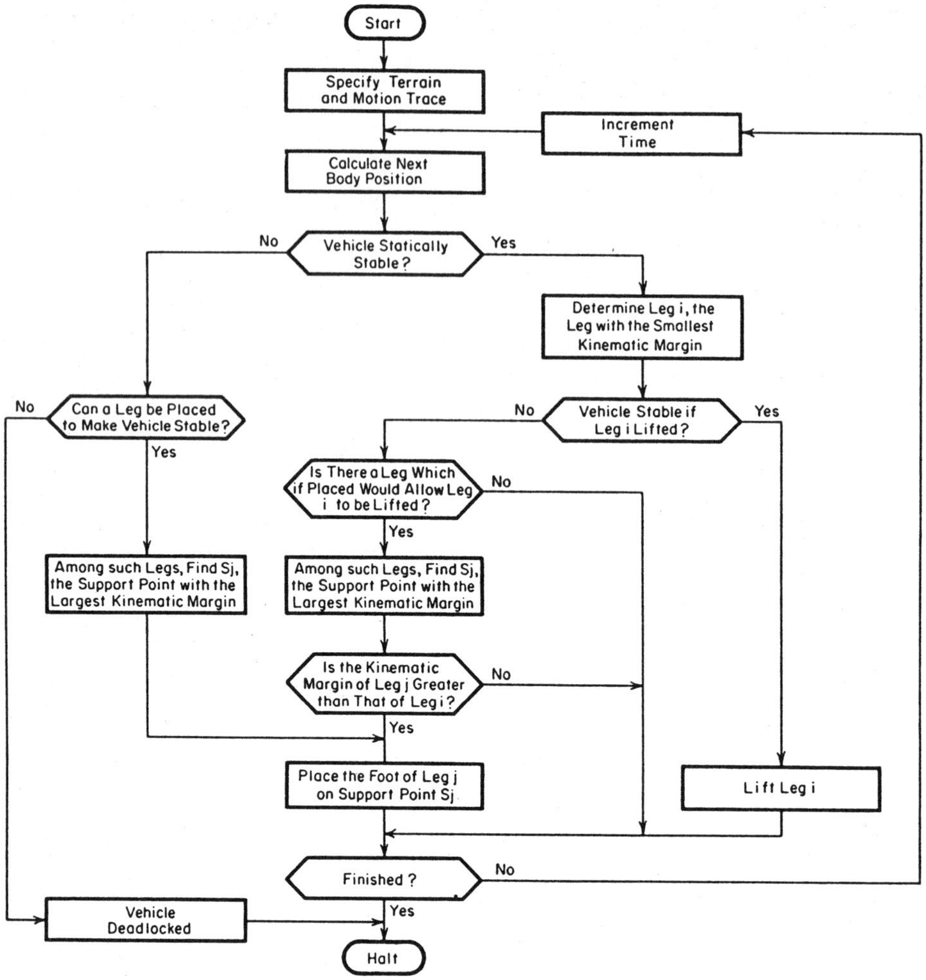

Fig. 4. Flowchart for free gait algorithm.

[17], although details differ. To understand its operation, it is useful to recognize that, at each iteration, one of four decisions can be made:

1) Maintain the current support state.
2) Lift a leg.
3) Place a leg.
4) Halt.

Careful examination of each of the corresponding paths in Fig. 4 shows that this algorithm tends to maximize the number of legs in the air and thus sacrifices stability for adaptability. It can also be seen that a halt state can occur prematurely. In such a circumstance, the algorithm is said to be *deadlocked*. In practical terms, the only solution for a deadlocked condition (cul-de-sac) is for the operator or navigation system to back up the vehicle for a suitable distance and then either alter the algorithm or specify a new motion trace.

Clearly, a simple change in the tests for static stability would allow modification of the algorithm of Fig. 4 to enforce a minimum value for stability margin. This would tend to increase the average number of legs in a supporting state and would certainly increase the probability of a deadlock. It is also evident that more complex changes in algorithm logic are possible such as, for example, placing legs whenever possible and lifting legs only when deadlock is imminent. These alternatives have not been explored as yet, but remain a subject for future research.

The underlying logic for the algorithm of Fig. 4 can be summarized as follows. Providing that the vehicle remains stable, legs are lifted so as to maximize the minimum value of the kinematic margin over all supporting legs. This in turn tends to extend forward the existence segment of each support state to a maximal extent, thereby increasing the likelihood that a new support state with an overlapping existence segment can be found. If the vehicle stability is imperiled, then the algorithm corrects this situation by placing a leg when this is possible. Whenever a leg is placed, among those legs which make the vehicle stable, the one with the largest kinematic margin is utilized. If leg placing is for the purpose of allowing a leg to be lifted in the *next* cycle of the algorithm, this is not done unless a net increase in the minimal value of the kinematic margin over all supporting legs results. This later feature appears to be missing from the Kugushev and Jaroshevskij algorithm [17]. Likewise, it is not apparent that in [17] the placing of legs to restore vehicle stability is governed by maximization of the minimum kinematic margin. Both of these features are believed to be

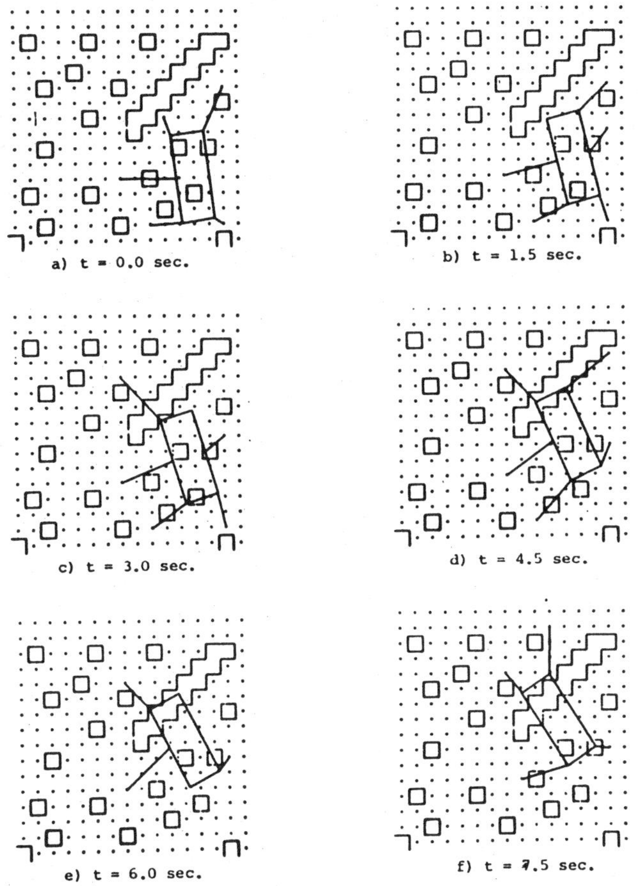

Fig. 5. Photographs of six representative phases of vehicle support as presented on display panel during simulated locomotion over terrain with forbidden cells.

quite important to the success of the free gait algorithm in finding feasible support state sequences over terrain containing a large proportion of forbidden cells.

As a final remark, it should be noted that leg placing requires an exhaustive search over every foothold of every leg not in a supporting state. This requirement places a lower bound on the size of terrain cells for real-time application of free gait algorithms.

Simulation Results

A simulation program including both a representation of the vehicle and terrain and the free gait algorithm of Fig. 4 has been written as part of this research. This program is implemented on a DEC-10 computer operating in a time-shared mode and utilizing an AG-60 plasma panel display as an output device. The display is equipped with a touch-panel sensor which allows the user to designate terrain cells as forbidden by simply touching them. A cell can be restored to the status of a permitted cell by simply touching it once again during the terrain specification portion of program execution. Permitted cells are displayed as a dot designating the cell center, while forbidden cells are outlined without the inclusion of such a dot. Fig. 5(a) is a photograph of a typical simulated terrain created by this process. This figure also shows a simplified depiction of the vehicle as seen from the top. The dimensions used to represent this vehicle and its reachable areas correspond to those of the hexapod system illustrated in Fig. 1. The convention used in the display is that only supporting legs are shown, since the position of legs in the air has no bearing on vehicle stability or kinematic limits.

After the terrain and vehicle turning radius have been specified by the operator, iterative execution of the free gait algorithm begins. Fig. 5(a)–(f) shows six representative phases of such a simulation experiment as presented on the plasma panel display. It can be seen, as expected, that the algorithm under consideration attempts to keep a maximal number of legs in the air with the minimum of three supporting legs occurring in Fig. 5(e). Many different terrain conditions have been investigated in this research, and the results presented in Fig. 5 are typical. In particular, support sequences are always nonperiodic and, in fact, are quite unpredictable when the terrain includes more than a very small percentage of forbidden cells. It is noteworthy that no terrain examples have been found as yet in which the system becomes deadlocked, unless the vehicle is required to traverse a region in which there is a set of contiguous forbidden cells constituting an area comparable in size to the vehicle body. Whenever such regions are avoided by proper path selection, the algorithm presented seems to be very effective, even when only substantially less than half of the terrain is composed of permitted cells. Similar results are reported in [17].

The program to realize the algorithm of Fig. 4 was written in Fortran and required about 250 statements. Execution was somewhat faster than real time for a vehicle speed of about 0.5 ft/s. Since the computer was used in a time-shared mode and execution also required display updating, it is believed that a small minicomputer (or possibly even a microprocessor) could solve the support state sequence problem on-line for a vehicle similar to the one illustrated in Fig. 1.

Summary and Conclusions

This paper has presented a formalization of the problem of choosing a sequence of support points for a legged vehicle to enable it to negotiate terrain in which some regions are unsuitable for support. The formal problem can be viewed as a rather complicated game somewhat akin to solving a maze, but with more complex constraints governing successive moves. The game is "won" if a sequence of support points is found which permits the vehicle to move from its starting point to its goal along a specified trajectory without stepping on forbidden cells and without becoming unstable. It is clear that biological systems are able to solve this problem with great efficiency as is evidenced, for example, by the behavior of cursorial quadrupeds in rapid motion over very irregular terrain.

Experience with other simulation studies and with actual vehicles has shown that, while human beings are naturally suited to solve the support sequence problem for biped locomotion, this capability is not readily extended to additional pairs of legs [2], [24], [25]. It appears that, instead, this task must be relegated to an automatic system if acceptable performance is to be attained in a legged vehicle when traversing very difficult terrain. One algorithm capable of solving this problem has been presented in this paper together with some typical results. The authors do not suggest that this algorithm is in any sense optimal, although it is believed that some improvements have been made over the only previous algorithm proposed for this problem [17]. Rather, the importance of this work lies in the complete formalization of the problem and the demonstration that automatic solution in real time is feasible with a modest-sized computer.

It is hoped that the presentation of these results will interest others sufficiently to stimulate further research, which surely should yield both better algorithms and better means for quantifying the performance of a given algorithm. It is the authors' belief that continued research on this problem in parallel with work on other aspects of the joint coordination control problem [2] can lead to an entirely new class of vehicles which may exhibit large performance advantages over existing vehicles in traversing very irregular terrain and in moving over soft soils. Although further basic research is needed before any practical machines can be constructed, the possible advantages of this type of vehicle in applications such as arctic transport, mining, agriculture, forestry, fire-fighting, explosive ordnance disposal, and even in unmanned ocean-floor or planetary exploration, provide incentives for such work.

References

[1] M. G. Bekker, *Introduction to Terrain-Vehicle Systems.* Ann Arbor, MI: Univ. Michigan Press, 1969.

[2] R. B. McGhee, "Control of legged locomotion systems," *Proc. Eighteenth Joint Automatic Control Conf.*, San Francisco, CA, pp. 205–215, June 1977.

[3] D. E. Orin, R. B. McGhee, and V. C. Jaswa, "Interactive computer-control of a six-legged robot vehicle with optimization of stability, terrain adaptability, and energy," *Proc. 1976 IEEE Conf. Decision and Control*, Clearwater Beach, FL, Dec. 1976.

[4] J. R. Buckett, "Design of an on-board electronic joint control system for a hexapod vehicle," M.S. thesis, Ohio State Univ., Mar. 1977.

[5] C. S. Chao, "A software system for on-line control of a hexapod vehicle utilizing a multiprocessor computing structure," M.S. thesis, Ohio State Univ., Aug. 1977.

[6] R. L. Briggs, "Real-time digital control of an electrically powered vehicle leg using vector force feedback," M.S. thesis, Ohio State Univ., Dec. 1977.

[7] D. E. Okhotsimski, V. S. Gurfinkel, E. A. Devyanin, and A. K. Platonov, "Integrated walking robot development," presented at *1977 Engineering Foundation Conf. Cybernetic Models of the Human Neuromuscular System*, Henniker, NH, Aug. 1977.

[8] D. E. Okhotsimski and A. K. Platonov, "Perceptive robot moving in 3D world," *Proc. IV Int. Joint Conf. Artificial Intelligence*, Tbilisi, Georgian SSR, USSR, Sept. 1975.

[9] S. S. Sun, "A theoretical study of gaits for legged locomotion systems," Ph.D. dissertation, Ohio State Univ., Mar. 1974.

[10] R. B. McGhee and S. S. Sun, "On the problem of selecting a gait for a legged vehicle," *Proc. VI IFAC Symp. Automatic Control in Space*, Armenian S.S.R., USSR, Aug. 1974.

[11] A. P. Bessonov and N. V. Umnov, "The analysis of gaits in six legged vehicles according to their static stability," *Proc. Symp. Theory and Practice of Robots and Manipulators*, International Centre for Mechanical Sciences, Udine, Italy, Sept. 1973.

[12] ——, "Choice of geometric parameters of walking machines," *On Theory and Practice of Robots and Manipulators*, Polish Scientific Publishers, Warsaw, 1976, pp. 63–74.

[13] R. Tomovic, "A general theoretical model of creeping displacement," *Cybernetica*, IV, pp. 98–107 (English translation), 1961.

[14] R. B. McGhee, "Some finite state aspects of legged locomotion," *Mathematical Biosciences*, vol. 2, no. 1/2, pp. 67–84, Feb. 1968.

[15] M. Hildebrand, "Symmetrical gaits of horses," *Science*, vol. 150, pp. 701–708, Nov. 1965.

[16] R. B. McGhee and A. A. Frank, "On the stability properties of quadruped creeping gaits," *Mathematical Biosciences*, vol. 3, no. 3/4, pp. 331–351, Oct. 1968.

[17] E. I. Kugushev and V. S. Jaroshevskij, "Problems of selecting a gait for an integrated locomotion robot," *Proc. Fourth Int. Conf. Artificial Intelligence*, Tbilisi, Georgian SSR, USSR, pp. 789–793, Sept. 1975.

[18] R. B. McGhee, "Robot locomotion," *Neural Control of Locomotion*, R. M. Herman *et al.*, Eds. New York: Plenum Press, 1976, pp. 237–264.

[19] D. M. Wilson, "Insect walking," *Annual Review of Entomology*, vol. 11, pp. 103–121, 1966.

[20] ——, "Stepping patterns in Tarantula spiders," *J. Experimental Biology*, vol. 47, pp. 133–151, 1967.

[21] P. E. Hart, N. J. Nilsson, and B. Raphael, "A formal basis for the heuristic determination of minimum cost paths," *IEEE Trans. Systems Sci. Cybern.*, vol. SSC-4, no. 2, pp. 100–107, July 1968.

[22] P. J. Blatman, "Environmental modeling and model preprocessing for a self-contained mobile robot," *Proc. 1975 IEEE SMC Conf.*, San Francisco, CA, Sept. 1975.

[23] G. Hadley, *Nonlinear and Dynamic Programming.* Reading, MA: Addison-Wesley, 1964.

[24] D. E. Okotsimski *et al.*, *Problems of Constructing and Modelling the Motion of an Operator-Controlled Walking Machine*, Preprint No. 125, Institute of Applied Mathematics, Academy of Sciences of the USSR, Moscow, 1974 (in Russian).

[25] R. S. Mosher, "Exploring the potential of quadruped," SAE Paper No. 690191, presented at the Int. Automotive Engineering Conf., Detroit, MI, Jan. 1969.

Adaptive Control of Left Ventricular Bypass Assist Devices

BAYLISS C. McINNIS, MEMBER, IEEE, ZHONG-WEI GUO, PO CHIEN LU, AND JIUN-CHUNG WANG

***Abstract*—This paper presents a computer-based adaptive control system for left ventricular bypass assist devices consisting of air driven diaphragm pumps. The system provides for 1) synchronization of pumping with ECG signals and 2) control of atrial pressure at desired levels. The system design includes an adaptive control algorithm which is a self-tuning PID-controller based on pole placement.**

The performance of the system has been demonstrated by *in vitro* experiments on a mock circulatory system. When there is an increase in atrial pressure, the system responds with an increase in stroke volume. Following major changes in the circulatory system, the control algorithm retunes itself and restores the system to the desired state.

I. Introduction

VENTRICULAR-assist pumps are designed to support the circulation of patients who cannot be separated from cardiopulmonary bypass after open-heart surgery. Intraaortic balloons presently are used to provide mechanical circulatory assistance in these patients. However, in the case of severe cardiogenic shock, counterpulsation with the intraaortic balloon is inadequate and many of these patients die. Hence, there is a need for mechanical circulatory assist devices that are capable of maintaining blood circulation and pressure until myocardial function returns.

Projects involving drive units for artificial ventricles (to be used as either assist devices or as total artificial hearts) are in progress by a number of research groups in the United States, as well as in Austria, Czechoslovakia, France, Germany, Japan, Peoples Republic of China, and Switzerland [1]–[3]. There are innovative features in the work of each group, but the potential of computer controlled assist devices to automatically maintain the blood pressures of the recipient at desired levels has not been fully realized.

The concept of synchronization with the natural ECG has been widely used [4]–[7]. Another widely used concept is that of termination of the systolic period (drive pressure) when the diaphragm or pusher plate reaches a desired end-position using a position sensor [5]–[9]. From the point of view of control, the drive system most similar to ours in overall concept and function is the one developed at Kyoto University, Kyoto, Japan [10].

The control systems of ventricular assist devices, as well as total artificial hearts, currently in use are still manual or semiautomatic with regard to regulation of the blood pressures of the recipient. The application of recent developments in adaptive control algorithms to stabilize the blood pressures of the circulatory system at desired levels by means of beat-by-beat adjustment of the pump stroke volume is a contribution of this work.

In order to realize beat-by-beat control it has been necessary to develop an improved drive system that utilizes a hydraulic servoloop to produce the pneumatic pressure that drives the assist pump. A description of our drive and control system design and the results of tests on a mock hydraulic circulatory system are presented in this paper.

Manuscript received February 25, 1983; revised March 24, 1983 and June 22, 1984. Paper recommended by Past Associate Editor, D. M. Wiberg. This work was supported in part by the National Heart, Lung and Blood Institute under Grant HL25029.

B. C. McInnis, P. C. Lu, and J.-C. Wang are with the Department of Electrical Engineering, University of Houston-University Park, Houston, TX 77044.

Z.-W. Guo was with the Department of Electrical Engineering, University of Houston-University Park, Houston, TX 77044. He is now with the Department of Automation, Tsinghua University, Beijing, China.

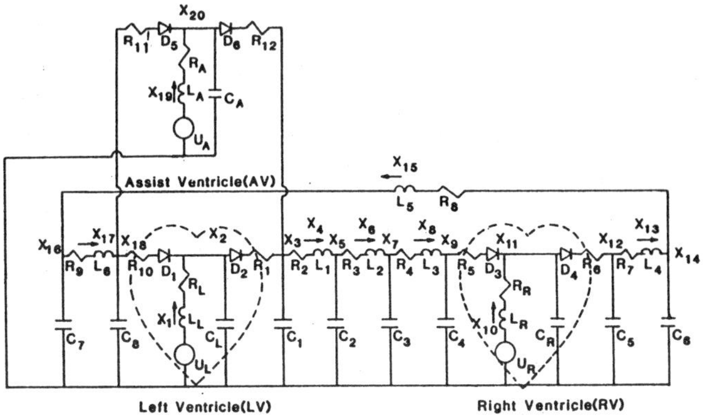

Fig. 1. Electrical analog model.

II. Model of the Circulatory System and Assist Devices

A. Electric Analog Model

An electric analog model similar to those of Noordergraaf [11] and Reul [12] of the circulatory system plus the assist pump is shown in Fig. 1. The RLC networks enclosed in the heart-shaped boundaries represent the natural ventricles. The voltage sources U_L and U_R represent the strength of these natural ventricles. Similarly, the voltage source U_A represents the driving force of the assist device. The diodes D_1, D_3, D_5 and resistors R_{10}, R_5, R_{11} model the inlet valves of the natural ventricles and assist device, while diodes D_2, D_4, D_6 and resistors R_1, R_6, R_{12} represent the outlet valves. The systemic and pulmonary loads are simulated by multiple stages of RLC networks. The inductors are used to model the inertance of the blood flow in the systemic and pulmonary circulations. During the systolic and diastolic periods both U_L and U_R, for the left and right ventricles, and U_A, for the assist device, have independent values. The states x_i as shown in the figure correspond to the blood flows and pressures.

The state equations for this model are the following:

$$\dot{X} = AX + BU \tag{1}$$

where $X \in R^{20}$, $U^T = [U_L U_R U_A]$.

The nonzero elements of the matrix A are listed in Table I. Elements of the matrix B are zero except for $B_{1,1} = 1/L_L$, $B_{10,2} = 1/L_R$, and $B_{19,3} = 1/L_A$.

B. Mock Circulatory System

The mock circulatory system (Fig. 2) used in our studies was designed and developed by H. Reul of the Helmholtz Institute of Biomedical Research, Aachen, Germany, based on the electric analog model mentioned previously. In the experimental apparatus, the natural ventricles are modeled by artificial ventricles

Reprinted from *IEEE Trans. Automat. Contr.*, vol. AC-30, pp. 322–329, Apr. 1985.

TABLE I
NONZERO ENTRIES OF THE SYSTEM MATRIX $A[ij]$

$$A_{1,1}=-\frac{R_L}{L_L}\quad A_{1,2}=-\frac{1}{L_L}$$

$$A_{2,1}=\frac{1}{C_L}\quad A_{2,2}=-\left[\frac{D(X_{18}-X_2)}{R_{10}C_L}+\frac{D(X_2-X_3)}{R_1C_2}\right]\quad A_{2,3}=\frac{D(X_2-X_3)}{R_1C_2}\quad A_{2,18}=\frac{D(X_{18}-X_2)}{R_{10}C_L}$$

$$A_{3,2}=\frac{D(X_2-X_3)}{R_1C_1}\quad A_{3,3}=-\left[\frac{D(X_2-X_3)}{R_1C_1}+\frac{D(X_{20}-X_3)}{R_{12}C_1}\right]\quad A_{3,4}=-\frac{1}{C_1}\quad A_{3,20}=\frac{D(X_{20}-X_3)}{R_{12}C_1}$$

$$A_{4,3}=\frac{1}{L_1}\quad A_{4,4}=-\frac{R_2}{L_1}\quad A_{4,5}=-\frac{1}{L_1}$$

$$A_{5,4}=\frac{1}{C_2}\quad A_{5,6}=-\frac{1}{C_2}$$

$$A_{6,5}=\frac{1}{L_2}\quad A_{6,6}=-\frac{R_3}{L_2}\quad A_{6,7}=-\frac{1}{L_2}$$

$$A_{7,6}=\frac{1}{C_3}\quad A_{7,8}=-\frac{1}{C_3}$$

$$A_{8,7}=\frac{1}{L_3}\quad A_{8,8}=-\frac{R_4}{L_3}\quad A_{8,9}=-\frac{1}{L_3}$$

$$A_{9,8}=\frac{1}{C_4}\quad A_{9,9}=-\frac{D(X_9-X_{11})}{R_5C_4}\quad A_{9,11}=\frac{D(X_9-X_{11})}{R_5C_4}$$

$$A_{10,10}=-\frac{R_R}{L_R}\quad A_{10,11}=-\frac{1}{L_R}$$

$$A_{11,9}=\frac{D(X_9-X_{11})}{R_5C_R}\quad A_{11,10}=\frac{1}{C_R}\quad A_{11,11}=-\left[\frac{D(X_9-X_{11})}{R_5C_R}+\frac{D(X_{11}-X_{12})}{R_6C_R}\right]\quad A_{11,12}=\frac{D(X_{11}-X_{12})}{R_6C_R}$$

$$A_{12,11}=\frac{D(X_{11}-X_{12})}{R_6C_5}\quad A_{12,12}=-\frac{D(X_{11}-X_{12})}{R_6C_5}\quad A_{12,13}=-\frac{1}{C_5}$$

$$A_{13,12}=\frac{1}{L_4}\quad A_{13,13}=-\frac{R_7}{L_4}\quad A_{13,14}=-\frac{1}{L_4}$$

$$A_{14,13}=\frac{1}{C_6}\quad A_{14,15}=-\frac{1}{C_6}$$

$$A_{15,14}=\frac{1}{L_5}\quad A_{15,15}=-\frac{R_8}{L_5}\quad A_{15,16}=-\frac{1}{L_5}$$

$$A_{16,15}=\frac{1}{C_7}\quad A_{16,17}=-\frac{1}{C_7}$$

$$A_{17,16}=\frac{1}{L_6}\quad A_{17,17}=-\frac{R_9}{L_6}\quad A_{17,18}=-\frac{1}{L_6}$$

$$A_{18,2}=\frac{D(X_{18}-X_2)}{R_{10}C_8}\quad A_{18,17}=\frac{1}{C_8}\quad A_{18,18}=-\left[\frac{D(X_{18}-X_{20})}{R_{11}C_8}+\frac{D(X_{18}-X_2)}{R_{10}C_8}\right]\quad A_{18,20}=\frac{D(X_{18}-X_{20})}{R_{11}C_8}$$

$$A_{19,19}=-\frac{R_A}{L_A}\quad A_{19,20}=-\frac{1}{L_A}$$

$$A_{20,3}=\frac{D(X_{20}-X_3)}{R_{12}C_A}\quad A_{20,18}=\frac{D(X_{18}-X_{20})}{R_{11}C_A}\quad A_{20,19}=\frac{1}{C_A}\quad A_{20,20}=-\left[\frac{D(X_{18}-X_{20})}{R_{11}C_A}+\frac{D(X_{20}-X_3)}{R_{12}C_A}\right]$$

$$D(y)=\begin{cases}1 & \text{if } y>0\\ o & \text{if } y\leq 0\end{cases}$$

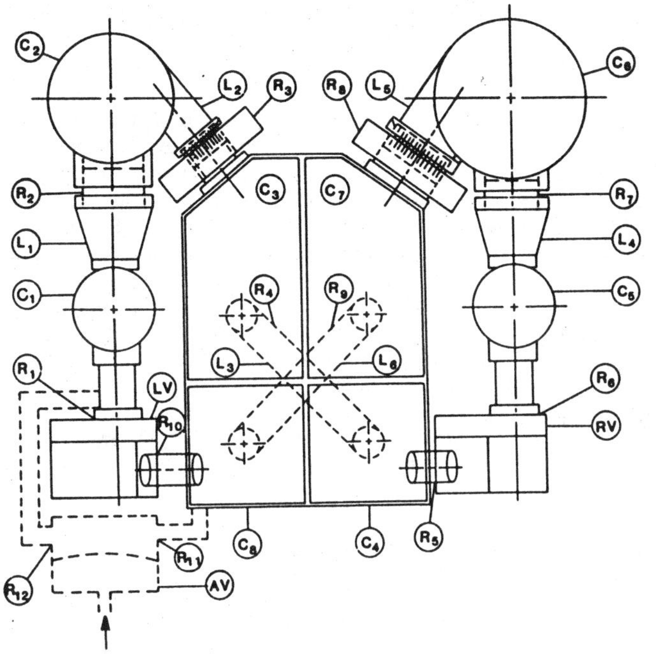

Fig. 2. Schematic diagram of the mock circulatory system.

designed by T. Akutsu at the Texas Heart Institute. A third ventricle, serving as an assist device, is a left ventricular bypass pump fabricated at the Baylor College of Medicine, Houston, TX. The afterload of each ventricle is simulated by a series of three resist units and four compliance chambers with three inertance components. Table II provides a set of nominal parameter values for the design of the systemic and pulmonary circulation. Each of the resistor units R_2, R_3, R_7, and R_8 consists of several thousand parallel nickel capillary tubes. The length and diameter of the tubes in each unit were calculated to ensure laminar flow over the range of flow rates desired for the experiment (4–15 l/min). These units represent, respectively, the characteristic resistance (R_2 or R_7) and the peripheral resistance (R_3 or R_8). The peripheral resistor is adjustable and regulates the magnitude of the arterial pressure. An adjustable compliance chamber (C_2 or C_6) is placed in series between the two resistors. By proper adjustment of the compliance and the peripheral resistor of the systemic and pulmonary circulation, a wide range of conditions can be produced in the mock circulatory system.

III. Control Strategy

Our objective is to develop an assist device system that will respond to atrial pressure in a manner that imitates the Frank–Starling mechanism of the natural circulatory system. The peripheral circulation is capable of determining cardiac output only if the heart is capable of pumping all the blood that flows into the atrium. Since a sick ventricle is unable to achieve this performance, the assist device should function so that the required pumping action is obtained.

It has been observed by Guyton [13] that, when the left heart fails, cardiac output falls proportionately as the left atrial pressure approaches the mean systemic pressure. Thus, the task of the left ventricular assist device is to restore cardiac output, and the extent to which flow has been restored is indicated by the level of atrial pressure.

In order for the assist device to assume the required load, we provide for beat-by-beat adjustment of stroke volume. This is accomplished by requiring full emptying of the assist device during systole and by adjusting the filling pressure during diastole. In this way we obtain a stroke volume that is proportional to atrial pressure.

TABLE II
MOCK LOOP PARAMETER VALUES FROM H. REUL [12]. COMPLIANCE (C) VALUES ARE IN UNITS OF cm^5/dyne, INERTANCE (L) IN dyne$\cdot s^2/cm^5$, AND RESISTANCE (R) IN dyne$\cdot s/cm^5$

Systemic Circulation	
$R_1 \sim 0$	outflow resistance
$C_1 = 0.165 \cdot 10^{-3}$	compliance of aorta
$R_2 = 90$	characteristic resistance
$L_1 = 1.1$	lumped tubing
$C_2 = 1.1 \cdot 10^{-3}$	compliance of arteries
$R_3 = 1300$	peripheral resistance
$L_2 = 3.1$	lumped tubing
$C_3 = 1.0 \cdot 10^{-2}$	venous compliance
$R_4 = 25$	venous resistance
$L_3 = 1.7$	lumped tubing
$C_4 = 5.0 \cdot 10^{-3}$	atrial compliance
$R_5 \sim 0$	inflow resistance
Pulmonary Circulation	
$R_6 \sim 0$	outflow resistance
$C_5 = 0.067 \cdot 10^{-3}$	compliance of pulmonary artery
$R_7 = 45$	characteristic resistance
$L_4 = 1.0$	lumped tubing
$C_6 = 2.0 \cdot 10^{-3}$	compliance of arteries
$R_8 = 110$	peripheral resistance
$L_5 = 2.4$	lumped tubing
$C_7 = 3.0 \cdot 10^{-2}$	pulmonary venous compliance
$R_9 = 25$	venous resistance
$L_6 = 1.7$	lumped tubing
$C_8 = 5.0 \cdot 10^{-3}$	atrial compliance
$R_{10} \sim 0$	inflow resistance

Since cardiac output is also a function of heart rate, we synchronize the assist device with the natural heart by means of the P or R wave [14]. This provides another link with the natural control loops of the circulatory system.

IV. Averaged State-Variable Model and Reduced-Order ARMA Model

For the system shown in Fig. 1 we define a discrete-time averaged state $\bar{X}(n)$ as the average of the continuous state $X(t)$ during the period Δ of a heartbeat (systolic plus diastolic periods) such that

$$\bar{X}(n) = \frac{1}{\Delta} \int_{t_{n-1}}^{t_n} X(t)\, dt \tag{2}$$

where

$$\Delta = t_n - t_{n-1} = \Delta_S + \Delta_D \text{ (see Fig. 3)}$$
$$\bar{X},\ X \in R^{20}.$$

Because of the open–closed characteristics of the heart valves (diodes, function $D(\cdot)$ in matrix A, Table I) both the systolic and diastolic periods can be divided into several stages in which the whole system is further decoupled into subsystems. Following the approach given in [15], we obtain the following averaged discrete state-variable model:

$$\bar{X}(n+1) = F\bar{X}(n) + GU(n) + HU(n-1) \tag{3}$$

where $F \in R^{20\times 20}$, $G,\ H \in R^{20\times 6}$

and

$$U^T = [U_N^T \vdots U_A^T] = [U_{LS}U_{LD}U_{RS}U_{RD} \vdots U_{AS}U_{AD}]$$

denotes the strength of the natural (left and right) and assist ventricles, respectively, where subscript S denotes the driving

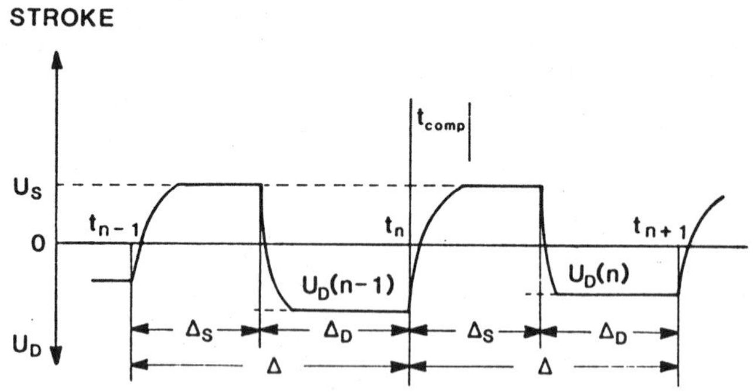

Fig. 3. Drive wave-form and timing diagram.

pressure during systole and subscript D the filling pressure during diastole. All of them are assumed to be constant during one heartbeat period.

Since U_N cannot be manipulated by the control system and U_{AS} is chosen to obtain full emptying of the assist ventricle, U_{AD} becomes the only control input. If we choose mean left atrial pressure $\overline{LAP}$ ($\bar{x}_{18}$) to be the controlled output, theoretically the averaged state-variable model (3) can be transformed to a canonical observer form and further to an ARMA model [16] as follows:

$$\bar{A}(q^{-1})y(t)=\bar{B}(q^{-1})u(t)+\bar{C}(q^{-1})W(t) \quad (4)$$

where

$$y=\overline{LAP}=\bar{x}_{18}=\text{controlled output (mean left atrial pressure)}$$
$$u=U_{AD}=\text{control input (diastolic stroke of assist device)}$$
$$W^T=[U_{LS}U_{LD}U_{RS}U_{RD}U_{AS}]=\text{unmanipulatable/fixed inputs}$$
$$\bar{A}(q^{-1})=1+\bar{a}_1q^{-1}+\cdots+\bar{a}_nq^{-n}$$
$$\bar{B}(q^{-1})=q^{-d}(\bar{b}_0+\bar{b}_1q^{-1}+\cdots+\bar{b}_mq^{-m})$$
$$\bar{C}(q^{-1})=q^{-d}(\bar{c}_0+\bar{c}_1q^{-1}+\cdots+\bar{c}_mq^{-m})$$

and $\bar{c}_i \in R^{1\times5}$, q^{-1} is backward shift operator, and d is time delay.

The question of error sensitivity in parameter estimation problems is particularly critical in biological systems such as the circulatory system because 1) such systems often involve many parameters whose values must be assumed known, 2) there is great variability, and 3) the data base tends to be very limited (see Bekey and Grove [17]). Thus a control design based on the full model (3) or (4) is impractical because of the large number of parameters and the variability of their values.

Since in this work the ultimate objective consists of the control of mean left atrial pressure ($\overline{LAP}$), a measurable output, we introduce the following deterministic second-order ARMA model:

$$A(q^{-1})y(t)=B(q^{-1})u(t)+C \quad (5)$$

where

$$y=\overline{LAP},\; u=U_{AD}$$
$$A(q^{-1})=1+a_1q^{-1}+a_2q^{-2}$$
$$B(q^{-1})=q^{-1}(b_0+b_1q^{-1}).$$

The term C is introduced to represent the effect of the strength of the natural ventricles and the drive pressure (for full emptying) of the assist ventricle, and it can be thought as a disturbance term. A second-order model is chosen because of its suitability for the development of an adaptive PID algorithm as shown below.

V. Adaptive PID-Control Algorithm

A. Motivation

Considering the complexity of the system and the large number of parameters involved, we should investigate the use of a conventional PI or PID controller. In our experimental studies, we found that these controllers worked under specific operating conditions. However, when major changes in the operating conditions occur (for example, changes in heart rate, strength of left natural ventricle U_{LD}, and systemic peripheral resistance R_3), the performance was not entirely satisfactory. Even though it is possible to determine the gains for a PID-controller (suitable for a range of operating conditions) by the classical tuning procedures, it is believed that these methods are not entirely satisfactory for clinical applications because of the disadvantages of being time consuming and requiring open-loop operation [18]. Therefore, the adaptive PID-controller based on pole placement was studied and adopted.

B. Pole Placement PID-Controller

The conventional digital PID-controller [19], [20] can be derived from the continuous-time form as

$$u(n)=\left[K_P+\frac{K_Iq^{-1}}{1-q^{-1}}+\frac{K_D(1-q^{-1})}{1+\nu q^{-1}}\right]\cdot[y_r(n)-y(n)] \quad (6)$$

where K_P, K_I, and K_D are, respectively, the proportional, integral, and derivative gains, ν is the filter constant ($-1<\nu<0$), and y_r is the set point (desired output). The delay in the integral part can be removed by using a backward-difference approximation.

There are many variations of the PID-controller (6). One alternative is to include the set-point signal y_r only in the integral part to avoid the so-called "set point and derivative kick" (sudden changes in the manipulated variable when the set point is changed) [21]. When the controller is further modified to include the filter $1/(1+\nu q^{-1})$ in the integral part, we obtain a structure that is desirable for pole placement because the two extra closed-loop zeros are at the origin [compared to one or two nonzero zeros in other structures of (6)] [19], [20].

The proposed PID structure can be expressed as

$$u(n)=\left[-K_P-\frac{K_D(1-q^{-1})}{1+\nu q^{-1}}\right]y(n)+\frac{K_I}{(1-q^{-1})(1+\nu q^{-1})}[y_r(n)-y(n)] \quad (7)$$

and rewritten in the form

$$P(q^{-1})u(n)=Ly_r(n)-S(q^{-1})y(n) \quad (8)$$

where

$$P(q^{-1})=(1-q^{-1})(1+\nu q^{-1})$$
$$S(q^{-1})=s_0+s_1q^{-1}+s_2q^{-2}$$
$$L=S(1)=K_I$$

and

$$s_0=K_P+K_I+K_D$$
$$s_1=K_P(\nu-1)-2K_D$$
$$s_2=K_D-\nu K_P.$$

Then for a given second-order process model

$$A(q^{-1})y(n)=B(q^{-1})u(n)+C \quad (5)$$

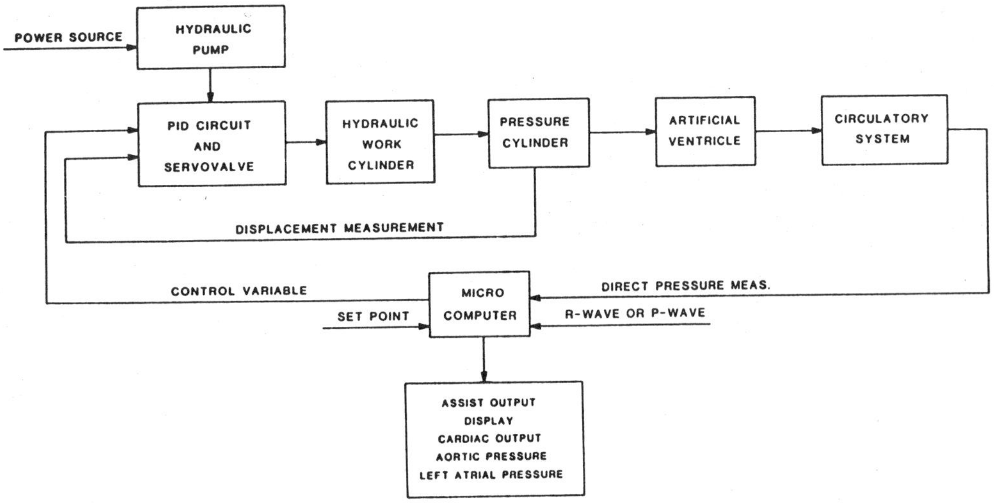

Fig. 4. Block diagram of the drive and control systems.

we can substitute the control law (8) to obtain the closed-loop equation

$$[A(q^{-1})P(q^{-1})+B(q^{-1})S(q^{-1})]y(n) = L \cdot B(q^{-1})y_r(n)+P(q^{-1})C. \quad (9)$$

Note that during steady state $y(n) = y_r(n)$ for a constant set point or disturbance. This is solely due to the integral action.

The order of the closed-loop system is increased to 4 with the system polynomial $A(q^{-1})P(q^{-1}) + B(q^{-1})S(q^{-1})$. Thus, under the assumption that $A(q^{-1})$ and $B(q^{-1})$ are coprime (i.e., no pole-zero cancellation) and $B(q^{-1})$ does not have the zero of $q = +1$, it can be shown that, for a desired closed-loop polynomial $A^*(q^{-1})$ all four poles can be arbitrarily assigned. The parameters, ν, s_0, s_1, and s_2 of the control law can be determined by equating the coefficients of $A^*(q^{-1})$ and $A(q^{-1})P(q^{-1}) + B(q^{-1})S(q^{-1})$ if the parameters a_i and b_i of the process model are known. In the adaptive case, a_i and b_i are obtained from the parameter estimation algorithm.

C. Parameter Estimation Algorithm

In order to estimate the parameters a_i and b_i, the process model (5) is rewritten as

$$y(n+1)=\phi^T(n)\theta \quad (10)$$

where

$$\phi^T(n)=[y(n)y(n-1)u(n)u(n-1)1]$$

$$\theta^T=[-a_1\ -a_2\ b_0\ b_1\ C].$$

For the system under study, the changes of operating conditions are mainly due to 1) heart rate, 2) strength of the natural ventricles, 3) peripheral resistance. While the effects of the latter two appear explicitly in the parameters of the full model (3) or (4) (note that the pumping ability of the natural ventricles is assumed to be neither manipulatable nor measurable, and its change can then be considered as a disturbance), the change of heart rate affects implicitly the parameters because it affects the systolic and diastolic periods and the decoupling of the whole model, which led to the derivation of the full model (3) from the continuous-time model (1) (Fig. 1). We conclude that we are dealing with a time-varying system.

The most commonly used algorithm for estimating time-varying parameters is recursive least squares with exponential forgetting. To avoid the severe problem of estimator windup and bursts [22] when the process is not well excited, the use of a variable forgetting factor has been suggested [23]. The ordinary least-square algorithm with covariance modification [24] or covariance resetting [25] has also been found to have excellent practical features when applied to time-varying problems. Since in this study the philosophy of the control design is to tune the PID-controller during the changes of the operating conditions and it is expected that the controlled output will track the constant set point in the steady state, it seems that the condition of persistent excitation cannot be fully satisfied. Also note that, to design the PID-controller by pole placement, a (reduced) second-order model was used to represent the process. Hence, there is always an associated modeling error in addition to measurement errors. Because of these considerations, an ordinary least-square algorithm with dead zone and covariance resetting was used for parameter estimation as follows:

$$\hat\theta(n)=\hat\theta(n-1) + \frac{g\cdot P(n-2)\phi(n-1)}{1+\phi^T(n-1)P(n-2)\phi(n-1)}[y(n)-\phi^T(n-1)\hat\theta(n-1)] \quad (11)$$

$$P(n-1)=P(n-2)-\frac{g\cdot P(n-2)\phi(n-1)\phi^T(n-1)P(n-2)}{1+\phi^T(n-1)P(n-2)\phi(n-1)} \quad (12)$$

$$g=\begin{cases}0 & \text{when } |y(n)-\phi^T(n-1)\hat\theta(n-1)|\le\epsilon \\ & \text{continually for more than } N \text{ heartbeats}\\ 1 & \text{otherwise}\end{cases}$$

$$P(-1)=\alpha_0 I \text{ and}$$

$$P(k)=\alpha I \text{ whenever } g(k+1)=0 \text{ and } g(k+2)=1.$$

The implication of this algorithm is that, when the prediction error is within a preset bound ϵ continually for a certain period (N heartbeats, $N = 4$ was used in this study), the estimation algorithm is frozen and the control algorithm behaves as a fixed-parameter PID-controller. Whenever the prediction error exceeds the preset bound, the covariance matrix P is reset, the estimation algorithm is reinitialized using the previous estimated parameter $\hat\theta$, and the control mode is back to adaptive control to tune the PID-controller.

VI. System Implementation

The hardware of the system is shown in the block diagram of Fig. 4. A linear variable differential transformer (LVDT) was used as the position feedback from the pressure cylinder to a PID analog circuit and a hydraulic servovalve. This loop was designed to produce the appropriate pneumatic pressure which is to drive the diaphragm of the assist ventricle. Thus, the control variable in this system is the stroke of the pressure cylinder. The PID analog circuit was adjusted to produce a smooth-rising waveform of stroke (and then driving or filling pressures) so that undesired hammering of the ventricle and the circulatory system is avoided (see Fig. 3). Similar servoloops were used to drive the simulated natural ventricles.

In addition to the mean atrial pressure ($\overline{LAP}$) we monitored the mean aortic pressure ($\overline{AOP}$) and the mean cardiac output ($\overline{CO}$). The pressures were measured using fluid-filled catheters and physiological pressure transducers while CO was measured using an electromagnetic blood flowmeter.

For the mock loop experiments a square-wave signal from a function generator was fed through a differentiator circuit to produce a signal which was used as the simulated P-wave.

The adaptive control algorithm and the data acquisition program were implemented in a DEC LSI-11/23 microcomputer, running under the DEC RSX-11S operating system.

LAP, AOP, and CO are measured every 16 ms, which is synchronized by a line clock interrupt. Whenever a simulated P-wave is sensed, the microcomputer sends out the D/A signal for systolic stroke U_{AS} (see Fig. 3) and starts computing the average of the outputs, the parameter estimates, and the values of the control variable. The length of the systolic period Δ_S of the current heartbeat is obtained from a table of optimal values [14] based on the average duration of the previous four heartbeats. After the systolic period is finished, the D/A signal of the control variable, the diastolic stroke U_{AD}, is then set out. The total computation time t_{comp} was estimated to be at most 0.1 s and the shortest systolic time interval is about 0.28 s for the heart rate at 120 beats/min. Therefore, the timing constraint is satisfied.

VII. Experimental Results

The results of experiments performed using the mock circulatory system to evaluate the performance of the controller in the presence of changes in heart rate, in systemic peripheral resistance, and in contractility of the simulated natural left ventricle, are given below. As stated earlier, our objective is to maintain mean left atrial pressure at a desired level.

In the experiments the initial values $\hat{\theta}^T(0) = [0.5\ 0.5\ 0.5\ 0.5\ 0.5]$, $P(-1) = 100 \cdot I$, $P(k) = 5 \cdot I$, and $\epsilon = 0.5$ were used in the parameter estimation algorithm (11) and (12). In the control law (8), the set point $y_r = \overline{LAP}_{\text{ref}} = -1$ mmHg was used, and due to the physical limitation of the hydraulic-driven pressure cylinder, an upper bound $(U_{AD})_{\max} = 22$ mm was added for diastolic stroke of the assist ventricle.

Since an overshoot response is not desired in the physiological system, all four poles were chosen on the right real axis as 0.1, 0.2, 0.5, 0.6. Thus, the closed-loop system can be thought as having two dominant poles located at $z = 0.5$ and $z = 0.6$. These poles correspond to an overdamped second-order continuous-time system with a damping ratio $\xi = 1.01$ and rise time $T_r = 7$ heartbeats.

Because there was little information available concerning $\hat{\theta}$, a large initial prediction error can be expected. To avoid undesirable transient response fixed gains $K_P = -1$, $K_I = -1$, $K_D = -1$, and filter constant $\nu = -0.7$ were used in the first heartbeat with only the estimation algorithm active. The negative sign of the gains K_P, K_I, and K_D was chosen because it is known that the input–output sensitivity $\Delta\overline{LAP}/\Delta U_{AD}$ is negative.

Experiment 1: Response in the presence of changes in heart rate.

The heart rate was increased in steps. Fig. 5 shows that increases in mean aortic pressure and mean cardiac output followed the increments in heart rate. The stroke volume of the assist device was adjusted automatically so that the atrial pressure was maintained at the desired level.

Experiment 2: Response in the presence of changes in systemic peripheral resistance.

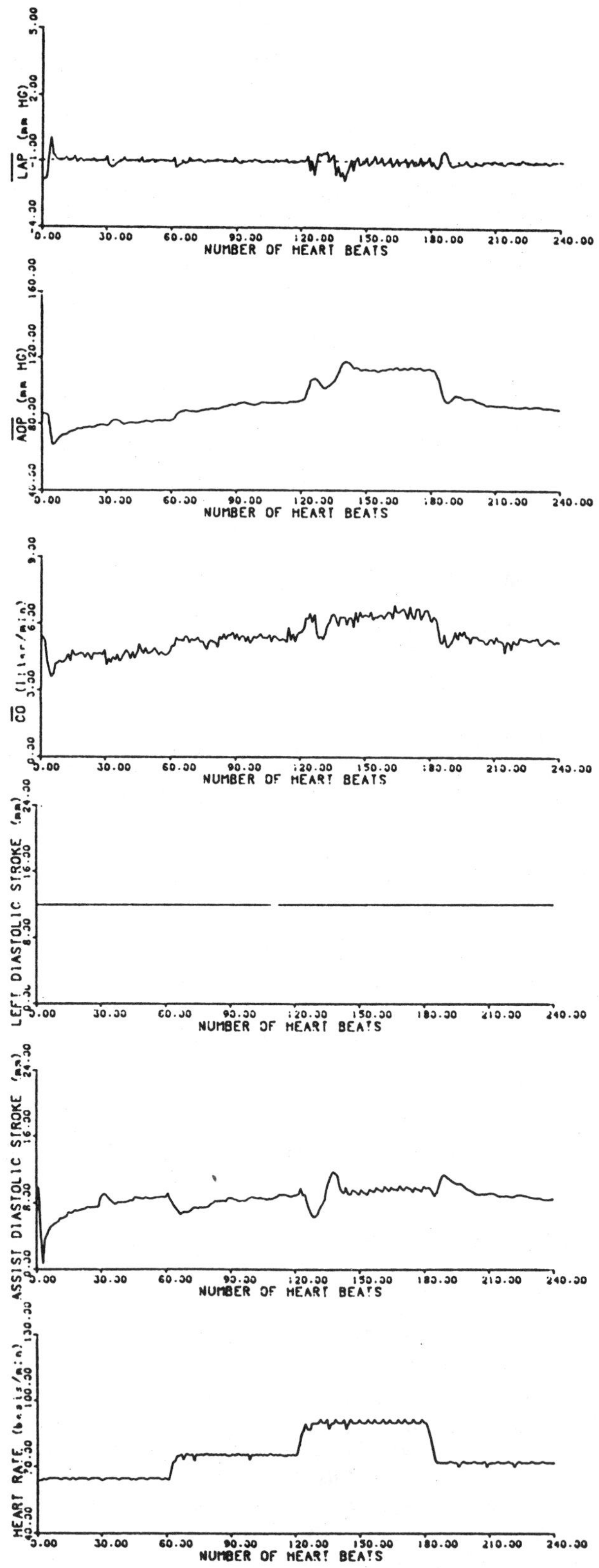

Fig. 5. Experiment 1: response in the presence of changes in heart rate.

The systemic peripheral resistance ($R3$ in Fig. 2) was altered manually at heartbeats 60, 120, and 180 (Fig. 6). These changes in peripheral resistance were followed immediately by proportionate changes in mean aortic pressure. In order to maintain mean atrial pressure at the desired level, the control system automatically adjusted the stroke volume.

Experiment 3: Response in the presence of changes in the contractility of the simulated left ventricle.

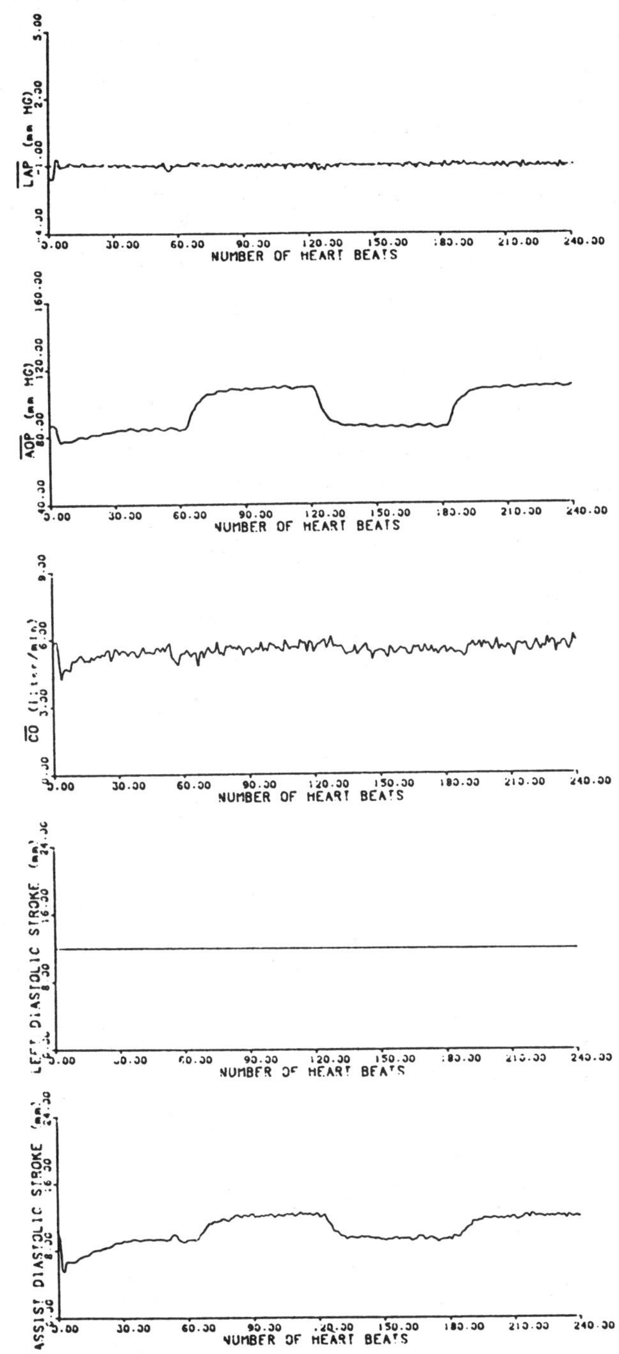

Fig. 6. Experiment 2: response in the presence of changes in peripheral resistance. (The heart rate was set at 75 beats/min.)

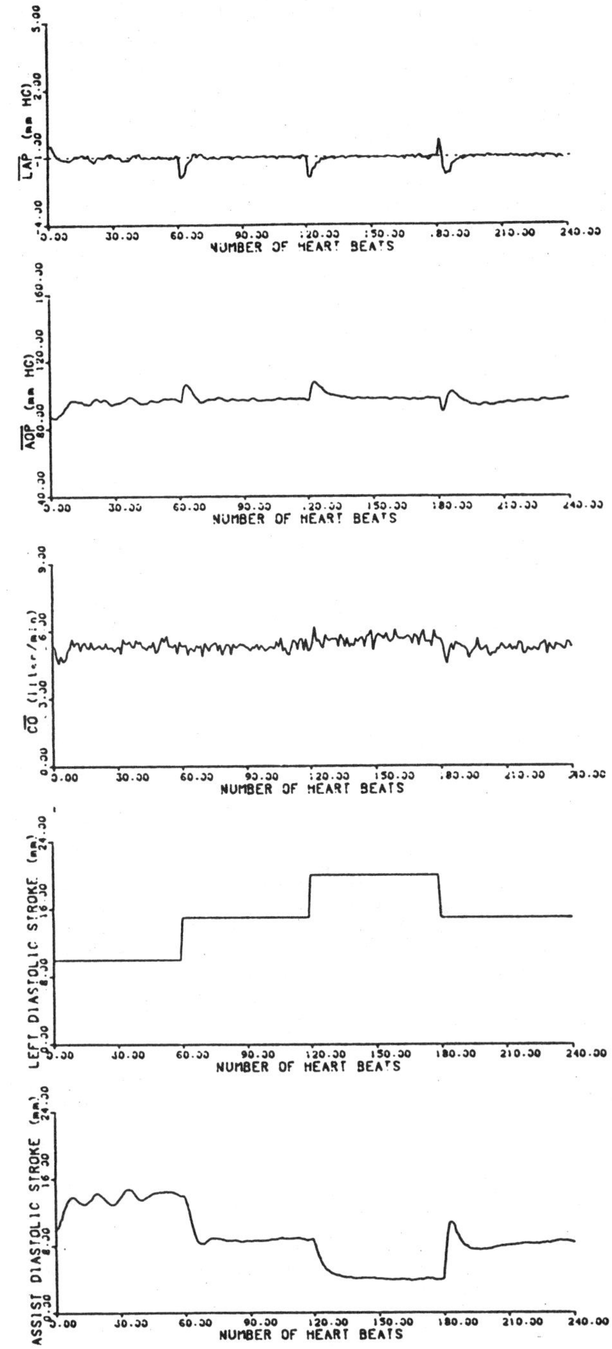

Fig. 7. Experiment 3: response in the presence of changes in the contractility of the simulated left ventricle. (The heart rate was set at 75 beats/min.)

When recovery of the left ventricle was simulated, the stroke volume of the assist device was adjusted by the control system so that the mean left atrial pressure remained at the desired level (Fig. 7). Similarly, when the pumping ability of the left ventricle underwent a step reduction, the assist device made the necessary adjustments to hold the mean left atrial pressure at the desired level. As a result of compensation by the assist device, the mean aortic pressure and cardiac output were unaffected by the changes in the strength of the left ventricle.

VIII. Discussion

For clinical applications we envision initiating control with a conventional (fixed parameter) PID controller. Later the adaptive PID controller can be brought on-line to fine tune the controller gains. After the prediction error has been reduced to an acceptable level, the self-tuning aspects of the controller can be shut off and remain inactive unless there is a major change in operating conditions.

A deterministic process model and a recursive least-squares estimation algorithm were used because only a low level of measurement noise was observed in the mean left atrial pressure $\overline{LAP}$ that was used as the output variable. If a significant level of colored noise exists, a stochastic ARMA model and an extended least-squares algorithm (pseudolinear regression algorithm) should be used [16].

The control of the fluid balance between the right and left sides of the circulatory system can be obtained by controlling the right and left atrial pressures. Since in the case of the use of a left ventricular assist device, we do not have control over right atrial pressure, unbalanced flow might occur. However, if the right heart is strong enough, it will adjust itself to the pumping of the left side and maintain the balance of the fluid between the left and right sides of the circulatory system. If not, a biventricular assist system is needed.

IX. Conclusion

In this paper, a drive system for the left ventricular bypass assist device and the design of a self-tuning PID controller based on pole placement are described. The mock loop experiments demonstrate that the control system including the servoloop of the drive system and the computer-based adaptive control algorithm produces desired performance. From an engineering point of view, this work shows that adaptive control is attractive for medical applications.

References

[1] F. Unger, Ed., *Assisted Circulation.* New York: Springer-Verlag, 1979.

[2] M. A. Fischetti, "The quest for the ultimate artificial heart," *IEEE Spectrum,* pp. 39–44, Mar. 1983.

[3] F. D. Altieri, "Status of implantable energy systems to actuate and control ventricular assist devices," *Artif. Organs,* vol. 7, no. 1, pp. 5–20, 1983.

[4] P. I. Singh, D. M. Lederman, D. C. de Sieyes, R. D. Cumming, B. B. Adams, and R. E. Lang, "A copulsating left heart assist system," in *Proc. IEEE Frontiers Eng. Health Care,* 1982, pp. 135–138.

[5] Y. Nose, G. Jacobs, R. J. Kiraly, L. Golding, H. Harosaki, S. Takantani, S. Murabayashi, R. W. Sukalac, H. Kambic, and J. Snow, "Experimental results for chronic left ventricular assist and total artificial heart development," *Artif. Organs,* vol. 7, no. 1, pp. 55–63, 1983.

[6] G. Rosenberg, A. Snyder, W. Weiss, D. L. Landis, D. B. Geselowitz, and W. S. Pierce, "A cam-type electric motor-driven left ventricular assist device," *J. Biomech. Eng.,* vol. 104, pp. 214–220, 1982.

[7] O. H. Frazier *et al.,* "Development of an implantable, integrated electrically-actuated left heart assist system (THI/GOULD LVAS)," NHLBI Rep. N01-HV-02915-2, Sept. 1982.

[8] J. Frank, K. Affeld, P. Baer, A. Mohnhaupt, F. Zartnack, and E. S Bucherl, "First experience with a mobile total artificial heart system," *Trans. Amer. Soc. Artif. Intern. Organs,* pp. 72–76, 1980.

[9] R. K. Jarvik, "The total artificial heart," *Sci. Amer.,* pp. 74–80, Jan. 1981.

[10] T. Kitamura and H. Akashi, "A design of an adaptive control system for left ventricular assist," in *Proc. IFAC 8th Triennial World Congress,* Kyoto, Japan, 1981.

[11] A. Noordergraaf, "Hemodynamics," in *Biological Engineering,* H. P. Schwan, Ed. New York: McGraw-Hill, 1969, pp. 391–545.

[12] H. Reul, H. Minamitani, and J. Runge, "A hydraulic analog of the systemic and pulmonary circulation for testing artificial hearts," in *Proc. ESAO II,* 1975, pp. 120–127.

[13] A. C. Guyton, *Cardiac Output and Its Regulation.* Philadelphia, PA: Saunders, 1973.

[14] B. C. McInnis, R. L. Everett, B. Vajapeyam, C. H. Lin, M. Iyer, and T. Akutsu, "Digital system for P-wave detection and synchronization of the artificial heart," *Int. J. Biomed. Comput.,* vol. 14, pp. 381–388, 1983.

[15] B. C. McInnis, J. C. Wang, and G. C. Goodwin, "Adaptive control systems for the artificial heart," in *Proc. IEEE Frontiers Eng. Health Care,* 1982, pp. 121–124.

[16] G. C. Goodwin and K. S. Sin, *Adaptive Filtering, Prediction and Control.* Englewood Cliffs, NJ: Prentice-Hall, 1984.

[17] G. A. Bekey and T. A. Grove, "Analysis of errors in the identification of biological systems," in *Proc. IEEE Int. Symp. Circuits Syst.,* 1980, pp. 444–447.

[18] M. Yuwana and D. E. Seborg, "A new method for on-line controller tuning," *A.I.Ch.E. J.,* vol. 28, no. 3, pp. 434–440, 1982.

[19] B. W. Wittenmark, "Self-tuning PID-controllers based on pole placement," Dept. Automat. Contr., Lund Inst. Technol., Lund, Sweden, Tech. Rep. CODEN: LUTFD/(TERT-7179)/1-037, 1979.

[20] K. J. Astrom and B. Wittenmark, *Computer Controlled Systems-Theory and Design.* Englewood Cliffs, NJ: Prentice-Hall, 1984.

[21] D. M. Auslander, Y. Takahashi, and M. Tomizuka, "Direct digital process control: Practice and algorithms for microprocessor application," *Proc. IEEE,* vol. 66, no. 2, pp. 199–208, 1978.

[22] K. J. Astrom, "Theory and applications of adaptive control—A survey," *Automatica,* vol. 19, no. 5, pp. 471–486, 1983.

[23] T. R. Fortescue, L. S. Kershenbaum, and B. E. Ydstie, "Implementation of self-tuning regulators with variable forgetting factors," *Automatica,* vol. 17, no. 6, pp. 831–835, 1981.

[24] E. F. Vogel and T. F. Edgar, "Application of an adaptive pole-zero placement controller to chemical process with variable dead time," in *Proc. Amer. Contr. Conf.,* Washington, DC, 1982.

[25] G. C. Goodwin, H. Elliott, and E. K. Teoh, "Deterministic convergence of a self-tuning regulator with covariance resetting," *Proc. Inst. Elec. Eng.,* vol. 130, pt. D, no. 1, pp. 6–8, 1983.

Adaptive Load-frequency Control of the Hungarian Power System*

I. VAJK,† M. VAJTA,+ L. KEVICZKY,‡ R. HABER,‡ J. HETTHÉSSY† and K. KOVÁCS§

An actually used regulator based on a self-tuning predictor of the area requirement and the required average for finite time controller of the area control error provides improved performance for load-frequency control.

Key Words—Adaptive control; computer applications; prediction; load-frequency control; power system control.

Abstract—The paper deals with the modelling of the power stations and the interconnected power systems for the design of the load-frequency controller of the Hungarian power system. It presents an adaptive regulator which uses the *a priori* known information and satisfies the multi-objective character of the control. The elaborated control strategy performs the following objectives:

—It eliminates the effect of the area load fluctuations to the tie-line power.

—It guarantees the scheduled value of the exported/imported energy.

—It reduces the commands sent to power stations.

—It satisfies the requirement with minimum cost.

The paper shows the real-time experiments with the implemented adaptive regulator which is presently applied for the load-frequency control of the Hungarian power system.

1. INTRODUCTION

IT MUST be admitted that the conventional PID regulator is the most widely used controller in the industry. This controller has well-known properties and its fundamental advantage is the ease of implementation.

Many theoretical and simulation studies have been carried out concerning the application of modern control theories to industrial processes. Criteria used for optimization do not always conform to the real system objectives. The weighting elements of the control algorithms are regularly chosen *a posteriori* to obtain a satisfactory compromise. Implementation of these regulators is quite complex. Often the relation between the physics of the process and the control strategy cannot be easily recognized. But modern control theory is valuable because it leads to a good algorithmic approach to the problem. It gives good ideas to modify the control algorithm for improving the performance of the system. On the other hand there are some methods in modern control theory which do not require too much *a priori* information in order to design a regulator. A certain class of these adaptive regulators are called self-tuning regulators (Åström, 1977). However, if certain information is known, it is useful to apply it for improving the performance of the system. The paper presents the load-frequency control of the Hungarian Electric Power System where all *a priori* information which is known about the system is used in the regulator design.

A project to implement a self-tuning regulator in the Hungarian load-frequency control system was started in 1978. The first efforts were directed to investigate the possibility of the application of such regulators using the model of the power system by simulation (Vajk *et al.*, 1978). Entering the field of practical problems a series of difficulties arose because of the multi-objective character of the task. The originally planned minimum variance type load-frequency regulator proved to be inconvenient because of the high control effort needed and a special half an hour accounting system working on the base of the integral of the area control error. It turned out that the limited increment rate at the generator control units makes the system dynamics practically known and only the area power requirement must be predicted. These facts caused a drastic change in the original control strategy and a new combined adaptive controller was developed. The adaptive regulator was implemented in the

† Department of Automation, Technical University of Budapest, Budapest, Hungary.

‡ Computer and Automation Research Institute, Hungarian Academy of Sciences, Budapest, Hungary.

§ Institute for Electric Power Research, Budapest, Hungary.

* Received 19 May 1982; revised 15 June 1983; revised 14 March 1984; revised 8 August 1984. The original version of this paper was presented at the 8th IFAC World Congress on Control Science and Technology for the Progress of Society which was held in Kyoto, Japan during August 1981. The Published Proceedings of this IFAC Meeting may be ordered from Pergamon Press Limited, Headington Hill Hall, Oxford, OX3 OBW, England. This paper was recommended for publication in revised form by associate editor J. H. Chow under the direction of editor H. A. Spang III.

Reprinted with permission from *Automatica*, vol. 21, pp. 129–137, Mar. 1985.

Copyright © 1985, International Federation of Automatic Control IFAC. Published by Pergamon Press Ltd.

central computer control system of the Hungarian power system and it is a standard facility now.

The main steps of the development of the applied adaptive control strategy are discussed in the paper and some records of the working system are also presented.

2. PROBLEM FORMULATION

The basic task of the power system is to satisfy the requirements of the consumers in respect of both quality and quantity continuously. This means that the power plants have to ensure the real and reactive power required by the consumers in such a way that the voltage and the frequency have to be kept between given limits. Since the electric power cannot be stored in relative large quantities, power stations have to immediately and continuously shift generations to fulfil the consumer's requirements.

A great number of controllers ensure the satisfaction of the requirements of the electric power system. An element of this hierarchical control system is the load-frequency control (LFC). The task of LFC is to keep the tie-line load exchange and the system frequency, as well as their integrals as close as possible to the setpoints, by acting through the speed control system of the generators (Zaborszky *et al.*, 1975).

Figure 1 shows the scheme of the control system. The real powers which flow on the individual tie-lines are summarized. The area interchange and the system frequency are used in order to adjust the power of the generators in the area. The applied controller should perform the following control objectives:

—The area regulates its own load fluctuations.
—It contributes to the control of system frequency.
—During the accounting intervals the exported or imported energy is of scheduled value.
—The regulator has to satisfy the requirement with minimum costs.
—The regulator reduces the commands sent to the power stations without compromising other control objectives.

These purposes are contradictory. The elimination of the disturbances caused by local fluctuation requires frequent interventions. Because of the contribution in the system frequency control the scheduled value of the exchange energy is not assured. It is evident from this that the synthesis of the optimal controller will be the result of a compromise.

3. CONTROLLED PROCESS

Power station

In the area to be controlled, most of the power plants consist of similar thermal units. Its functional block diagram is given in Fig. 2. The fuel burns in the boiler and produces steam, the turbine transforms the heat energy to mechanical energy, and then the generator is used to produce electrical energy for the consumers. For the realization of these energy transformations, several state variables of the unit

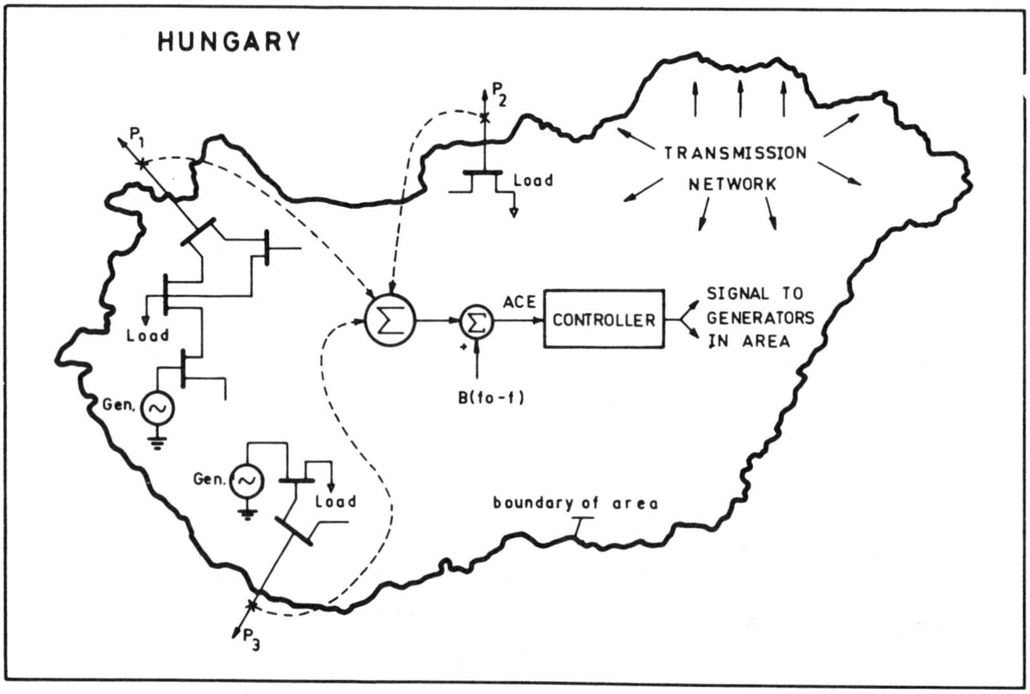

FIG. 1. Scheme of load-frequency control.

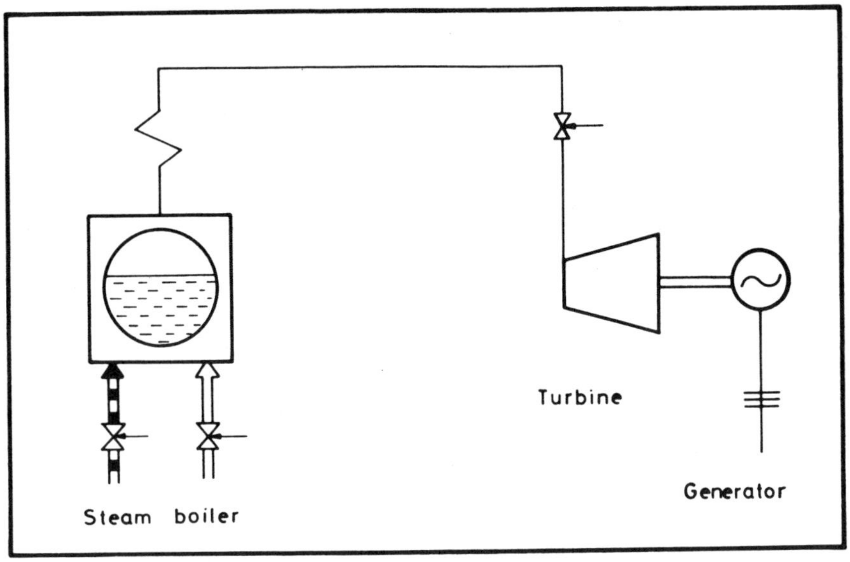

FIG. 2. Block diagram of power unit.

are controlled; for example the speed of generator, the power of the unit, pressure and the temperature of the steam, the water-level in the boiler, etc. The local control centre distributes the command sent from the Power Control Center among the parallel operating units. Indirectly, the Power Control Center regulates the position of the valve of the turbine in order to satisfy the power requirements of the consumers.

The investigations of the behaviour of the power unit is quite complex. Detailed studies can be found in (de Mello *et al.*, 1973; Handschin, 1972). Every turbine–generator system has a local power regulator, which has a unit rate limit controller. The limiter determines the permissible response rate. It has significant effect on the dynamics of the power plant. In practice, the time constants of the turbine–generator system can be neglected. Thus, if it is guaranteed that the increment rate of the commands does not exceed the maximal response rate, the power unit can be modelled by a dead-time lag. The block scheme of the power plants is given in Fig. 3., where U = command; P_g = generated electric power.

The maximal response rate of a power unit is about 3 MW/min. The delay time of the process is about 30 s caused by telecommunication and unit delay. Twelve power stations take place in the load frequency control of the country.

Interconnected power systems

During the model building of the electrical power system, two interconnected systems were assumed. It was considered that the power system consists of two generators interconnected by a 'lossless' network. The loads are modelled as real power sinks. This is illustrated in Fig. 4.

The simplified mathematical model of the interconnected system can be determined from the energy balance equations. These connections give the system frequency of the interconnected areas and the power interchange on the transmission lines. The difference of the produced and used electric power increases the rotating energy of the system by changing the system frequency. The frequency increase changes the load of the areas, because the power demand of the consumers depends on the frequency. From the energy balance of each area the tie-line power change can be obtained. For the subsystems

$$P_g - P_d(f) + P_t = H_1 \frac{df}{dt} \tag{1}$$

$$P_G - P_G(f) - P_t = H_2 \frac{df}{dt}$$

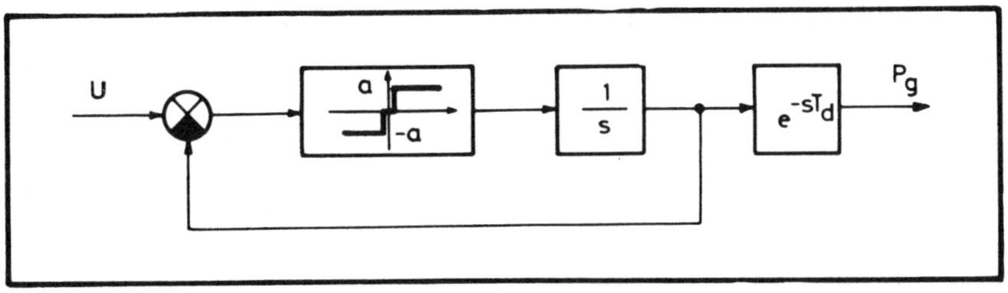

FIG. 3. Model of power unit.

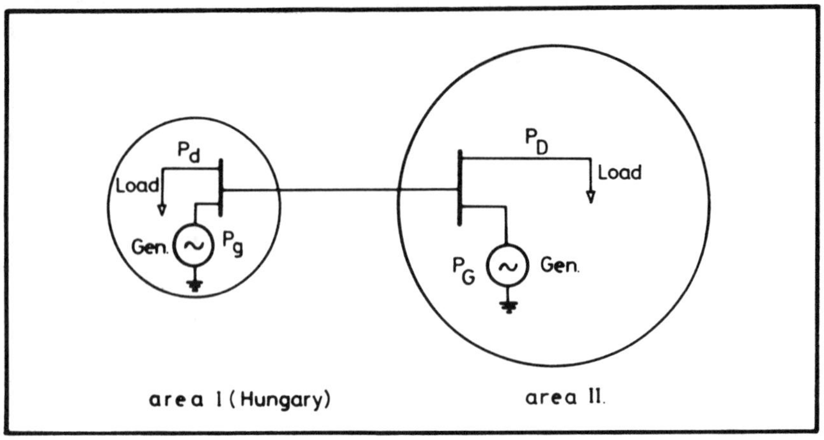

FIG. 4. Two interconnected power systems.

where:

f	system frequency
P_g, P_G	generated electric power of subsystems
P_d, P_D	load of areas
P_t	tie-line power
H_1, H_2	inertia constants.

In a large power system the individual areas do not play an important role and do not change significantly the system frequency. If it is considered that the cooperation power system is infinite, (1) is

$$P_g - P_d + P_t = 0 \tag{2}$$

i.e. the tie-line power is the difference of the power generation and the area load.

The Hungarian power system consists of 400, 220 and 120 kV distribution networks. At all these voltage levels and 750 kV level there is a permanent cooperation among Hungary and all her neighbours. The average load is about 3400 MW/h. Above 20% of energy consumption is imported.

4. A THEORETICAL CONTROL STRATEGY

Using the simplified model of the power system, it is not too difficult to determine the optimal control strategy. For the optimal operation, the commands sent to the power station have to be equal to the difference of the load and the scheduled tie-line power. Since the power station can realize this claim only with delay, the predicted value of the area requirement has to be sent, i.e.

$$U(t) = P_d(t + T_d) - P_0(t + T_d) \tag{3}$$

where

T_d	the dead time of the equivalent power unit
$U(t)$	the command sent to the equivalent power unit
$P_d(t + T_d)$	the load of the area at $t + T_d$
$P_0(t + T_d)$	the scheduled tie-line power.

Since the scheduled value of the tie-line power is given, the 'only' problem is the estimation of the load from the produced power of the plants and the measured tie-line power.

Let us consider that the required command $U(t)$ can be determined precisely. Applying the simplified model of the interconnected system (equation (2)) and the power units (Fig. 3) the control scheme can be created as Fig. 5 shows.

In the figure e^{sT_d} denotes an ideal predictor. The feed-back signal is the load. The scheme in Fig. 5 can

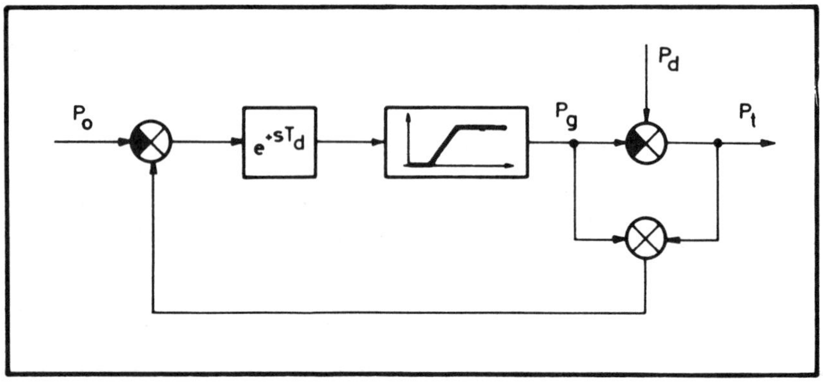

FIG. 5. Control on the basis of the simplified model.

be transformed to the block diagram presented in Fig. 6. Figure 6 shows that the optimal controller consists of a predictor, a comparator and an integral term. The advantage of the control strategy presented in Fig. 6 is that it is a closed loop control, while the control algorithm in Fig. 5 is an open loop. So it can reduce not only the load disturbances, but the characteristic changes of the power stations and the model inadequacy, too. This simple control algorithm cannot be directly applied for practical purposes. Some of the reasons are the following:

—The control is realized by a process computer applying sampled data.
—Each area has to take part in the system frequency control.
—During the accounting periods it is desired that the imported or exported energy should be of scheduled values.
—In the control scheme the area requirement is predicted in an ideal way.
—The above presented strategy guarantees a quick operation. It overburdens the power stations. It would be useful to modify the control algorithm in such a way that the control work of the plants could be reduced without compromising other control objectives.
—The Power Control Center regulates, not only one, but several power stations with different number of power units. The different power units have different cost characteristics. The area requirements have to be satisfied with minimum cost.

5. LOAD FREQUENCY CONTROL

Computer control

The application of a computer for process control causes some additional problems such as, sampling inaccuracy, approximate integration algorithms. These problems are quite common in digital control and will not be discussed here.

Frequency control

In order to keep the system frequency around the setpoint the area regulator operates by using the area control error, which is a linear combination of the tie-line exchange deviation from its setpoint $(P_t - P_0)$ and the frequency deviation from its setpoint $(f - f_0)$, i.e.

$$\mathrm{ACE} = (P_t - P_0) - B(f - f_0). \qquad (4)$$

The area droop is about 350 MW/Hz.

Using the modified measured $(P'_t = P_t - B(f - 50\,\mathrm{Hz}))$ and scheduled $(P'_0 = P_0 - B(f - 50\,\mathrm{Hz}))$ values of the tie line-power the area control error is

$$\mathrm{ACE} = P'_t - P'_0. \qquad (5)$$

In steady-state condition and when B is the area droop and the areas are controlled by similar method, the zero values of ACE implies that the generation and the load of the area are in balance, the frequency and the tie-line power are equal to their setpoints.

RAFT (Required Average for Finite Time) control

The export and the import of the electrical energy is settled on the base of half an hour accounting. The price of the extra energy obtained over the scheduled value depends on the average frequency in the current period. Therefore one of the tasks of the control system is to ensure that the integral of the area control error should be equal to zero at the accounting time instants. It means that the required energy is equal to the scheduled value at the end of accounting periods, if the correction caused by the frequency deviation is not taken into consideration.

The above mentioned control objectives can be satisfied by applying a modified reference value. The performance index for the determination of the new setpoint is

$$V(T_1) = E\left\{\int_{T_0}^{T_1} \mathrm{ACE}(\tau)\,\mathrm{d}\tau\right\} \qquad (6)$$

where E is the expectation operator, T_0 and T_1 are the beginning and the end of the investigated accounting interval. At the instant t the following conditional expected value can be minimized

$$V(T_1|t) = E\left\{\int_{T_0}^{T_1} \mathrm{ACE}(\tau)\,\mathrm{d}\tau|t\right\};\ T_0 < t < T_1 \qquad (7)$$

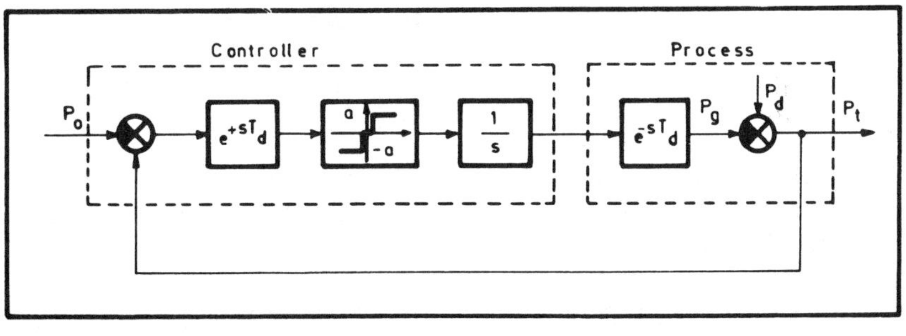

FIG. 6. Theoretical control strategy.

which can be estimated by

$$V(T_1|t) = \int_{T_0}^{t} \mathrm{ACE}(\tau)\,\mathrm{d}\tau + \int_{t}^{T_1} \widehat{\mathrm{ACE}}(\tau|t)\,\mathrm{d}\tau. \qquad (8)$$

Here the $\widehat{\ }$ sign refers to the fact of estimation. At time t the first integral of (8) is known. The area control error in the interval $t < \tau < t + T_d$ is determined by the control action $u(\tau)$, $\tau < t$. A possible estimation of $\widehat{\mathrm{ACE}}(\tau|t)$, $t + T_d < \tau < T_1$ would be

$$\widehat{\mathrm{ACE}}(\tau|t) = 0,\ t + T_d < \tau < T_1 \qquad (9)$$

i.e. there is no control error. In order to minimize the loss function, the reference value is changed in every step in such a way that

$$V(T_1|t) = 0. \qquad (10)$$

Therefore the reference signal of the controller is modified by

$$r(t) = \frac{-\int_{T_0}^{t} \mathrm{ACE}(\tau)\,\mathrm{d}\tau + \int_{t}^{t+T_d} r(\tau - T_d)\,\mathrm{d}\tau}{T_1 - T_d - t}. \qquad (11)$$

Here it was used that the expected value of the area control error at τ, $t < \tau < t + T_d$ is the value by means of which the reference value was modified.

At $t = T_1 - T_d$ $r(t)$ is infinite. But it does not happen, because $r(t)$ is limited. In practice, the total control consists of $T_1 - T_0$ interval long subprocesses. Because of the dead time, $u(\tau)$, $T_1 - T_d < \tau < T_1$ controls the next subprocesses. In this interval the control is based on the original reference value again.

The above, so-called RAFT (Required Average for Finite Time) algorithm was introduced for a composition control first in (Keviczky *et al.*, 1978).

Realization of an ideal predictor

The most difficult task is the realization of the e^{sT_d} lag; i.e. the prediction of area control error. The value of ACE at $t + T_d$ can be separated

$$\mathrm{ACE}(t + T_d) = \mathrm{ACE}(t) - [P'_0(t + T_d) - P'_0(t)] + [P'_d(t + T_d) - P'_d(t)] - [P_g(t + T_d) - P_g(t)]. \qquad (12)$$

It consists of the actual value of ACE, the increment of the generation power, the scheduled interchange and the load modified by the frequency.

The change of the scheduled tie-line power and of the scheduled frequency are *a priori* known. The increment of the load can be predicted. In order to eliminate the error committed at the realization of the ideal predictor and caused by the system constraints, an integral control action is used.

In the control of electric power system the difficult task is the following of the persistently changing consumption, the prediction of the short term load. The area load fluctuation is modelled by an autoregressive integrated moving average process. Since the control algorithm is realized by a computer, it can be considered as a discrete time stochastic process. For the prediction of the load an one step ahead predictor is used. It determines the value of $\delta(t + h|t)$

$$\delta(t + h|t) = P'_d(t + h) - P'_d(t) \qquad (13)$$

where h is the sampling time. Knowing the predictor equation, the increment of the load during the dead time

$$v(t) = P'_d(t + T_d) - P'_d(t) \qquad (14)$$

can be estimated by

$$\hat{v}(t) = \sum_{k=1}^{d} \delta(t + k\,h|t) \qquad (15)$$

where d is the dead time in sampling intervals. For the prediction of $\delta(t)$ the equation introduced in Wittenmark (1974)

$$\delta(\theta + 1|\theta) = \sum_{i=0}^{n} p_i\delta(\theta - i) + \sum_{i=1}^{m} q_i\delta(0 - i + 1|\theta - i) \qquad (16)$$

is used. In (16) θ denotes the discrete time, p_i and q_i are the parameters of the filter used for prediction. By applying vector notation

$$\mathbf{p}^{\mathrm{T}} = [p_0 p_1 \cdots p_n q_1 q_2 \cdots q_m]$$

and

$$\mathbf{x}^{\mathrm{T}}(\theta) = [\delta(\theta)\ \delta(\theta - 1)\ldots\delta(\theta - n)\ \ \delta(\theta|\theta - 1)\ldots\delta(\theta + 1 - m|\theta - m)].$$

Equation (16) can be written in the form

$$\delta(\theta + 1|\theta) = \mathbf{p}^{\mathrm{T}}\mathbf{x}(\theta). \qquad (17)$$

The parameters of this self-tuning (ST) predictor can be estimated by the recursive least-squares method

$$p(\theta) = p(\theta - 1) + R(\theta)\mathbf{x}(\theta - k)\,[\delta(\theta) - \mathbf{p}^{\mathrm{T}}(\theta - 1)\mathbf{x}(\theta - k)] \qquad (18)$$

where

$$R(\theta) = \frac{1}{w}\left[R(\theta-1) - \frac{R(\theta-1)\mathbf{x}(\theta-k)\mathbf{x}^{\mathrm{T}}(\theta-k)R(\theta-1)}{w + \mathbf{x}^{\mathrm{T}}(\theta-k)R(\theta-1)\mathbf{x}(\theta-k)}\right]. \tag{19}$$

In (19) w is the weighting factor.

The applied algorithm uses a second order autoregressive moving average process for modelling the load disturbance. The update of the parameter vector and the covariance matrix are realized by the UD filter (Thornton and Bierman, 1978). A stability test is used to prevent divergence.

Reduction of control effort

The presented control algorithm guarantees a quick and sensitive operation which cannot be permitted during the real-time work. In order to reduce the control action, a saturation characteristic is used instead of the relay one. So, if the area control error is not too large, the system response will be slower, but if the control error is greater than a prespecified value, the control remains as fast as before. Finally it can be summarized that the increment of the area requirement which has to be distributed among the power stations is

$$u(t) = \alpha\,\mathrm{ACE}(t) + \beta r(t) + \gamma\hat{v}(t) - (P'_0(t+T_d) - P'_0(t+T_d-h)) \tag{20}$$

where α, β and γ are the weighting components of the regulator. The first term provides an integral action, a feedback control effect. The second one is caused by the modification of the reference signal. It ensures the objective of the finite processes. The third one gives a feed-forward control action from the predicted load. The last ones are caused by the modification of the scheduled values. The control scheme is presented in Fig. 7.

Load dispatch

The Power Control Center allocates the total required generation among the power stations in such a way that the generation cost will be minimized. The working points of the power stations are chosen on the basis of the economic dispatch (ELD). The load-frequency controller moves the plants around these points.

For the economical use of the computer memory and to reduce the running time, the LFC takes the cost function of the plants into consideration in a linearized way. It uses the working point, the averaged gradient of the cost function upward and downward at the working point. These values are updated by the ELD.

The cost is minimal if all the plants are operated at the same incremental cost value and the constraints are satisfied.

$$\frac{dF_1}{dP_1} = \frac{dF_2}{dP_2} = \ldots = \frac{dF_p}{dP_p}. \tag{21}$$

Here F_i is the cost function of the plant which is a monotonous increasing function of the power. This method is known as the equal incremental cost method and it it widely applied for the load dispatch.

6. REAL TIME EXPERIMENTS

The Hungarian power system is controlled by a process computer made by Hitachi Ltd. (Japan). It consists of two HIDIC-80 CPUs with 64 kW private and 32 kW global capacities and two multi access discs. One of the CPUs executes the on-line tasks of the system, while the other CPU is a standby processor, and it can be used for any off-line purpose. In the case of any failure with the on-line CPU the on-line tasks are automatically taken over by the standy processor. This computer system realizes not only the load frequency control of the country, but many other functions of the power system, as well.

The task which realizes the above described control strategy, is only a little part of the entire control program system. Several other tasks are needed in order to ensure the real-time connection among the process, the computer and the dispatcher. The data for the operation of the LFC

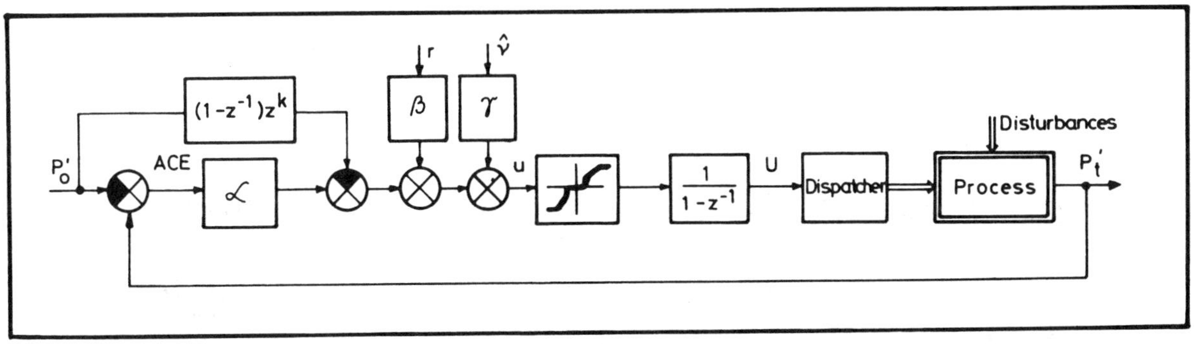

FIG. 7. The realized control scheme.

originate from off-line calculation, from telemetering and some are given by the dispatcher. For example, off-line programs determine the cost curves of the power plants, the daily schedule of the import energy and of the power plant generations.

The basic data from the telemetric system are called in cycles of 2 s. The measured signals are the tie-line measurements, the system frequency, the generation of individual power plants and the number of controllable blocks within a plant. All data are checked and they are stored with their status. The controller operates every minute. It uses prefiltered measurements. A separate program unit takes care of the check of the commands sent to the plants.

Several different diagrams on multicolor quasi-graphical display units and key-boards give effective connection between the dispatcher and the controlled system. The dispatcher can modify, among others, the power limits and the operational mode of the plants, or he can change the import schedule. If there is any unexpected failure, he can control the system manually.

At the realization the integral term of the controller was $\alpha = 0.4$. This value guarantees about 60° phase margin at crossover. The other parameters of the controller (β and γ) were chosen to be 1.

The initial values of the load predictor are $q_1 = 1.4$, $q_2 = -0.45$ and $p_1 = p_2 = 0.025$. The diagonal elements of the covariance matrix are started from $R_{ii} = 0.1$. The reasonable choice of the forgetting factor is 0.98.

Figure 8 shows the results of one of the introductory real-time experiments. It presents the 1 min average values of the area control error, the disturbances, and the commands sent to the controlled stations for a longer period. The area requirement is distributed among the plants in accordance with the optimal economical aspects and operation speed of the power stations.

7. CONCLUSIONS

The adaptive load frequency control of the Hungarian power system is discussed. The paper gives the problem formulation and a simplified model of the controlled process. Then a theoretical control strategy is presented as a base of the practical solution. A combined regulator based on the self-tuning predictor of the area requirement and the RAFT type regulator of the ACE is described for the load frequency control. Finally, real-time records are also shown.

The presented adaptive strategy of LFC leads to an improved dynamic performance, more efficient use of control action under a wide range of

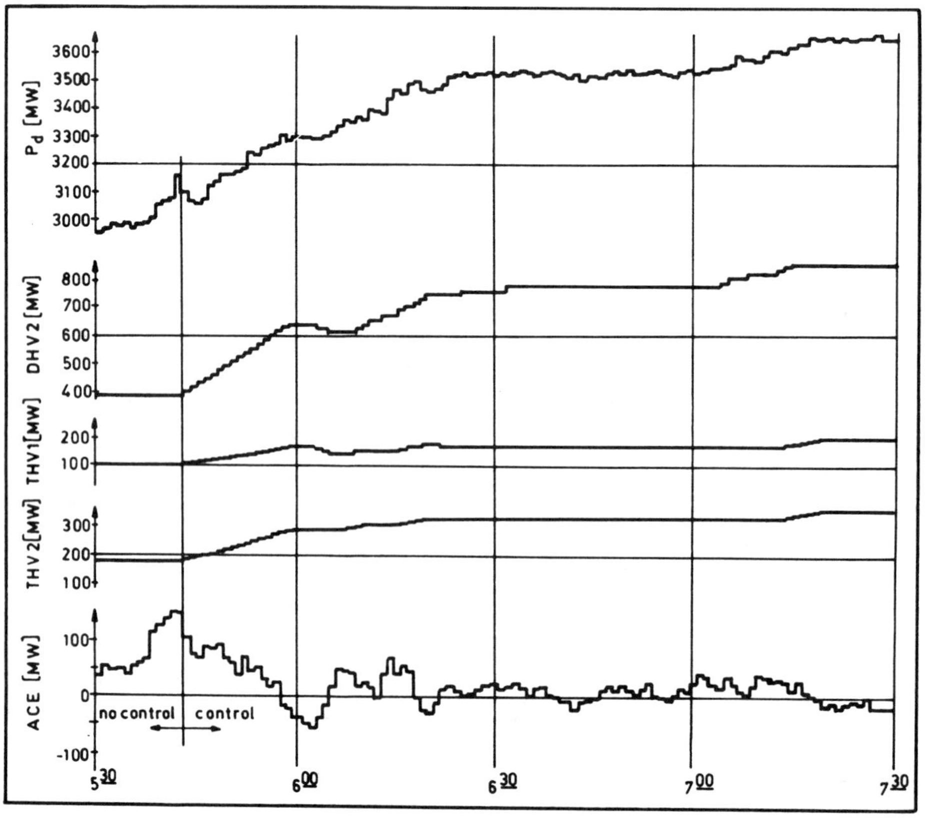

FIG. 8. Real-time records of load, commands of three power stations (DHV2, THV1, THV2) and area control error.

conditions. The controller guarantees the scheduled value of the exchange energy and the satisfaction of the area requirement with minimum cost.

The regulator proved to be good in practice. It has been applied for the control of the Hungarian power system without any problem since June 1981.

Acknowledgement—The authors would like to thank the Hungarian Electric Works and the Power Control Center for their support of this project and J. Paitz and J. Dobra for the technical help at the real-time installation of the control algorithm.

REFERENCES

Åström, K. J., U. Borisson, L. Ljung and B. Wittenmark (1977). Theory and applications of self-tuning regulators. *Automatica*, **13**, 457.

Haber, R., J. Hetthéssy, L. Keviczky, I. Vajk, M. Vajta and K. Kovács (1979). Some investigations on classical and adaptive load frequency control. *Intern. Workshop on Adaptive Control*, New Haven.

Handschin, E., (1972). *Real-time Control of Electric Power Systems*. Elsevier, Amsterdam.

Keviczky, L., J. Hetthéssy, M. Hilger and J. Kolostori (1978). Self-tuning adaptive control of cement raw material blending. *Automatica*, **14**, 525.

de Mello, F. P., R. J. Mills and W. F. B'rells (1973). Automatic generation control, Part I—Process modelling. Part II—Digital control techniques. *IEEE Trans. Power Appárat. Syst.* **PAS-92**, 710.

Thcrnton, C. L. and G. J. Bierman (1978). Filtering and error analysis via the UDU covariance factorization. *IEEE Trans. Aut. Control*, **AC-23**, 901.

Vajk, I., L. Keviczky, L. Kovács, and J. Hetthéssy (1978). Application of self-adjusting regulator in interconnected power system. *Proc.* 1978 *Summer Computer Simulation Conference*, Newport Beach, California.

Wittenmark, B. (1974). A self-tuning predictor. *IEEE Trans. Aut. Control*, **AC-19**, 848.

Zaborszky, J., A. K. Subramanian and K. M. Lu (1975). Control interfaces of generation allocation on the large interconnected power system. *Proc. IEEE Conf. on Decision and Control*, Houston, p. 283.

Author Index

Editor's Biography

Madan M. Gupta (M'63–SM'76) received the B.Eng. (Honors) in 1961, and the M.Sc. degree in 1962, both in electronics-communications engineering, from BITS Pilani, India. In 1964, he was awarded a British Commonwealth Fellowship for his higher studies in the U.K., first at the Queen's University of Belfast, and then at the University of Warwick. He was awarded the Ph.D. degree for his studies in adaptive control systems in 1967 by the University of Warwick, U.K.

From 1962 to 1964, he served as a lecturer in electrical-communication engineering at the University of Roorkee, India. He joined the faculty of the College of Engineering at the University of Saskatchewan in 1967 as a sessional lecturer, receiving a full professorship in July 1978. He has held several grants and research contracts since 1969. He has acted as a consultant to various agencies such as Atomic Energy of Canada Limited, Saskatchewan Power Corp., MacMillan Bloedel Research Limited, the Defence Research Board and the Honeywell of Canada. He was a visiting professor at the Florida State University during 1984. He has initiated research in the areas of Optimization and Adaptive Control Systems, Fuzzy Automata and Decision Processes, and Signal Processing. His present research interest also includes the Noninvasive Detection and Diagnosis of Early Cardiac Abnormalities and knowledge based expert systems and cognitive processes with application to intelligent systems. Professor Gupta is a coauthor, with A. Kaufmann, of the book entitled *Introduction to Fuzzy Arithmetic: Theory Applications* (Van Nostrand Reinhold, 1985) and is the editor of the books entitled *Fuzzy Automata and Decision Processes* (1977), *Advances in Fuzzy Set Theory and Applications* (1979), *Approximate Reasoning in Decision Analysis* (1982), *Fuzzy Information and Decision Processes* (1982), *Fuzzy Information, Knowledge Representation, and Decision Analysis* (1983), and *Approximate Reasoning in Expert Systems* (1985), all with North Holland. He is a subject editor for the *Encyclopedia of Systems and Control* (Pergamon Press, Oxford) and has authored or coauthored over 180 research papers. Dr. Gupta is an advisory editor for the *International Journal of Fuzzy Sets and Systems* (IFSA) and for other journals in the field.

Dr. Gupta is a member of many other professional organizations, and a recipient of a number of medals and awards including a Senior Industrial Fellowship of the Natural Science and Engineering Research Council of Canada during the academic year 1975–76. He was a program chairman of the special symposium on Fuzzy Set Theory and Applications held during the 1977 IEEE-CDC at New Orleans. Also, he was a cochairman of the IFAC symposium on Fuzzy Information, Knowledge Representation and Decision Analysis held at Marseille (France) in July 1983. He is a member of the Electrical Engineering Grant Selection Committee for the Natural Sciences and Engineering Research Council of Canada.